Available in MyMathLab® for your Algebra and Trigonometry Course

Achieve Your Potential

Success in math can make a difference in your life. MyMathLab is a learning experience with resources to help you Achieve Your Potential in this course and beyond. MyMathLab will help you learn the new skills required, also help you learn the bigger concepts, and make connections for future courses and careers.

Visualization and Conceptual Understanding

These MyMathLab resources will help you think visually and connect the concepts.

NEW! Guided Visualizations

These engaging interactive figures bring mathematical concepts to life, helping students visualize the concepts through directed explorations and purposeful manipulation. *Guided Visualizations* are assignable in MyMathLab and encourage active learning, critical thinking, and conceptual learning.

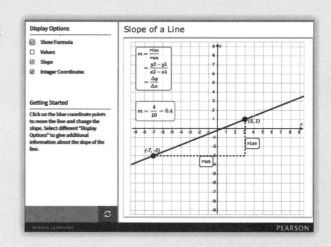

Given the function $P(x) = 5x^4 - 3x^3 + 7x^2 - 12x + 4$, what does Descartes' rule of signs tell you about the number of positive real zeros and the number of negative real zeros?

Video Assessment Exercises

Featuring authors Judy Beecher and Judy Penna, these Example Solutions videos walk students through the detailed solution process for nearly all examples from the text. Corresponding exercises check for conceptual understanding. Videos and exercises are assignable in MyMathLab.

Preparedness and Study Skills

MyMathLab® gives access to many learning resources that refresh knowledge of topics previously learned. Just-in-Time Review, Getting Ready material, and the Video Notebook are some of the tools available.

NEW! Just-in-Time Review

Review of prerequisite algebra topics is now presented when students need it most. References to the 28 *Just-in-Time* review topics are placed throughout the text as they become relevant allowing students to review just before learning a new skill. These efficient reviews of intermediate algebra topics are also assignable in MyMathLab.

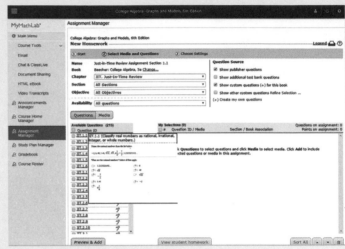

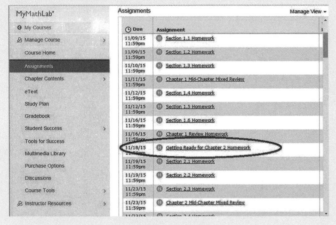

Getting Ready

With *Getting Ready* content students refresh prerequisite topics through assignable skill review quizzes and personalized homework.

NEW! Video Notebook

This author-created notebook contains fill-in-the-blank worksheets to accompany the authors' video examples. Key definitions, theorems, and procedures are also included. After filling in the worksheet while watching the video, the student has an excellent study guide for review and test preparation. This is available in print to accompany the text or as a download in MyMathLab.

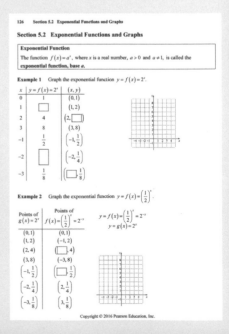

6TH EDITION

Precalculus

Graphs and Models
A Right Triangle Approach

Marvin L. Bittinger
Indiana University Purdue University Indianapolis

Judith A. Beecher

David J. Ellenbogen
Community College of Vermont

Judith A. Penna

PEARSON

Boston   Columbus   Indianapolis   New York   San Francisco
Amsterdam   Cape Town   Dubai   London   Madrid   Milan   Munich   Paris   Montréal   Toronto
Delhi   Mexico City   São Paulo   Sydney   Hong Kong   Seoul   Singapore   Taipei   Tokyo

Editorial Director	Chris Hoag
Editor in Chief	Anne Kelly
Acquisitions Editor	Chelsea Kharakozova
Editorial Assistant	Ashley Gordon
Project Manager	Rachel S. Reeve
Program Management Team Lead	Karen Wernholm
Project Management Team Lead	Christina Lepre
Media Producer	Erica Lange
TestGen Content Manager	Marty Wright
MathXL Content Manager	Kristina Evans
Product Marketing Manager	Rachel Ross
Field Marketing Manager	Peggy Lucas
Marketing Assistant	Fiona Murray
Senior Author Support/Technology Specialist	Joe Vetere
Rights and Permissions Project Manager	Gina M. Cheselka
Procurement Specialist	Carol Melville
Associate Director of Design	Andrea Nix
Program Design Lead	Barbara T. Atkinson
Text Design, Art Editing, and Photo Research	The Davis Group, Inc.
Editorial and Production Coordination	Martha Morong/Quadrata, Inc.
Composition	Lumina Datamatics, Inc.
Illustrations	Network Graphics, William Melvin, and Lumina Datamatics, Inc.
Cover Design	Jenny Willingham/Infiniti
Cover Image	hxdbzxy/Shutterstock

Library of Congress Cataloging-in-Publication Data

Precalculus : graphs and models : a right triangle approach / Marvin L. Bittinger, Indiana University-Perdue [i.e. Purdue] University, Indianapolis [and three others].—Sixth edition.
    pages cm
Includes index.
    ISBN 978-0-13-417905-6 (hardcover)—ISBN 0-13-417905-6 (hardcover)—ISBN 978-0-13-418196-7 (hardcover)—ISBN 0-13-418196-4 (hardcover)
    1. Precalculus—Textbooks. 2. Algebra—Graphic methods—Textbooks. 3. Trigonometry—Graphic methods—Textbooks. 4. Functional analysis—Textbooks. I. Bittinger, Marvin L.
    QA39.3.P74 2017
    512'.13—dc23
                                                                                    2015035066

1 2 3 4 5 6 7 8 9 10—RRD-W—19 18 17 16

www.pearsonhighered.com

ISBN 13: 978-0-13-417905-6
ISBN 10: 0-13-417905-6

Contents

Just-in-Time Review   J-1

1 Graphs, Functions, and Models   1

7

Trigonometric Identities, Inverse Functions, and Equations   **493**

10 ▷ Analytic Geometry Topics   719

Preface

Precalculus: Graphs and Models is known for enabling students to "see the math" through its

- focus on visualization,
- early introduction of functions,
- integration of technology, and
- connections between math concepts and the real world.

New!

With the new edition, we continue to innovate by positioning the review material as a more effective tool for teachers and students. Chapter R from the previous edition has been condensed into **28 Just-in-Time review topics** that are placed before Chapter 1. This new review feature is designed to give each student the opportunity to be successful in this course by providing a quick review of topics from intermediate algebra that will be built upon in new algebra and trigonometry topics. The review can be used in an individualized instruction format since some students will require more review than others. Treating the review in this manner will allow more time to cover the algebra and trigonometry topics in the syllabus.

On the other hand, some instructors might choose to review some or all of the topics with the entire class at the beginning of the course or in a just-in-time format as each is needed. We think that instructors will appreciate the flexibility that the Just-in-Time feature offers.

Additional resources in the MyMathLab course reflect the themes of just-in-time review and concept retention. For example, new Cumulative Review assignments allow students to synthesize and retain concepts learned throughout the course. The Just-in-Time review topics within MyMathLab allow for assignable Getting Ready review quizzes that lead to personalized Getting Ready homework focused on areas in which students need additional practice.

Our overarching goal is to provide students with a learning experience that will not only lead to success in this course, but also prepare them to be successful in the mathematics courses that they take in the future.

〉 Content Changes to the Sixth Edition

- **Just-in-Time Review** Review of prerequisite algebra topics is now presented when students need it most.

  New!
  - A set of 28 numbered, short review topics creates an efficient review of intermediate algebra topics:

    1. Real Numbers
    2. Properties of Real Numbers
    3. Order on the Number Line
    4. Absolute Value
    5. Operations with Real Numbers
    6. Interval Notation
    7. Integers as Exponents

    8. Scientific Notation
    9. Order of Operations
    10. Introduction to Polynomials
    11. Add and Subtract Polynomials
    12. Multiply Polynomials
    13. Special Products of Binomials
    14. Factor Polynomials; The FOIL Method

- This feature is placed before Chapter 1. Just-in-Time icons are positioned throughout the text next to the appropriate example where review of an intermediate algebra topic would be helpful.

- **Cumulative Reviews**   For enhanced concept review, cumulative reviews, assignable in MyMathLab, allow students to synthesize and retain concepts learned throughout the course.

- **Informed Exercises**   We have analyzed the MyMathLab usage data which has informed our revision of the exercises for this new edition. The goal is to ultimately improve the quality and quantity of exercises that are most relevant.

❯ Emphasis on Functions

Functions are the core of this course and are presented as a thread that runs throughout the course rather than as an isolated topic. We introduce functions in Chapter 1, whereas many traditional algebra and trigonometry textbooks cover equation-solving in Chapter 1. Our approach of introducing students to a relatively new concept at the beginning of the course, rather than requiring them to begin with a review of material that was previously covered in intermediate algebra, immediately engages them and serves to help them avoid the temptation to neglect studying early in the course because "I already know this."

The concept of a function can be challenging for students. By repeatedly exposing them to the language, notation, and use of functions, demonstrating visually how functions relate to equations and graphs, and also showing how functions can be used to model real data, we hope to ensure that students not only become comfortable with functions but also come to understand and appreciate them. You will see this emphasis on functions woven throughout the other themes that follow.

Classify the Function Exercises   With a focus on conceptual understanding, students are asked periodically to identify a number of functions by their type (linear, quadratic, rational, and so on). As students progress through the text, the variety of functions with which they are familiar increases and these exercises become more challenging. The "classifying the function" exercises appear with the review exercises in the Skill Maintenance portion of an exercise set. (See pp. 262 and 353–354.)

❯ Visual Emphasis

Our early introduction of functions allows graphs to be used to provide a visual aspect to solving equations and inequalities. For example, we are able to show students both algebraically and visually that the solutions of a quadratic equation $ax^2 + bx + c = 0$ are the zeros of the quadratic function $f(x) = ax^2 + bx + c$, as well as the first coordinates of the x-intercepts of the graph of that function. This makes it possible for students, particularly visual learners, to gain a quick understanding of these concepts. (See pp. 178, 181, 221, 281, and 342.)

Visualizing the Graph   Appearing at least once in every chapter, this feature provides students with an opportunity to match an equation with its graph by focusing on the characteristics of the equation and the corresponding attributes of the graph. (See pp. 138, 194, and 276.) In MyMathLab, animated Visualizing the Graph features for each chapter allow students to interact with graphs on an entirely new level. In addition to this full-page feature, many of the exercise sets include exercises in which the student is asked to match an equation with its graph or to find an equation of a function from its graph. (See pp. 140, 141, 230, and 326.)

Side-by-Side Examples   Many examples are presented in a side-by-side, two-column format in which the algebraic solution of an equation appears in the left column and a graphical solution appears in the right column. (See pp. 244, 284, and 357.) This enables students to visualize and comprehend the connections among the solutions of an equation, the zeros of a function, and the x-intercepts of the graph of a function.

New!

Guided Visualizations   These new figures help bring mathematical concepts to life. They are included in MyMathLab as both a teaching and a learning tool. Used as a lecture tool, the figures help engage students more fully and save the time that would otherwise be spent drawing figures by hand. Questions pertaining to each guided visualization are assignable in MyMathLab and reinforce active learning, critical thinking, and conceptual learning.

Integrated Technology   In order to increase students' understanding of the course content through a visual means, we integrate graphing calculator technology throughout. The use of the graphing calculator is woven throughout the text's exposition, exercise sets, and testing program without sacrificing algebraic skills. Graphing calculator technology is included in order to enhance—not replace—students' mathematical skills, and to alleviate the tedium associated with certain procedures. (See pp. 176, 273–274, and 355.) The graphing calculator windows enhance the visual element of the text, providing graphical interpretations of solutions of equations, zeros of functions, and x-intercepts of graphs of functions.

❯ Making Connections

Zeros, Solutions, and x-Intercepts   We find that when students understand the connections among the real zeros of a function, the solutions of its associated equation, and the first coordinates of the x-intercepts of its graph, a door opens to a new level of mathematical comprehension that increases the probability of success in this course. We emphasize zeros, solutions, and x-intercepts throughout the text by using consistent, precise terminology and including exceptional graphics. Seeing this theme repeated in different contexts leads to a better understanding and retention of these concepts. (See pp. 171 and 181.)

Connecting the Concepts   This feature highlights the importance of connecting concepts. When students are presented with concepts in visual form—using graphs, an outline, or a chart—rather than merely in paragraphs of text, comprehension is streamlined and retention is enhanced. The visual aspect of this feature invites students to stop and check their understanding of how concepts work together in one section or in several sections. This check in turn enhances student performance on homework assignments and exams. (See pp. 69, 181, and 249.)

Annotated Examples   We have included over 1070 annotated examples designed to fully prepare the student to work the exercises. Learning is carefully guided with the use of numerous color-coded art pieces and step-by-step annotations. Substitutions and annotations are highlighted in red for emphasis. (See pp. 175 and 349–350.)

Now Try Exercises   Now Try Exercises are found after nearly every example. This feature encourages active learning by asking students to do an exercise in the exercise set that is similar to the example that the student has just read. (See pp. 173, 268, and 322.)

Synthesis Exercises   These exercises appear at the end of each exercise set and encourage critical thinking by requiring students to synthesize concepts from several sections or to take a concept a step further than in the general exercises. For the Sixth Edition, these exercises are assignable in MyMathLab. (See pp. 251–252, 330, and 380.)

Real-Data Applications   We encourage students to see and interpret the mathematics that appears every day in the world around them. Throughout the writing process, we conducted an energetic search for real-data applications, and the result is a variety of examples and exercises that connect the mathematical content with everyday life. Most of these applications feature source lines and many include charts and graphs. Many are drawn from the fields of health, business and economics, life and physical sciences, social science, and areas of general interest such as sports and travel. (See pp. 37 ("Food Stamp Program"), 63 ("Industrial Robots"), 141 ("Medical Care Abroad"), 184 ("Funding for Afghan Security"), 231 ("Vinyl Album Sales"), 328 ("Alfalfa Imported by China"), 560 ("Vietnam Veterans Memorial"), 649 ("Cosmetic Surgery"), 658 ("Top Art Auction Sales"), 735 ("The Ellipse at the White House"), and 813 ("The Economic Multiplier; Super Bowl XLVII").)

〉 Ongoing Review

The most significant change to the Sixth Edition is the new Just-in-Time Review feature, designed to provide students with efficient and effective review of basic algebra skills.

New! Just-in-Time Review   Chapter R has been condensed into 28 numbered, short review topics to create an efficient review of intermediate algebra topics. This feature is placed before Chapter 1.

- Just-In-Time icons are placed throughout the text next to the example where review of an intermediate algebra topic would be helpful. (See pp. 33, 95, 164, 220, and 315.)
- The coverage of each topic contains worked-out examples and a short exercise set. Answers to all exercises appear at the end of the answers at the back of the book.
- Worked-out solutions to all Just-in-Time exercises are included in the *Student Solutions Manual*.

Just in Time

20

THE PRINCIPLE OF SQUARE ROOTS

The principle of square roots can be used to solve some quadratic equations.

THE PRINCIPLE OF SQUARE ROOTS

If $x^2 = k$, then $x = -\sqrt{k}$ or $x = \sqrt{k}$.

EXAMPLES Solve.

1.
$$s^2 - 144 = 0$$
$$s^2 = 144$$
$$s = -\sqrt{144} \quad or \quad s = \sqrt{144}$$
$$s = -12 \quad or \quad s = 12$$

The solutions are -12 and 12, or ± 12.

2.
$$3x^2 - 21 = 0$$
$$3x^2 = 21$$
$$x^2 = 7$$
$$x = -\sqrt{7} \quad or \quad x = \sqrt{7}$$

The solutions are $-\sqrt{7}$ and $\sqrt{7}$, or $\pm\sqrt{7}$.

Solve.

1. $x^2 - 36 = 0$

2. $2y^2 - 20 = 0$

3. $6z^2 = 18$

4. $3t^2 - 15 = 0$

5. $z^2 - 1 = 24$

6. $5x^2 - 75 = 0$

> **Do Exercises 1–6.**

Mid-Chapter Mixed Review   This review reinforces understanding of the mathematical concepts and skills covered in the first half of the chapter before students move on to new material in the second half of the chapter. Each review begins with at least three true/false exercises that require students to consider the concepts they have studied and also contains exercises that drill the skills from all prior sections of the chapter. They are available as assignments in MyMathLab. (See pp. 121–122 and 346–347.)

   Collaborative Discussion and Writing Exercises appear in the Mid-Chapter Mixed Review as well. These exercises can be discussed in small groups or by the class as a whole to encourage students to talk about the key mathematical concepts in the chapter. They can also be assigned to individual students to give them an opportunity to write about mathematics. (See pp. 199 and 253).

   A section reference (shown in red) is provided for each exercise in the Mid-Chapter Mixed Review. This tells the student which section to refer to if help is needed to work the exercise. Answers to all exercises in the Mid-Chapter Mixed Review are given at the back of the book.

Study Guide   This feature is found at the beginning of the **Summary and Review** near the end of each chapter. Presented in a two-column format and organized by section, this feature gives key concepts and terms in the left column and a worked-out example in the right column. It provides students with a concise and effective review of the chapter that is a solid basis for studying for a test. In MyMathLab, these Study Guides are accompanied by narrated examples to reinforce the key concepts and ideas. (See pp. 210–215 and 381–387.)

Exercise Sets   There are over 7060 exercises in this text. The exercise sets are enhanced with real-data applications and source lines, detailed art pieces, tables, graphs, and photographs. In addition to the exercises that provide students with concepts presented in the section, the exercise sets feature the following elements to provide ongoing review of topics presented earlier:

- **Skill Maintenance Exercises.** These exercises provide an ongoing review of concepts previously presented in the course, enhancing students' retention of these concepts. They include **Vocabulary Reinforcement**, described next, and **Classifying the Function** exercises, described earlier in the section "Emphasis on Functions." A section reference (shown in red) is provided for each exercise. This tells the student which section to refer to if help is needed to work the exercise. Answers to all Skill Maintenance exercises appear in the answer section at the back of the book. (See pp. 128, 206, 279–280, and 345.)
- **Enhanced Vocabulary Reinforcement Exercises.** This feature checks and reviews students' understanding of the vocabulary introduced throughout the text. It appears once in every chapter, in the Skill Maintenance portion of an exercise set, and is intended to provide a continuing review of the terms that students must know in order to be able to communicate effectively in the language of mathematics. (See pp. 149–150, 209, and 279.)
- **Enhanced Synthesis Exercises.** These exercises appear at the end of each exercise set and encourage critical thinking by requiring students to synthesize concepts from several sections or to take a concept a step further than in the general exercises. For the Sixth Edition, these exercises are assignable in MyMathLab.

Review Exercises   These exercises in the **Summary and Review** supplement the Study Guide by providing a thorough and comprehensive review of the skills taught in the chapter. A group of true/false exercises appears first, followed by a large number of exercises that drill the skills and concepts taught in the chapter. In addition, three multiple-choice exercises, one of which involves identifying the graph of a function, are included in the Review Exercises for every chapter. Each review exercise is accompanied by a section reference that, as in the Mid-Chapter Mixed Review, directs students to the section in which the material being reviewed can be found. Also included are Collaborative Discussion and Writing exercises. These exercises are described under the Mid-Chapter Mixed Review heading on p. xv. (See pp. 215–217 and 388–391.)

Chapter Test   The test at the end of each chapter allows students to test themselves and target areas that need further study before taking the in-class test. Each Chapter Test includes a multiple-choice exercise involving identifying the graph of a function. Answers to all questions in the Chapter Tests appear in the answer section at the back of the book, along with corresponding section references. (See pp. 218 and 391–392.)

DOMAIN
REVIEW SECTION 1.2.

Review Icons   Placed next to the concept that a student is currently studying, a review icon references a section of the text in which the student can find and review topics on which the current concept is built. (See pp. 263 and 308.)

❯ Acknowledgments

We wish to express our heartfelt thanks to a number of people who have contributed in special ways to the development of this textbook. Our editor, Chelsea Kharakozova, and Editor in Chief, Anne Kelly, encouraged and supported our vision. We are very appreciative of the marketing insight provided by Peggy Lucas and Rachel Ross, our marketing managers, and of the support that we received from the entire Pearson team, including Rachel Reeve, project manager, Barbara Atkinson, cover designer,

Ashley Gordon, editorial assistant, and Fiona Murray, marketing assistant. We also thank Erica Lange, media producer, for her creative work on the media products that accompany this text. And we are immensely grateful to Martha Morong for her editorial and production services, and to Geri Davis for her text design and art editing, and for the endless hours of hard work they have done to make this a book of which we are proud. We also thank Laurie Hurley and Holly Martinez for their meticulous accuracy checking and proofreading of the text.

The following reviewers made invaluable contributions to the development of the recent editions and we thank them for that:

Gerald Allen, *Howard College*
Robin Ayers, *Western Kentucky University*
Heidi Barrett, *Arapahoe Community College*
George Behr, *Coastline Community College*
Kimberly Bennekin, *Georgia Perimeter College*
*Nadine Bluett, *Front Range Community College*
Marc Campbell, *Daytona Beach Community College*
*Shawn Clift, *Eastern Kentucky University*
Mark A. Crawford, Jr., *Waubonsee Community College*
Brad Fcldscr, *Kennesaw State University*
Homa Ghaussi-Mujtaba, *Lansing Community College*
Bob Gravelle, *Colorado Technical University, Colorado Springs*
*Mako E. Haruta, *University of Hartford*
Judy Hayes, *Lake-Sumter Community College*
Michelle Hollis, *Bowling Green Community College*
*Patricia Ann Hussey, *Triton College*
Glenn Jablonski, *Triton College*
Bridgette Jacob, *Onondaga Community College*
Symon Kimitei, *Kennesaw State University*
Deanna Kindhart, *Illinois Central College*
Pamela Krompak, *Owens Community College*
Laud Kwaku, *Owens Community College*
Carol A. Lucas, *University of Central Oklahoma*
*Claude Moore, *Cape Fear Community College*
Daniel Olson, *Purdue University, North Central*
*Priti Patel, *Tarrant County Community College*
Cloyd A. Payne, *Owens Community College*
Randy K. Ross, *Morehead State University*
Daniel Russow, *Arizona Western College*
Brian Schworm, *Morehead State University*
*Nicholas Sedlock, *Framingham State University*
*Pavel Sikorskii, *Michigan State University*
Judith Staver, *Florida Community College at Jacksonville, South Campus*
*Laura Taylor, *Cape Fear Community College*
Jean Hunt Thorton, *Western Kentucky University*
*Pat Velicky, *Florence–Darlington Technical College*
*Jim Voss, *Front Range Community College*
Douglas Windham, *Tallahassee Community College*
Weicheng Xuan, *Arizona Western College*
*Cathleen Zucco-Teveloff, *Rider University*

* *Reviewers of the Sixth Edition*

M.L.B.
J.A.B.
D.J.E.
J.A.P.

Resources for Success

MyMathLab® Online Course for *Precalculus: Graphs and Models* by Bittinger/Beecher/Ellenbogen/Penna
(access code required)

MyMathLab is available to accompany Pearson's market-leading text offerings. To give students a consistent tone, voice, and teaching method, each text's flavor and approach is tightly integrated throughout the accompanying MyMathLab course, making learning the material as seamless as possible.

With this new edition, instructors will find even more assignable exercises in MyMathLab to match many of the in-text features so many instructors have grown to rely on, as well as a new Skills for Success Module.

Skill Maintenance Quizzes and Cumulative Reviews

Instructors can now assign MyMathLab quizzes generated from the *Skill Maintenance Exercises* found in the text. These quizzes support ongoing review to help students maintain essential skills.

For enhanced concept review throughout the course, *Cumulative Reviews* are now also assignable in MyMathLab. These assignments allow students to synthesize and retain concepts.

NEW! Guided Interactive Figures

These engaging interactive figures bring mathematical concepts to life, helping students visualize the concepts through directed explorations and purposeful manipulation. Guided Visualizations are assignable in MyMathLab and encourage active learning, critical thinking, and conceptual learning.

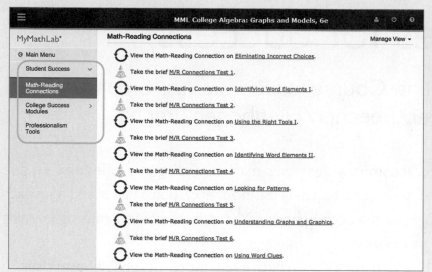

Skills for Success Modules

Skills for Success Modules help foster success in collegiate courses and prepare students for future professions. Topics such as "Time Management," "Stress Management" and "Financial Literacy" are available within the MyMathLab course. Instructors can integrate these media-rich activities with their traditional MyMathLab assignments.

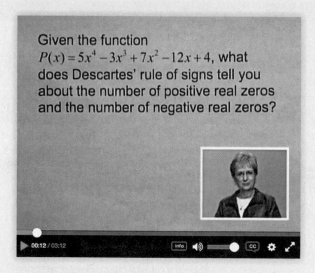

Video Assessment Exercises

Featuring authors Judy Beecher and Judy Penna, these Example Solutions videos walk students through the detailed solution process for nearly all examples from the text. Corresponding exercises check for conceptual understanding. Videos and exercises are assignable in MyMathLab.

NEW! Video Notebook

This author-created notebook contains fill-in-the-blank worksheets to accompany the authors' video examples. Key definitions, theorems, and procedures are also included. After filling in the worksheet while watching the video, the student has an excellent study guide for review and test preparation. This is available in print to accompany the text or as a download in MyMathLab.

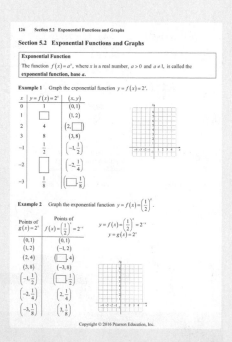

Instructor Resources

Additional resources can be downloaded from within MyMathLab and at www.pearsonhighered.com, or hardcopy resources can be ordered from your Pearson sales representative.

Annotated Instructor's Edition

The instructor's edition includes all answers to the exercise sets. Shorter answers are presented on the same page as the exercise; longer answers are in the back of the text. Sample homework assignments are indicated by a blue underline and may be assigned in MyMathLab®. Available upon request from your Pearson sales representative.

Instructor's Solutions Manual (Download Only)

Written by author Judy Penna, this resource contains worked-out solutions to all exercises in the exercise sets, Mid-Chapter Mixed Reviews, Chapter Reviews, and Chapter Tests, as well as solutions for all the Just-In-Time exercises.

PowerPoint® Lecture Slides

Feature presentations written and designed specifically for this text. These lecture slides provide an outline for presenting definitions, figures, and key examples from the text.

Online Test Bank (Download Only)

Contains four free-response test forms for each chapter following the same format and having the same level of difficulty as the test in the main text and two multiple-choice test forms for each chapter. It also provides six forms of the final examination, four with free-response questions and two with multiple-choice questions.

TestGen®

TestGen® (www.pearsoned.com/testgen) enables instructors to build, edit, print, and administer tests using a computerized bank of questions developed to cover all the objectives of the text.

Learning Catalytics Integration

MyMathLab now provides **Learning Catalytics**, an interactive student response tool that uses students' smartphones, tablets, or laptops to engage them in more sophisticated tasks and thinking. Learning Catalytics contains Pearson-created content for algebra and trigonometry that allows instructors to take advantage of this exciting technology immediately.

Student Resources

Additional resources to promote student success.

Author Example Videos

Ideal for distance learning or supplemental instruction, these videos feature authors Judy Beecher and Judy Penna working through and explaining examples in the text. Assignable in MyMathLab with new Video Assessment exercises.

Video Notebook

This notebook can accompany the text and/or MyMathLab course. It contains fill-in-the-blank worksheets to accompany the video examples presented by the authors. Key definitions, theorems, and procedures are also included. After filling in the worksheet while watching the video, the student has an excellent study guide for review and test preparation. This is available electronically for download within MyMathLab or as a printed resource.

Student's Solutions Manual

Written by author Judy Penna, this resource contains completely worked-out solutions with step-by-step annotations for all the odd-numbered exercises in the exercise sets, Mid-Chapter Mixed Reviews, and Chapter Reviews, as well as solutions for all the Chapter Test exercises and the Just-In-Time exercises.

Chapter R: Basic Concepts of Algebra

Available within MyMathLab, Chapter R supplements the prerequisite topics of the Just-in-Time review with more in-depth coverage and exercises.

- R.1 The Real-Number System
- R.2 Integer Exponents, Scientific Notation, and Order of Operations
- R.3 Addition, Subtraction, and Multiplication of Polynomials
- R.4 Factoring
- R.5 The Basics of Equation Solving
- R.6 Rational Expressions
- R.7 Radical Notation and Rational Exponents

To the Student

Guide to Success

Success can be planned. Combine goals and good study habits to create a plan for success that works for you. The following list contains study tips that your authors consider most helpful.

Skills for Success

❯ **Set goals and expect success.** Approach this class experience with a positive attitude.

❯ **Communicate with your instructor** when you need extra help.

❯ **Take your text with you to class and lab.** Each section in the text is designed with headings and boxed information that provide an outline for easy reference.

❯ **Ask questions in class, lab, and tutoring sessions.** Instructors encourage them, and other students probably have the same questions.

❯ **Begin each homework assignment as soon as possible.** If you have difficulty, you will then have the time to access supplementary resources.

❯ **Carefully read the instructions** before working homework exercises **and include all steps**.

❯ **Form a study group** with fellow students. Verbalizing questions about topics that you do not understand can clarify the material for you.

❯ After each quiz or test, **write out corrected step-by step solutions** to all missed questions. They will provide a valuable study guide for the midterm exam and the final exam.

❯ **MyMathLab has numerous tools to help you succeed.** Use MyMathLab to create a personalized study plan and practice skills with sample quizzes and tests.

❯ **Knowing math vocabulary is an important step toward success.** Review vocabulary using the Vocabulary Reinforcement exercises in the text and in MyMathLab.

❯ If you miss a lecture, **watch the video in the Multimedia Library** of MyMathLab that explains the concepts you missed.

In writing this textbook, we challenged ourselves to do everything possible to help you learn the concepts and skills contained between its covers so that you will be successful in this course and in the mathematics courses you take in the future. We realize that your time is both valuable and limited, so we communicate in a highly visual way that allows you to learn quickly and efficiently. We are confident that, if you invest an adequate amount of time in the learning process, this text will be of great value to you. We wish you a positive learning experience.

Marv Bittinger
Judy Beecher
David Ellenbogen
Judy Penna

JUST-IN-TIME Review

Throughout this text, there are Just-in-Time icons, numbered 1–28, that refer to brief reviews of the following 28 intermediate-algebra topics. Each mini-review lesson is accompanied by several exercises. All answers are provided in the answer section at the back of the text.

1. Real Numbers
2. Properties of Real Numbers
3. Order on the Number Line
4. Absolute Value
5. Operations with Real Numbers
6. Interval Notation
7. Integers as Exponents
8. Scientific Notation
9. Order of Operations
10. Introduction to Polynomials
11. Add and Subtract Polynomials
12. Multiply Polynomials
13. Special Products of Binomials
14. Factor Polynomials; The FOIL Method

15. Factor Polynomials: The *ac*-Method
16. Special Factorizations
17. Equation-Solving Principles
18. Inequality-Solving Principles
19. The Principle of Zero Products
20. The Principle of Square Roots
21. Simplify Rational Expressions
22. Multiply and Divide Rational Expressions
23. Add and Subtract Rational Expressions
24. Simplify Complex Rational Expressions
25. Simplify Radical Expressions
26. Rationalize Denominators
27. Rational Exponents
28. The Pythagorean Theorem

Just in Time

1

REAL NUMBERS

Some frequently used sets of real numbers and the relationships among them are shown below.

Real Numbers

Irrational Numbers
Examples: $\sqrt{42} = 6.480740698\ldots$, π, $-7.13133133313333\ldots$

Rational Numbers
Examples: $0, -9, \frac{7}{8}, 81.4, 9\frac{2}{3}$, $0.125, \frac{11}{6}, 0.\overline{27}$

Rational Numbers That Are Not Integers:
$\frac{2}{3}, -\frac{4}{5}, \frac{19}{-5}, \frac{-7}{8}, 8.3, 0.5\overline{6}, \ldots$

Integers:
$\ldots, -3, -2, -1, 0, 1, 2, 3, \ldots$

Negative Integers:
$-1, -2, -3, \ldots$

Whole Numbers:
$0, 1, 2, 3, \ldots$

Zero: 0

Natural Numbers:
(Positive Integers)
$1, 2, 3, \ldots$

(continued)

Numbers that can be expressed in the form p/q, where p and q are integers and $q \neq 0$, are **rational numbers**. Decimal notation for rational numbers either *terminates* (ends) or *repeats*. Each of the following is a rational number:

$$0, \qquad -17, \qquad \frac{13}{4}, \qquad \sqrt{25} = 5;$$

$$\frac{1}{4} = 0.25 \qquad \text{(terminating decimal)};$$

$$-\frac{5}{11} = -0.454545\ldots = -0.\overline{45} \qquad \text{(repeating decimal)};$$

$$\frac{5}{6} = 0.8333\ldots = 0.8\overline{3} \qquad \text{(repeating decimal)}.$$

The real numbers that are not rational are **irrational numbers**. Decimal notation for irrational numbers neither terminates nor repeats. Each of the following is an irrational number. Note in each case that there is no repeating block of digits.

$$\sqrt{2} = 1.414213562\ldots,$$
$$\sqrt{65} = 8.062257748\ldots,$$
$$-6.12122122212222\ldots,$$
$$\pi = 3.1415926535\ldots$$

$$\left(\tfrac{22}{7} \text{ and } 3.14 \text{ are } \textit{rational approximations} \text{ of the irrational number } \pi.\right)$$

The set of all rational numbers combined with the set of all irrational numbers gives us the set of **real numbers**.

> **Do Exercises 1–6.**

In Exercises 1–6, consider the numbers

$$\tfrac{2}{3}, \quad 6, \quad \sqrt{3}, \quad -2.45, \quad \sqrt[6]{26}, \quad 18.\overline{4},$$
$$-11, \quad \sqrt[3]{27}, \quad 5\tfrac{1}{6}, \quad 7.151551555\ldots,$$
$$-\sqrt{35}, \quad \sqrt[5]{3}, \quad -\tfrac{8}{7}, \quad 0, \quad \sqrt{16}.$$

1. Which are rational numbers?

2. Which are rational numbers but not integers?

3. Which are irrational numbers?

4. Which are integers?

5. Which are whole numbers?

6. Which are real numbers?

All answers to the Just-in-Time exercises appear at the end of the answers at the back of the book.

PROPERTIES OF REAL NUMBERS

For any real numbers a, b, and c:

$a + b = b + a$ and $ab = ba$	Commutative properties of addition and multiplication
$a + (b + c) = (a + b) + c$ and $a(bc) = (ab)c$	Associative properties of addition and multiplication
$a + 0 = 0 + a = a$	Additive identity property
$-a + a = a + (-a) = 0$	Additive inverse property
$a \cdot 1 = 1 \cdot a = a$	Multiplicative identity property
$a \cdot \dfrac{1}{a} = \dfrac{1}{a} \cdot a = 1 \ (a \neq 0)$	Multiplicative inverse property
$a(b + c) = ab + ac$ and $a(b - c) = ab - ac$	Distributive property

EXAMPLES Name the property illustrated in each sentence.

1. $8 \cdot 5 = 5 \cdot 8$ — Commutative property of multiplication

2. $14 + (-14) = 0$ — Additive inverse property

3. $2(a - b) = 2a - 2b$ — Distributive property

4. $5 + (m + n) = (5 + m) + n$ — Associative property of addition

5. $6 \cdot 1 = 1 \cdot 6 = 6$ — Multiplicative identity property

6. $q + t = t + q$ — Commutative property of addition

7. $\dfrac{7}{9} + 0 = \dfrac{7}{9}$ — Additive identity property

8. $-2(cd) = (-2c)d$ — Associative property of multiplication

9. $3(x + y) = 3x + 3y$ — Distributive property

10. $8 \cdot \dfrac{1}{8} = 1$ — Multiplicative inverse property

> **Do Exercises 1–10.**

Name the property illustrated by the sentence.

1. $-24 + 24 = 0$

2. $7(xy) = (7x)y$

3. $9(r - s) = 9r - 9s$

4. $11 + z = z + 11$

5. $-20 \cdot 1 = -20$

6. $5(x + y) = (x + y)5$

7. $q + 0 = q$

8. $75 \cdot \dfrac{1}{75} = 1$

9. $(x + y) + w = x + (y + w)$

10. $8(a + b) = 8a + 8b$

All answers to the Just-in-Time exercises appear at the end of the answers at the back of the book.

Just in Time 3

ORDER ON THE NUMBER LINE

The real numbers are modeled using a **number line,** as shown below. Each point on the line represents a real number, and every real number is represented by a point on the line.

$$-2.9 \qquad -\frac{3}{5} \qquad \sqrt{3} \qquad \pi \qquad \frac{17}{4}$$

⟵—+—+—+—+—•—+—•—+—+—+—•—+—•—•—+—⟶
-5 -4 -3 -2 -1 0 1 2 3 4 5

The order of the real numbers can be determined from the number line. If a number a is to the left of a number b, then a **is less than** b $(a < b)$. Similarly, a **is greater than** $b\,(a > b)$ if a is to the right of b on the number line. For example, we see from the number line above that $-2.9 < -\frac{3}{5}$, because -2.9 is to the left of $-\frac{3}{5}$. Also, $\frac{17}{4} > \sqrt{3}$, because $\frac{17}{4}$ is to the right of $\sqrt{3}$.

The statement $a \le b$, read "a is less than or equal to b," is true if either $a < b$ is true or $a = b$ is true. A similar statement holds for $a \ge b$.

▶ *Do Exercises 1–6.*

Classify the inequality as true or false.

1. $9 < -9$

2. $-10 \le -1$

3. $-\sqrt{26} < -5$

4. $\sqrt{6} \ge \sqrt{6}$

5. $-30 > -25$

6. $-\dfrac{4}{5} > -\dfrac{5}{4}$

Just in Time 4

ABSOLUTE VALUE

The **absolute value** of a number a, denoted $|a|$, is its distance from 0 on the number line. For example, $|-5| = 5$, because the distance of -5 from 0 is 5. For any real number a,

$$|a| = a \quad \text{if } a \ge 0 \qquad \text{and} \qquad |a| = -a \quad \text{if } a < 0.$$

EXAMPLES Simplify.

1. $|-10| = 10$ 2. $|0| = 0$ 3. $\left|\dfrac{4}{9}\right| = \dfrac{4}{9}$

Absolute value can be used to find the distance between two points on the number line. For any real numbers a and b, the distance between a and b is $|a - b|$ or, equivalently, $|b - a|$.

EXAMPLE 4 Find the distance between -2 and 3.

$$|-2 - 3| = |-5| = 5,$$

or equivalently,

$$|3 - (-2)| = |3 + 2| = |5| = 5$$

▶ *Do Exercises 1–8.*

Simplify.

1. $|-98|$ 2. $|0|$

3. $|4.7|$ 4. $\left|-\dfrac{2}{3}\right|$

Find the distance between the given pair of points on the number line.

5. $-7, \ 13$ 6. $2, \ 14.6$

7. $-39, \ -28$ 8. $-\dfrac{3}{4}, \dfrac{15}{8}$

RULES FOR OPERATIONS WITH REAL NUMBERS

Addition

- *Positive Numbers*: Add the same way that we add arithmetic numbers. The answer is positive.
- *Negative Numbers*: Add absolute values. The answer is negative.
- *A Positive Number and a Negative Number:* If the numbers have the same absolute value, the answer is 0. If the numbers have different absolute values, subtract the smaller absolute value from the larger. If the positive number has the greater absolute value, the answer is positive. If the negative number has the greater absolute value, the answer is negative.

Subtraction

- To subtract, add the opposite, or additive inverse, of the number being subtracted.

Multiplication and Division, where the divisor is nonzero

- Multiply or divide the absolute values. If the signs are the same, the answer is positive. If the signs are different, the answer is negative.

EXAMPLES Add.

1. $9 + (-29) = -20$

2. $-9 + (-29) = -38$

3. $-9 + 29 = 20$

EXAMPLES Subtract.

4. $15 - 6 = 15 + (-6) = 9$

5. $15 - (-6) = 15 + 6 = 21$

6. $-15 - 6 = -15 + (-6) = -21$

7. $-15 - (-6) = -15 + 6 = -9$

EXAMPLES Multiply or divide.

8. $-5 \cdot 20 = -100$ 9. $32 \div (-4) = -8$

10. $-32 \div 4 = -8$ 11. $-32 \div (-4) = 8$

12. $-5 \cdot (-20) = 100$ 13. $5 \cdot 20 = 100$

> **Do Exercises 1–15.**

Compute and simplify.

1. $8 - (-11)$

2. $-\dfrac{3}{10} \cdot \left(-\dfrac{1}{3}\right)$

3. $15 \div (-3)$

4. $-4 - (-1)$

5. $7 \cdot (-50)$

6. $-0.5 - 5$

7. $-3 + 27$

8. $-400 \div (-40)$

9. $4.2 \cdot (-3)$

10. $-13 - (-33)$

11. $-60 + 45$

12. $\dfrac{1}{2} - \dfrac{2}{3}$

13. $-24 \div 3$

14. $-6 + (-16)$

15. $-\dfrac{1}{2} \div \left(-\dfrac{5}{8}\right)$

JUST-IN-TIME Review

INTERVAL NOTATION

Sets of real numbers can be expressed using **interval notation**. For example, for real numbers a and b such that $a < b$, the **open interval** (a, b) is the set of real numbers between, but not including, a and b.

Some intervals extend without bound in one or both directions. The interval $[a, \infty)$, for example, begins at a and extends to the right without bound. The bracket indicates that a is included in the interval.

SET NOTATION	INTERVAL NOTATION	GRAPH
$\{x \mid a < x < b\}$	(a, b)	
$\{x \mid a \le x \le b\}$	$[a, b]$	
$\{x \mid a \le x < b\}$	$[a, b)$	
$\{x \mid a < x \le b\}$	$(a, b]$	
$\{x \mid x > a\}$	(a, ∞)	
$\{x \mid x \ge a\}$	$[a, \infty)$	
$\{x \mid x < b\}$	$(-\infty, b)$	
$\{x \mid x \le b\}$	$(-\infty, b]$	
$\{x \mid x \text{ is a real number}\}$	$(-\infty, \infty)$	

> **Do Exercises 1–10.**

Write interval notation.

1. $\{x \mid -5 \le x \le 5\}$

2. $\{x \mid -3 < x \le -1\}$

3. $\{x \mid x \le -2\}$

4. $\{x \mid x > 3.8\}$

5. $\{x \mid 7 < x\}$

6. $\{x \mid -2 < x < 2\}$

Write interval notation for the graph.

7. ←─(─┼─┼─┼─┼─┼─┼─┼─┼─)─→
   −5−4−3−2−1 0 1 2 3 4 5

8. 1.7
   ←─┼─┼─┼─┼─┼─┼─┼─█─┼─┼─┼─→
   −5−4−3−2−1 0 1 2 3 4 5

9. ←─(─┼─┼─┼─█─┼─┼─┼─┼─┼─┼─→
   −5−4−3−2−1 0 1 2 3 4 5

10. √5
    ←─┼─┼─┼─┼─┼─┼─┼─┼─)─┼─┼─→
    −5−4−3−2−1 0 1 2 3 4 5

INTEGERS AS EXPONENTS

When a positive integer is used as an *exponent*, it indicates the number of times that a factor appears in a product. For example, 7^3 means $7 \cdot 7 \cdot 7$, where 7 is the **base** and 3 is the **exponent**.

For any nonzero numbers a and b and any integers m and n,

$$a^0 = 1, \qquad a^{-m} = \frac{1}{a^m}, \quad \text{and} \quad \frac{a^{-m}}{b^{-n}} = \frac{b^n}{a^m}.$$

PROPERTIES OF EXPONENTS

For any real numbers a and b and any integers m and n, assuming 0 is not raised to a nonpositive power:

$a^m \cdot a^n = a^{m+n}$	Product rule
$\dfrac{a^m}{a^n} = a^{m-n} \ (a \neq 0)$	Quotient rule
$(a^m)^n = a^{mn}$	Power rule
$(ab)^m = a^m b^m$	Raising a product to a power
$\left(\dfrac{a}{b}\right)^m = \dfrac{a^m}{b^m} \ (b \neq 0)$	Raising a quotient to a power

EXAMPLES   Simplify each of the following.

1. $4^2 \cdot 4^{-5} = 4^{2+(-5)} = 4^{-3}$, or $\dfrac{1}{4^3}$

2. $\left(\dfrac{7}{9}\right)^0 = 1$

3. $(8^2)^{-5} = 8^{2(-5)} = 8^{-10}$, or $\dfrac{1}{8^{10}}$

4. $\dfrac{x^{11}}{x^4} = x^{11-4} = x^7$

5. $\left(\dfrac{a}{b}\right)^3 = \dfrac{a^3}{b^3}$

6. $(cd)^{-2} = c^{-2}d^{-2}$, or $\dfrac{1}{c^2 d^2}$

> **Do Exercises 1–10.**

Simplify.

1. 3^{-6}

2. $\dfrac{1}{(0.2)^{-5}}$

3. $\dfrac{w^{-4}}{z^{-9}}$

4. $\left(\dfrac{z}{y}\right)^2$

5. 100^0

6. $\dfrac{a^5}{a^{-3}}$

7. $(2xy^3)(-3x^{-5}y)$

8. $x^{-4} \cdot x^{-7}$

9. $(mn)^{-6}$

10. $(t^{-5})^4$

Just in Time 8   **SCIENTIFIC NOTATION**

We can use scientific notation to name both very large and very small positive numbers and to perform computations.

SCIENTIFIC NOTATION

Scientific notation for a number is an expression of the type

$$N \times 10^m,$$

where $1 \leq N < 10$, N is in decimal notation, and m is an integer.

EXAMPLES

Convert to scientific notation.

1. 9,460,000,000,000

   Since 9,460,000,000,000 is a large number, the exponent on 10 will be positive. We want to move the decimal point between 9 and 4. This is a move of 12 places, so the exponent will be 12.

   $$9,460,000,000,000 = 9.46 \times 10^{12}$$

2. 0.000073

   Since 0.000073 is a small number, the exponent on 10 will be negative. We want to move the decimal point between 7 and 3. This is a move of 5 places, so the exponent will be -5.

   $$0.000073 = 7.3 \times 10^{-5}$$

Convert to decimal notation.

3. 5.4×10^7

   The exponent is positive. We will move the decimal point 7 places to the right.

   $$5.4 \times 10^7 = 54,000,000$$

4. 3.819×10^{-3}

   The exponent is negative. We will move the decimal point 3 places to the left.

   $$3.819 \times 10^{-3} = 0.003819$$

> **Do Exercises 1–8.**

Convert to scientific notation.

1. 18,500,000

2. 0.000786

3. 0.0000000023

4. 8,927,000,000

Convert to decimal notation.

5. 4.3×10^{-8}

6. 5.17×10^6

7. 6.203×10^{11}

8. 2.94×10^{-5}

Recall that to simplify the expression $3 + 4 \cdot 5$, first we multiply 4 and 5 to get 20 and then we add 3 to get 23. Mathematicians have agreed on the following procedure, or rules for order of operations.

RULES FOR ORDER OF OPERATIONS

1. Do all calculations within grouping symbols before operations outside. When nested grouping symbols are present, work from the inside out.
2. Evaluate all exponential expressions.
3. Do all multiplications and divisions in order from left to right.
4. Do all additions and subtractions in order from left to right.

EXAMPLES

1. $8(5 - 3)^3 - 20$

$= 8 \cdot 2^3 - 20$

$= 8 \cdot 8 - 20$

$= 64 - 20$

$= 44$

2. $10[7 - 4(8 - 5)]$

$= 10[7 - 4 \cdot 3]$

$= 10[7 - 12]$

$= 10[-5]$

$= -50$

3. $-32 \div 2 \times (-4) \div (-2)$

$= -16 \times (-4) \div (-2)$

$= 64 \div (-2)$

$= -32$

4. $\dfrac{10 \div (8 - 6) + 9 \cdot 4}{2^5 + 3^2}$

$= \dfrac{10 \div 2 + 9 \cdot 4}{32 + 9}$

$= \dfrac{5 + 36}{41}$

$= \dfrac{41}{41}$

$= 1$

5. $3^{12} \cdot 3^{-4} \div 3^3 \cdot 3^{-1}$

$= 3^8 \div 3^3 \cdot 3^{-1}$

$= 3^5 \cdot 3^{-1}$

$= 3^4$

$= 81$

> **Do Exercises 1–8.**

Calculate.

1. $3 + 18 \div 6 - 3$

2. $5 \cdot 3 + 8 \cdot 3^2 + 4(6 - 2)$

3. $5(3 - 8 \cdot 3^2 + 4 \cdot 6 - 2)$

4. $16 \div 4 \cdot 4 \div 2 \cdot 256$

5. $2^6 \cdot 2^{-3} \div 2^{10} \div 2^{-8}$

6. $\dfrac{4(8 - 6)^2 - 4 \cdot 3 + 2 \cdot 8}{3^1 + 19^0}$

7. $64 \div [(-4) \div (-2)]$

8. $6[9 - (3 - 2)] + 4(2 - 3)$

Just in Time 10

INTRODUCTION TO POLYNOMIALS

Polynomials are a type of algebraic expression that you will often encounter in your study of algebra. Some examples of polynomials are

$$3x - 4y, \quad 5y^3 - \tfrac{7}{3}y^2 + 3y - 2, \quad -2.3a^4,$$
$$16, \quad \text{and} \quad z^6 - \sqrt{5}.$$

Algebraic expressions like $8x - 13$, $x^2 + 3x - 4$, and $3a^5 - 11 + a$ are **polynomials in one variable.** Algebraic expressions like $3ab^3 - 8$ and $5x^4y^2 - 3x^3y^8 + 7xy^2 + 6$ are **polynomials in several variables.** The terms of a polynomial are separated by $+$ signs. The terms of $2x^2 - 8x + 3 = 2x^2 + (-8x) + 6$ are $2x^2$, $-8x$, and 3. The **degree of a term** is the sum of the exponents of the variables in that term. The **degree of a polynomial** is the degree of the term of highest degree.

A polynomial with just one term, like $-9y^6$, is a **monomial.** If a polynomial has two terms, like $x^2 + 4$, it is a **binomial.** A polynomial with three terms, like $4x^2 - 4xy + 1$, is a **trinomial.**

EXAMPLES   Determine the degree of the polynomial.

1. $2x^3 - 1$ Degree: 3

2. $-5 \quad (-5 = -5x^0)$ Degree: 0

3. $w^2 - 3.5 + 4w^5 = 4w^5 + w^2 - 3.5$ Degree: 5

4. $7xy^3 - 16x^2y^4$ Degree: $2 + 4$, or 6

> *Do Exercises 1–8.*

Determine the degree of the polynomial.

1. $5 - x^6$

2. $x^2y^5 - x^7y + 4$

3. $2a^4 - 3 + a^2$

4. -41

5. $4x - x^3 + 0.1x^8 - 2x^5$

Classify the polynomial as a monomial, a binomial, or a trinomial.

6. $x - 3$

7. $14y^5$

8. $2y - \dfrac{1}{4}y^2 + 8$

Just in Time 11

ADD AND SUBTRACT POLYNOMIALS

If two terms of an expression have the same variables raised to the same powers, they are called **like terms**, or **similar terms**. We can **combine,** or **collect, like terms** using the distributive property. For example, $3y^2$ and $5y^2$ are like terms and $3y^2 + 5y^2 = (3 + 5)y^2 = 8y^2$. We add or subtract polynomials by combining like terms.

EXAMPLES   Add or subtract each of the following.

1. $(-5x^3 + 3x^2 - x) + (12x^3 - 7x^2 + 3)$
$= (-5x^3 + 12x^3) + (3x^2 - 7x^2) - x + 3$
$= (-5 + 12)x^3 + (3 - 7)x^2 - x + 3$
$= 7x^3 - 4x^2 - x + 3$

2. $(6x^2y^3 - 9xy) - (5x^2y^3 - 4xy)$
$= 6x^2y^3 - 9xy - 5x^2y^3 + 4xy$
$= x^2y^3 - 5xy$

> *Do Exercises 1–5.*

Add or subtract.

1. $(8y - 1) - (3 - y)$

2. $(3x^2 - 2x - x^3 + 2)$
$\quad - (5x^2 - 8x - x^3 + 4)$

3. $(2x + 3y + z - 7)$
$\quad + (4x - 2y - z + 8)$
$\quad + (-3x + y - 2z - 4)$

4. $(3ab^2 + 4a^2b - 2ab + 6)$
$\quad + (-ab^2 - 5a^2b + 8ab + 4)$

5. $(5x^2 + 4xy - 3y^2 + 2)$
$\quad - (9x^2 - 4xy + 2y^2 - 1)$

MULTIPLY POLYNOMIALS

To multiply monomials, we first multiply their coefficients, and then we multiply their variables.

EXAMPLES

1. $(-2x^3)(5x^4) = (-2 \cdot 5)(x^3 \cdot x^4) = -10x^7$

2. $(3yz^2)(8y^3z^5) = (3 \cdot 8)(y \cdot y^3)(z^2 \cdot z^5) = 24y^4z^7$

We can find the product of two binomials by multiplying the **F**irst terms, then the **O**uter terms, then the **I**nner terms, then the **L**ast terms. Then we combine like terms, if possible. This procedure is sometimes called **FOIL**.

EXAMPLE 3 Multiply: $(2x - 7)(3x + 4)$.

$$
\begin{array}{cccc}
\mathbf{F} & \mathbf{L} & \mathbf{F} \quad \mathbf{O} \quad \mathbf{I} \quad \mathbf{L} \\
(2x - 7)(3x + 4) & = & 6x^2 + 8x - 21x - 28 \\
& = & 6x^2 - 13x - 28 \\
\mathbf{I} & & \\
\mathbf{O} & &
\end{array}
$$

EXAMPLE 4 Multiply: $(5c - d)(4c - 9d)$.

$$(5c - d)(4c - 9d) = 20c^2 - 45cd - 4cd + 9d^2$$
$$= 20c^2 - 49cd + 9d^2$$

> **Do Exercises 1–6.**

Multiply.

1. $(3a^2)(-7a^4)$

2. $(y - 3)(y + 5)$

3. $(x + 6)(x + 3)$

4. $(2a + 3)(a + 5)$

5. $(2x + 3y)(2x + y)$

6. $(11t - 1)(3t + 4)$

SPECIAL PRODUCTS OF BINOMIALS

SPECIAL PRODUCTS OF BINOMIALS

$(A + B)^2 = A^2 + 2AB + B^2$ Square of a sum

$(A - B)^2 = A^2 - 2AB + B^2$ Square of a difference

$(A + B)(A - B) = A^2 - B^2$ Product of a sum and a difference

EXAMPLES

1. $(4x + 1)^2 = (4x)^2 + 2 \cdot 4x \cdot 1 + 1^2$
$$= 16x^2 + 8x + 1$$

2. $(3y^2 - 2)^2 = (3y^2)^2 - 2 \cdot 3y^2 \cdot 2 + 2^2$
$$= 9y^4 - 12y^2 + 4$$

3. $(x^2 + 3y)(x^2 - 3y) = (x^2)^2 - (3y)^2$
$$= x^4 - 9y^2$$

> **Do Exercises 1–6.**

Multiply.

1. $(x + 3)^2$

2. $(5x - 3)^2$

3. $(2x + 3y)^2$

4. $(a - 5b)^2$

5. $(n + 6)(n - 6)$

6. $(3y + 4)(3y - 4)$

FACTOR POLYNOMIALS; THE FOIL METHOD

When a polynomial is to be factored, we should always look first to factor out a factor that is common to all the terms using the distributive property. We generally look for the constant common factor with the largest absolute value and for variables with the largest exponent common to all the terms.

EXAMPLE 1   Factor: $15 + 10x - 5x^2$.

$$15 + 10x - 5x^2 = 5 \cdot 3 + 5 \cdot 2x - 5 \cdot x^2 = 5(3 + 2x - x^2)$$

In some polynomials, pairs of terms have a common binomial factor that can be removed in a process called **factoring by grouping**.

EXAMPLE 2   Factor: $x^3 + 3x^2 - 5x - 15$.

$$\begin{aligned}
x^3 + 3x^2 - 5x - 15 &= (x^3 + 3x^2) + (-5x - 15) \\
&= x^2(x + 3) - 5(x + 3) \\
&= (x + 3)(x^2 - 5)
\end{aligned}$$

Some trinomials can be factored into the product of two binomials. To factor a trinomial of the form $x^2 + bx + c$, we look for binomial factors of the form $(x + p)(x + q)$, where $p \cdot q = c$ and $p + q = b$. That is, we look for two numbers p and q whose sum is the coefficient of the middle term of the polynomial, b, and whose product is the constant term, c.

EXAMPLES   Factor.

3. $x^2 + 5x + 6 = (x + 2)(x + 3)$

4. $x^4 - 6x^3 + 8x^2 = x^2(x^2 - 6x + 8) = x^2(x - 2)(x - 4)$

To factor trinomials of the type $ax^2 + bx + c, a \neq 1$, using the **FOIL method:**

1. Factor out the largest common factor.
2. Find two First terms whose product is ax^2:

$$(\ \square x + \ )(\ \square x + \ ) = ax^2 + bx + c.$$
$$\underline{\hspace{2cm}}\text{ FOIL}$$

3. Find two Last terms whose product is c:

$$(\ x + \square)(\ x + \square) = ax^2 + bx + c$$
$$\underline{\hspace{2cm}}\text{ FOIL}$$

4. Repeat steps (2) and (3) until a combination is found for which the sum of the Outer product and the Inner product is bx:

$$(\ \square x + \square)(\ \square x + \square) = ax^2 + bx + c.$$
$$\underline{\hspace{2cm}}\text{ FOIL}$$
$$\text{I} \quad \text{O}$$

(continued)

EXAMPLES Factor.

5. $3x^2 - 10x - 8 = (3x + 2)(x - 4)$

6. $12y^2 + 44y - 45 = (2y + 9)(6y - 5)$

7. $r^2 - 7rs + 6s^2 = (r - 6s)(r - s)$

8. $y^4 - 3y^2 - 40 = (y^2 + 5)(y^2 - 8)$

> **Do Exercises 1–12.**

Factor out the largest common factor.

1. $3x + 18$

2. $2z^3 - 8z^2$

Factor by grouping.

3. $3x^3 - x^2 + 18x - 6$

4. $t^3 + 6t^2 - 2t - 12$

Factor the trinomial.

5. $w^2 - 7w + 10$

6. $t^2 + 8t + 15$

7. $2n^2 - 20n - 48$

8. $y^4 - 9y^3 + 14y^2$

9. $2n^2 + 9n - 56$

10. $2y^2 + y - 6$

11. $b^2 - 6bt + 5t^2$

12. $x^4 - 7x^2 - 30$

Just in Time

15

FACTOR POLYNOMIALS: THE *ac*-METHOD

A second method of factoring trinomials of the type
$ax^2 + bx + c, a \neq 1$, is known as the **ac-method**, or the
grouping method.

> ### THE *ac*-METHOD FOR FACTORING TRINOMIALS
>
> 1. Factor out the largest common factor. The remaining
>   trinomial is $ax^2 + bx + c$.
> 2. Multiply the leading coefficient a and the constant c.
> 3. Try to factor the product ac so that the sum of the factors is b.
>   That is, find integers p and q such that $pq = ac$ and $p + q = b$.
> 4. Split the middle term, writing it as a sum using the factors
>   found in step (3).
> 5. Factor by grouping.

EXAMPLE Factor: $6x^2 + 23x + 20$.

There is no common factor other than 1 or -1. We multiply the
leading coefficient, 6, and the constant, 20: $6 \cdot 20 = 120$. Then we
look for a factorization of 120 in which the sum of the factors is the
coefficient of the middle term, 23. That factorization is $8 \cdot 15$.

(continued)

Split the middle term: $23x = 8x + 15x$.

Factor by grouping:

$$6x^2 + 23x + 20 = 6x^2 + 8x + 15x + 20$$
$$= 2x(3x + 4) + 5(3x + 4)$$
$$= (3x + 4)(2x + 5).$$

▶ **Do Exercises 1–3.**

Just
in
Time

16

SPECIAL FACTORIZATIONS

SPECIAL FACTORIZATIONS

- Trinomial Squares:

$$A^2 + 2AB + B^2 = (A + B)^2;$$
$$A^2 - 2AB + B^2 = (A - B)^2$$

- Difference of Squares:

$$A^2 - B^2 = (A + B)(A - B)$$

- Sum or Difference of Cubes:

$$A^3 + B^3 = (A + B)(A^2 - AB + B^2);$$
$$A^3 - B^3 = (A - B)(A^2 + AB + B^2)$$

EXAMPLES

1. $x^2 - 16 = x^2 - 4^2 = (x + 4)(x - 4)$

2. $x^2 + 8x + 16 = x^2 + 2 \cdot x \cdot 4 + 4^2$
$$= (x + 4)^2$$

3. $25y^2 - 30y + 9 = (5y)^2 - 2 \cdot 5y \cdot 3 + 3^2$
$$= (5y - 3)^2$$

4. $x^3 + 27 = x^3 + 3^3 = (x + 3)(x^2 - 3x + 9)$

5. $16y^3 - 250 = 2(8y^3 - 125)$
$$= 2[(2y)^3 - 5^3]$$
$$= 2(2y - 5)(4y^2 + 10y + 25)$$

▶ **Do Exercises 1–10.**

Factor.

1. $8x^2 - 6x - 9$

2. $10t^2 + 4t - 6$

3. $18a^2 - 51a + 15$

Factor the difference of squares.

1. $z^2 - 81$

2. $16x^2 - 9$

3. $7pq^4 - 7py^4$

Factor the square of a binomial.

4. $x^2 + 12x + 36$

5. $9z^2 - 12z + 4$

6. $a^3 + 24a^2 + 144a$

Factor the sum or the difference of cubes.

7. $x^3 + 64$

8. $m^3 - 216$

9. $3a^5 - 24a^2$

10. $t^6 + 1$

For any real numbers a, b, and c,

THE ADDITION PRINCIPLE

If $a = b$ is true, then $a + c = b + c$ is true.

THE MULTIPLICATION PRINCIPLE

If $a = b$ is true, then $ac = bc$ is true.

EXAMPLES Solve.

1.

$$y - 11 = 12$$
$$y - 11 + 11 = 12 + 11$$
$$y = 23$$

Check:

$$\frac{y - 11 = 12}{23 - 11 \overset{?}{} 12}$$
$$12 \mid 12 \quad \text{TRUE}$$

The solution is 23.

2.

$$15c = 90$$
$$\frac{1}{15} \cdot 15c = \frac{1}{15} \cdot 90$$
$$c = 6$$

Check:

$$\frac{15c = 90}{15 \cdot 6 \overset{?}{} 90}$$
$$90 \mid 90 \quad \text{TRUE}$$

The solution is 6.

3.

$$\frac{1}{4}x + 5 = 8$$
$$\frac{1}{4}x + 5 - 5 = 8 - 5$$
$$\frac{1}{4}x = 3$$
$$4 \cdot \frac{1}{4}x = 4 \cdot 3$$
$$x = 12$$

Check:

$$\frac{1}{4}x + 5 = 8$$
$$\frac{}{\frac{1}{4} \cdot 12 + 5 \overset{?}{} 8}$$
$$3 + 5 $$
$$8 \mid 8 \quad \text{TRUE}$$

The solution is 12.

4.

$$2x + 3 = 1 - 6(x - 1)$$
$$2x + 3 = 1 - 6x + 6$$
$$2x + 3 = 7 - 6x$$
$$2x + 3 + 6x = 7 - 6x + 6x$$
$$8x + 3 = 7$$
$$8x + 3 - 3 = 7 - 3$$
$$8x = 4$$
$$\frac{8x}{8} = \frac{4}{8}$$
$$x = \frac{1}{2}$$

Check:

$$\frac{2x + 3 = 1 - 6(x - 1)}{2\left(\frac{1}{2}\right) + 3 \overset{?}{} 1 - 6\left(\frac{1}{2} - 1\right)}$$
$$1 + 3 \mid 1 + 3$$
$$4 \mid 4 \quad \text{TRUE}$$

The solution is $\frac{1}{2}$.

> **Do Exercises 1–8.**

Solve.

1. $7t = 70$

2. $x - 5 = 7$

3. $3x + 4 = -8$

4. $6x - 15 = 45$

5. $7y - 1 = 23 - 5y$

6. $3m - 7 = -13 + m$

7. $2(x + 7) = 5x + 14$

8. $5y - 4(2y - 10) = 25$

Just in Time

18

INEQUALITY-SOLVING PRINCIPLES

For any real numbers a, b, and c:

THE ADDITION PRINCIPLE FOR INEQUALITIES

If $a < b$ is true, then $a + c < b + c$ is true.

THE MULTIPLICATION PRINCIPLE FOR INEQUALITIES

a) If $a < b$ and $c > 0$ are true, then $ac < bc$ is true.

b) If $a < b$ and $c < 0$ are true, then $ac > bc$ is true.

(When both sides of an inequality are multiplied by a negative number, the inequality symbol must be reversed.)

Similar statements hold for $a \leq b$.

Solve.

1. $p + 25 \geq -100$

2. $-\dfrac{2}{3}x > 6$

3. $9x - 1 < 17$

4. $-x - 16 \geq 40$

5. $\dfrac{1}{3}y - 6 < 3$

6. $8 - 2w \leq -14$

EXAMPLES Solve.

1.
$$a + 9 \leq -50$$
$$a + 9 - 9 \leq -50 - 9$$
$$a \leq -59$$
The solution set is $(-\infty, -59]$.

2.
$$-5x < 4$$
$$-\tfrac{1}{5}(-5x) > -\tfrac{1}{5}(4)$$
$$x > -\tfrac{4}{5}$$
The solution set is $\left(-\tfrac{4}{5}, \infty\right)$.

3.
$$2y - 1 < 5$$
$$2y - 1 + 1 < 5 + 1$$
$$2y < 6$$
$$\tfrac{1}{2} \cdot 2y < \tfrac{1}{2} \cdot 6$$
$$y < 3$$
The solution set is $(-\infty, 3)$.

4.
$$4 - 3x \geq 13$$
$$-4 + 4 - 3x \geq -4 + 13$$
$$-3x \geq 9$$
$$\frac{-3x}{-3} \leq \frac{9}{-3}$$
$$x \leq -3$$
The solution set is $(-\infty, -3]$.

> **Do Exercises 1–6.**

The product of two numbers is 0 if one or both of the numbers is 0. Furthermore, *if any product is 0, then a factor must be* 0. For example:

If $7x = 0$, then we know that $x = 0$.

If $x(2x - 9) = 0$, then we know that $x = 0$ or $2x - 9 = 0$.

If $(x + 3)(x - 2) = 0$, then we know that $x + 3 = 0$ or $x - 2 = 0$.

THE PRINCIPLE OF ZERO PRODUCTS

If $ab = 0$ is true, then $a = 0$ or $b = 0$, and if $a = 0$ or $b = 0$, then $ab = 0$.

Some quadratic equations can be solved using the principle of zero products.

EXAMPLES Solve.

1. $(2q - 7)(q + 4) = 0$

$$2q - 7 = 0 \quad or \quad q + 4 = 0$$
$$2q = 7 \quad or \quad q = -4$$
$$q = \frac{7}{2} \quad or \quad q = -4$$

The solutions are -4 and $\frac{7}{2}$.

2. $5x^2 - 75x = 0$

$$5x(x - 15) = 0$$
$$5x = 0 \quad or \quad x - 15 = 0$$
$$x = 0 \quad or \quad x = 15$$

The solutions are 0 and 15.

3. $x^2 - 3x - 4 = 0$

$$(x + 1)(x - 4) = 0$$
$$x + 1 = 0 \quad or \quad x - 4 = 0$$
$$x = -1 \quad or \quad x = 4$$

The solutions are -1 and 4.

4. $2x^2 + 5x - 3 = 0$

$$(x + 3)(2x - 1) = 0$$
$$x + 3 = 0 \quad or \quad 2x - 1 = 0$$
$$x = -3 \quad or \quad 2x = 1$$
$$x = -3 \quad or \quad x = \tfrac{1}{2}$$

The solutions are -3 and $\tfrac{1}{2}$.

Solve.

1. $2y^2 + 42y = 0$

2. $(a + 7)(a - 1) = 0$

3. $(5y + 3)(y - 4) = 0$

4. $6x^2 + 7x - 5 = 0$

5. $t(t - 8) = 0$

6. $x^2 - 8x - 33 = 0$

7. $x^2 + 13x = 30$

8. $12x^2 - 7x - 12 = 0$

> **Do Exercises 1–8.**

THE PRINCIPLE OF SQUARE ROOTS

The principle of square roots can be used to solve some quadratic equations.

> ### THE PRINCIPLE OF SQUARE ROOTS
> If $x^2 = k$, then $x = -\sqrt{k}$ or $x = \sqrt{k}$.

EXAMPLES Solve.

1.
$$s^2 - 144 = 0$$
$$s^2 = 144$$
$$s = -\sqrt{144} \quad or \quad s = \sqrt{144}$$
$$s = -12 \quad\quad or \quad s = 12$$

The solutions are -12 and 12, or ± 12.

2.
$$3x^2 - 21 = 0$$
$$3x^2 = 21$$
$$x^2 = 7$$
$$x = -\sqrt{7} \quad or \quad x = \sqrt{7}$$

The solutions are $-\sqrt{7}$ and $\sqrt{7}$, or $\pm\sqrt{7}$.

> **Do Exercises 1–6.**

Solve.

1. $x^2 - 36 = 0$

2. $2y^2 - 20 = 0$

3. $6z^2 = 18$

4. $3t^2 - 15 = 0$

5. $z^2 - 1 = 24$

6. $5x^2 - 75 = 0$

A **rational expression** is the quotient of two polynomials. The **domain** of an algebraic expression is the set of all real numbers for which the expression is defined. Since division by 0 is not defined, any number that makes the denominator 0 is not in the domain of a rational expression.

EXAMPLE 1 Find the domain of

$$\frac{x^2 - 4}{x^2 - 4x - 5}.$$

We solve the equation $x^2 - 4x - 5 = 0$, or $(x + 1)(x - 5) = 0$, to find the numbers that are not in the domain. The solutions are -1 and 5. Since the denominator is 0 when $x = -1$ or $x = 5$, the domain is the set of all real numbers except -1 and 5.

EXAMPLE 2 Simplify: $\dfrac{9x^2 + 6x - 3}{12x^2 - 12}.$

$$\frac{9x^2 + 6x - 3}{12x^2 - 12} = \frac{3(3x^2 + 2x - 1)}{12(x^2 - 1)}$$

$$= \frac{3(x + 1)(3x - 1)}{3 \cdot 4(x + 1)(x - 1)}$$

$$= \frac{3(x + 1)}{3(x + 1)} \cdot \frac{3x - 1}{4(x - 1)}$$

$$= 1 \cdot \frac{3x - 1}{4(x - 1)}$$

$$= \frac{3x - 1}{4(x - 1)}$$

Canceling is a shortcut that is often used to remove a factor of 1.

EXAMPLE 3 Simplify: $\dfrac{2 - x}{x^2 + x - 6}.$

$$\frac{2 - x}{x^2 + x - 6} = \frac{2 - x}{(x + 3)(x - 2)}$$

$$= \frac{-1(x - 2)}{(x + 3)(x - 2)}$$

$$= \frac{-1(x - 2)}{(x + 3)(x - 2)}$$

$$= \frac{-1}{x + 3}, \text{ or } -\frac{1}{x + 3}$$

Find the domain of the rational expression.

1. $\dfrac{3x - 3}{x(x - 1)}$

2. $\dfrac{y + 6}{y^2 + 4y - 21}$

Simplify.

3. $\dfrac{x^2 - 4}{x^2 - 4x + 4}$

4. $\dfrac{x^2 + 2x - 3}{x^2 - 9}$

5. $\dfrac{x^3 - 6x^2 + 9x}{x^3 - 3x^2}$

6. $\dfrac{6y^2 + 12y - 48}{3y^2 - 9y + 6}$

> **Do Exercises 1–6.**

JUST-IN-TIME Review

MULTIPLY AND DIVIDE RATIONAL EXPRESSIONS

To multiply rational expressions, we multiply numerators and multiply denominators and, if possible, simplify the result. To divide rational expressions, we multiply the dividend by the reciprocal of the divisor and, if possible, simplify the result; that is,

$$\frac{a}{b} \cdot \frac{c}{d} = \frac{ac}{bd} \quad \text{and} \quad \frac{a}{b} \div \frac{c}{d} = \frac{a}{b} \cdot \frac{d}{c} = \frac{ad}{bc}.$$

EXAMPLES Multiply or divide.

1. $\dfrac{a^2 - 4}{16a} \cdot \dfrac{20a^2}{a + 2} = \dfrac{(a^2 - 4)(20a^2)}{16a(a + 2)}$

$\qquad = \dfrac{(a+2)(a-2) \cdot 4 \cdot 5 \cdot a \cdot a}{4 \cdot 4 \cdot a \cdot (a+2)}$

$\qquad = \dfrac{5a(a - 2)}{4}$

2. $\dfrac{x - 2}{12} \div \dfrac{x^2 - 4x + 4}{3x^3 + 15x^2} = \dfrac{x - 2}{12} \cdot \dfrac{3x^3 + 15x^2}{x^2 - 4x + 4}$

$\qquad = \dfrac{(x - 2)(3x^3 + 15x^2)}{12(x^2 - 4x + 4)}$

$\qquad = \dfrac{(x - 2)(3)(x^2)(x + 5)}{3 \cdot 4 \,(x - 2)(x - 2)}$

$\qquad = \dfrac{x^2(x + 5)}{4(x - 2)}$

> **Do Exercises 1–6.**

Multiply or divide and, if possible, simplify.

1. $\dfrac{r - s}{r + s} \cdot \dfrac{r^2 - s^2}{(r - s)^2}$

2. $\dfrac{m^2 - n^2}{r + s} \div \dfrac{m - n}{r + s}$

3. $\dfrac{4x^2 + 9x + 2}{x^2 + x - 2} \cdot \dfrac{x^2 - 1}{3x^2 + x - 2}$

4. $\dfrac{a^2 - a - 2}{a^2 - a - 6} \div \dfrac{a^2 - 2a}{2a + a^2}$

5. $\dfrac{3x + 12}{2x - 8} \div \dfrac{(x + 4)^2}{(x - 4)^2}$

6. $\dfrac{x^2 - y^2}{x^3 - y^3} \cdot \dfrac{x^2 + xy + y^2}{x^2 + 2xy + y^2}$

When rational expressions have the same denominator, we can add or subtract by adding or subtracting the numerators and retaining the common denominator. If the denominators differ, we must find equivalent rational expressions that have a common denominator before we can add or subtract. In general, it is most efficient to find the **least common denominator (LCD)** of the expressions.

To find the least common denominator of rational expressions, factor each denominator and form the product that uses each factor the greatest number of times it occurs in any factorization.

EXAMPLE 1 Add.

$$\frac{x^2 - 4x + 4}{2x^2 - 3x + 1} + \frac{x + 4}{2x - 2}$$

$$= \frac{x^2 - 4x + 4}{(2x - 1)(x - 1)} + \frac{x + 4}{2(x - 1)}$$

The LCD is $(2x - 1)(x - 1)(2)$, or $2(2x - 1)(x - 1)$.

$$= \frac{x^2 - 4x + 4}{(2x - 1)(x - 1)} \cdot \frac{2}{2} + \frac{x + 4}{2(x - 1)} \cdot \frac{2x - 1}{2x - 1}$$

$$= \frac{2x^2 - 8x + 8}{(2x - 1)(x - 1)(2)} + \frac{2x^2 + 7x - 4}{2(x - 1)(2x - 1)}$$

$$= \frac{4x^2 - x + 4}{2(2x - 1)(x - 1)}$$

EXAMPLE 2 Subtract.

$$\frac{x}{x^2 + 11x + 30} - \frac{5}{x^2 + 9x + 20}$$

$$= \frac{x}{(x + 5)(x + 6)} - \frac{5}{(x + 5)(x + 4)}$$

The LCD is $(x + 5)(x + 6)(x + 4)$.

$$= \frac{x}{(x + 5)(x + 6)} \cdot \frac{x + 4}{x + 4} - \frac{5}{(x + 5)(x + 4)} \cdot \frac{x + 6}{x + 6}$$

$$= \frac{x^2 + 4x}{(x + 5)(x + 6)(x + 4)} - \frac{5x + 30}{(x + 5)(x + 4)(x + 6)}$$

$$= \frac{x^2 + 4x - (5x + 30)}{(x + 5)(x + 6)(x + 4)} = \frac{x^2 + 4x - 5x - 30}{(x + 5)(x + 6)(x + 4)}$$

$$= \frac{x^2 - x - 30}{(x + 5)(x + 6)(x + 4)} = \frac{(x + 5)(x - 6)}{(x + 5)(x + 6)(x + 4)}$$

$$= \frac{x - 6}{(x + 6)(x + 4)}$$

> *Do Exercises 1–6.*

Add or subtract and, if possible, simplify.

1. $\dfrac{a - 3b}{a + b} + \dfrac{a + 5b}{a + b}$

2. $\dfrac{x^2 - 5}{3x^2 - 5x - 2} + \dfrac{x + 1}{3x - 6}$

3. $\dfrac{a^2 + 1}{a^2 - 1} - \dfrac{a - 1}{a + 1}$

4. $\dfrac{9x + 2}{3x^2 - 2x - 8} + \dfrac{7}{3x^2 + x - 4}$

5. $\dfrac{y}{y^2 - y - 20} - \dfrac{2}{y + 4}$

6. $\dfrac{3y}{y^2 - 7y + 10} - \dfrac{2y}{y^2 - 8y + 15}$

SIMPLIFY COMPLEX RATIONAL EXPRESSIONS

A **complex rational expression** has rational expressions in its numerator or its denominator or both.

EXAMPLE   Simplify: $\dfrac{\dfrac{1}{a} + \dfrac{1}{b}}{\dfrac{1}{a^3} + \dfrac{1}{b^3}}$.

Method 1:

$$\dfrac{\dfrac{1}{a} + \dfrac{1}{b}}{\dfrac{1}{a^3} + \dfrac{1}{b^3}} = \dfrac{\dfrac{1}{a} + \dfrac{1}{b}}{\dfrac{1}{a^3} + \dfrac{1}{b^3}} \cdot \dfrac{a^3 b^3}{a^3 b^3}$$

The LCD of the four rational expressions in the numerator and the denominator is $a^3 b^3$.

$$= \dfrac{\left(\dfrac{1}{a} + \dfrac{1}{b}\right)(a^3 b^3)}{\left(\dfrac{1}{a^3} + \dfrac{1}{b^3}\right)(a^3 b^3)} = \dfrac{a^2 b^3 + a^3 b^2}{b^3 + a^3}$$

$$= \dfrac{a^2 b^2(\cancel{b + a})}{(\cancel{b + a})(b^2 - ba + a^2)} = \dfrac{a^2 b^2}{b^2 - ba + a^2}$$

Method 2:

$$\dfrac{\dfrac{1}{a} + \dfrac{1}{b}}{\dfrac{1}{a^3} + \dfrac{1}{b^3}} = \dfrac{\dfrac{1}{a} \cdot \dfrac{b}{b} + \dfrac{1}{b} \cdot \dfrac{a}{a}}{\dfrac{1}{a^3} \cdot \dfrac{b^3}{b^3} + \dfrac{1}{b^3} \cdot \dfrac{a^3}{a^3}}$$

$\longleftarrow$ The LCD is ab.

$\longleftarrow$ The LCD is $a^3 b^3$.

$$= \dfrac{\dfrac{b}{ab} + \dfrac{a}{ab}}{\dfrac{b^3}{a^3 b^3} + \dfrac{a^3}{a^3 b^3}}$$

$$= \dfrac{\dfrac{b + a}{ab}}{\dfrac{b^3 + a^3}{a^3 b^3}} = \dfrac{b + a}{ab} \cdot \dfrac{a^3 b^3}{b^3 + a^3}$$

$$= \dfrac{(\cancel{b + a})(\cancel{ab})(a^2 b^2)}{(\cancel{ab})(\cancel{b + a})(b^2 - ba + a^2)}$$

$$= \dfrac{a^2 b^2}{b^2 - ba + a^2}$$

> **Do Exercises 1–5.**

Simplify.

1. $\dfrac{\dfrac{x}{y} - \dfrac{y}{x}}{\dfrac{1}{y} + \dfrac{1}{x}}$

2. $\dfrac{\dfrac{a - b}{b}}{\dfrac{a^2 - b^2}{ab}}$

3. $\dfrac{w + \dfrac{8}{w^2}}{1 + \dfrac{2}{w}}$

4. $\dfrac{\dfrac{x^2 - y^2}{xy}}{\dfrac{x - y}{y}}$

5. $\dfrac{\dfrac{a}{b} - \dfrac{b}{a}}{\dfrac{1}{a} - \dfrac{1}{b}}$

Note: $b - a = -1(a - b)$.

The symbol $\sqrt{a}$ denotes the nonnegative square root of a, and the symbol $\sqrt[3]{a}$ denotes the real-number cube root of a. The symbol $\sqrt[n]{a}$ denotes the nth root of a; that is, a number whose nth power is a. The symbol $\sqrt[n]{}$ is called a **radical**, and the expression under the radical is called the **radicand**. The number n (which is omitted when it is 2) is called the **index**.

Any positive number has two square roots, one positive and one negative. Similarly, for any even index, a positive number has two real-number roots. The positive root is called the **principal root**. Any real number has only one real-number odd root.

EXAMPLES Simplify.

1. $\sqrt{36} = 6$ because $6 \cdot 6 = 36$.

2. $-\sqrt{36} = -6$ **3.** $\sqrt[3]{-8} = -2$

4. $\sqrt[5]{\dfrac{32}{243}} = \dfrac{2}{3}$ **5.** $\sqrt[4]{-16}$ is not a real number.

PROPERTIES OF RADICALS

Let a and b be any real numbers or expressions for which the given roots exist. For any natural numbers m and n ($n \neq 1$):

1. If n is even, $\sqrt[n]{a^n} = |a|$.   **2.** If n is odd, $\sqrt[n]{a^n} = a$.

3. $\sqrt[n]{a} \cdot \sqrt[n]{b} = \sqrt[n]{ab}$.

4. $\sqrt[n]{\dfrac{a}{b}} = \dfrac{\sqrt[n]{a}}{\sqrt[n]{b}}$ ($b \neq 0$).

5. $\sqrt[n]{a^m} = \left(\sqrt[n]{a}\right)^m$.

> Here, we assume that no radicands are formed by raising negative quantities to even powers and, consequently, we will not use absolute-value notation when we simplify radical expressions involving variables.

EXAMPLES Simplify.

6. $\sqrt{(-5)^2} = |-5| = 5$ **7.** $\sqrt[3]{(-5)^3} = -5$

8. $\sqrt[4]{4} \cdot \sqrt[4]{5} = \sqrt[4]{4 \cdot 5} = \sqrt[4]{20}$ **9.** $\sqrt[3]{8^5} = \left(\sqrt[3]{8}\right)^5 = 2^5 = 32$

10. $\sqrt{50} = \sqrt{25 \cdot 2} = \sqrt{25} \cdot \sqrt{2} = 5\sqrt{2}$

11. $\dfrac{\sqrt{72}}{\sqrt{6}} = \sqrt{\dfrac{72}{6}} = \sqrt{12} = \sqrt{4 \cdot 3} = \sqrt{4} \cdot \sqrt{3} = 2\sqrt{3}$

12. $\sqrt{216x^5y^3} = \sqrt{36 \cdot 6 \cdot x^4 \cdot x \cdot y^2 \cdot y} = \sqrt{36x^4y^2}\sqrt{6xy}$
$= 6x^2y\sqrt{6xy}$

13. $8\sqrt{50} - 3\sqrt{8} = 8\sqrt{25 \cdot 2} - 3\sqrt{4 \cdot 2} = 8 \cdot 5\sqrt{2} - 3 \cdot 2\sqrt{2}$
$= 40\sqrt{2} - 6\sqrt{2} = (40 - 6)\sqrt{2} = 34\sqrt{2}$

14. $\left(5 - \sqrt{2}\right)\left(4 + 3\sqrt{2}\right) = 20 + 15\sqrt{2} - 4\sqrt{2} - 3\left(\sqrt{2}\right)^2$
$= 20 + 11\sqrt{2} - 6 = 14 + 11\sqrt{2}$

> **Do Exercises 1–20.**

Simplify. Assume that no radicands were formed by raising negative quantities to even powers.

1. $\sqrt{(-21)^2}$ **2.** $\sqrt{9y^2}$

3. $\sqrt{(a - 2)^2}$ **4.** $\sqrt[3]{-27x^3}$

5. $\sqrt[4]{81x^8}$ **6.** $\sqrt[5]{32}$

7. $\sqrt[4]{48x^6y^4}$ **8.** $\sqrt{15}\sqrt{35}$

9. $\dfrac{\sqrt{40xy}}{\sqrt{8x}}$ **10.** $\dfrac{\sqrt[3]{3x^2}}{\sqrt[3]{24x^5}}$

11. $\sqrt{x^2 - 4x + 4}$

12. $\sqrt{2x^3y}\sqrt{12xy}$

13. $\sqrt[3]{3x^2y}\sqrt[3]{36x}$

14. $5\sqrt{2} + 3\sqrt{32}$

15. $7\sqrt{12} - 2\sqrt{3}$

16. $2\sqrt{32} + 3\sqrt{8} - 4\sqrt{18}$

17. $6\sqrt{20} - 4\sqrt{45} + \sqrt{80}$

18. $\left(2 + \sqrt{3}\right)\left(5 + 2\sqrt{3}\right)$

19. $\left(\sqrt{8} + 2\sqrt{5}\right)\left(\sqrt{8} - 2\sqrt{5}\right)$

20. $\left(1 + \sqrt{3}\right)^2$

There are times when we need to remove the radicals in a denominator. This procedure is called **rationalizing the denominator**. It is done by multiplying by 1 in such a way as to obtain a perfect nth power in the denominator.

EXAMPLES Rationalize the denominator.

1. $\sqrt{\dfrac{3}{2}} = \sqrt{\dfrac{3}{2} \cdot \dfrac{2}{2}} = \sqrt{\dfrac{6}{4}} = \dfrac{\sqrt{6}}{\sqrt{4}} = \dfrac{\sqrt{6}}{2}$

2. $\dfrac{2}{\sqrt{3}} = \dfrac{2}{\sqrt{3}} \cdot \dfrac{\sqrt{3}}{\sqrt{3}} = \dfrac{2\sqrt{3}}{3}$

3. $\dfrac{\sqrt[3]{7}}{\sqrt[3]{9}} = \dfrac{\sqrt[3]{7}}{\sqrt[3]{9}} \cdot \dfrac{\sqrt[3]{3}}{\sqrt[3]{3}} = \dfrac{\sqrt[3]{21}}{\sqrt[3]{27}} = \dfrac{\sqrt[3]{21}}{3}$

Pairs of expressions of the form $a\sqrt{b} + c\sqrt{d}$ and $a\sqrt{b} - c\sqrt{d}$ are called **conjugates**. The product of such a pair contains no radicals and can be used to rationalize a denominator or a numerator.

EXAMPLE 4 Rationalize the denominator: $\dfrac{7}{3 + \sqrt{5}}$.

$$\dfrac{7}{3 + \sqrt{5}} = \dfrac{7}{3 + \sqrt{5}} \cdot \dfrac{3 - \sqrt{5}}{3 - \sqrt{5}}$$

$$= \dfrac{21 - 7\sqrt{5}}{3^2 - 3\sqrt{5} + 3\sqrt{5} - (\sqrt{5})^2}$$

$$= \dfrac{21 - 7\sqrt{5}}{9 - 5} = \dfrac{21 - 7\sqrt{5}}{4}$$

EXAMPLE 5 Rationalize the denominator: $\dfrac{1 + \sqrt{2}}{\sqrt{5} + \sqrt{10}}$.

$$\dfrac{1 + \sqrt{2}}{\sqrt{5} + \sqrt{10}} = \dfrac{1 + \sqrt{2}}{\sqrt{5} + \sqrt{10}} \cdot \dfrac{\sqrt{5} - \sqrt{10}}{\sqrt{5} - \sqrt{10}}$$

$$= \dfrac{\sqrt{5} - \sqrt{10} + \sqrt{10} - \sqrt{20}}{(\sqrt{5})^2 - (\sqrt{10})^2}$$

$$= \dfrac{\sqrt{5} - \sqrt{20}}{5 - 10}$$

$$= \dfrac{\sqrt{5} - 2\sqrt{5}}{-5}$$

$$= \dfrac{-\sqrt{5}}{-5}$$

$$= \dfrac{\sqrt{5}}{5}$$

Rationalize the denominator.

1. $\dfrac{4}{\sqrt{11}}$ **2.** $\sqrt{\dfrac{3}{7}}$

3. $\dfrac{\sqrt[3]{7}}{\sqrt[3]{2}}$ **4.** $\sqrt[3]{\dfrac{16}{9}}$

5. $\dfrac{3}{\sqrt{30} - 4}$

6. $\dfrac{4}{\sqrt{7} - \sqrt{3}}$

7. $\dfrac{6}{\sqrt{m} - \sqrt{n}}$

8. $\dfrac{1 - \sqrt{2}}{\sqrt{3} - \sqrt{6}}$

> **Do Exercises 1–8.**

For any real number a and any natural numbers m and n, $n \neq 1$, for which $\sqrt[n]{a}$ exists:

$$a^{1/n} = \sqrt[n]{a}, \qquad a^{m/n} = \left(\sqrt[n]{a}\right)^m = \sqrt[n]{a^m}, \quad \text{and} \quad a^{-m/n} = \frac{1}{a^{m/n}}.$$

EXAMPLES Convert to radical notation and, if possible, simplify.

1. $m^{1/6} = \sqrt[6]{m}$

2. $7^{3/4} = \sqrt[4]{7^3}$, or $\left(\sqrt[4]{7}\right)^3$

3. $8^{-5/3} = \dfrac{1}{8^{5/3}} = \dfrac{1}{\left(\sqrt[3]{8}\right)^5} = \dfrac{1}{2^5} = \dfrac{1}{32}$

EXAMPLES Convert to exponential notation.

4. $\left(\sqrt[4]{7xy}\right)^5 = (7xy)^{5/4}$ **5.** $\sqrt[6]{x^3} = x^{3/6} = x^{1/2}$

EXAMPLES Simplify and then, if appropriate, write radical notation.

6. $x^{5/6} \cdot x^{2/3} = x^{5/6+2/3} = x^{9/6} = x^{3/2} = \sqrt{x^3}$
      $= \sqrt{x^2}\,\sqrt{x} = x\sqrt{x}$

7. $(x+3)^{5/2}(x+3)^{-1/2} = (x+3)^{5/2-1/2} = (x+3)^2$

> **Do Exercises 1–11.**

Convert to radical notation and, if possible, simplify.

1. $y^{5/6}$ **2.** $x^{2/3}$

3. $16^{3/4}$ **4.** $4^{7/2}$

5. $125^{-1/3}$ **6.** $32^{-4/5}$

Convert to exponential notation.

7. $\sqrt[12]{y^4}$ **8.** $\sqrt{x^5}$

Simplify and then, if appropriate, write radical notation.

9. $x^{1/2} \cdot x^{2/3}$

10. $(a-2)^{9/4}(a-2)^{-1/4}$

11. $\left(m^{1/2}n^{5/2}\right)^{2/3}$

JUST-IN-TIME Review

Just in Time

28

THE PYTHAGOREAN THEOREM

A **right triangle** is a triangle with a 90° angle, as shown in the following figure. The small square in the corner indicates the 90° angle.

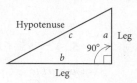

In a right triangle, the longest side is called the **hypotenuse**. It is also the side opposite the right angle. The other two sides are called **legs**. We generally use the letters a and b for the lengths of the legs and c for the length of the hypotenuse. They are related as follows.

THE PYTHAGOREAN THEOREM

In any right triangle, if a and b are the lengths of the legs and c is the length of the hypotenuse, then

$$a^2 + b^2 = c^2.$$

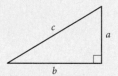

The equation $a^2 + b^2 = c^2$ is called the **Pythagorean equation**.

EXAMPLE 1 Find the length of the hypotenuse of this right triangle. Give an exact answer and an approximation to three decimal places.

$$4^2 + 5^2 = c^2$$
$$16 + 25 = c^2$$
$$41 = c^2$$
$$c = \sqrt{41}$$
$$c \approx 6.403$$

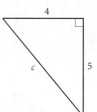

EXAMPLE 2 Find the length of leg b of this right triangle. Give an exact answer and an approximation to three decimal places.

$$10^2 + b^2 = 12^2$$
$$100 + b^2 = 144$$
$$b^2 = 144 - 100$$
$$b^2 = 44$$
$$b = \sqrt{44}$$
$$b \approx 6.633$$

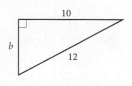

> **Do Exercises 1–5.**

Find the length of the third side of each right triangle. Where appropriate, give both an exact answer and an approximation to three decimal places.

1.

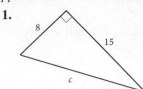

2.

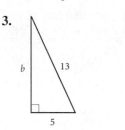

3.

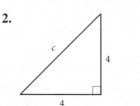

4.

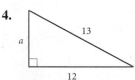

5.

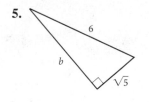

Graphs, Functions, and Models

APPLICATION

This problem appears as Exercise 60 in the Review Exercises.

The number of female medical school graduates has increased each year from 2006 to 2014. Model the data shown on p. 88 with a linear function where the number of female medical school graduates W is a function of the year x and where x is the number of years after 2006. Then, using this function, estimate the number of female graduates in 2013.

1.1   Introduction to Graphing

❯ Plot points.

❯ Determine whether an ordered pair is a solution of an equation.

❯ Find the *x*- and *y*-intercepts of an equation of the form $Ax + By = C$.

❯ Graph equations.

❯ Find the distance between two points in the plane, and find the midpoint of a segment.

❯ Find an equation of a circle with a given center and radius, and given an equation of a circle in standard form, find the center and the radius.

❯ Graph equations of circles.

❯ Graphs

Graphs provide a means of displaying, interpreting, and analyzing data in a visual format. It is not uncommon to open a newspaper or a magazine and encounter graphs. Examples of bar, line, and circle graphs are shown below.

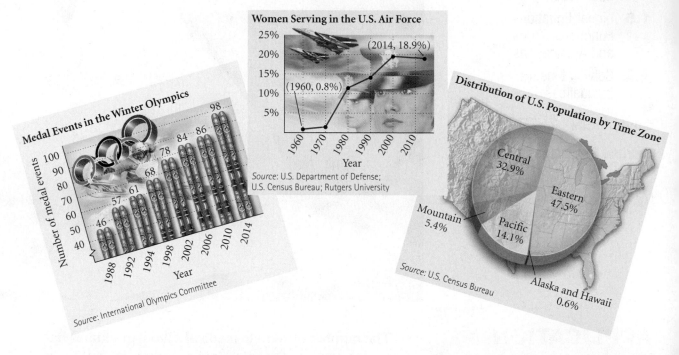

Many real-world situations can be modeled, or described mathematically, using equations in which two variables appear. We use a plane to graph a pair of numbers. To locate points on a plane, we use two perpendicular number lines, called **axes**, that intersect at $(0, 0)$. We call this point the **origin**. The horizontal axis is called the **x-axis**, and the vertical axis is called the **y-axis**. (Other variables, such as a and b, can also be used.) The axes divide the plane into four regions, called **quadrants**, denoted by Roman numerals and numbered counterclockwise from the upper right. Arrows show the positive direction of each axis.

Each point (x, y) in the plane is described by an **ordered pair**. The first number, x, indicates the point's horizontal location with respect to the y-axis, and the

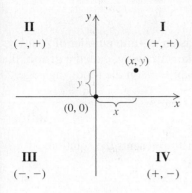

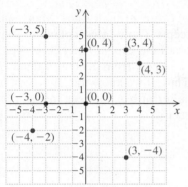

second number, *y*, indicates the point's vertical location with respect to the *x*-axis. We call *x* the **first coordinate**, the *x***-coordinate**, or the **abscissa**. We call *y* the **second coordinate**, the *y***-coordinate**, or the **ordinate**. Such a representation is called the **Cartesian coordinate system** in honor of the French mathematician and philosopher René Descartes (1596–1650).

In the first quadrant, both coordinates of a point are positive. In the second quadrant, the first coordinate is negative and the second is positive. In the third quadrant, both coordinates are negative, and in the fourth quadrant, the first coordinate is positive and the second is negative.

EXAMPLE 1   Graph and label the points $(-3, 5)$, $(4, 3)$, $(3, 4)$, $(-4, -2)$, $(3, -4)$, $(0, 4)$, $(-3, 0)$, and $(0, 0)$.

Solution   To graph or **plot** $(-3, 5)$, we note that the *x*-coordinate, -3, tells us to move from the origin 3 units horizontally in the negative direction, or 3 units to the left of the *y*-axis. Then we move 5 units up from the *x*-axis.* To graph the other points, we proceed in a similar manner. (See the graph at left.) Note that the point $(4, 3)$ is different from the point $(3, 4)$.

> *Now Try Exercise 3.*

❯ Solutions of Equations

Equations in two variables, like $2x + 3y = 18$, have solutions (x, y) that are ordered pairs such that when the first coordinate is substituted for *x* and the second coordinate is substituted for *y*, the result is a true equation. The first coordinate in an ordered pair generally represents the variable that occurs first alphabetically.

EXAMPLE 2   Determine whether each ordered pair is a solution of the equation $2x + 3y = 18$.

a) $(-5, 7)$ **b)** $(3, 4)$

Solution   We substitute the ordered pair into the equation and determine whether the resulting equation is true.

a)
$$2x + 3y = 18$$
$$\begin{array}{c|c} 2(-5) + 3(7) \;?\; 18 \\ -10 + 21 \\ 11 \;\big|\; 18 \end{array} \quad \text{FALSE}$$

We substitute -5 for *x* and 7 for *y* (alphabetical order).

The equation $11 = 18$ is false, so $(-5, 7)$ is not a solution.

b)
$$2x + 3y = 18$$
$$\begin{array}{c|c} 2(3) + 3(4) \;?\; 18 \\ 6 + 12 \\ 18 \;\big|\; 18 \end{array} \quad \text{TRUE}$$

We substitute 3 for *x* and 4 for *y*.

The equation $18 = 18$ is true, so $(3, 4)$ is a solution.

We can also perform these substitutions on a graphing calculator. When we substitute -5 for *x* and 7 for *y*, we get 11. Since $11 \neq 18$, $(-5, 7)$ is not a solution of the equation. When we substitute 3 for *x* and 4 for *y*, we get 18, so $(3, 4)$ is a solution.

> *Now Try Exercise 11.*

```
2(-5)+3*7
                        11
2*3+3*4
                        18
```

----

*Here the notation $(-3, 5)$ represents an ordered pair. This notation can also represent an open interval. See Just-in-Time 6 review on p. J-6. The context in which the notation appears usually makes the meaning clear.

❯ Graphs of Equations

The equation considered in Example 2 actually has an infinite number of solutions. Since we cannot list all the solutions, we will make a drawing, called a **graph**, that represents them. Some suggestions for drawing graphs are on the following page.

TO GRAPH AN EQUATION

To **graph an equation** is to make a drawing that represents the solutions of that equation.

Graphs of equations of the type $Ax + By = C$ are straight lines. Many such equations can be graphed conveniently using intercepts. The **x-intercept** of the graph of an equation is the point at which the graph crosses the x-axis. The **y-intercept** is the point at which the graph crosses the y-axis. We know from geometry that only one line can be drawn through two given points. Thus, if we know the intercepts, we can graph the line. To ensure that a computational error has not been made, it is a good idea to calculate and plot a third point as a check.

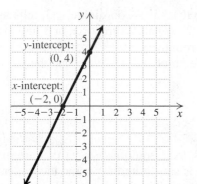

y-intercept: (0, 4)
x-intercept: (−2, 0)

---

x-INTERCEPT AND y-INTERCEPT

An **x-intercept** is a point $(a, 0)$. To find a, let $y = 0$ and solve for x.

A **y-intercept** is a point $(0, b)$. To find b, let $x = 0$ and solve for y.

EXAMPLE 3   Graph: $2x + 3y = 18$.

Solution   The graph is a line. To find ordered pairs that are solutions of this equation, we can replace either x or y with any number and then solve for the other variable. In this case, it is convenient to find the intercepts of the graph. For instance, if x is replaced with 0, then

$$2 \cdot 0 + 3y = 18$$
$$3y = 18$$
$$y = 6. \qquad \text{Dividing by 3 on both sides}$$

Thus, $(0, 6)$ is a solution. It is the *y-intercept* of the graph. If y is replaced with 0, then

$$2x + 3 \cdot 0 = 18$$
$$2x = 18$$
$$x = 9. \qquad \text{Dividing by 2 on both sides}$$

Thus, $(9, 0)$ is a solution. It is the *x-intercept* of the graph. We find a third solution as a check. If x is replaced with 3, then

$$2 \cdot 3 + 3y = 18$$
$$6 + 3y = 18$$
$$3y = 12 \qquad \text{Subtracting 6 on both sides}$$
$$y = 4. \qquad \text{Dividing by 3 on both sides}$$

Thus, $(3, 4)$ is a solution.

STUDY TIPS

Success can be planned. Combine goals and good study habits to create a plan for success that works for you. A list of study tips that your authors consider most helpful are included in the Guide to Success in the front of the text before Chapter 1.

We list the solutions in a table and then plot the points. Note that the points appear to lie on a straight line.

x	y	(x, y)
0	6	$(0, 6)$
9	0	$(9, 0)$
3	4	$(3, 4)$

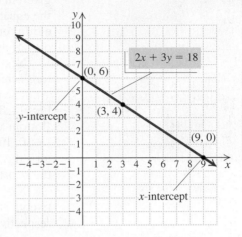

Were we to graph additional solutions of $2x + 3y = 18$, they would lie on the same straight line. Thus, to complete the graph, we use a straightedge to draw a line, as shown in the figure. This line represents all solutions of the equation. Every point on the line represents a solution; every solution is represented by a point on the line.

> **Now Try Exercise 17.**

When graphing some equations, it is convenient to first solve for y and then find ordered pairs. We can use the addition and multiplication principles to solve for y.

EXAMPLE 4 Graph: $3x - 5y = -10$.

Solution We first solve for y:

$$3x - 5y = -10$$
$$-5y = -3x - 10 \qquad \text{Subtracting } 3x \text{ on both sides}$$
$$y = \tfrac{3}{5}x + 2. \qquad \text{Multiplying by } -\tfrac{1}{5} \text{ on both sides}$$

By choosing multiples of 5 for x, we can avoid adding and subtracting fraction values when calculating y. For example, if we choose -5 for x, we get

$$y = \tfrac{3}{5}x + 2 = \tfrac{3}{5}(-5) + 2 = -3 + 2 = -1.$$

The following table lists a few points. We plot the points and draw the graph.

x	y	(x, y)
-5	-1	$(-5, -1)$
0	2	$(0, 2)$
5	5	$(5, 5)$

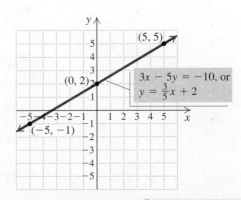

> **Now Try Exercise 29.**

In the equation $y = \tfrac{3}{5}x + 2$ in Example 4, the value of y *depends* on the value chosen for x, so x is said to be the **independent variable** and y the **dependent variable**.

Just in Time **17**

We can graph an equation on a graphing calculator. Many calculators require an equation to be entered in the form $y =$. In such a case, if the equation is not initially given in this form, it must be solved for y before it is entered in the calculator. For the equation $3x - 5y = -10$ in Example 4, we enter $y = \frac{3}{5}x + 2$ on the equation-editor, or $y =$, screen in the form $y = (3/5)x + 2$, which some calculators will display as shown in the window at left.

Next, we determine the portion of the xy-plane that will appear on the calculator's screen. That portion of the plane is called the **viewing window**.

The notation used in this text to denote a window setting consists of four numbers $[L, R, B, T]$, which represent the **L**eft and **R**ight endpoints of the x-axis and the **B**ottom and **T**op endpoints of the y-axis, respectively. The window with the settings $[-10, 10, -10, 10]$ is the **standard viewing window**. On some graphing calculators, the standard window can be selected quickly using the ZSTANDARD feature from the ZOOM menu.

Xmin and Xmax are used to set the left and right endpoints of the x-axis, respectively; Ymin and Ymax are used to set the bottom and top endpoints of the y-axis, respectively. The settings Xscl and Yscl give the scales for the axes. For example, Xscl $= 1$ and Yscl $= 1$ means that there is 1 unit between tick marks on each of the axes. In this text, scaling factors other than 1 will be listed by the window unless they are readily apparent.

After entering the equation $y = (3/5)x + 2$ and choosing a viewing window, we can then draw the graph.

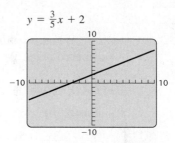

$y = \frac{3}{5}x + 2$

EXAMPLE 5   Graph: $y = x^2 - 9x - 12$.

Just in Time

7

Solution   Note that since this equation is not of the form $Ax + By = C$, its graph is not a straight line. We make a table of values, plot enough points to obtain an idea of the shape of the curve, and connect the points with a smooth curve. It is important to scale the axes to include most of the ordered pairs listed in the table. Here it is appropriate to use a larger scale on the y-axis than on the x-axis.

x	y	(x, y)
-3	24	$(-3, 24)$
-1	-2	$(-1, -2)$
0	-12	$(0, -12)$
2	-26	$(2, -26)$
4	-32	$(4, -32)$
5	-32	$(5, -32)$
10	-2	$(10, -2)$
12	24	$(12, 24)$

① **Select values for x.**

② **Compute values for y.**

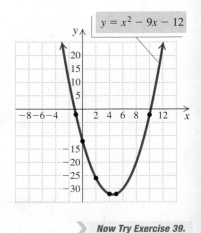

$y = x^2 - 9x - 12$

▶ *Now Try Exercise 39.*

FIGURE 1.

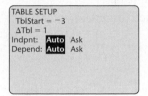

FIGURE 2.

A graphing calculator can be used to create a table of ordered pairs that are solutions of an equation. For the equation in Example 5, $y = x^2 - 9x - 12$, we first enter the equation on the equation-editor screen. (See Fig. 1.) Then we set up a table in AUTO mode by designating a value for TBLSTART and a value for ΔTBL. The calculator will produce a table beginning with the value of TBLSTART and continuing by adding ΔTBL to supply succeeding x-values. For the equation $y = x^2 - 9x - 12$, we let TBLSTART $= -3$ and ΔTBL $= 1$. (See Fig. 2.)

We can scroll up and down in the table to find values other than those shown in Fig. 3. We can also graph this equation on the graphing calculator, as shown in Fig. 4.

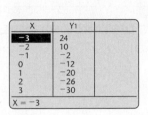

FIGURE 3.

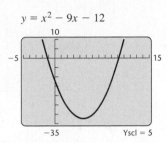

FIGURE 4.

❯ The Distance Formula

The $5.25 billion expansion of the Panama Canal doubled its capacity. A third canal lane is scheduled to open in 2016. (Source: Panama Canal Authority)

Suppose that a photographer must determine the distance between two points, A and B, on opposite sides of a lane of the Panama Canal. One way in which he or she might proceed is to measure two legs of a right triangle that is situated as shown below. The Pythagorean equation, $c^2 = a^2 + b^2$, where c is the length of the hypotenuse and a and b are the lengths of the legs, can then be used to find the length of the hypotenuse, which is the distance from A to B.

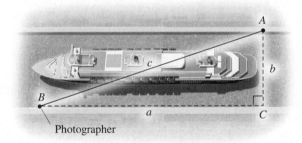

Photographer

A similar strategy is used to find the distance between two points in a plane. For two points (x_1, y_1) and (x_2, y_2), we can draw a right triangle in which the legs have lengths $|x_2 - x_1|$ and $|y_2 - y_1|$.

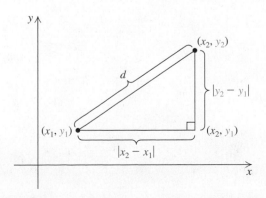

Just in Time
4, 28

Using the Pythagorean equation, $c^2 = a^2 + b^2$, we have

$$d^2 = |x_2 - x_1|^2 + |y_2 - y_1|^2.$$ Substituting d for c, $|x_2 - x_1|$ for a, and $|y_2 - y_1|$ for b in the Pythagorean equation

Because we are squaring, we can use parentheses to replace the absolute-value symbols:

$$d^2 = (x_2 - x_1)^2 + (y_2 - y_1)^2.$$

Taking the principal square root, we obtain the distance formula.

THE DISTANCE FORMULA

The **distance d** between any two points (x_1, y_1) and (x_2, y_2) is given by

$$d = \sqrt{(x_2 - x_1)^2 + (y_2 - y_1)^2}.$$

The subtraction of the x-coordinates can be done in any order, as can the subtraction of the y-coordinates. Although we derived the distance formula by considering two points not on a horizontal line or a vertical line, the distance formula holds for *any* two points.

Just in Time
5, 25

EXAMPLE 6 Find the distance between each pair of points.

a) $(-2, 2)$ and $(3, -6)$ **b)** $(-1, -5)$ and $(-1, 2)$

Solution We substitute into the distance formula.

a) $d = \sqrt{(x_2 - x_1)^2 + (y_2 - y_1)^2}$
$= \sqrt{[3 - (-2)]^2 + (-6 - 2)^2}$
$= \sqrt{5^2 + (-8)^2} = \sqrt{25 + 64}$
$= \sqrt{89} \approx 9.4$

b) $d = \sqrt{(x_2 - x_1)^2 + (y_2 - y_1)^2}$
$= \sqrt{[-1 - (-1)]^2 + (-5 - 2)^2}$
$= \sqrt{0^2 + (-7)^2} = \sqrt{0 + 49}$
$= \sqrt{49} = 7$

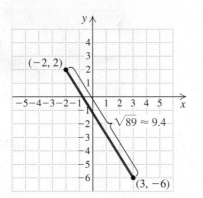

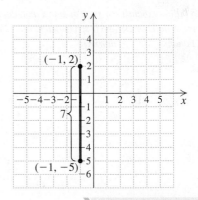

Now Try Exercises 63 and 71.

EXAMPLE 7   The point $(-2, 5)$ is on a circle that has $(3, -1)$ as its center. Find the length of the radius of the circle.

Solution   Since the length of the radius is the distance from the center to a point on the circle, we substitute into the distance formula:

$$d = \sqrt{(x_2 - x_1)^2 + (y_2 - y_1)^2};$$
$$r = \sqrt{[3 - (-2)]^2 + (-1 - 5)^2}$$

Substituting r for d, $(3, -1)$ for (x_2, y_2), and $(-2, 5)$ for (x_1, y_1). Either point can serve as (x_1, y_1).

$$= \sqrt{5^2 + (-6)^2} = \sqrt{25 + 36}$$
$$= \sqrt{61} \approx 7.8.$$

Rounding to the nearest tenth

The radius of the circle is approximately 7.8.

> *Now Try Exercise 77.*

❯ Midpoints of Segments

The distance formula can be used to develop a method of determining the *midpoint* of a segment when the endpoints are known. We state the formula and leave its proof to the exercises.

THE MIDPOINT FORMULA

If the endpoints of a segment are (x_1, y_1) and (x_2, y_2), then the coordinates of the **midpoint** of the segment are

$$\left(\frac{x_1 + x_2}{2}, \frac{y_1 + y_2}{2} \right).$$

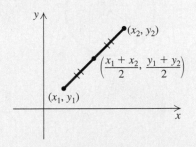

Note that we obtain the coordinates of the midpoint by averaging the coordinates of the endpoints. This is a good way to remember the midpoint formula.

EXAMPLE 8   Find the midpoint of the segment whose endpoints are $(-4, -2)$ and $(2, 5)$.

Solution   Using the midpoint formula, we obtain

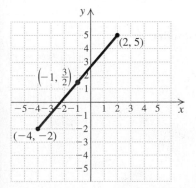

$$\left(\frac{-4 + 2}{2}, \frac{-2 + 5}{2} \right) = \left(\frac{-2}{2}, \frac{3}{2} \right) = \left(-1, \frac{3}{2} \right).$$

> *Now Try Exercise 83.*

EXAMPLE 9 The diameter of a circle connects the points $(2, -3)$ and $(6, 4)$ on the circle. Find the coordinates of the center of the circle.

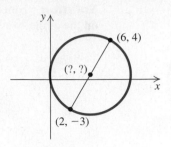

Solution Since the center of the circle is the midpoint of the diameter, we use the midpoint formula:

$$\left(\frac{2 + 6}{2}, \frac{-3 + 4}{2}\right), \quad \text{or} \quad \left(\frac{8}{2}, \frac{1}{2}\right), \quad \text{or} \quad \left(4, \frac{1}{2}\right).$$

The coordinates of the center are $\left(4, \frac{1}{2}\right)$.

> *Now Try Exercise 95.*

❯ Circles

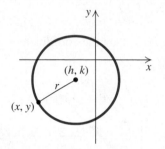

A **circle** is the set of all points in a plane that are a fixed distance r from a *center* (h, k). Thus if a point (x, y) is to be r units from the center, we must have

$$r = \sqrt{(x - h)^2 + (y - k)^2}. \qquad \text{Using the distance formula,}$$
$$d = \sqrt{(x_2 - x_1)^2 + (y_2 - y_1)^2}$$

Squaring both sides gives an equation of a circle. The distance r is the length of a *radius* of the circle.

THE EQUATION OF A CIRCLE

The standard form of the equation of a circle having center (h, k) and radius r is

$$(x - h)^2 + (y - k)^2 = r^2.$$

EXAMPLE 10 Find an equation of the circle having radius 5 and center $(3, -7)$.

Solution Using the standard form, we have

$$[x - 3]^2 + [y - (-7)]^2 = 5^2 \qquad \textbf{Substituting}$$
$$(x - 3)^2 + (y + 7)^2 = 25.$$

> *Now Try Exercise 99.*

EXAMPLE 11 Graph the circle $(x + 5)^2 + (y - 2)^2 = 16$.

Solution We write the equation in standard form to determine the center and the radius:

$$[x - (-5)]^2 + [y - 2]^2 = 4^2.$$

The center is $(-5, 2)$ and the radius is 4. We locate the center and draw the circle using a compass.

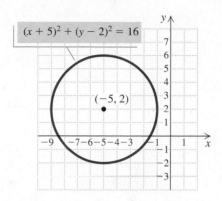

Now Try Exercise 111.

Circles can also be graphed using a graphing calculator. We show one method of doing so here.

When we graph a circle, we select a viewing window in which the distance between units is visually the same on both axes. This procedure is called **squaring the viewing window**. We do this so that the graph will not be distorted. A graph of the circle $x^2 + y^2 = 36$ in a nonsquared window is shown in Fig. 1.

On many graphing calculators, the ratio of the height to the width of the viewing screen is $\frac{2}{3}$. When we choose a window in which Xscl = Yscl and the length of the y-axis is $\frac{2}{3}$ the length of the x-axis, the window will be squared. The windows with dimensions $[-6, 6, -4, 4]$, $[-9, 9, -6, 6]$, and $[-12, 12, -8, 8]$ are examples of squared windows. A graph of the circle $x^2 + y^2 = 36$ in a squared window is shown in Fig. 2. Many graphing calculators have an option on the ZOOM menu that squares the window automatically.

$x^2 + y^2 = 36$

FIGURE 1.

$x^2 + y^2 = 36$

FIGURE 2.

EXAMPLE 12 Graph the circle $(x - 2)^2 + (y + 1)^2 = 16$.

Solution The circle $(x - 2)^2 + (y + 1)^2 = 16$ has center $(2, -1)$ and radius 4, so the viewing window $[-9, 9, -6, 6]$ is a good choice for the graph.

To graph a circle, we select the CIRCLE feature from the DRAW menu and enter the coordinates of the center and the length of the radius. The graph of the circle $(x - 2)^2 + (y + 1)^2 = 16$ is shown here. For more on graphing circles with a graphing calculator, see the material on conic sections in a later chapter.

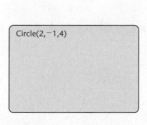

$(x - 2)^2 + (y + 1)^2 = 16$

Circle(2,−1,4)

Now Try Exercise 113.

Visualizing the Graph

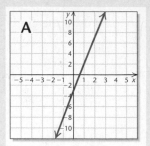

A

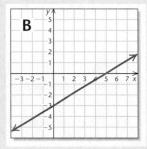

B

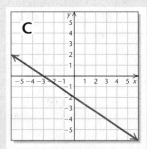

C

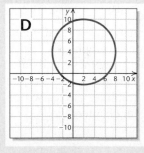

D

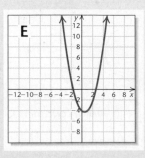

E

Match the equation with its graph.

1. $y = -x^2 + 5x - 3$

2. $3x - 5y = 15$

3. $(x - 2)^2 + (y - 4)^2 = 36$

4. $y - 5x = -3$

5. $x^2 + y^2 = \dfrac{25}{4}$

6. $15y - 6x = 90$

7. $y = -\dfrac{2}{3}x - 2$

8. $(x + 3)^2 + (y - 1)^2 = 16$

9. $3x + 5y = 15$

10. $y = x^2 - x - 4$

Answers on page A-1

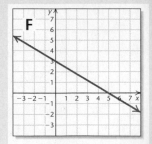

F

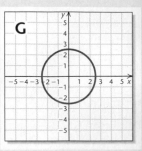

G

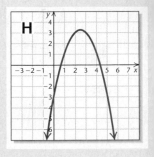

H

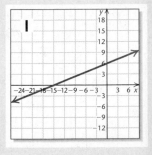

I

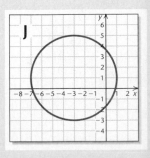

J

1.1   Exercise Set

Use the following graph for Exercises 1 and 2.

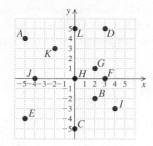

1. Find the coordinates of points A, B, C, D, E, and F.

2. Find the coordinates of points G, H, I, J, K, and L.

Graph and label the given points.

3. $(4, 0)$, $(-3, -5)$, $(-1, 4)$, $(0, 2)$, $(2, -2)$

4. $(1, 4)$, $(-4, -2)$, $(-5, 0)$, $(2, -4)$, $(0, 3)$

5. $(-5, 1)$, $(5, 1)$, $(2, 3)$, $(2, -1)$, $(0, 1)$

6. $(0, -1)$, $(4, -3)$, $(-5, 2)$, $(2, 0)$, $(-1, -5)$

Express the data pictured in the graph as ordered pairs, letting the first coordinate represent the year and the second coordinate the amount or percent.

7. Southwest Airlines: Number of Cities Served

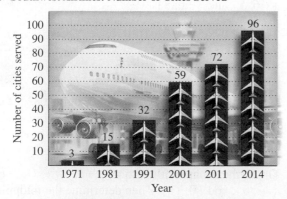

8. Women Serving in the Marines

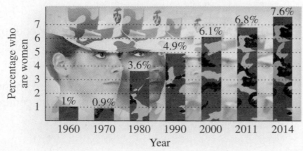

Source: U. S. Department of Veterans Affairs, Rutgers University

Use substitution to determine whether the given ordered pairs are solutions of the given equation.

9. $(-1, -9)$, $(0, 2)$; $y = 7x - 2$

10. $\left(\frac{1}{2}, 8\right)$, $(-1, 6)$; $y = -4x + 10$

11. $\left(\frac{2}{3}, \frac{3}{4}\right)$, $\left(1, \frac{3}{2}\right)$; $6x - 4y = 1$

12. $(1.5, 2.6)$, $(-3, 0)$; $x^2 + y^2 = 9$

13. $\left(-\frac{1}{2}, -\frac{4}{5}\right)$, $\left(0, \frac{3}{5}\right)$; $2a + 5b = 3$

14. $\left(0, \frac{3}{2}\right)$, $\left(\frac{2}{3}, 1\right)$; $3m + 4n = 6$

15. $(-0.75, 2.75)$, $(2, -1)$; $x^2 - y^2 = 3$

16. $(2, -4)$, $(4, -5)$; $5x + 2y^2 = 70$

Find the intercepts and then graph the line.

17. $5x - 3y = -15$ **18.** $2x - 4y = 8$

19. $2x + y = 4$ **20.** $3x + y = 6$

21. $4y - 3x = 12$ **22.** $3y + 2x = -6$

Graph the equation.

23. $y = 3x + 5$ **24.** $y = -2x - 1$

25. $x - y = 3$ **26.** $x + y = 4$

27. $y = -\frac{3}{4}x + 3$ **28.** $3y - 2x = 3$

29. $5x - 2y = 8$ **30.** $y = 2 - \frac{4}{3}x$

31. $x - 4y = 5$ **32.** $6x - y = 4$

33. $2x + 5y = -10$ **34.** $4x - 3y = 12$

35. $y = -x^2$ **36.** $y = x^2$

37. $y = x^2 - 3$ **38.** $y = 4 - x^2$

39. $y = -x^2 + 2x + 3$ **40.** $y = x^2 + 2x - 1$

In Exercises 41–44, use a graphing calculator to match the equation with one of the graphs (a)–(d) that follow.

a)
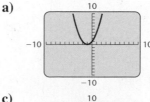

b)
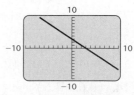

c)
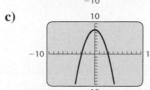

d)
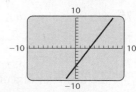

41. $y = 3 - x$ **42.** $2x - y = 6$

43. $y = x^2 + 2x + 1$ **44.** $y = 8 - x^2$

Use a graphing calculator to graph the equation in the standard window.

45. $y = 2x + 1$

46. $y = 3x - 4$

47. $4x + y = 7$

48. $5x + y = -8$

49. $y = \frac{1}{3}x + 2$

50. $y = \frac{3}{2}x - 4$

51. $2x + 3y = -5$

52. $3x + 4y = 1$

53. $y = x^2 + 6$

54. $y = x^2 - 8$

55. $y = 2 - x^2$

56. $y = 5 - x^2$

57. $y = x^2 + 4x - 2$

58. $y = x^2 - 5x + 3$

Graph the equation in the standard window and in the given window. Determine which window better shows the shape of the graph and the x- and y-intercepts.

59. $y = 3x^2 - 6$
$[-4, 4, -4, 4]$

60. $y = -2x + 24$
$[-15, 15, -10, 30]$, with Xscl = 3 and Yscl = 5

61. $y = -\frac{1}{6}x^2 + \frac{1}{12}$
$[-1, 1, -0.3, 0.3]$, with Xscl = 0.1 and Yscl = 0.1

62. $y = 6 - x^2$
$[-3, 3, -3, 3]$

Find the distance between the pair of points. Give an exact answer and, where appropriate, an approximation to three decimal places.

63. $(4, 6)$ and $(5, 9)$

64. $(-3, 7)$ and $(2, 11)$

65. $(-11, -8)$ and $(1, -13)$

66. $(-60, 5)$ and $(-20, 35)$

67. $(6, -1)$ and $(9, 5)$

68. $(-4, -7)$ and $(-1, 3)$

69. $\left(-8, \frac{7}{11}\right)$ and $\left(8, \frac{7}{11}\right)$

70. $\left(\frac{1}{2}, -\frac{4}{25}\right)$ and $\left(\frac{1}{2}, -\frac{13}{25}\right)$

71. $\left(-\frac{3}{5}, -4\right)$ and $\left(-\frac{3}{5}, \frac{2}{3}\right)$

72. $\left(-\frac{11}{3}, -\frac{1}{2}\right)$ and $\left(\frac{1}{3}, \frac{5}{2}\right)$

73. $(-4.2, 3)$ and $(2.1, -6.4)$

74. $(0.6, -1.5)$ and $(-8.1, -1.5)$

75. $(0, 0)$ and (a, b)

76. (r, s) and $(-r, -s)$

77. The points $(-3, -1)$ and $(9, 4)$ are the endpoints of the diameter of a circle. Find the length of the radius of the circle.

78. The point $(0, 1)$ is on a circle that has center $(-3, 5)$. Find the length of the diameter of the circle.

The converse of the Pythagorean theorem is also a true statement: If the sum of the squares of the lengths of two sides of a triangle is equal to the square of the length of the third side, then the triangle is a right triangle. Use the distance formula and the Pythagorean theorem to determine whether the set of points could be vertices of a right triangle.

79. $(-4, 5)$, $(6, 1)$, and $(-8, -5)$

80. $(-3, 1)$, $(2, -1)$, and $(6, 9)$

81. $(-4, 3)$, $(0, 5)$, and $(3, -4)$

82. The points $(-3, 4)$, $(2, -1)$, $(5, 2)$, and $(0, 7)$ are vertices of a quadrilateral. Show that the quadrilateral is a rectangle. (*Hint*: Show that the opposite sides of the quadrilateral are the same length and that the two diagonals are the same length.)

Find the midpoint of the segment having the given endpoints.

83. $(4, -9)$ and $(-12, -3)$

84. $(7, -2)$ and $(9, 5)$

85. $\left(0, \frac{1}{2}\right)$ and $\left(-\frac{2}{5}, 0\right)$

86. $(0, 0)$ and $\left(-\frac{7}{13}, \frac{2}{7}\right)$

87. $(6.1, -3.8)$ and $(3.8, -6.1)$

88. $(-0.5, -2.7)$ and $(4.8, -0.3)$

89. $(-6, 5)$ and $(-6, 8)$

90. $(1, -2)$ and $(-1, 2)$

91. $\left(-\frac{1}{6}, -\frac{3}{5}\right)$ and $\left(-\frac{2}{3}, \frac{5}{4}\right)$

92. $\left(\frac{2}{9}, \frac{1}{3}\right)$ and $\left(-\frac{2}{5}, \frac{4}{5}\right)$

93. Graph the rectangle described in Exercise 82. Then determine the coordinates of the midpoint of each of the four sides. Are the midpoints vertices of a rectangle?

94. Graph the square with vertices $(-5, -1)$, $(7, -6)$, $(12, 6)$, and $(0, 11)$. Then determine the midpoint of each of the four sides. Are the midpoints vertices of a square?

95. The points $\left(\sqrt{7}, -4\right)$ and $\left(\sqrt{2}, 3\right)$ are endpoints of the diameter of a circle. Determine the center of the circle.

96. The points $\left(-3, \sqrt{5}\right)$ and $\left(1, \sqrt{2}\right)$ are endpoints of the diagonal of a square. Determine the center of the square.

In Exercises 97 and 98, how would you change the window so that the circle is not distorted? Answers may vary.

97. $(x + 3)^2 + (y - 2)^2 = 36$

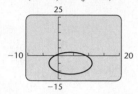

98. $(x - 4)^2 + (y + 5)^2 = 49$

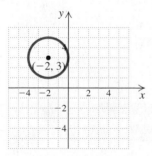

Find an equation for a circle satisfying the given conditions.

99. Center $(2, 3)$, radius of length $\frac{5}{3}$

100. Center $(4, 5)$, diameter of length 8.2

101. Center $(-1, 4)$, passes through $(3, 7)$

102. Center $(6, -5)$, passes through $(1, 7)$

103. The points $(7, 13)$ and $(-3, -11)$ are at the ends of a diameter.

104. The points $(-9, 4)$, $(-2, 5)$, $(-8, -3)$, and $(-1, -2)$ are vertices of an inscribed square.

105. Center $(-2, 3)$, tangent (touching at one point) to the y-axis

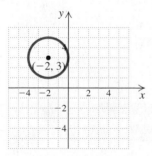

106. Center $(4, -5)$, tangent to the x-axis

Find the center and the radius of the circle. Then graph the circle by hand. Check your graph with a graphing calculator.

107. $x^2 + y^2 = 4$

108. $x^2 + y^2 = 81$

109. $x^2 + (y - 3)^2 = 16$

110. $(x + 2)^2 + y^2 = 100$

111. $(x - 1)^2 + (y - 5)^2 = 36$

112. $(x - 7)^2 + (y + 2)^2 = 25$

113. $(x + 4)^2 + (y + 5)^2 = 9$

114. $(x + 1)^2 + (y - 2)^2 = 64$

Find the equation of the circle.

115.

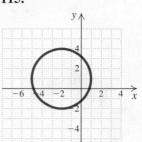

116.

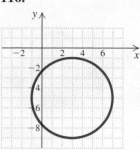

117.

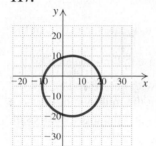

118.

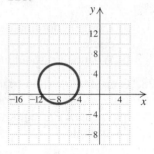

› Synthesis

To the student and the instructor: *The Synthesis exercises found at the end of every exercise set challenge students to combine concepts or skills studied in that section or in preceding parts of the text.*

119. If the point (p, q) is in the fourth quadrant, in which quadrant is the point $(q, -p)$?

Find the distance between the pair of points and find the midpoint of the segment having the given points as endpoints.

120. $\left(a, \dfrac{1}{a} \right)$ and $\left(a + h, \dfrac{1}{a + h} \right)$

121. $\left(a, \sqrt{a} \right)$ and $\left(a + h, \sqrt{a + h} \right)$

Find an equation of a circle satisfying the given conditions.

122. Center $(-5, 8)$ with a circumference of 10π units

123. Center $(2, -7)$ with an area of 36π square units

124. Find the point on the x-axis that is equidistant from the points $(-4, -3)$ and $(-1, 5)$.

125. Find the point on the y-axis that is equidistant from the points $(-2, 0)$ and $(4, 6)$.

126. Determine whether the points $(-1, -3)$, $(-4, -9)$, and $(2, 3)$ are collinear.

127. *An Arch of a Circle in Carpentry.* Matt is remodeling the front entrance to his home and needs to cut an arch for the top of an entranceway. The arch must be 8 ft wide and 2 ft high. To draw the arch, he will use a stretched string with chalk attached at one end as a compass.

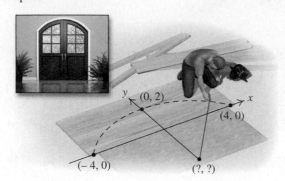

a) Using a coordinate system, locate the center of the circle.

b) What radius should Matt use to draw the arch?

128. Consider any right triangle with base *b* and height *h*, situated as shown. Show that the midpoint of the hypotenuse *P* is equidistant from the three vertices of the triangle.

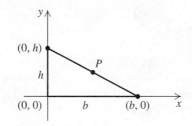

Determine whether each of the following points lies on the ***unit circle***, $x^2 + y^2 = 1$.

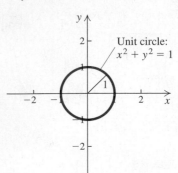

129. $\left(\dfrac{\sqrt{3}}{2}, -\dfrac{1}{2}\right)$

130. $(0, -1)$

131. $\left(-\dfrac{\sqrt{2}}{2}, \dfrac{\sqrt{2}}{2}\right)$

132. $\left(\dfrac{1}{2}, -\dfrac{\sqrt{3}}{2}\right)$

133. Prove the midpoint formula by showing that $\left(\dfrac{x_1 + x_2}{2}, \dfrac{y_1 + y_2}{2}\right)$ is equidistant from the points (x_1, y_1) and (x_2, y_2).

1.2 ▷ Functions and Graphs

> ❯ Determine whether a correspondence or a relation is a function.
> ❯ Find function values, or outputs, using a formula or a graph.
> ❯ Graph functions.
> ❯ Determine whether a graph is that of a function.
> ❯ Find the domain and the range of a function.
> ❯ Solve applied problems using functions.

We now focus our attention on a concept that is fundamental to many areas of mathematics—the idea of a *function*.

❯ Functions

We first consider an application.

Pay-As You-Go Exercise Program. A community center offers a pay-as you-go exercise program. The total cost is a membership fee of $20 plus $7.50 for every class attended. If a person attends 7 classes, the total cost is

$$\$7.50(7) + \$20, \quad \text{or} \quad \$72.50.$$

We can express this relationship with a set of ordered pairs, a graph, and an equation. A few ordered pairs are listed in the following table.

x	y	Ordered Pairs: (x, y)	Correspondence
1	27.50	$(1, 27.50)$	$1 \rightarrow 27.50$
2	35	$(2, 35)$	$2 \rightarrow 35$
4	50	$(4, 50)$	$4 \rightarrow 50$
7	72.50	$(7, 72.50)$	$7 \rightarrow 72.50$
10	95	$(10, 95)$	$10 \rightarrow 95$

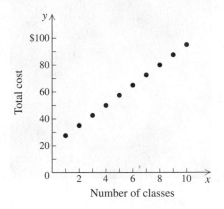

The ordered pairs express a relationship, or a correspondence, between the first coordinate and the second coordinate. We can see this relationship in the graph as well. The equation that describes the correspondence is

$$y = 7.50x + 20, \quad \text{where } x \text{ is a natural number.}$$

This is an example of a *function*. In this case, the total cost of the exercise classes y is a function of the number of classes attended x; that is, y is a function of x, where x is the independent variable and y is the dependent variable.

Let's consider some other correspondences before giving the definition of a function.

First Set	Correspondence	Second Set
To each person	there corresponds	that person's DNA.
To each truck sold	there corresponds	its price.
To each real number	there corresponds	the square of that number.

In each correspondence, the first set is called the **domain** and the second set is called the **range**. For each member, or **element**, in the domain, there is *exactly one* member in the range to which it corresponds. That is, each person has exactly *one* DNA, each truck has exactly *one* price, and each real number has exactly *one* square. Each correspondence is a *function*.

FUNCTION

A **function** is a correspondence between a first set, called the **domain**, and a second set, called the **range**, such that each member of the domain corresponds to *exactly one* member of the range.

It is important to note that not every correspondence between two sets is a function.

EXAMPLE 1   Determine whether each of the following correspondences is a function.

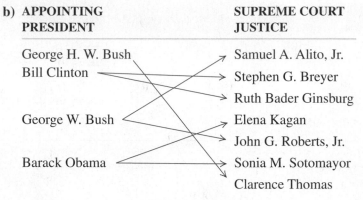

a)   −6 ⟶
     6 ⟶ 36

     −3 ⟶
     3 ⟶ 9

     0 ⟶ 0

b)

APPOINTING PRESIDENT	SUPREME COURT JUSTICE
George H. W. Bush	Samuel A. Alito, Jr.
Bill Clinton	Stephen G. Breyer
	Ruth Bader Ginsburg
George W. Bush	Elena Kagan
	John G. Roberts, Jr.
Barack Obama	Sonia M. Sotomayor
	Clarence Thomas

Solution

a) This correspondence *is* a function because each member of the domain corresponds to exactly one member of the range. Note that the definition of a function allows more than one member of the domain to correspond to the same member of the range.

b) This correspondence *is not* a function because there is at least one member of the domain that is paired with more than one member of the range (Bill Clinton with Stephen G. Breyer and Ruth Bader Ginsburg; George W. Bush with Samuel A. Alito, Jr., and John G. Roberts, Jr.; Barack Obama with Elena Kagan and Sonia M. Sotomayor).

> ⟩ *Now Try Exercises 5 and 7.*

EXAMPLE 2   Determine whether each of the following correspondences is a function.

DOMAIN	CORRESPONDENCE	RANGE
a) Years in which a presidential election occurs	The person elected	A set of presidents
b) All automobiles produced in 2016	Each automobile's VIN	A set of VINs
c) The set of all professional golfers who won a PGA tournament in 2014	The tournament won	The set of all PGA tournaments in 2014
d) The set of all PGA tournaments in 2014	The winner of the tournament	The set of all golfers who won a PGA tournament in 2014

Solution

a) This correspondence *is* a function because in each presidential election *exactly one* president is elected.

b) This correspondence *is* a function because each automobile has *exactly one* VIN.

c) This correspondence *is not* a function because a winning golfer could be paired with more than one tournament.

d) This correspondence *is* a function because each tournament has only one winning golfer.

> *Now Try Exercises 11 and 13.*

When a correspondence between two sets is not a function, it may still be an example of a **relation**.

RELATION

A **relation** is a correspondence between a first set, called the **domain**, and a second set, called the **range**, such that each member of the domain corresponds to *at least one* member of the range.

All the correspondences in Examples 1 and 2 are relations, but, as we have seen, not all are functions. Relations are sometimes written as sets of ordered pairs (as we saw earlier in the example on the total cost of attending a number of exercise classes) in which elements of the domain are the first coordinates of the ordered pairs and elements of the range are the second coordinates. For example, instead of writing $-3 \rightarrow 9$, as we did in Example 1(a), we could write the ordered pair $(-3, 9)$.

EXAMPLE 3 Determine whether each of the following relations is a function. Identify the domain and the range.

a) $\{(9, -5), (9, 5), (2, 4)\}$

b) $\{(-2, 5), (5, 7), (0, 1), (4, -2)\}$

c) $\{(-5, 3), (0, 3), (6, 3)\}$

Solution

a) The relation *is not* a function because the ordered pairs $(9, -5)$ and $(9, 5)$ have the same first coordinate and different second coordinates. (See Fig. 1.)

The domain is the set of all first coordinates: $\{9, 2\}$.

The range is the set of all second coordinates: $\{-5, 5, 4\}$.

b) The relation *is* a function because *no* two ordered pairs have the same first coordinate and different second coordinates. (See Fig. 2.)

The domain is the set of all first coordinates: $\{-2, 5, 0, 4\}$.

The range is the set of all second coordinates: $\{5, 7, 1, -2\}$.

c) The relation *is* a function because *no* two ordered pairs have the same first coordinate and different second coordinates. (See Fig. 3.)

The domain is $\{-5, 0, 6\}$.

The range is $\{3\}$.

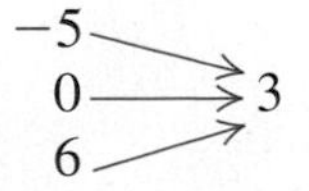

> *Now Try Exercises 17 and 19.*

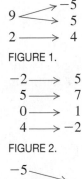

FIGURE 1.

FIGURE 2.

FIGURE 3.

❯ Notation for Functions

Functions used in mathematics are often given by equations. They generally require that certain calculations be performed in order to determine which member of the range is paired with each member of the domain. For example, in Section 1.1, we graphed the function $y = x^2 - 9x - 12$ by doing calculations like the following:

$$\text{for } x = -2, y = (-2)^2 - 9(-2) - 12 = 10,$$
$$\text{for } x = 0, y = 0^2 - 9 \cdot 0 - 12 = -12, \text{ and}$$
$$\text{for } x = 1, y = 1^2 - 9 \cdot 1 - 12 = -20.$$

A more concise notation is often used. For $y = x^2 - 9x - 12$, the **inputs** (members of the domain) are values of x substituted into the equation. The **outputs** (members of the range) are the resulting values of y. If we call the function f, we can use x to represent an arbitrary *input* and $f(x)$—read "f of x," or "f at x," or "the value of f at x"—to represent the corresponding *output*. In this notation, the function given by $y = x^2 - 9x - 12$ is written as $f(x) = x^2 - 9x - 12$, and the above calculations would be

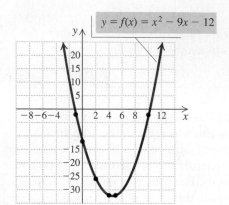

$y = f(x) = x^2 - 9x - 12$

$$f(-2) = (-2)^2 - 9(-2) - 12 = 10,$$
$$f(0) = 0^2 - 9 \cdot 0 - 12 = -12,$$
$$f(1) = 1^2 - 9 \cdot 1 - 12 = -20.$$

Keep in mind that $f(x)$ *does not* mean $f \cdot x$.

Thus, instead of writing "when $x = -2$, the value of y is 10," we can simply write "$f(-2) = 10$," which can be read as "f of -2 is 10" or "for the input -2, the output of f is 10." The letters g and h are also often used to name functions.

EXAMPLE 4   A function f is given by $f(x) = 2x^2 - x + 3$. Find each of the following.

a) $f(0)$ **b)** $f(-7)$
c) $f(5a)$ **d)** $f(a - 4)$

Solution   We can think of this formula as follows:

$$f(\,\blacksquare\,) = 2(\,\blacksquare\,)^2 - (\,\blacksquare\,) + 3.$$

Thus to find an output for a given input we think: "Whatever goes in the blank on the left goes in the blank(s) on the right." This gives us a "recipe" for finding outputs.

a) $f(0) = 2(0)^2 - 0 + 3 = 0 - 0 + 3 = 3$
b) $f(-7) = 2(-7)^2 - (-7) + 3 = 2 \cdot 49 + 7 + 3 = 108$
c) $f(5a) = 2(5a)^2 - 5a + 3 = 2 \cdot 25a^2 - 5a + 3 = 50a^2 - 5a + 3$
d) $f(a - 4) = 2(a - 4)^2 - (a - 4) + 3$
$$= 2(a^2 - 8a + 16) - a + 4 + 3$$
$$= 2a^2 - 16a + 32 - a + 4 + 3$$
$$= 2a^2 - 17a + 39$$

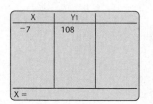

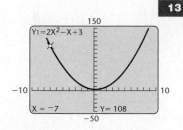

We can also find function values with a graphing calculator. Most calculators do not use function notation "$f(x) = \ldots$" to enter a function formula. Instead, we must enter the function using "$y = \ldots$." At left, we illustrate finding $f(-7)$ from part (b), first with the TABLE feature set in ASK mode and then with the VALUE feature from the CALC menu. We see on both screens that $f(-7) = 108$.

❯ **Now Try Exercise 21.**

❯ Graphs of Functions

We graph functions in the same way that we graph equations. We find ordered pairs (x, y), or $(x, f(x))$, plot points, and complete the graph.

EXAMPLE 5   Graph each of the following functions.

a) $f(x) = x^2 - 5$ **b)** $f(x) = x^3 - x$ **c)** $f(x) = \sqrt{x + 4}$

Solution   We select values for x and find the corresponding values of $f(x)$. Then we plot the points and connect them with a smooth curve.

a) $f(x) = x^2 - 5$

x	$f(x)$	$(x, f(x))$
-3	4	$(-3, 4)$
-2	-1	$(-2, -1)$
-1	-4	$(-1, -4)$
0	-5	$(0, -5)$
1	-4	$(1, -4)$
2	-1	$(2, -1)$
3	4	$(3, 4)$

b) $f(x) = x^3 - x$ **c)** $f(x) = \sqrt{x + 4}$

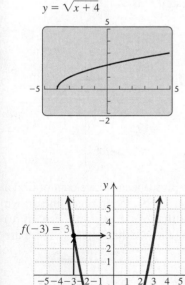

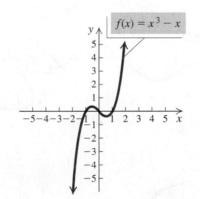

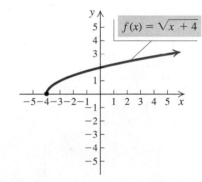

We can check the graphs with a graphing calculator. The checks for parts (b) and (c) are shown at left.

❯ *Now Try Exercise 33.*

Function values can also be determined from a graph.

EXAMPLE 6   For the function $f(x) = x^2 - 6$, use the graph at left to find each of the following function values.

a) $f(-3)$ **b)** $f(1)$

Solution

a) To find the function value $f(-3)$ from the graph, we locate the input -3 on the horizontal axis, move vertically to the graph of the function, and then move horizontally to find the output on the vertical axis. We see that $f(-3) = 3$.

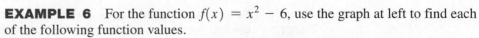

b) To find the function value $f(1)$, we locate the input 1 on the horizontal axis, move vertically to the graph, and then move horizontally to find the output on the vertical axis. We see that $f(1) = -5$.

> *Now Try Exercise 37.*

We know that when one member of the domain is paired with two or more different members of the range, the correspondence *is not* a function. Thus, when a graph contains two or more different points with the same first coordinate, the graph cannot represent a function. (See the graph at left. Note that 3 is paired with -1, 2, and 5.) Points sharing a common first coordinate are vertically above or below each other. This leads us to the *vertical-line test*.

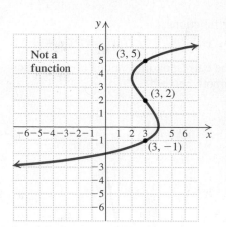

Since 3 is paired with more than one member of the range, the graph does not represent a function.

THE VERTICAL-LINE TEST

If it is possible for a vertical line to cross a graph more than once, then the graph *is not* the graph of a function.

To apply the vertical-line test, we try to find a vertical line that crosses the graph more than once. If we succeed, then the graph is not that of a function. If we do not, then the graph is that of a function.

EXAMPLE 7   Which of graphs (a)–(f) (in red) are graphs of functions? In graph (f), the solid dot shows that $(-1, 1)$ belongs to the graph. The open circle shows that $(-1, -2)$ does *not* belong to the graph.

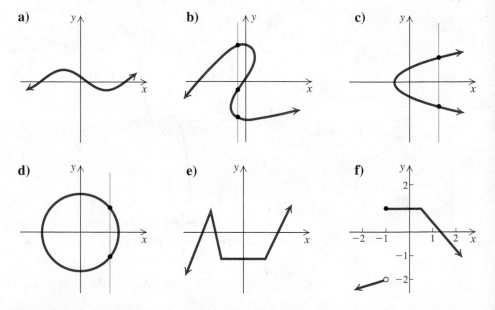

Solution   Graphs (a), (e), and (f) are graphs of functions because we cannot find a vertical line that crosses any of them more than once. In (b), the vertical line drawn crosses the graph at three points, so graph (b) is not that of a function. Also, in (c) and (d), we can find a vertical line that crosses the graph more than once, so these are not graphs of functions.

> *Now Try Exercises 45 and 49.*

❯ Finding Domains of Functions

When a function f whose inputs and outputs are real numbers is given by a formula, the *domain* is understood to be the set of all inputs for which the expression is defined as a real number. When an input results in an expression that is not defined as a real number, we say that the function value *does not exist* and that the number being substituted *is not* in the domain of the function.

EXAMPLE 8   Find the indicated function values, if possible, and determine whether the given values are in the domain of the function.

a) $f(1)$ and $f(3)$, for $f(x) = \dfrac{1}{x - 3}$

b) $g(16)$ and $g(-7)$, for $g(x) = \sqrt{x} + 5$

Solution

$y = 1/(x - 3)$

X	Y₁
1	−.5
3	ERROR

X =

$y = \sqrt{x} + 5$

X	Y₁
16	9
−7	ERROR

X =

a) $f(1) = \dfrac{1}{1 - 3} = \dfrac{1}{-2} = -\dfrac{1}{2}$

Since $f(1)$ is defined, 1 is in the domain of f.

$$f(3) = \frac{1}{3 - 3} = \frac{1}{0}$$

Since division by 0 is not defined, $f(3)$ does not exist and the number 3 is not in the domain of f. In a table from a graphing calculator, this is indicated with an ERROR message.

b) $g(16) = \sqrt{16} + 5 = 4 + 5 = 9$

Since $g(16)$ is defined, 16 is in the domain of g.

$$g(-7) = \sqrt{-7} + 5$$

Since $\sqrt{-7}$ is not defined as a real number, $g(-7)$ does not exist and the number -7 is not in the domain of g. Note the ERROR message in the table at left.

As we see in Example 8, inputs that make a denominator 0 or that yield a negative radicand in an even root are not in the domain of a function.

EXAMPLE 9   Find the domain of each of the following functions.

a) $f(x) = \dfrac{1}{x - 7}$

b) $h(x) = \dfrac{3x^2 - x + 7}{x^2 + 2x - 3}$

c) $f(x) = x^3 + |x|$

d) $g(x) = \sqrt[3]{x - 1}$

Just in Time
6, 14, 19

Solution

a) Because $x - 7 = 0$ when $x = 7$, the only input that results in a denominator of 0 is 7. The domain is $\{x \mid x \neq 7\}$. We can also write the solution using interval notation and the symbol $\cup$ for the **union**, or inclusion, of both sets: $(-\infty, 7) \cup (7, \infty)$.

b) We can substitute any real number in the numerator, but we must avoid inputs that make the denominator 0. To find those inputs, we solve $x^2 + 2x - 3 = 0$, or $(x + 3)(x - 1) = 0$. Since $x^2 + 2x - 3$ is 0 for -3 and 1, the domain consists of the set of all real numbers except -3 and 1, or $\{x \mid x \neq -3 \text{ and } x \neq 1\}$, or $(-\infty, -3) \cup (-3, 1) \cup (1, \infty)$.

c) We can substitute any real number for x. The domain is the set of all real numbers, or $(-\infty, \infty)$.

d) Because the index is odd, the radicand, $x - 1$, can be any real number. Thus x can be any real number. The domain is all real numbers, or $(-\infty, \infty)$.

Now Try Exercises 53, 55, and 59.

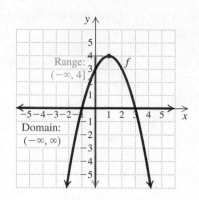

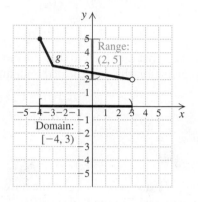

› Visualizing Domain and Range

Keep the following in mind regarding the *graph* of a function:

Domain = the set of a function's inputs, found on the horizontal *x*-axis;

Range = the set of a function's outputs, found on the vertical *y*-axis.

Consider the graph of function f, shown at left. To determine the domain of f, we look for the inputs on the *x*-axis that correspond to a point on the graph. We see that they include the entire set of real numbers, illustrated in red on the *x*-axis. Thus the domain is $(-\infty, \infty)$. To find the range, we look for the outputs on the *y*-axis that correspond to a point on the graph. We see that they include 4 and all real numbers less than 4, illustrated in blue on the *y*-axis. The bracket at 4 indicates that 4 is included in the interval. The range is $\{y \mid y \leq 4\}$, or $(-\infty, 4]$.

Let's now consider the graph of function g, shown at left. The solid dot shows that $(-4, 5)$ belongs to the graph. The open circle shows that $(3, 2)$ does *not* belong to the graph.

We see that the inputs of the function are -4 and all real numbers between -4 and 3, illustrated in red on the *x*-axis. The bracket at -4 indicates that -4 is included in the domain. The parenthesis at 3 indicates that 3 is not included in the domain. The domain is $\{x \mid -4 \leq x < 3\}$, or $[-4, 3)$. The outputs of the function are 5 and all real numbers between 2 and 5, illustrated in blue on the *y*-axis. The parenthesis at 2 indicates that 2 is not included in the range. The bracket at 5 indicates that 5 is included in the range. The range is $\{y \mid 2 < y \leq 5\}$, or $(2, 5]$.

EXAMPLE 10 Using the graph of the function, find the domain and the range of each of the following functions.

a) $f(x) = \frac{1}{2}x + 1$ **b)** $f(x) = \sqrt{x + 4}$

c) $f(x) = x^3 - x$ **d)** $f(x) = \dfrac{1}{x - 2}$

e) $f(x) = x^4 - 2x^2 - 3$ **f)** $f(x) = \sqrt{4 - (x - 3)^2}$

Solution

a)

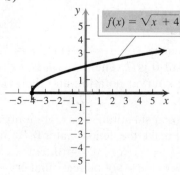

Domain = all real numbers, $(-\infty, \infty)$; range = all real numbers, $(-\infty, \infty)$

b)

Domain = $[-4, \infty)$; range = $[0, \infty)$

c)

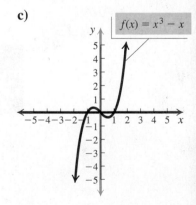

Domain = all real numbers, $(-\infty, \infty)$; range = all real numbers, $(-\infty, \infty)$

d)

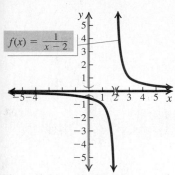

$f(x) = \dfrac{1}{x-2}$

Since the graph does not touch or cross either the vertical line $x = 2$ or the x-axis $y = 0$, 2 is excluded from the domain and 0 is excluded from the range. Domain $= (-\infty, 2) \cup (2, \infty)$; range $= (-\infty, 0) \cup (0, \infty)$

e)

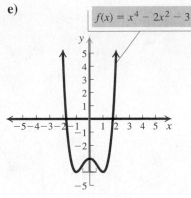

$f(x) = x^4 - 2x^2 - 3$

Domain $=$ all real numbers, $(-\infty, \infty)$; range $= [-4, \infty)$

f)

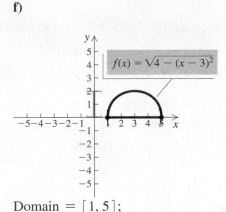

$f(x) = \sqrt{4 - (x - 3)^2}$

Domain $= [1, 5]$; range $= [0, 2]$

> *Now Try Exercises 75 and 83.*

Always consider adding the reasoning of Example 9 to a graphical analysis. Think, "What can I input?" to find the domain. Think, "What do I get out?" to find the range. Thus, in Examples 10(c) and 10(e), it might not appear as though the domain is all real numbers because the graph rises steeply, but by examining the equation we see that we can indeed substitute any real number for x.

> ## Applications of Functions

EXAMPLE 11 *Linear Expansion of a Bridge.* The linear expansion L of the steel center span of a suspension bridge that is 1420 m long is a function of the change in temperature t, in degrees Celsius, from winter to summer and is given by

$$L(t) = 0.000013 \cdot 1420 \cdot t,$$

where 0.000013 is the coefficient of linear expansion for steel and L is in meters. Find the linear expansion of the steel center span when the change in temperature from winter to summer is 30°, 42°, 50°, and 56° Celsius.

Solution We use a graphing calculator with the TABLE feature set in ASK mode to compute the function values. We find that

X	Y₁	
30	.5538	
42	.77532	
50	.923	
56	1.0338	

X =

$L(30) = 0.5538$ m,

$L(42) = 0.77532$ m,

$L(50) = 0.923$ m, and

$L(56) = 1.0338$ m.

> *Now Try Exercise 87.*

CONNECTING THE CONCEPTS

FUNCTION CONCEPTS

Formula for f: $f(x) = 5 + 2x^2 - x^4$.

For every input, there is exactly one output.

$(1, 6)$ is on the graph.

For the input 1, the output is 6.

$f(1) = 6$

Domain: set of all inputs $= (-\infty, \infty)$

Range: set of all outputs $= (-\infty, 6]$

GRAPH

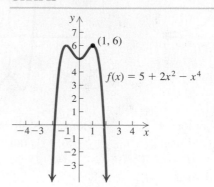

$f(x) = 5 + 2x^2 - x^4$

1.2 Exercise Set

In Exercises 1–14, determine whether the correspondence is a function.

1. $a \longrightarrow w$
$b \longrightarrow y$
$c \longrightarrow z$

2. $m \longrightarrow q$
$n \longrightarrow r$
$o \longrightarrow s$

3. $-6 \longrightarrow 36$
$-2 \longrightarrow 4$
$2 \nearrow$

4. $-3 \longrightarrow 2$
$1 \longrightarrow 4$
$5 \longrightarrow 6$
$9 \longrightarrow 8$

5. $m \longrightarrow A$
$n \searrow B$
$r \nearrow C$
$s \nearrow D$

6. $a \longrightarrow r$
$b \nearrow s$
$c \nearrow t$
d

7. PAINTING **ARTIST**

Night Watch
Old Guitarist
Irises, Saint-Remy
Starry Night
The Water-Lily Pond
Sunflowers
Mona Lisa
Woman with a Parasol
An Elephant

Vincent van Gogh
Claude Monet
Pablo Picasso
Rembrandt van Rijn
Leonardo da Vinci

8. COMPOSER **BROADWAY MUSICAL**

Marvin Hamlisch
Elton John
Claude-Michel Schönberg
Andrew Lloyd Webber

Billy Elliot the Musical
Cats
A Chorus Line
Evita
Les Misérables
The Lion King
The Phantom of the Opera
Miss Saigon

	DOMAIN	CORRESPONDENCE	RANGE
9.	A set of cars in a parking lot	Each car's license number	A set of letters and numbers
10.	A set of people in a town	A doctor a person uses	A set of doctors
11.	The integers less than 9	Five times the integer	A subset of integers
12.	A set of members of a rock band	An instrument each person plays	A set of instruments
13.	A set of students in a class	A student sitting in a neighboring seat	A set of students

14.

| A set of bags of chips on a shelf | Each bag's weight | A set of weights |

Determine whether the relation is a function. Identify the domain and the range.

15. $\{(2, 10), (3, 15), (4, 20)\}$

16. $\{(3, 1), (5, 1), (7, 1)\}$

17. $\{(-7, 3), (-2, 1), (-2, 4), (0, 7)\}$

18. $\{(1, 3), (1, 5), (1, 7), (1, 9)\}$

19. $\{(-2, 1), (0, 1), (2, 1), (4, 1), (-3, 1)\}$

20. $\{(5, 0), (3, -1), (0, 0), (5, -1), (3, -2)\}$

21. Given that $g(x) = 3x^2 - 2x + 1$, find each of the following.
   a) $g(0)$ **b)** $g(-1)$
   c) $g(3)$ **d)** $g(-x)$
   e) $g(1 - t)$

22. Given that $f(x) = 5x^2 + 4x$, find each of the following.
   a) $f(0)$ **b)** $f(-1)$
   c) $f(3)$ **d)** $f(t)$
   e) $f(t - 1)$

23. Given that $g(x) = x^3$, find each of the following.
   a) $g(2)$ **b)** $g(-2)$
   c) $g(-x)$ **d)** $g(3y)$
   e) $g(2 + h)$

24. Given that $f(x) = 2|x| + 3x$, find each of the following.
   a) $f(1)$ **b)** $f(-2)$
   c) $f(-x)$ **d)** $f(2y)$
   e) $f(2 - h)$

25. Given that
$$g(x) = \frac{x - 4}{x + 3},$$
find each of the following.
   a) $g(5)$ **b)** $g(4)$
   c) $g(-3)$ **d)** $g(-16.25)$
   e) $g(x + h)$

26. Given that
$$f(x) = \frac{x}{2 - x},$$
find each of the following.
   a) $f(2)$ **b)** $f(1)$
   c) $f(-16)$ **d)** $f(-x)$
   e) $f\left(-\frac{2}{3}\right)$

27. Find $g(0), g(-1), g(5)$, and $g\left(\frac{1}{2}\right)$ for
$$g(x) = \frac{x}{\sqrt{1 - x^2}}.$$

28. Find $h(0), h(2)$, and $h(-x)$ for
$$h(x) = x + \sqrt{x^2 - 1}.$$

In Exercises 29 and 30, use a graphing calculator and the TABLE *feature set in* ASK *mode.*

29. Given that
$$g(x) = 0.06x^3 - 5.2x^2 - 0.8x,$$
find $g(-2.1), g(5.08)$, and $g(10.003)$. Round answers to the nearest tenth.

30. Given that
$$h(x) = 3x^4 - 10x^3 + 5x^2 - x + 6,$$
find $h(-11), h(7)$, and $h(15)$.

Graph the function.

31. $f(x) = \frac{1}{2}x + 3$ **32.** $f(x) = \sqrt{x} - 1$

33. $f(x) = -x^2 + 4$ **34.** $f(x) = x^2 + 1$

35. $f(x) = \sqrt{x - 1}$ **36.** $f(x) = x - \frac{1}{2}x^3$

In Exercises 37–42, a graph of a function is shown. Using the graph, find the indicated function values; that is, given the inputs, find the outputs.

37. $h(1), h(3)$, and $h(4)$

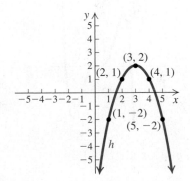

38. $t(-4), t(0)$, and $t(3)$

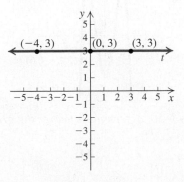

39. $s(-4)$, $s(-2)$, and $s(0)$

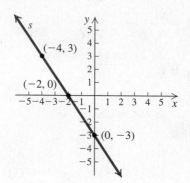

40. $g(-4)$, $g(-1)$, and $g(0)$

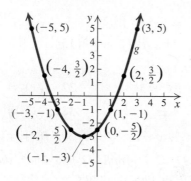

41. $f(-1)$, $f(0)$, and $f(1)$

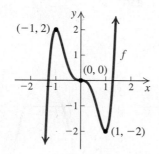

42. $g(-2)$, $g(0)$, and $g(2.4)$

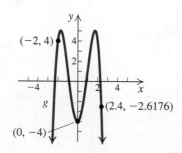

In Exercises 43–50, determine whether the graph is that of a function.

43.

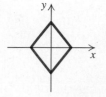

44.

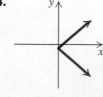

45.

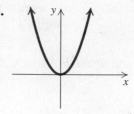

46.

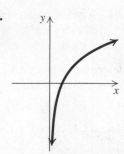

47.

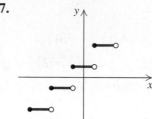

48.

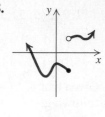

49.

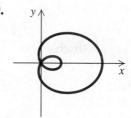

50.

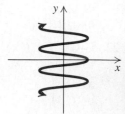

Find the domain of the function. Do not use a graphing calculator.

51. $f(x) = 7x + 4$

52. $f(x) = |3x - 2|$

53. $f(x) = |6 - x|$

54. $f(x) = \dfrac{1}{x^4}$

55. $f(x) = 4 - \dfrac{2}{x}$

56. $f(x) = \dfrac{1}{5}x^2 - 5$

57. $f(x) = \dfrac{x + 5}{2 - x}$

58. $f(x) = \dfrac{8}{x + 4}$

59. $f(x) = \dfrac{1}{x^2 - 4x - 5}$

60. $f(x) = \dfrac{(x - 2)(x + 9)}{x^3}$

61. $f(x) = \sqrt[3]{x + 10} - 1$

62. $f(x) = \sqrt[3]{4 - x}$

63. $f(x) = \dfrac{8 - x}{x^2 - 7x}$

64. $f(x) = \dfrac{x^4 - 2x^3 + 7}{3x^2 - 10x - 8}$

65. $f(x) = \frac{1}{10}|x|$

66. $f(x) = x^2 - 2x$

In Exercises 67–74, determine the domain and the range of the function.

67.

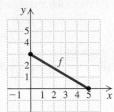

68.

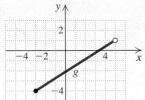

69.

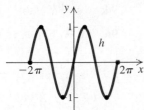

70.

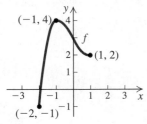

71.

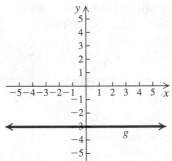

72.

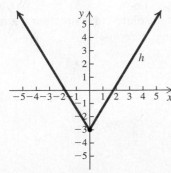

73.

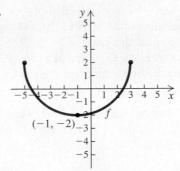

74.

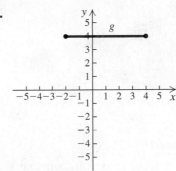

Graph the function with a graphing calculator. Then visually estimate the domain and the range.

75. $f(x) = |x|$

76. $f(x) = |x| - 2$

77. $f(x) = 3x - 2$

78. $f(x) = 5 - 3x$

79. $f(x) = \dfrac{1}{x - 3}$

80. $f(x) = \dfrac{1}{x + 1}$

81. $f(x) = (x - 1)^3 + 2$

82. $f(x) = (x - 2)^4 + 1$

83. $f(x) = \sqrt{7 - x}$

84. $f(x) = \sqrt{x + 8}$

85. $f(x) = -x^2 + 4x - 1$

86. $f(x) = 2x^2 - x^4 + 5$

87. *Boiling Point and Elevation.*   The elevation E, in meters, above sea level at which the boiling point of water is t degrees Celsius is given by the function

$$E(t) = 1000(100 - t) + 580(100 - t)^2.$$

At what elevation is the boiling point 99.5°? 100°?

88. *Windmill Power.*   Under certain conditions, the power P, in watts per hour, generated by a windmill with winds blowing v miles per hour is given by

$$P(v) = 0.015v^3.$$

Find the power generated by 15-mph winds and by 35-mph winds.

89. *Decreasing Value of the Dollar.*   In 2014, it took $23.63 to equal the value of $1 in 1913. In 2000, it took only $17.39 to equal the value of $1 in 1913. The amount that it takes to equal the value of $1 in 1913 can be estimated by the linear function V given by

$$V(x) = 0.4306x + 11.0043,$$

where x is the number of years since 1985. Thus, $V(10)$ gives the amount that it took in 1995 to equal the value of $1 in 1913.

Source: usinflationcalculator.com

a) Use this function to predict the amount that it will take in 2018 and in 2025 to equal the value of $1 in 1913.

b) When will it take approximately $32 to equal the value of $1 in 1913?

90. *Population of the United States.*   The population P of the United States in 1960 was 179,323,175. In 2015, the population was 320,400,215. The population of the United States can be estimated by the linear function P given by

$$P(x) = 2,511,040x + 151,143,509,$$

where x is the number of years after 1950. Thus, $P(20)$ gives the population in 1970.

a) Use this function to estimate the population in 1980 and in 2020.

b) When will the population be approximately 400,000,000?

❭ Skill Maintenance

To the student and the instructor: *The Skill Maintenance exercises review skills covered previously in the text. You can expect such exercises in every exercise set. They provide excellent review for a final examination. Answers to all skill maintenance exercises appear in the answer section at the back of the book. If you miss an exercise, restudy the objective shown in red next to the exercise or the instruction line that precedes it.*

Use substitution to determine whether the given ordered pairs are solutions of the given equation. **[1.1]**

91. $(-3, -2)$, $(2, -3)$; $y^2 - x^2 = -5$

92. $\left(\frac{4}{5}, -2\right)$, $\left(\frac{11}{5}, \frac{1}{10}\right)$; $15x - 10y = 32$

Graph the equation. **[1.1]**

93. $y = (x - 1)^2$

94. $y = \frac{1}{3}x - 6$

95. $-2x - 5y = 10$

96. $(x - 3)^2 + y^2 = 4$

❭ Synthesis

Find the domain of the function. Do not use a graphing calculator.

97. $f(x) = \sqrt[4]{2x + 5} + 3$

98. $f(x) = \dfrac{\sqrt{x + 1}}{x}$

99. $f(x) = \dfrac{\sqrt{x + 6}}{(x + 2)(x - 3)}$

100. $f(x) = \sqrt{x} - \sqrt{4 - x}$

101. Give an example of two different functions that have the same domain and the same range, but have no pairs in common. Answers may vary.

102. Draw a graph of a function for which the domain is $[-4, 4]$ and the range is $[1, 2] \cup [3, 5]$. Answers may vary.

103. Suppose that for some function g, $g(x + 3) = 2x + 1$. Find $g(-1)$.

104. Suppose that $f(x) = |x + 3| - |x - 4|$. Write $f(x)$ without using absolute-value notation if x is in each of the following intervals.

a) $(-\infty, -3)$

b) $[-3, 4)$

c) $[4, \infty)$

1.3 Linear Functions, Slope, and Applications

> ❯ Determine the slope of a line given two points on the line.
> ❯ Solve applied problems involving slope, or average rate of change.
> ❯ Find the slope and the *y*-intercept of a line given the equation $y = mx + b$, or $f(x) = mx + b$.
> ❯ Graph a linear equation using the slope and the *y*-intercept.
> ❯ Solve applied problems involving linear functions.

In real-life situations, we often need to make decisions on the basis of limited information. When the given information is used to formulate an equation or an inequality that at least approximates the situation mathematically, we have created a **model**. One of the most frequently used mathematical models is *linear*. The graph of a linear model is a straight line.

❯ Linear Functions

Let's examine the connections among equations, functions, and graphs that are *straight lines*. First, examine the graphs of linear functions and nonlinear functions shown here. Note that the graphs of the two types of functions are quite different.

Linear Functions

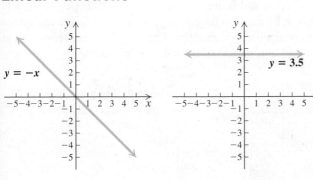

Nonlinear Functions

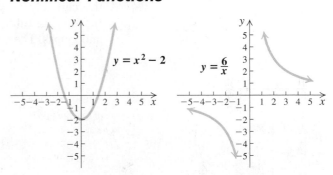

We begin with the definition of a linear function and related terminology, which are illustrated with graphs on the following page.

LINEAR FUNCTION

A function f is a **linear function** if it can be written as

$$f(x) = mx + b,$$

where m and b are constants.

If $m = 1$ and $b = 0$, then the function is the **identity function** $f(x) = x$.
If $m = 0$, then the function is a **constant function** $f(x) = b$.

Linear function:
$y = mx + b$

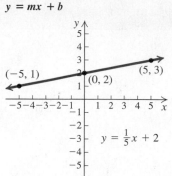

$$y = \frac{1}{5}x + 2$$

Identity function:
$y = 1 \cdot x + 0$, or $y = x$

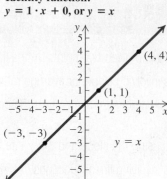

$y = x$

Constant function:
$y = 0 \cdot x + b$, or $y = b$ (Horizontal line)

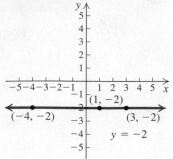

$y = -2$

Vertical line: $x = a$
(*not* a function)

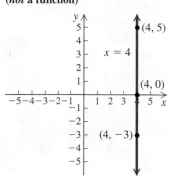

$x = 4$

HORIZONTAL LINES AND VERTICAL LINES

Horizontal lines are given by equations of the type $y = b$ or $f(x) = b$. (They *are* functions.)

Vertical lines are given by equations of the type $x = a$. (They *are not* functions.)

❯ The Linear Function $f(x) = mx + b$ and Slope

To attach meaning to the constant m in the equation $f(x) = mx + b$, we first consider an application. Suppose that Quality Foods is a wholesale supplier to restaurants that currently has stores in locations A and B in a large city. Their total operating costs for the same time period are given by the two functions shown in the following tables and graphs. The variable x represents time, in months. The variable y represents total costs, in thousands of dollars, over that period of time. Look for a pattern.

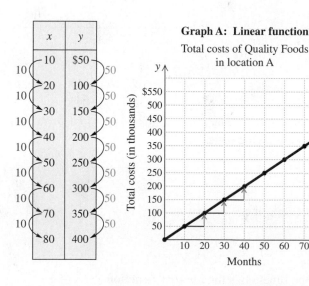

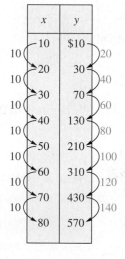

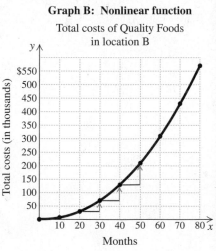

We see in graph A that *every* change of 10 months results in a $50 thousand change in total costs. But in graph B, changes of 10 months do *not* result in constant changes in total costs. This is a way to distinguish linear functions from nonlinear functions. The rate at which a linear function changes, or the steepness of its graph, is constant.

Mathematically, we define the steepness, or the **slope**, of a line as the ratio of its vertical change (*rise*) to the corresponding horizontal change (*run*). Slope represents the **rate of change** of y with respect to x.

SLOPE

The **slope m** of a line containing points (x_1, y_1) and (x_2, y_2) is given by

$$m = \frac{\text{rise}}{\text{run}}$$

$$= \frac{\text{the change in } y}{\text{the change in } x}$$

$$= \frac{y_2 - y_1}{x_2 - x_1} = \frac{y_1 - y_2}{x_1 - x_2}.$$

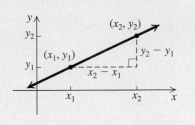

Just in Time

5

EXAMPLE 1   Graph the function $f(x) = -\frac{2}{3}x + 1$ and determine its slope.

Solution   Since the equation for f is in the form $f(x) = mx + b$, we know that it is a linear function. We can graph it by connecting two points on the graph with a straight line. We calculate two ordered pairs, plot the points, graph the function, and determine the slope:

$$f(3) = -\frac{2}{3} \cdot 3 + 1 = -2 + 1 = -1;$$

$$f(9) = -\frac{2}{3} \cdot 9 + 1 = -6 + 1 = -5;$$

Pairs: $(3, -1)$, $(9, -5)$;

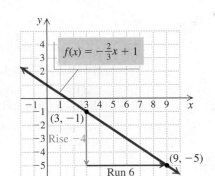

$$\text{Slope} = m = \frac{f(x_2) - f(x_1)}{x_2 - x_1} = \frac{y_2 - y_1}{x_2 - x_1}$$

$$= \frac{-5 - (-1)}{9 - 3} = \frac{-4}{6} = -\frac{2}{3}.$$

The slope is the same for any two points on a line. Thus, to check our work, note that $f(6) = -\frac{2}{3} \cdot 6 + 1 = -4 + 1 = -3$. Using the points $(6, -3)$ and $(3, -1)$, we have

$$m = \frac{-1 - (-3)}{3 - 6} = \frac{2}{-3} = -\frac{2}{3}.$$

EXPLORING WITH TECHNOLOGY

We can animate the effect of the slope m in linear functions of the type $f(x) = mx$ with a graphing calculator. Graph the equations

$$y_1 = x, \qquad y_2 = 2x,$$
$$y_3 = 5x, \quad \text{and} \quad y_4 = 10x$$

by entering them as $y_1 = \{1, 2, 5, 10\}x$. What do you think the graph of $y = 128x$ will look like?

Clear the screen and graph the equations

$$y_1 = -x, \qquad y_2 = -2x,$$
$$y_3 = -4x, \quad \text{and} \quad y_4 = -10x.$$

What do you think the graph of $y = -200x$ will look like?

We can also use the points in the opposite order when computing slope:

$$m = \frac{-3 - (-1)}{6 - 3} = \frac{-2}{3} = -\frac{2}{3}.$$

Note too that the slope of the line is the number m in the equation for the function $f(x) = -\frac{2}{3}x + 1$.

> ▶ *Now Try Exercises 7 and 31.*

The *slope* of the line given by $f(x) = mx + b$ is m.

If a line slants up from left to right, the change in x and the change in y have the same sign, so the line has a positive slope. The larger the slope, the steeper the line, as shown in Fig. 1. If a line slants down from left to right, the change in x and the change in y are of opposite signs, so the line has a negative slope. The larger the absolute value of the slope, the steeper the line, as shown in Fig. 2. Considering $y = mx$ when $m = 0$, we have $y = 0x$, or $y = 0$. Note that this horizontal line is the x-axis, as shown in Fig. 3.

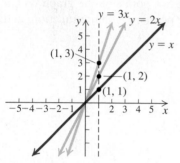

FIGURE 1.

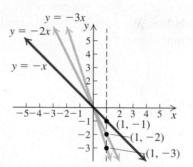

FIGURE 2.

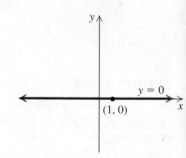

FIGURE 3.

HORIZONTAL LINES AND VERTICAL LINES

If a line is horizontal, the change in y for any two points is 0 and the change in x is nonzero. Thus a horizontal line has slope 0. (See Fig. 4.)

If a line is vertical, the change in x is 0. Thus the slope is *not defined* because we cannot divide by 0. (See Fig. 5.)

Horizontal lines

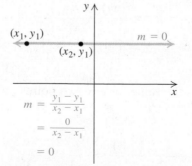

FIGURE 4.

Vertical lines

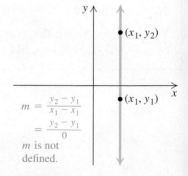

FIGURE 5.

Note that zero slope and an undefined slope are two very different concepts.

EXAMPLE 2 Graph each linear equation and determine its slope.

a) $x = -2$

b) $y = \frac{5}{2}$

Solution

a) Since y is missing in $x = -2$, any value for y will do.

x	y
-2	0
-2	3
-2	-4

Choose any number for y; x must be -2.

The graph is a *vertical line* 2 units to the left of the y-axis. (See Fig. 6.) The slope is not defined. The graph is *not* the graph of a function.

b) Since x is missing in $y = \frac{5}{2}$, any value for x will do.

x	y
0	$\frac{5}{2}$
-3	$\frac{5}{2}$
1	$\frac{5}{2}$

Choose any number for x; y must be $\frac{5}{2}$.

The graph is a *horizontal line* $\frac{5}{2}$, or $2\frac{1}{2}$, units above the x-axis. (See Fig. 7.) The slope is 0. The graph is the graph of a constant function.

> *Now Try Exercises 17 and 23.*

FIGURE 6.

FIGURE 7.

❯ Applications of Slope

Slope has many real-world applications. Numbers like 2%, 4%, and 7% are often used to represent the **grade** of a road. Such a number is meant to tell how steep a road is on a hill or a mountain. For example, a 4% grade means that the road rises (or falls) 4 ft for every horizontal distance of 100 ft.

Road grade $= \frac{a}{b}$
(Expressed as a percent)

The 2014 Olympic downhill course at Rosa Khutor Alpine Resort, located 40 km from Sochi, Russia, has the largest vertical drop ever built for an Olympic event. With a run of nearly 3500 m and a vertical drop of over 1075 m, the resulting grade, or slope, is approximately 31%. (*Source*: "Sochi's Gold Medal Ski Resort," by Brian Pinella, sochimagazine.com)

The concept of grade is also used with a treadmill. During a treadmill test, a cardiologist might change the slope, or grade, of the treadmill to measure its effect on a person's heart rate.

Another example occurs in hydrology. The strength or force of a river depends on how far the river falls vertically compared to how far it flows horizontally.

EXAMPLE 3 *Curb Ramps.* Curb ramps provide independent access to sidewalks for those who use wheelchairs. Guidelines for the grade of a curb ramp suggest a grade between 5.9% and 8.3%. A federal law states that every vertical rise of 1 ft requires a horizontal run of at least 12 ft. (*Source*: Federal Highway Administration, Office of Planning, Environment, and Realty) Find the grade of the curb ramp shown in the following figure.

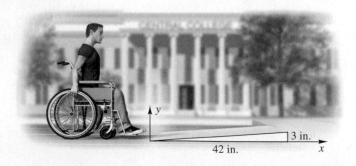

Solution The grade, or slope, is given by $m = \dfrac{3 \text{ in.}}{42 \text{ in.}} = \dfrac{1}{14} \approx 7.1\%$.

AVERAGE RATE OF CHANGE

Slope can also be considered as an **average rate of change**. To find the average rate of change between any two data points on a graph, we determine the slope of the line that passes through the two points.

EXAMPLE 4   *Food Stamp Program.*   The number of people participating in the federal Supplemental Nutrition Assistance Program has increased from 17.2 million in 2000 to 47.6 million in 2013. The following graph illustrates this upward trend. Find the average rate of change in the number of people using food stamps from 2000 to 2013.

Enrollment in the Federal Supplemental Nutrition Assistance Program

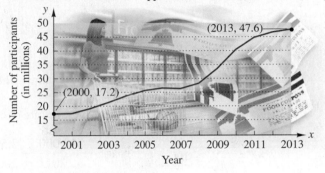

Source: U.S. Department of Agriculture

Solution   We use the coordinates of two points on the graph. In this case, we use (2000, 17.2) and (2013, 47.6). Then we compute the slope, or the average rate of change, as follows:

$$\text{Slope} = \text{Average rate of change}$$

$$= \frac{\text{Change in } y}{\text{Change in } x} = \frac{47.6 - 17.2}{2013 - 2000}$$

$$= \frac{30.4}{13} \approx 2.3.$$

The result tells us that each year from 2000 to 2013, the number of participants in the federal Supplemental Nutrition Assistance Program increased an average of 2.3 million. The average rate of change over this 13-year period was an increase of 2.3 million participants per year.

> *Now Try Exercise 41.*

EXAMPLE 5 *Oil Imports.* Increased oil production in the United States has resulted in decreased imports of crude oil. The total number of barrels imported in 2008 was 3,590,000. This number had decreased to 2,810,000 barrels in 2013. (*Source*: U.S. Census Bureau) Find the average rate of change in crude oil imports from 2008 to 2013.

Crude Oil Imports

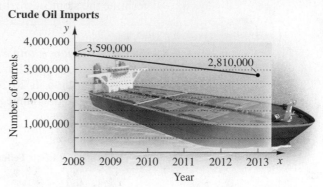

Source: U.S. Census Bureau

Solution Using the points (2008, 3,590,000) and (2013, 2,810,000), we compute the slope of the line containing these two points:

$$\text{Slope} = \text{Average rate of change} = \frac{\text{Change in } y}{\text{Change in } x}$$

$$= \frac{2,810,000 - 3,590,000}{2013 - 2008} = \frac{-780,000}{5} = -156,000.$$

The result tells us that each year from 2008 to 2013, the number of barrels of imported crude oil decreased on average 156,000 barrels. The average rate of change over the 5-year period was a decrease of 156,000 barrels per year.

❯ *Now Try Exercise 47.*

❯ Slope–Intercept Equations of Lines

y-INTERCEPT

REVIEW SECTION 1.1.

Compare the graphs of the equations

$$y = 3x \quad \text{and} \quad y = 3x - 2.$$

Note that the graph of $y = 3x - 2$ is a shift of the graph of $y = 3x$ down 2 units and that $y = 3x - 2$ has *y*-intercept $(0, -2)$. That is, the graph is parallel to $y = 3x$ and it crosses the *y*-axis at $(0, -2)$. The point $(0, -2)$ is the *y-intercept* of the graph.

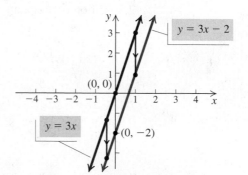

$y = -0.25x - 3.8$

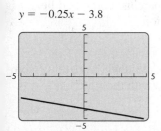

$y = \frac{1}{2}x - \frac{7}{6}$

Just in Time

2

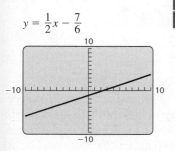

THE SLOPE–INTERCEPT EQUATION

The linear function f given by

$$f(x) = mx + b$$

is written in slope–intercept form. The graph of an equation in this form is a straight line parallel to $f(x) = mx$. The constant m is called the slope, and the y-intercept is $(0, b)$.

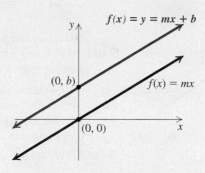

We can read the slope m and the y-intercept $(0, b)$ directly from the equation of a line written in slope–intercept form $y = mx + b$.

EXAMPLE 6 Find the slope and the y-intercept of the line with equation $y = -0.25x - 3.8$.

Solution

$$y = -0.25x - 3.8$$

Slope $= -0.25$; y-intercept $= (0, -3.8)$

> **Now Try Exercise 49.**

Any equation whose graph is a straight line is a **linear equation**. To find the slope and the y-intercept of the graph of a linear equation, we can solve for y, and then read the information from the equation.

EXAMPLE 7 Find the slope and the y-intercept of the line with equation $3x - 6y - 7 = 0$.

Solution We solve for y:

$$3x - 6y - 7 = 0$$
$$-6y = -3x + 7 \qquad \text{Adding } -3x \text{ and } 7 \text{ on both sides}$$
$$-\tfrac{1}{6}(-6y) = -\tfrac{1}{6}(-3x + 7) \qquad \text{Multiplying by } -\tfrac{1}{6} \text{ on both sides}$$
$$y = \tfrac{1}{2}x - \tfrac{7}{6}. \qquad \text{Using a distributive law}$$

Thus the slope is $\frac{1}{2}$, and the y-intercept is $\left(0, -\frac{7}{6}\right)$.

> **Now Try Exercise 61.**

❯ Graphing $f(x) = mx + b$ Using m and b

We can also graph a linear equation using its slope and y-intercept.

EXAMPLE 8   Graph: $y = -\frac{2}{3}x + 4$.

Solution   This equation is in slope–intercept form, $y = mx + b$. The y-intercept is $(0, 4)$. We plot this point. We can think of the slope $\left(m = -\frac{2}{3}\right)$ as $\frac{-2}{3}$.

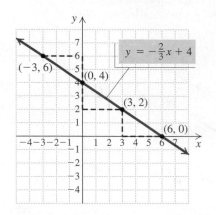

$$m = \frac{\text{rise}}{\text{run}} = \frac{\text{change in } y}{\text{change in } x} = \frac{-2}{3} \quad \begin{array}{l} \leftarrow \textbf{Move 2 units down.} \\ \leftarrow \textbf{Move 3 units to the right.} \end{array}$$

Starting at the y-intercept and using the slope, we find another point by moving 2 units down and 3 units to the right. We get a new point $(3, 2)$. In a similar manner, we can move from $(3, 2)$ to find another point, $(6, 0)$.

We could also think of the slope $\left(m = -\frac{2}{3}\right)$ as $\frac{2}{-3}$. Then we can start at $(0, 4)$ and move 2 units up and 3 units to the left. We get to another point on the graph, $(-3, 6)$. We now plot the points and draw the line. Note that we need only the y-intercept and one other point in order to graph the line, but it's a good idea to find a third point as a check that the first two points are correct.

❯ **Now Try Exercise 63.**

❯ Applications of Linear Functions

We now consider an application of linear functions.

EXAMPLE 9   *Estimating Adult Height.*   There is no *proven* way to predict a child's adult height, but a linear function can be used to *estimate* it, given the sum of the heights of the child's parents. The adult height M, in inches, of a male child whose parents' combined height is x, in inches, can be estimated with the function

$$M(x) = 0.5x + 2.5.$$

The adult height F, in inches, of a female child whose parents' combined height is x, in inches, can be estimated with the function

$$F(x) = 0.5x - 2.5.$$

(*Source*: Jay L. Hoecker, M.D., MayoClinic.com) Estimate the height of a female child whose parents' combined height is 135 in. What is the domain of this function?

Solution   We substitute in the function:

$$F(135) = 0.5(135) - 2.5 = 65.$$

Thus we can estimate the adult height of the female child as 65 in., or 5 ft 5 in.

Theoretically, the domain of the function is the set of all real numbers. However, the context of the problem dictates a different domain. Thus the domain consists of all positive real numbers—that is, the interval $(0, \infty)$. A more realistic domain might be 100 in. to 110 in.—that is, the interval $[100, 110]$.

❯ **Now Try Exercise 73.**

Visualizing the Graph

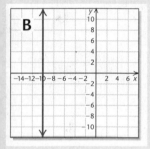

A

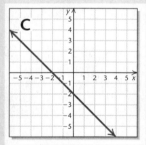

B

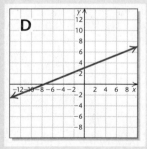

C

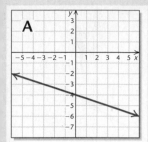

D

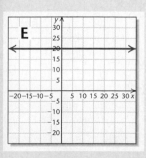

E

Match the equation with its graph.

1. $y = 20$

2. $5y = 2x + 15$

3. $y = -\dfrac{1}{3}x - 4$

4. $x = \dfrac{5}{3}$

5. $y = -x - 2$

6. $y = 2x$

7. $y = -3$

8. $3y = -4x$

9. $x = -10$

10. $y = x + \dfrac{7}{2}$

Answers on page A-3

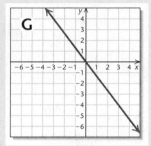

F

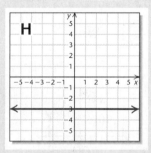

G

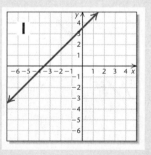

H

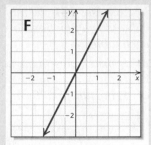

I

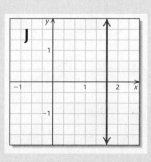

J

1.3 Exercise Set

In Exercises 1–4, the table of data contains input–output values for a function. Answer the following questions for each table.

a) *Is the change in the inputs x the same?*
b) *Is the change in the outputs y the same?*
c) *Is the function linear?*

1.

x	y
−3	7
−2	10
−1	13
0	16
1	19
2	22
3	25

2.

x	y
20	12.4
30	24.8
40	49.6
50	99.2
60	198.4
70	396.8
80	793.6

3.

x	y
11	3.2
26	5.7
41	8.2
56	9.3
71	11.3
86	13.7
101	19.1

4.

x	y
2	−8
4	−12
6	−16
8	−20
10	−24
12	−28
14	−32

Find the slope of the line containing the given points.

5.

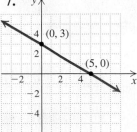

6.

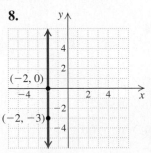

7.

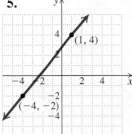

8.

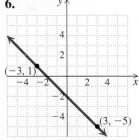

9.

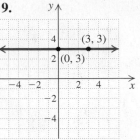

10.

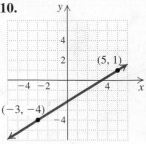

11. $(9, 4)$ and $(−1, 2)$

12. $(−3, 7)$ and $(5, −1)$

13. $(4, −9)$ and $(4, 6)$

14. $(−6, −1)$ and $(2, −13)$

15. $(0.7, −0.1)$ and $(−0.3, −0.4)$

16. $\left(−\frac{3}{4}, −\frac{1}{4}\right)$ and $\left(\frac{2}{7}, −\frac{5}{7}\right)$

17. $(2, −2)$ and $(4, −2)$

18. $(−9, 8)$ and $(7, −6)$

19. $\left(\frac{1}{2}, −\frac{3}{5}\right)$ and $\left(−\frac{1}{2}, \frac{3}{5}\right)$

20. $(−8.26, 4.04)$ and $(3.14, −2.16)$

21. $(16, −13)$ and $(−8, −5)$

22. $(\pi, −3)$ and $(\pi, 2)$

23. $(−10, −7)$ and $(−10, 7)$

24. $\left(\sqrt{2}, −4\right)$ and $(0.56, −4)$

25. $f(4) = 3$ and $f(−2) = 15$

26. $f(−4) = −5$ and $f(4) = 1$

27. $f\left(\frac{1}{5}\right) = \frac{1}{2}$ and $f(−1) = −\frac{11}{2}$

28. $f(8) = −1$ and $f\left(−\frac{2}{3}\right) = \frac{10}{3}$

29. $f(−6) = \frac{4}{5}$ and $f(0) = \frac{4}{5}$

30. $g\left(−\frac{9}{2}\right) = \frac{2}{9}$ and $g\left(\frac{2}{5}\right) = −\frac{5}{2}$

Determine the slope, if it exists, of the graph of the given linear equation.

31. $y = 1.3x − 5$

32. $y = −\frac{2}{5}x + 7$

33. $x = −2$

34. $f(x) = 4x − \frac{1}{4}$

35. $f(x) = −\frac{1}{2}x + 3$

36. $y = \frac{3}{4}$

37. $y = 9 − x$

38. $x = 8$

39. $y = 0.7$

40. $y = \frac{4}{5} − 2x$

41. *Price of a World Series Ticket.* In 1946, the lowest price of a World Series ticket was $1.20. By 2012, the lowest price of a ticket had increased to $110. (*Source*: AARP.com) Find the average rate of change in the lowest price of a World Series ticket from 1946 to 2012.

42. *Population Loss.* The population of Detroit, Michigan, decreased from 1,027,974 in 1990 to 688,701 in 2013 (*Source*: U.S. Census Bureau). Find the average rate of change in the population of Detroit, Michigan, over the 23-year period.

43. *Population Loss.* The population of Cleveland, Ohio, decreased from 478,403 in 2000 to 390,113 in 2013 (*Source*: U.S. Census Bureau). Find the average rate of change in the population of Cleveland, Ohio, over the 13-year period.

44. *Decreasing Size of Cattle Herd.* Drought has been the major reason for the decrease in the U.S. cattle herd in recent years. The number of cattle is at its lowest level since 1952. In 2006, there were 96.6 million head of cattle. This number had fallen to 87.7 million by 2014. (*Source*: U.S. Department of Agriculture) Find the average rate of change in the number of cattle from 2006 to 2014.

45. *Whole-Milk Consumption.* The annual per-capita consumption of whole milk in the United States was 25.3 gal in 1970. By 2011, this amount had decreased to 5.5 gal. Find the average rate of change in per-capita consumption of whole milk from 1970 to 2011.

46. *Chicken Consumption.* The annual per-capita consumption of chicken in the United States was 42.5 lb in 1990. By 2011, this amount had increased to 58.4 lb. (*Source*: Economic Research Service, U.S. Department of Agriculture) Find the average rate of change in per-capita consumption of chicken from 1990 to 2011.

47. *Growing Almonds.* In 2003, 550,000 acres of farmland in California were devoted to growing almonds. By 2012, the number of acres used to grow almonds had increased to 810,000. (*Source*: USDA National Agricultural Statistics Service) Find the average rate of change in the number of acres in California used to grow almonds from 2003 to 2012.

48. *ATM Fees.* The average fee to use an out-of-network ATM was $2.66 in 2004. By 2014, this fee had increased to $4.35. (*Source*: Bankrate.com) Find the average rate of change in out-of-network ATM fees from 2004 to 2014.

Find the slope and the y-intercept of the line with the given equation.

49. $y = \frac{3}{5}x - 7$

50. $f(x) = -2x + 3$

51. $x = -\frac{2}{5}$

52. $y = \frac{4}{7}$

53. $f(x) = 5 - \frac{1}{2}x$

54. $y = 2 + \frac{3}{7}x$

55. $3x + 2y = 10$

56. $2x - 3y = 12$

57. $y = -6$

58. $x = 10$

59. $5y - 4x = 8$

60. $5x - 2y + 9 = 0$

61. $4y - x + 2 = 0$

62. $f(x) = 0.3 + x$

Graph the equation using the slope and the y-intercept.

63. $y = -\frac{1}{2}x - 3$ **64.** $y = \frac{3}{2}x + 1$

65. $f(x) = 3x - 1$ **66.** $f(x) = -2x + 5$

67. $3x - 4y = 20$ **68.** $2x + 3y = 15$

69. $x + 3y = 18$ **70.** $5y - 2x = -20$

71. *Whales and Pressure at Sea Depth.* Whales can withstand extreme changes in atmospheric pressure because their bodies are flexible. Their rib cages and lungs can collapse safely under pressure. Sperm whales can hunt for squid at depths of 7000 ft or more. (*Sources*: National Ocean Service; National Oceanic and Atmospheric Administration) The function P, given by

$$P(d) = \frac{1}{33}d + 1,$$

gives the pressure, in atmospheres (atm), at a given depth d, in feet, under the sea.

a) Graph P.
b) Find $P(0), P(33), P(1000), P(5000),$ and $P(7000)$.

72. *Stopping Distance on Glare Ice.* The stopping distance (at some fixed speed) of regular tires on glare ice is a function of the air temperature F, in degrees Fahrenheit. This function is estimated by

$$D(F) = 2F + 115,$$

where $D(F)$ is the stopping distance, in feet, when the air temperature is F, in degrees Fahrenheit.

a) Graph D.
b) Find $D(-20°), D(0°), D(10°),$ and $D(32°)$.
c) Explain why the domain should be restricted to $[-57.5°, 32°]$.

73. *Reaction Time.* Suppose that while driving a car, you suddenly see a deer standing in the road. Your brain registers the information and sends a signal to your foot to hit the brake. The car travels a distance D, in feet, during this time, where D is a function of the speed r, in miles per hour, of the car when you see the deer. That reaction distance is a linear function given by

$$D(r) = \tfrac{11}{10}r + \tfrac{1}{2}.$$

a) Find the slope of this line and interpret its meaning in this application.
b) Graph D.
c) Find $D(5), D(10), D(20), D(50),$ and $D(65)$.
d) What is the domain of this function? Explain.

74. *Straight-Line Depreciation.* A contractor buys a new truck for $38,000. The truck is purchased on January 1 and is expected to last 5 years, at the end of which time its *trade-in*, or *salvage, value* will be $16,500. If the company figures the decline or depreciation in value to be the same each year, then the salvage value V, after t years, is given by the linear function

$$V(t) = \$38{,}000 - \$4300t, \quad \text{for } 0 \le t \le 5.$$

a) Find $V(0), V(1), V(2), V(3),$ and $V(5)$.
b) Find the domain and the range of this function.

75. *Total Cost.* Richard is considering relocating to an assisted living facility. He learns that there is an initial community fee of $2250 and a monthly charge of $3380 for level-one care. Write an equation that can be used to determine the total cost $C(t)$ for t months of level-one care. Then find the total cost for 20 months.

76. *Total Cost.* Superior Cable Television charges a $95 installation fee and $125 per month for the Star plan. Write an equation that can be used to determine the total cost $C(t)$ for t months of the Star plan. Then find the total cost for 18 months of service.

*In Exercises 77 and 78, the term **fixed costs** refers to the start-up costs of operating a business. This includes machinery and building costs. The term **variable costs** refers to what it costs a business to produce or service one item.*

77. Max's Custom Lacrosse Stringing experienced fixed costs of $750 and variable costs of $15 for each lacrosse stick that was restrung. Write an equation that can be used to determine the total cost when x sticks are restrung. Then determine the total cost of restringing 32 lacrosse sticks.

78. Soosie's Cookie Company had fixed costs of $1250 and variable costs of $4.25 per dozen gourmet cookies that were baked and packaged for sale. Write an equation that can be used to determine the total cost when x dozens of cookies are baked and sold. Then determine the total cost of baking and selling 85 dozen gourmet cookies.

❯ **Skill Maintenance**

If $f(x) = x^2 - 3x$, find each of the following. **[1.2]**

79. $f\left(\frac{1}{2}\right)$ **80.** $f(5)$

81. $f(-5)$ **82.** $f(-a)$

83. $f(a + h)$

❯ **Synthesis**

84. *Grade of Treadmills.* A treadmill is 5 ft long and is set at an 8% grade. How high is the end of the treadmill?

Find the slope of the line containing the given points.

85. (a, a^2) and $(a + h, (a + h)^2)$

86. $(r, s + t)$ and (r, s)

Suppose that f is a linear function. Determine whether each of the following statements is true or false.

87. $f(c - d) = f(c) - f(d)$

88. $f(kx) = kf(x)$

Let $f(x) = mx + b$. Find a formula for $f(x)$ given each of the following.

89. $f(x + 2) = f(x) + 2$

90. $f(3x) = 3f(x)$

Mid-Chapter Mixed Review

Determine whether each of the following statements is true or false.

1. The x-intercept of the line that passes through $\left(-\frac{2}{3}, \frac{3}{2}\right)$ and the origin is $\left(-\frac{2}{3}, 0\right)$. **[1.1]**

2. All functions are relations, but not all relations are functions. **[1.2]**

3. The line parallel to the y-axis that passes through $(-5, 25)$ is $y = -5$. **[1.3]**

4. Find the intercepts of the graph of the line $-8x + 5y = -40$. **[1.1]**

For each pair of points, find the distance between the points and the midpoint of the segment having the points as endpoints. **[1.1]**

5. $(-8, -15)$ and $(3, 7)$

6. $\left(-\frac{3}{4}, \frac{1}{5}\right)$ and $\left(\frac{1}{4}, -\frac{4}{5}\right)$

7. Find an equation for a circle having center $(-5, 2)$ and radius 13. **[1.1]**

8. Find the center and the radius of the circle given by the equation $(x - 3)^2 + (y + 1)^2 = 4$. **[1.1]**

Graph the equation.

9. $3x - 6y = 6$ **[1.1]**

10. $y = -\frac{1}{2}x + 3$ **[1.3]**

11. $y = 2 - x^2$ **[1.1]**

12. $(x + 4)^2 + y^2 = 4$ **[1.1]**

13. Given that $f(x) = x - 2x^2$, find $f(-4), f(0)$, and $f(1)$. **[1.2]**

14. Given that $g(x) = \dfrac{x + 6}{x - 3}$, find $g(-6), g(0)$, and $g(3)$. **[1.2]**

Find the domain of the function. **[1.2]**

15. $g(x) = x + 9$

16. $f(x) = \dfrac{-5}{x + 5}$

17. $h(x) = \dfrac{1}{x^2 + 2x - 3}$

Graph the function. **[1.2]**

18. $f(x) = -2x$

19. $g(x) = x^2 - 1$

20. Determine the domain and the range of the function shown in the figure at right. **[1.2]**

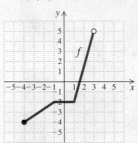

Find the slope of the line containing the given points. **[1.3]**

21. $(-2, 13)$ and $(-2, -5)$

22. $(10, -1)$ and $(-6, 3)$

23. $\left(\frac{5}{7}, \frac{1}{3}\right)$ and $\left(\frac{2}{7}, \frac{1}{3}\right)$

Determine the slope, if it exists, and the y-intercept of the line with the given equation. **[1.3]**

24. $f(x) = -\frac{1}{9}x + 12$

25. $y = -6$

26. $x = 2$

27. $3x - 16y + 1 = 0$

COLLABORATIVE DISCUSSION AND WRITING

To the student and the instructor: *The Collaborative Discussion and Writing exercises are meant to be answered with one or more sentences. They can be discussed and answered collaboratively by the entire class or by small groups.*

28. Explain as you would to a fellow student how the numerical value of the slope of a line can be used to describe the slant and the steepness of that line. **[1.3]**

29. Discuss why the graph of a vertical line $x = a$ cannot represent a function. **[1.3]**

30. Explain in your own words the difference between the domain of a function and the range of a function. **[1.2]**

31. Explain how you could find the coordinates of a point $\frac{7}{8}$ of the way from point A to point B. **[1.1]**

1.4 > Equations of Lines and Modeling

> Determine equations of lines.

> Given the equations of two lines, determine whether their graphs are parallel or perpendicular.

> Model a set of data with a linear function.

> Fit a regression line to a set of data. Then use the linear model to make predictions.

> Slope–Intercept Equations of Lines

In Section 1.3, we developed the slope–intercept equation $y = mx + b$, or $f(x) = mx + b$. If we know the slope and the y-intercept of a line, we can find an equation of the line using the slope–intercept equation.

EXAMPLE 1   A line has slope $-\frac{7}{9}$ and y-intercept $(0, 16)$. Find an equation of the line.

Solution   We use the slope–intercept equation and substitute $-\frac{7}{9}$ for m and 16 for b:

$$y = mx + b$$
$$y = -\frac{7}{9}x + 16, \quad \text{or}$$
$$f(x) = -\frac{7}{9}x + 16.$$

> **Now Try Exercise 7.**

EXAMPLE 2   A line has slope $-\frac{2}{3}$ and contains the point $(-3, 6)$. Find an equation of the line.

Solution   We use the slope–intercept equation, $y = mx + b$, and substitute $-\frac{2}{3}$ for m: $y = -\frac{2}{3}x + b$. Then, using the point $(-3, 6)$, we substitute -3 for x and 6 for y in $y = -\frac{2}{3}x + b$. Finally, we solve for b.

$$y = mx + b$$
$$y = -\frac{2}{3}x + b \qquad \text{Substituting } -\frac{2}{3} \text{ for } m$$
$$6 = -\frac{2}{3}(-3) + b \qquad \text{Substituting } -3 \text{ for } x \text{ and } 6 \text{ for } y$$
$$6 = 2 + b$$
$$4 = b \qquad \text{Solving for } b. \text{ The } y\text{-intercept is } (0, b).$$

The equation of the line is $y = -\frac{2}{3}x + 4$, or $f(x) = -\frac{2}{3}x + 4$.

> **Now Try Exercise 13.**

> Point–Slope Equations of Lines

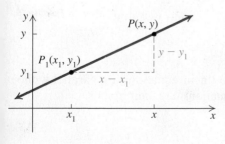

Another formula that can be used to determine an equation of a line is the *point–slope equation*. Suppose that we have a nonvertical line and that the coordinates of point P_1 on the line are (x_1, y_1). We can think of P_1 as fixed and imagine another point P on the line with coordinates (x, y). Thus the slope is given by

$$\frac{y - y_1}{x - x_1} = m.$$

Multiplying by $x - x_1$ on both sides, we get the *point–slope equation* of the line:

$$(x - x_1) \cdot \frac{y - y_1}{x - x_1} = m \cdot (x - x_1)$$
$$y - y_1 = m(x - x_1).$$

POINT–SLOPE EQUATION

The **point–slope equation** of the line with slope m passing through (x_1, y_1) is

$$y - y_1 = m(x - x_1).$$

If we know the slope of a line and the coordinates of one point on the line, we can find an equation of the line using either the point–slope equation,

$$y - y_1 = m(x - x_1),$$

or the slope–intercept equation,

$$y = mx + b.$$

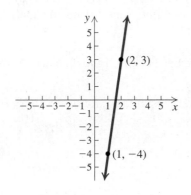

EXAMPLE 3   Find an equation of the line containing the points $(2, 3)$ and $(1, -4)$.

Solution   We first determine the slope:

$$m = \frac{-4 - 3}{1 - 2} = \frac{-7}{-1} = 7.$$

Using the Point–Slope Equation: We substitute 7 for m and either of the points $(2, 3)$ or $(1, -4)$ for (x_1, y_1) in the point–slope equation. In this case, we use $(2, 3)$.

$y - y_1 = m(x - x_1)$	**Point–slope equation**
$y - 3 = 7(x - 2)$	**Substituting**
$y - 3 = 7x - 14$	
$y = 7x - 11,$ or	
$f(x) = 7x - 11$	

Using the Slope–Intercept Equation: We substitute 7 for m and either of the points $(2, 3)$ or $(1, -4)$ for (x, y) in the slope–intercept equation and solve for b. Here we use $(1, -4)$.

$y = mx + b$	**Slope–intercept equation**
$-4 = 7 \cdot 1 + b$	**Substituting**
$-4 = 7 + b$	
$-11 = b$	**Solving for b**

We substitute 7 for m and -11 for b in $y = mx + b$ to get

$$y = 7x - 11,\quad \text{or}$$
$$f(x) = 7x - 11.$$

> *Now Try Exercise 19.*

❯ Parallel Lines

Can we determine whether the graphs of two linear equations are parallel without graphing them? Let's look at three pairs of equations and their graphs.

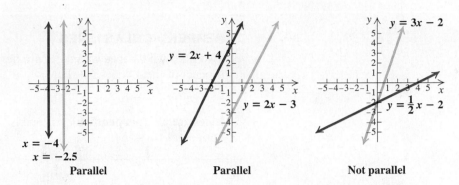

| Parallel | Parallel | Not parallel |

If two different lines, such as $x = -4$ and $x = -2.5$, are vertical, then they are parallel. Thus two equations such as $x = a_1$ and $x = a_2$, where $a_1 \neq a_2$, have graphs that are *parallel lines*. Two nonvertical lines, such as $y = 2x + 4$ and $y = 2x - 3$, or, in general, $y = mx + b_1$ and $y = mx + b_2$, where the slopes are the *same* and $b_1 \neq b_2$, also have graphs that are *parallel lines*.

PARALLEL LINES

Vertical lines are **parallel**. Nonvertical lines are **parallel** if and only if they have the same slope and different y-intercepts.

❯ Perpendicular Lines

Can we examine a pair of equations to determine whether their graphs are perpendicular without graphing the equations? Let's look at the following pairs of equations and their graphs.

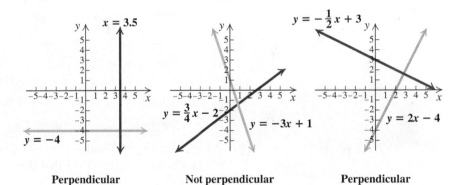

| Perpendicular | Not perpendicular | Perpendicular |

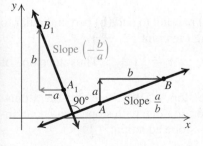

If one line is vertical and another is horizontal, they are perpendicular. For example, the lines $x = 3.5$ and $y = -4$ are perpendicular. Otherwise, how can we tell whether two lines are perpendicular? Consider a line $\overleftrightarrow{AB}$, as shown in the figure at left, with slope a/b. Then think of rotating the line 90° to get a line $\overleftrightarrow{A_1B_1}$ perpendicular to $\overleftrightarrow{AB}$. For the new line, the rise and the run are interchanged, but the run is now negative. Thus the slope of the new line is $-b/a$, which is the opposite of the reciprocal of the slope of the first line. Also note that when we multiply the slopes, we get

$$\frac{a}{b}\left(-\frac{b}{a}\right) = -1.$$

This is the condition under which lines will be perpendicular.

PERPENDICULAR LINES

Two lines with slopes m_1 and m_2 are **perpendicular** if and only if the product of their slopes is -1:

$$m_1 m_2 = -1.$$

Lines are also **perpendicular** if one is vertical $(x = a)$ and the other is horizontal $(y = b)$.

If a line has slope m_1, the slope m_2 of a line perpendicular to it is $-1/m_1$. The slope of one line is the *opposite of the reciprocal* of the other:

$$m_2 = -\frac{1}{m_1}, \quad \text{or} \quad m_1 = -\frac{1}{m_2}.$$

EXAMPLE 4   Determine whether each of the following pairs of lines is parallel, perpendicular, or neither.

a) $y + 2 = 5x$, $5y + x = -15$ **b)** $2y + 4x = 8$, $5 + 2x = -y$

c) $2x + 1 = y$, $y + 3x = 4$

Solution   We use the slopes of the lines to determine whether the lines are parallel or perpendicular.

a) We solve each equation for y:

$$y = 5x - 2, \qquad y = -\tfrac{1}{5}x - 3.$$

The slopes are 5 and $-\tfrac{1}{5}$. Their product is -1, so the lines are perpendicular. (See Fig. 1.) Note in the graphs at left that the graphing calculator windows have been squared to avoid distortion. (Review squaring windows in Section 1.1.)

b) Solving each equation for y, we get

$$y = -2x + 4, \qquad y = -2x - 5.$$

We see that $m_1 = -2$ and $m_2 = -2$. Since the slopes are the same and the y-intercepts, $(0, 4)$ and $(0, -5)$, are different, the lines are parallel. (See Fig. 2.)

c) Rewriting the first equation and solving the second equation for y, we have

$$y = 2x + 1, \qquad y = -3x + 4.$$

We see that $m_1 = 2$ and $m_2 = -3$. Since the slopes are not the same and their product is not -1, it follows that the lines are neither parallel nor perpendicular. (See Fig. 3.)

> *Now Try Exercises 35 and 39.*

EXAMPLE 5   Write equations of the lines **(a)** parallel to and **(b)** perpendicular to the graph of the line $4y - x = 20$ and containing the point $(2, -3)$.

Solution   We first solve $4y - x = 20$ for y to get $y = \tfrac{1}{4}x + 5$. Thus the slope of the given line is $\tfrac{1}{4}$.

a) The line parallel to the given line will have slope $\tfrac{1}{4}$. We use either the slope–intercept equation or the point–slope equation for a line with slope $\tfrac{1}{4}$ and containing the point $(2, -3)$. Here we use the point–slope equation:

$$
\begin{aligned}
y - y_1 &= m(x - x_1) \\
y - (-3) &= \tfrac{1}{4}(x - 2) \\
y + 3 &= \tfrac{1}{4}x - \tfrac{1}{2} \\
y &= \tfrac{1}{4}x - \tfrac{7}{2}.
\end{aligned}
$$

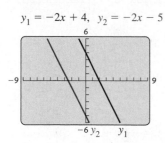

$y_1 = 5x - 2, \quad y_2 = -\tfrac{1}{5}x - 3$

FIGURE 1.

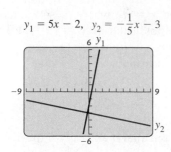

$y_1 = -2x + 4, \quad y_2 = -2x - 5$

FIGURE 2.

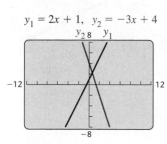

$y_1 = 2x + 1, \quad y_2 = -3x + 4$

FIGURE 3.

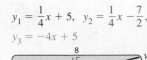

$$y_1 = \frac{1}{4}x + 5, \quad y_2 = \frac{1}{4}x - \frac{7}{2},$$
$$y_3 = -4x + 5$$

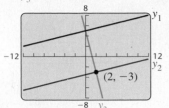

b) The slope of the perpendicular line is the opposite of the reciprocal of $\frac{1}{4}$, or -4. Again, we use the point–slope equation to write an equation for a line with slope -4 and containing the point $(2, -3)$:

$$y - y_1 = m(x - x_1)$$
$$y - (-3) = -4(x - 2)$$
$$y + 3 = -4x + 8$$
$$y = -4x + 5.$$

> **Now Try Exercise 43.**

SUMMARY OF TERMINOLOGY ABOUT LINES

TERMINOLOGY	MATHEMATICAL INTERPRETATION
Slope	$m = \dfrac{y_2 - y_1}{x_2 - x_1}$, or $\dfrac{y_1 - y_2}{x_1 - x_2}$
Slope–intercept equation	$y = mx + b$
Point–slope equation	$y - y_1 = m(x - x_1)$
Horizontal line	$y = b$
Vertical line	$x = a$
Parallel lines	$m_1 = m_2, b_1 \neq b_2$; or $x = a_1, x = a_2, a_1 \neq a_2$
Perpendicular lines	$m_1 m_2 = -1$, or $m_2 = -\dfrac{1}{m_1}$; or $x = a, y = b$

Creating a Mathematical Model

1. Recognize real-world problem.

2. Collect data.

3. Analyze data.

4. Construct model.

5. Test and refine model.

6. Explain and predict.

❯ Mathematical Models

When a real-world problem can be described in mathematical language, we have a **mathematical model**. For example, the natural numbers constitute a mathematical model for situations in which counting is essential. Situations in which algebra can be brought to bear often require the use of functions as models.

Mathematical models are abstracted from real-world situations. The mathematical model gives results that allow one to predict what will happen in that real-world situation. If the predictions are inaccurate or the results of experimentation do not conform to the model, the model must be changed or discarded.

Mathematical modeling can be an ongoing process. For example, finding a mathematical model that will provide an accurate prediction of population growth is not a simple task. Any population model that one might devise would need to be reshaped as further information is acquired.

❯ Curve Fitting

We will develop and use many kinds of mathematical models in this text. In this chapter, we have used *linear* functions as models. Other types of functions, such as quadratic, cubic, and exponential functions, can also model data. These functions are *nonlinear*. Modeling with quadratic functions and cubic functions is discussed in Chapter 4. Modeling with exponential functions is discussed in Chapter 5.

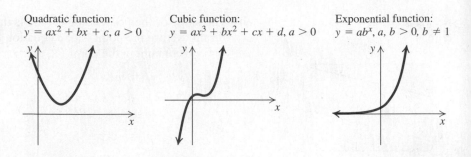

Quadratic function:
$y = ax^2 + bx + c, a > 0$

Cubic function:
$y = ax^3 + bx^2 + cx + d, a > 0$

Exponential function:
$y = ab^x, a, b > 0, b \neq 1$

In general, we try to find a function that fits, as well as possible, observations (data), theoretical reasoning, and common sense. We call this **curve fitting**; it is one aspect of mathematical modeling.

Let's look at some data and related graphs or **scatterplots** and determine whether a linear function seems to fit the set of data.

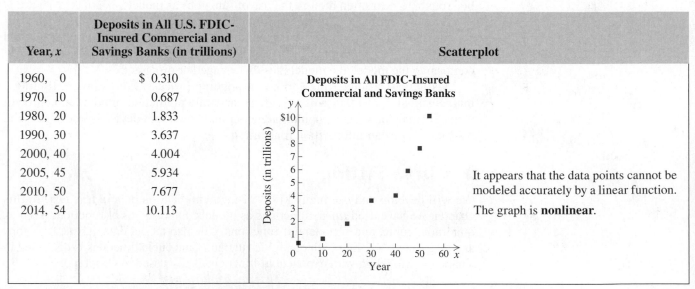

Year, x	Cost of a 30-Second Super Bowl Commercial (in millions)	Scatterplot
2010, 0	$2.9	
2011, 1	3.1	
2012, 2	3.5	It appears that the data points can be represented, or modeled, by a linear function.
2013, 3	3.8	
2014, 4	4.0	The graph is **linear**.
2015, 5	4.5	

Source: huffingtonpost.com

Year, x	Deposits in All U.S. FDIC-Insured Commercial and Savings Banks (in trillions)	Scatterplot
1960, 0	$ 0.310	
1970, 10	0.687	
1980, 20	1.833	
1990, 30	3.637	
2000, 40	4.004	
2005, 45	5.934	It appears that the data points cannot be modeled accurately by a linear function.
2010, 50	7.677	
2014, 54	10.113	The graph is **nonlinear**.

Source: *Summary of Deposits*, Federal Deposit Insurance Corporation

Looking at the scatterplots, we see that the data on the cost of a Super Bowl commercial seem to be rising in a manner to suggest that a *linear function* might fit, although a "perfect" straight line cannot be drawn through the data points. A linear function does not seem to fit the data on deposits in U.S. banks.

EXAMPLE 6   *Cost of a 30-Sec Super Bowl Commercial.*   The cost of a 30-sec Super Bowl commercial has increased $1.6 million from 2010 to 2015. Model the data in the table on p. 52 with a linear function. Then estimate the cost of a 30-sec commercial in 2018.

Solution   We can choose any two of the data points to determine an equation. Note that the first coordinate is the number of years since 2010 and the second coordinate is the corresponding cost of a 30-sec Super Bowl commercial in millions of dollars. Let's use $(1, 3.1)$ and $(4, 4.0)$.

We first determine the slope of the line:

$$m = \frac{4.0 - 3.1}{4 - 1} = \frac{0.9}{3} = 0.3.$$

Then we substitute 0.3 for m and either of the points $(1, 3.1)$ or $(4, 4.0)$ for (x_1, y_1) in the point–slope equation. In this case, let's use $(1, 3.1)$. We get

$$y - y_1 = m(x - x_1) \qquad \text{Point–slope equation}$$
$$y - 3.1 = 0.3(x - 1), \qquad \text{Substituting}$$

which simplifies to

$$y = 0.3x + 2.8,$$

where x is the number of years after 2010 and y is in millions of dollars.

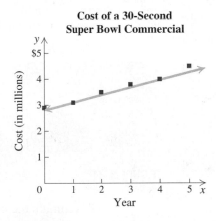

Cost of a 30-Second Super Bowl Commercial

Next, we estimate the cost of a 30-sec commercial in 2018 by substituting 8 $(2018 - 2010 = 8)$ for x in the model:

$$y = 0.3x + 2.8 \qquad \text{Model}$$
$$= 0.3(8) + 2.8 \qquad \text{Substituting}$$
$$= 5.2.$$

We estimate that the cost of a 30-sec Super Bowl commercial will be $5.2 million in 2018.

> **Now Try Exercise 61.**

A model and the estimates it produces are dependent on the data points used. In Example 6, if we were to use the data points $(0, 2.9)$ and $(5, 4.5)$, our model would be

$$y = 0.32x + 2.9,$$

and our estimate for the cost of a 30-sec commercial in 2018 would be $5.46 million, about $0.26 million more than the estimate provided by the first model.

Models that consider all the data points, not just two, are generally better models. The linear model that best fits the data can be found using a graphing calculator and a procedure called **linear regression**.

❯ Linear Regression

Although discussion leading to a complete understanding of linear regression belongs in a statistics course, we present the procedure here because we can carry it out easily using technology. The graphing calculator gives us the powerful capability to find linear models and to make predictions using them.

Consider the data presented before Example 6 on the cost of a 30-sec Super Bowl commercial. We can fit a regression line of the form $y = mx + b$ to the data using the LINEAR REGRESSION feature on a graphing calculator.

EXAMPLE 7 *Cost of a 30-Sec Super Bowl Commercial.* Fit a regression line to the data on the cost of a 30-sec Super Bowl commercial given in the table on p. 52. Then use the function to predict the cost of a 30-sec commercial in 2018.

Solution First, we enter the data in lists on the calculator. We enter the values of the independent variable x in list L1 and the corresponding values of the dependent variable y in list L2. (See Fig. 1.) The graphing calculator can then create a scatterplot of the data, as shown in Fig. 2.

When we select the LINEAR REGRESSION feature from the STAT CALC menu, we find the linear equation that best models the data. It is

$$y = 0.3142857143x + 2.847619048. \quad \textbf{Regression line}$$

(See Figs. 3 and 4.) We can then graph the regression line on the same graph as the scatterplot, as shown in Fig. 5.

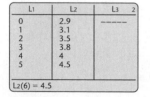

FIGURE 1.

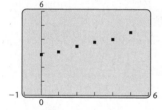

FIGURE 2.

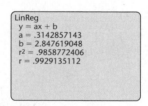

FIGURE 3.

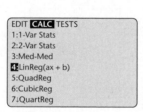

FIGURE 4.

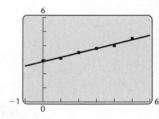

FIGURE 5.

To estimate the cost of a 30-sec Super Bowl commercial in 2018, we substitute 8 for x in the regression equation, which has been saved as Y1. Using this model, we see that the cost of a 30-sec commercial is predicted to be about $5.36 million in 2018. (See Fig. 6.)

FIGURE 6.

Note that $5.36 million is closer to the value $5.46 million that we found with data points $(0, 2.9)$ and $(5, 4.5)$ following Example 6 than to the value $5.2 million that we found with the data points $(1, 3.1)$ and $(4, 4.0)$ in Example 6.

❯ *Now Try Exercises 69(a) and 69(b).*

The Correlation Coefficient

On some graphing calculators with the DIAGNOSTIC feature turned on, a constant r between -1 and 1, called the **coefficient of linear correlation**, appears with the equation of the regression line. Though we cannot develop a formula for calculating r in this text, keep in mind that it is used to describe the strength of the linear relationship between x and y. The closer $|r|$ is to 1, the better the correlation. A positive value of r also indicates that the regression line has a positive slope, and a negative value of r indicates that the regression line has a negative slope. As shown in Fig. 4, for the data on the cost of a 30-sec Super Bowl commercial just discussed, $r = 0.9929135112$, which indicates a very good linear correlation.

The following scatterplots summarize the interpretation of a correlation coefficient.

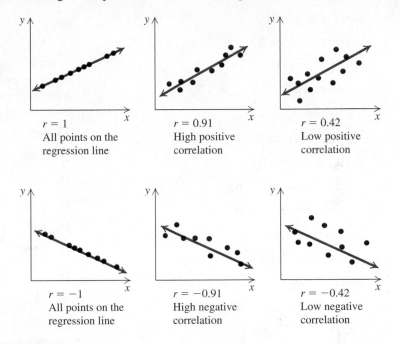

$r = 1$
All points on the regression line

$r = 0.91$
High positive correlation

$r = 0.42$
Low positive correlation

$r = -1$
All points on the regression line

$r = -0.91$
High negative correlation

$r = -0.42$
Low negative correlation

1.4   Exercise Set

Find the slope and the y-intercept of the graph of the linear equation. Then write the equation of the line in slope–intercept form.

1.

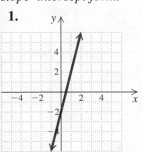

2.

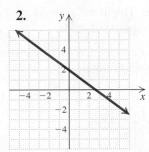

3.

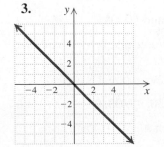

4.

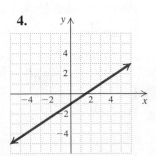

5.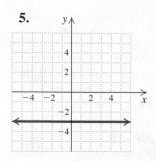

6.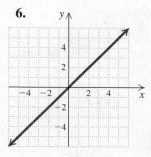

Write a slope–intercept equation for a line with the given characteristics.

7. $m = \frac{2}{9}$, y-intercept $(0, 4)$

8. $m = -\frac{3}{8}$, y-intercept $(0, 5)$

9. $m = -4$, y-intercept $(0, -7)$

10. $m = \frac{2}{7}$, y-intercept $(0, -6)$

11. $m = -4.2$, y-intercept $\left(0, \frac{3}{4}\right)$

12. $m = -4$, y-intercept $\left(0, -\frac{3}{2}\right)$

13. $m = \frac{2}{9}$, passes through $(3, 7)$

14. $m = -\frac{3}{8}$, passes through $(5, 6)$

15. $m = 0$, passes through $(-2, 8)$

16. $m = -2$, passes through $(-5, 1)$

17. $m = -\frac{3}{5}$, passes through $(-4, -1)$

18. $m = \frac{2}{3}$, passes through $(-4, -5)$

19. Passes through $(-1, 5)$ and $(2, -4)$

20. Passes through $\left(-3, \frac{1}{2}\right)$ and $\left(1, \frac{1}{2}\right)$

21. Passes through $(7, 0)$ and $(-1, 4)$

22. Passes through $(-3, 7)$ and $(-1, -5)$

23. Passes through $(0, -6)$ and $(3, -4)$

24. Passes through $(-5, 0)$ and $\left(0, \frac{4}{5}\right)$

25. Passes through $(-4, 7.3)$ and $(0, 7.3)$

26. Passes through $(-13, -5)$ and $(0, 0)$

Write equations of the horizontal line and the vertical line that pass through the given point.

27. $(0, -3)$　　　　**29.** $\left(\frac{2}{11}, -1\right)$

28. $\left(-\frac{1}{4}, 7\right)$　　　**30.** $(0.03, 0)$

31. Find a linear function h given $h(1) = 4$ and $h(-2) = 13$. Then find $h(2)$.

32. Find a linear function g given $g\left(-\frac{1}{4}\right) = -6$ and $g(2) = 3$. Then find $g(-3)$.

33. Find a linear function f given $f(5) = 1$ and $f(-5) = -3$. Then find $f(0)$.

34. Find a linear function h given $h(-3) = 3$ and $h(0) = 2$. Then find $h(-6)$.

Determine whether the pair of lines is parallel, perpendicular, or neither.

35. $y = \frac{26}{3}x - 11$,　　**36.** $y = -3x + 1$,
　　$y = -\frac{3}{26}x - 11$　　　$y = -\frac{1}{3}x + 1$

37. $y = \frac{2}{5}x - 4$,　　**38.** $y = \frac{3}{2}x - 8$,
　　$y = -\frac{2}{5}x + 4$　　　$y = 8 + 1.5x$

39. $x + 2y = 5$,　　**40.** $2x - 5y = -3$,
　　$2x + 4y = 8$　　　$2x + 5y = 4$

41. $y = 4x - 5$,　　**42.** $y = 7 - x$,
　　$4y = 8 - x$　　　$y = x + 3$

Write a slope–intercept equation for a line passing through the given point that is parallel to the given line. Then write a second equation for a line passing through the given point that is perpendicular to the given line.

43. $(3, 5)$, $y = \frac{2}{7}x + 1$

44. $(-1, 6)$, $f(x) = 2x + 9$

45. $(-7, 0)$, $y = -0.3x + 4.3$

46. $(-4, -5)$, $2x + y = -4$

47. $(3, -2)$, $3x + 4y = 5$

48. $(8, -2)$, $y = 4.2(x - 3) + 1$

49. $(3, -3)$, $x = -1$

50. $(4, -5)$, $y = -1$

Determine whether the statement is true or false.

51. The lines $x = -3$ and $y = 5$ are perpendicular.

52. The lines $y = 2x - 3$ and $y = -2x - 3$ are perpendicular.

53. The lines $y = \frac{2}{5}x + 4$ and $y = \frac{2}{5}x - 4$ are parallel.

54. The intersection of the lines $y = 2$ and $x = -\frac{3}{4}$ is $\left(-\frac{3}{4}, 2\right)$.

55. The lines $x = -1$ and $x = 1$ are perpendicular.

56. The lines $2x + 3y = 4$ and $3x - 2y = 4$ are perpendicular.

In Exercises 57–60, determine whether a linear model might fit the data.

57.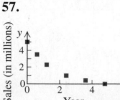

58.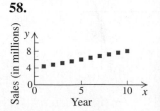

59.

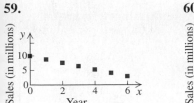

60.

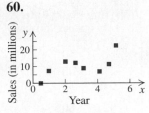

61. *Cost of Health Insurance.* The following table lists the average monthly cost to workers for family health insurance for various years.

Year, x	Average Monthly Cost to Workers for Family Health Insurance
2009, 0	$293
2010, 1	333
2011, 2	344
2012, 3	360
2013, 4	380
2014, 5	402

Source: Kaiser Family Foundation

a) Model the data with a linear function. Let the independent variable represent the number of years after 2009; that is, the data points are $(0, 293)$, $(3, 360)$, and so on. Answers may vary depending on the data points used.

b) Using the function found in part (a), predict the average cost to workers for family health insurance in 2018 and in 2023.

62. *U.S. Banks.* The following table lists the numbers of commercial and savings banks in the United States for various years.

Year, x	Number of U.S. Banks
2007, 0	8605
2008, 1	8441
2009, 2	8185
2010, 3	7821
2011, 4	7523
2012, 5	7255
2013, 6	6950
2014, 7	6669

Source: Federal Deposit Insurance Corporation

a) Model the data with a linear function. Let the independent variable represent the number of years after 2007. Answers may vary depending on the data points used.

b) Using the function found in part (a), predict the number of banks in the United States in 2017 and in 2020.

63. *Airline Add-On Fees.* Data on airline revenue from add-on fees are listed in the following table. Model the data with a linear function. Then, using that function, predict the revenue from airline add-on fees in 2019. Answers may vary depending on the data points used.

Year, x	Airline Revenue from Add-On Fees (in billions)
2010, 0	$22.6
2011, 1	32.5
2012, 2	36.1
2013, 3	42.6
2014, 4	49.9

Sources: U.S. airline consultancy, IdeaWorks Company; rental firm CarTrawler

64. *Licensed Drivers.* Data on the number of licensed drivers in the United States in selected years are listed in the following table. Model the data with a linear function, estimate the number of licensed drivers in 2005, and predict the number of licensed drivers in 2021. Answers may vary depending on the data points used.

Year, x	Number of Licensed Drivers (in millions)
1980, 0	145.295
1990, 10	167.015
2000, 20	190.625
2010, 30	210.115
2012, 32	211.815

Sources: Federal Highway Administration; U.S. Department of Transportation

65. *Bottled Water.* Data on the per-capita consumption, in gallons, of bottled water in the United States are given in the following table. Model the data with a linear function and predict the per-capita consumption of bottled water in 2017. Answers may vary depending on the data points used.

Year, x	Per-Capita Consumption of Bottled Water (in gallons)
2009, 0	27.6
2010, 1	28.3
2011, 2	29.2
2012, 3	30.8
2013, 4	32.0

Source: Beverage Marketing Corporation

66. *Electricity Use.* Data on the average annual household use of electricity, in kilowatt-hours, are listed in the following table. Model the data with a linear function and predict the average annual household electricity use in 2019. Answers may vary depending on the data points used.

Year, x	Annual Electricity Use (in kilowatt-hours)
2010, 0	11,504
2011, 1	11,280
2012, 2	10,837
2013, 3	10,819

Source: Energy Information Administration

67. a) Use a graphing calculator to fit a regression line to the data in Exercise 61.
  b) Predict the average monthly cost to workers for family health insurance in 2018 and compare the value with the result found in Exercise 61.
  c) Find the correlation coefficient for the regression line and determine whether the line fits the data closely.

68. a) Use a graphing calculator to fit a regression line to the data in Exercise 62.
  b) Predict the number of banks in the United States in 2020 and compare the result with the result found with the model in Exercise 62.
  c) Find the correlation coefficient for the regression line and determine whether the line fits the data closely.

69. a) Use a graphing calculator to fit a regression line to the data in Exercise 63.
  b) Predict airline revenue from add-on fees in 2019 and compare the value with the result found in Exercise 63.
  c) Find the correlation coefficient for the regression line and determine whether the line fits the data closely.

70. a) Use a graphing calculator to fit a regression line to the data in Exercise 64.
  b) Predict the number of licensed drivers in the United States in 2021 and compare the result with the prediction found with the model in Exercise 64.
  c) Find the correlation coefficient for the regression line and determine whether the line fits the data closely.

71. *Maximum Heart Rate.* A person who is exercising should not exceed his or her maximum heart rate, which is determined on the basis of that person's gender, age, and resting heart rate. The following table relates resting heart rate and maximum heart rate for a 20-year-old man.

Resting Heart Rate, H (in beats per minute)	Maximum Heart Rate, M (in beats per minute)
50	166
60	168
70	170
80	172

Source: American Heart Association

a) Use a graphing calculator to model the data with a linear function.
b) Estimate the maximum heart rate if the resting heart rate is 40, 65, 76, and 84.
c) What is the correlation coefficient? How confident are you about using the regression line to estimate function values?

72. *Study Time versus Grades.* A math instructor asked her students to keep track of how much time each spent studying a chapter on functions in her algebra–trigonometry course. She collected the information together with test scores from that chapter's test. The data are listed in the following table.

Study Time, x (in hours)	Test Grade, y (in percent)
23	81%
15	85
17	80
9	75
21	86
13	80
16	85
11	93

a) Use a graphing calculator to model the data with a linear function.
b) Predict a student's score if he or she studies 24 hr, 6 hr, and 18 hr.
c) What is the correlation coefficient? How confident are you about using the regression line to predict function values?

❯ Skill Maintenance

Find the slope of the line containing the given points. **[1.3]**

73. $(2, -8)$ and $(-5, -1)$

74. $(5, 7)$ and $(5, -7)$

Find an equation for a circle satisfying the given conditions. **[1.1]**

75. Center $(-7, -1)$, radius of length $\frac{9}{5}$

76. Center $(0, 3)$, diameter of length 5

❯ Synthesis

77. Find k so that the line containing the points $(-3, k)$ and $(4, 8)$ is parallel to the line containing the points $(5, 3)$ and $(1, -6)$.

78. *Road Grade.* Using the following figure, find the road grade and an equation giving the height y as a function of the horizontal distance x.

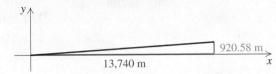

79. Find an equation of the line passing through the point $(4, 5)$ and perpendicular to the line passing through the points $(-1, 3)$ and $(2, 9)$.

1.5 ❯ Linear Equations, Functions, Zeros, and Applications

> ❯ Solve linear equations.

> ❯ Solve applied problems using linear models.

> ❯ Find zeros of linear functions.

An **equation** is a statement that two expressions are equal. To **solve** an equation in one variable is to find all the values of the variable that make the equation true. Each of these values is a **solution** of the equation. The set of all solutions of an equation is its **solution set**. Some examples of **equations in one variable** are

$$2x + 3 = 5, \qquad 3(x - 1) = 4x + 5,$$

$$x^2 - 3x + 2 = 0, \quad \text{and} \quad \frac{x - 3}{x + 4} = 1.$$

❯ Linear Equations

The first two equations above are *linear equations* in one variable. We define such equations as follows.

> A **linear equation in one variable** is an equation that can be expressed in the form $mx + b = 0$, where m and b are real numbers and $m \neq 0$.

Equations that have the same solution set are **equivalent equations**. For example, $2x + 3 = 5$ and $x = 1$ are equivalent equations because 1 is the solution of each equation. On the other hand, $x^2 - 3x + 2 = 0$ and $x = 1$ are not equivalent equations because 1 and 2 are both solutions of $x^2 - 3x + 2 = 0$ but 2 is not a solution of $x = 1$.

To solve an equation, we find an equivalent equation in which the variable is isolated. The following principles allow us to solve linear equations.

EQUATION-SOLVING PRINCIPLES

For any real numbers a, b, and c:

> **The Addition Principle:** If $a = b$ is true, then $a + c = b + c$ is true.
>
> **The Multiplication Principle:** If $a = b$ is true, then $ac = bc$ is true.

Note to the student and the instructor: We assume that students come to a College Algebra course with some equation-solving skills from their study of Intermediate Algebra. Thus a portion of the material in this section might be considered by some to be review in nature. We present this material here in order to use linear functions, with which students are familiar, to lay the groundwork for zeros of higher-order polynomial functions and their connection to solutions of equations and x-intercepts of graphs.

EXAMPLE 1 Solve: $\frac{3}{4}x - 1 = \frac{7}{5}$.

Solution When we have an equation that contains fractions, it is often convenient to multiply on both sides of the equation by the least common denominator (LCD) of the fractions in order to clear the equation of fractions. We have

$$\frac{3}{4}x - 1 = \frac{7}{5} \qquad \text{The LCD is } 4 \cdot 5, \text{ or } 20.$$

$$20\left(\frac{3}{4}x - 1\right) = 20 \cdot \frac{7}{5} \qquad \begin{array}{l}\text{Multiplying by the LCD on both sides}\\ \text{to clear fractions}\end{array}$$

$$20 \cdot \frac{3}{4}x - 20 \cdot 1 = 28$$

$$15x - 20 = 28$$

$$15x - 20 + 20 = 28 + 20 \qquad \begin{array}{l}\text{Using the addition principle to add}\\ \text{20 on both sides}\end{array}$$

$$15x = 48$$

$$\frac{15x}{15} = \frac{48}{15} \qquad \begin{array}{l}\text{Using the multiplication principle to}\\ \text{multiply by } \frac{1}{15}, \text{ or divide by 15, on}\\ \text{both sides}\end{array}$$

$$x = \frac{48}{15}$$

$$x = \frac{16}{5}. \qquad \begin{array}{l}\text{Simplifying. Note that } \frac{3}{4}x - 1 = \frac{7}{5} \text{ and}\\ x = \frac{16}{5} \text{ are equivalent equations.}\end{array}$$

Check:
$$\frac{3}{4}x - 1 = \frac{7}{5}$$
$$\frac{\frac{3}{4} \cdot \frac{16}{5} - 1 \ ? \ \frac{7}{5}}{\qquad} \qquad \text{Substituting } \frac{16}{5} \text{ for } x$$
$$\frac{12}{5} - \frac{5}{5} \quad\Big|$$
$$\frac{7}{5} \quad\Big|\quad \frac{7}{5} \quad \textbf{TRUE}$$

The solution is $\frac{16}{5}$.

> *Now Try Exercise 15.*

We can use the INTERSECT feature on a graphing calculator to solve equations. We call this the **Intersect method**. To use the Intersect method to solve the equation in Example 1, for instance, we graph $y_1 = \frac{3}{4}x - 1$ and $y_2 = \frac{7}{5}$. The value of x for which $y_1 = y_2$ is the solution of the equation $\frac{3}{4}x - 1 = \frac{7}{5}$. This value of x is the first coordinate of the point of intersection of the graphs of y_1 and y_2. Using the INTERSECT feature, we find that the first coordinate of this point is 3.2. We can find fraction notation for the solution by using the ▶FRAC feature. The solution is 3.2, or $\frac{16}{5}$.

$y_1 = \frac{3}{4}x - 1, \quad y_2 = \frac{7}{5}$

EXAMPLE 2   Solve: $2(5 - 3x) = 8 - 3(x + 2)$.

Algebraic Solution	**Graphical Solution**

Algebraic Solution

We have

$$2(5 - 3x) = 8 - 3(x + 2)$$
$$10 - 6x = 8 - 3x - 6 \qquad \text{Using the distributive property}$$
$$10 - 6x = 2 - 3x \qquad \text{Collecting like terms}$$
$$10 - 6x + 6x = 2 - 3x + 6x \qquad \text{Using the addition principle to add } 6x \text{ on both sides}$$
$$10 = 2 + 3x$$
$$10 - 2 = 2 + 3x - 2 \qquad \text{Using the addition principle to add } -2, \text{ or subtract 2, on both sides}$$
$$8 = 3x$$
$$\frac{8}{3} = \frac{3x}{3} \qquad \text{Using the multiplication principle to multiply by } \frac{1}{3}, \text{ or divide by 3, on both sides}$$
$$\frac{8}{3} = x.$$

Check:

$$\begin{array}{c|c} 2(5 - 3x) = 8 - 3(x + 2) \\ \hline 2\left(5 - 3 \cdot \frac{8}{3}\right) \ ? \ 8 - 3\left(\frac{8}{3} + 2\right) & \text{Substituting } \frac{8}{3} \text{ for } x \\ 2(5 - 8) \ \Big| \ 8 - 3\left(\frac{14}{3}\right) \\ 2(-3) \ \Big| \ 8 - 14 \\ -6 \ \Big| \ -6 & \text{TRUE} \end{array}$$

The solution is $\frac{8}{3}$.

Graphical Solution

We graph $y_1 = 2(5 - 3x)$ and $y_2 = 8 - 3(x + 2)$. The first coordinate of the point of intersection of the graphs is the value of x for which $2(5 - 3x) = 8 - 3(x + 2)$ and is thus the solution of the equation.

$y_1 = 2(5 - 3x), \ y_2 = 8 - 3(x + 2)$

y_2 5 y_1

Intersection
X = 2.6666667 Y = -6
-15

The solution is approximately 2.6666667.

We can find fraction notation for the exact solution by using the ▶ FRAC feature. The solution is $\frac{8}{3}$.

X ▶ Frac $\frac{8}{3}$

> **Now Try Exercise 27.**

X	Y1	Y2
2.6667	-6	-6

X =

We can use the TABLE feature on a graphing calculator, set in ASK mode, to check the solutions of equations. In Example 2, for instance, we let $y_1 = 2(5 - 3x)$ and $y_2 = 8 - 3(x + 2)$. When $\frac{8}{3}$ is entered for x, we see that $y_1 = y_2$, or $2(5 - 3x) = 8 - 3(x + 2)$. Thus, $\frac{8}{3}$ is the solution of the equation. (Note that the calculator converts $\frac{8}{3}$ to decimal notation in the table.)

Special Cases

Some equations have *no* solution.

EXAMPLE 3   Solve: $-24x + 7 = 17 - 24x$.

Solution   We have

$$-24x + 7 = 17 - 24x$$
$$24x - 24x + 7 = 24x + 17 - 24x \qquad \text{Adding } 24x$$
$$7 = 17. \qquad \text{We get a false equation.}$$

No matter what number we substitute for x, we get a false equation. Thus the equation has *no* solution.

> **Now Try Exercise 11.**

There are some equations for which *any* real number is a solution.

EXAMPLE 4   Solve: $3 - \frac{1}{3}x = -\frac{1}{3}x + 3$.

Solution   We have

$$3 - \frac{1}{3}x = -\frac{1}{3}x + 3$$
$$\frac{1}{3}x + 3 - \frac{1}{3}x = \frac{1}{3}x - \frac{1}{3}x + 3 \qquad \text{Adding } \tfrac{1}{3}x$$
$$3 = 3. \qquad\qquad \textbf{We get a true equation.}$$

Replacing x with any real number gives a true equation. Thus *any* real number is a solution. This equation has *infinitely* many solutions. The solution set is the set of real numbers, $\{x \mid x \text{ is a real number}\}$, or $(-\infty, \infty)$.

> **Now Try Exercise 3.**

〉 Applications Using Linear Models

Mathematical techniques can be used to answer questions arising from real-world situations. Linear equations and linear functions *model* many of these situations.

The following strategy is of great assistance in problem solving.

FIVE STEPS FOR PROBLEM SOLVING

1. **Familiarize** yourself with the problem situation. If the problem is presented in words, this means to read carefully. Some or all of the following can also be helpful.

   a) Make a drawing, if it makes sense to do so.
   b) Make a written list of the known facts and a list of what you wish to find out.
   c) Assign variables to represent unknown quantities.
   d) Organize the information in a chart or a table, if appropriate.
   e) Find further information. Look up a formula, consult a reference book or an expert in the field, or do research on the Internet.
   f) Guess or estimate the answer and check your guess or estimate.

2. **Translate** the problem situation to mathematical language or symbolism. For most of the problems you will encounter in algebra, this means to write one or more equations, but sometimes an inequality or some other mathematical symbolism may be appropriate.

3. **Carry out** some type of mathematical manipulation. Use your mathematical skills to find a possible solution. In algebra, this usually means to solve an equation, an inequality, or a system of equations or inequalities.

4. **Check** to see whether your possible solution actually fits the problem situation and is thus really a solution of the problem. Although you may have solved an equation, the solution(s) of the equation might not be solution(s) of the original problem.

5. **State** the answer clearly using a complete sentence.

EXAMPLE 5 *Industrial Robots.*   Companies in China purchased 36,560 industrial robots in 2013. This is 54% more than the number of industrial robots purchased by companies in the United States. (*Source*: International Federation of Robotics Statistical Department) How many industrial robots were bought by companies in the United States in 2013?

Solution

1. **Familiarize.** Let's estimate that 20,000 industrial robots were purchased by U.S. companies. Then the number of robots purchased by Chinese companies would be

$$20{,}000 + 54\% \cdot 20{,}000 = 1(20{,}000) + 0.54(20{,}000) = 1.54(20{,}000) = 30{,}800.$$

Since we know that Chinese companies actually bought 36,560 robots, our estimate of 20,000 is too low. Nevertheless, the calculations performed indicate how we can translate the problem to an equation. We let $x = $ the number of industrial robots purchased by U.S. companies in 2013. Then $x + 54\%x$, or $1 \cdot x + 0.54x$, or $1.54x$, is the number of robots purchased by Chinese companies.

2. **Translate.** We translate to an equation:

$$
\underbrace{\text{Number of robots purchased}\atop\text{by Chinese companies}}_{\textstyle 1.54x} \quad \underset{=}{\text{is}} \quad \underset{36{,}560.}{36{,}560}
$$

3. **Carry out.** We solve the equation, as follows:

$$1.54x = 36{,}560$$
$$x = \frac{36{,}560}{1.54} \approx 23{,}740.$$

4. **Check.** 54% of 23,740 is about 12,820, and $23{,}740 + 12{,}820 = 36{,}560$.

5. **State.** Companies in the United States purchased approximately 23,740 industrial robots in 2013.

> **Now Try Exercise 33.**

EXAMPLE 6 *Studying Abroad.*   In the 2012–2013 school year 47,058 U.S. students studied abroad in Italy and in France. There were 12,638 more students studying in Italy than in France. (*Source*: Institute of International Education 2014 Report) Find the number of U.S. students studying abroad in Italy and in France.

Solution

1. **Familiarize.** The number of U.S. students studying in Italy is described in terms of the number of students studying in France, so we let $x = $ the number of students studying in France. Then $x + 12{,}638 = $ the number of students studying in Italy.

2. **Translate.** We translate to an equation:

$$
\underbrace{\text{Number of students}\atop\text{studying in France}}_{\textstyle x} \quad \underset{+}{\text{plus}} \quad \underbrace{\text{Number of students}\atop\text{studying in Italy}}_{\textstyle x + 12{,}638} \quad \underset{=}{\text{is}} \quad \underset{47{,}058.}{47{,}058}
$$

3. **Carry out.** We solve the equation, as follows:

$$x + x + 12{,}638 = 47{,}058$$
$$2x = 34{,}420 \qquad \text{Collecting like terms and subtracting 12,638 on both sides}$$
$$x = 17{,}210. \qquad \text{Dividing by 2 on both sides}$$

If $x = 17{,}210$, then $x + 12{,}638 = 17{,}210 + 12{,}638 = 29{,}848$.

4. **Check.** If there were 29,848 U.S. students studying abroad in Italy and 17,210 in France, then the total number of U.S. students studying abroad in Italy and in France was 29,848 + 17,210, or 47,058. Also, 29,848 is 12,638 more than 17,210. The answer checks.

5. **State.** In the 2012–2013 school year, there were 29,848 U.S. students studying abroad in Italy and 17,210 students studying in France.

Now Try Exercise 57.

In some applications, we need to use a formula that describes the relationships among variables. When a situation involves distance, rate (also called speed or velocity), and time, for example, we use the following formula.

THE MOTION FORMULA

The distance d traveled by an object moving at rate r in time t is given by

$$d = r \cdot t.$$

EXAMPLE 7 *Airplane Speed.* Delta Airlines' fleet includes B737/800's, each with a cruising speed of 517 mph, and Saab 340B's, each with a cruising speed of 290 mph (*Source*: Delta Airlines). Suppose that a Saab 340B takes off and travels at its cruising speed. One hour later, a B737/800 takes off and follows the same route, traveling at its cruising speed. How long will it take the B737/800 to overtake the Saab 340B?

Solution

1. **Familiarize.** We make a drawing showing both the known information and the unknown information. We let t = the time, in hours, that the B737/800 travels before it overtakes the Saab 340B. Since the Saab 340B takes off 1 hr before the 737, it will travel for $t + 1$ hr before being overtaken. The planes will have traveled the same distance, d, when one overtakes the other.

We can also organize the information in a table, as follows.

$$d = r \cdot t$$

	Distance	Rate	Time	
B737/800	d	517	t	→ $d = 517t$
Saab 340B	d	290	$t + 1$	→ $d = 290(t + 1)$

2. **Translate.** Using the formula $d = rt$ in each row of the table, we get two expressions for d:

$$d = 517t \quad \text{and} \quad d = 290(t + 1).$$

Since the distances are the same, we have the following equation:

$$517t = 290(t + 1).$$

3. **Carry out.** We solve the equation, as follows:

$$517t = 290(t + 1)$$
$$517t = 290t + 290 \qquad \text{Using the distributive property}$$
$$227t = 290 \qquad \text{Subtracting } 290t \text{ on both sides}$$
$$t \approx 1.28. \qquad \text{Dividing by 227 on both sides and rounding to the nearest hundredth}$$

4. **Check.** If the B737/800 travels for about 1.28 hr, then the Saab 340B travels for about $1.28 + 1$, or 2.28 hr. In 2.28 hr, the Saab 340B travels $290(2.28)$, or 661.2 mi; and in 1.28 hr, the B737/800 travels $517(1.28)$, or 661.76 mi. Since $661.76 \text{ mi} \approx 661.2 \text{ mi}$, the answer checks. (Remember: We rounded the value of t.)

5. **State.** About 1.28 hr after the B737/800 has taken off, it will overtake the Saab 340B.

> *Now Try Exercise 41.*

For some applications, we need to use a formula to find the amount of interest earned by an investment or the amount of interest due on a loan.

THE SIMPLE-INTEREST FORMULA

The **simple interest** I on a principal of P dollars at interest rate r for t years is given by

$$I = Prt.$$

EXAMPLE 8 *Student Loans.* Damarion's two student loans total $28,000. One loan is at 5% simple interest, and the other is at 3% simple interest. After 1 year, Damarion owes $1040 in interest. What is the amount of each loan?

Solution

1. **Familiarize.** We let $x =$ the amount borrowed at 5% interest. Then the remainder of the $28,000, or $28,000 - x$, is borrowed at 3%. We organize the information in a table, keeping in mind the formula $I = Prt$.

	Amount Borrowed	Interest Rate	Time	Amount of Interest
5% Loan	x	5%, or 0.05	1 year	$x(0.05)(1)$, or $0.05x$
3% Loan	$28{,}000 - x$	3%, or 0.03	1 year	$(28{,}000 - x)(0.03)(1)$, or $0.03(28{,}000 - x)$
Total	28,000			1040

2. **Translate.** The total amount of interest on the two loans is $1040. Thus we can translate to the following equation:

$$\underbrace{\text{Interest on}}_{0.05x} \quad \underset{+}{\text{plus}} \quad \underbrace{\text{Interest on}}_{0.03(28{,}000 - x)} \quad \underset{=}{\text{is}} \quad \underset{1040.}{\$1040}$$

3. **Carry out.** We solve the equation, as follows:

$$0.05x + 0.03(28{,}000 - x) = 1040$$
$$0.05x + 840 - 0.03x = 1040 \qquad \text{Using the distributive property}$$
$$0.02x + 840 = 1040 \qquad \text{Collecting like terms}$$
$$0.02x = 200 \qquad \text{Subtracting 840 on both sides}$$
$$x = 10{,}000. \qquad \text{Dividing by 0.02 on both sides}$$

If $x = 10{,}000$, then $28{,}000 - x = 28{,}000 - 10{,}000 = 18{,}000$.

4. **Check.** The interest on $10,000 at 5% for 1 year is $10,000(0.05)(1), or $500. The interest on $18,000 at 3% for 1 year is $18,000(0.03)(1), or $540. Since $500 + $540 = $1040, the answer checks.

5. **State.** Damarion borrowed $10,000 at 5% interest and $18,000 at 3% interest.

> *Now Try Exercise 55.*

Sometimes we use formulas from geometry when solving applied problems. In the following example, we use the formula for the perimeter P of a rectangle with length l and width w: $P = 2l + 2w$.

EXAMPLE 9 *Solar Panels.* In December 2009, a solar energy farm was completed at the Denver International Airport. More than 9200 rectangular solar panels were installed (*Sources*: Woods Allee, Denver International Airport; www.solarpanelstore.com; *The Denver Post*). A solar panel, or photovoltaic panel, converts sunlight into electricity. The length of a panel is 13.6 in. less than twice the width, and the perimeter is 207.4 in. Find the length and the width.

Solution

1. **Familiarize.** We first make a drawing. Since the length of the panel is described in terms of the width, we let $w =$ the width, in inches. Then $2w - 13.6 =$ the length, in inches.

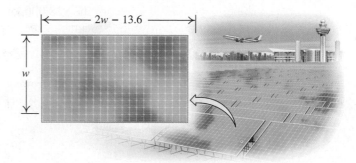

2. **Translate.** We use the formula for the perimeter of a rectangle:

$$P = 2l + 2w$$
$$207.4 = 2(2w - 13.6) + 2w. \qquad \text{Substituting 207.4 for } P \text{ and } 2w - 13.6 \text{ for } l$$

3. **Carry out.** We solve the equation:

$$207.4 = 2(2w - 13.6) + 2w$$
$$207.4 = 4w - 27.2 + 2w \qquad \text{Using the distributive property}$$
$$207.4 = 6w - 27.2 \qquad \text{Collecting like terms}$$
$$234.6 = 6w \qquad \text{Adding 27.2 on both sides}$$
$$39.1 = w. \qquad \text{Dividing by 6 on both sides}$$

If $w = 39.1$, then $2w - 13.6 = 2(39.1) - 13.6 = 78.2 - 13.6 = 64.6$.

4. **Check.** The length, 64.6 in., is 13.6 in. less than twice the width, 39.1 in. Also

$$2 \cdot 64.6 \text{ in.} + 2 \cdot 39.1 \text{ in.} = 129.2 \text{ in.} + 78.2 \text{ in.} = 207.4 \text{ in.}$$

The answer checks.

5. **State.** The length of the solar panel is 64.6 in., and the width is 39.1 in.

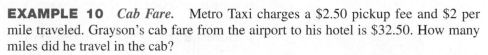 *Now Try Exercise 49.*

EXAMPLE 10 *Cab Fare.* Metro Taxi charges a $2.50 pickup fee and $2 per mile traveled. Grayson's cab fare from the airport to his hotel is $32.50. How many miles did he travel in the cab?

Solution

1. **Familiarize.** Let's guess that Grayson traveled 12 mi in the cab. Then his fare would be

$$\$2.50 + \$2 \cdot 12 = \$2.50 + \$24 = \$26.50.$$

We see that our guess is low, but the calculation shows us how to translate the problem to an equation. We let $m = $ the number of miles that Grayson traveled in the cab.

2. **Translate.** We translate to an equation:

Pickup fee	plus	Cost per mile	times	Number of miles traveled	is	Total charge
↓	↓	↓	↓	↓	↓	↓
2.50	+	2	·	m	=	32.50

3. **Carry out.** We solve the equation:

$$2.50 + 2 \cdot m = 32.50$$
$$2m = 30 \qquad \text{Subtracting 2.50 on both sides}$$
$$m = 15. \qquad \text{Dividing by 2 on both sides}$$

4. **Check.** If Grayson travels 15 mi in the cab, the mileage charge is $2 \cdot 15$, or $30. Then, with the $2.50 pickup fee included, his total charge is $2.50 + $30, or $32.50. The answer checks.

5. **State.** Grayson traveled 15 mi in the cab. *Now Try Exercise 65.*

❭ Zeros of Linear Functions

An input for which a function's output is 0 is called a **zero** of the function. We will restrict our attention in this section to zeros of linear functions. This allows us to become familiar with the concept of a zero, and it lays the groundwork for working with zeros of other types of functions in succeeding chapters.

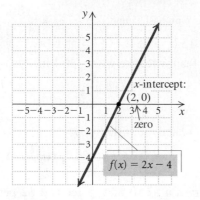

ZEROS OF FUNCTIONS

An input c of a function f is called a **zero** of the function if the output for the function is 0 when the input is c. That is, c is a zero of f if $f(c) = 0$.

Recall that a linear function is given by $f(x) = mx + b$, where m and b are constants. For the linear function $f(x) = 2x - 4$, we have $f(2) = 2 \cdot 2 - 4 = 0$, so 2 is a **zero** of the function. In fact, 2 is the *only* zero of this function. In general, a **linear function $f(x) = mx + b$, with $m \neq 0$, has exactly one zero**.

The zero, 2, is the first coordinate of the point at which the graph crosses the x-axis. This point, $(2, 0)$, is the *x-intercept* of the graph. Thus when we find the zero of a linear function, we are also finding the first coordinate of the x-intercept of the graph of the function.

For every linear function $f(x) = mx + b$, there is an associated linear equation $mx + b = 0$. When we find the zero of a function $f(x) = mx + b$, we are also finding the solution of the equation $mx + b = 0$.

EXAMPLE 11   Find the zero of $f(x) = 5x - 9$.

Algebraic Solution

We find the value of x for which $f(x) = 0$:

$$5x - 9 = 0 \qquad \text{Setting } f(x) = 0$$
$$5x = 9 \qquad \text{Adding 9 on both sides}$$
$$x = \tfrac{9}{5}, \text{ or } 1.8. \qquad \text{Dividing by 5 on both sides}$$

Using a table, set in ASK mode, we can check the solution. We enter $y = 5x - 9$ on the equation-editor screen and then enter the value $x = \tfrac{9}{5}$, or 1.8, in the table.

X	Y1	
1.8	0	
X =		

We see that $y = 0$ when $x = 1.8$, so the number 1.8 checks. The zero is $\tfrac{9}{5}$, or 1.8. This means that $f\left(\tfrac{9}{5}\right) = 0$, or $f(1.8) = 0$. Note that the *zero* of the function $f(x) = 5x - 9$ is the *solution* of the equation $5x - 9 = 0$.

Graphical Solution

The solution of $5x - 9 = 0$ is also the zero of $f(x) = 5x - 9$. Thus we can solve an equation by finding the zeros of the function associated with it. We call this the **Zero method**.

We graph $y = 5x - 9$ in the standard window and use the ZERO feature from the CALC menu to find the zero of $f(x) = 5x - 9$, as shown in the following figure. Note that the x-intercept must appear in the window when the ZERO feature is used.

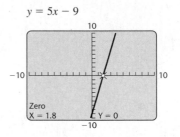

We can check algebraically by substituting 1.8 for x:

$$f(1.8) = 5(1.8) - 9 = 9 - 9 = 0.$$

The zero of $f(x) = 5x - 9$ is 1.8, or $\tfrac{9}{5}$.

Now Try Exercise 73.

CONNECTING THE CONCEPTS

The Intersect Method and the Zero Method

An equation such as $x - 1 = 2x - 6$ can be solved using the Intersect method by graphing $y_1 = x - 1$ and $y_2 = 2x - 6$ and using the INTERSECT feature to find the first coordinate of the point of intersection of the graphs.

The equation can also be solved using the Zero method by writing it with 0 on one side of the equals sign and then using the ZERO feature.

Solve: $x - 1 = 2x - 6$.

THE INTERSECT METHOD

Graph $y_1 = x - 1$ and $y_2 = 2x - 6$.
Point of intersection: $(5, 4)$
Solution: 5

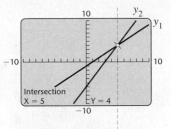

THE ZERO METHOD

First, add $-2x$ and 6 on both sides of the equation to get 0 on one side:

$$x - 1 = 2x - 6$$
$$x - 1 - 2x + 6 = 0.$$

Graph

$$y_3 = x - 1 - 2x + 6.$$

Zero: 5
Solution: 5

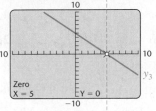

CONNECTING THE CONCEPTS

Zeros, Solutions, and Intercepts

The zero of a linear function $f(x) = mx + b$, with $m \neq 0$, is the solution of the linear equation $mx + b = 0$ and is the first coordinate of the x-intercept of the graph of $f(x) = mx + b$. To find the zero of $f(x) = mx + b$, we solve $f(x) = 0$, or $mx + b = 0$.

FUNCTION

Linear Function
$$f(x) = 2x - 4, \text{ or}$$
$$y = 2x - 4$$

ZERO OF THE FUNCTION; SOLUTION OF THE EQUATION

To find the **zero** of $f(x)$, we solve $f(x) = 0$:

$$2x - 4 = 0$$
$$2x = 4$$
$$x = 2.$$

The **solution** of $2x - 4 = 0$ is 2. This is the zero of the function $f(x) = 2x - 4$. That is, $f(2) = 0$.

ZERO OF THE FUNCTION; x-INTERCEPT OF THE GRAPH

The zero of $f(x)$ is the first coordinate of the **x-intercept** of the graph of $y = f(x)$.

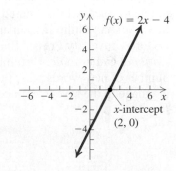

1.5   Exercise Set

Solve.

1. $4x + 5 = 21$

2. $2y - 1 = 3$

3. $23 - \frac{2}{5}x = -\frac{2}{5}x + 23$

4. $\frac{6}{5}y + 3 = \frac{3}{10}$

5. $4x + 3 = 0$

6. $3x - 16 = 0$

7. $3 - x = 12$

8. $4 - x = -5$

9. $3 - \frac{1}{4}x = \frac{3}{2}$

10. $10x - 3 = 8 + 10x$

11. $\frac{2}{11} - 4x = -4x + \frac{9}{11}$

12. $8 - \frac{2}{9}x = \frac{5}{6}$

13. $8 = 5x - 3$

14. $9 = 4x - 8$

15. $\frac{2}{5}y - 2 = \frac{1}{3}$

16. $-x + 1 = 1 - x$

17. $y + 1 = 2y - 7$

18. $5 - 4x = x - 13$

19. $2x + 7 = x + 3$

20. $5x - 4 = 2x + 5$

21. $3x - 5 = 2x + 1$

22. $4x + 3 = 2x - 7$

23. $4x - 5 = 7x - 2$

24. $5x + 1 = 9x - 7$

25. $5x - 2 + 3x = 2x + 6 - 4x$

26. $5x - 17 - 2x = 6x - 1 - x$

27. $7(3x + 6) = 11 - (x + 2)$

28. $4(5y + 3) = 3(2y - 5)$

29. $3(x + 1) = 5 - 2(3x + 4)$

30. $4(3x + 2) - 7 = 3(x - 2)$

31. $2(x - 4) = 3 - 5(2x + 1)$

32. $3(2x - 5) + 4 = 2(4x + 3)$

33. *New Words in the English Language.* During the nineteenth century, 75,029 new words entered the English language. This is about 46.9% more than the number of new words entered in the seventeenth century. (*Source*: Philip Durkin and Katherine Martin, Oxford University Press; "English by the Book," *National Geographic*, December 2013) Find the number of new words that appeared in the English language in the seventeenth century.

34. *Calorie Intake.* The average worldwide daily calorie intake per person has increased from 2200 to 2800 calories since the early 1960s. The average daily calorie intake per person in the United States is 3688. This is about 86.4% more than the average daily calorie intake per person in Haiti. (*Sources*: UN Food and Agriculture Organization; World Health Organization) Find the average daily calorie intake per person in Haiti.

35. *Amount Borrowed.* Kea borrowed money from her father at 5% simple interest to help pay her tuition at Wellington Community College. At the end of 1 year, she owed a total of $1365 in principal and interest. How much did she borrow?

36. *Amount of an Investment.* Miles makes an investment at 4% simple interest. At the end of 1 year, the total value of the investment is $1560. How much was originally invested?

37. *Angle Measure.* In triangle *ABC*, angle *B* is five times as large as angle *A*. The measure of angle *C* is 2° less than that of angle *A*. Find the measures of the angles. (*Hint*: The sum of the angle measures is 180°.)

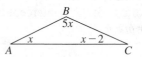

38. *Angle Measure.* In triangle *ABC*, angle *B* is twice as large as angle *A*. Angle *C* measures 20° more than angle *A*. Find the measures of the angles.

39. *Clothing Trade Deficit.* Imports of apparel and clothing accessories to the United States totaled $87.924 billion in 2013. This amount was $2.299 billion less than twenty-five times the apparel and clothing accessories exports that year. (*Source*: Bureau of Economic Analysis, U.S. Department of Commerce) Find the amount of apparel and clothing accessories exports from the United States in 2013.

40. *Foreign Trade.* In 2013, the total value of exports from the United States was $1,579,593,000,000. That year, exports were $445,432,000,000 more than half of the U.S. imports. (*Source*: U.S. Bureau of Economic Analysis, U.S. Department of Commerce) Find the value of imports to the United States in 2013.

41. *Train Speeds.* A Central Railway freight train leaves a station and travels due north at a speed of 60 mph. One hour later, an Amtrak passenger train leaves the same station and travels due north on a parallel track at a speed of 80 mph. How long will it take the passenger train to overtake the freight train?

42. *Distance Traveled.* A private airplane leaves Midway Airport and flies due east at a speed of 180 km/h. Two hours later, a jet leaves Midway and flies due east at a speed of 900 km/h. How far from the airport will the jet overtake the private plane?

43. *TV Tweets.* In 2014, there were 13.8 million tweets about the 56th Annual Grammy Awards. This number was 11.7 million more than the number of tweets about the State of the Union address. (*Source*: Nielsen) Find the number of tweets about the State of the Union address.

44. *Hourly Earnings.* The average hourly earnings of a medical records and health information technician is $33.05 less than the average hourly earnings of a purchasing manager. The average hourly earnings of a medical records and health information technician is $16.81. (*Sources*: Bureau of Labor Statistics; Economic Modeling Specialists International; Debra Auerbach, CareerBuilder.com) Find the average hourly earnings of a purchasing manager.

45. *Commission vs. Salary.* Samantha has a choice between receiving either a monthly salary of $1800 from Furniture by Design or a base salary of $1600 and a 4% commission on the amount of furniture that she sells during the month. For what amount of sales will the two choices be equal?

46. *Sales Commission.* Edward, a consumer electronics salesperson, earns a base salary of $1270 per month and a commission of 6% on the amount of sales that he makes. One month Edward received a paycheck for $3154. Find the amount of his sales for the month.

47. *Studying Abroad.* In the 2012–2013 school year, approximately 820,000 foreign students studied in the United States. The number of U.S. students who studied abroad that same year was about seven-twentieths of the number of foreign students who studied in the United States. (*Source*: Pew Research Center) Find the number of U.S. students who studied abroad during the 2012–2013 school year.

48. *Population Density.* The population density in China is 365.3 persons per square mile. The population density in the United States is approximately one-fourth of the density in China. (*Source*: *The World Almanac* 2014) Find the population density in the United States.

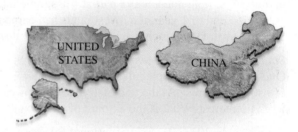

49. *Soccer-Field Dimensions.* The width of the soccer field recommended for players under the age of 12 is 35 yd less than the length. The perimeter of the field is 330 yd. (*Source*: U.S. Youth Soccer) Find the dimensions of the field.

50. *Poster Dimensions.* Marissa is designing a poster to promote the Talbot Street Art Fair. The width of the poster will be two-thirds of its height, and its perimeter will be 100 in. Find the dimensions of the poster.

51. *Test-Plot Dimensions.*   Morgan's Seeds has a rectangular test plot with a perimeter of 322 m. The length is 25 m more than the width. Find the dimensions of the plot.

52. *Garden Dimensions.*   The children at Tiny Tots Day Care plant a rectangular vegetable garden with a perimeter of 39 m. The length is twice the width. Find the dimensions of the garden.

53. *Flying into a Headwind.*   An airplane that travels 450 mph in still air encounters a 30-mph headwind. How long will it take the plane to travel 1050 mi into the wind?

54. *Flying with a Tailwind.*   An airplane that can travel 375 mph in still air is flying with a 25-mph tailwind. How long will it take the plane to travel 700 mi with the wind?

55. *Investment Income.*   Anton invested a total of $5000, part at 3% simple interest and part at 4% simple interest. At the end of 1 year, the investments had earned $176 interest. How much was invested at each rate?

56. *Student Loans.*   Vera's two student loans total $9000. One loan is at 5% simple interest, and the other is at 6% simple interest. At the end of 1 year, Vera owes $492 in interest. What is the amount of each loan?

57. *Patents.*   In 2013, IBM (International Business Machines) received 2133 more patents than Samsung. Together, they received 11,485 patents. (*Source*: IFI Claims Patent Services) How many patents did each company receive?

58. *Books about Presidents.*   There are 5493 print and e-books written about both George Washington and Abraham Lincoln. There are 1675 more books about Lincoln than about Washington. (*Source*: Bowker Books in Print) How many books have been written about each president?

59. *Estates of Celebrities.*   From October 2013 to October 2014, the estates of Michael Jackson and Elvis Presley together took in $195 million. The income to the estate of Michael Jackson was $30 million more than twice the income to the estate of Elvis Presley. (*Source*: 'Forbes' list of Top-Earning Dead Celebrities) Find the income to each estate.

60. *Ocean Depth.*   The average depth of the Pacific Ocean is 14,040 ft, and its depth is 8890 ft less than the sum of the average depths of the Atlantic Ocean and the Indian Ocean. The average depth of the Indian Ocean is 272 ft less than four-fifths of the average depth of the Atlantic Ocean. (*Source*: *Time Almanac* 2010) Find the average depth of the Indian Ocean.

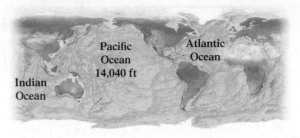

61. *Water Weight.*   Water accounts for 55% of a woman's weight (*Source*: ga.water.usgs.gov/edu). Lily weighs 135 lb. How much of her body weight is water?

62. *Water Weight.*   Water accounts for 60% of a man's weight (*Source*: ga.water.usgs.gov/edu). Jake weighs 186 lb. How much of his body weight is water?

63. *Traveling Upstream.*   A kayak moves at a rate of 12 mph in still water. If the river's current flows at a rate of 4 mph, how long does it take the boat to travel 36 mi upstream?

64. *Traveling Downstream.*   Angelo's kayak travels 14 km/h in still water. If the river's current flows at a rate of 2 km/h, how long will it take him to travel 20 km downstream?

65. *Cab Fare.*   City Cabs charges a $1.75 pickup fee and $1.50 per mile traveled. Diego's fare for a cross-town ride is $19.75. How far did he travel in the cab?

66. *Hourly Wage.*   Rosalyn worked 48 hr one week and earned a $1066 paycheck. She earns time and a half (1.5 times her regular hourly wage) for the number of hours that she works in excess of 40. What is Rosalyn's regular hourly wage?

67. *Olive Oil.* Together, Italy, Spain, and the United States consume 58% of the world's olive oil. The percentage consumed in Italy is $3\frac{3}{4}$ times the percentage consumed in the United States. The percentage consumed in Spain is $\frac{2}{3}$ of the percentage consumed in Italy. (*Source:* www.OliveOilEmporium.com) Find the percent of the world's olive oil consumed in each country.

68. *NFL Stadium Elevation.* The elevations of the 31 NFL stadiums range from 3 ft at Mercedes-Benz Superdome, New Orleans, Louisiana, to 5280 ft at Sports Authority Field at Mile High, Denver, Colorado. The elevation of Sports Authority Field at Mile High is 275 ft higher than seven times the elevation of Lucas Oil Stadium in Indianapolis, Indiana. What is the elevation of Lucas Oil Stadium?

Find the zero of the linear function.

69. $f(x) = x + 5$

70. $f(x) = 5x + 20$

71. $f(x) = -2x + 11$

72. $f(x) = 8 + x$

73. $f(x) = 16 - x$

74. $f(x) = -2x + 7$

75. $f(x) = x + 12$

76. $f(x) = 8x + 2$

77. $f(x) = -x + 6$

78. $f(x) = 4 + x$

79. $f(x) = 20 - x$

80. $f(x) = -3x + 13$

81. $f(x) = \frac{2}{5}x - 10$

82. $f(x) = 3x - 9$

83. $f(x) = -x + 15$

84. $f(x) = 4 - x$

In Exercises 85–90, use the given graph to find each of the following: (a) the x-intercept and (b) the zero of the function.

85.

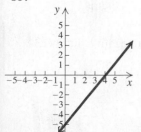

86.

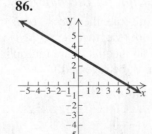

87.

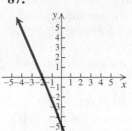

88.

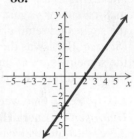

89.

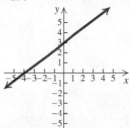

90.

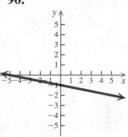

❯ Skill Maintenance

91. Write a slope–intercept equation for the line containing the point $(-1, 4)$ and parallel to the line $3x + 4y = 7$. **[1.4]**

92. Write an equation of the line containing the points $(-5, 4)$ and $(3, -2)$. **[1.4]**

93. Find the distance between $(2, 2)$ and $(-3, -10)$. **[1.1]**

94. Find the midpoint of the segment with endpoints $\left(-\frac{1}{2}, \frac{2}{5}\right)$ and $\left(-\frac{3}{2}, \frac{3}{5}\right)$. **[1.1]**

95. Given that $f(x) = \dfrac{x}{x - 3}$, find $f(-3), f(0)$, and $f(3)$.

96. Find the slope and the y-intercept of the line with the equation $7x - y = \frac{1}{2}$.

❯ Synthesis

State whether each of the following is a linear function.

97. $f(x) = 7 - \frac{3}{2}x$

98. $f(x) = \dfrac{3}{2x} + 5$

99. $f(x) = x^2 + 1$

100. $f(x) = \frac{3}{4}x - (2.4)^2$

Solve.

101. $2x - \{x - [3x - (6x + 5)]\} = 4x - 1$

102. $14 - 2[3 + 5(x - 1)] = 3\{x - 4[1 + 6(2 - x)]\}$

103. *Packaging and Price.*   Dannon recently replaced its 8-oz cup of yogurt with a 6-oz cup and reduced the suggested retail price from 89 cents to 71 cents (*Source*: IRI). Was the price per ounce reduced by the same percent as the size of the cup? If not, find the price difference per ounce in terms of a percent.

104. *Bestsellers.*   One week, 10 copies of the novel *Unbroken* by Laura Hillenbrand were sold for every 3.7 copies of *American Sniper* by Chris Kyle, Scott McEwen, and Jim DeFelice that were sold (*Source*: *USA Today* Best-Selling Books). If a total of 11,371 copies of the two books were sold, how many copies of each were sold?

105. *Running vs. Walking.*   A 150-lb person who runs at 6 mph for 1 hr burns about 720 calories. The same person, walking at 4 mph for 90 min, burns about 480 calories. (*Source*: FitSmart, *USA Weekend*, July 19–21, 2002) Suppose that a 150-lb person runs at 6 mph for 75 min. How far would the person have to walk at 4 mph in order to burn the same number of calories used running?

1.6 Solving Linear Inequalities

> Solve linear inequalities.

> Solve compound inequalities.

> Solve applied problems using inequalities.

An **inequality** is a sentence with $<$, $>$, $\leq$, or $\geq$ as its verb. An example is $3x - 5 < 6 - 2x$. To **solve** an inequality is to find all values of the variable that make the inequality true. Each of these values is a **solution** of the inequality, and the set of all such solutions is its **solution set**. Inequalities that have the same solution set are called **equivalent inequalities**.

Just in Time
18

> Linear Inequalities

The principles for solving inequalities are similar to those for solving equations.

PRINCIPLES FOR SOLVING INEQUALITIES

For any real numbers a, b, and c:

The Addition Principle for Inequalities:

If $a < b$ is true, then $a + c < b + c$ is true.

The Multiplication Principle for Inequalities:

a) If $a < b$ and $c > 0$ are true, then $ac < bc$ is true.
b) If $a < b$ and $c < 0$ are true, then $ac > bc$ is true.
(When both sides of an inequality are multiplied by a negative number, the inequality sign must be reversed.)

Similar statements hold for $a \leq b$.

First-degree inequalities with one variable, like those in Example 1 below, are **linear inequalities**.

Just in Time

3,6

EXAMPLE 1   Solve the inequality. Then graph the solution set.

a) $3x - 5 < 6 - 2x$ **b)** $13 - 7x \geq 10x - 4$

Solution

a) $3x - 5 < 6 - 2x$

$\quad 5x - 5 < 6$ Using the addition principle for inequalities; adding $2x$

$\qquad 5x < 11$ Using the addition principle for inequalities; adding 5

$\qquad\quad x < \frac{11}{5}$ Using the multiplication principle for inequalities; multiplying by $\frac{1}{5}$, or dividing by 5

Any number less than $\frac{11}{5}$ is a solution. The solution set is $\left\{x \mid x < \frac{11}{5}\right\}$, or $\left(-\infty, \frac{11}{5}\right)$. The graph of the solution set is shown below.

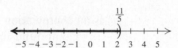

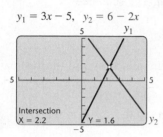

$y_1 = 3x - 5, \quad y_2 = 6 - 2x$

To check, we can graph $y_1 = 3x - 5$ and $y_2 = 6 - 2x$. The graph at left shows that for $x < 2.2$, or $x < \frac{11}{5}$, the graph of y_1 lies below the graph of y_2, or $y_1 < y_2$.

b) $\quad 13 - 7x \geq 10x - 4$

$\quad 13 - 17x \geq -4$ Subtracting $10x$

$\qquad\quad -17x \geq -17$ Subtracting 13

$\qquad\qquad x \leq 1$ Dividing by -17 and reversing the inequality sign

The solution set is $\{x \mid x \leq 1\}$, or $(-\infty, 1]$. The graph of the solution set is shown below.

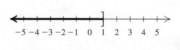

> **Now Try Exercises 1 and 3.**

EXAMPLE 2   Find the domain of the function.

a) $f(x) = \sqrt{x - 6}$ **b)** $h(x) = \dfrac{x}{\sqrt{3 - x}}$

Solution

a) The radicand, $x - 6$, must be greater than or equal to 0. We solve the inequality $x - 6 \geq 0$:

$\qquad x - 6 \geq 0$

$\qquad\quad x \geq 6.$

The domain is $\{x \mid x \geq 6\}$, or $[6, \infty)$.

b) Any real number can be an input for x in the numerator, but inputs for x must be restricted in the denominator. We must have $3 - x \geq 0$ and $\sqrt{3 - x} \neq 0$. Thus, $3 - x > 0$. We solve for x:

$$3 - x > 0$$
$$-x > -3 \qquad \text{Subtracting 3}$$
$$x < 3. \qquad \text{Multiplying by } -1 \text{ and reversing the inequality sign}$$

The domain is $\{x \mid x < 3\}$, or $(-\infty, 3)$.

> *Now Try Exercises 17 and 21.*

❯ Compound Inequalities

When two inequalities are joined by the word *and* or the word *or*, a **compound inequality** is formed. A compound inequality like

$$-3 < 2x + 5 \quad and \quad 2x + 5 \leq 7$$

is called a **conjunction**, because it uses the word *and*. The sentence

$$-3 < 2x + 5 \leq 7$$

is an abbreviation for the preceding conjunction.

Compound inequalities can be solved using the addition and multiplication principles for inequalities.

EXAMPLE 3   Solve $-3 < 2x + 5 \leq 7$. Then graph the solution set.

Solution   We have

$$-3 < 2x + 5 \leq 7$$
$$-8 < 2x \leq 2 \qquad \text{Subtracting 5}$$
$$-4 < x \leq 1. \qquad \text{Dividing by 2}$$

The solution set is $\{x \mid -4 < x \leq 1\}$, or $(-4, 1]$. The graph of the solution set is shown below.

> *Now Try Exercise 23.*

A compound inequality like $2x - 5 \leq -7 \ or \ 2x - 5 > 1$ is called a **disjunction**, because it contains the word *or*. Unlike some conjunctions, it cannot be abbreviated; that is, it cannot be written without the word *or*.

EXAMPLE 4   Solve: $2x - 5 \leq -7 \ or \ 2x - 5 > 1$. Then graph the solution set.

Solution   We have

$$2x - 5 \leq -7 \quad or \quad 2x - 5 > 1$$
$$2x \leq -2 \quad or \qquad 2x > 6 \qquad \text{Adding 5}$$
$$x \leq -1 \quad or \qquad x > 3. \qquad \text{Dividing by 2}$$

The solution set is $\{x \mid x \leq -1 \ or \ x > 3\}$. We can also write the solution set using interval notation and the symbol $\cup$ for the **union** or inclusion of both sets: $(-\infty, -1] \cup (3, \infty)$. The graph of the solution set is shown below.

$y_1 = 2x - 5, \quad y_2 = -7, \quad y_3 = 1$

To check, we graph $y_1 = 2x - 5$, $y_2 = -7$, and $y_3 = 1$. Note that for $\{x \mid x \le -1 \ or \ x > 3\}$, $y_1 \le y_2 \ or \ y_1 > y_3$.

> *Now Try Exercise 35.*

❯ An Application

EXAMPLE 5 *Income Plans.* For her interior decorating job, Natália can be paid in one of two ways:

    Plan A: $250 plus $10 per hour;

    Plan B: $20 per hour.

Suppose that a job takes n hours. For what values of n is plan B better for Natália?

Solution

1. **Familiarize.** Suppose that a job takes 20 hr. Then $n = 20$, and under plan A, Natália would earn $250 + $10 \cdot 20$, or $250 + 200, or $450. Her earnings under plan B would be $20 \cdot 20$, or $400. This shows that plan A is better for Natália if a job takes 20 hr. If a job takes 30 hr, then $n = 30$, and under plan A, Natália would earn $250 + $10 \cdot 30$, or $250 + 300, or $550. Under plan B, she would earn $20 \cdot 30$, or $600, so plan B is better in this case. To determine *all* values of n for which plan B is better for Natália, we solve an inequality. Our work in this step helps us write the inequality.

2. **Translate.** We translate to an inequality:

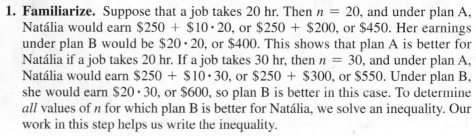

$$\underbrace{\text{Income from plan B}}_{20n} \quad \underbrace{\text{is greater than}}_{>} \quad \underbrace{\text{income from plan A}}_{250 + 10n.}$$

2. **Carry out.** We solve the inequality:

$$20n > 250 + 10n$$
$$10n > 250 \qquad \text{Subtracting } 10n \text{ on both sides}$$
$$n > 25. \qquad \text{Dividing by 10 on both sides}$$

4. **Check.** For $n = 25$, the income from plan A is $250 + $10 \cdot 25$, or $250 + 250, or $500, and the income from plan B is $20 \cdot 25$, or $500. This shows that for a job that takes 25 hr to complete, the income is the same under either plan. In the *Familiarize* step, we saw that plan B pays more for a 30-hr job. Since $30 > 25$, this provides a partial check of the result. We cannot check all values of n.

5. **State.** For values of n greater than 25 hr, plan B is better for Natália.

> *Now Try Exercise 45.*

1.6   Exercise Set

Solve and graph the solution set.

1. $4x - 3 > 2x + 7$

2. $8x + 1 \ge 5x - 5$

3. $x + 6 < 5x - 6$

4. $3 - x < 4x + 7$

5. $4 - 2x \le 2x + 16$

6. $3x - 1 > 6x + 5$

7. $14 - 5y \le 8y - 8$

8. $8x - 7 < 6x + 3$

9. $7x - 7 > 5x + 5$

10. $12 - 8y \ge 10y - 6$

11. $3x - 3 + 2x \ge 1 - 7x - 9$

12. $5y - 5 + y \leq 2 - 6y - 8$

13. $-\frac{3}{4}x \geq -\frac{5}{8} + \frac{2}{3}x$

14. $-\frac{5}{6}x \leq \frac{3}{4} + \frac{8}{3}x$

15. $4x(x - 2) < 2(2x - 1)(x - 3)$

16. $(x + 1)(x + 2) > x(x + 1)$

Find the domain of the function.

17. $h(x) = \sqrt{x - 7}$

18. $g(x) = \sqrt{x + 8}$

19. $f(x) = \sqrt{1 - 5x} + 2$

20. $f(x) = \sqrt{2x + 3} - 4$

21. $g(x) = \dfrac{5}{\sqrt{4 + x}}$

22. $h(x) = \dfrac{x}{\sqrt{8 - x}}$

Solve and write interval notation for the solution set. Then graph the solution set.

23. $-2 \leq x + 1 < 4$

24. $-3 < x + 2 \leq 5$

25. $5 \leq x - 3 \leq 7$

26. $-1 < x - 4 < 7$

27. $-3 \leq x + 4 \leq 3$

28. $-5 < x + 2 < 15$

29. $-2 < 2x + 1 < 5$

30. $-3 \leq 5x + 1 \leq 3$

31. $-4 \leq 6 - 2x < 4$

32. $-3 < 1 - 2x \leq 3$

33. $-5 < \frac{1}{2}(3x + 1) < 7$

34. $\frac{2}{3} \leq -\frac{4}{5}(x - 3) < 1$

35. $3x \leq -6 \; or \; x - 1 > 0$

36. $2x < 8 \; or \; x + 3 \geq 10$

37. $2x + 3 \leq -4 \; or \; 2x + 3 \geq 4$

38. $3x - 1 < -5 \; or \; 3x - 1 > 5$

39. $2x - 20 < -0.8 \; or \; 2x - 20 > 0.8$

40. $5x + 11 \leq -4 \; or \; 5x + 11 \geq 4$

41. $x + 14 \leq -\frac{1}{4} \; or \; x + 14 \geq \frac{1}{4}$

42. $x - 9 < -\frac{1}{2} \; or \; x - 9 > \frac{1}{2}$

43. *World Rice Production.*   The three countries with the most rice production are China, India, and Indonesia. The equation $y = 9.06x + 410.81$ provides a good estimate of world rice production in millions of metric tons, where x is the number of years after 1980. (*Source:* www.geohive.com) For what years will world rice production exceed 820 million metric tons?

44. *Social Security Disability.*   The equation $y = 0.326x + 7.148$ can be used to estimate the number of people, in millions, collecting Social Security disability payments, where x is the number of years after 2007 (*Source:* Social Security Administration). For what years will the number of people collecting disability payments be more than 12 million?

45. *Moving Costs.*   Acme Movers charges $200 plus $45 per hour to move a household across town. Leo's Movers charges $65 per hour. For what lengths of time does it cost less to hire Leo's Movers?

46. *Investment Income.*   Jalyn plans to invest $12,000, part at 4% simple interest and the rest at 6% simple interest. What is the most that she can invest at 4% and still be guaranteed at least $650 in interest per year?

47. *Investment Income.*   Dillon plans to invest $7500, part at 4% simple interest and the rest at 5% simple interest. What is the most that he can invest at 4% and still be guaranteed at least $325 in interest per year?

48. *Investment Income.*   A foundation invests $150,000 at simple interest, part at 7%, twice that amount at 4%, and the rest at 5.5%. What is the most that the foundation can invest at 4% and still be guaranteed at least $7575 in interest per year?

49. *Investment Income.*   A university invests $1,400,000 at simple interest, part at 5%, half that amount at 3.5%, and the rest at 5.5%. What is the most that the university can invest at 3.5% and still be guaranteed at least $68,000 in interest per year?

50. *Income Plans.* Karen can be paid in one of two ways for selling insurance policies:

Plan A: A salary of $750 per month, plus a commission of 10% of sales;

Plan B: A salary of $1000 per month, plus a commission of 8% of sales in excess of $2000.

For what amount of monthly sales is plan A better than plan B if we can assume that sales are always more than $2000?

51. *Income Plans.* Curt can be paid in one of two ways for the furniture he sells:

Plan A: A salary of $900 per month, plus a commission of 10% of sales;

Plan B: A salary of $1200 per month, plus a commission of 15% of sales in excess of $8000.

For what amount of monthly sales is plan B better than plan A if we can assume that Curt's sales are always more than $8000?

52. *Income Plans.* Jeanette can be paid in one of two ways for painting a house:

Plan A: $200 plus $12 per hour;

Plan B: $20 per hour.

Suppose that a job takes n hours to complete. For what values of n is plan A better for Jeanette?

❯ Skill Maintenance

Vocabulary Reinforcement

In each of Exercises 53–56, fill in the blank(s) with the correct term(s). Some of the given choices will not be used; others will be used more than once.

constant	domain
function	distance formula
any	exactly one
midpoint formula	identity
y-intercept	x-intercept
range	

53. A(n) _____ is a correspondence between a first set, called the _____, and a second set, called the _____, such that each member of the _____ corresponds to _____ member of the _____ . **[1.2]**

54. The _____ is $\left(\dfrac{x_1 + x_2}{2}, \dfrac{y_1 + y_2}{2} \right)$. **[1.1]**

55. A(n) _____ is a point $(a, 0)$. **[1.1]**

56. A function f is a linear function if it can be written as $f(x) = mx + b$, where m and b are constants. If $m = 0$, the function is a(n) _____ function $f(x) = b$. If $m = 1$ and $b = 0$, the function is the _____ function $f(x) = x$. **[1.3]**

❯ Synthesis

Solve.

57. $2x \leq 5 - 7x < 7 + x$

58. $x \leq 3x - 2 \leq 2 - x$

59. $3y < 4 - 5y < 5 + 3y$

60. $y - 10 < 5y + 6 \leq y + 10$

STUDY GUIDE

KEY TERMS AND CONCEPTS	EXAMPLES

SECTION 1.1: INTRODUCTION TO GRAPHING

Graphing Equations

To **graph** an equation is to make a drawing that represents the solutions of that equation. We can graph an equation by selecting values for one variable and finding the corresponding values for the other variable. We list the solutions (ordered pairs) in a table, plot the points, and draw the graph.

Graph: $y = 4 - x^2$.

x	$y = 4 - x^2$	(x, y)
0	4	$(0, 4)$
-1	3	$(-1, 3)$
1	3	$(1, 3)$
-2	0	$(-2, 0)$
2	0	$(2, 0)$

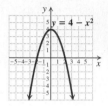

Intercepts

An **x-intercept** is a point $(a, 0)$.
To find a, let $y = 0$ and solve for x.

A **y-intercept** is a point $(0, b)$.
To find b, let $x = 0$ and solve for y.

We can graph a straight line by plotting the intercepts and drawing the line containing them.

Graph using intercepts: $2x - y = 4$.

We let $y = 0$:

$$2x - 0 = 4$$
$$2x = 4$$
$$x = 2.$$

The x-intercept is $(2, 0)$.
We let $x = 0$:

$$2 \cdot 0 - y = 4$$
$$-y = 4$$
$$y = -4.$$

The y-intercept is $(0, -4)$.

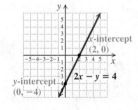

Distance Formula

The **distance** d between any two points (x_1, y_1) and (x_2, y_2) is given by

$$d = \sqrt{(x_2 - x_1)^2 + (y_2 - y_1)^2}.$$

Find the distance between $(-5, 7)$ and $(2, -3)$.

$$\begin{aligned} d &= \sqrt{[2 - (-5)]^2 + (-3 - 7)^2} \\ &= \sqrt{7^2 + (-10)^2} \\ &= \sqrt{49 + 100} \\ &= \sqrt{149} \approx 12.2 \end{aligned}$$

Midpoint Formula

If the endpoints of a segment are (x_1, y_1) and (x_2, y_2), then the coordinates of the **midpoint** of the segment are

$$\left(\frac{x_1 + x_2}{2}, \frac{y_1 + y_2}{2} \right).$$

Find the midpoint of the segment whose endpoints are $(-10, 4)$ and $(3, 8)$.

$$\begin{aligned} \left(\frac{x_1 + x_2}{2}, \frac{y_1 + y_2}{2} \right) &= \left(\frac{-10 + 3}{2}, \frac{4 + 8}{2} \right) \\ &= \left(-\frac{7}{2}, 6 \right) \end{aligned}$$

Circles

The **standard form** of the equation of a circle with center (h, k) and radius r is

$$(x - h)^2 + (y - k)^2 = r^2.$$

Find an equation of a circle with center $(1, -6)$ and radius 8.

$$(x - h)^2 + (y - k)^2 = r^2$$
$$(x - 1)^2 + [y - (-6)]^2 = 8^2$$
$$(x - 1)^2 + (y + 6)^2 = 64$$

Given the circle

$$(x + 9)^2 + (y - 2)^2 = 121,$$

determine the center and the radius.

Writing in standard form, we have

$$[x - (-9)]^2 + (y - 2)^2 = 11^2.$$

The center is $(-9, 2)$, and the radius is 11.

SECTION 1.2: FUNCTIONS AND GRAPHS

Functions

A **function** is a correspondence between a first set, called the **domain**, and a second set, called the **range**, such that each member of the domain corresponds to *exactly one* member of the range.

Consider the function given by

$$g(x) = |x| - 1.$$
$$g(-3) = |-3| - 1$$
$$= 3 - 1$$
$$= 2$$

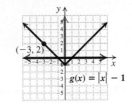

For the input -3, the output is 2: $g(-3) = 2$.
The point $(-3, 2)$ is on the graph.
Domain: Set of all inputs $= \{x \mid x \text{ is a real number}\}$, or $(-\infty, \infty)$.
Range: Set of all outputs: $\{y \mid y \geq -1\}$, or $[-1, \infty)$.

The Vertical-Line Test

If it is possible for a vertical line to cross a graph more than once, then the graph *is not* the graph of a function.

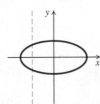

This *is not* the graph of a function because a vertical line can cross it more than once, as shown.

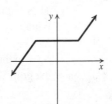

This *is* the graph of a function because no vertical line can cross it more than once.

Domain

When a function f whose inputs and outputs are real numbers is given by a formula, the **domain** is the set of all inputs for which the expression is defined as a real number.

Find the domain of the function given by

$$h(x) = \frac{x - 1}{(x + 5)(x - 10)}.$$

Division by 0 is not defined. Since $x + 5 = 0$ when $x = -5$ and $x - 10 = 0$ when $x = 10$, the domain of h is

$$\{x \mid x \text{ is a real number } and \ x \neq -5 \ and \ x \neq 10\},$$
$$\text{or} \quad (-\infty, -5) \cup (-5, 10) \cup (10, \infty).$$

SECTION 1.3: LINEAR FUNCTIONS, SLOPE, AND APPLICATIONS

Slope

$$m = \frac{\text{rise}}{\text{run}} = \frac{y_2 - y_1}{x_2 - x_1} = \frac{y_1 - y_2}{x_1 - x_2}$$

Slope can also be considered as an **average rate of change**. To find the average rate of change between two data points on a graph, determine the slope of the line that passes through the points.

The slope of the line containing the points $(3, -10)$ and $(-2, 6)$ is

$$m = \frac{y_2 - y_1}{x_2 - x_1} = \frac{6 - (-10)}{-2 - 3}$$

$$= \frac{16}{-5} = -\frac{16}{5}.$$

In 2000, the population of Flint, Michigan, was 124,943. By 2012, the population had decreased to 100,515. Find the average rate of change in population from 2000 to 2012.

$$\text{Average rate of change} = m = \frac{100{,}515 - 124{,}943}{2012 - 2000}$$

$$= \frac{-24{,}428}{12} \approx -2036$$

The average rate of change in population over the 12-year period was a decrease of about 2036 people per year.

Slope–Intercept Form of an Equation

$$f(x) = mx + b$$

The slope of the line is m.
The y-intercept of the line is $(0, b)$.

Determine the slope and the y-intercept of the line given by $5x - 7y = 14$.

We first find the slope–intercept form:

$$5x - 7y = 14$$
$$-7y = -5x + 14 \qquad \textbf{Adding } -5x$$
$$y = \frac{5}{7}x - 2. \qquad \textbf{Multiplying by } -\frac{1}{7}$$

The slope is $\frac{5}{7}$, and the y-intercept is $(0, -2)$.

To graph an equation written in slope–intercept form, plot the y-intercept and use the slope to find another point. Then draw the line.

Graph: $f(x) = -\frac{2}{3}x + 4$.

We plot the y-intercept, $(0, 4)$. Think of the slope as $\frac{-2}{3}$. From the y-intercept, we find another point by moving 2 units down and 3 units to the right to the point $(3, 2)$. We then draw the graph.

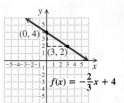

Horizontal Lines

The graph of $y = b$, or $f(x) = b$, is a horizontal line with y-intercept $(0, b)$. The slope of a horizontal line is 0.

Vertical Lines

The graph of $x = a$ is a vertical line with x-intercept $(a, 0)$. The slope of a vertical line is *not* defined.

Graph $y = -4$ and determine its slope.

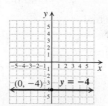

The slope is 0.

Graph $x = 3$ and determine its slope.

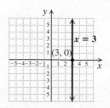

The slope is not defined.

SECTION 1.4: EQUATIONS OF LINES AND MODELING

Slope–Intercept Form of an Equation

$$y = mx + b, \text{ or } f(x) = mx + b$$

The slope of the line is m.

The y-intercept of the line is $(0, b)$.

Point–Slope Form of an Equation

$$y - y_1 = m(x - x_1)$$

The slope of the line is m.

The line passes through (x_1, y_1).

Write the slope–intercept equation for a line with slope $-\frac{2}{9}$ and y-intercept $(0, 4)$.

$$y = mx + b \qquad \textbf{Using the slope–intercept form}$$

$$y = -\frac{2}{9}x + 4 \qquad \textbf{Substituting } -\tfrac{2}{9} \textbf{ for } m \textbf{ and 4 for } b$$

Write the slope–intercept equation for a line that passes through $(-5, 7)$ and $(3, -9)$.

We first determine the slope:

$$m = \frac{-9 - 7}{3 - (-5)} = \frac{-16}{8} = -2.$$

Using the slope–intercept form: We substitute -2 for m and either $(-5, 7)$ or $(3, -9)$ for (x, y) and solve for b:

$$y = mx + b$$
$$7 = -2 \cdot (-5) + b \qquad \textbf{Using } (-5, 7)$$
$$7 = 10 + b$$
$$-3 = b.$$

The slope–intercept equation is $y = -2x - 3$.

Using the point–slope equation: We substitute -2 for m and either $(-5, 7)$ or $(3, -9)$ for (x_1, y_1):

$$y - y_1 = m(x - x_1)$$
$$y - (-9) = -2(x - 3) \qquad \textbf{Using } (3, -9)$$
$$y + 9 = -2x + 6$$
$$y = -2x - 3.$$

The slope–intercept equation is $y = -2x - 3$.

Parallel Lines

Vertical lines are parallel. Nonvertical lines are **parallel** if and only if they have the same slope and different y-intercepts.

Write the slope–intercept equation for a line passing through $(-3, 1)$ that is parallel to the line $y = \frac{2}{3}x + 5$.

The slope of $y = \frac{2}{3}x + 5$ is $\frac{2}{3}$, so the slope of a line parallel to this line is also $\frac{2}{3}$. We use either the slope–intercept equation or the point–slope equation for a line with slope $\frac{2}{3}$ and containing the point $(-3, 1)$. Here we use the point–slope equation and substitute $\frac{2}{3}$ for m, -3 for x_1, and 1 for y_1.

$$y - y_1 = m(x - x_1)$$

$$y - 1 = \frac{2}{3}[x - (-3)]$$

$$y - 1 = \frac{2}{3}(x + 3)$$

$$y - 1 = \frac{2}{3}x + 2$$

$$y = \frac{2}{3}x + 3 \qquad \textbf{Slope–intercept form}$$

Perpendicular Lines

Two lines are **perpendicular** if and only if the product of their slopes is -1 or if one line is vertical $(x = a)$ and the other is horizontal $(y = b)$.

Write the slope–intercept equation for a line that passes through $(-3, 1)$ and is perpendicular to the line $y = \frac{2}{3}x + 5$.

The slope of $y = \frac{2}{3}x + 5$ is $\frac{2}{3}$, so the slope of a line perpendicular to this line is the opposite of the reciprocal of $\frac{2}{3}$, or $-\frac{3}{2}$. Here we use the point–slope equation and substitute $-\frac{3}{2}$ for m, -3 for x_1, and 1 for y_1.

$$y - y_1 = m(x - x_1)$$

$$y - 1 = -\frac{3}{2}[x - (-3)]$$

$$y - 1 = -\frac{3}{2}(x + 3)$$

$$y - 1 = -\frac{3}{2}x - \frac{9}{2}$$

$$y = -\frac{3}{2}x - \frac{7}{2}$$

SECTION 1.5: LINEAR EQUATIONS, FUNCTIONS, ZEROS, AND APPLICATIONS

Equation–Solving Principles

Addition Principle: If $a = b$ is true, then $a + c = b + c$ is true.

Multiplication Principle: If $a = b$ is true, then $ac = bc$ is true.

Solve: $2(3x - 7) = 15 - (x + 1)$.

$$2(3x - 7) = 15 - (x + 1)$$

$6x - 14 = 15 - x - 1$ **Using the distributive property**

$6x - 14 = 14 - x$ **Collecting like terms**

$6x - 14 + x = 14 - x + x$ **Adding x on both sides**

$7x - 14 = 14$

$7x - 14 + 14 = 14 + 14$ **Adding 14 on both sides**

$7x = 28$

$\dfrac{7x}{7} = \dfrac{28}{7}$ **Dividing by 7 on both sides**

$x = 4$

Check:

$$\begin{array}{c|c} \multicolumn{2}{c}{2(3x - 7) = 15 - (x + 1)} \\ \hline 2(3 \cdot 4 - 7) \ ? \ 15 - (4 + 1) \\ 2(12 - 7) & 15 - 5 \\ 2 \cdot 5 & 10 \\ 10 & 10 \end{array}$$ **TRUE**

The solution is 4.

Special Cases

Some equations have *no* solution.

Solve: $2 + 17x = 17x - 9$.

$$2 + 17x = 17x - 9$$

$2 + 17x - 17x = 17x - 9 - 17x$ **Subtracting 17x on both sides**

$2 = -9$ **False equation**

We get a false equation; thus the equation has *no* solution.

(continued)

There are some equations for which *any* real number is a solution.

Solve: $5 - \dfrac{1}{2}x = -\dfrac{1}{2}x + 5$.

$$5 - \frac{1}{2}x = -\frac{1}{2}x + 5$$

$$5 - \frac{1}{2}x + \frac{1}{2}x = -\frac{1}{2}x + 5 + \frac{1}{2}x \qquad \text{Adding } \tfrac{1}{2}x \text{ on both sides}$$

$$5 = 5 \qquad \text{True equation}$$

We get a true equation. Thus any real number is a solution. The solution set is

$$\{x \mid x \text{ is a real number}\}, \quad \text{or} \quad (-\infty, \infty).$$

Zeros of Functions

An input c of a function f is called a **zero** of the function if the output for the function is 0 when the input is c. That is,

c is a zero of f if $f(c) = 0$.

A linear function $f(x) = mx + b$, with $m \neq 0$, has exactly one zero.

Find the zero of the linear function

$$f(x) = \frac{5}{8}x - 40.$$

We find the value of x for which $f(x) = 0$:

$$\frac{5}{8}x - 40 = 0 \qquad \text{Setting } f(x) = 0$$

$$\frac{5}{8}x = 40 \qquad \text{Adding 40 on both sides}$$

$$\frac{8}{5} \cdot \frac{5}{8}x = \frac{8}{5} \cdot 40 \qquad \text{Multiplying by } \tfrac{8}{5} \text{ on both sides}$$

$$x = 64.$$

We can check by substituting 64 for x:

$$f(64) = \tfrac{5}{8} \cdot 64 - 40 = 40 - 40 = 0.$$

The zero of $f(x) = \dfrac{5}{8}x - 40$ is 64.

SECTION 1.6: SOLVING LINEAR INEQUALITIES

Principles for Solving Linear Inequalities

Addition Principle:
If $a < b$ is true, then $a + c < b + c$ is true.

Multiplication Principle:
If $a < b$ and $c > 0$ are true, then $ac < bc$ is true.

If $a < b$ and $c < 0$ are true, then $ac > bc$ is true.

Similar statements hold for $a \leq b$.

Solve $3x - 2 \leq 22 - 5x$ and graph the solution set.

$$3x - 2 \leq 22 - 5x$$

$$3x - 2 + 5x \leq 22 - 5x + 5x \qquad \text{Adding 5x on both sides}$$

$$8x - 2 \leq 22$$

$$8x - 2 + 2 \leq 22 + 2 \qquad \text{Adding 2 on both sides}$$

$$8x \leq 24$$

$$\frac{8x}{8} \leq \frac{24}{8} \qquad \text{Dividing by 8 on both sides}$$

$$x \leq 3$$

The solution set is

$$\{x \mid x \leq 3\}, \quad \text{or} \quad (-\infty, 3].$$

The graph of the solution set is as follows.

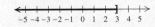

Compound Inequalities

When two inequalities are joined by the word *and* or the word *or*, a compound inequality is formed.

A Conjunction:

$1 < 3x - 20$ *and* $3x - 20 \le 40$,   or
$1 < 3x - 20 \le 40$

A Disjunction:

$8x - 1 \le -17$ *or* $8x - 1 > 7$

Solve: $1 < 3x - 20 \le 40$.

$$1 < 3x - 20 \le 40$$
$$21 < 3x \le 60 \quad \textbf{Adding 20}$$
$$7 < x \le 20 \quad \textbf{Dividing by 3}$$

The solution set is

$$\{x \mid 7 < x \le 20\}, \quad \text{or} \quad (7, 20].$$

Solve: $8x - 1 \le -17$ or $8x - 1 > 7$.

$$8x - 1 \le -17 \quad or \quad 8x - 1 > 7$$
$$8x \le -16 \quad or \quad 8x > 8 \quad \textbf{Adding 1}$$
$$x \le -2 \quad or \quad x > 1 \quad \textbf{Dividing by 8}$$

The solution set is

$$\{x \mid x \le -2 \ or \ x > 1\}, \quad \text{or} \quad (-\infty, -2] \cup (1, \infty).$$

REVIEW EXERCISES

Answers to all of the review exercises appear in the answer section at the back of the book. If you get an incorrect answer, restudy the objective indicated in red next to the exercise or the instruction line that precedes it.

Determine whether the statement is true or false.

1. If the line $ax + y = c$ is perpendicular to the line $x - by = d$, then $\dfrac{a}{b} = 1$. **[1.4]**

2. The intersection of the lines $y = \frac{1}{2}$ and $x = -5$ is $\left(-5, \frac{1}{2}\right)$. **[1.3]**

3. The domain of the function $f(x) = \dfrac{\sqrt{3 - x}}{x}$ does not contain -3 and 0. **[1.2]**

4. The line parallel to the *x*-axis that passes through $\left(-\frac{1}{4}, 7\right)$ is $x = -\frac{1}{4}$. **[1.3]**

5. The zero of a linear function *f* is the first coordinate of the *x*-intercept of the graph of $y = f(x)$. **[1.5]**

6. If $a < b$ is true and $c \ne 0$, then $ac < bc$ is true. **[1.6]**

Use substitution to determine whether the given ordered pairs are solutions of the given equation. **[1.1]**

7. $\left(3, \frac{24}{9}\right), (0, -9); \ 2x - 9y = -18$

8. $(0, 7), (7, 1); \ y = 7$

Find the intercepts and then graph the line. **[1.1]**

9. $2x - 3y = 6$ **10.** $10 - 5x = 2y$

Graph the equation. **[1.1]**

11. $y = -\frac{2}{3}x + 1$ **12.** $2x - 4y = 8$

13. $y = 2 - x^2$

14. Find the distance between $(3, 7)$ and $(-2, 4)$. **[1.1]**

15. Find the midpoint of the segment with endpoints $(3, 7)$ and $(-2, 4)$. **[1.1]**

16. Find the center and the radius of the circle with equation $(x + 1)^2 + (y - 3)^2 = 9$. Then graph the circle. **[1.1]**

Find an equation for a circle satisfying the given conditions. **[1.1]**

17. Center: $(0, -4)$, radius of length $\frac{3}{2}$

18. Center: $(-2, 6)$, radius of length $\sqrt{13}$

19. Diameter with endpoints $(-3, 5)$ and $(7, 3)$

Determine whether the correspondence is a function. **[1.2]**

20. $-6 \longrightarrow 1$
    $-1 \longrightarrow 3$
    $2 \longrightarrow 10$
    $7 \longrightarrow 12$

21. h $\longrightarrow$ r
    i $\longrightarrow$ s
    j $\nearrow$ t
    k $\nearrow$

Determine whether the relation is a function. Identify the domain and the range. **[1.2]**

22. $\{(3, 1), (5, 3), (7, 7), (3, 5)\}$

23. $\{(2, 7), (-2, -7), (7, -2), (0, 2), (1, -4)\}$

24. Given that $f(x) = x^2 - x - 3$, find each of the following. **[1.2]**

   a) $f(0)$ **b)** $f(-3)$
   c) $f(a - 1)$ **d)** $f(-x)$

25. Given that $f(x) = \dfrac{x - 7}{x + 5}$, find each of the following. **[1.2]**

   a) $f(7)$ **b)** $f(x + 1)$
   c) $f(-5)$ **d)** $f\left(-\frac{1}{2}\right)$

26. A graph of a function is shown below. Find $f(2)$, $f(-4)$, and $f(0)$. **[1.2]**

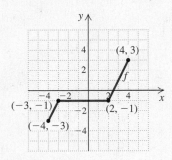

Determine whether the graph is that of a function. **[1.2]**

27.

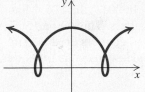

28.

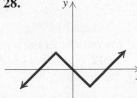

29.

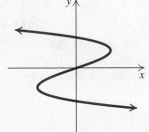

30.

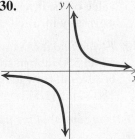

Find the domain of the function. **[1.2]**

31. $f(x) = 4 - 5x + x^2$

32. $f(x) = \dfrac{3}{x} + 2$

33. $f(x) = \dfrac{1}{x^2 - 6x + 5}$ **34.** $f(x) = \dfrac{-5x}{|16 - x^2|}$

Graph the function. Then visually estimate the domain and the range. **[1.2]**

35. $f(x) = \sqrt{16 - x^2}$

36. $g(x) = |x - 5|$

37. $f(x) = x^3 - 7$

38. $h(x) = x^4 + x^2$

In Exercises 39 and 40, the table of data contains input–output values for a function. Answer the following questions. **[1.3]**

   a) *Is the change in the inputs, x, the same?*
   b) *Is the change in the outputs, y, the same?*
   c) *Is the function linear?*

39.

x	y
−3	8
−2	11
−1	14
0	17
1	20
2	22
3	26

40.

x	y
20	11.8
30	24.2
40	36.6
50	49.0
60	61.4
70	73.8
80	86.2

Find the slope of the line containing the given points. **[1.3]**

41. $(2, -11), (5, -6)$

42. $(5, 4), (-3, 4)$

43. $\left(\frac{1}{2}, 3\right), \left(\frac{1}{2}, 0\right)$

44. *Coffee Consumption.* The U.S. annual per-capita consumption of coffee was 26.8 gal in 1990. By 2012, this amount had decreased to 24.7 gal. (*Source:* Economic Research Service, U.S. Department of Agriculture) Find the average rate of change in per-capita coffee consumption from 1990 to 2012. **[1.3]**

Find the slope and the y-intercept of the line with the given equation. **[1.3]**

45. $y = -\frac{7}{11}x - 6$ **46.** $-2x - y = 7$

47. Graph $y = -\frac{1}{4}x + 3$ using the slope and the y-intercept. **[1.3]**

48. *Total Cost.* Clear County Cable Television charges a $110 installation fee and $85 per month for basic service. Write an equation that can be used to determine the total cost $C(t)$ of t months of basic cable television service. Find the total cost of 1 year of service. **[1.3]**

49. *Temperature and Depth of the Earth.* The function T given by $T(d) = 10d + 20$ can be used to determine the temperature T, in degrees Celsius, at a depth d, in kilometers, inside the earth.

   a) Find $T(5)$, $T(20)$, and $T(1000)$. **[1.3]**
   b) The radius of the earth is about 5600 km. Use this fact to determine the domain of the function. **[1.3]**

Write a slope–intercept equation for a line with the following characteristics. **[1.4]**

50. $m = -\frac{2}{3}$, *y*-intercept $(0, -4)$

51. $m = 3$, passes through $(-2, -1)$

52. Passes through $(4, 1)$ and $(-2, -1)$

53. Write equations of the horizontal line and the vertical line that pass through $\left(-4, \frac{2}{5}\right)$. **[1.4]**

54. Find a linear function h given $h(-2) = -9$ and $h(4) = 3$. Then find $h(0)$. **[1.4]**

Determine whether the lines are parallel, perpendicular, or neither. **[1.4]**

55. $3x - 2y = 8$,
$6x - 4y = 2$

56. $y - 2x = 4$,
$2y - 3x = -7$

57. $y = \frac{3}{2}x + 7$,
$y = -\frac{2}{3}x - 4$

Given the point $(1, -1)$ *and the line* $2x + 3y = 4$:

58. Find an equation of the line containing the given point and parallel to the given line. **[1.4]**

59. Find an equation of the line containing the given point and perpendicular to the given line. **[1.4]**

60. *Female Medical School Graduates.* Data in the following table list the numbers of female medical school graduates for selected years 2006–2014.

Year, *x*	Female Medical School Graduates in the United States, *W*
2006, 0	7748
2008, 2	7969
2010, 4	8133
2012, 6	8291
2014, 8	8576

Source: Association of American Medical Colleges

a) Without using the regression feature on a graphing calculator, model the data with a linear function where the number of female medical school graduates W is a function of the year x and where x is the number of years after 2006. Then using this function, estimate the number of female graduates in 2013. Answers may vary depending on the data points used. **[1.4]**

b) Using a graphing calculator, fit a regression line to the data and use it to estimate the number of female medical school graduates in 2013. What is the correlation coefficient for the regression line? How close a fit is the regression line? **[1.4]**

Solve. **[1.5]**

61. $4y - 5 = 1$

62. $3x - 4 = 5x + 8$

63. $5(3x + 1) = 2(x - 4)$

64. $2(n - 3) = 3(n + 5)$

65. $\frac{3}{5}y - 2 = \frac{3}{8}$

66. $5 - 2x = -2x + 3$

67. $x - 13 = -13 + x$

68. *Production of Quarters.* In 2013, the U.S. Mint produced 1455 million quarters. This was a 156% increase over the number of quarters produced in 2012. (*Source*: U.S. Mint) How many quarters were produced in 2012? **[1.5]**

69. *Amount of Investment.* James makes an investment at 5.2% simple interest. At the end of 1 year, the total value of the investment is $2419.60. How much was originally invested? **[1.5]**

70. *Flying into a Headwind.* An airplane that can travel 550 mph in still air encounters a 20-mph headwind. How long will it take the plane to travel 1802 mi? **[1.5]**

Find the zero(s) of the function. **[1.5]**

71. $f(x) = 6x - 18$

72. $f(x) = x - 4$

73. $f(x) = 2 - 10x$

74. $f(x) = 8 - 2x$

Solve and write interval notation for the solution set. Then graph the solution set. **[1.6]**

75. $2x - 5 < x + 7$

76. $3x + 1 \geq 5x + 9$

77. $-3 \leq 3x + 1 \leq 5$

78. $-2 < 5x - 4 \leq 6$

79. $2x < -1$ *or* $x - 3 > 0$

80. $3x + 7 \leq 2$ *or* $2x + 3 \geq 5$

81. *Homeschooled Children in the United States.* The equation $y = 0.073x + 0.848$ can be used to estimate the number of homeschooled children in the United States, in millions, where x is the number of years after 1999 (*Source*: Department of Education's National Center for Education Statistics). For what years will the number of homeschooled children exceed 2.3 million? **[1.6]**

82. *Temperature Conversion.* The formula $C = \frac{5}{9}(F - 32)$ can be used to convert Fahrenheit temperatures F to Celsius temperatures C. For what Fahrenheit temperatures is the Celsius temperature lower than 45°C? **[1.6]**

83. The domain of the function
$$f(x) = \frac{x + 3}{8 - 4x}$$
is which of the following? **[1.2]**

A. $(-3, 2)$
B. $(-\infty, 2) \cup (2, \infty)$
C. $(-\infty, -3) \cup (-3, 2) \cup (2, \infty)$
D. $(-\infty, -3) \cup (-3, \infty)$

84. The center of the circle described by the equation $(x - 1)^2 + y^2 = 9$ is which of the following? **[1.1]**

A. $(-1, 0)$
B. $(1, 0)$
C. $(0, -3)$
D. $(-1, 3)$

85. The graph of $f(x) = -\frac{1}{2}x - 2$ is which of the following? **[1.3]**

A.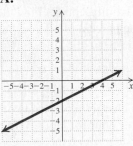

B.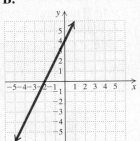

C.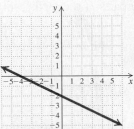

D.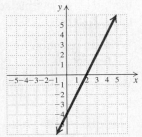

❯ Synthesis

86. Find the point on the x-axis that is equidistant from the points $(1, 3)$ and $(4, -3)$. **[1.1]**

Find the domain. **[1.2]**

87. $f(x) = \dfrac{\sqrt{1 - x}}{x - |x|}$

88. $f(x) = (x - 9x^{-1})^{-1}$

❯ Collaborative Discussion and Writing

89. Discuss why the graph of $f(x) = -\frac{3}{5}x + 4$ is steeper than the graph of $g(x) = \frac{1}{2}x - 6$. **[1.3]**

90. As the first step in solving
$$3x - 1 = 8,$$
Stella multiplies by $\frac{1}{3}$ on both sides. What advice would you give her about the procedure for solving equations? **[1.5]**

91. Is it possible for a disjunction to have no solution? Why or why not? **[1.6]**

92. Explain in your own words why a linear function $f(x) = mx + b$, with $m \neq 0$, has exactly one zero. **[1.5]**

93. Why can the conjunction $3 < x$ *and* $x < 4$ be written as $3 < x < 4$, but the disjunction $x < 3$ *or* $x > 4$ cannot be written as $3 > x > 4$? **[1.6]**

94. Explain in your own words what a function is. **[1.2]**

1. Determine whether the ordered pair $\left(\frac{1}{2}, \frac{9}{10}\right)$ is a solution of the equation $5y - 4 = x$.

2. Find the intercepts of $5x - 2y = -10$ and graph the line.

3. Find the distance between $(5, 8)$ and $(-1, 5)$.

4. Find the midpoint of the segment with endpoints $(-2, 6)$ and $(-4, 3)$.

5. Find the center and the radius of the circle
$$(x + 4)^2 + (y - 5)^2 = 36.$$

6. Find an equation of the circle with center $(-1, 2)$ and radius $\sqrt{5}$.

7. **a)** Determine whether the relation
$$\{(-4, 7), (3, 0), (1, 5), (0, 7)\}$$
   is a function. Answer yes or no.
   b) Find the domain of the relation.
   c) Find the range of the relation.

8. Given that $f(x) = 2x^2 - x + 5$, find each of the following.
   a) $f(-1)$
   b) $f(a + 2)$

9. Given that $f(x) = \dfrac{1 - x}{x}$, find each of the following.
   a) $f(0)$ **b)** $f(1)$

10. Using the graph below, find $f(-3)$.

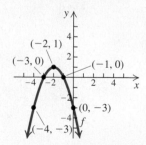

11. Determine whether the graph is that of a function. Answer yes or no.
   a)

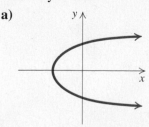

b)

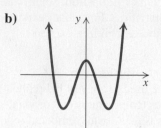

Find the domain of the function.

12. $f(x) = \dfrac{1}{x - 4}$

13. $g(x) = x^3 + 2$

14. $h(x) = \sqrt{25 - x^2}$

15. **a)** Graph: $f(x) = |x - 2| + 3$.
    b) Visually estimate the domain of $f(x)$.
    c) Visually estimate the range of $f(x)$.

Find the slope of the line containing the given points.

16. $\left(-2, \frac{2}{3}\right), (-2, 5)$

17. $(4, -10), (-8, 12)$

18. $(-5, 6), \left(\frac{3}{4}, 6\right)$

19. *Number of Married Adults.* The number of married adults in the United States is declining. In 1960, 72% of adults were married. This number had decreased to 51% by 2012. (*Source: AARP Magazine*, June–July 2014, "The New American Family," p. 32) Find the average rate of change in the percent of adults who are married for the years from 1960 to 2012.

20. Find the slope and the y-intercept of the line with equation $-3x + 2y = 5$.

21. *Total Cost.* An electrician charges a basic rate of $65 for a service call plus $48 per hour for labor. Write an equation that can be used to determine the cost $C(t)$ of hiring an electrician to do repair work. Then find the total cost, not including parts, if the repair work takes 2.25 hr.

22. Write an equation for the line with $m = -\frac{5}{8}$ and y-intercept $(0, -5)$.

23. Write an equation for the line that passes through $(-5, 4)$ and $(3, -2)$.

24. Write the equation of the vertical line that passes through $\left(-\frac{3}{8}, 11\right)$.

25. Determine whether the lines are parallel, perpendicular, or neither.

$$2x + 3y = -12,$$
$$2y - 3x = 8$$

26. Find an equation of the line containing the point $(-1, 3)$ and parallel to the line $x + 2y = -6$.

27. Find an equation of the line containing the point $(-1, 3)$ and perpendicular to the line $x + 2y = -6$.

28. *Weekly Earnings.* Data in the following table list the average weekly earnings of U.S. production workers from 2003 to 2013.

Year, x	Average Weekly Earnings of U.S. Production Workers
2003, 0	$517.82
2005, 2	544.05
2007, 4	589.27
2009, 6	616.01
2011, 8	653.19
2013, 10	677.67

Source: Bureau of Labor Statistics, U.S. Department of Labor

a) Without using the regression feature on a graphing calculator, model the data with a linear function. Then using this function, predict the average weekly earnings of U.S. production workers in 2017. Answers may vary depending on the data points used.

b) Using a graphing calculator, fit a regression line to the data and use it to predict the average weekly earnings of production workers in 2017. What is the correlation coefficient for the regression line?

Solve.

29. $6x + 7 = 1$

30. $2.5 - x = -x + 2.5$

31. $\frac{3}{2}y - 4 = \frac{5}{3}y + 6$

32. $2(4x + 1) = 8 - 3(x - 5)$

33. *Parking-Lot Dimensions.* The parking lot behind Kai's Kafé has a perimeter of 210 m. The width is three-fourths of the length. What are the dimensions of the parking lot?

34. *Pricing.* Kokona's Juice Bar prices its bottled juices by raising the wholesale price 50% and then adding 25¢. What is the wholesale price of a bottle of juice that sells for $2.95?

35. Find the zero(s) of the function

$$f(x) = 3x + 9.$$

Solve and write interval notation for the solution set. Then graph the solution set.

36. $5 - x \geq 4x + 20$

37. $-7 < 2x + 3 < 9$

38. $2x - 1 \leq 3 \ or \ 5x + 6 \geq 26$

39. *Moving Costs.* Morgan Movers charges $90 plus $25 per hour to move households across town. McKinley Movers charges $40 per hour for crosstown moves. For what lengths of time does it cost less to hire Morgan Movers?

40. The graph of $g(x) = 1 - \frac{1}{2}x$ is which of the following ?

A.
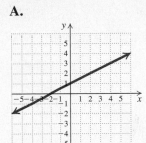

B.
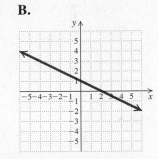

C.
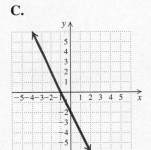

D.
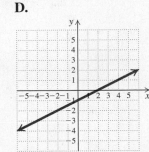

❯ Synthesis

41. Suppose that for some function h, $h(x + 2) = \frac{1}{2}x$. Find $h(-2)$.

More on Functions

APPLICATION

This problem appears as Exercise 27 in Section 2.1.

A wholesale nursery estimates that it will sell N fruit trees after spending a dollars on advertising, where
$$N(a) = -a^2 + 300a + 6, \quad 0 \le a \le 300,$$
and a is measured in thousands of dollars. For what advertising expenditure will the greatest number of fruit trees be sold? How many fruit trees will be sold for that amount?

2.1 ▷ Increasing, Decreasing, and Piecewise Functions; Applications

❯ Graph functions, looking for intervals on which the function is increasing, decreasing, or constant, and estimate relative maxima and minima.

❯ Given an application, find a function that models the application. Find the domain of the function and function values, and then graph the function.

❯ Graph functions defined piecewise.

Because functions occur in so many real-world situations, it is important to be able to analyze them carefully.

❯ Increasing, Decreasing, and Constant Functions

On a given interval, if the graph of a function rises from left to right, it is said to be **increasing** on that interval. If the graph drops from left to right, it is said to be **decreasing**. If the function values stay the same on the interval, the function is said to be **constant**.

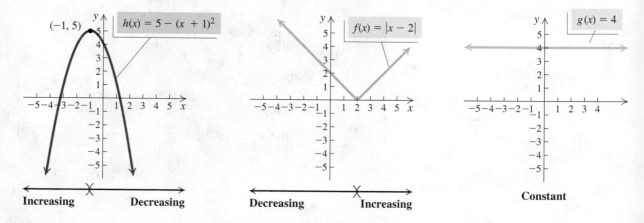

We are led to the following definitions.

INCREASING, DECREASING, AND CONSTANT FUNCTIONS

A function f is said to be **increasing** on an *open* interval I, if for all a and b in that interval, $a < b$ implies $f(a) < f(b)$. (See Fig. 1 on the following page.)

A function f is said to be **decreasing** on an *open* interval I, if for all a and b in that interval, $a < b$ implies $f(a) > f(b)$. (See Fig. 2.)

A function f is said to be **constant** on an *open* interval I, if for all a and b in that interval, $f(a) = f(b)$. (See Fig. 3.)

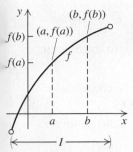

For $a < b$ in I, $f(a) < f(b)$; f is *increasing* on I.

FIGURE 1.

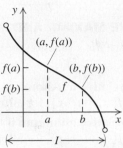

For $a < b$ in I, $f(a) > f(b)$; f is *decreasing* on I.

FIGURE 2.

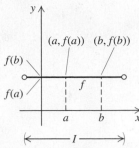

For all a and b in I, $f(a) = f(b)$; f is *constant* on I.

FIGURE 3.

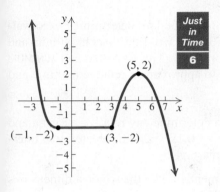

Just in Time

6

EXAMPLE 1 Determine the intervals on which the function in the figure at left is **(a)** increasing; **(b)** decreasing; **(c)** constant.

Solution When expressing interval(s) on which a function is increasing, decreasing, or constant, we consider only values in the *domain* of the function. Since the domain of this function is $(-\infty, \infty)$, we consider all real values of x.

a) As x-values (that is, values in the domain) increase from $x = 3$ to $x = 5$, y-values (that is, values in the range) increase from -2 to 2. Thus the function is increasing on the interval $(3, 5)$.

b) As x-values increase from negative infinity to -1, y-values decrease; y-values also decrease as x-values increase from 5 to positive infinity. Thus the function is decreasing on the intervals $(-\infty, -1)$ and $(5, \infty)$.

c) As x-values increase from -1 to 3, y remains -2. The function is constant on the interval $(-1, 3)$.

> *Now Try Exercise 5.*

In calculus, the slope of a line tangent to the graph of a function at a particular point is used to determine whether the function is increasing, decreasing, or neither. If the slope is positive, the function is increasing; if the slope is negative, the function is decreasing; if the slope is 0 over an interval, the function is constant. Since slope cannot be both positive and negative at the same point, a function cannot be both increasing and decreasing at a specific point. For this reason, increasing, decreasing, and constant intervals are expressed in *open interval* notation. In Example 1, if $[3, 5]$ had been used for the increasing interval and $[5, \infty)$ for a decreasing interval, the function would be both increasing and decreasing at $x = 5$. This is not possible.

❯ Relative Maximum and Minimum Values

Consider the graph shown below. Note the "peaks" and "valleys" at the x-values c_1, c_2, and c_3. The function value $f(c_2)$ is called a **relative maximum** (plural, **maxima**). Each of the function values $f(c_1)$ and $f(c_3)$ is called a **relative minimum** (plural, **minima**).

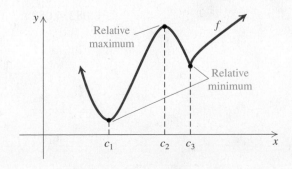

> **RELATIVE MAXIMA AND MINIMA**
>
> Suppose that f is a function for which $f(c)$ exists for some c in the domain of f. Then:
>
> $f(c)$ is a **relative maximum** if there exists an *open* interval I containing c such that $f(c) > f(x)$, for all x in I where $x \neq c$; and
>
> $f(c)$ is a **relative minimum** if there exists an *open* interval I containing c such that $f(c) < f(x)$, for all x in I where $x \neq c$.

Simply stated, $f(c)$ is a *relative maximum* if $(c, f(c))$ is the highest point in some *open* interval, and $f(c)$ is a *relative minimum* if $(c, f(c))$ is the lowest point in some *open* interval.

If you take a calculus course, you will learn a method for determining exact values of relative maxima and minima. In Section 3.3, we will find exact maximum and minimum values of quadratic functions algebraically. The MAXIMUM and MINIMUM features on a graphing calculator can be used to approximate relative maxima and minima.

EXAMPLE 2 Use a graphing calculator to determine any relative maxima and minima of the function $f(x) = 0.1x^3 - 0.6x^2 - 0.1x + 2$ and to determine intervals on which the function is increasing or decreasing.

Solution We first graph the function, experimenting with the window dimensions as needed. The curvature is seen fairly well with window settings of $[-4, 6, -3, 3]$. Using the MAXIMUM and MINIMUM features, we determine the relative maximum value and the relative minimum value of the function.

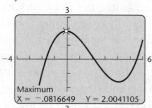

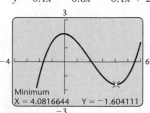

We see that the relative maximum value of the function is about 2.004. It occurs when $x \approx -0.082$. We also approximate the relative minimum, -1.604 at $x \approx 4.082$.

We note that the graph rises, or increases, from the left and stops increasing at the relative maximum. From this point, the graph decreases to the relative minimum and then begins to rise again. Thus the function is *increasing* on the intervals

$$(-\infty, -0.082) \quad \text{and} \quad (4.082, \infty)$$

and *decreasing* on the interval

$$(-0.082, 4.082).$$

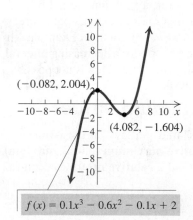

Now Try Exercise 23.

❯ Applications of Functions

Many real-world situations can be modeled by functions.

EXAMPLE 3   *Car Distance.*   Two nurses, Kiara and Matias, drive away from a hospital at right angles to each other. Kiara's speed is 35 mph and Matias's is 40 mph.

a) Express the distance between the cars as a function of time, $d(t)$.

b) Find the domain of the function.

Solution

a) Suppose that 1 hr goes by. At that time, Kiara has traveled 35 mi and Matias has traveled 40 mi. We can use the Pythagorean theorem to find the distance between them. This distance would be the length of the hypotenuse of a right triangle with legs measuring 35 mi and 40 mi. After 2 hr, the triangle's legs would measure $2 \cdot 35$, or 70 mi, and $2 \cdot 40$, or 80 mi. Noting that the distances will always be changing, we make a drawing and let $t =$ the time, in hours, that Kiara and Matias have been driving since leaving the hospital.

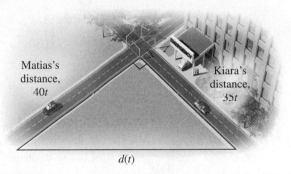

Matias's distance, $40t$

Kiara's distance, $35t$

$d(t)$

After t hours, Kiara has traveled $35t$ miles and Matias $40t$ miles. We now use the Pythagorean theorem:

$$[d(t)]^2 = (35t)^2 + (40t)^2.$$

Because distance must be nonnegative, we need consider only the positive square root when solving for $d(t)$:

$$
\begin{aligned}
d(t) &= \sqrt{(35t)^2 + (40t)^2} \\
&= \sqrt{1225t^2 + 1600t^2} \\
&= \sqrt{2825t^2} \\
&\approx 53.15|t| \quad \text{**Approximating the root to two decimal places**} \\
&\approx 53.15t. \quad \text{**Since $t \geq 0$, $|t| = t$.**}
\end{aligned}
$$

Thus, $d(t) = 53.15t, t \geq 0$.

b) Since the time traveled, t, must be nonnegative, the domain is the set of nonnegative real numbers $[0, \infty)$.

❯ *Now Try Exercise 35.*

$y = 53.15x$

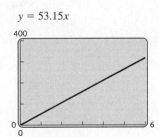

EXAMPLE 4   *Area of Office Space.*   A community college has 30 ft of dividers with which to set off a rectangular area for a student testing center. If a corner of the math lab is used for the testing center, the partition need only form two sides of a rectangle.

a) Express the floor area of the testing center as a function of the length of the partition.

b) Find the domain of the function.

c) Graph the function.

d) Find the dimensions that maximize the floor area.

Solution

a) Note that the dividers will form two sides of a rectangle. If, for example, 14 ft of dividers are used for the length of the rectangle, that would leave $30 - 14$, or 16 ft of dividers for the width. Thus if $x =$ the length, in feet, of the rectangle, then $30 - x =$ the width. We represent this information in a drawing, as shown below.

The area, $A(x)$, is given by

$$A(x) = x(30 - x) \quad \textbf{Area} = \textbf{length} \cdot \textbf{width}.$$
$$= 30x - x^2.$$

The function $A(x) = 30x - x^2$ can be used to express the area of the rectangle as a function of the length.

b) Because the length and the width of the rectangle must be positive and only 30 ft of dividers are available, we restrict the domain of A to $\{x \mid 0 < x < 30\}$—that is, the interval $(0, 30)$.

c) The graph is shown at left.

d) We use the MAXIMUM feature as shown on the graph at left. The maximum value of the area function on the interval $(0, 30)$ is 225 when $x = 15$. Thus the dimensions that maximize the area are

$$\text{Length} = x = 15 \text{ ft} \quad \text{and}$$
$$\text{Width} = 30 - x = 30 - 15 = 15 \text{ ft}.$$

▶ *Now Try Exercise 33.*

$y = 30x - x^2$
250

Maximum
0 X = 15   Y = 225   40
0

❭ Functions Defined Piecewise

Sometimes functions are defined **piecewise** using different output formulas for different pieces, or parts, of the domain.

Just in Time

5

EXAMPLE 5 For the function defined as

$$f(x) = \begin{cases} x + 1, & \text{for } x < -2, \\ 5, & \text{for } -2 \le x \le 3, \\ x^2, & \text{for } x > 3, \end{cases}$$

find $f(-5)$, $f(-3)$, $f(0)$, $f(3)$, $f(4)$, and $f(10)$.

Solution First, we determine which part of the domain contains the given input. Then we use the corresponding formula to find the output.

Since $-5 < -2$, we use the formula $f(x) = x + 1$:

$$f(-5) = -5 + 1 = -4.$$

Since $-3 < -2$, we use the formula $f(x) = x + 1$ again:

$$f(-3) = -3 + 1 = -2.$$

Since $-2 \le 0 \le 3$, we use the formula $f(x) = 5$:

$$f(0) = 5.$$

Since $-2 \le 3 \le 3$, we use the formula $f(x) = 5$ a second time:

$$f(3) = 5.$$

Since $4 > 3$, we use the formula $f(x) = x^2$:

$$f(4) = 4^2 = 16.$$

Since $10 > 3$, we once again use the formula $f(x) = x^2$:

$$f(10) = 10^2 = 100.$$

> **Now Try Exercise 47.**

EXAMPLE 6 Graph the function defined as

$$g(x) = \begin{cases} \frac{1}{3}x + 3, & \text{for } x < 3, \\ -x, & \text{for } x \ge 3. \end{cases}$$

Solution Since the function is defined in two pieces, or parts, we create the graph in two parts.

a) We graph $g(x) = \frac{1}{3}x + 3$ *only* for inputs x less than 3. That is, we use $g(x) = \frac{1}{3}x + 3$ only for x-values in the interval $(-\infty, 3)$. Some ordered pairs that are solutions of this piece of the function are shown in Table 1.

b) We graph $g(x) = -x$ *only* for inputs x greater than or equal to 3. That is, we use $g(x) = -x$ only for x-values in the interval $[3, \infty)$. Some ordered pairs that are solutions of this piece of the function are shown in Table 2.

TABLE 1

x $(x < 3)$	$g(x) = \frac{1}{3}x + 3$
-3	2
0	3
2	$3\frac{2}{3}$

TABLE 2

x $(x \ge 3)$	$g(x) = -x$
3	-3
4	-4
6	-6

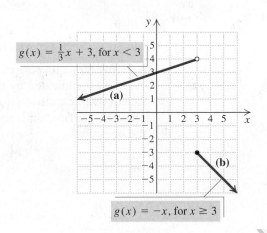

> **Now Try Exercise 51.**

TABLE 3

x $(x \le 0)$	$f(x) = 4$
-5	4
-2	4
0	4

TABLE 4

x $(0 < x \le 2)$	$f(x) = 4 - x^2$
$\frac{1}{2}$	$3\frac{3}{4}$
1	3
2	0

EXAMPLE 7 Graph the function defined as

$$f(x) = \begin{cases} 4, & \text{for } x \le 0, \\ 4 - x^2, & \text{for } 0 < x \le 2, \\ 2x - 6, & \text{for } x > 2. \end{cases}$$

Solution We create the graph in three pieces, or parts.

a) We graph $f(x) = 4$ *only* for inputs x less than or equal to 0. That is, we use $f(x) = 4$ only for x-values in the interval $(-\infty, 0]$. Some ordered pairs that are solutions of this piece of the function are shown in Table 3.

b) We graph $f(x) = 4 - x^2$ *only* for inputs x greater than 0 and less than or equal to 2. That is, we use $f(x) = 4 - x^2$ only for x-values in the interval $(0, 2]$. Some ordered pairs that are solutions of this piece of the function are shown in Table 4.

TABLE 5

x $(x > 2)$	$f(x) = 2x - 6$
$2\frac{1}{2}$	-1
3	0
5	4

c) We graph $f(x) = 2x - 6$ *only* for inputs x greater than 2. That is, we use $f(x) = 2x - 6$ only for x-values in the interval $(2, \infty)$. Some ordered pairs that are solutions of this piece of the function are shown in Table 5.

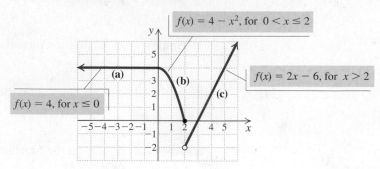

> **Now Try Exercise 55.**

EXAMPLE 8 Graph the function defined as

$$f(x) = \begin{cases} \dfrac{x^2 - 4}{x + 2}, & \text{for } x \neq -2, \\ 3, & \text{for } x = -2. \end{cases}$$

Solution When $x \neq -2$, the denominator of $(x^2 - 4)/(x + 2)$ is nonzero, so we can simplify:

$$\frac{x^2 - 4}{x + 2} = \frac{(x + 2)(x - 2)}{x + 2} = x - 2.$$

Thus,

$$f(x) = x - 2, \quad \text{for } x \neq -2.$$

The graph of this part of the function consists of a line with a "hole" at the point $(-2, -4)$, indicated by the open circle. The hole occurs because the piece of the function represented by $(x^2 - 4)/(x + 2)$ is not defined for $x = -2$. By the definition of the function, we see that $f(-2) = 3$, so we plot the point $(-2, 3)$ above the open circle.

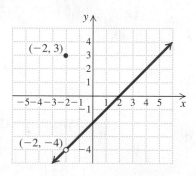

> **Now Try Exercise 59.**

$y = (x^2 - 4)/(x + 2)$

X	Y1
−2.3	−4.3
−2.2	−4.2
−2.1	−4.1
−2	**ERROR**
−1.9	−3.9
−1.8	−3.8
−1.7	−3.7

X = −2

The hand-drawn graph of the piece of the function in Example 8 represented by $y = (x^2 - 4)/(x + 2)$, $x \neq 2$, can be checked using a graphing calculator. When $y = (x^2 - 4)/(x + 2)$ is graphed, the hole may or may not be visible, depending on the window dimensions chosen. When we use the ZDECIMAL feature from the ZOOM menu, note that the hole will appear, as shown in the graph at left. If we examine a table of values for this function, we see that an ERROR message corresponds to the x-value -2. This indicates that -2 is *not* in the domain of the function $y = (x^2 - 4)/(x + 2)$. However, -2 *is* in the domain of the function f in Example 8 because $f(-2)$ is defined to be 3.

A piecewise function with importance in calculus and computer programming is the **greatest integer function**, denoted $f(x) = [\![x]\!]$, or $\text{int}(x)$.

GREATEST INTEGER FUNCTION

$f(x) = [\![x]\!] =$ the greatest integer *less than or equal to x.*

The greatest integer function pairs each input with the greatest integer *less than or equal to* that input. Thus, x-values 1, $1\frac{1}{2}$, and 1.8 are all paired with the y-value 1. Other pairings are shown below.

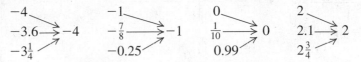

These values can be checked with a graphing calculator using the int(feature from the NUM submenu in the MATH menu.

EXAMPLE 9 Graph $f(x) = [\![x]\!]$ and determine its domain and range.

Solution The greatest integer function can also be defined as a piecewise function with an infinite number of statements. When plotting points by hand, it can be helpful to use the TABLE feature on a graphing calculator to find ordered pairs.

$$
f(x) = [\![x]\!] = \begin{cases}
\vdots \\
-3, & \text{for } -3 \le x < -2, \\
-2, & \text{for } -2 \le x < -1, \\
-1, & \text{for } -1 \le x < 0, \\
0, & \text{for } 0 \le x < 1, \\
1, & \text{for } 1 \le x < 2, \\
2, & \text{for } 2 \le x < 3, \\
3, & \text{for } 3 \le x < 4, \\
\vdots
\end{cases}
$$

We see that the domain of this function is the set of all real numbers, $(-\infty, \infty)$, and the range is the set of all integers, $\{\ldots, -3, -2, -1, 0, 1, 2, 3, \ldots\}$.

> ***Now Try Exercise 63.***

If we had used a calculator for Example 9, we would see the graph shown below.

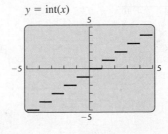

$y = \text{int}(x)$

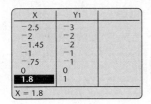

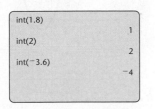

Just in Time

3

2.1   Exercise Set

Determine the intervals on which the function is
(a) *increasing,* **(b)** *decreasing, and* **(c)** *constant.*

1.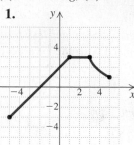

2.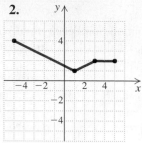

3.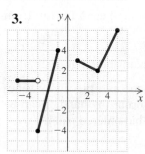

4.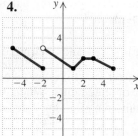

5.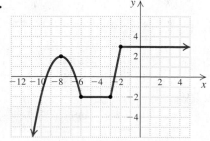

6.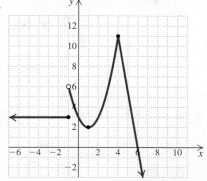

7.–12. Determine the domain and the range of each of the functions graphed in Exercises 1–6.

Using the graph, determine any relative maxima or minima of the function and the intervals on which the function is increasing or decreasing.

13. $f(x) = -x^2 + 5x - 3$

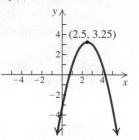

14. $f(x) = x^2 - 2x + 3$

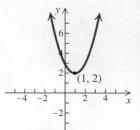

15. $f(x) = \frac{1}{4}x^3 - \frac{1}{2}x^2 - x + 2$

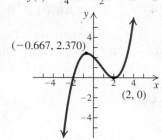

16. $f(x) = -0.09x^3 + 0.5x^2 - 0.1x + 1$

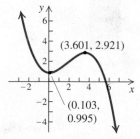

Graph the function. Estimate the intervals on which the function is increasing or decreasing and any relative maxima or minima.

17. $f(x) = x^2$

18. $f(x) = 4 - x^2$

19. $f(x) = 5 - |x|$

20. $f(x) = |x + 3| - 5$

21. $f(x) = x^2 - 6x + 10$

22. $f(x) = -x^2 - 8x - 9$

Graph the function using the given viewing window. Find the intervals on which the function is increasing or decreasing and find any relative maxima or minima. Change the viewing window if it seems appropriate for further analysis.

23. $f(x) = -x^3 + 6x^2 - 9x - 4$,
$[-3, 7, -20, 15]$

24. $f(x) = 0.2x^3 - 0.2x^2 - 5x - 4$,
$[-10, 10, -30, 20]$

25. $f(x) = 1.1x^4 - 5.3x^2 + 4.07$,
$[-4, 4, -4, 8]$

26. $f(x) = 1.2(x + 3)^4 + 10.3(x + 3)^2 + 9.78$,
$[-9, 3, -40, 100]$

27. *Advertising Effect.*   A wholesale nursery estimates that it will sell N fruit trees after spending a dollars on advertising, where

$$N(a) = -a^2 + 300a + 6, \quad 0 \le a \le 300,$$

and a is measured in thousands of dollars.

a) Graph the function using a graphing calculator.
b) Use the MAXIMUM feature to find the relative maximum.
c) For what advertising expenditure will the greatest number of fruit trees be sold? How many fruit trees will be sold for that amount?

28. *Temperature During an Illness.*   The temperature of a patient during an illness is given by the function

$$T(t) = -0.1t^2 + 1.2t + 98.6, \quad 0 \le t \le 12,$$

where T is the temperature, in degrees Fahrenheit, at time t, in days, after the onset of the illness.

a) Graph the function using a graphing calculator.
b) Use the MAXIMUM feature to determine at what time the patient's temperature was the highest. What was the highest temperature?

Use a graphing calculator to find the intervals on which the function is increasing or decreasing. Consider the entire set of real numbers if no domain is given.

29. $f(x) = \dfrac{8x}{x^2 + 1}$

30. $f(x) = \dfrac{-4}{x^2 + 1}$

31. $f(x) = x\sqrt{4 - x^2}$,   for $-2 \le x \le 2$

32. $f(x) = -0.8x\sqrt{9 - x^2}$,   for $-3 \le x \le 3$

33. *Lumberyard.*   Rick's lumberyard has 480 yd of fencing with which to enclose a rectangular area. If the enclosed area is x yards long, express its area as a function of its length.

34. *Triangular Flag.*   A seamstress is designing a triangular flag so that the length of the base of the triangle, in inches, is 7 less than twice the height h. Express the area of the flag as a function of the height.

35. *Blimp Distance.*   The Goodyear Blimp can be seen flying at an altitude of 3500 ft above the Motor Speedway during the Indianapolis 500 race. The slanted distance directly to the Pagoda at the start–finish line is d feet. Express the horizontal distance h as a function of d.

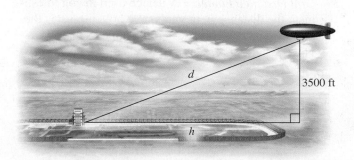

36. *Rising Balloon.*   A hot-air balloon rises straight up from the ground at a rate of 120 ft/min. The balloon is tracked from a rangefinder on the ground at point P, which is 400 ft from the release point Q of the balloon. Let $d =$ the distance from the balloon to the rangefinder and $t =$ the time, in minutes, since the balloon was released. Express d as a function of t.

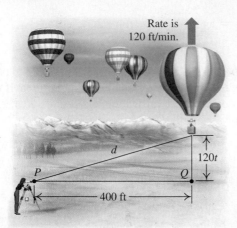

37. *Inscribed Rhombus.*   A rhombus is inscribed in a rectangle that is w meters wide with a perimeter of 40 m. Each vertex of the rhombus is a midpoint of a side of the rectangle. Express the area of the rhombus as a function of the width of the rectangle.

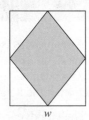

38. *Carpet Area.*   A carpet installer uses 46 ft of linen tape to bind the edges of a rectangular hall runner. If the runner is w feet wide, express its area as a function of the width.

39. *Golf Distance Finder.*   A device used in golf to estimate the distance d, in yards, to a hole measures the size s, in inches, that the 7-ft pin appears to be in a viewfinder. Express the distance d as a function of s.

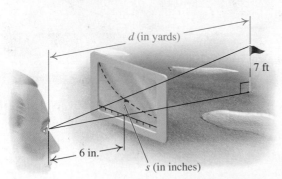

40. *Gas Tank Volume.*   A gas tank has ends that are hemispheres of radius r feet. The cylindrical midsection is 6 ft long. Express the volume of the tank as a function of r.

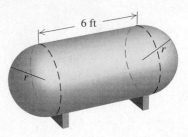

41. *Swimming Areas.*   A summer camp has 240 ft of float line with which to rope off three adjacent rectangular areas of a lake for swimming lessons, one for each of three levels of swimming ability. A beach forms one side of the swimming areas. Suppose that the width of each area is x yards.

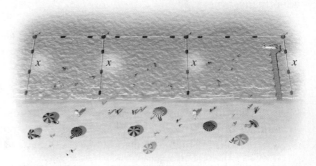

a) Express the total area of the three swimming areas as a function of x.
b) Find the domain of the function.
c) Using the graph of the function shown below, determine the dimensions that yield the maximum area.

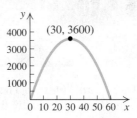

42. *Play Space.*   A car dealership has 24 ft of dividers with which to enclose a rectangular play space in a corner of a customer lounge. The sides against the wall require no partition. Suppose the play space is x feet long.

a) Express the area of the play space as a function of x.

b) Find the domain of the function.

c) Using the graph shown below, determine the dimensions that yield the maximum area.

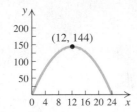

43. *Volume of a Box.* From a 12-cm by 12-cm piece of cardboard, square corners are cut out so that the sides can be folded up to make a box.

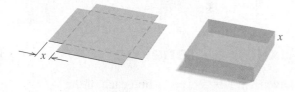

a) Express the volume of the box as a function of the side x, in centimeters, of a cut-out square.

b) Find the domain of the function.

c) Graph the function with a graphing calculator.

d) What dimensions yield the maximum volume?

44. *Office File.* Designs Unlimited plans to produce a one-component vertical file by bending the long side of an 8-in. by 14-in. sheet of plastic along two lines to form a ⊔ shape.

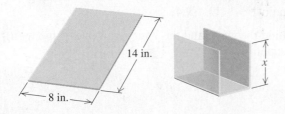

a) Express the volume of the file as a function of the height x, in inches, of the file.

b) Find the domain of the function.

c) Graph the function with a graphing calculator.

d) How tall should the file be in order to maximize the volume that the file can hold?

45. *Area of an Inscribed Rectangle.* A rectangle that is x feet wide is inscribed in a circle of radius 8 ft.

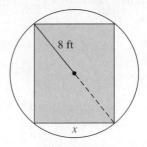

a) Express the area of the rectangle as a function of x.

b) Find the domain of the function.

c) Graph the function with a graphing calculator.

d) What dimensions maximize the area of the rectangle?

46. *Cost of Material.* A rectangular box with volume 320 ft^3 is built with a square base and top. The cost is $1.50/ft^2 for the bottom, $2.50/ft^2 for the sides, and $1/ft^2 for the top. Let $x =$ the length of the base, in feet.

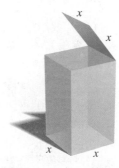

a) Express the cost of the box as a function of x.

b) Find the domain of the function.

c) Graph the function with a graphing calculator.

d) What dimensions minimize the cost of the box?

For each piecewise function, find the specified function values.

47. $g(x) = \begin{cases} x + 4, & \text{for } x \le 1, \\ 8 - x, & \text{for } x > 1 \end{cases}$

$g(-4), g(0), g(1),$ and $g(3)$

48. $f(x) = \begin{cases} 3, & \text{for } x \leq -2, \\ \frac{1}{2}x + 6, & \text{for } x > -2 \end{cases}$

$f(-5), f(-2), f(0),$ and $f(2)$

49. $h(x) = \begin{cases} -3x - 18, & \text{for } x < -5, \\ 1, & \text{for } -5 \leq x < 1, \\ x + 2, & \text{for } x \geq 1 \end{cases}$

$h(-6), h(0), h(1),$ and $h(4)$

50. $f(x) = \begin{cases} -5x - 8, & \text{for } x < -2, \\ \frac{1}{2}x + 5, & \text{for } -2 \leq x \leq 4, \\ 10 - 2x, & \text{for } x > 4 \end{cases}$

$f(-4), f(-2), f(4),$ and $f(6)$

Graph each of the following functions. Check your results using a graphing calculator.

51. $f(x) = \begin{cases} \frac{1}{2}x, & \text{for } x < 0, \\ x + 3, & \text{for } x \geq 0 \end{cases}$

52. $f(x) = \begin{cases} -\frac{1}{3}x + 2, & \text{for } x \leq 0, \\ x - 5, & \text{for } x > 0 \end{cases}$

53. $f(x) = \begin{cases} -\frac{3}{4}x + 2, & \text{for } x < 4, \\ -1, & \text{for } x \geq 4 \end{cases}$

54. $h(x) = \begin{cases} 2x - 1, & \text{for } x < 2, \\ 2 - x, & \text{for } x \geq 2 \end{cases}$

55. $f(x) = \begin{cases} x + 1, & \text{for } x \leq -3, \\ -1, & \text{for } -3 < x < 4, \\ \frac{1}{2}x, & \text{for } x \geq 4 \end{cases}$

56. $f(x) = \begin{cases} 4, & \text{for } x \leq -2, \\ x + 1, & \text{for } -2 < x < 3, \\ -x, & \text{for } x \geq 3 \end{cases}$

57. $g(x) = \begin{cases} \frac{1}{2}x - 1, & \text{for } x < 0, \\ 3, & \text{for } 0 \leq x \leq 1, \\ -2x, & \text{for } x > 1 \end{cases}$

58. $f(x) = \begin{cases} \dfrac{x^2 - 9}{x + 3}, & \text{for } x \neq -3, \\ 5, & \text{for } x = -3 \end{cases}$

59. $f(x) = \begin{cases} 2, & \text{for } x = 5, \\ \dfrac{x^2 - 25}{x - 5}, & \text{for } x \neq 5 \end{cases}$

60. $f(x) = \begin{cases} \dfrac{x^2 + 3x + 2}{x + 1}, & \text{for } x \neq -1, \\ 7, & \text{for } x = -1 \end{cases}$

61. $f(x) = [x]$ **62.** $f(x) = 2[x]$

63. $g(x) = 1 + [x]$ **64.** $h(x) = \frac{1}{2}[x] - 2$

65.–70. Find the domain and the range of each of the functions defined in Exercises 51–56.

Determine the domain and the range of the piecewise function. Then write an equation for the function.

71.
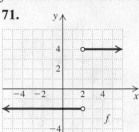

72.
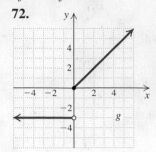

73.
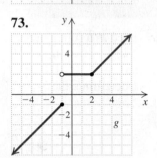

74.
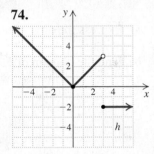

75.
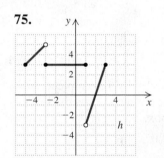

76.
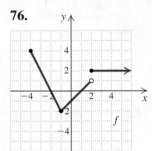

❭ Skill Maintenance

77. Given $f(x) = 5x^2 - 7$, find each of the following. **[1.2]**

  a) $f(-3)$ **b)** $f(3)$

  c) $f(a)$ **d)** $f(-a)$

78. Given $f(x) = 4x^3 - 5x$, find each of the following. **[1.2]**

  a) $f(2)$ **b)** $f(-2)$

  c) $f(a)$ **d)** $f(-a)$

79. Write an equation of the line perpendicular to the graph of the line $8x - y = 10$ and containing the point $(-1, 1)$. **[1.4]**

80. Find the slope and the y-intercept of the line with equation $2x - 9y + 1 = 0$. **[1.4]**

> # Synthesis

Using a graphing calculator, estimate the interval on which the function is increasing or decreasing and any relative maxima or minima.

81. $f(x) = x^4 + 4x^3 - 36x^2 - 160x + 400$

82. $f(x) = 3.22x^5 - 5.208x^3 - 11$

83. *Parking Costs.* A parking garage charges $3 for up to (but not including) 1 hr of parking, $6 for up to 2 hr of parking, $9 for up to 3 hr of parking, and so on. Let $C(t)$ = the cost of parking for t hours.

a) Graph the function.
b) Write an equation for $C(t)$ using the greatest integer notation $[\![t]\!]$.

84. If $[\![x + 2]\!] = -3$, what are the possible inputs for x?

85. If $([\![x]\!])^2 = 25$, what are the possible inputs for x?

86. *Minimizing Power Line Costs.* A power line is constructed from a power station at point A to an island at point I, which is 1 mi directly out in the water from a point B on the shore. Point B is 4 mi downshore from the power station at A. It costs $5000 per mile to lay the power line under water and $3000 per mile to lay the power line under ground. The line comes to the shore at point S downshore from A. Let x = the distance from B to S.

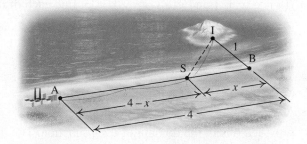

a) Express the cost C of laying the line as a function of x.
b) At what distance x from point B should the line come to shore in order to minimize cost?

87. *Volume of an Inscribed Cylinder.* A right circular cylinder of height h and radius r is inscribed in a right circular cone with a height of 10 ft and a base with radius 6 ft.

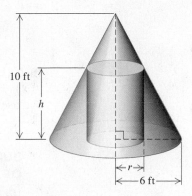

a) Express the height h of the cylinder as a function of r.
b) Express the volume V of the cylinder as a function of r.
c) Express the volume V of the cylinder as a function of h.

2.2 ▷ The Algebra of Functions

> ❯ Find the sum, the difference, the product, and the quotient of two functions, and determine the domains of the resulting functions.

> ❯ Find the difference quotient for a function.

> ## ❯ The Algebra of Functions: Sums, Differences, Products, and Quotients

We now use addition, subtraction, multiplication, and division to combine functions and obtain new functions.

Consider the following two functions f and g:

$$f(x) = x + 2 \quad \text{and} \quad g(x) = x^2 + 1.$$

Since $f(3) = 3 + 2 = 5$ and $g(3) = 3^2 + 1 = 10$, we have

$$f(3) + g(3) = 5 + 10 = 15,$$
$$f(3) - g(3) = 5 - 10 = -5,$$
$$f(3) \cdot g(3) = 5 \cdot 10 = 50,$$

and

$$\frac{f(3)}{g(3)} = \frac{5}{10} = \frac{1}{2}.$$

In fact, so long as x is in the domain of *both* f and g, we can easily compute $f(x) + g(x)$, $f(x) - g(x)$, $f(x) \cdot g(x)$, and, assuming $g(x) \neq 0$, $f(x)/g(x)$. We use the notation shown below.

SUMS, DIFFERENCES, PRODUCTS, AND QUOTIENTS OF FUNCTIONS

If f and g are functions and x is in the domain of each function, then:

$$(f + g)(x) = f(x) + g(x),$$
$$(f - g)(x) = f(x) - g(x),$$
$$(fg)(x) = f(x) \cdot g(x),$$
$$(f/g)(x) = f(x)/g(x), \text{ provided } g(x) \neq 0.$$

EXAMPLE 1   Given that $f(x) = x + 1$ and $g(x) = \sqrt{x + 3}$, find each of the following.

a) $(f + g)(x)$ **b)** $(f + g)(6)$ **c)** $(f + g)(-4)$

Solution

a) $(f + g)(x) = f(x) + g(x)$
$$= x + 1 + \sqrt{x + 3} \quad \textbf{This cannot be simplified.}$$

b) We can find $(f + g)(6)$ provided 6 is in the domain of *each* function. The domain of f is all real numbers. The domain of g is all real numbers x for which $x + 3 \geq 0$, or $x \geq -3$. This is the interval $[-3, \infty)$. We see that 6 is in both domains, so we have

$$f(6) = 6 + 1 = 7, \quad g(6) = \sqrt{6 + 3} = \sqrt{9} = 3,$$
$$(f + g)(6) = f(6) + g(6) = 7 + 3 = 10.$$

Another method is to use the formula found in part (a):

$$(f + g)(6) = 6 + 1 + \sqrt{6 + 3} = 7 + \sqrt{9} = 7 + 3 = 10.$$

We can check our work using a graphing calculator by entering

$$y_1 = x + 1, \quad y_2 = \sqrt{x + 3}, \quad \text{and} \quad y_3 = y_1 + y_2$$

on the $y =$ screen. Then on the home screen, we find $\text{Y}_3(6)$.

c) To find $(f + g)(-4)$, we must first determine whether -4 is in the domain of each function. We note that -4 is not in the domain of g, $[-3, \infty)$. That is, $\sqrt{-4 + 3}$ is not a real number. Thus, $(f + g)(-4)$ does not exist.

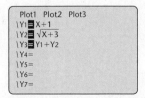

> *Now Try Exercise 15.*

It is useful to view the concept of the sum of two functions graphically. In the following graph, we see the graphs of two functions f and g and their sum, $f + g$. Consider finding $(f + g)(4)$, or $f(4) + g(4)$. We can locate $g(4)$ on the graph of g and measure it. Then we add that length on top of $f(4)$ on the graph of f. The sum gives us $(f + g)(4)$.

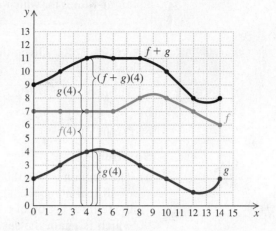

With this in mind, let's view Example 1 from a graphical perspective. Let's look at the graphs of

$$f(x) = x + 1, \qquad g(x) = \sqrt{x + 3}, \quad \text{and}$$
$$(f + g)(x) = x + 1 + \sqrt{x + 3}.$$

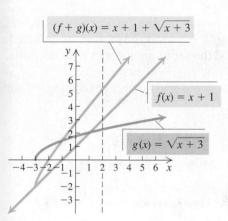

$(f + g)(x) = x + 1 + \sqrt{x + 3}$

$f(x) = x + 1$

$g(x) = \sqrt{x + 3}$

See the graph at left. Note that the domain of f is the set of all real numbers. The domain of g is $[-3, \infty)$. The domain of $f + g$ is the set of numbers in the intersection of the domains. This is the set of numbers in both domains.

Domain of f		$(-\infty, \infty)$
	$-5\ -4\ -3\ -2\ -1\ \ 0\ \ 1\ \ 2\ \ 3\ \ 4\ \ 5$	

Domain of g		$[-3, \infty)$
	$-5\ -4\ -3\ -2\ -1\ \ 0\ \ 1\ \ 2\ \ 3\ \ 4\ \ 5$	

Domain of $f + g$		$[-3, \infty)$
	$-5\ -4\ -3\ -2\ -1\ \ 0\ \ 1\ \ 2\ \ 3\ \ 4\ \ 5$	

Thus the domain of $f + g$ is $[-3, \infty)$.

We can confirm that the y-coordinates of the graph of $(f + g)(x)$ are the sums of the corresponding y-coordinates of the graphs of $f(x)$ and $g(x)$. Here we confirm it for $x = 2$.

$$f(x) = x + 1 \qquad\qquad g(x) = \sqrt{x + 3}$$
$$f(2) = 2 + 1 = 3; \qquad g(2) = \sqrt{2 + 3} = \sqrt{5};$$

$$(f + g)(x) = x + 1 + \sqrt{x + 3}$$
$$(f + g)(2) = 2 + 1 + \sqrt{2 + 3}$$
$$= 3 + \sqrt{5} = f(2) + g(2).$$

DOMAIN

REVIEW SECTION 1.2.

Let's also examine the domains of $f - g$, fg, and f/g for the functions $f(x) = x + 1$ and $g(x) = \sqrt{x + 3}$ of Example 1. The domains of $f - g$ and fg are the same as the domain of $f + g$, $[-3, \infty)$, because numbers in this interval are in the domains of *both* functions. For f/g, $g(x)$ cannot be 0. Since $\sqrt{x + 3} = 0$ when $x = -3$, we must exclude -3 so the domain of f/g is $(-3, \infty)$.

> **DOMAINS OF $f + g$, $f - g$, fg, AND f/g**
>
> If f and g are functions, then the domain of the functions $f + g$, $f - g$, and fg is the intersection of the domain of f and the domain of g. The domain of f/g is also the intersection of the domains of f and g with the exclusion of any x-values for which $g(x) = 0$.

Just in Time

11, 12

EXAMPLE 2 Given that $f(x) = x^2 - 4$ and $g(x) = x + 2$, find each of the following.

a) The domain of $f + g$, $f - g$, fg, and f/g

b) $(f + g)(x)$ **c)** $(f - g)(x)$ **d)** $(fg)(x)$

e) $(f/g)(x)$ **f)** $(gg)(x)$

Solution

a) The domain of f is the set of all real numbers. The domain of g is also the set of all real numbers. The domain of $f + g$, $f - g$, and fg is the set of numbers in the intersection of the domains—that is, the set of numbers in both domains, which is again the set of real numbers. For f/g, we must exclude -2, since $g(-2) = 0$. Thus the domain of f/g is the set of real numbers excluding -2, or $(-\infty, -2) \cup (-2, \infty)$.

b) $(f + g)(x) = f(x) + g(x) = (x^2 - 4) + (x + 2) = x^2 + x - 2$

c) $(f - g)(x) = f(x) - g(x) = (x^2 - 4) - (x + 2) = x^2 - x - 6$

d) $(fg)(x) = f(x) \cdot g(x) = (x^2 - 4)(x + 2) = x^3 + 2x^2 - 4x - 8$

e) $(f/g)(x) = \dfrac{f(x)}{g(x)} = \dfrac{x^2 - 4}{x + 2}$ Note that $g(x) = 0$ when $x = -2$, so $(f/g)(x)$ is *not defined* when $x = -2$.

$$= \frac{(x + 2)(x - 2)}{x + 2} \qquad \text{Factoring}$$

$$= x - 2 \qquad \text{Removing a factor of 1: } \frac{x + 2}{x + 2} = 1$$

Thus, $(f/g)(x) = x - 2$ with the added stipulation that $x \neq -2$ since -2 is not in the domain of $(f/g)(x)$.

f) $(gg)(x) = g(x) \cdot g(x) = [g(x)]^2 = (x + 2)^2 = x^2 + 4x + 4$

> **Now Try Exercise 21.**

❯ Difference Quotients

In Section 1.3, we learned that the slope of a line can be considered as an average *rate of change*. Here let's consider a nonlinear function f and draw a line through two points $(x, f(x))$ and $(x + h, f(x + h))$ as shown at left.

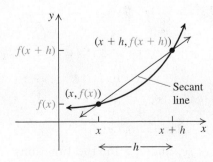

The slope of the line, called a **secant line**, is

$$\frac{f(x + h) - f(x)}{x + h - x},$$

which simplifies to

$$\frac{f(x + h) - f(x)}{h}. \qquad \textbf{Difference quotient}$$

This ratio is called the **difference quotient**, or the **average rate of change**. In calculus, it is important to be able to find and simplify difference quotients.

EXAMPLE 3   For the function f given by $f(x) = 2x - 3$, find and simplify the difference quotient

$$\frac{f(x + h) - f(x)}{h}.$$

Solution

$$\begin{aligned}
\frac{f(x + h) - f(x)}{h} &= \frac{2(x + h) - 3 - (2x - 3)}{h} && \text{Substituting} \\[2mm]
&= \frac{2x + 2h - 3 - 2x + 3}{h} && \text{Removing parentheses} \\[2mm]
&= \frac{2h}{h} = 2 && \text{Simplifying}
\end{aligned}$$

> *Now Try Exercise 49.*

EXAMPLE 4   For the function f given by $f(x) = \dfrac{1}{x}$, find and simplify the difference quotient

$$\frac{f(x + h) - f(x)}{h}.$$

Solution

$$\begin{aligned}
\frac{f(x + h) - f(x)}{h} &= \frac{\dfrac{1}{x + h} - \dfrac{1}{x}}{h} && \text{Substituting} \\[4mm]
&= \frac{\dfrac{1}{x + h} \cdot \dfrac{x}{x} - \dfrac{1}{x} \cdot \dfrac{x + h}{x + h}}{h} && \begin{array}{l}\text{The LCD of } \dfrac{1}{x + h} \\ \text{and } \dfrac{1}{x} \text{ is } x(x + h).\end{array} \\[4mm]
&= \frac{\dfrac{x}{x(x + h)} - \dfrac{x + h}{x(x + h)}}{h} \\[4mm]
&= \frac{\dfrac{x - (x + h)}{x(x + h)}}{h} && \text{Subtracting in the numerator} \\[4mm]
&= \frac{\dfrac{x - x - h}{x(x + h)}}{h} && \text{Removing parentheses} \\[4mm]
&= \frac{\dfrac{-h}{x(x + h)}}{h} && \text{Simplifying the numerator} \\[4mm]
&= \frac{-h}{x(x + h)} \cdot \frac{1}{h} && \begin{array}{l}\text{Multiplying by the reciprocal} \\ \text{of the divisor}\end{array} \\[4mm]
&= \frac{-h \cdot 1}{x \cdot (x + h) \cdot h} \\[4mm]
&= \frac{-1 \cdot \cancel{h}}{x \cdot (x + h) \cdot \cancel{h}} && \text{Rewriting } -h \cdot 1 \text{ as } -1 \cdot h \\[4mm]
&= \frac{-1}{x(x + h)}, \text{ or } -\frac{1}{x(x + h)}
\end{aligned}$$

> *Now Try Exercise 55.*

EXAMPLE 5   For the function f given by $f(x) = 2x^2 - x - 3$, find and simplify the difference quotient

$$\frac{f(x + h) - f(x)}{h}.$$

Solution   We first find $f(x + h)$:

$$f(x + h) = 2(x + h)^2 - (x + h) - 3 \qquad \begin{array}{l}\text{Substituting } x + h \text{ for } x \text{ in} \\ f(x) = 2x^2 - x - 3\end{array}$$

$$= 2[x^2 + 2xh + h^2] - (x + h) - 3$$

$$= 2x^2 + 4xh + 2h^2 - x - h - 3.$$

Then

$$\frac{f(x + h) - f(x)}{h} = \frac{[2x^2 + 4xh + 2h^2 - x - h - 3] - [2x^2 - x - 3]}{h}$$

$$= \frac{2x^2 + 4xh + 2h^2 - x - h - 3 - 2x^2 + x + 3}{h}$$

$$= \frac{4xh + 2h^2 - h}{h}$$

$$= \frac{h(4x + 2h - 1)}{h \cdot 1} = \frac{h}{h} \cdot \frac{4x + 2h - 1}{1} = 4x + 2h - 1.$$

▶ *Now Try Exercise 63.*

2.2   Exercise Set

Given that $f(x) = x^2 - 3$ and $g(x) = 2x + 1$, find each of the following, if it exists.

1. $(f + g)(5)$

2. $(fg)(0)$

3. $(f - g)(-1)$

4. $(fg)(2)$

5. $(f/g)\left(-\frac{1}{2}\right)$

6. $(f - g)(0)$

7. $(fg)\left(-\frac{1}{2}\right)$

8. $(f/g)\left(-\sqrt{3}\right)$

9. $(g - f)(-1)$

10. $(g/f)\left(-\frac{1}{2}\right)$

Given that $h(x) = x + 4$ and $g(x) = \sqrt{x - 1}$, find each of the following, if it exists.

11. $(h - g)(-4)$

12. $(gh)(10)$

13. $(g/h)(1)$

14. $(h/g)(1)$

15. $(g + h)(1)$

16. $(hg)(3)$

For each pair of functions in Exercises 17–34:

a) Find the domain of f, g, $f + g$, $f - g$, fg, ff, f/g, and g/f.

b) Find $(f + g)(x)$, $(f - g)(x)$, $(fg)(x)$, $(ff)(x)$, $(f/g)(x)$, and $(g/f)(x)$.

17. $f(x) = 2x + 3$, $g(x) = 3 - 5x$

18. $f(x) = -x + 1$, $g(x) = 4x - 2$

19. $f(x) = x - 3$, $g(x) = \sqrt{x + 4}$

20. $f(x) = x + 2$, $g(x) = \sqrt{x - 1}$

21. $f(x) = 2x - 1$, $g(x) = -2x^2$

22. $f(x) = x^2 - 1$, $g(x) = 2x + 5$

23. $f(x) = \sqrt{x - 3}$, $g(x) = \sqrt{x + 3}$

24. $f(x) = \sqrt{x}$, $g(x) = \sqrt{2 - x}$

25. $f(x) = x + 1$, $g(x) = |x|$

26. $f(x) = 4|x|$, $g(x) = 1 - x$

27. $f(x) = x^3$, $g(x) = 2x^2 + 5x - 3$

28. $f(x) = x^2 - 4$, $g(x) = x^3$

29. $f(x) = \dfrac{4}{x + 1}$, $g(x) = \dfrac{1}{6 - x}$

30. $f(x) = 2x^2$, $g(x) = \dfrac{2}{x - 5}$

31. $f(x) = \dfrac{1}{x}$, $g(x) = x - 3$

32. $f(x) = \sqrt{x+6}$, $g(x) = \dfrac{1}{x}$

33. $f(x) = \dfrac{3}{x-2}$, $g(x) = \sqrt{x-1}$

34. $f(x) = \dfrac{2}{4-x}$, $g(x) = \dfrac{5}{x-1}$

In Exercises 35–40, consider the functions F and G as shown in the following graph.

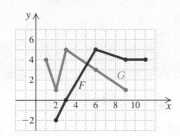

35. Find the domain of F, the domain of G, and the domain of $F + G$.

36. Find the domain of $F - G$, FG, and F/G.

37. Find the domain of G/F.

38. Graph $F + G$.

39. Graph $G - F$.

40. Graph $F - G$.

In Exercises 41–46, consider the functions F and G as shown in the following graph.

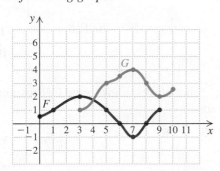

41. Find the domain of F, the domain of G, and the domain of $F + G$.

42. Find the domain of $F - G$, FG, and F/G.

43. Find the domain of G/F.

44. Graph $F + G$.

45. Graph $G - F$.

46. Graph $F - G$.

47. *Total Cost, Revenue, and Profit.* In economics, functions that involve revenue, cost, and profit are used. For example, suppose that $R(x)$ and $C(x)$ denote the total revenue and the total cost, respectively, of producing a new grocery cart for Ogata Wholesalers. Then the difference

$$P(x) = R(x) - C(x)$$

represents the total profit for producing x carts. Given

$$R(x) = 60x - 0.4x^2 \quad \text{and} \quad C(x) = 3x + 13,$$

find each of the following.

a) $P(x)$
b) $R(100)$, $C(100)$, and $P(100)$
c) Using a graphing calculator, graph the three functions in the viewing window $[0, 160, 0, 3000]$.

48. *Total Cost, Revenue, and Profit.* Given that

$$R(x) = 200x - x^2 \quad \text{and} \quad C(x) = 5000 + 8x$$

for a new tablet produced by Visual Communications, find each of the following. (See Exercise 47.)

a) $P(x)$
b) $R(175)$, $C(175)$, and $P(175)$
c) Using a graphing calculator, graph the three functions in the viewing window $[0, 200, 0, 10{,}000]$.

For each function f, construct and simplify the difference quotient

$$\frac{f(x+h) - f(x)}{h}.$$

49. $f(x) = 3x - 5$ **50.** $f(x) = 4x - 1$

51. $f(x) = 6x + 2$ **52.** $f(x) = 5x + 3$

53. $f(x) = \frac{1}{3}x + 1$ **54.** $f(x) = -\frac{1}{2}x + 7$

55. $f(x) = \dfrac{1}{3x}$ **56.** $f(x) = \dfrac{1}{2x}$

57. $f(x) = -\dfrac{1}{4x}$ **58.** $f(x) = -\dfrac{1}{x}$

59. $f(x) = x^2 + 1$ **60.** $f(x) = x^2 - 3$

61. $f(x) = 4 - x^2$ **62.** $f(x) = 2 - x^2$

63. $f(x) = 3x^2 - 2x + 1$ **64.** $f(x) = 5x^2 + 4x$

65. $f(x) = 4 + 5|x|$ **66.** $f(x) = 2|x| + 3x$

67. $f(x) = x^3$ **68.** $f(x) = x^3 - 2x$

69. $f(x) = \dfrac{x-4}{x+3}$ **70.** $f(x) = \dfrac{x}{2-x}$

❯ Skill Maintenance

Graph the equation. [1.1], [1.3]

71. $y = 3x - 1$ **72.** $2x + y = 4$

73. $x - 3y = 3$ **74.** $y = x^2 + 1$

❯ Synthesis

75. Write equations for two functions f and g such that the domain of $f - g$ is

$$\{x \mid x \neq -7 \text{ and } x \neq 3\}.$$

76. For functions h and f, find the domain of $h + f$, $h - f$, hf, and h/f if

$$h = \left\{(-4, 13), (-1, 7), (0, 5), \left(\tfrac{5}{2}, 0\right), (3, -5)\right\}, \quad \text{and}$$

$$f = \{(-4, -7), (-2, -5), (0, -3), (3, 0), (5, 2), (9, 6)\}.$$

77. Find the domain of $(h/g)(x)$ given that

$$h(x) = \frac{5x}{3x - 7} \quad \text{and} \quad g(x) = \frac{x^4 - 1}{5x - 15}.$$

2.3 ❯ The Composition of Functions

> ❯ Find the composition of two functions and the domain of the composition.
>
> ❯ Decompose a function as a composition of two functions.

❯ The Composition of Functions

In real-world situations, it is not uncommon for the output of a function to depend on some input that is itself an output of another function. For instance, the amount that a person pays as state income tax usually depends on the amount of adjusted gross income on the person's federal tax return, which, in turn, depends on his or her annual earnings. Such functions are called **composite functions**.

To see how composite functions work, suppose that a chemistry student needs a formula to convert Fahrenheit temperatures to Kelvin units. The formula

$$c(t) = \tfrac{5}{9}(t - 32)$$

gives the Celsius temperature $c(t)$ that corresponds to the Fahrenheit temperature t. The formula

$$k(c(t)) = c(t) + 273$$

gives the Kelvin temperature $k(c(t))$ that corresponds to the Celsius temperature $c(t)$. Thus, 50° Fahrenheit corresponds to

$$c(50) = \tfrac{5}{9}(50 - 32) = \tfrac{5}{9}(18) = 10° \text{ Celsius,}$$

and 10° Celsius corresponds to

$$k(c(50)) = k(10) = 10 + 273 = 283 \text{ Kelvin units,}$$

which is usually written 283K. We see that 50° Fahrenheit is the same as 283K. This two-step procedure can be used to convert any Fahrenheit temperature to Kelvin units.

	°F Fahrenheit	°C Celsius	K Kelvin
Boiling point of water	212°	100°	373 K
	50° ➡	10° ➡	283 K
Freezing point of water	32°	0°	273 K
Absolute zero	−460°	−273°	0 K

$y_1 = \dfrac{5}{9}(x - 32),\ \ y_2 = y_1 + 273$

X	Y₁	Y₂
50	10	283
59	15	288
68	20	293
77	25	298
86	30	303
95	35	308
104	40	313

X = 50

In the table shown at left, we use a graphing calculator to convert Fahrenheit temperatures x to Celsius temperatures y_1, using $y_1 = \frac{5}{9}(x - 32)$. We also convert Celsius temperatures to Kelvin units y_2, using $y_2 = y_1 + 273$.

A student making numerous conversions might look for a formula that converts directly from Fahrenheit to Kelvin. Such a formula can be found by substitution:

$$
\begin{aligned}
k(c(t)) &= c(t) + 273 \\
&= \frac{5}{9}(t - 32) + 273 && \textbf{Substituting } \frac{5}{9}(t - 32) \textbf{ for } c(t) \\
&= \frac{5}{9}t - \frac{160}{9} + 273 \\
&= \frac{5}{9}t - \frac{160}{9} + \frac{2457}{9} \\
&= \frac{5t + 2297}{9}. && \textbf{Simplifying}
\end{aligned}
$$

We can show on a graphing calculator that the same values that appear in the table for y_2 will appear when y_2 is entered as

$$ y_2 = \frac{5x + 2297}{9}. $$

Since the formula found above expresses the Kelvin temperature as a new function K of the Fahrenheit temperature t, we can write

$$ K(t) = \frac{5t + 2297}{9}, $$

where $K(t)$ is the Kelvin temperature corresponding to the Fahrenheit temperature, t. Here we have $K(t) = k(c(t))$. The new function K is called the **composition** of k and c and can be denoted $k \circ c$ (read "k composed with c," "the composition of k and c," or "k circle c").

COMPOSITION OF FUNCTIONS

The **composite function** $f \circ g$, the **composition** of f and g, is defined as

$$ (f \circ g)(x) = f(g(x)), $$

where x is in the domain of g and $g(x)$ is in the domain of f.

EXAMPLE 1   Given that $f(x) = 2x - 5$ and $g(x) = x^2 - 3x + 8$, find each of the following.

a) $(f \circ g)(x)$ and $(g \circ f)(x)$ 　　　　**b)** $(f \circ g)(7)$ and $(g \circ f)(7)$
c) $(g \circ g)(1)$ 　　　　　　　　　　　　**d)** $(f \circ f)(x)$

Solution   Consider each function separately:

$$ f(x) = 2x - 5 \qquad \textbf{This function multiplies each input by 2} $$
$$ \textbf{and then subtracts 5.} $$

and

$$ g(x) = x^2 - 3x + 8. \qquad \textbf{This function squares an input, subtracts three} $$
$$ \textbf{times the input from the result, and then adds 8.} $$

a) To find $(f \circ g)(x)$, we substitute $g(x)$ for x in the equation for $f(x)$:

$$(f \circ g)(x) = f(g(x)) = f(x^2 - 3x + 8) \qquad x^2 - 3x + 8 \text{ is the input for } f.$$

$$= 2(x^2 - 3x + 8) - 5 \qquad f \text{ multiplies the input by 2 and then subtracts 5.}$$

$$= 2x^2 - 6x + 16 - 5$$

$$= 2x^2 - 6x + 11.$$

To find $(g \circ f)(x)$, we substitute $f(x)$ for x in the equation for $g(x)$:

$$(g \circ f)(x) = g(f(x)) = g(2x - 5) \qquad 2x - 5 \text{ is the input for } g.$$

$$= (2x - 5)^2 - 3(2x - 5) + 8 \qquad g \text{ squares the input, subtracts three times the input, and then adds 8.}$$

$$= 4x^2 - 20x + 25 - 6x + 15 + 8$$

$$= 4x^2 - 26x + 48.$$

b) To find $(f \circ g)(7)$, we first find $g(7)$. Then we use $g(7)$ as an input for f:

$$(f \circ g)(7) = f(g(7)) = f(7^2 - 3 \cdot 7 + 8)$$

$$= f(36) = 2 \cdot 36 - 5$$

$$= 72 - 5 = 67.$$

To find $(g \circ f)(7)$, we first find $f(7)$. Then we use $f(7)$ as an input for g:

$$(g \circ f)(7) = g(f(7)) = g(2 \cdot 7 - 5)$$

$$= g(9) = 9^2 - 3 \cdot 9 + 8$$

$$= 81 - 27 + 8 = 62.$$

We could also find $(f \circ g)(7)$ and $(g \circ f)(7)$ by substituting 7 for x in the equations that we found in part (a):

$$(f \circ g)(x) = 2x^2 - 6x + 11$$

$$(f \circ g)(7) = 2 \cdot 7^2 - 6 \cdot 7 + 11 = 98 - 42 + 11 = 67;$$

$$(g \circ f)(x) = 4x^2 - 26x + 48$$

$$(g \circ f)(7) = 4 \cdot 7^2 - 26 \cdot 7 + 48 = 196 - 182 + 48 = 62.$$

$y_1 = 2x - 5, \quad y_2 = x^2 - 3x + 8$

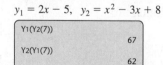

We can check our work using a graphing calculator. On the equation-editor screen, we enter $f(x)$ as $y_1 = 2x - 5$ and $g(x)$ as $y_2 = x^2 - 3x + 8$. Then, on the home screen, we find $(f \circ g)(7)$ and $(g \circ f)(7)$ using the function notations $Y_1(Y_2(7))$ and $Y_2(Y_1(7))$, respectively.

c) $(g \circ g)(1) = g(g(1)) = g(1^2 - 3 \cdot 1 + 8)$

$$= g(1 - 3 + 8) = g(6)$$

$$= 6^2 - 3 \cdot 6 + 8$$

$$= 36 - 18 + 8 = 26$$

d) $(f \circ f)(x) = f(f(x)) = f(2x - 5)$

$$= 2(2x - 5) - 5$$

$$= 4x - 10 - 5 = 4x - 15 \qquad \blacktriangleright \text{ **Now Try Exercises 1 and 15.**}$$

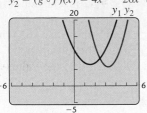

$y_1 = (f \circ g)(x) = 2x^2 - 6x + 11,$
$y_2 = (g \circ f)(x) = 4x^2 - 26x + 48$

Example 1 illustrates that, as a rule, $(f \circ g)(x) \neq (g \circ f)(x)$. We can see this graphically, as shown in the graphs at left.

EXAMPLE 2 Given that $f(x) = \sqrt{x}$ and $g(x) = x - 3$:

a) Find $f \circ g$ and $g \circ f$.

b) Find the domain of $f \circ g$ and the domain of $g \circ f$.

Solution

a) $(f \circ g)(x) = f(g(x)) = f(x - 3) = \sqrt{x - 3}$
$(g \circ f)(x) = g(f(x)) = g(\sqrt{x}) = \sqrt{x} - 3$

b) The domain of $f(x)$ is $\{x | x \geq 0\}$, or $[0, \infty)$. Any real number can be an input for $g(x)$, so the domain of $g(x)$ is $(-\infty, \infty)$.

The domain of $f \circ g$ consists of the values of x in the domain of g, $(-\infty, \infty)$, for which $g(x)$ is nonnegative. (Recall that the inputs of $f(x)$ must be nonnegative.) Thus we have

$$g(x) \geq 0$$
$$x - 3 \geq 0 \qquad \text{Substituting } x - 3 \text{ for } g(x)$$
$$x \geq 3.$$

We see that the domain of $f \circ g$ is $\{x | x \geq 3\}$, or $[3, \infty)$.

We can also find the domain of $f \circ g$ by examining the composite function itself, $(f \circ g)(x) = \sqrt{x - 3}$, keeping in mind any restrictions on the domain of g. Since any real number can be an input for g, the only restriction on $f \circ g$ is that the radicand must be nonnegative. We have

$$x - 3 \geq 0$$
$$x \geq 3.$$

Again, we see that the domain of $f \circ g$ is $\{x | x \geq 3\}$, or $[3, \infty)$. The graph in Fig. 1 confirms this.

The domain of $g \circ f$ consists of the values of x in the domain of f, $[0, \infty)$, for which $g(x)$ is defined. Since g can accept *any* real number as an input, any output from f is acceptable, so the entire domain of f is the domain of $g \circ f$. That is, the domain of $g \circ f$ is $\{x | x \geq 0\}$, or $[0, \infty)$.

We can also examine the composite function itself to find its domain, keeping in mind any restrictions on the domain of f. First, recall that the domain of f is $\{x | x \geq 0\}$, or $[0, \infty)$. Then consider $(g \circ f)(x) = \sqrt{x} - 3$. The radicand cannot be negative, so we have $x \geq 0$. As above, we see that the domain of $g \circ f$ is the domain of f, $\{x | x \geq 0\}$, or $[0, \infty)$. The graph in Fig. 2 confirms this.

> ***Now Try Exercise 27.***

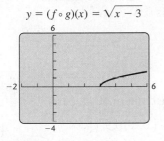

$y = (f \circ g)(x) = \sqrt{x - 3}$

FIGURE 1.

$y = (g \circ f)(x) = \sqrt{x} - 3$

FIGURE 2.

Just in Time 24

EXAMPLE 3 Given that $f(x) = \dfrac{1}{x - 2}$ and $g(x) = \dfrac{5}{x}$, find $f \circ g$ and $g \circ f$ and the domain of each.

Solution We have

$$(f \circ g)(x) = f(g(x)) = f\left(\frac{5}{x}\right) = \frac{1}{\dfrac{5}{x} - 2} = \frac{1}{\dfrac{5 - 2x}{x}} = \frac{x}{5 - 2x};$$

$$(g \circ f)(x) = g(f(x)) = g\left(\frac{1}{x - 2}\right) = \frac{5}{\dfrac{1}{x - 2}} = 5(x - 2).$$

Values of x that make the denominator 0 are not in the domains of these functions. Since $x - 2 = 0$ when $x = 2$, the domain of f is $\{x \mid x \neq 2\}$. The denominator of g is x, so the domain of g is $\{x \mid x \neq 0\}$.

The domain of $f \circ g$ consists of the values of x in the domain of g for which $g(x) \neq 2$. (Recall that 2 cannot be an input of f.) Since the domain of g is $\{x \mid x \neq 0\}$, 0 is not in the domain of $f \circ g$. In addition, we must find the value(s) of x for which $g(x) = 2$. We have

$$g(x) = 2$$

$$\frac{5}{x} = 2 \qquad \text{Substituting } \frac{5}{x} \text{ for } g(x)$$

$$5 = 2x$$

$$\frac{5}{2} = x.$$

This tells us that $\frac{5}{2}$ is also *not* in the domain of $f \circ g$. Then the domain of $f \circ g$ is

$$\left\{ x \mid x \neq 0 \text{ and } x \neq \tfrac{5}{2} \right\}, \quad \text{or} \quad (-\infty, 0) \cup \left(0, \tfrac{5}{2}\right) \cup \left(\tfrac{5}{2}, \infty\right).$$

We can also examine the composite function $f \circ g$ to find its domain, keeping in mind any restrictions on the domain of g. First, recall that 0 is not in the domain of g, so it cannot be in the domain of $(f \circ g)(x) = x/(5 - 2x)$. We must also exclude the value(s) of x for which the denominator of $f \circ g$ is 0. We have

$$5 - 2x = 0$$

$$5 = 2x$$

$$\tfrac{5}{2} = x.$$

Again, we see that $\frac{5}{2}$ is also not in the domain, so the domain of $f \circ g$ is

$$\left\{ x \mid x \neq 0 \text{ and } x \neq \tfrac{5}{2} \right\}, \quad \text{or} \quad (-\infty, 0) \cup \left(0, \tfrac{5}{2}\right) \cup \left(\tfrac{5}{2}, \infty\right).$$

The domain of $g \circ f$ consists of the values of x in the domain of f for which $f(x) \neq 0$. (Recall that 0 cannot be an input of g.) The domain of f is $\{x \mid x \neq 2\}$, so 2 is not in the domain of $g \circ f$. Next, we determine whether there are values of x for which $f(x) = 0$:

$$f(x) = 0$$

$$\frac{1}{x - 2} = 0 \qquad \text{Substituting } \frac{1}{x - 2} \text{ for } f(x)$$

$$(x - 2) \cdot \frac{1}{x - 2} = (x - 2) \cdot 0 \qquad \textbf{Multiplying by } x - 2$$

$$1 = 0. \qquad \textbf{False equation}$$

We see that there are no values of x for which $f(x) = 0$, so there are no additional restrictions on the domain of $g \circ f$. Thus the domain of $g \circ f$ is

$$\{x \mid x \neq 2\}, \quad \text{or} \quad (-\infty, 2) \cup (2, \infty).$$

We can also examine $g \circ f$ to find its domain. First, recall that 2 is not in the domain of f, so it cannot be in the domain of $(g \circ f)(x) = 5(x - 2)$. Since $5(x - 2)$ is defined for all real numbers, there are no additional restrictions on the domain of $g \circ f$. The domain is

$$\{x \mid x \neq 2\}, \quad \text{or} \quad (-\infty, 2) \cup (2, \infty).$$

> *Now Try Exercise 23.*

❱ Decomposing a Function as a Composition

In calculus, we often need to recognize how a function can be expressed as the composition of two functions. In this way, we are "decomposing" the function.

EXAMPLE 4   If $h(x) = (2x - 3)^5$, find $f(x)$ and $g(x)$ such that $h(x) = (f \circ g)(x)$.

Solution   The function $h(x)$ raises $(2x - 3)$ to the 5th power. Two functions that can be used for the composition are

$$f(x) = x^5 \quad \text{and} \quad g(x) = 2x - 3.$$

We can check by forming the composition:

$$h(x) = (f \circ g)(x) = f(g(x)) = f(2x - 3) = (2x - 3)^5.$$

This is the most "obvious" solution. There can be other less obvious solutions. For example, if

$$f(x) = (x + 7)^5 \quad \text{and} \quad g(x) = 2x - 10,$$

then

$$\begin{aligned}
h(x) = (f \circ g)(x) &= f(g(x)) \\
&= f(2x - 10) \\
&= [2x - 10 + 7]^5 = (2x - 3)^5.
\end{aligned}$$

❱ **Now Try Exercise 39.**

EXAMPLE 5   If $h(x) = \dfrac{1}{(x + 3)^3}$, find $f(x)$ and $g(x)$ such that $h(x) = (f \circ g)(x)$.

Solution   Two functions that can be used are

$$f(x) = \frac{1}{x^3} \quad \text{and} \quad g(x) = x + 3.$$

We check by forming the composition:

$$h(x) = (f \circ g)(x) = f(g(x)) = f(x + 3) = \frac{1}{(x + 3)^3}.$$

There are other functions that can be used as well. For example, if

$$f(x) = \frac{1}{x} \quad \text{and} \quad g(x) = (x + 3)^3,$$

then

$$h(x) = (f \circ g)(x) = f(g(x)) = f\big((x + 3)^3\big) = \frac{1}{(x + 3)^3}.$$

❱ **Now Try Exercise 41.**

2.3 Exercise Set

Given that $f(x) = 3x + 1$, $g(x) = x^2 - 2x - 6$, *and* $h(x) = x^3$, *find each of the following.*

1. $(f \circ g)(-1)$ **2.** $(g \circ f)(-2)$

3. $(h \circ f)(1)$ **4.** $(g \circ h)\left(\frac{1}{2}\right)$

5. $(g \circ f)(5)$ **6.** $(f \circ g)\left(\frac{1}{3}\right)$

7. $(f \circ h)(-3)$ **8.** $(h \circ g)(3)$

9. $(g \circ g)(-2)$ **10.** $(g \circ g)(3)$

11. $(h \circ h)(2)$ **12.** $(h \circ h)(-1)$

13. $(f \circ f)(-4)$ **14.** $(f \circ f)(1)$

15. $(h \circ h)(x)$ **16.** $(f \circ f)(x)$

Find $(f \circ g)(x)$ *and* $(g \circ f)(x)$ *and the domain of each.*

17. $f(x) = x + 3$, $g(x) = x - 3$

18. $f(x) = \frac{4}{5}x$, $g(x) = \frac{5}{4}x$

19. $f(x) = x + 1$, $g(x) = 3x^2 - 2x - 1$

20. $f(x) = 3x - 2$, $g(x) = x^2 + 5$

21. $f(x) = x^2 - 3$, $g(x) = 4x - 3$

22. $f(x) = 4x^2 - x + 10$, $g(x) = 2x - 7$

23. $f(x) = \dfrac{4}{1 - 5x}$, $g(x) = \dfrac{1}{x}$

24. $f(x) = \dfrac{6}{x}$, $g(x) = \dfrac{1}{2x + 1}$

25. $f(x) = 3x - 7$, $g(x) = \dfrac{x + 7}{3}$

26. $f(x) = \frac{2}{3}x - \frac{4}{5}$, $g(x) = 1.5x + 1.2$

27. $f(x) = 2x + 1$, $g(x) = \sqrt{x}$

28. $f(x) = \sqrt{x}$, $g(x) = 2 - 3x$

29. $f(x) = 20$, $g(x) = 0.05$

30. $f(x) = x^4$, $g(x) = \sqrt[4]{x}$

31. $f(x) = \sqrt{x + 5}$, $g(x) = x^2 - 5$

32. $f(x) = x^5 - 2$, $g(x) = \sqrt[5]{x + 2}$

33. $f(x) = x^2 + 2$, $g(x) = \sqrt{3 - x}$

34. $f(x) = 1 - x^2$, $g(x) = \sqrt{x^2 - 25}$

35. $f(x) = \dfrac{1 - x}{x}$, $g(x) = \dfrac{1}{1 + x}$

36. $f(x) = \dfrac{1}{x - 2}$, $g(x) = \dfrac{x + 2}{x}$

37. $f(x) = x^3 - 5x^2 + 3x + 7$, $g(x) = x + 1$

38. $f(x) = x - 1$, $g(x) = x^3 + 2x^2 - 3x - 9$

Find $f(x)$ *and* $g(x)$ *such that* $h(x) = (f \circ g)(x)$. *Answers may vary.*

39. $h(x) = (4 + 3x)^5$ **40.** $h(x) = \sqrt[3]{x^2 - 8}$

41. $h(x) = \dfrac{1}{(x - 2)^4}$ **42.** $h(x) = \dfrac{1}{\sqrt{3x + 7}}$

43. $h(x) = \dfrac{x^3 - 1}{x^3 + 1}$ **44.** $h(x) = |9x^2 - 4|$

45. $h(x) = \left(\dfrac{2 + x^3}{2 - x^3}\right)^6$ **46.** $h(x) = \left(\sqrt{x} - 3\right)^4$

47. $h(x) = \sqrt{\dfrac{x - 5}{x + 2}}$

48. $h(x) = \sqrt{1 + \sqrt{1 + x}}$

49. $h(x) = (x + 2)^3 - 5(x + 2)^2 + 3(x + 2) - 1$

50. $h(x) = 2(x - 1)^{5/3} + 5(x - 1)^{2/3}$

51. *Ripple Spread.* A stone is thrown into a pond, creating a circular ripple that spreads over the pond in such a way that the radius is increasing at a rate of 3 ft/sec.

a) Find a function $r(t)$ for the radius in terms of t.

b) Find a function $A(r)$ for the area of the ripple in terms of the radius r.

c) Find $(A \circ r)(t)$. Explain the meaning of this function.

52. The surface area S of a right circular cylinder is given by the formula $S = 2\pi rh + 2\pi r^2$. If the height is twice the radius, find each of the following.

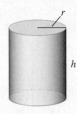

a) A function $S(r)$ for the surface area as a function of r

b) A function $S(h)$ for the surface area as a function of h

53. *Blouse Sizes.* A blouse that is size x in Japan is size $s(x)$ in the United States, where $s(x) = x - 3$. A blouse that is size x in the United States is size $t(x)$ in Australia, where $t(x) = x + 4$. (*Source:* www. onlineconversion.com) Find a function that will convert blouse sizes in Japan to blouse sizes in Australia.

54. A manufacturer of tools, selling rechargeable drills to a chain of home improvement stores, charges $6 more per drill than its manufacturing cost m. The stores then sell each drill for 150% of the price that it paid the manufacturer. Find a function $P(m)$ for the price at the home improvement stores.

❯ Skill Maintenance

Consider the following linear equations. Without graphing them, answer the questions below. **[1.3], [1.4]**

a) $y = x$ b) $y = -5x + 4$

c) $y = \frac{2}{3}x + 1$ d) $y = -0.1x + 6$

e) $y = 3x - 5$ f) $y = -x - 1$

g) $2x - 3y = 6$ h) $6x + 3y = 9$

55. Which, if any, have y-intercept $(0, 1)$?

56. Which, if any, have the same y-intercept?

57. Which slope down from left to right?

58. Which has the steepest slope?

59. Which pass(es) through the origin?

60. Which, if any, have the same slope?

61. Which, if any, are parallel?

62. Which, if any, are perpendicular?

❯ Synthesis

63. Let $p(a)$ represent the number of pounds of grass seed required to seed a lawn with area a. Let $c(s)$ represent the cost of s pounds of grass seed. Which composition makes sense: $(c \circ p)(a)$ or $(p \circ c)(s)$? What does it represent?

64. Write equations of two functions f and g such that $f \circ g = g \circ f = x$. (In Section 5.1, we will study inverse functions. If $f \circ g = g \circ f = x$, functions f and g are *inverses* of each other.)

Mid-Chapter Mixed Review

Determine whether the statement is true or false.

1. $f(c)$ is a relative maximum if $(c, f(c))$ is the highest point in some open interval containing c. **[2.1]**

2. If f and g are functions, then the domain of the functions $f + g$, $f - g$, fg, and f/g is the intersection of the domain of f and the domain of g. **[2.2]**

3. In general, $(f \circ g)(x) \neq (g \circ f)(x)$. **[2.3]**

4. Determine the intervals on which the function is
   (a) increasing; **(b)** decreasing; **(c)** constant. **[2.1]**

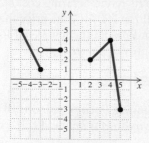

5. Using the graph, determine any relative maxima or
   minima of the function and the intervals on which the
   function is increasing or decreasing. **[2.1]**

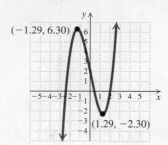

6. Determine the domain and the range of the function
   graphed in Exercise 4. **[2.1]**

7. *Window Design.* Lucas is designing a window for the
   peak of an A-frame house. The base is 4 ft more than
   the height h. Express the area of the window as a function of the height. **[2.1]**

8. For the function defined as

$$f(x) = \begin{cases} x - 5, & \text{for } x \leq -3, \\ 2x + 3, & \text{for } -3 < x \leq 0, \\ \frac{1}{2}x, & \text{for } x > 0, \end{cases}$$

   find $f(-5), f(-3), f(-1)$, and $f(6)$. **[2.1]**

9. Graph the function defined as

$$g(x) = \begin{cases} x + 2, & \text{for } x < -4, \\ -x, & \text{for } x \geq -4. \end{cases} \text{ [2.1]}$$

Given that $f(x) = 3x - 1$ and $g(x) = x^2 + 4$, find each of the following, if it exists. **[2.2]**

10. $(f + g)(-1)$

11. $(fg)(0)$

12. $(g - f)(3)$

13. $(g/f)\left(\frac{1}{3}\right)$

For each pair of functions in Exercises 14 and 15:

a) *Find the domains of $f, g, f + g, f - g, fg, ff, f/g$, and g/f.*
b) *Find $(f + g)(x), (f - g)(x), (fg)(x), (ff)(x), (f/g)(x)$, and $(g/f)(x)$.* **[2.2]**

14. $f(x) = 2x + 5, \ g(x) = -x - 4$

15. $f(x) = x - 1, \ g(x) = \sqrt{x + 2}$

For each function f in Exercises 16 and 17, construct and simplify the difference quotient

$$\frac{f(x + h) - f(x)}{h}. \text{ [2.2]}$$

16. $f(x) = 4x - 3$

17. $f(x) = 6 - x^2$

Given that $f(x) = 5x - 4, g(x) = x^3 + 1$, and $h(x) = x^2 - 2x + 3$, find each of the following. **[2.3]**

18. $(f \circ g)(1)$

19. $(g \circ h)(2)$

20. $(f \circ f)(0)$

21. $(h \circ f)(-1)$

Find $(f \circ g)(x)$ and $(g \circ f)(x)$ and the domain of each. **[2.3]**

22. $f(x) = \frac{1}{2}x, \ g(x) = 6x + 4$

23. $f(x) = 3x + 2, \ g(x) = \sqrt{x}$

COLLABORATIVE DISCUSSION AND WRITING

24. If $g(x) = b$, where b is a positive constant, describe how the graphs of $y = h(x)$ and $y = (h - g)(x)$ will differ. **[2.2]**

25. If the domain of a function f is the set of real numbers and the domain of a function g is also the set of real numbers, under what circumstances do $(f + g)(x)$ and $(f/g)(x)$ have different domains? **[2.2]**

26. If f and g are linear functions, what can you say about the domain of $f \circ g$ and the domain of $g \circ f$? **[2.3]**

27. Nora determines the domain of $f \circ g$ by examining only the formula for $(f \circ g)(x)$. Is her approach valid? Why or why not? **[2.3]**

2.4 ❭ Symmetry

❭ Determine whether a graph is symmetric with respect to the *x*-axis, the *y*-axis, and the origin.

❭ Determine whether a function is even, odd, or neither even nor odd.

❭ Symmetry

Symmetry occurs often in nature and in art. For example, when viewed from the front, the bodies of most animals are at least approximately symmetric. This means that each eye is the same distance from the center of the bridge of the nose, each shoulder is the same distance from the center of the chest, and so on. Architects have used symmetry for thousands of years to enhance the beauty of buildings.

A knowledge of symmetry in mathematics helps us graph and analyze equations and functions.

Consider the points $(4, 2)$ and $(4, -2)$ that appear on the graph of $x = y^2$. Points like these have the same x-value but opposite y-values and are **reflections** of each other across the x-axis. If, for any point (x, y) on a graph, the point $(x, -y)$ is also on the graph, then the graph is said to be **symmetric with respect to the x-axis**. If we fold the graph on the x-axis, the parts above and below the x-axis will coincide.

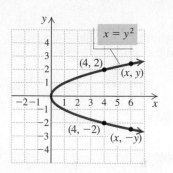

Consider the points $(3, 4)$ and $(-3, 4)$ that appear on the graph of $y = x^2 - 5$ shown below. Points like these have the same y-value but opposite x-values and are **reflections** of each other across the y-axis. If, for any point (x, y) on a graph, the point $(-x, y)$ is also on the graph, then the graph is said to be **symmetric with respect to the y-axis**. If we fold the graph on the y-axis, the parts to the left and to the right of the y-axis will coincide.

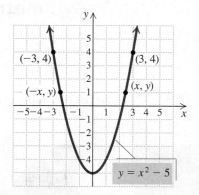

Consider the points $\left(-3, \sqrt{7}\right)$ and $\left(3, -\sqrt{7}\right)$ that appear on the graph of $x^2 = y^2 + 2$ shown below. Note that if we take the opposites of the coordinates of one pair, we get the other pair. If, for any point (x, y) on a graph, the point $(-x, -y)$ is also on the graph, then the graph is said to be **symmetric with respect to the origin**. Visually, if we rotate the graph $180°$ about the origin, the resulting figure coincides with the original.

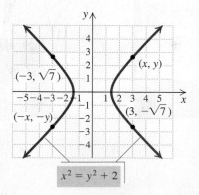

> **ALGEBRAIC TESTS OF SYMMETRY**
>
> *x-axis:* If replacing y with $-y$ produces an equivalent equation, then the graph is *symmetric with respect to the x-axis.*
>
> *y-axis:* If replacing x with $-x$ produces an equivalent equation, then the graph is *symmetric with respect to the y-axis.*
>
> *Origin:* If replacing x with $-x$ and y with $-y$ produces an equivalent equation, then the graph is *symmetric with respect to the origin.*

EXAMPLE 1 Test $y = x^2 + 2$ for symmetry with respect to the x-axis, the y-axis, and the origin.

Algebraic Solution

x-Axis:
We replace y with $-y$:

$$y = x^2 + 2$$
$$\downarrow$$
$$-y = x^2 + 2$$
$$y = -x^2 - 2. \quad \textbf{Multiplying by } -1 \textbf{ on both sides}$$

The resulting equation *is not* equivalent to the original equation, so the graph *is not* symmetric with respect to the x-axis.

y-Axis:
We replace x with $-x$:

$$y = x^2 + 2$$
$$y = (-x)^2 + 2$$
$$y = x^2 + 2. \quad \textbf{Simplifying}$$

The resulting equation *is* equivalent to the original equation, so the graph *is* symmetric with respect to the y-axis.

Origin:
We replace x with $-x$ and y with $-y$:

$$y = x^2 + 2$$
$$\downarrow$$
$$-y = (-x)^2 + 2$$
$$-y = x^2 + 2 \quad \textbf{Simplifying}$$
$$y = -x^2 - 2.$$

The resulting equation *is not* equivalent to the original equation, so the graph *is not* symmetric with respect to the origin.

Graphical Solution

We use a graphing calculator to graph the equation.

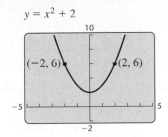

$y = x^2 + 2$

Note that if the graph were folded on the x-axis, the parts above and below the x-axis would not coincide so the graph *is not* symmetric with respect to the x-axis. If it were folded on the y-axis, the parts to the left and to the right of the y-axis would coincide so the graph *is* symmetric with respect to the y-axis. If we rotated it $180°$ around the origin, the resulting graph would not coincide with the original graph so the graph *is not* symmetric with respect to the origin.

▶ *Now Try Exercise 11.*

The algebraic method is often easier to apply than the graphical method, especially with equations that we may not be able to graph easily. It is also often more precise.

EXAMPLE 2   Test $x^2 + y^4 = 5$ for symmetry with respect to the x-axis, the y-axis, and the origin.

Algebraic Solution

x-Axis:
We replace y with $-y$:

$$x^2 + y^4 = 5$$
$$x^2 + (-y)^4 = 5$$
$$x^2 + y^4 = 5.$$

The resulting equation *is* equivalent to the original equation. Thus the graph *is* symmetric with respect to the x-axis.

y-Axis:
We replace x with $-x$:

$$x^2 + y^4 = 5$$
$$(-x)^2 + y^4 = 5$$
$$x^2 + y^4 = 5.$$

The resulting equation *is* equivalent to the original equation, so the graph *is* symmetric with respect to the y-axis.

Origin:
We replace x with $-x$ and y with $-y$:

$$x^2 + y^4 = 5$$
$$(-x)^2 + (-y)^4 = 5$$
$$x^2 + y^4 = 5.$$

The resulting equation *is* equivalent to the original equation, so the graph *is* symmetric with respect to the origin.

Graphical Solution

To graph $x^2 + y^4 = 5$ using a graphing calculator, we first solve the equation for y:

$$y = \pm\sqrt[4]{5 - x^2}.$$

Then on the Y= screen we enter the equations

$$y_1 = \sqrt[4]{5 - x^2} \quad \text{and}$$
$$y_2 = -\sqrt[4]{5 - x^2}.$$

$$x^2 + y^4 = 5$$
$$y_1 = \sqrt[4]{5 - x^2} \text{ and } y_2 = -\sqrt[4]{5 - x^2}$$

From the graph of the equation, we see symmetry with respect to both axes and with respect to the origin.

Now Try Exercise 21.

DETERMINING EVEN AND ODD FUNCTIONS

Given the function $f(x)$:

1. Find $f(-x)$ and simplify. If $f(x) = f(-x)$, then f is even.
2. Find $-f(x)$, simplify, and compare with $f(-x)$ from step (1). If $f(-x) = -f(x)$, then f is odd.

Except for the function $f(x) = 0$, a function cannot be *both* even and odd. Thus if $f(x) \neq 0$ and we see in step (1) that $f(x) = f(-x)$ (that is, f is even), we need not continue.

❯ Even Functions and Odd Functions

Now we relate symmetry to graphs of functions.

EVEN FUNCTIONS AND ODD FUNCTIONS

If the graph of a function f is symmetric with respect to the y-axis, we say that it is an **even function**. That is, for each x in the domain of f, $f(x) = f(-x)$.

If the graph of a function f is symmetric with respect to the origin, we say that it is an **odd function**. That is, for each x in the domain of f, $f(-x) = -f(x)$.

An algebraic procedure for determining even functions and odd functions is shown at left. In the next example, we show an even function and an odd function. Many functions are neither even nor odd.

EXAMPLE 3 Determine whether each of the following functions is even, odd, or neither.

a) $f(x) = 5x^7 - 6x^3 - 2x$ **b)** $h(x) = 5x^6 - 3x^2 - 7$

Algebraic Solution

a) $f(x) = 5x^7 - 6x^3 - 2x$

1. $f(-x) = 5(-x)^7 - 6(-x)^3 - 2(-x)$
$$= 5(-x^7) - 6(-x^3) + 2x$$
$$(-x)^7 = (-1 \cdot x)^7 = (-1)^7 x^7 = -x^7;$$
$$(-x)^3 = (-1 \cdot x)^3 = (-1)^3 x^3 = -x^3$$
$$= -5x^7 + 6x^3 + 2x$$

We see that $f(x) \neq f(-x)$. Thus, f is *not* even.

2. $-f(x) = -(5x^7 - 6x^3 - 2x)$
$$= -5x^7 + 6x^3 + 2x$$

We see that $f(-x) = -f(x)$. Thus, f is odd.

b) $h(x) = 5x^6 - 3x^2 - 7$

1. $h(-x) = 5(-x)^6 - 3(-x)^2 - 7$
$$= 5x^6 - 3x^2 - 7$$

We see that $h(x) = h(-x)$. Thus the function is even.

Graphical Solution

a) We see that the graph appears to be symmetric with respect to the origin. The function is odd.

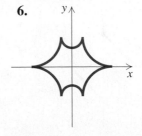

$y = 5x^7 - 6x^3 - 2x$

b) We see that the graph appears to be symmetric with respect to the y-axis. The function is even.

$y = 5x^6 - 3x^2 - 7$

Now Try Exercises 39 and 41.

2.4 Exercise Set

Determine visually whether the graph is symmetric with respect to the x-axis, the y-axis, and the origin.

1.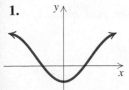

2.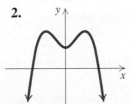

3.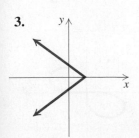

4.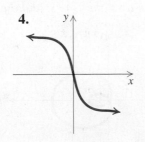

5.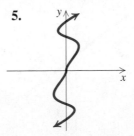

6.

First, graph the equation and determine visually whether it is symmetric with respect to the x-axis, the y-axis, and the origin. Then verify your assertion algebraically.

7. $y = |x| - 2$ **8.** $y = |x + 5|$

9. $5y = 4x + 5$ **10.** $2x - 5 = 3y$

11. $5y = 2x^2 - 3$ **12.** $x^2 + 4 = 3y$

13. $y = 1/x$ **14.** $y = -(4/x)$

Test algebraically whether the graph is symmetric with respect to the x-axis, the y-axis, and the origin. Then check your work graphically, if possible, using a graphing calculator.

15. $5x - 5y = 0$

16. $6x + 7y = 0$

17. $3x^2 - 2y^2 = 3$

18. $5y = 7x^2 - 2x$

19. $y = |2x|$

20. $y^3 = 2x^2$

21. $2x^4 + 3 = y^2$

22. $2y^2 = 5x^2 + 12$

23. $3y^3 = 4x^3 + 2$

24. $3x = |y|$

25. $xy = 12$

26. $xy - x^2 = 3$

Find the point that is symmetric to the given point with respect to the x-axis, the y-axis, and the origin.

27. $(-5, 6)$

28. $\left(\frac{7}{2}, 0\right)$

29. $(-10, -7)$

30. $\left(1, \frac{3}{8}\right)$

31. $(0, -4)$

32. $(8, -3)$

Determine visually whether the function is even, odd, or neither even nor odd.

33.

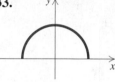

34.

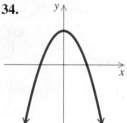

35.

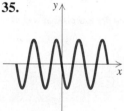

36.

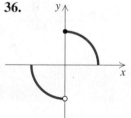

37.

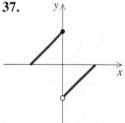

38.

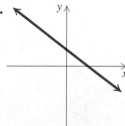

Determine algebraically whether the function is even, odd, or neither even nor odd. Then check your work graphically, where possible, using a graphing calculator.

39. $f(x) = -3x^3 + 2x$

40. $f(x) = 7x^3 + 4x - 2$

41. $f(x) = 5x^2 + 2x^4 - 1$

42. $f(x) = x + \frac{1}{x}$

43. $f(x) = x^{17}$

44. $f(x) = \sqrt[3]{x}$

45. $f(x) = x - |x|$

46. $f(x) = \frac{1}{x^2}$

47. $f(x) = 8$

48. $f(x) = \sqrt{x^2 + 1}$

❯ Skill Maintenance

49. Graph: $f(x) = \begin{cases} x - 2, & \text{for } x \le -1, \\ 3, & \text{for } -1 < x \le 2, \\ x, & \text{for } x > 2. \end{cases}$ [2.1]

50. *Peace Corps Volunteers.*   Since 1961, there has been a total of 6688 Peace Corps volunteers from the University of California–Berkeley and the University of Wisconsin–Madison. The number of volunteers from the University of California–Berkeley is 464 more than the number of volunteers from the University of Wisconsin–Madison. (*Source:* Peace Corps 2014) Find the number of Peace Corps volunteers from each university. [1.5]

❯ Synthesis

Determine whether the function is even, odd, or neither even nor odd.

51. $f(x) = x\sqrt{10 - x^2}$

52. $f(x) = \frac{x^2 + 1}{x^3 - 1}$

Determine whether the graph is symmetric with respect to the x-axis, the y-axis, and the origin.

53. $y^2 + 4xy^2 - y^4 = x^4 - 4x^3 + 3x^2 + 2x^2y^2$

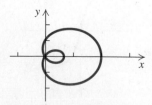

54. $(x^2 + y^2)^2 = 2xy$

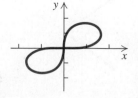

55. Show that if *f* is *any* function, then the function *E* defined by

$$E(x) = \frac{f(x) + f(-x)}{2}$$

is even.

56. Show that if *f* is *any* function, then the function *O* defined by

$$O(x) = \frac{f(x) - f(-x)}{2}$$

is odd.

57. Consider the functions *E* and *O* of Exercises 55 and 56.

a) Show that $f(x) = E(x) + O(x)$. This means that every function can be expressed as the sum of an even function and an odd function.

b) Let $f(x) = 4x^3 - 11x^2 + \sqrt{x} - 10$. Express *f* as a sum of an even function and an odd function.

Determine whether the statement is true or false.

58. The product of two odd functions is odd.

59. The sum of two even functions is even.

60. The product of an even function and an odd function is odd.

2.5 Transformations

> Given the graph of a function, graph its transformation under translations, reflections, stretchings, and shrinkings.

❯ Transformations of Functions

The graphs of some basic functions are shown below. Others can be seen on the inside back cover.

Identity function:
$y = x$

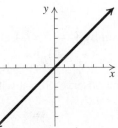

Squaring function:
$y = x^2$

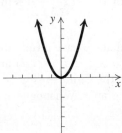

Square root function:
$y = \sqrt{x}$

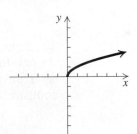

Cubing function:
$y = x^3$

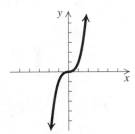

Cube root function:
$y = \sqrt[3]{x}$

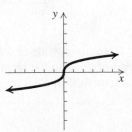

Reciprocal function:
$y = \frac{1}{x}$

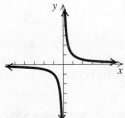

Absolute-value function:
$y = |x|$

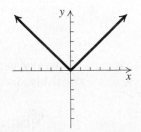

These functions can be considered building blocks for many other functions. We can create graphs of new functions by shifting them horizontally or vertically, stretching or shrinking them, and reflecting them across an axis. We now consider these **transformations**.

〉 Vertical Translations and Horizontal Translations

Suppose that we have a function given by $y = f(x)$. Let's explore the graphs of the new functions $y = f(x) + b$ and $y = f(x) - b$, for $b > 0$.

Consider the functions $y = \frac{1}{5}x^4$, $y = \frac{1}{5}x^4 + 5$, and $y = \frac{1}{5}x^4 - 3$ and compare their graphs. What pattern do you see? Test it with some other functions.

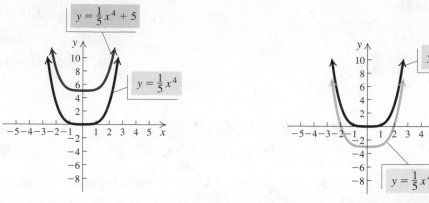

The effect of adding a constant to or subtracting a constant from $f(x)$ in $y = f(x)$ is a shift of the graph of $f(x)$ up or down, respectively. Such a shift is called a **vertical translation**.

VERTICAL TRANSLATION

For $b > 0$:

the graph of $y = f(x) + b$ is the graph of $y = f(x)$ shifted *up* b units;

the graph of $y = f(x) - b$ is the graph of $y = f(x)$ shifted *down* b units.

Suppose that we have a function given by $y = f(x)$. Let's explore the graphs of the new functions $y = f(x - d)$ and $y = f(x + d)$, for $d > 0$.

Consider the functions $y = \frac{1}{5}x^4$, $y = \frac{1}{5}(x - 3)^4$, and $y = \frac{1}{5}(x + 7)^4$ and compare their graphs. What pattern do you observe? Test it with some other functions.

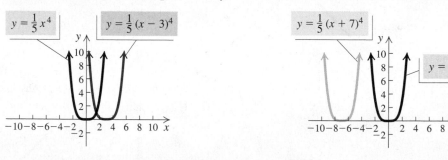

The effect of subtracting a constant from the x-value or adding a constant to the x-value in $y = f(x)$ is a shift of the graph of $f(x)$ to the right or to the left, respectively. Such a shift is called a **horizontal translation**.

HORIZONTAL TRANSLATION

For $d > 0$:

the graph of $y = f(x - d)$ is the graph of $y = f(x)$ shifted *to the right* d units;

the graph of $y = f(x + d)$ is the graph of $y = f(x)$ shifted *to the left* d units.

EXAMPLE 1 Graph each of the following. Before doing so, describe how each graph can be obtained from one of the basic graphs shown on the preceding pages.

a) $g(x) = x^2 - 6$ **b)** $h(x) = |x - 4|$

c) $g(x) = \sqrt{x + 2}$ **d)** $h(x) = \sqrt{x + 2} - 3$

Solution

a) To graph $g(x) = x^2 - 6$, first think of the graph of $f(x) = x^2$. Since $g(x) = f(x) - 6$, the graph of $g(x) = x^2 - 6$ is the graph of $f(x) = x^2$ shifted, or translated, *down* 6 units. (See Fig. 1.)

Let's compare some points on the graphs of f and g.

Points on f: $(-3, 9)$, $(0, 0)$, $(2, 4)$

Corresponding points on g: $(-3, 3)$, $(0, -6)$, $(2, -2)$

We note that the y-coordinate of a point on the graph of g is 6 less than the corresponding y-coordinate on the graph of f.

b) To graph $h(x) = |x - 4|$, first think of the graph of $f(x) = |x|$. Since $h(x) = f(x - 4)$, the graph of $h(x) = |x - 4|$ is the graph of $f(x) = |x|$ shifted *right* 4 units. (See Fig. 2.)

Let's again compare points on the two graphs.

Points on f: $(-4, 4)$, $(0, 4)$, $(6, 6)$

Corresponding points on h: $(0, 4)$, $(4, 0)$, $(10, 6)$

Noting points on f and h, we see that the x-coordinate of a point on the graph of h is 4 more than the x-coordinate of the corresponding point on f.

c) To graph $g(x) = \sqrt{x + 2}$, first think of the graph of $f(x) = \sqrt{x}$. Since $g(x) = f(x + 2)$, the graph of $g(x) = \sqrt{x + 2}$ is the graph of $f(x) = \sqrt{x}$ shifted *left* 2 units. (See Fig. 3.)

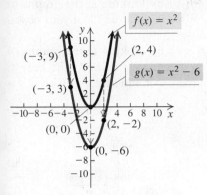

FIGURE 1.

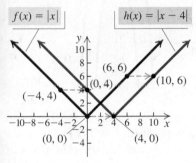

FIGURE 2.

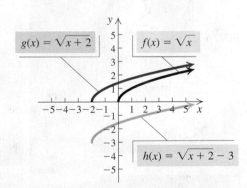

FIGURE 3.

FIGURE 4.

d) To graph $h(x) = \sqrt{x + 2} - 3$, first think of the graph of $f(x) = \sqrt{x}$. In part (c), we found that the graph of $g(x) = \sqrt{x + 2}$ is the graph of $f(x) = \sqrt{x}$ shifted *left* 2 units. Since $h(x) = g(x) - 3$, we shift the graph of $g(x) = \sqrt{x + 2}$ *down* 3 units. Together, the graph of $f(x) = \sqrt{x}$ is shifted *left* 2 units and *down* 3 units. (See Fig. 4.)

> **Now Try Exercises 3 and 15.**

❯ Reflections

Suppose that we have a function given by $y = f(x)$. Let's explore the graphs of the new functions $y = -f(x)$ and $y = f(-x)$.

Compare the functions $y = f(x)$ and $y = -f(x)$ by looking at the graphs of $y = \frac{1}{5}x^4$ and $y = -\frac{1}{5}x^4$ shown on the left below. What do you see? Test your observation with some other functions y_1 and y_2, where $y_2 = -y_1$.

Compare the functions $y = f(x)$ and $y = f(-x)$ by looking at the graphs of $y = 2x^3 - x^4 + 5$ and $y = 2(-x)^3 - (-x)^4 + 5$ shown on the right below. What do you see? Test your observation with some other functions in which x is replaced with $-x$.

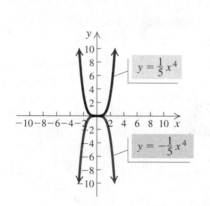

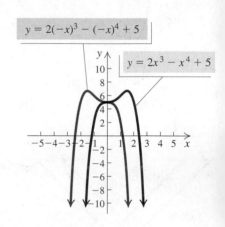

Given the graph of $y = f(x)$, we can reflect each point *across the x-axis* to obtain the graph of $y = -f(x)$. We can reflect each point of $y = f(x)$ *across the y-axis* to obtain the graph of $y = f(-x)$. The new graphs are called **reflections** of $y = f(x)$. The following photos illustrate reflection.

REFLECTIONS

The graph of $y = -f(x)$ is the **reflection** of the graph of $y = f(x)$ across the x-axis.

The graph of $y = f(-x)$ is the **reflection** of the graph of $y = f(x)$ across the y-axis.

If a point (x, y) is on the graph of $y = f(x)$, then $(x, -y)$ is on the graph of $y = -f(x)$, and $(-x, y)$ is on the graph of $y = f(-x)$.

$y_1 = x^3 - 4x^2$,
$y_2 = (-x)^3 - 4(-x)^2$

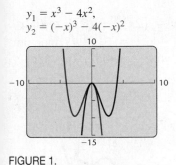

FIGURE 1.

$y_1 = x^3 - 4x^2$, $y_2 = -x^3 + 4x^2$

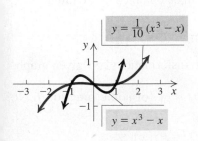

FIGURE 2.

EXAMPLE 2 Graph each of the following. Before doing so, describe how each graph can be obtained from the graph of $f(x) = x^3 - 4x^2$.

a) $g(x) = (-x)^3 - 4(-x)^2$ **b)** $h(x) = 4x^2 - x^3$

Solution

a) We first note that

$$f(-x) = (-x)^3 - 4(-x)^2 = g(x).$$

Thus the graph of g is a *reflection* of the graph of f across the y-axis. (See Fig. 1.) If (x, y) is on the graph of f, then $(-x, y)$ is on the graph of g. For example, $(2, -8)$ is on f and $(-2, -8)$ is on g.

b) We first note that

$$-f(x) = -(x^3 - 4x^2)$$
$$= -x^3 + 4x^2$$
$$= h(x).$$

Thus the graph of h is a *reflection* of the graph of f across the x-axis. (See Fig. 2.) If (x, y) is on the graph of f, then $(x, -y)$ is on the graph of h. For example, $(2, -8)$ is on f and $(2, 8)$ is on h. ❯

❯ Vertical and Horizontal Stretchings and Shrinkings

Suppose that we have a function given by $y = f(x)$. Let's explore the graphs of the new functions $y = af(x)$ and $y = f(cx)$.

Let's consider the functions $y = f(x) = x^3 - x$, $y = \frac{1}{10}(x^3 - x) = \frac{1}{10}f(x)$, $y = 2(x^3 - x) = 2f(x)$, and $y = -2(x^3 - x) = -2f(x)$ and compare their graphs. What pattern do you observe? Test it with some other functions.

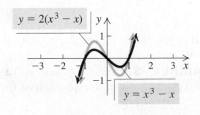

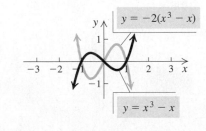

Consider any function f given by $y = f(x)$. Multiplying $f(x)$ by any constant a, where $|a| > 1$, to obtain $g(x) = af(x)$ will *stretch* the graph vertically away from the x-axis. If $0 < |a| < 1$, then the graph will be flattened or *shrunk* vertically toward the x-axis. If $a < 0$, the graph is also reflected across the x-axis.

Just
in
Time
4

VERTICAL STRETCHING AND SHRINKING

The graph of $y = af(x)$ can be obtained from the graph of $y = f(x)$ by

stretching vertically for $|a| > 1$, or

shrinking vertically for $0 < |a| < 1$.

For $a < 0$, the graph is also reflected across the *x*-axis.
(The *y*-coordinates of the graph of $y = af(x)$ can be obtained by multiplying
the *y*-coordinates of $y = f(x)$ by a.)

Let's consider the functions $y = f(x) = x^3 - x$, $y = (2x)^3 - (2x) = f(2x)$,
$y = \left(\frac{1}{2}x\right)^3 - \left(\frac{1}{2}x\right) = f\left(\frac{1}{2}x\right)$, and $y = \left(-\frac{1}{2}x\right)^3 - \left(-\frac{1}{2}x\right) = f\left(-\frac{1}{2}x\right)$ and compare their graphs. What pattern do you observe? Test it with some other functions.

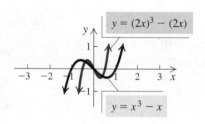

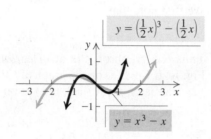

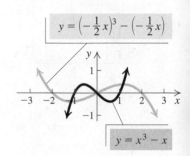

The constant c in the equation $g(x) = f(cx)$ will *shrink* the graph of $y = f(x)$
horizontally toward the *y*-axis if $|c| > 1$. If $0 < |c| < 1$, the graph will be
stretched horizontally away from the *y*-axis. If $c < 0$, the graph is also reflected
across the *y*-axis.

HORIZONTAL STRETCHING AND SHRINKING

The graph of $y = f(cx)$ can be obtained from the graph of $y = f(x)$ by

shrinking horizontally for $|c| > 1$, or

stretching horizontally for $0 < |c| < 1$.

For $c < 0$, the graph is also reflected across the *y*-axis.
(The *x*-coordinates of the graph of $y = f(cx)$ can be obtained by dividing
the *x*-coordinates of the graph of $y = f(x)$ by c.)

It is instructive to use these concepts to create transformations of a given graph.

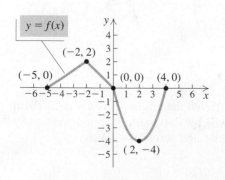

EXAMPLE 3   Shown at left is a graph of $y = f(x)$ for some function f. No
formula for f is given. Graph each of the following.

a) $g(x) = 2f(x)$ **b)** $h(x) = \frac{1}{2}f(x)$

c) $r(x) = f(2x)$ **d)** $s(x) = f\left(\frac{1}{2}x\right)$

e) $t(x) = f\left(-\frac{1}{2}x\right)$

Solution

a) Since $|2| > 1$, the graph of $g(x) = 2f(x)$ is a vertical stretching of the graph of
$y = f(x)$ by a factor of 2. We can consider the key points $(-5, 0)$, $(-2, 2)$, $(0, 0)$,
$(2, -4)$, and $(4, 0)$ on the graph of $y = f(x)$. The transformation multiplies each

y-coordinate by 2 to obtain the key points $(-5, 0)$, $(-2, 4)$, $(0, 0)$, $(2, -8)$, and $(4, 0)$ on the graph of $g(x) = 2f(x)$. The graph is shown below.

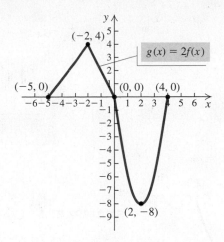

b) Since $\left|\frac{1}{2}\right| < 1$, the graph of $h(x) = \frac{1}{2}f(x)$ is a vertical shrinking of the graph of $y = f(x)$ by a factor of $\frac{1}{2}$. We again consider the key points $(-5, 0)$, $(-2, 2)$, $(0, 0)$, $(2, -4)$, and $(4, 0)$ on the graph of $y = f(x)$. The transformation multiplies each *y*-coordinate by $\frac{1}{2}$ to obtain the key points $(-5, 0)$, $(-2, 1)$, $(0, 0)$, $(2, -2)$, and $(4, 0)$ on the graph of $h(x) = \frac{1}{2}f(x)$. The graph is shown on the left below.

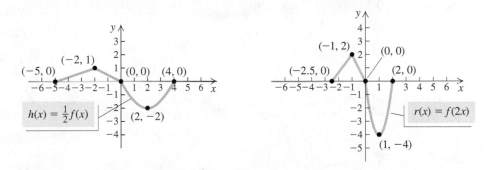

c) Since $|2| > 1$, the graph of $r(x) = f(2x)$ is a horizontal shrinking of the graph of $y = f(x)$. We consider the key points $(-5, 0)$, $(-2, 2)$, $(0, 0)$, $(2, -4)$, and $(4, 0)$ on the graph of $y = f(x)$. The transformation divides each *x*-coordinate by 2 to obtain the key points $(-2.5, 0)$, $(-1, 2)$, $(0, 0)$, $(1, -4)$, and $(2, 0)$ on the graph of $r(x) = f(2x)$. The graph is shown on the right above.

d) Since $\left|\frac{1}{2}\right| < 1$, the graph of $s(x) = f\left(\frac{1}{2}x\right)$ is a horizontal stretching of the graph of $y = f(x)$. We consider the key points $(-5, 0)$, $(-2, 2)$, $(0, 0)$, $(2, -4)$, and $(4, 0)$ on the graph of $y = f(x)$. The transformation divides each *x*-coordinate by $\frac{1}{2}$ (which is the same as multiplying by 2) to obtain the key points $(-10, 0)$, $(-4, 2)$, $(0, 0)$, $(4, -4)$, and $(8, 0)$ on the graph of $s(x) = f\left(\frac{1}{2}x\right)$. The graph is shown below.

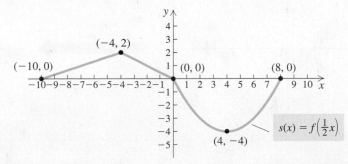

e) The graph of $t(x) = f\left(-\frac{1}{2}x\right)$ can be obtained by reflecting the graph in part (d) across the y-axis, as shown below.

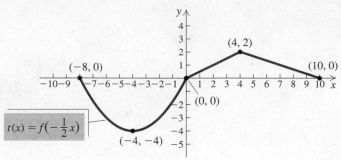

Now Try Exercises 59 and 61.

EXAMPLE 4 Use the graph of $y = f(x)$ shown at left to graph
$$y = -2f(x - 3) + 1.$$

Solution

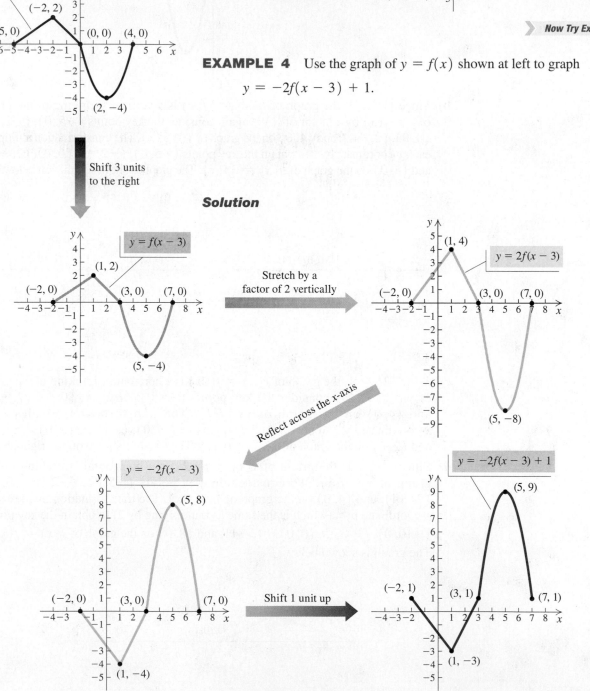

Now Try Exercise 63.

SUMMARY OF TRANSFORMATIONS OF $y = f(x)$

Vertical Translation: $y = f(x) \pm b$

For $b > 0$:

the graph of $y = f(x) + b$ is the graph of $y = f(x)$ shifted *up* b units;

the graph of $y = f(x) - b$ is the graph of $y = f(x)$ shifted *down* b units.

Horizontal Translation: $y = f(x \mp d)$

For $d > 0$:

the graph of $y = f(x - d)$ is the graph of $y = f(x)$ shifted *right* d units;

the graph of $y = f(x + d)$ is the graph of $y = f(x)$ shifted *left* d units.

Reflections

Across the x-axis:

The graph of $y = -f(x)$ is the reflection of the graph of $y = f(x)$ across the x-axis.

Across the y-axis:

The graph of $y = f(-x)$ is the reflection of the graph of $y = f(x)$ across the y-axis.

Vertical Stretching or Shrinking: $y = af(x)$

The graph of $y = af(x)$ can be obtained from the graph of $y = f(x)$ by

stretching vertically for $|a| > 1$, or

shrinking vertically for $0 < |a| < 1$.

For $a < 0$, the graph is also reflected across the x-axis.

Horizontal Stretching or Shrinking: $y = f(cx)$

The graph of $y = f(cx)$ can be obtained from the graph of $y = f(x)$ by

shrinking horizontally for $|c| > 1$, or

stretching horizontally for $0 < |c| < 1$.

For $c < 0$, the graph is also reflected across the y-axis.

Visualizing the Graph

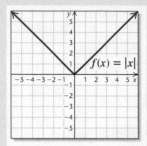

$f(x) = |x|$

Match the function with its graph. Use transformation graphing techniques to obtain the graph of g from the basic function $f(x) = |x|$ shown at top left.

1. $g(x) = -2|x|$

2. $g(x) = |x - 1| + 1$

3. $g(x) = -\left|\dfrac{1}{3}x\right|$

4. $g(x) = |2x|$

5. $g(x) = |x + 2|$

6. $g(x) = |x| + 3$

7. $g(x) = -\dfrac{1}{2}|x - 4|$

8. $g(x) = \dfrac{1}{2}|x| - 3$

9. $g(x) = -|x| - 2$

Answers on page A-10

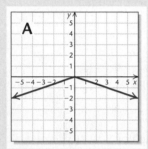

A

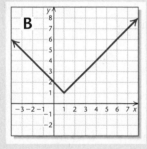

B

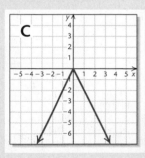

C

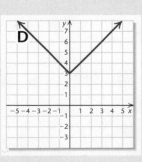

D

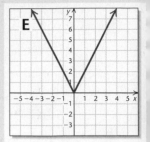

E

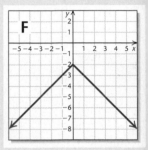

F

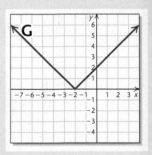

G

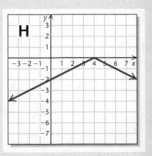

H

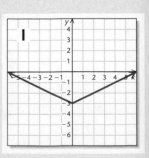

I

2.5   Exercise Set

Describe how the graph of the function can be obtained from one of the basic graphs on p. 129. Then graph the function by hand or with a graphing calculator.

1. $f(x) = (x - 3)^2$

2. $g(x) = x^2 + \frac{1}{2}$

3. $g(x) = x - 3$

4. $g(x) = -x - 2$

5. $h(x) = -\sqrt{x}$

6. $g(x) = \sqrt{x} - 1$

7. $h(x) = \frac{1}{x} + 4$

8. $g(x) = \frac{1}{x - 2}$

9. $h(x) = -3x + 3$

10. $f(x) = 2x + 1$

11. $h(x) = \frac{1}{2}|x| - 2$

12. $g(x) = -|x| + 2$

13. $g(x) = -(x - 2)^3$

14. $f(x) = (x + 1)^3$

15. $g(x) = (x + 1)^2 - 1$

16. $h(x) = -x^2 - 4$

17. $g(x) = \frac{1}{3}x^3 + 2$

18. $h(x) = (-x)^3$

19. $f(x) = \sqrt{x + 2}$

20. $f(x) = -\frac{1}{2}\sqrt{x} - 1$

21. $f(x) = \sqrt[3]{x} - 2$

22. $h(x) = \sqrt[3]{x + 1}$

Describe how the graph of the function can be obtained from one of the basic graphs on p. 129.

23. $g(x) = |3x|$

24. $f(x) = \frac{1}{2}\sqrt[3]{x}$

25. $h(x) = \frac{2}{x}$

26. $f(x) = |x - 3| - 4$

27. $f(x) = 3\sqrt{x} - 5$

28. $f(x) = 5 - \frac{1}{x}$

29. $g(x) = |\frac{1}{3}x| - 4$

30. $f(x) = \frac{2}{3}x^3 - 4$

31. $f(x) = -\frac{1}{4}(x - 5)^2$

32. $f(x) = (-x)^3 - 5$

33. $f(x) = \frac{1}{x + 3} + 2$

34. $g(x) = \sqrt{-x} + 5$

35. $h(x) = -(x - 3)^2 + 5$

36. $f(x) = 3(x + 4)^2 - 3$

The point $(-12, 4)$ is on the graph of $y = f(x)$. Find the corresponding point on the graph of $y = g(x)$.

37. $g(x) = \frac{1}{2}f(x)$

38. $g(x) = f(x - 2)$

39. $g(x) = f(-x)$

40. $g(x) = f(4x)$

41. $g(x) = f(x) - 2$

42. $g(x) = f(\frac{1}{2}x)$

43. $g(x) = 4f(x)$

44. $g(x) = -f(x)$

Given that $f(x) = x^2 + 3$, match the function g with a transformation of f from one of A–D.

45. $g(x) = x^2 + 4$ **A.** $f(x - 2)$

46. $g(x) = 9x^2 + 3$ **B.** $f(x) + 1$

47. $g(x) = (x - 2)^2 + 3$ **C.** $2f(x)$

48. $g(x) = 2x^2 + 6$ **D.** $f(3x)$

Write an equation for a function that has a graph with the given characteristics.

49. The shape of $y = x^2$, but reflected across the x-axis and shifted right 8 units

50. The shape of $y = \sqrt{x}$, but shifted left 6 units and down 5 units

51. The shape of $y = |x|$, but shifted left 7 units and up 2 units

52. The shape of $y = x^3$, but reflected across the x-axis and shifted right 5 units

53. The shape of $y = 1/x$, but shrunk horizontally by a factor of 2 and shifted down 3 units

54. The shape of $y = x^2$, but shifted right 6 units and up 2 units

55. The shape of $y = x^2$, but reflected across the x-axis and shifted right 3 units and up 4 units

56. The shape of $y = |x|$, but stretched horizontally by a factor of 2 and shifted down 5 units

57. The shape of $y = \sqrt{x}$, but reflected across the y-axis and shifted left 2 units and down 1 unit

58. The shape of $y = 1/x$, but reflected across the x-axis and shifted up 1 unit

A graph of $y = f(x)$ follows. No formula for f is given. In Exercises 59–66, graph the given equation.

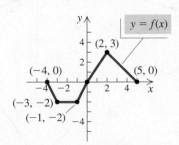

59. $g(x) = -2f(x)$ **60.** $g(x) = \frac{1}{2}f(x)$

61. $g(x) = f(-\frac{1}{2}x)$ **62.** $g(x) = f(2x)$

63. $g(x) = -\frac{1}{2}f(x - 1) + 3$

64. $g(x) = -3f(x + 1) - 4$

65. $g(x) = f(-x)$ **66.** $g(x) = -f(x)$

A graph of $y = g(x)$ follows. No formula for g is given. In Exercises 67–70, graph the given equation.

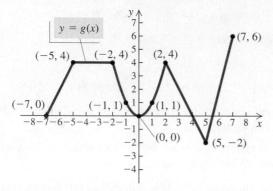

67. $h(x) = -g(x + 2) + 1$

68. $h(x) = \frac{1}{2}g(-x)$ **69.** $h(x) = g(2x)$

70. $h(x) = 2g(x - 1) - 3$

The graph of the function f is shown in figure (a). In Exercises 71–78, match the function g with one of the graphs (a)–(h) that follow. Some graphs may be used more than once and some may not be used at all.

a)

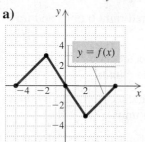

b)

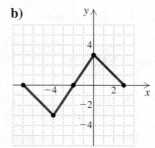

c)

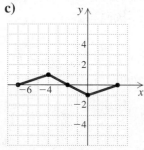

d)

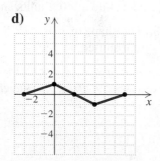

e)

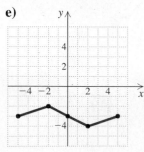

f)

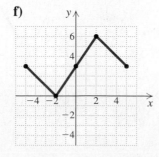

g)

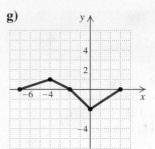

h)

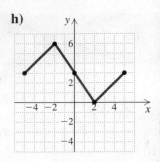

71. $g(x) = f(-x) + 3$

72. $g(x) = f(x) + 3$

73. $g(x) = -f(x) + 3$

74. $g(x) = -f(-x)$

75. $g(x) = \frac{1}{3}f(x - 2)$

76. $g(x) = \frac{1}{3}f(x) - 3$

77. $g(x) = \frac{1}{3}f(x + 2)$

78. $g(x) = -f(x + 2)$

For each pair of functions, determine whether $g(x) = f(-x)$.

79. $f(x) = 2x^4 - 35x^3 + 3x - 5,$
  $g(x) = 2x^4 + 35x^3 - 3x - 5$

80. $f(x) = \frac{1}{4}x^4 + \frac{1}{5}x^3 - 81x^2 - 17,$
  $g(x) = \frac{1}{4}x^4 + \frac{1}{5}x^3 + 81x^2 - 17$

A graph of the function $f(x) = x^3 - 3x^2$ is shown below. Exercises 81–84 show graphs of functions transformed from this one. Find a formula for each function.

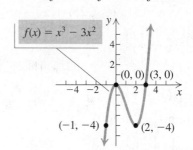

81.

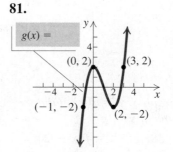

82.

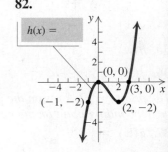

83.   **84.**

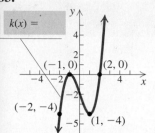   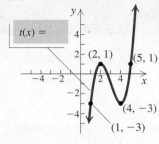

❯ Skill Maintenance

Determine algebraically whether the graph is symmetric, with respect to the x-axis, the y-axis, and the origin. [2.4]

85. $y = 3x^4 - 3$

86. $y^2 = x$

87. $2x - 5y = 0$

88. *Educational Level and Income.* The average annual wages of a 64-year-old person with a bachelor's degree is $67,735. This amount is approximately 53.7% more than the average wages of a 64-year-old person with only a high school diploma. (*Sources:* U.S. Census Bureau; Bureau of Labor Statistics, Current Population Survey, March Supplement) Find the average annual wages of a 64-year-old person with only a high school diploma. [1.5]

89. *Medical Care Abroad.* The cost of medical care abroad is often significantly less than in the United States. In 2014, the cost of knee replacement surgery in the United States was $34,000. This cost was $4000 more than four times the cost of knee replacement surgery in India. (*Sources:* aarp.org, Sarah Barchus and Beth Howard; Center for Medical Tourism Research in San Antonio) Find the cost of knee replacement surgery in India. [1.5]

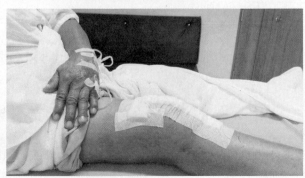

90. *Undergraduate Students from Foreign Countries.* In 2013–2014, Saudi Arabia and Canada sent 82,223 undergraduate students to the United States to study in universities. Saudi Arabia sent 25,615 more students than Canada sent. (*Source:* Institute for International Educators' "Open Doors" report) How many students did each country send to the United States to study? [1.5]

❯ Synthesis

Use the following graph of the function f for Exercises 91 and 92.

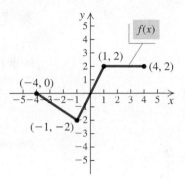

91. Graph: $y = |f(x)|$.   **92.** Graph: $y = f(|x|)$.

Use the following graph of the function g for Exercises 93 and 94.

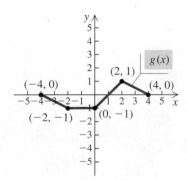

93. Graph: $y = g(|x|)$.   **94.** Graph: $y = |g(x)|$.

Graph each of the following using a graphing calculator. Before doing so, describe how the graph can be obtained from a more basic graph. Give the domain and the range of the function.

95. $f(x) = [\![x - \frac{1}{2}]\!]$ **96.** $f(x) = |\sqrt{x} - 1|$

97. If $(3, 4)$ is a point on the graph of $y = f(x)$, what point do you know is on the graph of $y = 2f(x)$? of $y = 2 + f(x)$? of $y = f(2x)$?

98. Find the zeros of $f(x) = 3x^5 - 20x^3$. Then, without using a graphing calculator, state the zeros of $f(x - 3)$ and $f(x + 8)$.

2.6   Variation and Applications

> Find equations of direct variation, inverse variation, and combined variation given values of the variables.

> Solve applied problems involving variation.

We now extend our study of formulas and functions by considering applications involving variation.

❯ Direct Variation

The median hourly wage for an elevator and escalator installer/repairer is $35 per hour (*Source:* U.S. Bureau of Labor Statistics). In 1 hr, $35 is earned; in 2 hr, $70 is earned; in 3 hr, $105 is earned; and so on. This gives rise to a set of ordered pairs:

$$(1, 35), \quad (2, 70), \quad (3, 105), \quad (4, 140), \quad \text{and so on.}$$

Note that the ratio of the second coordinate to the first coordinate is the same number for each pair:

$$\frac{35}{1} = 35, \quad \frac{70}{2} = 35, \quad \frac{105}{3} = 35, \quad \frac{140}{4} = 35, \quad \text{and so on.}$$

Earnings for Elevator and Escalator Installer/Repairer

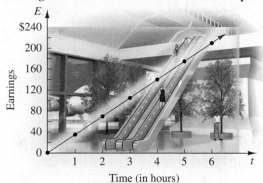

Whenever a situation produces pairs of numbers in which the *ratio is constant*, we say that there is **direct variation**. In this case, the amount earned E varies directly as the time worked t:

$$\frac{E}{t} = 35 \,(\text{a constant}), \quad \text{or} \quad E = 35t,$$

or, if we use function notation, $E(t) = 35t$. This equation is an equation of **direct variation**. The coefficient, 35, is called the **variation constant**. In this case, it is the rate of change of earnings with respect to time.

The graph of $y = kx$, $k > 0$, always goes through the origin and rises from left to right. Note that as x increases, y increases; that is, the function is increasing on the interval $(0, \infty)$. The constant k is also the slope of the line.

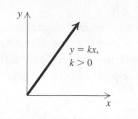

DIRECT VARIATION

If a situation gives rise to a linear function $f(x) = kx$, or $y = kx$, where k is a positive constant, we say that we have **direct variation**, or that **y varies directly as x**, or that **y is directly proportional to x**. The number k is called the **variation constant**, or the **constant of proportionality**.

EXAMPLE 1   Find the variation constant and an equation of variation in which y varies directly as x, and $y = 32$ when $x = 2$.

Solution   We know that $(2, 32)$ is a solution of $y = kx$. Thus,

$$y = kx$$
$$32 = k \cdot 2 \qquad \textbf{Substituting}$$
$$\frac{32}{2} = k \qquad \textbf{Solving for } k$$
$$16 = k. \qquad \textbf{Simplifying}$$

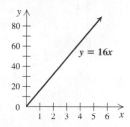

The variation constant, 16, is the rate of change of y with respect to x. The equation of variation is $y = 16x$.

> **Now Try Exercise 1.**

S cm of snow

W cm of water

EXAMPLE 2   *Water from Melting Snow.*   The number of centimeters of water W produced from melting snow varies directly as S, the number of centimeters of snow. Meteorologists have found that under certain conditions 150 cm of snow will melt to 16.8 cm of water. To how many centimeters of water will 200 cm of snow melt under the same conditions?

Solution   We can express the amount of water as a function of the amount of snow. Thus, $W(S) = kS$, where k is the variation constant. We first find k using the given data and then find an equation of variation:

$$W(S) = kS \qquad \textbf{W varies directly as S.}$$
$$W(150) = k \cdot 150 \qquad \textbf{Substituting 150 for S}$$
$$16.8 = k \cdot 150 \qquad \textbf{Replacing W(150) with 16.8}$$
$$\frac{16.8}{150} = k \qquad \textbf{Solving for } k$$
$$0.112 = k. \qquad \textbf{This is the variation constant.}$$

The equation of variation is $W(S) = 0.112S$.

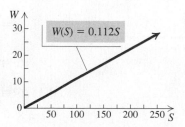

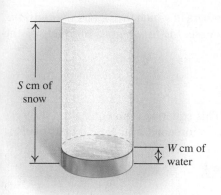

Next, we use the equation to find how many centimeters of water will result from melting 200 cm of snow:

$$W(S) = 0.112S$$
$$W(200) = 0.112(200) \qquad \textbf{Substituting}$$
$$W = 22.4.$$

Thus, 200 cm of snow will melt to 22.4 cm of water.

> *Now Try Exercise 17.*

❯ Inverse Variation

Suppose that a bus is traveling a distance of 20 mi. At a speed of 5 mph, the trip will take 4 hr; at 10 mph, it will take 2 hr; at 20 mph, it will take 1 hr; at 40 mph, it will take $\frac{1}{2}$ hr; and so on. We plot this information on a graph, using speed as the first coordinate and time as the second coordinate to determine a set of ordered pairs:

$$(5, 4), \qquad (10, 2), \qquad (20, 1), \qquad \left(40, \tfrac{1}{2}\right), \quad \text{and so on.}$$

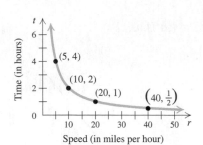

Time (in hours)
Speed (in miles per hour)

Note that the products of the coordinates are all the same number:

$$5 \cdot 4 = 20, \qquad 10 \cdot 2 = 20, \qquad 20 \cdot 1 = 20, \qquad 40 \cdot \tfrac{1}{2} = 20, \quad \text{and so on.}$$

Whenever a situation produces pairs of numbers in which the *product is constant*, we say that there is **inverse variation**. In this case, the time varies inversely as the speed, or rate:

$$rt = 20 \,(\text{a constant}), \quad \text{or} \quad t = \frac{20}{r},$$

or, if we use function notation, $t(r) = 20/r$. This equation is an equation of **inverse variation**. The coefficient, 20, is called the **variation constant**. Note that as the first number increases, the second number decreases.

The graph of $y = k/x, k > 0$, is like the one shown below. Note that as x increases, y decreases; that is, the function is decreasing on the interval $(0, \infty)$.

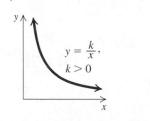

$$y = \frac{k}{x},$$
$$k > 0$$

INVERSE VARIATION

If a situation gives rise to a function $f(x) = k/x$, or $y = k/x$, where k is a positive constant, we say that we have **inverse variation**, or that **y varies inversely as x**, or that **y is inversely proportional to x**. The number k is called the **variation constant**, or the **constant of proportionality**.

EXAMPLE 3 Find the variation constant and an equation of variation in which y varies inversely as x, and $y = 16$ when $x = 0.3$.

Solution We know that $(0.3, 16)$ is a solution of $y = k/x$. We substitute:

$$y = \frac{k}{x}$$
$$16 = \frac{k}{0.3} \qquad \textbf{Substituting}$$
$$(0.3)16 = k \qquad \textbf{Solving for } k$$
$$4.8 = k. \qquad \textbf{Simplifying}$$

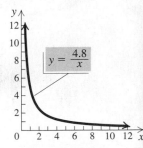

$$y = \frac{4.8}{x}$$

The variation constant is 4.8. The equation of variation is $y = 4.8/x$.

> *Now Try Exercise 3.*

There are many real-world problems that translate to an equation of inverse variation.

EXAMPLE 4   *Filling a Swimming Pool.*   The time t required to fill a swimming pool varies inversely as the rate of flow r of water into the pool. A tank truck can fill a pool in 90 min at a rate of 1500 L/min. How long would it take to fill the pool at a rate of 1800 L/min?

Solution   We can express the amount of time required as a function of the rate of flow. Thus we have $t(r) = k/r$. We first find k using the given information and then find an equation of variation:

$$t(r) = \frac{k}{r}$$ t varies inversely as r.

$$t(1500) = \frac{k}{1500}$$ **Substituting 1500 for r**

$$90 = \frac{k}{1500}$$ **Replacing $t(1500)$ with 90**

$$90 \cdot 1500 = k$$ **Solving for k**

$$135{,}000 = k.$$ **This is the variation constant.**

The equation of variation is

$$t(r) = \frac{135{,}000}{r}.$$

Next, we use the equation to find the time that it would take to fill the pool at a rate of 1800 L/min:

$$t(r) = \frac{135{,}000}{r}$$

$$t(1800) = \frac{135{,}000}{1800}$$ **Substituting**

$$t = 75.$$

Thus it would take 75 min to fill the pool at a rate of 1800 L/min.

> **Now Try Exercise 15.**

Let's summarize the procedure for solving variation problems.

SOLVING VARIATION PROBLEMS

1. Determine whether direct variation or inverse variation applies.
2. Write an equation of the form $y = kx$ (for direct variation) or $y = k/x$ (for inverse variation), substitute the known values, and solve for k.
3. Write the equation of variation, and use it to find the unknown value(s) in the problem.

❯ Combined Variation

We now look at other kinds of variation.

y varies **directly as the *n*th power of *x*** if there is some positive constant *k* such that

$$y = kx^n.$$

y varies **inversely as the *n*th power of *x*** if there is some positive constant *k* such that

$$y = \frac{k}{x^n}.$$

y varies **jointly as *x* and *z*** if there is some positive constant *k* such that

$$y = kxz.$$

There are other types of combined variation as well. Consider the formula for the volume of a right circular cylinder, $V = \pi r^2 h$, in which *V*, *r*, and *h* are variables and π is a constant. We say that *V* varies jointly as *h* and the square of *r*. In this formula, π is the variation constant.

EXAMPLE 5   Find an equation of variation in which *y* varies directly as the square of *x*, and $y = 12$ when $x = 2$.

Solution   We write an equation of variation and find *k*:

$$y = kx^2$$
$$12 = k \cdot 2^2 \qquad \textbf{Substituting}$$
$$12 = k \cdot 4$$
$$3 = k.$$

Thus, $y = 3x^2$.

> **Now Try Exercise 27.**

EXAMPLE 6   Find an equation of variation in which *y* varies jointly as *x* and *z*, and $y = 42$ when $x = 2$ and $z = 3$.

Solution   We have

$$y = kxz$$
$$42 = k \cdot 2 \cdot 3 \qquad \textbf{Substituting}$$
$$42 = k \cdot 6$$
$$7 = k.$$

Thus, $y = 7xz$.

> **Now Try Exercise 29.**

EXAMPLE 7   Find an equation of variation in which *y* varies jointly as *x* and *z* and inversely as the square of *w*, and $y = 105$ when $x = 3$, $z = 20$, and $w = 2$.

Solution   We have

$$y = k \cdot \frac{xz}{w^2}$$
$$105 = k \cdot \frac{3 \cdot 20}{2^2} \qquad \textbf{Substituting}$$
$$105 = k \cdot 15$$
$$7 = k.$$

Thus, $y = 7\frac{xz}{w^2}$, or $y = \frac{7xz}{w^2}$.

> **Now Try Exercise 33.**

Many applied problems can be modeled using equations of combined variation.

EXAMPLE 8   *Volume of a Tree.*   The volume of wood V in a tree varies jointly as the height h and the square of the girth g. (Girth is distance around.) If the volume of a redwood tree is 216 m^3 when the height is 30 m and the girth is 1.5 m, what is the height of a tree whose volume is 344 m^3 and whose girth is 1.6 m?

Solution   We first find k using the first set of data. Then we solve for h using the second set of data.

$$V = khg^2$$
$$216 = k \cdot 30 \cdot 1.5^2$$
$$216 = k \cdot 30 \cdot 2.25$$
$$216 = k \cdot 67.5$$
$$3.2 = k$$

Thus the equation of variation is $V = 3.2hg^2$. We substitute the second set of data into the equation:

$$344 = 3.2 \cdot h \cdot 1.6^2$$
$$344 = 3.2 \cdot h \cdot 2.56$$
$$344 = 8.192 \cdot h$$
$$42 \approx h.$$

The height of the tree is about 42 m.

Now Try Exercise 35.

2.6   Exercise Set

Find the variation constant and an equation of variation for the given situation.

1. y varies directly as x, and $y = 54$ when $x = 12$

2. y varies directly as x, and $y = 0.1$ when $x = 0.2$

3. y varies inversely as x, and $y = 3$ when $x = 12$

4. y varies inversely as x, and $y = 12$ when $x = 5$

5. y varies directly as x, and $y = 1$ when $x = \frac{1}{4}$

6. y varies inversely as x, and $y = 0.1$ when $x = 0.5$

7. y varies inversely as x, and $y = 32$ when $x = \frac{1}{8}$

8. y varies directly as x, and $y = 3$ when $x = 33$

9. y varies directly as x, and $y = \frac{3}{4}$ when $x = 2$

10. y varies inversely as x, and $y = \frac{1}{5}$ when $x = 35$

11. y varies inversely as x, and $y = 1.8$ when $x = 0.3$

12. y varies directly as x, and $y = 0.9$ when $x = 0.4$

13. *Child's Allowance.*   The Harrisons decide to give their children a weekly allowance that is directly proportional to each child's age. Their 6-year-old daughter receives an allowance of \$5.50. What is their 9-year-old son's allowance?

14. *Sales Tax.* The amount of sales tax paid on a product is directly proportional to the purchase price. In Iowa, the sales tax on a Nook Glowlight™ that sells for $119 is $7.14. What is the sales tax on an e-book that sells for $21?

15. *Rate of Travel.* The time *t* required to drive a fixed distance varies inversely as the speed *r*. It takes 5 hr at a speed of 80 km/h to drive a fixed distance. How long will it take to drive the same distance at a speed of 70 km/h?

16. *Beam Weight.* The weight *W* that a horizontal beam can support varies inversely as the length *L* of the beam. Suppose that an 8-m beam can support 1200 kg. How many kilograms can a 14-m beam support?

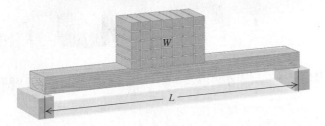

17. *Fat Intake.* The maximum number of grams of fat that should be in a diet varies directly as a person's weight. A person weighing 120 lb should consume no more than 60 g of fat per day. What is the maximum daily fat intake for a person weighing 180 lb?

18. *House of Representatives.* The number of representatives *N* that each state has varies directly as the number of people *P* living in the state. If California, with 38,333,000 residents, has 53 representatives, how many representatives does Texas, with a population of 26,448,000, have?

19. *Work Rate.* The time *T* required to do a job varies inversely as the number of people *P* working. It takes 5 hr for 7 bricklayers to build a park wall. How long will it take 10 bricklayers to complete the job?

20. *Pumping Rate.* The time *t* required to empty a tank varies inversely as the rate *r* of pumping. If a pump can empty a tank in 45 min at the rate of 600 kL/min, how long will it take the pump to empty the same tank at the rate of 1000 kL/min?

21. *Hooke's Law.* Hooke's law states that the distance *d* that a spring will stretch varies directly as the mass *m* of an object hanging from the spring. If a 3-kg mass stretches a spring 40 cm, how far will a 5-kg mass stretch the spring?

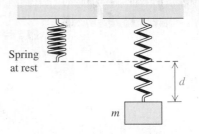

Spring at rest

22. *Relative Aperture.* The relative aperture, or f-stop, of a 23.5-mm diameter lens is directly proportional to the focal length *F* of the lens. If a 150-mm focal length has an f-stop of 6.3, find the f-stop of a 23.5-mm diameter lens with a focal length of 80 mm.

23. *Musical Pitch.* The pitch *P* of a musical tone varies inversely as its wavelength *W*. One tone has a pitch of 330 vibrations per second and a wavelength of 3.2 ft. Find the wavelength of another tone that has a pitch of 550 vibrations per second.

24. *Weight on Mars.* The weight *M* of an object on Mars varies directly as its weight *E* on Earth. A person who weighs 95 lb on Earth weighs 35.9 lb on Mars. How much would a 100-lb person weigh on Mars?

Find an equation of variation for the given situation.

25. y varies inversely as the square of x, and $y = 0.15$ when $x = 0.1$

26. y varies inversely as the square of x, and $y = 6$ when $x = 3$

27. y varies directly as the square of x, and $y = 0.15$ when $x = 0.1$

28. y varies directly as the square of x, and $y = 6$ when $x = 3$

29. y varies jointly as x and z, and $y = 56$ when $x = 7$ and $z = 8$

30. y varies directly as x and inversely as z, and $y = 4$ when $x = 12$ and $z = 15$

31. y varies jointly as x and the square of z, and $y = 105$ when $x = 14$ and $z = 5$

32. y varies jointly as x and z and inversely as w, and $y = \frac{3}{2}$ when $x = 2$, $z = 3$, and $w = 4$

33. y varies jointly as x and z and inversely as the product of w and p, and $y = \frac{3}{28}$ when $x = 3$, $z = 10$, $w = 7$, and $p = 8$

34. y varies jointly as x and z and inversely as the square of w, and $y = \frac{12}{5}$ when $x = 16$, $z = 3$, and $w = 5$

35. *Intensity of Light.*   The intensity I of light from a light bulb varies inversely as the square of the distance d from the bulb. Suppose that I is 90 W/m² (watts per square meter) when the distance is 5 m. How much *farther* would it be to a point where the intensity is 40 W/m²?

36. *Atmospheric Drag.*   Wind resistance, or atmospheric drag, tends to slow down moving objects. Atmospheric drag varies jointly as an object's surface area A and velocity v. If a car traveling at a speed of 40 mph with a surface area of 37.8 ft² experiences a drag of 222 N (Newtons), how fast must a car with 51 ft² of surface area travel in order to experience a drag force of 430 N?

37. *Braking Distance of a Car.*   The braking distance d of a car after the brakes have been applied varies directly as the square of the speed r. If a car traveling 60 mph can stop in 200 ft, how fast can a car travel and still stop in 72 ft?

38. *Weight of an Astronaut.*   The weight W of an object varies inversely as the square of the distance d from the center of the earth. At sea level (3978 mi from the center of the earth), an astronaut weighs 220 lb. Find his weight when he is 200 mi above the surface of the earth.

39. *Earned-Run Average.*   A pitcher's earned-run average E varies directly as the number R of earned runs allowed and inversely as the number I of innings pitched. In 2014, Clayton Kershaw of the Los Angeles Dodgers had an earned-run average of 1.77. He gave up 39 earned runs in 198.1 innings. How many earned runs would he have given up had he pitched 220 innings with the same average? Round to the nearest whole number.

40. *Boyle's Law.*   The volume V of a given mass of a gas varies directly as the temperature T and inversely as the pressure P. If $V = 231$ cm³ when $T = 42°$ and $P = 20$ kg/cm², what is the volume when $T = 30°$ and $P = 15$ kg/cm²?

❯ Skill Maintenance

Vocabulary Review

In each of Exercises 41–45, fill in the blank with the correct term. Some of the given choices will not be used.

even function	relative maximum
odd function	relative minimum
constant function	solution
composite function	zero
direct variation	perpendicular
inverse variation	parallel

41. Nonvertical lines are _____ if and only if they have the same slope and different y-intercepts. **[1.4]**

42. An input c of a function f is a(n) _____ of the function if $f(c) = 0$. **[1.5]**

43. For a function f for which $f(c)$ exists, $f(c)$ is a(n) _____ if $f(c)$ is the lowest point in some open interval. **[2.1]**

44. If the graph of a function is symmetric with respect to the origin, then f is a(n) _____. **[2.4]**

45. An equation $y = k/x$ is an equation of _____. **[2.6]**

❯ Synthesis

46. In each of the following equations, state whether y varies directly as x, inversely as x, or neither directly nor inversely as x.

a) $7xy = 14$ **b)** $x - 2y = 12$
c) $-2x + 3y = 0$ **d)** $x = \frac{3}{4}y$
e) $\dfrac{x}{y} = 2$

47. *Volume and Cost.* An 18-oz jar of peanut butter in the shape of a right circular cylinder is 5 in. high and 3 in. in diameter and sells for \$2.89. In the same store, a 22-oz jar of the same brand is $5\frac{1}{4}$ in. high and $3\frac{1}{4}$ in. in diameter. If the cost is directly proportional to volume, what should the price of the larger jar be? If the cost is directly proportional to weight, what should the price of the larger jar be?

48. Describe in words the variation given by the equation
$$Q = \frac{kp^2}{q^3}.$$

49. *Area of a Circle.* The area of a circle varies directly as the square of the length of a diameter. What is the variation constant?

Chapter 2 Summary and Review

STUDY GUIDE

KEY TERMS AND CONCEPTS	EXAMPLES

SECTION 2.1: INCREASING, DECREASING, AND PIECEWISE FUNCTIONS; APPLICATIONS

Increasing, Decreasing, and Constant Functions A function f is said to be **increasing** on an *open* interval I if for all a and b in that interval, $a < b$ implies $f(a) < f(b)$. A function f is said to be **decreasing** on an *open* interval I if for all a and b in that interval, $a < b$ implies $f(a) > f(b)$. A function f is said to be **constant** on an *open* interval I if for all a and b in that interval, $f(a) = f(b)$.	Determine the intervals on which the function is **(a)** increasing; **(b)** decreasing; **(c)** constant. 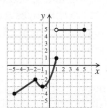 **a)** As x-values increase from -5 to -2, y-values increase from -4 to -2; y-values also increase as x-values increase from -1 to 1. Thus the function is increasing on the intervals $(-5, -2)$ and $(-1, 1)$. **b)** As x-values increase from -2 to -1, y-values decrease from -2 to -3, so the function is decreasing on the interval $(-2, -1)$. **c)** As x-values increase from 1 to 5, y remains 5, so the function is constant on the interval $(1, 5)$.

Relative Maxima and Minima

Suppose that f is a function for which $f(c)$ exists for some c in the domain of f. Then:

$f(c)$ is a **relative maximum** if there exists an *open* interval I containing c such that $f(c) > f(x)$ for all x in I, where $x \neq c$; and

$f(c)$ is a **relative minimum** if there exists an *open* interval I containing c such that $f(c) < f(x)$ for all x in I, where $x \neq c$.

Determine any relative maxima or minima of the function.

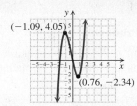

We see from the graph that the function has one relative maximum, 4.05. It occurs when $x = -1.09$. We also see that there is one relative minimum, -2.34. It occurs when $x = 0.76$.

Some applied problems can be modeled by functions.

See Examples 3 and 4 on pp. 97 and 98.

To graph a function that is defined **piecewise**, graph the function in parts as defined by its output formulas.

Graph the function defined as

$$f(x) = \begin{cases} 2x - 3, & \text{for } x < 1, \\ x + 1, & \text{for } x \geq 1. \end{cases}$$

We create the graph in two parts. First, we graph $f(x) = 2x - 3$ for inputs x less than 1. Then we graph $f(x) = x + 1$ for inputs x greater than or equal to 1.

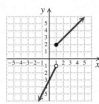

Greatest Integer Function

$f(x) = [\![x]\!] = $ the greatest integer less than or equal to x.

The graph of the greatest integer function is shown below. Each input is paired with the greatest integer less than or equal to that input.

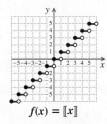

$$f(x) = [\![x]\!]$$

SECTION 2.2: THE ALGEBRA OF FUNCTIONS

Sums, Differences, Products, and Quotients of Functions

If f and g are functions and x is in the domain of each function, then:

$(f + g)(x) = f(x) + g(x),$
$(f - g)(x) = f(x) - g(x),$
$(fg)(x) = f(x) \cdot g(x),$
$(f/g)(x) = f(x)/g(x),$
provided $g(x) \neq 0.$

Given that $f(x) = x - 4$ and $g(x) = \sqrt{x + 5}$, find each of the following.

a) $(f + g)(x)$ b) $(f - g)(x)$
c) $(fg)(x)$ d) $(f/g)(x)$

a) $(f + g)(x) = f(x) + g(x) = x - 4 + \sqrt{x + 5}$
b) $(f - g)(x) = f(x) - g(x) = x - 4 - \sqrt{x + 5}$
c) $(fg)(x) = f(x) \cdot g(x) = (x - 4)\sqrt{x + 5}$
d) $(f/g)(x) = f(x)/g(x) = \dfrac{x - 4}{\sqrt{x + 5}}, x > -5$

Domains of $f + g$, $f - g$, fg, and f/g

If f and g are functions, then the domain of the functions $f + g$, $f - g$, and fg is the intersection of the domain of f and the domain of g. The domain of f/g is the intersection of the domain of f and the domain of g, with the exclusion of any x-values for which $g(x) = 0$.

For the functions f and g above, find the domains of $f + g$, $f - g$, fg, and f/g.

The domain of $f(x) = x - 4$ is the set of all real numbers. The domain of $g(x) = \sqrt{x + 5}$ is the set of all real numbers for which $x + 5 \geq 0$, or $x \geq -5$, or $[-5, \infty)$. Then the domain of $f + g$, $f - g$, and fg is the set of numbers in the intersection of these domains, or $[-5, \infty)$.

Since $g(-5) = 0$, we must exclude -5 from the domain of f/g. Thus the domain of f/g is $(-5, \infty)$.

The **difference quotient** for a function $f(x)$ is the ratio

$$\frac{f(x + h) - f(x)}{h}.$$

For the function $f(x) = x^2 - 4$, construct and simplify the difference quotient.

$$\frac{f(x + h) - f(x)}{h} = \frac{[(x + h)^2 - 4] - (x^2 - 4)}{h}$$

$$= \frac{x^2 + 2xh + h^2 - 4 - x^2 + 4}{h}$$

$$= \frac{2xh + h^2}{h} = \frac{h(2x + h)}{h}$$

$$= 2x + h$$

SECTION 2.3: THE COMPOSITION OF FUNCTIONS

The **composition of functions**, $f \circ g$, is defined as

$$(f \circ g)(x) = f(g(x)),$$

where x is in the domain of g and $g(x)$ is in the domain of f.

Given that $f(x) = 2x - 1$ and $g(x) = \sqrt{x}$, find each of the following.

a) $(f \circ g)(4)$ b) $(g \circ g)(625)$
c) $(f \circ g)(x)$ d) $(g \circ f)(x)$
e) The domain of $f \circ g$ and the domain of $g \circ f$

a) $(f \circ g)(4) = f(g(4)) = f(\sqrt{4}) = f(2) = 2 \cdot 2 - 1 = 4 - 1 = 3$
b) $(g \circ g)(625) = g(g(625)) = g(\sqrt{625}) = g(25) = \sqrt{25} = 5$
c) $(f \circ g)(x) = f(g(x)) = f(\sqrt{x}) = 2\sqrt{x} - 1$

(continued)

d) $(g \circ f)(x) = g(f(x)) = g(2x - 1) = \sqrt{2x - 1}$

e) The domain and the range of $f(x)$ are both $(-\infty, \infty)$, and the domain and the range of $g(x)$ are both $[0, \infty)$. Since f can accept any real number as an input, the domain of $f \circ g$ consists of all real numbers that are outputs of g, or $[0, \infty)$.

Any real number can be an input for f. Since $g(x)$ is not defined for negative radicands, we must have $2x - 1 \geq 0$, or $x \geq \frac{1}{2}$, so the domain of g and thus the domain of $g \circ f$ is $\left[\frac{1}{2}, \infty\right)$.

When we **decompose** a function, we write it as the composition of two functions.

If $h(x) = \sqrt{3x + 7}$, find $f(x)$ and $g(x)$ such that $h(x) = (f \circ g)(x)$.

This function finds the square root of $3x + 7$, so one decomposition is $f(x) = \sqrt{x}$ and $g(x) = 3x + 7$.

There are other correct answers, but this one is probably the most obvious.

SECTION 2.4: SYMMETRY

Algebraic Tests of Symmetry

x-axis: If replacing y with $-y$ produces an equivalent equation, then the graph is *symmetric with respect to the x-axis*.

y-axis: If replacing x with $-x$ produces an equivalent equation, then the graph is *symmetric with respect to the y-axis*.

Origin: If replacing x with $-x$ and y with $-y$ produces an equivalent equation, then the graph is *symmetric with respect to the origin*.

Test $y = 2x^3$ for symmetry with respect to the *x*-axis, the *y*-axis, and the origin.

x-axis: We replace y with $-y$:

$$-y = 2x^3$$
$$y = -2x^3. \quad \textbf{Multiplying by } -1$$

The resulting equation *is not* equivalent to the original equation, so the graph *is not* symmetric with respect to the *x*-axis.

y-axis: We replace x with $-x$:

$$y = 2(-x)^3 = -2x^3.$$

The resulting equation *is not* equivalent to the original equation, so the graph *is not* symmetric with respect to the *y*-axis.

Origin: We replace x with $-x$ and y with $-y$:

$$-y = 2(-x)^3$$
$$-y = -2x^3$$
$$y = 2x^3.$$

The resulting equation *is* equivalent to the original equation, so the graph *is* symmetric with respect to the origin.

Even Functions and Odd Functions

If the graph of a function is symmetric with respect to the *y*-axis, we say that it is an **even function**. That is, for each x in the domain of f, $f(x) = f(-x)$.

If the graph of a function is symmetric with respect to the origin, we say that it is an **odd function**. That is, for each x in the domain of f, $f(-x) = -f(x)$.

Determine whether each function is even, odd, or neither.

a) $g(x) = 2x^2 - 4$ **b)** $h(x) = x^5 - 3x^3 - x$

a) We first find $g(-x)$ and simplify:

$$g(-x) = 2(-x)^2 - 4 = 2x^2 - 4.$$

$g(x) = g(-x)$, so g *is* even. Since a function other than $f(x) = 0$ cannot be *both* even and odd and g is even, we need not test to see if it is an odd function.

(continued)

b) We first find $h(-x)$ and simplify:

$$h(-x) = (-x)^5 - 3(-x)^3 - (-x)$$
$$= -x^5 + 3x^3 + x.$$

$h(x) \neq h(-x)$, so *h is not* even.

Next, we find $-h(x)$ and simplify:

$$-h(x) = -(x^5 - 3x^3 - x)$$
$$= -x^5 + 3x^3 + x.$$

$h(-x) = -h(x)$, so *h is* odd.

SECTION 2.5: TRANSFORMATIONS

Vertical Translation

For $b > 0$:

the graph of $y = f(x) + b$ is the graph of $y = f(x)$ shifted *up b* units;

the graph of $y = f(x) - b$ is the graph of $y = f(x)$ shifted *down b* units.

Horizontal Translation

For $d > 0$:

the graph of $y = f(x - d)$ is the graph of $y = f(x)$ shifted *right d* units;

the graph of $y = f(x + d)$ is the graph of $y = f(x)$ shifted *left d* units.

Graph $g(x) = (x - 2)^2 + 1$. Before doing so, describe how the graph can be obtained from the graph of $f(x) = x^2$.

First, note that the graph of $h(x) = (x - 2)^2$ is the graph of $f(x) = x^2$ shifted right 2 units. Then the graph of $g(x) = (x - 2)^2 + 1$ is the graph of $h(x) = (x - 2)^2$ shifted up 1 unit. Thus the graph of g is obtained by shifting the graph of $f(x) = x^2$ right 2 units and up 1 unit.

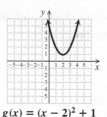

$g(x) = (x - 2)^2 + 1$

Reflections

The graph of $y = -f(x)$ is the **reflection** of $y = f(x)$ across the *x*-axis.

The graph of $y = f(-x)$ is the **reflection** of $y = f(x)$ across the *y*-axis.

If a point (x, y) is on the graph of $y = f(x)$, then $(x, -y)$ is on the graph of $y = -f(x)$, and $(-x, y)$ is on the graph of $y = f(-x)$.

Graph each of the following. Before doing so, describe how each graph can be obtained from the graph of $f(x) = x^2 - x$.

a) $g(x) = x - x^2$ **b)** $h(x) = (-x)^2 - (-x)$

a) Note that

$$-f(x) = -(x^2 - x)$$
$$= -x^2 + x$$
$$= x - x^2$$
$$= g(x).$$

Thus the graph is a reflection of the graph of $f(x) = x^2 - x$ across the *x*-axis.

$f(x) = x^2 - x$

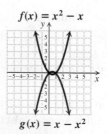

$g(x) = x - x^2$

(continued)

b) Note that

$$f(-x) = (-x)^2 - (-x) = h(x).$$

Thus the graph of $h(x) = (-x)^2 - (-x)$ is a reflection of the graph of $f(x) = x^2 - x$ across the y-axis.

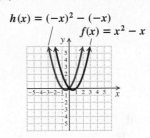

$h(x) = (-x)^2 - (-x)$

$f(x) = x^2 - x$

Vertical Stretching and Shrinking

The graph of $y = af(x)$ can be obtained from the graph of $y = f(x)$ by:

stretching vertically for $|a| > 1$, or

shrinking vertically for $0 < |a| < 1$.

For $a < 0$, the graph is also reflected across the x-axis.

(The y-coordinates of the graph of $y = af(x)$ can be obtained by multiplying the y-coordinates of $y = f(x)$ by a.)

Horizontal Stretching and Shrinking

The graph of $y = f(cx)$ can be obtained from the graph of $y = f(x)$ by:

shrinking horizontally for $|c| > 1$,

or

stretching horizontally for $0 < |c| < 1$.

For $c < 0$, the graph is also reflected across the y-axis.

(The x-coordinates of the graph of $y = f(cx)$ can be obtained by dividing the x-coordinates of $y = f(x)$ by c.)

A graph of $y = g(x)$ is shown below. Use this graph to graph each of the given equations.

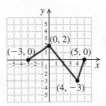

a) $f(x) = g(2x)$ **b)** $f(x) = -2g(x)$

c) $f(x) = \frac{1}{2}g(x)$ **d)** $f(x) = g\left(\frac{1}{2}x\right)$

a) Since $|2| > 1$, the graph of $f(x) = g(2x)$ is a horizontal shrinking of the graph of $y = g(x)$. The transformation divides each x-coordinate of g by 2.

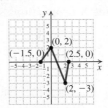

b) Since $|-2| > 1$, the graph of $f(x) = -2g(x)$ is a vertical stretching of the graph of $y = g(x)$. Since $-2 < 0$, the graph is also reflected across the x-axis. The transformation multiplies each y-coordinate of g by 2.

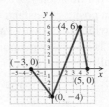

(continued)

c) Since $\left|\frac{1}{2}\right| < 1$, the graph of $f(x) = \frac{1}{2}g(x)$ is a vertical shrinking of the graph of $y = g(x)$. The transformation multiplies each y-coordinate of g by $\frac{1}{2}$.

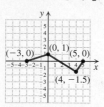

d) Since $\left|\frac{1}{2}\right| < 1$, the graph of $f(x) = g\left(\frac{1}{2}x\right)$ is a horizontal stretching of the graph of $y = g(x)$. The transformation divides each x-coordinate of g by $\frac{1}{2}$ (which is the same as multiplying by 2).

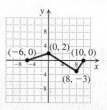

SECTION 2.6: VARIATION AND APPLICATIONS

Direct Variation

If a situation gives rise to a linear function $f(x) = kx$, or $y = kx$, where k is a positive constant, we say that we have **direct variation**, or that y **varies directly as** x, or that y **is directly proportional to** x. The number k is called the **variation constant**, or the **constant of proportionality**.

Find an equation of variation in which y varies directly as x, and $y = 24$ when $x = 8$. Then find the value of y when $x = 5$.

First, we have

$$y = kx \qquad \text{\textbf{\textit{y} varies directly as \textit{x}.}}$$
$$24 = k \cdot 8 \qquad \text{\textbf{Substituting}}$$
$$3 = k \qquad \text{\textbf{Variation constant}}$$

The equation of variation is $y = 3x$. Now we use the equation to find the value of y when $x = 5$:

$$y = 3x$$
$$= 3 \cdot 5 \qquad \text{\textbf{Substituting}}$$
$$= 15.$$

When $x = 5$, the value of y is 15.

Inverse Variation

If a situation gives rise to a function $f(x) = k/x$, or $y = k/x$, where k is a positive constant, we say that we have **inverse variation**, or that y **varies inversely as** x, or that y **is inversely proportional to** x. The number k is called the **variation constant**, or the **constant of proportionality**.

Find an equation of variation in which y varies inversely as x, and $y = 5$ when $x = 0.1$. Then find the value of y when $x = 10$.

First, we have

$$y = \frac{k}{x} \qquad \text{\textbf{\textit{y} varies inversely as \textit{x}.}}$$
$$5 = \frac{k}{0.1} \qquad \text{\textbf{Substituting}}$$
$$0.5 = k. \qquad \text{\textbf{Variation constant}}$$

The equation of variation is $y = \dfrac{0.5}{x}$.

(continued)

Now we use the equation to find the value of y when $x = 10$:

$$y = \frac{0.5}{x}$$

$$= \frac{0.5}{10} \qquad \text{Substituting}$$

$$= 0.05.$$

When $x = 10$, the value of y is 0.05.

Combined Variation

---

y varies **directly as the nth power of x** if there is some positive constant k such that

$$y = kx^n.$$

y varies **inversely as the nth power of x** if there is some positive constant k such that

$$y = \frac{k}{x^n}.$$

y varies **jointly as x and z** if there is some positive constant k such that

$$y = kxz.$$

Find an equation of variation in which y varies jointly as w and the square of x and inversely as z, and $y = 8$ when $w = 3$, $x = 2$, and $z = 6$.

First, we have

$$y = k \cdot \frac{wx^2}{z}$$

$$8 = k \cdot \frac{3 \cdot 2^2}{6} \qquad \text{Substituting}$$

$$8 = k \cdot \frac{3 \cdot 4}{6}$$

$$8 = 2k$$

$$4 = k. \qquad \text{Variation constant}$$

The equation of variation is $y = 4\,\dfrac{wx^2}{z}$, or $y = \dfrac{4wx^2}{z}$.

REVIEW EXERCISES

Determine whether the statement is true or false.

1. The greatest integer function pairs each input with the greatest integer less than or equal to that input. **[2.1]**

2. In general, for functions f and g, the domain of $f \circ g =$ the domain of $g \circ f$. **[2.3]**

3. The graph of $y = (x - 2)^2$ is the graph of $y = x^2$ shifted right 2 units. **[2.5]**

4. The graph of $y = -x^2$ is the reflection of the graph of $y = x^2$ across the x-axis. **[2.5]**

Determine the intervals on which the function is **(a)** *increasing;* **(b)** *decreasing; and* **(c)** *constant.* **[2.1]**

5.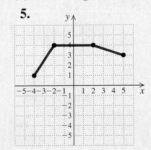

6.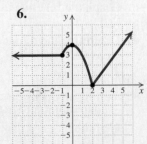

Graph the function. Estimate the intervals on which the function is increasing or decreasing, and estimate any relative maxima or minima. **[2.1]**

7. $f(x) = x^2 - 1$ **8.** $f(x) = 2 - |x|$

Use a graphing calculator to find the intervals on which the function is increasing or decreasing, and find any relative maxima or minima. **[2.1]**

9. $f(x) = x^2 - 4x + 3$

10. $f(x) = -x^2 + x + 6$

11. $f(x) = x^3 - 4x$

12. $f(x) = 2x - 0.5x^3$

13. *Fenced Patio.*   Syd has 48 ft of rolled bamboo fence with which to enclose a rectangular patio. The house forms one side of the patio. Suppose that two sides of the patio each measures x feet. Express the area of the patio as a function of x. **[2.1]**

14. *Inscribed Rectangle.*   A rectangle is inscribed in a semicircle of radius 2, as shown. The variable x = half the length of the rectangle. Express the area of the rectangle as a function of x. **[2.1]**

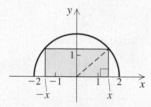

15. *Dog Pen.*   Mamie has 66 ft of fencing with which to enclose a rectangular dog pen. The side of her garage forms one side of the pen. Suppose that the side of the pen parallel to the garage is x feet long.

  a) Express the area of the dog pen as a function of x. **[2.1]**

  b) Find the domain of the function. **[2.1]**

  c) Graph the function using a graphing calculator. **[2.1]**

  d) Determine the dimensions that yield the maximum area. **[2.1]**

16. *Minimizing Surface Area.*   A container firm is designing an open-top rectangular box, with a square base, that will hold 108 in³. Let x = the length of a side of the base.

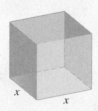

a) Express the surface area as a function of x. **[2.1]**

b) Find the domain of the function. **[2.1]**

c) Using the following graph, determine the dimensions that will minimize the surface area of the box. **[2.1]**

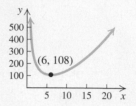

Graph each of the following. **[2.1]**

17. $f(x) = \begin{cases} -x, & \text{for } x \le -4, \\ \frac{1}{2}x + 1, & \text{for } x > -4 \end{cases}$

18. $f(x) = \begin{cases} x^3, & \text{for } x < -2, \\ |x|, & \text{for } -2 \le x \le 2, \\ \sqrt{x - 1}, & \text{for } x > 2 \end{cases}$

19. $f(x) = \begin{cases} \dfrac{x^2 - 1}{x + 1}, & \text{for } x \ne -1, \\ 3, & \text{for } x = -1 \end{cases}$

20. $f(x) = [\![x]\!]$

21. $f(x) = [\![x - 3]\!]$

22. For the function in Exercise 18, find $f(-1), f(5), f(-2)$, and $f(-3)$. **[2.1]**

23. For the function in Exercise 19, find $f(-2), f(-1), f(0)$, and $f(4)$. **[2.1]**

Given that $f(x) = \sqrt{x - 2}$ and $g(x) = x^2 - 1$, find each of the following, if it exists. **[2.2]**

24. $(f - g)(6)$

25. $(fg)(2)$

26. $(f + g)(-1)$

For each pair of functions in Exercises 27 and 28:

a) *Find the domain of $f, g, f + g, f - g, fg$, and f/g.* **[2.2]**

b) *Find $(f + g)(x), (f - g)(x), (fg)(x)$, and $(f/g)(x)$.* **[2.2]**

27. $f(x) = \dfrac{4}{x^2}, \ g(x) = 3 - 2x$

28. $f(x) = 3x^2 + 4x, \ g(x) = 2x - 1$

29. Given the total-revenue and total-cost functions $R(x) = 120x - 0.5x^2$ and $C(x) = 15x + 6$, find the total-profit function $P(x)$. **[2.2]**

For each function f, construct and simplify the difference quotient. **[2.2]**

30. $f(x) = 2x + 7$

31. $f(x) = 3 - x^2$

32. $f(x) = \dfrac{4}{x}$

Given that $f(x) = 2x - 1$, $g(x) = x^2 + 4$, and $h(x) = 3 - x^3$, find each of the following. **[2.3]**

33. $(f \circ g)(1)$

34. $(g \circ f)(1)$

35. $(h \circ f)(-2)$

36. $(g \circ h)(3)$

37. $(f \circ h)(-1)$

38. $(h \circ g)(2)$

39. $(f \circ f)(x)$

40. $(h \circ h)(x)$

For each pair of functions in Exercises 41 and 42:

a) *Find $(f \circ g)(x)$ and $(g \circ f)(x)$.* **[2.3]**

b) *Find the domain of $f \circ g$ and the domain of $g \circ f$.* **[2.3]**

41. $f(x) = \dfrac{4}{x^2}$, $g(x) = 3 - 2x$

42. $f(x) = 3x^2 + 4x$, $g(x) = 2x - 1$

Find $f(x)$ and $g(x)$ such that $h(x) = (f \circ g)(x)$. **[2.3]**

43. $h(x) = \sqrt{5x + 2}$

44. $h(x) = 4(5x - 1)^2 + 9$

Graph the given equation and determine visually whether it is symmetric with respect to the x-axis, the y-axis, and the origin. Then verify your assertion algebraically. **[2.4]**

45. $x^2 + y^2 = 4$

46. $y^2 = x^2 + 3$

47. $x + y = 3$

48. $y = x^2$

49. $y = x^3$

50. $y = x^4 - x^2$

Determine visually whether the function is even, odd, or neither even nor odd. **[2.4]**

51.

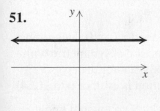

52.

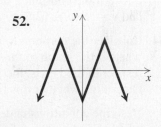

53.

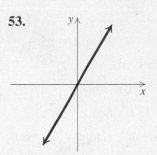

54.

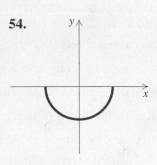

In Exercises 55–60, test whether the function is even, odd, or neither even nor odd. **[2.4]**

55. $f(x) = 9 - x^2$

56. $f(x) = x^3 - 2x + 4$

57. $f(x) = x^7 - x^5$

58. $f(x) = |x|$

59. $f(x) = \sqrt{16 - x^2}$

60. $f(x) = \dfrac{10x}{x^2 + 1}$

Write an equation for a function that has a graph with the given characteristics. **[2.5]**

61. The shape of $y = x^2$, but shifted left 3 units

62. The shape of $y = \sqrt{x}$, but reflected across the x-axis and shifted right 3 units and up 4 units

63. The shape of $y = |x|$, but stretched vertically by a factor of 2 and shifted right 3 units

A graph of $y = f(x)$ is shown below. No formula for f is given. Graph each of the following. **[2.5]**

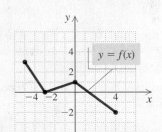

64. $y = f(x - 1)$

65. $y = f(2x)$

66. $y = -2f(x)$

67. $y = 3 + f(x)$

Find an equation of variation for the given situation. **[2.6]**

68. y varies directly as x, and $y = 100$ when $x = 25$.

69. y varies directly as x, and $y = 6$ when $x = 9$.

70. y varies inversely as x, and $y = 100$ when $x = 25$.

71. y varies inversely as x, and $y = 6$ when $x = 9$.

72. y varies inversely as the square of x, and $y = 12$ when $x = 2$.

73. y varies jointly as x and the square of z and inversely as w, and $y = 2$ when $x = 16$, $w = 0.2$, and $z = \frac{1}{2}$.

74. *Pumping Time.*   The time t required to empty a tank varies inversely as the rate r of pumping. If a pump can empty a tank in 35 min at the rate of 800 kL/min, how long will it take the pump to empty the same tank at the rate of 1400 kL/min? **[2.6]**

75. *Test Score.*   The score N on a test varies directly as the number of correct responses a. Sam answers 29 questions correctly and earns a score of 87. What would Sam's score have been if he had answered 25 questions correctly? **[2.6]**

76. *Power of Electric Current.*   The power P expended by heat in an electric circuit of fixed resistance varies directly as the square of the current C in the circuit. A circuit expends 180 watts when a current of 6 amperes is flowing. What is the amount of heat expended when the current is 10 amperes? **[2.6]**

77. For $f(x) = x + 1$ and $g(x) = \sqrt{x}$, the domain of $(g \circ f)(x)$ is which of the following? **[2.3]**

A. $[-1, \infty)$ **B.** $[-1, 0)$

C. $[0, \infty)$ **D.** $(-\infty, \infty)$

78. For $b > 0$, the graph of $y = f(x) + b$ is the graph of $y = f(x)$ shifted in which of the following ways? **[2.5]**

A. Right b units   **B.** Left b units

C. Up b units   **D.** Down b units

79. The graph of the function f is shown below.

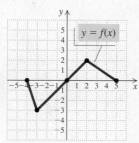

The graph of $g(x) = -\frac{1}{2} f(x) + 1$ is which of the following? **[2.5]**

A.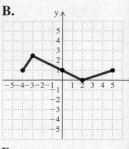

B.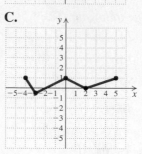

C.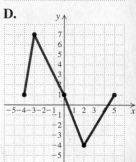

D.

› Synthesis

80. Prove that the sum of two odd functions is odd. **[2.2], [2.4]**

81. Describe how the graph of $y = -f(-x)$ is obtained from the graph of $y = f(x)$. **[2.5]**

› Collaborative Discussion and Writing

82. Given that $f(x) = 4x^3 - 2x + 7$, find each of the following. Then discuss how each expression differs from the other. **[1.2], [2.5]**

a) $f(x) + 2$
b) $f(x + 2)$
c) $f(x) + f(2)$

83. Given the graph of $y = f(x)$, explain and contrast the effect of the constant c on the graphs of $y = f(cx)$ and $y = cf(x)$. **[2.5]**

84. Consider the constant function $f(x) = 0$. Determine whether the graph of this function is symmetric with respect to the x-axis, the y-axis, and/or the origin. Determine whether this function is even or odd. **[2.4]**

85. Describe conditions under which you would know whether a polynomial function

$$f(x) = a_n x^n + a_{n-1} x^{n-1} + \cdots + a_2 x^2 + a_1 x + a_0$$

is even or odd without using an algebraic procedure. Explain. **[2.4]**

86. If y varies directly as x^2, explain why doubling x would not cause y to be doubled as well. **[2.6]**

87. If y varies directly as x and x varies inversely as z, how does y vary with regard to z? Why? **[2.6]**

2 Chapter Test

1. Determine the intervals on which the function is (a) increasing; (b) decreasing; and (c) constant.

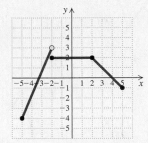

2. Graph the function $f(x) = 2 - x^2$. Estimate the intervals on which the function is increasing or decreasing, and estimate any relative maxima or minima.

3. Use a graphing calculator to find the intervals on which the function $f(x) = x^3 + 4x^2$ is increasing or decreasing, and find any relative maxima or minima.

4. *Triangular Pennant.* A softball team is designing a triangular pennant such that the height is 6 in. less than four times the length of the base b. Express the area of the pennant as a function of b.

5. Graph:
$$f(x) = \begin{cases} x^2, & \text{for } x < -1, \\ |x|, & \text{for } -1 \le x \le 1, \\ \sqrt{x - 1}, & \text{for } x > 1. \end{cases}$$

6. For the function in Exercise 5, find $f\left(-\frac{7}{8}\right), f(5)$, and $f(-4)$.

Given that $f(x) = x^2 - 4x + 3$ and $g(x) = \sqrt{3 - x}$, find each of the following, if it exists.

7. $(f + g)(-6)$

8. $(f - g)(-1)$

9. $(fg)(2)$

10. $(f/g)(1)$

For $f(x) = x^2$ and $g(x) = \sqrt{x - 3}$, find each of the following.

11. The domain of f

12. The domain of g

13. The domain of $f + g$

14. The domain of $f - g$

15. The domain of fg

16. The domain of f/g

17. $(f + g)(x)$

18. $(f - g)(x)$

19. $(fg)(x)$

20. $(f/g)(x)$

For each function, construct and simplify the difference quotient.

21. $f(x) = \frac{1}{2}x + 4$

22. $f(x) = 2x^2 - x + 3$

Given that $f(x) = x^2 - 1, g(x) = 4x + 3$, and $h(x) = 3x^2 + 2x + 4$, find each of the following.

23. $(g \circ h)(2)$

24. $(f \circ g)(-1)$

25. $(h \circ f)(1)$

26. $(g \circ g)(x)$

For $f(x) = \sqrt{x - 5}$ and $g(x) = x^2 + 1$:

27. Find $(f \circ g)(x)$ and $(g \circ f)(x)$.

28. Find the domain of $(f \circ g)(x)$ and the domain of $(g \circ f)(x)$.

29. Find $f(x)$ and $g(x)$ such that
$$h(x) = (f \circ g)(x) = (2x - 7)^4.$$

30. Determine whether the graph of $y = x^4 - 2x^2$ is symmetric with respect to the x-axis, the y-axis, and the origin.

31. Determine whether the function
$$f(x) = \frac{2x}{x^2 + 1}$$
is even, odd, or neither even nor odd. Show your work.

32. Write an equation for a function that has the shape of $y = x^2$, but shifted right 2 units and down 1 unit.

33. Write an equation for a function that has the shape of $y = x^2$, but reflected across the x-axis and shifted left 2 units and up 3 units.

34. The graph of a function $y = f(x)$ is shown below. No formula for f is given. Graph $y = -\frac{1}{2}f(x)$.

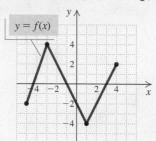

35. Find an equation of variation in which y varies inversely as x, and $y = 5$ when $x = 6$.

36. Find an equation of variation in which y varies directly as x, and $y = 60$ when $x = 12$.

37. Find an equation of variation in which y varies jointly as x and the square of z and inversely as w, and $y = 100$ when $x = 0.1$, $z = 10$, and $w = 5$.

38. The stopping distance d of a car after the brakes have been applied varies directly as the square of the speed r. If a car traveling 60 mph can stop in 200 ft, how long will it take a car traveling 30 mph to stop?

39. The graph of the function f is shown below.

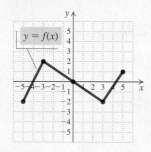

The graph of $g(x) = 2f(x) - 1$ is which of the following?

A.
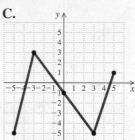

B.
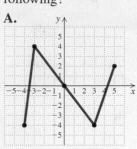

C.
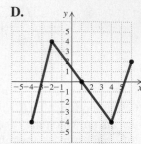

D.
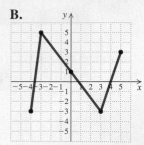

❭ Synthesis

40. If $(-3, 1)$ is a point on the graph of $y = f(x)$, what point do you know is on the graph of $y = f(3x)$?

Quadratic Functions and Equations; Inequalities

APPLICATION

This problem appears as Example 10 in Section 3.2.

The numbers of both magazine launches and magazine closures have increased in recent years. The function $m(x) = 34x^2 - 59x + 81$ can be used to estimate the number of magazine closures after 2012 (*Source*: MediaFinder.com, online database of U.S. and Canadian print and digital publications). In what year was the number of magazine closures about 99?

3.1   The Complex Numbers

❯ Perform computations involving complex numbers.

Some functions have zeros that are not real numbers. In order to find the zeros of such functions, we must consider the **complex-number system**.

❯ The Complex-Number System

We know that the square root of a negative number is not a real number. For example, $\sqrt{-1}$ is not a real number because there is no real number x such that $x^2 = -1$. This means that certain equations, like $x^2 = -1$ or $x^2 + 1 = 0$, do not have real-number solutions, and certain functions, like $f(x) = x^2 + 1$, do not have real-number zeros. Consider the graph of $f(x) = x^2 + 1$.

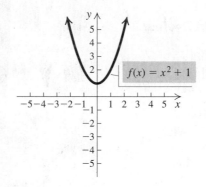

We see that the graph does not cross the x-axis and thus has no x-intercepts. This illustrates that the function $f(x) = x^2 + 1$ has no real-number zeros. Thus there are no real-number solutions of the corresponding equation $x^2 + 1 = 0$.

We can define a nonreal number that is a solution of the equation $x^2 + 1 = 0$.

THE NUMBER i

The number i is defined such that

$$i = \sqrt{-1} \quad \text{and} \quad i^2 = -1.$$

To express roots of negative numbers in terms of i, we can use the fact that

$$\sqrt{-p} = \sqrt{-1 \cdot p} = \sqrt{-1} \cdot \sqrt{p} = i\sqrt{p}$$

when p is a positive real number.

EXAMPLE 1   Express each number in terms of i.

a) $\sqrt{-7}$ b) $\sqrt{-16}$ c) $-\sqrt{-13}$

d) $-\sqrt{-64}$ e) $\sqrt{-48}$

Solution

a) $\sqrt{-7} = \sqrt{-1 \cdot 7} = \sqrt{-1} \cdot \sqrt{7}$
$= i\sqrt{7}, \text{ or } \sqrt{7}i$ ←

> *i is not under the radical.*

b) $\sqrt{-16} = \sqrt{-1 \cdot 16} = \sqrt{-1} \cdot \sqrt{16}$
$= i \cdot 4 = 4i$

c) $-\sqrt{-13} = -\sqrt{-1 \cdot 13} = -\sqrt{-1} \cdot \sqrt{13}$
$= -i\sqrt{13}, \text{ or } -\sqrt{13}i$ ←

d) $-\sqrt{-64} = -\sqrt{-1 \cdot 64} = -\sqrt{-1} \cdot \sqrt{64}$
$= -i \cdot 8 = -8i$

e) $\sqrt{-48} = \sqrt{-1 \cdot 48} = \sqrt{-1} \cdot \sqrt{48}$
$= i\sqrt{16 \cdot 3}$
$= i \cdot 4\sqrt{3}$
$= 4i\sqrt{3}, \text{ or } 4\sqrt{3}i$ ←

> **Now Try Exercises 1, 7, and 9.**

The complex numbers are formed by adding real numbers and multiples of *i*.

COMPLEX NUMBERS

A **complex number** is a number of the form $a + bi$, where a and b are real numbers. The number a is said to be the **real part** of $a + bi$, and the number b is said to be the **imaginary part** of $a + bi$.*

Note that either a or b or both can be 0. When $b = 0$, $a + bi = a + 0i = a$, so every real number is a complex number. A complex number like $3 + 4i$ or $17i$, in which $b \neq 0$, is called an **imaginary number**. A complex number like $17i$ or $-4i$, in which $a = 0$ and $b \neq 0$, is sometimes called a **pure imaginary number**. The relationships among various types of complex numbers are shown in the figure below.

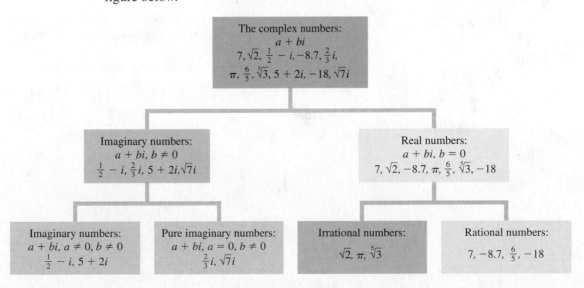

---

*Sometimes bi is considered to be the imaginary part.

❯ Addition and Subtraction

The complex numbers obey the commutative, associative, and distributive laws. Thus we can add and subtract them as we do binomials. We collect the real parts and the imaginary parts of complex numbers just as we collect like terms in binomials.

EXAMPLE 2   Add or subtract and simplify each of the following.

a) $(8 + 6i) + (3 + 2i)$ **b)** $(4 + 5i) - (6 - 3i)$

Solution

a) $(8 + 6i) + (3 + 2i) = (8 + 3) + (6i + 2i)$

                 Collecting the real parts and the imaginary parts

               $= 11 + (6 + 2)i = 11 + 8i$

b) $(4 + 5i) - (6 - 3i) = (4 - 6) + [5i - (-3i)]$

                 Note that 6 and $-3i$ are both being subtracted.

              $= -2 + 8i$

                         ❯ *Now Try Exercises 11 and 21.*

When set in $a + bi$ mode, most graphing calculators can perform operations on complex numbers. The operations in Example 2 are shown in the window at left. Some calculators will express a complex number in the form (a, b) rather than $a + bi$.

```
(8+6i)+(3+2i)
                    11+8i
(4+5i)−(6−3i)
                    −2+8i
```

❯ Multiplication

When $\sqrt{a}$ and $\sqrt{b}$ are real numbers, $\sqrt{a} \cdot \sqrt{b} = \sqrt{ab}$, but this is not true when $\sqrt{a}$ and $\sqrt{b}$ are not real numbers. Thus,

$$\sqrt{-2} \cdot \sqrt{-5} = \sqrt{-1} \cdot \sqrt{2} \cdot \sqrt{-1} \cdot \sqrt{5}$$
$$= i\sqrt{2} \cdot i\sqrt{5}$$
$$= i^2\sqrt{10} = -1\sqrt{10} = -\sqrt{10} \quad \text{is correct!}$$

But

$$\sqrt{-2} \cdot \sqrt{-5} = \sqrt{(-2)(-5)} = \sqrt{10} \quad \text{is wrong!}$$

Keeping this and the fact that $i^2 = -1$ in mind, we multiply with imaginary numbers in much the same way that we do with real numbers.

Just in Time
12, 13

EXAMPLE 3   Multiply and simplify each of the following.

a) $\sqrt{-16} \cdot \sqrt{-25}$ **b)** $(1 + 2i)(1 + 3i)$ **c)** $(3 - 7i)^2$

Solution

a) $\sqrt{-16} \cdot \sqrt{-25} = \sqrt{-1} \cdot \sqrt{16} \cdot \sqrt{-1} \cdot \sqrt{25}$
$$= i \cdot 4 \cdot i \cdot 5$$
$$= i^2 \cdot 20$$
$$= -1 \cdot 20 \quad i^2 = -1$$
$$= -20$$

b) $(1 + 2i)(1 + 3i) = 1 + 3i + 2i + 6i^2$ **Multiplying each term of one number by every term of the other (FOIL)**

                       $= 1 + 3i + 2i - 6$ $i^2 = -1$

                       $= -5 + 5i$ **Collecting like terms**

c) $(3 - 7i)^2 = 3^2 - 2 \cdot 3 \cdot 7i + (7i)^2$ **Recall that** $(A - B)^2 = A^2 - 2AB + B^2.$

$\qquad\qquad = 9 - 42i + 49i^2$

$\qquad\qquad = 9 - 42i - 49$ $i^2 = -1$

$\qquad\qquad = -40 - 42i$

> *Now Try Exercises 31, 39, and 55.*

We can multiply complex numbers on a graphing calculator set in $a + bi$ mode. The products found in Example 3 are shown below.

$$\boxed{\begin{array}{l} \sqrt{-16}\,\sqrt{-25} \\ \qquad\qquad\qquad -20 \\ (1+2i)(1+3i) \\ \qquad\qquad\qquad -5+5i \\ (3-7i)^2 \\ \qquad\qquad\qquad -40-42i \end{array}}$$

Just in Time 7

Recall that -1 raised to an *even* power is 1, and -1 raised to an *odd* power is -1. Simplifying powers of i can then be done by using the fact that $i^2 = -1$ and expressing the given power of i in terms of i^2. Consider the following:

$i = \sqrt{-1},$

$i^2 = -1,$

$i^3 = i^2 \cdot i = (-1)i = -i,$

$i^4 = (i^2)^2 = (-1)^2 = 1,$

$i^5 = i^4 \cdot i = (i^2)^2 \cdot i = (-1)^2 \cdot i = 1 \cdot i = i,$

$i^6 = (i^2)^3 = (-1)^3 = -1,$

$i^7 = i^6 \cdot i = (i^2)^3 \cdot i = (-1)^3 \cdot i = -1 \cdot i = -i,$

$i^8 = (i^2)^4 = (-1)^4 = 1.$

Note that the powers of i cycle through the values i, -1, $-i$, and 1.

EXAMPLE 4   Simplify each of the following.

a) i^{37} b) i^{58}

c) i^{75} d) i^{80}

Solution

a) $i^{37} = i^{36} \cdot i = (i^2)^{18} \cdot i = (-1)^{18} \cdot i = 1 \cdot i = i$

b) $i^{58} = (i^2)^{29} = (-1)^{29} = -1$

c) $i^{75} = i^{74} \cdot i = (i^2)^{37} \cdot i = (-1)^{37} \cdot i = -1 \cdot i = -i$

d) $i^{80} = (i^2)^{40} = (-1)^{40} = 1$

> *Now Try Exercises 79 and 83.*

These powers of i can also be simplified in terms of i^4 rather than i^2. Consider i^{37} in Example 4(a), for instance. When we divide 37 by 4, we get 9 with a remainder of 1. Then $37 = 4 \cdot 9 + 1$, so

$$i^{37} = (i^4)^9 \cdot i = 1^9 \cdot i = 1 \cdot i = i.$$

The other examples shown above can be done in a similar manner.

❯ Conjugates and Division

Conjugates of complex numbers are defined as follows.

> **CONJUGATE OF A COMPLEX NUMBER**
>
> The **conjugate** of a complex number $a + bi$ is $a - bi$. The numbers $a + bi$ and $a - bi$ are **complex conjugates**.

Each of the following pairs of numbers are complex conjugates:

$$-3 + 7i \text{ and } -3 - 7i; \quad 14 - 5i \text{ and } 14 + 5i; \quad \text{and} \quad 8i \text{ and } -8i.$$

The product of a complex number and its conjugate is a real number.

EXAMPLE 5 Multiply each of the following.

a) $(5 + 7i)(5 - 7i)$ **b)** $(8i)(-8i)$

Solution

a) $(5 + 7i)(5 - 7i) = 5^2 - (7i)^2$ Using $(A + B)(A - B) = A^2 - B^2$

$$= 25 - 49i^2$$
$$= 25 - 49(-1)$$
$$= 25 + 49$$
$$= 74$$

b) $(8i)(-8i) = -64i^2$
$$= -64(-1)$$
$$= 64$$

> **Now Try Exercise 49.**

```
(5+7i)(5−7i)
                74
(8i)(−8i)
                64
```

Conjugates are used when we divide complex numbers.

EXAMPLE 6 Divide $2 - 5i$ by $1 - 6i$.

Solution We write fraction notation and then multiply by 1, using the conjugate of the denominator to form the symbol for 1.

$$\frac{2 - 5i}{1 - 6i} = \frac{2 - 5i}{1 - 6i} \cdot \frac{1 + 6i}{1 + 6i} \quad \begin{array}{l}\text{Note that } 1 + 6i \text{ is the conjugate} \\ \text{of the divisor, } 1 - 6i.\end{array}$$

$$= \frac{(2 - 5i)(1 + 6i)}{(1 - 6i)(1 + 6i)}$$

$$= \frac{2 + 12i - 5i - 30i^2}{1 - 36i^2}$$

$$= \frac{2 + 7i + 30}{1 + 36} \quad i^2 = -1$$

$$= \frac{32 + 7i}{37}$$

$$= \frac{32}{37} + \frac{7}{37}i. \qquad \textbf{Writing the quotient in the form } a + bi$$

> **Now Try Exercise 69.**

```
(2−5i)/(1−6i)▶Frac
           32     7
           ── + ── i
           37    37
```

With a graphing calculator set in $a + bi$ mode, we can divide complex numbers and express the real parts and the imaginary parts in fraction form, just as we did in Example 6.

3.1   Exercise Set

Express the number in terms of i.

1. $\sqrt{-3}$

2. $\sqrt{-21}$

3. $\sqrt{-25}$

4. $\sqrt{-100}$

5. $-\sqrt{-33}$

6. $-\sqrt{-59}$

7. $-\sqrt{-81}$

8. $-\sqrt{-9}$

9. $\sqrt{-98}$

10. $\sqrt{-28}$

Simplify. Write answers in the form a + bi, where a and b are real numbers.

11. $(-5 + 3i) + (7 + 8i)$

12. $(-6 - 5i) + (9 + 2i)$

13. $(4 - 9i) + (1 - 3i)$

14. $(7 - 2i) + (4 - 5i)$

15. $(12 + 3i) + (-8 + 5i)$

16. $(-11 + 4i) + (6 + 8i)$

17. $(-1 - i) + (-3 - i)$

18. $(-5 - i) + (6 + 2i)$

19. $\left(3 + \sqrt{-16}\right) + \left(2 + \sqrt{-25}\right)$

20. $\left(7 - \sqrt{-36}\right) + \left(2 + \sqrt{-9}\right)$

21. $(10 + 7i) - (5 + 3i)$

22. $(-3 - 4i) - (8 - i)$

23. $(13 + 9i) - (8 + 2i)$

24. $(-7 + 12i) - (3 - 6i)$

25. $(6 - 4i) - (-5 + i)$

26. $(8 - 3i) - (9 - i)$

27. $(-5 + 2i) - (-4 - 3i)$

28. $(-6 + 7i) - (-5 - 2i)$

29. $(4 - 9i) - (2 + 3i)$

30. $(10 - 4i) - (8 + 2i)$

31. $\sqrt{-4} \cdot \sqrt{-36}$

32. $\sqrt{-49} \cdot \sqrt{-9}$

33. $\sqrt{-81} \cdot \sqrt{-25}$

34. $\sqrt{-16} \cdot \sqrt{-100}$

35. $7i(2 - 5i)$

36. $3i(6 + 4i)$

37. $-2i(-8 + 3i)$

38. $-6i(-5 + i)$

39. $(1 + 3i)(1 - 4i)$

40. $(1 - 2i)(1 + 3i)$

41. $(2 + 3i)(2 + 5i)$

42. $(3 - 5i)(8 - 2i)$

43. $(-4 + i)(3 - 2i)$

44. $(5 - 2i)(-1 + i)$

45. $(8 - 3i)(-2 - 5i)$

46. $(7 - 4i)(-3 - 3i)$

47. $\left(3 + \sqrt{-16}\right)\left(2 + \sqrt{-25}\right)$

48. $\left(7 - \sqrt{-16}\right)\left(2 + \sqrt{-9}\right)$

49. $(5 - 4i)(5 + 4i)$

50. $(5 + 9i)(5 - 9i)$

51. $(3 + 2i)(3 - 2i)$

52. $(8 + i)(8 - i)$

53. $(7 - 5i)(7 + 5i)$

54. $(6 - 8i)(6 + 8i)$

55. $(4 + 2i)^2$

56. $(5 - 4i)^2$

57. $(-2 + 7i)^2$

58. $(-3 + 2i)^2$

59. $(1 - 3i)^2$

60. $(2 - 5i)^2$

61. $(-1 - i)^2$

62. $(-4 - 2i)^2$

63. $(3 + 4i)^2$

64. $(6 + 5i)^2$

65. $\dfrac{3}{5 - 11i}$

66. $\dfrac{i}{2 + i}$

67. $\dfrac{5}{2 + 3i}$

68. $\dfrac{-3}{4 - 5i}$

69. $\dfrac{4 + i}{-3 - 2i}$

70. $\dfrac{5 - i}{-7 + 2i}$

71. $\dfrac{5 - 3i}{4 + 3i}$

72. $\dfrac{6 + 5i}{3 - 4i}$

73. $\dfrac{2 + \sqrt{3}i}{5 - 4i}$

74. $\dfrac{\sqrt{5} + 3i}{1 - i}$

75. $\dfrac{1 + i}{(1 - i)^2}$

76. $\dfrac{1 - i}{(1 + i)^2}$

77. $\dfrac{4 - 2i}{1 + i} + \dfrac{2 - 5i}{1 + i}$

78. $\dfrac{3 + 2i}{1 - i} + \dfrac{6 + 2i}{1 - i}$

Simplify.

79. i^{11}

80. i^7

81. i^{35}

82. i^{24}

83. i^{64}

84. i^{42}

85. $(-i)^{71}$

86. $(-i)^6$

87. $(5i)^4$

88. $(2i)^5$

❯ Skill Maintenance

89. Write a slope–intercept equation for the line containing the point $(3, -5)$ and perpendicular to the line $3x - 6y = 7$. **[1.4]**

Given that $f(x) = x^2 + 4$ *and* $g(x) = 3x + 5$, *find each of the following.* **[2.2]**

90. The domain of $f - g$

91. The domain of f/g

92. $(f - g)(x)$

93. $(f/g)(2)$

94. For the function $f(x) = x^2 - 3x + 4$, construct and simplify the difference quotient

$$\frac{f(x + h) - f(x)}{h}.\ \text{[2.2]}$$

96. The conjugate of a sum is the sum of the conjugates of the individual complex numbers.

97. The conjugate of a product is the product of the conjugates of the individual complex numbers.

Let $z = a + bi$ *and* $\bar{z} = a - bi$.

98. Find a general expression for $1/z$.

99. Find a general expression for $z\bar{z}$.

100. Solve $z + 6\bar{z} = 7$ for z.

101. Multiply and simplify:

$$[x - (3 + 4i)][x - (3 - 4i)].$$

❯ Synthesis

Determine whether the statement is true or false.

95. The sum of two numbers that are complex conjugates of each other is always a real number.

3.2 ❯ **Quadratic Equations, Functions, Zeros, and Models**

❯ Find zeros of quadratic functions and solve quadratic equations by using the principle of zero products, by using the principle of square roots, by completing the square, and by using the quadratic formula.

❯ Solve equations that are reducible to quadratic.

❯ Solve applied problems using quadratic equations.

❯ Quadratic Equations and Quadratic Functions

In this section, we will explore the relationship between the solutions of quadratic equations and the zeros of quadratic functions. We define quadratic equations and quadratic functions as follows.

QUADRATIC EQUATIONS

A **quadratic equation** is an equation that can be written in the form

$$ax^2 + bx + c = 0, \quad a \neq 0,$$

where a, b, and c are real numbers.

QUADRATIC FUNCTIONS

A **quadratic function** f is a function that can be written in the form

$$f(x) = ax^2 + bx + c, \quad a \neq 0,$$

where a, b, and c are real numbers.

A quadratic equation written in the form $ax^2 + bx + c = 0$ is said to be in **standard form**.

ZEROS OF A FUNCTION

REVIEW SECTION 1.5.

The *zeros* of a quadratic function $f(x) = ax^2 + bx + c$ are the *solutions* of the associated quadratic equation $ax^2 + bx + c = 0$. (These solutions are sometimes called *roots* of the equation.) Quadratic functions can have real-number zeros or imaginary-number zeros and quadratic equations can have real-number solutions or imaginary-number solutions. If the zeros or solutions are real numbers, they are also the first coordinates of the x-intercepts of the graph of the quadratic function.

The following principles allow us to solve many quadratic equations.

EQUATION-SOLVING PRINCIPLES

The Principle of Zero Products: If $ab = 0$ is true, then $a = 0$ or $b = 0$, and if $a = 0$ or $b = 0$, then $ab = 0$.

The Principle of Square Roots: If $x^2 = k$, then $x = \sqrt{k}$ or $x = -\sqrt{k}$.

Just in Time

14, 19

EXAMPLE 1 Solve: $2x^2 - x = 3$.

Algebraic Solution

We factor and use the principle of zero products:

$$2x^2 - x = 3$$
$$2x^2 - x - 3 = 0$$
$$(x + 1)(2x - 3) = 0$$
$$x + 1 = 0 \quad or \quad 2x - 3 = 0$$
$$x = -1 \quad or \quad 2x = 3$$
$$x = -1 \quad or \quad x = \tfrac{3}{2}.$$

Check:

For $x = -1$:

$$\begin{array}{c|c} 2x^2 - x = 3 \\ \hline 2(-1)^2 - (-1) \,?\, 3 \\ 2 \cdot 1 + 1 \\ 2 + 1 \\ 3 \,\big|\, 3 \quad \text{TRUE} \end{array}$$

For $x = \tfrac{3}{2}$:

$$\begin{array}{c|c} 2x^2 - x = 3 \\ \hline 2\left(\tfrac{3}{2}\right)^2 - \tfrac{3}{2} \,?\, 3 \\ 2 \cdot \tfrac{9}{4} - \tfrac{3}{2} \\ \tfrac{9}{2} - \tfrac{3}{2} \\ \tfrac{6}{2} \\ 3 \,\big|\, 3 \quad \text{TRUE} \end{array}$$

The solutions are -1 and $\tfrac{3}{2}$.

Graphical Solution

The solutions of the equation $2x^2 - x = 3$, or the equivalent equation $2x^2 - x - 3 = 0$, are the zeros of the function $f(x) = 2x^2 - x - 3$. They are also the first coordinates of the x-intercepts of the graph of $f(x) = 2x^2 - x - 3$.

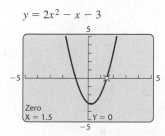

$y = 2x^2 - x - 3$

Zero
X = -1   Y = 0

$y = 2x^2 - x - 3$

Zero
X = 1.5   Y = 0

The solutions are -1 and 1.5, or -1 and $\tfrac{3}{2}$.

> *Now Try Exercise 3.*

Just in Time
20

EXAMPLE 2   Solve: $2x^2 - 10 = 0$.

Solution   We have

$$2x^2 - 10 = 0$$
$$2x^2 = 10 \qquad \text{Adding 10 on both sides}$$
$$x^2 = 5 \qquad \text{Dividing by 2 on both sides}$$
$$x = \sqrt{5} \quad or \quad x = -\sqrt{5}. \qquad \text{Using the principle of square roots}$$

Check:

$$\begin{array}{c|c} 2x^2 - 10 = 0 \\ \hline 2(\pm\sqrt{5})^2 - 10 \;?\; 0 & \text{We can check both solutions at once.} \\ 2 \cdot 5 - 10 & \\ 10 - 10 & \\ 0 \;\big|\; 0 & \text{TRUE} \end{array}$$

The solutions are $\sqrt{5}$ and $-\sqrt{5}$, or $\pm\sqrt{5}$.

> *Now Try Exercise 7.*

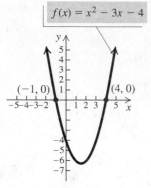

$$f(x) = x^2 - 3x - 4$$

$(-1, 0)$ $(4, 0)$

Two real-number zeros
Two x-intercepts

FIGURE 1.

We have seen that some quadratic equations can be solved by factoring and using the principle of zero products. For example, consider the equation $x^2 - 3x - 4 = 0$:

$$x^2 - 3x - 4 = 0$$
$$(x + 1)(x - 4) = 0 \qquad \text{Factoring}$$
$$x + 1 = 0 \quad or \quad x - 4 = 0 \qquad \text{Using the principle of zero products}$$
$$x = -1 \quad or \qquad x = 4.$$

The equation $x^2 - 3x - 4 = 0$ has *two real-number* solutions, -1 and 4. These are the zeros of the associated quadratic function $f(x) = x^2 - 3x - 4$ and the first coordinates of the x-intercepts of the graph of this function. (See Fig. 1.)

Next, consider the equation $x^2 - 6x + 9 = 0$. Again, we factor and use the principle of zero products:

$$x^2 - 6x + 9 = 0$$
$$(x - 3)(x - 3) = 0 \qquad \text{Factoring}$$
$$x - 3 = 0 \quad or \quad x - 3 = 0 \qquad \text{Using the principle of zero products}$$
$$x = 3 \quad or \qquad x = 3.$$

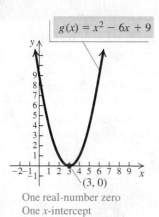

$$g(x) = x^2 - 6x + 9$$

$(3, 0)$

One real-number zero
One x-intercept

FIGURE 2.

The equation $x^2 - 6x + 9 = 0$ has *one real-number* solution, 3. It is the zero of the quadratic function $g(x) = x^2 - 6x + 9$ and the first coordinate of the x-intercept of the graph of this function. (See Fig. 2.)

The principle of square roots can be used to solve quadratic equations like $x^2 + 13 = 0$:

$$x^2 + 13 = 0$$
$$x^2 = -13$$
$$x = \pm\sqrt{-13} \qquad \text{Using the principle of square roots}$$
$$x = \pm\sqrt{13}\,i. \qquad \sqrt{-13} = \sqrt{-1} \cdot \sqrt{13} = i \cdot \sqrt{13} = \sqrt{13}\,i$$

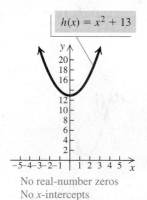

$$h(x) = x^2 + 13$$

No real-number zeros
No x-intercepts

FIGURE 3.

The equation has *two imaginary-number* solutions, $-\sqrt{13}\,i$ and $\sqrt{13}\,i$. These are the zeros of the associated quadratic function $h(x) = x^2 + 13$. Since the zeros are not real numbers, the graph of the function has no x-intercepts. (See Fig. 3.)

❯ Completing the Square

Neither the principle of zero products nor the principle of square roots would yield the *exact* zeros of a function like $f(x) = x^2 - 6x - 10$ or the *exact* solutions of the associated equation $x^2 - 6x - 10 = 0$. If we wish to find exact zeros or solutions, we can use a procedure called **completing the square** and then use the principle of square roots.

EXAMPLE 3 Find the zeros of $f(x) = x^2 - 6x - 10$ by completing the square.

Solution   We find the values of x for which $f(x) = 0$; that is, we solve the associated equation $x^2 - 6x - 10 = 0$. Our goal is to find an equivalent equation of the form $x^2 + bx + c = d$ in which $x^2 + bx + c$ is a perfect square. Since

$$x^2 + bx + \left(\frac{b}{2}\right)^2 = \left(x + \frac{b}{2}\right)^2,$$

the number c is found by taking half the coefficient of the x-term and squaring it. Thus for the equation $x^2 - 6x - 10 = 0$, we have

$$
\begin{aligned}
x^2 - 6x - 10 &= 0 \\
x^2 - 6x &= 10 \qquad && \text{Adding 10} \\
x^2 - 6x + 9 &= 10 + 9 \qquad && \text{Adding 9 on both sides to complete the square:} \\
&&& \left(\frac{b}{2}\right)^2 = \left(\frac{-6}{2}\right)^2 = (-3)^2 = 9 \\
x^2 - 6x + 9 &= 19.
\end{aligned}
$$

Because $x^2 - 6x + 9$ is a perfect square, we are able to write it as $(x - 3)^2$, the square of a binomial. We can then use the principle of square roots to finish the solution:

$$
\begin{aligned}
(x - 3)^2 &= 19 && \text{Factoring} \\
x - 3 &= \pm\sqrt{19} && \text{Using the principle of square roots} \\
x &= 3 \pm \sqrt{19}. && \text{Adding 3}
\end{aligned}
$$

Therefore, the solutions of the equation are $3 + \sqrt{19}$ and $3 - \sqrt{19}$, or simply $3 \pm \sqrt{19}$. The zeros of $f(x) = x^2 - 6x - 10$ are also $3 + \sqrt{19}$ and $3 - \sqrt{19}$, or $3 \pm \sqrt{19}$.

We can find decimal approximations for $3 \pm \sqrt{19}$ using a calculator:

$$3 + \sqrt{19} \approx 7.359 \quad \text{and} \quad 3 - \sqrt{19} \approx -1.359.$$

The zeros are approximately 7.359 and -1.359.

❯ **Now Try Exercise 31.**

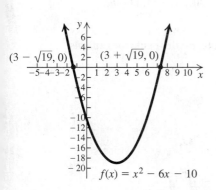

$(3 - \sqrt{19}, 0)$ $(3 + \sqrt{19}, 0)$

$f(x) = x^2 - 6x - 10$

Approximations for the zeros of the quadratic function $f(x) = x^2 - 6x - 10$ in Example 3 can be found using the Zero method.

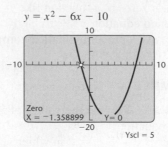

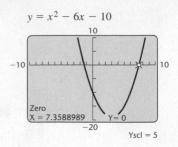

Before we can complete the square, the coefficient of the x^2-term must be 1. When it is not, we divide by the x^2-coefficient on both sides of the equation.

EXAMPLE 4   Solve: $2x^2 - 1 = 3x$.

Solution   We have

$$2x^2 - 1 = 3x$$

$$2x^2 - 3x - 1 = 0 \qquad \text{Subtracting } 3x. \text{ We are unable to factor the result.}$$

$$2x^2 - 3x = 1 \qquad \text{Adding 1}$$

$$x^2 - \frac{3}{2}x = \frac{1}{2} \qquad \text{Dividing by 2 to make the } x^2\text{-coefficient 1}$$

$$x^2 - \frac{3}{2}x + \frac{9}{16} = \frac{1}{2} + \frac{9}{16} \qquad \text{Completing the square: } \tfrac{1}{2}\left(-\tfrac{3}{2}\right) = -\tfrac{3}{4} \text{ and } \left(-\tfrac{3}{4}\right)^2 = \tfrac{9}{16}; \text{ adding } \tfrac{9}{16}$$

$$\left(x - \frac{3}{4}\right)^2 = \frac{17}{16} \qquad \text{Factoring and simplifying}$$

$$x - \frac{3}{4} = \pm\frac{\sqrt{17}}{4} \qquad \text{Using the principle of square roots and the quotient rule for radicals}$$

$$x = \frac{3}{4} \pm \frac{\sqrt{17}}{4} \qquad \text{Adding } \tfrac{3}{4}$$

$$x = \frac{3 \pm \sqrt{17}}{4}.$$

The solutions are

$$\frac{3 + \sqrt{17}}{4} \quad \text{and} \quad \frac{3 - \sqrt{17}}{4}, \quad \text{or} \quad \frac{3 \pm \sqrt{17}}{4}.$$

> **Now Try Exercise 35.**

To solve a quadratic equation by completing the square:

1. Isolate the terms with variables on one side of the equation and arrange them in descending order.
2. Divide by the coefficient of the squared term if that coefficient is not 1.
3. Complete the square by finding half the coefficient of the first-degree term and adding its square on both sides of the equation.
4. Express one side of the equation as the square of a binomial.
5. Use the principle of square roots.
6. Solve for the variable.

❯ Using the Quadratic Formula

Because completing the square works for *any* quadratic equation, it can be used to solve the general quadratic equation $ax^2 + bx + c = 0$ for x. The result will be a formula that can be used to solve any quadratic equation quickly.

Consider any quadratic equation in standard form:

$$ax^2 + bx + c = 0, \quad a \neq 0.$$

For now, we assume that $a > 0$ and solve by completing the square. As the steps are carried out, compare them with those of Example 4.

$$ax^2 + bx + c = 0 \qquad \text{Standard form}$$
$$ax^2 + bx = -c \qquad \text{Adding } -c$$
$$x^2 + \frac{b}{a}x = -\frac{c}{a} \qquad \text{Dividing by } a$$

Half of $\frac{b}{a}$ is $\frac{b}{2a}$, and $\left(\frac{b}{2a}\right)^2 = \frac{b^2}{4a^2}$. Thus we add $\frac{b^2}{4a^2}$:

$$x^2 + \frac{b}{a}x + \frac{b^2}{4a^2} = -\frac{c}{a} + \frac{b^2}{4a^2} \qquad \text{Adding } \frac{b^2}{4a^2} \text{ to complete the square}$$

$$\left(x + \frac{b}{2a}\right)^2 = -\frac{4ac}{4a^2} + \frac{b^2}{4a^2} \qquad \begin{array}{l}\text{Factoring on the left; finding a}\\\text{common denominator on the}\\\text{right: } -\frac{c}{a} = -\frac{c}{a} \cdot \frac{4a}{4a} = -\frac{4ac}{4a^2}\end{array}$$

$$\left(x + \frac{b}{2a}\right)^2 = \frac{b^2 - 4ac}{4a^2}$$

$$x + \frac{b}{2a} = \pm\frac{\sqrt{b^2 - 4ac}}{2a} \qquad \begin{array}{l}\text{Using the principle of square}\\\text{roots and the quotient rule}\\\text{for radicals. Since } a > 0,\\\sqrt{4a^2} = 2a.\end{array}$$

$$x = -\frac{b}{2a} \pm \frac{\sqrt{b^2 - 4ac}}{2a} \qquad \text{Adding } -\frac{b}{2a}$$

$$x = \frac{-b \pm \sqrt{b^2 - 4ac}}{2a}.$$

It can also be shown that this result holds if $a < 0$.

THE QUADRATIC FORMULA

The solutions of $ax^2 + bx + c = 0$, $a \neq 0$, are given by

$$x = \frac{-b \pm \sqrt{b^2 - 4ac}}{2a}.$$

EXAMPLE 5 Solve $3x^2 + 2x = 7$. Find exact solutions and approximate solutions rounded to three decimal places.

Algebraic Solution

After writing the equation in standard form, we are unable to factor, so we identify a, b, and c in order to use the quadratic formula:

$$3x^2 + 2x = 7$$
$$3x^2 + 2x - 7 = 0;$$
$$a = 3, \quad b = 2, \quad c = -7.$$

We then use the quadratic formula:

$$x = \frac{-b \pm \sqrt{b^2 - 4ac}}{2a}$$

$$= \frac{-2 \pm \sqrt{2^2 - 4(3)(-7)}}{2(3)} \quad \textbf{Substituting}$$

$$= \frac{-2 \pm \sqrt{4 + 84}}{6} = \frac{-2 \pm \sqrt{88}}{6}$$

$$= \frac{-2 \pm \sqrt{4 \cdot 22}}{6} = \frac{-2 \pm 2\sqrt{22}}{6}$$

$$= \frac{2(-1 \pm \sqrt{22})}{2 \cdot 3}$$

$$= \frac{2}{2} \cdot \frac{-1 \pm \sqrt{22}}{3}$$

$$= \frac{-1 \pm \sqrt{22}}{3}.$$

The exact solutions are

$$\frac{-1 - \sqrt{22}}{3} \quad \text{and} \quad \frac{-1 + \sqrt{22}}{3}.$$

Using a calculator, we approximate the solutions to be -1.897 and 1.230.

Graphical Solution

Using the Intersect method, we graph $y_1 = 3x^2 + 2x$ and $y_2 = 7$ and use the INTERSECT feature to find the coordinates of the points of intersection. The first coordinates of these points are the solutions of the equation $y_1 = y_2$, or $3x^2 + 2x = 7$.

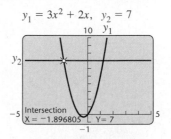

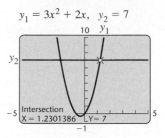

The solutions of $3x^2 + 2x = 7$ are approximately -1.897 and 1.230. We could also write the equation in standard form, $3x^2 + 2x - 7 = 0$, and use the Zero method.

Now Try Exercise 41.

Not all quadratic equations can be solved graphically.

EXAMPLE 6 Solve: $x^2 + 5x + 8 = 0$.

Algebraic Solution	**Graphical Solution**
To find the solutions, we use the quadratic formula. For $x^2 + 5x + 8 = 0$, we have $$a = 1, \quad b = 5, \quad c = 8;$$ $$x = \frac{-b \pm \sqrt{b^2 - 4ac}}{2a}$$ $$= \frac{-5 \pm \sqrt{5^2 - 4(1)(8)}}{2 \cdot 1} \quad \text{Substituting}$$ $$= \frac{-5 \pm \sqrt{25 - 32}}{2}$$ $$= \frac{-5 \pm \sqrt{-7}}{2} = \frac{-5 \pm \sqrt{7}i}{2}.$$ The solutions are $-\dfrac{5}{2} - \dfrac{\sqrt{7}}{2}i$ and $-\dfrac{5}{2} + \dfrac{\sqrt{7}}{2}i$.	The graph of the function $f(x) = x^2 + 5x + 8$ shows no x-intercepts. $y = x^2 + 5x + 8$ Thus the function has no real-number zeros and there are no real-number solutions of the associated equation $x^2 + 5x + 8 = 0$. This is a quadratic equation that cannot be solved graphically.

> **Now Try Exercise 47.**

❯ The Discriminant

From the quadratic formula, we know that the solutions x_1 and x_2 of a quadratic equation are given by

$$x_1 = \frac{-b + \sqrt{b^2 - 4ac}}{2a} \quad \text{and} \quad x_2 = \frac{-b - \sqrt{b^2 - 4ac}}{2a}.$$

The expression $b^2 - 4ac$ shows the nature of the solutions. This expression is called the **discriminant**. If it is 0, then it makes no difference whether we choose the plus sign or the minus sign in the formula. That is, $x_1 = -\dfrac{b}{2a} = x_2$, so there is just one solution. In this case, we sometimes say that there is one repeated real solution. If the discriminant is positive, there will be two different real solutions. If it is negative, we will be taking the square root of a negative number; hence there will be two imaginary-number solutions, and they will be complex conjugates.

DISCRIMINANT

For $ax^2 + bx + c = 0$, where a, b, and c are real numbers, $a \neq 0$:

$b^2 - 4ac = 0 \longrightarrow$ One real-number solution;

$b^2 - 4ac > 0 \longrightarrow$ Two different real-number solutions;

$b^2 - 4ac < 0 \longrightarrow$ Two different imaginary-number solutions, complex conjugates.

In Example 5, the discriminant, 88, is positive, indicating that there are two different real-number solutions. The negative discriminant, -7, in Example 6 indicates that there are two different imaginary-number solutions.

❯ Equations Reducible to Quadratic

Some equations can be treated as quadratic, provided we make a suitable substitution. For example, consider the following:

$$x^4 - 5x^2 + 4 = 0$$
$$(x^2)^2 - 5x^2 + 4 = 0 \qquad x^4 = (x^2)^2$$
$$u^2 - 5u + 4 = 0. \qquad \text{Substituting } u \text{ for } x^2$$

The equation $u^2 - 5u + 4 = 0$ can be solved for u by factoring or using the quadratic formula. Then we can reverse the substitution, replacing u with x^2, and solve for x. Equations like the one above are said to be **reducible to quadratic**, or **quadratic in form**.

EXAMPLE 7   Solve: $x^4 - 5x^2 + 4 = 0$.

Algebraic Solution

We let $u = x^2$ and substitute:

$$u^2 - 5u + 4 = 0 \qquad \text{Substituting } u \text{ for } x^2$$
$$(u - 1)(u - 4) = 0 \qquad \text{Factoring}$$
$$u - 1 = 0 \quad or \quad u - 4 = 0 \qquad \begin{array}{l}\text{Using the} \\ \text{principle of zero} \\ \text{products}\end{array}$$
$$u = 1 \quad or \qquad u = 4.$$

Don't stop here! We must solve for the original variable. We substitute x^2 for u and solve for x:

$$x^2 = 1 \quad or \quad x^2 = 4$$
$$x = \pm 1 \quad or \quad x = \pm 2. \qquad \begin{array}{l}\text{Using the principle of} \\ \text{square roots}\end{array}$$

The solutions are -1, 1, -2, and 2.

Graphical Solution

Using the Zero method, we graph the function $y = x^4 - 5x^2 + 4$ and use the ZERO feature to find the zeros.

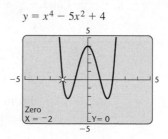

The leftmost zero is -2. Using the ZERO feature three more times, we find that the other zeros are -1, 1, and 2. Thus the solutions of $x^4 - 5x^2 + 4 = 0$ are -2, -1, 1, and 2.

❯ *Now Try Exercise 91.*

Just in Time

27

EXAMPLE 8   Solve: $t^{2/3} - 2t^{1/3} - 3 = 0$.

Solution   We let $u = t^{1/3}$ and substitute:

$$t^{2/3} - 2t^{1/3} - 3 = 0$$
$$(t^{1/3})^2 - 2t^{1/3} - 3 = 0$$
$$u^2 - 2u - 3 = 0 \qquad \text{Substituting } u \text{ for } t^{1/3}$$
$$(u + 1)(u - 3) = 0 \qquad \text{Factoring}$$
$$u + 1 = 0 \quad or \quad u - 3 = 0 \qquad \text{Using the principle of zero products}$$
$$u = -1 \quad or \qquad u = 3.$$

Now we must solve for the original variable, t. We substitute $t^{1/3}$ for u and solve for t:

$$t^{1/3} = -1 \qquad or \qquad t^{1/3} = 3$$
$$(t^{1/3})^3 = (-1)^3 \quad or \quad (t^{1/3})^3 = 3^3 \qquad \text{Cubing on both sides}$$
$$t = -1 \qquad or \qquad t = 27.$$

The solutions are -1 and 27.

❯ *Now Try Exercise 99.*

❯ Applications

Some applied problems can be translated to quadratic equations.

EXAMPLE 9 *Museums in China.* The number of museums in China increased from approximately 2000 in the year 2000 to over 3500 by the end of 2012. In 2012, a record 451 new museums opened. For comparison, in the United States, only 20–40 new museums were opened per year from 2000 to 2008. The function

$$h(x) = 30.992x^2 + 4.108x + 2294.594$$

can be used to estimate the number of museums in China, *x* years after 2005. (*Source:* The Economist/www.economist.com).

a) Estimate the number of museums that will be in China in 2017 if the number of new museums that open per year continues growing at the same rate.

b) In what year was the number of museums in China 2600?

Solution

a) For 2017, $x = 2017 - 2005 = 12$. We substitute 12 for *x* and find $h(12)$:

$$h(x) = 30.992x^2 + 4.108x + 2294.594$$
$$h(12) = 30.992(12)^2 + 4.108(12) + 2294.594$$
$$h(12) = 4462.848 + 49.296 + 2294.594 \approx 6807.$$

In 2017, there will be approximately 6807 museums in China.

b) We substitute 2600 for $h(x)$ and solve for *x*:

$$h(x) = 30.992x^2 + 4.108x + 2294.594$$
$$2600 = 30.992x^2 + 4.108x + 2294.594$$
$$0 = 30.992x^2 + 4.108x - 305.406.$$

We then use the quadratic formula, with $a = 30.992$, $b = 4.108$, and $c = -305.406$:

$$x = \frac{-b \pm \sqrt{b^2 - 4ac}}{2a}$$

$$x = \frac{-4.108 \pm \sqrt{(4.108)^2 - 4(30.992)(-305.406)}}{2(30.992)}$$

$$x = \frac{-4.108 \pm \sqrt{37{,}877.44667}}{61.984}$$

$$x = 3.074 \quad or \quad x = -3.206.$$

Because we are looking for a year after 2005, we use the positive solution. Thus there were about 2600 museums in China 3 years after 2005, or in 2008.

❯ **Now Try Exercise 107.**

EXAMPLE 10 *Magazine Closures.* The numbers of both magazine launches and magazine closures have increased in recent years. The function

$$m(x) = 34x^2 - 59x + 81$$

can be used to estimate the number of magazine closures after 2012 (*Source*: Media Finder.com, online database of U.S. and Canadian print and digital publications). In what year was the number of magazine closures 99?

Solution   We substitute 99 for $m(x)$ and solve for x:

$$99 = 34x^2 - 59x + 81$$
$$0 = 34x^2 - 59x - 18.$$

We then use the quadratic formula, with $a = 34$, $b = -59$, and $c = -18$:

$$x = \frac{-(-59) \pm \sqrt{(-59)^2 - 4 \cdot 34 \cdot (-18)}}{2 \cdot 34}$$ **Substituting**

$$x = \frac{59 \pm \sqrt{5929}}{68}$$

$$x = 2 \quad \text{or} \quad x \approx -0.3.$$

Because we are looking for a year after 2012, we use the positive solution. Thus there were 99 magazine closures 2 years after 2012, or in 2014.

> **Now Try Exercise 109.**

EXAMPLE 11   *Train Speeds.*   Two trains leave a station at the same time. One train travels due west, and the other travels due south. The train traveling west travels 20 km/h faster than the train traveling south. After 2 hr, the trains are 200 km apart. Find the speed of each train.

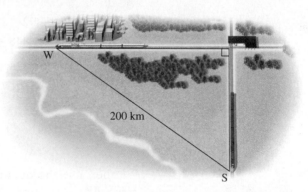

Solution

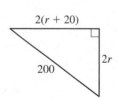

Just in Time

28

1. **Familiarize.**   First, we make a drawing. We let $r =$ the speed of the train traveling south, in kilometers per hour. Then $r + 20 =$ the speed of the train traveling west, in kilometers per hour. We use the motion formula $d = rt$, where d is the distance, r is the rate (or speed), and t is the time. After 2 hr, the train traveling south has traveled $2r$ kilometers, and the train traveling west has traveled $2(r + 20)$ kilometers. We add these distances to the drawing.

2. **Translate.**   We use the Pythagorean theorem, $a^2 + b^2 = c^2$, where a and b are the lengths of the legs of a right triangle and c is the length of the hypotenuse:

$$[2(r + 20)]^2 + (2r)^2 = 200^2.$$

3. **Carry out.**   We solve the equation:

$$\begin{aligned}
[2(r + 20)]^2 + (2r)^2 &= 200^2 \\
4(r^2 + 40r + 400) + 4r^2 &= 40{,}000 \\
4r^2 + 160r + 1600 + 4r^2 &= 40{,}000 \\
8r^2 + 160r + 1600 &= 40{,}000 \qquad \text{\textbf{Collecting like terms}} \\
8r^2 + 160r - 38{,}400 &= 0 \qquad \text{\textbf{Subtracting 40,000}} \\
r^2 + 20r - 4800 &= 0 \qquad \text{\textbf{Dividing by 8}} \\
(r + 80)(r - 60) &= 0 \qquad \text{\textbf{Factoring}} \\
r + 80 = 0 \quad \text{or} \quad r - 60 &= 0 \qquad \text{\textbf{Principle of zero products}} \\
r = -80 \quad \text{or} \qquad r &= 60.
\end{aligned}$$

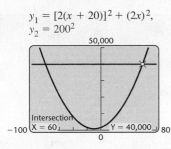

$y_1 = [2(x + 20)]^2 + (2x)^2,$
$y_2 = 200^2$

We also can solve this equation with a graphing calculator using the Intersect method, as shown in the window at left.

4. **Check.** Since speed cannot be negative, we check only 60. If the speed of the train traveling south is 60 km/h, then the speed of the train traveling west is 60 + 20, or 80 km/h. In 2 hr, the train heading south travels 60·2, or 120 km, and the train heading west travels 80·2, or 160 km. Then they are $\sqrt{120^2 + 160^2}$, or $\sqrt{40,000}$, or 200 km apart. The answer checks.

5. **State.** The speed of the train heading south is 60 km/h, and the speed of the train heading west is 80 km/h.

> **Now Try Exercise 113.**

CONNECTING THE CONCEPTS

Zeros, Solutions, and Intercepts

The zeros of a function $y = f(x)$ are also the solutions of the equation $f(x) = 0$, and the real-number zeros are the first coordinates of the *x*-intercepts of the graph of the function.

FUNCTION	**ZEROS OF THE FUNCTION; SOLUTIONS OF THE EQUATION**	***x*-INTERCEPTS OF THE GRAPH**

Linear Function

$f(x) = 2x - 4,$ or
$\quad y = 2x - 4$

To find the **zero** of $f(x)$, we solve $f(x) = 0$:

$$2x - 4 = 0$$
$$2x = 4$$
$$x = 2.$$

The **solution** of the equation $2x - 4 = 0$ is 2. This is the zero of the function $f(x) = 2x - 4$; that is, $f(2) = 0$.

The zero of $f(x)$ is the first coordinate of the *x*-intercept of the graph of $y = f(x)$.

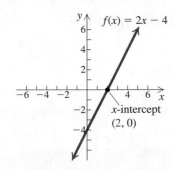

Quadratic Function

$g(x) = x^2 - 3x - 4,$ or
$\quad y = x^2 - 3x - 4$

To find the **zeros** of $g(x)$, we solve $g(x) = 0$:

$$x^2 - 3x - 4 = 0$$
$$(x + 1)(x - 4) = 0$$
$$x + 1 = 0 \quad or \quad x - 4 = 0$$
$$x = -1 \quad or \quad x = 4.$$

The **solutions** of the equation $x^2 - 3x - 4 = 0$ are −1 and 4. They are the zeros of the function $g(x) = x^2 - 3x - 4$; that is, $g(-1) = 0$ and $g(4) = 0$.

The real-number zeros of $g(x)$ are the first coordinates of the ***x*-intercepts** of the graph of $y = g(x)$.

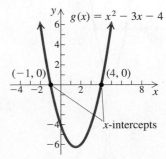

3.2 | Exercise Set

Solve.

1. $(2x - 3)(3x - 2) = 0$

2. $(5x - 2)(2x + 3) = 0$

3. $x^2 - 8x - 20 = 0$

4. $x^2 + 6x + 8 = 0$

5. $3x^2 + x - 2 = 0$

6. $10x^2 - 16x + 6 = 0$

7. $4x^2 - 12 = 0$ **8.** $6x^2 = 36$

9. $3x^2 = 21$ **10.** $2x^2 - 20 = 0$

11. $5x^2 + 10 = 0$

12. $4x^2 + 12 = 0$

13. $x^2 + 16 = 0$

14. $x^2 + 25 = 0$

15. $2x^2 = 6x$

16. $18x + 9x^2 = 0$

17. $3y^3 - 5y^2 - 2y = 0$

18. $3t^3 + 2t = 5t^2$

19. $7x^3 + x^2 - 7x - 1 = 0$
(*Hint*: Factor by grouping.)

20. $3x^3 + x^2 - 12x - 4 = 0$
(*Hint*: Factor by grouping.)

In Exercises 21–28, use the given graph to find (**a**) *the x-intercepts and* (**b**) *the zeros of the function.*

21.

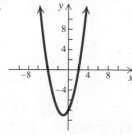

22.

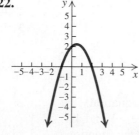

23.

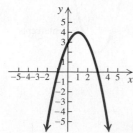

24.

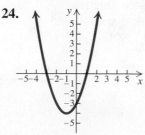

25.

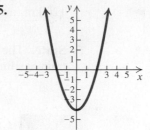

26.

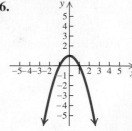

27.

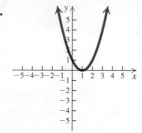

28.

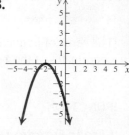

Solve by completing the square to obtain exact solutions.

29. $x^2 + 6x = 7$ **30.** $x^2 + 8x = -15$

31. $x^2 = 8x - 9$ **32.** $x^2 = 22 + 10x$

33. $x^2 + 8x + 25 = 0$ **34.** $x^2 + 6x + 13 = 0$

35. $3x^2 + 5x - 2 = 0$ **36.** $2x^2 - 5x - 3 = 0$

Use the quadratic formula to find exact solutions.

37. $x^2 - 2x = 15$ **38.** $x^2 + 4x = 5$

39. $5m^2 + 3m = 2$ **40.** $2y^2 - 3y - 2 = 0$

41. $3x^2 + 6 = 10x$ **42.** $3t^2 + 8t + 3 = 0$

43. $x^2 + x + 2 = 0$ **44.** $x^2 + 1 = x$

45. $5t^2 - 8t = 3$ **46.** $5x^2 + 2 = x$

47. $3x^2 + 4 = 5x$ **48.** $2t^2 - 5t = 1$

49. $x^2 - 8x + 5 = 0$ **50.** $x^2 - 6x + 3 = 0$

51. $3x^2 + x = 5$ **52.** $5x^2 + 3x = 1$

53. $2x^2 + 1 = 5x$ **54.** $4x^2 + 3 = x$

55. $5x^2 + 2x = -2$ **56.** $3x^2 + 3x = -4$

For each of the following, find the discriminant, $b^2 - 4ac$, and then determine whether one real-number solution, two different real-number solutions, or two different imaginary-number solutions exist.

57. $4x^2 = 8x + 5$ **58.** $4x^2 - 12x + 9 = 0$

59. $x^2 + 3x + 4 = 0$ **60.** $x^2 - 2x + 4 = 0$

61. $5t^2 - 7t = 0$ **62.** $5t^2 - 4t = 11$

Solve graphically. Round solutions to three decimal places, where appropriate.

63. $x^2 - 8x + 12 = 0$

64. $5x^2 + 42x + 16 = 0$

65. $7x^2 - 43x + 6 = 0$

66. $10x^2 - 23x + 12 = 0$

67. $6x + 1 = 4x^2$

68. $3x^2 + 5x = 3$

69. $2x^2 - 4 = 5x$

70. $4x^2 - 2 = 3x$

Find the zeros of the function algebraically. Give exact answers.

71. $f(x) = x^2 + 6x + 5$

72. $f(x) = x^2 - x - 2$

73. $f(x) = x^2 - 3x - 3$

74. $f(x) = 3x^2 + 8x + 2$

75. $f(x) = x^2 - 5x + 1$

76. $f(x) = x^2 - 3x - 7$

77. $f(x) = x^2 + 2x - 5$

78. $f(x) = x^2 - x - 4$

79. $f(x) = 2x^2 - x + 4$

80. $f(x) = 2x^2 + 3x + 2$

81. $f(x) = 3x^2 - x - 1$

82. $f(x) = 3x^2 + 5x + 1$

83. $f(x) = 5x^2 - 2x - 1$

84. $f(x) = 4x^2 - 4x - 5$

85. $f(x) = 4x^2 + 3x - 3$

86. $f(x) = x^2 + 6x - 3$

Use a graphing calculator to find the zeros of the function. Round to three decimal places.

87. $f(x) = 3x^2 + 2x - 4$

88. $f(x) = 9x^2 - 8x - 7$

89. $f(x) = 5.02x^2 - 4.19x - 2.057$

90. $f(x) = 1.21x^2 - 2.34x - 5.63$

Solve.

91. $x^4 - 3x^2 + 2 = 0$

92. $x^4 + 3 = 4x^2$

93. $x^4 + 3x^2 = 10$

94. $x^4 - 8x^2 = 9$

95. $y^4 + 4y^2 - 5 = 0$

96. $y^4 - 15y^2 - 16 = 0$

97. $x - 3\sqrt{x} - 4 = 0$
(*Hint*: Let $u = \sqrt{x}$.)

98. $2x - 9\sqrt{x} + 4 = 0$

99. $m^{2/3} - 2m^{1/3} - 8 = 0$
(*Hint*: Let $u = m^{1/3}$.)

100. $t^{2/3} + t^{1/3} - 6 = 0$

101. $x^{1/2} - 3x^{1/4} + 2 = 0$

102. $x^{1/2} - 4x^{1/4} = -3$

103. $(2x - 3)^2 - 5(2x - 3) + 6 = 0$
(*Hint*: Let $u = 2x - 3$.)

104. $(3x + 2)^2 + 7(3x + 2) - 8 = 0$

105. $(2t^2 + t)^2 - 4(2t^2 + t) + 3 = 0$

106. $12 = (m^2 - 5m)^2 + (m^2 - 5m)$

Multigenerational Households. *After declining between 1940 and 1980, the number of multigenerational American households has been increasing since 1980. The function*

$$h(x) = 0.012x^2 - 0.583x + 35.727$$

can be used to estimate the number of multigenerational households in the United States, in millions, x years after 1940 (Source: *Pew Research Center*). *Use this function for Exercises 107 and 108.*

107. In what year were there 40 million multigenerational households?

108. In what year were there 55 million multigenerational households?

Funding for Afghan Security. The number of U.S. forces in Afghanistan decreased to approximately 34,000 in 2014 from a high of about 100,000 in 2010. The amount of U.S. funding for Afghan security forces also decreased during this period. The function

$$f(x) = -1.321x^2 + 5.156x + 5.517$$

can be used to estimate the amount of U.S. funding for Afghan security forces, in billions of dollars, x years after 2009 (Sources: *U.S. Department of Defense; Brookings Institution; International Security Assistance Force; ESRI*). Use this function for Exercises 109 and 110.

109. In what year was the amount of U.S. funding for Afghan security forces about $10.5 billion?

110. In what year was the amount of U.S. funding for Afghan security forces about $5.0 billion?

Time of a Free Fall. The formula $s = 16t^2$ is used to approximate the distance s, in feet, that an object falls freely from rest in t seconds. Use this formula for Exercises 111 and 112.

111. The Taipei 101 Tower, also known as the Taipei Financial Center, in Taipei, Taiwan, is 1670 ft tall. How long would it take an object dropped from the top to reach the ground?

112. At 630 ft, the Gateway Arch in St. Louis is the tallest man-made monument in the United States. How long would it take an object dropped from the top to reach the ground?

113. The length of a rectangular poster is 1 ft more than the width, and a diagonal of the poster is 5 ft. Find the length and the width.

114. The length of one leg of a right triangle is 7 cm less than the length of the other leg. The length of the hypotenuse is 13 cm. Find the lengths of the legs.

115. One number is 5 greater than another. The product of the numbers is 36. Find the numbers.

116. One number is 6 less than another. The product of the numbers is 72. Find the numbers.

117. *Box Construction.* An open box is made from a 10-cm by 20-cm piece of tin by cutting a square from each corner and folding up the edges. The area of the resulting base is 96 cm². What is the length of the sides of the squares?

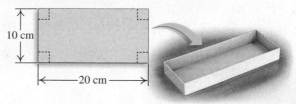

118. *Petting Zoo Dimensions.* At the Glen Island Zoo, 170 m of fencing was used to enclose a rectangular petting area of 1750 m². Find the dimensions of the petting area.

119. *Dimensions of a Rug.* Find the dimensions of a rectangular Persian rug whose perimeter is 28 ft and whose area is 48 ft².

120. *Picture Frame Dimensions.* The rectangular frame on a picture is 8 in. by 10 in. outside and is of uniform width. What is the width of the frame if 48 in² of the picture shows?

State whether the function is linear or quadratic.

121. $f(x) = 4 - 5x$

122. $f(x) = 4 - 5x^2$

123. $f(x) = 7x^2$

124. $f(x) = 23x + 6$

125. $f(x) = 1.2x - (3.6)^2$

126. $f(x) = 2 - x - x^2$

〉 Skill Maintenance

Cost of a Super Bowl Ad. The cost of a 30-sec Super Bowl ad has increased more than 70% since 2004. The function

$$C(x) = 0.17x + 2.25$$

can be used to estimate the cost of a 30-sec ad, in millions of dollars, x years after 2004 (Source: *Katar Media*). Use this function for Exercises 127 and 128. **[1.2]**

127. Estimate the cost of a 30-sec Super Bowl ad in 2014.

128. When will the cost of a 30-sec Super Bowl ad reach $5.0 million?

Determine whether the graph is symmetric with respect to the x-axis, the y-axis, and the origin. **[2.4]**

129. $3x^2 + 4y^2 = 5$

130. $y^3 = 6x^2$

Determine whether the function is even, odd, or neither even nor odd. **[2.4]**

131. $f(x) = 2x^3 - x$

132. $f(x) = 4x^2 + 2x - 3$

❯ Synthesis

For each equation in Exercises 133–136, under the given condition: **(a)** *Find k and* **(b)** *find a second solution.*

133. $kx^2 - 17x + 33 = 0$; one solution is 3

134. $kx^2 - 2x + k = 0$; one solution is -3

135. $x^2 - kx + 2 = 0$; one solution is $1 + i$

136. $x^2 - (6 + 3i)x + k = 0$; one solution is 3

Solve.

137. $(x - 2)^3 = x^3 - 2$

138. $(x + 1)^3 = (x - 1)^3 + 26$

139. $(6x^3 + 7x^2 - 3x)(x^2 - 7) = 0$

140. $\left(x - \frac{1}{5}\right)\left(x^2 - \frac{1}{4}\right) + \left(x - \frac{1}{5}\right)\left(x^2 + \frac{1}{8}\right) = 0$

141. $x^2 + x - \sqrt{2} = 0$

142. $x^2 + \sqrt{5}x - \sqrt{3} = 0$

143. $2t^2 + (t - 4)^2 = 5t(t - 4) + 24$

144. $9t(t + 2) - 3t(t - 2) = 2(t + 4)(t + 6)$

145. $\sqrt{x - 3} - \sqrt[4]{x - 3} = 2$

146. $x^2 + 3x + 1 - \sqrt{x^2 + 3x + 1} = 8$

147. $\left(y + \frac{2}{y}\right)^2 + 3y + \frac{6}{y} = 4$

148. Solve $\frac{1}{2}at^2 + v_0t + x_0 = 0$ for t.

3.3 ❯ **Analyzing Graphs of Quadratic Functions**

> ❯ Find the vertex, the axis of symmetry, and the maximum or minimum value of a quadratic function using the method of completing the square.
>
> ❯ Graph quadratic functions.
>
> ❯ Solve applied problems involving maximum and minimum function values.

❯ Graphing Quadratic Functions of the Type $f(x) = a(x - h)^2 + k$

The graph of a quadratic function is called a **parabola**. The graph of every parabola evolves from the graph of the squaring function $f(x) = x^2$ using transformations.

Exploring with Technology

Think of transformations and look for patterns. Consider the following functions:

$$y_1 = x^2, \quad y_2 = -0.4x^2,$$
$$y_3 = -0.4(x - 2)^2, \quad y_4 = -0.4(x - 2)^2 + 3.$$

Graph y_1 and y_2. How do you get from the graph of y_1 to y_2?

Graph y_2 and y_3. How do you get from the graph of y_2 to y_3?

Graph y_3 and y_4. How do you get from the graph of y_3 to y_4?

Consider the following functions:

$$y_1 = x^2, \qquad y_2 = 2x^2,$$
$$y_3 = 2(x + 3)^2, \qquad y_4 = 2(x + 3)^2 - 5.$$

Graph y_1 and y_2. How do you get from the graph of y_1 to y_2?

Graph y_2 and y_3. How do you get from the graph of y_2 to y_3?

Graph y_3 and y_4. How do you get from the graph of y_3 to y_4?

TRANSFORMATIONS

REVIEW SECTION 2.5.

We get the graph of $f(x) = a(x - h)^2 + k$ from the graph of $f(x) = x^2$ as follows:

$$f(x) = x^2$$
$$\downarrow$$
$$f(x) = ax^2 \qquad \text{**Vertical stretching or shrinking with a**}$$
$$\qquad\qquad\qquad \text{**reflection across the x-axis if $a < 0$**}$$
$$\downarrow$$
$$f(x) = a(x - h)^2 \qquad \text{**Horizontal translation**}$$
$$\downarrow$$
$$f(x) = a(x - h)^2 + k. \qquad \text{**Vertical translation**}$$

Consider the following graphs of the form $f(x) = a(x - h)^2 + k$. The point (h, k) at which the graph turns is called the **vertex**. The maximum or minimum value of $f(x)$ occurs at the vertex. Each graph has a line $x = h$ that is called the **axis of symmetry**.

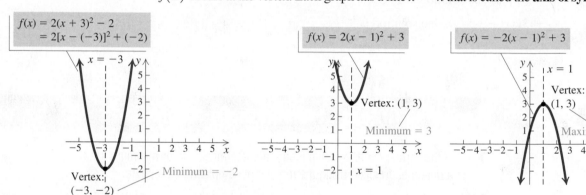

$$f(x) = 2(x + 3)^2 - 2$$
$$= 2[x - (-3)]^2 + (-2)$$

$x = -3$

Vertex: $(-3, -2)$

Minimum = -2

$$f(x) = 2(x - 1)^2 + 3$$

Vertex: $(1, 3)$

Minimum = 3

$x = 1$

$$f(x) = -2(x - 1)^2 + 3$$

$x = 1$

Vertex: $(1, 3)$

Maximum = 3

CONNECTING THE CONCEPTS   **Graphing Quadratic Functions**

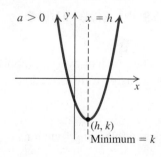

$a > 0$

$x = h$

(h, k)

Minimum = k

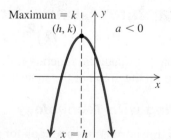

Maximum = k

(h, k)

$a < 0$

$x = h$

The graph of the function $f(x) = a(x - h)^2 + k$ is a parabola that

- opens up if $a > 0$ and down if $a < 0$;
- has (h, k) as the vertex;
- has $x = h$ as the axis of symmetry;
- has k as a minimum value (output) if $a > 0$;
- has k as a maximum value if $a < 0$.

As we saw in Section 2.5, the constant a serves to stretch or shrink the graph vertically. As a parabola is stretched vertically, it becomes narrower, and as it is shrunk vertically, it becomes wider. That is, as $|a|$ increases, the graph becomes narrower, and as $|a|$ gets close to 0, the graph becomes wider.

If the equation is in the form $f(x) = a(x - h)^2 + k$, we can learn a great deal about the graph without actually graphing the function.

Function	$f(x) = 3\left(x - \frac{1}{4}\right)^2 - 2$ $= 3\left(x - \frac{1}{4}\right)^2 + (-2)$	$g(x) = -3(x + 5)^2 + 7$ $= -3[x - (-5)]^2 + 7$
Vertex	$\left(\frac{1}{4}, -2\right)$	$(-5, 7)$
Axis of Symmetry	$x = \frac{1}{4}$	$x = -5$
Maximum	None ($3 > 0$, so the graph opens up.)	7 ($-3 < 0$, so the graph opens down.)
Minimum	-2 ($3 > 0$, so the graph opens up.)	None ($-3 < 0$, so the graph opens down.)

Note that the vertex (h, k) is used to find the maximum or minimum value of the function. The maximum or minimum value is the number k, *not* the ordered pair (h, k).

❯ Graphing Quadratic Functions of the Type $f(x) = ax^2 + bx + c, a \neq 0$

We now use a modification of the method of completing the square as an aid in graphing and analyzing quadratic functions of the form $f(x) = ax^2 + bx + c, a \neq 0$.

EXAMPLE 1 Find the vertex, the axis of symmetry, and the maximum or minimum value of $f(x) = x^2 + 10x + 23$. Then graph the function.

Solution To express

$$f(x) = x^2 + 10x + 23$$

in the form

$$f(x) = a(x - h)^2 + k,$$

we complete the square on the terms involving x. To do so, we take half the coefficient of x and square it, obtaining $(10/2)^2$, or 25. We now add and subtract that number on the *right side*:

$$f(x) = x^2 + 10x + 23 = x^2 + 10x + 25 - 25 + 23.$$

Since $25 - 25 = 0$, the new expression for the function is equivalent to the original expression. Note that this process differs from the one we used to complete the square in order to solve a quadratic equation, where we added the same number on both sides of the equation to obtain an equivalent equation. Instead, when we complete the square to write a function in the form $f(x) = a(x - h)^2 + k$, we add and subtract the same number on one side. The entire process is shown below:

$$f(x) = x^2 + 10x + 23 \qquad \text{Note that 25 completes the square for } x^2 + 10x.$$

$$= x^2 + 10x + 25 - 25 + 23 \qquad \text{Adding } 25 - 25, \text{ or 0, to the right side}$$

$$= (x^2 + 10x + 25) - 25 + 23 \qquad \text{Regrouping}$$

$$= (x + 5)^2 - 2 \qquad \text{Factoring and simplifying}$$

$$= [x - (-5)]^2 + (-2). \qquad \text{Writing in the form } f(x) = a(x - h)^2 + k$$

Keeping in mind that this function will have a minimum value since $a > 0$ ($a = 1$), from this form of the function we know the following:

Vertex: $(-5, -2)$;

Axis of symmetry: $x = -5$;

Minimum value of the function: -2.

To graph the function by hand, we first plot the vertex, $(-5, -2)$, and find several points on either side of $x = -5$. Then we plot these points and connect them with a smooth curve. We see that the points $(-4, -1)$ and $(-3, 2)$ are reflections of the points $(-6, -1)$ and $(-7, 2)$, respectively, across the axis of symmetry, $x = -5$.

x	$f(x)$	
-5	-2	$\leftarrow$ Vertex
-6	-1	
-4	-1	
-7	2	
-3	2	

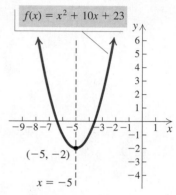

The graph of $f(x) = x^2 + 10x + 23$, or

$$f(x) = [x - (-5)]^2 + (-2),$$

shown above, is a shift of the graph of $y = x^2$ left 5 units and down 2 units.

> *Now Try Exercise 3.*

Keep in mind that the axis of symmetry is not part of the graph; it is a characteristic of the graph. If you fold the graph on its axis of symmetry, the two halves of the graph will coincide.

EXAMPLE 2 Find the vertex, the axis of symmetry, and the maximum or minimum value of $g(x) = x^2/2 - 4x + 8$. Then graph the function.

Solution We complete the square in order to write the function in the form $g(x) = a(x - h)^2 + k$. First, we factor $\frac{1}{2}$ out of the first two terms. This makes the coefficient of x^2 within the parentheses 1:

$$g(x) = \frac{x^2}{2} - 4x + 8$$

$$= \frac{1}{2}(x^2 - 8x) + 8. \qquad \text{Factoring } \tfrac{1}{2} \text{ out of the first two terms:} \atop x^2/2 - 4x = \tfrac{1}{2}\cdot x^2 - \tfrac{1}{2}\cdot 8x$$

Next, we complete the square inside the parentheses: Half of -8 is -4, and $(-4)^2 = 16$. We add and subtract 16 inside the parentheses:

$$g(x) = \tfrac{1}{2}(x^2 - 8x + 16 - 16) + 8$$
$$= \tfrac{1}{2}(x^2 - 8x + 16) - \tfrac{1}{2}\cdot 16 + 8 \qquad \text{\textbf{Using the distributive law to remove} } \atop -16 \text{ from within the parentheses}$$
$$= \tfrac{1}{2}(x^2 - 8x + 16) - 8 + 8$$
$$= \tfrac{1}{2}(x - 4)^2 + 0, \text{ or } \tfrac{1}{2}(x - 4)^2. \qquad \text{\textbf{Factoring and simplifying}}$$

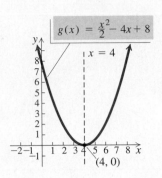

$$g(x) = \frac{x^2}{2} - 4x + 8$$

We know the following:

Vertex: $(4, 0)$;

Axis of symmetry: $x = 4$;

Minimum value of the function: 0.

Finally, we plot the vertex and several points on either side of it and draw the graph of the function. The graph of g is a vertical shrinking of the graph of $y = x^2$ along with a shift right 4 units.

> **Now Try Exercise 9.**

EXAMPLE 3 Find the vertex, the axis of symmetry, and the maximum or minimum value of $f(x) = -2x^2 + 10x - \frac{23}{2}$. Then graph the function.

Solution We have

$$
\begin{aligned}
f(x) &= -2x^2 + 10x - \frac{23}{2} \\
&= -2(x^2 - 5x) - \frac{23}{2} && \text{Factoring } -2 \text{ out of the first two terms} \\
&= -2\left(x^2 - 5x + \frac{25}{4} - \frac{25}{4}\right) - \frac{23}{2} && \text{Completing the square inside the parentheses} \\
&= -2\left(x^2 - 5x + \frac{25}{4}\right) - 2\left(-\frac{25}{4}\right) - \frac{23}{2} && \text{Using the distributive law to remove } -\frac{25}{4} \text{ from within the parentheses} \\
&= -2\left(x^2 - 5x + \frac{25}{4}\right) + \frac{25}{2} - \frac{23}{2} \\
&= -2\left(x - \frac{5}{2}\right)^2 + 1.
\end{aligned}
$$

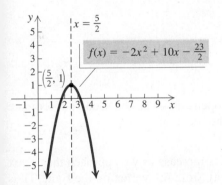

$$x = \frac{5}{2}$$

$$f(x) = -2x^2 + 10x - \frac{23}{2}$$

$$\left(\frac{5}{2}, 1\right)$$

This form of the function yields the following:

Vertex: $\left(\frac{5}{2}, 1\right)$;

Axis of symmetry: $x = \frac{5}{2}$;

Maximum value of the function: 1.

The graph is found by shifting the graph of $f(x) = x^2$ right $\frac{5}{2}$ units, reflecting it across the x-axis, stretching it vertically, and shifting it up 1 unit.

> **Now Try Exercise 13.**

Just in Time

24

In many situations, we want to use a formula to find the coordinates of the vertex directly from the equation $f(x) = ax^2 + bx + c$. One way to develop such a formula is to first note that the x-coordinate of the vertex is centered between the x-intercepts, or zeros, of the function. By averaging the two solutions of $ax^2 + bx + c = 0$, we find a formula for the x-coordinate of the vertex:

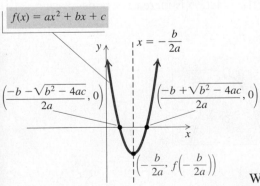

$$f(x) = ax^2 + bx + c$$

$$x = -\frac{b}{2a}$$

$$\left(\frac{-b - \sqrt{b^2 - 4ac}}{2a}, 0\right)$$

$$\left(\frac{-b + \sqrt{b^2 - 4ac}}{2a}, 0\right)$$

$$\left(-\frac{b}{2a}, f\left(-\frac{b}{2a}\right)\right)$$

$$
\begin{aligned}
x\text{-coordinate of vertex} &= \frac{\dfrac{-b - \sqrt{b^2 - 4ac}}{2a} + \dfrac{-b + \sqrt{b^2 - 4ac}}{2a}}{2} \\
&= \frac{\dfrac{-2b}{2a}}{2} = \frac{-\dfrac{b}{a}}{2} \\
&= -\frac{b}{a} \cdot \frac{1}{2} = -\frac{b}{2a}.
\end{aligned}
$$

We use this value of x to find the y-coordinate of the vertex, $f\left(-\dfrac{b}{2a}\right)$.

THE VERTEX OF A PARABOLA

The **vertex** of the graph of $f(x) = ax^2 + bx + c$ is

$$\left(-\frac{b}{2a}, f\left(-\frac{b}{2a}\right)\right).$$

We calculate the   We substitute to
x-coordinate.   find the y-coordinate.

EXAMPLE 4 For the function $f(x) = -x^2 + 14x - 47$:

a) Find the vertex.

b) Determine whether there is a maximum or a minimum value and find that value.

c) Find the range.

d) On what intervals is the function increasing? decreasing?

Solution There is no need to graph the function.

a) The x-coordinate of the vertex is

$$-\frac{b}{2a} = -\frac{14}{2(-1)} = -\frac{14}{-2} = 7.$$

Since

$$f(7) = -7^2 + 14 \cdot 7 - 47 = -49 + 98 - 47 = 2,$$

the vertex is $(7, 2)$.

b) Since a is negative ($a = -1$), the graph opens down so the second coordinate of the vertex, 2, is the maximum value of the function.

c) The range is $(-\infty, 2]$.

d) Since the graph opens down, function values increase as we approach the vertex from the left and decrease as we move away from the vertex on the right. Thus the function is increasing on the interval $(-\infty, 7)$ and decreasing on $(7, \infty)$.

> *Now Try Exercise 31.*

We can use a graphing calculator to work Example 4. Once we have graphed $y = -x^2 + 14x - 47$, we see that the graph opens down and thus has a maximum value. We can use the MAXIMUM feature to find the coordinates of the vertex. Using these coordinates, we can then find the maximum value and the range of the function along with the intervals on which the function is increasing or decreasing.

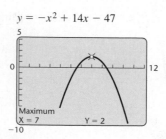

$y = -x^2 + 14x - 47$

Maximum
X = 7   Y = 2

❯ Applications

Many real-world situations involve finding the maximum or the minimum value of a quadratic function.

EXAMPLE 5 *Maximizing Area.*   A landscaper has enough stone to enclose a rectangular koi pond next to an existing garden wall of the Englemans' house with 24 ft of stone wall. If the garden wall forms one side of the rectangle, what is the maximum area that the landscaper can enclose? What dimensions of the koi pond will yield this area?

PROBLEM-SOLVING STRATEGY

REVIEW SECTION 1.5.

Solution   We will use the five-step problem-solving strategy.

1. **Familiarize.** We first make a drawing of the situation, using w to represent the width of the koi pond, in feet. Then $(24 - 2w)$ feet of stone is available for the length. Suppose that the koi pond were 1 ft wide. Then its length would be $24 - 2 \cdot 1 = 22$ ft, and its area would be $(22 \text{ ft})(1 \text{ ft}) = 22 \text{ ft}^2$. If the koi pond were 2 ft wide, its length would be $24 - 2 \cdot 2 = 20$ ft, and its area would be $(20 \text{ ft})(2 \text{ ft}) = 40 \text{ ft}^2$. This is larger than the first area we found, but we do not know if it is the maximum possible area. To find the maximum area, we will find a function that represents the area and then determine its maximum value.

$24 - 2w$ w

2. **Translate.** We write a function for the area of the koi pond. We have

$$A(w) = (24 - 2w)w \qquad A = lw; l = 24 - 2w$$
$$= -2w^2 + 24w,$$

where $A(w)$ is the area of the koi pond, in square feet, as a function of the width w.

3. **Carry out.** To solve this problem, we need to determine the maximum value of $A(w)$ and find the dimensions for which that maximum occurs. Since A is a quadratic function and w^2 has a negative coefficient, we know that the function has a maximum value that occurs at the vertex of the graph of the function. The first coordinate of the vertex, $(w, A(w))$, is

$$w = -\frac{b}{2a} = -\frac{24}{2(-2)} = -\frac{24}{-4} = 6.$$

Thus, if $w = 6$ ft, then the length $l = 24 - 2 \cdot 6 = 12$ ft, and the area is $(12 \text{ ft})(6 \text{ ft}) = 72 \text{ ft}^2$.

4. **Check.** As a partial check, we note that $72 \text{ ft}^2 > 40 \text{ ft}^2$, which is the larger area that we found in a guess in the *Familiarize* step. As a more complete check, assuming that the function $A(W)$ is correct, we could examine a table of values for $A(w) = (24 - 2w)w$ and/or examine its graph.

5. **State.** The maximum possible area is 72 ft^2 when the koi pond is 6 ft wide and 12 ft long.

> *Now Try Exercise 45.*

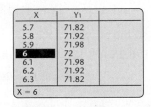

X	Y₁
5.7	71.82
5.8	71.92
5.9	71.98
6	72
6.1	71.98
6.2	71.92
6.3	71.82
X = 6	

$y = (24 - 2x)x$

EXAMPLE 6 *Height of a Rocket.*   A model rocket is launched with an initial velocity of 100 ft/sec from the top of a hill that is 20 ft high. Its height, in feet, *t* seconds after it has been launched is given by the function $s(t) = -16t^2 + 100t + 20$. Determine the time at which the rocket reaches its maximum height and find the maximum height.

Solution

1., 2. Familiarize and **Translate.** We are given the function in the statement of the problem: $s(t) = -16t^2 + 100t + 20$.

3. Carry out. We need to find the maximum value of the function and the value of *t* for which it occurs. Since $s(t)$ is a quadratic function and t^2 has a negative coefficient, we know that the maximum value of the function occurs at the vertex of the graph of the function. The first coordinate of the vertex gives the time *t* at which the rocket reaches its maximum height. It is

$$t = -\frac{b}{2a} = -\frac{100}{2(-16)} = -\frac{100}{-32} = 3.125.$$

The second coordinate of the vertex gives the maximum height of the rocket. We substitute in the function to find it:

$$s(3.125) = -16(3.125)^2 + 100(3.125) + 20 = 176.25.$$

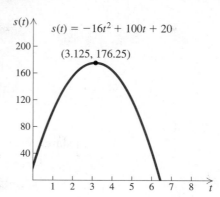

4. Check. As a check, we can complete the square to write the function in the form $s(t) = a(t - h)^2 + k$ and determine the coordinates of the vertex from this form of the function. We get

$$s(t) = -16(t - 3.125)^2 + 176.25.$$

This confirms that the vertex is $(3.125, 176.25)$, so the answer checks.

5. State. The rocket reaches a maximum height of 176.25 ft 3.125 sec after it has been launched.

> *Now Try Exercise 41.*

EXAMPLE 7 *Determining the Height of an Elevator Shaft.*   Jared drops a screwdriver from the top of an elevator shaft. Exactly 5 sec later, he hears the sound of the screwdriver hitting the bottom of the shaft. The speed of sound is 1100 ft/sec. How tall is the elevator shaft?

Solution

1. Familiarize. We first make a drawing and label it with known and unknown information. We let s = the height of the elevator shaft, in feet, t_1 = the time, in seconds, that it takes for the screwdriver to hit the bottom of the elevator shaft, and t_2 = the time, in seconds, that it takes for the sound to reach the top of the elevator shaft. This gives us the equation

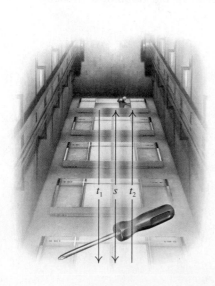

$$t_1 + t_2 = 5. \tag{1}$$

2. Translate. Can we find any relationship between the two times and the distance s? Often in problem solving you may need to look up related formulas in a physics book, in another mathematics book, or on the Internet. We find that the formula

$$s = 16t^2$$

gives the distance, in feet, that a dropped object falls in t seconds. The time t_1 that it takes the screwdriver to hit the bottom of the elevator shaft can be found as follows:

$$s = 16t_1^2, \quad \text{or} \quad \frac{s}{16} = t_1^2, \quad \text{so} \quad t_1 = \frac{\sqrt{s}}{4}. \quad \text{Taking the positive square root} \tag{2}$$

To find an expression for t_2, the time that it takes the sound to travel to the top of the well, recall that *Distance = Rate · Time*. Thus,

$$s = 1100t_2, \quad \text{or} \quad t_2 = \frac{s}{1100}. \tag{3}$$

We now have expressions for t_1 and t_2, both in terms of s. Substituting into equation (1), we obtain

$$t_1 + t_2 = 5, \quad \text{or} \quad \frac{\sqrt{s}}{4} + \frac{s}{1100} = 5. \tag{4}$$

3. Carry out.

Algebraic Solution

We solve equation (4) for s. Multiplying by 1100, we get

$$275\sqrt{s} + s = 5500, \quad \text{or} \quad s + 275\sqrt{s} - 5500 = 0.$$

This equation is reducible to quadratic with $u = \sqrt{s}$. Substituting, we get

$$u^2 + 275u - 5500 = 0.$$

Using the quadratic formula, we can solve for u:

$$u = \frac{-b \pm \sqrt{b^2 - 4ac}}{2a}$$

$$= \frac{-275 + \sqrt{275^2 - 4 \cdot 1 \cdot (-5500)}}{2 \cdot 1} \quad \begin{array}{l}\text{We want only the}\\\text{positive solution.}\end{array}$$

$$= \frac{-275 + \sqrt{97{,}625}}{2} \approx 18.725.$$

Since $u \approx 18.725$, we have

$$\sqrt{s} \approx 18.725$$

$$s \approx 350.6. \quad \begin{array}{l}\text{Squaring both sides and rounding}\\\text{to the nearest tenth}\end{array}$$

Graphical Solution

We use the Intersect method. It will probably require some trial and error to determine an appropriate window.

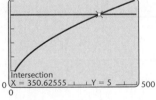

4. Check. To check, we can substitute 350.6 for s in equation (4) and see that $t_1 + t_2 \approx 5$. We leave the computation to the student.

5. State. The height of the elevator shaft is about 350.6 ft.

▷ *Now Try Exercise 55.*

Visualizing the Graph

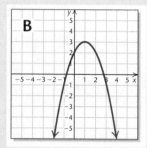

A

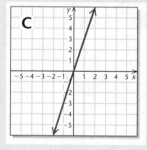

B

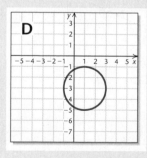

C

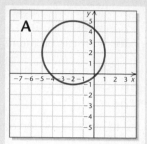

D

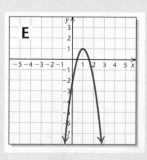

E

Match the equation with its graph.

1. $y = 3x$

2. $y = -(x - 1)^2 + 3$

3. $(x + 2)^2 + (y - 2)^2 = 9$

4. $y = 3$

5. $2x - 3y = 6$

6. $(x - 1)^2 + (y + 3)^2 = 4$

7. $y = -2x + 1$

8. $y = 2x^2 - x - 4$

9. $x = -2$

10. $y = -3x^2 + 6x - 2$

Answers on page A-14

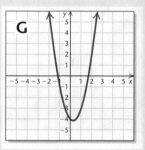

F

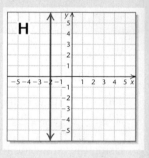

G

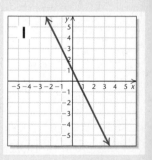

H

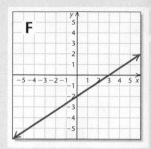

I

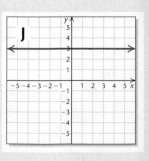

J

3.3   Exercise Set

In Exercises 1 and 2, use the given graph to find the following.

a) *The vertex.*
b) *The axis of symmetry.*
c) *The maximum or the minimum value of the function.*

1.

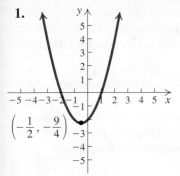

$\left(-\frac{1}{2}, -\frac{9}{4}\right)$

2.

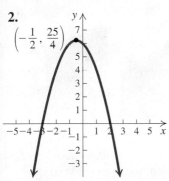

$\left(-\frac{1}{2}, \frac{25}{4}\right)$

In Exercises 3–16:

a) *Find the vertex.*
b) *Find the axis of symmetry.*
c) *Determine whether there is a maximum or a minimum value and find that value.*
d) *Graph the function.*

3. $f(x) = x^2 - 8x + 12$

4. $g(x) = x^2 + 7x - 8$

5. $f(x) = x^2 - 7x + 12$

6. $g(x) = x^2 - 5x + 6$

7. $f(x) = x^2 + 4x + 5$

8. $f(x) = x^2 + 2x + 6$

9. $g(x) = \dfrac{x^2}{2} + 4x + 6$

10. $g(x) = \dfrac{x^2}{3} - 2x + 1$

11. $g(x) = 2x^2 + 6x + 8$

12. $f(x) = 2x^2 - 10x + 14$

13. $f(x) = -x^2 - 6x + 3$

14. $f(x) = -x^2 - 8x + 5$

15. $g(x) = -2x^2 + 2x + 1$

16. $f(x) = -3x^2 - 3x + 1$

In Exercises 17–24, match the equation with one of the graphs (a)–(h) that follow.

a)

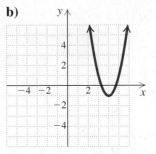

b)

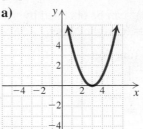

c)

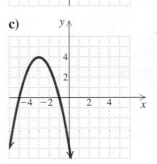

d)

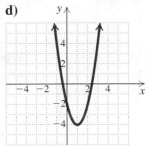

e)

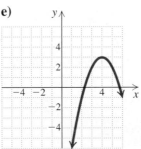

f)

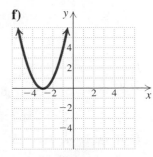

g)

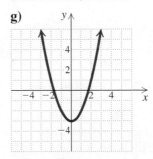

h)

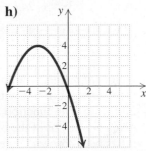

17. $y = (x + 3)^2$

18. $y = -(x - 4)^2 + 3$

19. $y = 2(x - 4)^2 - 1$

20. $y = x^2 - 3$

21. $y = -\frac{1}{2}(x + 3)^2 + 4$

22. $y = (x - 3)^2$

23. $y = -(x + 3)^2 + 4$

24. $y = 2(x - 1)^2 - 4$

Determine whether the statement is true or false.

25. The function $f(x) = -3x^2 + 2x + 5$ has a maximum value.

26. The vertex of the graph of $f(x) = ax^2 + bx + c$ is $-\dfrac{b}{2a}$.

27. The graph of $h(x) = (x + 2)^2$ can be obtained by translating the graph of $h(x) = x^2$ right 2 units.

28. The vertex of the graph of the function $g(x) = 2(x - 4)^2 - 1$ is $(-4, -1)$.

29. The axis of symmetry of the function $f(x) = -(x + 2)^2 - 4$ is $x = -2$.

30. The minimum value of the function $f(x) = 3(x - 1)^2 + 5$ is 5.

In Exercises 31–40:

a) *Find the vertex.*

b) *Determine whether there is a maximum or a minimum value and find that value.*

c) *Find the range.*

d) *Find the intervals on which the function is increasing and the intervals on which the function is decreasing.*

31. $f(x) = x^2 - 6x + 5$

32. $f(x) = x^2 + 4x - 5$

33. $f(x) = 2x^2 + 4x - 16$

34. $f(x) = \frac{1}{2}x^2 - 3x + \frac{5}{2}$

35. $f(x) = -\frac{1}{2}x^2 + 5x - 8$

36. $f(x) = -2x^2 - 24x - 64$

37. $f(x) = 3x^2 + 6x + 5$

38. $f(x) = -3x^2 + 24x - 49$

39. $g(x) = -4x^2 - 12x + 9$

40. $g(x) = 2x^2 - 6x + 5$

41. *Height of a Ball.* A ball is thrown directly upward from a height of 6 ft with an initial velocity of 20 ft/sec. The function $s(t) = -16t^2 + 20t + 6$ gives the height of the ball, in feet, t seconds after it has been thrown. Determine the time at which the ball reaches its maximum height and find the maximum height.

42. *Height of a Projectile.* A stone is thrown directly upward from a height of 30 ft with an initial velocity of 60 ft/sec. The height of the stone, in feet, t seconds after it has been thrown is given by the function $s(t) = -16t^2 + 60t + 30$. Determine the time at which the stone reaches its maximum height and find the maximum height.

43. *Height of a Rocket.* A model rocket is launched with an initial velocity of 120 ft/sec from a height of 80 ft. The height of the rocket, in feet, t seconds after it has been launched is given by the function $s(t) = -16t^2 + 120t + 80$. Determine the time at which the rocket reaches its maximum height and find the maximum height.

44. *Height of a Rocket.* A model rocket is launched with an initial velocity of 150 ft/sec from a height of 40 ft. The function $s(t) = -16t^2 + 150t + 40$ gives the height of the rocket, in feet, t seconds after it has been launched. Determine the time at which the rocket reaches its maximum height and find the maximum height.

45. *Maximizing Volume.* Mendoza Manufacturing plans to produce a one-compartment vertical file by bending the long side of a 10-in. by 18-in. sheet of plastic along two lines to form a ⊔-shape. How tall should the file be in order to maximize the volume that it can hold?

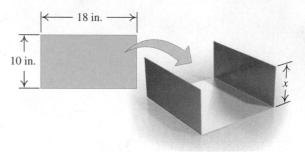

46. *Maximizing Area.* A fourth-grade class decides to enclose a rectangular garden, using the side of the school as one side of the rectangle. What is the maximum area that the class can enclose using 32 ft of fence? What should the dimensions of the garden be in order to yield this area?

47. *Maximizing Area.* The sum of the base and the height of a triangle is 20 cm. Find the dimensions for which the area is a maximum.

48. *Maximizing Area.* The sum of the base and the height of a parallelogram is 69 cm. Find the dimensions for which the area is a maximum.

49. *Minimizing Cost.*   Designs for #1 Canines has determined that when x hundred portable doghouses are produced, the average cost per doghouse is given by

$$C(x) = 0.1x^2 - 4.2x + 72.4,$$

where $C(x)$ is in hundreds of dollars. How many doghouses should be produced in order to minimize the average cost per doghouse?

Portable doghouse   Carry bag

Maximizing Profit.   *In business, profit is the difference between revenue and cost; that is,*

$$Total\ profit = Total\ revenue - Total\ cost,$$
$$P(x) = R(x) - C(x),$$

where x is the number of units sold. Find the maximum profit and the number of units that must be sold in order to yield the maximum profit for each of the following.

50. $R(x) = 5x,\ C(x) = 0.001x^2 + 1.2x + 60$

51. $R(x) = 50x - 0.5x^2,\ C(x) = 10x + 3$

52. $R(x) = 20x - 0.1x^2,\ C(x) = 4x + 2$

53. *Maximizing Area.*   A berry farmer needs to separate and enclose two adjacent rectangular fields, one for blueberries and one for strawberries. If a lake forms one side of the fields and 240 yd of fencing is available, what is the largest total area that can be enclosed?

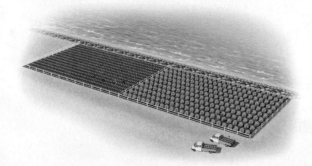

54. *Norman Window.*   A Norman window is a rectangle with a semicircle on top. Sky Blue Windows is designing a Norman window that will require 24 ft of trim on the outer edges. What dimensions will allow the maximum amount of light to enter a house?

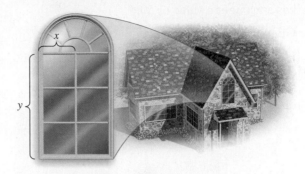

55. *Finding the Depth of a Well.*   Two seconds after a chlorine tablet has been dropped into a well, a splash is heard. The speed of sound is 1100 ft/sec. How far is the top of the well from the water? (*Hint*: See Example 7.)

56. *Finding the Height of a Cliff.*   A water balloon is dropped from a cliff. Exactly 3 sec later, the sound of the balloon hitting the ground reaches the top of the cliff. How high is the cliff? (*Hint*: See Example 7.)

❯ Skill Maintenance

For each function f, construct and simplify the difference quotient

$$\frac{f(x + h) - f(x)}{h}.\ \ \textbf{[2.2]}$$

57. $f(x) = 3x - 7$ **58.** $f(x) = 2x^2 - x + 4$

A graph of $y = f(x)$ follows. No formula is given for f. Make a hand-drawn graph of each of the following. **[2.5]**

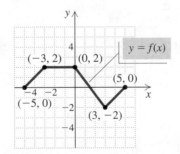

59. $g(x) = -2f(x)$ **60.** $g(x) = f(2x)$

❯ Synthesis

61. Find c such that
$$f(x) = -0.2x^2 - 3x + c$$
has a maximum value of -225.

62. Find b such that
$$f(x) = -4x^2 + bx + 3$$
has a maximum value of 50.

63. Graph: $f(x) = (|x| - 5)^2 - 3$.

64. Find a quadratic function with vertex $(4, -5)$ and containing the point $(-3, 1)$.

65. *Minimizing Area.*   A 24-in. piece of string is cut into two pieces. One piece is used to form a circle while the other is used to form a square. How should the string be cut so that the sum of the areas is a minimum?

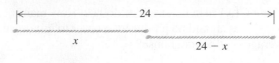

Mid-Chapter Mixed Review

Determine whether the statement is true or false.

1. The product of a complex number and its conjugate is a real number. **[3.1]**

2. Every quadratic equation has at least one x-intercept. **[3.2]**

3. If a quadratic equation has two different real-number solutions, then its discriminant is positive. **[3.2]**

4. The vertex of the graph of the function $f(x) = 3(x + 4)^2 + 5$ is $(4, 5)$. **[3.3]**

Express the number in terms of i. **[3.1]**

5. $\sqrt{-36}$

6. $\sqrt{-5}$

7. $-\sqrt{-16}$

8. $\sqrt{-32}$

Simplify. Write answers in the form $a + bi$, where a and b are real numbers. **[3.1]**

9. $(3 - 2i) + (-4 + 3i)$

10. $(-5 + i) - (2 - 4i)$

11. $(2 + 3i)(4 - 5i)$

12. $\dfrac{3 + i}{-2 + 5i}$

Simplify. **[3.1]**

13. i^{13}

14. i^{44}

15. $(-i)^5$

16. $(2i)^6$

Solve. **[3.2]**

17. $x^2 + 3x - 4 = 0$

18. $2x^2 + 6 = -7x$

19. $4x^2 = 24$

20. $x^2 + 100 = 0$

21. Find the zeros of $f(x) = 4x^2 - 8x - 3$ by completing the square. Show your work. **[3.2]**

In Exercises 22–24:

a) *Find the discriminant $b^2 - 4ac$, and then determine whether one real-number solution, two different real-number solutions, or two different imaginary-number solutions exist.* **[3.2]**

b) *Solve the equation, finding exact solutions and approximate solutions rounded to three decimal places, where appropriate.* **[3.2]**

22. $x^2 - 3x - 5 = 0$

23. $4x^2 - 12x + 9 = 0$

24. $3x^2 + 2x = -1$

Solve. [3.2]

25. $x^4 + 5x^2 - 6 = 0$

26. $2x - 5\sqrt{x} + 2 = 0$

27. One number is 2 more than another. The product of the numbers is 35. Find the numbers. [3.2]

In Exercises 28 and 29:

a) *Find the vertex.* [3.3]
b) *Find the axis of symmetry.* [3.3]
c) *Determine whether there is a maximum or a minimum value, and find that value.* [3.3]
d) *Find the range.* [3.3]
e) *Find the intervals on which the function is increasing and the intervals on which the function is decreasing.* [3.3]
f) *Graph the function.* [3.3]

28. $f(x) = x^2 - 6x + 7$

29. $f(x) = -2x^2 - 4x - 5$

30. The sum of the base and the height of a triangle is 16 in. Find the dimensions for which the area is a maximum. [3.3]

COLLABORATIVE DISCUSSION AND WRITING

31. Is the sum of two imaginary numbers always an imaginary number? Explain your answer. [3.1]

32. The graph of a quadratic function can have 0, 1, or 2 x-intercepts. How can you predict the number of x-intercepts without drawing the graph or (completely) solving an equation? [3.2]

33. Discuss two ways in which we used completing the square in this chapter. [3.2], [3.3]

34. Suppose that the graph of $f(x) = ax^2 + bx + c$ has x-intercepts $(x_1, 0)$ and $(x_2, 0)$. What are the x-intercepts of $g(x) = -ax^2 - bx - c$? Explain. [3.3]

3.4 ❯ **Solving Rational Equations and Radical Equations**

> ❯ Solve rational equations.
> ❯ Solve radical equations.

❯ Rational Equations

Equations containing rational expressions are called **rational equations**. Solving such equations involves multiplying on both sides by the least common denominator (LCD) of all the rational expressions to *clear the equation of fractions*.

EXAMPLE 1 Solve: $\dfrac{x - 8}{3} + \dfrac{x - 3}{2} = 0.$

Algebraic Solution

We have

$$\frac{x-8}{3} + \frac{x-3}{2} = 0 \qquad \text{The LCD is } 3 \cdot 2, \text{ or } 6.$$

$$6\left(\frac{x-8}{3} + \frac{x-3}{2}\right) = 6 \cdot 0 \qquad \text{Multiplying by the LCD on both sides to clear fractions}$$

$$6 \cdot \left(\frac{x-8}{3}\right) + 6 \cdot \left(\frac{x-3}{2}\right) = 0$$

$$2(x-8) + 3(x-3) = 0$$

$$2x - 16 + 3x - 9 = 0$$

$$5x - 25 = 0$$

$$5x = 25$$

$$x = 5.$$

The possible solution is 5. We check using a table in ASK mode.

$$y = \frac{x-8}{3} + \frac{x-3}{2}$$

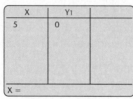

Since the value of $\dfrac{x-8}{3} + \dfrac{x-3}{2}$ is 0 when $x = 5$, the number 5 is the solution.

Graphical Solution

We use the Zero method. The solution of the equation

$$\frac{x-8}{3} + \frac{x-3}{2} = 0$$

is the zero of the function

$$f(x) = \frac{x-8}{3} + \frac{x-3}{2}.$$

$$y = \frac{x-8}{3} + \frac{x-3}{2}$$

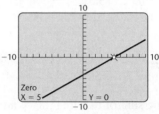

The zero of the function is 5. Thus the solution of the equation is 5.

> *Now Try Exercise 3.*

CAUTION! Clearing fractions is a valid procedure when solving rational equations but not when adding, subtracting, multiplying, or dividing rational expressions. A rational expression may have operation signs but it will have no equals sign. A rational equation *always* has an equals sign. For example, $\dfrac{x-8}{3} + \dfrac{x-3}{2}$ is a rational expression but $\dfrac{x-8}{3} + \dfrac{x-3}{2} = 0$ is a rational equation.

To *simplify* the rational *expression* $\dfrac{x-8}{3} + \dfrac{x-3}{2}$, we first find the LCD and write each fraction with that denominator. The final result is usually a rational expression.

To *solve* the rational *equation* $\dfrac{x-8}{3} + \dfrac{x-3}{2} = 0$, we first multiply on both sides by the LCD to clear fractions. The final result is one or more numbers. As we will see in Example 2, these numbers must be checked in the original equation.

When we use the multiplication principle to multiply (or divide) on both sides of an equation by an expression with a variable, we might not obtain an equivalent equation. We must check the possible solutions obtained in this manner by substituting them in the original equation. The next example illustrates this.

EXAMPLE 2 Solve: $\dfrac{x^2}{x-3} = \dfrac{9}{x-3}$.

Solution The LCD is $x - 3$.

$$(x-3) \cdot \frac{x^2}{x-3} = (x-3) \cdot \frac{9}{x-3}$$

$$x^2 = 9$$

$$x = -3 \quad or \quad x = 3 \qquad \text{\textbf{Using the principle of square roots}}$$

The possible solutions are -3 and 3. We check.

Check: For -3:

$$\dfrac{x^2}{x-3} = \dfrac{9}{x-3}$$

$$\dfrac{(-3)^2}{-3-3} \;\overset{?}{\vert}\; \dfrac{9}{-3-3}$$

$$\dfrac{9}{-6} \;\bigg\vert\; \dfrac{9}{-6} \qquad \text{\textbf{TRUE}}$$

For 3:

$$\dfrac{x^2}{x-3} = \dfrac{9}{x-3}$$

$$\dfrac{3^2}{3-3} \;\overset{?}{\vert}\; \dfrac{9}{3-3}$$

$$\dfrac{9}{0} \;\bigg\vert\; \dfrac{9}{0} \qquad \text{\textbf{NOT DEFINED}}$$

The number -3 checks, so it is a solution. Since division by 0 is not defined, 3 is not a solution. Note that 3 is not in the domain of either $x^2/(x-3)$ or $9/(x-3)$.

We can also use a table on a graphing calculator to check the possible solutions. When $x = -3$, we see that $y_1 = -1.5 = y_2$, so -3 is a solution. When $x = 3$, we get ERROR messages. This indicates that 3 is not in the domain of y_1 or y_2 and thus is not a solution.

> *Now Try Exercise 9.*

$$y_1 = \frac{x^2}{x-3}, \quad y_2 = \frac{9}{x-3}$$

X	Y₁	Y₂
-3	-1.5	-1.5
3	ERROR	ERROR

X =

EXAMPLE 3 Solve: $\dfrac{2}{3x+6} + \dfrac{1}{x^2-4} = \dfrac{4}{x-2}$.

Solution We first factor the denominators in order to determine the LCD:

$$\frac{2}{3(x+2)} + \frac{1}{(x+2)(x-2)} = \frac{4}{x-2} \qquad \begin{array}{l}\textbf{The LCD is}\\ \textbf{3}(x+2)(x-2).\end{array}$$

$$3(x+2)(x-2)\left(\frac{2}{3(x+2)} + \frac{1}{(x+2)(x-2)}\right) = 3(x+2)(x-2) \cdot \frac{4}{x-2}$$

$$\text{\textbf{Multiplying by the LCD to clear fractions}}$$

$$2(x-2) + 3 = 3 \cdot 4(x+2)$$

$$2x - 4 + 3 = 12x + 24$$

$$2x - 1 = 12x + 24$$

$$-10x = 25$$

$$x = -\tfrac{5}{2}.$$

The possible solution is $-\frac{5}{2}$. We check this on a graphing calculator.

$$y_1 = \frac{2}{3x + 6} + \frac{1}{x^2 - 4}, \quad y_2 = \frac{4}{x - 2}$$

X	Y₁	Y₂
−2.5	−.8889	−.8889

X =

We see that $y_1 = y_2$ when $x = -\frac{5}{2}$, or -2.5, so $-\frac{5}{2}$ is the solution.

> *Now Try Exercise 21.*

❯ Radical Equations

A **radical equation** is an equation in which variables appear in one or more radicands. For example,

$$\sqrt{2x - 5} - \sqrt{x - 3} = 1$$

is a radical equation. The following principle is used to solve such equations.

THE PRINCIPLE OF POWERS

For any positive integer n:

If $a = b$ is true, then $a^n = b^n$ is true.

EXAMPLE 4 Solve: $\sqrt{3x + 1} = 4$.

Algebraic Solution

We use the principle of powers and square both sides:

$$\sqrt{3x + 1} = 4$$
$$\left(\sqrt{3x + 1}\right)^2 = 4^2$$
$$3x + 1 = 16$$
$$3x = 15$$
$$x = 5.$$

Check:

$$\begin{array}{c|c} \sqrt{3x + 1} = 4 \\ \hline \sqrt{3 \cdot 5 + 1} & 4 \\ \sqrt{15 + 1} & \\ \sqrt{16} & \\ 4 & 4 \quad \text{TRUE} \end{array}$$

The solution is 5.

Graphical Solution

We graph $y_1 = \sqrt{3x + 1}$ and $y_2 = 4$ and then use the INTERSECT feature. We see that the solution is 5. The check shown in the following table confirms that the solution is 5.

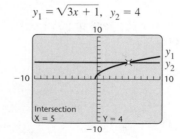

$$y_1 = \sqrt{3x + 1}, \quad y_2 = 4$$

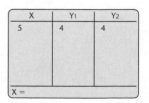

X	Y₁	Y₂
5	4	4

X =

> *Now Try Exercise 31.*

In Example 4, the radical was isolated on one side of the equation. If this had not been the case, our first step would have been to isolate the radical. We do so in the next example.

EXAMPLE 5   Solve: $5 + \sqrt{x + 7} = x$.

Algebraic Solution

We first isolate the radical and then use the principle of powers:

$$5 + \sqrt{x + 7} = x$$
$$\sqrt{x + 7} = x - 5 \qquad \text{Subtracting 5 on both sides to isolate the radical}$$
$$\left(\sqrt{x + 7}\right)^2 = (x - 5)^2 \qquad \text{Using the principle of powers; squaring both sides}$$
$$x + 7 = x^2 - 10x + 25$$
$$0 = x^2 - 11x + 18 \qquad \text{Subtracting } x \text{ and 7}$$
$$0 = (x - 9)(x - 2) \qquad \text{Factoring}$$
$$x - 9 = 0 \quad \text{or} \quad x - 2 = 0$$
$$x = 9 \quad \text{or} \qquad x = 2.$$

The possible solutions are 9 and 2.

Check: For 9:

$$5 + \sqrt{x + 7} = x$$

$$5 + \sqrt{9 + 7} \; ? \; 9$$
$$5 + \sqrt{16}$$
$$5 + 4$$
$$9 \;\big|\; 9 \quad \text{TRUE}$$

For 2:

$$5 + \sqrt{x + 7} = x$$

$$5 + \sqrt{2 + 7} \; ? \; 2$$
$$5 + \sqrt{9}$$
$$5 + 3$$
$$8 \;\big|\; 2 \quad \text{FALSE}$$

Since 9 checks but 2 does not, the only solution is 9.

Graphical Solution

We graph $y_1 = 5 + \sqrt{x + 7}$ and $y_2 = x$. Using the INTERSECT feature, we see that the solution is 9.

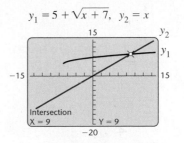

We can also use the ZERO feature to get this result. To do so, we first write the equivalent equation $5 + \sqrt{x + 7} - x = 0$. The zero of the function $f(x) = 5 + \sqrt{x + 7} - x$ is 9, so the solution of the original equation is 9.

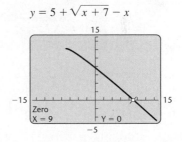

Note that the graphs show that the equation has only one solution.

Now Try Exercise 55.

When we raise both sides of an equation to an even power, the resulting equation can have solutions that the original equation does not. This is because the converse of the principle of powers is not necessarily true. That is, if $a^n = b^n$ is true, we do not know that $a = b$ is true. For example, $(-2)^2 = 2^2$, but $-2 \neq 2$. Thus, as we saw in Example 5, it is necessary to check the possible solutions in the original equation when the principle of powers is used to raise both sides of an equation to an even power.

When a radical equation has two radical terms on one side, we isolate one of them and then use the principle of powers. If, after doing so, a radical term remains, we repeat these steps.

EXAMPLE 6 Solve: $\sqrt{x - 3} + \sqrt{x + 5} = 4$.

Solution We have

$$\sqrt{x - 3} = 4 - \sqrt{x + 5} \qquad \text{Isolating one radical}$$

$$\left(\sqrt{x - 3}\right)^2 = \left(4 - \sqrt{x + 5}\right)^2 \qquad \begin{array}{l}\text{Using the principle of powers;}\\\text{squaring both sides}\end{array}$$

$$x - 3 = 16 - 8\sqrt{x + 5} + (x + 5)$$

$$x - 3 = 21 - 8\sqrt{x + 5} + x \qquad \text{Collecting like terms}$$

$$-24 = -8\sqrt{x + 5} \qquad \begin{array}{l}\text{Isolating the remaining radical;}\\\text{subtracting } x \text{ and 21 on both sides}\end{array}$$

$$3 = \sqrt{x + 5} \qquad \text{Dividing by } -8 \text{ on both sides}$$

$$3^2 = \left(\sqrt{x + 5}\right)^2 \qquad \begin{array}{l}\text{Using the principle of powers;}\\\text{squaring both sides}\end{array}$$

$$9 = x + 5$$

$$4 = x. \qquad \text{Subtracting 5 on both sides}$$

We check the possible solution, 4, on a graphing calculator.

$$y_1 = \sqrt{x - 3} + \sqrt{x + 5}, \quad y_2 = 4$$

X	Y₁	Y₂
4	4	4

X =

Since $y_1 = y_2$ when $x = 4$, the number 4 checks. It is the solution.

> *Now Try Exercise 65.*

3.4 Exercise Set

Solve.

1. $\dfrac{1}{4} + \dfrac{1}{5} = \dfrac{1}{t}$

2. $\dfrac{1}{3} - \dfrac{5}{6} = \dfrac{1}{x}$

3. $\dfrac{x + 2}{4} - \dfrac{x - 1}{5} = 15$

4. $\dfrac{t + 1}{3} - \dfrac{t - 1}{2} = 1$

5. $\dfrac{1}{2} + \dfrac{2}{x} = \dfrac{1}{3} + \dfrac{3}{x}$

6. $\dfrac{1}{t} + \dfrac{1}{2t} + \dfrac{1}{3t} = 5$

7. $\dfrac{5}{3x + 2} = \dfrac{3}{2x}$

8. $\dfrac{2}{x - 1} = \dfrac{3}{x + 2}$

9. $\dfrac{y^2}{y + 4} = \dfrac{16}{y + 4}$

10. $\dfrac{49}{w - 7} = \dfrac{w^2}{w - 7}$

11. $x + \dfrac{6}{x} = 5$

12. $x - \dfrac{12}{x} = 1$

13. $\dfrac{6}{y + 3} + \dfrac{2}{y} = \dfrac{5y - 3}{y^2 - 9}$

14. $\dfrac{3}{m + 2} + \dfrac{2}{m} = \dfrac{4m - 4}{m^2 - 4}$

15. $\dfrac{2x}{x - 1} = \dfrac{5}{x - 3}$

16. $\dfrac{2x}{x + 7} = \dfrac{5}{x + 1}$

17. $\dfrac{2}{x + 5} + \dfrac{1}{x - 5} = \dfrac{16}{x^2 - 25}$

18. $\dfrac{2}{x^2 - 9} + \dfrac{5}{x - 3} = \dfrac{3}{x + 3}$

19. $\dfrac{3x}{x + 2} + \dfrac{6}{x} = \dfrac{12}{x^2 + 2x}$

20. $\dfrac{3y + 5}{y^2 + 5y} + \dfrac{y + 4}{y + 5} = \dfrac{y + 1}{y}$

21. $\dfrac{1}{5x+20} - \dfrac{1}{x^2-16} = \dfrac{3}{x-4}$

22. $\dfrac{1}{4x+12} - \dfrac{1}{x^2-9} = \dfrac{5}{x-3}$

23. $\dfrac{2}{5x+5} - \dfrac{3}{x^2-1} = \dfrac{4}{x-1}$

24. $\dfrac{1}{3x+6} - \dfrac{1}{x^2-4} = \dfrac{3}{x-2}$

25. $\dfrac{8}{x^2-2x+4} = \dfrac{x}{x+2} + \dfrac{24}{x^3+8}$

26. $\dfrac{18}{x^2-3x+9} - \dfrac{x}{x+3} = \dfrac{81}{x^3+27}$

27. $\dfrac{x}{x-4} - \dfrac{4}{x+4} = \dfrac{32}{x^2-16}$

28. $\dfrac{x}{x-1} - \dfrac{1}{x+1} = \dfrac{2}{x^2-1}$

29. $\dfrac{1}{x-6} - \dfrac{1}{x} = \dfrac{6}{x^2-6x}$

30. $\dfrac{1}{x-15} - \dfrac{1}{x} = \dfrac{15}{x^2-15x}$

31. $\sqrt{3x-4} = 1$ **32.** $\sqrt{4x+1} = 3$

33. $\sqrt{2x-5} = 2$ **34.** $\sqrt{3x+2} = 6$

35. $\sqrt{7-x} = 2$ **36.** $\sqrt{5-x} = 1$

37. $\sqrt{1-2x} = 3$ **38.** $\sqrt{2-7x} = 2$

39. $\sqrt[3]{5x-2} = -3$ **40.** $\sqrt[3]{2x+1} = -5$

41. $\sqrt[4]{x^2-1} = 1$ **42.** $\sqrt[5]{3x+4} = 2$

43. $\sqrt{y-1} + 4 = 0$ **44.** $\sqrt{m+1} - 5 = 8$

45. $\sqrt{b+3} - 2 = 1$ **46.** $\sqrt{x-4} + 1 = 5$

47. $\sqrt{z+2} + 3 = 4$ **48.** $\sqrt{y-5} - 2 = 3$

49. $\sqrt{2x+1} - 3 = 3$ **50.** $\sqrt{3x-1} + 2 = 7$

51. $\sqrt{2-x} - 4 = 6$ **52.** $\sqrt{5-x} + 2 = 8$

53. $\sqrt[3]{6x+9} + 8 = 5$ **54.** $\sqrt[5]{2x-3} - 1 = 1$

55. $\sqrt{x+4} + 2 = x$ **56.** $\sqrt{x+1} + 1 = x$

57. $\sqrt{x-3} + 5 = x$ **58.** $\sqrt{x+3} - 1 = x$

59. $\sqrt{x+7} = x+1$ **60.** $\sqrt{6x+7} = x+2$

61. $\sqrt{3x+3} = x+1$ **62.** $\sqrt{2x+5} = x-5$

63. $\sqrt{5x+1} = x-1$ **64.** $\sqrt{7x+4} = x+2$

65. $\sqrt{x-3} + \sqrt{x+2} = 5$

66. $\sqrt{x} - \sqrt{x-5} = 1$

67. $\sqrt{3x-5} + \sqrt{2x+3} + 1 = 0$

68. $\sqrt{2m-3} = \sqrt{m+7} - 2$

69. $\sqrt{x} - \sqrt{3x-3} = 1$

70. $\sqrt{2x+1} - \sqrt{x} = 1$

71. $\sqrt{2y-5} - \sqrt{y-3} = 1$

72. $\sqrt{4p+5} + \sqrt{p+5} = 3$

73. $\sqrt{y+4} - \sqrt{y-1} = 1$

74. $\sqrt{y+7} + \sqrt{y+16} = 9$

75. $\sqrt{x+5} + \sqrt{x+2} = 3$

76. $\sqrt{6x+6} = 5 + \sqrt{21-4x}$

77. $x^{1/3} = -2$ **78.** $t^{1/5} = 2$

79. $t^{1/4} = 3$ **80.** $m^{1/2} = -7$

Solve.

81. $\dfrac{P_1V_1}{T_1} = \dfrac{P_2V_2}{T_2}$, for T_1
(A chemistry formula for gases)

82. $\dfrac{1}{F} = \dfrac{1}{m} + \dfrac{1}{p}$, for F
(A formula from optics)

83. $W = \sqrt{\dfrac{1}{LC}}$, for C
(An electricity formula)

84. $s = \sqrt{\dfrac{A}{6}}$, for A
(A geometry formula)

85. $\dfrac{1}{R} = \dfrac{1}{R_1} + \dfrac{1}{R_2}$, for R_2
(A formula for resistance)

86. $\dfrac{1}{t} = \dfrac{1}{a} + \dfrac{1}{b}$, for t
(A formula for work rate)

87. $I = \sqrt{\dfrac{A}{P}} - 1$, for P
(A compound-interest formula)

88. $T = 2\pi\sqrt{\dfrac{1}{g}}$, for g
(A pendulum formula)

89. $\dfrac{1}{F} = \dfrac{1}{m} + \dfrac{1}{p}$, for p
(A formula from optics)

90. $\dfrac{V^2}{R^2} = \dfrac{2g}{R+h}$, for h
(A formula for escape velocity)

❯ Skill Maintenance

Find the zero of the function. **[1.5]**

91. $f(x) = 15 - 2x$

92. $f(x) = -3x + 9$

Solve. **[1.5]**

93. *Pork Production.* Together, China and the United States, the top two pork producers worldwide, produced 64,308,000 metric tons of pork in 2013. China produced 1,260,000 metric tons more than five times the number of metric tons produced by the United States. (*Source*: United Nations Food and Agriculture Organization) How many metric tons of pork did each country produce in 2013?

94. *Student Loan Debt.* In 2014, the average student loan debt per college graduate was about $33,050. This was about 77.6% more than the average student loan debt per college graduate in 2004. (*Source*: Analysis by Mark Kantrowitz, publisher of Edvisors.com) What was the average student loan debt per college graduate in 2004?

❯ Synthesis

Solve.

95. $(x - 3)^{2/3} = 2$

96. $\dfrac{x + 3}{x + 2} - \dfrac{x + 4}{x + 3} = \dfrac{x + 5}{x + 4} - \dfrac{x + 6}{x + 5}$

97. $\sqrt{x + 5} + 1 = \dfrac{6}{\sqrt{x + 5}}$

98. $\sqrt{15 + \sqrt{2x + 80}} = 5$

99. $x^{2/3} = x$

3.5 ❯ Solving Equations and Inequalities with Absolute Value

> ❯ Solve equations with absolute value.

> ❯ Solve inequalities with absolute value.

Just in Time

4

❯ Equations with Absolute Value

Recall that the absolute value of a number is its distance from 0 on the number line. We use this concept to solve equations with absolute value.

> For $a > 0$ and an algebraic expression X:
>
> $|X| = a$ is equivalent to $X = -a$ or $X = a$.

EXAMPLE 1 Solve: $|x| = 5$.

Algebraic Solution	**Graphical Solution**

Algebraic Solution

We have

$$|x| = 5$$
$$x = -5 \quad or \quad x = 5. \qquad \text{Writing an equivalent statement}$$

The solutions are −5 and 5.

To check, note that −5 and 5 are both 5 units from 0 on the number line.

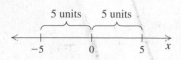

Graphical Solution

Using the Intersect method, we graph $y_1 = |x|$ and $y_2 = 5$ and find the first coordinates of the points of intersection.

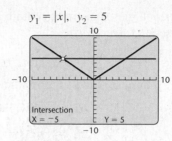

The solutions are −5 and 5.

We could also have used the Zero method to get this result, graphing $y = |x| - 5$ and using the ZERO feature twice.

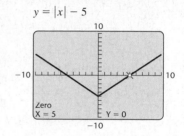

The zeros of $f(x) = |x| - 5$ are −5 and 5, so the solutions of the original equation are −5 and 5.

> **Now Try Exercise 1.**

EXAMPLE 2 Solve: $|x - 3| - 1 = 4$.

Solution First, we add 1 on both sides to get an expression of the form $|X| = a$:

$$|x - 3| - 1 = 4$$
$$|x - 3| = 5$$
$$x - 3 = -5 \quad or \quad x - 3 = 5 \qquad \begin{array}{l}|X| = a \text{ is equivalent to } X = -a \\ \text{or } X = a.\end{array}$$
$$x = -2 \quad or \qquad x = 8. \qquad \textbf{Adding 3}$$

Check: For −2:

$$|x - 3| - 1 = 4$$
$$\overline{|-2 - 3| - 1 \; ? \; 4}$$
$$|-5| - 1$$
$$5 - 1$$
$$4 \; | \; 4 \quad \text{TRUE}$$

For 8:

$$|x - 3| - 1 = 4$$
$$\overline{|8 - 3| - 1 \; ? \; 4}$$
$$|5| - 1$$
$$5 - 1$$
$$4 \; | \; 4 \quad \text{TRUE}$$

The solutions are −2 and 8.

> **Now Try Exercise 21.**

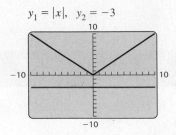

$y_1 = |x|, \; y_2 = -3$

When $a = 0$, $|X| = a$ is equivalent to $X = 0$. Note that for $a < 0$, $|X| = a$ has no solution, because the absolute value of an expression is never negative. We can use a graph to illustrate the last statement for a specific value of a. For example, if we let $a = -3$ and graph $y_1 = |x|$ and $y_2 = -3$, we see that the graphs do not intersect, as shown at left. Thus the equation $|x| = -3$ has no solution. The solution set is the **empty set**, denoted $\varnothing$.

❯ Inequalities with Absolute Value

Inequalities sometimes contain absolute-value notation. The following properties are used to solve them.

For $a > 0$ and an algebraic expression X:

$|X| < a$ is equivalent to $-a < X < a$.

$|X| > a$ is equivalent to $X < -a \; or \; X > a$.

Similar statements hold for $|X| \le a$ and $|X| \ge a$.

For example,

$|x| < 3$ is equivalent to $-3 < x < 3$;

$|y| \ge 1$ is equivalent to $y \le -1 \; or \; y \ge 1$; and

$|2x + 3| \le 4$ is equivalent to $-4 \le 2x + 3 \le 4$.

EXAMPLE 3   Solve and graph the solution set: $|3x + 2| < 5$.

Solution   We have

$$|3x + 2| < 5$$
$$-5 < 3x + 2 < 5 \qquad \textbf{Writing an equivalent inequality}$$
$$-7 < 3x < 3 \qquad \textbf{Subtracting 2}$$
$$-\tfrac{7}{3} < x < 1. \qquad \textbf{Dividing by 3}$$

The solution set is $\left\{ x \,\middle|\, -\tfrac{7}{3} < x < 1 \right\}$, or $\left(-\tfrac{7}{3}, 1 \right)$. The graph of the solution set is shown below.

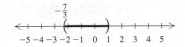

$-\tfrac{7}{3}$

❯ **Now Try Exercise 45.**

EXAMPLE 4   Solve and graph the solution set: $|5 - 2x| \ge 1$.

Solution   We have

$$|5 - 2x| \ge 1$$
$$5 - 2x \le -1 \quad or \quad 5 - 2x \ge 1 \qquad \textbf{Writing an equivalent inequality}$$
$$-2x \le -6 \quad or \qquad -2x \ge -4 \qquad \textbf{Subtracting 5}$$
$$x \ge 3 \quad or \qquad \quad x \le 2. \qquad \textbf{Dividing by } -\textbf{2 and reversing the inequality signs}$$

The solution set is $\{ x \,|\, x \le 2 \; or \; x \ge 3 \}$, or $(-\infty, 2] \cup [3, \infty)$. The graph of the solution set is shown below.

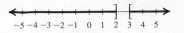

❯ **Now Try Exercise 47.**

3.5 | Exercise Set

Solve.

1. $|x| = 7$

2. $|x| = 4.5$

3. $|x| = 0$

4. $|x| = \frac{3}{2}$

5. $|x| = \frac{5}{6}$

6. $|x| = -\frac{3}{5}$

7. $|x| = -10.7$

8. $|x| = 12$

9. $|3x| = 1$

10. $|5x| = 4$

11. $|8x| = 24$

12. $|6x| = 0$

13. $|x - 1| = 4$

14. $|x - 7| = 5$

15. $|x + 2| = 6$

16. $|x + 5| = 1$

17. $|3x + 2| = 1$

18. $|7x - 4| = 8$

19. $\left|\frac{1}{2}x - 5\right| = 17$

20. $\left|\frac{1}{3}x - 4\right| = 13$

21. $|x - 1| + 3 = 6$

22. $|x + 2| - 5 = 9$

23. $|x + 3| - 2 = 8$

24. $|x - 4| + 3 = 9$

25. $|3x + 1| - 4 = -1$

26. $|2x - 1| - 5 = -3$

27. $|4x - 3| + 1 = 7$

28. $|5x + 4| + 2 = 5$

29. $12 - |x + 6| = 5$

30. $9 - |x - 2| = 7$

31. $7 - |2x - 1| = 6$

32. $5 - |4x + 3| = 2$

Solve and write interval notation for the solution set. Then graph the solution set.

33. $|x| < 7$

34. $|x| \le 4.5$

35. $|x| \le 2$

36. $|x| < 3$

37. $|x| \ge 4.5$

38. $|x| > 7$

39. $|x| > 3$

40. $|x| \ge 2$

41. $|3x| < 1$

42. $|5x| \le 4$

43. $|2x| \ge 6$

44. $|4x| > 20$

45. $|x + 8| < 9$

46. $|x + 6| \le 10$

47. $|x + 8| \ge 9$

48. $|x + 6| > 10$

49. $\left|x - \frac{1}{4}\right| < \frac{1}{2}$

50. $|x - 0.5| \le 0.2$

51. $|2x + 3| \le 9$

52. $|3x + 4| < 13$

53. $|x - 5| > 0.1$

54. $|x - 7| \ge 0.4$

55. $|6 - 4x| \ge 8$

56. $|5 - 2x| > 10$

57. $\left|x + \frac{2}{3}\right| \le \frac{5}{3}$

58. $\left|x + \frac{3}{4}\right| < \frac{1}{4}$

59. $\left|\frac{2x + 1}{3}\right| > 5$

60. $\left|\frac{2x - 1}{3}\right| \ge \frac{5}{6}$

61. $|2x - 4| < -5$

62. $|3x + 5| < 0$

63. $|7 - x| \ge -4$

64. $|2x + 1| > -\frac{1}{2}$

❯ Skill Maintenance

Vocabulary Reinforcement

In each of Exercises 65–72, fill in the blank with the correct term. Some of the given choices will not be used.

distance formula
midpoint formula
function
relation
x-intercept
y-intercept
perpendicular
parallel
horizontal lines
vertical lines

symmetric with respect
  to the x-axis
symmetric with respect
  to the y-axis
symmetric with respect
  to the origin
increasing
decreasing
constant

65. A(n) _____ is a point $(0, b)$. **[1.1]**

66. The _____ is
$d = \sqrt{(x_2 - x_1)^2 + (y_2 - y_1)^2}$. **[1.1]**

67. A(n) _____ is a correspondence such that each member of the domain corresponds to at least one member of the range. **[1.2]**

68. A(n) _____ is a correspondence such that each member of the domain corresponds to exactly one member of the range. **[1.2]**

69. _____ are given by equations of the type $y = b$, or $f(x) = b$. **[1.3]**

70. Nonvertical lines are _____ if and only if they have the same slope and different y-intercepts. **[1.4]**

71. A function f is said to be _____ on an open interval I if, for all a and b in that interval, $a < b$ implies $f(a) > f(b)$. **[2.1]**

72. For an equation $y = f(x)$, if replacing x with $-x$ produces an equivalent equation, then the graph is _____ . **[2.4]**

❯ Synthesis

Solve.

73. $|3x - 1| > 5x - 2$

74. $|x + 2| \le |x - 5|$

75. $|p - 4| + |p + 4| < 8$

76. $|x| + |x + 1| < 10$

77. $|x - 3| + |2x + 5| > 6$

STUDY GUIDE

KEY TERMS AND CONCEPTS	EXAMPLE

SECTION 3.1: THE COMPLEX NUMBERS

The number i is defined such that $i = \sqrt{-1}$ and $i^2 = -1$.	Express each number in terms of i. $$\sqrt{-5} = \sqrt{-1 \cdot 5} = \sqrt{-1} \cdot \sqrt{5} = i\sqrt{5}, \text{ or } \sqrt{5}i;$$ $$-\sqrt{-36} = -\sqrt{-1 \cdot 36} = -\sqrt{-1} \cdot \sqrt{36} = -i \cdot 6 = -6i$$
A **complex number** is a number of the form $a + bi$, where a and b are real numbers. The number a is said to be the **real part** of $a + bi$, and the number b is said to be the **imaginary part** of $a + bi$. To **add** or **subtract complex numbers**, we add or subtract the real parts, and we add or subtract the imaginary parts.	Add or subtract. $$(-3 + 4i) + (5 - 8i) = (-3 + 5) + (4i - 8i)$$ $$= 2 - 4i;$$ $$(6 - 7i) - (10 + 3i) = (6 - 10) + (-7i - 3i)$$ $$= -4 - 10i$$
When we **multiply complex numbers**, we must keep in mind the fact that $i^2 = -1$. Note that $\sqrt{a} \cdot \sqrt{b} \neq \sqrt{ab}$ when $\sqrt{a}$ and $\sqrt{b}$ are not real numbers.	Multiply. $$\sqrt{-4} \cdot \sqrt{-100} = \sqrt{-1} \cdot \sqrt{4} \cdot \sqrt{-1} \cdot \sqrt{100}$$ $$= i \cdot 2 \cdot i \cdot 10$$ $$= i^2 \cdot 20$$ $$= -1 \cdot 20 \qquad i^2 = -1$$ $$= -20;$$ $$(2 - 5i)(3 + i) = 6 + 2i - 15i - 5i^2$$ $$= 6 - 13i - 5(-1)$$ $$= 6 - 13i + 5$$ $$= 11 - 13i$$
The **conjugate of a complex number** $a + bi$ is $a - bi$. The numbers $a + bi$ and $a - bi$ are **complex conjugates**. Conjugates are used when we **divide complex numbers**.	Divide. $$\frac{5 - 2i}{3 + i} = \frac{5 - 2i}{3 + i} \cdot \frac{3 - i}{3 - i} \qquad \begin{array}{l} 3 - i \text{ is the conjugate} \\ \text{of the divisor, } 3 + i. \end{array}$$ $$= \frac{15 - 5i - 6i + 2i^2}{9 - i^2}$$ $$= \frac{15 - 11i - 2}{9 + 1} \qquad i^2 = -1$$ $$= \frac{13 - 11i}{10}$$ $$= \frac{13}{10} - \frac{11}{10}i$$

SECTION 3.2: QUADRATIC EQUATIONS, FUNCTIONS, ZEROS, AND MODELS

A **quadratic equation** is an equation that can be written in the form

$$ax^2 + bx + c = 0, \quad a \neq 0,$$

where a, b, and c are real numbers.

A **quadratic function** f is a function that can be written in the form

$$f(x) = ax^2 + bx + c, \quad a \neq 0,$$

where a, b, and c are real numbers.

The **zeros** of a quadratic function $f(x) = ax^2 + bx + c$ are the *solutions* of the associated quadratic equation $ax^2 + bx + c = 0$.

$3x^2 - 2x + 4 = 0$ and $5 - 4x = x^2$ are examples of quadratic equations. The equation $3x^2 - 2x + 4 = 0$ is written in **standard form**.

The functions $f(x) = 2x^2 + x + 1$ and $f(x) = 5x^2 - 4$ are examples of quadratic functions.

The Principle of Zero Products

If $ab = 0$ is true, then $a = 0$ *or* $b = 0$, and if $a = 0$ *or* $b = 0$, then $ab = 0$.

Solve: $3x^2 - 4 = 11x$.

$$3x^2 - 4 = 11x$$

$$3x^2 - 11x - 4 = 0 \qquad \text{Subtracting } 11x \text{ on both sides to get 0 on one side of the equation}$$

$$(3x + 1)(x - 4) = 0 \qquad \text{Factoring}$$

$$3x + 1 = 0 \quad or \quad x - 4 = 0 \qquad \text{Using the principle of zero products}$$

$$3x = -1 \quad or \qquad x = 4$$

$$x = -\frac{1}{3} \quad or \qquad x = 4$$

The solutions are $-\dfrac{1}{3}$ and 4.

The Principle of Square Roots

If $x^2 = k$, then $x = \sqrt{k}$ *or* $x = -\sqrt{k}$.

Solve: $3x^2 - 18 = 0$.

$$3x^2 - 18 = 0$$

$$3x^2 = 18 \qquad \text{Adding 18 on both sides}$$

$$x^2 = 6 \qquad \text{Dividing by 3 on both sides}$$

$$x = \sqrt{6} \quad or \quad x = -\sqrt{6} \qquad \text{Using the principle of square roots}$$

The solutions are $\sqrt{6}$ and $-\sqrt{6}$, or $\pm\sqrt{6}$.

To solve a quadratic equation by **completing the square**:

1. Isolate the terms with variables on one side of the equation and arrange them in descending order.
2. Divide by the coefficient of the squared term if that coefficient is not 1.
3. Complete the square by taking half the coefficient of the first-degree term and adding its square on both sides of the equation.
4. Express one side of the equation as the square of a binomial.
5. Use the principle of square roots.
6. Solve for the variable.

Solve: $2x^2 - 3 = 6x$.

$$2x^2 - 3 = 6x$$

$$2x^2 - 6x - 3 = 0 \qquad \text{Subtracting } 6x$$

$$2x^2 - 6x \phantom{{}- 3} = 3 \qquad \text{Adding 3}$$

$$x^2 - 3x \phantom{{}- 3} = \frac{3}{2} \qquad \text{Dividing by 2 to make the } x^2\text{-coefficient 1}$$

$$x^2 - 3x + \frac{9}{4} = \frac{3}{2} + \frac{9}{4} \qquad \begin{array}{l}\text{Completing the square:}\\ \frac{1}{2}(-3) = -\frac{3}{2} \text{ and}\\ \left(-\frac{3}{2}\right)^2 = \frac{9}{4}; \text{ adding } \frac{9}{4}\end{array}$$

$$\left(x - \frac{3}{2}\right)^2 = \frac{15}{4} \qquad \text{Factoring and simplifying}$$

$$x - \frac{3}{2} = \pm\frac{\sqrt{15}}{2} \qquad \begin{array}{l}\text{Using the principle of square}\\ \text{roots and the quotient rule}\\ \text{for radicals}\end{array}$$

$$x = \frac{3}{2} \pm \frac{\sqrt{15}}{2} = \frac{3 \pm \sqrt{15}}{2}$$

The solutions are $\dfrac{3 + \sqrt{15}}{2}$ and $\dfrac{3 - \sqrt{15}}{2}$, or $\dfrac{3 \pm \sqrt{15}}{2}$.

The solutions of $ax^2 + bx + c = 0$, $a \neq 0$, can be found using the **quadratic formula**:

$$x = \frac{-b \pm \sqrt{b^2 - 4ac}}{2a}.$$

Solve: $x^2 - 6 = 3x$.

$$x^2 - 6 = 3x$$

$$x^2 - 3x - 6 = 0 \qquad \text{Standard form}$$

$$a = 1, b = -3, c = -6$$

$$x = \frac{-b \pm \sqrt{b^2 - 4ac}}{2a}$$

$$= \frac{-(-3) \pm \sqrt{(-3)^2 - 4(1)(-6)}}{2 \cdot 1}$$

$$= \frac{3 \pm \sqrt{9 + 24}}{2}$$

$$= \frac{3 \pm \sqrt{33}}{2} \qquad \text{Exact solutions}$$

Using a calculator, we approximate the solutions to be 4.372 and −1.372.

Discriminant

For $ax^2 + bx + c = 0$, where a, b, and c are real numbers:

$b^2 - 4ac = 0 \longrightarrow$ One real-number solution;

$b^2 - 4ac > 0 \longrightarrow$ Two different real-number solutions;

$b^2 - 4ac < 0 \longrightarrow$ Two different imaginary-number solutions, complex conjugates.

For the equation above, $x^2 - 6 = 3x$, we see that $b^2 - 4ac$ is 33. Since 33 is positive, there are two different real-number solutions.

For $2x^2 - x + 4 = 0$, with $a = 2$, $b = -1$, and $c = 4$, the discriminant, $(-1)^2 - 4 \cdot 2 \cdot 4 = 1 - 32 = -31$, is negative, so there are two different imaginary-number (or nonreal) solutions.

For $x^2 - 6x + 9 = 0$, with $a = 1$, $b = -6$, and $c = 9$, the discriminant, $(-6)^2 - 4 \cdot 1 \cdot 9 = 36 - 36 = 0$, is 0 so there is one real-number solution.

Equations **reducible to quadratic**, or **quadratic in form**, can be treated as quadratic equations if a suitable substitution is made.

Solve: $x^4 - x^2 - 12 = 0$.

$$x^4 - x^2 - 12 = 0 \qquad \text{Let } u = x^2. \text{ Then } u^2 = (x^2)^2 = x^4.$$

$$u^2 - u - 12 = 0 \qquad \textbf{Substituting}$$

$$(u - 4)(u + 3) = 0$$

$$u - 4 = 0 \quad \textit{or} \quad u + 3 = 0$$

$$u = 4 \quad \textit{or} \quad u = -3 \qquad \textbf{Solving for } u$$

$$x^2 = 4 \quad \textit{or} \quad x^2 = -3$$

$$x = \pm 2 \quad \textit{or} \quad x = \pm\sqrt{3}i \qquad \textbf{Solving for } x$$

The solutions are 2, -2, $\sqrt{3}i$, and $-\sqrt{3}i$.

SECTION 3.3: ANALYZING GRAPHS OF QUADRATIC FUNCTIONS

Graphing Quadratic Equations

The graph of the function $f(x) = a(x - h)^2 + k$ is a parabola that:

- opens up if $a > 0$ and down if $a < 0$;
- has (h, k) as the vertex;
- has $x = h$ as the axis of symmetry;
- has k as a minimum value (output) if $a > 0$;
- has k as a maximum value if $a < 0$.

We can use a modification of the technique of completing the square as an aid in analyzing and graphing quadratic functions.

Find the vertex, the axis of symmetry, and the maximum or minimum value of $f(x) = 2x^2 + 12x + 12$.

$$
\begin{aligned}
f(x) &= 2x^2 + 12x + 12 \\
&= 2(x^2 + 6x) + 12 && \text{\textbf{Note that 9} } \\
& && \text{\textbf{completes the square}} \\
& && \text{\textbf{for } } x^2 + 6x. \\
&= 2(x^2 + 6x + 9 - 9) + 12 && \text{\textbf{Adding } } 9 - 9, \\
& && \text{\textbf{or 0, inside the}} \\
& && \text{\textbf{parentheses}} \\
&= 2(x^2 + 6x + 9) - 2 \cdot 9 + 12 && \text{\textbf{Using the}} \\
& && \text{\textbf{distributive law}} \\
& && \text{\textbf{to remove } } -9 \text{ \textbf{from}} \\
& && \text{\textbf{within the}} \\
& && \text{\textbf{parentheses}} \\
&= 2(x + 3)^2 - 6 \\
&= 2[x - (-3)]^2 + (-6)
\end{aligned}
$$

The function is now written in the form

$$f(x) = a(x - h)^2 + k$$

with $a = 2$, $h = -3$, and $k = -6$. Because $a > 0$, we know the graph opens up and thus the function has a minimum value. We also know the following:

Vertex (h, k): $(-3, -6)$;

Axis of symmetry $x = h$: $x = -3$;

Minimum value of the function k: -6.

To graph the function, we first plot the vertex and then find several points on either side of it. We plot these points and connect them with a smooth curve.

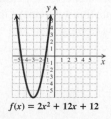

$f(x) = 2x^2 + 12x + 12$

The Vertex of a Parabola

The **vertex** of the graph of $f(x) = ax^2 + bx + c$ is

$$\left(-\frac{b}{2a}, f\left(-\frac{b}{2a}\right)\right).$$

We calculate the   We substitute to
x-coordinate.   find the y-coordinate.

Find the vertex of the function $f(x) = -3x^2 + 6x + 1$.

$$-\frac{b}{2a} = -\frac{6}{2(-3)} = 1$$

$$f(1) = -3 \cdot 1^2 + 6 \cdot 1 + 1 = 4$$

The vertex is $(1, 4)$.

Some applied problems can be solved by finding the maximum or minimum value of a quadratic function.

See Examples 5–7 on pp. 191–193.

SECTION 3.4: SOLVING RATIONAL EQUATIONS AND RADICAL EQUATIONS

A **rational equation** is an equation containing one or more rational expressions. When we solve a rational equation, we usually first multiply by the least common denominator (LCD) of all the rational expressions to clear the fractions.

CAUTION! When we multiply by an expression containing a variable, we might not obtain an equation equivalent to the original equation, so we *must* check the possible solutions obtained by substituting them in the original equation.

Solve: $\dfrac{5}{x+2} - \dfrac{4}{x^2-4} = \dfrac{x-3}{x-2}$.

$$\frac{5}{x+2} - \frac{4}{x^2-4} = \frac{x-3}{x-2}$$

$$\frac{5}{x+2} - \frac{4}{(x+2)(x-2)} = \frac{x-3}{x-2} \qquad \begin{array}{l}\textbf{The LCD is}\\ \textbf{(x + 2)(x - 2).}\end{array}$$

$$(x+2)(x-2)\left(\frac{5}{x+2} - \frac{4}{(x+2)(x-2)}\right)$$

$$= (x+2)(x-2) \cdot \frac{x-3}{x-2}$$

$$5(x-2) - 4 = (x+2)(x-3)$$

$$5x - 10 - 4 = x^2 - x - 6$$

$$5x - 14 = x^2 - x - 6$$

$$0 = x^2 - 6x + 8$$

$$0 = (x-2)(x-4)$$

$$x - 2 = 0 \quad or \quad x - 4 = 0$$

$$x = 2 \quad or \qquad x = 4$$

The number 2 does not check, but 4 does. The solution is 4.

A **radical equation** is an equation that contains one or more radicals. We use the **principle of powers** to solve radical equations.

For any positive integer n:

If $a = b$ is true, then $a^n = b^n$ is true.

CAUTION! If $a^n = b^n$ is true, it is not necessarily true that $a = b$, so we *must* check the possible solutions obtained by substituting them in the original equation.

Solve: $\sqrt{x+2} + \sqrt{x-1} = 3$.

$$\sqrt{x+2} + \sqrt{x-1} = 3$$

$$\sqrt{x+2} = 3 - \sqrt{x-1} \qquad \textbf{Isolating one radical}$$

$$\left(\sqrt{x+2}\right)^2 = \left(3 - \sqrt{x-1}\right)^2$$

$$x + 2 = 9 - 6\sqrt{x-1} + (x-1)$$

$$x + 2 = 8 - 6\sqrt{x-1} + x$$

$$-6 = -6\sqrt{x-1} \qquad \begin{array}{l}\textbf{Isolating the}\\ \textbf{remaining radical}\end{array}$$

$$1 = \sqrt{x-1} \qquad \textbf{Dividing by } -6$$

$$1^2 = \left(\sqrt{x-1}\right)^2$$

$$1 = x - 1$$

$$2 = x$$

The number 2 checks. It is the solution.

SECTION 3.5: SOLVING EQUATIONS AND INEQUALITIES WITH ABSOLUTE VALUE

We use the following property to **solve equations with absolute value**.

For $a > 0$ and an algebraic expression X:

$|X| = a$ is equivalent to
$X = -a$ or $X = a$.

Solve: $|x + 1| = 4$.

$$|x + 1| = 4$$
$$x + 1 = -4 \quad or \quad x + 1 = 4$$
$$x = -5 \quad or \quad x = 3$$

Both numbers check. The solutions are -5 and 3.

The following properties are used to **solve inequalities with absolute value**.

For $a > 0$ and an algebraic expression X:

$|X| < a$ is equivalent to $-a < X < a$.
$|X| > a$ is equivalent to
$X < -a$ or $X > a$.

Similar statements hold for
$|X| \leq a$ and $|X| \geq a$.

Solve: $|x - 2| < 3$.

$$|x - 2| < 3$$
$$-3 < x - 2 < 3$$
$$-1 < x < 5 \quad \textbf{Adding 2}$$

The solution set is $\{x | -1 < x < 5\}$, or $(-1, 5)$.

$$|3x| \geq 6$$
$$3x \leq -6 \quad or \quad 3x \geq 6$$
$$x \leq -2 \quad or \quad x \geq 2 \quad \textbf{Dividing by 3}$$

The solution set is $\{x | x \leq -2 \text{ or } x \geq 2\}$, or $(-\infty, -2] \cup [2, \infty)$.

REVIEW EXERCISES

Determine whether the statement is true or false.

1. We can use the quadratic formula to solve any quadratic equation. **[3.2]**

2. The function $f(x) = -3(x + 4)^2 - 1$ has a maximum value. **[3.3]**

3. For any positive integer n, if $a^n = b^n$ is true, then $a = b$ is true. **[3.4]**

4. An equation with absolute value cannot have two negative-number solutions. **[3.5]**

Solve. **[3.2]**

5. $(2y + 5)(3y - 1) = 0$

6. $x^2 + 4x - 5 = 0$

7. $3x^2 + 2x = 8$

8. $5x^2 = 15$

9. $x^2 + 10 = 0$

Find the zero(s) of the function. **[3.2]**

10. $f(x) = x^2 - 2x + 1$

11. $f(x) = x^2 + 2x - 15$

12. $f(x) = 2x^2 - x - 5$

13. $f(x) = 3x^2 + 2x + 3$

Solve.

14. $\dfrac{5}{2x + 3} + \dfrac{1}{x - 6} = 0$ **[3.4]**

15. $\dfrac{3}{8x + 1} + \dfrac{8}{2x + 5} = 1$ **[3.4]**

16. $\sqrt{5x + 1} - 1 = \sqrt{3x}$ **[3.4]**

17. $\sqrt{x - 1} - \sqrt{x - 4} = 1$ **[3.4]**

18. $|x - 4| = 3$ **[3.5]**

19. $|2y + 7| = 9$ **[3.5]**

Solve and write interval notation for the solution set. Then graph the solution set. **[3.5]**

20. $|5x| \geq 15$ **21.** $|3x + 4| < 10$

22. $|6x - 1| < 5$ **23.** $|x + 4| \geq 2$

24. Solve $\dfrac{1}{M} + \dfrac{1}{N} = \dfrac{1}{P}$ for P. **[3.4]**

Express in terms of i. **[3.1]**

25. $-\sqrt{-40}$

26. $\sqrt{-12} \cdot \sqrt{-20}$

27. $\dfrac{\sqrt{-49}}{-\sqrt{-64}}$

Simplify each of the following. Write the answer in the form $a + bi$, where a and b are real numbers. **[3.1]**

28. $(6 + 2i) + (-4 - 3i)$

29. $(3 - 5i) - (2 - i)$

30. $(6 + 2i)(-4 - 3i)$

31. $\dfrac{2 - 3i}{1 - 3i}$

32. i^{23}

Solve by completing the square to obtain exact solutions. Show your work. **[3.2]**

33. $x^2 - 3x = 18$

34. $3x^2 - 12x - 6 = 0$

Solve. Give exact solutions. **[3.2]**

35. $3x^2 + 10x = 8$

36. $r^2 - 2r + 10 = 0$

37. $x^2 = 10 + 3x$

38. $x = 2\sqrt{x} - 1$

39. $y^4 - 3y^2 + 1 = 0$

40. $(x^2 - 1)^2 - (x^2 - 1) - 2 = 0$

41. $(p - 3)(3p + 2)(p + 2) = 0$

42. $x^3 + 5x^2 - 4x - 20 = 0$

In Exercises 43 and 44, complete the square to:

a) *find the vertex;*

b) *find the axis of symmetry;*

c) *determine whether there is a maximum or a minimum value and find that value;*

d) *find the range; and*

e) *graph the function.* **[3.3]**

43. $f(x) = -4x^2 + 3x - 1$

44. $f(x) = 5x^2 - 10x + 3$

In Exercises 45–48, match the equation with one of the figures (a)–(d) that follow. **[3.3]**

a)

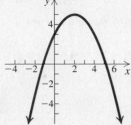

b)

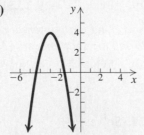

c)

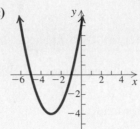

d)

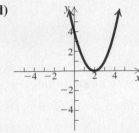

45. $y = (x - 2)^2$

46. $y = (x + 3)^2 - 4$

47. $y = -2(x + 3)^2 + 4$

48. $y = -\frac{1}{2}(x - 2)^2 + 5$

49. *Legs of a Right Triangle.*   The hypotenuse of a right triangle is 50 ft. One leg is 10 ft longer than the other. What are the lengths of the legs? **[3.2]**

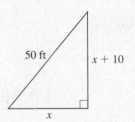

50. *Bicycling Speed.*   Harry and Rebecca leave a campsite, Harry biking due north and Rebecca biking due east. Harry bikes 7 km/h slower than Rebecca. After 4 hr, they are 68 km apart. Find the speed of each bicyclist. **[3.2]**

51. *Sidewalk Width.*   A 60-ft by 80-ft parking lot is torn up to install a sidewalk of uniform width around its perimeter. The new area of the parking lot is two-thirds of the old area. How wide is the sidewalk? **[3.2]**

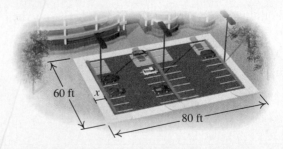

52. *Maximizing Volume.*   The Garcias have 24 ft of flexible fencing with which to build a rectangular "toy corral." If the fencing is 2 ft high, what dimensions should the corral have in order to maximize its volume? **[3.3]**

53. *Dimensions of a Box.* An open box is made from a 10-cm by 20-cm piece of aluminum by cutting a square from each corner and folding up the edges. The area of the resulting base is 90 cm². What is the length of the sides of the squares? **[3.2]**

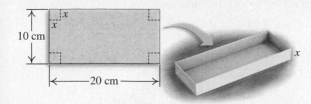

54. What are the zeros of $f(x) = 2x^2 - 5x + 1$? **[3.2]**

A. $\dfrac{5 \pm \sqrt{17}}{2}$ B. $\dfrac{5 \pm \sqrt{17}}{4}$

C. $\dfrac{5 \pm \sqrt{33}}{4}$ D. $\dfrac{-5 \pm \sqrt{17}}{4}$

55. Solve: $\sqrt{4x + 1} + \sqrt{2x} = 1$. **[3.4]**

A. There are two solutions.
B. There is only one solution. It is less than 1.
C. There is only one solution. It is greater than 1.
D. There is no solution.

56. The graph of $f(x) = (x - 2)^2 - 3$ is which of the following? **[3.3]**

A.
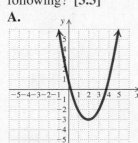

B.
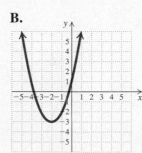

C.
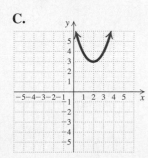

D.
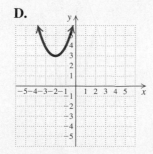

› Synthesis

Solve.

57. $\sqrt{\sqrt{\sqrt{\sqrt{x}}}} = 2$ **[3.4]**

58. $(t - 4)^{4/5} = 3$ **[3.4]**

59. $(x - 1)^{2/3} = 4$ **[3.4]**

60. $(2y - 2)^2 + y - 1 = 5$ **[3.2]**

61. $\sqrt{x + 2} + \sqrt[4]{x + 2} - 2 = 0$ **[3.2]**

62. At the beginning of the year, $3500 was deposited in a savings account. One year later, $4000 was deposited in another account. The interest rate was the same for both accounts. At the end of the second year, there was a total of $8518.35 in the accounts. What was the annual interest rate? **[3.2]**

63. Find b such that $f(x) = -3x^2 + bx - 1$ has a maximum value of 2. **[3.3]**

› Collaborative Discussion and Writing

64. Is the product of two imaginary numbers always an imaginary number? Explain your answer. **[3.1]**

65. Is it possible for a quadratic function to have one real zero and one imaginary zero? Why or why not? **[3.2]**

66. If the graphs of
$$f(x) = a_1(x - h_1)^2 + k_1$$
and
$$g(x) = a_2(x - h_2)^2 + k_2$$
have the same shape, what, if anything, can you conclude about the a's, the h's, and the k's? Explain your answer. **[3.3]**

67. Explain why it is necessary to check the possible solutions of a rational equation. **[3.4]**

68. Explain why it is necessary to check the possible solutions when the principle of powers is used to solve an equation. **[3.4]**

69. Explain why $|x| < p$ has no solution for $p \le 0$. **[3.5]**

70. Explain why all real numbers are solutions of $|x| > p$, for $p < 0$. **[3.5]**

3 Chapter Test

Solve. Find exact solutions.

1. $(2x - 1)(x + 5) = 0$

2. $6x^2 - 36 = 0$

3. $x^2 + 4 = 0$

4. $x^2 - 2x - 3 = 0$

5. $x^2 - 5x + 3 = 0$

6. $2t^2 - 3t + 4 = 0$

7. $x + 5\sqrt{x} - 36 = 0$

8. $\dfrac{3}{3x + 4} + \dfrac{2}{x - 1} = 2$

9. $\sqrt{x + 4} - 2 = 1$

10. $\sqrt{x + 4} - \sqrt{x - 4} = 2$

11. $|x + 4| = 7$

12. $|4y - 3| = 5$

Solve and write interval notation for the solution set. Then graph the solution set.

13. $|x + 3| \leq 4$

14. $|2x - 1| < 5$

15. $|x + 5| > 2$

16. $|3x - 5| \geq 7$

17. Solve $\dfrac{1}{A} + \dfrac{1}{B} = \dfrac{1}{C}$ for B.

18. Solve $R = \sqrt{3np}$ for n.

19. Solve $x^2 + 4x = 1$ by completing the square. Find the exact solutions. Show your work.

20. The tallest structure in the United States, at 2063 ft, is the KTHI-TV tower in Blanchard, North Dakota (*Source*: *The Cambridge Fact Finder*). How long would it take an object falling freely from the top to reach the ground? (Use the formula $s = 16t^2$, where s is the distance, in feet, that an object falls freely from rest in t seconds.)

Express in terms of i.

21. $\sqrt{-43}$

22. $-\sqrt{-25}$

Simplify.

23. $(5 - 2i) - (2 + 3i)$

24. $(3 + 4i)(2 - i)$

25. $\dfrac{1 - i}{6 + 2i}$

26. i^{33}

Find the zeros of each function.

27. $f(x) = 4x^2 - 11x - 3$

28. $f(x) = 2x^2 - x - 7$

29. For the graph of the function
$$f(x) = -x^2 + 2x + 8:$$
a) Find the vertex.
b) Find the axis of symmetry.
c) State whether there is a maximum or a minimum value and find that value.
d) Find the range.
e) Graph the function.

30. *Maximizing Area.* A homeowner wants to fence a rectangular play yard using 80 ft of fencing. The side of the house will be used as one side of the rectangle. Find the dimensions for which the area is a maximum.

31. The graph of $f(x) = (x - 1)^2 - 2$ is which of the following?

A.
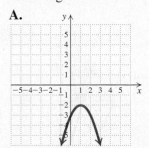

B.
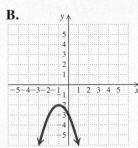

C.
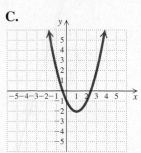

D.
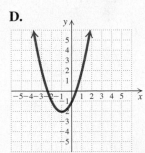

❯ Synthesis

32. Find a such that $f(x) = ax^2 - 4x + 3$ has a maximum value of 12.

Polynomial Functions and Rational Functions

APPLICATION

This problem appears as Example 9 in Section 4.1.

U.S. Army personnel on active duty make up approximately 37% of the total number in the U.S. military. Midyear numbers, in thousands, of U.S. Army personnel for selected years are listed in the table on p. 228.

Model the data with a quadratic function, a cubic function, and a quartic function. Then using R^2, the coefficient of determination, decide which function is the best fit.

219

4.1 Polynomial Functions and Modeling

> ❯ Determine the behavior of the graph of a polynomial function using the leading-term test.

> ❯ Factor polynomial functions and find their zeros and their multiplicities.

> ❯ Use a graphing calculator to graph a polynomial function and find its real-number zeros, its relative maximum and minimum values, and its domain and range.

> ❯ Solve applied problems using polynomial models; fit quadratic, cubic, and quartic polynomial functions to data.

There are many different kinds of functions. The constant, linear, and quadratic functions that we studied in Chapters 1 and 3 are part of a larger group of functions called *polynomial functions*.

POLYNOMIAL FUNCTION

A **polynomial function** P is given by

$$P(x) = a_n x^n + a_{n-1}x^{n-1} + a_{n-2}x^{n-2} + \cdots + a_1 x + a_0,$$

where the coefficients $a_n, a_{n-1}, \ldots, a_1, a_0$ are real numbers and the exponents are whole numbers.

Just in Time 10

The first nonzero coefficient, a_n, is called the **leading coefficient**. The term $a_n x^n$ is called the **leading term**. The **degree** of the polynomial function is n. Some examples of polynomial functions follow.

POLYNOMIAL FUNCTION	EXAMPLE	DEGREE	LEADING TERM	LEADING COEFFICIENT
Constant	$f(x) = 3 \quad (f(x) = 3 = 3x^0)$	0	3	3
Linear	$f(x) = \frac{2}{3}x + 5 \quad \left(f(x) = \frac{2}{3}x + 5 = \frac{2}{3}x^1 + 5\right)$	1	$\frac{2}{3}x$	$\frac{2}{3}$
Quadratic	$f(x) = 4x^2 - x + 3$	2	$4x^2$	4
Cubic	$f(x) = x^3 + 2x^2 + x - 5$	3	x^3	1
Quartic	$f(x) = -x^4 - 1.1x^3 + 0.3x^2 - 2.8x - 1.7$	4	$-x^4$	-1

The function $f(x) = 0$ can be described in many ways:

$$f(x) = 0 = 0x^2 = 0x^{15} = 0x^{48},$$

and so on. For this reason, we say that the constant function $f(x) = 0$ has no degree. Functions such as

$$f(x) = \frac{2}{x} + 5, \text{ or } 2x^{-1} + 5, \quad \text{and} \quad g(x) = \sqrt{x} - 6, \text{ or } x^{1/2} - 6,$$

are *not* polynomial functions because the exponents -1 and $\frac{1}{2}$ are *not* whole numbers.

From our study of functions in Chapters 1–3, we know how to find or at least estimate many characteristics of a polynomial function. Let's consider two examples for review.

Quadratic Function

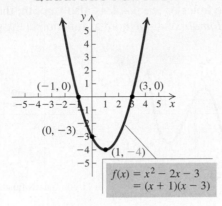

Function: $f(x) = x^2 - 2x - 3$
$= (x + 1)(x - 3)$

Zeros: $-1, 3$

x-intercepts: $(-1, 0), (3, 0)$

y-intercept: $(0, -3)$

Minimum: -4 at $x = 1$

Maximum: None

Domain: All real numbers, $(-\infty, \infty)$

Range: $[-4, \infty)$

Cubic Function

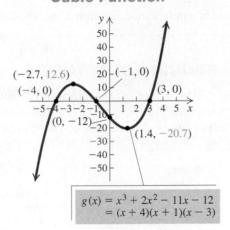

Function: $g(x) = x^3 + 2x^2 - 11x - 12$
$= (x + 4)(x + 1)(x - 3)$

Zeros: $-4, -1, 3$

x-intercepts: $(-4, 0), (-1, 0), (3, 0)$

y-intercept: $(0, -12)$

Relative minimum: -20.7 at $x = 1.4$

Relative maximum: 12.6 at $x = -2.7$

Domain: All real numbers, $(-\infty, \infty)$

Range: All real numbers, $(-\infty, \infty)$

All graphs of polynomial functions have some characteristics in common. Compare the following graphs. How do the graphs of polynomial functions differ from the graphs of nonpolynomial functions? Describe some characteristics of the graphs of polynomial functions that you observe.

Polynomial Functions

$f(x) = x^2 + 3x + 1$

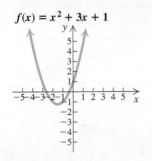

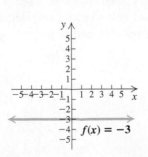

$f(x) = 2x^3 + x^2 + x - 1$

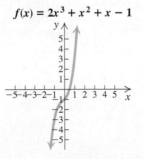

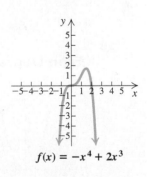

$f(x) = -x^4 + 2x^3$

Nonpolynomial Functions

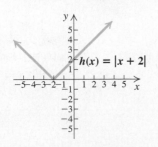

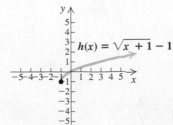

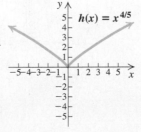

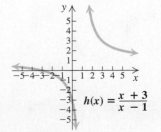

DOMAIN OF A FUNCTION

REVIEW SECTION 1.2.

You probably noted that the graph of a polynomial function is *continuous*; that is, it has no holes or breaks. It is also smooth; there are no sharp corners. Furthermore, the *domain* of a polynomial function is the set of all real numbers, $(-\infty, \infty)$.

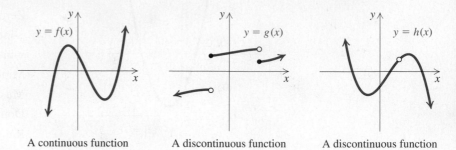

A continuous function   A discontinuous function   A discontinuous function

The *domain* of a polynomial function is the set of all real numbers, $(-\infty, \infty)$.

❯ The Leading-Term Test

The behavior of the graph of a polynomial function as x becomes very large $(x \rightarrow \infty)$ or very small $(x \rightarrow -\infty)$ is referred to as the end behavior of the graph. The leading term of a polynomial function determines its end behavior.

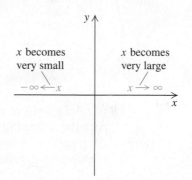

Using the graphs shown below, let's see if we can discover some general patterns by comparing the end behavior of even-degree and odd-degree functions. We also observe the effect of positive and negative leading coefficients.

Even Degree

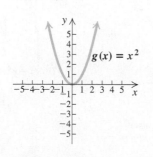

$g(x) = x^2$

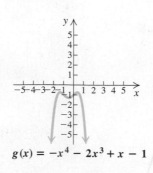

$g(x) = -x^4 - 2x^3 + x - 1$

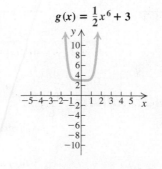

$g(x) = \frac{1}{2}x^6 + 3$

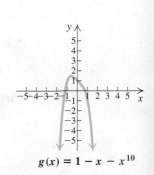

$g(x) = 1 - x - x^{10}$

Odd Degree

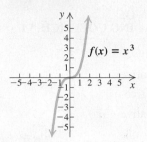

$$f(x) = x^3$$

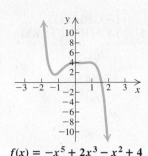

$$f(x) = -x^5 + 2x^3 - x^2 + 4$$

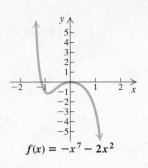

$$f(x) = -x^7 - 2x^2$$

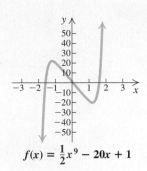

$$f(x) = \tfrac{1}{2}x^9 - 20x + 1$$

We can summarize our observations as follows.

THE LEADING-TERM TEST

If $a_n x^n$ is the leading term of a polynomial function, then the behavior of the graph as $x \to \infty$ or as $x \to -\infty$ can be described in one of the four following ways.

n	$a_n > 0$	$a_n < 0$
Even	⌣	⌢
Odd	∿	∿

The ∿ portion of the graph is not determined by this test.

EXAMPLE 1 Using the leading-term test, match each of the following functions with one of the graphs A–D that follow.

a) $f(x) = 3x^4 - 2x^3 + 3$

b) $f(x) = -5x^3 - x^2 + 4x + 2$

c) $f(x) = x^5 + \tfrac{1}{4}x + 1$

d) $f(x) = -x^6 + x^5 - 4x^3$

A.

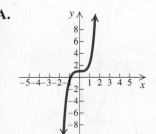

B.

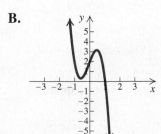

C.

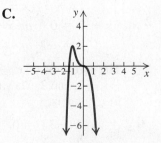

D.

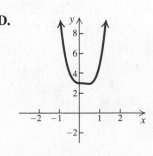

Solution

	LEADING TERM	DEGREE OF LEADING TERM	SIGN OF LEADING COEFFICIENT	GRAPH
a)	$3x^4$	4, even	Positive	D
b)	$-5x^3$	3, odd	Negative	B
c)	x^5	5, odd	Positive	A
d)	$-x^6$	6, even	Negative	C

> **Now Try Exercise 19.**

❯ Finding Zeros of Factored Polynomial Functions

Let's review the meaning of the real zeros of a function and their connection to the *x*-intercepts of the function's graph.

CONNECTING THE CONCEPTS   Zeros, Solutions, and Intercepts

FUNCTION

ZEROS OF THE FUNCTION; SOLUTIONS OF THE EQUATION

ZEROS OF THE FUNCTION; *x*-INTERCEPTS OF THE GRAPH

Quadratic Polynomial

$$g(x) = x^2 - 2x - 8$$
$$= (x + 2)(x - 4),$$
or
$$y = (x + 2)(x - 4)$$

To find the **zeros** of $g(x)$, we solve $g(x) = 0$:

$$x^2 - 2x - 8 = 0$$
$$(x + 2)(x - 4) = 0$$
$$x + 2 = 0 \quad or \quad x - 4 = 0$$
$$x = -2 \quad or \qquad x = 4.$$

The **solutions** of $x^2 - 2x - 8 = 0$ are -2 and 4. They are the zeros of the function $g(x)$; that is,

$$g(-2) = 0 \quad and \quad g(4) = 0.$$

The real-number zeros of $g(x)$ are the *x*-coordinates of the **x-intercepts** of the graph of $y = g(x)$.

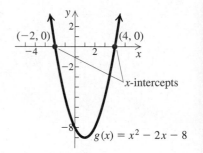

Cubic Polynomial

$$h(x)$$
$$= x^3 + 2x^2 - 5x - 6$$
$$= (x + 3)(x + 1)(x - 2),$$
or
$$y = (x + 3)(x + 1)(x - 2)$$

To find the **zeros** of $h(x)$, we solve $h(x) = 0$:

$$x^3 + 2x^2 - 5x - 6 = 0$$
$$(x + 3)(x + 1)(x - 2) = 0$$
$$x + 3 = 0 \quad or \: x + 1 = 0 \quad or \: x - 2 = 0$$
$$x = -3 \: or \qquad x = -1 \: or \qquad x = 2.$$

The **solutions** of $x^3 + 2x^2 - 5x - 6 = 0$ are $-3, -1$, and 2. They are the zeros of the function $h(x)$; that is,

$$h(-3) = 0,$$
$$h(-1) = 0, \quad and$$
$$h(2) = 0.$$

The real-number zeros of $h(x)$ are the *x*-coordinates of the **x-intercepts** of the graph of $y = h(x)$.

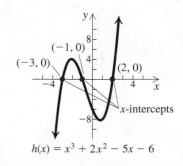

The connection between the real-number zeros of a function and the x-intercepts of the graph of the function is easily seen in the preceding examples. If c is a real zero of a function (that is, if $f(c) = 0$), then $(c, 0)$ is an x-intercept of the graph of the function.

EXAMPLE 2   Consider $P(x) = x^3 + x^2 - 17x + 15$. Determine whether each of the numbers 2 and -5 is a zero of $P(x)$.

Solution   We first evaluate $P(2)$:

$$P(2) = (2)^3 + (2)^2 - 17(2) + 15 = -7. \qquad \text{Substituting 2 into the polynomial}$$

Since $P(2) \neq 0$, we know that 2 is *not* a zero of the polynomial function.
    We then evaluate $P(-5)$:

$$P(-5) = (-5)^3 + (-5)^2 - 17(-5) + 15 = 0. \qquad \text{Substituting } -5 \text{ into the polynomial}$$

Since $P(-5) = 0$, we know that -5 is a zero of $P(x)$.   ❯ **Now Try Exercise 23.**

Let's take a closer look at the polynomial function

$$h(x) = x^3 + 2x^2 - 5x - 6$$

(see Connecting the Concepts on p. 224). The factors of $h(x)$ are

$$x + 3, \qquad x + 1, \quad \text{and} \quad x - 2,$$

and the zeros are

$$-3, \qquad -1, \quad \text{and} \quad 2.$$

We note that when the polynomial is expressed as a product of linear factors, each factor determines a zero of the function. Thus if we know the linear factors of a polynomial function $f(x)$, we can easily find the zeros of $f(x)$ by solving the equation $f(x) = 0$ using the principle of zero products.

EXAMPLE 3   Find the zeros of

$$f(x) = 5(x - 2)(x - 2)(x - 2)(x + 1) = 5(x - 2)^3(x + 1).$$

Solution   To solve the equation $f(x) = 0$, we use the principle of zero products, solving $x - 2 = 0$ and $x + 1 = 0$. The zeros of $f(x)$ are 2 and -1. (See Fig. 1.)   ❯

EXAMPLE 4   Find the zeros of

$$g(x) = -(x - 1)(x - 1)(x + 2)(x + 2) = -(x - 1)^2(x + 2)^2.$$

Solution   To solve the equation $g(x) = 0$, we use the principle of zero products, solving $x - 1 = 0$ and $x + 2 = 0$. The zeros of $g(x)$ are 1 and -2. (See Fig. 2.)   ❯

Let's consider the occurrences of the zeros in the functions in Examples 3 and 4 and their relationship to the graphs of those functions. In Example 3, the factor $x - 2$ occurs three times. In a case like this, we say that the zero we obtain from this factor, 2, has a **multiplicity** of 3. The factor $x + 1$ occurs one time. The zero we obtain from this factor, -1, has a *multiplicity* of 1.

In Example 4, the factors $x - 1$ and $x + 2$ each occur two times. Thus both zeros, 1 and -2, have a *multiplicity* of 2.

Note, in Example 3, that the zeros have odd multiplicities and the graph crosses the x-axis at both -1 and 2. But in Example 4, the zeros have even multiplicities and the graph is tangent to (touches but does not cross) the x-axis at -2 and 1. This leads us to the following generalization.

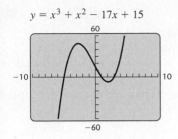

$y = x^3 + x^2 - 17x + 15$

Just in Time

19

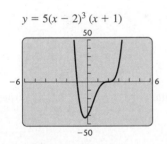

$y = 5(x - 2)^3 (x + 1)$

FIGURE 1

$y = -(x - 1)^2 (x + 2)^2$

FIGURE 2

> **EVEN MULTIPLICITY AND ODD MULTIPLICITY**
>
> If $(x - c)^k$, $k \geq 1$, is a factor of a polynomial function $P(x)$ and $(x - c)^{k+1}$ is not a factor and:
>
> - k is odd, then the graph crosses the x-axis at $(c, 0)$;
> - k is even, then the graph is tangent to the x-axis at $(c, 0)$.

Just in Time

14

Some polynomials can be factored by grouping. Then we use the principle of zero products to find their zeros.

EXAMPLE 5   Find the zeros of

$$f(x) = x^3 - 2x^2 - 9x + 18.$$

Solution   We factor by grouping, as follows:

$$
\begin{aligned}
f(x) &= x^3 - 2x^2 - 9x + 18 \\
&= x^2(x - 2) - 9(x - 2) \quad &&\text{Grouping } x^3 \text{ with } -2x^2 \text{ and } -9x \text{ with } \\
& &&\text{18 and factoring each group} \\
&= (x - 2)(x^2 - 9) \quad &&\text{Factoring out } x - 2 \\
&= (x - 2)(x + 3)(x - 3). \quad &&\text{Factoring } x^2 - 9
\end{aligned}
$$

$y = x^3 - 2x^2 - 9x + 18$

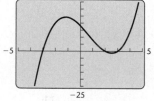

Then, by the principle of zero products, the solutions of the equation $f(x) = 0$ are 2, -3, and 3. These are the zeros of $f(x)$.

▶ **Now Try Exercise 39.**

Other factoring techniques can also be used.

EXAMPLE 6   Find the zeros of

$$f(x) = x^4 + 4x^2 - 45.$$

Solution   We factor as follows:

$$f(x) = x^4 + 4x^2 - 45 = (x^2 - 5)(x^2 + 9).$$

We now solve the equation $f(x) = 0$ to determine the zeros. We use the principle of zero products:

$$
\begin{aligned}
(x^2 - 5)(x^2 + 9) &= 0 \\
x^2 - 5 = 0 \quad &or \quad x^2 + 9 = 0 \\
x^2 = 5 \quad &or \quad x^2 = -9 \\
x = \pm\sqrt{5} \quad &or \quad x = \pm\sqrt{-9} = \pm 3i.
\end{aligned}
$$

The solutions are $\pm\sqrt{5}$ and $\pm 3i$. These are the zeros of $f(x)$.

▶ **Now Try Exercise 37.**

$y = x^4 + 4x^2 - 45$

Only the real-number zeros of a function correspond to the x-intercepts of its graph. For instance, the real-number zeros of the function in Example 6, $-\sqrt{5}$ and $\sqrt{5}$, can be seen on the graph of the function at left, but the nonreal zeros, $-3i$ and $3i$, cannot.

> Every polynomial function of degree n, with $n \geq 1$, has at least one zero and at most n zeros.

This is often stated as follows: "Every polynomial function of degree n, with $n \geq 1$, has *exactly* n zeros." This statement is compatible with the preceding statement, if one takes multiplicities into account.

❯ Finding Real Zeros on a Calculator

Finding exact values of the real zeros of a function can be difficult. We can find approximations using a graphing calculator.

EXAMPLE 7 Find the real zeros of the function f given by

$$f(x) = 0.1x^3 - 0.6x^2 - 0.1x + 2.$$

Approximate the zeros to three decimal places.

Solution We use a graphing calculator, trying to create a graph that clearly shows the curvature and the intercepts. Then we look for points where the graph crosses the x-axis. We know that there are no more than 3 because the degree of the polynomial is 3. It appears that there are three different zeros, one near -2, one near 2, and one near 6. We use the ZERO feature to find them.

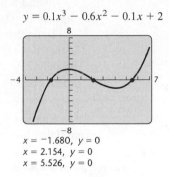

$$y = 0.1x^3 - 0.6x^2 - 0.1x + 2$$

$x = -1.680, \ y = 0$
$x = 2.154, \ y = 0$
$x = 5.526, \ y = 0$

The zeros are approximately -1.680, 2.154, and 5.526.   ❯ *Now Try Exercise 43.*

❯ Polynomial Models

Polynomial functions have many uses as models in science, engineering, and business. The simplest use of polynomial functions in applied problems occurs when we merely evaluate a polynomial function. In such cases, a model has already been developed.

EXAMPLE 8 *Ibuprofen in the Bloodstream.* The polynomial function

$$M(t) = 0.5t^4 + 3.45t^3 - 96.65t^2 + 347.7t$$

can be used to estimate the number of milligrams of the pain relief medication ibuprofen in the bloodstream t hours after 400 mg of the medication has been taken.

a) Find the number of milligrams in the bloodstream at $t = 0, 0.5, 1, 1.5$, and so on, up to 6 hr. Round the function values to the nearest tenth.

b) Find the domain, the relative maximum and where it occurs, and the range.

Solution

a) We can evaluate the function with the TABLE feature of a graphing calculator set in AUTO mode. We start at 0 and use a step-value of 0.5.

$M(0) = 0,$	$M(3.5) = 255.9,$
$M(0.5) = 150.2,$	$M(4) = 193.2,$
$M(1) = 255,$	$M(4.5) = 126.9,$
$M(1.5) = 318.3,$	$M(5) = 66,$
$M(2) = 344.4,$	$M(5.5) = 20.2,$
$M(2.5) = 338.6,$	$M(6) = 0.$
$M(3) = 306.9,$	

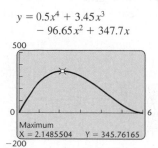

$y = 0.5x^4 + 3.45x^3 - 96.65x^2 + 347.7x$

b) Recall that the domain of a polynomial function, unless restricted by a statement of the function, is $(-\infty, \infty)$. The implications of this application restrict the domain of the function. If we assume that a patient had not taken any of the medication before, it seems reasonable that $M(0) = 0$; that is, at time 0, there is 0 mg of the medication in the bloodstream. After the medication has been taken, $M(t)$ will be positive for a period of time and eventually decrease back to 0 when $t = 6$ and not increase again (unless another dose is taken). Thus the restricted domain is $[0, 6]$.

To determine the range, we find the relative maximum value of the function using the MAXIMUM feature. The maximum is about 345.8 mg. It occurs approximately 2.15 hr, or 2 hr 9 min, after the initial dose has been taken. The range is about $[0, 345.8]$.

> ▶ *Now Try Exercise 63.*

In Chapter 1, we used regression to model data with linear functions. We now expand that procedure to include quadratic, cubic, and quartic models.

EXAMPLE 9 *Army Personnel on Active Duty.* U.S. Army personnel on active duty make up approximately 37% of the total number in the U.S. military. Midyear numbers, in thousands, of U.S. Army personnel for selected years are listed in the following table.

Year, x	Number of Active-Duty Army Personnel (in thousands)
2005, 0	492.7
2007, 2	522.0
2008, 3	539.2
2010, 5	566.0
2012, 7	550.1
2014, 9	512.1

Source: Department of the Army, U.S. Department of Defense

Looking at the table above, we note that the data, in thousands of active-duty Army personnel, could be modeled with a quadratic function, a cubic function, or a quartic function.

a) Model the data with a quadratic function, a cubic function, and a quartic function. Let the first coordinate of each data point be the number of years after 2005; that is, enter the data as $(0, 492.7)$, $(2, 522.0)$, $(3, 539.2)$, and so on. Then using R^2, the **coefficient of determination**, decide which function is the best fit.

b) Graph the function with the scatterplot of the data.

c) Use the answer to part (a) to estimate the number of active-duty Army personnel in 2006, in 2009, in 2013, and in 2020.

Solution

a) Using the REGRESSION feature with DIAGNOSTIC turned on, we get the following.

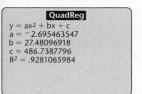

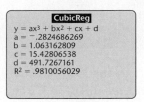

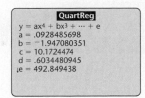

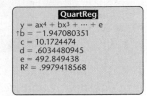

The R^2-value for the quartic function is closer to 1 than the R^2-values for the quadratic function and the cubic function. Thus the quartic function is the best fit:

$$f(x) = 0.0928485698x^4 - 1.947080351x^3$$
$$+ 10.1724474x^2 + 0.6034480945x$$
$$+ 492.849438.$$

b) The scatterplot and the graph are shown below.

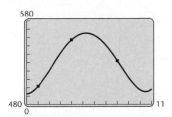

c) We evaluate the function found in part (a).

X	Y₁	
1	501.77	
4	557.18	
8	532.12	
15	919.76	
X =		

Using this function, we can estimate the number of active-duty Army personnel as 501,800 in 2006, 557,200 in 2009, and 532,100 in 2013. Looking at the given data in the table above, we see that these estimates appear to be fairly accurate.

If we use the function to estimate the number of active-duty Army personnel in 2020, we find an answer of about 919,800. This estimate is not realistic since it is not reasonable to expect the number of active-duty Army personnel to increase by 407,700 from 2014 to 2020.

The quartic model has a higher value for R^2 than the quadratic model or the cubic model over the domain of the data, but this number does not reflect the degree of accuracy for extended values. It is always important when using regression to analyze predictions with common sense and knowledge of current trends.

> **Now Try Exercise 77.**

4.1 Exercise Set

Determine the leading term, the leading coefficient, and the degree of the polynomial. Then classify the polynomial function as constant, linear, quadratic, cubic, or quartic.

1. $g(x) = \frac{1}{2}x^3 - 10x + 8$

2. $f(x) = 15x^2 - 10 + 0.11x^4 - 7x^3$

3. $h(x) = 0.9x - 0.13$

4. $f(x) = -6$

5. $g(x) = 305x^4 + 4021$

6. $h(x) = 2.4x^3 + 5x^2 - x + \frac{7}{8}$

7. $h(x) = -5x^2 + 7x^3 + x^4$

8. $f(x) = 2 - x^2$

9. $g(x) = 4x^3 - \frac{1}{2}x^2 + 8$

10. $f(x) = 12 + x$

In Exercises 11–18, select one of the four sketches (a)–(d) that follow to describe the end behavior of the graph of the function.

a)

b)

c)

d)

11. $f(x) = -3x^3 - x + 4$

12. $f(x) = \frac{1}{4}x^4 + \frac{1}{2}x^3 - 6x^2 + x - 5$

13. $f(x) = -x^6 + \frac{3}{4}x^4$

14. $f(x) = \frac{2}{5}x^5 - 2x^4 + x^3 - \frac{1}{2}x + 3$

15. $f(x) = -3.5x^4 + x^6 + 0.1x^7$

16. $f(x) = -x^3 + x^5 - 0.5x^6$

17. $f(x) = 10 + \frac{1}{10}x^4 - \frac{2}{5}x^3$

18. $f(x) = 2x + x^3 - x^5$

In Exercises 19–22, use the leading-term test to match the function with one of the graphs (a)–(d) that follow.

a)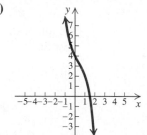

b)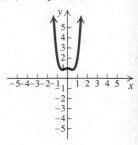

c)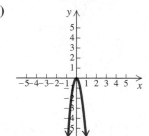

d)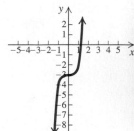

19. $f(x) = -x^6 + 2x^5 - 7x^2$

20. $f(x) = 2x^4 - x^2 + 1$

21. $f(x) = x^5 + \frac{1}{10}x - 3$

22. $f(x) = -x^3 + x^2 - 2x + 4$

23. Use substitution to determine whether 4, 5, and -2 are zeros of
$$f(x) = x^3 - 9x^2 + 14x + 24.$$

24. Use substitution to determine whether 2, 3, and -1 are zeros of
$$f(x) = 2x^3 - 3x^2 + x + 6.$$

25. Use substitution to determine whether 2, 3, and -1 are zeros of
$$g(x) = x^4 - 6x^3 + 8x^2 + 6x - 9.$$

26. Use substitution to determine whether 1, -2, and 3 are zeros of
$$g(x) = x^4 - x^3 - 3x^2 + 5x - 2.$$

Find the zeros of the polynomial function and state the multiplicity of each.

27. $f(x) = (x + 3)^2(x - 1)$

28. $f(x) = (x + 5)^3(x - 4)(x + 1)^2$

29. $f(x) = -2(x - 4)(x - 4)(x - 4)(x + 6)$

30. $f(x) = \left(x + \frac{1}{2}\right)(x + 7)(x + 7)(x + 5)$

31. $f(x) = (x^2 - 9)^3$

32. $f(x) = (x^2 - 4)^2$

33. $f(x) = x^3(x - 1)^2(x + 4)$

34. $f(x) = x^2(x + 3)^2(x - 4)(x + 1)^4$

35. $f(x) = -8(x - 3)^2(x + 4)^3x^4$

36. $f(x) = (x^2 - 5x + 6)^2$

37. $f(x) = x^4 - 4x^2 + 3$

38. $f(x) = x^4 - 10x^2 + 9$

39. $f(x) = x^3 + 3x^2 - x - 3$

40. $f(x) = x^3 - x^2 - 2x + 2$

41. $f(x) = 2x^3 - x^2 - 8x + 4$

42. $f(x) = 3x^3 + x^2 - 48x - 16$

Using a graphing calculator, find the real zeros of the function. Approximate the zeros to three decimal places.

43. $f(x) = x^3 - 3x - 1$

44. $f(x) = x^3 + 3x^2 - 9x - 13$

45. $f(x) = x^4 - 2x^2$

46. $f(x) = x^4 - 2x^3 - 5.6$

47. $f(x) = x^3 - x$

48. $f(x) = 2x^3 - x^2 - 14x - 10$

49. $f(x) = x^8 + 8x^7 - 28x^6 - 56x^5 + 70x^4$
$\quad\quad + 56x^3 - 28x^2 - 8x + 1$

50. $f(x) = x^6 - 10x^5 + 13x^3 - 4x^2 - 5$

Using a graphing calculator, estimate the real zeros, the relative maxima and minima, and the range of the polynomial function.

51. $g(x) = x^3 - 1.2x + 1$

52. $h(x) = -\frac{1}{2}x^4 + 3x^3 - 5x^2 + 3x + 6$

53. $f(x) = x^6 - 3.8$

54. $h(x) = 2x^3 - x^4 + 20$

55. $f(x) = x^2 + 10x - x^5$

56. $f(x) = 2x^4 - 5.6x^2 + 10$

Determine whether the statement is true or false.

57. If $P(x) = (x - 3)^4(x + 1)^3$, then the graph of the polynomial function $y = P(x)$ crosses the x-axis at $(3, 0)$.

58. If $P(x) = (x + 2)^2\left(x - \frac{1}{4}\right)^5$, then the graph of the polynomial function $y = P(x)$ crosses the x-axis at $\left(\frac{1}{4}, 0\right)$.

59. If $P(x) = (x - 2)^3(x + 5)^6$, then the graph of $y = P(x)$ is tangent to the x-axis at $(-5, 0)$.

60. If $P(x) = (x + 4)^2(x - 1)^2$, then the graph of $y = P(x)$ is tangent to the x-axis at $(4, 0)$.

61. *Vinyl Album Sales.* Vinyl record albums are making a comeback. Sales of vinyl albums rose 51% from 2013 to 2014. The sales data over the years 2001 to 2014 are modeled by the quartic function

$$f(x) = 0.000553x^4 - 0.00735x^3$$
$$+ 0.06496x^2 - 0.24441x$$
$$+ 1.41345,$$

where x is the number of years after 2001 and $f(x)$ is the number of albums in millions (*Source:* Nielsen SoundScan). Find the number of vinyl albums sold in 2008, in 2012, and in 2016.

62. *Railroad Miles.* The greatest combined length of U.S.-owned operating railroad track existed in 1916, when industrial activity increased during World War I. The total length has decreased ever since. The data over the years 1900 to 2011 are modeled by the quartic function

$$f(x) = -0.002391x^4 + 0.949686x^3$$
$$- 123.648199x^2 + 4729.3635x$$
$$+ 198,846.4097,$$

where x is the number of years after 1900 and $f(x)$ is in miles (*Source:* Association of American Railroads). Find the number of miles of operating railroad track in the United States in 1916, in 1960, in 2000, and in 2016.

63. *Print Book Sales.* The sales of print books increased about 7% from 2012 to 2014. The sales data over the years 2008 to 2014 are modeled by the cubic function

$$s(x) = 4.958x^3 - 38.125x^2 + 26.417x + 778,$$

where x is the number of years after 2008 and $s(x)$ is the number of print books, in millions (*Source:* Nielsen). Find the number of print books sold in 2009, in 2014, and in 2015.

64. *Dog Years.* A dog's life span is typically much shorter than that of a human. The cubic function

$$d(x) = 0.010255x^3 - 0.340119x^2 + 7.397499x + 6.618361,$$

where x is the dog's age, in years, approximates the equivalent human age in years. Estimate the equivalent human age for dogs that are 3, 12, and 16 years old.

65. *Projectile Motion.* A stone thrown downward with an initial velocity of 34.3 m/sec will travel a distance of s meters, where

$$s(t) = 4.9t^2 + 34.3t$$

and t is in seconds. If a stone is thrown downward at 34.3 m/sec from a height of 294 m, how long will it take the stone to hit the ground?

66. *Games in a Sports League.* If there are x teams in a sports league and all the teams play each other twice, a total of $N(x)$ games are played, where

$$N(x) = x^2 - x.$$

A softball league has 9 teams, each of which plays the others twice. If the league pays $110 per game for the field and the umpires, how much will it cost to play the entire schedule?

67. *Prison Admissions.* Since 2006, total admissions to state and federal prisons have been declining (*Source:* Bureau of Justice Statistics). The quartic function

$$p(x) = 6.213x^4 - 432.347x^3 + 1922.987x^2 + 20,503.912x + 638,684.984,$$

where x is the number of years after 2001, can be used to estimate the number of admissions to state and federal prisons from 2001 to 2012. Estimate the number of prison admissions in 2003, in 2006, and in 2011.

68. *Obesity.* The percentage of adults in the United States who are obese is increasing (*Source:* Gallup–Healthways Well-Being Index). The cubic function

$$f(x) = 0.102x^3 - 0.764x^2 + 1.595x + 25.494,$$

where x is the number of years after 2008, can be used to estimate the percentage of adults who are obese. Using this function, estimate the percentage of adults who were obese in 2009 and in 2013.

69. *Interest Compounded Annually.* When P dollars is invested at interest rate i, compounded annually, for t years, the investment grows to A dollars, where

$$A = P(1 + i)^t.$$

Trevor's parents deposit $8000 in a savings account when Trevor is 16 years old. The principal plus interest is to be used for a truck when Trevor is 18 years old. Find the interest rate i if the $8000 grows to $9039.75 in 2 years.

70. *Interest Compounded Annually.* When P dollars is invested at interest rate i, compounded annually, for t years, the investment grows to A dollars, where

$$A = P(1 + i)^t.$$

When Sara enters the 11th grade, her grandparents deposit $10,000 in a college savings account. Find the interest rate i if the $10,000 grows to $11,193.64 in 2 years.

For the scatterplots and graphs in Exercises 71–76, determine which, if any, of the following functions might be used as a model for the data.

a) *Linear,* $f(x) = mx + b$
b) *Quadratic,* $f(x) = ax^2 + bx + c, a > 0$
c) *Quadratic,* $f(x) = ax^2 + bx + c, a < 0$
d) *Polynomial, not linear or quadratic*

71.

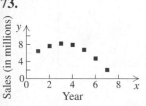

72.

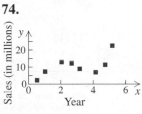

73.

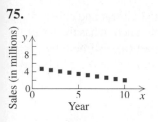

74.

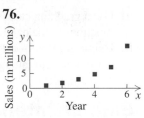

75.

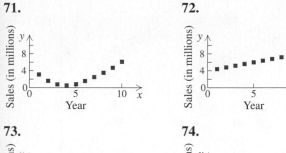

76.

77. *Foreign Adoptions.*   The number of foreign adoptions in the United States has declined in recent years, as shown in the following table.

Year, x	Number of U.S. Foreign Adoptions from Top 15 Countries, y
2003, 0	21,320
2004, 1	22,911
2005, 2	22,710
2006, 3	20,705
2007, 4	19,741
2008, 5	17,229
2009, 6	12,782
2010, 7	11,059
2011, 8	9,320
2012, 9	8,668
2013, 10	7,094

Sources: Office of Immigration Statistics; Department of Homeland Security

a) Use a graphing calculator to fit quadratic, cubic, and quartic functions to the data. Let x represent the number of years after 2003. Using R^2-values, determine which function is the best fit.
b) Using the function found in part (a), estimate the number of U.S. foreign adoptions in 2014.

78. *U.S. Farm Acreage.*   As the number of farms has decreased in the United States, the average size of the remaining farms has grown larger, as shown in the following table.

Year, x	Average Acreage per Farm, y
1900, 0	147
1910, 10	139
1920, 20	149
1930, 30	157
1940, 40	174
1950, 50	213
1960, 60	297
1970, 70	374
1980, 80	426
1990, 90	460
1995, 95	438
2000, 100	436
2003, 103	441
2005, 105	445
2009, 109	418
2013, 113	435

Sources: National Agricultural Statistics Service; U.S. Department of Agriculture

a) Use a graphing calculator to fit quadratic, cubic, and quartic functions to the data. Let x represent the number of years since 1900. Using R^2-values, determine which function is the best fit.

b) Using the function found in part (a), estimate the average acreage in 1955, in 1998, and in 2011.

79. *Classified Ad Revenue.* The following table lists the newspaper revenue, in billions of dollars, from classified ads for selected years from 1975 to 2013.

Year, x	Newspaper Revenue from Classified Ads, y (in billions of dollars)
1975, 0	$ 2.159
1980, 5	4.222
1985, 10	8.375
1990, 15	11.506
1995, 20	13.742
2000, 25	19.608
2005, 30	17.312
2006, 31	16.986
2007, 32	14.186
2008, 33	9.975
2009, 34	6.179
2010, 35	5.648
2011, 36	5.028
2012, 37	4.626
2013, 38	4.140

Source: Research Department, Newspaper Association of America

a) Use a graphing calculator to fit cubic and quartic functions to the data. Let x represent the number of years after 1975. Using R^2-values, determine which function is the better fit.

b) Using the function found in part (a), estimate the newspaper revenue from classified ads in 1988 and in 2002.

80. *Dog Years.* A dog's life span is typically much shorter than that of a human. Age equivalents for dogs and humans are listed in the following table.

Age of Dog, x (in years)	Human Age, $h(x)$ (in years)
0.25	5
0.5	10
1	15
2	24
4	32
6	40
8	48
10	56
14	72
18	91
21	106

Source: Based on "How to Determine a Dog's Age in Human Years," by Melissa Maroff, eHow Contributor, www.ehow.com

a) Use a graphing calculator to fit linear and cubic functions to the data. Which function has the better fit?

b) Using the function from part (a), estimate the equivalent human age for dogs that are 5, 11, and 15 years old.

❯ Skill Maintenance

Find the distance between the pair of points. **[1.1]**

81. $(3, -5)$ and $(0, -1)$

82. $(4, 2)$ and $(-2, -4)$

83. Find the center and the radius of the circle
$$(x - 3)^2 + (y + 5)^2 = 49. \ [1.1]$$

84. The diameter of a circle connects the points $(-6, 5)$ and $(-2, 1)$ on the circle. Find the coordinates of the center of the circle and the length of the radius. **[1.1]**

Solve.

85. $2y - 3 \geq 1 - y + 5$ **[1.6]**

86. $(x - 2)(x + 5) > x(x - 3)$ **[1.6]**

87. $|x + 6| \geq 7$ **[3.5]**

88. $|x + \frac{1}{4}| \leq \frac{2}{3}$ **[3.5]**

❯ Synthesis

Determine the degree and the leading term of the polynomial function.

89. $f(x) = (x^5 - 1)^2(x^2 + 2)^3$

90. $f(x) = (10 - 3x^5)^2(5 - x^4)^3(x + 4)$

4.2

Graphing Polynomial Functions

❯ Graph polynomial functions.

❯ Use the intermediate value theorem to determine whether a function has a real zero between two given real numbers.

❯ Graphing Polynomial Functions

In addition to using the leading-term test and finding the zeros of a polynomial function, it is helpful to consider the following facts when graphing the function.

If $P(x)$ is a polynomial function of degree n, then the graph of the function has:

- at most n real zeros, and thus at most n x-intercepts;
- at most $n - 1$ turning points.

(Turning points on a graph, also called relative maxima and minima, occur when the function changes from decreasing to increasing or from increasing to decreasing.)

EXAMPLE 1 Graph the polynomial function $h(x) = -2x^4 + 3x^3$.

Solution

1. First, we use the leading-term test to determine the end behavior of the graph. The leading term is $-2x^4$. The degree, 4, is even, and the coefficient, -2, is negative. Thus the end behavior of the graph as $x \to \infty$ and as $x \to -\infty$ can be sketched as follows.

2. The zeros of the function are the first coordinates of the x-intercepts of the graph. To find the zeros, we solve $h(x) = 0$ by factoring and using the principle of zero products.

$$-2x^4 + 3x^3 = 0$$
$$-x^3(2x - 3) = 0 \qquad \text{Factoring}$$
$$-x^3 = 0 \quad or \quad 2x - 3 = 0 \qquad \text{Using the principle of zero products}$$
$$x = 0 \quad or \qquad x = \tfrac{3}{2}.$$

The zeros of the function are 0 and $\frac{3}{2}$. Note that the multiplicity of 0 is 3 and the multiplicity of $\frac{3}{2}$ is 1. The x-intercepts are $(0, 0)$ and $\left(\frac{3}{2}, 0\right)$.

3. The zeros divide the x-axis into three intervals:

$$(-\infty, 0), \qquad \left(0, \tfrac{3}{2}\right), \quad \text{and} \quad \left(\tfrac{3}{2}, \infty\right).$$

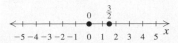

The sign of $h(x)$ is the same for all values of x in an interval. That is, $h(x)$ is positive for all x-values in an interval or $h(x)$ is negative for all x-values in an interval. To determine which, we choose a test value for x from each interval and find $h(x)$.

$y_1 = -2x^4 + 3x^3$

X	Y1	
−1	−5	
1	1	
2	−8	

X =

Interval	$(-\infty, 0)$	$\left(0, \frac{3}{2}\right)$	$\left(\frac{3}{2}, \infty\right)$
Test Value, x	−1	1	2
Function Value, $h(x)$	−5	1	−8
Sign of $h(x)$	−	+	−
Location of Points on Graph	Below x-axis	Above x-axis	Below x-axis

This test-point procedure also gives us three points to plot. In this case, we have $(-1, -5)$, $(1, 1)$, and $(2, -8)$.

4. We found the y-intercept in step (2) but if we hadn't, we would find $h(0)$ to determine it:

$$h(x) = -2x^4 + 3x^3$$
$$h(0) = -2 \cdot 0^4 + 3 \cdot 0^3 = 0.$$

The y-intercept is $(0, 0)$.

5. A few additional points are helpful when completing the graph.

x	$h(x)$
−1.5	−20.25
−0.5	−0.5
0.5	0.25
2.5	−31.25

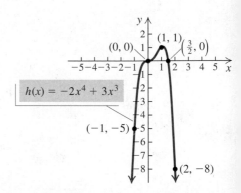

$h(x) = -2x^4 + 3x^3$

6. The degree of h is 4. The graph of h can have at most 4 x-intercepts and at most 3 turning points. In fact, it has 2 x-intercepts and 1 turning point. The zeros, 0 and $\frac{3}{2}$, each have odd multiplicities: 3 for 0 and 1 for $\frac{3}{2}$. Since the multiplicities are odd, the graph crosses the x-axis at 0 and $\frac{3}{2}$. The end behavior of the graph is what we described in step (1). As $x \to \infty$ and also as $x \to -\infty$, $h(x) \to -\infty$. The graph appears to be correct.

▶ *Now Try Exercise 23.*

The following is a procedure for graphing polynomial functions.

> To graph a polynomial function:
>
> 1. Use the leading-term test to determine the end behavior.
> 2. Find the zeros of the function by solving $f(x) = 0$. Any real zeros are the first coordinates of the x-intercepts.
> 3. Use the x-intercepts (zeros) to divide the x-axis into intervals, and choose a test point in each interval to determine the sign of all function values in that interval.
> 4. Find $f(0)$. This gives the y-intercept of the function.
> 5. If necessary, find additional function values to determine the general shape of the graph and then draw the graph.
> 6. As a partial check, use the facts that the graph has at most n x-intercepts and at most $n - 1$ turning points. Multiplicity of zeros can also be considered in order to check where the graph crosses or is tangent to the x-axis. We can also check the graph with a graphing calculator.

EXAMPLE 2 Graph the polynomial function

$$f(x) = 2x^3 + x^2 - 8x - 4.$$

Solution

1. The leading term is $2x^3$. The degree, 3, is odd, and the coefficient, 2, is positive. Thus the end behavior of the graph will appear as follows.

Just in Time

14

2. To find the zeros, we solve $f(x) = 0$. Here we can use factoring by grouping.

$$2x^3 + x^2 - 8x - 4 = 0$$
$$x^2(2x + 1) - 4(2x + 1) = 0 \qquad \textbf{Factoring by grouping}$$
$$(2x + 1)(x^2 - 4) = 0$$
$$(2x + 1)(x + 2)(x - 2) = 0 \qquad \textbf{Factoring a difference of squares}$$

The zeros are $-\frac{1}{2}$, -2, and 2. Each is of multiplicity 1. The x-intercepts are $(-2, 0)$, $\left(-\frac{1}{2}, 0\right)$, and $(2, 0)$.

3. The zeros divide the x-axis into four intervals:

$$(-\infty, -2), \qquad \left(-2, -\tfrac{1}{2}\right), \qquad \left(-\tfrac{1}{2}, 2\right), \quad \text{and} \quad (2, \infty).$$

We choose a test value for x from each interval and find $f(x)$.

$y_1 = 2x^3 + x^2 - 8x - 4$

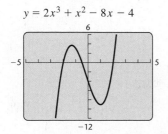

Interval	$(-\infty, -2)$	$\left(-2, -\frac{1}{2}\right)$	$\left(-\frac{1}{2}, 2\right)$	$(2, \infty)$
Test Value, x	-3	-1	1	3
Function Value, $f(x)$	-25	3	-9	35
Sign of $f(x)$	$-$	$+$	$-$	$+$
Location of Points on Graph	Below x-axis	Above x-axis	Below x-axis	Above x-axis

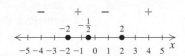

The test values and the corresponding function values also give us four points on the graph: $(-3, -25)$, $(-1, 3)$, $(1, -9)$, and $(3, 35)$.

4. To determine the y-intercept, we find $f(0)$:

$$f(x) = 2x^3 + x^2 - 8x - 4$$
$$f(0) = 2 \cdot 0^3 + 0^2 - 8 \cdot 0 - 4 = -4.$$

The y-intercept is $(0, -4)$.

5. We find a few additional points and complete the graph.

x	$f(x)$
-2.5	-9
-1.5	3.5
0.5	-7.5
1.5	-7

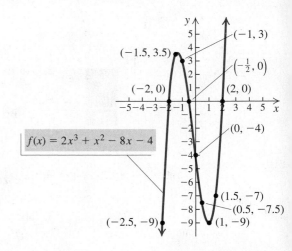

$y = 2x^3 + x^2 - 8x - 4$

6. The degree of f is 3. The graph of f can have at most 3 x-intercepts and at most 2 turning points. It has 3 x-intercepts and 2 turning points. Each zero has a multiplicity of 1. Thus the graph crosses the x-axis at -2, $-\frac{1}{2}$, and 2. The graph has the end behavior described in step (1). As $x \to -\infty$, $f(x) \to -\infty$, and as $x \to \infty$, $f(x) \to \infty$. The graph appears to be correct.

> **Now Try Exercise 33.**

Some polynomials are difficult to factor. In the next example, the polynomial is given in factored form. In Sections 4.3 and 4.4, we will learn methods that facilitate determining factors of such polynomials.

EXAMPLE 3 Graph the polynomial function

$$g(x) = x^4 - 7x^3 + 12x^2 + 4x - 16$$
$$= (x + 1)(x - 2)^2(x - 4).$$

Solution

1. The leading term is x^4. The degree, 4, is even, and the coefficient, 1, is positive. The following sketch shows the end behavior.

2. To find the zeros, we solve $g(x) = 0$:

$$(x + 1)(x - 2)^2(x - 4) = 0.$$

The zeros are -1, 2, and 4; 2 is of multiplicity 2; the others are of multiplicity 1. The x-intercepts are $(-1, 0)$, $(2, 0)$, and $(4, 0)$.

3. The zeros divide the x-axis into four intervals:

$$(-\infty, -1), \quad (-1, 2), \quad (2, 4), \quad \text{and} \quad (4, \infty).$$

We choose a test value for x from each interval and find $g(x)$.

$y_1 = x^4 - 7x^3 + 12x^2 + 4x - 16$

X	Y₁
−1.25	13.863
1	−6
3	−4
4.25	6.6445

X =

Interval	$(-\infty, -1)$	$(-1, 2)$	$(2, 4)$	$(4, \infty)$
Test Value, x	-1.25	1	3	4.25
Function Value, $g(x)$	≈ 13.9	-6	-4	≈ 6.6
Sign of $g(x)$	$+$	$-$	$-$	$+$
Location of Points on Graph	Above x-axis	Below x-axis	Below x-axis	Above x-axis

The test values and the corresponding function values also give us four points on the graph: $(-1.25, 13.9)$, $(1, -6)$, $(3, -4)$, and $(4.25, 6.6)$.

4. To determine the y-intercept, we find $g(0)$:

$$g(x) = x^4 - 7x^3 + 12x^2 + 4x - 16$$
$$g(0) = 0^4 - 7 \cdot 0^3 + 12 \cdot 0^2 + 4 \cdot 0 - 16 = -16.$$

The y-intercept is $(0, -16)$.

5. We find a few additional points and draw the graph.

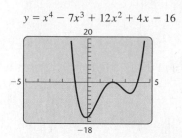

$y = x^4 - 7x^3 + 12x^2 + 4x - 16$

x	$g(x)$
-0.5	-14.1
0.5	-11.8
1.5	-1.6
2.5	-1.3
3.5	-5.1

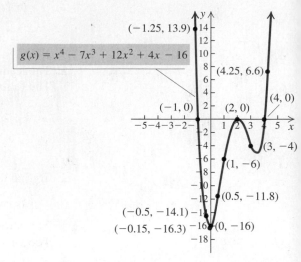

6. The degree of g is 4. The graph of g can have at most 4 x-intercepts and at most 3 turning points. It has 3 x-intercepts and 3 turning points. One of the zeros, 2, has a multiplicity of 2, so the graph is tangent to the x-axis at 2. The other zeros, -1 and 4, each have a multiplicity of 1 so the graph crosses the x-axis at -1 and 4. The graph has the end behavior described in step (1). As $x \to \infty$ and as $x \to -\infty$, $g(x) \to \infty$. The graph appears to be correct.

> *Now Try Exercise 19.*

❯ The Intermediate Value Theorem

Polynomial functions are continuous, hence their graphs are unbroken. The domain of a polynomial function, unless restricted by the statement of the function, is $(-\infty, \infty)$. Suppose that two polynomial function values $P(a)$ and $P(b)$ have opposite signs. Since P is continuous, its graph must be a curve from $(a, P(a))$ to $(b, P(b))$ without a break. Then it follows that the curve must cross the x-axis at some point c between a and b; that is, the function has a zero at c between a and b.

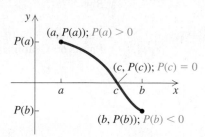

THE INTERMEDIATE VALUE THEOREM

For any polynomial function $P(x)$ with real coefficients, suppose that for $a \neq b$, $P(a)$ and $P(b)$ are of opposite signs. Then the function has a real zero between a and b.

The intermediate value theorem *cannot* be used to determine whether there is or is not a real zero between a and b when $P(a)$ and $P(b)$ have the *same* sign.

EXAMPLE 4 Using the intermediate value theorem, determine, if possible, whether the function has a real zero between a and b.

a) $f(x) = x^3 + x^2 - 6x$; $a = -4$, $b = -2$

b) $f(x) = x^3 + x^2 - 6x$; $a = -1$, $b = 3$

c) $g(x) = \frac{1}{3}x^4 - x^3$; $a = -\frac{1}{2}$, $b = \frac{1}{2}$

d) $g(x) = \frac{1}{3}x^4 - x^3$; $a = 1$, $b = 2$

Solution We find $f(a)$ and $f(b)$ or $g(a)$ and $g(b)$ and determine whether they differ in sign. The graphs of $f(x)$ and $g(x)$ at left provide a visual check of the conclusions.

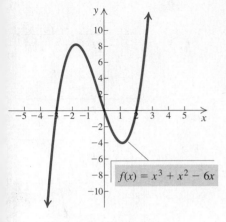

$f(x) = x^3 + x^2 - 6x$

a) $f(-4) = (-4)^3 + (-4)^2 - 6(-4) = -24$,

$f(-2) = (-2)^3 + (-2)^2 - 6(-2) = 8$

Note that $f(-4)$ is negative and $f(-2)$ is positive. By the intermediate value theorem, since $f(-4)$ and $f(-2)$ have opposite signs, then $f(x)$ has a real zero between -4 and -2. The graph at left confirms this.

b) $f(-1) = (-1)^3 + (-1)^2 - 6(-1) = 6$,

$f(3) = 3^3 + 3^2 - 6(3) = 18$

Both $f(-1)$ and $f(3)$ are positive. Thus the intermediate value theorem *does not allow* us to determine whether there is a real zero between -1 and 3. Note that the graph of $f(x)$ shows that there are two zeros between -1 and 3.

c) $g\left(-\frac{1}{2}\right) = \frac{1}{3}\left(-\frac{1}{2}\right)^4 - \left(-\frac{1}{2}\right)^3 = \frac{7}{48}$,

$g\left(\frac{1}{2}\right) = \frac{1}{3}\left(\frac{1}{2}\right)^4 - \left(\frac{1}{2}\right)^3 = -\frac{5}{48}$

Since $g\left(-\frac{1}{2}\right)$ and $g\left(\frac{1}{2}\right)$ have opposite signs, $g(x)$ has a real zero between $-\frac{1}{2}$ and $\frac{1}{2}$. The graph at left confirms this.

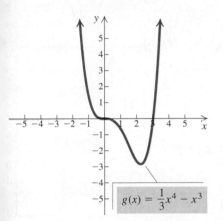

$g(x) = \frac{1}{3}x^4 - x^3$

d) $g(1) = \frac{1}{3}(1)^4 - 1^3 = -\frac{2}{3}$,

$g(2) = \frac{1}{3}(2)^4 - 2^3 = -\frac{8}{3}$

Both $g(1)$ and $g(2)$ are negative. Thus the intermediate value theorem *does not allow* us to determine whether there is a real zero between 1 and 2. Note that the graph of $g(x)$ at left shows that there are no real zeros between 1 and 2.

> ***Now Try Exercises 39 and 43.***

Visualizing the Graph

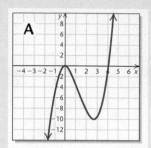

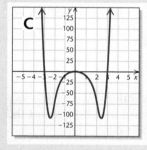

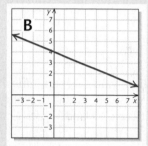

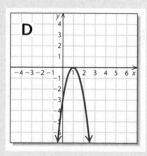

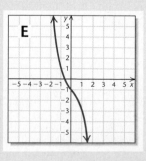

Match the function with its graph.

1. $f(x) = -x^4 - x + 5$

2. $f(x) = -3x^2 + 6x - 3$

3. $f(x) = x^4 - 4x^3 + 3x^2 + 4x - 4$

4. $f(x) = -\dfrac{2}{5}x + 4$

5. $f(x) = x^3 - 4x^2$

6. $f(x) = x^6 - 9x^4$

7. $f(x) = x^5 - 3x^3 + 2$

8. $f(x) = -x^3 - x - 1$

9. $f(x) = x^2 + 7x + 6$

10. $f(x) = \dfrac{7}{2}$

Answers on page A-18

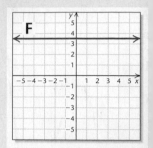

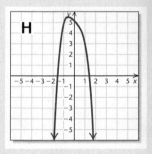

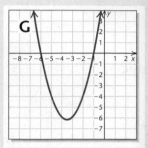

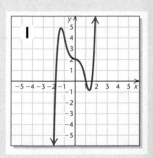

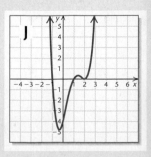

4.2   Exercise Set

For each function in Exercises 1–6, state:

a) *the maximum number of real zeros that the function can have;*

b) *the maximum number of x-intercepts that the graph of the function can have; and*

c) *the maximum number of turning points that the graph of the function can have.*

1. $f(x) = x^5 - x^2 + 6$

2. $f(x) = -x^2 + x^4 - x^6 + 3$

3. $f(x) = x^{10} - 2x^5 + 4x - 2$

4. $f(x) = \frac{1}{4}x^3 + 2x^2$

5. $f(x) = -x - x^3$

6. $f(x) = -3x^4 + 2x^3 - x - 4$

In Exercises 7–12, use the leading-term test and your knowledge of y-intercepts to match the function with one of the graphs (a)–(f) that follow.

a)
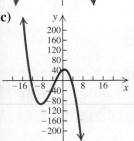

b)
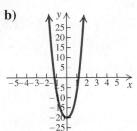

c)
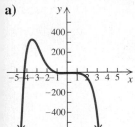

d)
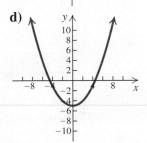

e)
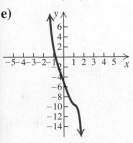

f)
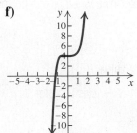

7. $f(x) = \frac{1}{4}x^2 - 5$

8. $f(x) = -0.5x^6 - x^5 + 4x^4 - 5x^3 - 7x^2 + x - 3$

9. $f(x) = x^5 - x^4 + x^2 + 4$

10. $f(x) = -\frac{1}{3}x^3 - 4x^2 + 6x + 42$

11. $f(x) = x^4 - 2x^3 + 12x^2 + x - 20$

12. $f(x) = -0.3x^7 + 0.11x^6 - 0.25x^5 + x^4 + x^3 - 6x - 5$

Graph the polynomial function. Follow the steps outlined in the procedure on p. 237.

13. $f(x) = -x^3 - 2x^2$

14. $g(x) = x^4 - 4x^3 + 3x^2$

15. $h(x) = x^2 + 2x - 3$

16. $f(x) = x^2 - 5x + 4$

17. $h(x) = x^5 - 4x^3$

18. $f(x) = x^3 - x$

19. $h(x) = x(x - 4)(x + 1)(x - 2)$

20. $f(x) = x(x - 1)(x + 3)(x + 5)$

21. $g(x) = -\frac{1}{4}x^3 - \frac{3}{4}x^2$

22. $f(x) = \frac{1}{2}x^3 + \frac{5}{2}x^2$

23. $g(x) = -x^4 - 2x^3$

24. $h(x) = x^3 - 3x^2$

25. $f(x) = -\frac{1}{2}(x - 2)(x + 1)^2(x - 1)$

26. $g(x) = (x - 2)^3(x + 3)$

27. $g(x) = -x(x - 1)^2(x + 4)^2$

28. $h(x) = -x(x - 3)(x - 3)(x + 2)$

29. $f(x) = (x - 2)^2(x + 1)^4$

30. $g(x) = x^4 - 9x^2$

31. $g(x) = -(x - 1)^4$

32. $h(x) = (x + 2)^3$

33. $h(x) = x^3 + 3x^2 - x - 3$

34. $g(x) = -x^3 + 2x^2 + 4x - 8$

35. $f(x) = 6x^3 - 8x^2 - 54x + 72$

36. $h(x) = x^5 - 5x^3 + 4x$

Graph the piecewise function.

37. $g(x) = \begin{cases} -x + 3, & \text{for } x \leq -2, \\ 4, & \text{for } -2 < x < 1, \\ \frac{1}{2}x^3, & \text{for } x \geq 1 \end{cases}$

38. $h(x) = \begin{cases} -x^2, & \text{for } x < -2, \\ x + 1, & \text{for } -2 \le x < 0, \\ x^3 - 1, & \text{for } x \ge 0 \end{cases}$

Using the intermediate value theorem, determine, if possible, whether the function f has a real zero between a and b.

39. $f(x) = x^3 + 3x^2 - 9x - 13; \ a = -5, b = -4$

40. $f(x) = x^3 + 3x^2 - 9x - 13; \ a = 1, b = 2$

41. $f(x) = 3x^2 - 2x - 11; \ a = -3, b = -2$

42. $f(x) = 3x^2 - 2x - 11; \ a = 2, b = 3$

43. $f(x) = x^4 - 2x^2 - 6; \ a = 2, b = 3$

44. $f(x) = 2x^5 - 7x + 1; \ a = 1, b = 2$

45. $f(x) = x^3 - 5x^2 + 4; \ a = 4, b = 5$

46. $f(x) = x^4 - 3x^2 + x - 1; \ a = -3, b = -2$

❯ Skill Maintenance

Match the equation with one of the graphs (a)–(f) that follow.

a)

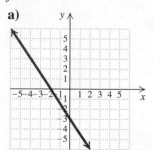

b)

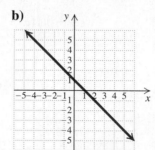

c)

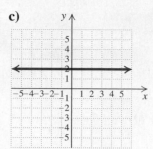

d)

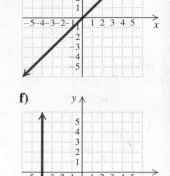

e)

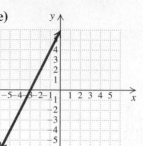

f)

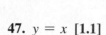

47. $y = x$ **[1.1]**

48. $x = -4$ **[1.3]**

49. $y - 2x = 6$ **[1.1]**

50. $3x + 2y = -6$ **[1.1]**

51. $y = 1 - x$ **[1.1]**

52. $y = 2$ **[1.3]**

Solve.

53. $2x - \frac{1}{2} = 4 - 3x$ **[1.5]**

54. $x^3 - x^2 - 12x = 0$ **[4.1]**

55. $6x^2 - 23x - 55 = 0$ **[3.2]**

56. $\frac{3}{4}x + 10 = \frac{1}{5} + 2x$ **[1.5]**

4.3 Polynomial Division; The Remainder Theorem and the Factor Theorem

> ❯ Perform long division with polynomials and determine whether one polynomial is a factor of another.

> ❯ Use synthetic division to divide a polynomial by $x - c$.

> ❯ Use the remainder theorem to find a function value $f(c)$.

> ❯ Use the factor theorem to determine whether $x - c$ is a factor of $f(x)$.

In general, finding exact zeros of many polynomial functions is neither easy nor straightforward. We now develop concepts that help us find exact zeros of certain polynomial functions with degree 3 or greater. Consider the polynomial

$$h(x) = x^3 + 2x^2 - 5x - 6 = (x + 3)(x + 1)(x - 2).$$

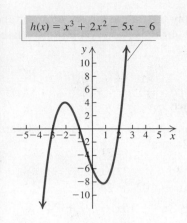

$h(x) = x^3 + 2x^2 - 5x - 6$

The factors are

$$x + 3, \quad x + 1, \quad \text{and} \quad x - 2,$$

and the zeros are

$$-3, \quad -1, \quad \text{and} \quad 2.$$

When a polynomial is expressed in factored form, each factor determines a zero of the function. Thus if we know the factors of a polynomial, we can easily find the zeros. The "reverse" is also true: If we know the zeros of a polynomial function, we can find the factors of the polynomial.

〉 Division and Factors

When we divide one polynomial by another, we obtain a quotient and a remainder. If the remainder is 0, then the divisor is a factor of the dividend.

EXAMPLE 1   Divide to determine whether $x + 1$ and $x - 3$ are factors of

$$x^3 + 2x^2 - 5x - 6.$$

Solution   We divide $x^3 + 2x^2 - 5x - 6$ by $x + 1$.

$$
\begin{array}{r}
\text{Quotient} \\
x^2 + x - 6 \\
x + 1 \overline{\smash{)}\, x^3 + 2x^2 - 5x - 6} \quad \leftarrow \textbf{Dividend} \\
\underline{x^3 + x^2} \\
x^2 - 5x \\
\underline{x^2 + x} \\
-6x - 6 \\
\underline{-6x - 6} \\
0 \quad \leftarrow \text{Remainder}
\end{array}
$$

Divisor

Since the remainder is 0, we know that $x + 1$ is a factor of $x^3 + 2x^2 - 5x - 6$. In fact, we know that

$$x^3 + 2x^2 - 5x - 6 = (x + 1)(x^2 + x - 6).$$

We divide $x^3 + 2x^2 - 5x - 6$ by $x - 3$.

$$
\begin{array}{r}
x^2 + 5x + 10 \\
x - 3 \overline{\smash{)}\, x^3 + 2x^2 - 5x - 6} \\
\underline{x^3 - 3x^2} \\
5x^2 - 5x \\
\underline{5x^2 - 15x} \\
10x - 6 \\
\underline{10x - 30} \\
24 \quad \leftarrow \text{Remainder}
\end{array}
$$

The remainder, 24, is not 0, so we know that $x - 3$ is *not* a factor of $x^3 + 2x^2 - 5x - 6$.

〉 **Now Try Exercise 3.**

When we divide a polynomial $P(x)$ by a divisor $d(x)$, a polynomial $Q(x)$ is the quotient and a polynomial $R(x)$ is the remainder. The quotient $Q(x)$ must have degree less than that of the dividend $P(x)$. The remainder $R(x)$ must either be 0 or have degree less than that of the divisor $d(x)$.

As in arithmetic, to check division, we multiply the quotient by the divisor and add the remainder, to see if we get the dividend. Thus these polynomials are related as follows:

$$P(x) = d(x) \cdot Q(x) + R(x)$$

Dividend   Divisor   Quotient   Remainder

For instance, if $P(x) = x^3 + 2x^2 - 5x - 6$ and $d(x) = x - 3$, as in Example 1, then $Q(x) = x^2 + 5x + 10$ and $R(x) = 24$, and

$$
\begin{aligned}
P(x) &= d(x) \cdot Q(x) + R(x) \\
x^3 + 2x^2 - 5x - 6 &= (x - 3) \cdot (x^2 + 5x + 10) + 24 \\
&= x^3 + 5x^2 + 10x - 3x^2 - 15x - 30 + 24 \\
&= x^3 + 2x^2 - 5x - 6.
\end{aligned}
$$

❯ The Remainder Theorem and Synthetic Division

Consider the function

$$h(x) = x^3 + 2x^2 - 5x - 6.$$

When we divided $h(x)$ by $x + 1$ and $x - 3$ in Example 1, the remainders were 0 and 24, respectively. Let's now find the function values $h(-1)$ and $h(3)$:

$$
\begin{aligned}
h(-1) &= (-1)^3 + 2(-1)^2 - 5(-1) - 6 = 0; \\
h(3) &= (3)^3 + 2(3)^2 - 5(3) - 6 = 24.
\end{aligned}
$$

Note that the function values are the same as the remainders. This suggests the following theorem.

THE REMAINDER THEOREM

If a number c is substituted for x in the polynomial $f(x)$, then the result $f(c)$ is the remainder that would be obtained by dividing $f(x)$ by $x - c$. That is, if $f(x) = (x - c) \cdot Q(x) + R$, then $f(c) = R$.

Proof (Optional).   The equation $f(x) = d(x) \cdot Q(x) + R(x)$, where $d(x) = x - c$, is the basis of this proof. If we divide $f(x)$ by $x - c$, we obtain a quotient $Q(x)$ and a remainder $R(x)$ related as follows:

$$f(x) = (x - c) \cdot Q(x) + R(x).$$

The remainder $R(x)$ must either be 0 or have degree less than $x - c$. Thus, $R(x)$ must be a constant. Let's call this constant R. The equation above is true for any replacement of x, so we replace x with c. We get

$$
\begin{aligned}
f(c) &= (c - c) \cdot Q(c) + R \\
&= 0 \cdot Q(c) + R \\
&= R.
\end{aligned}
$$

Thus the function value $f(c)$ is the remainder obtained when we divide $f(x)$ by $x - c$.   ■

The remainder theorem motivates us to find a rapid way of dividing by $x - c$ in order to find function values. To streamline division, we can arrange the work so that duplicate and unnecessary writing is avoided. Consider the following:

$$(4x^3 - 3x^2 + x + 7) \div (x - 2).$$

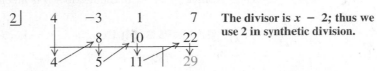

A.
$$
\begin{array}{r}
4x^2 + 5x + 11 \\
x - 2 \overline{)\, 4x^3 - 3x^2 + x + 7} \\
\underline{4x^3 - 8x^2} \\
5x^2 + x \\
\underline{5x^2 - 10x} \\
11x + 7 \\
\underline{11x - 22} \\
29
\end{array}
$$

B.
$$
\begin{array}{r}
4 \quad 5 \quad 11 \\
1 - 2 \overline{)\, 4 - 3 + 1 + 7} \\
\underline{4 - 8} \\
5 + 1 \\
\underline{5 - 10} \\
11 + 7 \\
\underline{11 - 22} \\
29
\end{array}
$$

The division in (B) is the same as that in (A), but we wrote only the coefficients. The red numerals are duplicated, so we look for an arrangement in which they are not duplicated. In place of the divisor in the form $x - c$, we can simply use c and then add rather than subtract. When the procedure is "collapsed," we have the algorithm known as **synthetic division**.

C. *Synthetic Division*

$$
\begin{array}{c|cccc}
2 & 4 & -3 & 1 & 7 \\
 & & 8 & 10 & 22 \\
\hline
 & 4 & 5 & 11 & 29
\end{array}
$$
The divisor is $x - 2$; thus we use 2 in synthetic division.

We "bring down" the 4. Then we multiply it by the 2 to get 8 and add to get 5. We then multiply 5 by 2 to get 10, add, and so on. The last number, 29, is the remainder. The others, 4, 5, and 11, are the coefficients of the quotient, $4x^2 + 5x + 11$. (Note that the degree of the quotient is 1 less than the degree of the dividend when the degree of the divisor is 1.)

When using synthetic division, we write a 0 for a missing term in the dividend.

Just in Time **5**

EXAMPLE 2 Use synthetic division to find the quotient and the remainder:

$$(2x^3 + 7x^2 - 5) \div (x + 3).$$

Solution First, we note that $x + 3 = x - (-3)$.

$$
\begin{array}{c|cccc}
-3 & 2 & 7 & 0 & -5 \\
 & & -6 & -3 & 9 \\
\hline
 & 2 & 1 & -3 & 4
\end{array}
$$
Note: **We must write a 0 for the missing x-term.**

The quotient is $2x^2 + x - 3$. The remainder is 4.

> **Now Try Exercise 13.**

We can now use synthetic division to find polynomial function values.

EXAMPLE 3 Given that $f(x) = 2x^5 - 3x^4 + x^3 - 2x^2 + x - 8$, find $f(10)$.

Solution By the remainder theorem, $f(10)$ is the remainder when $f(x)$ is divided by $x - 10$. We use synthetic division to find that remainder.

$$
\begin{array}{c|cccccc}
10 & 2 & -3 & 1 & -2 & 1 & -8 \\
 & & 20 & 170 & 1710 & 17{,}080 & 170{,}810 \\
\hline
 & 2 & 17 & 171 & 1708 & 17{,}081 & 170{,}802
\end{array}
$$

Thus, $f(10) = 170{,}802$.

> **Now Try Exercise 25.**

Compare the computations in Example 3 with those in a direct substitution:

$$f(10) = 2(10)^5 - 3(10)^4 + (10)^3 - 2(10)^2 + 10 - 8$$
$$= 2 \cdot 100,000 - 3 \cdot 10,000 + 1000 - 2 \cdot 100 + 10 - 8$$
$$= 200,000 - 30,000 + 1000 - 200 + 10 - 8$$
$$= 170,802.$$

The computations in synthetic division are less complicated than those involved in substituting. The easiest way to find $f(10)$ is to use one of the methods for evaluating a function on a graphing calculator. In the figure at left, we show the result when we enter $y_1 = 2x^5 - 3x^4 + x^3 - 2x^2 + x - 8$ and then use function notation on the home screen.

```
Y1(10)
              170802
```

EXAMPLE 4   Determine whether 5 is a zero of $g(x)$, where

$$g(x) = x^4 - 26x^2 + 25.$$

Solution   We use synthetic division and the remainder theorem to find $g(5)$.

$$
\begin{array}{r|rrrrr}
5 & 1 & 0 & -26 & 0 & 25 \\
  &   & 5 & 25 & -5 & -25 \\
\hline
  & 1 & 5 & -1 & -5 & 0
\end{array}
$$

Writing 0's for missing terms:
$x^4 + 0x^3 - 26x^2 + 0x + 25$

Since $g(5) = 0$, the number 5 is a zero of $g(x)$.

> **Now Try Exercise 31.**

EXAMPLE 5   Determine whether i is a zero of $f(x)$, where

$$f(x) = x^3 - 3x^2 + x - 3.$$

Solution   We use synthetic division and the remainder theorem to find $f(i)$.

$$
\begin{array}{r|rrrr}
i & 1 & -3 & 1 & -3 \\
  &   & i & -3i - 1 & 3 \\
\hline
  & 1 & -3 + i & -3i & 0
\end{array}
$$

$i(-3 + i) = -3i + i^2 = -3i - 1$
$i(-3i) = -3i^2 = 3$

Since $f(i) = 0$, the number i is a zero of $f(x)$.

> **Now Try Exercise 35.**

❯ Finding Factors of Polynomials

We now consider a useful result that follows from the remainder theorem.

THE FACTOR THEOREM

For a polynomial $f(x)$, if $f(c) = 0$, then $x - c$ is a factor of $f(x)$.

Proof (Optional).   If we divide $f(x)$ by $x - c$, we obtain a quotient and a remainder, related as follows:

$$f(x) = (x - c) \cdot Q(x) + f(c).$$

Then if $f(c) = 0$, we have

$$f(x) = (x - c) \cdot Q(x),$$

so $x - c$ is a factor of $f(x)$. ∎

The factor theorem is very useful in factoring polynomials and hence in solving polynomial equations and finding zeros of polynomial functions. If we know a zero of a polynomial function, we know a factor.

EXAMPLE 6   Let $f(x) = x^3 - 3x^2 - 6x + 8$. Factor $f(x)$ and solve the equation $f(x) = 0$.

Solution   We look for linear factors of the form $x - c$. Let's try $x + 1$, or $x - (-1)$. (In the next section, we will learn a method for choosing the numbers to try for c.) We use synthetic division to determine whether $f(-1) = 0$.

$$\begin{array}{r|rrrr} -1 & 1 & -3 & -6 & 8 \\ & & -1 & 4 & 2 \\ \hline & 1 & -4 & -2 & | \quad 10 \end{array}$$

Since $f(-1) \neq 0$, we know that $x + 1$ *is not a factor* of $f(x)$. We now try $x - 1$.

$$\begin{array}{r|rrrr} 1 & 1 & -3 & -6 & 8 \\ & & 1 & -2 & -8 \\ \hline & 1 & -2 & -8 & | \quad 0 \end{array}$$

Since $f(1) = 0$, we know that $x - 1$ *is one factor* of $f(x)$ and the quotient, $x^2 - 2x - 8$, is another. Thus,

$$f(x) = (x - 1)(x^2 - 2x - 8).$$

The trinomial $x^2 - 2x - 8$ is easily factored, so we have

$$f(x) = (x - 1)(x - 4)(x + 2).$$

We now solve the equation $f(x) = 0$. To do so, we use the principle of zero products:

$$(x - 1)(x - 4)(x + 2) = 0$$
$$x - 1 = 0 \quad or \quad x - 4 = 0 \quad or \quad x + 2 = 0$$
$$x = 1 \quad or \qquad x = 4 \quad or \qquad x = -2.$$

The solutions of the equation $x^3 - 3x^2 - 6x + 8 = 0$ are -2, 1, and 4. They are also the zeros of the function $f(x) = x^3 - 3x^2 - 6x + 8$. We can use a table set in ASK mode to check the solutions.

> **Now Try Exercise 41.**

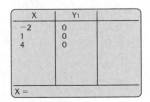

CONNECTING THE CONCEPTS

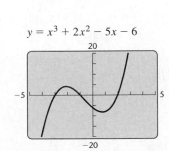

$y = x^3 + 2x^2 - 5x - 6$

Consider the function

$$f(x) = (x - 2)(x + 3)(x + 1), \quad or \quad f(x) = x^3 + 2x^2 - 5x - 6,$$

and its graph.

We can make the following statements:

- -3 is a zero of f.
- $f(-3) = 0$.
- -3 is a solution of $f(x) = 0$.
- $(-3, 0)$ is an x-intercept of the graph of f.
- 0 is the remainder when $f(x)$ is divided by $x + 3$, or $x - (-3)$.
- $x - (-3)$, or $x + 3$, is a factor of f.

Similar statements are also true for -1 and 2.

4.3 Exercise Set

1. For the function
$$f(x) = x^4 - 6x^3 + x^2 + 24x - 20,$$
use long division to determine whether each of the following is a factor of $f(x)$.
   a) $x + 1$ b) $x - 2$ c) $x + 5$

2. For the function
$$h(x) = x^3 - x^2 - 17x - 15,$$
use long division to determine whether each of the following is a factor of $h(x)$.
   a) $x + 5$ b) $x + 1$ c) $x + 3$

3. For the function
$$g(x) = x^3 - 2x^2 - 11x + 12,$$
use long division to determine whether each of the following is a factor of $g(x)$.
   a) $x - 4$ b) $x - 3$ c) $x - 1$

4. For the function
$$f(x) = x^4 + 8x^3 + 5x^2 - 38x + 24,$$
use long division to determine whether each of the following is a factor of $f(x)$.
   a) $x + 6$ b) $x + 1$ c) $x - 4$

In each of the following, a polynomial $P(x)$ and a divisor $d(x)$ are given. Use long division to find the quotient $Q(x)$ and the remainder $R(x)$ when $P(x)$ is divided by $d(x)$. Express $P(x)$ in the form $d(x) \cdot Q(x) + R(x)$.

5. $P(x) = x^3 - 8$,
   $d(x) = x + 2$

6. $P(x) = 2x^3 - 3x^2 + x - 1$,
   $d(x) = x - 3$

7. $P(x) = x^3 + 6x^2 - 25x + 18$,
   $d(x) = x + 9$

8. $P(x) = x^3 - 9x^2 + 15x + 25$,
   $d(x) = x - 5$

9. $P(x) = x^4 - 2x^2 + 3$,
   $d(x) = x + 2$

10. $P(x) = x^4 + 6x^3$,
    $d(x) = x - 1$

Use synthetic division to find the quotient and the remainder.

11. $(2x^4 + 7x^3 + x - 12) \div (x + 3)$

12. $(x^3 - 7x^2 + 13x + 3) \div (x - 2)$

13. $(x^3 - 2x^2 - 8) \div (x + 2)$

14. $(x^3 - 3x + 10) \div (x - 2)$

15. $(3x^3 - x^2 + 4x - 10) \div (x + 1)$

16. $(4x^4 - 2x + 5) \div (x + 3)$

17. $(x^5 + x^3 - x) \div (x - 3)$

18. $(x^7 - x^6 + x^5 - x^4 + 2) \div (x + 1)$

19. $(x^4 - 1) \div (x - 1)$

20. $(x^5 + 32) \div (x + 2)$

21. $(2x^4 + 3x^2 - 1) \div \left(x - \frac{1}{2}\right)$

22. $(3x^4 - 2x^2 + 2) \div \left(x - \frac{1}{4}\right)$

Use synthetic division to find the function values. Then check your work using a graphing calculator.

23. $f(x) = x^3 - 6x^2 + 11x - 6$; find $f(1)$, $f(-2)$, and $f(3)$.

24. $f(x) = x^3 + 7x^2 - 12x - 3$; find $f(-3)$, $f(-2)$, and $f(1)$.

25. $f(x) = x^4 - 3x^3 + 2x + 8$; find $f(-1)$, $f(4)$, and $f(-5)$.

26. $f(x) = 2x^4 + x^2 - 10x + 1$; find $f(-10)$, $f(2)$, and $f(3)$.

27. $f(x) = 2x^5 - 3x^4 + 2x^3 - x + 8$; find $f(20)$ and $f(-3)$.

28. $f(x) = x^5 - 10x^4 + 20x^3 - 5x - 100$; find $f(-10)$ and $f(5)$.

29. $f(x) = x^4 - 16$; find $f(2)$, $f(-2)$, $f(3)$, and $f(1 - \sqrt{2})$.

30. $f(x) = x^5 + 32$; find $f(2)$, $f(-2)$, $f(3)$, and $f(2 + 3i)$.

Using synthetic division, determine whether the numbers are zeros of the polynomial function.

31. $-3, 2$; $f(x) = 3x^3 + 5x^2 - 6x + 18$

32. $-4, 2$; $f(x) = 3x^3 + 11x^2 - 2x + 8$

33. $-3, 1$; $h(x) = x^4 + 4x^3 + 2x^2 - 4x - 3$

34. $2, -1$; $g(x) = x^4 - 6x^3 + x^2 + 24x - 20$

35. $i, -2i$; $g(x) = x^3 - 4x^2 + 4x - 16$

36. $\frac{1}{3}, 2$; $h(x) = x^3 - x^2 - \frac{1}{9}x + \frac{1}{9}$

37. $-3, \frac{1}{2}; \ f(x) = x^3 - \frac{7}{2}x^2 + x - \frac{3}{2}$

38. $i, -i, -2; \ f(x) = x^3 + 2x^2 + x + 2$

Factor the polynomial function $f(x)$. Then solve the equation $f(x) = 0$.

39. $f(x) = x^3 + 4x^2 + x - 6$

40. $f(x) = x^3 + 5x^2 - 2x - 24$

41. $f(x) = x^3 - 6x^2 + 3x + 10$

42. $f(x) = x^3 + 2x^2 - 13x + 10$

43. $f(x) = x^3 - x^2 - 14x + 24$

44. $f(x) = x^3 - 3x^2 - 10x + 24$

45. $f(x) = x^4 - 7x^3 + 9x^2 + 27x - 54$

46. $f(x) = x^4 - 4x^3 - 7x^2 + 34x - 24$

47. $f(x) = x^4 - x^3 - 19x^2 + 49x - 30$

48. $f(x) = x^4 + 11x^3 + 41x^2 + 61x + 30$

Sketch the graph of the polynomial function. Follow the procedure outlined on p. 237. Use synthetic division and the remainder theorem to find the zeros.

49. $f(x) = x^4 - x^3 - 7x^2 + x + 6$

50. $f(x) = x^4 + x^3 - 3x^2 - 5x - 2$

51. $f(x) = x^3 - 7x + 6$

52. $f(x) = x^3 - 12x + 16$

53. $f(x) = -x^3 + 3x^2 + 6x - 8$

54. $f(x) = -x^4 + 2x^3 + 3x^2 - 4x - 4$

❯ Skill Maintenance

Solve. Find exact solutions. **[3.2]**

55. $2x^2 + 12 = 5x$

56. $7x^2 + 4x = 3$

In Exercises 57–59, consider the function
$$g(x) = x^2 + 5x - 14.$$

57. What are the inputs if the output is -14? **[3.2]**

58. What is the output if the input is 3? **[1.2]**

59. Given an output of -20, find the corresponding inputs. **[3.2]**

60. *Disruptive Airline Passengers.* The number of reported disruptive airline passengers has increased linearly over the years, rising from 339 cases in 2007 to 8217 cases in 2013 (*Source*: International Air Transportation Association). Using these two data points, find a linear function, $f(x) = mx + b$, that models the data. Let x represent the number of years after

2007. Then use this function to estimate the number of reported cases of disruptive airline passengers in 2011 and in 2017. **[1.4]**

61. The sum of the base and the height of a triangle is 30 in. Find the dimensions for which the area is a maximum. **[3.3]**

❯ Synthesis

In Exercises 62 and 63, a graph of a polynomial function is given. On the basis of the graph:

a) *Find as many factors of the polynomial as you can.*

b) *Construct a polynomial function with the zeros shown in the graph.*

c) *Can you find any other polynomial functions with the given zeros?*

d) *Can you find more than one polynomial function with the given zeros and the same graph?*

62.

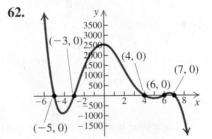

63.

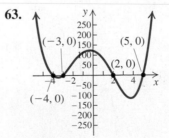

64. For what values of k will the remainder be the same when $x^2 + kx + 4$ is divided by $x - 1$ and by $x + 1$?

65. Find k such that $x + 2$ is a factor of $x^3 - kx^2 + 3x + 7k$.

66. *Beam Deflection.* A beam rests at two points A and B and has a concentrated load applied to its center. Let $y =$ the deflection, in feet, of the beam at a distance of x feet from A. Under certain conditions, this deflection is given by

$$y = \tfrac{1}{13}x^3 - \tfrac{1}{14}x.$$

Find the zeros of the polynomial on the interval $[0, 2]$.

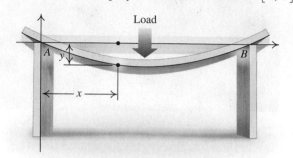

Solve.

67. $\dfrac{2x^2}{x^2 - 1} + \dfrac{4}{x + 3} = \dfrac{12x - 4}{x^3 + 3x^2 - x - 3}$

68. $\dfrac{6x^2}{x^2 + 11} + \dfrac{60}{x^3 - 7x^2 + 11x - 77} = \dfrac{1}{x - 7}$

69. Find a 15th-degree polynomial for which $x - 1$ is a factor. Answers may vary.

Use synthetic division to divide.

70. $(x^4 - y^4) \div (x - y)$

71. $(x^3 + 3ix^2 - 4ix - 2) \div (x + i)$

72. $(x^2 - 4x - 2) \div [x - (3 + 2i)]$

73. $(x^2 - 3x + 7) \div (x - i)$

Mid-Chapter Mixed Review

Determine whether the statement is true or false.

1. The y-intercept of the graph of the function $P(x) = 5 - 2x^3$ is $(5, 0)$. **[4.2]**

2. The degree of the polynomial $x - \tfrac{1}{2}x^4 - 3x^6 + x^5$ is 6. **[4.1]**

3. If $f(x) = (x + 7)(x - 8)$, then $f(8) = 0$. **[4.3]**

4. If $f(12) = 0$, then $x + 12$ is a factor of $f(x)$. **[4.3]**

Find the zeros of the polynomial function and state the multiplicity of each. **[4.1]**

5. $f(x) = (x^2 - 10x + 25)^3$

6. $h(x) = 2x^3 + x^2 - 50x - 25$

7. $g(x) = x^4 - 3x^2 + 2$

8. $f(x) = -6(x - 3)^2(x + 4)$

In Exercises 9–12, match the function with one of the graphs (a)–(d) that follow. **[4.2]**

a)

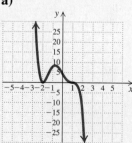

b)

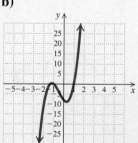

c)

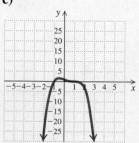

d)

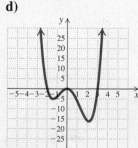

9. $f(x) = x^4 - x^3 - 6x^2$

10. $f(x) = -(x - 1)^3(x + 2)^2$

11. $f(x) = 6x^3 + 8x^2 - 6x - 8$

12. $f(x) = -(x - 1)^3(x + 1)$

Using the intermediate value theorem, determine, if possible, whether the function has at least one real zero between a and b. **[4.2]**

13. $f(x) = x^3 - 2x^2 + 3$; $a = -2, b = 0$

14. $f(x) = x^3 - 2x^2 + 3$; $a = -\tfrac{1}{2}, b = 1$

15. For the polynomial $P(x) = x^4 - 6x^3 + x - 2$ and the divisor $d(x) = x - 1$, use long division to find the quotient $Q(x)$ and the remainder $R(x)$ when $P(x)$ is divided by $d(x)$. Express $P(x)$ in the form $d(x) \cdot Q(x) + R(x)$. **[4.3]**

Use synthetic division to find the quotient and the remainder. **[4.3]**

16. $(3x^4 - x^3 + 2x^2 - 6x + 6) \div (x - 2)$

17. $(x^5 - 5) \div (x + 1)$

Use synthetic division to find the function values. **[4.3]**

18. $g(x) = x^3 - 9x^2 + 4x - 10$; find $g(-5)$

19. $f(x) = 20x^2 - 40x$; find $f\left(\frac{1}{2}\right)$

20. $f(x) = 5x^4 + x^3 - x$; find $f(-\sqrt{2})$

Using synthetic division, determine whether the numbers are zeros of the polynomial function. **[4.3]**

21. $-3i, 3$; $f(x) = x^3 - 4x^2 + 9x - 36$

22. $-1, 5$; $f(x) = x^6 - 35x^4 + 259x^2 - 225$

Factor the polynomial function $f(x)$. Then solve the equation $f(x) = 0$. **[4.3]**

23. $f(x) = x^3 - 2x^2 - 55x + 56$

24. $f(x) = x^4 - 2x^3 - 13x^2 + 14x + 24$

COLLABORATIVE DISCUSSION AND WRITING

25. How is the range of a polynomial function related to the degree of the polynomial? **[4.1]**

26. Is it possible for the graph of a polynomial function to have no y-intercept? no x-intercepts? Explain your answer. **[4.2]**

27. Explain why values of a function must be all positive or all negative between consecutive zeros. **[4.2]**

28. In synthetic division, why is the degree of the quotient 1 less than that of the dividend? **[4.3]**

4.4 ⟩ Theorems about Zeros of Polynomial Functions

> ❯ Find a polynomial with specified zeros.

> ❯ For a polynomial function with integer coefficients, find the rational zeros and the other zeros, if possible.

> ❯ Use Descartes' rule of signs to find information about the number of real zeros of a polynomial function with real coefficients.

Just in Time
1

We will now allow the coefficients of a polynomial to be complex numbers. In certain cases, we will restrict the coefficients to be real numbers, rational numbers, or integers, as shown in the following examples.

Polynomial	Type of Coefficient
$5x^3 - 3x^2 + (2 + 4i)x + i$	Complex
$5x^3 - 3x^2 + \sqrt{2}x - \pi$	Real
$5x^3 - 3x^2 + \frac{2}{3}x - \frac{7}{4}$	Rational
$5x^3 - 3x^2 + 8x - 11$	Integer

❯ The Fundamental Theorem of Algebra

A linear, or first-degree, polynomial function $f(x) = mx + b$ (where $m \neq 0$) has just one zero, $-b/m$. It can be shown that any quadratic polynomial function $f(x) = ax^2 + bx + c$ with complex numbers for coefficients has at least one, and at most two, complex zeros. The following theorem is a generalization. No proof is given in this text.

THE FUNDAMENTAL THEOREM OF ALGEBRA

Every polynomial function of degree n, with $n \geq 1$, has at least one zero in the set of complex numbers.

Note that although the fundamental theorem of algebra guarantees that a zero exists, it does not tell how to find it. Recall that the zeros of a polynomial function $f(x)$ are the solutions of the polynomial equation $f(x) = 0$. We now develop some concepts that can help in finding zeros. First, we consider one of the results of the fundamental theorem of algebra.

Every polynomial function f of degree n, with $n \geq 1$, can be factored into n linear factors (not necessarily unique); that is,

$$f(x) = a_n(x - c_1)(x - c_2) \cdots (x - c_n).$$

❯ Finding Polynomials with Given Zeros

Given several numbers, we can find a polynomial function with those numbers as its zeros.

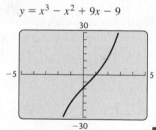

$y = x^3 - x^2 + 9x - 9$

EXAMPLE 1 Find a polynomial function of degree 3, having the zeros 1, $3i$, and $-3i$.

Solution Such a function has factors $x - 1$, $x - 3i$, and $x + 3i$, so we have

$$f(x) = a_n(x - 1)(x - 3i)(x + 3i).$$

The number a_n can be any nonzero number. The simplest polynomial function will be obtained if we let it be 1. If we then multiply the factors, we obtain

$$\begin{aligned}
f(x) &= (x - 1)(x^2 - 9i^2) &&\text{Multiplying } (x - 3i)(x + 3i) \\
&= (x - 1)(x^2 + 9) &&-9i^2 = -9(-1) = 9 \\
&= x^3 - x^2 + 9x - 9.
\end{aligned}$$

Just in Time **12**

❯ *Now Try Exercise 3.*

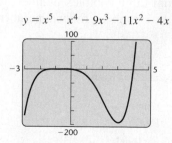

$y = x^5 - x^4 - 9x^3 - 11x^2 - 4x$

EXAMPLE 2 Find a polynomial function of degree 5 with -1 as a zero of multiplicity 3, 4 as a zero of multiplicity 1, and 0 as a zero of multiplicity 1.

Solution Proceeding as in Example 1, letting $a_n = 1$, we obtain

$$\begin{aligned}
f(x) &= [x - (-1)]^3(x - 4)(x - 0) \\
&= (x + 1)^3(x - 4)x \\
&= (x^3 + 3x^2 + 3x + 1)(x^2 - 4x) \\
&= x^5 - x^4 - 9x^3 - 11x^2 - 4x.
\end{aligned}$$

❯ *Now Try Exercise 13.*

❯ Zeros of Polynomial Functions with Real Coefficients

Consider the quadratic equation $x^2 - 2x + 2 = 0$, with real coefficients. Its solutions are $1 + i$ and $1 - i$. Note that they are complex conjugates. This generalizes to any polynomial equation with real coefficients.

NONREAL ZEROS: $a + bi$ AND $a - bi$, $b \neq 0$

If a complex number $a + bi$, $b \neq 0$, is a zero of a polynomial function $f(x)$ with *real* coefficients, then its conjugate, $a - bi$, is also a zero. For example, if $2 + 7i$ is a zero of a polynomial function $f(x)$ with real coefficients, then its conjugate, $2 - 7i$, is also a zero. (Nonreal zeros occur in conjugate pairs.)

In order for the preceding to be true, it is essential that the coefficients be *real* numbers.

❯ Rational Coefficients

When a polynomial function has rational numbers for coefficients, certain irrational zeros also occur in pairs, as described in the following theorem.

IRRATIONAL ZEROS: $a + c\sqrt{b}$ AND $a - c\sqrt{b}$, b IS NOT A PERFECT SQUARE

If $a + c\sqrt{b}$, where a, b, and c are rational and b is not a perfect square, is a zero of a polynomial function $f(x)$ with *rational* coefficients, then its conjugate, $a - c\sqrt{b}$, is also a zero. For example, if $-3 + 5\sqrt{2}$ is a zero of a polynomial function $f(x)$ with rational coefficients, then its conjugate, $-3 - 5\sqrt{2}$, is also a zero. (Irrational zeros occur in conjugate pairs.)

EXAMPLE 3   Suppose that a polynomial function of degree 6 with rational coefficients has

$$-2 + 5i, \quad -2i, \quad \text{and} \quad 1 - \sqrt{3}$$

as three of its zeros. Find the other zeros.

Solution   Since the coefficients are rational, the other zeros are the conjugates of the given zeros,

$$-2 - 5i, \quad 2i, \quad \text{and} \quad 1 + \sqrt{3}.$$

There are no other zeros because a polynomial function of degree 6 can have at most 6 zeros.

❯ *Now Try Exercise 19.*

EXAMPLE 4   Find a polynomial function of lowest degree with rational coefficients that has $-\sqrt{3}$ and $1 + i$ as two of its zeros.

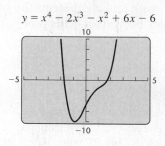

$y = x^4 - 2x^3 - x^2 + 6x - 6$

Solution   The function must also have the zeros $\sqrt{3}$ and $1 - i$. Because we want to find the polynomial function of lowest degree with the given zeros, we will not include additional zeros; that is, we will write a polynomial function of degree 4. Thus if we let $a_n = 1$, the polynomial function is

$$
\begin{aligned}
f(x) &= [x - (-\sqrt{3})][x - \sqrt{3}][x - (1 + i)][x - (1 - i)] \\
&= (x + \sqrt{3})(x - \sqrt{3})[(x - 1) - i][(x - 1) + i] \\
&= (x^2 - 3)[(x - 1)^2 - i^2] \\
&= (x^2 - 3)[x^2 - 2x + 1 + 1] \\
&= (x^2 - 3)(x^2 - 2x + 2) \\
&= x^4 - 2x^3 - x^2 + 6x - 6.
\end{aligned}
$$

> **Now Try Exercise 39.**

❯ Integer Coefficients and the Rational Zeros Theorem

It is not always easy to find the zeros of a polynomial function. However, if a polynomial function has integer coefficients, there is a procedure that will yield all the rational zeros.

THE RATIONAL ZEROS THEOREM

Let

$$P(x) = a_n x^n + a_{n-1}x^{n-1} + \cdots + a_1 x + a_0,$$

where all the coefficients are integers. Consider a rational number denoted by p/q, where p and q are relatively prime (having no common factor besides -1 and 1). If p/q is a zero of $P(x)$, then p is a factor of a_0 and q is a factor of a_n.

EXAMPLE 5   Given $f(x) = 3x^4 - 11x^3 + 10x - 4$:

a) Find the rational zeros and then the other zeros; that is, solve $f(x) = 0$.

b) Factor $f(x)$ into linear factors.

Solution

a) Because the degree of $f(x)$ is 4, there are at most 4 distinct zeros. All of the coefficients are integers. The rational zeros theorem says that if a rational number p/q is a zero of $f(x)$, then p must be a factor of -4 and q must be a factor of 3. Thus the possibilities for p/q are

$$\frac{Possibilities\ for\ p}{Possibilities\ for\ q}: \quad \frac{\pm 1, \pm 2, \pm 4}{\pm 1, \pm 3};$$

$$Possibilities\ for\ p/q: \quad 1, -1, 2, -2, 4, -4, \tfrac{1}{3}, -\tfrac{1}{3}, \tfrac{2}{3}, -\tfrac{2}{3}, \tfrac{4}{3}, -\tfrac{4}{3}.$$

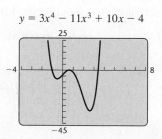

$y = 3x^4 - 11x^3 + 10x - 4$

We could use the TABLE feature or some other method to find function values. However, if we use synthetic division, the quotient polynomial becomes a beneficial by-product if a zero is found. Rather than use synthetic division to check *each* of these possibilities, we graph the function and inspect the graph for zeros that appear to be near any of the possible rational zeros. (See the graph at left.)

From the graph, we see that of the possibilities in the list, only the numbers $-1, \tfrac{1}{3}$, and $\tfrac{2}{3}$ might be rational zeros.

We try -1.

$$
\begin{array}{r|rrrrr}
-1 & 3 & -11 & 0 & 10 & -4 \\
   &   & -3 & 14 & -14 & 4 \\
\hline
   & 3 & -14 & 14 & -4 & \,0
\end{array}
$$

We have $f(-1) = 0$, so -1 is a zero. Thus, $x + 1$ is a factor of $f(x)$. Using the results of the synthetic division, we can express $f(x)$ as

$$f(x) = (x + 1)(3x^3 - 14x^2 + 14x - 4).$$

We now consider the factor $3x^3 - 14x^2 + 14x - 4$ and check the other possible zeros. We try $\frac{1}{3}$.

$$
\begin{array}{r|rrrr}
1/3 & 3 & -14 & 14 & -4 \\
    &   & 1 & -\frac{13}{3} & \frac{29}{9} \\
\hline
    & 3 & -13 & \frac{29}{3} & -\frac{7}{9}
\end{array}
$$

Since $f\left(\frac{1}{3}\right) \neq 0$, we know that $\frac{1}{3}$ is not a zero.
Let's now try $\frac{2}{3}$.

$$
\begin{array}{r|rrrr}
2/3 & 3 & -14 & 14 & -4 \\
    &   & 2 & -8 & 4 \\
\hline
    & 3 & -12 & 6 & 0
\end{array}
$$

Since the remainder is 0, we know that $x - \frac{2}{3}$ is a factor of $3x^3 - 14x^2 + 14x - 4$ and is also a factor of $f(x)$. Thus, $\frac{2}{3}$ is a zero of $f(x)$.

We can check the zeros with the TABLE feature. (See the window at left.) Note that the graphing calculator converts $\frac{1}{3}$ and $\frac{2}{3}$ to decimal notation. Since $f(-1) = 0$ and $f\left(\frac{2}{3}\right) = 0$, -1 and $\frac{2}{3}$ are zeros. Since $f\left(\frac{1}{3}\right) \neq 0$, $\frac{1}{3}$ is not a zero.

Using the results of the synthetic division, we can factor further:

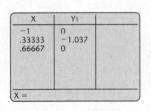

$$f(x) = (x + 1)\left(x - \tfrac{2}{3}\right)(3x^2 - 12x + 6) \qquad \text{Using the results of the last synthetic division}$$

$$= (x + 1)\left(x - \tfrac{2}{3}\right) \cdot 3 \cdot (x^2 - 4x + 2). \qquad \text{Removing a factor of 3}$$

The quadratic formula can be used to find the values of x for which $x^2 - 4x + 2 = 0$. Those values are also zeros of $f(x)$:

$$x = \frac{-b \pm \sqrt{b^2 - 4ac}}{2a}$$

$$= \frac{-(-4) \pm \sqrt{(-4)^2 - 4 \cdot 1 \cdot 2}}{2 \cdot 1} \qquad a = 1, b = -4, \text{ and } c = 2$$

$$= \frac{4 \pm \sqrt{8}}{2} = \frac{4 \pm 2\sqrt{2}}{2} = \frac{2(2 \pm \sqrt{2})}{2}$$

$$= 2 \pm \sqrt{2}.$$

The rational zeros are -1 and $\frac{2}{3}$. The other zeros are $2 \pm \sqrt{2}$.

b) The complete factorization of $f(x)$ is

$$f(x) = 3(x + 1)\left(x - \tfrac{2}{3}\right)[x - (2 - \sqrt{2})][x - (2 + \sqrt{2})], \quad \text{or}$$
$$f(x) = (x + 1)(3x - 2)[x - (2 - \sqrt{2})][x - (2 + \sqrt{2})].$$

Replacing $3\left(x - \tfrac{2}{3}\right)$ with $(3x - 2)$

> **Now Try Exercise 55.**

EXAMPLE 6   Given $f(x) = 2x^5 - x^4 - 4x^3 + 2x^2 - 30x + 15$:

a) Find the rational zeros and then the other zeros; that is, solve $f(x) = 0$.

b) Factor $f(x)$ into linear factors.

Solution

a) Because the degree of $f(x)$ is 5, there are at most 5 distinct zeros. All of the coefficients are integers. According to the rational zeros theorem, any rational zero of f must be of the form p/q, where p is a factor of 15 and q is a factor of 2. The possibilities are

$$\frac{Possibilities\ for\ p}{Possibilities\ for\ q}: \qquad \frac{\pm 1,\ \pm 3,\ \pm 5,\ \pm 15}{\pm 1,\ \pm 2};$$

$Possibilities\ for\ p/q:$ $1, -1, 3, -3, 5, -5, 15, -15, \frac{1}{2}, -\frac{1}{2}, \frac{3}{2}, -\frac{3}{2}, \frac{5}{2}, -\frac{5}{2},$

$\frac{15}{2}, -\frac{15}{2}.$

Rather than use synthetic division to check each of these possibilities, we graph $y = 2x^5 - x^4 - 4x^3 + 2x^2 - 30x + 15$. (See the graph at left.) We can then inspect the graph for zeros that appear to be near any of the possible rational zeros.

From the graph, we see that of the possibilities in the list, only the numbers $-\frac{5}{2}, \frac{1}{2},$ and $\frac{5}{2}$ might be rational zeros. By synthetic division or using the TABLE feature of a graphing calculator (see the table at left), we see that only $\frac{1}{2}$ is actually a rational zero.

$$\begin{array}{r|rrrrrr}
1/2 & 2 & -1 & -4 & 2 & -30 & 15 \\
    &   & 1 & 0 & -2 & 0 & -15 \\
\hline
    & 2 & 0 & -4 & 0 & -30 & \bigm| \ \ 0
\end{array}$$

This means that $x - \frac{1}{2}$ is a factor of $f(x)$. We write the factorization and try to factor further:

$$\begin{aligned}
f(x) &= \left(x - \tfrac{1}{2}\right)\left(2x^4 - 4x^2 - 30\right) \\
&= \left(x - \tfrac{1}{2}\right) \cdot 2 \cdot \left(x^4 - 2x^2 - 15\right) \qquad \textbf{Factoring out the 2} \\
&= \left(x - \tfrac{1}{2}\right) \cdot 2 \cdot \left(x^2 - 5\right)\left(x^2 + 3\right). \qquad \textbf{Factoring the trinomial}
\end{aligned}$$

We now solve the equation $f(x) = 0$ to determine the zeros. We use the principle of zero products:

$$\left(x - \tfrac{1}{2}\right) \cdot 2 \cdot \left(x^2 - 5\right)\left(x^2 + 3\right) = 0$$

$$\begin{array}{lllll}
x - \tfrac{1}{2} = 0 & or & x^2 - 5 = 0 & or & x^2 + 3 = 0 \\
x = \tfrac{1}{2} & or & x^2 = 5 & or & x^2 = -3 \\
x = \tfrac{1}{2} & or & x = \pm\sqrt{5} & or & x = \pm\sqrt{3}\,i.
\end{array}$$

There is only one rational zero, $\frac{1}{2}$. The other zeros are $\pm\sqrt{5}$ and $\pm\sqrt{3}\,i$.

b) The factorization into linear factors is

$$f(x) = 2\left(x - \tfrac{1}{2}\right)\left(x + \sqrt{5}\right)\left(x - \sqrt{5}\right)\left(x + \sqrt{3}\,i\right)\left(x - \sqrt{3}\,i\right), \quad or$$
$$f(x) = (2x - 1)\left(x + \sqrt{5}\right)\left(x - \sqrt{5}\right)\left(x + \sqrt{3}\,i\right)\left(x - \sqrt{3}\,i\right).$$

Replacing $2\left(x - \tfrac{1}{2}\right)$ **with** $(2x - 1)$

▶ *Now Try Exercise 61.*

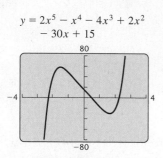

$y = 2x^5 - x^4 - 4x^3 + 2x^2 - 30x + 15$

X	Y1
-2.5	-69.38
.5	0
2.5	46.25

X =

❯ Descartes' Rule of Signs

The development of a rule that helps determine the number of positive real zeros and the number of negative real zeros of a polynomial function is credited to the French mathematician René Descartes. To use the rule, we must have the polynomial arranged in descending order or ascending order, with no zero terms written in and the constant term not 0. Then we determine the number of *variations of sign*, that is, the number of times, in reading through the polynomial, that successive coefficients are of different signs.

EXAMPLE 7   Determine the number of variations of sign in the polynomial function $P(x) = 2x^5 - 3x^2 + x + 4$.

Solution   We have

$$P(x) = \underbrace{2x^5 - 3x^2 + x + 4}$$

From positive to
negative; a variation

Both positive; no variation

From negative to positive;
a variation

The number of variations of sign is 2.

Note the following:

$$P(-x) = 2(-x)^5 - 3(-x)^2 + (-x) + 4$$
$$= -2x^5 - 3x^2 - x + 4.$$

Just in Time 7

We see that the number of variations of sign in $P(-x)$ is 1. It occurs as we go from $-x$ to 4.

We now state Descartes' rule, without proof.

DESCARTES' RULE OF SIGNS

Let $P(x)$, written in descending order or ascending order, be a polynomial function with real coefficients and a nonzero constant term. The number of positive real zeros of $P(x)$ is either:

1. The same as the number of variations of sign in $P(x)$, or
2. Less than the number of variations of sign in $P(x)$ by a positive even integer.

The number of negative real zeros of $P(x)$ is either:

3. The same as the number of variations of sign in $P(-x)$, or
4. Less than the number of variations of sign in $P(-x)$ by a positive even integer.

A zero of multiplicity m must be counted m times.

In each of Examples 8–10, what does Descartes' rule of signs tell you about the number of positive real zeros and the number of negative real zeros?

EXAMPLE 8 $P(x) = 2x^5 - 5x^2 - 3x + 6$

Solution   The number of variations of sign in $P(x)$ is 2. Therefore, the number of positive real zeros is either 2 or less than 2 by 2, 4, 6, and so on. Thus the number of positive real zeros is either 2 or 0, since a negative number of zeros has no meaning.

$$P(-x) = -2x^5 - 5x^2 + 3x + 6$$

The number of variations of sign in $P(-x)$ is 1. Thus there is exactly 1 negative real zero. Since nonreal, complex conjugates occur in pairs, we also know the possible ways in which nonreal zeros might occur. The table at left summarizes all the possibilities for real zeros and nonreal zeros of $P(x)$.

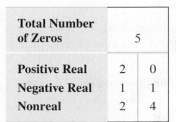

Total Number of Zeros	5	
Positive Real	2	0
Negative Real	1	1
Nonreal	2	4

Now Try Exercise 93.

EXAMPLE 9 $P(x) = 5x^4 - 3x^3 + 7x^2 - 12x + 4$

Solution There are 4 variations of sign. Thus the number of positive real zeros is either

$$4 \quad or \quad 4 - 2 \quad or \quad 4 - 4;$$

that is, the number of positive real zeros is 4, 2, or 0.

$$P(-x) = 5x^4 + 3x^3 + 7x^2 + 12x + 4$$

There are 0 variations of sign, so there are no negative real zeros. The table at left summarizes all the possibilities for real zeros and nonreal zeros of $P(x)$.

> *Now Try Exercise 81.*

Total Number of Zeros	4		
Positive Real	4	2	0
Negative Real	0	0	0
Nonreal	0	2	4

EXAMPLE 10 $P(x) = 6x^6 - 2x^2 - 5x$

Solution As written, the polynomial does not satisfy the conditions of Descartes' rule of signs because the constant term is 0. But because x is a factor of every term, we know that the polynomial has 0 as a zero. We can then factor as follows:

$$P(x) = x(6x^5 - 2x - 5).$$

Now we analyze $Q(x) = 6x^5 - 2x - 5$ and $Q(-x) = -6x^5 + 2x - 5$. The number of variations of sign in $Q(x)$ is 1. Therefore, there is exactly 1 positive real zero. The number of variations of sign in $Q(-x)$ is 2. Thus the number of negative real zeros is 2 or 0. The same results apply to $P(x)$. Since nonreal, complex conjugates occur in pairs, we know the possible ways in which nonreal zeros might occur. The table at left summarizes all the possibilities for real zeros and nonreal zeros of $P(x)$.

> *Now Try Exercise 95.*

Total Number of Zeros	6	
0 as a Zero	1	1
Positive Real	1	1
Negative Real	2	0
Nonreal	2	4

4.4 Exercise Set

Find a polynomial function of degree 3 with the given numbers as zeros.

1. $-2, 3, 5$

2. $-1, 0, 4$

3. $-3, 2i, -2i$

4. $2, i, -i$

5. $\sqrt{2}, -\sqrt{2}, 3$

6. $-5, \sqrt{3}, -\sqrt{3}$

7. $1 - \sqrt{3}, 1 + \sqrt{3}, -2$

8. $-4, 1 - \sqrt{5}, 1 + \sqrt{5}$

9. $1 + 6i, 1 - 6i, -4$

10. $1 + 4i, 1 - 4i, -1$

11. $-\frac{1}{3}, 0, 2$

12. $-3, 0, \frac{1}{2}$

13. Find a polynomial function of degree 5 with -1 as a zero of multiplicity 3, 0 as a zero of multiplicity 1, and 1 as a zero of multiplicity 1.

14. Find a polynomial function of degree 4 with -2 as a zero of multiplicity 1, 3 as a zero of multiplicity 2, and -1 as a zero of multiplicity 1.

15. Find a polynomial function of degree 4 with -1 as a zero of multiplicity 3 and 0 as a zero of multiplicity 1.

16. Find a polynomial function of degree 5 with $-\frac{1}{2}$ as a zero of multiplicity 2, 0 as a zero of multiplicity 1, and 1 as a zero of multiplicity 2.

Suppose that a polynomial function of degree 4 with rational coefficients has the given numbers as zeros. Find the other zero(s).

17. $-1, \sqrt{3}, \frac{11}{3}$

18. $-\sqrt{2}, -1, \frac{4}{5}$

19. $-i, 2 - \sqrt{5}$

20. $i, -3 + \sqrt{3}$

21. $3i, 0, -5$

22. $3, 0, -2i$

23. $-4 - 3i, 2 - \sqrt{3}$

24. $6 - 5i, -1 + \sqrt{7}$

Suppose that a polynomial function of degree 5 with rational coefficients has the given numbers as zeros. Find the other zero(s).

25. $-\frac{1}{2}, \sqrt{5}, -4i$

26. $\frac{3}{4}, -\sqrt{3}, 2i$

27. $-5, 0, 2 - i, 4$

28. $-2, 3, 4, 1 - i$

29. $6, -3 + 4i, 4 - \sqrt{5}$

30. $-3 - 3i, 2 + \sqrt{13}, 6$

31. $-\frac{3}{4}, \frac{3}{4}, 0, 4 - i$

32. $-0.6, 0, 0.6, -3 + \sqrt{2}$

Find a polynomial function of lowest degree with rational coefficients that has the given numbers as some of its zeros.

33. $1 + i, 2$

34. $2 - i, -1$

35. $4i$

36. $-5i$

37. $-4i, 5$

38. $3, -i$

39. $1 - i, -\sqrt{5}$

40. $2 - \sqrt{3}, 1 + i$

41. $\sqrt{5}, -3i$

42. $-\sqrt{2}, 4i$

Given that the polynomial function has the given zero, find the other zeros.

43. $f(x) = x^3 + 5x^2 - 2x - 10; \ -5$

44. $f(x) = x^3 - x^2 + x - 1; \ 1$

45. $f(x) = x^4 - 5x^3 + 7x^2 - 5x + 6; \ -i$

46. $f(x) = x^4 - 16; \ 2i$

47. $f(x) = x^3 - 6x^2 + 13x - 20; \ 4$

48. $f(x) = x^3 - 8; \ 2$

List all possible rational zeros of the function.

49. $f(x) = x^5 - 3x^2 + 1$

50. $f(x) = x^7 + 37x^5 - 6x^2 + 12$

51. $f(x) = 2x^4 - 3x^3 - x + 8$

52. $f(x) = 3x^3 - x^2 + 6x - 9$

53. $f(x) = 15x^6 + 47x^2 + 2$

54. $f(x) = 10x^{25} + 3x^{17} - 35x + 6$

For each polynomial function:

a) *Find the rational zeros and then the other zeros; that is, solve $f(x) = 0$.*

b) *Factor $f(x)$ into linear factors.*

55. $f(x) = x^3 + 3x^2 - 2x - 6$

56. $f(x) = x^3 - x^2 - 3x + 3$

57. $f(x) = 3x^3 - x^2 - 15x + 5$

58. $f(x) = 4x^3 - 4x^2 - 3x + 3$

59. $f(x) = x^3 - 3x + 2$

60. $f(x) = x^3 - 2x + 4$

61. $f(x) = 2x^3 + 3x^2 + 18x + 27$

62. $f(x) = 2x^3 + 7x^2 + 2x - 8$

63. $f(x) = 5x^4 - 4x^3 + 19x^2 - 16x - 4$

64. $f(x) = 3x^4 - 4x^3 + x^2 + 6x - 2$

65. $f(x) = x^4 - 3x^3 - 20x^2 - 24x - 8$

66. $f(x) = x^4 + 5x^3 - 27x^2 + 31x - 10$

67. $f(x) = x^3 - 4x^2 + 2x + 4$

68. $f(x) = x^3 - 8x^2 + 17x - 4$

69. $f(x) = x^3 + 8$

70. $f(x) = x^3 - 8$

71. $f(x) = \frac{1}{3}x^3 - \frac{1}{2}x^2 - \frac{1}{6}x + \frac{1}{6}$

72. $f(x) = \frac{2}{3}x^3 - \frac{1}{2}x^2 + \frac{2}{3}x - \frac{1}{2}$

Find only the rational zeros of the function. If there are none, state this.

73. $f(x) = x^4 + 2x^3 - 5x^2 - 4x + 6$

74. $f(x) = x^4 - 3x^3 - 9x^2 - 3x - 10$

75. $f(x) = x^3 - x^2 - 4x + 3$

76. $f(x) = 2x^3 + 3x^2 + 2x + 3$

77. $f(x) = x^4 + 2x^3 + 2x^2 - 4x - 8$

78. $f(x) = x^4 + 6x^3 + 17x^2 + 36x + 66$

79. $f(x) = x^5 - 5x^4 + 5x^3 + 15x^2 - 36x + 20$

80. $f(x) = x^5 - 3x^4 - 3x^3 + 9x^2 - 4x + 12$

What does Descartes' rule of signs tell you about the number of positive real zeros and the number of negative real zeros of the function?

81. $f(x) = 3x^5 - 2x^2 + x - 1$

82. $g(x) = 5x^6 - 3x^3 + x^2 - x$

83. $h(x) = 6x^7 + 2x^2 + 5x + 4$

84. $P(x) = -3x^5 - 7x^3 - 4x - 5$

85. $F(p) = 3p^{18} + 2p^4 - 5p^2 + p + 3$

86. $H(t) = 5t^{12} - 7t^4 + 3t^2 + t + 1$

87. $C(x) = 7x^6 + 3x^4 - x - 10$

88. $g(z) = -z^{10} + 8z^7 + z^3 + 6z - 1$

89. $h(t) = -4t^5 - t^3 + 2t^2 + 1$

90. $P(x) = x^6 + 2x^4 - 9x^3 - 4$

91. $f(y) = y^4 + 13y^3 - y + 5$

92. $Q(x) = x^4 - 2x^2 + 12x - 8$

93. $r(x) = x^4 - 6x^2 + 20x - 24$

94. $f(x) = x^5 - 2x^3 - 8x$

95. $R(x) = 3x^5 - 5x^3 - 4x$

96. $f(x) = x^4 - 9x^2 - 6x + 4$

Sketch the graph of the polynomial function. Follow the procedure outlined on p. 237. Use the rational zeros theorem when finding the zeros.

97. $f(x) = 4x^3 + x^2 - 8x - 2$

98. $f(x) = 3x^3 - 4x^2 - 5x + 2$

99. $f(x) = 2x^4 - 3x^3 - 2x^2 + 3x$

100. $f(x) = 4x^4 - 37x^2 + 9$

> ## Skill Maintenance

For Exercises 101 and 102, complete the square to:

a) *find the vertex;*

b) *find the axis of symmetry; and*

c) *determine whether there is a maximum or a minimum function value and find that value.* **[3.3]**

101. $f(x) = x^2 - 8x + 10$

102. $f(x) = 3x^2 - 6x - 1$

Find the zeros of the function.

103. $f(x) = -\frac{4}{5}x + 8$ **[1.5]**

104. $g(x) = x^2 - 8x - 33$ **[3.2]**

Determine the leading term, the leading coefficient, and the degree of the polynomial. Then describe the end behavior of the function's graph and classify the polynomial function as constant, linear, quadratic, cubic, or quartic.

105. $g(x) = -x^3 - 2x^2$ **[4.1]**

106. $f(x) = -x^2 - 3x + 6$ **[3.3]**

107. $f(x) = -\frac{4}{9}$ **[1.3], [4.1]**

108. $h(x) = x - 2$ **[1.3], [4.1]**

109. $g(x) = x^4 - 2x^3 + x^2 - x + 2$ **[4.1]**

110. $h(x) = x^3 + \frac{1}{2}x^2 - 4x - 3$ **[4.1]**

> ## Synthesis

111. Consider $f(x) = 2x^3 - 5x^2 - 4x + 3$. Find the solutions of each equation.

   a) $f(x) = 0$

   b) $f(x - 1) = 0$

   c) $f(x + 2) = 0$

   d) $f(2x) = 0$

112. Use the rational zeros theorem and the equation $x^4 - 12 = 0$ to show that $\sqrt[4]{12}$ is irrational.

Find the rational zeros of the function.

113. $P(x) = 2x^5 - 33x^4 - 84x^3 + 2203x^2 - 3348x - 10{,}080$

114. $P(x) = x^6 - 6x^5 - 72x^4 - 81x^2 + 486x + 5832$

4.5 ❯ Rational Functions

> ❯ For a rational function, find the domain and graph the function, identifying all of the asymptotes.
>
> ❯ Solve applied problems involving rational functions.

Now we turn our attention to functions that represent the quotient of two polynomials. Whereas the sum, the difference, or the product of two polynomials is a polynomial, in general the quotient of two polynomials is *not* itself a polynomial.

A *rational number* can be expressed as the quotient of two integers, p/q, where $q \neq 0$. A *rational function* is formed by the quotient of two polynomials, $p(x)/q(x)$, where $q(x) \neq 0$. Here are some examples of rational functions and their graphs.

$$f(x) = \frac{1}{x}$$

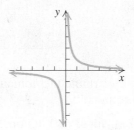

$$f(x) = \frac{1}{x^2}$$

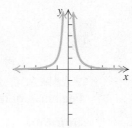

$$f(x) = \frac{x-3}{x^2 + x - 2}$$

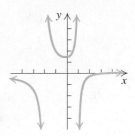

$$f(x) = \frac{2x+5}{2x-6}$$

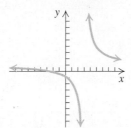

$$f(x) = \frac{x^2 + 2x - 3}{x^2 - x - 2}$$

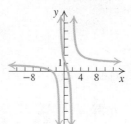

$$f(x) = \frac{-x^2}{x+1}$$

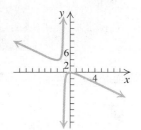

RATIONAL FUNCTION

A **rational function** is a function f that is a quotient of two polynomials. That is,

$$f(x) = \frac{p(x)}{q(x)},$$

where $p(x)$ and $q(x)$ are polynomials and where $q(x)$ is not the zero polynomial. The domain of f consists of all inputs x for which $q(x) \neq 0$.

❯ The Domain of a Rational Function

EXAMPLE 1 Consider

$$f(x) = \frac{1}{x-3}.$$

Find the domain and graph f.

DOMAINS OF FUNCTIONS

REVIEW SECTION 1.2.

Solution When the denominator $x - 3$ is 0, we have $x = 3$, so the only input that results in a denominator of 0 is 3. Thus the domain is

$$\{x | x \neq 3\}, \text{ or } (-\infty, 3) \cup (3, \infty).$$

The graph of this function is the graph of $y = 1/x$ translated 3 units to the right.

$$y = \frac{1}{x - 3}$$

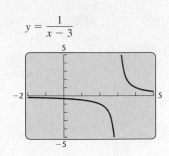

EXAMPLE 2   Determine the domain of each of the functions illustrated at the beginning of this section.

Solution   The domain of each rational function will be the set of all real numbers except those values that make the denominator 0. To determine those exceptions, we set the denominator equal to 0 and solve for x.

Just in Time

6

FUNCTION	DOMAIN
$f(x) = \dfrac{1}{x}$	$\{x \mid x \neq 0\}$, or $(-\infty, 0) \cup (0, \infty)$
$f(x) = \dfrac{1}{x^2}$	$\{x \mid x \neq 0\}$, or $(-\infty, 0) \cup (0, \infty)$
$f(x) = \dfrac{x - 3}{x^2 + x - 2} = \dfrac{x - 3}{(x + 2)(x - 1)}$	$\{x \mid x \neq -2 \text{ and } x \neq 1\}$, or $(-\infty, -2) \cup (-2, 1) \cup (1, \infty)$
$f(x) = \dfrac{2x + 5}{2x - 6} = \dfrac{2x + 5}{2(x - 3)}$	$\{x \mid x \neq 3\}$, or $(-\infty, 3) \cup (3, \infty)$
$f(x) = \dfrac{x^2 + 2x - 3}{x^2 - x - 2} = \dfrac{x^2 + 2x - 3}{(x + 1)(x - 2)}$	$\{x \mid x \neq -1 \text{ and } x \neq 2\}$, or $(-\infty, -1) \cup (-1, 2) \cup (2, \infty)$
$f(x) = \dfrac{-x^2}{x + 1}$	$\{x \mid x \neq -1\}$, or $(-\infty, -1) \cup (-1, \infty)$

As a partial check of the domains, we can observe the discontinuities (breaks) in the graphs of these functions. (See p. 263.)

❯ Asymptotes

Vertical Asymptotes

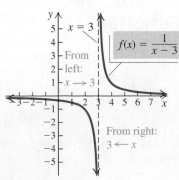

Vertical asymptote: $x = 3$

Look at the graph of $f(x) = 1/(x - 3)$, shown at left. (Also see Example 1.) Let's explore what happens as x-values get closer and closer to 3 from the left. We then explore what happens as x-values get closer and closer to 3 from the right.

From the left:

x	2	$2\frac{1}{2}$	$2\frac{99}{100}$	$2\frac{9999}{10,000}$	$2\frac{999,999}{1,000,000}$	$\longrightarrow 3$
$f(x)$	-1	-2	-100	$-10,000$	$-1,000,000$	$\longrightarrow -\infty$

From the right:

x	4	$3\frac{1}{2}$	$3\frac{1}{100}$	$3\frac{1}{10,000}$	$3\frac{1}{1,000,000}$	$\longrightarrow 3$
$f(x)$	1	2	100	10,000	1,000,000	$\longrightarrow \infty$

We see that as x-values get closer and closer to 3 from the left, the function values (y-values) decrease without bound (that is, they approach negative infinity, $-\infty$). Similarly, as the x-values approach 3 from the right, the function values increase without bound (that is, they approach positive infinity, ∞). We write this as

$$f(x) \to -\infty \text{ as } x \to 3^{-} \quad \text{and} \quad f(x) \to \infty \text{ as } x \to 3^{+}.$$

We read "$f(x) \to -\infty$ as $x \to 3^{-}$" as "$f(x)$ decreases without bound as x approaches 3 from the left." We read "$f(x) \to \infty$ as $x \to 3^{+}$" as "$f(x)$ increases without bound as x approaches 3 from the right." The notation $x \to 3$ means that x gets as close to 3 as possible without being equal to 3. The vertical line $x = 3$ is said to be a *vertical asymptote* for this curve.

In general, the line $x = a$ is a **vertical asymptote** for the graph of f if any of the following is true:

$$f(x) \to \infty \text{ as } x \to a^{-}, \quad \text{or} \quad f(x) \to -\infty \text{ as } x \to a^{-}, \quad \text{or}$$
$$f(x) \to \infty \text{ as } x \to a^{+}, \quad \text{or} \quad f(x) \to -\infty \text{ as } x \to a^{+}.$$

The following figures show the four ways in which a vertical asymptote can occur.

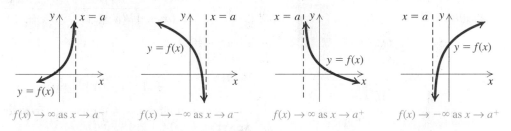

$f(x) \to \infty \text{ as } x \to a^{-}$ $f(x) \to -\infty \text{ as } x \to a^{-}$ $f(x) \to \infty \text{ as } x \to a^{+}$ $f(x) \to -\infty \text{ as } x \to a^{+}$

The vertical asymptotes of a rational function $f(x) = p(x)/q(x)$ are found by determining the zeros of $q(x)$ that are not also zeros of $p(x)$. If $p(x)$ and $q(x)$ are polynomials with no common factors other than constants, we need determine only the zeros of the denominator $q(x)$.

DETERMINING VERTICAL ASYMPTOTES

For a rational function $f(x) = p(x)/q(x)$, where $p(x)$ and $q(x)$ are polynomials with no common factors other than constants, if a is a zero of the denominator, then the line $x = a$ is a vertical asymptote for the graph of the function.

EXAMPLE 3   Determine the vertical asymptotes for the graph of each of the following functions.

a) $f(x) = \dfrac{2x - 11}{x^2 + 2x - 8}$ **b)** $h(x) = \dfrac{x^2 - 4x}{x^3 - x}$

c) $g(x) = \dfrac{x - 2}{x^3 - 5x}$

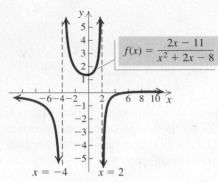

FIGURE 1.

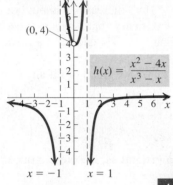

FIGURE 2.

Just in Time

21

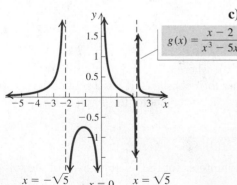

FIGURE 3.

Solution

a) First, we factor the denominator:

$$f(x) = \frac{2x - 11}{x^2 + 2x - 8} = \frac{2x - 11}{(x + 4)(x - 2)}.$$

The numerator and the denominator have no common factors. The zeros of the denominator are -4 and 2. Thus the vertical asymptotes for the graph of $f(x)$ are the lines $x = -4$ and $x = 2$. (See Fig. 1.)

b) We factor the numerator and the denominator:

$$h(x) = \frac{x^2 - 4x}{x^3 - x} = \frac{x(x - 4)}{x(x^2 - 1)} = \frac{x(x - 4)}{x(x + 1)(x - 1)}.$$

The domain of the function is $\{x \mid x \neq -1 \text{ and } x \neq 0 \text{ and } x \neq 1\}$, or $(-\infty, -1) \cup (-1, 0) \cup (0, 1) \cup (1, \infty)$. Note that the numerator and the denominator share a common factor, x. The vertical asymptotes of $h(x)$ are found by determining the zeros of the denominator, $x(x + 1)(x - 1)$, that are *not* also zeros of the numerator, $x(x - 4)$. The zeros of $x(x + 1)(x - 1)$ are $0, -1,$ and 1. The zeros of $x(x - 4)$ are 0 and 4. Thus, although the denominator has three zeros, the graph of $h(x)$ has only two vertical asymptotes, $x = -1$ and $x = 1$. (See Fig. 2.)

The rational expression $[x(x - 4)]/[x(x + 1)(x - 1)]$ can be simplified. Thus,

$$h(x) = \frac{x(x - 4)}{x(x + 1)(x - 1)} = \frac{x - 4}{(x + 1)(x - 1)},$$

where $x \neq 0, x \neq -1,$ and $x \neq 1$. The graph of $h(x)$ is the graph of

$$h(x) = \frac{x - 4}{(x + 1)(x - 1)}$$

with the point where $x = 0$ missing. To determine the y-coordinate of the "hole," we substitute 0 for x:

$$h(0) = \frac{0 - 4}{(0 + 1)(0 - 1)} = \frac{-4}{1 \cdot (-1)} = 4.$$

Thus the "hole" is located at $(0, 4)$.

c) We factor the denominator:

$$g(x) = \frac{x - 2}{x^3 - 5x} = \frac{x - 2}{x(x^2 - 5)}.$$

The numerator and the denominator have no common factors. We find the zeros of the denominator, $x(x^2 - 5)$. Solving $x(x^2 - 5) = 0$, we get

$$x = 0 \quad or \quad x^2 - 5 = 0$$
$$x = 0 \quad or \quad x^2 = 5$$
$$x = 0 \quad or \quad x = \pm\sqrt{5}.$$

The zeros of the denominator are $0, \sqrt{5},$ and $-\sqrt{5}$. Thus the vertical asymptotes are the lines $x = 0, x = \sqrt{5},$ and $x = -\sqrt{5}$. (See Fig. 3.)

> *Now Try Exercises 15 and 19.*

Horizontal Asymptotes

Looking again at the graph of $f(x) = 1/(x - 3)$ (also see Example 1), let's explore what happens to $f(x) = 1/(x - 3)$ as x increases without bound (approaches positive infinity, ∞) and as x decreases without bound (approaches negative infinity, $-\infty$).

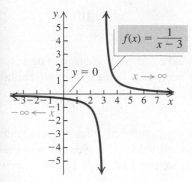

Horizontal asymptote: $y = 0$

x increases without bound:

x	100	5000	1,000,000	$\longrightarrow \infty$
$f(x)$	≈ 0.0103	≈ 0.0002	≈ 0.000001	$\longrightarrow 0$

x decreases without bound:

x	-300	-8000	$-1,000,000$	$\longrightarrow -\infty$
$f(x)$	≈ -0.0033	≈ -0.0001	≈ -0.000001	$\longrightarrow 0$

We see that

$$\frac{1}{x - 3} \to 0 \text{ as } x \to \infty \quad \text{and} \quad \frac{1}{x - 3} \to 0 \text{ as } x \to -\infty.$$

Since $y = 0$ is the equation of the x-axis, we say that the curve approaches the x-axis asymptotically and that the x-axis is a *horizontal asymptote* for the curve.

In general, the line $y = b$ is a **horizontal asymptote** for the graph of f if either or both of the following are true:

$$f(x) \to b \text{ as } x \to \infty \quad \text{or} \quad f(x) \to b \text{ as } x \to -\infty.$$

The following figures illustrate four ways in which horizontal asymptotes can occur. In each case, the curve gets close to the line $y = b$ either as $x \to \infty$ or as $x \to -\infty$. Keep in mind that the symbols ∞ and $-\infty$ convey the idea of increasing without bound and decreasing without bound, respectively.

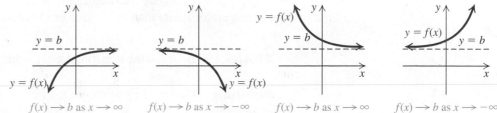

$f(x) \to b \text{ as } x \to \infty$ $f(x) \to b \text{ as } x \to -\infty$ $f(x) \to b \text{ as } x \to \infty$ $f(x) \to b \text{ as } x \to -\infty$

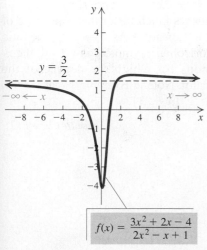

$$f(x) = \frac{3x^2 + 2x - 4}{2x^2 - x + 1}$$

How can we determine a horizontal asymptote? As x gets very large or very small, the value of a polynomial function $p(x)$ is dominated by the function's leading term. Because of this, if $p(x)$ and $q(x)$ have the *same* degree, the value of $p(x)/q(x)$ as $x \to \infty$ or as $x \to -\infty$ is dominated by the ratio of the numerator's leading coefficient to the denominator's leading coefficient.

For $f(x) = (3x^2 + 2x - 4)/(2x^2 - x + 1)$, we see that the numerator, $3x^2 + 2x - 4$, is dominated by $3x^2$ and the denominator, $2x^2 - x + 1$, is dominated by $2x^2$, so $f(x)$ approaches $3x^2/2x^2$, or $3/2$ as x gets very large or very small:

$$\frac{3x^2 + 2x - 4}{2x^2 - x + 1} \to \frac{3}{2}, \text{ or } 1.5, \text{ as } x \to \infty, \quad \text{and}$$

$$\frac{3x^2 + 2x - 4}{2x^2 - x + 1} \to \frac{3}{2}, \text{ or } 1.5, \text{ as } x \to -\infty.$$

We say that the curve approaches the horizontal line $y = \frac{3}{2}$ asymptotically and that $y = \frac{3}{2}$ is a *horizontal asymptote* for the curve.

It follows that when the numerator and the denominator of a rational function have the same degree, the line $y = a/b$ is the horizontal asymptote, where a and b are the leading coefficients of the numerator and the denominator, respectively.

EXAMPLE 4 Find the horizontal asymptote: $f(x) = \dfrac{-7x^4 - 10x^2 + 1}{11x^4 + x - 2}$.

Solution The numerator and the denominator have the same degree. The ratio of the leading coefficients is $-\frac{7}{11}$, so the line $y = -\frac{7}{11}$, or $-0.\overline{63}$, is the horizontal asymptote.

> *Now Try Exercise 21.*

X	Y₁	
100000	−.6364	
−80000	−.6364	
X =		

As a partial check of Example 4, we could use a graphing calculator to evaluate the function for a very large value of x and a very small value of x. (See the window at left.) It is useful in calculus to multiply by 1. Here we use $(1/x^4)/(1/x^4)$:

$$f(x) = \frac{-7x^4 - 10x^2 + 1}{11x^4 + x - 2} \cdot \frac{\dfrac{1}{x^4}}{\dfrac{1}{x^4}} = \frac{\dfrac{-7x^4}{x^4} - \dfrac{10x^2}{x^4} + \dfrac{1}{x^4}}{\dfrac{11x^4}{x^4} + \dfrac{x}{x^4} - \dfrac{2}{x^4}}$$

$$= \frac{-7 - \dfrac{10}{x^2} + \dfrac{1}{x^4}}{11 + \dfrac{1}{x^3} - \dfrac{2}{x^4}}.$$

As $|x|$ becomes very large, each expression whose denominator is a power of x tends toward 0. Specifically, as $x \to \infty$ or as $x \to -\infty$, we have

$$f(x) \to \frac{-7 - 0 + 0}{11 + 0 - 0}, \quad \text{or} \quad f(x) \to -\frac{7}{11}.$$

The horizontal asymptote is $y = -\frac{7}{11}$, or $-0.\overline{63}$.

We now investigate the occurrence of a horizontal asymptote when the degree of the numerator is less than the degree of the denominator.

EXAMPLE 5 Find the horizontal asymptote: $f(x) = \dfrac{2x + 3}{x^3 - 2x^2 + 4}$.

Solution We let $p(x) = 2x + 3$, $q(x) = x^3 - 2x^2 + 4$, and $f(x) = p(x)/q(x)$. Note that as $x \to \infty$, the value of $q(x)$ grows much faster than the value of $p(x)$. Because of this, the ratio $p(x)/q(x)$ shrinks toward 0. As $x \to -\infty$, the ratio $p(x)/q(x)$ behaves in a similar manner. The horizontal asymptote is $y = 0$, the x-axis. This is the case for all rational functions for which the degree of the numerator is less than the degree of the denominator. Note in Example 1 that $y = 0$, the x-axis, is the horizontal asymptote of $f(x) = 1/(x - 3)$.

> *Now Try Exercise 23.*

The following statements describe the two ways in which a horizontal asymptote occurs.

DETERMINING A HORIZONTAL ASYMPTOTE

- When the numerator and the denominator of a rational function have the same degree, the line $y = a/b$ is the horizontal asymptote, where a and b are the leading coefficients of the numerator and the denominator, respectively.
- When the degree of the numerator of a rational function is less than the degree of the denominator, the x-axis, or $y = 0$, is the horizontal asymptote.
- When the degree of the numerator of a rational function is greater than the degree of the denominator, there is no horizontal asymptote.

EXAMPLE 6   Graph

$$g(x) = \frac{2x^2 + 1}{x^2}.$$

Include and label all asymptotes.

Solution   Since 0 is the zero of the denominator and is not a zero of the numerator, the y-axis, $x = 0$, is the vertical asymptote. Note also that the degree of the numerator is the same as the degree of the denominator. Thus, $y = 2/1$, or 2, is the horizontal asymptote.

To draw the graph, we first draw the asymptotes with dashed lines. Then we compute and plot some ordered pairs and draw the two branches of the curve. We can check the graph with a graphing calculator.

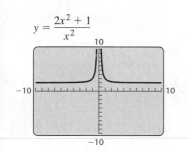

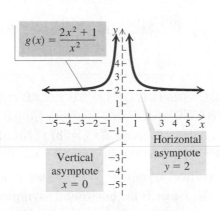

x	$g(x)$
-2	2.25
-1.5	$2.\overline{4}$
-1	3
-0.5	6
0.5	6
1	3
1.5	$2.\overline{4}$
2	2.25

> **Now Try Exercise 41.**

Oblique Asymptotes

Sometimes a line that is neither horizontal nor vertical is an asymptote. Such a line is called an **oblique asymptote**, or a **slant asymptote**.

EXAMPLE 7   Find all the asymptotes of

$$f(x) = \frac{2x^2 - 3x - 1}{x - 2}.$$

Solution   The line $x = 2$ is the vertical asymptote because 2 is the zero of the denominator and is not a zero of the numerator. There is no horizontal asymptote because the degree of the numerator is greater than the degree of the denominator. When the degree of the numerator is 1 greater than the degree of the denominator, we divide to find an equivalent expression:

$$\frac{2x^2 - 3x - 1}{x - 2} = (2x + 1) + \frac{1}{x - 2}.$$

$$
\require{enclose}
\begin{array}{r}
2x + 1 \\
x - 2 \enclose{longdiv}{2x^2 - 3x - 1} \\
\underline{2x^2 - 4x } \\
x - 1 \\
\underline{x - 2} \\
1
\end{array}
$$

We see that when $x \to \infty$ or $x \to -\infty$, $1/(x - 2) \to 0$ and the value of $f(x) \to 2x + 1$. This means that as $|x|$ becomes very large, the graph of $f(x)$ gets very close to the graph of $y = 2x + 1$. Thus the line $y = 2x + 1$ is the oblique asymptote.

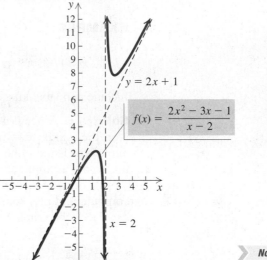

> **Now Try Exercise 59.**

OCCURRENCE OF LINES AS ASYMPTOTES OF RATIONAL FUNCTIONS

For a rational function $f(x) = p(x)/q(x)$, where $p(x)$ and $q(x)$ have no common factors other than constants:

> **Vertical asymptotes** occur at any x-values that make the denominator 0.
>
> **The x-axis is the horizontal asymptote** when the degree of the numerator is less than the degree of the denominator.
>
> **A horizontal asymptote other than the x-axis** occurs when the numerator and the denominator have the same degree.
>
> **An oblique asymptote** occurs when the degree of the numerator is 1 greater than the degree of the denominator.

There can be only one horizontal asymptote or one oblique asymptote and never both.

An asymptote is *not* part of the graph of the function.

The following statements are also true.

CROSSING AN ASYMPTOTE

- The graph of a rational function *never crosses* a vertical asymptote.
- The graph of a rational function *might cross* a horizontal asymptote but does not necessarily do so.

Shown below is an outline of a procedure that we can follow to create accurate graphs of rational functions.

To graph a rational function $f(x) = p(x)/q(x)$, where $p(x)$ and $q(x)$ have no common factor other than constants:

1. Find any real zeros of the denominator. Determine the domain of the function and sketch any vertical asymptotes.
2. Find the horizontal asymptote or the oblique asymptote, if there is one, and sketch it.
3. Find any zeros of the function. The zeros are found by determining the zeros of the numerator. These are the first coordinates of the *x*-intercepts of the graph.
4. Find $f(0)$. This gives the *y*-intercept, $(0, f(0))$, of the function.
5. Find other function values to determine the general shape. Then draw the graph.

EXAMPLE 8   Graph: $f(x) = \dfrac{2x + 3}{3x^2 + 7x - 6}$.

Solution

1. We find the zeros of the denominator by solving $3x^2 + 7x - 6 = 0$. Since

$$3x^2 + 7x - 6 = (3x - 2)(x + 3),$$

the zeros are $\frac{2}{3}$ and -3. Thus the domain excludes $\frac{2}{3}$ and -3 and is

$$(-\infty, -3) \cup \left(-3, \tfrac{2}{3}\right) \cup \left(\tfrac{2}{3}, \infty\right).$$

Since neither zero of the denominator is a zero of the numerator, the graph has vertical asymptotes $x = -3$ and $x = \frac{2}{3}$. We sketch these as dashed lines.

2. Because the degree of the numerator is less than the degree of the denominator, the *x*-axis, $y = 0$, is the horizontal asymptote.

3. To find the zeros of the numerator, we solve $2x + 3 = 0$ and get $x = -\frac{3}{2}$. Thus, $-\frac{3}{2}$ is the zero of the function, and the pair $\left(-\frac{3}{2}, 0\right)$ is the *x*-intercept.

4. We find $f(0)$:

$$f(0) = \frac{2 \cdot 0 + 3}{3 \cdot 0^2 + 7 \cdot 0 - 6}$$

$$= \frac{3}{-6} = -\frac{1}{2}.$$

The point $\left(0, -\frac{1}{2}\right)$ is the *y*-intercept.

5. We find other function values to determine the general shape. We choose values in each interval of the domain, as shown in the following table, and then draw the graph. Note that the graph of this function crosses its horizontal asymptote at $x = -\frac{3}{2}$.

x	y
-4.5	-0.26
-3.25	-1.19
-2.5	0.42
-0.5	-0.23
0.5	-2.29
0.75	4.8
1.5	0.53
3.5	0.18

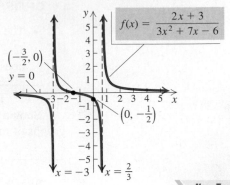

▷ *Now Try Exercise 65.*

EXAMPLE 9   Graph: $g(x) = \dfrac{x^2 - 1}{x^2 + x - 6}$.

Solution

1. We find the zeros of the denominator by solving $x^2 + x - 6 = 0$. Since

$$x^2 + x - 6 = (x + 3)(x - 2),$$

the zeros are -3 and 2. Thus the domain excludes the x-values -3 and 2 and is

$$(-\infty, -3) \cup (-3, 2) \cup (2, \infty).$$

Since neither zero of the denominator is a zero of the numerator, the graph has vertical asymptotes $x = -3$ and $x = 2$. We sketch these as dashed lines.

2. The numerator and the denominator have the same degree, so the horizontal asymptote is determined by the ratio of the leading coefficients: $1/1$, or 1. Thus, $y = 1$ is the horizontal asymptote. We sketch it with a dashed line.

3. To find the zeros of the numerator, we solve $x^2 - 1 = 0$. The solutions are -1 and 1. Thus, -1 and 1 are the zeros of the function and the pairs $(-1, 0)$ and $(1, 0)$ are the x-intercepts.

4. We find $g(0)$:

$$g(0) = \frac{0^2 - 1}{0^2 + 0 - 6} = \frac{-1}{-6} = \frac{1}{6}.$$

Thus, $\left(0, \frac{1}{6}\right)$ is the y-intercept.

5. We find other function values to determine the general shape and draw the graph.

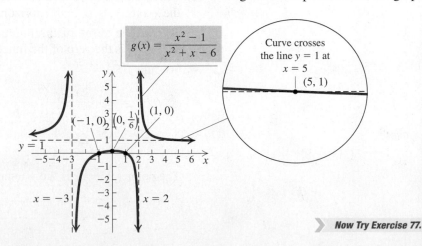

Curve crosses the line $y = 1$ at $x = 5$

$(5, 1)$

▷ *Now Try Exercise 77.*

The magnified portion of the graph in Example 9 above shows another situation in which a graph can cross its horizontal asymptote. The point where $g(x)$ crosses $y = 1$ can be found by setting $g(x) = 1$ and solving for x:

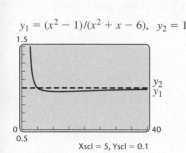

$y_1 = (x^2 - 1)/(x^2 + x - 6), \quad y_2 = 1$

Xscl = 5, Yscl = 0.1

$$\frac{x^2 - 1}{x^2 + x - 6} = 1$$

$$x^2 - 1 = x^2 + x - 6$$

$$-1 = x - 6 \qquad \text{Subtracting } x^2$$

$$5 = x. \qquad \text{Adding 6}$$

The point of intersection is $(5, 1)$. Let's look at the behavior of the curve after it crosses the horizontal asymptote at $x = 5$. (See the graph at left.) It continues to decrease for a short interval and then begins to increase, getting closer and closer to $y = 1$ as $x \rightarrow \infty$.

Graphs of rational functions can also cross an oblique asymptote. The graph of

$$f(x) = \frac{2x^3}{x^2 + 1}$$

shown below crosses its oblique asymptote $y = 2x$. **Remember: Graphs can cross horizontal asymptotes or oblique asymptotes, but they cannot cross vertical asymptotes.**

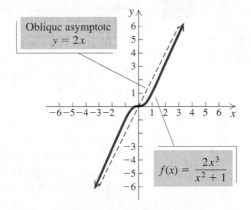

Let's now graph a rational function $f(x) = p(x)/q(x)$, where $p(x)$ and $q(x)$ have a common factor, $x - c$. The graph of such a function has a "hole" in it. We first saw this situation in Example 3(b), where the common factor was x.

EXAMPLE 10   Graph: $g(x) = \dfrac{x - 2}{x^2 - x - 2}$.

Solution   We first express the denominator in factored form:

$$g(x) = \frac{x - 2}{x^2 - x - 2} = \frac{x - 2}{(x + 1)(x - 2)}.$$

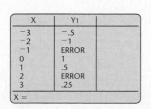

The domain of the function is $\{x \mid x \neq -1 \text{ and } x \neq 2\}$, or $(-\infty, -1) \cup (-1, 2) \cup (2, \infty)$. Note that both the numerator and the denominator have the factor $x - 2$. The zeros of the denominator are -1 and 2, and the zero of the numerator is 2. Since -1 is the only zero of the denominator that is *not* a zero of the numerator, the graph of the function has $x = -1$ as its only vertical asymptote. The degree of the numerator is less than the degree of the denominator, so $y = 0$ is the horizontal asymptote. There are no zeros of the function and thus no x-intercepts, because 2 is the only zero of the numerator and 2 is not in the domain of the function. Since $g(0) = 1$, $(0, 1)$ is the y-intercept. We draw the graph indicating the "hole" when $x = 2$ with an open circle.

The rational expression $(x - 2)/[(x + 1)(x - 2)]$ can be simplified. Thus,

$$g(x) = \frac{x - 2}{(x + 1)(x - 2)} = \frac{1}{x + 1}, \quad \text{where } x \neq -1 \text{ and } x \neq 2.$$

The graph of $g(x)$ is the graph of $y = 1/(x + 1)$ with the point where $x = 2$ missing. To determine the coordinates of the "hole," we substitute 2 for x in $g(x) = 1/(x + 1)$:

$$g(2) = \frac{1}{2 + 1} = \frac{1}{3}.$$

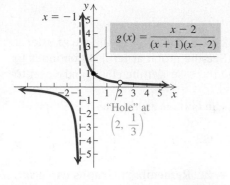

Thus the hole is located at $\left(2, \frac{1}{3}\right)$. With certain window dimensions, the hole is visible on a graphing calculator, as shown at right.

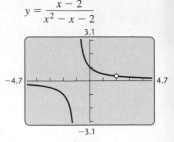

> **Now Try Exercise 49.**

Just in Time

21

EXAMPLE 11 Graph: $f(x) = \dfrac{-2x^2 - x + 15}{x^2 - x - 12}$.

Solution We first express the numerator and the denominator in factored form:

$$f(x) = \frac{-2x^2 - x + 15}{x^2 - x - 12} = \frac{-(2x^2 + x - 15)}{x^2 - x - 12} = \frac{-(2x - 5)(x + 3)}{(x - 4)(x + 3)}.$$

The domain of the function is $\{x \mid x \neq -3 \text{ and } x \neq 4\}$, or $(-\infty, -3) \cup (-3, 4) \cup (4, \infty)$. The numerator and the denominator have the common factor $x + 3$. The zeros of the denominator are -3 and 4, and the zeros of the numerator are -3 and $\frac{5}{2}$. Since 4 is the only zero of the denominator that is *not* a zero of the numerator, the graph of the function has $x = 4$ as its *only* vertical asymptote.

The degrees of the numerator and the denominator are the same, so the line $y = \frac{-2}{1} = -2$ is the horizontal asymptote. The zeros of the numerator are $\frac{5}{2}$ and -3. Because -3 is not in the domain of the function, the only x-intercept is $\left(\frac{5}{2}, 0\right)$. Since $f(0) = \frac{15}{-12} = -\frac{5}{4}$, then $\left(0, -\frac{5}{4}\right)$ is the y-intercept.

The rational function

$$\frac{-(2x - 5)(x + 3)}{(x - 4)(x + 3)}$$

can be simplified. Thus,

$$f(x) = \frac{-(2x - 5)(x + 3)}{(x - 4)(x + 3)} = \frac{-(2x - 5)}{x - 4}, \quad \text{where } x \neq -3 \text{ and } x \neq 4.$$

The graph of $f(x)$ is the graph of $y = -(2x - 5)/(x - 4)$ with the point where $x = -3$ missing. To determine the coordinates of the hole, we substitute -3 for x in $f(x) = -(2x - 5)/(x - 4)$:

$$f(-3) = \frac{-[2(-3) - 5]}{-3 - 4}$$

$$= \frac{-[-11]}{-7} = \frac{11}{-7} = -\frac{11}{7}.$$

Thus the hole is located at $\left(-3, -\frac{11}{7}\right)$. We draw the graph indicating the hole when $x = -3$ with an open circle.

x	y
-5	-1.67
-4	-1.63
-3	Not defined
-2	-1.5
-1	-1.4
0	-1.25
1	-1
2	-0.5

x	y
3	1
3.5	4
4	Not defined
5	-5
6	-3.5
7	-3
8	-2.75

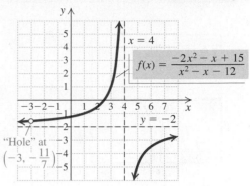

$$f(x) = \frac{-2x^2 - x + 15}{x^2 - x - 12}$$

"Hole" at $\left(-3, -\frac{11}{7}\right)$

> *Now Try Exercise 67.*

❯ Applications

EXAMPLE 12 *Temperature During an Illness.* A person's temperature T, in degrees Fahrenheit, during an illness is given by the function

$$T(t) = \frac{4t}{t^2 + 1} + 98.6,$$

where time t is given in hours since the onset of the illness.

a) Graph the function on the interval $[0, 48]$.

b) Find the temperature at $t = 0, 1, 2, 5, 12,$ and 24.

c) Find the horizontal asymptote of the graph of $T(t)$. Complete:

$$T(t) \rightarrow \boxed{} \text{ as } t \rightarrow \infty.$$

d) Give the meaning of the answer to part (c) in terms of the application.

e) Find the maximum temperature during the illness.

Solution

a) The graph is shown at left.

b) We have

$$T(0) = 98.6, \quad T(1) = 100.6, \quad T(2) = 100.2,$$
$$T(5) \approx 99.369, \quad T(12) \approx 98.931, \quad \text{and} \quad T(24) \approx 98.766.$$

c) Since

$$T(t) = \frac{4t}{t^2 + 1} + 98.6,$$

$$= \frac{98.6t^2 + 4t + 98.6}{t^2 + 1},$$

the horizontal asymptote is $y = 98.6/1$, or 98.6. Then it follows that $T(t) \rightarrow 98.6$ as $t \rightarrow \infty$.

d) As time goes on, the temperature returns to "normal," which is $98.6°F$.

e) Using the MAXIMUM feature on a graphing calculator, we find the maximum temperature to be $100.6°F$ at $t = 1$ hr.

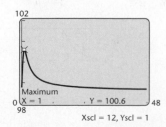

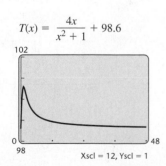

$$T(x) = \frac{4x}{x^2 + 1} + 98.6$$

Xscl = 12, Yscl = 1

Maximum
X = 1   Y = 100.6
Xscl = 12, Yscl = 1

> *Now Try Exercise 83.*

Visualizing the Graph

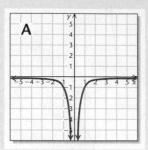

A

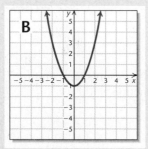

B

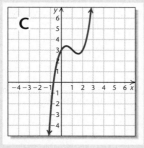

C

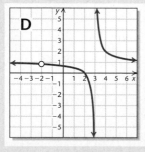

D

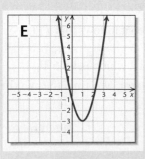

E

Match the function with its graph.

1. $f(x) = -\dfrac{1}{x^2}$

2. $f(x) = x^3 - 3x^2 + 2x + 3$

3. $f(x) = \dfrac{x^2 - 4}{x^2 - x - 6}$

4. $f(x) = -x^2 + 4x - 1$

5. $f(x) = \dfrac{x - 3}{x^2 + x - 6}$

6. $f(x) = \dfrac{3}{4}x + 2$

7. $f(x) = x^2 - 1$

8. $f(x) = x^4 - 2x^2 - 5$

9. $f(x) = \dfrac{8x - 4}{3x + 6}$

10. $f(x) = 2x^2 - 4x - 1$

Answers on page A-20

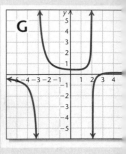

F

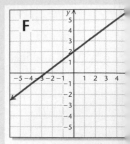

G

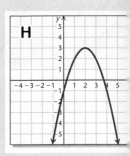

H

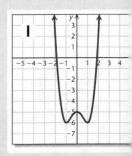

I

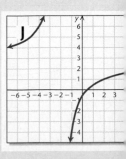

J

4.5   Exercise Set

Determine the domain of the function.

1. $f(x) = \dfrac{x^2}{2 - x}$

2. $f(x) = \dfrac{1}{x^3}$

3. $f(x) = \dfrac{x + 1}{x^2 - 6x + 5}$

4. $f(x) = \dfrac{(x + 4)^2}{4x - 3}$

5. $f(x) = \dfrac{3x - 4}{3x + 15}$

6. $f(x) = \dfrac{x^2 + 3x - 10}{x^2 + 2x}$

In Exercises 7–12, use your knowledge of asymptotes and intercepts to match the equation with one of the graphs (a)–(f) that follow. List all asymptotes. Check your work using a graphing calculator.

a)
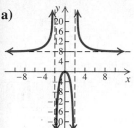

b)
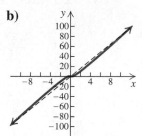

c)
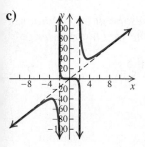

d)
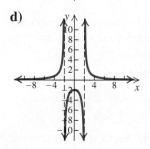

e)
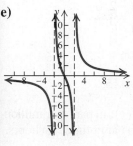

f)
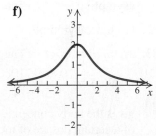

7. $f(x) = \dfrac{8}{x^2 - 4}$

8. $f(x) = \dfrac{8}{x^2 + 4}$

9. $f(x) = \dfrac{8x}{x^2 - 4}$

10. $f(x) = \dfrac{8x^2}{x^2 - 4}$

11. $f(x) = \dfrac{8x^3}{x^2 - 4}$

12. $f(x) = \dfrac{8x^3}{x^2 + 4}$

Determine the vertical asymptotes of the graph of the function.

13. $g(x) = \dfrac{1}{x^2}$

14. $f(x) = \dfrac{4x}{x^2 + 10x}$

15. $h(x) = \dfrac{x + 7}{2 - x}$

16. $g(x) = \dfrac{x^4 + 2}{x}$

17. $f(x) = \dfrac{3 - x}{(x - 4)(x + 6)}$

18. $h(x) = \dfrac{x^2 - 4}{x(x + 5)(x - 2)}$

19. $g(x) = \dfrac{x^3}{2x^3 - x^2 - 3x}$

20. $f(x) = \dfrac{x + 5}{x^2 + 4x - 32}$

Determine the horizontal asymptote of the graph of the function.

21. $f(x) = \dfrac{3x^2 + 5}{4x^2 - 3}$

22. $g(x) = \dfrac{x + 6}{x^3 + 2x^2}$

23. $h(x) = \dfrac{x^2 - 4}{2x^4 + 3}$

24. $f(x) = \dfrac{x^5}{x^5 + x}$

25. $g(x) = \dfrac{x^3 - 2x^2 + x - 1}{x^2 - 16}$

26. $h(x) = \dfrac{8x^4 + x - 2}{2x^4 - 10}$

Determine the oblique asymptote of the graph of the function.

27. $g(x) = \dfrac{x^2 + 4x - 1}{x + 3}$

28. $f(x) = \dfrac{x^2 - 6x}{x - 5}$

29. $h(x) = \dfrac{x^4 - 2}{x^3 + 1}$

30. $g(x) = \dfrac{12x^3 - x}{6x^2 + 4}$

31. $f(x) = \dfrac{x^3 - x^2 + x - 4}{x^2 + 2x - 1}$

32. $h(x) = \dfrac{5x^3 - x^2 + x - 1}{x^2 - x + 2}$

Make a hand-drawn graph for each of Exercises 33–78. Be sure to label all the asymptotes. List the domain and the x-intercepts and the y-intercepts. Check your work using a graphing calculator.

33. $f(x) = \dfrac{1}{x}$

34. $g(x) = \dfrac{1}{x^2}$

35. $h(x) = -\dfrac{4}{x^2}$

36. $f(x) = -\dfrac{6}{x}$

37. $g(x) = \dfrac{x^2 - 4x + 3}{x + 1}$

38. $h(x) = \dfrac{2x^2 - x - 3}{x - 1}$

39. $f(x) = \dfrac{-2}{x - 5}$

40. $f(x) = \dfrac{1}{x - 5}$

41. $f(x) = \dfrac{2x + 1}{x}$

42. $f(x) = \dfrac{3x - 1}{x}$

43. $f(x) = \dfrac{x + 3}{x^2 - 9}$

44. $f(x) = \dfrac{x - 1}{x^2 - 1}$

45. $f(x) = \dfrac{x}{x^2 + 3x}$

46. $f(x) = \dfrac{3x}{3x - x^2}$

47. $f(x) = \dfrac{1}{(x - 2)^2}$

48. $f(x) = \dfrac{-2}{(x - 3)^2}$

49. $f(x) = \dfrac{x^2 + 2x - 3}{x^2 + 4x + 3}$

50. $f(x) = \dfrac{x^2 - x - 2}{x^2 - 5x - 6}$

51. $f(x) = \dfrac{1}{x^2 + 3}$

52. $f(x) = \dfrac{-1}{x^2 + 2}$

53. $f(x) = \dfrac{x^2 - 4}{x - 2}$

54. $f(x) = \dfrac{x^2 - 9}{x + 3}$

55. $f(x) = \dfrac{x - 1}{x + 2}$

56. $f(x) = \dfrac{x - 2}{x + 1}$

57. $f(x) = \dfrac{x^2 + 3x}{2x^3 - 5x^2 - 3x}$

58. $f(x) = \dfrac{3x}{x^2 + 5x + 4}$

59. $f(x) = \dfrac{x^2 - 9}{x + 1}$

60. $f(x) = \dfrac{x^3 - 4x}{x^2 - x}$

61. $f(x) = \dfrac{x^2 + x - 2}{2x^2 + 1}$

62. $f(x) = \dfrac{x^2 - 2x - 3}{3x^2 + 2}$

63. $g(x) = \dfrac{3x^2 - x - 2}{x - 1}$

64. $f(x) = \dfrac{2x^2 - 5x - 3}{2x + 1}$

65. $f(x) = \dfrac{x - 1}{x^2 - 2x - 3}$

66. $f(x) = \dfrac{x + 2}{x^2 + 2x - 15}$

67. $f(x) = \dfrac{3x^2 + 11x - 4}{x^2 + 2x - 8}$

68. $f(x) = \dfrac{2x^2 - 3x - 9}{x^2 - 2x - 3}$

69. $f(x) = \dfrac{x - 3}{(x + 1)^3}$

70. $f(x) = \dfrac{x + 2}{(x - 1)^3}$

71. $f(x) = \dfrac{x^3 + 1}{x}$

72. $f(x) = \dfrac{x^3 - 1}{x}$

73. $f(x) = \dfrac{x^3 + 2x^2 - 15x}{x^2 - 5x - 14}$

74. $f(x) = \dfrac{x^3 + 2x^2 - 3x}{x^2 - 25}$

75. $f(x) = \dfrac{5x^4}{x^4 + 1}$

76. $f(x) = \dfrac{x + 1}{x^2 + x - 6}$

77. $f(x) = \dfrac{x^2}{x^2 - x - 2}$

78. $f(x) = \dfrac{x^2 - x - 2}{x + 2}$

Find a rational function that satisfies the given conditions. Answers may vary, but try to give the simplest answer possible.

79. Vertical asymptotes $x = -4$, $x = 5$

80. Vertical asymptotes $x = -4$, $x = 5$; x-intercept $(-2, 0)$

81. Vertical asymptotes $x = -4$, $x = 5$; horizontal asymptote $y = \frac{3}{2}$; x-intercept $(-2, 0)$

82. Oblique asymptote $y = x - 1$

83. *Medical Dosage.*   The function

$$N(t) = \frac{0.8t + 1000}{5t + 4}, \quad t \geq 15,$$

gives the body concentration $N(t)$, in parts per million, of a certain dosage of medication after time t, in hours.

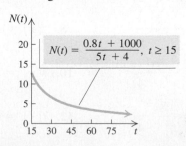

a) Find the horizontal asymptote of the graph and complete the following:

$$N(t) \longrightarrow \boxed{} \text{ as } t \longrightarrow \infty.$$

b) Explain the meaning of the answer to part (a) in terms of the application.

84. *Average Cost.* The average cost per light, in dollars, for a company to produce x roadside emergency lights is given by the function

$$A(x) = \frac{2x + 100}{x}, \quad x > 0.$$

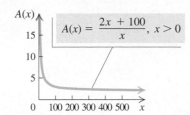

a) Find the horizontal asymptote of the graph and complete the following:

$$A(x) \longrightarrow \boxed{} \text{ as } x \longrightarrow \infty.$$

b) Explain the meaning of the answer to part (a) in terms of the application.

85. *Population Growth.* The population P, in thousands, of a resort community is given by

$$P(t) = \frac{500t}{2t^2 + 9},$$

where t is the time, in months, since the city council raised the property taxes.

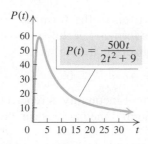

a) Find the population at $t = 0, 1, 3,$ and 8 months.
b) Find the horizontal asymptote of the graph and complete the following:

$$P(t) \longrightarrow \boxed{} \text{ as } t \longrightarrow \infty.$$

c) Explain the meaning of the answer to part (b) in terms of the application.

86. *Minimizing Surface Area.* The Hold-It Container Co. is designing an open-top rectangular box, with a square base, that will hold 108 cubic centimeters.

a) Express the surface area S as a function of the length x of a side of the base.
b) Use a graphing calculator to graph the function on the interval $(0, \infty)$.
c) Estimate the minimum surface area and the value of x that will yield it.

❯ Skill Maintenance

Vocabulary Reinforcement

In each of Exercises 87–95, fill in the blank with the correct term. Some of the given choices will not be used. Others will be used more than once.

x-intercept	midpoint formula
y-intercept	horizontal lines
odd function	vertical lines
even function	point–slope equation
domain	slope–intercept equation
range	difference quotient
slope	$f(x) = f(-x)$
distance formula	$f(-x) = -f(x)$

87. A function is a correspondence between a first set, called the _____ , and a second set, called the _____ , such that each member of the _____ corresponds to exactly one member of the _____ . **[1.2]**

88. The _____ of a line containing (x_1, y_1) and (x_2, y_2) is given by $(y_2 - y_1)/(x_2 - x_1)$. **[1.3]**

89. The _____ of the line with slope m and y-intercept $(0, b)$ is $y = mx + b$. **[1.3]**

90. The _____ of the line with slope m passing through (x_1, y_1) is $y - y_1 = m(x - x_1)$. **[1.4]**

91. A(n) _____ is a point $(a, 0)$. **[1.1]**

92. For each x in the domain of an odd function f, _____ . **[2.4]**

93. _____ are given by equations of the type $x = a$. **[1.3]**

94. The _____ is $\left(\dfrac{x_1 + x_2}{2}, \dfrac{y_1 + y_2}{2}\right)$. **[1.1]**

95. A(n) _____ is a point $(0, b)$. **[1.1]**

❯ Synthesis

96. Graph

$$y_1 = \frac{x^3 + 4}{x} \quad \text{and} \quad y_2 = x^2$$

using the same viewing window. Explain how the parabola $y_2 = x^2$ can be thought of as a nonlinear asymptote for y_1.

Find the nonlinear asymptote of the function.

97. $f(x) = \dfrac{x^5 + 2x^3 + 4x^2}{x^2 + 2}$

98. $f(x) = \dfrac{x^4 + 3x^2}{x^2 + 1}$

Graph the function.

99. $f(x) = \dfrac{2x^3 + x^2 - 8x - 4}{x^3 + x^2 - 9x - 9}$

100. $f(x) = \dfrac{x^3 + 4x^2 + x - 6}{x^2 - x - 2}$

4.6 ❯ Polynomial Inequalities and Rational Inequalities

❯ Solve polynomial inequalities.

❯ Solve rational inequalities.

We will use a combination of algebraic methods and graphical methods to solve polynomial inequalities and rational inequalities.

❯ Polynomial Inequalities

Just as a quadratic equation can be written in the form $ax^2 + bx + c = 0$, a **quadratic inequality** can be written in the form $ax^2 + bx + c \ \square \ 0$, where $\square$ is $<$, $>$, $\le$, or $\ge$. Here are some examples of quadratic inequalities:

$$x^2 - 4x - 5 < 0 \quad \text{and} \quad -\tfrac{1}{2}x^2 + 4x - 7 \ge 0.$$

When the inequality symbol in a polynomial inequality is replaced with an equals sign, a **related equation** is formed. Polynomial inequalities can be solved once the related equation has been solved.

EXAMPLE 1 Solve: $x^2 - 4x - 5 > 0$.

Solution We are asked to find all x-values for which $x^2 - 4x - 5 > 0$. To locate these values, we graph $f(x) = x^2 - 4x - 5$. Then we note that whenever the graph passes through an x-intercept, the function changes sign. Thus to solve $x^2 - 4x - 5 > 0$, we first solve the *related equation* $x^2 - 4x - 5 = 0$ to find all zeros of the function:

$$x^2 - 4x - 5 = 0$$
$$(x + 1)(x - 5) = 0.$$

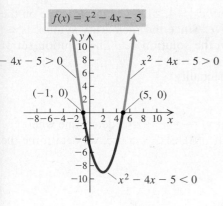

$f(x) = x^2 - 4x - 5$

$-4x - 5 > 0$

$x^2 - 4x - 5 > 0$

$(-1, 0)$

$(5, 0)$

$x^2 - 4x - 5 < 0$

The zeros are -1 and 5. Thus the x-intercepts of the graph are $(-1, 0)$ and $(5, 0)$, as shown at left. The zeros divide the x-axis into three intervals:

$$(-\infty, -1), \qquad (-1, 5), \quad \text{and} \quad (5, \infty).$$

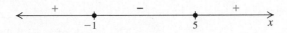

The sign of $x^2 - 4x - 5$ is the same for all values of x in a given interval. Thus we choose a test value for x from each interval and find $f(x)$. We can also determine the sign of $f(x)$ in each interval by simply looking at the graph of the function.

Interval	$(-\infty, -1)$	$(-1, 5)$	$(5, \infty)$
Test Value	$f(-2) = 7$	$f(0) = -5$	$f(7) = 16$
Sign of $f(x)$	Positive	Negative	Positive

Just in Time

3

Since we are solving $x^2 - 4x - 5 > 0$, the solution set consists of only two of the three intervals, those in which the sign of $f(x)$ is positive. Since the inequality sign is $>$, we do not include the endpoints of the intervals in the solution set. The solution set is

$$(-\infty, -1) \cup (5, \infty), \quad \text{or} \quad \{x | x < -1 \ or \ x > 5\}.$$

> **Now Try Exercise 27.**

EXAMPLE 2   Solve: $x^2 + 3x - 5 \leq x + 3$.

Solution   By subtracting $x + 3$ on both sides, we form an equivalent inequality:

$$x^2 + 3x - 5 - x - 3 \leq 0$$
$$x^2 + 2x - 8 \leq 0.$$

We need to find all x-values for which $x^2 + 2x - 8 \leq 0$. To visualize these values, we first graph $f(x) = x^2 + 2x - 8$ and then determine the zeros of the function.

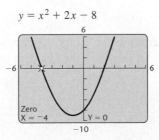

$y = x^2 + 2x - 8$

Zero
X = -4   Y = 0

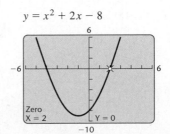

$y = x^2 + 2x - 8$

Zero
X = 2   Y = 0

X	Y1	
-5	7	
0	-8	
4	16	
X =		

Using the ZERO feature, we see that the zeros are -4 and 2.

The intervals to be considered are $(-\infty, -4)$, $(-4, 2)$, and $(2, \infty)$. Using test values for $f(x)$, we determine the sign of $f(x)$ in each interval. (See the table at left.)

Function values are negative in the interval $(-4, 2)$. We can also note on the graph where the function values are negative. Since the inequality symbol is $\leq$, we include the endpoints of the interval in the solution set. The solution set of $x^2 + 3x - 5 \leq x + 3$ is $[-4, 2]$, or $\{x \mid -4 \leq x \leq 2\}$.

An alternative approach to solving the inequality

$$x^2 + 3x - 5 \leq x + 3$$

is to graph both sides, $y_1 = x^2 + 3x - 5$ and $y_2 = x + 3$, and determine where the graph of y_1 is below the graph of y_2. Using the INTERSECT feature, we determine the points of intersection, $(-4, -1)$ and $(2, 5)$.

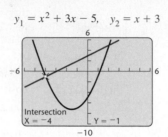

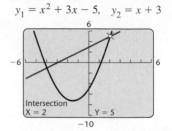

The graph of y_1 is below the graph of y_2 over the interval $(-4, 2)$. Since the inequality symbol is $\leq$, we include the endpoints of the interval in the solution. Thus the solution set of $x^2 + 3x - 5 \leq x + 3$ is $[-4, 2]$.

> *Now Try Exercise 29.*

Quadratic inequalities are one type of **polynomial inequality**. Other examples of polynomial inequalities are

$$-2x^4 + x^2 - 3 < 7, \qquad \tfrac{2}{3}x + 4 \geq 0, \quad \text{and} \quad 4x^3 - 2x^2 > 5x + 7.$$

EXAMPLE 3   Solve: $x^3 - x > 0$.

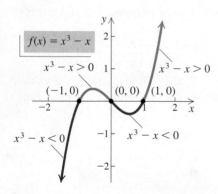

Solution   We are asked to find all x-values for which $x^3 - x > 0$. To locate these values, we graph $f(x) = x^3 - x$. Then we note that whenever the function changes sign, its graph passes through an x-intercept. Thus to solve $x^3 - x > 0$, we first solve the related equation $x^3 - x = 0$ to find all zeros of the function:

$$x^3 - x = 0$$
$$x(x^2 - 1) = 0$$
$$x(x + 1)(x - 1) = 0.$$

The zeros are $-1, 0$, and 1. Thus the x-intercepts of the graph are $(-1, 0), (0, 0)$, and $(1, 0)$, as shown in the figure at left. The zeros divide the x-axis into four intervals:

$$(-\infty, -1), \qquad (-1, 0), \qquad (0, 1), \quad \text{and} \quad (1, \infty).$$

The sign of $x^3 - x$ is the same for all values of x in a given interval. Thus we choose a test value for x from each interval and find $f(x)$. We can use the TABLE feature set in ASK mode to determine the sign of $f(x)$ in each interval.

X	Y₁
−2	−6
−.5	.375
.5	−.375
2	6

X =

We can also determine the sign of $f(x)$ in each interval by simply looking at the graph of the function.

Interval	$(-\infty, -1)$	$(-1, 0)$	$(0, 1)$	$(1, \infty)$
Test Value	$f(-2) = -6$	$f(-0.5) = 0.375$	$f(0.5) = -0.375$	$f(2) = 6$
Sign of $f(x)$	Negative	Positive	Negative	Positive

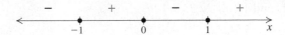

Since we are solving $x^3 - x > 0$, the solution set consists of only two of the four intervals, those in which the sign of $f(x)$ is *positive*. We see that the solution set is $(-1, 0) \cup (1, \infty)$, or $\{x \mid -1 < x < 0 \; or \; x > 1\}$.

> **Now Try Exercise 39.**

To solve a polynomial inequality:

1. Find an equivalent inequality with $P(x)$ on one side and 0 on the other.
2. Change the inequality symbol to an equals sign and solve the related equation; that is, solve $P(x) = 0$.
3. Use the solutions to divide the x-axis into intervals. Then select a test value from each interval and determine the sign of the polynomial on the interval.
4. Determine the intervals for which the inequality is satisfied and write interval notation or set-builder notation for the solution set. Include the endpoints of the intervals in the solution set if the inequality symbol is $\leq$ or $\geq$.

EXAMPLE 4 Solve: $3x^4 + 10x \le 11x^3 + 4$.

Solution By subtracting $11x^3 + 4$, we form the equivalent inequality

$$3x^4 - 11x^3 + 10x - 4 \le 0.$$

Algebraic Solution

To solve the related equation

$$3x^4 - 11x^3 + 10x - 4 = 0,$$

we need to use the theorems of Section 4.4. We solved this equation in Example 5 in Section 4.4. The solutions are

$$-1, \quad 2 - \sqrt{2}, \quad \tfrac{2}{3}, \quad \text{and} \quad 2 + \sqrt{2},$$

or approximately

$$-1, \quad 0.586, \quad 0.667, \quad \text{and} \quad 3.414.$$

These numbers divide the x-axis into five intervals:

$$(-\infty, -1), \ \left(-1, 2 - \sqrt{2}\right), \ \left(2 - \sqrt{2}, \tfrac{2}{3}\right),$$
$$\left(\tfrac{2}{3}, 2 + \sqrt{2}\right), \text{ and } \left(2 + \sqrt{2}, \infty\right).$$

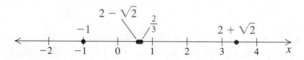

We then let $f(x) = 3x^4 - 11x^3 + 10x - 4$ and, using test values for x, determine the sign of $f(x)$ in each interval.

X	Y1
−2	112
0	−4
.6	.0128
1	−2
4	100

X =

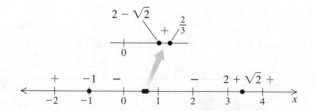

Function values are negative in the intervals $\left(-1, 2 - \sqrt{2}\right)$ and $\left(\tfrac{2}{3}, 2 + \sqrt{2}\right)$. Since the inequality sign is $\le$, we include the endpoints of the intervals in the solution set. The solution set is

$$\left[-1, 2 - \sqrt{2}\right] \cup \left[\tfrac{2}{3}, 2 + \sqrt{2}\right], \quad \text{or}$$
$$\left\{x \mid -1 \le x \le 2 - \sqrt{2} \ or \ \tfrac{2}{3} \le x \le 2 + \sqrt{2}\right\}.$$

Graphical Solution

We graph $y = 3x^4 - 11x^3 + 10x - 4$ using a viewing window that reveals the curvature of the graph.

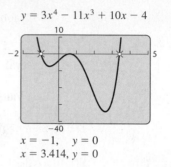

$x = -1, \quad y = 0$
$x = 3.414, \ y = 0$

Using the ZERO feature, we see that two of the zeros are -1 and approximately 3.414 $\left(2 + \sqrt{2} \approx 3.414\right)$. However, this window leaves us uncertain about the number of zeros of the function on the interval $[0, 1]$. The following window shows another view of the zeros on the interval $[0, 1]$. Those zeros are about 0.586 and 0.667 $\left(2 - \sqrt{2} \approx 0.586; \tfrac{2}{3} \approx 0.667\right)$.

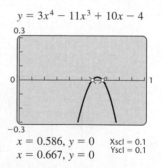

$x = 0.586, \ y = 0$ Xscl = 0.1
$x = 0.667, \ y = 0$ Yscl = 0.1

The intervals to be considered are $(-\infty, -1)$, $(-1, 0.586)$, $(0.586, 0.667)$, $(0.667, 3.414)$, and $(3.414, \infty)$. We note on the graph where the function is negative. Then including appropriate endpoints, we find that the solution set is approximately

$$[-1, 0.586] \cup [0.667, 3.414], \quad \text{or}$$
$$\{x \mid -1 \le x \le 0.586 \ or \ 0.667 \le x \le 3.414\}.$$

Now Try Exercise 45.

❭ Rational Inequalities

Some inequalities involve rational expressions and functions. These are called **rational inequalities**. To solve rational inequalities, we need to make some adjustments to the preceding method.

EXAMPLE 5   Solve: $\dfrac{3x}{x+6} < 0$.

Solution   We look for all values of x for which the related function

$$f(x) = \frac{3x}{x+6}$$

is not defined or is 0. These are called **critical values**.

The denominator tells us that $f(x)$ is not defined when $x = -6$. Next, we solve $f(x) = 0$:

$$\frac{3x}{x+6} = 0$$

$$(x+6) \cdot \frac{3x}{x+6} = (x+6) \cdot 0 \quad \textbf{Multiplying by } x+6$$

$$3x = 0$$

$$x = 0.$$

The critical values are -6 and 0. These values divide the x-axis into three intervals:

$$(-\infty, -6), \qquad (-6, 0), \quad \text{and} \quad (0, \infty).$$

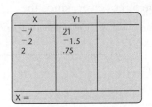

We then use a test value to determine the sign of $f(x)$ on each interval.

Function values are negative on only the interval $(-6, 0)$. Although $f(0) = 0$, the inequality symbol is $<$, so we know that 0 is not included in the solution set. Note that since -6 is not in the domain of f, -6 cannot be part of the solution set. The solution set is

$$(-6, 0), \quad \text{or} \quad \{x \mid -6 < x < 0\}.$$

The graph of $f(x)$ shows where $f(x)$ is positive and where it is negative.

❭ **Now Try Exercise 53.**

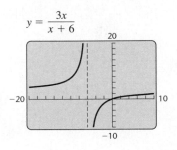

EXAMPLE 6   Solve: $\dfrac{x+1}{2x-4} \le 1$.

Solution   We first subtract 1 on both sides in order to find an equivalent inequality with 0 on one side:

$$\frac{x+1}{2x-4} - 1 \le 0.$$

Algebraic Solution

We look for all values of x for which the related function

$$f(x) = \frac{x+1}{2x-4} - 1$$

is not defined or is 0. These are called **critical values**.

A look at the denominator shows that $f(x)$ is not defined for $x = 2$. Next, we solve $f(x) = 0$:

$$\frac{x+1}{2x-4} - 1 = 0$$

$$(2x-4)\left(\frac{x+1}{2x-4} - 1\right) = (2x-4) \cdot 0$$

$$x + 1 - (2x-4) \cdot 1 = 0$$

$$x + 1 - 2x + 4 = 0$$

$$-x = -5$$

$$x = 5.$$

The critical values are 2 and 5. These values divide the x-axis into three intervals:

$$(-\infty, 2), \quad (2, 5), \quad \text{and} \quad (5, \infty).$$

We then use a test value to determine the sign of $f(x)$ in each interval.

X	Y₁	
0	−1.25	
3	1	
6	−.125	
X =		

Function values are negative on the intervals $(-\infty, 2)$ and $(5, \infty)$. Since $f(5) = 0$ and the inequality symbol is $\leq$, we know that 5 is in the solution set. Note that since 2 is not in the domain of f, it cannot be part of the solution set. The solution set is $(-\infty, 2) \cup [5, \infty)$.

Graphical Solution

The graph of

$$y = \frac{x+1}{2x-4} - 1$$

confirms the two critical values found algebraically: 2 where $f(x)$ is not defined and 5 where $f(x) = 0$.

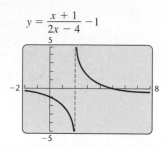

The graph shows where y is negative. Note that 2 cannot be in the solution set since y is not defined for this value. We do include 5, however, since the inequality symbol is $\leq$ and $f(5) = 0$.

The solution set is

$$(-\infty, 2) \cup [5, \infty).$$

> *Now Try Exercise 59.*

EXAMPLE 7 Solve: $\dfrac{x-3}{x+4} \geq \dfrac{x+2}{x-5}$.

Solution We first subtract $(x+2)/(x-5)$ on both sides in order to find an equivalent inequality with 0 on one side:

$$\frac{x-3}{x+4} - \frac{x+2}{x-5} \geq 0.$$

Algebraic Solution

We look for all values of x for which the related function

$$f(x) = \frac{x-3}{x+4} - \frac{x+2}{x-5}$$

is not defined or is 0. These are the critical values.

A look at the denominators shows that $f(x)$ is not defined for $x = -4$ and $x = 5$. Next, we solve $f(x) = 0$:

$$\frac{x-3}{x+4} - \frac{x+2}{x-5} = 0$$

$$(x+4)(x-5)\left(\frac{x-3}{x+4} - \frac{x+2}{x-5}\right) = (x+4)(x-5)\cdot 0$$

$$(x-5)(x-3) - (x+4)(x+2) = 0$$

$$x^2 - 8x + 15 - (x^2 + 6x + 8) = 0$$

$$x^2 - 8x + 15 - x^2 - 6x - 8 = 0$$

$$-14x + 7 = 0$$

$$-14x = -7$$

$$x = \tfrac{1}{2}.$$

The critical values are -4, $\tfrac{1}{2}$, and 5. These values divide the x-axis into four intervals:

$$(-\infty, -4), \qquad \left(-4, \tfrac{1}{2}\right), \qquad \left(\tfrac{1}{2}, 5\right), \quad \text{and} \quad (5, \infty).$$

We then use a test value to determine the sign of $f(x)$ in each interval.

X	Y₁	
-5	7.7	
-2	-2.5	
3	2.5	
6	-7.7	

X =

Function values are positive on the intervals $(-\infty, -4)$ and $\left(\tfrac{1}{2}, 5\right)$. Since $f\left(\tfrac{1}{2}\right) = 0$ and the inequality symbol is $\geq$, we know that $\tfrac{1}{2}$ must be in the solution set. Note that since neither -4 nor 5 is in the domain of f, they cannot be part of the solution set.

The solution set is $(-\infty, -4) \cup \left[\tfrac{1}{2}, 5\right)$.

Graphical Solution

We graph

$$y = \frac{x-3}{x+4} - \frac{x+2}{x-5}$$

in the standard window, which shows the curvature of the function.

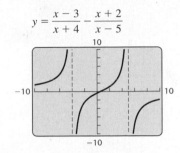

$$y = \frac{x-3}{x+4} - \frac{x+2}{x-5}$$

Using the ZERO feature, we find that 0.5 is a zero.

We then look for values where the function is not defined. By examining the denominators $x + 4$ and $x - 5$, we see that $f(x)$ is not defined for $x = -4$ and $x = 5$.

The critical values are -4, 0.5, and 5.

The graph shows where y is positive and where it is negative. Note that -4 and 5 cannot be in the solution set since y is not defined for these values. We do include 0.5, however, since the inequality symbol is $\geq$ and $f(0.5) = 0$. The solution set is

$$(-\infty, -4) \cup [0.5, 5).$$

▶ **Now Try Exercise 61.**

The following is a method for solving rational inequalities.

To solve a rational inequality:

1. Find an equivalent inequality with 0 on one side.
2. Change the inequality symbol to an equals sign and solve the related equation.
3. Find values of the variable for which the related rational function is not defined.
4. The numbers found in steps (2) and (3) are called critical values. Use the critical values to divide the *x*-axis into intervals. Then determine the function's sign in each interval using an *x*-value from the interval or using the graph of the equation.
5. Select the intervals for which the inequality is satisfied and write interval notation or set-builder notation for the solution set. If the inequality symbol is ≤ or ≥, then the solutions to step (2) should be included in the solution set. The *x*-values found in step (3) are never included in the solution set.

It works well to use a combination of algebraic methods and graphical methods to solve polynomial inequalities and rational inequalities. The algebraic methods give exact numbers for the critical values, and the graphical methods usually allow us to see easily what intervals satisfy the inequality.

4.6   Exercise Set

For the function $f(x) = x^2 + 2x - 15$, solve each of the following.

1. $f(x) = 0$ **2.** $f(x) < 0$

3. $f(x) \leq 0$ **4.** $f(x) > 0$

5. $f(x) \geq 0$

For the function

$$g(x) = \frac{x - 2}{x + 4},$$

solve each of the following.

6. $g(x) = 0$ **7.** $g(x) > 0$

8. $g(x) \leq 0$ **9.** $g(x) \geq 0$

10. $g(x) < 0$

For the function

$$h(x) = \frac{7x}{(x - 1)(x + 5)},$$

solve each of the following.

11. $h(x) = 0$

12. $h(x) \leq 0$

13. $h(x) \geq 0$

14. $h(x) > 0$

15. $h(x) < 0$

For the function $g(x) = x^5 - 9x^3$, solve each of the following.

16. $g(x) = 0$

17. $g(x) < 0$

18. $g(x) \leq 0$

19. $g(x) > 0$

20. $g(x) \geq 0$

In Exercises 21–24, a related function is graphed. Solve the given inequality.

21. $x^3 + 6x^2 < x + 30$

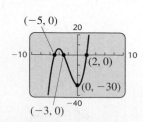

22. $x^4 - 27x^2 - 14x + 120 \geq 0$

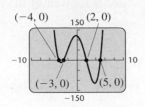

23. $\dfrac{8x}{x^2 - 4} \geq 0$

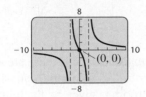

24. $\dfrac{8}{x^2 - 4} < 0$

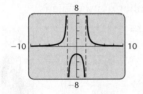

Solve.

25. $(x - 1)(x + 4) < 0$

26. $(x + 3)(x - 5) < 0$

27. $x^2 + x - 2 > 0$

28. $x^2 - x - 6 > 0$

29. $x^2 - x - 5 \geq x - 2$

30. $x^2 + 4x + 7 \geq 5x + 9$

31. $x^2 > 25$

32. $x^2 \leq 1$

33. $4 - x^2 \leq 0$

34. $11 - x^2 \geq 0$

35. $6x - 9 - x^2 < 0$

36. $x^2 + 2x + 1 \leq 0$

37. $x^2 + 12 < 4x$

38. $x^2 - 8 > 6x$

39. $4x^3 - 7x^2 \leq 15x$

40. $2x^3 - x^2 < 5x$

41. $x^3 + 3x^2 - x - 3 \geq 0$

42. $x^3 + x^2 - 4x - 4 \geq 0$

43. $x^3 - 2x^2 < 5x - 6$

44. $x^3 + x \leq 6 - 4x^2$

45. $x^5 + x^2 \geq 2x^3 + 2$

46. $x^5 + 24 > 3x^3 + 8x^2$

47. $2x^3 + 6 \leq 5x^2 + x$

48. $2x^3 + x^2 < 10 + 11x$

49. $x^3 + 5x^2 - 25x \leq 125$

50. $x^3 - 9x + 27 \geq 3x^2$

51. $0.1x^3 - 0.6x^2 - 0.1x + 2 < 0$

52. $19.2x^3 + 12.8x^2 + 144 \geq 172.8x + 3.2x^4$

List the critical values of the related function. Then solve the inequality.

53. $\dfrac{1}{x + 4} > 0$

54. $\dfrac{1}{x - 3} \leq 0$

55. $\dfrac{-4}{2x + 5} < 0$

56. $\dfrac{-2}{5 - x} \geq 0$

57. $\dfrac{2x}{x - 4} \geq 0$

58. $\dfrac{5x}{x + 1} < 0$

59. $\dfrac{x + 1}{x - 2} \geq 3$

60. $\dfrac{x}{x - 5} < 2$

61. $\dfrac{x - 4}{x + 3} - \dfrac{x + 2}{x - 1} \leq 0$

62. $\dfrac{x + 1}{x - 2} - \dfrac{x - 3}{x - 1} < 0$

63. $\dfrac{x + 6}{x - 2} > \dfrac{x - 8}{x - 5}$

64. $\dfrac{x - 7}{x + 2} \geq \dfrac{x - 9}{x + 3}$

65. $x - 2 > \dfrac{1}{x}$

66. $4 \geq \dfrac{4}{x} + x$

67. $\dfrac{2}{x^2 - 4x + 3} \leq \dfrac{5}{x^2 - 9}$

68. $\dfrac{3}{x^2 - 4} \leq \dfrac{5}{x^2 + 7x + 10}$

69. $\dfrac{3}{x^2 + 1} \geq \dfrac{6}{5x^2 + 2}$

70. $\dfrac{4}{x^2 - 9} < \dfrac{3}{x^2 - 25}$

71. $\dfrac{5}{x^2 + 3x} < \dfrac{3}{2x + 1}$

72. $\dfrac{2}{x^2 + 3} > \dfrac{3}{5 + 4x^2}$

73. $\dfrac{5x}{7x - 2} > \dfrac{x}{x + 1}$

74. $\dfrac{x^2 - x - 2}{x^2 + 5x + 6} < 0$

75. $\dfrac{x}{x^2 + 4x - 5} + \dfrac{3}{x^2 - 25} \leq \dfrac{2x}{x^2 - 6x + 5}$

76. $\dfrac{2x}{x^2 - 9} + \dfrac{x}{x^2 + x - 12} \geq \dfrac{3x}{x^2 + 7x + 12}$

77. *Temperature During an Illness.* A person's temperature T, in degrees Fahrenheit, during an illness is given by the function

$$T(t) = \dfrac{4t}{t^2 + 1} + 98.6,$$

where t is the time since the onset of the illness, in hours. Find the interval on which the temperature was over 100°F. (See Example 12 in Section 4.5.)

78. *Population Growth.* The population P, in thousands, of a resort community is given by

$$P(t) = \dfrac{500t}{2t^2 + 9},$$

where t is the time, in months, since the city council raised the property taxes. Find the interval on which the population was 40,000 or greater. (See Exercise 85 in Exercise Set 4.5.)

79. *Total Profit.* Flexl, Inc., determines that its total profit is given by the function

$$P(x) = -3x^2 + 630x - 6000.$$

a) Flexl makes a profit for those nonnegative values of x for which $P(x) > 0$. Find the values of x for which Flexl makes a profit.

b) Flexl loses money for those nonnegative values of x for which $P(x) < 0$. Find the values of x for which Flexl loses money.

80. *Height of a Thrown Object.* The function

$$S(t) = -16t^2 + 32t + 1920$$

gives the height S, in feet, of an object thrown upward with a velocity of 32 ft/sec from a cliff that is 1920 ft high. Here t is the time, in seconds, that the object is in the air.

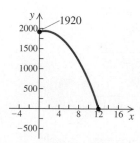

a) For what times is the height greater than 1920 ft?

b) For what times is the height less than 640 ft?

81. *Number of Diagonals.* A polygon with n sides has D diagonals, where D is given by the function

$$D(n) = \dfrac{n(n - 3)}{2}.$$

Find the number of sides n if

$$27 \leq D \leq 230.$$

82. *Number of Handshakes.* If there are n people in a room, the number N of possible handshakes by all the people in the room is given by the function

$$N(n) = \dfrac{n(n - 1)}{2}.$$

For what number n of people is

$$66 \leq N \leq 300?$$

❯ Skill Maintenance

Find an equation for a circle satisfying the given conditions. [1.1]

83. Center: $(-2, 4)$; radius of length 3

84. Center: $(0, -3)$; diameter of length $\frac{7}{2}$

In Exercises 85 and 86:

a) *Find the vertex.*

b) *Determine whether there is a maximum or a minimum value and find that value.*

c) *Find the range.* [3.3]

85. $h(x) = -2x^2 + 3x - 8$

86. $g(x) = x^2 - 10x + 2$

❯ Synthesis

Solve.

87. $|x^2 - 5| = 5 - x^2$

88. $x^4 - 6x^2 + 5 > 0$

89. $2|x|^2 - |x| + 2 \leq 5$

90. $(7 - x)^{-2} < 0$

91. $\left| 1 + \dfrac{1}{x} \right| < 3$

92. $\left| 2 - \dfrac{1}{x} \right| \leq 2 + \left| \dfrac{1}{x} \right|$

93. Write a quadratic inequality for which the solution set is $(-4, 3)$.

94. Write a polynomial inequality for which the solution set is $[-4, 3] \cup [7, \infty)$.

Find the domain of the function.

95. $f(x) = \sqrt{\dfrac{72}{x^2 - 4x - 21}}$

96. $f(x) = \sqrt{x^2 - 4x - 21}$

Chapter 4 Summary and Review

STUDY GUIDE

KEY TERMS AND CONCEPTS

EXAMPLES

SECTION 4.1: POLYNOMIAL FUNCTIONS AND MODELING

Polynomial Function

$$P(x) = a_n x^n + a_{n-1}x^{n-1} + a_{n-2}x^{n-2} + \cdots + a_1 x + a_0,$$

where the coefficients $a_n, a_{n-1}, \ldots, a_1, a_0$ are real numbers and the exponents are whole numbers.

The first nonzero coefficient, a_n, is called the **leading coefficient**. The term $a_n x^n$ is called the **leading term**. The **degree** of the polynomial function is n.

Classifying polynomial functions by degree:

Type	Degree
Constant	0
Linear	1
Quadratic	2
Cubic	3
Quartic	4

Consider the polynomial

$$P(x) = \frac{1}{3}x^2 + x - 4x^5 + 2.$$

Leading term: $-4x^5$

Leading coefficient: -4

Degree of polynomial: 5

Classify the following polynomial functions:

Function	Type
$f(x) = -2$	Constant
$f(x) = 0.6x - 11$	Linear
$f(x) = 5x^2 + x - 4$	Quadratic
$f(x) = 5x^3 - x + 10$	Cubic
$f(x) = -x^4 + 8x^3 + x$	Quartic

The Leading-Term Test

If $a_n x^n$ is the leading term of a polynomial function, then the behavior of the graph as $x \to \infty$ and as $x \to -\infty$ can be described in one of the following four ways.

a) If n is even, and $a_n > 0$:

b) If n is even, and $a_n < 0$:

c) If n is odd, and $a_n > 0$:

d) If n is odd, and $a_n < 0$:

Using the leading-term test, describe the end behavior of the graph of each of the following functions by selecting one of (a)–(d) shown at left.

$$h(x) = -2x^6 + x^4 - 3x^2 + x$$

The leading term $a_n x^n$ is $-2x^6$. Since 6 is even and $-2 < 0$, the shape is shown in (b).

$$g(x) = 4x^3 - 8x + 1$$

The leading term, $a_n x^n$, is $4x^3$. Since 3 is odd and $4 > 0$, the shape is shown in (c).

Zeros of Functions

If c is a real zero of a function $f(x)$ (that is, $f(c) = 0$), then $x - c$ is a factor of $f(x)$ and $(c, 0)$ is an x-intercept of the graph of the function.

  If we know the linear factors of a polynomial function $f(x)$, we can find the zeros of $f(x)$ by solving the equation $f(x) = 0$ using the principle of zero products.

  Every function of degree n, with $n \geq 1$, has at least one zero and at most n zeros.

To find the zeros of

$$f(x) = -2(x - 3)(x + 8)^2,$$

solve $-2(x - 3)(x + 8)^2 = 0$ using the principle of zero products:

$$x - 3 = 0 \quad or \quad x + 8 = 0$$
$$x = 3 \quad or \quad x = -8.$$

The zeros of $f(x)$ are 3 and -8.

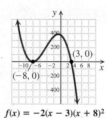

$$f(x) = -2(x - 3)(x + 8)^2$$

To find the zeros of

$$h(x) = x^4 - 12x^2 - 64,$$

solve $h(x) = 0$:

$$x^4 - 12x^2 - 64 = 0$$
$$(x^2 - 16)(x^2 + 4) = 0$$
$$(x + 4)(x - 4)(x^2 + 4) = 0$$
$$x + 4 = 0 \quad or \quad x - 4 = 0 \quad or \quad x^2 + 4 = 0$$
$$x = -4 \quad or \quad x = 4 \quad or \quad x^2 = -4$$
$$x = -4 \quad or \quad x = 4 \quad or \quad x = \pm\sqrt{-4}$$
$$x = -4 \quad or \quad x = 4 \quad or \quad x = \pm 2i.$$

The zeros of $h(x)$ are $-4, 4, -2i$, and $2i$.

Even and Odd Multiplicity

If $(x - c)^k, k \geq 1$, is a factor of a polynomial function $P(x)$ and $(x - c)^{k+1}$ is not a factor and:

- k is odd, then the graph crosses the x-axis at $(c, 0)$;

- k is even, then the graph is tangent to the x-axis at $(c, 0)$.

For $f(x) = -2(x - 3)(x + 8)^2$ graphed above, note that for the factor $x - 3$, or $(x - 3)^1$, the exponent 1 is odd and the graph crosses the x-axis at $(3, 0)$. For the factor $(x + 8)^2$, the exponent 2 is even and the graph is tangent to the x-axis at $(-8, 0)$.

SECTION 4.2: GRAPHING POLYNOMIAL FUNCTIONS

If $P(x)$ is a polynomial function of degree n, the graph of the function has:

- at most n real zeros, and thus at most n x-intercepts, and

- at most $n - 1$ turning points.

To Graph a Polynomial Function

1. Use the leading-term test to determine the end behavior.

2. Find the zeros of the function by solving $f(x) = 0$. Any real zeros are the first coordinates of the x-intercepts.

3. Use the x-intercepts (zeros) to divide the x-axis into intervals and choose a test point in each interval to determine the sign of all function values in that interval. For all x-values in an interval, $f(x)$ is either always positive for all values or always negative for all values.

4. Find $f(0)$. This gives the y-intercept of the function.

5. If necessary, find additional function values to determine the general shape of the graph and then draw the graph.

Graph: $h(x) = x^4 - 12x^2 - 16x = x(x - 4)(x + 2)^2$.

1. The leading term is x^4. Since 4 is even and $1 > 0$, the end behavior of the graph can be sketched as follows.

2. Solve $x(x - 4)(x + 2)^2 = 0$. The solutions are 0, 4, and -2. The zeros of $h(x)$ are 0, 4, and -2. The x-intercepts are $(0, 0)$, $(4, 0)$, and $(-2, 0)$. The multiplicity of 0 and 4 is 1. The graph will cross the x-axis at 0 and 4. The multiplicity of -2 is 2. The graph is tangent to the x-axis at -2.

3. The zeros divide the x-axis into four intervals.

Interval	$(-\infty, -2)$	$(-2, 0)$	$(0, 4)$	$(4, \infty)$
Test Value	-3	-1	1	5
Function Value, $h(x)$	21	5	-27	245
Sign of $h(x)$	$+$	$+$	$-$	$+$
Location of Points on Graph	Above x-axis	Above x-axis	Below x-axis	Above x-axis

Four points on the graph are $(-3, 21)$, $(-1, 5)$, $(1, -27)$, and $(5, 245)$.

4. Find $h(0)$:

$$h(0) = 0(0 - 4)(0 + 2)^2 = 0.$$

The y-intercept is $(0, 0)$.

(continued)

5. Find additional points and draw the graph.

x	$h(x)$
-2.5	4.1
-1.5	2.1
-0.5	5.1
0.5	-10.9
2	-64
3	-75

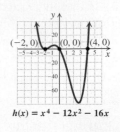

$$h(x) = x^4 - 12x^2 - 16x$$

The Intermediate Value Theorem

For any polynomial function $P(x)$ with real coefficients, suppose that for $a \neq b$, $P(a)$ and $P(b)$ are of opposite signs. Then the function has at least one real zero between a and b.

The intermediate value theorem *cannot* be used to determine whether there is a real zero between a and b when $P(a)$ and $P(b)$ have the *same* sign.

Use the intermediate value theorem to determine, if possible, whether each of the following functions has a real zero between a and b.

$$f(x) = 2x^3 - 5x^2 + x - 2, \quad a = 2, b = 3;$$
$$f(2) = -4; \ f(3) = 10$$

Since $f(2)$ and $f(3)$ have opposite signs, $f(x)$ has at least one real zero between 2 and 3.

$$f(x) = 2x^3 - 5x^2 + x - 2, \quad a = -2, b = -1;$$
$$f(-2) = -40; \ f(-1) = -10$$

Both $f(-2)$ and $f(-1)$ are negative. Thus the intermediate value theorem does not allow us to determine whether there is a real zero between -2 and -1.

SECTION 4.3: POLYNOMIAL DIVISION; THE REMAINDER THEOREM AND THE FACTOR THEOREM

Polynomial Division

$$P(x) = d(x) \cdot Q(x) + R(x)$$

Dividend   Divisor   Quotient   Remainder

When we divide a polynomial $P(x)$ by a divisor $d(x)$, a polynomial $Q(x)$ is the quotient and a polynomial $R(x)$ is the remainder. The quotient $Q(x)$ must have degree less than that of the dividend $P(x)$. The remainder $R(x)$ must either be 0 or have degree less than that of the divisor $d(x)$. If $R(x) = 0$, then the divisor $d(x)$ is a factor of the dividend.

Given $P(x) = x^4 - 6x^3 + 9x^2 + 4x - 12$ and $d(x) = x + 2$, use long division to find the quotient and the remainder when $P(x)$ is divided by $d(x)$. Express $P(x)$ in the form $d(x) \cdot Q(x) + R(x)$.

$$
\begin{array}{r}
x^3 - 8x^2 + 25x - 46 \\
x + 2 \overline{\smash{)}\ x^4 - 6x^3 + 9x^2 + 4x - 12} \\
\underline{x^4 + 2x^3} \\
-8x^3 + 9x^2 \\
\underline{-8x^3 - 16x^2} \\
25x^2 + 4x \\
\underline{25x^2 + 50x} \\
-46x - 12 \\
\underline{-46x - 92} \\
80
\end{array}
$$

$Q(x) = x^3 - 8x^2 + 25x - 46$ and $R(x) = 80$. Thus, $P(x) = (x + 2)(x^3 - 8x^2 + 25x - 46) + 80$. Since $R(x) \neq 0$, $x + 2$ is not a factor of $P(x)$.

The Remainder Theorem

If a number c is substituted for x in the polynomial $f(x)$, then the result $f(c)$ is the remainder that would be obtained by dividing $f(x)$ by $x - c$. That is, if $f(x) = (x - c) \cdot Q(x) + R$, then $f(c) = R$.

The long-division process can be streamlined with synthetic division. Synthetic division can also be used to find polynomial function values.

The Factor Theorem

For a polynomial $f(x)$, if $f(c) = 0$, then $x - c$ is a factor of $f(x)$.

Repeat the division shown above using synthetic division. Note that the divisor $x + 2 = x - (-2)$.

$$
\begin{array}{r|rrrrr}
-2 & 1 & -6 & 9 & 4 & -12 \\
   &   & -2 & 16 & -50 & 92 \\
\hline
   & 1 & -8 & 25 & -46 & \mid \; 80
\end{array}
$$

Again, note that $Q(x) = x^3 - 8x^2 + 25x - 46$ and $R(x) = 80$. Since $R(x) \neq 0$, $x - (-2)$, or $x + 2$, is not a factor of $P(x)$.

Now divide $P(x)$ by $x - 3$.

$$
\begin{array}{r|rrrrr}
3 & 1 & -6 & 9 & 4 & -12 \\
  &   & 3 & -9 & 0 & 12 \\
\hline
  & 1 & -3 & 0 & 4 & \mid \; 0
\end{array}
$$

$Q(x) = x^3 - 3x^2 + 4$ and $R(x) = 0$. Since $R(x) = 0$, $x - 3$ is a factor of $P(x)$.

For $f(x) = 2x^5 - x^3 - 3x^2 - 4x + 15$, find $f(-2)$.

$$
\begin{array}{r|rrrrrr}
-2 & 2 & 0 & -1 & -3 & -4 & 15 \\
   &   & -4 & 8 & -14 & 34 & -60 \\
\hline
   & 2 & -4 & 7 & -17 & 30 & \mid \; -45
\end{array}
$$

Thus, $f(-2) = -45$.

Let $g(x) = x^4 + 8x^3 + 6x^2 - 40x + 25$. Factor $g(x)$ and solve $g(x) = 0$.

Use synthetic division to look for factors of the form $x - c$. Let's try $x + 5$.

$$
\begin{array}{r|rrrrr}
-5 & 1 & 8 & 6 & -40 & 25 \\
   &   & -5 & -15 & 45 & -25 \\
\hline
   & 1 & 3 & -9 & 5 & \mid \; 0
\end{array}
$$

Since $g(-5) = 0$, the number -5 is a zero of $g(x)$ and $x - (-5)$, or $x + 5$, is a factor of $g(x)$. This gives us

$$g(x) = (x + 5)(x^3 + 3x^2 - 9x + 5).$$

Let's try $x + 5$ again with the factor $x^3 + 3x^2 - 9x + 5$.

$$
\begin{array}{r|rrrr}
-5 & 1 & 3 & -9 & 5 \\
   &   & -5 & 10 & -5 \\
\hline
   & 1 & -2 & 1 & \mid \; 0
\end{array}
$$

Now we have

$$g(x) = (x + 5)^2(x^2 - 2x + 1).$$

The trinomial $x^2 - 2x + 1$ easily factors, so

$$g(x) = (x + 5)^2(x - 1)^2.$$

Solve $g(x) = 0$. The solutions of $(x + 5)^2(x - 1)^2 = 0$ are -5 and 1. They are also the zeros of $g(x)$.

SECTION 4.4: THEOREMS ABOUT ZEROS OF POLYNOMIAL FUNCTIONS

The Fundamental Theorem of Algebra

Every polynomial function of degree n, $n \geq 1$, with complex coefficients has at least one zero in the system of complex numbers.

Every polynomial function f of degree n, with $n \geq 1$, can be factored into n linear factors (not necessarily unique); that is,

$$f(x) = a_n(x - c_1)(x - c_2) \cdots (x - c_n).$$

Nonreal Zeros

$a + bi$ and $a - bi, b \neq 0$

If a complex number $a + bi$, $b \neq 0$, is a zero of a polynomial function $f(x)$ with *real* coefficients, then its conjugate, $a - bi$, is also a zero. (Nonreal zeros occur in conjugate pairs.)

Irrational Zeros

$a + c\sqrt{b}$ and $a - c\sqrt{b},$
b not a perfect square

If $a + c\sqrt{b}$, where a, b, and c are rational and b is not a perfect square, is a zero of a polynomial function $f(x)$ with *rational* coefficients, then its conjugate, $a - c\sqrt{b}$, is also a zero. (Irrational zeros occur in conjugate pairs.)

The Rational Zeros Theorem

Consider the polynomial function

$$P(x) = a_n x^n + a_{n-1}x^{n-1} + a_{n-2}x^{n-2}$$
$$+ \cdots + a_1 x + a_0,$$

where all the coefficients are integers and $n \geq 1$. Also, consider a rational number p/q, where p and q have no common factor other than -1 and 1. If p/q is a zero of $P(x)$, then p is a factor of a_0 and q is a factor of a_n.

Find a polynomial function of degree 5 with -4 and 2 as zeros of multiplicity 1, and -1 as a zero of multiplicity 3.

$$\begin{aligned} f(x) &= [x - (-4)]\,[x - 2]\,[x - (-1)]^3 \\ &= (x + 4)(x - 2)(x + 1)^3 \\ &= x^5 + 5x^4 + x^3 - 17x^2 - 22x - 8 \end{aligned}$$

Find a polynomial function with rational coefficients of lowest degree with $1 - i$ and $\sqrt{7}$ as two of its zeros.

If $1 - i$ is a zero, then $1 + i$ is also a zero. If $\sqrt{7}$ is a zero, then $-\sqrt{7}$ is also a zero.

$$\begin{aligned} f(x) &= [x - (1 - i)][x - (1 + i)][x - \sqrt{7}] \times \\ &\quad [x - (-\sqrt{7})] \\ &= [(x - 1) + i][(x - 1) - i](x - \sqrt{7})(x + \sqrt{7}) \\ &= [(x - 1)^2 - i^2](x^2 - 7) \\ &= (x^2 - 2x + 1 + 1)(x^2 - 7) \\ &= (x^2 - 2x + 2)(x^2 - 7) \\ &= x^4 - 2x^3 - 5x^2 + 14x - 14 \end{aligned}$$

For $f(x) = 2x^4 - 9x^3 - 16x^2 - 9x - 18$, solve $f(x) = 0$ and factor $f(x)$ into linear factors.

There are at most 4 distinct zeros. Any rational zeros of f must be of the form p/q, where p is a factor of -18 and q is a factor of 2.

$\dfrac{\textit{Possibilities for } p}{\textit{Possibilities for } q}$: $\dfrac{\pm 1,\ \pm 2,\ \pm 3,\ \pm 6,\ \pm 9,\ \pm 18}{\pm 1,\ \pm 2}$

Possibilities for p/q: $1, -1, 2, -2, 3, -3, 6, -6, 9, -9,$

$$18, -18, \frac{1}{2}, -\frac{1}{2}, \frac{3}{2}, -\frac{3}{2}, \frac{9}{2}, -\frac{9}{2}$$

Use synthetic division to check the possibilities. We leave it to the student to verify that ± 1, ± 2, and ± 3 are not zeros. Let's try 6.

$$\begin{array}{r|rrrrr} 6 & 2 & -9 & -16 & -9 & -18 \\ & & 12 & 18 & 12 & 18 \\ \hline & 2 & 3 & 2 & 3 & 0 \end{array}$$

Since $f(6) = 0$, 6 is a zero and $x - 6$ is a factor of $f(x)$. Now we express $f(x)$ as

$$f(x) = (x - 6)(2x^3 + 3x^2 + 2x + 3).$$

(continued)

Let's consider the factor $2x^3 + 3x^2 + 2x + 3$ and check the other possibilities. Let's try $-\frac{3}{2}$.

$$
\begin{array}{r|rrrr}
-\frac{3}{2} & 2 & 3 & 2 & 3 \\
  &   & -3 & 0 & -3 \\
\hline
  & 2 & 0 & 2 & \;0
\end{array}
$$

Since $f\left(-\frac{3}{2}\right) = 0$, $-\frac{3}{2}$ is also a zero and $x + \frac{3}{2}$ is a factor of $f(x)$. We express $f(x)$ as

$$
\begin{aligned}
f(x) &= (x - 6)\left(x + \tfrac{3}{2}\right)\left(2x^2 + 2\right) \\
     &= 2(x - 6)\left(x + \tfrac{3}{2}\right)\left(x^2 + 1\right).
\end{aligned}
$$

Now we solve the equation $f(x) = 0$ to determine the zeros. We see that the only rational zeros are 6 and $-\frac{3}{2}$. The other zeros are $\pm i$.

The factorization into linear factors is

$$
\begin{aligned}
f(x) = {}& 2(x - 6)\left(x + \tfrac{3}{2}\right)(x - i)(x + i), \quad \text{or} \\
& (x - 6)(2x + 3)(x - i)(x + i).
\end{aligned}
$$

Descartes' Rule of Signs

Let $P(x)$, written in descending order or ascending order, be a polynomial function with real coefficients and a nonzero constant term. The number of positive real zeros of $P(x)$ is either:

1. The same as the number of variations of sign in $P(x)$, or

2. Less than the number of variations of sign in $P(x)$ by a positive even integer.

The number of negative real zeros of $P(x)$ is either:

3. The same as the number of variations of sign in $P(-x)$, or

4. Less than the number of variations of sign in $P(-x)$ by a positive even integer.

A zero of multiplicity m must be counted m times.

Determine the number of positive real zeros and the number of negative real zeros of

$$
P(x) = 4x^5 - x^4 - 2x^3 + 8x - 10.
$$

There are 3 variations of sign in $P(x)$. Thus the number of positive real zeros is 3 or 1.

$$
P(-x) = -4x^5 - x^4 + 2x^3 - 8x - 10
$$

There are 2 variations of sign in $P(-x)$. Thus the number of negative real zeros is 2 or 0.

SECTION 4.5: RATIONAL FUNCTIONS

Rational Function

A **rational function** is a function f that is a quotient of two polynomials. That is,

$$f(x) = \frac{p(x)}{q(x)},$$

where $p(x)$ and $q(x)$ are polynomials and $q(x)$ is not the zero polynomial. The domain of $f(x)$ consists of all x for which $q(x) \neq 0$.

Determine the domain of each function.

FUNCTION	DOMAIN
$f(x) = \dfrac{1}{x^5}$	$(-\infty, 0) \cup (0, \infty)$
$f(x) = \dfrac{x + 6}{x^2 + 2x - 8}$	
$\quad = \dfrac{x + 6}{(x - 2)(x + 4)}$	$(-\infty, -4) \cup (-4, 2) \cup (2, \infty)$

Vertical Asymptotes

For a rational function $f(x) = p(x)/q(x)$, where $p(x)$ and $q(x)$ are polynomials with *no common factors* other than constants, if a is a zero of the denominator, then the line $x = a$ is a vertical asymptote for the graph of the function.

Horizontal Asymptotes

When the numerator and the denominator of a rational function have the same degree, the line $y = a/b$ is the horizontal asymptote, where a and b are the leading coefficients of the numerator and the denominator, respectively.

When the degree of the numerator is less than the degree of the denominator, the x-axis, or $y = 0$, is the horizontal asymptote.

When the degree of the numerator is greater than the degree of the denominator, there is *no* horizontal asymptote.

Oblique Asymptotes

When the degree of the numerator is 1 greater than the degree of the denominator, there is an oblique asymptote.

Determine the vertical, horizontal, and oblique asymptotes of the graph of the function.

FUNCTION	ASYMPTOTES
$f(x) = \dfrac{x^2 - 2}{x - 1}$	Vertical: $x = 1$ Horizontal: None Oblique: $y = x + 1$
$f(x) = \dfrac{3x - 4}{x^2 + 6x - 7}$	Vertical: $x = -7$; $x = 1$ Horizontal: $y = 0$ Oblique: None
$f(x) = \dfrac{2x^2 + 9x - 5}{3x^2 + 13x + 12}$	Vertical: $x = -\dfrac{4}{3}$; $x = -3$ Horizontal: $y = \dfrac{2}{3}$ Oblique: None

To Graph a Rational Function

$f(x) = p(x)/q(x)$, where $p(x)$ and $q(x)$ have no common factor other than constants:

1. Find any real zeros of the denominator. Determine the domain of the function and sketch any vertical asymptotes.

2. Find the horizontal asymptote or the oblique asymptote, if there is one, and sketch it.

Graph: $g(x) = \dfrac{x^2 - 4}{x^2 + 4x - 5}$.

Domain: The zeros of the denominator are -5 and 1. The domain is $(-\infty, -5) \cup (-5, 1) \cup (1, \infty)$.

Vertical asymptotes: Since neither zero of the denominator is a zero of the numerator, the graph has vertical asymptotes at $x = -5$ and $x = 1$.

Horizontal asymptote: The degree of the numerator is the same as the degree of the denominator, so the horizontal asymptote is determined by the ratio of the leading coefficients: $1/1$, or 1. The horizontal asymptote is $y = 1$.

(continued)   *(continued)*

3. Find any zeros of the function. The zeros are found by determining the zeros of the numerator. These are the first coordinates of the *x*-intercepts of the graph.

4. Find $f(0)$. This gives the *y*-intercept, $(0, f(0))$, of the graph.

5. Find other function values to determine the general shape. Then draw the graph.

Crossing an Asymptote

The graph of a rational function never crosses a vertical asymptote.

The graph of a rational function might cross a horizontal asymptote but does not necessarily do so.

Oblique asymptote: None

Zeros of g: Solving $g(x) = 0$ gives us -2 and 2, so the zeros are -2 and 2.

x-intercepts: $(-2, 0)$ and $(2, 0)$

y-intercept: $\left(0, \dfrac{4}{5}\right)$, because $g(0) = \dfrac{4}{5}$

Other values:

x	y
-8	2.22
-6	4.57
-4	-2.4
-3	-0.63
-1	0.38
0.5	1.36
1.5	-0.54
3	0.31
4	0.44

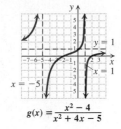

$$g(x) = \frac{x^2 - 4}{x^2 + 4x - 5}$$

Graph: $f(x) = \dfrac{x + 3}{x^2 - x - 12}$.

Domain: The zeros of the denominator are -3 and 4. The domain is $(-\infty, -3) \cup (-3, 4) \cup (4, \infty)$.

Vertical asymptote: Since 4 is the only zero of the denominator that is not a zero of the numerator, the only vertical asymptote is $x = 4$.

Horizontal asymptote: Because the degree of the numerator is less than the degree of the denominator, the *x*-axis, $y = 0$, is the horizontal asymptote.

Oblique asymptote: None

Zeros of f: The equation $f(x) = 0$ has no solutions because -3 is not in the domain of the function. Thus there are no zeros of f.

x-intercepts: None

y-intercept: $\left(0, -\dfrac{1}{4}\right)$ because $f(0) = -\dfrac{1}{4}$

Hole in the graph:

$$f(x) = \frac{x + 3}{(x + 3)(x - 4)} = \frac{1}{x - 4},$$

where $x \neq -3$ and $x \neq 4$.

(continued)

To determine the coordinates of the hole, substitute -3 for x in $f(x) = 1/(x - 4)$:

$$f(-3) = \frac{1}{-3 - 4} = -\frac{1}{7}.$$

The hole is located at $\left(-3, -\frac{1}{7}\right)$.

Other values:

x	y
-4	-0.13
-2	-0.17
1	-0.33
3	-1
5	1
7	0.33

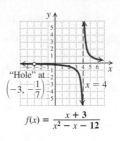

"Hole" at $\left(-3, -\frac{1}{7}\right)$

$x = 4$

$$f(x) = \frac{x + 3}{x^2 - x - 12}$$

SECTION 4.6: POLYNOMIAL INEQUALITIES AND RATIONAL INEQUALITIES

To Solve a Polynomial Inequality

1. Find an equivalent inequality with 0 on one side.

2. Change the inequality symbol to an equals sign and solve the related equation.

3. Use the solutions to divide the x-axis into intervals. Then select a test value from each interval and determine the polynomial's sign on the interval.

4. Determine the intervals for which the inequality is satisfied and write interval notation or set-builder notation for the solution set. Include the endpoints of the intervals in the solution set if the inequality symbol is $\leq$ or $\geq$.

Solve: $x^3 - 3x^2 \leq 6x - 8$.

Equivalent inequality: $x^3 - 3x^2 - 6x + 8 \leq 0$.

First, we solve the related equation:

$$x^3 - 3x^2 - 6x + 8 = 0.$$

The solutions are -2, 1, and 4. These numbers divide the x-axis into 4 intervals. Next, we let $f(x) = x^3 - 3x^2 - 6x + 8$ and, using test values for $f(x)$, determine the sign of $f(x)$ in each interval.

INTERVAL	TEST VALUE	SIGN OF $f(x)$
$(-\infty, -2)$	$f(-3) = -28$	$-$
$(-2, 1)$	$f(0) = 8$	$+$
$(1, 4)$	$f(2) = -8$	$-$
$(4, \infty)$	$f(6) = 80$	$+$

Test values are negative in the intervals $(-\infty, -2)$ and $(1, 4)$. Since the inequality sign is $\leq$, we include the endpoints of the intervals in the solution set.

The solution set is

$$(-\infty, -2] \cup [1, 4].$$

To Solve a Rational Inequality

1. Find an equivalent inequality with 0 on one side.

2. Change the inequality symbol to an equals sign and solve the related equation.

3. Find values of the variable for which the related rational function is not defined.

4. The numbers found in steps (2) and (3) are called *critical values*. Use the critical values to divide the x-axis into intervals. Then determine the function's sign in each interval using an x-value from the interval or using the graph of the equation.

5. Select the intervals for which the inequality is satisfied and write interval notation or set-builder notation for the solution set. If the inequality symbol is $\leq$ or $\geq$, then the solutions to step (2) should be included in the solution set. The x-values found in step (3) are never included in the solution set.

Solve: $\dfrac{x-1}{x+5} > \dfrac{x+3}{x-2}$.

Equivalent inequality: $\dfrac{x-1}{x+5} - \dfrac{x+3}{x-2} > 0$

Related function: $f(x) = \dfrac{x-1}{x+5} - \dfrac{x+3}{x-2}$

The function is not defined for $x = -5$ and $x = 2$. Solving $f(x) = 0$, we get $x = -\frac{13}{11}$. The critical values are -5, $-\frac{13}{11}$, and 2. These divide the x-axis into four intervals.

INTERVAL	TEST VALUE	SIGN OF $f(x)$
$(-\infty, -5)$	$f(-6) = 6.63$	$+$
$\left(-5, -\frac{13}{11}\right)$	$f(-2) = -0.75$	$-$
$\left(-\frac{13}{11}, 2\right)$	$f(0) = 1.3$	$+$
$(2, \infty)$	$f(3) = -5.75$	$-$

Test values are positive in the intervals $(-\infty, -5)$ and $\left(-\frac{13}{11}, 2\right)$. Since $f\left(-\frac{13}{11}\right) = 0$ and -5 and 2 are not in the domain of f, -5, $-\frac{13}{11}$, and 2 cannot be part of the solution set. The solution set is

$$(-\infty, -5) \cup \left(-\tfrac{13}{11}, 2\right).$$

REVIEW EXERCISES

Determine whether the statement is true or false.

1. If $f(x) = (x+a)(x+b)(x-c)$, then $f(-b) = 0$. **[4.3]**

2. The graph of a rational function never crosses a vertical asymptote. **[4.5]**

3. For the function $g(x) = x^4 - 8x^2 - 9$, the only possible rational zeros are $1, -1, 3$, and -3. **[4.4]**

4. The graph of $P(x) = x^6 - x^8$ has at most 6 x-intercepts. **[4.2]**

5. The domain of the function

$$f(x) = \frac{x-4}{(x+2)(x-3)}$$

is $(-\infty, -2) \cup (3, \infty)$. **[4.5]**

Use a graphing calculator to graph the polynomial function. Then estimate the function's (**a**) *zeros,* (**b**) *relative maxima,* (**c**) *relative minima, and* (**d**) *domain and range.* **[4.1]**

6. $f(x) = -2x^2 - 3x + 6$

7. $f(x) = x^3 + 3x^2 - 2x - 6$

8. $f(x) = x^4 - 3x^3 + 2x^2$

Determine the leading term, the leading coefficient, and the degree of the polynomial. Then classify the polynomial function as constant, linear, quadratic, cubic, or quartic. **[4.1]**

9. $f(x) = 7x^2 - 5 + 0.45x^4 - 3x^3$

10. $h(x) = -25$

11. $g(x) = 6 - 0.5x$

12. $f(x) = \frac{1}{3}x^3 - 2x + 3$

Use the leading-term test to describe the end behavior of the graph of the function. **[4.1]**

13. $f(x) = -\frac{1}{2}x^4 + 3x^2 + x - 6$

14. $f(x) = x^5 + 2x^3 - x^2 + 5x + 4$

Find the zeros of the polynomial function and state the multiplicity of each. **[4.1]**

15. $g(x) = \left(x - \frac{2}{3}\right)(x + 2)^3(x - 5)^2$

16. $f(x) = x^4 - 26x^2 + 25$

17. $h(x) = x^3 + 4x^2 - 9x - 36$

18. *Interest Compounded Annually.*   When P dollars is invested at interest rate i, compounded annually, for t years, the investment grows to A dollars, where

$$A = P(1 + i)^t.$$

a) Find the interest rate i if $6250 grows to $6760 in 2 years. **[4.1]**

b) Find the interest rate i if $1,000,000 grows to $1,215,506.25 in 4 years. **[4.1]**

19. *Cholesterol Level and the Risk of Heart Attack.* The following table lists data concerning the relationship of cholesterol level in men to the risk of a heart attack.

Cholesterol Level	Number of Men per 100,000 Who Suffer a Heart Attack
100	30
200	65
250	100
275	130

Source: Nutrition Action Newsletter

a) Use regression on a graphing calculator to fit linear, quadratic, and cubic functions to the data. **[4.1]**

b) It is also known that 180 of 100,000 men with a cholesterol level of 300 have a heart attack. Which function in part (a) would best make this prediction? **[4.1]**

c) Use the answer to part (b) to predict the heart attack rate for men with cholesterol levels of 350 and of 400. **[4.1]**

Sketch the graph of the polynomial function.

20. $f(x) = -x^4 + 2x^3$ **[4.2]**

21. $g(x) = (x - 1)^3(x + 2)^2$ **[4.2]**

22. $h(x) = x^3 + 3x^2 - x - 3$ **[4.2]**

23. $f(x) = x^4 - 5x^3 + 6x^2 + 4x - 8$
[4.2], [4.3], [4.4]

24. $g(x) = 2x^3 + 7x^2 - 14x + 5$ **[4.2], [4.4]**

Using the intermediate value theorem, determine, if possible, whether the function f has a zero between a and b. **[4.2]**

25. $f(x) = 4x^2 - 5x - 3;\ a = 1, b = 2$

26. $f(x) = x^3 - 4x^2 + \frac{1}{2}x + 2;\ a = -1, b = 1$

In each of the following, a polynomial $P(x)$ and a divisor $d(x)$ are given. Use long division to find the quotient $Q(x)$ and the remainder $R(x)$ when $P(x)$ is divided by $d(x)$. Express $P(x)$ in the form $d(x) \cdot Q(x) + R(x)$. **[4.3]**

27. $P(x) = 6x^3 - 2x^2 + 4x - 1,$
$d(x) = x - 3$

28. $P(x) = x^4 - 2x^3 + x + 5,$
$d(x) = x + 1$

Use synthetic division to find the quotient and the remainder. **[4.3]**

29. $(x^3 + 2x^2 - 13x + 10) \div (x - 5)$

30. $(x^4 + 3x^3 + 3x^2 + 3x + 2) \div (x + 2)$

31. $(x^5 - 2x) \div (x + 1)$

Use synthetic division to find the indicated function value. **[4.3]**

32. $f(x) = x^3 + 2x^2 - 13x + 10;\ f(-2)$

33. $f(x) = x^4 - 16;\ f(-2)$

34. $f(x) = x^5 - 4x^4 + x^3 - x^2 + 2x - 100;$
$f(-10)$

Using synthetic division, determine whether the given numbers are zeros of the polynomial function. **[4.3]**

35. $-i, -5;\ f(x) = x^3 - 5x^2 + x - 5$

36. $-1, -2;\ f(x) = x^4 - 4x^3 - 3x^2 + 14x - 8$

37. $\frac{1}{3}, 1;\ f(x) = x^3 - \frac{4}{3}x^2 - \frac{5}{3}x + \frac{2}{3}$

38. $2, -\sqrt{3};\ f(x) = x^4 - 5x^2 + 6$

Factor the polynomial $f(x)$. Then solve the equation $f(x) = 0$. **[4.3], [4.4]**

39. $f(x) = x^3 + 2x^2 - 7x + 4$

40. $f(x) = x^3 + 4x^2 - 3x - 18$

41. $f(x) = x^4 - 4x^3 - 21x^2 + 100x - 100$

42. $f(x) = x^4 - 3x^2 + 2$

Find a polynomial function of degree 3 with the given numbers as zeros. **[4.4]**

43. $-4, -1, 2$

44. $-3, 1 - i, 1 + i$

45. $\frac{1}{2}, 1 - \sqrt{2}, 1 + \sqrt{2}$

46. Find a polynomial function of degree 4 with -5 as a zero of multiplicity 3 and $\frac{1}{2}$ as a zero of multiplicity 1. **[4.4]**

47. Find a polynomial function of degree 5 with -3 as a zero of multiplicity 2, 2 as a zero of multiplicity 1, and 0 as a zero of multiplicity 2. **[4.4]**

Suppose that a polynomial function of degree 5 with rational coefficients has the given zeros. Find the other zero(s). **[4.4]**

48. $-\frac{2}{3}, \sqrt{5}, 4 + i$

49. $0, 1 + \sqrt{3}, -\sqrt{3}$

50. $-\sqrt{2}, \frac{1}{2}, 1, 2$

Find a polynomial function of lowest degree with rational coefficients and the following as some of its zeros. **[4.4]**

51. $\sqrt{11}$

52. $-i, 6$

53. $-1, 4, 1 + i$

54. $\sqrt{5}, -2i$

55. $\frac{1}{3}, 0, -3$

List all possible rational zeros. **[4.4]**

56. $h(x) = 4x^5 - 2x^3 + 6x - 12$

57. $g(x) = 3x^4 - x^3 + 5x^2 - x + 1$

58. $f(x) = x^3 - 2x^2 + x - 24$

For each polynomial function:

a) *Find the rational zeros and then the other zeros; that is, solve $f(x) = 0$.* **[4.4]**
b) *Factor $f(x)$ into linear factors.* **[4.4]**

59. $f(x) = 3x^5 + 2x^4 - 25x^3 - 28x^2 + 12x$

60. $f(x) = x^3 - 2x^2 - 3x + 6$

61. $f(x) = x^4 - 6x^3 + 9x^2 + 6x - 10$

62. $f(x) = x^3 + 3x^2 - 11x - 5$

63. $f(x) = 3x^3 - 8x^2 + 7x - 2$

64. $f(x) = x^5 - 8x^4 + 20x^3 - 8x^2 - 32x + 32$

65. $f(x) = x^6 + x^5 - 28x^4 - 16x^3 + 192x^2$

66. $f(x) = 2x^5 - 13x^4 + 32x^3 - 38x^2 + 22x - 5$

What does Descartes' rule of signs tell you about the number of positive real zeros and the number of negative real zeros of each of the following polynomial functions? **[4.4]**

67. $f(x) = 2x^6 - 7x^3 + x^2 - x$

68. $h(x) = -x^8 + 6x^5 - x^3 + 2x - 2$

69. $g(x) = 5x^5 - 4x^2 + x - 1$

Graph the function. Be sure to label all the asymptotes. List the domain and the x- and y-intercepts. **[4.5]**

70. $f(x) = \dfrac{x^2 - 5}{x + 2}$

71. $f(x) = \dfrac{5}{(x - 2)^2}$

72. $f(x) = \dfrac{x^2 + x - 6}{x^2 - x - 20}$

73. $f(x) = \dfrac{x - 2}{x^2 - 2x - 15}$

In Exercises 74 and 75, find a rational function that satisfies the given conditions. Answers may vary, but try to give the simplest answer possible. **[4.5]**

74. Vertical asymptotes $x = -2, x = 3$

75. Vertical asymptotes $x = -2, x = 3$; horizontal asymptote $y = 4$; x-intercept $(-3, 0)$

76. *Medical Dosage.* The function

$$N(t) = \frac{0.7t + 2000}{8t + 9}, \quad t \geq 5,$$

gives the body concentration $N(t)$, in parts per million, of a certain dosage of medication after time t, in hours.

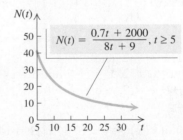

a) Find the horizontal asymptote of the graph and complete the following:

$N(t) \rightarrow$ _____ as $t \rightarrow \infty$. **[4.5]**

b) Explain the meaning of the answer to part (a) in terms of the application. **[4.5]**

Solve. **[4.6]**

77. $x^2 - 9 < 0$

78. $2x^2 > 3x + 2$

79. $(1 - x)(x + 4)(x - 2) \leq 0$

80. $\dfrac{x - 2}{x + 3} < 4$

81. *Height of a Rocket.* The function

$$S(t) = -16t^2 + 80t + 224$$

gives the height S, in feet, of a model rocket launched with a velocity of 80 ft/sec from a hill that is 224 ft high, where t is the time, in seconds.

a) Determine when the rocket reaches the ground. [4.1]

b) On what interval is the height greater than 320 ft? [4.1], [4.6]

82. *Population Growth.* The population P, in thousands, of Novi is given by

$$P(t) = \frac{8000t}{4t^2 + 10},$$

where t is the time, in months. Find the interval on which the population was 400,000 or greater. [4.6]

83. Which of the following is the domain of the function

$$g(x) = \frac{x^2 + 2x - 3}{x^2 - 5x + 6}? \quad [4.5]$$

A. $(-\infty, 2) \cup (2, 3) \cup (3, \infty)$
B. $(-\infty, -3) \cup (-3, 1) \cup (1, \infty)$
C. $(-\infty, 2) \cup (3, \infty)$
D. $(-\infty, -3) \cup (1, \infty)$

84. Which of the following are the vertical asymptotes of the function

$$f(x) = \frac{x - 4}{(x + 1)(x - 2)(x + 4)}? \quad [4.5]$$

A. $x = 1$, $x = -2$, and $x = 4$
B. $x = -1$, $x = 2$, $x = -4$, and $x = 4$
C. $x = -1$, $x = 2$, and $x = -4$
D. $x = 4$

85. The graph of $f(x) = -\frac{1}{2}x^4 + x^3 + 1$ is which of the following? [4.2]

A.
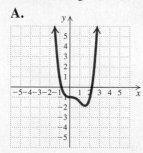

B.
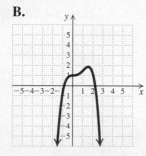

C.
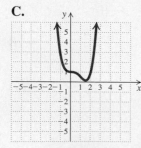

D.
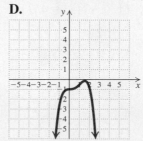

› Synthesis

Solve.

86. $x^2 \geq 5 - 2x$ [4.6]

87. $\left| 1 - \dfrac{1}{x^2} \right| < 3$ [4.6]

88. $x^4 - 2x^3 + 3x^2 - 2x + 2 = 0$ [4.4]

89. $(x - 2)^{-3} < 0$ [4.6]

90. Express $x^3 - 1$ as a product of linear factors. [4.4]

91. Find k such that $x + 3$ is a factor of $x^3 + kx^2 + kx - 15$. [4.3]

92. When $x^2 - 4x + 3k$ is divided by $x + 5$, the remainder is 33. Find the value of k. [4.3]

Find the domain of the function. [4.6]

93. $f(x) = \sqrt{x^2 + 3x - 10}$

94. $f(x) = \sqrt{x^2 - 3.1x + 2.2} + 1.75$

95. $f(x) = \dfrac{1}{\sqrt{5 - |7x + 2|}}$

› Collaborative Discussion and Writing

96. Explain the difference between a polynomial function and a rational function. [4.1], [4.5]

97. Is it possible for a third-degree polynomial with rational coefficients to have no real zeros? Why or why not? [4.4]

98. Explain and contrast the three types of asymptotes considered for rational functions. [4.5]

99. If $P(x)$ is an even function, and by Descartes' rule of signs, $P(x)$ has one positive real zero, how many negative real zeros does $P(x)$ have? Explain. [4.4]

100. Explain why the graph of a rational function cannot have both a horizontal asymptote and an oblique asymptote. [4.5]

101. Under what circumstances would a quadratic inequality have a solution set that is a closed interval? [4.6]

4 Chapter Test

Determine the leading term, the leading coefficient, and the degree of the polynomial. Then classify the polynomial as constant, linear, quadratic, cubic, or quartic.

1. $f(x) = 2x^3 + 6x^2 - x^4 + 11$

2. $h(x) = -4.7x + 29$

3. Find the zeros of the polynomial function and state the multiplicity of each:
$$f(x) = x(3x - 5)(x - 3)^2(x + 1)^3.$$

4. *Hybrid Automobiles.* In 2005, only 205,828 hybrid automobiles were sold, while in 2013, 498,054 were sold (*Source*: WardsAuto Group, a division of Penton). The cubic function
$$f(x) = 3707.968x^3 - 40{,}437.526x^2$$
$$+ 126{,}421.240x + 197{,}407.131,$$
where x is the number of years after 2005, can be used to estimate the number of hybrid automobiles sold in years 2005 to 2013. Use this function to estimate the number of hybrid automobiles sold in 2008 and in 2012.

Sketch the graph of the polynomial function.

5. $f(x) = x^3 - 5x^2 + 2x + 8$

6. $f(x) = -2x^4 + x^3 + 11x^2 - 4x - 12$

Using the intermediate value theorem, determine, if possible, whether the function has a zero between a and b.

7. $f(x) = -5x^2 + 3;\ a = 0, b = 2$

8. $g(x) = 2x^3 + 6x^2 - 3;\ a = -2, b = -1$

9. Use long division to find the quotient $Q(x)$ and the remainder $R(x)$ when $P(x)$ is divided by $d(x)$. Express $P(x)$ in the form $d(x) \cdot Q(x) + R(x)$. Show your work.
$$P(x) = x^4 + 3x^3 + 2x - 5,$$
$$d(x) = x - 1$$

10. Use synthetic division to find the quotient and the remainder. Show your work.
$$(3x^3 - 12x + 7) \div (x - 5)$$

11. Use synthetic division to find $P(-3)$ for $P(x) = 2x^3 - 6x^2 + x - 4$. Show your work.

12. Use synthetic division to determine whether -2 is a zero of $f(x) = x^3 + 4x^2 + x - 6$. Answer yes or no. Show your work.

13. Find a polynomial function of degree 4 with -3 as a zero of multiplicity 2 and 0 and 6 as zeros of multiplicity 1.

14. Suppose that a polynomial function of degree 5 with rational coefficients has 1, $\sqrt{3}$, and $2 - i$ as zeros. Find the other zeros.

Find a polynomial function of lowest degree with rational coefficients and the following as some of its zeros.

15. $-10, 3i$

16. $0, -\sqrt{3}, 1 - i$

List all possible rational zeros.

17. $f(x) = 2x^3 + x^2 - 2x + 12$

18. $h(x) = 10x^4 - x^3 + 2x - 5$

For each polynomial function:

a) *Find the rational zeros and then the other zeros; that is, solve $f(x) = 0$.*

b) *Factor $f(x)$ into linear factors.*

19. $f(x) = x^3 + x^2 - 5x - 5$

20. $f(x) = 2x^4 - 11x^3 + 16x^2 - x - 6$

21. $f(x) = x^3 + 4x^2 + 4x + 16$

22. $f(x) = 3x^4 - 11x^3 + 15x^2 - 9x + 2$

23. What does Descartes' rule of signs tell you about the number of positive real zeros and the number of negative real zeros of the following function?
$$g(x) = -x^8 + 2x^6 - 4x^3 - 1$$

Graph the function. Be sure to label all the asymptotes. List the domain and the x- and y-intercepts.

24. $f(x) = \dfrac{2}{(x - 3)^2}$

25. $f(x) = \dfrac{x + 3}{x^2 - 3x - 4}$

26. Find a rational function that has vertical asymptotes $x = -1$ and $x = 2$ and x-intercept $(-4, 0)$.

Solve.

27. $2x^2 > 5x + 3$

28. $\dfrac{x + 1}{x - 4} \le 3$

29. The function $S(t) = -16t^2 + 64t + 192$ gives the height S, in feet, of a model rocket launched with a velocity of 64 ft/sec from a hill that is 192 ft high.

a) Determine how long it will take the rocket to reach the ground.

b) Find the interval on which the height of the rocket is greater than 240 ft.

30. The graph of $f(x) = x^3 - x^2 - 2$ is which of the following?

A.

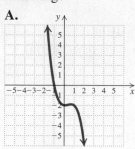

B.

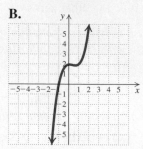

C.

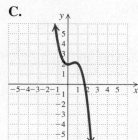

D.

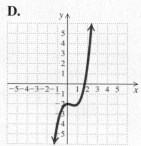

❯ Synthesis

31. Find the domain of $f(x) = \sqrt{x^2 + x - 12}$.

Exponential Functions and Logarithmic Functions

APPLICATION

This problem appears as Exercise 65 in Section 5.2.

Annual milk consumption per capita in China has increased greatly in recent years. Consequently, the demand for imported alfalfa has increased exponentially. In 2008, approximately 20,000 tons of U.S. alfalfa were imported by China. This number increased to approximately 650,000 tons by 2013. (*Source*: *National Geographic*, May 2015). The increase in imported U.S. alfalfa can be modeled by the exponential function $A(x) = 22{,}611.008(1.992)^x$, where x is the number of years after 2008. Find the number of tons of U.S. alfalfa imported by China in 2012 and in 2016.

5.1 > Inverse Functions

> ❯ Determine whether a function is one-to-one, and if it is, find a formula for its inverse.
> ❯ Simplify expressions of the type $(f \circ f^{-1})(x)$ and $(f^{-1} \circ f)(x)$.

❯ Inverses

When we go from an output of a function back to its input or inputs, we get an inverse relation. When that relation is a function, we have an inverse function.

Consider the relation h given as follows:

$$h = \{(-8, 5), (4, -2), (-7, 1), (3.8, 6.2)\}.$$

Suppose that we *interchange* the first and second coordinates. The relation we obtain is called the **inverse** of the relation h and is given as follows:

$$\text{Inverse of } h = \{(5, -8), (-2, 4), (1, -7), (6.2, 3.8)\}.$$

RELATIONS

REVIEW SECTION 1.2.

INVERSE RELATION

Interchanging the first and second coordinates of each ordered pair in a relation produces the **inverse relation**.

EXAMPLE 1 Consider the relation g given by

$$g = \{(2, 4), (-1, 3), (-2, 0)\}.$$

Graph the relation in blue. Find the inverse and graph it in red.

Solution The relation g is shown in blue in the figure at left. The inverse of the relation is

$$\{(4, 2), (3, -1), (0, -2)\}$$

and is shown in red. The pairs in the inverse are reflections of the pairs in g across the line $y = x$.

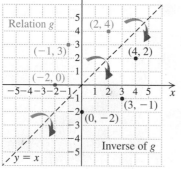

> *Now Try Exercise 1.*

INVERSE RELATION

If a relation is defined by an equation, interchanging the variables produces an equation of the **inverse relation.**

EXAMPLE 2 Find an equation for the inverse of the relation

$$y = x^2 - 5x.$$

Solution We interchange x and y and obtain an equation of the inverse:

$$x = y^2 - 5y.$$

> *Now Try Exercise 9.*

If a relation is given by an equation, then the solutions of the inverse can be found from those of the original equation by interchanging the first and second coordinates of each ordered pair. Thus the graphs of a relation and its inverse are always reflections of each other across the line $y = x$. This is illustrated with the equations of Example 2 in the following tables and graph. We will explore inverses and their graphs later in this section.

$x = y^2 - 5y$	y
6	−1
0	0
−6	2
−4	4

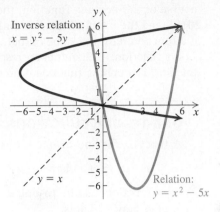

x	$y = x^2 - 5x$
−1	6
0	0
2	−6
4	−4

› Inverses and One-to-One Functions

Let's consider the following two functions.

Year (domain)	First-Class Postage Cost, in cents (range)
2006 ⟶	39
2007 ⟶	41
2008 ⟶	42
2009 ⟶	
2010 ⟶	44
2011 ⟶	
2012 ⟶	45
2013 ⟶	46
2014 ⟶	
2015 ⟶	49

Source: U.S. Postal Service

Number (domain)	Cube (range)
−3 ⟶	−27
−2 ⟶	−8
−1 ⟶	−1
0 ⟶	0
1 ⟶	1
2 ⟶	8
3 ⟶	27

Suppose that we reverse the arrows. Are these inverse relations functions?

Year (range)	First-Class Postage Cost, in cents (domain)
2006 ⟵	39
2007 ⟵	41
2008 ⟵	42
2009 ⟵	
2010 ⟵	44
2011 ⟵	
2012 ⟵	45
2013 ⟵	46
2014 ⟵	
2015 ⟵	49

Source: U.S. Postal Service

Number (range)	Cube (domain)
−3 ⟵	−27
−2 ⟵	−8
−1 ⟵	−1
0 ⟵	0
1 ⟵	1
2 ⟵	8
3 ⟵	27

We see that the inverse of the postage function is not a function. Like all functions, each input in the postage function has exactly one output. However, the output for 2009, 2010, and 2011 is 44. Also, the output for 2014 and 2015 is 49. Thus in the inverse of the postage function, the input 44 has *three* outputs—2009, 2010, and 2011—and the input 49 has *two* outputs—2014 and 2015. When two or more inputs of a function have the same output, the inverse relation cannot be a function. In the cubing function, each output corresponds to exactly one input, so its inverse is also a function. The cubing function is an example of a **one-to-one function**.

ONE-TO-ONE FUNCTIONS

A function f is **one-to-one** if different inputs have different outputs—that is,

$$\text{if} \quad a \neq b, \quad \text{then} \quad f(a) \neq f(b).$$

Or, a function f is **one-to-one** if when the outputs are the same, the inputs are the same—that is,

$$\text{if} \quad f(a) = f(b), \quad \text{then} \quad a = b.$$

If the inverse of a function f is also a function, it is named f^{-1} (read "f-inverse").

The -1 in f^{-1} is *not* an exponent!

Do *not* misinterpret the -1 in f^{-1} as a negative exponent: f^{-1} does *not* mean the reciprocal of f and $f^{-1}(x)$ is *not* equal to $\dfrac{1}{f(x)}$.

ONE-TO-ONE FUNCTIONS AND INVERSES

- If a function f is one-to-one, then its inverse f^{-1} is a function.
- The domain of a one-to-one function f is the range of the inverse f^{-1}.
- The range of a one-to-one function f is the domain of the inverse f^{-1}.
- A function that is increasing over its entire domain or is decreasing over its entire domain is a one-to-one function.

$$\begin{array}{cc} D_f & D_{f^{-1}} \\ R_f & R_{f^{-1}} \end{array}$$

EXAMPLE 3 Given the function f described by $f(x) = 2x - 3$, prove that f is one-to-one (that is, it has an inverse that is a function).

Solution To show that f is one-to-one, we show that if $f(a) = f(b)$, then $a = b$. Assume that $f(a) = f(b)$ for a and b in the domain of f. Since $f(a) = 2a - 3$ and $f(b) = 2b - 3$, we have

$$2a - 3 = 2b - 3$$
$$2a = 2b \qquad \text{Adding 3}$$
$$a = b. \qquad \text{Dividing by 2}$$

Thus, if $f(a) = f(b)$, then $a = b$. This shows that f is one-to-one.

> *Now Try Exercise 17.*

EXAMPLE 4 Given the function g described by $g(x) = x^2$, prove that g is not one-to-one.

Solution We can prove that g is not one-to-one by finding two numbers a and b for which $a \neq b$ and $g(a) = g(b)$. Two such numbers are -3 and 3, because $-3 \neq 3$ and $g(-3) = g(3) = 9$. Thus g is not one-to-one.

> *Now Try Exercise 21.*

The following graphs show a function, in blue, and its inverse, in red. To determine whether the inverse is a function, we can apply the vertical-line test to its graph. By reflecting each such vertical line across the line $y = x$, we obtain an equivalent **horizontal-line test** for the original function.

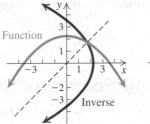

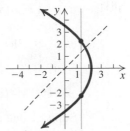
The vertical-line test shows that the inverse is not a function.

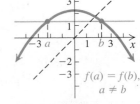
The horizontal-line test shows that the function is not one-to-one.

HORIZONTAL-LINE TEST

If it is possible for a horizontal line to intersect the graph of a function more than once, then the function is *not* one-to-one and its inverse is *not* a function.

EXAMPLE 5 From the graph shown in each of (a)–(d), determine whether the function is one-to-one and thus has an inverse that is a function.

a)
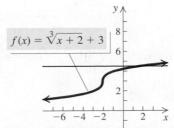

b)
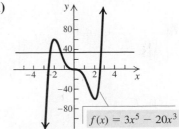

c)
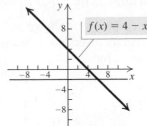

d)
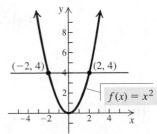

Solution   For each function, we apply the horizontal-line test.

RESULT	REASON
a) One-to-one; inverse is a function	No horizontal line intersects the graph more than once.
b) Not one-to-one; inverse is not a function	There are many horizontal lines that intersect the graph more than once. Note that where the line $y = 4$ intersects the graph, the first coordinates are -2 and 2. Although these are different inputs, they have the same output, 4.
c) One-to-one; inverse is a function	No horizontal line intersects the graph more than once.
d) Not one-to-one; inverse is not a function	There are many horizontal lines that intersect the graph more than once.

> *Now Try Exercises 25 and 27.*

❯ Finding Formulas for Inverses

Suppose that a function is described by a formula. If it has an inverse that is a function, we proceed as follows to find a formula for f^{-1}.

OBTAINING A FORMULA FOR AN INVERSE

If a function f is one-to-one, a formula for its inverse can generally be found as follows:

1. Replace $f(x)$ with y.
2. Interchange x and y.
3. Solve for y.
4. Replace y with $f^{-1}(x)$.

EXAMPLE 6   Determine whether the function $f(x) = 2x - 3$ is one-to-one, and if it is, find a formula for $f^{-1}(x)$.

Solution   The graph of f is shown at left. It passes the horizontal-line test. Thus it is one-to-one and its inverse is a function. We also proved that f is one-to-one in Example 3. We find a formula for $f^{-1}(x)$.

$y = 2x - 3$

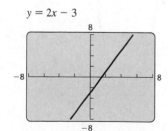

1. Replace $f(x)$ with y: $\qquad y = 2x - 3$
2. Interchange x and y: $\qquad x = 2y - 3$
3. Solve for y: $\qquad x + 3 = 2y$

$$\frac{x + 3}{2} = y$$

4. Replace y with $f^{-1}(x)$: $\quad f^{-1}(x) = \dfrac{x + 3}{2}.$

> *Now Try Exercise 57.*

Consider

$$f(x) = 2x - 3 \quad \text{and} \quad f^{-1}(x) = \frac{x + 3}{2}$$

from Example 6. For the input 5, we have

$$f(5) = 2 \cdot 5 - 3 = 10 - 3 = 7.$$

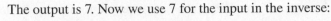

The output is 7. Now we use 7 for the input in the inverse:

$$f^{-1}(7) = \frac{7 + 3}{2} = \frac{10}{2} = 5.$$

The function f takes the number 5 to 7. The inverse function f^{-1} takes the number 7 back to 5.

EXAMPLE 7   Graph

$$f(x) = 2x - 3 \quad \text{and} \quad f^{-1}(x) = \frac{x + 3}{2}$$

using the same set of axes. Then compare the two graphs.

Solution   The graphs of f and f^{-1} are shown at left. The solutions of the inverse function can be found from those of the original function by interchanging the first and second coordinates of each ordered pair.

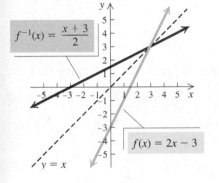

x	$f(x) = 2x - 3$
-1	-5
0	-3 ← *y*-intercept
2	1
3	3

x	$f^{-1}(x) = \dfrac{x + 3}{2}$
-5	-1
-3	0 ← *x*-intercept
1	2
3	3

On some graphing calculators, we can graph the inverse of a function after graphing the function itself by accessing a drawing feature.

When we interchange x and y in finding a formula for the inverse of $f(x) = 2x - 3$, we are in effect reflecting the graph of that function across the line $y = x$. For example, when the coordinates of the y-intercept of the graph of f, $(0, -3)$, are reversed, we get the x-intercept of the graph of $f^{-1}, (-3, 0)$. If we were to graph $f(x) = 2x - 3$ in wet ink and fold along the line $y = x$, the graph of $f^{-1}(x) = (x + 3)/2$ would be formed by the ink transferred from f.

> The graph of f^{-1} is a reflection of the graph of f across the line $y = x$.

EXAMPLE 8   Consider $g(x) = x^3 + 2$.

a) Determine whether the function is one-to-one.

b) If it is one-to-one, find a formula for its inverse.

c) Graph the function and its inverse.

Solution

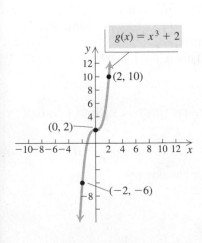

a) The graph of $g(x) = x^3 + 2$ is shown at left. It passes the horizontal-line test and thus has an inverse that is a function. We also know that $g(x)$ is one-to-one because it is an increasing function over its entire domain.

b) We follow the procedure for finding an inverse.

1. Replace $g(x)$ with y: $y = x^3 + 2$

2. Interchange x and y: $x = y^3 + 2$

3. Solve for y: $x - 2 = y^3$

$$\sqrt[3]{x - 2} = y$$

4. Replace y with $g^{-1}(x)$: $g^{-1}(x) = \sqrt[3]{x - 2}$.

We can test a point as a partial check:

$$g(x) = x^3 + 2$$
$$g(3) = 3^3 + 2 = 27 + 2 = 29.$$

Will $g^{-1}(29) = 3$? We have

$$g^{-1}(x) = \sqrt[3]{x - 2}$$
$$g^{-1}(29) = \sqrt[3]{29 - 2} = \sqrt[3]{27} = 3.$$

Since $g(3) = 29$ and $g^{-1}(29) = 3$, we can be reasonably certain that the formula for $g^{-1}(x)$ is correct.

c) To find the graph of the inverse function, we reflect the graph of $g(x) = x^3 + 2$ across the line $y = x$. This can be done by plotting points or by using a graphing calculator.

x	$g(x)$
-2	-6
-1	1
0	2
1	3
2	10

x	$g^{-1}(x)$
-6	-2
1	-1
2	0
3	1
10	2

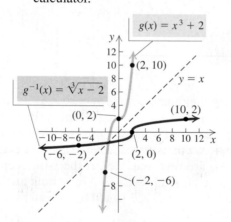

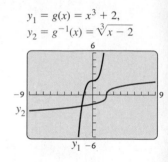

> *Now Try Exercises 63 and 79.*

❯ Inverse Functions and Composition

Suppose that we were to use some input a for a one-to-one function f and find its output, $f(a)$. The function f^{-1} would then take that output back to a. Similarly, if we began with an input b for the function f^{-1} and found its output, $f^{-1}(b)$, the original function f would then take that output back to b. This is summarized as follows.

If a function f is one-to-one, then f^{-1} is the unique function such that each of the following holds:

$$(f^{-1} \circ f)(x) = f^{-1}(f(x)) = x, \quad \text{for each } x \text{ in the domain of } f, \text{ and}$$

$$(f \circ f^{-1})(x) = f(f^{-1}(x)) = x, \quad \text{for each } x \text{ in the domain of } f^{-1}.$$

EXAMPLE 9 Given that $f(x) = 5x + 8$, use composition of functions to show that

$$f^{-1}(x) = \frac{x - 8}{5}.$$

Solution We find $(f^{-1} \circ f)(x)$ and $(f \circ f^{-1})(x)$ and check to see that each is x:

$$(f^{-1} \circ f)(x) = f^{-1}(f(x))$$
$$= f^{-1}(5x + 8) = \frac{(5x + 8) - 8}{5} = \frac{5x}{5} = x;$$

$$(f \circ f^{-1})(x) = f(f^{-1}(x))$$
$$= f\left(\frac{x - 8}{5}\right) = 5\left(\frac{x - 8}{5}\right) + 8 = x - 8 + 8 = x.$$

> **Now Try Exercise 87.**

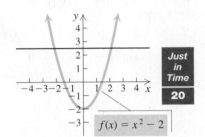

❯ Restricting a Domain

In the case in which the inverse of a function is not a function, the domain of the function can be restricted to allow the inverse to be a function. Let's consider the function $f(x) = x^2 - 2$. It is not one-to-one. The graph is shown at left.

Suppose that we had tried to find a formula for the inverse as follows:

$$y = x^2 - 2 \qquad \textbf{Replacing } f(x) \textbf{ with } y$$
$$x = y^2 - 2 \qquad \textbf{Interchanging } x \textbf{ and } y$$
$$x + 2 = y^2$$
$$\pm\sqrt{x + 2} = y. \qquad \textbf{Solving for } y$$

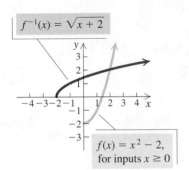

This is not the equation of a function. An input of, say, 2 would yield two outputs, -2 and 2. In such cases, it is convenient to consider "part" of the function by restricting the domain of $f(x)$. For example, if we restrict the domain of $f(x) = x^2 - 2$ to nonnegative numbers, then its inverse is a function, as shown at left by the graphs of $f(x) = x^2 - 2$, $x \geq 0$, and $f^{-1}(x) = \sqrt{x + 2}$.

5.1   Exercise Set

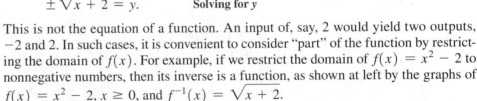

Find the inverse of the relation.

1. $\{(7, 8), (-2, 8), (3, -4), (8, -8)\}$

2. $\{(0, 1), (5, 6), (-2, -4)\}$

3. $\{(-1, -1), (-3, 4)\}$

4. $\{(-1, 3), (2, 5), (-3, 5), (2, 0)\}$

Find an equation of the inverse relation.

5. $y = 4x - 5$

6. $2x^2 + 5y^2 = 4$

7. $x^3y = -5$

8. $y = 3x^2 - 5x + 9$

9. $x = y^2 - 2y$

10. $x = \frac{1}{2}y + 4$

Graph the equation by substituting and plotting points. Then reflect the graph across the line $y = x$ to obtain the graph of its inverse.

11. $x = y^2 - 3$ **12.** $y = x^2 + 1$

13. $y = 3x - 2$ **14.** $x = -y + 4$

15. $y = |x|$ **16.** $x + 2 = |y|$

Given the function f, prove that f is one-to-one using the definition of a one-to-one function on p. 310.

17. $f(x) = \frac{1}{3}x - 6$ **18.** $f(x) = 4 - 2x$

19. $f(x) = x^3 + \frac{1}{2}$ **20.** $f(x) = \sqrt[3]{x}$

Given the function g, prove that g is not one-to-one using the definition of a one-to-one function on p. 310.

21. $g(x) = 1 - x^2$ **22.** $g(x) = 3x^2 + 1$

23. $g(x) = x^4 - x^2$ **24.** $g(x) = \frac{1}{x^6}$

Using the horizontal-line test, determine whether the function is one-to-one.

25. $f(x) = 2.7^x$ **26.** $f(x) = 2^{-x}$

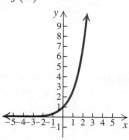   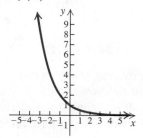

27. $f(x) = 4 - x^2$ **28.** $f(x) = x^3 - 3x + 1$

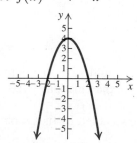   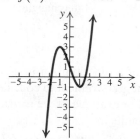

29. $f(x) = \frac{8}{x^2 - 4}$ **30.** $f(x) = \sqrt{\frac{10}{4 + x}}$

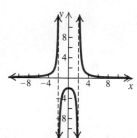   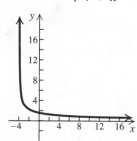

31. $f(x) = \sqrt[3]{x + 2} - 2$ **32.** $f(x) = \frac{8}{x}$

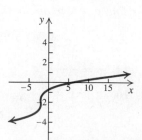

Graph the function and determine whether the function is one-to-one using the horizontal-line test.

33. $f(x) = 5x - 8$ **34.** $f(x) = 3 + 4x$

35. $f(x) = 1 - x^2$ **36.** $f(x) = |x| - 2$

37. $f(x) = |x + 2|$ **38.** $f(x) = -0.8$

39. $f(x) = -\frac{4}{x}$ **40.** $f(x) = \frac{2}{x + 3}$

41. $f(x) = \frac{2}{3}$ **42.** $f(x) = \frac{1}{2}x^2 + 3$

43. $f(x) = \sqrt{25 - x^2}$ **44.** $f(x) = -x^3 + 2$

Graph the function and its inverse using a graphing calculator. Use an inverse drawing feature, if available. Find the domain and the range of f and of f^{-1}.

45. $f(x) = 0.8x + 1.7$ **46.** $f(x) = 2.7 - 1.08x$

47. $f(x) = \frac{1}{2}x - 4$ **48.** $f(x) = x^3 - 1$

49. $f(x) = \sqrt{x - 3}$ **50.** $f(x) = -\frac{2}{x}$

51. $f(x) = x^2 - 4, x \geq 0$

52. $f(x) = 3 - x^2, x \geq 0$

53. $f(x) = (3x - 9)^3$

54. $f(x) = \sqrt[3]{\frac{x - 3.2}{1.4}}$

In Exercises 55–70, for each function:

a) *Determine whether it is one-to-one.*

b) *If the function is one-to-one, find a formula for the inverse.*

55. $f(x) = x + 4$ **56.** $f(x) = 7 - x$

57. $f(x) = 2x - 1$ **58.** $f(x) = 5x + 8$

59. $f(x) = \frac{4}{x + 7}$ **60.** $f(x) = -\frac{3}{x}$

61. $f(x) = \frac{x + 4}{x - 3}$

62. $f(x) = \frac{5x - 3}{2x + 1}$

63. $f(x) = x^3 - 1$

64. $f(x) = (x + 5)^3$

65. $f(x) = x\sqrt{4 - x^2}$

66. $f(x) = 2x^2 - x - 1$

67. $f(x) = 5x^2 - 2, x \geq 0$

68. $f(x) = 4x^2 + 3, x \geq 0$

69. $f(x) = \sqrt{x + 1}$

70. $f(x) = \sqrt[3]{x - 8}$

Find the inverse by thinking about the operations of the function and then reversing, or undoing, them. Check your work algebraically.

FUNCTION	INVERSE
71. $f(x) = 3x$	$f^{-1}(x) = $ ▢
72. $f(x) = \frac{1}{4}x + 7$	$f^{-1}(x) = $ ▢
73. $f(x) = -x$	$f^{-1}(x) = $ ▢
74. $f(x) = \sqrt[3]{x} - 5$	$f^{-1}(x) = $ ▢
75. $f(x) = \sqrt[3]{x - 5}$	$f^{-1}(x) = $ ▢
76. $f(x) = x^{-1}$	$f^{-1}(x) = $ ▢

Each graph in Exercises 77–82 is the graph of a one-to-one function f. Sketch the graph of the inverse function f^{-1}.

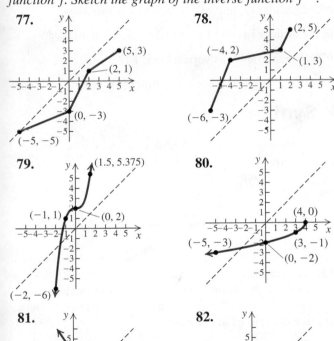

77.

78.

79.

80.

81.

82.

For the function f, use composition of functions to show that f^{-1} is as given.

83. $f(x) = \frac{7}{8}x$, $f^{-1}(x) = \frac{8}{7}x$

84. $f(x) = \dfrac{x + 5}{4}$, $f^{-1}(x) = 4x - 5$

85. $f(x) = \dfrac{1 - x}{x}$, $f^{-1}(x) = \dfrac{1}{x + 1}$

86. $f(x) = \sqrt[3]{x + 4}$, $f^{-1}(x) = x^3 - 4$

87. $f(x) = \dfrac{2}{5}x + 1$, $f^{-1}(x) = \dfrac{5x - 5}{2}$

88. $f(x) = \dfrac{x + 6}{3x - 4}$, $f^{-1}(x) = \dfrac{4x + 6}{3x - 1}$

Find the inverse of the given one-to-one function f. Give the domain and the range of f and of f^{-1}, and then graph both f and f^{-1} on the same set of axes.

89. $f(x) = 5x - 3$

90. $f(x) = 2 - x$

91. $f(x) = \dfrac{2}{x}$

92. $f(x) = -\dfrac{3}{x + 1}$

93. $f(x) = \frac{1}{3}x^3 - 2$

94. $f(x) = \sqrt[3]{x} - 1$

95. $f(x) = \dfrac{x + 1}{x - 3}$

96. $f(x) = \dfrac{x - 1}{x + 2}$

97. Find $f(f^{-1}(5))$ and $f^{-1}(f(a))$:

$f(x) = x^3 - 4$.

98. Find $f^{-1}(f(p))$ and $f(f^{-1}(1253))$:

$$f(x) = \sqrt[5]{\dfrac{2x - 7}{3x + 4}}.$$

99. *Hitting Lessons.*   A summer little-league baseball team determines that the cost per player of a group hitting lesson is given by the formula

$$C(x) = \dfrac{72 + 2x}{x},$$

where x is the number of players in the group and $C(x)$ is in dollars.

a) Determine the cost per player of a group hitting lesson when there are 2, 5, and 8 players in the group.

b) Find a formula for the inverse of the function and explain what it represents.

c) Use the inverse function to determine the number of players in the group lesson when the cost per player is $74, $20, and $11.

100. *Women's Shoe Sizes.* A function that will convert women's shoe sizes in the United States to those in Australia is

$$s(x) = \frac{2x - 3}{2}$$

(*Source:* OnlineConversion.com).

a) Determine the women's shoe sizes in Australia that correspond to sizes 5, $7\frac{1}{2}$, and 8 in the United States.

b) Find a formula for the inverse of the function and explain what it represents.

c) Use the inverse function to determine the women's shoe sizes in the United States that correspond to sizes 3, $5\frac{1}{2}$, and 7 in Australia.

101. *E-Commerce Holiday Sales.* Retail e-commerce holiday season sales (November and December), in billions of dollars, x years after 2009 is given by the function

$$H(x) = 4.798\,x + 28.1778$$

(*Source:* statista.com).

a) Determine the total amount of e-commerce holiday sales in 2012 and in 2016.

b) Find $H^{-1}(x)$ and explain what it represents.

102. *Converting Temperatures.* The following formula can be used to convert Fahrenheit temperatures x to Celsius temperatures $T(x)$:

$$T(x) = \frac{5}{9}(x - 32).$$

a) Find $T(-13°)$ and $T(86°)$.

b) Find $T^{-1}(x)$ and explain what it represents.

› **Skill Maintenance**

Consider quadratic functions (a)–(h) that follow. Without graphing them, answer the questions below. [3.3]

a) $f(x) = 2x^2$
b) $f(x) = -x^2$
c) $f(x) = \frac{1}{4}x^2$
d) $f(x) = -5x^2 + 3$
e) $f(x) = \frac{2}{3}(x - 1)^2 - 3$
f) $f(x) = -2(x + 3)^2 + 1$
g) $f(x) = (x - 3)^2 + 1$
h) $f(x) = -4(x + 1)^2 - 3$

103. Which functions have a maximum value?

104. Which graphs open up?

105. Consider (a) and (c). Which graph is narrower?

106. Consider (d) and (e). Which graph is narrower?

107. Which graph has vertex $(-3, 1)$?

108. For which is the line of symmetry $x = 0$?

› **Synthesis**

Using only a graphing calculator, determine whether the functions are inverses of each other.

109. $f(x) = \sqrt[3]{\frac{x - 3.2}{1.4}}$, $g(x) = 1.4x^3 + 3.2$

110. $f(x) = \frac{2x - 5}{4x + 7}$, $g(x) = \frac{7x - 4}{5x + 2}$

111. The function $f(x) = x^2 - 3$ is not one-to-one. Restrict the domain of f so that its inverse is a function. Find the inverse and state the restriction on the domain of the inverse.

112. Consider the function f given by

$$f(x) = \begin{cases} x^3 + 2, & x \le -1, \\ x^2, & -1 < x < 1, \\ x + 1, & x \ge 1. \end{cases}$$

Does f have an inverse that is a function? Why or why not?

113. Find three examples of functions that are their own inverses; that is, $f = f^{-1}$.

114. Given the function $f(x) = ax + b$, $a \ne 0$, find the values of a and b for which $f^{-1}(x) = f(x)$.

5.2 Exponential Functions and Graphs

> Graph exponential equations and exponential functions.

> Solve applied problems involving exponential functions and their graphs.

We now turn our attention to the study of a set of functions that are very rich in application. Consider the following graphs. Each one illustrates an *exponential function*. In this section, we consider such functions and some important applications.

Skype Users Online at Same Time

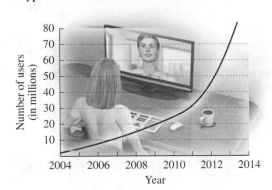

Source: Skype Numerology Blog

Postseason Bowl Games

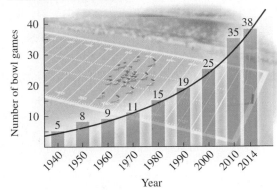

Source: USA TODAY research, College Football Data Warehouse

> Graphing Exponential Functions

We now define exponential functions. We assume that a^x has meaning for any real number x and any positive real number a and that the laws of exponents still hold, though we will not prove them here.

EXPONENTIAL FUNCTION

The function $f(x) = a^x$, where x is a real number, $a > 0$ and $a \neq 1$, is called the **exponential function, base a**.

We require the **base** to be positive in order to avoid the imaginary numbers that would occur by taking even roots of negative numbers: An example is $(-1)^{1/2}$, the square root of -1, which is not a real number. The restriction $a \neq 1$ is made to exclude the constant function $f(x) = 1^x = 1$, which does not have an inverse that is a function because it is not one-to-one.

The following are examples of exponential functions:

$$f(x) = 2^x, \qquad f(x) = \left(\tfrac{1}{2}\right)^x, \qquad f(x) = 3.57^{(x-2)}.$$

Note that, in contrast to functions like $f(x) = x^5$ and $f(x) = x^{1/2}$ in which the variable is the base of an exponential expression, the variable in an exponential function is *in the exponent*.

Let's now consider graphs of exponential functions.

Just in Time

7

EXAMPLE 1   Graph the exponential function

$$y = f(x) = 2^x.$$

Solution   We compute some function values and list the results in a table.

$$f(0) = 2^0 = 1; \qquad f(-1) = 2^{-1} = \frac{1}{2^1} = \frac{1}{2};$$
$$f(1) = 2^1 = 2;$$
$$f(2) = 2^2 = 4; \qquad f(-2) = 2^{-2} = \frac{1}{2^2} = \frac{1}{4};$$
$$f(3) = 2^3 = 8;$$
$$f(-3) = 2^{-3} = \frac{1}{2^3} = \frac{1}{8}.$$

x	$y = f(x) = 2^x$	(x, y)
0	1	$(0, 1)$
1	2	$(1, 2)$
2	4	$(2, 4)$
3	8	$(3, 8)$
-1	$\frac{1}{2}$	$(-1, \frac{1}{2})$
-2	$\frac{1}{4}$	$(-2, \frac{1}{4})$
-3	$\frac{1}{8}$	$(-3, \frac{1}{8})$

Next, we plot these points and connect them with a smooth curve. Be sure to plot enough points to determine how steeply the curve rises.

The curve comes very close to the *x*-axis, but does not touch or cross it.

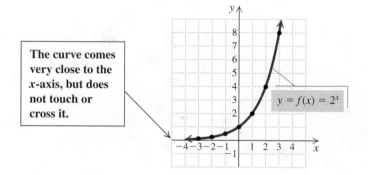

Note that as x increases, the function values increase without bound. As x decreases, the function values decrease, getting close to 0. That is, as $x \to -\infty$, $y \to 0$. Thus the *x*-axis, or the line $y = 0$, is a horizontal asymptote. As the *x*-inputs decrease, the curve gets closer and closer to this line, but does not cross it.

> **Now Try Exercise 11.**

HORIZONTAL ASYMPTOTES

REVIEW SECTION 4.5.

EXAMPLE 2   Graph the exponential function $y = f(x) = \left(\frac{1}{2}\right)^x$.

Solution   Before we plot points and draw the curve, note that

$$y = f(x) = \left(\frac{1}{2}\right)^x = (2^{-1})^x = 2^{-x}.$$

This tells us that this graph is a reflection of the graph of $y = 2^x$ across the *y*-axis. For example, if $(3, 8)$ is a point of the graph of $f(x) = 2^x$, then $(-3, 8)$ is a point of the graph of $f(x) = 2^{-x}$. Selected points are listed in the table at left.

Next, we plot points and connect them with a smooth curve.

Points of $f(x) = 2^x$	Points of $f(x) = \left(\frac{1}{2}\right)^x = 2^{-x}$
$(0, 1)$	$(0, 1)$
$(1, 2)$	$(-1, 2)$
$(2, 4)$	$(-2, 4)$
$(3, 8)$	$(-3, 8)$
$(-1, \frac{1}{2})$	$(1, \frac{1}{2})$
$(-2, \frac{1}{4})$	$(2, \frac{1}{4})$
$(-3, \frac{1}{8})$	$(3, \frac{1}{8})$

EXPLORING WITH TECHNOLOGY

Use the TRACE and TABLE features to confirm that the graphs of $y = f(x) = 2^x$ and $y = f(x) = \left(\frac{1}{2}\right)^x$ never cross the *x*-axis.

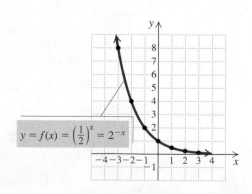

Note that as x increases, the function values decrease, getting close to 0. The x-axis, $y = 0$, is the horizontal asymptote. As x decreases, the function values increase without bound.

> **Now Try Exercise 15.**

Note the patterns in the following graphs of exponential functions.

CONNECTING THE CONCEPTS   **Properties of Exponential Functions**

Let's list and compare some characteristics of exponential functions, keeping in mind that the definition of an exponential function, $f(x) = a^x$, requires that a be positive and different from 1.

$f(x) = a^x, a > 0, a \neq 1$

Continuous

One-to-one

Domain: $(-\infty, \infty)$

Range: $(0, \infty)$

Increasing if $a > 1$

Decreasing if $0 < a < 1$

Horizontal asymptote is x-axis

y-intercept: $(0, 1)$

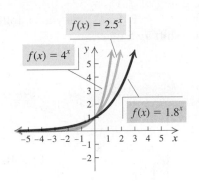

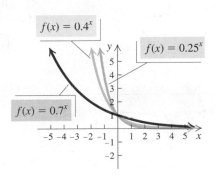

TRANSFORMATIONS

REVIEW SECTION 2.5.

To graph other types of exponential functions, keep in mind the ideas of translation, stretching, and reflection. All these concepts allow us to visualize the graph before drawing it.

EXAMPLE 3   Graph each of the following by hand. Before doing so, describe how each graph can be obtained from the graph of $f(x) = 2^x$. Then check your graph with a graphing calculator.

a) $f(x) = 2^{x-2}$ **b)** $f(x) = 2^x - 4$ **c)** $f(x) = 5 - 0.5^x$

Solution

a) The graph of $f(x) = 2^{x-2}$ is the graph of $y = 2^x$ shifted *right* 2 units.

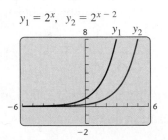

x	$f(x)$
-1	$\frac{1}{8}$
0	$\frac{1}{4}$
1	$\frac{1}{2}$
2	1
3	2
4	4
5	8

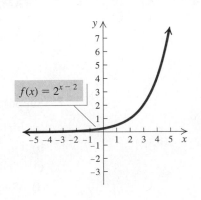

b) The graph of $f(x) = 2^x - 4$ is the graph of $y = 2^x$ shifted *down* 4 units.

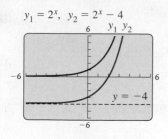

$y_1 = 2^x, \ y_2 = 2^x - 4$

x	$f(x)$
-2	$-3\frac{3}{4}$
-1	$-3\frac{1}{2}$
0	-3
1	-2
2	0
3	4

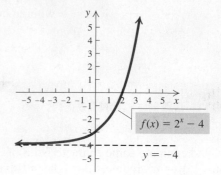

c) The graph of $f(x) = 5 - 0.5^x = 5 - \left(\frac{1}{2}\right)^x = 5 - 2^{-x}$ is a reflection of the graph of $y = 2^x$ across the y-axis, followed by a reflection across the x-axis and then a shift *up* 5 units.

$y_1 = 2^x, \ y_2 = 5 - 0.5^x$

x	$f(x)$
-3	-3
-2	1
-1	3
0	4
1	$4\frac{1}{2}$
2	$4\frac{3}{4}$

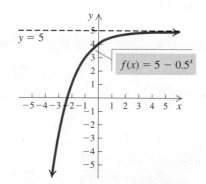

> *Now Try Exercises 27 and 33.*

❯ Applications

Graphing calculators are especially helpful when we are working with exponential functions. They not only facilitate computations but they also allow us to visualize the functions. One of the most frequent applications of exponential functions occurs with compound interest.

EXAMPLE 4 *Compound Interest.* The amount of money A to which a principal P will grow after t years at interest rate r (in decimal form), compounded n times per year, is given by the function

$$A(t) = P\left(1 + \frac{r}{n}\right)^{nt}.$$

Suppose that \$100,000 is invested at 6.5% interest, compounded semiannually.

a) Find a function for the amount to which the investment grows after t years.

b) Graph the function.

c) Find the amount of money in the account at $t = 0$, 4, 8, and 10 years.

d) When will the amount of money in the account reach \$400,000?

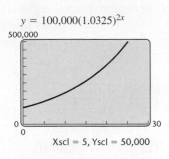

$y = 100,000(1.0325)^{2x}$

Xscl = 5, Yscl = 50,000

$y = 100,000(1.0325)^{2x}$

Y₁(0)	
	100000
Y₁(4)	
	129157.7535
Y₁(8)	
	166817.253

Solution

a) Since $P = \$100,000$, $r = 6.5\% = 0.065$, and $n = 2$, we can substitute these values and write the following function:

$$A(t) = 100,000\left(1 + \frac{0.065}{2}\right)^{2\cdot t} = \$100,000(1.0325)^{2t}.$$

b) For the graph shown at left, we use the viewing window $[0, 30, 0, 500,000]$ because of the large numbers and the fact that negative time values have no meaning in this application.

c) We can compute function values using function notation on the home screen of a graphing calculator. (See the window at left.) We can also calculate the values directly on a graphing calculator by substituting in the expression for $A(t)$:

$$A(0) = 100,000(1.0325)^{2\cdot 0} = \$100,000;$$
$$A(4) = 100,000(1.0325)^{2\cdot 4} \approx \$129,157.75;$$
$$A(8) = 100,000(1.0325)^{2\cdot 8} \approx \$166,817.25;$$
$$A(10) = 100,000(1.0325)^{2\cdot 10} \approx \$189,583.79.$$

d) To find the amount of time it takes for the account to grow to \$400,000, we set

$$100,000(1.0325)^{2t} = 400,000$$

and solve for t. One way we can do this is by graphing the equations

$$y_1 = 100,000(1.0325)^{2x} \quad \text{and} \quad y_2 = 400,000.$$

Then we can use the Intersect method to estimate the first coordinate of the point of intersection. (See Fig. 1 below.)

We can also use the Zero method to estimate the zero of the function $y = 100,000(1.0325)^{2x} - 400,000$. (See Fig. 2 below.)

Regardless of the method we use, we see that the account grows to \$400,000 after about 21.67 years, or about 21 years and 8 months.

INTERSECT METHOD

REVIEW SECTION 1.5.

ZERO METHOD

REVIEW SECTION 1.5.

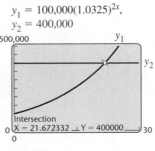

$y_1 = 100,000(1.0325)^{2x}$,
$y_2 = 400,000$

FIGURE 1.

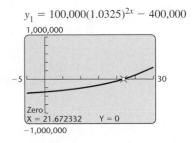

$y_1 = 100,000(1.0325)^{2x} - 400,000$

FIGURE 2.

> *Now Try Exercise 51.*

❯ The Number e

We now consider a very special number in mathematics. In 1741, Leonhard Euler named this number e. Though you may not have encountered it before, you will see here and in future mathematics courses that it has many important applications. To explain this number, we use the compound interest formula $A = P(1 + r/n)^{nt}$ discussed in Example 4. Suppose that \$1 is invested at 100% interest for 1 year. Since $P = 1$, $r = 100\% = 1$, and $t = 1$, the formula above becomes a function A defined in terms of the number of compounding periods n:

$$A = P\left(1 + \frac{r}{n}\right)^{nt} = 1\left(1 + \frac{1}{n}\right)^{n\cdot 1} = \left(1 + \frac{1}{n}\right)^{n}.$$

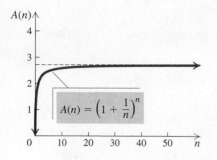

Let's use the graph of this function shown at left to explore the values of $A(n)$ as $n \to \infty$. Consider the graph for larger and larger values of n. Does this function have a horizontal asymptote?

Let's find some function values using a calculator.

n, Number of Compounding Periods	$A(n) = \left(1 + \dfrac{1}{n}\right)^n$
1 (compounded annually)	$2.00
2 (compounded semiannually)	2.25
3	2.3704
4 (compounded quarterly)	2.4414
5	2.4883
100	2.7048
365 (compounded daily)	2.7146
8760 (compounded hourly)	2.7181

It appears from these values that the graph does have a horizontal asymptote, $y \approx 2.7$. As the values of n get larger and larger, the function values get closer and closer to the number Euler named e. Its decimal representation does not terminate or repeat; it is irrational.

$$e = 2.7182818284\ldots$$

EXAMPLE 5 Find each value of e^x, to four decimal places, using the $\boxed{e^x}$ key on a calculator.

a) e^3 **b)** $e^{-0.23}$ **c)** e^0

Solution

FUNCTION VALUE	READOUT	ROUNDED
a) e^3	e³ 20.08553692	20.0855
b) $e^{-0.23}$	e⁻.²³ .7945336025	0.7945
c) e^0	e⁰ 1	1

> **Now Try Exercises 1 and 3.**

〉 Graphs of Exponential Functions, Base e

We demonstrate ways in which to graph exponential functions.

EXAMPLE 6 Graph $f(x) = e^x$ and $g(x) = e^{-x}$.

Solution We can compute points for each equation using a table or the $\boxed{e^x}$ key on a calculator. Then we plot these points and draw the graphs of the functions.

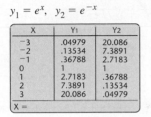

$y_1 = e^x, \quad y_2 = e^{-x}$

X	Y₁	Y₂
−3	.04979	20.086
−2	.13534	7.3891
−1	.36788	2.7183
0	1	1
1	2.7183	.36788
2	7.3891	.13534
3	20.086	.04979
X =		

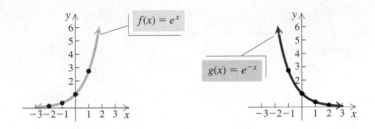

Note that the graphs are reflections of each other across the *y*-axis.

〉 *Now Try Exercise 23.*

EXAMPLE 7 Graph each of the following by hand. Before doing so, describe how each graph can be obtained from the graph of $y = e^x$.

a) $f(x) = e^{x+3}$ **b)** $f(x) = e^{-0.5x}$ **c)** $f(x) = 1 - e^{-2x}$

Solution

a) The graph of $f(x) = e^{x+3}$ is a translation of the graph of $y = e^x$ left 3 units.

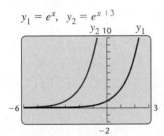

$y_1 = e^x, \quad y_2 = e^{x+3}$

x	f(x)
−7	0.018
−5	0.135
−3	1
−1	7.389
0	20.086

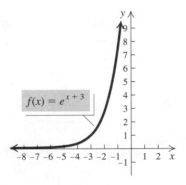

b) We note that the graph of $f(x) = e^{-0.5x}$ is a horizontal stretching of the graph of $y = e^x$ followed by a reflection across the *y*-axis.

$y_1 = e^x, \quad y_2 = e^{-0.5x}$

x	f(x)
−2	2.718
−1	1.649
0	1
1	0.607
2	0.368

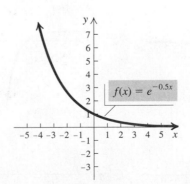

c) The graph of $f(x) = 1 - e^{-2x}$ is a horizontal shrinking of the graph of $y = e^x$ followed by a reflection across the y-axis, then across the x-axis, followed by a translation up 1 unit.

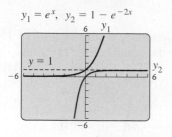

$y_1 = e^x, \ y_2 = 1 - e^{-2x}$

x	$f(x)$
-1	-6.389
0	0
1	0.865
2	0.982
3	0.998

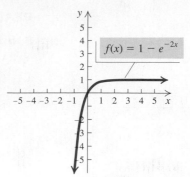

$f(x) = 1 - e^{-2x}$

Now Try Exercises 41 and 47.

5.2   Exercise Set

Find each of the following, to four decimal places, using a calculator.

1. e^4

2. e^{10}

3. $e^{-2.458}$

4. $\left(\dfrac{1}{e^3}\right)^2$

In Exercises 5–10, match the function with one of the graphs (a)–(f) that follow.

a)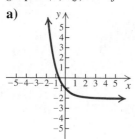

b)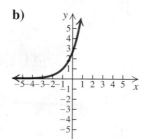

c)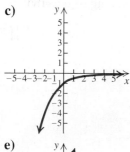

d)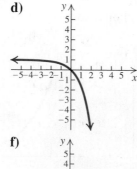

e)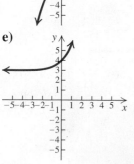

f)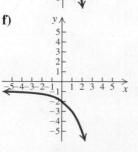

5. $f(x) = -2^x - 1$

6. $f(x) = -\left(\frac{1}{2}\right)^x$

7. $f(x) = e^x + 3$

8. $f(x) = e^{x+1}$

9. $f(x) = 3^{-x} - 2$

10. $f(x) = 1 - e^x$

Graph the function by substituting and plotting points. Then check your work using a graphing calculator.

11. $f(x) = 3^x$

12. $f(x) = 5^x$

13. $f(x) = 6^x$

14. $f(x) = 3^{-x}$

15. $f(x) = \left(\frac{1}{4}\right)^x$

16. $f(x) = \left(\frac{2}{3}\right)^x$

17. $y = -2^x$

18. $y = 3 - 3^x$

19. $f(x) = -0.25^x + 4$

20. $f(x) = 0.6^x - 3$

21. $f(x) = 1 + e^{-x}$

22. $f(x) = 2 - e^{-x}$

23. $y = \frac{1}{4}e^x$

24. $y = 2e^{-x}$

25. $f(x) = 1 - e^{-x}$

26. $f(x) = e^x - 2$

Sketch the graph of the function and check the graph with a graphing calculator. Describe how each graph can be obtained from the graph of a basic exponential function.

27. $f(x) = 2^{x+1}$

28. $f(x) = 2^{x-1}$

29. $f(x) = 2^x - 3$

30. $f(x) = 2^x + 1$

31. $f(x) = 2^{1-x} + 2$

32. $f(x) = 5 - 2^{-x}$

33. $f(x) = 4 - 3^{-x}$

34. $f(x) = 2^{x-1} - 3$

35. $f(x) = \left(\frac{3}{2}\right)^{x-1}$

36. $f(x) = 3^{4-x}$

37. $f(x) = 2^{x+3} - 5$

38. $f(x) = -3^{x-2}$

39. $f(x) = 3 \cdot 2^{x-1} + 1$

40. $f(x) = 2 \cdot 3^{x+1} - 2$

41. $f(x) = e^{2x}$

42. $f(x) = e^{-0.2x}$

43. $f(x) = \frac{1}{2}(1 - e^x)$

44. $f(x) = 3(1 + e^x) - 2$

45. $y = e^{-x+1}$ **46.** $y = e^{2x} + 1$

47. $f(x) = 2(1 - e^{-x})$ **48.** $f(x) = 1 - e^{-0.01x}$

Graph the piecewise function.

49. $f(x) = \begin{cases} e^{-x} - 4, & \text{for } x < -2, \\ x + 3, & \text{for } -2 \le x < 1, \\ x^2, & \text{for } x \ge 1 \end{cases}$

50. $g(x) = \begin{cases} 4, & \text{for } x \le -3, \\ x^2 - 6, & \text{for } -3 < x < 0, \\ e^x, & \text{for } x \ge 0 \end{cases}$

51. *Compound Interest.* Suppose that $82,000 is invested at $4\frac{1}{2}\%$ interest, compounded quarterly.

   a) Find the function for the amount to which the investment grows after t years.

   b) Graph the function.

   c) Find the amount of money in the account at $t = 0, 2,$ 5, and 10 years.

   d) When will the amount of money in the account reach $100,000?

52. *Compound Interest.* Suppose that $750 is invested at 7% interest, compounded semiannually.

   a) Find the function for the amount to which the investment grows after t years.

   b) Graph the function.

   c) Find the amount of money in the account at $t = 1, 6,$ 10, 15, and 25 years.

   d) When will the amount of money in the account reach $3000?

53. *Interest on a CD.* On Elizabeth's sixth birthday, her grandparents present her with a $3000 certificate of deposit (CD) that earns 5% interest, compounded quarterly. If the CD matures on her sixteenth birthday, what amount will be available then?

54. *Interest in a College Trust Fund.* Following the birth of his child, Benjamin deposits $10,000 in a college trust fund where interest is 3.9%, compounded semiannually.

   a) Find a function for the amount in the account after t years.

   b) Find the amount of money in the account at $t = 0, 4,$ 8, 10, 18, and 21 years.

In Exercises 55–62, use the compound-interest formula to find the account balance A with the given conditions:

   $P = $ principal,
   $r = $ interest rate,
   $n = $ number of compounding periods per year,
   $t = $ time, in years,
   $A = $ account balance.

	P	r	Compounded	n	t	A
55.	$3,000	4%	Semiannually		2	
56.	$12,500	3%	Quarterly		3	
57.	$120,000	2.5%	Annually		10	
58.	$120,000	2.5%	Quarterly		10	
59.	$53,500	$5\frac{1}{2}\%$	Quarterly		$6\frac{1}{2}$	
60.	$6,250	$6\frac{3}{4}\%$	Semiannually		$4\frac{1}{2}$	
61.	$17,400	8.1%	Daily		5	
62.	$900	7.3%	Daily		$7\frac{1}{4}$	

63. *Alternative-Fuel Vehicles.* The sales of alternative-fuel vehicles have more than tripled since 1995 (*Source:* Energy Information Administration). The exponential function

$$A(x) = 246{,}855(1.0931)^x,$$

where x is the number of years after 1995, can be used to estimate the number of alternative-fuel vehicles sold in a given year. Find the number of alternative-fuel vehicles sold in 2000 and in 2013. Then project the number of alternative-fuel vehicles sold in 2018.

64. *Increasing CPU Power.* The central processing unit (CPU) power in computers has increased significantly over the years. The CPU power in Macintosh computers has grown exponentially from 8 MHz in 1984 to 3400 MHz in 2013 (*Source:* Apple). The exponential function

$$M(t) = 7.91477(1.26698)^t,$$

where t is the number of years after 1984, can be used to estimate the CPU power in a Macintosh computer in a given year. Find the CPU power of a Macintosh Performa 5320CD in 1995 and of an iMac G6 in 2009. Round to the nearest one MHz.

65. *Alfalfa Imported by China.* The amount and quality of milk produced increases when dairy cows are fed high-quality alfalfa. Although annual milk consumption per capita in China has been one of the lowest in the world, it has increased greatly in recent years. Consequently, the demand for imported alfalfa has increased exponentially. In 2008, approximately 20,000 tons of U.S. alfalfa were imported by China. This number increased to approximately 650,000 tons by 2013. (*Sources: National Geographic*, May 2015, Arjen Hoekstra, University of Twente; USDA Economic Research Service; Shefali Sharma and Zhang Rou, Institute for Agriculture and Trade Policy; FAO; Ministry of Agriculture, People's Republic of China) The increase in imported U.S. alfalfa can be modeled by the exponential function

$$A(x) = 22{,}611.008(1.992)^x,$$

where x is the number of years after 2008. Find the number of tons of U.S. alfalfa imported by China in 2012 and in 2016.

66. *U.S. Imports.* The amount of imports to the United States has increased exponentially since 1980 (*Sources*: U.S. Census Bureau; U.S. Bureau of Economic Analysis; U.S. Department of Commerce). The exponential function

$$I(x) = 297.539(1.075)^x,$$

where x is the number of years after 1980, can be used to estimate the total amount of U.S. imports, in billions of dollars. Find the total amount of imports to the United States in 1995, in 2005, in 2010, and in 2013. Round to the nearest billion dollars.

67. *E-Cigarette Sales.* The electronic cigarette was launched in 2007, and since then sales have increased from about $20 million in 2008 to about $500 million

in 2012 (*Sources*: UBS; forbes.com). The exponential function

$$S(x) = 20.913(2.236)^x,$$

where x is the number of years after 2008, models the sales, in millions of dollars. Use this function to estimate the sales of e-cigarettes in 2011 and in 2015. Round to the nearest million dollars.

68. *Earthquakes in Oklahoma.* The number of earthquakes in Oklahoma has increased dramatically since 2008. Researchers are noting that as the number of wastewater wells that have resulted from oil and gas operations increases, the number of earthquakes also increases (*Sources*: U.S. Geological Survey; Oklahoma Geological Survey). The number of earthquakes with a 3.0 or greater magnitude in Oklahoma can be modeled by the exponential function

$$E(x) = 20.279(1.785)^x,$$

where x is the number of years after 2009. Find the number of earthquakes with a magnitude greater than 3.0 in 2010, in 2013, and in 2016.

69. *Centenarian Population.* The centenarian population in the United States has grown over 65% in the last 30 years. In 1980, there were only 32,194 residents ages 100 and over. This number had grown to 53,364 by 2010. (*Sources*: Population Projections Program; U.S. Census Bureau; U.S. Department of Commerce; "What People Who Live to 100 Have in Common," by Emily Brandon, *U.S. News and World Report*, January 7, 2013) The exponential function

$$H(t) = 80{,}040.68(1.0481)^t,$$

where t is the number of years after 2015, can be used to project the number of centenarians. Use this function to project the centenarian population in 2020 and in 2050.

70. *Bachelor's Degrees Earned.*   The exponential function

$$D(t) = 347(1.024)^t$$

gives the number of bachelor's degrees, in thousands, earned in the United States t years after 1970 (*Sources*: National Center for Educational Statistics; U.S. Department of Education). Find the number of bachelor's degrees earned in 1985, in 2000, and in 2014. Then estimate the number of bachelor's degrees that will be earned in 2020. Round to the nearest thousand degrees.

71. *Charitable Giving.*   Over the last four decades, the amount of charitable giving in the United States has grown exponentially from approximately $20.7 billion in 1969 to approximately $316.2 billion in 2012 (*Sources*: Giving USA Foundation; Volunteering in America by the Corporation for National & Community Service; National Philanthropic Trust; School of Philanthropy, Indiana University Purdue University Indianapolis). The exponential function

$$G(x) = 20.7(1.066)^x,$$

where x is the number of years after 1969, can be used to estimate the amount of charitable giving, in billions of dollars, in a given year. Find the amount of charitable giving in 1982, in 1995, and in 2010. Then use this function to estimate the amount of charitable giving in 2017. Round to the nearest billion dollars.

72. *Price of Admission to the Magic Kingdom.*   In 2015, the price of a one-day, one-park admission to Disney's Magic Kingdom in Florida rose to $105. The exponential function

$$D(x) = 4.532(1.078)^x,$$

where x is the number of years after 1971, models the price of a ticket. (*Source*: AllEars.net, an independent Disney consumer website) Find the price of a ticket in 1980, in 2000, and in 2012. Then use the function to project the price of a ticket in 2020.

73. *Salvage Value.*   A restaurant purchased a 72-in. range with six burners for $6982. The value of the range each year is 85% of the value of the preceding year. After t years, its value, in dollars, is given by the exponential function

$$V(t) = 6982(0.85)^t.$$

a) Graph the function.
b) Find the value of the range after 0, 1, 2, 5, and 8 years.
c) The restaurant decides to replace the range when its value has declined to $1000. After how long will the range be replaced?

74. *Salvage Value.*   A landscape company purchased a backhoe for $56,395. The value of the backhoe each year is 90% of the value of the preceding year. After t years, its value, in dollars, is given by the exponential function

$$V(t) = 56,395(0.9)^t.$$

a) Graph the function.
b) Find the value of the backhoe after 0, 1, 3, 6, and 10 years. Round to the nearest dollar.

75. *Advertising.*   A company begins an Internet advertising campaign to market a new telephone. The percentage of the target market that buys a product is generally a function of the length of the advertising campaign. The estimated percentage is given by

$$f(t) = 100(1 - e^{-0.04t}),$$

where t is the number of days of the campaign.

a) Graph the function.
b) Find $f(25)$, the percentage of the target market that has bought the phone after a 25-day advertising campaign.
c) After how long will 90% of the target market have bought the phone?

76. *Growth of a Stock.*   The value of a stock is given by the function

$$V(t) = 58(1 - e^{-1.1t}) + 20,$$

where V is the value of the stock after time t, in months.

a) Graph the function.
b) Find $V(1)$, $V(2)$, $V(4)$, $V(6)$, and $V(12)$.
c) After how long will the value of the stock be $75?

In Exercises 77–90, use a graphing calculator to match the equation with one of the figures (a)–(n) that follow.

a)
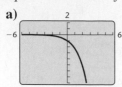

b)
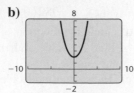

c)
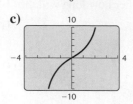

d)
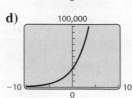

e)
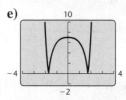

f)
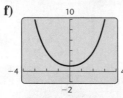

g)
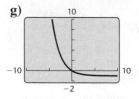

h)
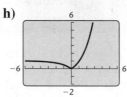

i)
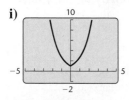

j)
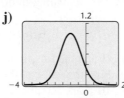

k)
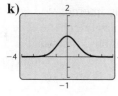

l)
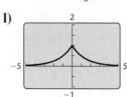

m)
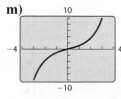

n)
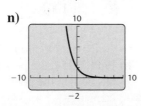

77. $y = 3^x - 3^{-x}$

78. $y = 3^{-(x+1)^2}$

79. $f(x) = -2.3^x$

80. $f(x) = 30{,}000(1.4)^x$

81. $y = 2^{-|x|}$

82. $y = 2^{-(x-1)}$

83. $f(x) = (0.58)^x - 1$

84. $y = 2^x + 2^{-x}$

85. $g(x) = e^{|x|}$

86. $f(x) = |2^x - 1|$

87. $y = 2^{-x^2}$

88. $y = |2^{x^2} - 8|$

89. $g(x) = \dfrac{e^x - e^{-x}}{2}$

90. $f(x) = \dfrac{e^x + e^{-x}}{2}$

Use a graphing calculator to find the point(s) of intersection of the graphs of each of the following pairs of equations.

91. $y = |1 - 3^x|, \quad y = 4 + 3^{-x^2}$

92. $y = 4^x + 4^{-x}, \quad y = 8 - 2x - x^2$

93. $y = 2e^x - 3, \quad y = \dfrac{e^x}{x}$

94. $y = \dfrac{1}{e^x + 1}, \quad y = 0.3x + \dfrac{7}{9}$

Solve graphically.

95. $5.3^x - 4.2^x = 1073$

96. $e^x = x^3$

97. $2^x > 1$

98. $3^x \leq 1$

99. $2^x + 3^x = x^2 + x^3$

100. $31{,}245e^{-3x} = 523{,}467$

❯ Skill Maintenance

Simplify. **[3.1]**

101. $(1 - 4i)(7 + 6i)$

102. $\dfrac{2 - i}{3 + i}$

Find the x-intercepts and the zeros of the function.

103. $f(x) = 2x^2 - 13x - 7$ **[3.2]**

104. $h(x) = x^3 - 3x^2 + 3x - 1$ **[4.3]**

105. $h(x) = x^4 - x^2$ **[4.1]**

106. $g(x) = x^3 + x^2 - 12x$ **[4.1]**

Solve.

107. $x^3 + 6x^2 - 16x = 0$ **[4.1]**

108. $3x^2 - 6 = 5x$ **[3.2]**

❯ Synthesis

109. Which is larger, 7^π or π^7? 70^{80} or 80^{70}?

In Exercises 110 and 111:

a) *Graph using a graphing calculator.*

b) *Approximate the zeros.*

c) *Approximate the relative maximum and minimum values. If your graphing calculator has a* MAX–MIN *feature, use it.*

110. $f(x) = x^2 e^{-x}$

111. $f(x) = e^{-x^2}$

112. Graph $f(x) = x^{1/(x-1)}$ for $x > 0$. Use a graphing calculator and the TABLE feature to identify the horizontal asymptote.

113. For the function f, construct and simplify the difference quotient. (See Section 2.2 for review.)

$$f(x) = 2e^x - 3$$

5.3

Logarithmic Functions and Graphs

❯ Find common logarithms and natural logarithms with and without a calculator.

❯ Convert between exponential equations and logarithmic equations.

❯ Change logarithm bases.

❯ Graph logarithmic functions.

❯ Solve applied problems involving logarithmic functions.

We now consider *logarithmic*, or *logarithm*, *functions*. These functions are inverses of exponential functions and have many applications.

❯ Logarithmic Functions

We have noted that every exponential function (with $a > 0$ and $a \neq 1$) is one-to-one. Thus such a function has an inverse that is a function. In this section, we will study these inverse functions, called logarithmic functions, and use them in applications. We can draw the graph of the inverse of an exponential function by interchanging x and y.

EXAMPLE 1 Graph: $x = 2^y$.

Solution Note that x is alone on one side of the equation. We can find ordered pairs that are solutions by choosing values for y and then computing the corresponding x-values.

For $y = 0$, $x = 2^0 = 1$.

For $y = 1$, $x = 2^1 = 2$.

For $y = 2$, $x = 2^2 = 4$.

For $y = 3$, $x = 2^3 = 8$.

For $y = -1$, $x = 2^{-1} = \dfrac{1}{2^1} = \dfrac{1}{2}$.

For $y = -2$, $x = 2^{-2} = \dfrac{1}{2^2} = \dfrac{1}{4}$.

For $y = -3$, $x = 2^{-3} = \dfrac{1}{2^3} = \dfrac{1}{8}$.

x		
$x = 2^y$	y	(x, y)
1	0	$(1, 0)$
2	1	$(2, 1)$
4	2	$(4, 2)$
8	3	$(8, 3)$
$\dfrac{1}{2}$	-1	$\left(\dfrac{1}{2}, -1\right)$
$\dfrac{1}{4}$	-2	$\left(\dfrac{1}{4}, -2\right)$
$\dfrac{1}{8}$	-3	$\left(\dfrac{1}{8}, -3\right)$

↑ **(1) Choose values for y.**

↑ **(2) Compute values for x.**

We plot the points and connect them with a smooth curve. Note that the curve does not touch or cross the y-axis. The y-axis is a vertical asymptote.

Note too that this curve is the graph of $y = 2^x$ reflected across the line $y = x$ as we would expect for an inverse. The inverse of $y = 2^x$ is $x = 2^y$.

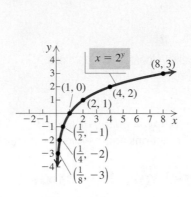
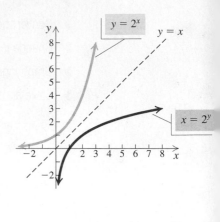

Now Try Exercise 1.

To find a formula for f^{-1} when $f(x) = 2^x$, we use the method discussed in Section 5.1:

1. Replace $f(x)$ with y: $y = 2^x$

2. Interchange x and y: $x = 2^y$

3. Solve for y: $y = $ the power to which we raise 2 to get x

4. Replace y with $f^{-1}(x)$: $f^{-1}(x) = $ the power to which we raise 2 to get x.

Mathematicians have defined a new symbol to replace the words "the power to which we raise 2 to get x." That symbol is "$\log_2 x$," read "the logarithm, base 2, of x."

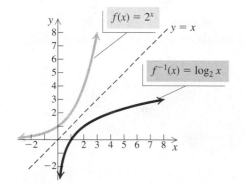

LOGARITHMIC FUNCTION, BASE 2

"$\log_2 x$," read "the logarithm, base 2, of x," means "the power to which we raise 2 to get x."

Thus if $f(x) = 2^x$, then $f^{-1}(x) = \log_2 x$. For example,

$$f^{-1}(8) = \log_2 8 = 3,$$

because

3 is the power to which we raise 2 to get 8.

Similarly, $\log_2 13$ is the power to which we raise 2 to get 13. As yet, we have no simpler way to say this other than

"$\log_2 13$ is the power to which we raise 2 to get 13."

Later, however, we will learn how to approximate this expression using a calculator.

For any exponential function $f(x) = a^x$, its inverse is called a **logarithmic function, base a**. The graph of the inverse can be obtained by reflecting the graph of $y = a^x$ across the line $y = x$, to obtain $x = a^y$. Then $x = a^y$ is equivalent to $y = \log_a x$. We read $\log_a x$ as "the logarithm, base a, of x."

The inverse of $f(x) = a^x$ is given by $f^{-1}(x) = \log_a x$.

> **LOGARITHMIC FUNCTION, BASE *a***
>
> We define $y = \log_a x$ as that number y such that $x = a^y$, where $x > 0$ and a is a positive constant other than 1.

Let's look at the graphs of $f(x) = a^x$ and $f^{-1}(x) = \log_a x$ for $a > 1$ and for $0 < a < 1$.

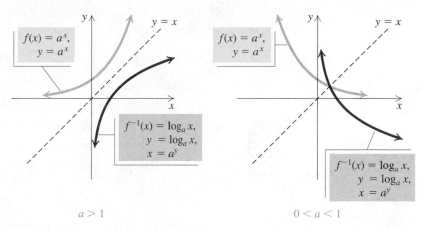

Note that the graphs of $f(x)$ and $f^{-1}(x)$ are reflections of each other across the line $y = x$.

CONNECTING THE CONCEPTS

Comparing Exponential Functions and Logarithmic Functions

In the following table, we compare exponential functions and logarithmic functions with bases *a* greater than 1. Similar statements could be made for *a*, where $0 < a < 1$. It is helpful to visualize the differences by carefully observing the graphs.

EXPONENTIAL FUNCTION

$y = a^x$
$f(x) = a^x$
$a > 1$
Continuous
One-to-one
Domain: All real
  numbers, $(-\infty, \infty)$
Range: All positive
  real numbers, $(0, \infty)$
Increasing
Horizontal asymptote is *x*-axis:
  $(a^x \to 0$ as $x \to -\infty)$
y-intercept: $(0, 1)$
There is no *x*-intercept.

LOGARITHMIC FUNCTION

$x = a^y$
$f^{-1}(x) = \log_a x$
$a > 1$
Continuous
One-to-one
Domain: All positive
  real numbers, $(0, \infty)$
Range: All real
  numbers, $(-\infty, \infty)$
Increasing
Vertical asymptote is *y*-axis:
  $(\log_a x \to -\infty$ as $x \to 0^+)$
x-intercept: $(1, 0)$
There is no *y*-intercept.

❭ Finding Certain Logarithms

Let's use the definition of logarithms to find some logarithmic values.

EXAMPLE 2   Find each of the following logarithms.

a) $\log_{10} 10{,}000$ **b)** $\log_{10} 0.01$ **c)** $\log_5 125$

d) $\log_9 3$ **e)** $\log_6 1$ **f)** $\log_5 5$

Solution

a) The exponent to which we raise 10 to obtain 10,000 is 4; thus, we have $\log_{10} 10{,}000 = 4$.

b) We have $0.01 = \dfrac{1}{100} = \dfrac{1}{10^2} = 10^{-2}$. The exponent to which we raise 10 to get 0.01 is -2, so $\log_{10} 0.01 = -2$.

c) $125 = 5^3$. The exponent to which we raise 5 to get 125 is 3, so $\log_5 125 = 3$.

d) $3 = \sqrt{9} = 9^{1/2}$. The exponent to which we raise 9 to get 3 is $\frac{1}{2}$, so $\log_9 3 = \frac{1}{2}$.

e) $1 = 6^0$. The exponent to which we raise 6 to get 1 is 0, so $\log_6 1 = 0$.

f) $5 = 5^1$. The exponent to which we raise 5 to get 5 is 1, so $\log_5 5 = 1$.

> ❭ *Now Try Exercises 9 and 15.*

Examples 2(e) and 2(f) illustrate two important properties of logarithms. The property $\log_a 1 = 0$ follows from the fact that $a^0 = 1$. Thus, $\log_5 1 = 0$, $\log_{10} 1 = 0$, and so on. The property $\log_a a = 1$ follows from the fact that $a^1 = a$. Thus, $\log_5 5 = 1$, $\log_{10} 10 = 1$, and so on.

$$\log_a 1 = 0 \quad \text{and} \quad \log_a a = 1, \quad \text{for any logarithmic base } a.$$

❭ Converting Between Exponential Equations and Logarithmic Equations

It is helpful in dealing with logarithmic functions to remember that a logarithm of a number is an *exponent*. It is the exponent y in $x = a^y$. You might think to yourself, "the logarithm, base a, of a number x is the power to which a must be raised to get x."

We are led to the following. (The symbol $\longleftrightarrow$ means that the two statements are equivalent; that is, when one is true, the other is true. The words "if and only if" can be used in place of $\longleftrightarrow$.)

$$\log_a x = y \longleftrightarrow x = a^y \qquad \text{A logarithm is an exponent!}$$

EXAMPLE 3   Convert each of the following to a logarithmic equation.

a) $16 = 2^x$ **b)** $10^{-3} = 0.001$ **c)** $e^t = 70$

Solution

The exponent is the logarithm.

a) $16 = 2^x \qquad \log_2 16 = x$

The base remains the same.

b) $10^{-3} = 0.001 \rightarrow \log_{10} 0.001 = -3$

c) $e^t = 70 \rightarrow \log_e 70 = t$

> ❭ *Now Try Exercise 37.*

EXAMPLE 4   Convert each of the following to an exponential equation.

a) $\log_2 32 = 5$ **b)** $\log_a Q = 8$ **c)** $x = \log_t M$

Solution

The logarithm is the exponent.

a) $\log_2 32 = 5 \qquad 2^5 = 32$

The base remains the same.

b) $\log_a Q = 8 \rightarrow a^8 = Q$

c) $x = \log_t M \rightarrow t^x = M$

> **Now Try Exercise 45.**

❯ Finding Logarithms on a Calculator

Before calculators became so widely available, base-10 logarithms, or **common logarithms**, were used extensively to simplify complicated calculations. In fact, that is why logarithms were invented. The abbreviation **log**, with no base written, is used to represent common logarithms, or base-10 logarithms. Thus,

$\log x$ means $\log_{10} x.$

For example, $\log 29$ means $\log_{10} 29$. Let's compare $\log 29$ with $\log 10$ and $\log 100$:

$$\left.\begin{array}{l} \log 10 = \log_{10} 10 = 1 \\ \log 29 = ? \\ \log 100 = \log_{10} 100 = 2 \end{array}\right\}$$ Since 29 is between 10 and 100, it seems reasonable that $\log 29$ is between 1 and 2.

On a calculator, the key for common logarithms is generally marked **LOG**. Using that key, we find that

$$\log 29 \approx 1.462397998 \approx 1.4624$$

rounded to four decimal places. Since $1 < 1.4624 < 2$, our answer seems reasonable. This also tells us that $10^{1.4624} \approx 29$.

EXAMPLE 5   Find each of the following common logarithms on a calculator. If you are using a graphing calculator, set the calculator in REAL mode. Round to four decimal places.

a) $\log 645,778$ **b)** $\log 0.0000239$ **c)** $\log (-3)$

Solution

FUNCTION VALUE	READOUT	ROUNDED
a) $\log 645,778$	log(645778) 5.810083246	5.8101
b) $\log 0.0000239$	log(0.0000239) −4.621602099	−4.6216
c) $\log (-3)$	ERR:NONREAL ANS **1:** Quit 2: Goto *	Does not exist as a real number

---

*If the graphing calculator is set in $a + bi$ mode, the readout is $.4771212547 + 1.364376354i$.

log(645778)
　　　　　　5.810083246
10^Ans
　　　　　　　645778

A check for part (a) is shown at left. Since 5.810083246 is the power to which we raise 10 to get 645,778, we can check part (a) by finding $10^{5.810083246}$. We can check part (b) in a similar manner. In part (c), log (-3) does not exist as a real number because there is no real-number power to which we can raise 10 to get -3. The number 10 raised to any real-number power is positive. The common logarithm of a negative number does not exist as a real number. Recall that logarithmic functions are inverses of exponential functions, and since the range of an exponential function is $(0, \infty)$, the domain of $f(x) = \log_a x$ is $(0, \infty)$.

> *Now Try Exercises 57 and 61.*

❯ Natural Logarithms

Logarithms, base e, are called **natural logarithms**. The abbreviation "ln" is generally used for natural logarithms. Thus,

ln x　means　$\log_e x$.

For example, ln 53 means $\log_e 53$. On a calculator, the key for natural logarithms is generally marked **LN**. Using that key, we find that

$$\ln 53 \approx 3.970291914$$
$$\approx 3.9703$$

rounded to four decimal places. This also tells us that $e^{3.9703} \approx 53$.

EXAMPLE 6　Find each of the following natural logarithms on a calculator. If you are using a graphing calculator, set the calculator in REAL mode. Round to four decimal places.

a) ln 645,778　　　　b) ln 0.0000239　　　　c) ln (-5)

d) ln e　　　　　　　e) ln 1

Solution

FUNCTION VALUE	READOUT	ROUNDED
a) ln 645,778	ln(645778)　　　13.37821107	13.3782
b) ln 0.0000239	ln(0.0000239)　　　−10.6416321	−10.6416
c) ln (-5)	ERR:NONREAL ANS *　**1:** Quit　2: Goto	Does not exist as a real number
d) ln e	ln(e)　　　1	1
e) ln 1	ln(1)　　　0	0

---

*If the graphing calculator is set in $a + bi$ mode, the readout is $1.609437912 + 3.141592654i$.

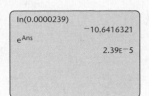

Since 13.37821107 is the power to which we raise e to get 645,778, we can check part (a) by finding $e^{13.37821107}$. We can check parts (b), (d), and (e) in a similar manner. A check for part (b) is shown at left. In parts (d) and (e), note that $\ln e = \log_e e = 1$ and $\ln 1 = \log_e 1 = 0$.

> **Now Try Exercises 65 and 67.**

$\ln 1 = 0$ and $\ln e = 1$, for the logarithmic base e.

❯ Changing Logarithmic Bases

Most calculators give the values of both common logarithms and natural logarithms. To find a logarithm with a base other than 10 or e, we can use the following conversion formula.

THE CHANGE-OF-BASE FORMULA

For any logarithmic bases a and b, and any positive number M,

$$\log_b M = \frac{\log_a M}{\log_a b}.$$

With some calculators, it is possible to find a logarithm with any logarithmic base using the logBASE operation from the MATH MATH menu. The computation of $\log_4 15$ is shown in the window below.

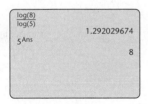

We will prove this result in the next section.

EXAMPLE 7 Find $\log_5 8$ using common logarithms.

Solution First, we let $a = 10$, $b = 5$, and $M = 8$. Then we substitute into the change-of-base formula:

$$\log_5 8 = \frac{\log_{10} 8}{\log_{10} 5} \qquad \textbf{Substituting}$$

$$\approx 1.2920. \qquad \textbf{Using a calculator}$$

Since $\log_5 8$ is the power to which we raise 5 to get 8, we would expect this power to be greater than 1 ($5^1 = 5$) and less than 2 ($5^2 = 25$), so the result is reasonable. The check is shown in the window at left.

> **Now Try Exercise 69.**

We can also use base e for a conversion.

EXAMPLE 8 Find $\log_5 8$ using natural logarithms.

Solution Substituting e for a, 5 for b, and 8 for M, we have

$$\log_5 8 = \frac{\log_e 8}{\log_e 5}$$

$$= \frac{\ln 8}{\ln 5} \approx 1.2920.$$

The check is shown at left. Note that we get the same value using base e for the conversion that we did using base 10 in Example 7.

> **Now Try Exercise 75.**

❯ Graphs of Logarithmic Functions

We demonstrate several ways to graph logarithmic functions.

EXAMPLE 9   Graph: $y = f(x) = \log_5 x$.

Solution   The equation $y = \log_5 x$ is equivalent to $x = 5^y$. We can find ordered pairs that are solutions by choosing values for y and computing the corresponding x-values. We then plot points, remembering that x is still the first coordinate.

To use a graphing calculator to graph $y = \log_5 x$ in Example 9, we first change the base or use the logBASE operation. Here we change from base 5 to base e:

$$y = \log_5 x = \frac{\ln x}{\ln 5}.$$

$$y = \log_5 x = \frac{\ln x}{\ln 5} = \frac{\ln (x)}{\ln (5)}$$

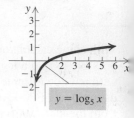

For $y = 0$, $x = 5^0 = 1$.
For $y = 1$, $x = 5^1 = 5$.
For $y = 2$, $x = 5^2 = 25$.
For $y = 3$, $x = 5^3 = 125$.
For $y = -1$, $x = 5^{-1} = \dfrac{1}{5}$.
For $y = -2$, $x = 5^{-2} = \dfrac{1}{25}$.

x, or 5^y	y
1	0
5	1
25	2
125	3
$\dfrac{1}{5}$	-1
$\dfrac{1}{25}$	-2

↑ (1) Select y.
↑ (2) Compute x.

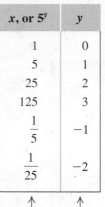

Some graphing calculators have a feature that graphs inverses automatically. If we begin with $y_1 = 5^x$, the graphs of both y_1 and its inverse $y_2 = \log_5 x$ will be drawn.

$$y_1 = 5^x, \quad y_2 = \log_5 x$$

Two other methods for graphing this function are shown at left.

> *Now Try Exercise 5.*

EXAMPLE 10   Graph: $g(x) = \ln x$.

Solution   To graph $y = g(x) = \ln x$, we select values for x and use a calculator to find the corresponding values of $\ln x$. We then plot points and draw the curve. We can graph the function $y = \ln x$ directly using a graphing calculator.

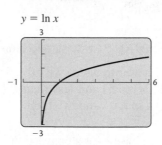

$y = \ln x$

x	$g(x)$
0.5	-0.7
1	0
2	0.7
3	1.1
4	1.4
5	1.6

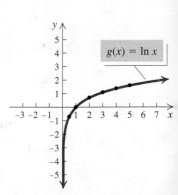

We could also write $g(x) = \ln x$, or $y = \ln x$, as $x = e^y$, select values for y, and use a calculator to find the corresponding values of x.

> *Now Try Exercise 7.*

Recall that the graph of $f(x) = \log_a x$, for any base a, has the x-intercept $(1, 0)$. The domain is the set of positive real numbers, and the range is the set of all real numbers. The y-axis is the vertical asymptote.

EXAMPLE 11 Graph each of the following by hand. Before doing so, describe how each graph can be obtained from the graph of $y = \ln x$. Give the domain and the vertical asymptote of each function. Then check your graph with a graphing calculator.

a) $f(x) = \ln(x + 3)$

b) $f(x) = 3 - \frac{1}{2}\ln x$

c) $f(x) = |\ln(x - 1)|$

Solution

a) The graph of $f(x) = \ln(x + 3)$ is a shift of the graph of $y = \ln x$ left 3 units. The domain is the set of all real numbers greater than -3, $(-3, \infty)$. The line $x = -3$ is the vertical asymptote.

$y_1 = \ln x, \quad y_2 = \ln(x + 3)$

x	$f(x)$
-2.9	-2.303
-2	0
0	1.099
2	1.609
4	1.946

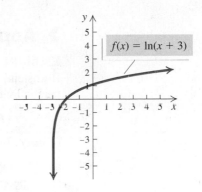

b) The graph of $f(x) = 3 - \frac{1}{2}\ln x$ is a vertical shrinking of the graph of $y = \ln x$, followed by a reflection across the x-axis, and then a translation up 3 units. The domain is the set of all positive real numbers, $(0, \infty)$. The y-axis is the vertical asymptote.

$y_1 = \ln x, \quad y_2 = 3 - \frac{1}{2}\ln x$

x	$f(x)$
0.1	4.151
1	3
3	2.451
6	2.104
9	1.901

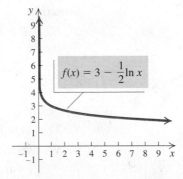

c) The graph of $f(x) = |\ln (x - 1)|$ is a translation of the graph of $y = \ln x$ right 1 unit. Then the absolute value has the effect of reflecting negative outputs across the x-axis. The domain is the set of all real numbers greater than 1, $(1, \infty)$. The line $x = 1$ is the vertical asymptote.

x	$f(x)$
1.1	2.303
2	0
4	1.099
6	1.609
8	1.946

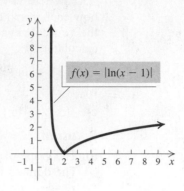

> **Now Try Exercise 89.**

❯ Applications

EXAMPLE 12 *Walking Speed.* In a study by psychologists Bornstein and Bornstein, it was found that the average walking speed w, in feet per second, of a person living in a city of population P, in thousands, is given by the function

$$w(P) = 0.37 \ln P + 0.05$$

(*Source*: *International Journal of Psychology*).

a) The population of Billings, Montana, is 106,954. Find the average walking speed of people living in Billings.

b) The population of Chicago, Illinois, is 2,714,856. Find the average walking speed of people living in Chicago.

c) Graph the function.

d) A sociologist computes the average walking speed in a city to be approximately 2.0 ft/sec. Use this information to estimate the population of the city.

Solution

a) Since P is in thousands and $106,954 = 106.954$ thousand, we substitute 106.954 for P:

$$w(106.954) = 0.37 \ln 106.954 + 0.05 \qquad \textbf{Substituting}$$
$$\approx 1.8. \qquad \textbf{Finding the natural logarithm and simplifying}$$

The average walking speed of people living in Billings is about 1.8 ft/sec.

b) We substitute 2714.856 for P:

$$w(2714.856) = 0.37 \ln 2714.856 + 0.05 \qquad \textbf{Substituting}$$
$$\approx 3.0.$$

The average walking speed of people living in Chicago is about 3.0 ft/sec.

c) We graph with a viewing window of $[0, 600, 0, 4]$ because inputs are very large and outputs are very small by comparison.

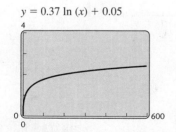

$y = 0.37 \ln (x) + 0.05$

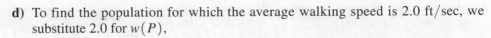

$y_1 = 0.37 \ln(x) + 0.05, \ y_2 = 2$

Intersection
X = 194.46851   Y = 2

d) To find the population for which the average walking speed is 2.0 ft/sec, we substitute 2.0 for $w(P)$,

$$2.0 = 0.37 \ln P + 0.05,$$

and solve for P.

We will use the Intersect method. We graph the equations $y_1 = 0.37 \ln x + 0.05$ and $y_2 = 2$ and use the INTERSECT feature to approximate the point of intersection.

We see that in a city in which the average walking speed is 2.0 ft/sec, the population is about 194.5 thousand, or 194,500.

> *Now Try Exercise 95(d).*

EXAMPLE 13 *Earthquake Magnitude.* Measured on the Richter scale, the magnitude R of an earthquake of intensity I is defined as

$$R = \log \frac{I}{I_0},$$

where I_0 is a minimum intensity used for comparison. We can think of I_0 as a threshold intensity that is the weakest earthquake that can be recorded on a seismograph. If one earthquake is 10 times as intense as another, its magnitude on the Richter scale is 1 greater than that of the other. If one earthquake is 100 times as intense as another, its magnitude on the Richter scale is 2 higher, and so on. Thus an earthquake whose magnitude is 5 on the Richter scale is 10 times as intense as an earthquake whose magnitude is 4. Earthquake intensities can be interpreted as multiples of the minimum intensity I_0.

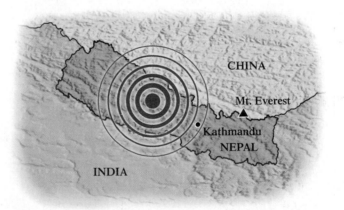

The Nepal region earthquake in South Asia on April 25, 2015, had an intensity of $10^{7.8} \cdot I_0$ (*Source*: earthquake.usgs.gov). It caused extensive loss of life and severe structural damage to buildings, railways, and roads. What was the magnitude of this earthquake on the Richter scale?

Solution We substitute into the formula:

$$R = \log \frac{I}{I_0} = \log \frac{10^{7.8} \cdot I_0}{I_0} = \log 10^{7.8} = 7.8.$$

The magnitude of the earthquake was 7.8 on the Richter scale.

> *Now Try Exercise 97(a).*

Visualizing the Graph

A

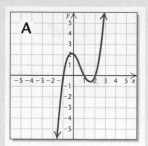

B

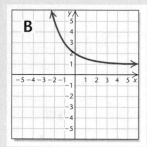

C

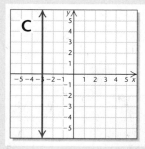

D

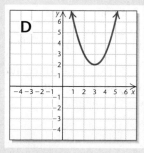

E

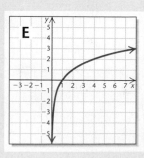

Match the equation or function with its graph.

1. $f(x) = 4^x$

2. $f(x) = \ln x - 3$

3. $(x + 3)^2 + y^2 = 9$

4. $f(x) = 2^{-x} + 1$

5. $f(x) = \log_2 x$

6. $f(x) = x^3 - 2x^2 - x + 2$

7. $x = -3$

8. $f(x) = e^x - 4$

9. $f(x) = (x - 3)^2 + 2$

10. $3x = 6 + y$

Answers on page A-28

F

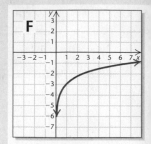

G

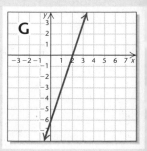

H

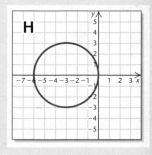

I

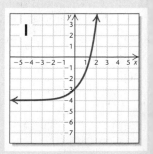

J

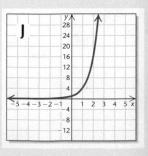

5.3 Exercise Set

Make a hand-drawn graph of each of the following. Then check your work using a graphing calculator.

1. $x = 3^y$

2. $x = 4^y$

3. $x = \left(\frac{1}{2}\right)^y$

4. $x = \left(\frac{4}{3}\right)^y$

5. $y = \log_3 x$

6. $y = \log_4 x$

7. $f(x) = \log x$

8. $f(x) = \ln x$

Find each of the following. Do not use a calculator.

9. $\log_2 16$

10. $\log_3 9$

11. $\log_5 125$

12. $\log_2 64$

13. $\log 0.001$

14. $\log 100$

15. $\log_2 \frac{1}{4}$

16. $\log_8 2$

17. $\ln 1$

18. $\ln e$

19. $\log 10$

20. $\log 1$

21. $\log_5 5^4$

22. $\log \sqrt{10}$

23. $\log_3 \sqrt[4]{3}$

24. $\log 10^{8/5}$

25. $\log 10^{-7}$

26. $\log_5 1$

27. $\log_{49} 7$

28. $\log_3 3^{-2}$

29. $\ln e^{3/4}$

30. $\log_2 \sqrt{2}$

31. $\log_4 1$

32. $\ln e^{-5}$

33. $\ln \sqrt{e}$

34. $\log_{64} 4$

Convert to a logarithmic equation.

35. $10^3 = 1000$

36. $5^{-3} = \frac{1}{125}$

37. $8^{1/3} = 2$

38. $10^{0.3010} = 2$

39. $e^3 = t$

40. $Q^t = x$

41. $e^2 = 7.3891$

42. $e^{-1} = 0.3679$

43. $p^k = 3$

44. $e^{-t} = 4000$

Convert to an exponential equation.

45. $\log_5 5 = 1$

46. $t = \log_4 7$

47. $\log 0.01 = -2$

48. $\log 7 = 0.845$

49. $\ln 30 = 3.4012$

50. $\ln 0.38 = -0.9676$

51. $\log_a M = -x$

52. $\log_t Q = k$

53. $\log_a T^3 = x$

54. $\ln W^5 = t$

Find each of the following using a calculator. Round to four decimal places.

55. $\log 3$

56. $\log 8$

57. $\log 532$

58. $\log 93{,}100$

59. $\log 0.57$

60. $\log 0.082$

61. $\log (-2)$

62. $\ln 50$

63. $\ln 2$

64. $\ln (-4)$

65. $\ln 809.3$

66. $\ln 0.00037$

67. $\ln (-1.32)$

68. $\ln 0$

Find the logarithm using common logarithms and the change-of-base formula.

69. $\log_4 100$

70. $\log_3 20$

71. $\log_{100} 0.3$

72. $\log_\pi 100$

73. $\log_{200} 50$

74. $\log_{5.3} 1700$

Find the logarithm using natural logarithms and the change-of-base formula.

75. $\log_3 12$

76. $\log_4 25$

77. $\log_{100} 15$

78. $\log_9 100$

Graph the function and its inverse using the same set of axes. Use any method.

79. $f(x) = 3^x, \ f^{-1}(x) = \log_3 x$

80. $f(x) = \log_4 x, \ f^{-1}(x) = 4^x$

81. $f(x) = \log x, \ f^{-1}(x) = 10^x$

82. $f(x) = e^x, \ f^{-1}(x) = \ln x$

For each of the following functions, briefly describe how the graph can be obtained from the graph of a basic logarithmic function. Then graph the function using a graphing calculator. Give the domain and the vertical asymptote of each function.

83. $f(x) = \log_2 (x + 3)$

84. $f(x) = \log_3 (x - 2)$

85. $y = \log_3 x - 1$

86. $y = 3 + \log_2 x$

87. $f(x) = 4 \ln x$

88. $f(x) = \frac{1}{2} \ln x$

89. $y = 2 - \ln x$

90. $y = \ln (x + 1)$

91. $f(x) = \frac{1}{2} \log (x - 1) - 2$

92. $f(x) = 5 - 2 \log (x + 1)$

Graph the piecewise function.

93. $g(x) = \begin{cases} 5, & \text{for } x \le 0, \\ \log x + 1, & \text{for } x > 0 \end{cases}$

94. $f(x) = \begin{cases} 1 - x, & \text{for } x \le -1, \\ \ln (x + 1), & \text{for } x > -1 \end{cases}$

95. *Walking Speed.* Refer to Example 12. The data on various cities and their populations are given below. Find the average walking speed in each city. Round to the nearest tenth of a foot per second.

a) El Paso, Texas: 672,538
b) Phoenix, Arizona: 1,488,750
c) Birmingham, Alabama: 212,038
d) Milwaukee, Wisconsin: 598,916
e) Honolulu, Hawaii: 345,610
f) Charlotte, North Carolina: 775,202
g) Omaha, Nebraska: 421,570
h) Sydney, Australia: 3,908,643

96. *Forgetting.* Students in a political science class took a final exam and then took equivalent forms of the exam at monthly intervals thereafter. The average score $S(t)$, as a percent, after t months was found to be given by the function

$$S(t) = 78 - 15 \log (t + 1), \quad t \geq 0.$$

a) What was the average score when the students initially took the test, $t = 0$?
b) What was the average score after 4 months? after 24 months?
c) Graph the function.
d) After what time t was the average score 50%?

97. *Earthquake Magnitude.* Refer to Example 13. The data on various locations of earthquakes and their intensities are given below. Find the magnitude of each earthquake on the Richter scale.

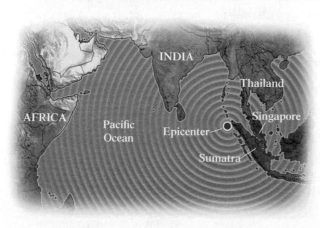

a) San Francisco, California, 1906: $10^{7.7} \cdot I_0$
b) Chile, 1960: $10^{9.5} \cdot I_0$
c) Iran, 2003: $10^{6.6} \cdot I_0$
d) Turkey, 1999: $10^{7.6} \cdot I_0$
e) Chile, 2014: $10^{8.2} \cdot I_0$
f) China, 2008: $10^{7.9} \cdot I_0$
g) Spain, 2011: $10^{5.1} \cdot I_0$
h) Sumatra, 2004: $10^{9.3} \cdot I_0$

98. *pH of Substances in Chemistry.* In chemistry, the pH of a substance is defined as

$$pH = -\log [H^+],$$

where H^+ is the hydrogen ion concentration, in moles per liter. Find the pH of each substance.

SUBSTANCE	HYDROGEN ION CONCENTRATION
a) Pineapple juice	1.6×10^{-4}
b) Hair conditioner	0.0013
c) Mouthwash	6.3×10^{-7}
d) Eggs	1.6×10^{-8}
e) Tomatoes	6.3×10^{-5}

99. Find the hydrogen ion concentration of each substance, given the pH. (See Exercise 98.) Express the answer in scientific notation.

SUBSTANCE	pH
a) Tap water	7
b) Rainwater	5.4
c) Orange juice	3.2
d) Wine	4.8

100. *Advertising.* A model for advertising response is given by the function

$$N(a) = 1000 + 200 \ln a, \quad a \geq 1,$$

where $N(a)$ is the number of units sold when a is the amount spent on advertising, in thousands of dollars.

a) How many units were sold after spending $1000 ($a = 1$) on advertising?
b) How many units were sold after spending $5000?
c) Graph the function.
d) How much would have to be spent in order to sell 2000 units?

101. *Loudness of Sound.* The **loudness** L, in bels (after Alexander Graham Bell), of a sound of intensity I is defined to be

$$L = \log \frac{I}{I_0},$$

where I_0 is the minimum intensity detectable by the human ear (such as the tick of a watch at 20 ft under quiet conditions). If a sound is 10 times as intense as another, its loudness is 1 bel greater than that of the other. If a sound is 100 times as intense as another, its loudness is 2 bels greater, and so on. The bel is a large unit, so a subunit, the **decibel**, is generally used. For L, in decibels, the formula is

$$L = 10 \log \frac{I}{I_0}.$$

Find the loudness, in decibels, of each sound with the given intensity.

SOUND	INTENSITY
a) Jet engine at 100 ft	$10^{14} \cdot I_0$
b) Loud rock concert	$10^{11.5} \cdot I_0$
c) Bird calls	$10^{4} \cdot I_0$
d) Normal conversation	$10^{6.5} \cdot I_0$
e) Thunder	$10^{12} \cdot I_0$
f) Loudest sound possible	$10^{19.4} \cdot I_0$

❭ Skill Maintenance

Find the slope and the y-intercept of the line. **[1.4]**

102. $3x - 10y = 14$

103. $y = 6$

104. $x = -4$

Use synthetic division to find the function values. **[4.3]**

105. $g(x) = x^3 - 6x^2 + 3x + 10$; find $g(-5)$

106. $f(x) = x^4 - 2x^3 + x - 6$; find $f(-1)$

Find a polynomial function of degree 3 with the given numbers as zeros. Answers may vary. **[4.4]**

107. $\sqrt{7}, -\sqrt{7}, 0$

108. $4i, -4i, 1$

❭ Synthesis

Simplify.

109. $\dfrac{\log_5 8}{\log_5 2}$

110. $\dfrac{\log_3 64}{\log_3 16}$

Find the domain of the function.

111. $f(x) = \log_5 x^3$

112. $f(x) = \log_4 x^2$

113. $f(x) = \ln |x|$

114. $f(x) = \log (3x - 4)$

Solve.

115. $\log_2 (2x + 5) < 0$

116. $\log_2 (x - 3) \geq 4$

In Exercises 117–120, match the equation with one of the figures (a)–(d) that follow.

a)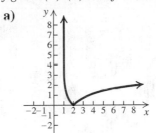

b)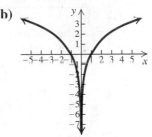

c)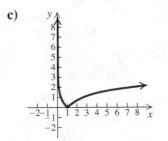

d)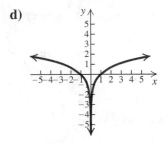

117. $f(x) = \ln |x|$

118. $f(x) = |\ln x|$

119. $f(x) = \ln x^2$

120. $g(x) = |\ln (x - 1)|$

For Exercises 121–124:

a) *Graph the function.*

b) *Estimate the zeros.*

c) *Estimate the relative maximum values and the relative minimum values.*

121. $f(x) = x \ln x$

122. $f(x) = x^2 \ln x$

123. $f(x) = \dfrac{\ln x}{x^2}$

124. $f(x) = e^{-x} \ln x$

125. Using a graphing calculator, find the point(s) of intersection of the graphs of the following.

$$y = 4 \ln x, \qquad y = \frac{4}{e^x + 1}$$

Determine whether the statement is true or false.

1. The domain of all logarithmic functions is $[1, \infty)$. **[5.3]**

2. The range of a one-to-one function f is the domain of its inverse f^{-1}. **[5.1]**

3. The y-intercept of $f(x) = e^{-x}$ is $(0, -1)$. **[5.2]**

For each function, determine whether it is one-to-one, and if the function is one-to-one, find a formula for its inverse. **[5.1]**

4. $f(x) = -\dfrac{2}{x}$

5. $f(x) = 3 + x^2$

6. $f(x) = \dfrac{5}{x - 2}$

7. Given the function $f(x) = \sqrt{x - 5}$, use composition of functions to show that $f^{-1}(x) = x^2 + 5$. **[5.1]**

8. Given the one-to-one function $f(x) = x^3 + 2$, find the inverse and give the domain and the range of f and of f^{-1}. **[5.1]**

Match the function with one of the graphs (a)–(h) that follow. **[5.2], [5.3]**

a)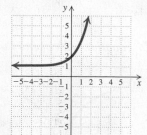

b)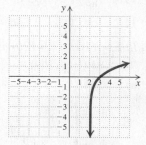

c)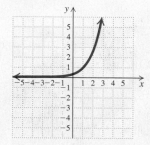

d)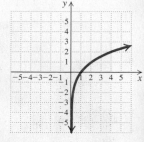

e)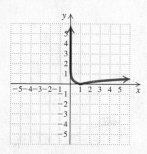

f)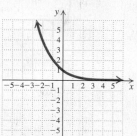

g)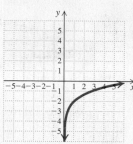

h)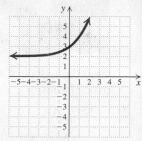

9. $y = \log_2 x$

10. $f(x) = 2^x + 2$

11. $f(x) = e^{x-1}$

12. $f(x) = \ln x - 2$

13. $f(x) = \ln (x - 2)$

14. $y = 2^{-x}$

15. $f(x) = |\log x|$

16. $f(x) = e^x + 1$

17. Suppose that \$3200 is invested at $4\frac{1}{2}\%$ interest, compounded quarterly. Find the amount of money in the account in 6 years. **[5.2]**

Find each of the following without a calculator. **[5.3]**

18. $\log_4 1$

19. $\ln e^{-4/5}$

20. $\log 0.01$

21. $\ln e^2$

22. $\ln 1$

23. $\log_2 \dfrac{1}{16}$

24. $\log 1$

25. $\log_3 27$

26. $\log \sqrt[4]{10}$

27. $\ln e$

28. Convert $e^{-6} = 0.0025$ to a logarithmic equation. **[5.3]**

29. Convert $\log T = r$ to an exponential equation. **[5.3]**

Find the logarithm using the change-of-base formula. **[5.3]**

30. $\log_3 20$

31. $\log_\pi 10$

COLLABORATIVE DISCUSSION AND WRITING

32. Explain why an even function f does not have an inverse f^{-1} that is a function. **[5.1]**

34. Describe the differences between the graphs of $f(x) = x^3$ and $g(x) = 3^x$. **[5.2]**

33. Suppose that \$10,000 is invested for 8 years at 6.4% interest, compounded annually. In what year will the most interest be earned? Why? **[5.2]**

35. If $\log b < 0$, what can you say about b? **[5.3]**

5.4 ❯ **Properties of Logarithmic Functions**

> ❯ Convert from logarithms of products, powers, and quotients to expressions in terms of individual logarithms, and conversely.
> ❯ Simplify expressions of the type $\log_a a^x$ and $a^{\log_a x}$.

We now establish some properties of logarithmic functions. These properties are based on the corresponding rules for exponents.

Just in Time

7

❯ Logarithms of Products

The first property of logarithms corresponds to the product rule for exponents: $a^m \cdot a^n = a^{m+n}$.

THE PRODUCT RULE

For any positive numbers M and N and any logarithmic base a,

$$\log_a MN = \log_a M + \log_a N.$$

(The logarithm of a product is the sum of the logarithms of the factors.)

EXAMPLE 1 Express as a sum of logarithms: $\log_3 (9 \cdot 27)$.

Solution We have

$$\log_3 (9 \cdot 27) = \log_3 9 + \log_3 27. \quad \textbf{Using the product rule}$$

As a check, note that

$$\log_3 (9 \cdot 27) = \log_3 243 = 5 \qquad \mathbf{3^5 = 243}$$
$$\text{and} \quad \log_3 9 + \log_3 27 = 2 + 3 = 5. \qquad \mathbf{3^2 = 9;\ 3^3 = 27}$$

❯ *Now Try Exercise 1.*

EXAMPLE 2 Express as a single logarithm: $\log_2 p^3 + \log_2 q$.

Solution We have

$$\log_2 p^3 + \log_2 q = \log_2 (p^3 q).$$

❯ *Now Try Exercise 35.*

A Proof of the Product Rule: Let $\log_a M = x$ and $\log_a N = y$. Converting to exponential equations, we have $a^x = M$ and $a^y = N$. Then

$$MN = a^x \cdot a^y = a^{x+y}.$$

Converting back to a logarithmic equation, we get

$$\log_a MN = x + y.$$

Remembering what x and y represent, we know it follows that

$$\log_a MN = \log_a M + \log_a N.$$

› Logarithms of Powers

The second property of logarithms corresponds to the power rule for exponents: $(a^m)^n = a^{mn}$.

THE POWER RULE

For any positive number M, any logarithmic base a, and any real number p,

$$\log_a M^p = p \log_a M.$$

(The logarithm of a power of M is the exponent times the logarithm of M.)

EXAMPLE 3   Express each of the following as a product.

a) $\log_a 11^{-3}$

b) $\log_a \sqrt[4]{7}$

c) $\ln x^6$

Solution

a) $\log_a 11^{-3} = -3 \log_a 11$ | Using the power rule

b) $\log_a \sqrt[4]{7} = \log_a 7^{1/4}$ | Writing exponential notation

$\qquad\qquad\quad = \frac{1}{4} \log_a 7$ | Using the power rule

c) $\ln x^6 = 6 \ln x$ | Using the power rule

> *Now Try Exercises 13 and 15.*

A Proof of the Power Rule: Let $x = \log_a M$. The equivalent exponential equation is $a^x = M$. Raising both sides to the power p, we obtain

$$(a^x)^p = M^p, \quad \text{or} \quad a^{xp} = M^p.$$

Converting back to a logarithmic equation, we get

$$\log_a M^p = xp.$$

But $x = \log_a M$, so substituting gives us

$$\log_a M^p = (\log_a M)p = p \log_a M.$$

› Logarithms of Quotients

The third property of logarithms corresponds to the quotient rule for exponents: $a^m / a^n = a^{m-n}$.

THE QUOTIENT RULE

For any positive numbers M and N, and any logarithmic base a,

$$\log_a \frac{M}{N} = \log_a M - \log_a N.$$

(The logarithm of a quotient is the logarithm of the numerator minus the logarithm of the denominator.)

EXAMPLE 4   Express as a difference of logarithms: $\log_t \dfrac{8}{w}$.

Solution   We have

$$\log_t \frac{8}{w} = \log_t 8 - \log_t w. \qquad \text{\textbf{Using the quotient rule}}$$ > **Now Try Exercise 17.**

EXAMPLE 5   Express as a single logarithm: $\log_b 64 - \log_b 16$.

Solution   We have

$$\log_b 64 - \log_b 16 = \log_b \frac{64}{16} = \log_b 4. \qquad$$ > **Now Try Exercise 37.**

A Proof of the Quotient Rule: The proof follows from both the product rule and the power rule:

$$\log_a \frac{M}{N} = \log_a MN^{-1} = \log_a M + \log_a N^{-1} \qquad \text{\textbf{Using the product rule}}$$

$$= \log_a M + (-1) \log_a N \qquad \text{\textbf{Using the power rule}}$$

$$= \log_a M - \log_a N.$$ ∎

COMMON ERRORS

$\log_a MN \neq (\log_a M)(\log_a N)$ The logarithm of a product is *not* the product of the logarithms.

$\log_a (M + N) \neq \log_a M + \log_a N$ The logarithm of a sum is *not* the sum of the logarithms.

$\log_a \dfrac{M}{N} \neq \dfrac{\log_a M}{\log_a N}$ The logarithm of a quotient is *not* the quotient of the logarithms.

$(\log_a M)^p \neq p \log_a M$ The power of a logarithm is *not* the exponent times the logarithm.

❯ Applying the Properties

EXAMPLE 6   Express each of the following in terms of sums and differences of logarithms.

a) $\log_a \dfrac{x^2 y^5}{z^4}$ **b)** $\log_a \sqrt[3]{\dfrac{a^2 b}{c^5}}$ **c)** $\log_b \dfrac{ay^5}{m^3 n^4}$

Solution

a) $\log_a \dfrac{x^2 y^5}{z^4} = \log_a (x^2 y^5) - \log_a z^4$ **Using the quotient rule**

$$= \log_a x^2 + \log_a y^5 - \log_a z^4 \qquad \text{\textbf{Using the product rule}}$$

$$= 2 \log_a x + 5 \log_a y - 4 \log_a z \qquad \text{\textbf{Using the power rule}}$$

b) $\log_a \sqrt[3]{\dfrac{a^2 b}{c^5}} = \log_a \left(\dfrac{a^2 b}{c^5}\right)^{1/3}$ Writing exponential notation

$\qquad = \dfrac{1}{3} \log_a \dfrac{a^2 b}{c^5}$ Using the power rule

$\qquad = \dfrac{1}{3} \left(\log_a a^2 b - \log_a c^5\right)$ Using the quotient rule. The parentheses are necessary.

$\qquad = \dfrac{1}{3} \left(2 \log_a a + \log_a b - 5 \log_a c\right)$ Using the product rule and the power rule

$\qquad = \dfrac{1}{3} \left(2 + \log_a b - 5 \log_a c\right)$ $\log_a a = 1$

$\qquad = \dfrac{2}{3} + \dfrac{1}{3} \log_a b - \dfrac{5}{3} \log_a c$ Multiplying to remove parentheses

c) $\log_b \dfrac{a y^5}{m^3 n^4} = \log_b a y^5 - \log_b m^3 n^4$ Using the quotient rule

$\qquad = \left(\log_b a + \log_b y^5\right) - \left(\log_b m^3 + \log_b n^4\right)$ Using the product rule

$\qquad = \log_b a + \log_b y^5 - \log_b m^3 - \log_b n^4$ Removing parentheses

$\qquad = \log_b a + 5 \log_b y - 3 \log_b m - 4 \log_b n$ Using the power rule

> *Now Try Exercises 25 and 31.*

EXAMPLE 7 Express as a single logarithm:

$$5 \log_b x - \log_b y + \tfrac{1}{4} \log_b z.$$

Solution   We have

$5 \log_b x - \log_b y + \tfrac{1}{4} \log_b z = \log_b x^5 - \log_b y + \log_b z^{1/4}$

$\qquad\qquad$ Using the power rule

$\qquad = \log_b \dfrac{x^5}{y} + \log_b z^{1/4}$ Using the quotient rule

$\qquad = \log_b \dfrac{x^5 z^{1/4}}{y},\ \text{or}\ \log_b \dfrac{x^5 \sqrt[4]{z}}{y}.$

$\qquad\qquad$ Using the product rule

> *Now Try Exercise 41.*

EXAMPLE 8 Express as a single logarithm:

$$\ln (3x + 1) - \ln (3x^2 - 5x - 2).$$

Solution   We have

$\ln (3x + 1) - \ln (3x^2 - 5x - 2) = \ln \dfrac{3x + 1}{3x^2 - 5x - 2}$ Using the quotient rule

$\qquad = \ln \dfrac{3x + 1}{(3x + 1)(x - 2)}$ Factoring

$\qquad = \ln \dfrac{1}{x - 2}.$ Simplifying

> *Now Try Exercise 45.*

EXAMPLE 9 Given that $\log_a 2 \approx 0.301$ and $\log_a 3 \approx 0.477$, find each of the following, if possible.

a) $\log_a 6$

b) $\log_a \frac{2}{3}$

c) $\log_a 81$

d) $\log_a \frac{1}{4}$

e) $\log_a 5$

f) $\dfrac{\log_a 3}{\log_a 2}$

Solution

a) $\log_a 6 = \log_a (2 \cdot 3) = \log_a 2 + \log_a 3$ Using the product rule
$$\approx 0.301 + 0.477$$
$$\approx 0.778$$

b) $\log_a \frac{2}{3} = \log_a 2 - \log_a 3$ Using the quotient rule
$$\approx 0.301 - 0.477 \approx -0.176$$

c) $\log_a 81 = \log_a 3^4 = 4 \log_a 3$ Using the power rule
$$\approx 4(0.477) \approx 1.908$$

d) $\log_a \frac{1}{4} = \log_a 1 - \log_a 4$ Using the quotient rule
$$= 0 - \log_a 2^2 \qquad \log_a 1 = 0; 4 = 2^2$$
$$= -2 \log_a 2 \qquad \text{Using the power rule}$$
$$\approx -2(0.301) \approx -0.602$$

e) $\log_a 5$ *cannot* be found using these properties and the given information.
$$\log_a 5 \neq \log_a 2 + \log_a 3 \qquad \log_a 2 + \log_a 3 = \log_a (2 \cdot 3) = \log_a 6$$

f) $\dfrac{\log_a 3}{\log_a 2} \approx \dfrac{0.477}{0.301} \approx 1.585$ **We simply divide, not using any of the properties.**

> *Now Try Exercises 53 and 55.*

❭ Simplifying Expressions of the Type $\log_a a^x$ and $a^{\log_a x}$

We have two final properties of logarithms to consider. The first follows from the product rule: Since $\log_a a^x = x \log_a a = x \cdot 1 = x$, we have $\log_a a^x = x$. This property also follows from the definition of a logarithm: x is the power to which we raise a in order to get a^x.

THE LOGARITHM OF A BASE TO A POWER

For any logarithmic base a and any real number x,

$$\log_a a^x = x.$$

(The logarithm, base a, of a to a power is the power.)

EXAMPLE 10 Simplify each of the following.

a) $\log_a a^8$

b) $\ln e^{-t}$

c) $\log 10^{3k}$

Solution

a) $\log_a a^8 = 8$ **8 is the power to which we raise a in order to get a^8.**

b) $\ln e^{-t} = \log_e e^{-t} = -t$ $\ln e^x = x$

c) $\log 10^{3k} = \log_{10} 10^{3k} = 3k$

> *Now Try Exercises 65 and 73.*

Let $M = \log_a x$. Then $a^M = x$. Substituting $\log_a x$ for M, we obtain $a^{\log_a x} = x$. This also follows from the definition of a logarithm: $\log_a x$ is the power to which a is raised in order to get x.

A BASE TO A LOGARITHMIC POWER

For any logarithmic base a and any positive real number x,

$$a^{\log_a x} = x.$$

(The number a raised to the power $\log_a x$ is x.)

EXAMPLE 11   Simplify each of the following.

a) $4^{\log_4 k}$ b) $e^{\ln 5}$ c) $10^{\log 7t}$

Solution

a) $4^{\log_4 k} = k$

b) $e^{\ln 5} = e^{\log_e 5} = 5$

c) $10^{\log 7t} = 10^{\log_{10} 7t} = 7t$

> **Now Try Exercises 69 and 71.**

CHANGE-OF-BASE FORMULA

REVIEW SECTION 5.3.

A Proof of the Change-of-Base Formula: We close this section by proving the change-of-base formula and summarizing the properties of logarithms considered thus far in this chapter. In Section 5.3, we used the change-of-base formula,

$$\log_b M = \frac{\log_a M}{\log_a b},$$

to make base conversions in order to find logarithmic values using a calculator. Let $x = \log_b M$. Then

$b^x = M$	**Definition of logarithm**
$\log_a b^x = \log_a M$	**Taking the logarithm on both sides**
$x \log_a b = \log_a M$	**Using the power rule**
$x = \dfrac{\log_a M}{\log_a b},$	**Dividing by $\log_a b$**

so

$$x = \log_b M = \frac{\log_a M}{\log_a b}.$$

Following is a summary of the properties of logarithms.

SUMMARY OF THE PROPERTIES OF LOGARITHMS

The Product Rule: $\log_a MN = \log_a M + \log_a N$

The Power Rule: $\log_a M^p = p \log_a M$

The Quotient Rule: $\log_a \dfrac{M}{N} = \log_a M - \log_a N$

The Change-of-Base Formula: $\log_b M = \dfrac{\log_a M}{\log_a b}$

Other Properties: $\log_a a = 1, \quad \log_a 1 = 0,$

$\log_a a^x = x, \quad a^{\log_a x} = x$

5.4   Exercise Set

Express as a sum of logarithms and simplify, if possible.

1. $\log_3 (81 \cdot 27)$

2. $\log_2 (8 \cdot 64)$

3. $\log_5 (5 \cdot 125)$

4. $\log_4 (64 \cdot 4)$

5. $\log_t 8Y$

6. $\log 0.2x$

7. $\ln xy$

8. $\ln ab$

Express as a product.

9. $\log_b t^3$

10. $\log_a x^4$

11. $\log y^8$

12. $\ln y^5$

13. $\log_c K^{-6}$

14. $\log_b Q^{-8}$

15. $\ln \sqrt[3]{4}$

16. $\ln \sqrt{a}$

Express as a difference of logarithms.

17. $\log_t \dfrac{M}{8}$

18. $\log_a \dfrac{76}{13}$

19. $\log \dfrac{x}{y}$

20. $\ln \dfrac{a}{b}$

21. $\ln \dfrac{r}{s}$

22. $\log_b \dfrac{3}{w}$

Express in terms of sums and differences of logarithms.

23. $\log_a 6xy^5z^4$

24. $\log_a x^3y^2z$

25. $\log_b \dfrac{p^2q^5}{m^4b^9}$

26. $\log_b \dfrac{x^2y}{b^3}$

27. $\ln \dfrac{2}{3x^3y}$

28. $\log \dfrac{5a}{4b^2}$

29. $\log \sqrt{r^3t}$

30. $\ln \sqrt[3]{5x^5}$

31. $\log_a \sqrt{\dfrac{x^6}{p^5q^8}}$

32. $\log_c \sqrt[3]{\dfrac{y^3z^2}{x^4}}$

33. $\log_a \sqrt[4]{\dfrac{m^8n^{12}}{a^3b^5}}$

34. $\log_a \sqrt{\dfrac{a^6b^8}{a^2b^5}}$

Express as a single logarithm and, if possible, simplify.

35. $\log_a 75 + \log_a 2$

36. $\log 0.01 + \log 1000$

37. $\log 10{,}000 - \log 100$

38. $\ln 54 - \ln 6$

39. $\frac{1}{2} \log n + 3 \log m$

40. $\frac{1}{2} \log a - \log 2$

41. $\frac{1}{2} \log_a x + 4 \log_a y - 3 \log_a x$

42. $\frac{2}{5} \log_a x - \frac{1}{3} \log_a y$

43. $\ln x^2 - 2 \ln \sqrt{x}$

44. $\ln 2x + 3(\ln x - \ln y)$

45. $\ln (x^2 - 4) - \ln (x + 2)$

46. $\log (x^3 - 8) - \log (x - 2)$

47. $\log (x^2 - 5x - 14) - \log (x^2 - 4)$

48. $\log_a \dfrac{a}{\sqrt{x}} - \log_a \sqrt{ax}$

49. $\ln x - 3[\ln (x - 5) + \ln (x + 5)]$

50. $\frac{2}{3}[\ln (x^2 - 9) - \ln (x + 3)] + \ln (x + y)$

51. $\frac{3}{2} \ln 4x^6 - \frac{4}{5} \ln 2y^{10}$

52. $120(\ln \sqrt[5]{x^3} + \ln \sqrt[3]{y^2} - \ln \sqrt[4]{16z^5})$

Given that $\log_a 2 \approx 0.301$, $\log_a 7 \approx 0.845$, and $\log_a 11 \approx 1.041$, find each of the following, if possible. Round the answer to the nearest thousandth.

53. $\log_a \frac{2}{11}$

54. $\log_a 14$

55. $\log_a 98$

56. $\log_a \frac{1}{7}$

57. $\dfrac{\log_a 2}{\log_a 7}$

58. $\log_a 9$

Given that $\log_b 2 \approx 0.693$, $\log_b 3 \approx 1.099$, and $\log_b 5 \approx 1.609$, find each of the following, if possible. Round the answer to the nearest thousandth.

59. $\log_b 125$

60. $\log_b \frac{5}{3}$

61. $\log_b \frac{1}{6}$

62. $\log_b 30$

63. $\log_b \dfrac{3}{b}$

64. $\log_b 15b$

Simplify.

65. $\log_p p^3$

66. $\log_t t^{2713}$

67. $\log_e e^{|x-4|}$

68. $\log_q q^{\sqrt{3}}$

69. $3^{\log_3 4x}$

70. $5^{\log_5 (4x - 3)}$

71. $10^{\log w}$

72. $e^{\ln x^3}$

73. $\ln e^{8t}$

74. $\log 10^{-k}$

75. $\log_b \sqrt{b}$

76. $\log_b \sqrt{b^3}$

❯ Skill Maintenance

In each of Exercises 77–86, classify the function as linear, quadratic, cubic, quartic, rational, exponential, or logarithmic.

77. $f(x) = 5 - x^2 + x^4$ **[4.1]**

78. $f(x) = 2^x$ **[5.2]**

79. $f(x) = -\frac{3}{4}$ **[1.3]**

80. $f(x) = 4^x - 8$ **[5.2]**

81. $f(x) = -\frac{3}{x}$ **[4.5]**

82. $f(x) = \log x + 6$ **[5.3]**

83. $f(x) = -\frac{1}{3}x^3 - 4x^2 + 6x + 42$ **[4.1]**

84. $f(x) = \frac{x^2 - 1}{x^2 + x - 6}$ **[4.5]**

85. $f(x) = \frac{1}{2}x + 3$ **[1.4]**

86. $f(x) = 2x^2 - 6x + 3$ **[3.2]**

› Synthesis

Solve for x.

87. $5^{\log_5 8} = 2x$

88. $\ln e^{3x-5} = -8$

Express as a single logarithm and, if possible, simplify.

89. $\log_a (x^2 + xy + y^2) + \log_a (x - y)$

90. $\log_a (a^{10} - b^{10}) - \log_a (a + b)$

Express as a sum or a difference of logarithms.

91. $\log_a \dfrac{x - y}{\sqrt{x^2 - y^2}}$

92. $\log_a \sqrt{9 - x^2}$

93. Given that $\log_a x = 2$, $\log_a y = 3$, and $\log_a z = 4$, find

$$\log_a \frac{\sqrt[4]{y^2 z^5}}{\sqrt[4]{x^3 z^{-2}}}.$$

Determine whether each of the following is true or false. Assume that a, x, M, and N are positive.

94. $\log_a M + \log_a N = \log_a (M + N)$

95. $\log_a M - \log_a N = \log_a \dfrac{M}{N}$

96. $\dfrac{\log_a M}{\log_a N} = \log_a M - \log_a N$

97. $\dfrac{\log_a M}{x} = \log_a M^{1/x}$

98. $\log_a x^3 = 3 \log_a x$

99. $\log_a 8x = \log_a x + \log_a 8$

100. $\log_N (MN)^x = x \log_N M + x$

Suppose that $\log_a x = 2$. Find each of the following.

101. $\log_a \left(\dfrac{1}{x} \right)$

102. $\log_{1/a} x$

103. Simplify:

$$\log_{10} 11 \cdot \log_{11} 12 \cdot \log_{12} 13 \cdots \log_{998} 999 \cdot \log_{999} 1000.$$

Write each of the following without using logarithms.

104. $\log_a x + \log_a y - mz = 0$

105. $\ln a - \ln b + xy = 0$

Prove each of the following for any base a and any positive number x.

106. $\log_a \left(\dfrac{1}{x} \right) = -\log_a x = \log_{1/a} x$

107. $\log_a \left(\dfrac{x + \sqrt{x^2 - 5}}{5} \right) = -\log_a \left(x - \sqrt{x^2 - 5} \right)$

5.5 Solving Exponential Equations and Logarithmic Equations

› Solve exponential equations.

› Solve logarithmic equations.

› Solving Exponential Equations

Equations with variables in the exponents, such as

$$3^x = 20 \quad \text{and} \quad 2^{5x} = 64,$$

are called **exponential equations**.

Sometimes, as is the case with the equation $2^{5x} = 64$, we can write each side as a power of the same number:

$$2^{5x} = 2^6.$$

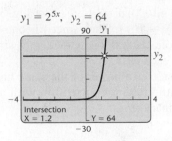

We can then set the exponents equal and solve:

$$5x = 6$$
$$x = \frac{6}{5}, \text{ or } 1.2.$$

We use the following property to solve exponential equations.

BASE–EXPONENT PROPERTY

For any $a > 0$, $a \neq 1$,

$$a^x = a^y \longleftrightarrow x = y.$$

ONE-TO-ONE FUNCTIONS

REVIEW SECTION 5.1.

This property follows from the fact that for any $a > 0$, $a \neq 1$, $f(x) = a^x$ is a one-to-one function. If $a^x = a^y$, then $f(x) = f(y)$. Then since f is one-to-one, it follows that $x = y$. Conversely, if $x = y$, it follows that $a^x = a^y$, since we are raising a to the same power in each case.

EXAMPLE 1 Solve: $2^{3x-7} = 32$.

Algebraic Solution

Note that $32 = 2^5$. Thus we can write each side as a power of the same number:

$$2^{3x-7} = 2^5.$$

Since the bases are the same number, 2, we can use the base–exponent property and set the exponents equal:

$$3x - 7 = 5$$
$$3x = 12$$
$$x = 4.$$

Check:

$$\begin{array}{c|c} 2^{3x-7} = 32 \\ \hline 2^{3(4)-7} \overset{?}{\,} 32 \\ 2^{12-7} \\ 2^5 \\ 32 & 32 \quad \text{TRUE} \end{array}$$

The solution is 4.

Graphical Solution

We will use the Intersect method. We graph

$$y_1 = 2^{3x-7} \quad \text{and} \quad y_2 = 32$$

to find the coordinates of the point of intersection. The first coordinate of this point is the solution of the equation $y_1 = y_2$, or $2^{3x-7} = 32$.

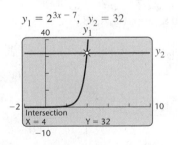

The solution is 4.

We could also write the equation in the form $2^{3x-7} - 32 = 0$ and use the Zero method.

> *Now Try Exercise 7.*

Another property that is used when solving some exponential equations and logarithmic equations is as follows.

PROPERTY OF LOGARITHMIC EQUALITY

For any $M > 0$, $N > 0$, $a > 0$, and $a \neq 1$,

$$\log_a M = \log_a N \longleftrightarrow M = N.$$

This property follows from the fact that for any $a > 0$, $a \neq 1$, $f(x) = \log_a x$ is a one-to-one function. If $\log_a x = \log_a y$, then $f(x) = f(y)$. Then since f is one-to-one, it follows that $x = y$. Conversely, if $x = y$, it follows that $\log_a x = \log_a y$, since we are taking the logarithm of the same number in each case.

When it does not seem possible to write each side as a power of the same base, we can use the property of logarithmic equality and take the logarithm with any base on each side and then use the power rule for logarithms.

EXAMPLE 2 Solve: $3^x = 20$.

Algebraic Solution

We have
$$3^x = 20$$
$$\log 3^x = \log 20 \quad \text{Taking the common logarithm on both sides}$$
$$x \log 3 = \log 20 \quad \text{Using the power rule}$$
$$x = \frac{\log 20}{\log 3}. \quad \text{Dividing by log 3}$$

This is an exact answer. We cannot simplify further, but we can approximate using a calculator:
$$x = \frac{\log 20}{\log 3} \approx 2.7268.$$

We can check this by finding $3^{2.7268}$:
$$3^{2.7268} \approx 20.$$

The solution is about 2.7268.

Graphical Solution

We will use the Intersect method. We graph
$$y_1 = 3^x \quad \text{and} \quad y_2 = 20$$

to find the x-coordinate of the point of intersection. That x-coordinate is the value of x for which $3^x = 20$ and is thus the solution of the equation.

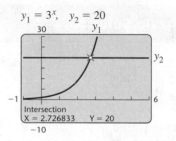

The solution is approximately 2.7268.
We could also write the equation in the form $3^x - 20 = 0$ and use the Zero method.

Now Try Exercise 11.

In Example 2, we took the common logarithm on both sides of the equation. Any base will give the same result. Let's try base 3. We have
$$3^x = 20$$
$$\log_3 3^x = \log_3 20$$
$$x = \log_3 20 \quad \log_a a^x = x$$
$$x = \frac{\log 20}{\log 3} \quad \text{Using the change-of-base formula}$$
$$x \approx 2.7268.$$

log₃(20)
2.726833028

Note that we changed the base to do the final calculation. We also could find $\log_3 20$ directly using the logBASE operation.

EXAMPLE 3 Solve: $100e^{0.08t} = 2500$.

Algebraic Solution

It will make our work easier if we take the natural logarithm when working with equations that have e as a base.

We have

$$100e^{0.08t} = 2500$$

$$e^{0.08t} = 25 \qquad \textbf{Dividing by 100}$$

$$\ln e^{0.08t} = \ln 25 \qquad \begin{array}{l}\textbf{Taking the natural}\\ \textbf{logarithm on both sides}\end{array}$$

$$0.08t = \ln 25 \qquad \begin{array}{l}\textbf{Finding the logarithm of a}\\ \textbf{base to a power: } \log_a a^x = x\end{array}$$

$$t = \frac{\ln 25}{0.08} \qquad \textbf{Dividing by 0.08}$$

$$\approx 40.2.$$

The solution is about 40.2.

Graphical Solution

Using the Intersect method, we graph the equations

$$y_1 = 100e^{0.08x} \quad \text{and} \quad y_2 = 2500$$

and determine the point of intersection. The first coordinate of the point of intersection is the solution of the equation $100e^{0.08x} = 2500$.

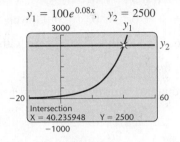

The solution is approximately 40.2.

> *Now Try Exercise 19.*

EXAMPLE 4 Solve: $4^{x+3} = 3^{-x}$.

Algebraic Solution

We have

$$4^{x+3} = 3^{-x}$$

$$\log 4^{x+3} = \log 3^{-x} \qquad \begin{array}{l}\textbf{Taking the}\\ \textbf{common logarithm}\\ \textbf{on both sides}\end{array}$$

$$(x+3)\log 4 = -x\log 3 \qquad \textbf{Using the power rule}$$

$$x\log 4 + 3\log 4 = -x\log 3 \qquad \textbf{Removing parentheses}$$

$$x\log 4 + x\log 3 = -3\log 4 \qquad \begin{array}{l}\textbf{Adding } x \log 3 \textbf{ and}\\ \textbf{subtracting } 3 \log 4\end{array}$$

$$x(\log 4 + \log 3) = -3\log 4 \qquad \textbf{Factoring on the left}$$

$$x = \frac{-3\log 4}{\log 4 + \log 3} \qquad \begin{array}{l}\textbf{Dividing by}\\ \log 4 + \log 3\end{array}$$

$$\approx -1.6737.$$

The solution is about -1.6737.

Graphical Solution

We will use the Intersect method. We graph

$$y_1 = 4^{x+3} \quad \text{and} \quad y_2 = 3^{-x}$$

to find the x-coordinate of the point of intersection. That x-coordinate is the value of x for which $4^{x+3} = 3^{-x}$ and is the solution of the equation.

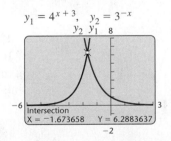

The solution is approximately -1.6737.

We could also write the equation in the form $4^{x+3} - 3^{-x} = 0$ and use the Zero method.

> *Now Try Exercise 21.*

EQUATIONS REDUCIBLE TO
QUADRATIC

REVIEW SECTION 3.2.

EXAMPLE 5   Solve: $e^x + e^{-x} - 6 = 0$.

Algebraic Solution

In this case, we have more than one term with x in the exponent:

$$e^x + e^{-x} - 6 = 0$$

$$e^x + \frac{1}{e^x} - 6 = 0 \qquad \text{Rewriting } e^{-x} \text{ with a positive exponent}$$

$$e^{2x} + 1 - 6e^x = 0. \qquad \text{Multiplying by } e^x \text{ on both sides}$$

This equation is reducible to quadratic with $u = e^x$:

$$u^2 - 6u + 1 = 0.$$

We use the quadratic formula with $a = 1$, $b = -6$, and $c = 1$:

$$u = \frac{-b \pm \sqrt{b^2 - 4ac}}{2a}$$

$$u = \frac{-(-6) \pm \sqrt{(-6)^2 - 4 \cdot 1 \cdot 1}}{2 \cdot 1}$$

$$u = \frac{6 \pm \sqrt{32}}{2} = \frac{6 \pm 4\sqrt{2}}{2}$$

$$u = \frac{2(3 \pm 2\sqrt{2})}{2}$$

$$u = 3 \pm 2\sqrt{2}$$

$$e^x = 3 \pm 2\sqrt{2}. \qquad \text{Replacing } u \text{ with } e^x$$

We now take the natural logarithm on both sides:

$$\ln e^x = \ln (3 \pm 2\sqrt{2})$$

$$x = \ln (3 \pm 2\sqrt{2}). \qquad \text{Using } \ln e^x = x$$

Approximating each of the solutions, we obtain 1.76 and -1.76.

Graphical Solution

Using the Zero method, we begin by graphing the function

$$y = e^x + e^{-x} - 6.$$

Then we find the zeros of the function.

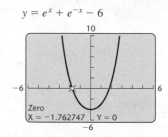

$y = e^x + e^{-x} - 6$

The leftmost zero is about -1.76. Using the ZERO feature one more time, we find that the other zero is about 1.76.

The solutions are about -1.76 and 1.76.

Now Try Exercise 25.

It is possible that when encountering an equation like the one in Example 5, you might not recognize that it could be solved in the algebraic manner shown. This points out the value of the graphical solution.

❯ Solving Logarithmic Equations

Equations containing variables in logarithmic expressions, such as $\log_2 x = 4$ and $\log x + \log (x + 3) = 1$, are called **logarithmic equations**. To solve logarithmic equations algebraically, we first try to obtain a single logarithmic expression on one side and then write an equivalent exponential equation.

EXAMPLE 6 Solve: $\log_3 x = -2$.

Algebraic Solution	**Graphical Solution**
We have $\log_3 x = -2$ $3^{-2} = x$ **Converting to an exponential equation** $\dfrac{1}{3^2} = x$ $\dfrac{1}{9} = x.$ *Check:* $\dfrac{\log_3 x = -2}{\log_3 \frac{1}{9} \ ? \ -2}$ $\log_3 3^{-2}$ | $\qquad\qquad -2$ | -2 **TRUE** The solution is $\frac{1}{9}$.	We use the change-of-base formula and graph the equations $$y_1 = \log_3 x = \frac{\ln x}{\ln 3} \quad \text{and} \quad y_2 = -2.$$ We could also graph $y_1 = \log_3 x$ directly with the logBASE operation. Then we use the Intersect method with a graphing calculator. 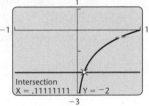  If the solution is a rational number, we can usually find fraction notation for the exact solution by using the FRAC feature from the MATH submenu of the MATH menu. The solution is $\frac{1}{9}$.

❯ *Now Try Exercise 33.*

EXAMPLE 7 Solve: $\log x + \log (x + 3) = 1$.

Just in Time

19

Algebraic Solution	**Graphical Solution**

Algebraic Solution

In this case, we have common logarithms. Writing the base of 10 will help us understand the problem:

$$\log_{10} x + \log_{10} (x + 3) = 1$$

$$\log_{10} [x(x + 3)] = 1 \qquad \textbf{Using the product rule to obtain a single logarithm}$$

$$x(x + 3) = 10^1 \qquad \textbf{Writing an equivalent exponential equation}$$

$$x^2 + 3x = 10$$

$$x^2 + 3x - 10 = 0$$

$$(x - 2)(x + 5) = 0 \qquad \textbf{Factoring}$$

$$x - 2 = 0 \quad or \quad x + 5 = 0$$

$$x = 2 \quad or \quad x = -5.$$

Check: For 2:

$$\frac{\log x + \log (x + 3) = 1}{\log 2 + \log (2 + 3) \;\; ? \;\; 1}$$
$$\log 2 + \log 5$$
$$\log (2 \cdot 5)$$
$$\log 10$$
$$1 \;\;\Big|\;\; 1 \quad \text{TRUE}$$

For -5:

$$\frac{\log x + \log (x + 3) = 1}{\log (-5) + \log (-5 + 3) \;\; ? \;\; 1} \quad \text{FALSE}$$

The number -5 is not a solution because negative numbers do not have real-number logarithms. The solution is 2.

Graphical Solution

We can graph the equations

$$y_1 = \log x + \log (x + 3)$$

and

$$y_2 = 1$$

and use the Intersect method. The first coordinate of the point of intersection is the solution of the equation.

$$y_1 = \log x + \log (x + 3), \; y_2 = 1$$

Intersection
X = 2   Y = 1

We could also graph the function

$$y = \log x + \log (x + 3) - 1$$

and use the Zero method. The zero of the function is the solution of the equation.

$$y_1 = \log x + \log (x + 3) - 1$$

Zero
X = 2   Y = 0

The solution of the equation is 2. From the graph, we can easily see that there is only one solution.

> *Now Try Exercise 41.*

EXAMPLE 8   Solve: $\log_3 (2x - 1) - \log_3 (x - 4) = 2$.

Algebraic Solution

We have

$$\log_3 (2x - 1) - \log_3 (x - 4) = 2$$

$$\log_3 \frac{2x - 1}{x - 4} = 2 \qquad \textbf{Using the quotient rule}$$

$$\frac{2x - 1}{x - 4} = 3^2 \qquad \textbf{Writing an equivalent exponential equation}$$

$$\frac{2x - 1}{x - 4} = 9$$

$$(x - 4) \cdot \frac{2x - 1}{x - 4} = (x - 4) \cdot 9 \qquad \textbf{Multiplying by the LCD, } x - 4$$

$$2x - 1 = 9x - 36$$

$$35 = 7x$$

$$5 = x.$$

Check:

$$\frac{\log_3 (2x - 1) - \log_3 (x - 4) = 2}{\log_3 (2 \cdot 5 - 1) - \log_3 (5 - 4) \ \overset{?}{\mid} \ 2}$$

$$\log_3 9 - \log_3 1$$

$$2 - 0$$

$$2 \ \mid \ 2 \quad \text{TRUE}$$

The solution is 5.

Graphical Solution

Here we use the Intersect method to find the solution of the equation. We use the change-of-base formula and graph the equations

$$y_1 = \frac{\ln (2x - 1)}{\ln 3} - \frac{\ln (x - 4)}{\ln 3}$$

and

$$y_2 = 2.$$

We could also graph y_1 directly with the logBASE operation.

$$y_1 = \frac{\ln (2x - 1)}{\ln 3} - \frac{\ln (x - 4)}{\ln 3}, \quad y_2 = 2$$

```
Intersection
X = 5          Y = 2
```

The solution is 5.

> *Now Try Exercise 45.*

EXAMPLE 9 Solve: $\ln(4x + 6) - \ln(x + 5) = \ln x$.

Algebraic Solution

We have

$$\ln(4x + 6) - \ln(x + 5) = \ln x$$

$$\ln\frac{4x + 6}{x + 5} = \ln x \qquad \text{Using the quotient rule}$$

$$\frac{4x + 6}{x + 5} = x \qquad \text{Using the property of logarithmic equality}$$

$$(x + 5) \cdot \frac{4x + 6}{x + 5} = (x + 5)x \qquad \text{Multiplying by } x + 5$$

$$4x + 6 = x^2 + 5x$$

$$0 = x^2 + x - 6$$

$$0 = (x + 3)(x - 2) \qquad \text{Factoring}$$

$$x + 3 = 0 \quad or \quad x - 2 = 0$$

$$x = -3 \quad or \qquad x = 2.$$

The number -3 is not a solution because $\ln[4(-3) + 6]$, or $\ln(-6)$, on the left side of the equation and $\ln(-3)$ on the right side of the equation are not real numbers. The value 2 checks and is the solution.

Graphical Solution

The solution of the equation

$$\ln(4x + 6) - \ln(x + 5) = \ln x$$

is the zero of the function

$$f(x) = \ln(4x + 6) - \ln(x + 5) - \ln x.$$

The solution is also the first coordinate of the x-intercept of the graph of the function. Here we use the Zero method.

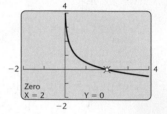

$y_1 = \ln(4x + 6) - \ln(x + 5) - \ln x$

The solution of the equation is 2. From the graph, we can easily see that there is only one solution.

> *Now Try Exercise 43.*

Sometimes we encounter equations for which an algebraic solution seems difficult or impossible.

EXAMPLE 10 Solve: $e^{0.5x} - 7.3 = 2.08x + 6.2$.

Graphical Solution We graph the equations

$$y_1 = e^{0.5x} - 7.3 \quad \text{and} \quad y_2 = 2.08x + 6.2$$

on a graphing calculator and use the Intersect method. (See the window at left.)

We can also consider the equation

$$y = e^{0.5x} - 7.3 - 2.08x - 6.2, \quad \text{or} \quad y = e^{0.5x} - 2.08x - 13.5,$$

and use the Zero method. The approximate solutions are -6.471 and 6.610.

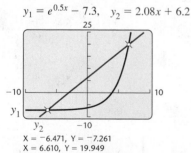

$y_1 = e^{0.5x} - 7.3, \quad y_2 = 2.08x + 6.2$

X = −6.471, Y = −7.261
X = 6.610, Y = 19.949

> *Now Try Exercise 67.*

5.5 Exercise Set

Solve the exponential equation algebraically. Then check using a graphing calculator. Round to three decimal places, if appropriate.

1. $3^x = 81$

2. $2^x = 32$

3. $2^{2x} = 8$

4. $3^{7x} = 27$

5. $2^x = 33$

6. $2^x = 40$

7. $5^{4x-7} = 125$

8. $4^{3x-5} = 16$

9. $27 = 3^{5x} \cdot 9^{x^2}$

10. $3^{x^2+4x} = \frac{1}{27}$

11. $84^x = 70$

12. $28^x = 10^{-3x}$

13. $10^{-x} = 5^{2x}$

14. $15^x = 30$

15. $e^{-c} = 5^{2c}$

16. $e^{4t} = 200$

17. $e^t = 1000$

18. $e^{-t} = 0.04$

19. $e^{-0.03t} = 0.08$

20. $1000e^{0.09t} = 5000$

21. $3^x = 2^{x-1}$

22. $5^{x+2} = 4^{1-x}$

23. $(3.9)^x = 48$

24. $250 - (1.87)^x = 0$

25. $e^x + e^{-x} = 5$

26. $e^x - 6e^{-x} = 1$

27. $3^{2x-1} = 5^x$

28. $2^{x+1} = 5^{2x}$

29. $2e^x = 5 - e^{-x}$

30. $e^x + e^{-x} = 4$

Solve the logarithmic equation algebraically. Then check using a graphing calculator.

31. $\log_5 x = 4$

32. $\log_2 x = -3$

33. $\log x = -4$

34. $\log x = 1$

35. $\ln x = 1$

36. $\ln x = -2$

37. $\log_{64} \frac{1}{4} = x$

38. $\log_{125} \frac{1}{25} = x$

39. $\log_2 (10 + 3x) = 5$

40. $\log_5 (8 - 7x) = 3$

41. $\log x + \log (x - 9) = 1$

42. $\log_2 (x + 1) + \log_2 (x - 1) = 3$

43. $\log_2 (x + 20) - \log_2 (x + 2) = \log_2 x$

44. $\log (x + 5) - \log (x - 3) = \log 2$

45. $\log_8 (x + 1) - \log_8 x = 2$

46. $\log x - \log (x + 3) = -1$

47. $\log x + \log (x + 4) = \log 12$

48. $\log_3 (x + 14) - \log_3 (x + 6) = \log_3 x$

49. $\log (x + 8) - \log (x + 1) = \log 6$

50. $\ln x - \ln (x - 4) = \ln 3$

51. $\log_4 (x + 3) + \log_4 (x - 3) = 2$

52. $\ln (x + 1) - \ln x = \ln 4$

53. $\log (2x + 1) - \log (x - 2) = 1$

54. $\log_5 (x + 4) + \log_5 (x - 4) = 2$

55. $\ln (x + 8) + \ln (x - 1) = 2 \ln x$

56. $\log_3 x + \log_3 (x + 1) = \log_3 2 + \log_3 (x + 3)$

Solve.

57. $\log_6 x = 1 - \log_6 (x - 5)$

58. $2^{x^2-9x} = \frac{1}{256}$

59. $9^{x-1} = 100(3^x)$

60. $2 \ln x - \ln 5 = \ln (x + 10)$

61. $e^x - 2 = -e^{-x}$

62. $2 \log 50 = 3 \log 25 + \log (x - 2)$

Use a graphing calculator to find the approximate solutions of the equation.

63. $2^x - 5 = 3x + 1$

64. $0.082e^{0.05x} = 0.034$

65. $xe^{3x} - 1 = 3$

66. $4x - 3^x = -6$

67. $5e^{5x} + 10 = 3x + 40$

68. $4 \ln (x + 3.4) = 2.5$

69. $\log_8 x + \log_8 (x + 2) = 2$

70. $\ln x^2 = -x^2$

71. $\log_5 (x + 7) - \log_5 (2x - 3) = 1$

72. $\log_3 x + 7 = 4 - \log_5 x$

Approximate the point(s) of intersection of the pair of equations.

73. $2.3x + 3.8y = 12.4, \ y = 1.1 \ln (x - 2.05)$

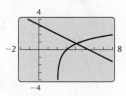

74. $y = \ln 3x, \ y = 3x - 8$

75. $y = 2.3 \ln (x + 10.7), \ y = 10e^{-0.007x^2}$

76. $y = 2.3 \ln (x + 10.7)$, $y = 10e^{-0.07x^2}$

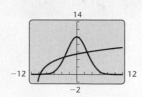

❯ Skill Maintenance

In Exercises 77–80:

a) *Find the vertex.*
b) *Find the axis of symmetry.*
c) *Determine whether there is a maximum or a minimum value and find that value.* **[3.3]**

77. $g(x) = x^2 - 6$

78. $f(x) = -x^2 + 6x - 8$

79. $G(x) = -2x^2 - 4x - 7$

80. $H(x) = 3x^2 - 12x + 16$

❯ Synthesis

Solve using any method.

81. $\dfrac{e^x + e^{-x}}{e^x - e^{-x}} = 3$

82. $\ln (\ln x) = 2$

83. $\sqrt{\ln x} = \ln \sqrt{x}$

84. $\ln \sqrt[4]{x} = \sqrt{\ln x}$

85. $(\log_3 x)^2 - \log_3 x^2 = 3$

86. $\log_3 (\log_4 x) = 0$

87. $\ln x^2 = (\ln x)^2$

88. $x \left(\ln \frac{1}{6} \right) = \ln 6$

89. $5^{2x} - 3 \cdot 5^x + 2 = 0$

90. $x^{\log x} = \dfrac{x^3}{100}$

91. $\ln x^{\ln x} = 4$

92. $|2^{x^2} - 8| = 3$

93. $\dfrac{\sqrt{(e^{2x} \cdot e^{-5x})^{-4}}}{e^x \div e^{-x}} = e^7$

94. Given that $a = (\log_{125} 5)^{\log_5 125}$, find the value of $\log_3 a$.

95. Given that $a = \log_8 225$ and $b = \log_2 15$, express a as a function of b.

96. Given that $f(x) = e^x - e^{-x}$, find $f^{-1}(x)$ if it exists.

5.6 ❯ Applications and Models: Growth and Decay; Compound Interest

> ❯ Solve applied problems involving exponential growth and exponential decay.

> ❯ Solve applied problems involving compound interest.

> ❯ Find models involving exponential functions and logarithmic functions.

Exponential functions and logarithmic functions with base e are rich in applications to many fields such as business, science, psychology, and sociology.

❯ Population Growth

The function

$$P(t) = P_0 e^{kt}, \quad k > 0,$$

is a model of many kinds of population growth, whether it be a population of people, bacteria, smartphones, or money. In this function, P_0 is the population at time 0, P is the population after time t, and k is called the **exponential growth rate**. The graph of such an equation is shown at left.

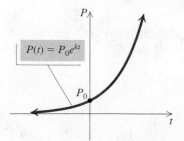

EXAMPLE 1 *Population Growth of Ghana.*   In 2014, the population of Ghana, located on the west coast of Africa, was about 25.8 million, and the exponential growth rate was 2.19% per year (*Source: CIA World Factbook*, 2015).

a) Find the exponential growth function.

b) Graph the exponential growth function.

c) Estimate the population in 2018.

d) At this growth rate, when will the population be 40 million?

Solution

a) At $t = 0$ (2014), the population was 25.8 million, and the exponential growth rate was 2.19% per year. We substitute 25.8 for P_0 and 2.19%, or 0.0219, for k to obtain the exponential growth function

$$P(t) = 25.8e^{0.0219t},$$

where t is the number of years after 2014 and $P(t)$ is in millions.

b) Using a graphing calculator, we obtain the graph of the exponential growth function.

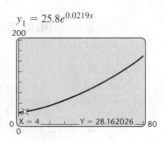

$y_1 = 25.8e^{0.0219x}$

$X = 4$ $Y = 28.162026$

c) In 2018, $t = 4$; that is, 4 years have passed since 2014. To find the population in 2018, we substitute 4 for t:

$$P(4) = 25.8e^{0.0219(4)} = 25.8e^{0.0876} \approx 28.2.$$

We can also use the VALUE feature from the CALC menu on a graphing calculator to find $P(4)$. We estimate the population to be about 28.2 million, or 28,200,000, in 2018.

d) We are looking for the time t for which $P(t) = 40$. To find t, we solve the equation

$$40 = 25.8e^{0.0219t}$$

using both an algebraic method and a graphical method.

Algebraic Solution	**Graphical Solution**
We have	Using the Intersect method, we graph the equations

Algebraic Solution

We have

$$40 = 25.8e^{0.0219t} \qquad \text{Substituting 40 for } P(t)$$

$$\frac{40}{25.8} = e^{0.0219t} \qquad \text{Dividing by 25.8}$$

$$\ln \frac{40}{25.8} = \ln e^{0.0219t} \qquad \begin{array}{l}\text{Taking the natural}\\\text{logarithm on both sides}\end{array}$$

$$\ln \frac{40}{25.8} = 0.0219t \qquad \ln e^x = x$$

$$\frac{\ln \dfrac{40}{25.8}}{0.0219} = t \qquad \text{Dividing by 0.0219}$$

$$20 \approx t.$$

The population of Ghana will be 40 million about 20 years after 2014.

Graphical Solution

Using the Intersect method, we graph the equations

$$y_1 = 25.8e^{0.0219t} \quad \text{and} \quad y_2 = 40$$

and find the first coordinate of their point of intersection.

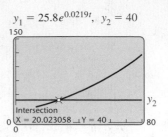

$$y_1 = 25.8e^{0.0219t}, \ y_2 = 40$$

Intersection
X = 20.023058   Y = 40

The solution is about 20, so the population of Ghana will be 40 million about 20 years after 2014.

> *Now Try Exercise 1.*

》 Interest Compounded Continuously

When interest is paid on interest, we call it **compound interest**. Suppose that an amount P_0 is invested in a savings account at interest rate k **compounded continuously**. The amount $P(t)$ in the account after t years is given by the exponential function

$$P(t) = P_0e^{kt}.$$

EXAMPLE 2 *Interest Compounded Continuously.*   Suppose that $2000 is invested at interest rate k, compounded continuously, and grows to $2504.65 in 5 years.

a) What is the interest rate?

b) Find the exponential growth function.

c) What will the balance be after 10 years?

d) After how long will the $2000 have doubled?

Solution

a) At $t = 0$, $P(0) = P_0 = \$2000$. Thus the exponential growth function is of the form

$$P(t) = 2000e^{kt}.$$

We know that $P(5) = \$2504.65$. We substitute and solve for k:

$$2504.65 = 2000e^{k(5)} \qquad \text{Substituting 2504.65 for } P(t) \text{ and 5 for } t$$

$$2504.65 = 2000e^{5k}$$

$$\frac{2504.65}{2000} = e^{5k}. \qquad \text{Dividing by 2000}$$

Then

$$\ln \frac{2504.65}{2000} = \ln e^{5k} \qquad \textbf{Taking the natural logarithm}$$

$$\ln \frac{2504.65}{2000} = 5k \qquad \textbf{Using } \ln e^x = x$$

$$\frac{\ln \dfrac{2504.65}{2000}}{5} = k \qquad \textbf{Dividing by 5}$$

$$0.045 \approx k.$$

The interest rate is about 0.045, or 4.5%.

We can also find k by graphing the equations

$$y_1 = 2000e^{5x} \quad \text{and} \quad y_2 = 2504.65$$

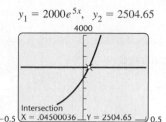

$y_1 = 2000e^{5x}, \quad y_2 = 2504.65$

and using the Intersect feature to approximate the first coordinate of the point of intersection. The interest rate is about 0.045, or 4.5%.

b) Substituting 0.045 for k in the function $P(t) = 2000e^{kt}$, we see that the exponential growth function is

$$P(t) = 2000e^{0.045t}.$$

c) The balance after 10 years is

$$P(10) - 2000e^{0.045(10)} = 2000e^{0.45} \approx \$3136.62.$$

d) To find the doubling time T, we set $P(T) = 2 \cdot P_0 = 2 \cdot \$2000 = \$4000$ and solve for T. We solve

$$4000 = 2000e^{0.045T}$$

using both an algebraic method and a graphical method.

Algebraic Solution

We have

$$4000 = 2000e^{0.045T}$$

$$2 = e^{0.045T} \qquad \textbf{Dividing by 2000}$$

$$\ln 2 = \ln e^{0.045T} \qquad \textbf{Taking the natural logarithm}$$

$$\ln 2 = 0.045T \qquad \textbf{ln } e^x = x$$

$$\frac{\ln 2}{0.045} = T \qquad \textbf{Dividing by 0.045}$$

$$15.4 \approx T.$$

Thus the original investment of $2000 will double in about 15.4 years.

Graphical Solution

We use the Zero method. We graph the function

$$y = 2000e^{0.045x} - 4000$$

and find the zero of the function. The zero of the function is the solution of the equation.

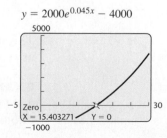

$y = 2000e^{0.045x} - 4000$

The solution is about 15.4, so the original investment of $2000 will double in about 15.4 years.

Now Try Exercise 7.

We can find a general expression relating the growth rate k and the doubling time T by solving the following equation:

$$2P_0 = P_0 e^{kT} \qquad \text{Substituting } 2P_0 \text{ for } P \text{ and } T \text{ for } t$$
$$2 = e^{kT} \qquad \text{Dividing by } P_0$$
$$\ln 2 = \ln e^{kT} \qquad \text{Taking the natural logarithm}$$
$$\ln 2 = kT \qquad \text{Using } \ln e^x = x$$
$$\frac{\ln 2}{k} = T.$$

GROWTH RATE AND DOUBLING TIME

The **growth rate** k and the **doubling time** T are related by

$$kT = \ln 2, \quad \text{or} \quad k = \frac{\ln 2}{T}, \quad \text{or} \quad T = \frac{\ln 2}{k}.$$

Note that the relationship between k and T does not depend on P_0.

EXAMPLE 3 *Population Growth.* The population of the Philippines is now doubling every 37.7 years (*Source: CIA World Factbook*, 2014). What is the exponential growth rate?

Solution We have

$$k = \frac{\ln 2}{T} = \frac{\ln 2}{37.7} \approx 0.0184 \approx 1.84\%.$$

The growth rate of the population of the Philippines is about 1.84% per year.

> *Now Try Exercise 3(e).*

Philippines

❯ Models of Limited Growth

The model $P(t) = P_0 e^{kt}, k > 0$, has many applications involving unlimited population growth. However, in some populations, there can be factors that prevent a population from exceeding some limiting value—perhaps a limitation on food, living space, or other natural resources. One model of such growth is

$$P(t) = \frac{a}{1 + be^{-kt}}.$$

This is called a **logistic function**. This function increases toward a *limiting value a* as $t \to \infty$. Thus, $y = a$ is the horizontal asymptote of the graph of $P(t)$.

EXAMPLE 4 *Limited Population Growth in a Lake.* A lake is stocked with 400 fish of a new variety. The size of the lake, the availability of food, and the number of other fish restrict the growth of that type of fish in the lake to a limiting value of 2500. The population gets closer and closer to this limiting value, but never reaches it. The population of fish in the lake after time t, in months, is given by the logistic function

$$P(t) = \frac{2500}{1 + 5.25e^{-0.32t}}.$$

a) Graph the function.

b) Find the population after 0, 1, 5, 10, 15, and 20 months.

Solution

a) We use a graphing calculator to graph the function. The graph is shown below. Note that this function increases toward a limiting value of 2500. The graph has $y = 2500$ as a horizontal asymptote.

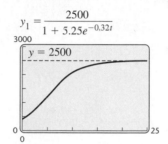

$$y_1 = \frac{2500}{1 + 5.25e^{-0.32t}}$$

X	Y₁
0	400
1	519.5
5	1213.6
10	2059.3
15	2396.5
20	2478.4

X =

b) We can use the TABLE feature on a graphing calculator set in ASK mode to find the function values. (See the window on the right above.) Thus the population will be about 400 after 0 months, 520 after 1 month, 1214 after 5 months, 2059 after 10 months, 2397 after 15 months, and 2478 after 20 months.

> **Now Try Exercise 17.**

Another model of limited growth is provided by the function

$$P(t) = L(1 - e^{-kt}), \quad k > 0,$$

which is shown graphed at right. This function also increases toward a limiting value L as $t \to \infty$, so $y = L$ is the horizontal asymptote of the graph of $P(t)$.

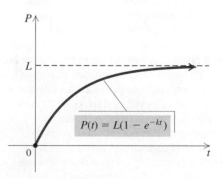

❯ Exponential Decay

The function

$$P(t) = P_0 e^{-kt}, \quad k > 0,$$

is an effective model of the decline, or decay, of a population. An example is the decay of a radioactive substance. In this case, P_0 is the amount of the substance at time $t = 0$, and $P(t)$ is the amount of the substance left after time t, where k is a positive constant that depends on the situation. The constant k is called the **decay rate**.

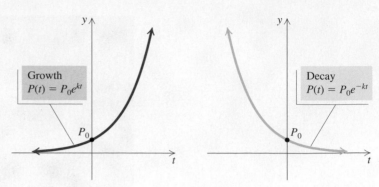

The **half-life** of bismuth (Bi-210) is 5 days. This means that half of an amount of radioactive bismuth will cease to be radioactive in 5 days. The effect of half-life T for nonnegative inputs is shown in the following graph. The exponential function gets close to 0, but never reaches 0, as t gets very large. Thus, according to an exponential decay model, a radioactive substance never completely decays.

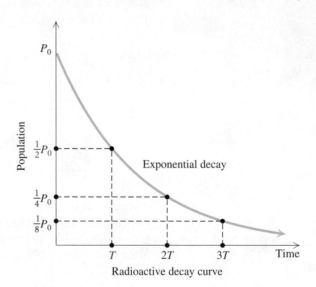

Radioactive decay curve

How can scientists determine that an animal bone has lost 30% of its carbon-14? The assumption is that the percentage of carbon-14 in the atmosphere is the same as that in living plants and animals. When a plant or an animal dies, the amount of carbon-14 that it contains decays exponentially. A scientist can burn an animal bone and use a Geiger counter to determine the percentage of the smoke that is carbon-14. The amount by which this varies from the percentage in the atmosphere tells how much carbon-14 has been lost.

The process of carbon-14 dating was developed by the American chemist Willard E. Libby in 1952. It is known that the radioactivity in a living plant is 16 disintegrations per gram per minute. Since the half-life of carbon-14 is 5750 years, an object with an activity of 8 disintegrations per gram per minute is 5750 years old, one with an activity of 4 disintegrations per gram per minute is 11,500 years old, and so on. Carbon-14 dating can be used to measure the age of objects up to 40,000 years old. Beyond such an age, it is too difficult to measure the radioactivity and some other method would need to be used.

Carbon-14 dating was used to find the age of the Dead Sea Scrolls. It was also used to refute the authenticity of the Shroud of Turin, presumed to have covered the body of Christ.

We can find a general expression relating the decay rate k and the half-life time T by solving the following equation:

$$\frac{1}{2} P_0 = P_0 e^{-kT} \qquad \textbf{Substituting } \frac{1}{2} P_0 \textbf{ for } P \textbf{ and } T \textbf{ for } t$$

$$\frac{1}{2} = e^{-kT} \qquad \textbf{Dividing by } P_0$$

$$\ln \frac{1}{2} = \ln e^{-kT} \qquad \textbf{Taking the natural logarithm on both sides}$$

$$\ln 2^{-1} = -kT \qquad \frac{1}{2} = 2^{-1}; \ln e^x = x$$

$$-\ln 2 = -kT \qquad \textbf{Using the power rule}$$

$$\frac{\ln 2}{k} = T. \qquad \textbf{Dividing by } -k$$

DECAY RATE AND HALF-LIFE

The **decay rate** k and the **half-life** T are related by

$$kT = \ln 2, \quad \text{or} \quad k = \frac{\ln 2}{T}, \quad \text{or} \quad T = \frac{\ln 2}{k}.$$

Note that the relationship between decay rate and half-life is the same as that between growth rate and doubling time.

EXAMPLE 5 *Carbon Dating.* The radioactive element carbon-14 has a half-life of 5750 years. The percentage of carbon-14 present in the remains of organic matter can be used to determine the age of that organic matter. Archaeologists discovered that the linen wrapping from one of the Dead Sea Scrolls had lost 22.3% of its carbon-14 at the time it was found. How old was the linen wrapping?

In 1947, a Bedouin youth looking for a stray goat climbed into a cave at Kirbet Qumran on the shores of the Dead Sea near Jericho and came upon earthenware jars containing an incalculable treasure of ancient manuscripts. Shown here are fragments of those Dead Sea Scrolls, a portion of some 600 or so texts found so far and which concern the Jewish books of the Bible. Officials date them before 70 A.D., making them the oldest Biblical manuscripts by 1000 years.

Solution We first find k when the half-life T is 5750 years:

$$k = \frac{\ln 2}{T} = \frac{\ln 2}{5750} = 0.00012.$$

Now we have the function

$$P(t) = P_0 e^{-0.00012t}.$$

(This function can be used for any subsequent carbon-dating problem.) If the linen wrapping has lost 22.3% of its carbon-14 from an initial amount P_0, then $77.7\% P_0$ is the amount present. To find the age t of the wrapping, we solve the equation for t:

$77.7\% P_0 = P_0 e^{-0.00012t}$	**Substituting $77.7\% P_0$ for P**
$0.777 = e^{-0.00012t}$	**Dividing by P_0 and writing 77.7% as 0.777**
$\ln 0.777 = \ln e^{-0.00012t}$	**Taking the natural logarithm on both sides**
$\ln 0.777 = -0.00012t$	$\ln e^x = x$
$\dfrac{\ln 0.777}{-0.00012} = t$	**Dividing by -0.00012**
$2103 \approx t.$	

Thus the linen wrapping on the Dead Sea Scrolls was about 2103 years old when it was found.

Now Try Exercise 9.

› Exponential and Logarithmic Curve Fitting

We have added several new functions that can be considered when we fit curves to data. Let's review some of them.

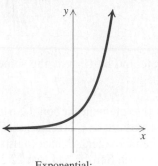

Exponential:
$f(x) = ab^x$, or ae^{kx}
$a > 0, b > 1, k > 0$

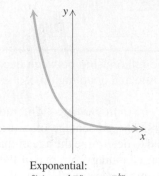

Exponential:
$f(x) = ab^{-x}$, or ae^{-kx}
$a > 0, b > 1, k > 0$

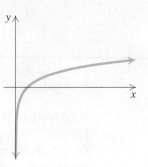

Logarithmic:
$f(x) = a + b \ln x$
$b > 0$

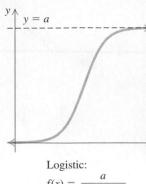

Logistic:
$f(x) = \dfrac{a}{1 + be^{-kx}}$

$a, b, k > 0$

Now, when we analyze a set of data for curve fitting, these models can be considered as well as polynomial functions (such as linear, quadratic, cubic, and quartic functions) and rational functions.

EXAMPLE 6 *Health Expenditures.* Health costs have increased exponentially over the past few decades. The following table lists national health expenditures in selected years from 1960 to 2012.

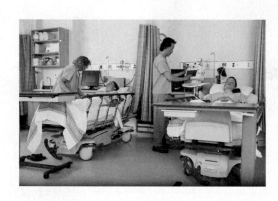

Year, x	National Health Expenditures (in billions)
1960, 0	$ 27
1970, 10	75
1980, 20	256
1990, 30	724
2000, 40	1400
2010, 50	2600
2012, 52	2800

Sources: Centers for Medicare and Medicaid Services

a) Use a graphing calculator to fit an exponential function to the data.

b) Graph the function with the scatterplot of the data.

c) Use the function to estimate national health expenditures in 1985, in 2003, and in 2011. Then use the function to estimate health expenditures in 2016.

Solution

a) We will fit an equation of the type $y = a \cdot b^x$ to the data, where x is the number of years after 1960. Entering the data into the calculator and carrying out the regression procedure, we find that the equation is

$$y = 34.11812301(1.093629292)^x.$$

(See Fig. 1.) The correlation coefficient, r, is very close to 1. This gives us an indication that the exponential function fits the data well.

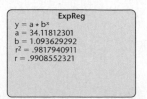

ExpReg
y = a∗b^x
a = 34.11812301
b = 1.093629292
r² = .9817940911
r = .9908552321

FIGURE 1.

b) The scatterplot and the graph are shown in Fig. 2 on the left below.

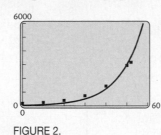

FIGURE 2.

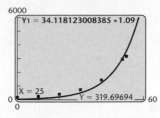

FIGURE 3.

c) Using the VALUE feature in the CALC menu (see Fig. 3 on the right above), we evaluate the function found in part (a) for $x = 25$ ($1985 - 1960 = 25$) and estimate national health expenditures in 1985 to be about \$320 billion. We continue using the VALUE feature to find health expenditures in additional years—2003: \$1601 billion, or \$1.601 trillion; 2011: \$3276 billion, or \$3.276 trillion; 2016: \$5125 billion, or \$5.125 trillion.

> **Now Try Exercise 29.**

On some graphing calculators, there may be a REGRESSION feature that yields an exponential function, base e. If not, and you wish to find such a function, a conversion can be done using the following.

CONVERTING FROM BASE b TO BASE e

$$b^x = e^{x(\ln b)}$$

Then, for the equation in Example 6, we have

$$
\begin{aligned}
y &= 34.11812301(1.093629292)^x \\
&= 34.11812301 e^{x(\ln 1.093629292)} \\
&= 34.11812301 e^{0.089501791x}.
\end{aligned}
$$

We can prove this conversion formula using properties of logarithms, as follows:

$$e^{x(\ln b)} = e^{\ln b^x} = b^x.$$

5.6 Exercise Set

1. *Population Growth of Raleigh.* Raleigh, North Carolina, is one of the fastest growing cities in the United States. In 2014, the population was approximately 438,000, and the growth rate was about 3.4% per year.

NORTH CAROLINA

Raleigh

a) Find the exponential growth function.
b) Estimate the population of Raleigh in 2017.
c) When will the population of Raleigh be 500,000?
d) Find the doubling time.

2. *Population Growth of Rabbits.* Under ideal conditions, a population of rabbits has an exponential growth rate of 11.7% per day. Consider an initial population of 100 rabbits.

a) Find the exponential growth function.
b) Graph the function.
c) What will the population be after 7 days? after 2 weeks?
d) Find the doubling time.

3. *Population Growth.*   Complete the following table.

Country	Growth Rate, k	Doubling Time, T
a) United States		99.0 years
b) Bolivia		42.5 years
c) Uganda	3.32%	
d) Australia	1.11%	
e) Sweden		385 years
f) Laos	2.32%	
g) India	1.28%	
h) China		150.7 years
i) Guinea		26.3 years
j) Hong Kong	0.39%	

4. *E-Book Sales.*   The revenue from e-book sales (not including educational textbooks) accounted for only 0.5% of U.S. publishing sales in 2006. This percentage grew to 22.6% in 2012. (*Source*: Association of American Publishers) Assuming that the exponential growth model applies:

a) Find the value of k and write the function.
b) Estimate the percentage of U.S. publishing sales that were e-book sales in 2009 and in 2010. Round to the nearest tenth.

5. *Population Growth of Haiti.*   The population of Haiti has a growth rate of 1.08% per year. In 2015, the population was 9,996,731, and the land area of Haiti is 32,961,561,600 square yards. (*Source*: U.S. Census Bureau)

Assuming that this growth rate continues and is exponential, after how long will there be one person for every square yard of land?

6. *Georgia O'Keeffe Painting.*   On November 20, 2014, Georgia O'Keeffe's painting, "Jimson Weed/White Flower No. 1," sold at Sotheby's auction in New York City for $44.4 million, a record price for any female artist's work that was sold at auction. The painting had previously sold for $1 million in 1994. (*Sources*: theguardian.com; sothebys.com) Assuming that the value A of the painting has grown exponentially:

a) Find the value of k, and determine the exponential growth function, assuming that $A_0 =$ $1 million and t is the number of years after 1994.
b) Estimate the value of the painting in 2019.
c) What is the doubling time for the value of the painting?
d) After how long will the value of the painting be $280 million, assuming that there is no change in the growth rate?

7. *Interest Compounded Continuously.* Suppose that $10,000 is invested at an interest rate of 5.4% per year, compounded continuously.

a) Find the exponential function that describes the amount in the account after time t, in years.

b) What is the balance after 1 year? 2 years? 5 years? 10 years?

c) What is the doubling time?

8. *Interest Compounded Continuously.* Complete the following table.

Initial Investment at $t = 0$, P_0	Interest Rate, k	Doubling Time, T	Amount After 5 Years
a) $35,000	3.2%		
b) $5000			$7,130.90
c)	5.6%		$9,923.47
d)		11 years	$17,539.32
e) $109,000			$136,503.18
f)		46.2 years	$19,552.82

9. *Carbon Dating.* In 1970, Amos Flora of Flora, Indiana, discovered teeth and jawbones while dredging a creek. Scientists discovered that the bones were from a mastodon and that they had lost 77.2% of their carbon-14. How old were the bones at the time that they were discovered? (*Sources*: "Farm Yields Bones Thousands of Years Old," by Dan McFeely, *Indianapolis Star*, October 20, 2008; Field Museum of Chicago, Bill Turnbull, anthropologist)

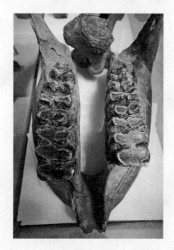

10. *Tomb in the Valley of the Kings.* In February 2006, in the Valley of the Kings in Egypt, a team of archaeologists uncovered the first tomb since King Tut's tomb was found in 1922. The tomb contained five wooden sarcophagi that contained mummies. The archaeologists believe that the mummies are from the 18th Dynasty, about 3300 to 3500 years ago. Determine the amount of carbon-14 that the mummies have lost.

11. *Radioactive Decay.* Complete the following table.

Radioactive Substance	Decay Rate, k	Half-Life T
a) Polonium (Po-218)		3.1 min
b) Lead (Pb-210)		22.3 years
c) Iodine (I-125)	1.15% per day	
d) Krypton (Kr-85)	6.5% per year	
e) Strontium (Sr-90)		29.1 years
f) Uranium (U-232)		70.0 years
g) Plutonium (Pu-239)		24,100 years

12. *Advertising Revenue.* The amount of advertising revenue in U.S. newspapers has declined continually since 2006. In 2006, the advertising revenue was $49.3 billion, and by 2013 that amount had decreased to $20.7 billion (*Source*: Newspaper Association of America). Assuming that the amount of newspaper advertising revenue decreased according to the exponential decay model:

a) Find the value of k, and write an exponential function that describes the advertising revenue after time t, in years, where t is the number of years after 2006.

b) Estimate the advertising revenue in 2008 and in 2012.

c) At this decay rate, when will the advertising revenue be $16 billion?

13. *Porsche 928.* The market value of the 1993–1995 Porsche 928 has had a recent upswing, increasing from $8000 in 2000 to $15,400 in 2015 (*Sources*: Haggerty; *Hemmings Motor News*, May 2015, p. 36) Assuming that the value *V* of the car has grown exponentially:

a) Find the value of *k*, and determine the exponential growth function, assuming that $V_0 = 8000$ and *t* is the number of years after 2000.

b) Estimate the value of the car in 2011.

c) What is the doubling time for the value of the car?

d) After how long will the value of the car be $25,000, assuming that there is no change in the growth rate?

14. *Married Adults.* The data in the following table show that the percentage of adults in the United States who are currently married is declining.

Year	Percent of Adults Who Are Married
1960	72.2%
1980	62.3
2000	57.4
2010	51.4
2012	50.5

Sources: Pew Research Center;
U.S. Census Bureau

Assuming that the percentage of adults who are married will continue to decrease according to the exponential decay model:

a) Use the data for 1960 and 2012 to find the value of *k* and to write an exponential function that describes the percent of adults married after time *t*, in years, where *t* is the number of years after 1960.

b) Estimate the percent of adults who are married in 2015 and in 2018.

c) At this decay rate, in which year will the percent of adults who are married be 40%?

15. *British Guiana 1c Magenta Stamp.* The British Guiana 1c magenta stamp is considered the world's most famous stamp. It was issued in 1856, and only one of these stamps is known to exist. The sale of this stamp has broken the record for a single auction price four times. In 1980, John E. DuPont bought the stamp for $935,000. On June 17, 2014, the DuPont estate sold the stamp at Sotheby's auction in New York for a record price of $9,480,000 (*Source*: www.sothebys.com). Assuming that the value *S* of the stamp has grown exponentially:

a) Find the value of *k*, and determine the exponential growth function, assuming that $S_0 = \$935,000$ and *t* is the number of years after 1980.

b) Estimate the value of the stamp in 2000.

c) What is the doubling time for the value of the stamp?

d) After how long will the value of the stamp be $12 million, assuming that there is no change in growth rate?

16. *Oil Consumption.* In 1980, China consumed 1.85 million barrels of oil per day. By 2012, that consumption had grown to 10.28 million barrels per day. (*Sources*: U.S. Energy Information Administration; NextBigThingInvestor.com) Assuming that the consumption of oil C_0 in China has grown exponentially:

a) Find the value of k, and determine the exponential growth function, assuming that $C_0 = 1.85$ and t is the number of years after 1980.
b) Estimate the consumption of oil in 2005.
c) What is the doubling time for the consumption of oil in China?
d) After how long will the consumption of oil in China be 13 million barrels per day, assuming that there is no change in the growth rate?

17. *Spread of an Epidemic.* In a town whose population is 3500, a disease creates an epidemic. The number of people N infected t days after the disease has begun is given by the function

$$N(t) = \frac{3500}{1 + 19.9e^{-0.6t}}.$$

a) Graph the function.
b) How many are initially infected with the disease $(t = 0)$?
c) Find the number infected after 2 days, 5 days, 8 days, 12 days, and 16 days.
d) Using this model, can you say whether all 3500 people will ever be infected? Explain.

18. *Limited Population Growth in a Lake.* A lake is stocked with 640 fish of a new variety. The size of the lake, the availability of food, and the number of other fish restrict the growth of that type of fish in the lake to a limiting value of 3040. The population of fish in the lake after time t, in months, is given by the function

$$P(t) = \frac{3040}{1 + 3.75e^{-0.32t}}.$$

a) Graph the function.
b) Find the population after 0, 1, 5, 10, 15, and 20 months.

Newton's Law of Cooling. *Suppose that a body with temperature T_1 is placed in surroundings with temperature T_0 different from that of T_1. The body will either cool or warm to temperature $T(t)$ after time t, in minutes, where*

$$T(t) = T_0 + (T_1 - T_0)e^{-kt}.$$

Use this law in Exercises 19–22.

19. A cup of coffee with temperature 105°F is placed in a freezer with temperature 0°F. After 5 min, the temperature of the coffee is 70°F. What will its temperature be after 10 min?

20. A dish of lasagna baked at 375°F is taken out of the oven at 11:15 A.M. into a kitchen that is 72°F. After 3 min, the temperature of the lasagna is 365°F. What will the temperature of the lasagna be at 11:30 A.M.?

21. A chilled gelatin salad that has a temperature of 43°F is taken from the refrigerator and placed on the dining room table in a room that is 68°F. After 12 min, the temperature of the salad is 55°F. What will the temperature of the salad be after 20 min?

22. *When Was the Murder Committed?* The police discover the body of a murder victim. Critical to solving the crime is determining when the murder was committed. The coroner arrives at the murder scene at 12:00 P.M. She immediately takes the temperature of the body and finds it to be 94.6°F. She then takes the temperature 1 hr later and finds it to be 93.4°F. The temperature of the room is 70°F. When was the murder committed?

In Exercises 23–28, determine which, if any, of these functions might be used as a model for the data in the scatterplot.

a) *Quadratic, $f(x) = ax^2 + bx + c$*
b) *Polynomial, not quadratic*
c) *Exponential, $f(x) = ab^x$, or P_0e^{kx}, $k > 0$*
d) *Exponential, $f(x) = ab^{-x}$, or P_0e^{-kx}, $k > 0$*
e) *Logarithmic, $f(x) = a + b \ln x$*
f) *Logistic, $f(x) = \dfrac{a}{1 + be^{-kx}}$*

23.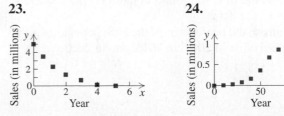

24.

25.   **26.**   **27.**   **28.**

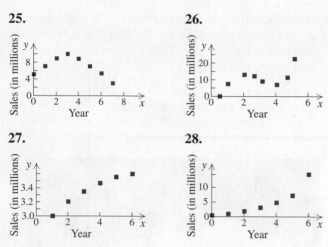

29. *Percent of Americans Ages 85 and Older.*   In 1900, 0.2% of the U.S. population, or 122,000 people, were ages 85 and older. This number grew to 5,751,000 in 2010. The following table lists data regarding the percentage of the U.S. population ages 85 and older in selected years from 1900 to 2010.

Year, x	Percent of U.S. Population Ages 85 and Older, y
1900, 0	0.2%
1910, 10	0.2
1920, 20	0.2
1930, 30	0.2
1940, 40	0.3
1950, 50	0.4
1960, 60	0.5
1970, 70	0.7
1980, 80	1.0
1990, 90	1.2
1995, 95	1.4
2000, 100	1.5
2005, 105	1.7
2010, 110	1.9

Sources: U.S. Census Bureau; U.S. Department of Commerce

a) Use a graphing calculator to fit an exponential function to the data, where x is the number of years after 1900. Determine whether the function is a good fit.
b) Graph the function found in part (a) with a scatterplot of the data.
c) Estimate the percentage of the U.S. population ages 85 and older in 2007, in 2015, and in 2020.

30. *Forgetting.*   In an economics class, students were given a final exam at the end of the course. Then they were retested with an equivalent test at subsequent time intervals. Their scores after time x, in months, are listed in the following table.

Time, x (in months)	Score, y
1	84.9%
2	84.6
3	84.4
4	84.2
5	84.1
6	83.9

a) Use a graphing calculator to fit a logarithmic function $y = a + b \ln x$ to the data.
b) Use the function to predict test scores after 8, 10, 24, and 36 months.
c) After how long will the test scores fall below 82%?

31. *Costs of Alzheimer's Disease.*   The costs of caring for Americans with Alzheimer's disease have surpassed the costs of caring for cancer patients. The following table lists the costs and projected costs of caring for Alzheimer's patients in selected years.

Year, x	Costs of Alzheimer's Disease (in billions)
2010, 0	$ 172
2014, 4	214
2020, 10	276
2025, 15	363
2030, 20	486
2035, 25	640
2040, 30	824
2045, 35	1009
2050, 40	1205

Sources: aarp.org/bulletin, "Where's the War on Alzheimer's," by T. R. Reid, January–February 2015; Alzheimer's Association

a) Use a graphing calculator to fit an exponential function to the data, where x is the number of years after 2010.
b) Graph the function found in part (a).
c) Use the function found in part (a) to estimate the costs of Alzheimer's disease in 2016 and in 2023.

32. *Wind Power Capacity.* U.S. wind power capacity has been increasing exponentially in recent years. The following table lists, for selected years, the U.S. wind power capacity, in megawatts (MW).

Year, x	U.S. Wind Power Capacity (in megawatts, MW)
2002, 0	4,557
2004, 2	6,619
2006, 4	11,450
2008, 6	25,065
2010, 8	40,283
2012, 10	60,012
2014, 12	65,879

Source: American Wind Energy Association

a) Use a graphing calculator to fit an exponential function to the data, where x is the number of years after 2002.
b) Graph the function in part (a).
c) Use the function in part (a) to estimate the wind capacity in 2011 and in 2015.

33. *Architects in Indiana.* The number of architects employed in Indiana in 2013 is more than 500 less than the number employed in 2007. The following table lists the number of architects employed in selected years.

Year, x	Number of Architects Employed In Indiana
2007, 0	1460
2009, 2	1280
2011, 4	1010
2013, 6	950

Source: Indiana Department of Workforce Development

a) Use a graphing calculator to model the data with an exponential function, where x is the number of years after 2007.
b) Use the function found in part (a) to estimate the number of architects employed in Indiana in 2008 and in 2012.
c) If the number of architects continues to decrease at the same rate, in what year will the number of architects employed in Indiana be 900?

34. *Effect of Advertising.* A company introduced a new software product on a trial run in a city. They advertised the product on television and found the following data regarding the percent P of people who bought the product after x ads were run.

Number of Ads, x	Percentage Who Bought, P
0	0.2%
10	0.7
20	2.7
30	9.2
40	27.0
50	57.6
60	83.3
70	94.8
80	98.5
90	99.6

a) Use a graphing calculator to fit a logistic function

$$P(x) = \frac{a}{1 + be^{-kx}}$$

to the data.
b) What percent of people bought the product when 55 ads were run? 100 ads?
c) Find the horizontal asymptote for the graph. Interpret the asymptote in terms of the advertising situation.

❯ **Skill Maintenance**

Vocabulary Reinforcement

In Exercises 35–40, choose the correct name of the principle or rule from the given choices.

principle of zero products
multiplication principle for equations
product rule
addition principle for inequalities
power rule
multiplication principle for inequalities
principle of square roots
quotient rule

35. For any real numbers a, b, and c: If $a < b$ and $c > 0$ are true, then $ac < bc$ is true. If $a < b$ and $c < 0$ are true, then $ac > bc$ is true. **[1.6]**

———————

36. For any positive numbers M and N and any logarithm base a, $\log_a MN = \log_a M + \log_a N$. **[5.4]**

———————

37. If $ab = 0$ is true, then $a = 0$ or $b = 0$, and if $a = 0$ or $b = 0$, then $ab = 0$. **[3.2]**

———————

38. If $x^2 = k$, then $x = \sqrt{k}$ or $x = -\sqrt{k}$. **[3.2]**

———————

39. For any positive number M, any logarithm base a, and any real number p, $\log_a M^p = p \log_a M$. **[5.4]**

———————

40. For any real numbers a, b, and c: If $a = b$ is true, then $ac = bc$ is true. **[1.5]**

———————

❯ **Synthesis**

41. *Supply and Demand.* The supply function and the demand function for the sale of a certain type of DVD player are given by
$$S(p) = 150e^{0.004p} \quad \text{and} \quad D(p) = 480e^{-0.003p},$$
respectively, where $S(p)$ is the number of DVD players that the company is willing to sell at price p and $D(p)$ is the quantity that the public is willing to buy at price p. Find p such that $D(p) = S(p)$. This is called the **equilibrium price**.

42. *Carbon Dating.* Recently, while digging in Chaco Canyon, New Mexico, archaeologists found corn pollen that was 4000 years old (*Source*: *American Anthropologist*). This was evidence that Native Americans had been cultivating crops in the Southwest centuries earlier than scientists had thought. What percent of the carbon-14 had been lost from the pollen?

43. *Present Value.* Following the birth of a child, a grandparent wants to make an initial investment P_0 that will grow to $50,000 for the child's education at age 18. Interest is compounded continuously at 5.2%. What should the initial investment be? Such an amount is called the **present value** of $50,000 due 18 years from now.

44. *Present Value.*
  a) Solve $P = P_0 e^{kt}$ for P_0.
  b) Referring to Exercise 43, find the present value of $50,000 due 18 years from now at interest rate 6.4%, compounded continuously.

45. *The Beer–Lambert Law.* A beam of light enters a medium such as water or smog with initial intensity I_0. Its intensity decreases depending on the thickness (or concentration) of the medium. The intensity I at a depth (or concentration) of x units is given by
$$I = I_0 e^{-\mu x}.$$
The constant μ (the Greek letter "mu") is called the **coefficient of absorption**, and it varies with the medium. For sea water, $\mu = 1.4$.
  a) What percentage of light intensity I_0 remains in sea water at a depth of 1 m? 3 m? 5 m? 50 m?
  b) Plant life cannot exist below 10 m. What percentage of I_0 remains at 10 m?

46. Given that $y = ax^b$, take the natural logarithm on both sides. Let $Y = \ln y$ and $X = \ln x$. Consider Y as a function of X. What kind of function is Y?

47. Given that $y = ae^x$, take the natural logarithm on both sides. Let $Y = \ln y$. Consider Y as a function of x. What kind of function is Y?

STUDY GUIDE

KEY TERMS AND CONCEPTS	EXAMPLES

SECTION 5.1: INVERSE FUNCTIONS

Inverse Relation

If a relation is defined by an equation, then interchanging the variables produces an equation of the inverse relation.

Given $y = -5x + 7$, find an equation of the inverse relation.

$$y = -5x + 7 \quad \textbf{Relation}$$
$$\downarrow \qquad \downarrow$$
$$x = -5y + 7 \quad \textbf{Inverse relation}$$

One-to-One Functions

A function f is one-to-one if different inputs have different outputs—that is,

$$\text{if } a \neq b, \quad \text{then} \quad f(a) \neq f(b).$$

Or a function f is one-to-one if when the outputs are the same, the inputs are the same—that is,

$$\text{if } f(a) = f(b), \quad \text{then} \quad a = b.$$

Prove that $f(x) = 16 - 3x$ is one-to-one.

Show that if $f(a) = f(b)$, then $a = b$. Assume that $f(a) = f(b)$. Since $f(a) = 16 - 3a$ and $f(b) = 16 - 3b$,

$$16 - 3a = 16 - 3b$$
$$-3a = -3b$$
$$a = b.$$

Thus, if $f(a) = f(b)$, then $a = b$ and f is one-to-one.

Horizontal-Line Test

If it is possible for a horizontal line to intersect the graph of a function more than once, then the function is *not* one-to-one and its inverse is *not* a function.

One-to-One Functions and Inverses

- If a function f is one-to-one, then its inverse f^{-1} is a function.
- The domain of a one-to-one function f is the range of the inverse f^{-1}.
- The range of a one-to-one function f is the domain of the inverse f^{-1}.
- A function that is increasing over its entire domain or is decreasing over its entire domain is a one-to-one function.

The -1 in f^{-1} is *not* an exponent.

Using its graph, determine whether each function is one-to-one.

a)

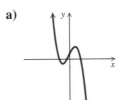

b)

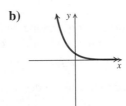

a) There are many horizontal lines that intersect the graph more than once. Thus the function is *not* one-to-one and its inverse is *not* a function.

b) No horizontal line intersects the graph more than once. Thus the function is one-to-one and its inverse is a function.

(continued)

Obtaining a Formula for an Inverse

If a function f is one-to-one, a formula for its inverse can generally be found as follows:

1. Replace $f(x)$ with y.
2. Interchange x and y.
3. Solve for y.
4. Replace y with $f^{-1}(x)$.

The graph of f^{-1} is a reflection of the graph of f across the line $y = x$.

Given the one-to-one function $f(x) = 2 - x^3$, find a formula for its inverse. Then graph the function and its inverse on the same set of axes.

$$f(x) = 2 - x^3$$
$$\downarrow$$

1. $y = 2 - x^3$ **Replacing $f(x)$ with y**
$$\downarrow \qquad\qquad \downarrow$$
2. $x = 2 - y^3$ **Interchanging x and y**

3. Solve for y:

$$y^3 = 2 - x \qquad \textbf{Adding } y^3 \textbf{ and subtracting } x$$
$$y = \sqrt[3]{2 - x}.$$
$$\downarrow$$

4. $f^{-1}(x) = \sqrt[3]{2 - x}$ **Replacing y with $f^{-1}(x)$**

$$f(x) = 2 - x^3 \text{ and } f^{-1}(x) = \sqrt[3]{2 - x}$$

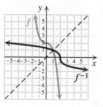

If a function f is one-to-one, then f^{-1} is the unique function such that each of the following holds:

$$(f^{-1} \circ f)(x) = f^{-1}(f(x)) = x,$$
for each x in the domain of f, and

$$(f \circ f^{-1})(x) = f(f^{-1}(x)) = x,$$
for each x in the domain of f^{-1}.

Given $f(x) = \dfrac{3 + x}{x}$, use composition of functions to show that

$$f^{-1}(x) = \frac{3}{x - 1}.$$

$$(f^{-1} \circ f)(x) = f^{-1}(f(x))$$
$$= f^{-1}\left(\frac{3 + x}{x}\right) = \frac{3}{\dfrac{3 + x}{x} - 1}$$

$$= \frac{3}{\dfrac{3 + x - x}{x}} = \frac{3}{\dfrac{3}{x}} = 3 \cdot \frac{x}{3} = x;$$

$$(f \circ f^{-1})(x) = f(f^{-1}(x)) = f\left(\frac{3}{x - 1}\right)$$

$$= \frac{3 + \dfrac{3}{x - 1}}{\dfrac{3}{x - 1}} = \frac{\dfrac{3(x - 1) + 3}{x - 1}}{\dfrac{3}{x - 1}}$$

$$= \frac{3(x - 1) + 3}{x - 1} \cdot \frac{x - 1}{3}$$

$$= \frac{3x - 3 + 3}{3} = \frac{3x}{3} = x$$

SECTION 5.2: EXPONENTIAL FUNCTIONS AND GRAPHS

Exponential Function

$y = a^x$, or $f(x) = a^x$, $\quad a > 0, a \neq 1$

Continuous

One-to-one

Domain: $(-\infty, \infty)$

Range: $(0, \infty)$

Increasing if $a > 1$

Decreasing if $0 < a < 1$

Horizontal asymptote is x-axis

y-intercept: $(0, 1)$

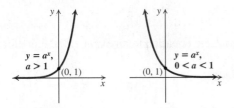

Graph: $f(x) = 2^x$, $g(x) = 2^{-x}$, $h(x) = 2^{x-1}$, and $t(x) = 2^x - 1$.

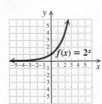

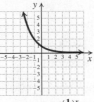

$$g(x) = 2^{-x} = \left(\tfrac{1}{2}\right)^x$$

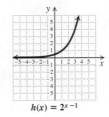

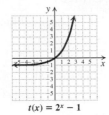

$h(x) = 2^{x-1}$

$t(x) = 2^x - 1$

Compound Interest

The amount of money A to which a principal P will grow after t years at interest rate r (in decimal form), compounded n times per year, is given by the formula

$$A = P\left(1 + \frac{r}{n}\right)^{nt}.$$

Suppose that \$5000 is invested at 3.5% interest, compounded quarterly. Find the money in the account after 3 years.

$$A = P\left(1 + \frac{r}{n}\right)^{nt} = 5000\left(1 + \frac{0.035}{4}\right)^{4 \cdot 3}$$
$$\approx \$5551.02$$

The Number e

$e = 2.7182818284\ldots$

Find each of the following, to four decimal places, using a calculator.

$e^{-3} \approx 0.0498;$

$e^{4.5} \approx 90.0171$

Graph: $f(x) = e^x$ and $g(x) = e^{-x+2} - 4$.

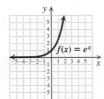

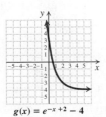

$g(x) = e^{-x+2} - 4$

SECTION 5.3: LOGARITHMIC FUNCTIONS AND GRAPHS

Logarithmic Function

$y = \log_a x, \quad x > 0, a > 0, a \neq 1$

Continuous

One-to-one

Domain: $(0, \infty)$

Range: $(-\infty, \infty)$

Increasing if $a > 1$

Vertical asymptote is y-axis

x-intercept: $(1, 0)$

The inverse of an exponential function $f(x) = a^x$ is given by $f^{-1}(x) = \log_a x$.

Graph: $f(x) = \log_2 x$ and $g(x) = \ln(x - 1) + 2$.

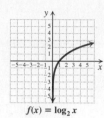

$f(x) = \log_2 x$

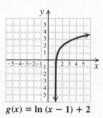

$g(x) = \ln(x - 1) + 2$

A logarithm is an exponent:

$$\log_a x = y \longleftrightarrow x = a^y.$$

$\log x$ means $\log_{10} x$ **Common logarithms**

$\ln x$ means $\log_e x$ **Natural logarithms**

Convert each logarithmic equation to an exponential equation.

$$\log_4 \frac{1}{16} = -2 \longleftrightarrow 4^{-2} = \frac{1}{16};$$

$$\ln R = 3 \longleftrightarrow e^3 = R$$

Convert each exponential equation to a logarithmic equation.

$$e^{-5} = 0.0067 \longleftrightarrow \ln 0.0067 = -5;$$

$$7^2 = 49 \longleftrightarrow \log_7 49 = 2$$

For any logarithm base a,

$$\log_a 1 = 0 \quad \text{and} \quad \log_a a = 1.$$

For the logarithm base e,

$$\ln 1 = 0 \quad \text{and} \quad \ln e = 1.$$

Find each of the following without using a calculator.

$\log 100 = 2$; $\log 10^{-5} = -5$;

$\ln 1 = 0$; $\log_9 9 = 1$;

$\ln \sqrt[3]{e} = \frac{1}{3}$; $\log_2 64 = 6$;

$\log_8 1 = 0$; $\ln e = 1$

Find each of the following using a calculator and rounding to four decimal places.

$\ln 223 = 5.4072$; $\log \frac{2}{9} = -0.6532$;

$\log(-8)$ Does not exist; $\ln 0.06 = -2.8134$

The Change-of-Base Formula

For any logarithmic bases a and b, and any positive number M,

$$\log_b M = \frac{\log_a M}{\log_a b}.$$

Find $\log_3 11$ using common logarithms:

$$\log_3 11 = \frac{\log 11}{\log 3} \approx 2.1827.$$

Find $\log_3 11$ using natural logarithms:

$$\log_3 11 = \frac{\ln 11}{\ln 3} \approx 2.1827.$$

Earthquake Magnitude

The magnitude R, measured on the Richter scale, of an earthquake of intensity I is defined as

$$R = \log \frac{I}{I_0},$$

where I_0 is a minimum intensity used for comparison.

What is the magnitude on the Richter scale of an earthquake of intensity $10^{6.8} \cdot I_0$?

$$R = \log \frac{I}{I_0} = \log \frac{10^{6.8} \cdot I_0}{I_0} = \log 10^{6.8} = 6.8$$

SECTION 5.4: PROPERTIES OF LOGARITHMIC FUNCTIONS

The Product Rule

For any positive numbers M and N, and any logarithmic base a,

$$\log_a MN = \log_a M + \log_a N.$$

The Power Rule

For any positive number M, any logarithmic base a, and any real number p,

$$\log_a M^p = p \log_a M.$$

The Quotient Rule

For any positive numbers M and N, and any logarithmic base a,

$$\log_a \frac{M}{N} = \log_a M - \log_a N.$$

Express $\log_c \sqrt{\dfrac{c^2 r}{b^3}}$ in terms of sums and differences of logarithms.

$$\begin{aligned}
\log_c \sqrt{\frac{c^2 r}{b^3}} &= \log_c \left(\frac{c^2 r}{b^3} \right)^{1/2} \\
&= \tfrac{1}{2} \log_c \left(\frac{c^2 r}{b^3} \right) \\
&= \tfrac{1}{2} (\log_c c^2 r - \log_c b^3) \\
&= \tfrac{1}{2} (\log_c c^2 + \log_c r - \log_c b^3) \\
&= \tfrac{1}{2} (2 + \log_c r - 3 \log_c b) \\
&= 1 + \tfrac{1}{2} \log_c r - \tfrac{3}{2} \log_c b
\end{aligned}$$

Express $\ln (3x^2 + 5x - 2) - \ln (x + 2)$ as a single logarithm.

$$\begin{aligned}
\ln (3x^2 + 5x - 2) - \ln (x + 2) &= \ln \frac{3x^2 + 5x - 2}{x + 2} \\
&= \ln \frac{(3x - 1)(x + 2)}{x + 2} \\
&= \ln (3x - 1)
\end{aligned}$$

Given $\log_a 7 \approx 0.8451$ and $\log_a 5 \approx 0.6990$, find $\log_a \frac{1}{7}$ and $\log_a 35$.

$$\log_a \frac{1}{7} = \log_a 1 - \log_a 7 \approx 0 - 0.8451 \approx -0.8451;$$

$$\begin{aligned}
\log_a 35 = \log_a (7 \cdot 5) &= \log_a 7 + \log_a 5 \\
&\approx 0.8451 + 0.6990 \\
&\approx 1.5441
\end{aligned}$$

For any base a and any real number x,

$$\log_a a^x = x.$$

For any base a and any positive real number x,

$$a^{\log_a x} = x.$$

Simplify each of the following.

$$8^{\log_8 k} = k; \qquad \log 10^{43} = 43;$$

$$\log_a a^4 = 4; \qquad e^{\ln 2} = 2$$

SECTION 5.5: SOLVING EXPONENTIAL EQUATIONS AND LOGARITHMIC EQUATIONS

The Base–Exponent Property

For any $a > 0$, $a \neq 1$,

$$a^x = a^y \longleftrightarrow x = y.$$

Solve: $3^{2x-3} = 81$.

$$3^{2x-3} = 3^4 \qquad \mathbf{81 = 3^4}$$
$$2x - 3 = 4$$
$$2x = 7$$
$$x = \tfrac{7}{2}$$

The solution is $\frac{7}{2}$.

The Property of Logarithmic Equality

For any $M > 0$, $N > 0$, $a > 0$, and $a \neq 1$,

$$\log_a M = \log_a N \longleftrightarrow M = N.$$

Solve: $6^{x-2} = 2^{-3x}$.

$$\log 6^{x-2} = \log 2^{-3x}$$
$$(x - 2)\log 6 = -3x \log 2$$
$$x \log 6 - 2 \log 6 = -3x \log 2$$
$$x \log 6 + 3x \log 2 = 2 \log 6$$
$$x(\log 6 + 3 \log 2) = 2 \log 6$$
$$x = \frac{2 \log 6}{\log 6 + 3 \log 2}$$
$$x \approx 0.9257$$

Solve: $\log_3 (x - 2) + \log_3 x = 1$.

$$\log_3 [x(x - 2)] = 1$$
$$x(x - 2) = 3^1$$
$$x^2 - 2x - 3 = 0$$
$$(x - 3)(x + 1) = 0$$
$$x - 3 = 0 \quad or \quad x + 1 = 0$$
$$x = 3 \quad or \quad x = -1$$

The number -1 is not a solution because negative numbers do not have real-number logarithms. The value 3 checks and is the solution.

Solve: $\ln (x + 10) - \ln (x + 4) = \ln x$.

$$\ln \frac{x + 10}{x + 4} = \ln x$$
$$\frac{x + 10}{x + 4} = x$$
$$x + 10 = x(x + 4)$$
$$x + 10 = x^2 + 4x$$
$$0 = x^2 + 3x - 10$$
$$0 = (x + 5)(x - 2)$$
$$x + 5 = 0 \quad or \quad x - 2 = 0$$
$$x = -5 \quad or \quad x = 2$$

The number -5 is not a solution because $-5 + 4 = -1$ and $\ln (-1)$ is not a real number. The value 2 checks and is the solution.

SECTION 5.6: APPLICATIONS AND MODELS: GROWTH AND DECAY; COMPOUND INTEREST

Exponential Growth Model

$P(t) = P_0e^{kt}, \quad k > 0$

Doubling Time

$kT = \ln 2, \quad \text{or} \quad k = \dfrac{\ln 2}{T},$

$\text{or} \quad T = \dfrac{\ln 2}{k}$

In April 2015, the population of the United States was 320.8 million, and the exponential growth rate was 0.7% per year (*Source: CIA World Factbook* 2015). After how long will the population be double what it was in 2015? Estimate the population in 2020.

With a population growth rate of 0.7%, or 0.007, the doubling time T is

$$T = \frac{\ln 2}{k} = \frac{\ln 2}{0.007} \approx 99.$$

The population of the United States will be double what it was in 2015 in about 99 years after 2015.

The exponential growth function is

$$P(t) = 320.8e^{0.007t},$$

where t is the number of years after 2015 and $P(t)$ is in millions. Since $t = 5$ in 2020, we substitute 5 for t:

$$P(5) = 320.8e^{0.007(5)} = 320.8e^{0.035} \approx 332.2.$$

The population will be about 332.2 million, or 332,200,000 in 2020.

Interest Compounded Continuously

$P(t) = P_0e^{kt}, \quad k > 0$

Suppose that $20,000 is invested at interest rate k, compounded continuously, and grows to $23,236.68 in 3 years. What is the interest rate? What will the balance be in 8 years?

The exponential growth function is of the form $P(t) = 20{,}000e^{kt}$. Given that $P(3) = \$23{,}236.68$, substituting 3 for t and 23,236.68 for $P(t)$ gives

$$23{,}236.68 = 20{,}000e^{k(3)}$$

to get $k \approx 0.05$, or 5%.

We then substitute 0.05 for k and 8 for t and determine $P(8)$:

$$P(8) = 20{,}000e^{0.05(8)} = 20{,}000e^{0.4} \approx \$29{,}836.49.$$

Exponential Decay Model

$P(t) = P_0e^{-kt}, \quad k > 0$

Half-Life

$kT = \ln 2, \quad \text{or} \quad k = \dfrac{\ln 2}{T},$

$\text{or} \quad T = \dfrac{\ln 2}{k}$

Archaeologists discovered an animal bone that had lost 65.2% of its carbon-14 at the time it was found. How old was the bone?

The decay rate for carbon-14 is 0.012%, or 0.00012. If the bone has lost 65.2% of its carbon-14 from an initial amount P_0, then 34.8%P_0 is the amount present. We substitute 34.8%P_0 for $P(t)$ and solve:

$$34.8\%P_0 = P_0e^{-0.00012t}$$
$$0.348 = e^{-0.00012t}$$
$$\ln 0.348 = -0.00012t$$
$$\frac{\ln 0.348}{-0.00012} = t$$
$$8796 \approx t.$$

The bone was about 8796 years old when it was found.

REVIEW EXERCISES

Determine whether the statement is true or false.

1. The domain of a one-to-one function f is the range of the inverse f^{-1}. **[5.1]**

2. The x-intercept of $f(x) = \log x$ is $(0, 1)$. **[5.3]**

3. The graph of f^{-1} is a reflection of the graph of f across $y = 0$. **[5.1]**

4. If it is not possible for a horizontal line to intersect the graph of a function more than once, then the function is one-to-one and its inverse is a function. **[5.1]**

5. The range of all exponential functions is $[0, \infty)$. **[5.2]**

6. The horizontal asymptote of $y = 2^x$ is $y = 0$. **[5.2]**

7. Find the inverse of the relation

$$\{(1.3, -2.7), (8, -3), (-5, 3), (6, -3), (7, -5)\}.$$

   [5.1]

8. Find an equation of the inverse relation. **[5.1]**

   a) $y = -2x + 3$
   b) $y = 3x^2 + 2x - 1$
   c) $0.8x^3 - 5.4y^2 = 3x$

Graph the function and determine whether the function is one-to-one using the horizontal-line test. **[5.1]**

9. $f(x) = -|x| + 3$

10. $f(x) = x^2 + 1$

11. $f(x) = 2x - \frac{3}{4}$

12. $f(x) = -\dfrac{6}{x + 1}$

In Exercises 13–18, given the function:

a) *Sketch the graph and determine whether the function is one-to-one.* **[5.1], [5.3]**

b) *If it is one-to-one, find a formula for the inverse.* **[5.1], [5.3]**

13. $f(x) = 2 - 3x$

14. $f(x) = \dfrac{x + 2}{x - 1}$

15. $f(x) = \sqrt{x - 6}$

16. $f(x) = x^3 - 8$

17. $f(x) = 3x^2 + 2x - 1$

18. $f(x) = e^x$

For the function f, use composition of functions to show that f^{-1} is as given. **[5.1]**

19. $f(x) = 6x - 5$, $f^{-1}(x) = \dfrac{x + 5}{6}$

20. $f(x) = \dfrac{x + 1}{x}$, $f^{-1}(x) = \dfrac{1}{x - 1}$

Find the inverse of the given one-to-one function f. Give the domain and the range of f and of f^{-1} and then graph both f and f^{-1} on the same set of axes. **[5.1]**

21. $f(x) = 2 - 5x$

22. $f(x) = \dfrac{x - 3}{x + 2}$

23. Find $f(f^{-1}(657))$:

$$f(x) = \frac{4x^5 - 16x^{37}}{119x}, \quad x > 1. \ \textbf{[5.1]}$$

24. Find $f(f^{-1}(a))$: $f(x) = \sqrt[3]{3x - 4}$. **[5.1]**

Graph the function.

25. $f(x) = \left(\frac{1}{3}\right)^x$ **[5.2]**

26. $f(x) = 1 + e^x$ **[5.2]**

27. $f(x) = -e^{-x}$ **[5.2]**

28. $f(x) = \log_2 x$ **[5.3]**

29. $f(x) = \frac{1}{2} \ln x$ **[5.3]**

30. $f(x) = \log x - 2$ **[5.3]**

In Exercises 31–36, match the equation with one of the figures (a)–(f) that follow.

a)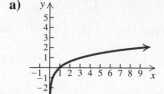

b)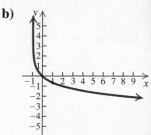

c)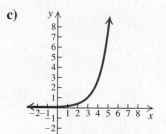

d)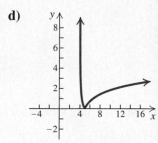

e)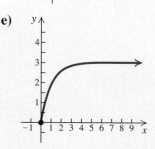

f)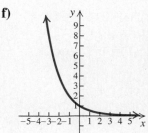

31. $f(x) = e^{x-3}$ **[5.2]**

32. $f(x) = \log_3 x$ **[5.3]**

33. $y = -\log_3 (x + 1)$ **[5.3]**

34. $y = \left(\frac{1}{2}\right)^x$ **[5.2]**

35. $f(x) = 3(1 - e^{-x}), x \geq 0$ **[5.2]**

36. $f(x) = |\ln (x - 4)|$ **[5.3]**

Find each of the following. Do not use a calculator. **[5.3]**

37. $\log_5 125$ **38.** $\log 100{,}000$

39. $\ln e$ **40.** $\ln 1$

41. $\log 10^{1/4}$ **42.** $\log_3 \sqrt{3}$

43. $\log 1$ **44.** $\log 10$

45. $\log_2 \sqrt[3]{2}$ **46.** $\log 0.01$

Convert to an exponential equation. **[5.3]**

47. $\log_4 x = 2$ **48.** $\log_a Q = k$

Convert to a logarithmic equation. **[5.3]**

49. $4^{-3} = \frac{1}{64}$ **50.** $e^x = 80$

Find each of the following using a calculator. Round to four decimal places. **[5.3]**

51. $\log 11$ **52.** $\log 0.234$

53. $\ln 3$ **54.** $\ln 0.027$

55. $\log (-3)$ **56.** $\ln 0$

Find the logarithm using the change-of-base formula. **[5.3]**

57. $\log_5 24$ **58.** $\log_8 3$

Express as a single logarithm and, if possible, simplify. **[5.4]**

59. $3 \log_b x - 4 \log_b y + \frac{1}{2} \log_b z$

60. $\ln (x^3 - 8) - \ln (x^2 + 2x + 4) + \ln (x + 2)$

Express in terms of sums and differences of logarithms. **[5.4]**

61. $\ln \sqrt[4]{wr^2}$ **62.** $\log \sqrt[3]{\dfrac{M^2}{N}}$

Given that $\log_a 2 = 0.301$, $\log_a 5 = 0.699$, and $\log_a 6 = 0.778$, find each of the following. **[5.4]**

63. $\log_a 3$ **64.** $\log_a 50$

65. $\log_a \frac{1}{5}$ **66.** $\log_a \sqrt[3]{5}$

Simplify. **[5.4]**

67. $\ln e^{-5k}$ **68.** $\log_5 5^{-6t}$

Solve. **[5.5]**

69. $\log_4 x = 2$ **70.** $3^{1-x} = 9^{2x}$

71. $e^x = 80$ **72.** $4^{2x-1} - 3 = 61$

73. $\log_{16} 4 = x$ **74.** $\log_x 125 = 3$

75. $\log_2 x + \log_2 (x - 2) = 3$

76. $\log (x^2 - 1) - \log (x - 1) = 1$

77. $\log x^2 = \log x$ **78.** $e^{-x} = 0.02$

79. *Saving for College.* Following the birth of triplets, the grandparents deposit $30,000 in a college trust fund that earns 4.2% interest, compounded quarterly.

  a) Find a function for the amount in the account after t years. **[5.2]**

  b) Find the amount in the account at $t = 0, 6, 12$, and 18 years. **[5.2]**

80. *Breweries.* The number of breweries in the United States has increased in recent years. The total number of breweries can be estimated using the exponential function

$$B(t) = 2456.1(1.188)^t,$$

where t is the number of years after 2012 (*Source:* Brewers Association). Find the number of breweries in 2014. Then use this function to estimate the number of breweries in 2018. **[5.2]**

81. How long will it take an investment to double if it is invested at 4.5%, compounded continuously? **[5.6]**

82. The population of a metropolitan area consisting of 8 counties doubled in 26 years. What was the exponential growth rate? **[5.6]**

83. How old is a skeleton that has lost 27% of its carbon-14? **[5.6]**

84. The hydrogen ion concentration of milk is 2.3×10^{-6}. What is the pH? (See Exercise 98 in Exercise Set 5.3.) **[5.3]**

85. *Earthquake Magnitude.* The earthquake in Haiti on January 12, 2010, had an intensity of $10^{7.0} \cdot I_0$ (*Source:* U.S. Geological Survey). What is the magnitude of the earthquake on the Richter scale? **[5.3]**

86. What is the loudness, in decibels, of a sound whose intensity is $1000I_0$? (See Exercise 101 in Exercise Set 5.3.) **[5.3]**

87. *Walking Speed.* The average walking speed w, in feet per second, of a person living in a city of population P, in thousands, is given by the function

$$w(P) = 0.37 \ln P + 0.05.$$

  a) The population of Wichita, Kansas, is 353,823. Find the average walking speed. **[5.3]**

  b) A city's population has an average walking speed of 3.4 ft/sec. Find the population. **[5.6]**

88. *Social Security Distributions.* Cash Social Security distributions were $35 million, or $0.035 billion, in 1940. This amount has increased exponentially to $786 billion in 2012. (*Source*: Pew Research Center) Assuming that the exponential growth model applies:

a) Find the exponential growth rate k. **[5.6]**
b) Find the exponential growth function. **[5.6]**
c) Estimate the total cash distributions in 1970, in 2000, and in 2015. **[5.6]**
d) In what year will the cash benefits reach $2 trillion? **[5.6]**

89. *The Population of Cambodia.* The population of Cambodia was 15.2 million in 2013, and the exponential growth rate was 1.67% per year (*Source*: U.S. Census Bureau, World Population Profile).

a) Find the exponential growth function. **[5.6]**
b) What will the population be in 2017? in 2020? **[5.6]**
c) When will the population be 18 million? **[5.6]**
d) What is the doubling time? **[5.6]**

Cambodia

90. *Atmospheric Carbon as Carbon Dioxide (CO_2).* Scientists are of the opinion that the world's cumulative emissions of carbon as CO_2 should not exceed one trillion metric tons. The following table lists the cumulative atmospheric carbon as CO_2, in billions of metric tons, in selected years.

Year, x	Cumulative Atmospheric Carbon as CO_2 (in billions of metric tons)
1933, 0	100
1967, 34	200
1983, 50	300
1997, 64	400
2008, 75	500
2012, 79	545

Sources: *National Geographic*, April 2014, "Can Goal Ever Be Clean?," p. 38, by Michelle Nijhuis; Thomas Boden, Carbon Dioxide Information Analysis Center/Oak Ridge Laboratory, U.S. Department of Energy; EPA

a) Use a graphing calculator to fit an exponential function to the data, where x is the number of years after 1933. **[5.6]**
b) Graph the function with a scatterplot of the data. **[5.6]**
c) Estimate the cumulative atmospheric carbon in 1970, in 2000, and in 2015. **[5.6]**
d) In what year will the cumulative atmospheric carbon as CO_2 reach one trillion metric tons? **[5.6]**

91. Using only a graphing calculator, determine whether the following functions are inverses of each other.

$$f(x) = \frac{4 + 3x}{x - 2}, \qquad g(x) = \frac{x + 4}{x - 3} \quad \textbf{[5.1]}$$

92. a) Use a graphing calculator to graph
$$f(x) = 5e^{-x} \ln x$$
in the viewing window $[-1, 10, -5, 5]$. **[5.2], [5.3]**
b) Estimate the relative maximum and minimum values of the function. **[5.2], [5.3]**

93. Which of the following is the horizontal asymptote of the graph of $f(x) = e^{x-3} + 2$? **[5.2]**

A. $y = -2$ B. $y = -3$
C. $y = 3$ D. $y = 2$

94. Which of the following is the domain of the logarithmic function $f(x) = \log(2x - 3)$? **[5.3]**

A. $\left(\frac{3}{2}, \infty\right)$ B. $\left(-\infty, \frac{3}{2}\right)$
C. $(3, \infty)$ D. $(-\infty, \infty)$

95. The graph of $f(x) = 2^{x-2}$ is which of the following? **[5.2]**

A.   B.

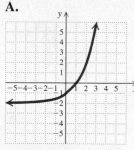 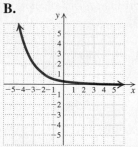

C.   D.

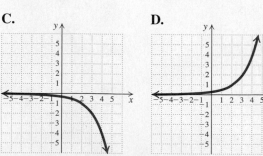

96. The graph of $f(x) = \log_2 x$ is which of the following? **[5.3]**

A.

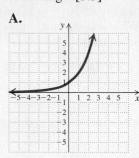

B.

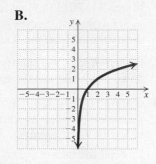

C.

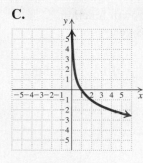

D.

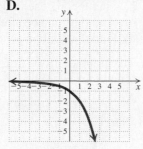

❯ Synthesis

Solve. **[5.5]**

97. $|\log_4 x| = 3$

98. $\log x = \ln x$

99. $5^{\sqrt{x}} = 625$

100. Find the domain: $f(x) = \log_3 (\ln x)$. **[5.3]**

❯ Collaborative Discussion and Writing

101. Explain how the graph of $f(x) = \ln x$ can be used to obtain the graph of $g(x) = e^{x-2}$. **[5.3]**

102. *Atmospheric Pressure.*   Atmospheric pressure P at an altitude a is given by
$$P = P_0 e^{-0.00005a},$$
where P_0 is the pressure at sea level, approximately $14.7 \, \text{lb/in}^2$ (pounds per square inch). Explain how a barometer, or some device for measuring atmospheric pressure, can be used to find the height of a skyscraper. **[5.6]**

103. Explain the errors, if any, in the following: **[5.4]**
$$\log_a ab^3 = (\log_a a)(\log_a b^3) = 3 \log_a b.$$

104. Describe the difference between $f^{-1}(x)$ and $[f(x)]^{-1}$. **[5.1]**

---

5 | Chapter Test

1. Find the inverse of the relation
$$\{(-2, 5), (4, 3), (0, -1), (-6, -3)\}.$$

Determine whether the function is one-to-one. Answer yes or no.

2.

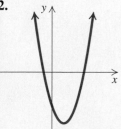

3.

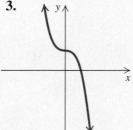

In Exercises 4–7, given the function:

a) *Sketch the graph and determine whether the function is one-to-one.*

b) *If it is one-to-one, find a formula for the inverse.*

4. $f(x) = x^3 + 1$

5. $f(x) = 1 - x$

6. $f(x) = \dfrac{x}{2 - x}$

7. $f(x) = x^2 + x - 3$

8. Use composition of functions to show that f^{-1} is as given:
$$f(x) = -4x + 3, \qquad f^{-1}(x) = \frac{3 - x}{4}.$$

9. Find the inverse of the one-to-one function

$$f(x) = \frac{1}{x - 4}.$$

Give the domain and the range of f and of f^{-1} and then graph both f and f^{-1} on the same set of axes.

Graph the function.

10. $f(x) = 4^{-x}$

11. $f(x) = \log x$

12. $f(x) = e^x - 3$

13. $f(x) = \ln (x + 2)$

Find each of the following. Do not use a calculator.

14. $\log 0.00001$ **15.** $\ln e$

16. $\ln 1$ **17.** $\log_4 \sqrt[5]{4}$

18. Convert to an exponential equation: $\ln x = 4$.

19. Convert to a logarithmic equation: $3^x = 5.4$.

Find each of the following using a calculator. Round to four decimal places.

20. $\ln 16$ **21.** $\log 0.293$

22. Find $\log_6 10$ using the change-of-base formula.

23. Express as a single logarithm:

$$2 \log_a x - \log_a y + \tfrac{1}{2} \log_a z.$$

24. Express $\ln \sqrt[5]{x^2 y}$ in terms of sums and differences of logarithms.

25. Given that $\log_a 3 = 1.585$ and $\log_a 15 = 3.907$, find $\log_a 5$.

26. Simplify: $\ln e^{-4t}$.

Solve.

27. $\log_{25} 5 = x$

28. $\log_3 x + \log_3 (x + 8) = 2$

29. $3^{4-x} = 27^x$

30. $e^x = 65$

31. *Earthquake Magnitude.*   The Tohoku earthquake near the northeast coast of Honshu, Japan, on March 11, 2011, had an intensity of $10^{9.0} \cdot I_0$ (*Source:* U.S. Geological Survey). What was the magnitude of the earthquake on the Richter scale?

32. *Growth Rate.*   A country's population doubled in 45 years. What was the exponential growth rate?

33. *Compound Interest.*   Suppose that $1000 is invested at interest rate k, compounded continuously, and grows to $1144.54 in 3 years.

  a) Find the interest rate.
  b) Find the exponential growth function.
  c) Find the balance after 8 years.
  d) Find the doubling time.

34. The graph of $f(x) = 2^{x-1} + 1$ is which of the following?

A.

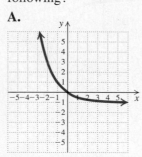

B.

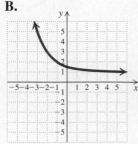

C.

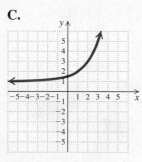

D.

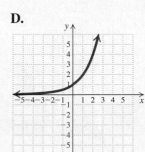

❯ Synthesis

35. Solve: $4^{\sqrt[3]{x}} = 8$.

The Trigonometric Functions

APPLICATION

This problem appears as Exercise 23 in Section 6.2.

The longest escalator in the world is in the subway system in St. Petersburg, Russia. The escalator is 1084.6 ft long and drops a vertical distance of 195.8 ft. What is its angle of depression?

6.1   Trigonometric Functions of Acute Angles

> Determine the six trigonometric ratios for a given acute angle of a right triangle.

> Determine the trigonometric function values of 30°, 45°, and 60°.

> Using a calculator, find function values for any acute angle, and given a function value of an acute angle, find the angle.

> Given the function values of an acute angle, find the function values of its complement.

❯ The Trigonometric Ratios

We begin our study of trigonometry by considering right triangles and acute angles measured in degrees. An **acute angle** is an angle with measure greater than 0° and less than 90°. Greek letters such as α (alpha), β (beta), γ (gamma), θ (theta), and ϕ (phi) are often used to denote an angle. Consider a right triangle with one of its acute angles labeled θ. The side opposite the right angle is called the **hypotenuse**. The other sides of the triangle are referenced by their position relative to the acute angle θ. One side is opposite θ and one is adjacent to θ.

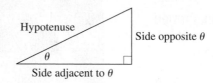

The *lengths* of the sides of the triangle are used to define the six trigonometric ratios:

sine (sin),	cosecant (csc),
cosine (cos),	secant (sec),
tangent (tan),	cotangent (cot).

The **sine of θ** is the length of the side opposite θ divided by the length of the hypotenuse (see Fig. 1):

$$\sin \theta = \frac{\text{length of side opposite } \theta}{\text{length of hypotenuse}}.$$

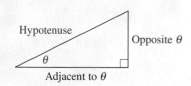

FIGURE 1.

The ratio depends on the measure of angle θ and thus the ratio is a function of θ. The notation $\sin \theta$ actually means $\sin (\theta)$, where sin, or sine, is the name of the function.

The **cosine of θ** is the length of the side adjacent to θ divided by the length of the hypotenuse (see Fig. 2):

$$\cos \theta = \frac{\text{length of side adjacent to } \theta}{\text{length of hypotenuse}}.$$

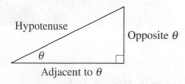

FIGURE 2.

The **tangent of θ** is the length of the side opposite θ divided by the length of the side adjacent to θ (see Fig. 3):

$$\tan \theta = \frac{\text{length of side opposite } \theta}{\text{length of side adjacent to } \theta}.$$

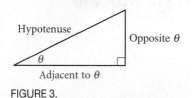

FIGURE 3.

The six trigonometric ratios, or trigonometric functions, are defined as follows. Here the *domain* of each function is the *set of acute angles*. Later in this chapter, the domain will be extended first to the set of all angles, or rotations, and then to the set of real numbers.

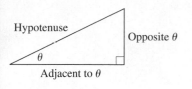

Hypotenuse / Opposite θ / Adjacent to θ / θ

TRIGONOMETRIC FUNCTION VALUES OF AN ACUTE ANGLE θ

Let θ be an acute angle of a right triangle. Then the six trigonometric functions of θ are as follows:

$$\sin \theta = \frac{\text{side opposite } \theta}{\text{hypotenuse}}, \qquad \csc \theta = \frac{\text{hypotenuse}}{\text{side opposite } \theta},$$

$$\cos \theta = \frac{\text{side adjacent to } \theta}{\text{hypotenuse}}, \qquad \sec \theta = \frac{\text{hypotenuse}}{\text{side adjacent to } \theta},$$

$$\tan \theta = \frac{\text{side opposite } \theta}{\text{side adjacent to } \theta}, \qquad \cot \theta = \frac{\text{side adjacent to } \theta}{\text{side opposite } \theta}.$$

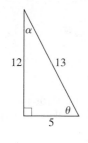

α / 12 / 13 / 5 / θ

EXAMPLE 1 In the right triangle shown at left, find the six trigonometric function values of **(a)** θ and **(b)** α.

Solution We use the definitions.

a) $\sin \theta = \dfrac{\text{opp}}{\text{hyp}} = \dfrac{12}{13}, \qquad \csc \theta = \dfrac{\text{hyp}}{\text{opp}} = \dfrac{13}{12},$

$\cos \theta = \dfrac{\text{adj}}{\text{hyp}} = \dfrac{5}{13}, \qquad \sec \theta = \dfrac{\text{hyp}}{\text{adj}} = \dfrac{13}{5},$

$\tan \theta = \dfrac{\text{opp}}{\text{adj}} = \dfrac{12}{5}, \qquad \cot \theta = \dfrac{\text{adj}}{\text{opp}} = \dfrac{5}{12}$

The references to opposite, adjacent, and hypotenuse are relative to θ.

b) $\sin \alpha = \dfrac{\text{opp}}{\text{hyp}} = \dfrac{5}{13}, \qquad \csc \alpha = \dfrac{\text{hyp}}{\text{opp}} = \dfrac{13}{5},$

$\cos \alpha = \dfrac{\text{adj}}{\text{hyp}} = \dfrac{12}{13}, \qquad \sec \alpha = \dfrac{\text{hyp}}{\text{adj}} = \dfrac{13}{12},$

$\tan \alpha = \dfrac{\text{opp}}{\text{adj}} = \dfrac{5}{12}, \qquad \cot \alpha = \dfrac{\text{adj}}{\text{opp}} = \dfrac{12}{5}$

The references to opposite, adjacent, and hypotenuse are relative to α.

> **Now Try Exercise 1.**

In Example 1(a), we note that the value of $\csc \theta$, $\frac{13}{12}$, is the reciprocal of $\frac{12}{13}$, the value of $\sin \theta$. Likewise, we see the same reciprocal relationship between the values of $\sec \theta$ and $\cos \theta$ and between the values of $\cot \theta$ and $\tan \theta$. For any angle, the cosecant, secant, and cotangent function values are the reciprocals of the sine, cosine, and tangent function values, respectively.

RECIPROCAL FUNCTIONS

$$\csc \theta = \frac{1}{\sin \theta}, \qquad \sec \theta = \frac{1}{\cos \theta}, \qquad \cot \theta = \frac{1}{\tan \theta}$$

If we know the values of the sine, cosine, and tangent functions of an angle, we can use these reciprocal relationships to find the values of the cosecant, secant, and cotangent functions of that angle.

EXAMPLE 2   Given that $\sin \phi = \frac{4}{5}$, $\cos \phi = \frac{3}{5}$, and $\tan \phi = \frac{4}{3}$, find $\csc \phi$, $\sec \phi$, and $\cot \phi$.

Solution   Using the reciprocal relationships, we have

$$\csc \phi = \frac{1}{\sin \phi} = \frac{1}{\dfrac{4}{5}} = \frac{5}{4}, \qquad \sec \phi = \frac{1}{\cos \phi} = \frac{1}{\dfrac{3}{5}} = \frac{5}{3},$$

and $\quad \cot \phi = \dfrac{1}{\tan \phi} = \dfrac{1}{\dfrac{4}{3}} = \dfrac{3}{4}.$

Just in Time

26

EXAMPLE 3   Given that $\sin \beta = \dfrac{\sqrt{21}}{5}$, $\cos \beta = \dfrac{2}{5}$, and $\tan \beta = \dfrac{\sqrt{21}}{2}$, find $\csc \beta$, $\sec \beta$, and $\cot \beta$.

Solution   Using the reciprocal relationships, we have

$$\csc \beta = \frac{1}{\sin \beta} = \frac{1}{\dfrac{\sqrt{21}}{5}} = \frac{5}{\sqrt{21}}$$

$$= \frac{5}{\sqrt{21}} \cdot \frac{\sqrt{21}}{\sqrt{21}} = \frac{5\sqrt{21}}{21}, \qquad \textbf{Rationalizing the denominator}$$

$$\sec \beta = \frac{1}{\cos \beta} = \frac{1}{\dfrac{2}{5}} = \frac{5}{2},$$

and $\quad \cot \beta = \dfrac{1}{\tan \beta} = \dfrac{1}{\dfrac{\sqrt{21}}{2}} = \dfrac{2}{\sqrt{21}} = \dfrac{2}{\sqrt{21}} \cdot \dfrac{\sqrt{21}}{\sqrt{21}} = \dfrac{2\sqrt{21}}{21}.$

> **Now Try Exercise 7.**

Triangles are said to be **similar** if their corresponding angles have the *same* measure. In similar triangles, the lengths of corresponding sides are in the same ratio. The right triangles shown below are similar. Note that the corresponding angles are equal, and the length of each side of the second triangle is four times the length of the corresponding side of the first triangle.

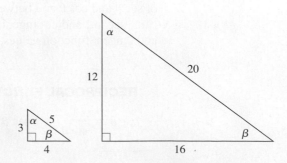

Let's take a look at the sine, cosine, and tangent values of β in each triangle. Can we expect corresponding function values to be the same?

FIRST TRIANGLE	**SECOND TRIANGLE**
$\sin \beta = \dfrac{3}{5}$	$\sin \beta = \dfrac{12}{20} = \dfrac{3}{5}$
$\cos \beta = \dfrac{4}{5}$	$\cos \beta = \dfrac{16}{20} = \dfrac{4}{5}$
$\tan \beta = \dfrac{3}{4}$	$\tan \beta = \dfrac{12}{16} = \dfrac{3}{4}$

For the two triangles, the values of $\sin \beta$, $\cos \beta$, and $\tan \beta$ are the same. The lengths of the sides are proportional—thus the *ratios* are the same. This must be the case because in order for the sine, cosine, and tangent to be functions, there must be only one output (the ratio) for each input (the angle β).

> The trigonometric function values of θ depend only on the measure of the angle, *not* on the size of the triangle.

❯ The Six Related Functions

We can find the other five trigonometric function values of an acute angle when one of the function-value ratios is known.

EXAMPLE 4   If $\sin \beta = \frac{6}{7}$ and β is an acute angle, find the other five trigonometric function values of β.

Solution   We know from the definition of the sine function that the ratio

$$\frac{6}{7} \quad \text{is} \quad \frac{\text{opp}}{\text{hyp}}.$$

PYTHAGOREAN EQUATION

REVIEW SECTION 1.1.

Using this information, let's consider a right triangle in which the hypotenuse has length 7 and the side opposite β has length 6. To find the length of the side adjacent to β, we use the *Pythagorean equation*:

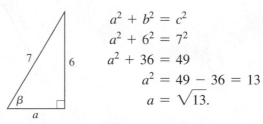

$$a^2 + b^2 = c^2$$
$$a^2 + 6^2 = 7^2$$
$$a^2 + 36 = 49$$
$$a^2 = 49 - 36 = 13$$
$$a = \sqrt{13}.$$

We now use the lengths of the three sides to find the other five ratios:

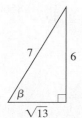

$$\sin \beta = \frac{6}{7}, \qquad\qquad\qquad \csc \beta = \frac{7}{6},$$

$$\cos \beta = \frac{\sqrt{13}}{7}, \qquad\qquad \sec \beta = \frac{7}{\sqrt{13}}, \quad \text{or} \quad \frac{7\sqrt{13}}{13},$$

$$\tan \beta = \frac{6}{\sqrt{13}}, \quad \text{or} \quad \frac{6\sqrt{13}}{13}, \qquad \cot \beta = \frac{\sqrt{13}}{6}.$$

> *Now Try Exercise 9.*

❯ Function Values of 30°, 45°, and 60°

In Examples 1 and 4, we found the trigonometric function values of an acute angle of a right triangle when the lengths of the three sides were known. In most situations, we are asked to find the function values when the measure of the acute angle is given. For certain special angles such as 30°, 45°, and 60°, which are frequently seen in applications, we can use geometry to determine the function values.

A right triangle with a 45° angle actually has two 45° angles. Thus the triangle is *isosceles*, and the legs are the same length. Let's consider such a triangle whose legs have length 1. Then we can find the length of its hypotenuse, c, using the Pythagorean equation as follows:

$$1^2 + 1^2 = c^2, \quad \text{or} \quad c^2 = 2, \quad \text{or} \quad c = \sqrt{2}.$$

Such a triangle is shown below. From this diagram, we can easily determine the trigonometric function values of 45°.

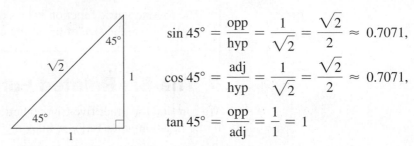

$$\sin 45° = \frac{\text{opp}}{\text{hyp}} = \frac{1}{\sqrt{2}} = \frac{\sqrt{2}}{2} \approx 0.7071,$$

$$\cos 45° = \frac{\text{adj}}{\text{hyp}} = \frac{1}{\sqrt{2}} = \frac{\sqrt{2}}{2} \approx 0.7071,$$

$$\tan 45° = \frac{\text{opp}}{\text{adj}} = \frac{1}{1} = 1$$

It is sufficient to find only the function values of the sine, cosine, and tangent, since the other three function values are their reciprocals.

It is also possible to determine the function values of 30° and 60°. A right triangle with 30° and 60° acute angles is half of an equilateral triangle, as shown in the following figure. Thus if we choose an equilateral triangle whose sides have length 2 and take half of it, we obtain a right triangle that has a hypotenuse of length 2 and a leg of length 1. The other leg has length a, which can be found as follows:

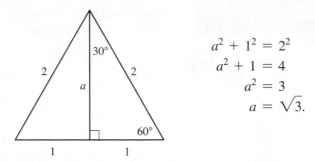

$$a^2 + 1^2 = 2^2$$
$$a^2 + 1 = 4$$
$$a^2 = 3$$
$$a = \sqrt{3}.$$

We can now determine the function values of 30° and 60°:

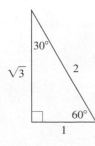

$$\sin 30° = \frac{1}{2} = 0.5, \qquad\qquad \sin 60° = \frac{\sqrt{3}}{2} \approx 0.8660,$$

$$\cos 30° = \frac{\sqrt{3}}{2} \approx 0.8660, \qquad\qquad \cos 60° = \frac{1}{2} = 0.5,$$

$$\tan 30° = \frac{1}{\sqrt{3}} = \frac{\sqrt{3}}{3} \approx 0.5774, \qquad \tan 60° = \frac{\sqrt{3}}{1} = \sqrt{3} \approx 1.7321.$$

Since we will often use the function values of 30°, 45°, and 60°, either the triangles that yield them or the values themselves should be memorized.

	30°	45°	60°
sin	$1/2$	$\sqrt{2}/2$	$\sqrt{3}/2$
cos	$\sqrt{3}/2$	$\sqrt{2}/2$	$1/2$
tan	$\sqrt{3}/3$	1	$\sqrt{3}$

Let's now use what we have learned about trigonometric functions of special angles to solve problems. We will consider such applications in greater detail in Section 6.2.

EXAMPLE 5 *Height of a Fireworks Display.* Trajectories for fireworks involve variables such as the launch angle, the launch velocity, and the size of the shell. Using physics to calculate the trajectory, fireworks technicians know exactly the path of the shell. Launch angles vary from 45° to 90°. For every inch of the diameter of the shell, a firework can travel about 100 ft vertically and 70 ft horizontally. These distances can vary depending on air resistance. (*Sources*: www.thinkquest.org, Pyrotechnico; www.dispatch.com, Aaron Harden) A 6-in. shell launched at an angle of 60° travels a horizontal distance of 390 ft. Approximate the height of the fireworks display. Round the answer to the nearest foot.

Solution We begin with a diagram of the situation. We know the measure of an acute angle and the length of the adjacent side.

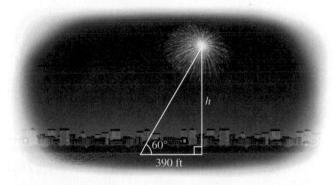

Since we want to determine the length of the opposite side, we can use either the tangent ratio or the cotangent ratio. In this case, we use the tangent ratio:

$$\tan 60° = \frac{\text{opp}}{\text{adj}} = \frac{h}{390}$$

$$390 \cdot \tan 60° = h \qquad \text{Multiplying by 390 on both sides}$$

$$390 \cdot \sqrt{3} = h \qquad \text{Substituting; } \tan 60° = \sqrt{3}$$

$$675 \approx h.$$

The fireworks display is approximately 675 ft high.

▶ **Now Try Exercise 29.**

❯ Function Values of Any Acute Angle

Historically, the measure of an angle has been expressed in degrees, minutes, and seconds. One minute, denoted $1'$, is such that $60' = 1°$, or $1' = \frac{1}{60} \cdot (1°)$. One second, denoted $1''$, is such that $60'' = 1'$, or $1'' = \frac{1}{60} \cdot (1')$. Then 61 degrees, 27 minutes, 4 seconds can be written as $61°27'4''$. The use of this **D°M′S″ form** was common before the widespread use of calculators. Now the preferred notation is to express fraction parts of degrees in **decimal degree form**. For example, $61°27'4'' \approx 61.45°$ when written in decimal degree form. Although the D°M′S″ notation is still widely used in navigation, we will most often use the decimal form in this text.

Most calculators can convert D°M′S″ notation to decimal degree notation and vice versa. Procedures among calculators vary.

EXAMPLE 6   Convert $5°42'30''$ to decimal degree notation.

Solution   We can use a graphing calculator set in DEGREE mode (see the window at left) to convert between D°M′S″ form and decimal degree form.

To convert D°M′S″ form to decimal degree form, we enter $5°42'30''$ using the ANGLE menu for the degree and minute symbols and **ALPHA** **(+)** for the third symbol. Pressing **ENTER** gives us

$$5°42'30'' \approx 5.71°,$$

rounded to the nearest hundredth of a degree.

| NORMAL SCI ENG |
| FLOAT 0123456789 |
| RADIAN **DEGREE** |
| FUNC PAR POL SEQ |
| CONNECTED DOT |
| SEQUENTIAL SIMUL |
| REAL a+bi re^θi |
| FULL HORIZ G-T |

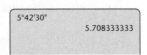

Without a calculator, we can convert as follows:

$$
\begin{aligned}
5°42'30'' &= 5° + 42' + 30'' \\
&= 5° + 42' + \frac{30'}{60} \qquad && 1'' = \frac{1}{60}'; 30'' = \frac{30'}{60} \\
&= 5° + 42.5' \qquad && \frac{30'}{60} = 0.5' \\
&= 5° + \frac{42.5°}{60} \qquad && 1' = \frac{1}{60}°; 42.5' = \frac{42.5°}{60} \\
&\approx 5.71°. \qquad && \frac{42.5°}{60} \approx 0.71°
\end{aligned}
$$

> **Now Try Exercise 37.**

EXAMPLE 7   Convert $72.18°$ to D°M′S″ notation.

Solution   To convert decimal degree form to D°M′S″ form, we enter 72.18 and access the ▶DMS feature in the ANGLE menu.

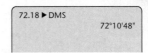

The result is

$$72.18° = 72°10'48''.$$

Without a calculator, we can convert as follows:

$$
\begin{aligned}
72.18° &= 72° + 0.18 \times 1° \\
&= 72° + 0.18 \times 60' \qquad && 1° = 60' \\
&= 72° + 10.8' \\
&= 72° + 10' + 0.8 \times 1' \\
&= 72° + 10' + 0.8 \times 60'' \qquad && 1' = 60'' \\
&= 72° + 10' + 48'' \\
&= 72°10'48''.
\end{aligned}
$$

> **Now Try Exercise 45.**

So far we have measured angles using degrees. Another useful unit for angle measure is the radian, which we will study in Section 6.4. Most calculators work with either degrees or radians. Be sure to use whichever mode is appropriate. In this section, we use the DEGREE mode.

Keep in mind the difference between an exact answer and an approximation. For example,

$$\sin 60° = \frac{\sqrt{3}}{2}. \quad \textbf{This is exact.}$$

But using a calculator, you get an answer like

$$\sin 60° \approx 0.8660254038. \quad \textbf{This is an approximation.}$$

Calculators generally provide values only of the sine, cosine, and tangent functions. You can find values of the cosecant, secant, and cotangent by taking reciprocals of the sine, cosine, and tangent functions, respectively.

EXAMPLE 8   Using a calculator, find the trigonometric function value, rounded to four decimal places, of each of the following.

a) $\tan 29.7°$ **b)** $\sec 48°$ **c)** $\sin 84°10'39''$

Solution

a) We check to be sure that the calculator is in DEGREE mode. The function value is

$$\tan 29.7° \approx 0.5703899297$$

$$\approx 0.5704. \quad \textbf{Rounded to four decimal places}$$

b) The secant function value can be found by taking the reciprocal of the cosine function value:

$$\sec 48° = \frac{1}{\cos 48°} \approx 1.49447655 \approx 1.4945.$$

c) We enter $\sin 84°10'39''$. The result is

$$\sin 84°10'39'' \approx 0.9948409474 \approx 0.9948.$$

> *Now Try Exercises 61 and 69.*

We can use the TABLE feature on a graphing calculator to find an angle for which we know a trigonometric function value.

EXAMPLE 9   Find the acute angle, to the nearest tenth of a degree, whose sine value is approximately 0.20113—that is, given $\sin \theta = 0.20113$, find θ.

Solution   With a graphing calculator set in DEGREE mode, we first enter the equation $y = \sin x$. With a minimum value of 0 and a step-value of 0.1, we scroll through the table of values looking for the y-value closest to 0.20113.

X	Y₁
11.1	.19252
11.2	.19423
11.3	.19595
11.4	.19766
11.5	.19937
11.6	.20108
11.7	.20279
X = 11.6	

sin 11.6° ≈ 0.20108

We find that $11.6°$ is the angle whose sine value is about 0.20113.

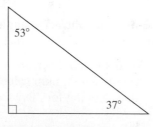

The quickest way to find the angle with a calculator is to use an inverse function key. (We first studied inverse functions in the chapter on exponential functions and logarithmic functions.) First, check to be sure that your calculator is in DEGREE mode. Usually two keys must be pressed in sequence. For this example, if we press

 **2ND** **SIN** .20113 **ENTER**,

we find that the acute angle whose sine is 0.20113 is approximately 11.60304613°, or 11.6°.

> **Now Try Exercise 75.**

EXAMPLE 10 *Ladder Safety.* A window-washing crew has purchased new 30-ft extension ladders. The manufacturer states that the safest placement on a wall is to extend the ladder to 25 ft and to position the base 6.5 ft from the wall (*Source*: R. D. Werner Co., Inc.). What angle does the ladder make with the ground in this position?

Solution We make a drawing and then use the most convenient trigonometric function. Because we know the length of the side adjacent to θ and the length of the hypotenuse, we choose the cosine function.

From the definition of the cosine function, we have

$$\cos\theta = \frac{\text{adj}}{\text{hyp}} = \frac{6.5\text{ ft}}{25\text{ ft}} = 0.26.$$

Using a calculator, we find the acute angle whose cosine is 0.26:

$\theta \approx 74.92993786°.$ Pressing **2ND** **COS** 0.26 **ENTER**

Thus when the ladder is in its safest position, it makes an angle of about 75° with the ground.

❯ Cofunctions and Complements

Two angles are **complementary** whenever the sum of their measures is 90°. Each is the complement of the other. In a right triangle, the acute angles are complementary, because the sum of all three angle measures is 180° and the right angle accounts for 90° of this total. Thus if one acute angle of a right triangle is θ, the other is $90° - \theta$.

The six trigonometric function values of each of the acute angles in the following triangle are listed on the right. Note that 53° and 37° are complementary angles because $53° + 37° = 90°$.

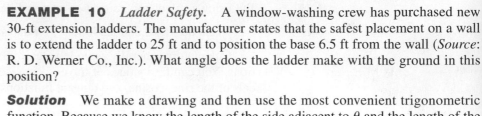

$\sin 37° \approx 0.6018$	$\csc 37° \approx 1.6616$
$\cos 37° \approx 0.7986$	$\sec 37° \approx 1.2521$
$\tan 37° \approx 0.7536$	$\cot 37° \approx 1.3270$
$\sin 53° \approx 0.7986$	$\csc 53° \approx 1.2521$
$\cos 53° \approx 0.6018$	$\sec 53° \approx 1.6616$
$\tan 53° \approx 1.3270$	$\cot 53° \approx 0.7536$

For these angles, we note that

$$\sin 37° = \cos 53°, \qquad \cos 37° = \sin 53°,$$
$$\tan 37° = \cot 53°, \qquad \cot 37° = \tan 53°,$$
$$\sec 37° = \csc 53°, \qquad \csc 37° = \sec 53°.$$

The sine of an angle is also the cosine of the angle's complement. Similarly, the tangent of an angle is the cotangent of the angle's complement, and the secant of an angle is the cosecant of the angle's complement. These pairs of functions are called **cofunctions**. A list of cofunction identities follows.

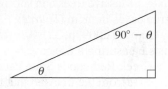

COFUNCTION IDENTITIES

$\sin \theta = \cos (90° - \theta), \qquad \cos \theta = \sin (90° - \theta),$

$\tan \theta = \cot (90° - \theta), \qquad \cot \theta = \tan (90° - \theta),$

$\sec \theta = \csc (90° - \theta), \qquad \csc \theta = \sec (90° - \theta)$

EXAMPLE 11   Given that $\sin 18° \approx 0.3090$, $\cos 18° \approx 0.9511$, and $\tan 18° \approx 0.3249$, find the six trigonometric function values of $72°$.

Solution   Using reciprocal relationships, we know that

$$\csc 18° = \frac{1}{\sin 18°} \approx 3.2361,$$

$$\sec 18° = \frac{1}{\cos 18°} \approx 1.0515,$$

and $$\cot 18° = \frac{1}{\tan 18°} \approx 3.0777.$$

Since $72°$ and $18°$ are complementary, we have

$$\sin 72° = \cos 18° \approx 0.9511, \qquad \cos 72° = \sin 18° \approx 0.3090,$$
$$\tan 72° = \cot 18° \approx 3.0777, \qquad \cot 72° = \tan 18° \approx 0.3249,$$
$$\sec 72° = \csc 18° \approx 3.2361, \qquad \csc 72° = \sec 18° \approx 1.0515.$$

> **Now Try Exercise 97.**

6.1   Exercise Set

In Exercises 1–6, find the six trigonometric function values of the specified angle.

1.

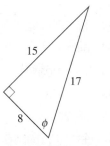

2.

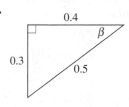

3.

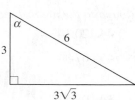

4.

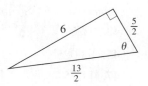

5.

6.

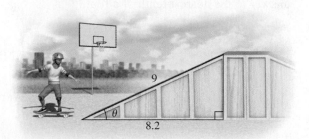

7. Given that $\sin \alpha = \dfrac{\sqrt{5}}{3}$, $\cos \alpha = \dfrac{2}{3}$, and

$\tan \alpha = \dfrac{\sqrt{5}}{2}$, find $\csc \alpha$, $\sec \alpha$, and $\cot \alpha$.

8. Given that $\sin \beta = \dfrac{2\sqrt{2}}{3}$, $\cos \beta = \dfrac{1}{3}$, and

$\tan \beta = 2\sqrt{2}$, find $\csc \beta$, $\sec \beta$, and $\cot \beta$.

Given a function value of an acute angle, find the other five trigonometric function values.

9. $\sin \theta = \frac{24}{25}$

10. $\cos \sigma = 0.7$

11. $\tan \phi = 2$

12. $\cot \theta = \frac{1}{3}$

13. $\csc \theta = 1.5$

14. $\sec \beta = \sqrt{17}$

15. $\cos \beta = \dfrac{\sqrt{5}}{5}$

16. $\sin \sigma = \frac{10}{11}$

Find the exact function value.

17. $\cos 45°$ **18.** $\tan 30°$ **19.** $\sec 60°$

20. $\sin 45°$ **21.** $\cot 60°$ **22.** $\csc 45°$

23. $\sin 30°$ **24.** $\cos 60°$ **25.** $\tan 45°$

26. $\sec 30°$ **27.** $\csc 30°$ **28.** $\tan 60°$

29. *Four Square.* The game Four Square is making a comeback on college campuses. The game is played on a 16-ft square court divided into four smaller squares that meet in the center (*Source:* www.squarefour.org/rules).

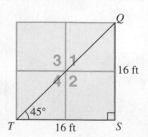

If a line is drawn diagonally from one corner to another corner, then a right triangle *QTS* is formed, where $\angle QTS$ is 45°. Using the cosecant function, find the length of the diagonal. Round the answer to the nearest tenth of a foot.

30. *Distance to a Fire Cave.* Massive trees can survive wildfires that leave large caves in them (*Source: National Geographic,* October 2009, p. 32). A hiker observes scientists measuring a fire cave in a redwood tree in Prairie Creek Redwoods State Park. He estimates that he is 80 ft from the tree and that the angle between the ground and the line of sight to the scientists is 60°. Approximate how high the fire cave is. Round the answer to the nearest foot.

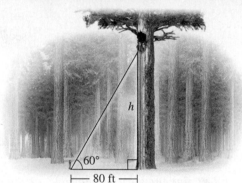

Convert to decimal degree notation. Round to two decimal places.

31. $9°43'$ **32.** $52°15'$

33. $35°50''$ **34.** $64°53'$

35. $3°2'$ **36.** $19°47'23''$

37. $49°38'46''$ **38.** $76°11'34''$

39. $15'5''$ **40.** $68°2''$

41. $5°53''$ **42.** $44'10''$

Convert to D°M′S″ notation. Round to the nearest second.

43. $17.6°$ **44.** $20.14°$

45. $83.025°$ **46.** $67.84°$

47. $11.75°$ **48.** $29.8°$

49. $47.8268°$ **50.** $0.253°$

51. $0.9°$ **52.** $30.2505°$

53. $39.45°$ **54.** $2.4°$

Find the function value. Round to four decimal places.

55. cos 51°

56. cot 17°

57. tan 4°13′

58. sin 26.1°

59. sec 38.43°

60. cos 74°10′40″

61. cos 40.35°

62. csc 45.2°

63. sin 69°

64. tan 63°48′

65. tan 85.4°

66. cos 4°

67. csc 89.5°

68. sec 35.28°

69. cot 30°25′6″

70. sin 59.2°

Find the acute angle θ, to the nearest tenth of a degree, for the given function value.

71. $\sin \theta = 0.5125$

72. $\tan \theta = 2.032$

73. $\tan \theta = 0.2226$

74. $\cos \theta = 0.3842$

75. $\sin \theta = 0.9022$

76. $\tan \theta = 3.056$

77. $\cos \theta = 0.6879$

78. $\sin \theta = 0.4005$

79. $\cot \theta = 2.127$

80. $\csc \theta = 1.147$

$\left(Hint: \tan \theta = \dfrac{1}{\cot \theta}. \right)$

81. $\sec \theta = 1.279$

82. $\cot \theta = 1.351$

Find the exact acute angle θ for the given function value.

83. $\sin \theta = \dfrac{\sqrt{2}}{2}$

84. $\cot \theta = \dfrac{\sqrt{3}}{3}$

85. $\cos \theta = \dfrac{1}{2}$

86. $\sin \theta = \dfrac{1}{2}$

87. $\tan \theta = 1$

88. $\cos \theta = \dfrac{\sqrt{3}}{2}$

89. $\csc \theta = \dfrac{2\sqrt{3}}{3}$

90. $\tan \theta = \sqrt{3}$

91. $\cot \theta = \sqrt{3}$

92. $\sec \theta = \sqrt{2}$

Use the cofunction and reciprocal identities to complete each of the following.

93. $\cos 20° = $ _____ $70° = \dfrac{1}{\underline{\quad} 20°}$

94. $\sin 64° = $ _____ $26° = \dfrac{1}{\underline{\quad} 64°}$

95. $\tan 52° = \cot$ _____ $= \dfrac{1}{\underline{\quad} 52°}$

96. $\sec 13° = \csc$ _____ $= \dfrac{1}{\underline{\quad} 13°}$

97. Given that

$$\sin 65° \approx 0.9063, \qquad \cos 65° \approx 0.4226,$$
$$\tan 65° \approx 2.1445, \qquad \cot 65° \approx 0.4663,$$
$$\sec 65° \approx 2.3662, \qquad \csc 65° \approx 1.1034,$$

find the six function values of 25°.

98. Given that

$$\sin 8° \approx 0.1392, \qquad \cos 8° \approx 0.9903,$$
$$\tan 8° \approx 0.1405, \qquad \cot 8° \approx 7.1154,$$
$$\sec 8° \approx 1.0098, \qquad \csc 8° \approx 7.1853,$$

find the six function values of 82°.

99. Given that $\sin 71°10′5″ \approx 0.9465$, $\cos 71°10′5″ \approx 0.3228$, and $\tan 71°10′5″ \approx 2.9321$, find the six function values of 18°49′55″.

100. Given that $\sin 38.7° \approx 0.6252$, $\cos 38.7° \approx 0.7804$, and $\tan 38.7° \approx 0.8012$, find the six function values of 51.3°.

101. Given that $\sin 82° = p$, $\cos 82° = q$, and $\tan 82° = r$, find the six function values of 8° in terms of p, q, and r.

❯ Skill Maintenance

Make a hand-drawn graph of the function. Then check your work using a graphing calculator.

102. $f(x) = 2^{-x}$ **[5.2]**

103. $f(x) = e^{x/2}$ **[5.2]**

104. $g(x) = \log_2 x$ **[5.3]**

105. $h(x) = \ln x$ **[5.3]**

Solve. **[5.5]**

106. $e^t = 10,000$

107. $5^x = 625$

108. $\log(3x + 1) - \log(x - 1) = 2$

109. $\log_7 x = 3$

❯ Synthesis

110. Given that $\cos \theta = 0.9651$, find $\csc(90° - \theta)$.

111. Given that $\sec \beta = 1.5304$, find $\sin(90° - \beta)$.

112. Find the six trigonometric function values of α.

113. Show that the area of this triangle is $\frac{1}{2}ab \sin \theta$.

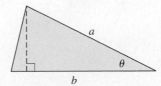

6.2 > Applications of Right Triangles

> Solve right triangles.

> Solve applied problems involving right triangles and trigonometric functions.

> Solving Right Triangles

Now that we can find function values for any acute angle, it is possible to *solve* right triangles. To **solve** a triangle means to find the lengths of *all* sides and the measures of *all* angles.

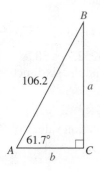

EXAMPLE 1 In $\triangle ABC$ (shown at left), find a, b, and B, where a and b represent lengths of sides and B represents the measure of $\angle B$. Here we use standard lettering for naming the sides and angles of a right triangle: Side a is opposite angle A, side b is opposite angle B, where a and b are the legs, and side c, the hypotenuse, is opposite angle C, the right angle.

Solution In $\triangle ABC$, we know three of the measures:

$$A = 61.7°, \qquad a = \text{?},$$
$$B = \text{?}, \qquad b = \text{?},$$
$$C = 90°, \qquad c = 106.2.$$

Since the sum of the angle measures of any triangle is 180° and $C = 90°$, the sum of A and B is 90°. Thus,

$$B = 90° - A = 90° - 61.7° = 28.3°.$$

We are given an acute angle and the hypotenuse. This suggests that we can use the sine and cosine ratios to find a and b, respectively:

$$\sin 61.7° = \frac{\text{opp}}{\text{hyp}} = \frac{a}{106.2} \quad \text{and} \quad \cos 61.7° = \frac{\text{adj}}{\text{hyp}} = \frac{b}{106.2}.$$

Solving for a and b, we get

$$a = 106.2 \sin 61.7° \quad \text{and} \quad b = 106.2 \cos 61.7°$$
$$a \approx 93.5 \qquad\qquad\qquad b \approx 50.3.$$

Thus,

$$A = 61.7°, \qquad a \approx 93.5,$$
$$B = 28.3°, \qquad b \approx 50.3,$$
$$C = 90°, \qquad c = 106.2.$$

> *Now Try Exercise 1.*

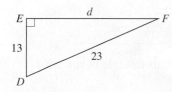

EXAMPLE 2 In $\triangle DEF$ (shown at left), find D and F. Then find d.

Solution In $\triangle DEF$, we know three of the measures:

$$D = \text{?}, \qquad d = \text{?},$$
$$E = 90°, \qquad e = 23,$$
$$F = \text{?}, \qquad f = 13.$$

We know the side adjacent to D and the hypotenuse. This suggests the use of the cosine ratio:

$$\cos D = \frac{\text{adj}}{\text{hyp}} = \frac{13}{23}.$$

We now find the angle whose cosine is $\frac{13}{23}$. To the nearest hundredth of a degree,

$$D \approx 55.58°.$$ **Pressing** **2ND** **COS** (13/23) **ENTER**

Since the sum of D and F is $90°$, we can find F by subtracting:

$$F = 90° - D \approx 90° - 55.58° \approx 34.42°.$$

We could use the Pythagorean equation to find d, but we will use a trigonometric function here. We could use $\cos F$, $\sin D$, or the tangent or cotangent ratios for either D or F. Let's use $\tan D$:

$$\tan D = \frac{\text{opp}}{\text{adj}} = \frac{d}{13}, \quad \text{or} \quad \tan 55.58° \approx \frac{d}{13}.$$

Then

$$d \approx 13 \tan 55.58° \approx 19.$$

The six measures are

$$D \approx 55.58°, \qquad d \approx 19,$$
$$E = 90°, \qquad e = 23,$$
$$F \approx 34.42°, \qquad f = 13.$$

> **Now Try Exercise 9.**

❯ Applications

Right triangles can be used to model and solve many applied problems in the real world.

EXAMPLE 3 *Walking at Niagara Falls.* While visiting Niagara Falls, a tourist walking toward Horseshoe Falls on a walkway next to Niagara Parkway notices the entrance to the Cave of the Winds attraction directly across the Niagara River. She continues walking for another 1000 ft and finds that the entrance is still visible but at approximately a 50° angle to the walkway.

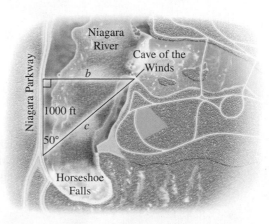

a) How many feet is she from the entrance to the Cave of the Winds?

b) What is the approximate width of the Niagara River at the point directly across from the entrance to the Cave of the Winds?

Solution

a) We know the side adjacent to the 50° angle and want to find the hypotenuse. We can use the cosine function:

$$\cos 50° = \frac{1000 \text{ ft}}{c}$$

$$c \cos 50° = 1000 \text{ ft} \qquad \textbf{Multiplying by } c$$

$$c = \frac{1000 \text{ ft}}{\cos 50°} \qquad \textbf{Dividing by } \cos 50°$$

$$c \approx 1556 \text{ ft}.$$

After walking 1000 ft, she is approximately 1556 ft from the entrance to the Cave of the Winds.

b) We know the side adjacent to the 50° angle and want to find the opposite side. We can use the tangent function:

$$\tan 50° = \frac{b}{1000 \text{ ft}}$$

$$b = 1000 \text{ ft} \cdot \tan 50° \approx 1192 \text{ ft}.$$

The width of the Niagara River directly across from the Cave of the Winds is approximately 1192 ft.

> **Now Try Exercise 21.**

EXAMPLE 4 *Rafters for a House.* House framers can use trigonometric functions to determine the lengths of rafters for a house. They first choose the pitch of the roof, or the ratio of the rise over the run. Then using a triangle with that ratio, they calculate the length of the rafter needed for the house. José is constructing rafters for a roof with a 10/12 pitch on a house that is 42 ft wide. Find the length x of the rafter of the house to the nearest tenth of a foot.

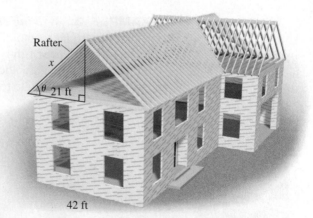

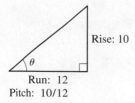

Pitch: 10/12

Solution We first find the angle θ that the rafter makes with the side wall. We know the rise, 10, and the run, 12, so we can use the tangent function to determine the angle that corresponds to the pitch of 10/12:

$$\tan \theta = \frac{10}{12} \approx 0.8333.$$

Using a calculator, we find that $\theta \approx 39.8°$. Since trigonometric function values of θ depend only on the measure of the angle and not on the size of the triangle, the angle for the rafter is also 39.8°.

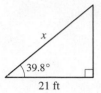

21 ft

To determine the length x of the rafter, we can use the cosine function. (See the figure at left.) Note that the width of the house is 42 ft, and a leg of this triangle is half that length, 21 ft.

$$\cos 39.8° = \frac{21 \text{ ft}}{x}$$

$$x \cos 39.8° = 21 \text{ ft} \qquad \textbf{Multiplying by } x$$

$$x = \frac{21 \text{ ft}}{\cos 39.8°} \qquad \textbf{Dividing by } \cos 39.8°$$

$$x \approx 27.3 \text{ ft}$$

The length of the rafter for this house is approximately 27.3 ft.

> *Now Try Exercise 33.*

Many applications with right triangles involve an *angle of elevation* or an *angle of depression*. The angle between the horizontal and a line of sight above the horizontal is called an **angle of elevation**. The angle between the horizontal and a line of sight below the horizontal is called an **angle of depression**. For example, suppose that you are looking straight ahead and then you move your eyes up to look at an approaching airplane. The angle that your eyes pass through is an angle of elevation. If the pilot of the plane is looking forward and then looks down, the pilot's eyes pass through an angle of depression.

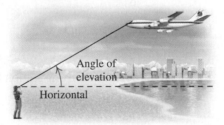
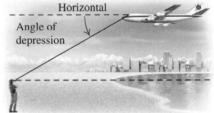

EXAMPLE 5 *Gondola Aerial Lift.* In Telluride, Colorado, there is a free gondola ride that provides a spectacular view of the town and the surrounding mountains. The gondolas that begin in the town at an elevation of 8725 ft travel 5750 ft to Station St. Sophia, whose altitude is 10,550 ft. They then continue 3913 ft to Mountain Village, whose elevation is 9500 ft.

a) What is the angle of elevation from the town to Station St. Sophia?

b) What is the angle of depression from Station St. Sophia to Mountain Village?

Solution We begin by labeling a drawing with the given information.

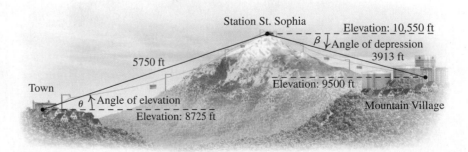

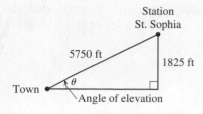

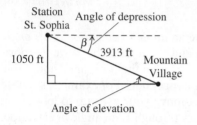

a) The difference in the elevation of Station St. Sophia and the elevation of the town is 10,550 ft − 8725 ft, or 1825 ft. This measure is the length of the side opposite the angle of elevation, θ, in the right triangle shown at left. Since we know the side opposite θ and the hypotenuse, we can find θ by using the sine function. We first find $\sin \theta$:

$$\sin \theta = \frac{1825 \text{ ft}}{5750 \text{ ft}} \approx 0.3174.$$

Using a calculator, we find that

$$\theta \approx 18.5°. \qquad \textbf{Pressing } \boxed{\text{2ND}} \ \boxed{\text{SIN}} \ 0.3174 \ \boxed{\text{ENTER}}$$

Thus the angle of elevation from the town to Station St. Sophia is approximately 18.5°.

b) When parallel lines are cut by a transversal, alternate interior angles are equal. Thus the angle of depression, β, from Station St. Sophia to Mountain Village is equal to the angle of elevation from Mountain Village to Station St. Sophia, so we can use the right triangle shown at left.

The difference in the elevation of Station St. Sophia and the elevation of Mountain Village is 10,550 ft − 9500 ft, or 1050 ft. Since we know the side opposite the angle of elevation and the hypotenuse, we can again use the sine function:

$$\sin \beta = \frac{1050 \text{ ft}}{3913 \text{ ft}} \approx 0.2683.$$

Using a calculator, we find that

$$\beta \approx 15.6°.$$

The angle of depression from Station St. Sophia to Mountain Village is approximately 15.6°.

> *Now Try Exercise 17.*

EXAMPLE 6 *Height of a Bamboo Plant.* Bamboo is the fastest growing land plant in the world and is becoming a popular wood for hardwood flooring. It can grow up to 46 in. per day and reaches its maximum height and girth in one season of growth. (*Sources: Farm Show*, Vol. 34, No. 4, 2010, p. 7; *U-Cut Bamboo Business*; American Bamboo Society) To estimate the height of a bamboo shoot, a farmer walks off 27 ft from the base and estimates the angle of elevation to the top of the shoot to be 70°. What is the approximate height h of the bamboo shoot?

Solution From the figure, we have

$$\tan 70° = \frac{h}{27 \text{ ft}}$$

$$h = 27 \text{ ft} \cdot \tan 70° \approx 74 \text{ ft}.$$

The height of the bamboo shoot is approximately 74 ft.

>

Some applications of trigonometry involve the concept of direction, or bearing. In this text, we present two ways of giving direction, the first below and the second in Section 6.3.

Bearing: First-Type

One method of giving direction, or **bearing**, involves reference to a north–south line using an acute angle. For example, N55°W means 55° west of north and S67°E means 67° east of south.

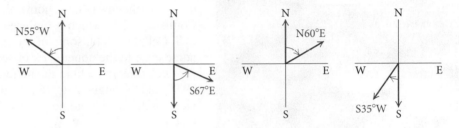

A second-type of bearing that gives directions in degrees from north is covered in Section 6.3.

EXAMPLE 7 *Distance to a Forest Fire.* A forest ranger at point *A* sights a fire directly south. A second ranger at point *B*, 7.5 mi east of the first ranger, sights the same fire at a bearing of S27°23′W. How far from *A* is the fire?

Solution   We first find the complement of 27°23′:

$$B = 90° - 27°23'$$ **Angle B is opposite side d in triangle BAF.**

$$= 62°37'$$

$$\approx 62.62°.$$

From the figure shown on the preceding page, we see that the desired distance d is a side of right triangle BAF. We have

$$\frac{d}{7.5 \text{ mi}} \approx \tan 62.62°$$

$$d \approx 7.5 \text{ mi} \cdot \tan 62.62° \approx 14.5 \text{ mi}.$$

The forest ranger at point A is about 14.5 mi from the fire.  **Now Try Exercise 37.**

EXAMPLE 8   *U.S. Cellular Field.*   In U.S. Cellular Field, the home of the Chicago White Sox baseball team, the first row of seats in the upper deck is farther away from home plate than the last row of seats in the original Comiskey Park, which it replaced. Although there is no obstructed view in U.S. Cellular Field, some of the fans still complain about the distance from home plate to the upper deck of seats. From a seat in the last row of the upper deck directly behind the batter, the angle of depression to home plate is 29.9°, and the angle of depression to the pitcher's mound is 24.2°. Find **(a)** the viewing distance to home plate and **(b)** the viewing distance to the pitcher's mound.

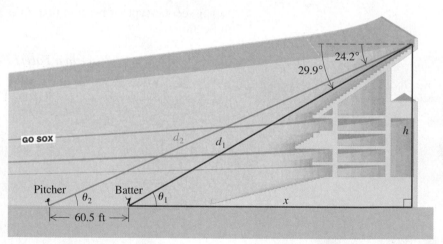

Solution   From geometry, we know that $\theta_1 = 29.9°$ and $\theta_2 = 24.2°$. The standard distance from home plate to the pitcher's mound is 60.5 ft. In the drawing, we let d_1 be the viewing distance to home plate, d_2 the viewing distance to the pitcher's mound, h the elevation of the last row, and x the horizontal distance from the batter to a point directly below the seat in the last row of the upper deck.

We begin by determining the distance x. We use the tangent function with $\theta_1 = 29.9°$ and $\theta_2 = 24.2°$:

$$\tan 29.9° = \frac{h}{x} \qquad \text{and} \quad \tan 24.2° = \frac{h}{x + 60.5}$$

or

$$h = x \tan 29.9° \quad \text{and} \qquad h = (x + 60.5) \tan 24.2°.$$

Then substituting $x \tan 29.9°$ for h in the second equation, we obtain

$$x \tan 29.9° = (x + 60.5) \tan 24.2°.$$

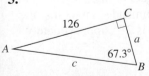

Just in Time 17

Solving for x, we get

$$x \tan 29.9° = x \tan 24.2° + 60.5 \tan 24.2°$$

$$x \tan 29.9° - x \tan 24.2° = x \tan 24.2° + 60.5 \tan 24.2° - x \tan 24.2°$$

$$x(\tan 29.9° - \tan 24.2°) = 60.5 \tan 24.2°$$

$$x = \frac{60.5 \tan 24.2°}{\tan 29.9° - \tan 24.2°}$$

$$x \approx 216.5.$$

We can then find d_1 and d_2 using the cosine function:

$$\cos 29.9° = \frac{216.5}{d_1} \quad \text{and} \quad \cos 24.2° = \frac{216.5 + 60.5}{d_2}$$

or

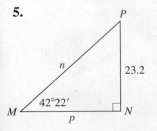

$$d_1 = \frac{216.5}{\cos 29.9°} \quad \text{and} \quad d_2 = \frac{277}{\cos 24.2°}$$

$$d_1 \approx 249.7 \qquad\qquad d_2 \approx 303.7.$$

The distance to home plate is about 250 ft,[*] and the distance to the pitcher's mound is about 304 ft.

> **Now Try Exercise 25.**

_____

[*]In the original Comiskey Park, the distance to home plate was only 150 ft.

6.2 | Exercise Set

In Exercises 1–6, solve the right triangle.

1.

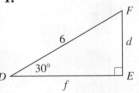

2.

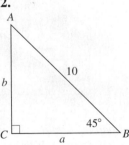

3.

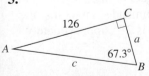

4.

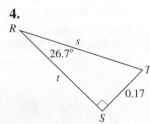

5.

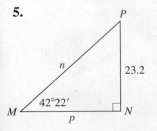

6.

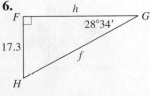

In Exercises 7–16, solve the right triangle. (Standard lettering has been used.)

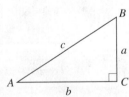

7. $A = 87°43'$, $a = 9.73$

8. $a = 12.5$, $b = 18.3$

9. $b = 100$, $c = 450$

10. $B = 56.5°$, $c = 0.0447$

11. $A = 47.58°$, $c = 48.3$

12. $B = 20.6°$, $a = 7.5$

13. $A = 35°$, $b = 40$

14. $B = 69.3°$, $b = 93.4$

15. $b = 1.86$, $c = 4.02$

16. $a = 10.2$, $c = 20.4$

17. *Aerial Photography.* An aerial photographer who photographs farm properties for a real estate company has determined from experience that the best photo is taken at a height of approximately 475 ft and a distance of 850 ft from the farmhouse. What is the angle of depression from the plane to the house?

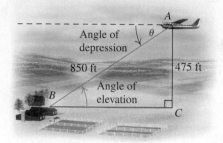

18. *Memorial Flag Case.* A tradition in the United States is to drape an American flag over the casket of a deceased U.S. Forces veteran. At the burial, the flag is removed, folded into a triangle, and presented to the family. The folded flag will fit in a case in the form of an isosceles right triangle, as shown below. The inside dimension across the bottom of the case is $21\frac{1}{2}$ in. (*Source:* Bruce Kieffer, *Woodworker's Journal,* August 2006). Using trigonometric functions, find the length x and round the answer to the nearest tenth of an inch.

19. *Zip Line.* The ZipRider®, a zip line at Icy Straight Point, Alaska, is 5495 ft long, and has a vertical drop of 1320 ft (*Source:* www.ziprider.com). Find its angle of depression.

20. *Setting a Fishing Reel Line Counter.* A fisherman who is fishing 50 ft directly out from a visible tree stump near the shore wants to position his line and bait approximately N35°W of the boat and west of the stump. Using the right triangle shown in the drawing, determine the reel's line counter setting, to the nearest foot, to position the line directly west of the stump.

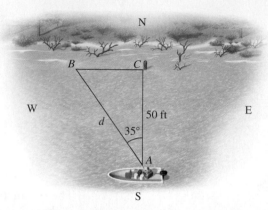

21. *Framing a Closet.* Sam is framing a closet under a stairway. The stairway is 16 ft 3 in. long, and its angle of elevation is 38°. Find the depth of the closet to the nearest inch.

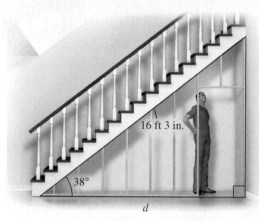

22. *Loading Ramp.* Charles needs to purchase a custom ramp to use while loading and unloading a garden tractor. When down, the tailgate of his truck is 38 in. from the ground. If the recommended angle that the ramp makes with the ground is 28°, approximately how long must the ramp be?

23. *Longest Escalator.* The longest escalator in the world is in the subway system in St. Petersburg, Russia. The escalator is 1084.6 ft long and drops a vertical distance of 195.8 ft. What is its angle of depression?

24. *Cloud Height.* To measure cloud height at night, a vertical beam of light is directed on a spot on the cloud. From a point 135 ft away from the light source, the angle of elevation to the spot is found to be 67.35°. Find the height of the cloud to the nearest foot.

25. *Mount Rushmore National Memorial.* While visiting Mount Rushmore in Rapid City, South Dakota, Landon approximated the angle of elevation to the top of George Washington's head to be 35°. After walking 250 ft closer, he guessed that the angle of elevation had increased by 15°. Approximate the height of the Mount Rushmore memorial, to the top of George Washington's head. Round the answer to the nearest foot.

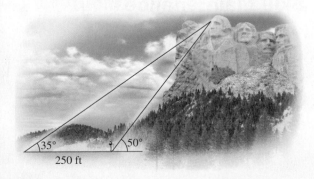

26. *Golden Gate Bridge.* The Golden Gate Bridge has two main towers of equal height that support the two main cables. A visitor on a tour boat passing through San Francisco Bay views the top of one of the towers and estimates the angle of elevation to be 30°. After sailing 670 ft closer, he estimates the angle of elevation to this same tower to be 50°. Approximate the height of the tower to the nearest foot.

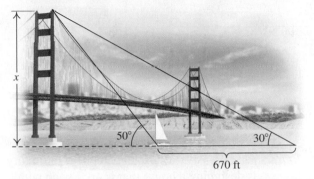

27. *Inscribed Pentagon.* A regular pentagon is inscribed in a circle of radius 15.8 cm. Find the perimeter of the pentagon.

28. *Height of a Weather Balloon.* A weather balloon is directly west of two observing stations that are 10 mi apart. The angles of elevation of the balloon from the two stations are 17.6° and 78.2°. How high is the balloon?

29. *Height of a Building.* A window washer on a ladder looks at a nearby building 100 ft away, noting that the angle of elevation to the top of the building is 18.7° and the angle of depression to the bottom of the building is 6.5°. How tall is the nearby building?

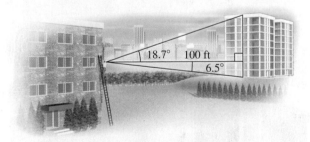

30. *Height of a Kite.* For a science fair project, a group of students tested different materials used to construct kites. Their instructor provided an instrument that accurately measures the angle of elevation. In one of the tests, the angle of elevation was 63.4° with 670 ft of string out. Assuming the string was taut, how high was the kite?

31. *Quilt Design.* Nancy is designing a quilt that she will enter in the quilt competition at the State Fair. The quilt consists of twelve identical squares with 4 rows of 3 squares each. Each square is to have a regular octagon inscribed in a circle, as shown in the figure.

Each side of the octagon is to be 7 in. long. Find the radius of the circumscribed circle and the dimensions of the quilt. Round the answers to the nearest hundredth of an inch.

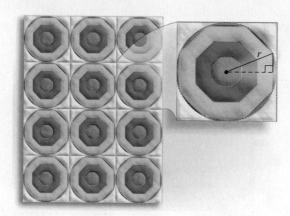

32. *Rafters for a House.* Blaise, an architect for luxury homes, is designing a house that is 46 ft wide with a roof whose pitch is 11/12. Determine the length of the rafters needed for this house. Round the answer to the nearest tenth of a foot.

33. *Rafters for a Medical Office.* The pitch of the roof for a medical office needs to be 5/12. If the building is 33 ft wide, how long must the rafters be?

34. *Angle of Elevation.* The Millau Viaduct in southern France is the tallest cable-stayed bridge in the world (*Source*: www.abelard.org). What is the angle of elevation of the sun when a pylon with height 343 m casts a shadow of 186 m?

35. *Distance Between Towns.* From a hot-air balloon 2 km high, the angles of depression to two towns in line with the balloon and on the same side of the balloon are 81.2° and 13.5°. How far apart are the towns?

36. *Distance from a Lighthouse.* From the top of a lighthouse 55 ft above sea level, the angle of depression to a small boat is 11.3°. How far from the foot of the lighthouse is the boat?

37. *Lightning Detection.* In extremely large forests, it is not cost-effective to position forest rangers in towers or to use small aircraft to continually watch for fires. Since lightning is a frequent cause of fire, lightning detectors are now commonly used instead. These devices not only give a bearing on the location but also measure the intensity of the lightning. A detector at point Q is situated 15 mi west of a central fire station at point R. The bearing from Q to where lightning hits due south of R is S37.6°E. How far is the hit from point R?

38. *Length of an Antenna.* A vertical antenna is mounted atop a 50-ft pole. From a point on level ground 75 ft from the base of the pole, the antenna subtends an angle of 10.5°. Find the length of the antenna.

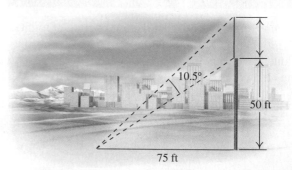

39. *Lobster Boat.* A lobster boat is situated due west of a lighthouse. A barge is 12 km south of the lobster boat. From the barge, the bearing to the lighthouse is N63°20′E. How far is the lobster boat from the lighthouse?

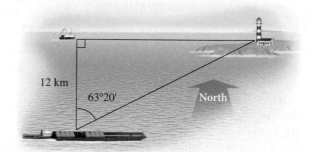

❯ Skill Maintenance

Find the distance between the points. **[1.1]**

40. $(-9, 3)$ and $(0, 0)$

41. $(8, -2)$ and $(-6, -4)$

42. Convert to a logarithmic equation: $e^4 = t$. **[5.3]**

43. Convert to an exponential equation:
$\log 0.001 = -3$. **[5.3]**

> ## Synthesis

44. *Diameter of a Pipe.* A V-gauge is used to find the diameter of a pipe. The advantage of such a device is that it is rugged, it is accurate, and it has no moving parts to break down. In the figure, the measure of angle *AVB* is 54°. A pipe is placed in the V-shaped slot and the distance *VP* is used to estimate the diameter. The line *VP* is calibrated by listing as its units the corresponding diameters. This, in effect, establishes a function between *VP* and *d*.

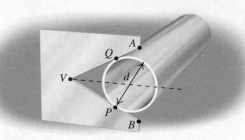

a) Suppose that the diameter of a pipe is 2 cm. What is the distance *VP*?
b) Suppose that the distance *VP* is 3.93 cm. What is the diameter of the pipe?
c) Find a formula for *d* in terms of *VP*.
d) Find a formula for *VP* in terms of *d*.

45. Find *h*, to the nearest tenth.

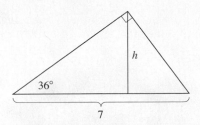

46. *Sound of an Airplane.* It is common experience to hear the sound of a low-flying airplane and look at the wrong place in the sky to see the plane. Suppose that a plane is traveling directly at you at a speed of 200 mph and an altitude of 3000 ft, and you hear the sound at what seems to be an angle of elevation of 20°. At what angle θ should you actually look in order to see the plane? Consider the speed of sound to be 1100 ft/sec.

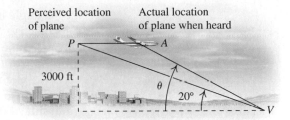

6.3 ▶ Trigonometric Functions of Any Angle

> ❯ Find angles that are coterminal with a given angle, and find the complement and the supplement of a given angle.

> ❯ Determine the six trigonometric function values for any angle in standard position when the coordinates of a point on the terminal side are given.

> ❯ Find the function values for any angle whose terminal side lies on an axis.

> ❯ Find the function values for an angle whose terminal side makes an angle of 30°, 45°, or 60° with the *x*-axis.

> ❯ Use a calculator to find function values and angles.

❯ Angles, Rotations, and Degree Measure

An *angle* is a familiar figure in the world around us.

Horizon
Canadian border

Horizon
Mexican border

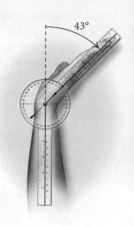

An **angle** is the union of two rays with a common endpoint called the **vertex**. In trigonometry, we often think of an angle as a **rotation**. To do so, think of locating a ray along the positive *x*-axis with its endpoint at the origin. This ray is called the **initial side** of the angle. Though we leave that ray fixed, think of making a copy of it and rotating it. A rotation *counterclockwise* is a **positive rotation**, and a rotation *clockwise* is a **negative rotation**. The ray at the end of the rotation is called the **terminal side** of the angle. The angle formed is said to be in **standard position**.

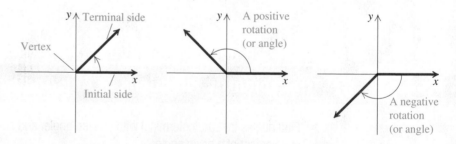

The measure of an angle or rotation may be given in degrees. The Babylonians developed the idea of dividing the circumference of a circle into 360 equal parts, or degrees. If we let the measure of one of these parts be 1°, then one complete positive revolution or rotation has a measure of 360°. One half of a revolution has a measure of 180°, one fourth of a revolution has a measure of 90°, and so on. We can also speak of an angle of measure 60°, 135°, 330°, or 420°. The terminal sides of these angles lie in quadrants I, II, IV, and I, respectively. The negative rotations −30°, −110°, and −225° represent angles with terminal sides in quadrants IV, III, and II, respectively.

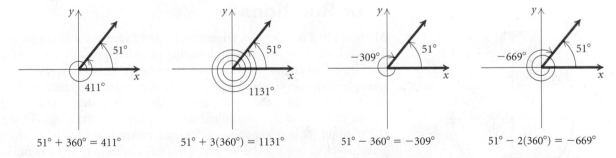

If two or more angles have the same terminal side, the angles are said to be **coterminal**. To find angles coterminal with a given angle, we add or subtract multiples of 360°. For example, 420°, shown above, has the same terminal side as 60°, since 420° = 360° + 60°. Thus we say that angles of measure 60° and 420° are coterminal. The negative rotation that measures −300° is also coterminal with 60° because 60° − 360° = −300°. The set of all angles coterminal with 60° can be expressed as 60° + n · 360°, where n is an integer. Other examples of coterminal angles shown above are 90° and −270°, −90° and 270°, 135° and −225°, −30° and 330°, and −110° and 610°.

Just in Time

5

EXAMPLE 1 Find two positive angles and two negative angles that are coterminal with **(a)** 51° and **(b)** −7°.

Solution

a) We add and subtract multiples of 360°. Many answers are possible.

$$51° + 360° = 411° \qquad 51° + 3(360°) = 1131° \qquad 51° - 360° = -309° \qquad 51° - 2(360°) = -669°$$

Thus angles of measure 411°, 1131°, −309°, and −669° are coterminal with 51°.

b) We have the following:

$$-7° + 360° = 353°, \qquad -7° + 2(360°) = 713°,$$
$$-7° - 360° = -367°, \qquad -7° - 10(360°) = -3607°.$$

Thus angles of measure 353°, 713°, −367°, and −3607° are coterminal with −7°.

> **Now Try Exercise 13.**

Angles can be classified by their measures, as we see in the following figures.

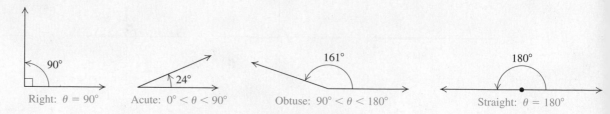

Right: $\theta = 90°$ Acute: $0° < \theta < 90°$ Obtuse: $90° < \theta < 180°$ Straight: $\theta = 180°$

Recall that two acute angles are **complementary** if the sum of their measures is 90°. For example, angles that measure 10° and 80° are complementary because $10° + 80° = 90°$. Two positive angles are **supplementary** if the sum of their measures is 180°. For example, angles that measure 45° and 135° are supplementary because $45° + 135° = 180°$.

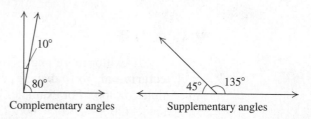

Complementary angles Supplementary angles

EXAMPLE 2 Find the complement and the supplement of 71.46°.

Solution We have

$$90° - 71.46° = 18.54°,$$
$$180° - 71.46° = 108.54°.$$

Thus the complement of 71.46° is 18.54° and the supplement is 108.54°.

> *Now Try Exercise 19.*

❯ Trigonometric Functions of Angles or Rotations

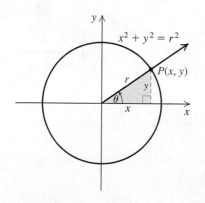

Many applied problems in trigonometry involve the use of angles that are not acute. Thus we need to extend the domains of the trigonometric functions defined in Section 6.1 to angles, or rotations, of *any* size. To do this, we first consider a right triangle with one vertex at the origin of a coordinate system and one vertex *on the positive x-axis*. (See the figure at left.) The other vertex is at *P*, a point on the circle whose center is at the origin and whose radius *r* is the length of the hypotenuse of the triangle. This triangle is a **reference triangle** for angle θ, which is in standard position. Note that *y* is the length of the side opposite θ, and *x* is the length of the side adjacent to θ.

Recalling the definitions from Section 6.1, we note that three of the trigonometric functions of angle θ are defined as follows:

$$\sin \theta = \frac{\text{opp}}{\text{hyp}} = \frac{y}{r}, \qquad \cos \theta = \frac{\text{adj}}{\text{hyp}} = \frac{x}{r}, \qquad \tan \theta = \frac{\text{opp}}{\text{adj}} = \frac{y}{x}.$$

Since *x* and *y* are the coordinates of the point *P* and the length of the radius is the length of the hypotenuse, we can also define these functions as follows:

$$\sin \theta = \frac{y\text{-coordinate}}{\text{radius}}, \qquad \cos \theta = \frac{x\text{-coordinate}}{\text{radius}}, \qquad \tan \theta = \frac{y\text{-coordinate}}{x\text{-coordinate}}.$$

We will use these definitions for functions of angles of any measure. The following figures show angles whose terminal sides lie in quadrants II, III, and IV.

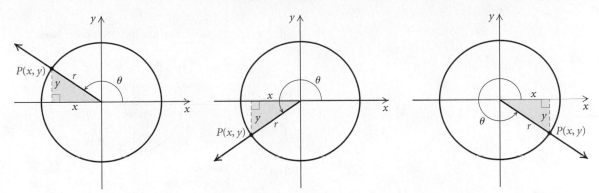

A reference triangle can be drawn for angles in any quadrant, as shown. Note that the angle is in standard position; that is, it is always measured from the positive half of the *x*-axis. The point $P(x, y)$ is a point, other than the vertex, on the terminal side of the angle. Each of its two coordinates may be positive, negative, or zero, depending on the location of the terminal side. *The length of the radius, which is also the length of the hypotenuse of the reference triangle, is always considered positive.* (Note that $x^2 + y^2 = r^2$, or $r = \sqrt{x^2 + y^2}$.) Regardless of the location of *P*, we have the following definitions. We now extend the *domain* of the six trigonometric functions from acute angles (see p. 395) to include *all angles, or rotations.*

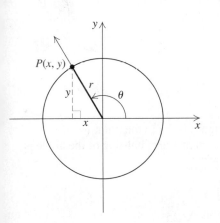

TRIGONOMETRIC FUNCTIONS OF ANY ANGLE θ

Suppose that $P(x, y)$ is any point other than the vertex on the terminal side of any angle θ in standard position, and r is the radius, or distance from the origin to $P(x, y)$. Then the trigonometric functions are defined as follows:

$$\sin \theta = \frac{y\text{-coordinate}}{\text{radius}} = \frac{y}{r}, \qquad \csc \theta = \frac{\text{radius}}{y\text{-coordinate}} = \frac{r}{y} \ (y \neq 0),$$

$$\cos \theta = \frac{x\text{-coordinate}}{\text{radius}} = \frac{x}{r}, \qquad \sec \theta = \frac{\text{radius}}{x\text{-coordinate}} = \frac{r}{x} \ (x \neq 0),$$

$$\tan \theta = \frac{y\text{-coordinate}}{x\text{-coordinate}} = \frac{y}{x} \ (x \neq 0), \quad \cot \theta = \frac{x\text{-coordinate}}{y\text{-coordinate}} = \frac{x}{y} \ (y \neq 0).$$

Values of the trigonometric functions can be positive, negative, or zero, depending on where the terminal side of the angle lies. Since the length of the radius is always positive, the signs of the function values depend only on the coordinates of the point *P* on the terminal side of the angle. In the first quadrant, all function values are positive because both coordinates are positive. In the second quadrant, first coordinates are negative and second coordinates are positive; thus only the sine and the cosecant values are positive. Similarly, we can determine the signs of the function values in the third and the fourth quadrants. *Because of the reciprocal relationships, we need to learn only the signs for the sine, cosine, and tangent functions.*

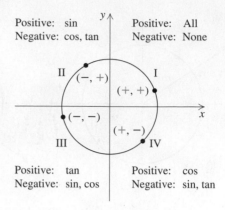

EXAMPLE 3 Find the six trigonometric function values for each angle shown.

a)

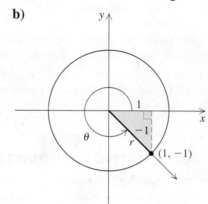

b)

c)

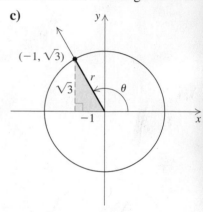

Solution

a) We first determine r, the distance from the origin $(0, 0)$ to the point $(-4, -3)$. The distance between $(0, 0)$ and any point (x, y) on the terminal side of the angle is

$$r = \sqrt{(x - 0)^2 + (y - 0)^2} = \sqrt{x^2 + y^2}.$$

Substituting -4 for x and -3 for y, we find

$$r = \sqrt{(-4)^2 + (-3)^2} = \sqrt{16 + 9} = \sqrt{25} = 5.$$

Just in Time

25

Using the definitions of the trigonometric functions, we can now find the function values of θ. We substitute -4 for x, -3 for y, and 5 for r:

$$\sin \theta = \frac{y}{r} = \frac{-3}{5} = -\frac{3}{5}, \qquad \csc \theta = \frac{r}{y} = \frac{5}{-3} = -\frac{5}{3},$$

$$\cos \theta = \frac{x}{r} = \frac{-4}{5} = -\frac{4}{5}, \qquad \sec \theta = \frac{r}{x} = \frac{5}{-4} = -\frac{5}{4},$$

$$\tan \theta = \frac{y}{x} = \frac{-3}{-4} = \frac{3}{4}, \qquad \cot \theta = \frac{x}{y} = \frac{-4}{-3} = \frac{4}{3}.$$

As expected, the tangent value and the cotangent value are positive and the other four values are negative. This is true for all angles in quadrant III.

b) We first determine r, the distance from the origin to the point $(1, -1)$:

$$r = \sqrt{1^2 + (-1)^2} = \sqrt{1 + 1} = \sqrt{2}.$$

Substituting 1 for x, -1 for y, and $\sqrt{2}$ for r, we find

$$\sin\theta = \frac{y}{r} = \frac{-1}{\sqrt{2}} = -\frac{\sqrt{2}}{2}, \qquad \csc\theta = \frac{r}{y} = \frac{\sqrt{2}}{-1} = -\sqrt{2},$$

$$\cos\theta = \frac{x}{r} = \frac{1}{\sqrt{2}} = \frac{\sqrt{2}}{2}, \qquad \sec\theta = \frac{r}{x} = \frac{\sqrt{2}}{1} = \sqrt{2},$$

$$\tan\theta = \frac{y}{x} = \frac{-1}{1} = -1, \qquad \cot\theta = \frac{x}{y} = \frac{1}{-1} = -1.$$

c) We determine r, the distance from the origin to the point $\left(-1, \sqrt{3}\right)$:

$$r = \sqrt{(-1)^2 + \left(\sqrt{3}\right)^2} = \sqrt{1+3} = \sqrt{4} = 2.$$

Substituting -1 for x, $\sqrt{3}$ for y, and 2 for r, we find the trigonometric function values of θ are

$$\sin\theta = \frac{\sqrt{3}}{2}, \qquad \csc\theta = \frac{2}{\sqrt{3}} = \frac{2\sqrt{3}}{3},$$

$$\cos\theta = \frac{-1}{2} = -\frac{1}{2}, \qquad \sec\theta = \frac{2}{-1} = -2,$$

$$\tan\theta = \frac{\sqrt{3}}{-1} = -\sqrt{3}, \qquad \cot\theta = \frac{-1}{\sqrt{3}} = -\frac{\sqrt{3}}{3}.$$

> **Now Try Exercise 29.**

Any point other than the origin on the terminal side of an angle in standard position can be used to determine the trigonometric function values of that angle. The function values are the same regardless of which point is used. To illustrate this, let's consider an angle θ in standard position whose terminal side lies on the line $y = -\frac{1}{2}x$. We can determine two second-quadrant solutions of the equation, find the length r for each point, and then compare the sine, cosine, and tangent function values using each point.

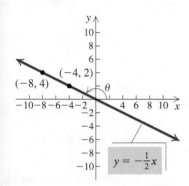

If $x = -4$, then $y = -\frac{1}{2}(-4) = 2$.

If $x = -8$, then $y = -\frac{1}{2}(-8) = 4$.

For $(-4, 2)$, $r = \sqrt{(-4)^2 + 2^2} = \sqrt{20} = 2\sqrt{5}$.

For $(-8, 4)$, $r = \sqrt{(-8)^2 + 4^2} = \sqrt{80} = 4\sqrt{5}$.

Using $(-4, 2)$ and $r = 2\sqrt{5}$, we find that

$$\sin\theta = \frac{y}{r} = \frac{2}{2\sqrt{5}} = \frac{1}{\sqrt{5}} = \frac{\sqrt{5}}{5},$$

$$\cos\theta = \frac{x}{r} = \frac{-4}{2\sqrt{5}} = \frac{-2}{\sqrt{5}} = -\frac{2\sqrt{5}}{5},$$

and $\quad \tan\theta = \dfrac{y}{x} = \dfrac{2}{-4} = -\dfrac{1}{2}.$

Using $(-8, 4)$ and $r = 4\sqrt{5}$, we find that

$$\sin\theta = \frac{y}{r} = \frac{4}{4\sqrt{5}} = \frac{1}{\sqrt{5}} = \frac{\sqrt{5}}{5},$$

$$\cos\theta = \frac{x}{r} = \frac{-8}{4\sqrt{5}} = \frac{-2}{\sqrt{5}} = -\frac{2\sqrt{5}}{5},$$

and $\quad \tan\theta = \dfrac{y}{x} = \dfrac{4}{-8} = -\dfrac{1}{2}.$

We see that the function values are the same using either point. This illustrates that any point other than the origin on the terminal side of an angle can be used to determine the trigonometric function values.

The trigonometric function values of θ depend only on the angle, not on the choice of the point on the terminal side that is used to compute them.

❯ The Six Related Functions

When we know one of the function values of an angle, we can find the other five if we know the quadrant in which the terminal side lies. The procedure is to sketch a reference triangle in the appropriate quadrant, use the Pythagorean equation as needed to find the lengths of its sides, and then find the ratios of the sides.

EXAMPLE 4 Given that $\tan \theta = -\frac{2}{3}$ and θ is in the second quadrant, find the other function values.

Solution We first sketch a second-quadrant angle. Since

$$\tan \theta = \frac{y}{x} = -\frac{2}{3} = \frac{2}{-3}, \qquad \text{Expressing } -\frac{2}{3} \text{ as } \frac{2}{-3} \text{ since } \theta \text{ is in quadrant II}$$

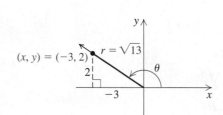

we make the legs' lengths 3 and 2. We measure off the 3 units in the negative direction since θ is in quadrant II. The hypotenuse must then have length $\sqrt{3^2 + 2^2}$, or $\sqrt{13}$. Now we read off the appropriate ratios:

$$\sin \theta = \frac{y}{r} = \frac{2}{\sqrt{13}}, \quad \text{or} \quad \frac{2\sqrt{13}}{13}, \qquad \csc \theta = \frac{r}{y} = \frac{\sqrt{13}}{2},$$

$$\cos \theta = \frac{x}{r} = -\frac{3}{\sqrt{13}}, \quad \text{or} \quad -\frac{3\sqrt{13}}{13}, \qquad \sec \theta = \frac{r}{x} = -\frac{\sqrt{13}}{3},$$

$$\tan \theta = \frac{y}{x} = -\frac{2}{3}, \qquad \cot \theta = \frac{x}{y} = -\frac{3}{2}.$$

❯ *Now Try Exercise 35.*

❯ Terminal Side on an Axis

An angle whose terminal side falls on one of the axes is a **quadrantal angle**. One of the coordinates of any point on that side is 0. The definitions of the trigonometric functions still apply, but in some cases, function values will not be defined because a denominator will be 0.

EXAMPLE 5 Find the sine, cosine, and tangent values for $90°$, $180°$, $270°$, and $360°$.

Solution We first make a drawing of each angle in standard position and label a point on the terminal side. Since the function values are the same for all points on the terminal side, we choose $(0, 1)$, $(-1, 0)$, $(0, -1)$, and $(1, 0)$ for convenience. Note that $r = 1$ for each choice.

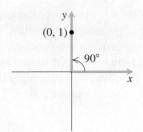
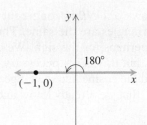
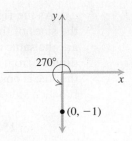
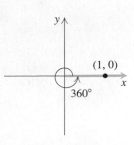

Then by the definitions we get

$$\sin 90° = \frac{1}{1} = 1,$$

$$\cos 90° = \frac{0}{1} = 0,$$

$$\tan 90° = \frac{1}{0}, \quad \textbf{Not defined}$$

$$\sin 180° = \frac{0}{1} = 0,$$

$$\cos 180° = \frac{-1}{1} = -1,$$

$$\tan 180° = \frac{0}{-1} = 0,$$

$$\sin 270° = \frac{-1}{1} = -1,$$

$$\cos 270° = \frac{0}{1} = 0,$$

$$\tan 270° = \frac{-1}{0}, \quad \textbf{Not defined}$$

$$\sin 360° = \frac{0}{1} = 0,$$

$$\cos 360° = \frac{1}{1} = 1,$$

$$\tan 360° = \frac{0}{1} = 0.$$

In Example 5, all the values can be found using a calculator, but you will find that it is convenient to be able to compute them mentally. It is also helpful to note that coterminal angles have the same function values. For example, 0° and 360° are coterminal; thus, sin 0° = 0, cos 0° = 1, and tan 0° = 0.

EXAMPLE 6 Find each of the following.

a) $\sin(-90°)$ **b)** csc 540°

Solution

a) We note that −90° is coterminal with 270°. Thus,

$$\sin(-90°) = \sin 270° = -1.$$

b) Since 540° = 180° + 360°, 540° and 180° are coterminal. Thus,

$$\csc 540° = \csc 180° = \frac{1}{\sin 180°} = \frac{1}{0}, \quad \text{which is not defined.}$$

> *Now Try Exercises 45 and 55.*

ERR: DIVIDE BY 0
1: Quit
2: Goto

or

ERR: Domain
1: Quit
2: Goto

Trigonometric values can always be checked using a calculator. When the value is not defined, the calculator will display an ERROR message, as shown at left.

❯ Reference Angles: 30°, 45°, and 60°

We can also mentally determine trigonometric function values whenever the terminal side makes a 30°, 45°, or 60° angle with the *x*-axis. Consider, for example, an angle of 150°. The terminal side makes a 30° angle with the *x*-axis, since 180° − 150° = 30°.

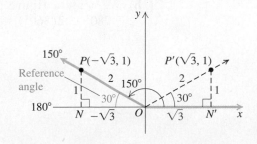

As the figure shows, $\triangle ONP$ is congruent to $\triangle ON'P'$; therefore, the ratios of the sides of the two triangles are the same. Thus the trigonometric function values are the same except perhaps for the sign. We could determine the function values directly from $\triangle ONP$, but this is not necessary. If we remember that in quadrant II, the sine is positive and the cosine and the tangent are negative, we can simply use the function values of 30° that we already know and prefix the appropriate sign. Thus,

$$\sin 150° = \sin 30° = \frac{1}{2},$$

$$\cos 150° = -\cos 30° = -\frac{\sqrt{3}}{2},$$

and $\tan 150° = -\tan 30° = -\frac{1}{\sqrt{3}},$ or $-\frac{\sqrt{3}}{3}.$

Triangle ONP is the reference triangle and the acute angle $\angle NOP$ is called a *reference angle*.

REFERENCE ANGLE

The **reference angle** for an angle is the acute angle formed by the terminal side of the angle and the *x*-axis.

EXAMPLE 7 Find the sine, cosine, and tangent function values for each of the following.

a) 225°   **b)** −780°

Solution

a) We draw a figure showing the terminal side of a 225° angle. The reference angle is 225° − 180°, or 45°.

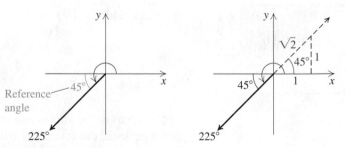

Recall from Section 6.1 that $\sin 45° = \sqrt{2}/2$, $\cos 45° = \sqrt{2}/2$, and $\tan 45° = 1$. Also note that in the third quadrant, the sine and the cosine are negative and the tangent is positive. Thus we have

$$\sin 225° = -\frac{\sqrt{2}}{2}, \quad \cos 225° = -\frac{\sqrt{2}}{2}, \quad \text{and} \quad \tan 225° = 1.$$

b) We draw a figure showing the terminal side of a −780° angle. Since −780° + 2(360°) = −60°, we know that −780° and −60° are coterminal.

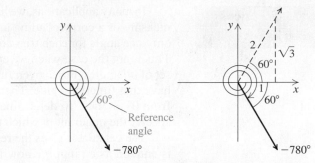

The reference angle for $-60°$ is the acute angle formed by the terminal side of the angle and the x-axis. Thus the reference angle for $-60°$ is $60°$. We know that since $-780°$ is a fourth-quadrant angle, the cosine is positive and the sine and the tangent are negative. Recalling that $\sin 60° = \sqrt{3}/2$, $\cos 60° = 1/2$, and $\tan 60° = \sqrt{3}$, we have

$$\sin(-780°) = -\frac{\sqrt{3}}{2},$$

$$\cos(-780°) = \frac{1}{2},$$

and $\quad \tan(-780°) = -\sqrt{3}.$

> *Now Try Exercises 47 and 51.*

❯ Function Values for Any Angle

When the terminal side of an angle falls on one of the axes or makes a 30°, 45°, or 60° angle with the x-axis, we can find exact function values without the use of a calculator. But this group is only a small subset of *all* angles. Using a calculator, we can approximate the trigonometric function values of *any* angle. In fact, we can approximate or find exact function values of all angles without using a reference angle.

EXAMPLE 8   Find each of the following function values using a calculator and round the answer to four decimal places, where appropriate.

a) $\cos 112°$ **b)** $\sec 500°$

c) $\tan(-83.4°)$ **d)** $\csc 351.75°$

e) $\cos 2400°$ **f)** $\sin 175°40'9''$

g) $\cot(-135°)$

Solution   Using a calculator set in DEGREE mode, we find the values.

a) $\cos 112° \approx -0.3746$

b) $\sec 500° = \dfrac{1}{\cos 500°} \approx -1.3054$

c) $\tan(-83.4°) \approx -8.6427$

d) $\csc 351.75° = \dfrac{1}{\sin 351.75°} \approx -6.9690$

e) $\cos 2400° = -0.5$

f) $\sin 175°40'9'' \approx 0.0755$

g) $\cot(-135°) = \dfrac{1}{\tan(-135°)} = 1$

> *Now Try Exercises 85 and 91.*

```
cos(112)
               -.3746065934
  1
cos(500)
               -1.305407289
tan(-83.4)
               -8.642747461
```

```
   1
sin(351.75)
               -6.968999424
cos(2400)
                         -.5
sin(175°40'9")
                .0755153443
```

INVERSE FUNCTIONS

REVIEW SECTION 5.1.

In many applications, we have a trigonometric function value and want to find the measure of a corresponding angle. When only acute angles are considered, there is only one angle for each trigonometric function value. (See Example 9 in Section 6.1.) This is not the case when we extend the domain of the trigonometric functions to the set of *all* angles. For a given function value, there is an infinite number of angles that have that function value. There can be two such angles for each value in the range from 0° to 360°. To determine a unique answer in the interval $(0°, 360°)$, we must specify the quadrant in which the terminal side lies.

The calculator gives the reference angle as an output for each function value that is entered as an input. Knowing the reference angle and the quadrant in which the terminal side lies, we can find the specified angle.

EXAMPLE 9 Given the function value and the quadrant restriction, find θ.

a) $\sin \theta = 0.2812, \ 90° < \theta < 180°$

b) $\cot \theta = -0.1611, \ 270° < \theta < 360°$

Solution

a) We first sketch the angle in the second quadrant. We use the calculator to find the acute angle (reference angle) whose sine is 0.2812. The reference angle is approximately 16.33°. We find the angle θ by subtracting 16.33° from 180°:

$$180° - 16.33° = 163.67°.$$

Thus, $\theta \approx 163.67°$.

b) We begin by sketching the angle in the fourth quadrant. Because the tangent and cotangent values are reciprocals, we know that

$$\tan \theta \approx \frac{1}{-0.1611} \approx -6.2073.$$

We use the calculator to find the acute angle (reference angle) whose tangent is 6.2073, ignoring the fact that $\tan \theta$ is negative. The reference angle is approximately 80.85°. We find angle θ by subtracting 80.85° from 360°:

$$360° - 80.85° = 279.15°.$$

Thus, $\theta \approx 279.15°$.

> *Now Try Exercise 97.*

BEARING: FIRST-TYPE

REVIEW SECTION 6.2.

Bearing: Second-Type

In aerial navigation, directions are given in degrees clockwise from north. Thus east is 90°, south is 180°, and west is 270°. Several aerial directions, or **bearings**, are given below.

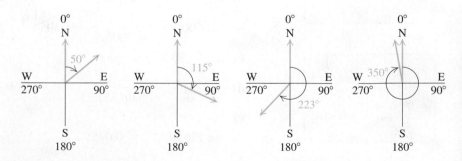

EXAMPLE 10 *Aerial Navigation.* An airplane flies 218 mi from an airport in a direction of 245°. How far south of the airport is the plane then? How far west?

Solution We first find the measure of $\angle ABC$:

$$B = 270° - 245° = 25°. \quad \textbf{Angle } B \textbf{ is opposite side } b \textbf{ in the right triangle.}$$

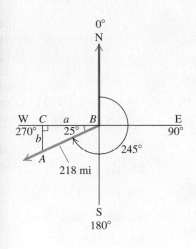

From the figure shown at left, we see that the distance south of the airport *b* and the distance west of the airport *a* are legs of a right triangle. We have

$$\frac{b}{218} = \sin 25°$$

$$b = 218 \sin 25° \approx 92 \text{ mi}$$

and

$$\frac{a}{218} = \cos 25°$$

$$a = 218 \cos 25° \approx 198 \text{ mi}.$$

The airplane is about 92 mi south and about 198 mi west of the airport.

> **Now Try Exercise 105.**

6.3 Exercise Set

For angles of the following measures, state in which quadrant the terminal side lies. It helps to sketch the angle in standard position.

1. 187°

2. −14.3°

3. 245°15′

4. −120°

5. 800°

6. 1075°

7. −460.5°

8. 315°

9. −912°

10. 13°15′58″

11. 537°

12. −345.14°

Find two positive angles and two negative angles that are coterminal with the given angle. Answers may vary.

13. 74°

14. −81°

15. 115.3°

16. 275°10′

17. −180°

18. −310°

Find the complement and the supplement.

19. 17.11°

20. 47°38′

21. 12°3′14″

22. 9.038°

23. 45.2°

24. 67.31°

Find the six trigonometric function values for the angle shown.

25.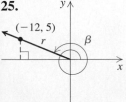

26.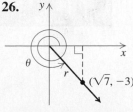

27.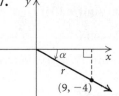

28.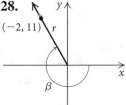

29.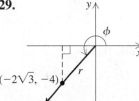

30.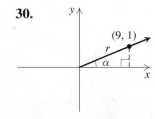

The terminal side of angle θ in standard position lies on the given line in the given quadrant. Find sin θ, cos θ, and tan θ.

31. $2x + 3y = 0$; quadrant IV

32. $4x + y = 0$; quadrant II

33. $5x − 4y = 0$; quadrant I

34. $y = 0.8x$; quadrant III

A function value and a quadrant are given. Find the other five trigonometric function values. Give exact answers.

35. $\sin \theta = -\frac{1}{3}$; quadrant III

36. $\tan \beta = 5$; quadrant I

37. $\cot \theta = -2$; quadrant IV

38. $\cos \alpha = -\frac{4}{5}$; quadrant II

39. $\cos \phi = \frac{3}{5}$; quadrant IV

40. $\sin \theta = -\frac{5}{13}$; quadrant III

Find the reference angle and the exact function value if they exist.

41. $\cos 150°$

42. $\sec (-225°)$

43. $\tan (-135°)$

44. $\sin (-45°)$

45. $\sin 7560°$

46. $\tan 270°$

47. $\cos 495°$

48. $\tan 675°$

49. $\csc (-210°)$

50. $\sin 300°$

51. $\cot 570°$

52. $\cos (-120°)$

53. $\tan 330°$

54. $\cot 855°$

55. $\sec (-90°)$

56. $\sin 90°$

57. $\cos (-180°)$

58. $\csc 90°$

59. $\tan 240°$

60. $\cot (-180°)$

61. $\sin 495°$

62. $\sin 1050°$

63. $\csc 225°$

64. $\sin (-450°)$

65. $\cos 0°$

66. $\tan 480°$

67. $\cot (-90°)$

68. $\sec 315°$

69. $\cos 90°$

70. $\sin (-135°)$

71. $\cos 270°$

72. $\tan 0°$

In Exercises 73–80, find the signs of the six trigonometric function values for the given angles.

73. $319°$

74. $-57°$

75. $194°$

76. $-620°$

77. $-215°$

78. $290°$

79. $-272°$

80. $91°$

Use a calculator in Exercises 81–84, but do not use the trigonometric function keys.

81. Given that
$$\sin 41° = 0.6561,$$
$$\cos 41° = 0.7547,$$
$$\tan 41° = 0.8693,$$
find the trigonometric function values for $319°$.

82. Given that
$$\sin 27° = 0.4540,$$
$$\cos 27° = 0.8910,$$
$$\tan 27° = 0.5095,$$
find the trigonometric function values for $333°$.

83. Given that
$$\sin 65° = 0.9063,$$
$$\cos 65° = 0.4226,$$
$$\tan 65° = 2.1445,$$
find the trigonometric function values for $115°$.

84. Given that
$$\sin 35° = 0.5736,$$
$$\cos 35° = 0.8192,$$
$$\tan 35° = 0.7002,$$
find the trigonometric function values for $215°$.

Find the function value. Round to four decimal places.

85. $\tan 310.8°$

86. $\cos 205.5°$

87. $\cot 146.15°$

88. $\sin (-16.4°)$

89. $\sin 118°42'$

90. $\cos 273°45'$

91. $\cos (-295.8°)$

92. $\tan 1086.2°$

93. $\cos 5417°$

94. $\sec 240°55'$

95. $\csc 520°$

96. $\sin 3824°$

Given the function value and the quadrant restriction, find θ.

	FUNCTION VALUE	INTERVAL	θ
97.	$\sin \theta = -0.9956$	$(270°, 360°)$	_____
98.	$\tan \theta = 0.2460$	$(180°, 270°)$	_____
99.	$\cos \theta = -0.9388$	$(180°, 270°)$	_____
100.	$\sec \theta = -1.0485$	$(90°, 180°)$	_____
101.	$\tan \theta = -3.0545$	$(270°, 360°)$	_____
102.	$\sin \theta = -0.4313$	$(180°, 270°)$	_____
103.	$\csc \theta = 1.0480$	$(0°, 90°)$	_____
104.	$\cos \theta = -0.0990$	$(90°, 180°)$	_____

105. *Aerial Navigation.* An airplane flies 150 km from an airport in a direction of 120°. How far east of the airport is the plane then? How far south?

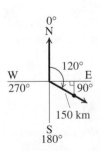

106. *Aerial Navigation.*   An airplane leaves an airport and travels for 100 mi in a direction of 300°. How far north of the airport is the plane then? How far west?

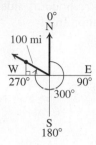

107. *Aerial Navigation.*   An airplane travels at 150 km/h for 2 hr in a direction of 138° from Omaha. At the end of this time, how far south of Omaha is the plane?

108. *Aerial Navigation.*   An airplane travels at 120 km/h for 2 hr in a direction of 319° from Chicago. At the end of this time, how far north of Chicago is the plane?

❯ Skill Maintenance

Graph the function. Sketch and label any vertical asymptotes.

109. $f(x) = \dfrac{1}{x^2 - 25}$ [4.5]

110. $g(x) = x^3 - 2x + 1$ [4.2]

Determine the domain and the range of the function.

111. $f(x) = \dfrac{x - 4}{x + 2}$ [1.2], [4.5]

112. $g(x) = \dfrac{x^2 - 9}{2x^2 - 7x - 15}$ [1.2], [4.1], [4.5]

Find the zeros of the function.

113. $f(x) = 12 - x$ [1.5]

114. $g(x) = x^2 - x - 6$ [3.2]

Find the x-intercept(s) of the graph of the function.

115. $f(x) = 12 - x$ [1.5]

116. $g(x) = x^2 - x - 6$ [3.2]

❯ Synthesis

117. *Tallest Ferris Wheel.*   On March 31, 2014, the world's tallest ferris wheel, the High Roller, opened in Las Vegas at the LINQ. Each cabin, which can fit up to 40 passengers, is 220 ft from the center of the wheel. The wheel makes one revolution in approximately 30 min. (*Source*: Nancy Trejos, *USA Today*, April 7, 2014). When you board, you are 6 ft above the ground. After you have rotated through an angle of 315°, how far above the ground are you?

118. *Valve Cap on a Bicycle.*   The valve cap on a bicycle wheel is 12.5 in. from the center of the wheel. From the position shown, the wheel starts to roll. After the wheel has turned 390°, how far above the ground is the valve cap? Assume that the outer radius of the tire is 13.375 in.

Determine whether the statement is true or false.

1. If $\sin \alpha > 0$ and $\cot \alpha > 0$, then α is in the first quadrant. **[6.3]**

2. The lengths of corresponding sides in similar triangles are in the same ratio. **[6.1]**

3. If θ is an acute angle and $\csc \theta \approx 1.5539$, then $\cos (90° - \theta) \approx 0.6435$. **[6.1]**

Solve the right triangle. **[6.2]**

4.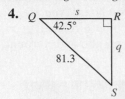

5.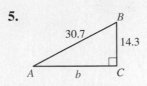

Find two positive angles and two negative angles that are coterminal with the given angle. Answers may vary. **[6.3]**

6. $-75°$

7. $214°30'$

Find the complement and the supplement of the given angle. **[6.3]**

8. $18.2°$

9. $87°15'10''$

10. Given that $\sin 25° = 0.4226$, $\cos 25° = 0.9063$, and $\tan 25° = 0.4663$, find the six trigonometric function values for $155°$. Use a calculator, but do not use the trigonometric function keys. **[6.3]**

11. Find the six trigonometric function values for the angle shown. **[6.3]**

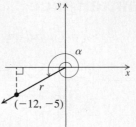

12. Given $\cot \theta = 2$ and θ in quadrant III, find the other five trigonometric function values. **[6.3]**

13. Given $\cos \alpha = \frac{2}{9}$ and $0° < \alpha < 90°$, find the other five trigonometric function values. **[6.1]**

14. Convert $42°8'50''$ to decimal degree notation. Round to four decimal places. **[6.1]**

15. Convert $51.18°$ to degrees, minutes, and seconds. **[6.1]**

16. Given that $\sin 9° \approx 0.1564$, $\cos 9° \approx 0.9877$, and $\tan 9° \approx 0.1584$, find the six trigonometric function values of $81°$. **[6.1]**

17. If $\tan \theta = 2.412$ and θ is acute, find the angle to the nearest tenth of a degree. **[6.1]**

18. *Aerial Navigation.* An airplane travels at 200 mph for $1\frac{1}{2}$ hr in a direction of $285°$ from Atlanta. At the end of this time, how far west of Atlanta is the plane? **[6.3]**

Without a calculator, find the exact function value. **[6.1], [6.3]**

19. $\tan 210°$

20. $\sin 45°$

21. $\cot 30°$

22. $\sec 135°$

23. $\cos 45°$

24. $\csc (-30°)$

25. $\sin 90°$

26. $\cos 270°$

27. $\sin 120°$

28. $\sec 180°$

29. $\tan (-240°)$

30. $\cot (-315°)$

31. $\sin 750°$

32. $\csc 45°$

33. $\cos 210°$

34. $\cot 0°$

35. $\csc 150°$

36. $\tan 90°$

37. $\sec 3600°$

38. $\cos 495°$

Find the function value. Round the answer to four decimal places. **[6.1], [6.3]**

39. $\cos 39.8°$

40. $\sec 50°$

41. $\tan 2183°$

42. $\sin 10°28'3''$

43. $\csc (-74°)$

44. $\cot 142.7°$

45. $\sin (-40.1°)$

46. $\cos 87°15'$

COLLABORATIVE DISCUSSION AND WRITING

47. Why do the function values of θ depend only on the angle and not on the choice of a point on the terminal side? **[6.3]**

48. Explain the difference between reciprocal functions and cofunctions. **[6.1]**

49. In Section 6.1, the trigonometric functions are defined as functions of acute angles. What appear to be the ranges for the sine, cosine, and tangent functions given the restricted domain as the set of angles whose measures are greater than 0° and less than 90°? **[6.1]**

50. Why is the domain of the tangent function different from the domains of the sine function and the cosine function? **[6.3]**

6.4 ⟩ Radians, Arc Length, and Angular Speed

> ❯ Find points on the unit circle that are determined by real numbers.

> ❯ Convert between radian measure and degree measure; find coterminal, complementary, and supplementary angles.

> ❯ Find the length of an arc of a circle; find the measure of a central angle of a circle.

> ❯ Convert between linear speed and angular speed.

Another useful unit of angle measure is called a **radian**. To introduce radian measure, we use a circle centered at the origin with a radius of length 1. Such a circle is called a **unit circle**. Its equation is $x^2 + y^2 = 1$.

CIRCLES

REVIEW SECTION 1.1.

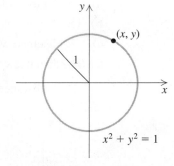

$x^2 + y^2 = 1$

❯ Distances on the Unit Circle

The circumference of a circle of radius r is $2\pi r$. Thus for the unit circle, where $r = 1$, the circumference is 2π. If a point starts at A and travels around the circle (Fig. 1 on the following page), it will travel a distance of 2π. If it travels halfway around the circle (Fig. 2), it will travel a distance of $\frac{1}{2} \cdot 2\pi$, or π.

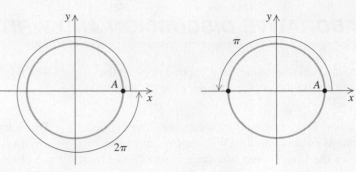

FIGURE 1.   FIGURE 2.

If a point C travels $\frac{1}{8}$ of the way around the circle (Fig. 3), it will travel a distance of $\frac{1}{8} \cdot 2\pi$, or $\pi/4$. Note that C is $\frac{1}{4}$ of π the distance from A to B. If a point D travels $\frac{1}{6}$ of the way around the circle (Fig. 4), it will travel a distance of $\frac{1}{6} \cdot 2\pi$, or $\pi/3$. Note that D is $\frac{1}{3}$ of π the distance from A to B.

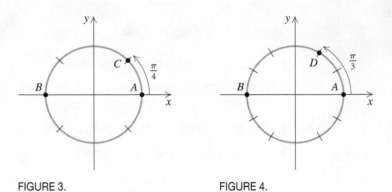

FIGURE 3.   FIGURE 4.

EXAMPLE 1   How far will a point travel if it goes **(a)** $\frac{1}{4}$, **(b)** $\frac{1}{12}$, **(c)** $\frac{3}{8}$, and **(d)** $\frac{5}{6}$ of the way around the unit circle?

Solution

a) $\frac{1}{4}$ of the total distance around the circle is $\frac{1}{4} \cdot 2\pi$, which is $\frac{1}{2}\pi$, or $\pi/2$.

b) The distance will be $\frac{1}{12} \cdot 2\pi$, which is $\frac{1}{6}\pi$, or $\pi/6$.

c) The distance will be $\frac{3}{8} \cdot 2\pi$, which is $\frac{3}{4}\pi$, or $3\pi/4$.

d) The distance will be $\frac{5}{6} \cdot 2\pi$, which is $\frac{5}{3}\pi$, or $5\pi/3$. Think of $5\pi/3$ as $\pi + \frac{2}{3}\pi$.

These distances are illustrated in the following figures.

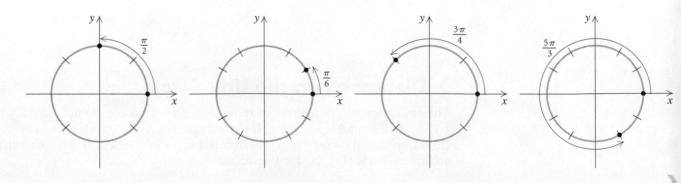

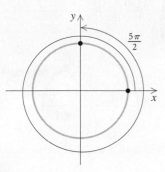

FIGURE 5.

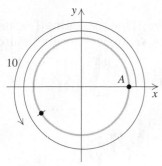

FIGURE 6.

A point may travel completely around the circle and then continue. For example, if it goes around once and then continues $\frac{1}{4}$ of the way around, it will have traveled a distance of $2\pi + \frac{1}{4} \cdot 2\pi$, or $5\pi/2$ (Fig. 5). *Every* real number determines a point on the unit circle. For the positive number 10, for example, we start at A and travel counterclockwise a distance of 10. The point at which we stop is the point "determined" by the number 10. Note that $2\pi \approx 6.28$ and that $10 \approx 1.6(2\pi)$. Thus the point for 10 travels around the unit circle about 1.6, or $1\frac{3}{5}$, times (Fig. 6).

For a negative number, we move clockwise around the circle. Points for $-\pi/4$ and $-3\pi/2$ are shown in the following figures. The number 0 determines the point A.

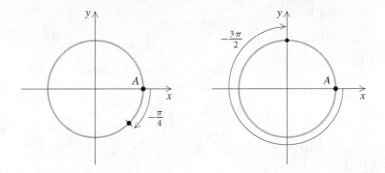

EXAMPLE 2 On the unit circle, mark the point determined by each of the following real numbers.

a) $\dfrac{9\pi}{4}$ **b)** $-\dfrac{7\pi}{6}$

Solution

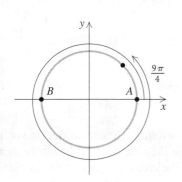

a) Think of $9\pi/4$ as $2\pi + \frac{1}{4}\pi$. (See the figure at left.) Since $9\pi/4 > 0$, the point moves counterclockwise. The point goes completely around once and then continues $\frac{1}{4}$ of the way from A to B.

b) The number $-7\pi/6$ is negative, so the point moves clockwise. From A to B, the distance is π, or $\frac{6}{6}\pi$, so we need to go beyond B another distance of $\pi/6$, clockwise. (See the figure below.)

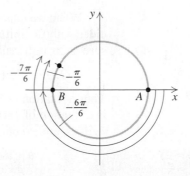

> **Now Try Exercise 1.**

❯ Radian Measure

Degree measure is a common unit of angle measure in many everyday applications. But in many scientific fields and in mathematics (calculus, in particular), there is another commonly used unit of measure called the *radian*.

RADIANS

Consider the unit circle. Recall that this circle has radius 1. Suppose we measure, moving counterclockwise, an arc of length 1, and mark a point T on the circle. If we draw a ray from the origin through T, we have formed an angle. The measure of that angle is 1 **radian**.

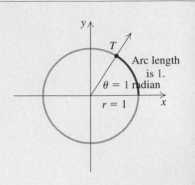

The word radian is derived from the word *radius*. Thus measuring 1 "radius" along the circumference of the circle determines an angle whose measure is 1 *radian*. One radian is about 57.3°.

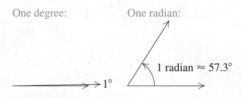

Angles that measure 2 radians, 3 radians, and 6 radians are shown below.

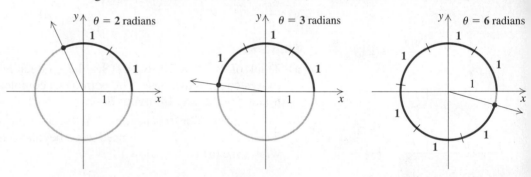

When we make a complete (counterclockwise) revolution, the terminal side coincides with the initial side on the positive x-axis. We then have an angle whose measure is 2π radians, or about 6.28 radians, which is the circumference of the circle:

$$2\pi r = 2\pi(1) = 2\pi.$$

Thus a rotation of 360° (1 revolution) has a measure of 2π radians. A half revolution is a rotation of 180°, or π radians. A quarter revolution is a rotation of 90°, or $\pi/2$ radians, and so on.

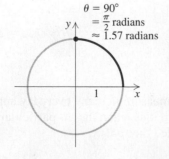

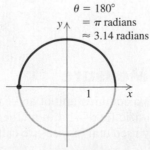

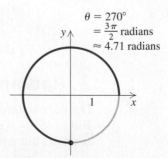

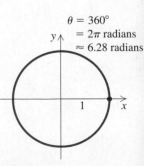

To convert between degrees and radians, we first note that

$$360° = 2\pi \text{ radians}.$$

It follows that

$$180° = \pi \text{ radians.}$$

To make conversions, we multiply by 1, noting that:

CONVERTING BETWEEN DEGREE MEASURE AND RADIAN MEASURE

$$\frac{\pi \text{ radians}}{180°} = \frac{180°}{\pi \text{ radians}} = 1.$$

To convert from degree to radian measure, multiply by $\dfrac{\pi \text{ radians}}{180°}$.

To convert from radian to degree measure, multiply by $\dfrac{180°}{\pi \text{ radians}}$.

EXAMPLE 3 Convert each of the following to radians.

a) 120°   **b)** −297.25°

Solution

a) $120° = 120° \cdot \dfrac{\pi \text{ radians}}{180°}$ **Multiplying by 1**

$\qquad = \dfrac{120°}{180°}\pi \text{ radians}$

$\qquad = \dfrac{2\pi}{3}\text{radians, or about 2.09 radians}$

b) $-297.25° = -297.25° \cdot \dfrac{\pi \text{ radians}}{180°}$

$\qquad = -\dfrac{297.25°}{180°}\pi \text{ radians}$

$\qquad = -\dfrac{297.25\pi}{180}\text{radians}$

$\qquad \approx -5.19 \text{ radians}$

We also can use a calculator set in RADIAN mode to convert the angle measures. We enter the angle measure followed by ° (degrees) from the ANGLE menu. Finally, we press **ENTER** to see radian measure.

```
120°
                        2.094395102
-297.25°
                       -5.187991202
```

> *Now Try Exercises 11 and 23.*

EXAMPLE 4 Convert each of the following to degrees.

a) $\dfrac{3\pi}{4} \text{ radians}$ **b)** 8.5 radians

Solution

a) $\dfrac{3\pi}{4}$ radians $= \dfrac{3\pi}{4}$ radians $\cdot \dfrac{180°}{\pi \text{ radians}}$ **Multiplying by 1**

$$= \dfrac{3\pi}{4\pi} \cdot 180° = \dfrac{3}{4} \cdot 180° = 135°$$

b) 8.5 radians $= 8.5$ radians $\cdot \dfrac{180°}{\pi \text{ radians}}$

$$= \dfrac{8.5(180°)}{\pi} \approx 487.01°$$

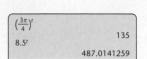

With a calculator set in DEGREE mode, we can enter the angle measure followed by ^{r} (radians) from the ANGLE menu. Finally, we press **ENTER** to see degree measure.

> *Now Try Exercises 35 and 43.*

The radian–degree equivalents of the most commonly used angle measures are illustrated in the following figures.

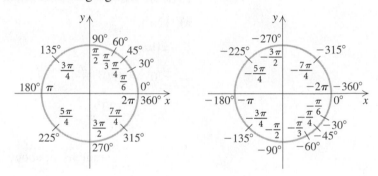

> It is also helpful to visualize radian–degree equivalents separately with unit circles divided into 8 and 12 sections.
>
>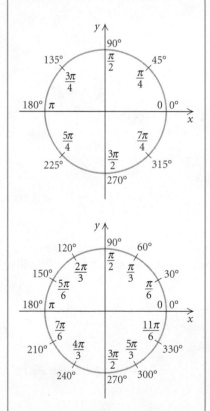

When a rotation is given in radians, the word "radians" is optional and is most often omitted. **Thus if no unit is given for a rotation, the rotation is understood to be in radians.**

We can also find coterminal, complementary, and supplementary angles in radian measure just as we did for degree measure in Section 6.3.

EXAMPLE 5 Find a positive angle and a negative angle that are coterminal with $2\pi/3$. Many answers are possible.

Solution To find angles coterminal with a given angle, we add or subtract multiples of 2π:

$$\dfrac{2\pi}{3} + 2\pi = \dfrac{2\pi}{3} + \dfrac{6\pi}{3} = \dfrac{8\pi}{3},$$

$$\dfrac{2\pi}{3} - 3(2\pi) = \dfrac{2\pi}{3} - \dfrac{18\pi}{3} = -\dfrac{16\pi}{3}.$$

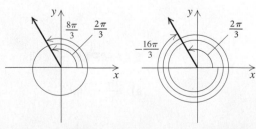

Thus, $8\pi/3$ and $-16\pi/3$ are two of the many angles coterminal with $2\pi/3$.

> *Now Try Exercise 51.*

EXAMPLE 6   Find the complement and the supplement of $\pi/6$.

Solution   Since $90°$ equals $\pi/2$ radians, the complement of $\pi/6$ is

$$\frac{\pi}{2} - \frac{\pi}{6} = \frac{3\pi}{6} - \frac{\pi}{6} = \frac{2\pi}{6}, \quad \text{or} \quad \frac{\pi}{3}.$$

Since $180°$ equals π radians, the supplement of $\pi/6$ is

$$\pi - \frac{\pi}{6} = \frac{6\pi}{6} - \frac{\pi}{6} = \frac{5\pi}{6}.$$

Thus the complement of $\pi/6$ is $\pi/3$ and the supplement is $5\pi/6$.

> *Now Try Exercise 55.*

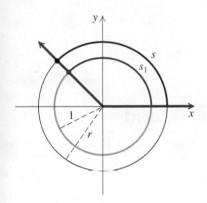

❯ Arc Length and Central Angles

Radian measure can be determined using a circle other than a unit circle. In the figure at left, a unit circle (with radius 1) is shown along with another circle (with radius r, $r \neq 1$). The angle shown is a **central angle** of both circles.

From geometry, we know that the arcs that the angle subtends have their lengths in the same ratio as the radii of the circles. The radii of the circles are r and 1. The corresponding arc lengths are s and s_1. Thus we have the proportion

$$\frac{s}{s_1} = \frac{r}{1},$$

which also can be written as

$$\frac{s_1}{1} = \frac{s}{r}.$$

Now s_1 is the *radian measure* of the rotation in question. It is common to use a Greek letter, such as θ, for the measure of an angle or a rotation and the letter s for arc length. Adopting this convention, we rewrite the proportion above as

$$\theta = \frac{s}{r}.$$

In any circle, the measure (in radians) of a central angle, the arc length the angle subtends, and the length of the radius are related in this fashion. Or, in general, the following is true.

RADIAN MEASURE

The **radian measure** θ of a rotation is the **ratio** of the distance s, traveled by a point at a radius r from the center of rotation, to the length of the radius r:

$$\theta = \frac{s}{r}.$$

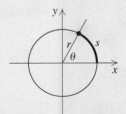

When we are using the formula $\theta = s/r$, θ must be in radians and s and r must be expressed in the same unit.

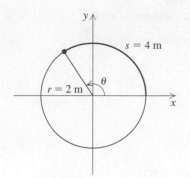

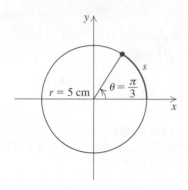

EXAMPLE 7 Find the measure of a rotation in radians when a point 2 m from the center of rotation travels 4 m.

Solution We have

$$\theta = \frac{s}{r}$$

$$= \frac{4 \text{ m}}{2 \text{ m}} = 2.$$ **The unit is understood to be radians.**

▶ *Now Try Exercise 65.*

EXAMPLE 8 Find the length of an arc of a circle of radius 5 cm associated with an angle of $\pi/3$ radians.

Solution We have

$$\theta = \frac{s}{r}, \quad \text{or} \quad s = r\theta.$$

Thus, $s = 5 \text{ cm} \cdot \pi/3$, or about 5.24 cm.

▶ *Now Try Exercise 63.*

❯ Linear Speed and Angular Speed

Linear speed is defined to be distance traveled per unit of time. If we use v for linear speed, s for distance, and t for time, then

$$v = \frac{s}{t}.$$

Similarly, **angular speed** is defined to be amount of rotation per unit of time. For example, we might speak of the angular speed of a bicycle wheel as 150 revolutions per minute or the angular speed of the earth as 2π radians per day. The Greek letter ω (omega) is generally used for angular speed. Thus for a rotation θ and time t, angular speed is defined as

$$\omega = \frac{\theta}{t}.$$

As an example of how these definitions can be applied, let's consider the refurbished carousel at the Children's Museum in Indianapolis, Indiana. It consists of three circular rows of animals. All animals, regardless of the row, travel at the same angular speed. But the animals in the outer row travel at a greater linear speed than those in the inner rows. What is the relationship between the linear speed v and the angular speed ω?

To develop the relationship we seek, recall that, for rotations measured in radians, $\theta = s/r$. This is equivalent to

$$s = r\theta.$$

We divide by time, t, to obtain

$$\frac{s}{t} = \frac{r\theta}{t} \qquad \textbf{Dividing by } t$$

$$\frac{s}{t} = r \cdot \frac{\theta}{t}$$
$$\underset{v}{\downarrow} \qquad \underset{\omega}{\downarrow}.$$

Now s/t is linear speed v, and θ/t is angular speed ω. Thus we have the relationship we seek,

$$v = r\omega.$$

© 2006, The Children's Museum of Indianapolis

LINEAR SPEED IN TERMS OF ANGULAR SPEED

The **linear speed** v of a point a distance r from the center of rotation is given by

$$v = r\omega,$$

where ω is the **angular speed**, in radians, per unit of time.

For the formula $v = r\omega$, the units of distance for v and r must be the same, ω must be in radians per unit of time, and the units of time for v and ω must be the same.

EXAMPLE 9 *Linear Speed of an Earth Satellite.* An earth satellite in circular orbit 1200 km high makes one complete revolution every 90 min. What is its linear speed? Use 6400 km for the length of a radius of the earth.

Solution To use the formula $v = r\omega$, we need to know r and ω:

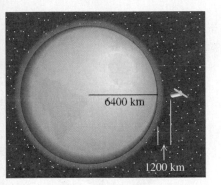

6400 km

1200 km

$$r = 6400 \text{ km} + 1200 \text{ km} \qquad \textbf{Radius of earth plus height of satellite}$$
$$= 7600 \text{ km};$$

$$\omega = \frac{\theta}{t} = \frac{2\pi}{90 \text{ min}} = \frac{\pi}{45 \text{ min}}. \qquad \textbf{We have, as usual, omitted the word radians.}$$

Now, using $v = r\omega$, we have

$$v = 7600 \text{ km} \cdot \frac{\pi}{45 \text{ min}} = \frac{7600\pi}{45} \cdot \frac{\text{km}}{\text{min}} \approx 531 \frac{\text{km}}{\text{min}}.$$

Thus the linear speed of the satellite is approximately 531 km/min.

> **Now Try Exercise 71.**

EXAMPLE 10 *Angular Speed of a Capstan.* An anchor on a Navy vessel is hoisted at a rate of 2 ft/sec as the line is wound around a capstan with a 1.4-yd diameter. What is the angular speed of the capstan?

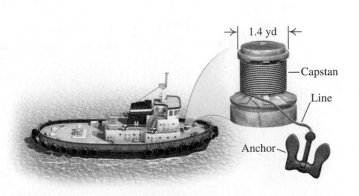

1.4 yd

Capstan

Line

Anchor

Solution   We will use the formula $v = r\omega$ in the form $\omega = v/r$, taking care to use the proper units. Since v is given in feet per second, we need to give r in feet:

$$r = \frac{d}{2} = \frac{1.4}{2}\,\text{yd} \cdot \frac{3\,\text{ft}}{1\,\text{yd}} = 2.1\,\text{ft}.$$

Then ω will be in radians per second:

$$\omega = \frac{v}{r} = \frac{2\,\text{ft/sec}}{2.1\,\text{ft}} = \frac{2\,\text{ft}}{\text{sec}} \cdot \frac{1}{2.1\,\text{ft}} \approx 0.952/\text{sec}.$$

Thus the angular speed is approximately 0.952 radian/sec.   ▸ *Now Try Exercise 77.*

The formulas $\theta = \omega t$ and $v = r\omega$ can be used in combination to find distances and angles in various situations involving rotational motion.

EXAMPLE 11   *Angle of Revolution.*   A 2014 Toyota FJ Cruiser is traveling at a speed of 70 mph. Its tires have an outside diameter of 30.875 in. Find the angle through which a tire turns in 10 sec.

30.875 in.

Solution   Recall that $\omega = \theta/t$, or $\theta = \omega t$. Thus we can find θ if we know ω and t. To find ω, we use the formula $v = r\omega$. The linear speed v of a point on the outside of the tire is the speed of the FJ Cruiser, 70 mph. For convenience, we first convert 70 mph to feet per second:

$$v = 70\,\frac{\text{mi}}{\text{hr}} \cdot \frac{1\,\text{hr}}{60\,\text{min}} \cdot \frac{1\,\text{min}}{60\,\text{sec}} \cdot \frac{5280\,\text{ft}}{1\,\text{mi}}$$

$$\approx 102.667\,\frac{\text{ft}}{\text{sec}}.$$

The radius of the tire is half the diameter. Now $r = d/2 = 30.875/2 = 15.4375$ in. We will convert to feet, since v is in feet per second:

$$r = 15.4375\,\text{in.} \cdot \frac{1\,\text{ft}}{12\,\text{in.}}$$

$$= \frac{15.4375}{12}\,\text{ft}$$

$$\approx 1.29\,\text{ft}.$$

Using $v = r\omega$, we have

$$102.667\,\frac{\text{ft}}{\text{sec}} = 1.29\,\text{ft} \cdot \omega,$$

so

$$\omega = \frac{102.667 \text{ ft/sec}}{1.29 \text{ ft}} \approx \frac{79.59}{\text{sec}}.$$

Then in 10 sec,

$$\theta = \omega t = \frac{79.59}{\text{sec}} \cdot 10 \text{ sec} \approx 796.$$

Thus the angle, in radians, through which a tire turns in 10 sec is 796.

> *Now Try Exercise 79.*

EXAMPLE 12 *Angular Speed of a Gear Wheel.* One gear wheel turns another, the teeth being on the rims. The wheels have 9-in. and 5-in. radii, and the smaller wheel rotates at 48 rpm (revolutions per minute). Find the angular speed of the larger wheel, in radians per second.

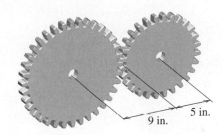

Solution Let $\omega_1 =$ the angular speed of the smaller wheel and $\omega_2 =$ the angular speed of the larger wheel. The wheels have the same linear speed, so we have

$$v = 5\omega_1 = 9\omega_2.$$

We first convert the angular speed of the smaller wheel, 48 rpm, to radians per second:

$$\omega_1 = 48 \text{ rpm} = \frac{48 \cdot 2\pi}{1 \text{ min}} \qquad \begin{array}{l} \text{1 revolution} = 2\pi; \\ \text{48 revolutions} = 48 \cdot 2\pi = 96\pi \end{array}$$

$$= \frac{96\pi}{1 \text{ min}} \cdot \frac{1 \text{ min}}{60 \text{ sec}}$$

$$= 1.6\pi/\text{sec}$$

$$\approx 5.027/\text{sec}.$$

Next, we substitute 5.027/sec for ω_1 and solve for ω_2:

$$5\omega_1 = 9\omega_2$$
$$5(5.027/\text{sec}) = 9\omega_2$$
$$25.135/\text{sec} = 9\omega_2$$
$$2.793/\text{sec} \approx \omega_2.$$

The angular speed of the larger wheel is about 2.793 radians/sec, or 2.793/sec.

> *Now Try Exercise 81.*

6.4   Exercise Set

For each of Exercises 1–4, sketch a unit circle and mark the points determined by the given real numbers.

1. a) $\dfrac{\pi}{4}$ **b)** $\dfrac{3\pi}{2}$ **c)** $\dfrac{3\pi}{4}$

  d) π **e)** $\dfrac{11\pi}{4}$ **f)** $\dfrac{17\pi}{4}$

2. a) $\dfrac{\pi}{2}$ **b)** $\dfrac{5\pi}{4}$ **c)** 2π

  d) $\dfrac{9\pi}{4}$ **e)** $\dfrac{13\pi}{4}$ **f)** $\dfrac{23\pi}{4}$

3. a) $\dfrac{\pi}{6}$ **b)** $\dfrac{2\pi}{3}$ **c)** $\dfrac{7\pi}{6}$

  d) $\dfrac{10\pi}{6}$ **e)** $\dfrac{14\pi}{6}$ **f)** $\dfrac{23\pi}{4}$

4. a) $-\dfrac{\pi}{2}$ **b)** $-\dfrac{3\pi}{4}$ **c)** $-\dfrac{5\pi}{6}$

  d) $-\dfrac{5\pi}{2}$ **e)** $-\dfrac{17\pi}{6}$ **f)** $-\dfrac{9\pi}{4}$

Find two real numbers between -2π and 2π that determine each of the points on the unit circle.

5.

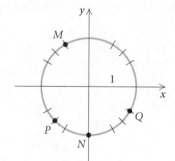

6.

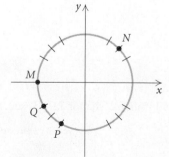

For Exercises 7 and 8, sketch a unit circle and mark the approximate location of the point determined by the given real number.

7. a) 2.4   **b)** 7.5
  c) 32   **d)** 320

8. a) 0.25   **b)** 1.8
  c) 47   **d)** 500

Convert to radian measure. Leave the answer in terms of π.

9. 75°   **10.** 30°

11. 200°   **12.** −135°

13. −214.6°   **14.** 37.71°

15. −180°   **16.** 90°

17. 12.5°   **18.** 6.3°

19. −340°   **20.** −60°

Convert to radian measure. Round the answer to two decimal places.

21. 240°   **22.** 15°

23. −60°   **24.** 145°

25. 117.8°   **26.** −231.2°

27. 1.354°   **28.** 584°

29. 345°   **30.** −75°

31. 95°   **32.** 24.8°

Convert to degree measure. Round the answer to two decimal places where appropriate.

33. $-\dfrac{3\pi}{4}$ **34.** $\dfrac{7\pi}{6}$

35. 8π **36.** $-\dfrac{\pi}{3}$

37. 1   **38.** −17.6

39. 2.347   **40.** 25

41. $\dfrac{5\pi}{4}$ **42.** -6π

43. −90   **44.** 37.12

45. $\dfrac{2\pi}{7}$ **46.** $\dfrac{\pi}{9}$

47. Certain positive angles are marked here in degrees. Find the corresponding radian measures.

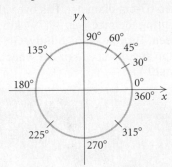

48. Certain negative angles are marked here in degrees. Find the corresponding radian measures.

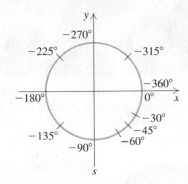

Find a positive angle and a negative angle that are co-terminal with the given angle. Answers may vary.

49. $\dfrac{\pi}{4}$

50. $\dfrac{5\pi}{3}$

51. $\dfrac{7\pi}{6}$

52. π

53. $-\dfrac{2\pi}{3}$

54. $-\dfrac{3\pi}{4}$

Find the complement and the supplement.

55. $\dfrac{\pi}{3}$ **56.** $\dfrac{5\pi}{12}$ **57.** $\dfrac{3\pi}{8}$

58. $\dfrac{\pi}{4}$ **59.** $\dfrac{\pi}{12}$ **60.** $\dfrac{\pi}{6}$

Complete the following table. Round the answers to two decimal places where appropriate.

DISTANCE, s (ARC LENGTH)	RADIUS, r	ANGLE, θ
61. 8 ft	$3\frac{1}{2}$ ft	_____
62. 200 cm	_____	45°
63. _____	4.2 in.	$\dfrac{5\pi}{12}$
64. 16 yd	_____	5

65. In a circle with a 120-cm radius, an arc 132 cm long subtends an angle of how many radians? how many degrees, to the nearest degree?

66. In a circle with a 10-ft diameter, an arc 20 ft long subtends an angle of how many radians? how many degrees, to the nearest degree?

67. In a circle with a 2-yd radius, how long is an arc associated with an angle of 1.6 radians?

68. In a circle with a 5-m radius, how long is an arc associated with an angle of 2.1 radians?

69. *Angle of Revolution.* A tire on a 2014 Dodge Durango SUV has an outside diameter of 36.32 in. Through what angle (in radians) does the tire turn while traveling 1 mi?

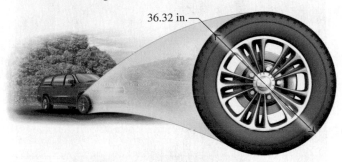

36.32 in.

70. *Angle of Revolution.* Through how many radians does the minute hand of a wristwatch rotate from 12:40 P.M. to 1:30 P.M.?

71. *Linear Speed.* A flywheel with a 15-cm diameter is rotating at a rate of 7 radians/sec. What is the linear speed of a point on its rim, in centimeters per minute?

72. *Linear Speed.* A wheel with a 30-cm radius is rotating at a rate of 3 radians/sec. What is the linear speed of a point on its rim, in meters per minute?

73. *Linear Speeds on a Carousel.* When Brett and Will ride the carousel described earlier in this section, Brett always selects a horse on the outside row, whereas Will prefers the row closest to the center. These rows are 19 ft 3 in. and 13 ft 11 in. from the center, respectively. The angular speed of the carousel is 2.4 revolutions per minute. (*Source*: The Children's Museum,

Indianapolis, IN) What is the difference, in miles per hour, in the linear speeds of Brett and Will?

74. *Angular Speed of a Printing Press.* This text was printed on a four-color web heatset offset press. A cylinder on this press has a 21-in. diameter. The linear speed of a point on the cylinder's surface is 18.33 feet per second. What is the angular speed of the cylinder, in revolutions per hour? Printers often refer to the angular speed as impressions per hour (IPH). (*Source*: R. R. Donnelley, Willard, Ohio)

75. *Linear Speed at the Equator.* The earth has a 4000-mi radius and rotates one revolution every 24 hr. What is the linear speed of a point on the equator, in miles per hour?

76. *Linear Speed of the Earth.* The earth is about 93,000,000 mi from the sun and traverses its orbit, which is nearly circular, every 365.25 days. What is the linear velocity of the earth in its orbit, in miles per hour?

77. *Determining the Speed of a River.* A water wheel has a 10-ft radius. To get a good approximation of the speed of the river, you count the revolutions of the wheel and find that it makes 14 revolutions per minute (rpm). What is the speed of the river, in miles per hour?

78. *The Tour de France.* Cadel Evans of Australia won the 2015 Tour de France bicycle race. The wheel of his bicycle had a 67-cm diameter. His overall average linear speed during the race was 39.794 km/h (*Source*: Preston Green, Bicycle Garage Indy, Greenwood, IN). What was the angular speed of the wheel, in revolutions per hour?

79. *John Deere Tractor.* A rear wheel on a John Deere 8300 farm tractor has a 23-in. radius. Find the angle (in radians) through which a wheel rotates in 12 sec if the tractor is traveling at a speed of 22 mph.

80. *Angular Speed of a Pulley.* Two pulleys, 50 cm and 30 cm in diameter, respectively, are connected by a belt. The larger pulley makes 12 revolutions per minute. Find the angular speed of the smaller pulley, in radians per second.

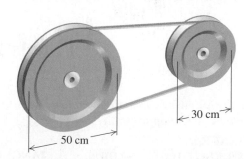

81. *Angular Speed of a Gear Wheel.*   One gear wheel turns another, the teeth being on the rims. The wheels have 40-cm and 50-cm radii, and the smaller wheel rotates at 20 rpm. Find the angular speed of the larger wheel, in radians per second.

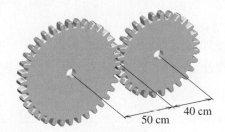

❯ Skill Maintenance

Vocabulary Reinforcement

In each of Exercises 82–89, fill in the blanks with the correct terms. Some of the given choices will not be used.

inverse	relation
horizontal line	vertical asymptote
vertical line	horizontal asymptote
exponential function	even function
logarithmic function	odd function
natural	sine of θ
common	cosine of θ
logarithm	tangent of θ
one-to-one	

82. The domain of a(n) _____ function f is the range of the inverse f^{-1}. **[5.1]**

83. The _____ is the length of the side adjacent to θ divided by the length of the hypotenuse. **[6.1]**

84. The function $f(x) = a^x$, where x is a real number, $a > 0$ and $a \neq 1$, is called the _____, base a. **[5.2]**

85. The graph of a rational function may or may not cross a(n) _____. **[4.5]**

86. If the graph of a function f is symmetric with respect to the origin, we say that it is a(n) _____. **[2.4]**

87. Logarithms, base e, are called _____ logarithms. **[5.3]**

88. If it is possible for a(n) _____ to intersect the graph of a function more than once, then the function is not one-to-one and its _____ is not a function. **[5.1]**

89. A(n) _____ is an exponent. **[5.3]**

❯ Synthesis

90. A point on the unit circle has y-coordinate $-\sqrt{21}/5$. What is its x-coordinate? Check using a calculator.

91. On the earth, one degree of latitude is how many kilometers? how many miles? (Assume that the radius of the earth is 6400 km, or 4000 mi, approximately.)

92. A **grad** is a unit of angle measure similar to a degree. A right angle has a measure of 100 grads. Convert each of the following to grads.

a) 48°   **b)** $\dfrac{5\pi}{7}$

93. A **mil** is a unit of angle measure. A right angle has a measure of 1600 mils. Convert each of the following to degrees, minutes, and seconds.

a) 100 mils   **b)** 350 mils

94. *Hands of a Clock.*   At what time between noon and 1:00 P.M. are the hands of a clock perpendicular?

95. *Distance Between Points on the Earth.*   To find the distance between two points on the earth when their latitude and longitude are known, we can use a right triangle for an excellent approximation if the points are not too far apart. Point A is at latitude 38°27′30″ N, longitude 82°57′15″ W, and point B is at latitude 38°28′45″ N, longitude 82°56′30″ W. Find the distance from A to B in nautical miles. (One minute of latitude is one nautical mile.)

6.5 ▷ Circular Functions: Graphs and Properties

❯ Given the coordinates of a point on the unit circle, find its reflections across the *x*-axis, the *y*-axis, and the origin.

❯ Determine the six trigonometric function values for a real number when the coordinates of the point on the unit circle determined by that real number are given.

❯ Find trigonometric function values for any real number using a calculator.

❯ Graph the six circular functions and state their properties.

The domains of the trigonometric functions, defined in Sections 6.1 and 6.3, have been sets of angles or rotations measured in a *real number* of degree units. We can also consider the domains to be sets of **real numbers**, or **radians**, introduced in Section 6.4. Many applications in calculus that use the trigonometric functions refer only to radians.

Let's again consider radian measure and the unit circle. We defined radian measure for θ as

$$\theta = \frac{s}{r}. \qquad \textbf{\textit{θ} is a real number without units.}$$

When $r = 1$,

$$\theta = \frac{s}{1}, \quad \text{or} \quad \theta = s.$$

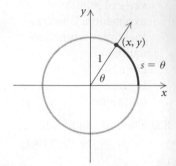

The arc length s on the unit circle is the same as the radian measure of the angle θ.

In the figure above, the point (x, y) is the point where the terminal side of the angle with radian measure s intersects the *unit circle*. We can now extend our definitions of the trigonometric functions using *domains* composed of *real numbers*, or *radians*. Trigonometric functions with domains composed of real numbers are called **circular functions**.

In the definitions,

• *s can be considered the radian measure of an angle or*
• *the measure of an arc length on the unit circle.*

Either way, s is a real number. To each real number s, there corresponds an arc length s on the unit circle.

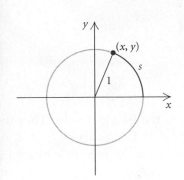

BASIC CIRCULAR FUNCTIONS

For a real number s that determines a point (x, y) on the unit circle:

$$\sin s = \text{second coordinate} = \frac{y}{1} = y,$$

$$\cos s = \text{first coordinate} = \frac{x}{1} = x,$$

$$\tan s = \frac{\text{second coordinate}}{\text{first coordinate}} = \frac{y}{x} \ (x \neq 0),$$

$$\csc s = \frac{1}{\text{second coordinate}} = \frac{1}{y} \ (y \neq 0),$$

$$\sec s = \frac{1}{\text{first coordinate}} = \frac{1}{x} \ (x \neq 0),$$

$$\cot s = \frac{\text{first coordinate}}{\text{second coordinate}} = \frac{x}{y} \ (y \neq 0).$$

We can consider the domains of trigonometric functions to be real numbers rather than angles. We can determine these values for a specific real number if we know the coordinates of the point on the unit circle determined by that number. As with degree measure, we can also find these function values directly using a calculator.

❭ Reflections on the Unit Circle

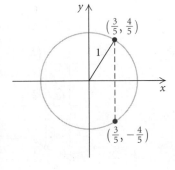

Let's consider the unit circle and a few of its points. For any point (x, y) on the unit circle, $x^2 + y^2 = 1$, we know that $-1 \leq x \leq 1$ and $-1 \leq y \leq 1$. If we know the x- or y-coordinate of a point on the unit circle, we can find the other coordinate. If $x = \frac{3}{5}$, then

$$\left(\tfrac{3}{5}\right)^2 + y^2 = 1$$
$$y^2 = 1 - \tfrac{9}{25} = \tfrac{16}{25}$$
$$y = \pm\tfrac{4}{5}.$$

Thus, $\left(\frac{3}{5}, \frac{4}{5}\right)$ and $\left(\frac{3}{5}, -\frac{4}{5}\right)$ are points on the unit circle. There are two points with an x-coordinate of $\frac{3}{5}$.

Now let's consider the radian measure $\pi/3$ and determine the coordinates of the point on the unit circle determined by $\pi/3$. We construct a right triangle by dropping a perpendicular segment from the point to the x-axis.

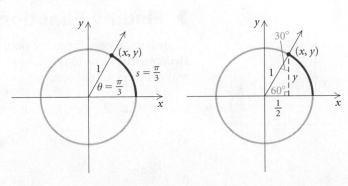

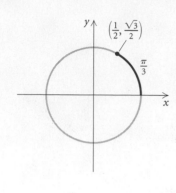

Since $\pi/3 = 60°$, we have a 30°–60° right triangle in which the side opposite the 30° angle is one half of the hypotenuse. The hypotenuse, or radius, is 1, so the side opposite the 30° angle is $\frac{1}{2} \cdot 1$, or $\frac{1}{2}$. Using the Pythagorean equation, we can find the other side:

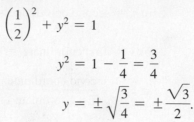

$$\left(\frac{1}{2}\right)^2 + y^2 = 1$$

$$y^2 = 1 - \frac{1}{4} = \frac{3}{4}$$

$$y = \pm\sqrt{\frac{3}{4}} = \pm\frac{\sqrt{3}}{2}.$$

We know that y is positive since the point is in the first quadrant. Thus the coordinates of the point determined by $\pi/3$ are $x = 1/2$ and $y = \sqrt{3}/2$, or $\left(1/2, \sqrt{3}/2\right)$. We can always check to see if a point is on the unit circle by substituting into the equation $x^2 + y^2 = 1$:

$$\left(\frac{1}{2}\right)^2 + \left(\frac{\sqrt{3}}{2}\right)^2 = \frac{1}{4} + \frac{3}{4} = 1.$$

Because a unit circle is symmetric with respect to the x-axis, the y-axis, and the origin, we can use the coordinates of one point on the unit circle to find coordinates of its reflections.

EXAMPLE 1 Each of the following points lies on the unit circle. Find their reflections across the x-axis, the y-axis, and the origin.

a) $\left(\dfrac{3}{5}, \dfrac{4}{5}\right)$ **b)** $\left(\dfrac{\sqrt{2}}{2}, \dfrac{\sqrt{2}}{2}\right)$ **c)** $\left(\dfrac{1}{2}, \dfrac{\sqrt{3}}{2}\right)$

Solution

a)

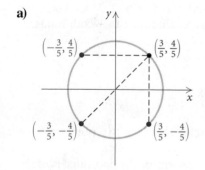

b)

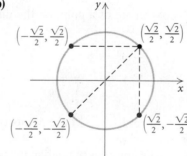

c)

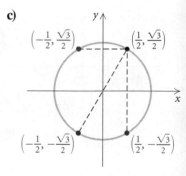

> **Now Try Exercise 1.**

❯ Finding Function Values

Knowing the coordinates of only a few points on the unit circle along with their reflections allows us to find trigonometric function values of the most frequently used real numbers, or radians.

EXAMPLE 2 Find each of the following function values.

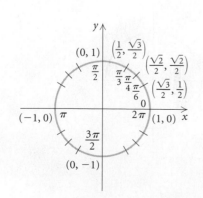

a) $\tan\dfrac{\pi}{3}$ **b)** $\cos\dfrac{3\pi}{4}$ **c)** $\sin\left(-\dfrac{\pi}{6}\right)$

d) $\cos\dfrac{4\pi}{3}$ **e)** $\cot\pi$ **f)** $\csc\left(-\dfrac{7\pi}{2}\right)$

Solution We locate the point on the unit circle determined by the rotation, and then find its coordinates using reflection if necessary.

a) The coordinates of the point determined by $\pi/3$ are $\left(1/2, \sqrt{3}/2\right)$.

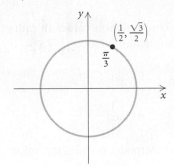

Thus, $\tan\dfrac{\pi}{3} = \dfrac{y}{x} = \dfrac{\sqrt{3}/2}{1/2} = \sqrt{3}$.

b) The reflection of $\left(\sqrt{2}/2, \sqrt{2}/2\right)$ across the y-axis is $\left(-\sqrt{2}/2, \sqrt{2}/2\right)$.

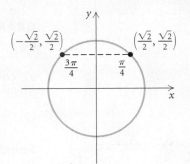

Thus, $\cos\dfrac{3\pi}{4} = x = -\dfrac{\sqrt{2}}{2}$.

c) The reflection of $\left(\sqrt{3}/2, 1/2\right)$ across the x-axis is $\left(\sqrt{3}/2, -1/2\right)$.

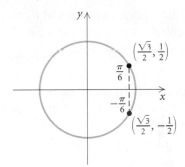

Thus, $\sin\left(-\dfrac{\pi}{6}\right) = y = -\dfrac{1}{2}$.

d) The reflection of $\left(1/2, \sqrt{3}/2\right)$ across the origin is $\left(-1/2, -\sqrt{3}/2\right)$.

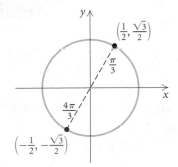

Thus, $\cos\dfrac{4\pi}{3} = x = -\dfrac{1}{2}$.

e) The coordinates of the point determined by π are $(-1, 0)$.

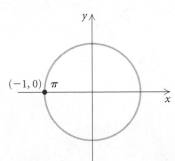

Thus, $\cot\pi = \dfrac{x}{y} = \dfrac{-1}{0}$, which is not defined.

We can also think of $\cot\pi$ as the reciprocal of $\tan\pi$. Since $\tan\pi = y/x = 0/-1 = 0$ and the reciprocal of 0 is not defined, we know that $\cot\pi$ is not defined.

f) The coordinates of the point determined by $-7\pi/2$ are $(0, 1)$.

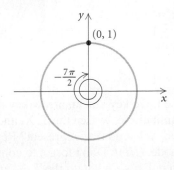

Thus, $\csc\left(-\dfrac{7\pi}{2}\right) = \dfrac{1}{y} = \dfrac{1}{1} = 1$.

> *Now Try Exercises 9 and 11.*

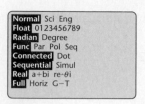

Using a calculator, we can find trigonometric function values of any real number without knowing the coordinates of the point that it determines on the unit circle. Most calculators have both degree mode and radian mode. When finding function values of radian measures, or real numbers, we *must* set the calculator in RADIAN mode. (See the window at left.)

EXAMPLE 3   Find each of the following function values of radian measures using a calculator. Round the answers to four decimal places.

a) $\cos \dfrac{2\pi}{5}$ **b)** $\tan(-3)$

c) $\sin 24.9$ **d)** $\sec \dfrac{\pi}{7}$

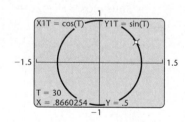

Solution   Using a calculator set in RADIAN mode, we find the values.

a) $\cos \dfrac{2\pi}{5} \approx 0.3090$ **b)** $\tan(-3) \approx 0.1425$

c) $\sin 24.9 \approx -0.2306$ **d)** $\sec \dfrac{\pi}{7} = \dfrac{1}{\cos \dfrac{\pi}{7}} \approx 1.1099$

Note in part (d) that the secant function value can be found by taking the reciprocal of the cosine value. Thus we can enter cos $\pi/7$ and use the reciprocal key.

> *Now Try Exercises 25 and 33.*

EXPLORING WITH TECHNOLOGY

We can graph the unit circle using a graphing calculator. We use PARAMETRIC mode with the following window and let $X_{1T} = \cos T$ and $Y_{1T} = \sin T$. Here we use DEGREE mode.

WINDOW

    Tmin = 0
    Tmax = 360
    Tstep = 15
    Xmin = -1.5
    Xmax = 1.5
    Xscl = 1
    Ymin = -1
    Ymax = 1
    Yscl = 1

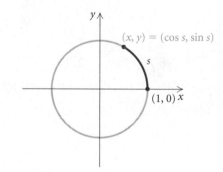

Using the trace key and an arrow key to move the cursor around the unit circle, we see the T, X, and Y values appear on the screen. What do they represent? Repeat this exercise in RADIAN mode. (*Hint*: Don't forget to convert Tmin, Tmax, and Tstep to radians.) What do the T, X, and Y values represent? (For more on parametric equations, see Section 10.7.)

From the definitions on p. 449, we can relabel any point (x, y) on the unit circle as $(\cos s, \sin s)$, where s is any real number.

❯ Graphs of the Sine and Cosine Functions

Properties of functions can be observed from their graphs. We begin by graphing the sine and cosine functions. We make a table of values, plot the points, and then connect those points with a smooth curve. It is helpful to first draw a unit circle and label a few points with coordinates. We can either use the coordinates as the function values or find approximate sine and cosine values directly with a calculator.

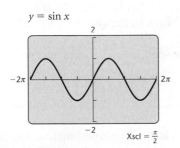

s	$\sin s$	$\cos s$
0	0	1
$\pi/6$	0.5	0.8660
$\pi/4$	0.7071	0.7071
$\pi/3$	0.8660	0.5
$\pi/2$	1	0
$3\pi/4$	0.7071	−0.7071
π	0	−1
$5\pi/4$	−0.7071	−0.7071
$3\pi/2$	−1	0
$7\pi/4$	−0.7071	0.7071
2π	0	1

s	$\sin s$	$\cos s$
0	0	1
$-\pi/6$	−0.5	0.8660
$-\pi/4$	−0.7071	0.7071
$-\pi/3$	−0.8660	0.5
$-\pi/2$	−1	0
$-3\pi/4$	−0.7071	−0.7071
$-\pi$	0	−1
$-5\pi/4$	0.7071	−0.7071
$-3\pi/2$	1	0
$-7\pi/4$	0.7071	0.7071
-2π	0	1

The graphs are as follows.

$y = \sin x$

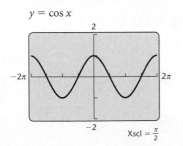

$y = \cos x$

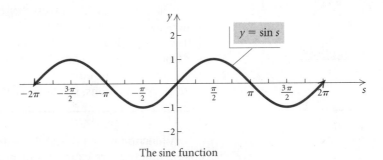

The sine function

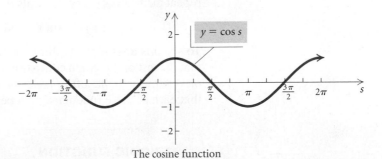

The cosine function

The sine and cosine functions are continuous functions. Note in the graph of the sine function that function values increase from 0 at $s = 0$ to 1 at $s = \pi/2$, then decrease to 0 at $s = \pi$, decrease further to −1 at $s = 3\pi/2$, and increase to 0 at 2π. The reverse pattern follows when s decreases from 0 to -2π. Note in the graph of the cosine function that function values start at 1 when $s = 0$, and decrease to 0 at $s = \pi/2$. They decrease further to −1 at $s = \pi$, then increase to 0 at $s = 3\pi/2$, and increase further to 1 at $s = 2\pi$. An identical pattern follows when s decreases from 0 to -2π.

From the unit circle and the graphs of the functions, we know that the domain of both the sine and cosine functions is the entire set of real numbers, $(-\infty, \infty)$. The range of each function is the set of all real numbers from -1 to 1, $[-1, 1]$.

DOMAIN AND RANGE OF THE SINE FUNCTION AND THE COSINE FUNCTION

The *domain* of the sine function and the cosine function is $(-\infty, \infty)$.

The *range* of the sine function and the cosine function is $[-1, 1]$.

EXPLORING WITH TECHNOLOGY

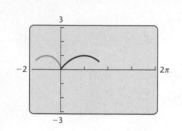

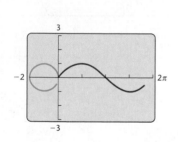

Another way to construct the sine and cosine graphs is by considering the unit circle and transferring vertical distances for the sine function and horizontal distances for the cosine function. Using a graphing calculator, we can visualize the transfer of these distances. We use the calculator set in PARAMETRIC mode and RADIAN mode and let $X_{1T} = \cos T - 1$ and $Y_{1T} = \sin T$ for the unit circle centered at $(-1, 0)$ and $X_{2T} = T$ and $Y_{2T} = \sin T$ for the sine curve. Use the following window settings.

Tmin = 0	Xmin = -2	Ymin = -3
Tmax = 2π	Xmax = 2π	Ymax = 3
Tstep = .1	Xscl = $\pi/2$	Yscl = 1

With the calculator set in SIMULTANEOUS mode, we can actually watch the sine function (in red) "unwind" from the unit circle (in blue). In the two screens at left, we partially illustrate this animated procedure.

Consult your calculator's instruction manual for specific keystrokes and graph both the sine curve and the cosine curve in this manner. (For more on parametric equations, see Section 10.7.)

A function with a repeating pattern is called **periodic**. The sine function and the cosine function are examples of periodic functions. The values of these functions repeat themselves every 2π units. In other words, for any s, we have

$$\sin(s + 2\pi) = \sin s \quad \text{and} \quad \cos(s + 2\pi) = \cos s.$$

To see this another way, think of the part of the graph between 0 and 2π and note that the rest of the graph consists of copies of it. If we translate the graph of $y = \sin x$ or $y = \cos x$ to the left or right 2π units, we will obtain the original graph. We say that each of these functions has a period of 2π.

PERIODIC FUNCTION

A function f is said to be **periodic** if there exists a positive constant p such that

$$f(s + p) = f(s)$$

for all s in the domain of f. The smallest such positive number p is called the period of the function.

The period p can be thought of as the length of the shortest recurring interval.

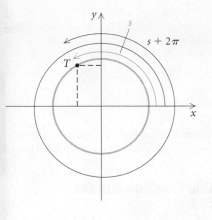

We can also use the unit circle to verify that the period of the sine and cosine functions is 2π. Consider any real number s and the point T that it determines on a unit circle, as shown at left. If we increase s by 2π, the point determined by $s + 2\pi$ is again the point T. Hence for any real number s,

$$\sin(s + 2\pi) = \sin s \quad \text{and} \quad \cos(s + 2\pi) = \cos s.$$

It is also true that $\sin(s + 4\pi) = \sin s$, $\sin(s + 6\pi) = \sin s$, and so on. In fact, for *any* integer k, the following equations are identities:

$$\sin[s + k(2\pi)] = \sin s \quad \text{and} \quad \cos[s + k(2\pi)] = \cos s,$$

or $\quad \sin s = \sin(s + 2k\pi) \quad \text{and} \quad \cos s = \cos(s + 2k\pi)$.

The **amplitude** of a periodic function is defined as one half of the distance between its maximum and minimum function values. It is always positive. Both the graphs and the unit circle verify that the maximum value of the sine and cosine functions is 1, whereas the minimum value of each is -1. Thus,

$$\text{the amplitude of the sine function} = \tfrac{1}{2}\left|1 - (-1)\right| = 1$$

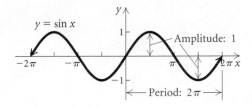

and the amplitude of the cosine function $= \tfrac{1}{2}\left|1 - (-1)\right| = 1$.

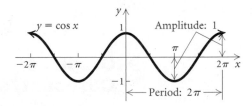

Consider any real number s and its opposite, $-s$. These numbers determine points T and T_1 on a unit circle that are symmetric with respect to the x-axis.

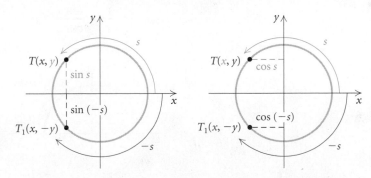

Because their second coordinates are opposites of each other, we know that for any number s,

$$\sin(-s) = -\sin s.$$

Because their first coordinates are the same, we know that for any number s,

$$\cos(-s) = \cos s.$$

EXPLORING WITH TECHNOLOGY

Using the TABLE feature on a graphing calculator, compare the y-values for $y_1 = \sin x$ and $y_2 = \sin(-x)$ and for $y_3 = \cos x$ and $y_4 = \cos(-x)$. Set TblStart $= 0$ and $\triangle$Tbl $= \pi/12$.

X	Y₁	Y₂
0	0	0
.2618	.25882	−.2588
.5236	.5	−.5
.7854	.70711	−.7071
1.0472	.86603	−.866
1.309	.96593	−.9659
1.5708	1	−1
X = 0		

X	Y₃	Y₄
0	1	1
.2618	.96593	.96593
.5236	.86603	.86603
.7854	.70711	.70711
1.0472	.5	.5
1.309	.25882	.25882
1.5708	0	0
X = 0		

What appears to be the relationship between $\sin x$ and $\sin(-x)$ and that between $\cos x$ and $\cos(-x)$?

Thus we have shown the following.

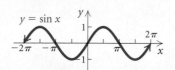

> The sine function is *odd*.
> The cosine function is *even*.

A summary of the properties of the sine function and the cosine function follows.

CONNECTING THE CONCEPTS

Comparing the Sine Function and the Cosine Function

SINE FUNCTION

$y = \sin x$

1. Continuous
2. Period: 2π
3. Domain: All real numbers
4. Range: $[-1, 1]$
5. Amplitude: 1
6. Odd: $\sin(-s) = -\sin s$

COSINE FUNCTION

$y = \cos x$

1. Continuous
2. Period: 2π
3. Domain: All real numbers
4. Range: $[-1, 1]$
5. Amplitude: 1
6. Even: $\cos(-s) = \cos s$

❯ Graphs of the Tangent, Cotangent, Cosecant, and Secant Functions

To graph the tangent function, we could make a table of values using a calculator, but in this case it is easier to begin with the definition of tangent and the coordinates of a few points on the unit circle. We recall that

$$\tan s = \frac{y}{x} = \frac{\sin s}{\cos s}.$$

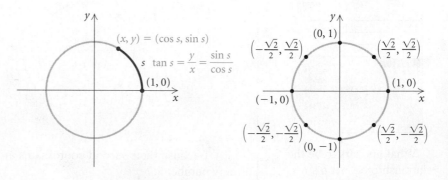

The tangent function is not defined when x, the first coordinate, is 0. That is, it is not defined for any number s whose cosine is 0:

$$s = \pm\frac{\pi}{2}, \pm\frac{3\pi}{2}, \pm\frac{5\pi}{2}, \ldots.$$

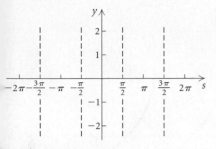

FIGURE 1.

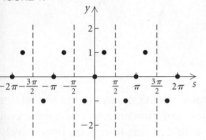

FIGURE 2.

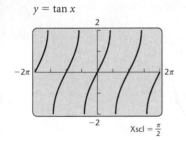

$y = \tan x$

We draw vertical asymptotes at these locations (see Fig. 1 at left).

We also note that

$$\tan s = 0 \text{ at } s = 0, \pm\pi, \pm 2\pi, \pm 3\pi, \ldots,$$

$$\tan s = 1 \text{ at } s = \ldots -\frac{7\pi}{4}, -\frac{3\pi}{4}, \frac{\pi}{4}, \frac{5\pi}{4}, \frac{9\pi}{4}, \ldots,$$

$$\tan s = -1 \text{ at } s = \ldots -\frac{9\pi}{4}, -\frac{5\pi}{4}, -\frac{\pi}{4}, \frac{3\pi}{4}, \frac{7\pi}{4}, \ldots.$$

We can add these ordered pairs to the graph (see Fig. 2 at left) and investigate the values in $(-\pi/2, \pi/2)$ using a calculator. Note that the function value is 0 when $s = 0$, and the values increase without bound as s increases toward $\pi/2$. The graph gets closer and closer to the vertical asymptote as s gets closer to $\pi/2$, but it never touches the line. As s decreases from 0 to $-\pi/2$, the values decrease without bound. Again the graph gets closer and closer to a vertical asymptote, but it never touches it. We now complete the graph.

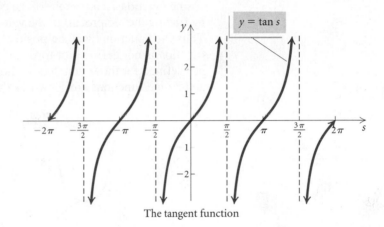

The tangent function

From the graph, we see that the tangent function is continuous except where it is not defined. The period of the tangent function is π. Note that although there is a period, there is no amplitude because there are no maximum and minimum values. When $\cos s = 0$, $\tan s$ is not defined ($\tan s = \sin s / \cos s$). Thus the domain of the tangent function is the set of all real numbers except $(\pi/2) + k\pi$, where k is an integer. The range of the function is the set of all real numbers.

THE DOMAIN AND THE RANGE OF THE TANGENT FUNCTION

The domain of the tangent function is all real numbers except $\dfrac{\pi}{2} + k\pi$, where k is an integer.

The range of the tangent function is $(-\infty, \infty)$.

The cotangent function ($\cot s = \cos s / \sin s$) is not defined when y, the second coordinate, is 0—that is, it is not defined for any number s whose sine is 0. Thus the cotangent is not defined for $s = 0, \pm\pi, \pm 2\pi, \pm 3\pi, \ldots$.

The graph of the function is shown below.

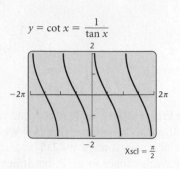

$$y = \cot x = \frac{1}{\tan x}$$

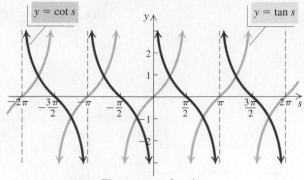

The cotangent function

The cosecant and sine functions are reciprocal functions, as are the secant and cosine functions. The graphs of the cosecant and secant functions can be constructed by finding the reciprocals of the values of the sine and cosine functions, respectively. Thus the functions will be positive together and negative together. The cosecant function is not defined for those numbers s whose sine is 0. The secant function is not defined for those numbers s whose cosine is 0. In the following graphs, the sine and cosine functions are shown by the gray curves for reference.

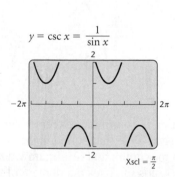

$$y = \csc x = \frac{1}{\sin x}$$

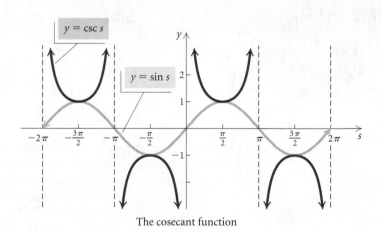

The cosecant function

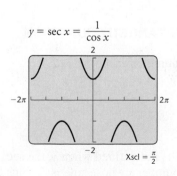

$$y = \sec x = \frac{1}{\cos x}$$

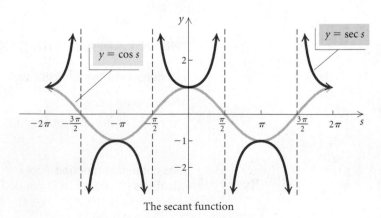

The secant function

The following is a summary of the basic properties of the tangent, cotangent, cosecant, and secant functions. These functions are continuous except where they are not defined.

Now Try Exercises 49.

CONNECTING THE CONCEPTS

Comparing the Tangent, Cotangent, Cosecant, and Secant Functions

TANGENT FUNCTION

1. Period: π
2. Domain: All real numbers except $(\pi/2) + k\pi$, where k is an integer
3. Range: All real numbers

COSECANT FUNCTION

1. Period: 2π
2. Domain: All real numbers except $k\pi$, where k is an integer
3. Range: $(-\infty, -1] \cup [1, \infty)$

COTANGENT FUNCTION

1. Period: π
2. Domain: All real numbers except $k\pi$, where k is an integer
3. Range: All real numbers

SECANT FUNCTION

1. Period: 2π
2. Domain: All real numbers except $(\pi/2) + k\pi$, where k is an integer
3. Range: $(-\infty, -1] \cup [1, \infty)$

In this chapter, we have used the letter s for arc length and have avoided the letters x and y, which generally represent first and second coordinates, respectively. Nevertheless, we can represent the arc length on a unit circle by any variable, such as s, t, x, or θ. Each arc length determines a point that can be labeled with an ordered pair. The first coordinate of that ordered pair is the cosine of the arc length, and the second coordinate is the sine of the arc length. The identities we have developed hold no matter what symbols are used for variables—for example, $\cos(-s) = \cos s$, $\cos(-x) = \cos x$, $\cos(-\theta) = \cos \theta$, and $\cos(-t) = \cos t$.

6.5 Exercise Set

*The following points are on the unit circle. Find the coordinates of their reflections across **(a)** the x-axis, **(b)** the y-axis, and **(c)** the origin.*

1. $\left(-\dfrac{3}{4}, \dfrac{\sqrt{7}}{4}\right)$

2. $\left(\dfrac{2}{3}, \dfrac{\sqrt{5}}{3}\right)$

3. $\left(\dfrac{2}{5}, -\dfrac{\sqrt{21}}{5}\right)$

4. $\left(-\dfrac{\sqrt{3}}{2}, -\dfrac{1}{2}\right)$

5. The number $\pi/4$ determines a point on the unit circle with coordinates $\left(\sqrt{2}/2, \sqrt{2}/2\right)$. What are the coordinates of the point determined by $-\pi/4$?

6. A number β determines a point on the unit circle with coordinates $\left(-2/3, \sqrt{5}/3\right)$. What are the coordinates of the point determined by $-\beta$?

Find the function value using coordinates of points on the unit circle. Give exact answers.

7. $\sin \pi$

8. $\cos\left(-\dfrac{\pi}{3}\right)$

9. $\cot \dfrac{7\pi}{6}$

10. $\tan \dfrac{11\pi}{4}$

11. $\sin(-3\pi)$

12. $\csc \dfrac{3\pi}{4}$

13. $\cos \dfrac{5\pi}{6}$

14. $\tan\left(-\dfrac{\pi}{4}\right)$

15. $\sec \dfrac{\pi}{2}$

16. $\cos 10\pi$

17. $\cos \dfrac{\pi}{6}$

18. $\sin \dfrac{2\pi}{3}$

19. $\sin \dfrac{5\pi}{4}$

20. $\cos \dfrac{11\pi}{6}$

21. $\sin (-5\pi)$

22. $\tan \dfrac{3\pi}{2}$

23. $\cot \dfrac{5\pi}{2}$

24. $\tan \dfrac{5\pi}{3}$

Find the function value using a calculator set in RADIAN *mode. Round the answer to four decimal places, where appropriate.*

25. $\tan \dfrac{\pi}{7}$

26. $\cos \left(-\dfrac{2\pi}{5}\right)$

27. $\sec 37$

28. $\sin 11.7$

29. $\cot 342$

30. $\tan 1.3$

31. $\cos 6\pi$

32. $\sin \dfrac{\pi}{10}$

33. $\csc 4.16$

34. $\sec \dfrac{10\pi}{7}$

35. $\tan \dfrac{7\pi}{4}$

36. $\cos 2000$

37. $\sin \left(-\dfrac{\pi}{4}\right)$

38. $\cot 7\pi$

39. $\sin 0$

40. $\cos (-29)$

41. $\tan \dfrac{2\pi}{9}$

42. $\sin \dfrac{8\pi}{3}$

In Exercises 43–46, recall that the graph of $y = f(-x)$ is a reflection of the graph of $y = f(x)$ across the y-axis and that the graph of $y = -f(x)$ is a reflection of the graph of $y = f(x)$ across the x-axis.

43. a) Sketch a graph of $y = \sin x$.
  b) By reflecting the graph in part (a), sketch a graph of $y = \sin (-x)$.
  c) By reflecting the graph in part (a), sketch a graph of $y = -\sin x$.
  d) How do the graphs in parts (b) and (c) compare?

44. a) Sketch a graph of $y = \cos x$.
  b) By reflecting the graph in part (a), sketch a graph of $y = \cos (-x)$.
  c) By reflecting the graph in part (a), sketch a graph of $y = -\cos x$.
  d) How do the graphs in parts (a) and (b) compare?

45. a) Sketch a graph of $y = \tan x$.
  b) By reflecting the graph of part (a), sketch a graph of $y = \tan (-x)$.
  c) By reflecting the graph of part (a), sketch a graph of $y = -\tan x$.
  d) How do the graphs of parts (b) and (c) compare?

46. a) Sketch a graph of $y = \sec x$.
  b) By reflecting the graph of part (a), sketch a graph of $y = \sec (-x)$.
  c) By reflecting the graph of part (a), sketch a graph of $y = -\sec x$.
  d) How do the graphs of parts (a) and (b) compare?

In Exercises 47–50, recall that the graph of $y = f(x - d)$ is the graph of $y = f(x)$ shifted right d units and that the graph of $y = f(x + d)$ is the graph of $y = f(x)$ shifted left d units.

47. a) Sketch a graph of $y = \sin x$.
  b) By translating, sketch a graph of $y = \sin (x + \pi)$.
  c) By reflecting the graph of part (a), sketch a graph of $y = -\sin x$.
  d) How do the graphs of parts (b) and (c) compare?

48. a) Sketch a graph of $y = \sin x$.
  b) By translating, sketch a graph of $y = \sin (x - \pi)$.
  c) By reflecting the graph of part (a), sketch a graph of $y = -\sin x$.
  d) How do the graphs of parts (b) and (c) compare?

49. a) Sketch a graph of $y = \cos x$.
  b) By translating, sketch a graph of $y = \cos (x + \pi)$.
  c) By reflecting the graph of part (a), sketch a graph of $y = -\cos x$.
  d) How do the graphs of parts (b) and (c) compare?

50. a) Sketch a graph of $y = \cos x$.
  b) By translating, sketch a graph of $y = \cos (x - \pi)$.
  c) By reflecting the graph of part (a), sketch a graph of $y = -\cos x$.
  d) How do the graphs of parts (b) and (c) compare?

51. Of the six circular functions, which are even? Which are odd?

52. Of the six circular functions, which have period π? Which have period 2π?

Consider the coordinates on the unit circle for Exercises 53–56.

53. In which quadrants is the tangent function positive? negative?

54. In which quadrants is the sine function positive? negative?

55. In which quadrants is the cosine function positive? negative?

56. In which quadrants is the cosecant function positive? negative?

> Skill Maintenance

Graph both functions in the same viewing window and describe how g is a transformation of f. **[2.4]**

57. $f(x) = x^2$, $g(x) = 2x^2 - 3$

58. $f(x) = x^2$, $g(x) = (x - 2)^2$

59. $f(x) = |x|$, $g(x) = \frac{1}{2}|x - 4| + 1$

60. $f(x) = x^3$, $g(x) = -x^3$

Write an equation for a function that has a graph with the given characteristics. Check using a graphing calculator.

61. The shape of $y = x^3$, but reflected across the x-axis, shifted right 2 units, and shifted down 1 unit **[2.4]**

62. The shape of $y = 1/x$, but shrunk vertically by a factor of $\frac{1}{4}$ and shifted up 3 units **[2.4]**

> Synthesis

Complete. (For example, $\sin(x + 2\pi) = \sin x$.)

63. $\sin(x + 2k\pi)$, k an integer $=$ _____

64. $\cos(x + 2k\pi)$, k an integer $=$ _____

65. $\sin(\pi - x) =$ _____

66. $\cos(\pi - x) =$ _____

67. $\cos(x - \pi) =$ _____

68. $\cos(x + \pi) =$ _____

69. $\sin(x + \pi) =$ _____

70. $\sin(x - \pi) =$ _____

71. Find all numbers x that satisfy the following. Check using a graphing calculator.

a) $\sin x = 1$
b) $\cos x = -1$
c) $\sin x = 0$

72. Find $f \circ g$ and $g \circ f$, where $f(x) = x^2 + 2x$ and $g(x) = \cos x$.

Use a graphing calculator to determine the domain, the range, the period, and the amplitude of the function.

73. $y = (\sin x)^2$

74. $y = |\cos x| + 1$

Determine the domain of the function.

75. $f(x) = \sqrt{\cos x}$

76. $g(x) = \dfrac{1}{\sin x}$

77. $f(x) = \dfrac{\sin x}{\cos x}$

78. $g(x) = \log(\sin x)$

Graph.

79. $y = 3 \sin x$

80. $y = \sin |x|$

81. $y = \sin x + \cos x$

82. $y = |\cos x|$

83. One of the motivations for developing trigonometry with a unit circle is that you can actually "see" $\sin \theta$ and $\cos \theta$ on the circle. Note in the following figure that $AP = \sin \theta$ and $OA = \cos \theta$. It turns out that you can also "see" the other four trigonometric functions. Prove each of the following.

a) $BD = \tan \theta$ b) $OD = \sec \theta$
c) $OE = \csc \theta$ d) $CE = \cot \theta$

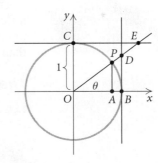

84. Using graphs, determine all numbers x that satisfy
$$\sin x < \cos x.$$

85. Using a calculator, consider $(\sin x)/x$, where x is between 0 and $\pi/2$. As x approaches 0, this function approaches a limiting value. What is it?

6.6 ▷ **Graphs of Transformed Sine Functions and Cosine Functions**

> ❯ Graph transformations of $y = \sin x$ and $y = \cos x$ in the form
> $$y = A \sin (Bx - C) + D$$
> and
> $$y = A \cos (Bx - C) + D$$
> and determine the amplitude, the period, and the phase shift.
>
> ❯ Graph sums of functions.
>
> ❯ Graph functions (damped oscillations) found by multiplying trignometric functions by other functions.

❯ Variations of Basic Graphs

In Section 6.5, we graphed all six trigonometric functions. In this section, we will consider variations of the graphs of the sine and the cosine functions. For example, we will graph equations like the following:

$$y = 5 \sin \tfrac{1}{2}x, \qquad y = \cos (2x - \pi), \quad \text{and} \quad y = \tfrac{1}{2}\sin x - 3.$$

In particular, we are interested in graphs of functions in the form

$$y = A \sin (Bx - C) + D$$

and

$$y = A \cos (Bx - C) + D,$$

**TRANSFORMATIONS
OF FUNCTIONS**

REVIEW SECTION 2.5.

where A, B, C, and D are constants. These constants have the effect of translating, reflecting, stretching, and shrinking the basic graphs. Let's first examine the effect of each constant individually. Then we will consider the combined effects of more than one constant.

The Constant D

Let's observe the effect of the constant D in the following graphs.

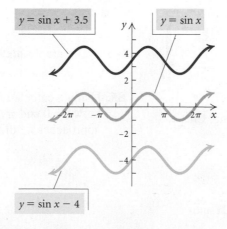

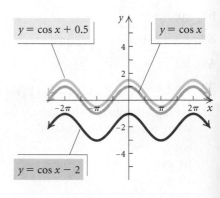

THE CONSTANT *D*: TRANSLATING VERTICALLY
The constant *D* in $y = A \sin(Bx - C) + D$ and $y = A \cos(Bx - C) + D$ *translates* the graphs *vertically* up *D* units if $D > 0$ or down $|D|$ units if $D < 0$.

EXAMPLE 1   Sketch a graph of $y = \sin x + 3$.

Solution   The graph of $y = \sin x + 3$ is a *vertical* translation of the graph of $y = \sin x$ up 3 units. One way to sketch the graph is to first consider $y = \sin x$ on an interval of length 2π, say, $[0, 2\pi]$. The zeros of the function and the maximum and minimum values can be considered key points. These are

$$(0, 0), \quad \left(\frac{\pi}{2}, 1\right), \quad (\pi, 0), \quad \left(\frac{3\pi}{2}, -1\right), \quad (2\pi, 0).$$

These key points are translated up 3 units to obtain the key points of the graph of $y = \sin x + 3$. These are

$$(0, 3), \quad \left(\frac{\pi}{2}, 4\right), \quad (\pi, 3), \quad \left(\frac{3\pi}{2}, 2\right), \quad (2\pi, 3).$$

The graph of $y = \sin x + 3$ can be sketched on the interval $[0, 2\pi]$ and extended to obtain the rest of the graph by repeating the graph on intervals of length 2π.

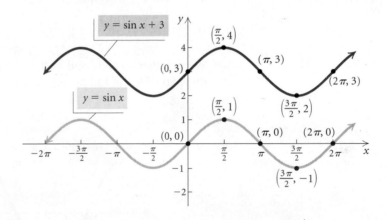

The Constant *A*

Next, we consider the effect of the constant *A*. What can we observe in the following graphs? What is the effect of the constant *A* on the graph of the basic function when **(a)** $0 < A < 1$? **(b)** $A > 1$? **(c)** $-1 < A < 0$? **(d)** $A < -1$?

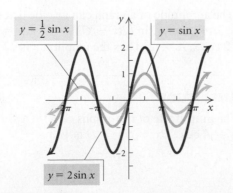

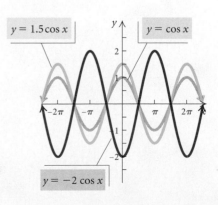

THE CONSTANT *A*: VERTICAL STRETCHING AND SHRINKING

The constant A in $y = A \sin (Bx - C) + D$ and $y = A \cos (Bx - C) + D$ *stretches* or *shrinks* the graphs *vertically*. If $|A| > 1$, then there will be a vertical stretching. If $|A| < 1$, then there will be a vertical shrinking. If $A < 0$, the graph is also *reflected* across the *x*-axis.

EXAMPLE 2   Sketch a graph of $y = 2 \cos x$. What is the amplitude?

Solution   The constant 2 in $y = 2 \cos x$ has the effect of stretching the graph of $y = \cos x$ vertically by a factor of 2. Since the function values of $y = \cos x$ are such that $-1 \le \cos x \le 1$, the function values of $y = 2 \cos x$ are such that $-2 \le 2 \cos x \le 2$. The maximum value of $y = 2 \cos x$ is 2, and the minimum value is -2. Thus the **amplitude** is

AMPLITUDE

REVIEW SECTION 6.5.

$$\tfrac{1}{2}|2 - (-2)|, \text{ or } 2.$$

We draw the graph of $y = \cos x$ and consider its key points,

$$(0, 1), \quad \left(\frac{\pi}{2}, 0\right), \quad (\pi, -1), \quad \left(\frac{3\pi}{2}, 0\right), \quad (2\pi, 1),$$

on the interval $[0, 2\pi]$.

We then multiply the second coordinates by 2 to obtain the key points of $y = 2 \cos x$. These are

$$(0, 2), \quad \left(\frac{\pi}{2}, 0\right), \quad (\pi, -2), \quad \left(\frac{3\pi}{2}, 0\right), \quad (2\pi, 2).$$

We plot these points and sketch the graph on the interval $[0, 2\pi]$. Then we repeat this part of the graph on adjacent intervals of length 2π.

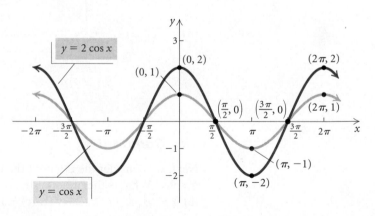

The amplitude of a graph can also be determined by finding $|A|$ from equations of the form $y = A \sin (Bx - C) + D$ or $y = A \cos (Bx - C) + D$. In Example 2, $y = 2 \cos x$, $A = 2$. Thus the amplitude is $|2|$, or 2.

AMPLITUDE

The **amplitude** of the graphs of $y = A \sin (Bx - C) + D$ and $y = A \cos (Bx - C) + D$ is $|A|$.

EXAMPLE 3 Sketch a graph of $y = -\frac{1}{2}\sin x$. What is the amplitude?

Solution The amplitude of the graph is $\left|-\frac{1}{2}\right|$, or $\frac{1}{2}$. The graph of $y = -\frac{1}{2}\sin x$ is a vertical shrinking and a reflection of the graph of $y = \sin x$ across the x-axis. In graphing, the key points of $y = \sin x$,

$$(0, 0), \quad \left(\frac{\pi}{2}, 1\right), \quad (\pi, 0), \quad \left(\frac{3\pi}{2}, -1\right), \quad (2\pi, 0),$$

are transformed to

$$(0, 0), \quad \left(\frac{\pi}{2}, -\frac{1}{2}\right), \quad (\pi, 0), \quad \left(\frac{3\pi}{2}, \frac{1}{2}\right), \quad (2\pi, 0).$$

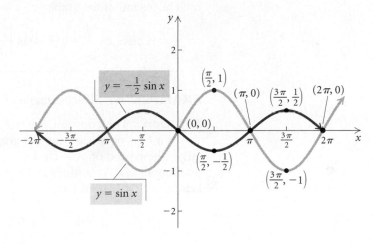

The Constant B

Next, let's consider the effect of the constant B. Changes in the constants A and D *do not* change the period. But what effect, if any, does a change in B have on the period of the function? Let's observe the period of each of the following graphs.

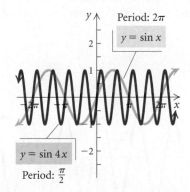

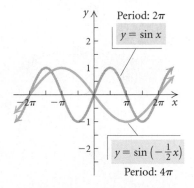

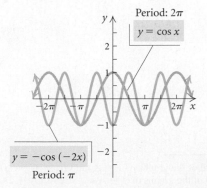

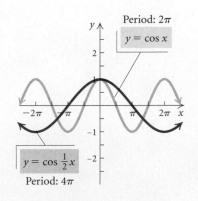

THE CONSTANT *B*: STRETCHING OR SHRINKING HORIZONTALLY

The constant *B* in $y = A \sin (Bx - C) + D$ and $y = A \cos (Bx - C) + D$ *stretches* or *shrinks* the graph *horizontally*. If $|B| < 1$, then there will be a horizontal stretching. If $|B| > 1$, then there will be a horizontal shrinking. If $B < 0$, the graph is also *reflected* across the *y*-axis.

PERIOD

The **period** of the graphs of $y = A \sin (Bx - C) + D$ and

$y = A \cos (Bx - C) + D$ is $\left| \dfrac{2\pi}{B} \right|$.*

EXAMPLE 4   Sketch a graph of $y = \sin 4x$. What is the period?

Solution   The constant *B* has the effect of changing the period. The graph of $y = \sin 4x$ is obtained from the graph of $y = \sin x$ by shrinking the graph horizontally. The period of $y = \sin 4x$ is $|2\pi/4|$, or $\pi/2$. The new graph is obtained by dividing the first coordinate of each ordered-pair solution of $y = \sin 4x$ by 4. The key points of $y = \sin x$ are

$$(0, 0), \quad \left(\frac{\pi}{2}, 1 \right), \quad (\pi, 0), \quad \left(\frac{3\pi}{2}, -1 \right), \quad (2\pi, 0).$$

These are transformed to the key points of $y = \sin 4x$, which are

$$(0, 0), \quad \left(\frac{\pi}{8}, 1 \right), \quad \left(\frac{\pi}{4}, 0 \right), \quad \left(\frac{3\pi}{8}, -1 \right), \quad \left(\frac{\pi}{2}, 0 \right).$$

We plot these key points and sketch in the graph on the shortened interval $[0, \pi/2]$. Then we repeat the graph on other intervals of length $\pi/2$.

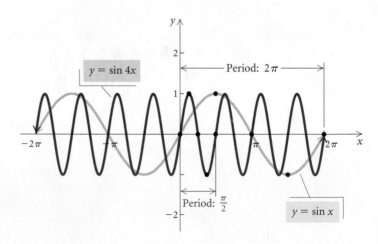

---

*The period of the graphs of $y = A \tan (Bx - C) + D$ and $y = A \cot (Bx - C) + D$ is $|\pi/B|$.
The period of the graphs of $y = A \sec (Bx - C) + D$ and $y = A \csc (Bx - C) + D$ is $|2\pi/B|$.

The Constant C

Next, we examine the effect of the constant C. The curve in each of the following graphs has an amplitude of 1 and a period of 2π, but there are six distinct graphs. What is the effect of the constant C?

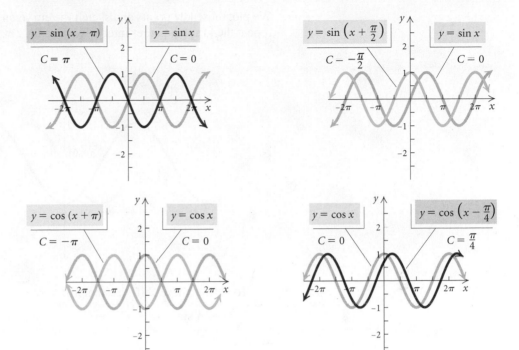

THE CONSTANT C: TRANSLATING HORIZONTALLY $\dfrac{C}{B}$

For each of the functions of the form

$$y = A \sin (Bx - C) + D \quad \text{and} \quad y = A \cos (Bx - C) + D:$$

- If $B = 1$, then the constant C translates the graph horizontally to the right C units if $C > 0$ and to the left $|C|$ units if $C < 0$.

- If $B \neq 1$ and $B > 0$, then $\dfrac{C}{B}$ translates the graph horizontally to the right $\dfrac{C}{B}$ units if $\dfrac{C}{B} > 0$ and to the left $\left|\dfrac{C}{B}\right|$ units if $\dfrac{C}{B} < 0$.

EXAMPLE 5   Sketch a graph of $y = \sin \left(x - \dfrac{\pi}{2} \right)$.

Solution   The amplitude is 1, and the period is 2π. The graph of $y = \sin (x - \pi/2)$ is obtained from the graph of $y = \sin x$ by translating the graph horizontally—to the right C units if $C > 0$ and to the left $|C|$ units if $C < 0$. The graph of $y = \sin (x - \pi/2)$ is a translation of the graph of $y = \sin x$ to the right $\pi/2$ units. The value $\pi/2$ is called the **phase shift**. The key points of $y = \sin x$,

$$(0,0), \quad \left(\frac{\pi}{2}, 1 \right), \quad (\pi, 0), \quad \left(\frac{3\pi}{2}, -1 \right), \quad (2\pi, 0),$$

are transformed by adding $\pi/2$ to each of the first coordinates to obtain the following key points of $y = \sin(x - \pi/2)$:

$$\left(\frac{\pi}{2}, 0\right), \quad (\pi, 1), \quad \left(\frac{3\pi}{2}, 0\right), \quad (2\pi, -1), \quad \left(\frac{5\pi}{2}, 0\right).$$

We plot these key points and sketch the curve on the interval $[\pi/2, 5\pi/2]$. Then we repeat the graph on other intervals of length 2π.

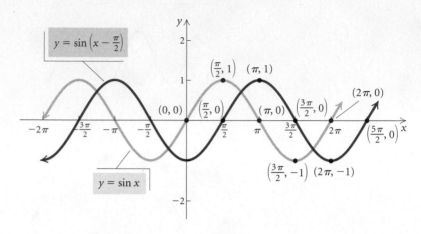

To define phase shift, it is helpful to rewrite

$$y = A \sin(Bx - C) + D \qquad \text{and} \quad y = A \cos(Bx - C) + D$$

as

$$y = A \sin\left[B\left(x - \frac{C}{B}\right)\right] + D \quad \text{and} \quad y = A \cos\left[B\left(x - \frac{C}{B}\right)\right] + D.$$

PHASE SHIFT

The **phase shift** of the graphs

$$y = A \sin(Bx - C) + D = A \sin\left[B\left(x - \frac{C}{B}\right)\right] + D$$

and

$$y = A \cos(Bx - C) + D = A \cos\left[B\left(x - \frac{C}{B}\right)\right] + D \text{ is } \frac{C}{B}.$$

- If $\dfrac{C}{B} > 0$, then the graph is translated to the right $\dfrac{C}{B}$ units.

- If $\dfrac{C}{B} < 0$, then the graph is translated to the left $\left|\dfrac{C}{B}\right|$ units.

EXAMPLE 6 Sketch a graph of $y = \cos(2x - \pi)$.

Solution The graph of

$$y = \cos(2x - \pi)$$

is the same as the graph of

$$y = 1 \cdot \cos\left[2\left(x - \frac{\pi}{2}\right)\right] + 0.$$

The amplitude is 1. The factor 2 shrinks the period by half, making the period $|2\pi/2|$, or π. The phase shift $\pi/2$ translates the graph of $y = \cos 2x$ to the right $\pi/2$ units. Because $D = 0$, there is no vertical translation. Thus, to form the graph, we first graph $y = \cos x$, followed by $y = \cos 2x$ and then $y = \cos[2(x - \pi/2)]$.

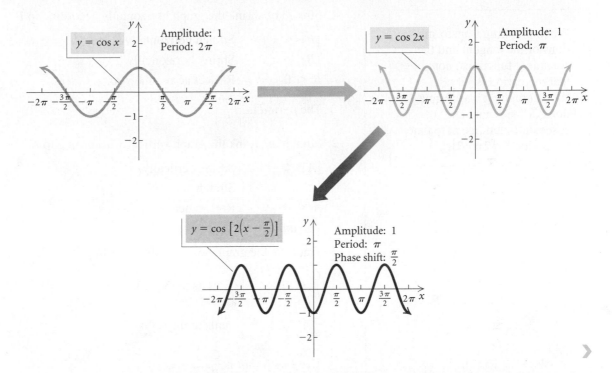

Let's now summarize the effect of the constants. When graphing, we carry out the procedures *in the order* listed. Be sure that the horizontal stretching or shrinking based on the constant B is done before the translation based on the phase shift C/B.

When graphing transformations of the tangent and the cotangent functions, note that the period is $|\pi/B|$. When graphing transformations of the secant and the cosecant functions, note that the period is $|2\pi/B|$.

TRANSFORMATIONS OF SINE FUNCTIONS AND COSINE FUNCTIONS

To graph

$$y = A \sin (Bx - C) + D = A \sin \left[B \left(x - \frac{C}{B} \right) \right] + D$$

and

$$y = A \cos (Bx - C) + D = A \cos \left[B \left(x - \frac{C}{B} \right) \right] + D,$$

follow these steps in the order in which they are listed.

1. Stretch or shrink the graph horizontally according to B.

$|B| < 1$	Stretch horizontally
$|B| > 1$	Shrink horizontally
$B < 0$	Reflect across the y-axis

The *period* is $\left| \dfrac{2\pi}{B} \right|$.

2. Stretch or shrink the graph vertically according to A.

$|A| < 1$	Shrink vertically
$|A| > 1$	Stretch vertically
$A < 0$	Reflect across the x-axis

The *amplitude* is $|A|$.

3. Translate the graph horizontally according to C/B.

$\dfrac{C}{B} < 0$	$\left| \dfrac{C}{B} \right|$ units to the left
$\dfrac{C}{B} > 0$	$\dfrac{C}{B}$ units to the right

The *phase shift* is $\dfrac{C}{B}$.

4. Translate the graph vertically according to D.

$D < 0$	$|D|$ units down
$D > 0$	D units up

EXAMPLE 7   Sketch a graph of $y = 3 \sin (2x + \pi/2) + 1$. Find the amplitude, the period, and the phase shift.

Solution   We first note that

$$y = 3 \sin \left(2x + \frac{\pi}{2} \right) + 1$$

$$= 3 \sin \left[2x - \left(-\frac{\pi}{2} \right) \right] + 1 \qquad y = A \sin (Bx - C) + D; C = -\pi/2$$

$$= 3 \sin \left[2 \left(x - \left(-\frac{\pi}{4} \right) \right) \right] + 1. \qquad y = A \sin \left[B \left(x - \frac{C}{B} \right) \right] + D;$$

$$\frac{C}{B} = \frac{-\pi/2}{2} = -\frac{\pi}{4}$$

Then we have the following:

$$\text{Amplitude} = |A| = |3| = 3,$$

$$\text{Period} = \left| \frac{2\pi}{B} \right| = \left| \frac{2\pi}{2} \right| = \pi,$$

$$\text{Phase shift} = \frac{C}{B} = \frac{-\pi/2}{2} = -\frac{\pi}{4}.$$

To create the final graph, we begin with the basic sine curve, $y = \sin x$. Then we sketch graphs of each of the following equations in sequence.

1. $y = \sin 2x$

2. $y = 3 \sin 2x$

3. $y = 3 \sin \left[2 \left(x - \left(-\frac{\pi}{4} \right) \right) \right]$

4. $y = 3 \sin \left[2 \left(x - \left(-\frac{\pi}{4} \right) \right) \right] + 1$

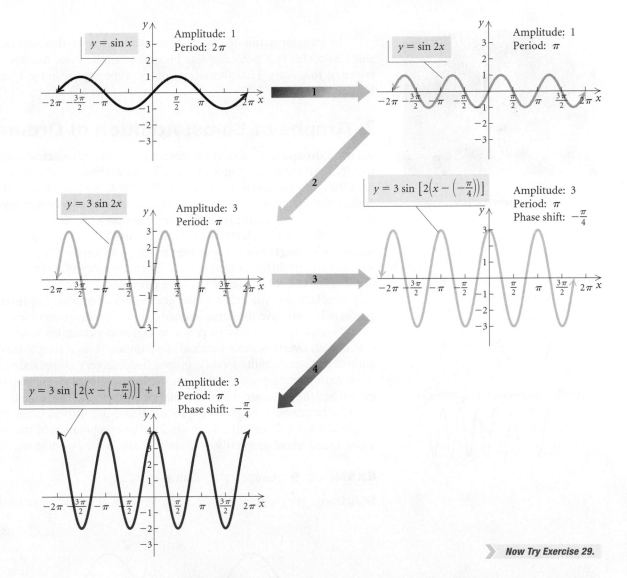

Now Try Exercise 29.

All the graphs in Examples 1–7 can be checked using a graphing calculator. Even though it is faster and more accurate to graph using a calculator, graphing by hand gives us a greater understanding of the effect of changing the constants A, B, C, and D.

Graphing calculators are especially convenient when a period or a phase shift is not a multiple of $\pi/4$.

EXAMPLE 8   Graph $y = 3 \cos 2\pi x - 1$. Find the amplitude, the period, and the phase shift.

Solution   First we note the following:

$$\text{Amplitude} = |A| = |3| = 3,$$

$$\text{Period} = \left|\frac{2\pi}{B}\right| = \left|\frac{2\pi}{2\pi}\right| = |1| = 1,$$

$$\text{Phase shift} = \frac{C}{B} = \frac{0}{2\pi} = 0.$$

There is no phase shift in this case because the constant $C = 0$. The graph has a vertical translation of the graph of the cosine function down 1 unit, an amplitude of 3, and a period of 1, so we can use $[-4, 4, -5, 5]$ as the viewing window.

> *Now Try Exercise 31.*

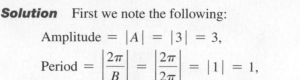

$y = 3 \cos (2\pi x) - 1$

The transformation techniques that we learned in this section for graphing the sine and cosine functions can also be applied in the same manner to the other trigonometric functions. Transformations of this type appear in the synthesis exercises in Exercise Set 6.6.

❯ Graphs of Sums: Addition of Ordinates

220 vibrations/second: 1 octave lower

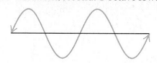

440 vibrations/second

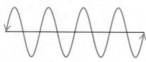

880 vibrations/second: 1 octave higher

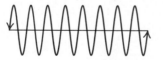

An **oscilloscope** is an electronic device that converts electrical signals into graphs like those in the preceding examples. These graphs are often called sine waves. By manipulating the controls of the oscilloscope, we can change the amplitude, the period, and the phase of sine waves. The oscilloscope has many applications, and the trigonometric functions play a major role in many of them.

The output of an electronic synthesizer used in the recording and playing of music can be converted into sine waves by an oscilloscope. The graphs at left illustrate simple tones of different frequencies. The frequency of a simple tone is the number of vibrations in the signal of the tone per second. The loudness or intensity of the tone is reflected in the height of the graph (its amplitude). The three tones in the diagrams at left all have the same intensity but different frequencies.

Musical instruments can generate extremely complex sine waves. On a single instrument, overtones can become superimposed on a simple tone. When multiple notes are played simultaneously, graphs become very complicated. This can happen when multiple notes are played on a single instrument or a group of instruments, or even when the same simple note is played on different instruments.

Combinations of simple tones produce interesting curves. Consider two tones whose graphs are $y_1 = 2 \sin x$ and $y_2 = \sin 2x$. The combination of the two tones produces a new sound whose graph is $y = 2 \sin x + \sin 2x$, as shown in the following example.

EXAMPLE 9   Graph: $y = 2 \sin x + \sin 2x$.

Solution   We graph $y = 2 \sin x$ and $y = \sin 2x$ using the same set of axes.

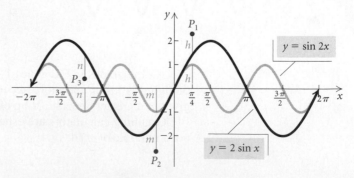

Next, we graphically add some y-coordinates, or ordinates, to obtain points on the graph that we seek. At $x = \pi/4$, we transfer the distance h, which is the value of $\sin 2x$, up to add it to the value of $2 \sin x$. Point P_1 is on the graph that we seek. At $x = -\pi/4$, we use a similar procedure, but this time both ordinates are negative. Point P_2 is on the graph. At $x = -5\pi/4$, we add the negative ordinate of $\sin 2x$ to the positive ordinate of $2 \sin x$. Point P_3 is also on the graph. We continue to plot points in this fashion and then connect them to get the desired graph, shown below. This method is called **addition of ordinates**, because we add the y-values (ordinates) of $y = \sin 2x$ to the y-values (ordinates) of $y = 2 \sin x$. Note that the period of $2 \sin x$ is 2π and the period of $\sin 2x$ is π. The period of the sum $2 \sin x + \sin 2x$ is 2π, the least common multiple of 2π and π.

$y_1 = 2 \sin x, \quad y_2 = \sin 2x, \quad y_3 = y_1 + y_2$

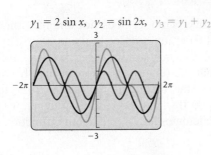

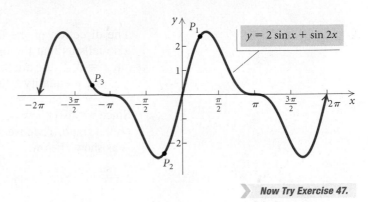

$y = 2 \sin x + \sin 2x$

> **Now Try Exercise 47.**

Using a graphing calculator, we can quickly determine the period of a trigonometric function that is a combination of sine and cosine functions.

EXAMPLE 10   Graph $y = 2 \cos x - \sin 3x$ and determine its period.

Solution   We graph $y = 2 \cos x - \sin 3x$ with appropriate dimensions. The period appears to be 2π.

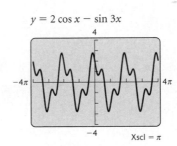

$y = 2 \cos x - \sin 3x$

Xscl $= \pi$

> **Now Try Exercise 55.**

❯ Damped Oscillation: Multiplication of Ordinates

Suppose that a weight is attached to a spring and the spring is stretched and put into motion. The weight oscillates up and down. If we could assume falsely that the weight will bob up and down forever, then its height h after time t, in seconds, might be approximated by a function like

$$h(t) = 5 + 2 \sin (6\pi t).$$

Over a short time period, this might be a valid model, but experience tells us that eventually the spring will come to rest. A more appropriate model is provided by the following example, which illustrates **damped oscillation**.

EXAMPLE 11 Sketch a graph of $f(x) = e^{-x/2}\sin x$.

Solution The function f is the product of two functions g and h, where

$$g(x) = e^{-x/2} \quad \text{and} \quad h(x) = \sin x.$$

Thus, to find function values, we can **multiply ordinates.** Let's do more analysis before graphing. Note that for any real number x,

$$-1 \le \sin x \le 1.$$

Recall from Chapter 5 that all values of the exponential function are positive. Thus we can multiply by $e^{-x/2}$ and obtain the inequality

$$-e^{-x/2} \le e^{-x/2}\sin x \le e^{-x/2}.$$

The direction of the inequality symbols does not change since $e^{-x/2} > 0$. This also tells us that the original function crosses the x-axis only at values for which $\sin x = 0$. These are the numbers $k\pi$, for any integer k.

The inequality tells us that the function f is constrained between the graphs of $y = -e^{-x/2}$ and $y = e^{-x/2}$. We start by graphing these functions using dashed lines. Since we also know that $f(x) = 0$ when $x = k\pi$, k an integer, we mark these points on the graph. Then we use a calculator and compute other function values. The graph is as shown below.

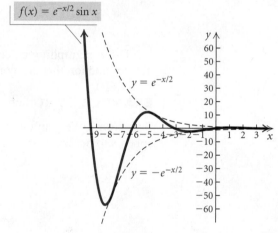

Now Try Exercise 63.

Visualizing the Graph

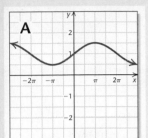

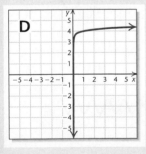

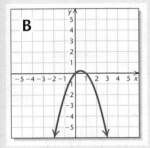

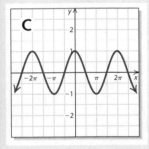

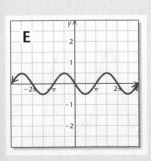

Match the function with its graph.

1. $f(x) = -\sin x$

2. $f(x) = 2x^3 - x + 1$

3. $y = \dfrac{1}{2}\cos\left(x + \dfrac{\pi}{2}\right)$

4. $f(x) = \cos\left(\dfrac{1}{2}x\right)$

5. $y = -x^2 + x$

6. $y = \dfrac{1}{2}\log x + 4$

7. $f(x) = 2^{x-1}$

8. $f(x) = \dfrac{1}{2}\sin\left(\dfrac{1}{2}x\right) + 1$

9. $f(x) = -\cos(x - \pi)$

10. $f(x) = -\dfrac{1}{2}x^4$

Answers on page A-36

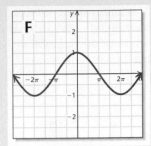

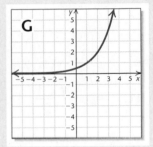

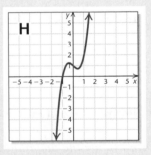

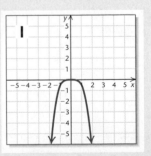

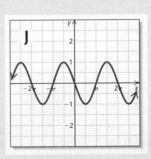

6.6   Exercise Set

Determine the amplitude, the period, and the phase shift of the function and, without a graphing calculator, sketch the graph of the function by hand. Then check the graph using a graphing calculator.

1. $y = \sin x + 1$

2. $y = \frac{1}{4}\cos x$

3. $y = -3\cos x$

4. $y = \sin(-2x)$

5. $y = \frac{1}{2}\cos x$

6. $y = \sin\left(\frac{1}{2}x\right)$

7. $y = \sin(2x)$

8. $y = \cos x - 1$

9. $y = 2\sin\left(\frac{1}{2}x\right)$

10. $y = \cos\left(x - \frac{\pi}{2}\right)$

11. $y = \cos\left(-\frac{1}{2}x\right)$

12. $y = \sin\left(-\frac{1}{4}x\right)$

13. $y = \frac{1}{2}\sin\left(x + \frac{\pi}{2}\right)$

14. $y = \cos x - \frac{1}{2}$

15. $y = 3\cos(x - \pi)$

16. $y = -\sin\left(\frac{1}{4}x\right) + 1$

17. $y = \frac{1}{3}\sin x - 4$

18. $y = \cos\left(\frac{1}{2}x + \frac{\pi}{2}\right)$

19. $y = -\cos(-x) + 2$

20. $y = \frac{1}{2}\sin\left(2x - \frac{\pi}{4}\right)$

Determine the amplitude, the period, and the phase shift of the function. Then check by graphing the function using a graphing calculator. Try to visualize the graph before creating it.

21. $y = 2\cos\left(\frac{1}{2}x - \frac{\pi}{2}\right)$

22. $y = 4\sin\left(\frac{1}{4}x + \frac{\pi}{8}\right)$

23. $y = -\frac{1}{2}\sin\left(2x + \frac{\pi}{2}\right)$

24. $y = -3\cos(4x - \pi) + 2$

25. $y = 2 + 3\cos(\pi x - 3)$

26. $y = 5 - 2\cos\left(\frac{\pi}{2}x + \frac{\pi}{2}\right)$

27. $y = -\frac{1}{2}\cos(2\pi x) + 2$

28. $y = -2\sin(-2x + \pi) - 2$

29. $y = -\sin\left(\frac{1}{2}x - \frac{\pi}{2}\right) + \frac{1}{2}$

30. $y = \frac{1}{3}\cos(-3x) + 1$

31. $y = \cos(-2\pi x) + 2$

32. $y = \frac{1}{2}\sin(2\pi x + \pi)$

33. $y = -\frac{1}{4}\cos(\pi x - 4)$

34. $y = 2\sin(2\pi x + 1)$

In Exercises 35–42, without a graphing calculator, match the function with one of the graphs (a)–(h) that follow. Then check your work using a graphing calculator.

a)

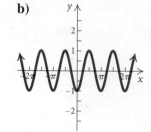

b)

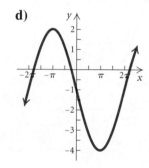

c)

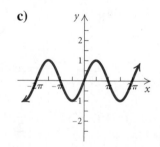

d)

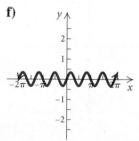

e)

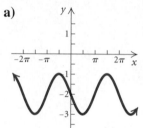

f)

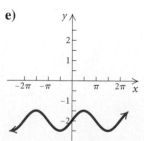

g)

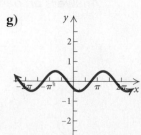

h)

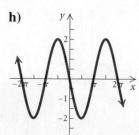

35. $y = -\cos 2x$

36. $y = \frac{1}{2}\sin x - 2$

37. $y = 2\cos\left(x + \frac{\pi}{2}\right)$

38. $y = -3\sin\frac{1}{2}x - 1$

39. $y = \sin(x - \pi) - 2$

40. $y = -\frac{1}{2}\cos\left(x - \frac{\pi}{4}\right)$

41. $y = \frac{1}{3}\sin 3x$

42. $y = \cos\left(x - \frac{\pi}{2}\right)$

In Exercises 43–46, determine the equation of the function that is graphed. Answers may vary.

43.

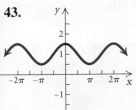

44.

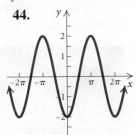

45.

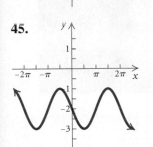

46.

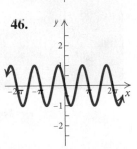

Graph using addition of ordinates. Then check your work using a graphing calculator.

47. $y = 2\cos x + \cos 2x$

48. $y = 3\cos x + \cos 3x$

49. $y = \sin x + \cos 2x$

50. $y = 2\sin x + \cos 2x$

51. $y = \sin x - \cos x$

52. $y = 3\cos x - \sin x$

53. $y = 3\cos x + \sin 2x$

54. $y = 3\sin x - \cos 2x$

Use a graphing calculator to graph the function.

55. $y = x + \sin x$

56. $y = -x - \sin x$

57. $y = \cos x - x$

58. $y = -(\cos x - x)$

59. $y = \cos 2x + 2x$

60. $y = \cos 3x + \sin 3x$

61. $y = 4\cos 2x - 2\sin x$

62. $y = 7.5\cos x + \sin 2x$

Graph each of the following.

63. $f(x) = e^{-x/2}\cos x$

64. $f(x) = e^{-0.4x}\sin x$

65. $f(x) = 0.6x^2\cos x$

66. $f(x) = e^{-x/4}\sin x$

67. $f(x) = x\sin x$

68. $f(x) = |x|\cos x$

69. $f(x) = 2^{-x}\sin x$

70. $f(x) = 2^{-x}\cos x$

❯ Skill Maintenance

Classify the function as linear, quadratic, cubic, quartic, rational, exponential, logarithmic, or trigonometric.

71. $f(x) = \dfrac{x + 4}{x}$ **[4.5]**

72. $y = \frac{1}{2}\log x - 4$ **[5.3]**

73. $y = x^4 - x - 2$ **[4.1]**   **74.** $\frac{3}{4}x + \frac{1}{2}y = -5$ **[1.3]**

75. $f(x) = \sin x - 3$ **[6.6]**

76. $f(x) = 0.5e^{x-2}$ **[5.2]**

77. $y = \frac{2}{5}$ **[1.3]**

78. $y = \sin x + \cos x$ **[6.6]**

79. $y = x^2 - x^3$ **[4.1]**   **80.** $f(x) = \left(\frac{1}{2}\right)^x$ **[5.2]**

❯ Synthesis

Find the maximum and minimum values of the function.

81. $y = 2\cos\left[3\left(x - \dfrac{\pi}{2}\right)\right] + 6$

82. $y = \frac{1}{2}\sin(2x - 6\pi) - 4$

The transformation techniques that we learned in this section for graphing the sine and cosine functions can also be applied to the other trigonometric functions. Sketch a graph of each of the following. Then check your work using a graphing calculator.

83. $y = -\tan x$

84. $y = \tan(-x)$

85. $y = \csc(-x)$

86. $y = -\cot x$

87. $y = \frac{1}{2}\sec\left(\frac{1}{2}x\right)$

88. $y = -\frac{3}{2}\csc x$

89. $y = -2 + \cot x$

90. $y = -\sec x + 2$

91. $y = 2\tan\left(\frac{1}{2}x\right)$

92. $y = \cot(2x)$

93. $y = 2\sec(x - \pi)$

94. $y = 4\tan\left(\dfrac{1}{4}x + \dfrac{\pi}{8}\right)$

95. $y = \cot\left(x + \dfrac{\pi}{2}\right) - 1$

96. $y = \sec(x + \pi) + 2$

97. $y = 2\csc\left(\dfrac{1}{2}x - \dfrac{3\pi}{4}\right)$ **98.** $y = 4\sec(2x - \pi)$

Use a graphing calculator to graph each of the following on the given interval and approximate the zeros.

99. $f(x) = \dfrac{\sin x}{x}; \ [-12, 12]$

100. $f(x) = \dfrac{\cos x - 1}{x}; \ [-12, 12]$

101. $f(x) = x^3 \sin x; \ [-5, 5]$

102. $f(x) = \dfrac{(\sin x)^2}{x}; \ [-4, 4]$

103. *Temperature During an Illness.* The temperature T of a patient during a 12-day illness is given by

$$T(t) = 101.6° + 3° \sin\left(\dfrac{\pi}{8}t\right).$$

a) Graph the function on the interval $[0, 12]$.
b) What are the maximum and minimum temperatures during the illness?

104. *Periodic Sales.* A company in a northern climate has sales of skis as given by

$$S(t) = 10\left(1 - \cos \dfrac{\pi}{6}t\right),$$

where t is the time, in months ($t = 0$ corresponds to July 1), and $S(t)$ is in thousands of dollars.

a) Graph the function on a 12-month interval $[0, 12]$.
b) What is the period of the function?
c) What is the minimum amount of sales and when does it occur?
d) What is the maximum amount of sales and when does it occur?

105. *Satellite Location.* A satellite circles the earth in such a way that it is y miles from the equator (north or south, height not considered) t minutes after its launch, where

$$y(t) = 3000\left[\cos \dfrac{\pi}{45}(t - 10)\right].$$

What are the amplitude, the period, and the phase shift?

106. *Water Wave.* The cross-section of a water wave is given by

$$y = 3 \sin\left(\dfrac{\pi}{4}x + \dfrac{\pi}{4}\right),$$

where y is the vertical height of the water wave and x is the distance from the origin to the wave.

$$y = 3\sin\left(\tfrac{\pi}{4}x + \tfrac{\pi}{4}\right)$$

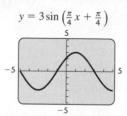

What are the amplitude, the period, and the phase shift?

107. *Damped Oscillations.* Suppose that the motion of a spring is given by

$$d(t) = 6e^{-0.8t}\cos(6\pi t) + 4,$$

where d is the distance, in inches, of a weight from the point at which the spring is attached to a ceiling, after t seconds. How far do you think the spring is from the ceiling when the spring stops bobbing?

STUDY GUIDE

KEY TERMS AND CONCEPTS	EXAMPLES

SECTION 6.1: TRIGONOMETRIC FUNCTIONS OF ACUTE ANGLES

Trigonometric Function Values of an Acute Angle θ

Let θ be an acute angle of a right triangle. The six trigonometric functions of θ are as follows:

$$\sin \theta = \frac{\text{opp}}{\text{hyp}}, \qquad \csc \theta = \frac{\text{hyp}}{\text{opp}},$$

$$\cos \theta = \frac{\text{adj}}{\text{hyp}}, \qquad \sec \theta = \frac{\text{hyp}}{\text{adj}},$$

$$\tan \theta = \frac{\text{opp}}{\text{adj}}, \qquad \cot \theta = \frac{\text{adj}}{\text{opp}}.$$

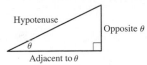

If $\cos \alpha = \frac{3}{8}$ and α is an acute angle, find the other five trigonometric function values of α.

$$\cos \alpha = \frac{3}{8} \quad \leftarrow \quad \text{adj} \\ \qquad\qquad \leftarrow \quad \text{hyp}$$

We find the missing length using the Pythagorean equation: $a^2 + b^2 = c^2$.

$$a^2 + 3^2 = 8^2$$
$$a^2 = 64 - 9$$
$$a = \sqrt{55}$$

$$\sin \alpha = \frac{\sqrt{55}}{8}, \qquad \csc \alpha = \frac{8}{\sqrt{55}}, \text{ or } \frac{8\sqrt{55}}{55},$$

$$\cos \alpha = \frac{3}{8}, \qquad \sec \alpha = \frac{8}{3},$$

$$\tan \alpha = \frac{\sqrt{55}}{3}, \qquad \cot \alpha = \frac{3}{\sqrt{55}}, \text{ or } \frac{3\sqrt{55}}{55}$$

Reciprocal Functions

$$\csc \theta = \frac{1}{\sin \theta}, \qquad \sec \theta = \frac{1}{\cos \theta}, \qquad \cot \theta = \frac{1}{\tan \theta}$$

Given that $\sin \beta = \frac{5}{13}$, $\cos \beta = \frac{12}{13}$, and $\tan \beta = \frac{5}{12}$, find $\csc \beta$, $\sec \beta$, and $\cot \beta$.

$$\csc \beta = \frac{13}{5}, \qquad \sec \beta = \frac{13}{12}, \qquad \cot \beta = \frac{12}{5}$$

Function Values of Special Angles

We often use the function values of 30°, 45°, and 60°. Either the following triangles or the values themselves should be memorized.

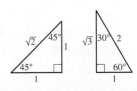

	30°	45°	60°
sin	$1/2$	$\sqrt{2}/2$	$\sqrt{3}/2$
cos	$\sqrt{3}/2$	$\sqrt{2}/2$	$1/2$
tan	$\sqrt{3}/3$	1	$\sqrt{3}$

Find the exact function value.

$$\csc 45° = \frac{2}{\sqrt{2}}, \text{ or } \sqrt{2}, \qquad \tan 60° = \sqrt{3},$$

$$\sin 30° = \frac{1}{2}, \qquad\qquad \cot 45° = 1,$$

$$\cos 60° = \frac{1}{2}, \qquad\qquad \sec 30° = \frac{2}{\sqrt{3}}, \text{ or } \frac{2\sqrt{3}}{3},$$

$$\cot 30° = \frac{3}{\sqrt{3}}, \text{ or } \sqrt{3}, \qquad \sin 60° = \frac{\sqrt{3}}{2},$$

$$\tan 45° = 1, \qquad\qquad \sec 45° = \frac{2}{\sqrt{2}}, \text{ or } \sqrt{2},$$

$$\csc 60° = \frac{2}{\sqrt{3}}, \text{ or } \frac{2\sqrt{3}}{3}, \qquad \cos 45° = \frac{\sqrt{2}}{2}$$

Most calculators can convert D°M′S″ notation to decimal degree notation and vice versa. Procedures among calculators vary. We also can convert without using a calculator.

Convert 17°42′35″ to decimal degree notation, rounding the answer to the nearest hundredth of a degree.

$$17°42′35″ = 17° + 42′ + \frac{35′}{60} \approx 17° + 42.5833′$$

$$\approx 17° + \frac{42.5833°}{60} \approx 17.71°$$

Convert 23.12° to D°M′S″ notation.

$$23.12° = 23° + 0.12 \times 1° = 23° + 0.12 \times 60′$$
$$= 23° + 7.2′ = 23° + 7′ + 0.2 \times 1′$$
$$= 23° + 7′ + 0.2 \times 60″ = 23° + 7′ + 12″$$
$$= 23°\,7′12″$$

Cofunction Identities

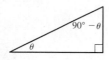

$\sin\theta = \cos(90° - \theta),$
$\cos\theta = \sin(90° - \theta),$
$\tan\theta = \cot(90° - \theta),$
$\cot\theta = \tan(90° - \theta),$
$\sec\theta = \csc(90° - \theta),$
$\csc\theta = \sec(90° - \theta)$

Given that $\sin 47° \approx 0.7314$, $\cos 47° \approx 0.6820$, and $\tan 47° \approx 1.0724$, find the six trigonometric function values for 43°.

First, we find $\csc 47°$, $\sec 47°$, and $\cot 47°$:

$$\csc 47° = \frac{1}{\sin 47°} \approx 1.3672,$$

$$\sec 47° = \frac{1}{\cos 47°} \approx 1.4663,$$

$$\cot 47° = \frac{1}{\tan 47°} \approx 0.9325.$$

We know that $43° = 90° - 47°$, so we have

$\sin 43° = \cos 47° \approx 0.6820,$
$\cos 43° = \sin 47° \approx 0.7314,$
$\tan 43° = \cot 47° \approx 0.9325,$
$\csc 43° = \sec 47° \approx 1.4663,$
$\sec 43° = \csc 47° \approx 1.3672,$
$\cot 43° = \tan 47° \approx 1.0724.$

SECTION 6.2: APPLICATIONS OF RIGHT TRIANGLES

Solving a Triangle

To **solve** a triangle means to find the lengths of *all* sides and the measures of *all* angles.

Solve this right triangle.

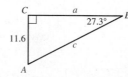

$$A = ?, \qquad a = ?,$$
$$B = 27.3°, \qquad b = 11.6,$$
$$C = 90°, \qquad c = ?$$

First, we find A: $A = 90° - 27.3° = 62.7°$.
Then we use the tangent and the cosine functions to find a and c:

$$\tan 62.7° = \frac{a}{11.6} \qquad \cos 62.7° = \frac{11.6}{c}$$

$$11.6 \tan 62.7° = a$$
$$22.5 \approx a, \qquad\qquad c = \frac{11.6}{\cos 62.7°}$$

$$c \approx 25.3.$$

SECTION 6.3: TRIGONOMETRIC FUNCTIONS OF ANY ANGLE

Coterminal Angles

If two or more angles have the same terminal side, the angles are said to be **coterminal**.

To find angles coterminal with a given angle, we add or subtract multiples of 360°.

Find two positive angles and two negative angles that are coterminal with 123°.

$$123° + 360° = 483°,$$
$$123° + 3(360°) = 1203°,$$
$$123° - 360° = -237°,$$
$$123° - 2(360°) = -597°$$

The angles 483°, 1203°, −237°, and −597° are coterminal with 123°.

Complementary Angles and Supplementary Angles

Two acute angles are **complementary** if the sum of their measures is 90°.

Two positive angles are **supplementary** if the sum of their measures is 180°.

Find the complement and the supplement of 83.5°.

$$90° - 83.5° = 6.5°,$$
$$180° - 83.5° = 96.5°$$

The complement of 83.5° is 6.5°, and the supplement of 83.5° is 96.5°.

Trigonometric Functions of Any Angle θ

If $P(x, y)$ is any point on the terminal side of any angle θ in standard position, and r is the distance from the origin to $P(x, y)$, where $r = \sqrt{x^2 + y^2}$, then

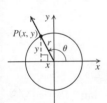

$$\sin \theta = \frac{y}{r}, \quad \csc \theta = \frac{r}{y},$$
$$\cos \theta = \frac{x}{r}, \quad \sec \theta = \frac{r}{x},$$
$$\tan \theta = \frac{y}{x}, \quad \cot \theta = \frac{x}{y}.$$

The trigonometric function values of θ depend only on the angle, not on the choice of the point on the terminal side that is used to compute them.

Signs of Function Values

The signs of the function values depend only on the coordinates of the point P on the terminal side of an angle.

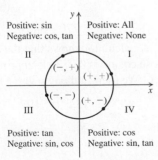

Find the six trigonometric function values for the angle shown.

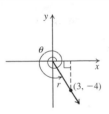

We first determine r:

$$r = \sqrt{x^2 + y^2} = \sqrt{3^2 + (-4)^2} = \sqrt{25} = 5.$$

$$\sin \theta = -\frac{4}{5}, \quad \csc \theta = -\frac{5}{4},$$
$$\cos \theta = \frac{3}{5}, \quad \sec \theta = \frac{5}{3},$$
$$\tan \theta = -\frac{4}{3}, \quad \cot \theta = -\frac{3}{4}$$

Given that $\cos \alpha = -\frac{1}{5}$ and α is in the third quadrant, find the other function values.

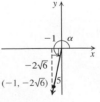

One leg of the reference triangle has length 1, and the length of the hypotenuse is 5. The length of the other leg is $\sqrt{5^2 - 1^2}$, or $\sqrt{24}$, or $2\sqrt{6}$.

$$\sin \alpha = -\frac{2\sqrt{6}}{5}, \quad \csc \alpha = -\frac{5}{2\sqrt{6}}, \text{ or } -\frac{5\sqrt{6}}{12},$$
$$\cos \alpha = -\frac{1}{5}, \quad \sec \alpha = -5,$$
$$\tan \alpha = 2\sqrt{6}, \quad \cot \alpha = \frac{1}{2\sqrt{6}}, \text{ or } \frac{\sqrt{6}}{12}$$

Trigonometric Function Values of Quadrantal Angles

An angle whose terminal side falls on one of the axes is a **quadrantal angle**.

	0° 360°	90°	180°	270°
sin	0	1	0	−1
cos	1	0	−1	0
tan	0	Not defined	0	Not defined

Find the exact function value.

$\tan(-90°)$ is not defined,
$\sin 450° = 1$,
$\csc 270° = -1$,
$\cos 720° = 1$,
$\sec(-180°) = -1$,
$\cot(-360°)$ is not defined

Reference Angles

The **reference angle** for an angle is the acute angle formed by the terminal side of the angle and the *x*-axis.

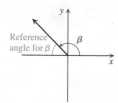

When the reference angle is 30°, 45°, or 60°, we can mentally determine trigonometric function values.

Find the sine, cosine, and tangent values for 240°.

The reference angle is 240° − 180°, or 60°. Recall that $\sin 60° = \dfrac{\sqrt{3}}{2}$, $\cos 60° = \dfrac{1}{2}$, and $\tan 60° = \sqrt{3}$.

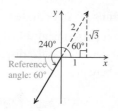

In the third quadrant, the sine and the cosine functions are negative, and the tangent function is positive. Thus,

$$\sin 240° = -\frac{\sqrt{3}}{2}, \qquad \cos 240° = -\frac{1}{2}, \quad \text{and} \quad \tan 240° = \sqrt{3}.$$

Trigonometric Function Values of Any Angle

Using a calculator, we can approximate the trigonometric function values of any angle.

Find each of the following function values using a calculator set in DEGREE mode. Round the values to four decimal places, where appropriate.

$\csc 285° \approx -1.0353$, $\cos 51° \approx 0.6293$,
$\sin 25°14'38'' \approx 0.4265$, $\sec(-45°) \approx 1.4142$,
$\tan(-1020°) \approx 1.7321$, $\sin 810° = 1$

Given $\cos \theta \approx -0.9724$, $180° < \theta < 270°$, find θ.

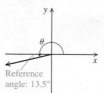

A calculator shows that the acute angle whose cosine is 0.9724 is approximately 13.5°. We then find angle θ:

$$180° + 13.5° = 193.5°.$$

Thus, $\theta \approx 193.5°$.

Aerial Navigation

In aerial navigation, directions are given in degrees clockwise from *north*. For example, a direction, or bearing, of 195° is shown below.

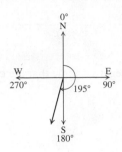

An airplane flies 320 mi from an airport in a direction of 305°. How far north of the airport is the plane then? How far west?

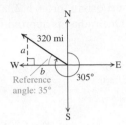

The distance north of the airport a and the distance west of the airport b are legs of a right triangle. The reference angle is $305° - 270° = 35°$. Thus,

$$\frac{a}{320} = \sin 35°$$
$$a = 320 \sin 35° \approx 184;$$
$$\frac{b}{320} = \cos 35°$$
$$b = 320 \cos 35° \approx 262.$$

The airplane is about 184 mi north and about 262 mi west of the airport.

SECTION 6.4: RADIANS, ARC LENGTH, AND ANGULAR SPEED

The Unit Circle

A circle centered at the origin with a radius of length 1 is called a **unit circle**. Its equation is $x^2 + y^2 = 1$.

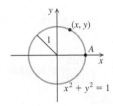

The circumference of a circle of radius r is $2\pi r$. For a unit circle, where $r = 1$, the circumference is 2π. If a point starts at A and travels around the circle, it travels a distance of 2π.

Find two real numbers between -2π and 2π that determine each of the labeled points.

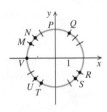

$M:\ \dfrac{5\pi}{6},\ -\dfrac{7\pi}{6}$ $S:\ \dfrac{7\pi}{4},\ -\dfrac{\pi}{4}$

$N:\ \dfrac{3\pi}{4},\ -\dfrac{5\pi}{4}$ $T:\ \dfrac{4\pi}{3},\ -\dfrac{2\pi}{3}$

$P:\ \dfrac{\pi}{2},\ -\dfrac{3\pi}{2}$ $U:\ \dfrac{5\pi}{4},\ -\dfrac{3\pi}{4}$

$Q:\ \dfrac{\pi}{3},\ -\dfrac{5\pi}{3}$ $V:\ \pi,\ -\pi$

$R:\ \dfrac{11\pi}{6},\ -\dfrac{\pi}{6}$

Radian Measure

Consider the unit circle ($r = 1$) and arc length 1. If a ray is drawn from the origin through T, an angle of 1 radian is formed. One radian is approximately 57.3°.

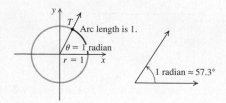

A complete counterclockwise revolution is an angle whose measure is 2π radians, or about 6.28 radians. Thus a rotation of 360° (1 revolution) has a measure of 2π radians.

Radian–Degree Equivalents

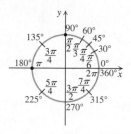

Converting Between Degree Measure and Radian Measure

To convert from degree measure to radian measure, multiply by $\dfrac{\pi \text{ radians}}{180°}$.

To convert from radian measure to degree measure, multiply by $\dfrac{180°}{\pi \text{ radians}}$.

If no unit is given for a rotation, the rotation is understood to be in radians.

Convert 150° and $-63.5°$ to radian measure. Leave answers in terms of π.

$$150° = 150° \cdot \frac{\pi \text{ radians}}{180°} = \frac{150°}{180°}\,\pi \text{ radians} = \frac{5\pi}{6};$$

$$-63.5° = -63.5° \cdot \frac{\pi \text{ radians}}{180°} = -\frac{63.5°}{180°}\,\pi \text{ radians} \approx -0.35\pi$$

Find a positive angle and a negative angle that are coterminal with $\dfrac{7\pi}{4}$.

$$\frac{7\pi}{4} + 2\pi = \frac{7\pi}{4} + \frac{8\pi}{4} = \frac{15\pi}{4};$$

$$\frac{7\pi}{4} - 3(2\pi) = \frac{7\pi}{4} - 6\pi = \frac{7\pi}{4} - \frac{24\pi}{4} = -\frac{17\pi}{4}$$

Two angles coterminal with $\dfrac{7\pi}{4}$ are $\dfrac{15\pi}{4}$ and $-\dfrac{17\pi}{4}$.

Find the complement and the supplement of $\dfrac{\pi}{8}$.

$$\frac{\pi}{2} - \frac{\pi}{8} = \frac{4\pi}{8} - \frac{\pi}{8} = \frac{3\pi}{8}; \qquad 90° = \frac{\pi}{2}$$

$$\pi - \frac{\pi}{8} = \frac{8\pi}{8} - \frac{\pi}{8} = \frac{7\pi}{8} \qquad 180° = \pi$$

The complement of $\dfrac{\pi}{8}$ is $\dfrac{3\pi}{8}$, and the supplement of $\dfrac{\pi}{8}$ is $\dfrac{7\pi}{8}$.

Convert $-328°$ and 29.2° to radian measure. Round the answers to two decimal places.

$$-328° = -328° \cdot \frac{\pi \text{ radians}}{180°} = -\frac{328°}{180°}\,\pi \text{ radians} \approx -5.72;$$

$$29.2° = 29.2° \cdot \frac{\pi \text{ radians}}{180°} = \frac{29.2°}{180°}\,\pi \text{ radians} \approx 0.51$$

Convert $-\dfrac{2\pi}{3}$, 5π, and -1.3 to degree measure. Round the answers to two decimal places.

$$-\frac{2\pi}{3} = -\frac{2\pi}{3} \cdot \frac{180°}{\pi \text{ radians}} = -\frac{2}{3} \cdot 180° = -120°;$$

$$5\pi = 5\pi \cdot \frac{180°}{\pi \text{ radians}} = 5 \cdot 180° = 900°;$$

$$-1.3 = -1.3 \cdot \frac{180°}{\pi \text{ radians}} = \frac{-1.3(180°)}{\pi} \approx -74.48°$$

Radian Measure

The **radian measure** θ of a rotation is the ratio of the distance s traveled by a point at a radius r from the center of rotation to the length of the radius r:

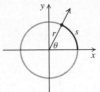

$$\theta = \frac{s}{r}.$$

When the formula $\theta = s/r$ is used, θ must be in radians and s and r must be expressed in the same unit.

Find the measure of a rotation in radians when a point 6 cm from the center of rotation travels 13 cm.

$$\theta = \frac{s}{r} = \frac{13\text{ cm}}{6\text{ cm}} = \frac{13}{6}\text{ radians}$$

Find the length of an arc of a circle of radius 10 yd associated with an angle of $5\pi/4$ radians.

$$\theta = \frac{s}{r}, \quad \text{or} \quad s = r\theta;$$

$$s = r\theta = 10\text{ yd} \cdot \frac{5\pi}{4} \approx 39.3\text{ yd}$$

Linear Speed and Angular Speed

Linear speed v is the distance s traveled per unit of time t:

$$v = \frac{s}{t}.$$

Angular speed ω is the amount of rotation θ per unit of time t:

$$\omega = \frac{\theta}{t}.$$

Linear Speed in Terms of Angular Speed

The linear speed v of a point a distance r from the center of rotation is given by

$$v = r\omega,$$

where ω is the angular speed, in radians, per unit of time. The unit of distance for v and r must be the same, ω must be in radians per unit of time, and v and ω must be expressed in the same unit of time.

A wheel with a 40-cm radius is rotating at a rate of 2.5 radians/sec. What is the linear speed of a point on its rim, in meters per minute?

We first express r in meters and w in radians per minute.

$$r = 40\text{ cm} \cdot \frac{1\text{ m}}{100\text{ cm}} = 0.4\text{ m};$$

$$\omega = \frac{2.5\text{ radians}}{1\text{ sec}} \cdot \frac{60\text{ sec}}{1\text{ min}} = \frac{150\text{ radians}}{1\text{ min}}$$

Then we find linear speed.

$$v = r\omega = 0.4\text{ m} \cdot \frac{150}{1\text{ min}} = \frac{60\text{ m}}{\text{min}}$$

SECTION 6.5: CIRCULAR FUNCTIONS: GRAPHS AND PROPERTIES

Domains of the Trigonometric Functions

In Sections 6.1 and 6.3, the domains of the trigonometric functions were defined as a set of angles or rotations measured in a real number of degree units. In Section 6.4, the domains were considered to be sets of real numbers, or radians. Radian measure for θ is defined as $\theta = s/r$. When $r = 1$, $\theta = s$. The arc length s on the unit circle is the same as the radian measure of the angle θ.

Basic Circular Functions

On the unit circle, s can be considered the radian measure of an angle or the measure of an arc length. In either case, it is a real number. Trigonometric functions with domains composed of real numbers are called **circular functions**.

For a real number s that determines a point (x, y) on the unit circle:

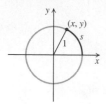

$$\sin s = y,$$
$$\cos s = x,$$
$$\tan s = \frac{y}{x}, x \neq 0,$$
$$\csc s = \frac{1}{y}, y \neq 0,$$
$$\sec s = \frac{1}{x}, x \neq 0,$$
$$\cot s = \frac{x}{y}, y \neq 0.$$

Find each function value using coordinates of a point on the unit circle.

$$\sin(-5\pi) = 0, \qquad \csc\frac{\pi}{3} = \frac{2}{\sqrt{3}}, \text{ or } \frac{2\sqrt{3}}{3},$$

$$\cos\left(-\frac{3\pi}{4}\right) = -\frac{\sqrt{2}}{2}, \qquad \sec\frac{\pi}{6} = \frac{2}{\sqrt{3}}, \text{ or } \frac{2\sqrt{3}}{3},$$

$$\tan\frac{5\pi}{2} \text{ is not defined}, \qquad \cot\frac{23\pi}{6} = -\sqrt{3},$$

$$\cos\left(-\frac{5\pi}{6}\right) = -\frac{\sqrt{3}}{2}, \qquad \tan\frac{7\pi}{4} = -1$$

Find each function value using a calculator set in RADIAN mode. Round the answers to four decimal places, where appropriate.

$$\cos(-14.7) \approx -0.5336, \qquad \tan\frac{3\pi}{2} \text{ is not defined},$$

$$\sin\frac{9\pi}{5} \approx -0.5878, \qquad \sec 214 \approx 1.0733$$

Reflections

Because a unit circle is symmetric with respect to the x-axis, the y-axis, and the origin, the coordinates of one point on the unit circle can be used to find coordinates of its reflections.

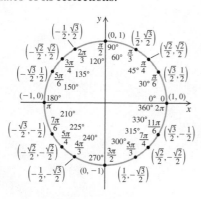

The point $\left(\frac{3}{5}, \frac{4}{5}\right)$ is on the unit circle. Find the coordinates of its reflection across **(a)** the x-axis, **(b)** the y-axis, and **(c)** the origin.

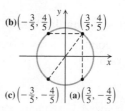

Periodic Function

A function f is said to be **periodic** if there exists a positive constant p such that

$$f(s + p) = f(s)$$

for all s in the domain of f. The smallest such positive number p is called the period of the function.

Graph the sine, the cosine, and the tangent functions. For graphs of the cosecant, the secant, and the cotangent functions, see p. 458.

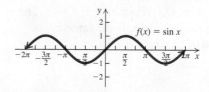

(continued)

Amplitude

The **amplitude** of a periodic function is defined as one half of the distance between its maximum and minimum function values. It is always positive.

Sine Function

1. Continuous
2. Period: 2π
3. Domain: All real numbers
4. Range: $[-1, 1]$
5. Amplitude: 1
6. Odd: $\sin(-s) = -\sin s$

Cosine Function

1. Continuous
2. Period: 2π
3. Domain: All real numbers
4. Range: $[-1, 1]$
5. Amplitude: 1
6. Even: $\cos(-s) = \cos s$

Tangent Function

1. Period: π
2. Domain: All real numbers except $(\pi/2) + k\pi$, where k is an integer
3. Range: All real numbers
4. Odd: $\tan(-s) = -\tan s$

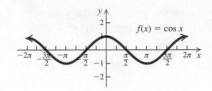

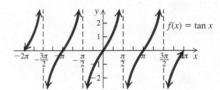

Compare the domains of the sine, the cosine, and the tangent functions.

FUNCTION	DOMAIN
sine	All real numbers
cosine	All real numbers
tangent	All real numbers except $\pi/2 + k\pi$, where k is an integer

Compare the ranges of the sine, the cosine, and the tangent functions.

FUNCTION	RANGE
sine	$[-1, 1]$
cosine	$[-1, 1]$
tangent	All real numbers

Compare the periods of the six trigonometric functions.

FUNCTION	PERIOD
sine, cosine, cosecant, secant	2π
tangent, cotangent	π

SECTION 6.6: GRAPHS OF TRANSFORMED SINE FUNCTIONS AND COSINE FUNCTIONS

Transformations of the Sine Function and the Cosine Function

To graph $y = A \sin(Bx - C) + D$ and $y = A \cos(Bx - C) + D$:

1. Stretch or shrink the graph horizontally according to B. Reflect across the y-axis if $B < 0$. $\left(\text{Period} = \left|\dfrac{2\pi}{B}\right|\right)$

Determine the amplitude, the period, and the phase shift of

$$y = -\frac{1}{2}\sin\left(2x - \frac{\pi}{2}\right) + 1$$

and sketch the graph of the function.

$$y = -\frac{1}{2}\sin\left(2x - \frac{\pi}{2}\right) + 1$$

$$= -\frac{1}{2}\sin\left[2\left(x - \frac{\pi}{4}\right)\right] + 1$$

(continued)

2. Stretch or shrink the graph vertically according to A. Reflect across the x-axis if $A < 0$. (Amplitude = $|A|$)

3. Translate the graph horizontally according to C/B.

$\left(\text{Phase shift} = \dfrac{C}{B}\right)$

4. Translate the graph vertically according to D.

Period: $\left|\dfrac{2\pi}{2}\right| = \pi$

Amplitude: $\left|-\dfrac{1}{2}\right| = \dfrac{1}{2}$

Phase shift: $\dfrac{\pi/2}{2} = \dfrac{\pi}{4}$

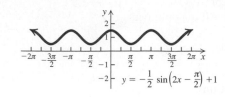

$y = -\dfrac{1}{2}\sin\left(2x - \dfrac{\pi}{2}\right) + 1$

REVIEW EXERCISES

Determine whether the statement is true or false.

1. Given that $(-a, b)$ is a point on the unit circle and θ is in the second quadrant, then $\cos\theta$ is a. **[6.4], [6.5]**

2. Given that $(-c, -d)$ is a point on the unit circle and θ is in the second quadrant, then $\tan\theta = -\dfrac{c}{d}$. **[6.4], [6.5]**

3. The measure $300°$ is greater than the measure 5 radians. **[6.4]**

4. If $\sec\theta > 0$ and $\cot\theta < 0$, then θ is in the fourth quadrant. **[6.3]**

5. The amplitude of $y = \frac{1}{2}\sin x$ is twice as large as the amplitude of $y = \sin\frac{1}{2}x$. **[6.6]**

6. The supplement of $\frac{9}{13}\pi$ is greater than the complement of $\dfrac{\pi}{6}$. **[6.4]**

7. Find the six trigonometric function values of θ. **[6.1]**

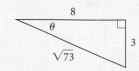

8. Given that β is acute and $\sin\beta = \dfrac{\sqrt{91}}{10}$, find the other five trigonometric function values. **[6.1]**

Find the exact function value, if it exists.

9. $\cos 45°$ **[6.1]**

10. $\cot 60°$ **[6.1]**

11. $\cos 495°$ **[6.3]**

12. $\sin 150°$ **[6.3]**

13. $\sec(-270°)$ **[6.3]**

14. $\tan(-600°)$ **[6.3]**

15. $\csc 60°$ **[6.1]**

16. $\cot(-45°)$ **[6.3]**

17. Convert $22.27°$ to degrees, minutes, and seconds. Round to the nearest second. **[6.1]**

18. Convert $47°33'27''$ to decimal degree notation. Round to two decimal places. **[6.1]**

Find the function value. Round to four decimal places. **[6.3]**

19. $\tan 2184°$

20. $\sec 27.9°$

21. $\cos 18°13'42''$

22. $\sin 245°24'$

23. $\cot(-33.2°)$

24. $\sin 556.13°$

Find θ in the interval indicated. Round the answer to the nearest tenth of a degree. **[6.3]**

25. $\cos\theta = -0.9041, \ (180°, 270°)$

26. $\tan\theta = 1.0799, \ (0°, 90°)$

Find the exact acute angle θ, in degrees, given the function value. **[6.1]**

27. $\sin\theta = \dfrac{\sqrt{3}}{2}$

28. $\tan\theta = \sqrt{3}$

29. $\cos \theta = \dfrac{\sqrt{2}}{2}$

30. $\sec \theta = \dfrac{2\sqrt{3}}{3}$

31. Given that $\sin 59.1° \approx 0.8581$, $\cos 59.1° \approx 0.5135$, and $\tan 59.1° \approx 1.6709$, find the six function values for $30.9°$. **[6.1]**

Solve each of the following right triangles. Standard lettering has been used. **[6.2]**

32. $a = 7.3$, $c = 8.6$

33. $a = 30.5$, $B = 51.17°$

34. One leg of a right triangle bears east. The hypotenuse is 734 m long and bears N57°23′E. Find the perimeter of the triangle.

35. An observer's eye is 6 ft above the floor. A mural is being viewed. The bottom of the mural is at floor level. The observer looks down 13° to see the bottom and up 17° to see the top. How tall is the mural?

For angles of the following measures, state in which quadrant the terminal side lies. **[6.3]**

36. $142°11′5″$

37. $-635.2°$

38. $-392°$

Find a positive angle and a negative angle that are coterminal with the given angle. Answers may vary.

39. $65°$ **[6.3]**

40. $\dfrac{7\pi}{3}$ **[6.4]**

Find the complement and the supplement.

41. $13.4°$ **[6.3]**

42. $\dfrac{\pi}{6}$ **[6.4]**

43. Find the six trigonometric function values for the angle θ shown. **[6.3]**

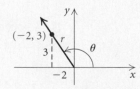

44. Given that $\tan \theta = 2/\sqrt{5}$ and that the terminal side is in quadrant III, find the other five trigonometric function values. **[6.3]**

45. An airplane travels at 530 mph for $3\frac{1}{2}$ hr in a direction of 160° from Minneapolis, Minnesota. At the end of that time, how far south of Minneapolis is the airplane? **[6.3]**

46. On a unit circle, mark and label the points determined by $7\pi/6$, $-3\pi/4$, $-\pi/3$, and $9\pi/4$. **[6.4]**

For angles of the following measures, convert to radian measure in terms of π, and convert to radian measure not in terms of π. Round the answer to two decimal places. **[6.4]**

47. $145.2°$

48. $-30°$

Convert to degree measure. Round the answer to two decimal places where appropriate. **[6.4]**

49. $\dfrac{3\pi}{2}$

50. 3

51. -4.5

52. 11π

53. Find the length of an arc of a circle, given a central angle of $\pi/4$ and a radius of 7 cm. **[6.4]**

54. An arc 18 m long on a circle of radius 8 m subtends an angle of how many radians? how many degrees, to the nearest degree? **[6.4]**

55. A waterwheel in a watermill has a radius of 7 ft and makes a complete revolution in 70 sec. What is the linear speed, in feet per minute, of a point on the rim? **[6.4]**

56. An automobile wheel has a diameter of 14 in. If the car travels at a speed of 55 mph, what is the angular velocity, in radians per hour, of a point on the edge of the wheel? **[6.4]**

57. The point $\left(\frac{3}{5}, -\frac{4}{5}\right)$ is on a unit circle. Find the coordinates of its reflections across the x-axis, the y-axis, and the origin. **[6.5]**

Find the exact function value, if it exists. **[6.5]**

58. $\cos \pi$

59. $\tan \dfrac{5\pi}{4}$

60. $\sin \dfrac{5\pi}{3}$

61. $\sin \left(-\dfrac{7\pi}{6}\right)$

62. $\tan \dfrac{\pi}{6}$

63. $\cos(-13\pi)$

Find the function value, if it exists. Round to four decimal places. **[6.5]**

64. $\sin 24$

65. $\cos(-75)$

66. $\cot 16\pi$

67. $\tan \dfrac{3\pi}{7}$

68. $\sec 14.3$

69. $\cos \left(-\dfrac{\pi}{5}\right)$

70. Graph each of the six trigonometric functions from -2π to 2π. **[6.5]**

71. What is the period of each of the six trigonometric functions? **[6.5]**

72. Complete the following table. **[6.5]**

Function	Domain	Range
sine		
cosine		
tangent		

73. Complete the following table with the sign of the specified trigonometric function value in each of the four quadrants. **[6.3]**

Function	I	II	III	IV
sine				
cosine				
tangent				

Determine the amplitude, the period, and the phase shift of the function, and sketch the graph of the function. Then check the graph using a graphing calculator. **[6.6]**

74. $y = \sin \left(x + \dfrac{\pi}{2}\right)$

75. $y = 3 + \dfrac{1}{2}\cos \left(2x - \dfrac{\pi}{2}\right)$

In Exercises 76–79, without using a graphing calculator, match the function with one of the graphs (a)–(d) that follow. Then check your work using a graphing calculator. **[6.6]**

a)

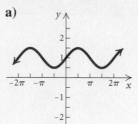

b)

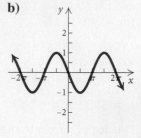

c)

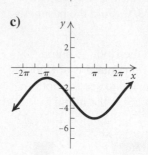

d)

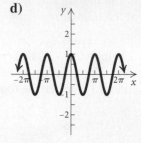

76. $y = \cos 2x$

77. $y = \dfrac{1}{2}\sin x + 1$

78. $y = -2\sin \dfrac{1}{2}x - 3$

79. $y = -\cos \left(x - \dfrac{\pi}{2}\right)$

80. Sketch a graph of $y = 3\cos x + \sin x$ for values of x between 0 and 2π. **[6.6]**

81. Graph: $f(x) = e^{-0.7x}\cos x$. **[6.6]**

82. Which of the following is the reflection of $\left(-\dfrac{1}{2}, \dfrac{\sqrt{3}}{2}\right)$ across the y-axis? **[6.5]**

A. $\left(\dfrac{1}{2}, -\dfrac{\sqrt{3}}{2}\right)$

B. $\left(\dfrac{\sqrt{3}}{2}, \dfrac{1}{2}\right)$

C. $\left(\dfrac{1}{2}, \dfrac{\sqrt{3}}{2}\right)$

D. $\left(\dfrac{\sqrt{3}}{2}, -\dfrac{1}{2}\right)$

83. Which of the following is the domain of the cosine function? **[6.5]**

A. $(-1, 1)$

B. $(-\infty, \infty)$

C. $[0, \infty)$

D. $[-1, 1]$

84. The graph of $f(x) = -\cos(-x)$ is which of the following? **[6.6]**

A.

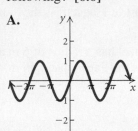

B.

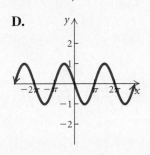

C.

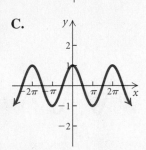

D.

Synthesis

85. Graph $y = 3 \sin(x/2)$, and determine the domain, the range, and the period. **[6.6]**

86. In the following graph, $y_1 = \sin x$ is shown in black and y_2 is shown in red. Express y_2 as a transformation of the graph of y_1. **[6.6]**

$$y_1 = \sin x, \quad y_2 = ?$$

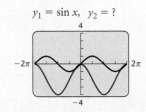

87. Find the domain of $y = \log(\cos x)$. **[6.6]**

Collaborative Discussion and Writing

88. Compare the terms radian and degree. **[6.1], [6.4]**

89. In circular motion with a fixed angular speed, the length of the radius is directly proportional to the linear speed. Explain why using an example. **[6.4]**

90. Explain why both the sine function and the cosine function are continuous, but the tangent function is not continuous. **[6.5]**

91. In the transformation steps listed in Section 6.6, why must step (1) precede step (3)? Give an example that illustrates this. **[6.6]**

92. In the equations $y = A \sin(Bx - C) + D$ and $y = A \cos(Bx - C) + D$, which constants translate the graphs and which constants stretch and shrink the graphs? Describe in your own words the effect of each constant. **[6.6]**

93. Two new cars are each driven at an average speed of 60 mph for an extended highway test drive of 2000 mi. The diameters of the wheels of the two cars are 15 in. and 16 in., respectively. If the cars use tires of equal durability and profile, differing only by the diameter, which car will probably need new tires first? Explain your answer. **[6.4]**

6 Chapter Test

1. Find the six trigonometric function values of θ.

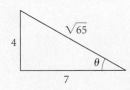

Find the exact function value, if it exists.

2. $\sin 120°$

3. $\tan(-45°)$

4. $\cos 3\pi$

5. $\sec \dfrac{5\pi}{4}$

6. Convert $38°27'56''$ to decimal degree notation. Round to two decimal places.

Find the function values. Round to four decimal places.

7. $\tan 526.4°$

8. $\sin(-12°)$

9. $\sec \dfrac{5\pi}{9}$

10. $\cos 76.07$

11. Find the exact acute angle θ, in degrees, for which $\sin \theta = \frac{1}{2}$.

12. Given that $\sin 28.4° \approx 0.4756$, $\cos 28.4° \approx 0.8796$, and $\tan 28.4° \approx 0.5407$, find the six trigonometric function values for $61.6°$.

13. Solve the right triangle with $b = 45.1$ and $A = 35.9°$. Standard lettering has been used.

14. Find a positive angle and a negative angle coterminal with a 112° angle.

15. Find the supplement of $\dfrac{5\pi}{6}$.

16. Given that $\sin \theta = -4/\sqrt{41}$ and that the terminal side is in quadrant IV, find the other five trigonometric function values.

17. Convert 210° to radian measure in terms of π.

18. Convert $\dfrac{3\pi}{4}$ to degree measure.

19. Find the length of an arc of a circle given a central angle of $\pi/3$ and a radius of 16 cm.

Consider the function $y = -\sin(x - \pi/2) + 1$ *for Exercises 20–23.*

20. Find the amplitude.

21. Find the period.

22. Find the phase shift.

23. Which of the following is the graph of the function?

A.
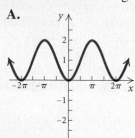

B.
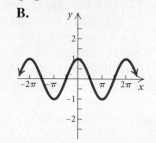

C.
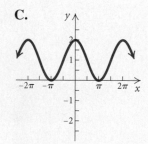

D.
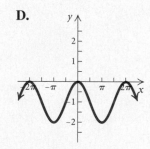

24. *Ski Dubai Resort.* Ski Dubai is the first indoor ski resort in the Middle East. Its longest ski run drops 60 ft and has an angle of depression of approximately 8.6° (*Source*: www.SkiDubai.com). Find the length of the ski run. Round the answer to the nearest foot.

25. *Location.* A motor home travels at 50 mph for 6 hr in a direction of 115° from Flagstaff, Arizona. At the end of that time, how far east of Flagstaff is the motor home?

26. *Linear Speed.* A ferris wheel has a radius of 6 m and revolves at 1.5 rpm. What is the linear speed, in meters per minute?

27. Graph: $f(x) = \frac{1}{2} x^2 \sin x$.

28. The graph of $f(x) = -\sin(-x)$ is which of the following?

A.
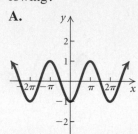

B.
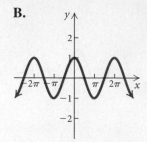

C.
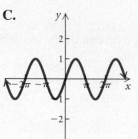

D.
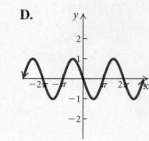

❯ Synthesis

29. Determine the domain of $f(x) = \dfrac{-3}{\sqrt{\cos x}}$.

Trigonometric Identities, Inverse Functions, and Equations

APPLICATION

This problem appears as Exercise 51 in Section 7.5.

Sales of fishing boats fluctuate in cycles. The following cosine function can be used to estimate the total amount of sales of fishing boats, y, in thousands of dollars, in month x, for a business:

$$y = 15.328 \cos(0.475x - 1.728) + 87.223.$$

Approximate the total amount of sales to the nearest dollar for November and for March. (*Hint*: 1 represents January, 2 represents February, and so on.)

7.1 ▷ Identities: Pythagorean and Sum and Difference

❯ State the Pythagorean identities.

❯ Simplify and manipulate expressions containing trigonometric expressions.

❯ Use the sum and difference identities to find function values.

An **identity** is an equation that is true for all *possible* replacements of the variables. The following is a list of the identities studied in Chapter 6.

BASIC IDENTITIES

$$\sin x = \frac{1}{\csc x}, \qquad \csc x = \frac{1}{\sin x}, \qquad \begin{aligned} \sin(-x) &= -\sin x, \\ \cos(-x) &= \cos x, \end{aligned}$$

$$\cos x = \frac{1}{\sec x}, \qquad \sec x = \frac{1}{\cos x}, \qquad \tan(-x) = -\tan x,$$

$$\tan x = \frac{1}{\cot x}, \qquad \cot x = \frac{1}{\tan x}, \qquad \tan x = \frac{\sin x}{\cos x},$$

$$\cot x = \frac{\cos x}{\sin x}$$

In this section, we will develop some other important identities.

❯ Pythagorean Identities

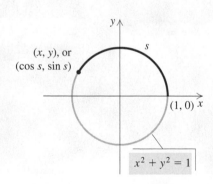

We now consider three other identities that are fundamental to a study of trigonometry. They are called the *Pythagorean identities*. Recall that the equation of a unit circle in the *xy*-plane is

$$x^2 + y^2 = 1.$$

For any point on the unit circle, the coordinates x and y satisfy this equation. Suppose that a real number s determines a point on the unit circle with coordinates (x, y), or $(\cos s, \sin s)$. Then $x = \cos s$ and $y = \sin s$. Substituting $\cos s$ for x and $\sin s$ for y in the equation of the unit circle gives us the identity

$$(\cos s)^2 + (\sin s)^2 = 1, \qquad \textbf{Substituting cos } \textbf{\textit{s}} \textbf{ for } \textbf{\textit{x}} \textbf{ and sin } \textbf{\textit{s}} \textbf{ for } \textbf{\textit{y}}$$

which can be expressed as

$$\mathbf{\sin^2 s + \cos^2 s = 1.}$$

It is conventional in trigonometry to use the notation $\sin^2 s$ rather than $(\sin s)^2$. Note that $\sin^2 s \neq \sin s^2$.

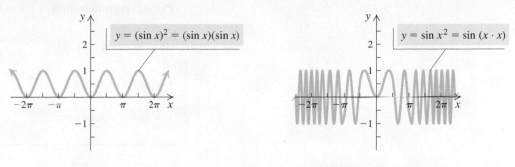

The identity $\sin^2 s + \cos^2 s = 1$ gives a relationship between the sine and the cosine of any real number s. It is an important **Pythagorean identity**.

We can divide by $\sin^2 s$ on both sides of the preceding identity:

$$\frac{\sin^2 s}{\sin^2 s} + \frac{\cos^2 s}{\sin^2 s} = \frac{1}{\sin^2 s}. \qquad \textbf{Dividing by } \sin^2 s$$

Simplifying gives us a second Pythagorean identity:

$$\mathbf{1 + \cot^2 s = \csc^2 s.}$$

This equation is true for any replacement of s with a real number for which $\sin^2 s \neq 0$, since we divided by $\sin^2 s$. But the numbers for which $\sin^2 s = 0$ (or $\sin s = 0$) are exactly those for which the cotangent function and the cosecant function are not defined. Thus our new equation holds for all real numbers s for which $\cot s$ and $\csc s$ are defined and is thus an identity.

The third Pythagorean identity can be obtained by dividing by $\cos^2 s$ on both sides of the first Pythagorean identity:

$$\frac{\sin^2 s}{\cos^2 s} + \frac{\cos^2 s}{\cos^2 s} = \frac{1}{\cos^2 s} \qquad \textbf{Dividing by } \cos^2 s$$

$$\mathbf{\tan^2 s + 1 = \sec^2 s.} \qquad \textbf{Simplifying}$$

This equation is true for any replacement of s with a real number for which $\cos^2 s \neq 0$, since we divided by $\cos^2 s$. But the numbers for which $\cos^2 s = 0$ (or $\cos s = 0$) are exactly those for which the tangent function and the secant function are not defined. Thus our new equation holds for all real numbers s for which $\tan s$ and $\sec s$ are defined and is thus an identity.

The identities that we have developed hold no matter what symbols are used for the variables. For example, we could write

$$\sin^2 s + \cos^2 s = 1, \qquad \sin^2 \theta + \cos^2 \theta = 1, \quad \text{or} \quad \sin^2 x + \cos^2 x = 1.$$

PYTHAGOREAN IDENTITIES

$\sin^2 x + \cos^2 x = 1,$

$1 + \cot^2 x = \csc^2 x,$

$1 + \tan^2 x = \sec^2 x$

It is often helpful to express the Pythagorean identities in equivalent forms.

Pythagorean Identities	Equivalent Forms
$\sin^2 x + \cos^2 x = 1$	$\sin^2 x = 1 - \cos^2 x$ $\cos^2 x = 1 - \sin^2 x$
$1 + \cot^2 x = \csc^2 x$	$1 = \csc^2 x - \cot^2 x$ $\cot^2 x = \csc^2 x - 1$
$1 + \tan^2 x = \sec^2 x$	$1 = \sec^2 x - \tan^2 x$ $\tan^2 x = \sec^2 x - 1$

❯ Simplifying Trigonometric Expressions

We can factor, simplify, and manipulate trigonometric expressions in the same way that we manipulate strictly algebraic expressions.

EXAMPLE 1 Multiply and simplify: $\cos x \, (\tan x - \sec x)$.

Solution

$$\cos x(\tan x - \sec x)$$
$$= \cos x \tan x - \cos x \sec x \qquad \text{Multiplying}$$
$$= \cos x \, \frac{\sin x}{\cos x} - \cos x \, \frac{1}{\cos x} \qquad \begin{array}{l}\text{Recalling the identities } \tan x = \sin x/\cos x \\ \text{and } \sec x = 1/\cos x \text{ and substituting}\end{array}$$
$$= \sin x - 1 \qquad \text{Simplifying}$$

> *Now Try Exercise 3.*

There is no general procedure for simplifying trigonometric expressions, but it is often helpful to write everything in terms of sines and cosines, as we did in Example 1. We also look for a Pythagorean identity within a trigonometric expression.

EXAMPLE 2 Factor and simplify: $\sin^2 x \cos^2 x + \cos^4 x$.

Solution

$$\sin^2 x \cos^2 x + \cos^4 x$$
$$= \cos^2 x \, (\sin^2 x + \cos^2 x) \qquad \text{Factoring out the common factor}$$
$$= \cos^2 x \cdot (1) \qquad \text{Using } \sin^2 x + \cos^2 x = 1$$
$$= \cos^2 x$$

> *Now Try Exercises 9 and 13.*

A graphing calculator can be used to perform a partial check of an identity. First, we graph the expression on the left side of the equals sign. Then we graph the expression on the right side using the same screen. If the two graphs are indistinguishable, then we have a partial verification that the equation is an identity. Of course, we can never see the entire graph, so there can always be some doubt. Also, the graphs may not overlap precisely, but you may not be able to tell because the difference between the graphs may be less than the width of a pixel. However, if the graphs are obviously different, we know that a mistake has been made.

Consider the identity in Example 1:

$$\cos x \, (\tan x - \sec x) = \sin x - 1.$$

Recalling that $\sec x = 1/\cos x$, we enter

$$y_1 = \cos x \, (\tan x - 1/\cos x) \quad \text{and} \quad y_2 = \sin x - 1.$$

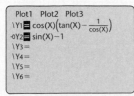
```
Plot1  Plot2  Plot3
\Y1 ■ cos(X)(tan(X)− 1/cos(X))
•0Y2 ■ sin(X)−1
\Y3=
\Y4=
\Y5=
\Y6=
```

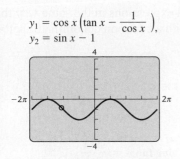

$y_1 = \cos x \left(\tan x - \dfrac{1}{\cos x} \right),$
$y_2 = \sin x - 1$

X	Y1	Y2
−6.283	−1	−1
−5.498	−.2929	−.2929
−4.712	ERROR	0
−3.927	−.2929	−.2929
−3.142	−1	−1
−2.356	−1.707	−1.707
−1.571	ERROR	−2

X = −6.28318530718

TblStart = −2π
ΔTbl = π/4

To graph, we first select SEQUENTIAL mode. Then we select the "line"-graph style for y_1 and the "path"-graph style, denoted by $-\bigcirc$, for y_2. The calculator will graph y_1 first. Then it will graph y_2 as the circular cursor traces the leading edge of the graph, allowing us to determine whether the graphs coincide. As you can see in the graph screen at left, the graphs appear to be identical. Thus, $\cos x (\tan x - \sec x) = \sin x - 1$ is most likely an identity.

The TABLE feature can also be used to check identities. Note in the table at left that the function values are the same except for those values of x for which $\cos x = 0$. The domain of y_1 excludes these values. The domain of y_2 is the set of all real numbers. Thus all real numbers except $\pm\pi/2, \pm 3\pi/2, \pm 5\pi/2, \ldots$ are possible replacements for x in the identity. Recall that an identity is an equation that is true for all *possible* replacements.

EXAMPLE 3   Simplify each of the following trigonometric expressions.

a) $\dfrac{\cot(-\theta)}{\csc(-\theta)}$

b) $\dfrac{2\sin^2 t + \sin t - 3}{1 - \cos^2 t - \sin t}$

Solution

a) $\dfrac{\cot(-\theta)}{\csc(-\theta)} = \dfrac{\dfrac{\cos(-\theta)}{\sin(-\theta)}}{\dfrac{1}{\sin(-\theta)}}$ **Rewriting in terms of sines and cosines**

$= \dfrac{\cos(-\theta)}{\sin(-\theta)} \cdot \dfrac{\sin(-\theta)}{1}$ **Multiplying by the reciprocal, $\sin(-\theta)/1$**

$= \cos(-\theta)$ **Removing a factor of 1, $\sin(-\theta)/\sin(-\theta)$**

$= \cos\theta$ **The cosine function is even.**

Recall that the sine function is odd, $\sin(-\theta) = -\sin\theta$, and the cosine function is even, $\cos(-\theta) = \cos\theta$. It can be shown that the tangent, the cotangent, and the cosecant functions are odd and the secant function is even. We can also simplify this expression using those identities:

$$\frac{\cot(-\theta)}{\csc(-\theta)} = \frac{-\cot\theta}{-\csc\theta} = \frac{\cot\theta}{\csc\theta} = \frac{\dfrac{\cos\theta}{\sin\theta}}{\dfrac{1}{\sin\theta}}$$

$$= \frac{\cos\theta}{\sin\theta} \cdot \frac{\sin\theta}{1} = \cos\theta.$$

b) $\dfrac{2\sin^2 t + \sin t - 3}{1 - \cos^2 t - \sin t}$

$= \dfrac{2\sin^2 t + \sin t - 3}{\sin^2 t - \sin t}$ **Substituting $\sin^2 t$ for $1 - \cos^2 t$**

$= \dfrac{(2\sin t + 3)(\sin t - 1)}{\sin t (\sin t - 1)}$ **Factoring in both the numerator and the denominator**

$= \dfrac{2\sin t + 3}{\sin t}$ **Simplifying**

$= \dfrac{2\sin t}{\sin t} + \dfrac{3}{\sin t}$

$= 2 + \dfrac{3}{\sin t},$ or $2 + 3\csc t$

Just in Time 21

Just in Time 14

> **Now Try Exercises 17 and 19.**

We can add and subtract trigonometric rational expressions in the same way that we do algebraic expressions, writing expressions with a common denominator before adding and subtracting numerators.

> **Just in Time**
> **23**

EXAMPLE 4 Add and simplify: $\dfrac{\cos x}{1 + \sin x} + \tan x$.

Solution

$$\dfrac{\cos x}{1 + \sin x} + \tan x = \dfrac{\cos x}{1 + \sin x} + \dfrac{\sin x}{\cos x} \qquad \text{Using } \tan x = \dfrac{\sin x}{\cos x}$$

$$= \dfrac{\cos x}{1 + \sin x} \cdot \dfrac{\cos x}{\cos x} + \dfrac{\sin x}{\cos x} \cdot \dfrac{1 + \sin x}{1 + \sin x}$$

Multiplying by forms of 1

$$= \dfrac{\cos^2 x + \sin x + \sin^2 x}{\cos x\,(1 + \sin x)} \qquad \text{Adding}$$

$$= \dfrac{1 + \sin x}{\cos x\,(1 + \sin x)} \qquad \text{Using } \sin^2 x + \cos^2 x = 1$$

$$= \dfrac{1}{\cos x}, \quad \text{or} \quad \sec x \qquad \text{Simplifying}$$

> ❯ *Now Try Exercise 27.*

When radicals occur, the use of absolute value is sometimes necessary, but it can be difficult to determine when to use it. In Examples 5 and 6, we will assume that all radicands are nonnegative. This means that the identities are meant to be confined to certain quadrants.

> **Just in Time**
> **25**

EXAMPLE 5 Multiply and simplify: $\sqrt{\sin^3 x \cos x} \cdot \sqrt{\cos x}$.

Solution

$$\sqrt{\sin^3 x \, \cos x} \cdot \sqrt{\cos x} = \sqrt{\sin^3 x \, \cos^2 x}$$

$$= \sqrt{\sin^2 x \, \cos^2 x \, \sin x}$$

$$= \sin x \, \cos x \sqrt{\sin x}$$

> ❯ *Now Try Exercise 31.*

> **Just in Time**
> **26**

EXAMPLE 6 Rationalize the denominator: $\sqrt{\dfrac{2}{\tan x}}$.

Solution

$$\sqrt{\dfrac{2}{\tan x}} = \sqrt{\dfrac{2}{\tan x} \cdot \dfrac{\tan x}{\tan x}}$$

$$= \sqrt{\dfrac{2 \tan x}{\tan^2 x}}$$

$$= \dfrac{\sqrt{2 \tan x}}{\tan x}$$

> ❯ *Now Try Exercise 37.*

Often in calculus, a substitution is a useful manipulation, as we show in the following example.

EXAMPLE 7   Express $\sqrt{9 + x^2}$ as a trigonometric function of θ without using radicals by letting $x = 3 \tan \theta$. Assume that $0 < \theta < \pi/2$. Then find $\sin \theta$ and $\cos \theta$.

Solution   We have

$$\sqrt{9 + x^2} = \sqrt{9 + (3 \tan \theta)^2} \qquad \text{Substituting } 3 \tan \theta \text{ for } x$$

$$= \sqrt{9 + 9 \tan^2 \theta}$$

$$= \sqrt{9(1 + \tan^2 \theta)} \qquad \text{Factoring}$$

$$= \sqrt{9 \sec^2 \theta} \qquad \text{Using } 1 + \tan^2 x = \sec^2 x$$

$$= 3|\sec \theta| = 3 \sec \theta. \qquad \text{For } 0 < \theta < \pi/2, \sec \theta > 0,$$
$$\text{so } |\sec \theta| = \sec \theta.$$

We can express $\sqrt{9 + x^2} = 3 \sec \theta$ as

$$\sec \theta = \frac{\sqrt{9 + x^2}}{3}.$$

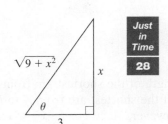

Just in Time

28

In a right triangle, we know that $\sec \theta$ is hypotenuse/adjacent, when θ is one of the acute angles. Using the Pythagorean equation, we can determine that the side opposite θ is x. Then from the right triangle, we see that

$$\sin \theta = \frac{x}{\sqrt{9 + x^2}} \quad \text{and} \quad \cos \theta = \frac{3}{\sqrt{9 + x^2}}.$$

> *Now Try Exercise 45.*

❭ Sum and Difference Identities

We now develop some important identities involving sums or differences of two numbers (or angles), beginning with an identity for the cosine of the difference of two numbers. We use the letters u and v for these numbers.

Let's consider a real number u in the interval $[\pi/2, \pi]$ and a real number v in the interval $[0, \pi/2]$. These determine points A and B on the unit circle, as shown below. The arc length s is $u - v$, and we know that $0 \le s \le \pi$. Recall that the coordinates of A are $(\cos u, \sin u)$ and the coordinates of B are $(\cos v, \sin v)$.

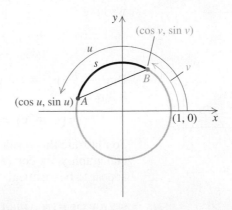

DISTANCE FORMULA

REVIEW SECTION 1.1.

Using the distance formula, we can write an expression for the distance AB:

$$AB = \sqrt{(\cos u - \cos v)^2 + (\sin u - \sin v)^2}.$$

This can be simplified as follows:

$$AB = \sqrt{\cos^2 u - 2 \cos u \cos v + \cos^2 v + \sin^2 u - 2 \sin u \sin v + \sin^2 v}$$

$$= \sqrt{(\sin^2 u + \cos^2 u) + (\sin^2 v + \cos^2 v) - 2(\cos u \cos v + \sin u \sin v)}$$

$$= \sqrt{2 - 2(\cos u \cos v + \sin u \sin v)}.$$

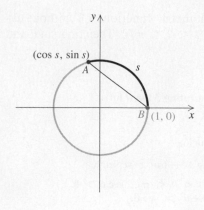

Now let's imagine rotating the circle on p. 499 so that point B is at $(1, 0)$, as shown at left. Although the coordinates of point A are now $(\cos s, \sin s)$, the distance AB has not changed.

Again we use the distance formula to write an expression for the distance AB:

$$AB = \sqrt{(\cos s - 1)^2 + (\sin s - 0)^2}.$$

This can be simplified as follows:

$$
\begin{aligned}
AB &= \sqrt{\cos^2 s - 2\cos s + 1 + \sin^2 s} \\
&= \sqrt{(\sin^2 s + \cos^2 s) + 1 - 2\cos s} \\
&= \sqrt{2 - 2\cos s}.
\end{aligned}
$$

Equating our two expressions for AB, we obtain

$$\sqrt{2 - 2(\cos u \cos v + \sin u \sin v)} = \sqrt{2 - 2\cos s}.$$

Solving this equation for $\cos s$ gives

$$\cos s = \cos u \cos v + \sin u \sin v. \qquad \textbf{(1)}$$

But $s = u - v$, so we have the equation

$$\cos(u - v) = \cos u \cos v + \sin u \sin v. \qquad \textbf{(2)}$$

Formula (1) above holds when s is the length of the shortest arc from A to B. Given any real numbers u and v, the length of the shortest arc from A to B is not always $u - v$. In fact, it could be $v - u$. However, since $\cos(-x) = \cos x$, we know that $\cos(v - u) = \cos(u - v)$. Thus, $\cos s$ is always equal to $\cos(u - v)$. Formula (2) holds for all real numbers u and v. That formula is thus the identity we sought:

$$\cos(u - v) = \cos u \cos v + \sin u \sin v.$$

To illustrate this result using a graphing calculator, we replace u with x and v with 3 and graph $y_1 = \cos(x - 3)$ and $y_2 = \cos x \cos 3 + \sin x \sin 3$. The graphs appear to be identical.

$y_1 = \cos(x - 3),$
$y_2 = \cos x \cos 3 + \sin x \sin 3$

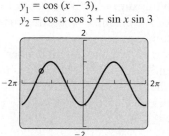

The cosine sum formula follows easily from the one we have just derived. Let's consider $\cos(u + v)$. This is equal to $\cos[u - (-v)]$, and by the identity above, we have

$$
\begin{aligned}
\cos(u + v) &= \cos[u - (-v)] \\
&= \cos u \cos(-v) + \sin u \sin(-v).
\end{aligned}
$$

But $\cos(-v) = \cos v$ and $\sin(-v) = -\sin v$, so the identity we seek is the following:

$$\cos(u + v) = \cos u \cos v - \sin u \sin v.$$

To illustrate this result using a graphing calculator, we replace u with x and v with 2 and graph $y_1 = \cos(x + 2)$ and $y_2 = \cos x \cos 2 - \sin x \sin 2$. Again, the graphs appear to be identical.

$y_1 = \cos(x + 2),$
$y_2 = \cos x \cos 2 - \sin x \sin 2$

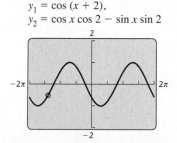

EXAMPLE 8 Find $\cos(5\pi/12)$ exactly.

Solution We can express $5\pi/12$ as a difference of two numbers whose exact sine and cosine values are known:

$$\frac{5\pi}{12} = \frac{9\pi}{12} - \frac{4\pi}{12}, \quad \text{or} \quad \frac{3\pi}{4} - \frac{\pi}{3}.$$

Then, using $\cos (u - v) = \cos u \cos v + \sin u \sin v$, we have

$$\cos \frac{5\pi}{12} = \cos \left(\frac{3\pi}{4} - \frac{\pi}{3} \right) = \cos \frac{3\pi}{4} \cos \frac{\pi}{3} + \sin \frac{3\pi}{4} \sin \frac{\pi}{3}$$

$$= -\frac{\sqrt{2}}{2} \cdot \frac{1}{2} + \frac{\sqrt{2}}{2} \cdot \frac{\sqrt{3}}{2}$$

$$= -\frac{\sqrt{2}}{4} + \frac{\sqrt{6}}{4}$$

$$= \frac{\sqrt{6} - \sqrt{2}}{4}.$$

We can check using a graphing calculator set in RADIAN mode.

> **Now Try Exercise 51.**

$$\begin{array}{l} \cos\left(\frac{5\pi}{12}\right) \\ \qquad\qquad\qquad .2588190451 \\ \frac{\sqrt{6}-\sqrt{2}}{4} \\ \qquad\qquad\qquad .2588190451 \end{array}$$

Consider $\cos (\pi/2 - \theta)$. We can use the identity for the cosine of a difference to simplify as follows:

$$\cos \left(\frac{\pi}{2} - \theta \right) = \cos \frac{\pi}{2} \cos \theta + \sin \frac{\pi}{2} \sin \theta$$

$$= 0 \cdot \cos \theta + 1 \cdot \sin \theta$$

$$= \sin \theta.$$

Thus we have developed the identity

$$\sin \theta = \cos \left(\frac{\pi}{2} - \theta \right). \qquad \begin{array}{l} \textbf{This cofunction identity first} \\ \textbf{appeared in Section 6.1.} \end{array} \qquad (3)$$

This identity holds for any real number θ. From it, we can obtain an identity for the cosine function. We first let α be any real number. Then we replace θ in $\sin \theta = \cos (\pi/2 - \theta)$ with $\pi/2 - \alpha$. This gives us

$$\sin \left(\frac{\pi}{2} - \alpha \right) = \cos \left[\frac{\pi}{2} - \left(\frac{\pi}{2} - \alpha \right) \right] = \cos \alpha,$$

which yields the identity

$$\cos \alpha = \sin \left(\frac{\pi}{2} - \alpha \right). \qquad\qquad\qquad\qquad (4)$$

Using identities (3) and (4) and the identity for the cosine of a difference, we can obtain an identity for the sine of a sum. We start with identity (3) and substitute $u + v$ for θ:

$$\sin \theta = \cos \left(\frac{\pi}{2} - \theta \right) \qquad\qquad \textbf{Identity (3)}$$

$$\sin (u + v) = \cos \left[\frac{\pi}{2} - (u + v) \right] \qquad\qquad \textbf{Substituting } u + v \textbf{ for } \theta$$

$$= \cos \left[\left(\frac{\pi}{2} - u \right) - v \right]$$

$$= \cos \left(\frac{\pi}{2} - u \right) \cos v + \sin \left(\frac{\pi}{2} - u \right) \sin v$$

Using the identity for the cosine of a difference

$$= \sin u \cos v + \cos u \sin v. \qquad\qquad \textbf{Using identities (3) and (4)}$$

Thus the identity we seek is

$$\sin (u + v) = \sin u \cos v + \cos u \sin v.$$

To find a formula for the sine of a difference, we can use the identity just derived, substituting $-v$ for v:

$$\sin (u + (-v)) = \sin u \cos (-v) + \cos u \sin (-v).$$

Simplifying gives us

$$\sin (u - v) = \sin u \cos v - \cos u \sin v.$$

EXAMPLE 9 Find sin 105° exactly.

Solution We express 105° as the sum of two measures:

$$105° = 45° + 60°.$$

Then

$$\sin 105° = \sin (45° + 60°)$$
$$= \sin 45° \cos 60° + \cos 45° \sin 60°$$

Using $\sin (u + v) = \sin u \cos v + \cos u \sin v$

$$= \frac{\sqrt{2}}{2} \cdot \frac{1}{2} + \frac{\sqrt{2}}{2} \cdot \frac{\sqrt{3}}{2}$$
$$= \frac{\sqrt{2} + \sqrt{6}}{4}.$$

We can check this result using a graphing calculator set in DEGREE mode.

```
sin(105)
              .9659258263
√2+√6
―――
 4
              .9659258263
```

> *Now Try Exercise 55.*

Formulas for the tangent of a sum or a difference can be derived using identities already established. A summary of the sum and difference identities follows.

SUM AND DIFFERENCE IDENTITIES

$$\sin (u \pm v) = \sin u \cos v \pm \cos u \sin v,$$
$$\cos (u \pm v) = \cos u \cos v \mp \sin u \sin v,$$
$$\tan (u \pm v) = \frac{\tan u \pm \tan v}{1 \mp \tan u \tan v}$$

There are six identities here, half of them obtained by using the signs shown in color.

Just in Time

24

EXAMPLE 10 Find tan 15° exactly.

Solution We rewrite 15° as 45° − 30° and use the identity for the tangent of a difference:

$$\tan 15° = \tan (45° - 30°)$$
$$= \frac{\tan 45° - \tan 30°}{1 + \tan 45° \tan 30°}$$
$$= \frac{1 - \sqrt{3}/3}{1 + 1 \cdot \sqrt{3}/3}$$
$$= \frac{3 - \sqrt{3}}{3 + \sqrt{3}}.$$

> *Now Try Exercise 53.*

EXAMPLE 11 Assume that $\sin \alpha = \frac{2}{3}$ and $\sin \beta = \frac{1}{3}$ and that α and β are between 0 and $\pi/2$. Then evaluate $\sin(\alpha + \beta)$.

Solution Using the identity for the sine of a sum, we have

$$\sin(\alpha + \beta) = \sin \alpha \cos \beta + \cos \alpha \sin \beta$$
$$= \tfrac{2}{3} \cos \beta + \tfrac{1}{3} \cos \alpha.$$

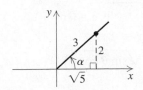

To finish, we need to know the values of $\cos \beta$ and $\cos \alpha$. Using reference triangles and the Pythagorean equation, we can determine these values from the diagrams:

$$\cos \alpha = \frac{\sqrt{5}}{3} \quad \text{and} \quad \cos \beta = \frac{2\sqrt{2}}{3}.$$

Cosine values are positive in the first quadrant.

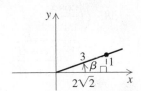

Substituting these values gives us

$$\sin(\alpha + \beta) = \frac{2}{3} \cdot \frac{2\sqrt{2}}{3} + \frac{1}{3} \cdot \frac{\sqrt{5}}{3}$$
$$= \frac{4\sqrt{2}}{9} + \frac{\sqrt{5}}{9}, \quad \text{or} \quad \frac{4\sqrt{2} + \sqrt{5}}{9}.$$

> *Now Try Exercise 65.*

EXAMPLE 12 Assume that $\cos \alpha = -\frac{4}{5}$ with α between π and $3\pi/2$ and that $\cos \beta = -\frac{2}{5}$ with β between $\pi/2$ and π. Then evaluate $\cos(\alpha - \beta)$.

Solution Using the identity for the cosine of a difference, we have

$$\cos(\alpha - \beta) = \cos \alpha \cos \beta + \sin \alpha \sin \beta$$

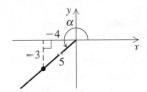

$$= \left(-\frac{4}{5}\right) \cdot \left(-\frac{2}{5}\right) + \sin \alpha \sin \beta$$
$$= \frac{8}{25} + \sin \alpha \sin \beta.$$

We need to know the values of $\sin \alpha$ and $\sin \beta$. Using reference triangles and the Pythagorean theorem, we can determine these values from the diagrams:

$$\sin \alpha = -\frac{3}{5} \quad \text{and} \quad \sin \beta = \frac{\sqrt{21}}{5}.$$

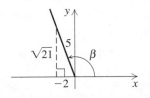

Substituting these values gives us

$$\cos(\alpha - \beta) = \frac{8}{25} + \left(-\frac{3}{5}\right) \cdot \left(\frac{\sqrt{21}}{5}\right)$$
$$= \frac{8}{25} - \frac{3\sqrt{21}}{25} = \frac{8 - 3\sqrt{21}}{25}.$$

> *Now Try Exercise 69.*

7.1   Exercise Set

Multiply and simplify. Check your result using a graphing calculator.

1. $(\sin x - \cos x)(\sin x + \cos x)$

2. $\tan x (\cos x - \csc x)$

3. $\cos y \sin y (\sec y + \csc y)$

4. $(\sin x + \cos x)(\sec x + \csc x)$

5. $(\sin \phi - \cos \phi)^2$

6. $(1 + \tan x)^2$

7. $(\sin x + \csc x)(\sin^2 x + \csc^2 x - 1)$

8. $(1 - \sin t)(1 + \sin t)$

Factor and simplify, if possible. Check your result using a graphing calculator.

9. $\sin x \cos x + \cos^2 x$

10. $\tan^2 \theta - \cot^2 \theta$

11. $\sin^4 x - \cos^4 x$

12. $4 \sin^2 y + 8 \sin y + 4$

13. $2 \cos^2 x + \cos x - 3$

14. $3 \cot^2 \beta + 6 \cot \beta + 3$

15. $\sin^3 x + 27$

16. $1 - 125 \tan^3 s$

Simplify and check using a graphing calculator.

17. $\dfrac{\sin^2 x \cos x}{\cos^2 x \sin x}$

18. $\dfrac{30 \sin^3 x \cos x}{6 \cos^2 x \sin x}$

19. $\dfrac{\sin^2 x + 2 \sin x + 1}{\sin x + 1}$

20. $\dfrac{\cos^2 \alpha - 1}{\cos \alpha + 1}$

21. $\dfrac{4 \tan t \sec t + 2 \sec t}{6 \tan t \sec t + 2 \sec t}$

22. $\dfrac{\csc (-x)}{\cot (-x)}$

23. $\dfrac{\sin^4 x - \cos^4 x}{\sin^2 x - \cos^2 x}$

24. $\dfrac{4 \cos^3 x}{\sin^2 x} \cdot \left(\dfrac{\sin x}{4 \cos x} \right)^2$

25. $\dfrac{5 \cos \phi}{\sin^2 \phi} \cdot \dfrac{\sin^2 \phi - \sin \phi \cos \phi}{\sin^2 \phi - \cos^2 \phi}$

26. $\dfrac{\tan^2 y}{\sec y} \div \dfrac{3 \tan^3 y}{\sec y}$

27. $\dfrac{1}{\sin^2 s - \cos^2 s} - \dfrac{2}{\cos s - \sin s}$

28. $\left(\dfrac{\sin x}{\cos x} \right)^2 - \dfrac{1}{\cos^2 x}$

29. $\dfrac{\sin^2 \theta - 9}{2 \cos \theta + 1} \cdot \dfrac{10 \cos \theta + 5}{3 \sin \theta + 9}$

30. $\dfrac{9 \cos^2 \alpha - 25}{2 \cos \alpha - 2} \cdot \dfrac{\cos^2 \alpha - 1}{6 \cos \alpha - 10}$

Simplify and check using a graphing calculator. Assume that all radicands are nonnegative.

31. $\sqrt{\sin^2 x \cos x} \cdot \sqrt{\cos x}$

32. $\sqrt{\cos^2 x \sin x} \cdot \sqrt{\sin x}$

33. $\sqrt{\cos \alpha \sin^2 \alpha} - \sqrt{\cos^3 \alpha}$

34. $\sqrt{\tan^2 x - 2 \tan x \sin x + \sin^2 x}$

35. $(1 - \sqrt{\sin y})(\sqrt{\sin y} + 1)$

36. $\sqrt{\cos \theta}(\sqrt{2 \cos \theta} + \sqrt{\sin \theta \cos \theta})$

Rationalize the denominator.

37. $\sqrt{\dfrac{\sin x}{\cos x}}$

38. $\sqrt{\dfrac{\cos x}{\tan x}}$

39. $\sqrt{\dfrac{\cos^2 y}{2 \sin^2 y}}$

40. $\sqrt{\dfrac{1 - \cos \beta}{1 + \cos \beta}}$

Rationalize the numerator.

41. $\sqrt{\dfrac{\cos x}{\sin x}}$

42. $\sqrt{\dfrac{\sin x}{\cot x}}$

43. $\sqrt{\dfrac{1 + \sin y}{1 - \sin y}}$

44. $\sqrt{\dfrac{\cos^2 x}{2 \sin^2 x}}$

Use the given substitution to express the given radical expression as a trigonometric function without radicals. Assume that $a > 0$ and $0 < \theta < \pi/2$. Then find expressions for the indicated trigonometric functions.

45. Let $x = a \sin \theta$ in $\sqrt{a^2 - x^2}$. Then find $\cos \theta$ and $\tan \theta$.

46. Let $x = 2 \tan \theta$ in $\sqrt{4 + x^2}$. Then find $\sin \theta$ and $\cos \theta$.

47. Let $x = 3 \sec \theta$ in $\sqrt{x^2 - 9}$. Then find $\sin \theta$ and $\cos \theta$.

48. Let $x = a \sec \theta$ in $\sqrt{x^2 - a^2}$. Then find $\sin \theta$ and $\cos \theta$.

Use the given substitution to express the given radical expression as a trigonometric function without radicals. Assume that $0 < \theta < \pi/2$.

49. Let $x = \sin \theta$ in $\dfrac{x^2}{\sqrt{1 - x^2}}$.

50. Let $x = 4 \sec \theta$ in $\dfrac{\sqrt{x^2 - 16}}{x^2}$.

Use the sum and difference identities to evaluate exactly. Then check using a graphing calculator.

51. $\sin \dfrac{\pi}{12}$

52. $\cos 75°$

53. $\tan 105°$

54. $\tan \dfrac{5\pi}{12}$

55. $\cos 15°$

56. $\sin \dfrac{7\pi}{12}$

First write each of the following as a trigonometric function of a single angle. Then evaluate.

57. $\sin 37° \cos 22° + \cos 37° \sin 22°$

58. $\cos 83° \cos 53° + \sin 83° \sin 53°$

59. $\cos 19° \cos 5° - \sin 19° \sin 5°$

60. $\sin 40° \cos 15° - \cos 40° \sin 15°$

61. $\dfrac{\tan 20° + \tan 32°}{1 - \tan 20° \tan 32°}$

62. $\dfrac{\tan 35° - \tan 12°}{1 + \tan 35° \tan 12°}$

63. Derive the formula for the tangent of a sum.

64. Derive the formula for the tangent of a difference.

Assuming that $\sin u = \frac{3}{5}$ and $\sin v = \frac{4}{5}$ and that u and v are between 0 and $\pi/2$, evaluate each of the following exactly.

65. $\cos (u + v)$

66. $\tan (u - v)$

67. $\sin (u - v)$

68. $\cos (u - v)$

Assuming that $\cos \alpha = -\frac{3}{7}$ with α between $\pi/2$ and π and that $\cos \beta = \frac{8}{9}$ with β between $3\pi/2$ and 2π, evaluate each of the following exactly.

69. $\cos (\alpha + \beta)$

70. $\sin (\alpha - \beta)$

Assuming that $\sin \theta = 0.6249$ and $\cos \phi = 0.1102$ and that both θ and ϕ are first-quadrant angles, evaluate each of the following.

71. $\tan (\theta + \phi)$

72. $\sin (\theta - \phi)$

73. $\cos (\theta - \phi)$

74. $\cos (\theta + \phi)$

Simplify.

75. $\sin (\alpha + \beta) + \sin (\alpha - \beta)$

76. $\cos (\alpha + \beta) - \cos (\alpha - \beta)$

77. $\cos (u + v) \cos v + \sin (u + v) \sin v$

78. $\sin (u - v) \cos v + \cos (u - v) \sin v$

❯ Skill Maintenance

Solve. [1.5]

79. $2x - 3 = 2\left(x - \frac{3}{2}\right)$

80. $x - 7 = x + 3.4$

Given that $\sin 31° = 0.5150$ and $\cos 31° = 0.8572$, find the specified function value. [6.1]

81. $\sec 59°$

82. $\tan 59°$

❯ Synthesis

Angles Between Lines. *One of the identities gives an easy way to find an angle formed by two lines. Consider two lines with equations l_1: $y = m_1 x + b_1$ and l_2: $y = m_2 x + b_2$.*

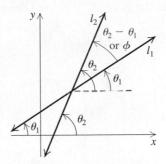

The slopes m_1 and m_2 are the tangents of the angles θ_1 and θ_2 that the lines form with the positive direction of the x-axis. Thus we have $m_1 = \tan \theta_1$ and $m_2 = \tan \theta_2$. To find the measure of $\theta_2 - \theta_1$, or ϕ, we proceed as follows:

$$\tan \phi = \tan (\theta_2 - \theta_1)$$
$$= \frac{\tan \theta_2 - \tan \theta_1}{1 + \tan \theta_2 \tan \theta_1} = \frac{m_2 - m_1}{1 + m_2 m_1}.$$

This formula also holds when the lines are taken in the reverse order. When ϕ is acute, $\tan \phi$ will be positive. When ϕ is obtuse, $\tan \phi$ will be negative.

Find the measure of the angle from l_1 to l_2.

83. l_1: $2x = 3 - 2y$,
l_2: $x + y = 5$

84. l_1: $3y = \sqrt{3}x + 3$,
l_2: $y = \sqrt{3}x + 2$

85. l_1: $y = 3$,
l_2: $x + y = 5$

86. l_1: $2x + y - 4 = 0$,
l_2: $y - 2x + 5 = 0$

87. *Circus Guy Wire.* In a circus, a guy wire A is attached to the top of a 30-ft pole. Wire B is used for performers to walk up to the tight wire, 10 ft above the ground. Find the angle ϕ between the wires if they are attached to the ground 40 ft from the pole.

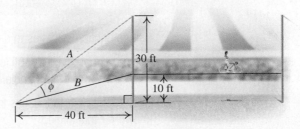

88. *Rope Course and Climbing Wall.* For a rope course and climbing wall, a guy wire R is attached 47 ft high on a vertical pole. Another guy wire S is attached 40 ft above the ground on the same pole. (*Source:* Experiential Resources, Inc., Todd Domeck, Owner) Find the angle α between the wires if they are attached to the ground 50 ft from the pole.

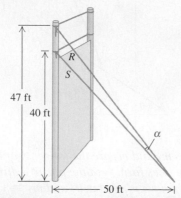

89. Given that $f(x) = \cos x$, show that

$$\frac{f(x + h) - f(x)}{h} = \cos x \left(\frac{\cos h - 1}{h} \right) - \sin x \left(\frac{\sin h}{h} \right).$$

90. Given that $f(x) = \sin x$, show that

$$\frac{f(x + h) - f(x)}{h} = \sin x \left(\frac{\cos h - 1}{h} \right) + \cos x \left(\frac{\sin h}{h} \right).$$

Show that each of the following is not an identity by finding a replacement or replacements for which the sides of the equation do not name the same number. Then use a graphing calculator to show that the equation is not an identity.

91. $\dfrac{\sin 5x}{x} = \sin 5$

92. $\sqrt{\sin^2 \theta} = \sin \theta$

93. $\cos (2\alpha) = 2 \cos \alpha$

94. $\sin (-x) = \sin x$

95. $\dfrac{\cos 6x}{\cos x} = 6$

96. $\tan^2 \theta + \cot^2 \theta = 1$

Find the slope of line l_1, where m_2 is the slope of line l_2 and ϕ is the smallest positive angle from l_1 to l_2.

97. $m_2 = \frac{2}{3}$, $\phi = 30°$

98. $m_2 = \frac{4}{3}$, $\phi = 45°$

99. Line l_1 contains the points $(-3, 7)$ and $(-3, -2)$. Line l_2 contains $(0, -4)$ and $(2, 6)$. Find the smallest positive angle from l_1 to l_2.

100. Line l_1 contains the points $(-2, 4)$ and $(5, -1)$. Find the slope of line l_2 such that the angle from l_1 to l_2 is 45°.

101. Find an identity for $\cos 2\theta$. (*Hint:* $2\theta = \theta + \theta$.)

102. Find an identity for $\sin 2\theta$. (*Hint:* $2\theta = \theta + \theta$.)

Derive the identity. Check using a graphing calculator.

103. $\tan \left(x + \dfrac{\pi}{4} \right) = \dfrac{1 + \tan x}{1 - \tan x}$

104. $\sin \left(x - \dfrac{3\pi}{2} \right) = \cos x$

105. $\sin (\alpha + \beta) + \sin (\alpha - \beta) = 2 \sin \alpha \cos \beta$

106. $\dfrac{\sin (\alpha + \beta)}{\cos (\alpha - \beta)} = \dfrac{\tan \alpha + \tan \beta}{1 + \tan \alpha \tan \beta}$

7.2 Identities: Cofunction, Double-Angle, and Half-Angle

> ❯ Use cofunction identities to derive other identities.

> ❯ Use the double-angle identities to find function values of twice an angle when one function value is known for that angle.

> ❯ Use the half-angle identities to find function values of half an angle when one function value is known for that angle.

> ❯ Simplify trigonometric expressions using the double-angle identities and the half-angle identities.

❯ Cofunction Identities

Each of the identities listed below yields a conversion to a *cofunction*. For this reason, we call them cofunction identities.

COFUNCTION IDENTITIES

$$\sin\left(\frac{\pi}{2} - x\right) = \cos x, \qquad \cos\left(\frac{\pi}{2} - x\right) = \sin x,$$

$$\tan\left(\frac{\pi}{2} - x\right) = \cot x, \qquad \cot\left(\frac{\pi}{2} - x\right) = \tan x,$$

$$\sec\left(\frac{\pi}{2} - x\right) = \csc x, \qquad \csc\left(\frac{\pi}{2} - x\right) = \sec x$$

We verified the first two of these identities in Section 7.1. The other four can be proved using the first two and the definitions of the trigonometric functions. These identities hold for all real numbers, and thus, for all angle measures, but if we restrict θ to values such that $0° < \theta < 90°$, or $0 < \theta < \pi/2$, then we have a special application to the acute angles of a right triangle.

Comparing graphs can lead to possible identities. On the left below, we see that the graph of $y = \sin(x + \pi/2)$ is a translation of the graph of $y = \sin x$ to the left $\pi/2$ units. On the right, we see the graph of $y = \cos x$.

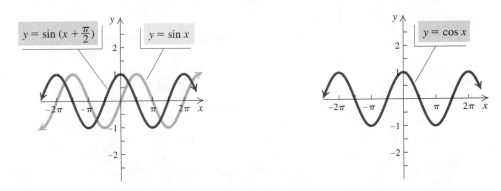

Comparing the graphs, we note a possible identity:

$$\sin\left(x + \frac{\pi}{2}\right) = \cos x.$$

The identity can be proved using the identity for the sine of a sum developed in Section 7.1.

EXAMPLE 1   Prove the identity $\sin(x + \pi/2) = \cos x$.

Solution

$$\sin\left(x + \frac{\pi}{2}\right) = \sin x \cos\frac{\pi}{2} + \cos x \sin\frac{\pi}{2} \qquad \text{Using } \sin(u + v) = \\ \sin u \cos v + \cos u \sin v$$

$$= \sin x \cdot 0 + \cos x \cdot 1$$

$$= \cos x$$

We now state four additional cofunction identities. These new identities that involve the sine and cosine functions can be verified using previously established identities as seen in Example 1.

COFUNCTION IDENTITIES FOR THE SINE AND THE COSINE

$$\sin\left(x \pm \frac{\pi}{2}\right) = \pm\cos x, \qquad \cos\left(x \pm \frac{\pi}{2}\right) = \mp\sin x$$

EXAMPLE 2   Find an identity for each of the following.

a) $\tan\left(x + \dfrac{\pi}{2}\right)$

b) $\sec(x - 90°)$

Solution

a) We have

$$\tan\left(x + \frac{\pi}{2}\right) = \frac{\sin\left(x + \dfrac{\pi}{2}\right)}{\cos\left(x + \dfrac{\pi}{2}\right)} \qquad \text{Using } \tan x = \frac{\sin x}{\cos x}$$

$$= \frac{\cos x}{-\sin x} \qquad \text{Using cofunction identities}$$

$$= -\cot x.$$

Thus the identity we seek is

$$\tan\left(x + \frac{\pi}{2}\right) = -\cot x.$$

b) We have

$$\sec(x - 90°) = \frac{1}{\cos(x - 90°)} = \frac{1}{\sin x} = \csc x.$$

Thus, $\sec(x - 90°) = \csc x$.

> **Now Try Exercises 5 and 7.**

❯ Double-Angle Identities

If we double an angle of measure x, the new angle will have measure $2x$. **Double-angle identities** give trigonometric function values of $2x$ in terms of function values of x. To develop these identities, we will use the sum formulas from the preceding section. We first develop a formula for $\sin 2x$. Recall that

$$\sin(u + v) = \sin u \cos v + \cos u \sin v.$$

We will consider a number x and substitute it for both u and v in this identity. Doing so gives us

$$\sin(x + x) = \sin 2x$$
$$= \sin x \cos x + \cos x \sin x$$
$$= 2\sin x \cos x.$$

Our first double-angle identity is thus

$$\textbf{sin } 2x = 2\,\textbf{sin } x\,\textbf{cos } x.$$

$y_1 = \sin 2x, \quad y_2 = 2 \sin x \cos x$

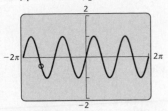

X = -1.57079632679

TblStart = -2π
ΔTbl = $\pi/4$

As a partial verification of this identity, we can graph

$$y_1 = \sin 2x \quad \text{and} \quad y_2 = 2 \sin x \cos x$$

using the "line"-graph style for y_1 and the "path"-graph style for y_2 and see that they appear to have the same graph, as shown at left. We can also use the TABLE feature.

Double-angle identities for the cosine and tangent functions can be derived in much the same way as the identity above:

$$\cos 2x = \cos^2 x - \sin^2 x, \quad \tan 2x = \frac{2 \tan x}{1 - \tan^2 x}.$$

EXAMPLE 3 Given that $\tan \theta = -\frac{3}{4}$ and θ is in quadrant II, find each of the following.

a) $\sin 2\theta$ **b)** $\cos 2\theta$

c) $\tan 2\theta$ **d)** The quadrant in which 2θ lies

Solution By drawing a reference triangle as shown, we find that

$$\sin \theta = \frac{3}{5}$$

and

$$\cos \theta = -\frac{4}{5}.$$

Thus we have the following.

a) $\sin 2\theta = 2 \sin \theta \cos \theta = 2 \cdot \frac{3}{5} \cdot \left(-\frac{4}{5}\right) = -\frac{24}{25}$

b) $\cos 2\theta = \cos^2 \theta - \sin^2 \theta = \left(-\frac{4}{5}\right)^2 - \left(\frac{3}{5}\right)^2 = \frac{16}{25} - \frac{9}{25} = \frac{7}{25}$

c) $\tan 2\theta = \dfrac{2 \tan \theta}{1 - \tan^2 \theta} = \dfrac{2 \cdot \left(-\frac{3}{4}\right)}{1 - \left(-\frac{3}{4}\right)^2} = \dfrac{-\frac{3}{2}}{1 - \frac{9}{16}} = -\frac{3}{2} \cdot \frac{16}{7} = -\frac{24}{7}$

Note that $\tan 2\theta$ could have been found more easily in this case by simply dividing:

$$\tan 2\theta = \frac{\sin 2\theta}{\cos 2\theta} = \frac{-\frac{24}{25}}{\frac{7}{25}} = -\frac{24}{7}.$$

d) Since $\sin 2\theta$ is negative and $\cos 2\theta$ is positive, we know that 2θ is in quadrant IV.

> *Now Try Exercise 9.*

Two other useful identities for $\cos 2x$ can be derived easily, as follows.

$\cos 2x = \cos^2 x - \sin^2 x$ $\cos 2x = \cos^2 x - \sin^2 x$

$\quad\quad = (1 - \sin^2 x) - \sin^2 x$ $\quad\quad = \cos^2 x - (1 - \cos^2 x)$

$\quad\quad = 1 - 2 \sin^2 x$ $\quad\quad = 2 \cos^2 x - 1$

DOUBLE-ANGLE IDENTITIES

$\sin 2x = 2 \sin x \cos x,$ $\cos 2x = \cos^2 x - \sin^2 x$

$\quad\quad\quad\quad\quad\quad\quad\quad\quad\quad\quad\quad\quad = 1 - 2 \sin^2 x$

$\tan 2x = \dfrac{2 \tan x}{1 - \tan^2 x}$ $\quad\quad = 2 \cos^2 x - 1$

Solving the last two cosine double-angle identities for $\sin^2 x$ and $\cos^2 x$, respectively, we obtain two more identities:

$$\sin^2 x = \frac{1 - \cos 2x}{2} \quad \text{and} \quad \cos^2 x = \frac{1 + \cos 2x}{2}.$$

Using division and these two identities gives us the following useful identity:

$$\tan^2 x = \frac{1 - \cos 2x}{1 + \cos 2x}.$$

EXAMPLE 4 Find an equivalent expression for each of the following.

a) $\sin 3\theta$ in terms of function values of θ

b) $\cos^3 x$ in terms of function values of x or $2x$, raised only to the first power

Solution

a) $\sin 3\theta = \sin(2\theta + \theta)$

$\qquad = \sin 2\theta \cos \theta + \cos 2\theta \sin \theta$

$\qquad = (2 \sin \theta \cos \theta) \cos \theta + (2 \cos^2 \theta - 1) \sin \theta$

$\qquad\qquad$ Using $\sin 2\theta = 2 \sin \theta \cos \theta$ and $\cos 2\theta = 2 \cos^2 \theta - 1$

$\qquad = 2 \sin \theta \cos^2 \theta + 2 \sin \theta \cos^2 \theta - \sin \theta$

$\qquad = 4 \sin \theta \cos^2 \theta - \sin \theta$

We could also substitute $\cos^2 \theta - \sin^2 \theta$ or $1 - 2 \sin^2 \theta$ for $\cos 2\theta$. Each substitution leads to a different result, but all results are equivalent.

b) $\cos^3 x = \cos^2 x \cos x$

$\qquad = \dfrac{1 + \cos 2x}{2} \cos x$

$\qquad = \dfrac{\cos x + \cos x \cos 2x}{2}$

> *Now Try Exercise 15.*

❯ Half-Angle Identities

If we take half of an angle of measure x, the new angle will have measure $x/2$. **Half-angle identities** give trigonometric function values of $x/2$ in terms of function values of x. To develop these identities, we replace x with $x/2$ and take square roots. For example,

$$\sin^2 x = \frac{1 - \cos 2x}{2} \qquad \begin{array}{l}\text{Solving the identity}\\ \cos 2x = 1 - 2 \sin^2 x \text{ for } \sin^2 x\end{array}$$

$$\sin^2 \frac{x}{2} = \frac{1 - \cos 2 \cdot \dfrac{x}{2}}{2} \qquad \text{Substituting } \frac{x}{2} \text{ for } x$$

$$\sin^2 \frac{x}{2} = \frac{1 - \cos x}{2}$$

$$\sin \frac{x}{2} = \pm\sqrt{\frac{1 - \cos x}{2}}. \qquad \text{Taking square roots}$$

The formula is called a *half-angle formula*. The use of $+$ and $-$ depends on the quadrant in which the angle $x/2$ lies. Half-angle identities for the cosine and tangent functions can be derived in a similar manner. Two additional formulas for the half-angle tangent identity are also listed on the following page.

HALF-ANGLE IDENTITIES

$$\sin \frac{x}{2} = \pm \sqrt{\frac{1 - \cos x}{2}},$$

$$\cos \frac{x}{2} = \pm \sqrt{\frac{1 + \cos x}{2}},$$

$$\tan \frac{x}{2} = \pm \sqrt{\frac{1 - \cos x}{1 + \cos x}}$$

$$= \frac{\sin x}{1 + \cos x} = \frac{1 - \cos x}{\sin x}$$

EXAMPLE 5 Find $\tan(\pi/8)$ exactly. Then check the answer using a graphing calculator in RADIAN mode.

Solution We have

$$\tan \frac{\pi}{8} = \tan \frac{\frac{\pi}{4}}{2} = \frac{\sin \frac{\pi}{4}}{1 + \cos \frac{\pi}{4}} = \frac{\frac{\sqrt{2}}{2}}{1 + \frac{\sqrt{2}}{2}} = \frac{\frac{\sqrt{2}}{2}}{\frac{2 + \sqrt{2}}{2}} = \frac{\sqrt{2}}{2 + \sqrt{2}}$$

$$= \frac{\sqrt{2}}{2 + \sqrt{2}} \cdot \frac{2 - \sqrt{2}}{2 - \sqrt{2}} = \frac{2\sqrt{2} - 2}{4 - 2} = \frac{2(\sqrt{2} - 1)}{2} = \sqrt{2} - 1.$$

> **Now Try Exercise 21.**

The identities that we have developed are also useful for simplifying trigonometric expressions.

EXAMPLE 6 Simplify each of the following.

a) $\dfrac{\sin x \cos x}{\frac{1}{2}\cos 2x}$ **b)** $2 \sin^2 \dfrac{x}{2} + \cos x$

Solution

a) We can obtain $2 \sin x \cos x$ in the numerator by multiplying the expression by $\frac{2}{2}$:

$$\frac{\sin x \cos x}{\frac{1}{2}\cos 2x} = \frac{2}{2} \cdot \frac{\sin x \cos x}{\frac{1}{2}\cos 2x} = \frac{2 \sin x \cos x}{\cos 2x}$$

$$= \frac{\sin 2x}{\cos 2x} \quad \text{Using } \sin 2x = 2 \sin x \cos x$$

$$= \tan 2x.$$

$y_1 = 2 \sin^2 \dfrac{x}{2} + \cos x, \quad y_2 = 1$

b) We have

$$2 \sin^2 \frac{x}{2} + \cos x = 2\left(\frac{1 - \cos x}{2}\right) + \cos x$$

$$\text{Using } \sin \frac{x}{2} = \pm \sqrt{\frac{1 - \cos x}{2}}, \text{ or } \sin^2 \frac{x}{2} = \frac{1 - \cos x}{2}$$

$$= 1 - \cos x + \cos x$$

$$= 1.$$

X	Y1	Y2
−6.283	1	1
−5.498	1	1
−4.712	1	1
−3.927	1	1
−3.142	1	1
−2.356	1	1
−1.571	1	1
X = −6.28318530718		

TblStart = −2π
ΔTbl = π/4

We can check this result using a graph or a table, as shown at left.

> **Now Try Exercise 31.**

7.2   Exercise Set

1. Given that $\sin(3\pi/10) \approx 0.8090$ and $\cos(3\pi/10) \approx 0.5878$, find each of the following.

   a) The other four function values for $3\pi/10$

   b) The six function values for $\pi/5$

2. Given that

$$\sin\frac{\pi}{12} = \frac{\sqrt{2-\sqrt{3}}}{2} \quad \text{and} \quad \cos\frac{\pi}{12} = \frac{\sqrt{2+\sqrt{3}}}{2},$$

find exact answers for each of the following.

   a) The other four function values for $\pi/12$

   b) The six function values for $5\pi/12$

3. Given that $\sin\theta = \frac{1}{3}$ and that the terminal side is in quadrant II, find exact answers for each of the following.

   a) The other function values for θ

   b) The six function values for $\pi/2 - \theta$

   c) The six function values for $\theta - \pi/2$

4. Given that $\cos\phi = \frac{4}{5}$ and that the terminal side is in quadrant IV, find exact answers for each of the following.

   a) The other function values for ϕ

   b) The six function values for $\pi/2 - \phi$

   c) The six function values for $\phi + \pi/2$

Find an equivalent expression for each of the following.

5. $\sec\left(x + \dfrac{\pi}{2}\right)$ **6.** $\cot\left(x - \dfrac{\pi}{2}\right)$

7. $\tan\left(x - \dfrac{\pi}{2}\right)$ **8.** $\csc\left(x + \dfrac{\pi}{2}\right)$

Find the exact value of $\sin 2\theta$, $\cos 2\theta$, $\tan 2\theta$, and the quadrant in which 2θ lies.

9. $\sin\theta = \frac{4}{5}$, θ in quadrant I

10. $\cos\theta = \frac{5}{13}$, θ in quadrant I

11. $\cos\theta = -\frac{3}{5}$, θ in quadrant III

12. $\tan\theta = -\frac{15}{8}$, θ in quadrant II

13. $\tan\theta = -\frac{5}{12}$, θ in quadrant II

14. $\sin\theta = -\dfrac{\sqrt{10}}{10}$, θ in quadrant IV

15. Find an equivalent expression for $\cos 4x$ in terms of function values of x.

16. Find an equivalent expression for $\sin^4\theta$ in terms of function values of θ, 2θ, or 4θ, raised only to the first power.

Use the half-angle identities to evaluate exactly.

17. $\cos 15°$ **18.** $\tan 67.5°$

19. $\sin 112.5°$ **20.** $\cos\dfrac{\pi}{8}$

21. $\tan 75°$ **22.** $\sin\dfrac{5\pi}{12}$

Given that $\sin\theta = 0.3416$ and θ is in quadrant I, find each of the following using identities.

23. $\sin 2\theta$ **24.** $\cos\dfrac{\theta}{2}$

25. $\sin\dfrac{\theta}{2}$ **26.** $\sin 4\theta$

In Exercises 27–30, use a graphing calculator to determine which of the following expressions asserts an identity. Then derive the identity algebraically.

27. $\dfrac{\cos 2x}{\cos x - \sin x} = \cdots$

   a) $1 + \cos x$ **b)** $\cos x - \sin x$

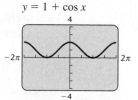

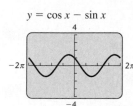

   c) $-\cot x$ **d)** $\sin x (\cot x + 1)$

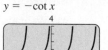

28. $2\cos^2\dfrac{x}{2} = \cdots$

   a) $\sin x (\csc x + \tan x)$ **b)** $\sin x - 2\cos x$

   c) $2(\cos^2 x - \sin^2 x)$ **d)** $1 + \cos x$

29. $\dfrac{\sin 2x}{2\cos x} = \cdots$

   a) $\cos x$ **b)** $\tan x$

   c) $\cos x + \sin x$ **d)** $\sin x$

30. $2 \sin \dfrac{\theta}{2} \cos \dfrac{\theta}{2} = \cdots$

a) $\cos^2 \theta$

b) $\sin \dfrac{\theta}{2}$

c) $\sin \theta$

d) $\sin \theta - \cos \theta$

Simplify. Check your results using a graphing calculator.

31. $2 \cos^2 \dfrac{x}{2} - 1$

32. $\cos^4 x - \sin^4 x$

33. $\dfrac{2 - \sec^2 x}{\sec^2 x}$

34. $(\sin x + \cos x)^2$

35. $(\sin x - \cos x)^2 + \sin 2x$

36. $\dfrac{1 + \sin 2x + \cos 2x}{1 + \sin 2x - \cos 2x}$

37. $(-4 \cos x \sin x + 2 \cos 2x)^2 +$ $(2 \cos 2x + 4 \sin x \cos x)^2$

38. $2 \sin x \cos^3 x - 2 \sin^3 x \cos x$

> ### Skill Maintenance

Complete the identity. [7.1]

39. $1 - \cos^2 x =$

40. $\sec^2 x - \tan^2 x =$

41. $\sin^2 x - 1 =$

42. $1 + \cot^2 x =$

43. $\csc^2 x - \cot^2 x$

44. $1 + \tan^2 x =$

45. $1 - \sin^2 x$

46. $\sec^2 x - 1$

Consider the following functions (a)–(f). Without graphing them, answer questions 47–50 below. [6.5]

a) $f(x) = 2 \sin \left(\dfrac{1}{2} x - \dfrac{\pi}{2} \right)$

b) $f(x) = \dfrac{1}{2} \cos \left(2x - \dfrac{\pi}{4} \right) + 2$

c) $f(x) = -\sin \left[2 \left(x - \dfrac{\pi}{2} \right) \right] + 2$

d) $f(x) = \sin (x + \pi) - \dfrac{1}{2}$

e) $f(x) = -2 \cos (4x - \pi)$

f) $f(x) = -\cos \left[2 \left(x - \dfrac{\pi}{8} \right) \right]$

47. Which functions have a graph with an amplitude of 2?

48. Which functions have a graph with a period of π?

49. Which functions have a graph with a period of 2π?

50. Which functions have a graph with a phase shift of $\dfrac{\pi}{4}$?

> ### Synthesis

51. Given that $\cos 51° \approx 0.6293$, find the six function values for $141°$.

52. Find $\sin 15°$ first using a difference identity and then using a half-angle identity. Then compare the results.

Simplify. Check your results using a graphing calculator.

53. $\cos (\pi - x) + \cot x \sin \left(x - \dfrac{\pi}{2} \right)$

54. $\sin \left(\dfrac{\pi}{2} - x \right) [\sec x - \cos x]$

55. $\dfrac{\cos^2 y \sin \left(y + \dfrac{\pi}{2} \right)}{\sin^2 y \sin \left(\dfrac{\pi}{2} - y \right)}$

56. $\dfrac{\cos x - \sin \left(\dfrac{\pi}{2} - x \right) \sin x}{\cos x - \cos (\pi - x) \tan x}$

Find $\sin \theta$, $\cos \theta$, and $\tan \theta$ under the given conditions.

57. $\tan \dfrac{\theta}{2} = -\dfrac{5}{3}, \ \pi < \theta \le \dfrac{3\pi}{2}$

58. $\cos 2\theta = \dfrac{7}{12}, \ \dfrac{3\pi}{2} \le 2\theta \le 2\pi$

59. *Acceleration Due to Gravity.* The acceleration due to gravity is often denoted by g in a formula such as $S = \frac{1}{2} g t^2$, where S is the distance that an object falls in time t. The number g relates to motion near the earth's surface and is generally considered constant. In fact, however, g is not constant, but varies slightly with latitude. *Latitude* is used to measure north–south location on the earth between the equator and the poles. If ϕ stands for latitude, in degrees, g is given with good approximation by the formula

$g = 9.78049(1 + 0.005288 \sin^2 \phi - 0.000006 \sin^2 2\phi)$,

where g is measured in meters per second per second at sea level.

a) Chicago has latitude $42°$N. Find g.

b) Philadelphia has latitude $40°$N. Find g.

c) Express g in terms of $\sin \phi$ only. That is, eliminate the double angle.

60. *Nautical Mile.* Sydney, Australia, has latitude $34°$S. (See the figure.) In Great Britain, the *nautical mile* is defined as the length of a minute of arc of the earth's radius. Since the earth is flattened slightly at the poles, a British nautical mile varies with latitude. In fact, it is given, in feet, by the function

$$N(\phi) = 6066 - 31\cos 2\phi,$$

where ϕ is the latitude in degrees.

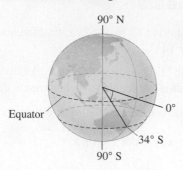

a) What is the length of a British nautical mile at Sydney?

b) What is the length of a British nautical mile at the North Pole?

c) Express $N(\phi)$ in terms of $\cos\phi$ only; that is, do not use the double angle.

7.3 ▶ Proving Trigonometric Identities

> ❯ Prove identities using other identities.

> ❯ Use the product-to-sum identities and the sum-to-product identities to derive other identities.

❯ The Logic of Proving Identities

We outline two algebraic methods for proving identities.

Method 1. Start with either the left side or the right side of the equation and obtain the other side. For example, suppose that you are trying to prove that the equation $P = Q$ is an identity. You might try to produce a string of statements $(R_1, R_2, \ldots$ or $T_1, T_2, \ldots)$ like the following, which start with P and end with Q or start with Q and end with P:

$$
\begin{aligned}
P &= R_1 \quad \text{or} \quad & Q &= T_1 \\
&= R_2 & &= T_2 \\
&\;\;\vdots & &\;\;\vdots \\
&= Q & &= P.
\end{aligned}
$$

Method 2. Work with each side separately until you obtain the same expression. For example, suppose that you are trying to prove that $P = Q$ is an identity. You might be able to produce two strings of statements like the following, each ending with the same statement S.

$$
\begin{aligned}
P &= R_1 & Q &= T_1 \\
&= R_2 & &= T_2 \\
&\;\;\vdots & &\;\;\vdots \\
&= S & &= S.
\end{aligned}
$$

The number of steps in each string might be different, but in each case the result is S.

A first step in learning to prove identities is to have at hand a list of the identities that you have already learned. Such a list is on the inside back cover of this text. Ask your instructor which ones you are expected to memorize. The more identities you prove, the easier it will be to prove new ones. A list of helpful hints is shown at left.

❯ Proving Identities

EXAMPLE 1 Prove the identity $1 + \sin 2\theta = (\sin \theta + \cos \theta)^2$.

Solution Let's use method 1. We begin with the right side and obtain the left side:

$$(\sin \theta + \cos \theta)^2 = \sin^2\theta + 2 \sin \theta \cos \theta + \cos^2\theta \qquad \text{Squaring}$$
$$= 1 + 2 \sin \theta \cos \theta \qquad \text{Recalling the identity } \sin^2 x + \cos^2 x = 1 \text{ and substituting}$$
$$= 1 + \sin 2\theta. \qquad \text{Using } \sin 2x = 2 \sin x \cos x$$

We could also begin with the left side and obtain the right side:

$$1 + \sin 2\theta = 1 + 2 \sin \theta \cos \theta \qquad \text{Using } \sin 2x = 2 \sin x \cos x$$
$$= \sin^2\theta + 2 \sin \theta \cos \theta + \cos^2\theta \qquad \text{Replacing 1 with } \sin^2\theta + \cos^2\theta$$
$$= (\sin \theta + \cos \theta)^2. \qquad \text{Factoring}$$

❯ *Now Try Exercises 13 and 19.*

EXAMPLE 2 Prove the identity

$$\sin^2 x \tan^2 x = \tan^2 x - \sin^2 x.$$

Solution For this proof, we are going to use method 2, working with each side separately using method 2. We try to obtain the same expression on each side. In actual practice, you might work on one side for a while, then work on the other side, and then go back to the first side. In other words, you work back and forth until you arrive at the same expression. Let's start with the right side.

$$\tan^2 x - \sin^2 x = \frac{\sin^2 x}{\cos^2 x} - \sin^2 x \qquad \text{Recalling the identity } \tan x = \frac{\sin x}{\cos x} \text{ and substituting}$$

$$= \frac{\sin^2 x}{\cos^2 x} - \sin^2 x \cdot \frac{\cos^2 x}{\cos^2 x} \qquad \text{Multiplying by 1 in order to subtract}$$

$$= \frac{\sin^2 x - \sin^2 x \cos^2 x}{\cos^2 x} \qquad \text{Carrying out the subtraction}$$

$$= \frac{\sin^2 x (1 - \cos^2 x)}{\cos^2 x} \qquad \text{Factoring}$$

$$= \frac{\sin^2 x \sin^2 x}{\cos^2 x} \qquad \text{Recalling the identity } 1 - \cos^2 x = \sin^2 x \text{ and substituting}$$

$$= \frac{\sin^4 x}{\cos^2 x}$$

At this point, we stop and work with the left side of the original identity, $\sin^2 x \tan^2 x$, and try to end with the same expression that we ended with on the right side:

$$\sin^2 x \tan^2 x = \sin^2 x \frac{\sin^2 x}{\cos^2 x} \qquad \text{Recalling the identity } \tan x = \frac{\sin x}{\cos x} \text{ and substituting}$$

$$= \frac{\sin^4 x}{\cos^2 x}.$$

We have obtained the same expression from each side, so the proof is complete.

❯ *Now Try Exercise 25.*

HINTS FOR PROVING IDENTITIES

1. Use either method 1 or method 2.
2. Work with the more complex side first.
3. Carry out any algebraic manipulations, such as adding, subtracting, multiplying, or factoring.
4. When rational expressions are involved, it can be helpful to multiply by forms of 1 to obtain a common denominator.
5. Converting all expressions to sines and cosines is often helpful.
6. Try something! Put your pencil to work and get involved. You will be amazed at how often this leads to success.

EXAMPLE 3 Prove the identity

$$\frac{\sin 2x}{\sin x} - \frac{\cos 2x}{\cos x} = \sec x.$$

Solution We will use method 1, starting with the left side since it is the more complex side of the identity.

$$\frac{\sin 2x}{\sin x} - \frac{\cos 2x}{\cos x} = \frac{2 \sin x \cos x}{\sin x} - \frac{\cos^2 x - \sin^2 x}{\cos x} \qquad \text{Using double-angle identities}$$

$$= 2 \cos x - \frac{\cos^2 x - \sin^2 x}{\cos x} \qquad \text{Simplifying}$$

$$= \frac{2 \cos^2 x}{\cos x} - \frac{\cos^2 x - \sin^2 x}{\cos x} \qquad \begin{array}{l}\text{Multiplying } 2 \cos x \\ \text{by 1, or } \cos x / \cos x\end{array}$$

$$= \frac{2 \cos^2 x - \cos^2 x + \sin^2 x}{\cos x} \qquad \text{Subtracting}$$

$$= \frac{\cos^2 x + \sin^2 x}{\cos x}$$

$$= \frac{1}{\cos x} \qquad \text{Using } \cos^2 x + \sin^2 x = 1$$

$$= \sec x \qquad \text{Recalling a basic identity}$$

> *Now Try Exercise 15.*

EXAMPLE 4 Prove the identity

$$\frac{\sec t - 1}{t \sec t} = \frac{1 - \cos t}{t}.$$

Solution We use method 1, starting with the left side. Note that the left side involves sec t, whereas the right side involves cos t, so it might be wise to make use of a basic identity that involves these two expressions: sec $t = 1/\cos t$.

$$\frac{\sec t - 1}{t \sec t} = \frac{\dfrac{1}{\cos t} - 1}{t \cdot \dfrac{1}{\cos t}} \qquad \text{Substituting } 1/\cos t \text{ for sec } t$$

$$= \left(\frac{1}{\cos t} - 1\right) \cdot \frac{\cos t}{t}$$

$$= \frac{1}{t} - \frac{\cos t}{t} \qquad \text{Multiplying}$$

$$= \frac{1 - \cos t}{t}$$

We started with the left side and obtained the right side, so the proof is complete.

> *Now Try Exercise 5.*

EXAMPLE 5   Prove the identity

$$\cot \phi + \csc \phi = \frac{\sin \phi}{1 - \cos \phi}.$$

Solution   We are again using method 2, beginning with the left side:

$$\cot \phi + \csc \phi = \frac{\cos \phi}{\sin \phi} + \frac{1}{\sin \phi} \qquad \textbf{Using basic identities}$$

$$= \frac{1 + \cos \phi}{\sin \phi}. \qquad \textbf{Adding}$$

At this point, we stop and work with the right side of the original identity:

$$\frac{\sin \phi}{1 - \cos \phi} = \frac{\sin \phi}{1 - \cos \phi} \cdot \frac{1 + \cos \phi}{1 + \cos \phi} \qquad \textbf{Multiplying by 1}$$

$$= \frac{\sin \phi \, (1 + \cos \phi)}{1 - \cos^2 \phi}$$

$$= \frac{\sin \phi \, (1 + \cos \phi)}{\sin^2 \phi} \qquad \textbf{Using } \sin^2 x = 1 - \cos^2 x$$

$$= \frac{1 + \cos \phi}{\sin \phi}. \qquad \textbf{Simplifying}$$

The proof is complete since we obtained the same expression from each side.

> *Now Try Exercise 29.*

❯ Product-to-Sum and Sum-to-Product Identities

On occasion, it is convenient to convert a product of trigonometric expressions to a sum, or the reverse. The following identities are useful in this connection.

PRODUCT-TO-SUM IDENTITIES

$$\sin x \cdot \sin y = \tfrac{1}{2}[\cos (x - y) - \cos (x + y)] \qquad \textbf{(1)}$$

$$\cos x \cdot \cos y = \tfrac{1}{2}[\cos (x - y) + \cos (x + y)] \qquad \textbf{(2)}$$

$$\sin x \cdot \cos y = \tfrac{1}{2}[\sin (x + y) + \sin (x - y)] \qquad \textbf{(3)}$$

$$\cos x \cdot \sin y = \tfrac{1}{2}[\sin (x + y) - \sin (x - y)] \qquad \textbf{(4)}$$

We can derive product-to-sum identities (1) and (2) using the sum and difference identities for the cosine function:

$$\cos (x + y) = \cos x \cos y - \sin x \sin y, \qquad \textbf{Sum identity}$$

$$\cos (x - y) = \cos x \cos y + \sin x \sin y. \qquad \textbf{Difference identity}$$

Subtracting the sum identity from the difference identity, we have

$$\cos (x - y) - \cos (x + y) = 2 \sin x \sin y \qquad \textbf{Subtracting}$$

$$\tfrac{1}{2}[\cos (x - y) - \cos (x + y)] = \sin x \sin y. \qquad \textbf{Multiplying by } \tfrac{1}{2}$$

Thus, $\sin x \sin y = \tfrac{1}{2}[\cos (x - y) - \cos (x + y)]$.

Adding the cosine sum and difference identities, we have

$$\cos (x - y) + \cos (x + y) = 2 \cos x \cos y \qquad \text{Adding}$$

$$\tfrac{1}{2}[\cos (x - y) + \cos (x + y)] = \cos x \cos y. \qquad \text{Multiplying by } \tfrac{1}{2}$$

Thus, $\cos x \cos y = \tfrac{1}{2}[\cos (x - y) + \cos (x + y)]$.

Identities (3) and (4) can be derived in a similar manner using the sum and difference identities for the sine function.

EXAMPLE 6 Find an identity for $2 \sin 3\theta \cos 7\theta$.

Solution We will use the identity

$$\sin x \cdot \cos y = \tfrac{1}{2}[\sin (x + y) + \sin (x - y)].$$

Here $x = 3\theta$ and $y = 7\theta$. Thus,

$$2 \sin 3\theta \cos 7\theta = 2 \cdot \tfrac{1}{2}[\sin (3\theta + 7\theta) + \sin (3\theta - 7\theta)]$$

$$= \sin 10\theta + \sin (-4\theta)$$

$$= \sin 10\theta - \sin 4\theta. \qquad \text{Using } \sin (-x) = -\sin x$$

> *Now Try Exercise 37.*

SUM-TO-PRODUCT IDENTITIES

$$\sin x + \sin y = 2 \sin \frac{x + y}{2} \cos \frac{x - y}{2} \qquad\qquad (5)$$

$$\sin x - \sin y = 2 \cos \frac{x + y}{2} \sin \frac{x - y}{2} \qquad\qquad (6)$$

$$\cos y + \cos x = 2 \cos \frac{x + y}{2} \cos \frac{x - y}{2} \qquad\qquad (7)$$

$$\cos y - \cos x = 2 \sin \frac{x + y}{2} \sin \frac{x - y}{2} \qquad\qquad (8)$$

The sum-to-product identities (5)–(8) can be derived using the product-to-sum identities. Proofs are left to the exercises.

EXAMPLE 7 Find an identity for $\cos \theta + \cos 5\theta$.

Solution We will use the identity

$$\cos y + \cos x = 2 \cos \frac{x + y}{2} \cos \frac{x - y}{2}.$$

Here $x = 5\theta$ and $y = \theta$. Thus,

$$\cos \theta + \cos 5\theta = 2 \cos \frac{5\theta + \theta}{2} \cos \frac{5\theta - \theta}{2}$$

$$= 2 \cos 3\theta \cos 2\theta.$$

> *Now Try Exercise 35.*

7.3   Exercise Set

Prove the identity.

1. $\sec x - \sin x \tan x = \cos x$

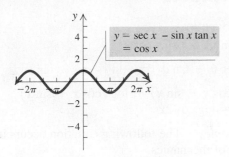

$y = \sec x - \sin x \tan x$
$= \cos x$

2. $\dfrac{1 + \cos \theta}{\sin \theta} + \dfrac{\sin \theta}{\cos \theta} = \dfrac{\cos \theta + 1}{\sin \theta \cos \theta}$

3. $\dfrac{1 - \cos x}{\sin x} = \dfrac{\sin x}{1 + \cos x}$

4. $\dfrac{1 + \tan y}{1 + \cot y} = \dfrac{\sec y}{\csc y}$

5. $\dfrac{1 + \tan \theta}{1 - \tan \theta} + \dfrac{1 + \cot \theta}{1 - \cot \theta} = 0$

6. $\dfrac{\sin x + \cos x}{\sec x + \csc x} = \dfrac{\sin x}{\sec x}$

7. $\dfrac{\cos^2 \alpha + \cot \alpha}{\cos^2 \alpha - \cot \alpha} = \dfrac{\cos^2 \alpha \tan \alpha + 1}{\cos^2 \alpha \tan \alpha - 1}$

8. $\sec 2\theta = \dfrac{\sec^2 \theta}{2 - \sec^2 \theta}$

9. $\dfrac{2 \tan \theta}{1 + \tan^2 \theta} = \sin 2\theta$

10. $\dfrac{\cos (u - v)}{\cos u \sin v} = \tan u + \cot v$

11. $1 - \cos 5\theta \cos 3\theta - \sin 5\theta \sin 3\theta = 2 \sin^2 \theta$

12. $\cos^4 x - \sin^4 x = \cos 2x$

13. $2 \sin \theta \cos^3 \theta + 2 \sin^3 \theta \cos \theta = \sin 2\theta$

14. $\dfrac{\tan 3t - \tan t}{1 + \tan 3t \ \tan t} = \dfrac{2 \tan t}{1 - \tan^2 t}$

15. $\dfrac{\tan x - \sin x}{2 \tan x} = \sin^2 \dfrac{x}{2}$

16. $\dfrac{\cos^3 \beta - \sin^3 \beta}{\cos \beta - \sin \beta} = \dfrac{2 + \sin 2\beta}{2}$

17. $\sin (\alpha + \beta) \sin (\alpha - \beta) = \sin^2 \alpha - \sin^2 \beta$

18. $\cos^2 x (1 - \sec^2 x) = -\sin^2 x$

19. $\tan \theta (\tan \theta + \cot \theta) = \sec^2 \theta$

20. $\dfrac{\cos \theta + \sin \theta}{\cos \theta} = 1 + \tan \theta$

21. $\dfrac{1 + \cos^2 x}{\sin^2 x} = 2 \csc^2 x - 1$

22. $\dfrac{\tan y + \cot y}{\csc y} = \sec y$

23. $\dfrac{1 + \sin x}{1 - \sin x} + \dfrac{\sin x - 1}{1 + \sin x} = 4 \sec x \tan x$

24. $\tan \theta - \cot \theta = (\sec \theta - \csc \theta)(\sin \theta + \cos \theta)$

25. $\cos^2 \alpha \cot^2 \alpha = \cot^2 \alpha - \cos^2 \alpha$

26. $\dfrac{\tan x + \cot x}{\sec x + \csc x} = \dfrac{1}{\cos x + \sin x}$

27. $2 \sin^2 \theta \cos^2 \theta + \cos^4 \theta = 1 - \sin^4 \theta$

28. $\dfrac{\cot \theta}{\csc \theta - 1} = \dfrac{\csc \theta + 1}{\cot \theta}$

29. $\dfrac{1 + \sin x}{1 - \sin x} = (\sec x + \tan x)^2$

30. $\sec^4 s - \tan^2 s = \tan^4 s + \sec^2 s$

31. Derive the product-to-sum identities (3) and (4) using the sine sum and difference identities.

32. Derive the sum-to-product identities (5)–(8) using the product-to-sum identities (1)–(4).

Use the product-to-sum identities and the sum-to-product identities to find identities for each of the following.

33. $\sin 3\theta - \sin 5\theta$ **34.** $\sin 7x - \sin 4x$

35. $\sin 8\theta + \sin 5\theta$ **36.** $\cos \theta - \cos 7\theta$

37. $\sin 7u \sin 5u$ **38.** $2 \sin 7\theta \cos 3\theta$

39. $7 \cos \theta \sin 7\theta$ **40.** $\cos 2t \sin t$

41. $\cos 55° \sin 25°$ **42.** $7 \cos 5\theta \cos 7\theta$

Use the product-to-sum identities and the sum-to-product identities to prove each of the following.

43. $\sin 4\theta + \sin 6\theta = \cot \theta (\cos 4\theta - \cos 6\theta)$

44. $\tan 2x (\cos x + \cos 3x) = \sin x + \sin 3x$

45. $\cot 4x (\sin x + \sin 4x + \sin 7x)$
$= \cos x + \cos 4x + \cos 7x$

46. $\tan \dfrac{x+y}{2} = \dfrac{\sin x + \sin y}{\cos x + \cos y}$

47. $\cot \dfrac{x+y}{2} = \dfrac{\sin y - \sin x}{\cos x - \cos y}$

48. $\tan \dfrac{\theta+\phi}{2} \tan \dfrac{\phi-\theta}{2} = \dfrac{\cos \theta - \cos \phi}{\cos \theta + \cos \phi}$

49. $\tan \dfrac{\theta+\phi}{2} (\sin \theta - \sin \phi)$

$\qquad = \tan \dfrac{\theta-\phi}{2} (\sin \theta + \sin \phi)$

50. $\sin 2\theta + \sin 4\theta + \sin 6\theta = 4\cos \theta \cos 2\theta \sin 3\theta$

In Exercises 51–56, use a graphing calculator to determine which expression (A)–(F) on the right can be used to complete the identity. Then try to prove that identity algebraically.

51. $\dfrac{\cos x + \cot x}{1 + \csc x}$

52. $\cot x + \csc x$

53. $\sin x \cos x + 1$

54. $2 \cos^2 x - 1$

55. $\dfrac{1}{\cot x \ \sin^2 x}$

56. $(\cos x + \sin x)(1 - \sin x \ \cos x)$

A. $\dfrac{\sin^3 x - \cos^3 x}{\sin x - \cos x}$

B. $\cos x$

C. $\tan x + \cot x$

D. $\cos^3 x + \sin^3 x$

E. $\dfrac{\sin x}{1 - \cos x}$

F. $\cos^4 x - \sin^4 x$

❯ Skill Maintenance

For each function:
a) *Graph the function.* **[5.1]**
b) *Determine whether the function is one-to-one.* **[5.1]**
c) *If the function is one-to-one, find an equation for its inverse.* **[5.1]**
d) *Graph the inverse of the function.* **[5.1]**

57. $f(x) = 3x - 2$

58. $f(x) = x^3 + 1$

59. $f(x) = x^2 - 4, \ x \geq 0$

60. $f(x) = \sqrt{x+2}$

Solve.

61. $2x^2 = 5x$ **[3.2]**

62. $3x^2 + 5x - 10 = 18$ **[3.2]**

63. $x^4 + 5x^2 - 36 = 0$ **[3.2]**

64. $x^2 - 10x + 1 = 0$ **[3.2]**

65. $\sqrt{x-2} = 5$ **[3.4]**

66. $x = \sqrt{x+7} + 5$ **[3.4]**

❯ Synthesis

Prove the identity.

67. $\ln |\tan x| = -\ln |\cot x|$

68. $\ln |\sec \theta + \tan \theta| = -\ln |\sec \theta - \tan \theta|$

69. $\log (\cos x - \sin x) + \log (\cos x + \sin x)$
$\qquad = \log \cos 2x$

70. *Mechanics.* The following equation occurs in the study of mechanics:

$$\sin \theta = \dfrac{I_1 \cos \phi}{\sqrt{(I_1 \cos \phi)^2 + (I_2 \sin \phi)^2}}.$$

It can happen that $I_1 = I_2$. Assuming that this happens, simplify the equation.

71. *Alternating Current.* In the theory of alternating current, the following equation occurs:

$$R = \dfrac{1}{\omega C (\tan \theta + \tan \phi)}.$$

Show that this equation is equivalent to

$$R = \dfrac{\cos \theta \cos \phi}{\omega C \sin (\theta + \phi)}.$$

72. *Electrical Theory.* In electrical theory, the following equations occur:

$$E_1 = \sqrt{2} E_t \cos \left(\theta + \dfrac{\pi}{P} \right)$$

and

$$E_2 = \sqrt{2} E_t \cos \left(\theta - \dfrac{\pi}{P} \right).$$

Assuming that these equations hold, show that

$$\dfrac{E_1 + E_2}{2} = \sqrt{2} E_t \cos \theta \cos \dfrac{\pi}{P}$$

and

$$\dfrac{E_1 - E_2}{2} = -\sqrt{2} E_t \sin \theta \sin \dfrac{\pi}{P}.$$

Determine whether the statement is true or false.

1. $\sin x \, (\csc x - \cot x) = 1 - \cos x$ **[7.1]**

2. $\sin 42° = \sqrt{\dfrac{1 + \cos 84°}{2}}$ **[7.2]**

3. $\sin \dfrac{\pi}{9} = \cos \dfrac{7\pi}{18}$ **[7.2]**

4. $\cos^2 x \neq \cos x^2$ **[7.1]**

For Exercises 5–14, choose one of expressions A–J to complete the identity. **[7.1], [7.2]**

5. $\cos (-x) =$

6. $\cos (u + v) =$

7. $\tan 2x =$

8. $\tan \left(\dfrac{\pi}{2} - x \right) =$

9. $1 + \cot^2 x =$

10. $\sin \dfrac{x}{2} =$

11. $\sin 2x =$

12. $\sin (u - v) =$

13. $\csc \left(\dfrac{\pi}{2} - x \right) =$

14. $\cos \dfrac{x}{2} =$

A. $2 \sin x \cos x$

B. $\pm \sqrt{\dfrac{1 + \cos x}{2}}$

C. $\csc^2 x$

D. $\dfrac{2 \tan x}{1 - \tan^2 x}$

E. $\pm \sqrt{\dfrac{1 - \cos x}{2}}$

F. $\sec x$

G. $\sin u \cos v - \cos u \sin v$

H. $\cos u \cos v - \sin u \sin v$

I. $\cot x$

J. $\cos x$

Simplify.

15. $\sqrt{\dfrac{\cot x}{\sin x}}$ **[7.1]**

16. $\dfrac{1}{\sin^2 x} - \left(\dfrac{\cos x}{\sin x} \right)^2$ **[7.1]**

17. $\dfrac{2 \cos^2 x - 5 \cos x - 3}{\cos x - 3}$ **[7.1]**

18. $\dfrac{\sin x}{\tan (-x)}$ **[7.1]**

19. $(\cos x - \sin x)^2$ **[7.2]**

20. $1 - 2 \sin^2 \dfrac{x}{2}$ **[7.2]**

21. Rationalize the denominator:

$\sqrt{\dfrac{\sec x}{1 - \cos x}}.$ **[7.1]**

22. Write $\cos 41° \cos 29° + \sin 41° \sin 29°$ as a trigonometric function of a single angle and then evaluate. **[7.1]**

23. Evaluate $\cos \dfrac{3\pi}{8}$ exactly. **[7.1]**

24. Evaluate $\sin 105°$ exactly. **[7.1]**

25. Assuming that $\sin \alpha = \frac{5}{13}$ and $\sin \beta = \frac{12}{13}$ and that α and β are between 0 and $\pi/2$, evaluate $\tan (\alpha - \beta)$. **[7.1]**

26. Find the exact value of $\sin 2\theta$ and the quadrant in which 2θ lies if $\cos \theta = -\frac{4}{5}$ and θ is in quadrant II. **[7.2]**

Prove the identity. **[7.3]**

27. $\cos^2 \dfrac{x}{2} = \dfrac{\tan x + \sin x}{2 \tan x}$

28. $\dfrac{1 - \sin x}{\cos x} = \dfrac{\cos x}{1 + \sin x}$

29. $\dfrac{\sin^3 x - \cos^3 x}{\sin x - \cos x} = \dfrac{2 + \sin 2x}{2}$

30. $\sin 6\theta - \sin 2\theta = \tan 2\theta \, (\cos 2\theta + \cos 6\theta)$

COLLABORATIVE DISCUSSION AND WRITING

31. Explain why $\tan (x + 450°)$ cannot be simplified using the tangent sum formula, but can be simplified using the sine and cosine sum formulas. **[7.1]**

32. Discuss and compare the graphs of $y = \sin x$, $y = \sin 2x$, and $y = \sin (x/2)$. **[7.2]**

33. What restrictions must be placed on the variable in each of the following identities? Why? **[7.3]**

a) $\sin 2x = \dfrac{2 \tan x}{1 + \tan^2 x}$

b) $\dfrac{1 - \cos x}{\sin x} = \dfrac{\sin x}{1 + \cos x}$

34. Find all errors in the following:

$2 \sin^2 2x + \cos 4x$
$= 2(2 \sin x \cos x)^2 + 2 \cos 2x$
$= 8 \sin^2 x \cos^2 x + 2(\cos^2 x + \sin^2 x)$
$= 8 \sin^2 x \cos^2 x + 2.$ **[7.2]**

521

7.4 Inverses of the Trigonometric Functions

> ❯ Find values of the inverse trigonometric functions.

> ❯ Simplify expressions involving compositions such as $\sin(\sin^{-1} x)$ and $\sin^{-1}(\sin x)$.

> ❯ Simplify expressions involving compositions such as $\sin\left(\cos^{-1}\frac{1}{2}\right)$ without using a calculator.

> ❯ Simplify expressions involving compositions such as $\sin \arctan(a/b)$ by making a drawing and reading off appropriate ratios.

The graphs of the sine, cosine, and tangent functions are shown below. Are the inverses of these functions also functions? We learned earlier that a function has an inverse that is a function if it is one-to-one, which we can check with the horizontal-line test.

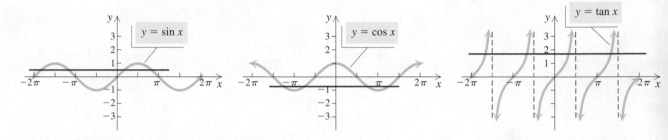

INVERSE FUNCTIONS

REVIEW SECTION 5.1.

Note that for each function, a horizontal line (shown in red) crosses the graph more than once. Therefore, none of them has an inverse that is a function.

The graphs of an equation and its inverse are reflections of each other across the line $y = x$. Let's examine the graphs of the inverses of each of the three functions graphed above.

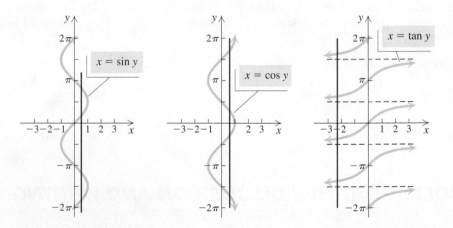

We can check again to see whether these are graphs of functions by using the vertical-line test. In each case, there is a vertical line (shown in red) that crosses the graph more than once, so each *fails* to be a function.

❯ Restricting Ranges to Define Inverse Functions

Recall that a function like $f(x) = x^2$ does not have an inverse that is a function, but by restricting the domain of f to nonnegative numbers, we have a new squaring function, $f(x) = x^2, x \geq 0$, that has an inverse that is a function, $f^{-1}(x) = \sqrt{x}$. This is equivalent to restricting the range of the inverse relation to exclude ordered pairs that contain negative numbers.

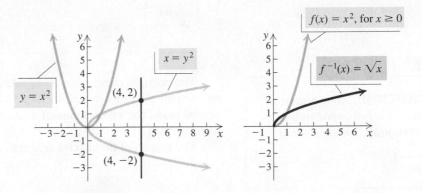

In a similar manner, we can define new trigonometric functions whose inverses are functions. We can do this by restricting either the domains of the basic trigonometric functions or the ranges of their inverse relations. This can be done in many ways, but the restrictions illustrated below with solid red curves are fairly standard in mathematics.

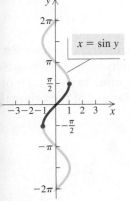

FIGURE 1.

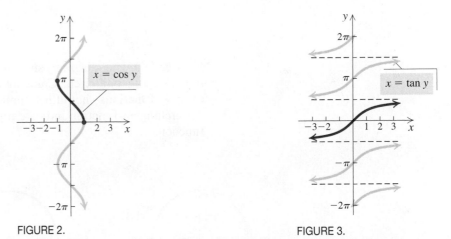

FIGURE 2.

FIGURE 3.

Just in Time 6

For the inverse sine function, we choose a range close to the origin that allows all inputs on the interval $[-1, 1]$ to have function values. Thus we choose the interval $[-\pi/2, \pi/2]$ for the range (Fig. 1). For the inverse cosine function, we choose a range close to the origin that allows all inputs on the interval $[-1, 1]$ to have function values. We choose the interval $[0, \pi]$ (Fig. 2). For the inverse tangent function, we choose a range close to the origin that allows all real numbers to have function values. The interval $(-\pi/2, \pi/2)$ satisfies this requirement (Fig. 3).

INVERSE TRIGONOMETRIC FUNCTIONS

FUNCTION	DOMAIN	RANGE
$y = \sin^{-1} x$ $= \arcsin x$, where $x = \sin y$	$[-1, 1]$	$[-\pi/2, \pi/2]$
$y = \cos^{-1} x$ $= \arccos x$, where $x = \cos y$	$[-1, 1]$	$[0, \pi]$
$y = \tan^{-1} x$ $= \arctan x$, where $x = \tan y$	$(-\infty, \infty)$	$(-\pi/2, \pi/2)$

> **CAUTION!**   The notation $\sin^{-1} x$ is *not* exponential notation. It does *not* mean $\dfrac{1}{\sin x}$!

The notation arcsin x arises because the function value, y, is the length of an arc on the unit circle for which the sine is x. Either of the two kinds of notation above can be read "the inverse sine of x" or "the arc sine of x" or "the number (or angle) whose sine is x."

The graphs of the inverse trigonometric functions are as follows.

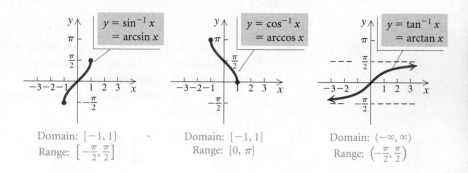

Domain: $[-1, 1]$ Domain: $[-1, 1]$ Domain: $(-\infty, \infty)$
Range: $\left[-\frac{\pi}{2}, \frac{\pi}{2}\right]$ Range: $[0, \pi]$ Range: $\left(-\frac{\pi}{2}, \frac{\pi}{2}\right)$

The following diagrams show the restricted ranges for the inverse trigonometric functions on a unit circle. Compare these graphs with the graphs above. The ranges of these functions should be memorized. The missing endpoints in the graph of the arctangent function indicate inputs that are not in the domain of the original function.

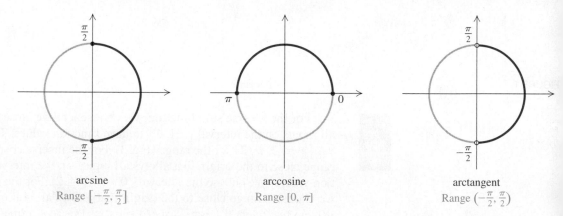

arcsine   arccosine   arctangent
Range $\left[-\frac{\pi}{2}, \frac{\pi}{2}\right]$ Range $[0, \pi]$ Range $\left(-\frac{\pi}{2}, \frac{\pi}{2}\right)$

The following is a summary of the domains and ranges of the trigonometric functions together with a summary of the domains and ranges of the inverse trigonometric functions. For completeness, we have included the arccosecant, the arcsecant, and the arccotangent, though there is a lack of uniformity in their definitions in mathematical literature.

FUNCTION	DOMAIN	RANGE
sin	All reals, $(-\infty, \infty)$	$[-1, 1]$
cos	All reals, $(-\infty, \infty)$	$[-1, 1]$
tan	All reals except $k\pi/2$, k odd	All reals, $(-\infty, \infty)$
csc	All reals except $k\pi$	$(-\infty, -1] \cup [1, \infty)$
sec	All reals except $k\pi/2$, k odd	$(-\infty, -1] \cup [1, \infty)$
cot	All reals except $k\pi$	All reals, $(-\infty, \infty)$

INVERSE FUNCTION	DOMAIN	RANGE
$\sin^{-1}$	$[-1, 1]$	$\left[-\dfrac{\pi}{2}, \dfrac{\pi}{2}\right]$
$\cos^{-1}$	$[-1, 1]$	$[0, \pi]$
$\tan^{-1}$	All reals, $(-\infty, \infty)$	$\left(-\dfrac{\pi}{2}, \dfrac{\pi}{2}\right)$
$\csc^{-1}$	$(-\infty, -1] \cup [1, \infty)$	$\left[-\dfrac{\pi}{2}, 0\right) \cup \left(0, \dfrac{\pi}{2}\right]$
$\sec^{-1}$	$(-\infty, -1] \cup [1, \infty)$	$\left[0, \dfrac{\pi}{2}\right) \cup \left(\dfrac{\pi}{2}, \pi\right]$
$\cot^{-1}$	All reals, $(-\infty, \infty)$	$(0, \pi)$

EXAMPLE 1 Find each of the following function values. For parts (d) and (e), see the preceding Connecting the Concepts for the restricted ranges.

a) $\sin^{-1}\dfrac{\sqrt{2}}{2}$ b) $\cos^{-1}\left(-\dfrac{1}{2}\right)$ c) $\tan^{-1}\left(-\dfrac{\sqrt{3}}{3}\right)$

d) $\cot^{-1} 0$ e) $\sec^{-1}\left(-\sqrt{2}\right)$

Solution

a) Another way to state "find $\sin^{-1}\sqrt{2}/2$" is to say "find β such that $\sin \beta = \sqrt{2}/2$." In the restricted range $[-\pi/2, \pi/2]$, the only number with a sine of $\sqrt{2}/2$ is $\pi/4$. Thus, $\sin^{-1}\left(\sqrt{2}/2\right) = \pi/4$, or 45°. (See Fig. 4 on the following page.)

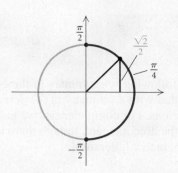

FIGURE 4.

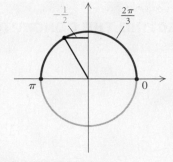

FIGURE 5.

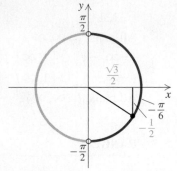

FIGURE 6.

b) The only number with a cosine of $-\frac{1}{2}$ in the restricted range $[0, \pi]$ is $2\pi/3$. Thus, $\cos^{-1}\left(-\frac{1}{2}\right) = 2\pi/3$, or $120°$. (See Fig. 5 above.)

c) The only number in the restricted range $(-\pi/2, \pi/2)$ with a tangent of $-\sqrt{3}/3$ is $-\pi/6$. Thus, $\tan^{-1}\left(-\sqrt{3}/3\right)$ is $-\pi/6$, or $-30°$. (See Fig. 6 at left.)

d) The only number in the restricted range $(0, \pi)$ with a cotangent of 0 is $\pi/2$. Thus, $\cot^{-1} 1 = \pi/2$, or $90°$.

e) The only number in the restricted range $[0, \pi/2) \cup (\pi/2, \pi]$ with a secant of $-\sqrt{2}$ is $3\pi/4$. Thus, $\sec^{-1}\left(-\sqrt{2}\right) = 3\pi/4$, or $135°$.

> *Now Try Exercises 1 and 11.*

We can also use a calculator to find inverse trigonometric function values. The cosecant, the secant, and the cotangent functions are the reciprocals of the sine, the cosine, and the tangent functions, respectively. Thus we have

$$\csc^{-1}(x) = \sin^{-1}\left(\frac{1}{x}\right),$$

$$\sec^{-1}(x) = \cos^{-1}\left(\frac{1}{x}\right), \quad \text{and}$$

$$\cot^{-1}(x) = \tan^{-1}\left(\frac{1}{x}\right).$$

On most graphing calculators, we can find inverse function values in either radians or degrees simply by selecting the appropriate mode. The keystrokes involved in finding inverse function values vary with the calculator. Be sure to read the instructions for the particular calculator that you are using.

EXAMPLE 2   Approximate each of the following function values in both radians and degrees. Round radian measure to four decimal places and degree measure to the nearest tenth of a degree.

a) $\cos^{-1}(-0.2689)$ **b)** $\tan^{-1}(-0.2623)$
c) $\sin^{-1} 0.20345$ **d)** $\cos^{-1} 1.318$
e) $\csc^{-1} 8.205$

Solution

FUNCTION VALUE	MODE	READOUT	ROUNDED
a) $\cos^{-1}(-0.2689)$	Radian	1.843047111	1.8430
	Degree	105.5988209	105.6°
b) $\tan^{-1}(-0.2623)$	Radian	-.2565212141	-0.2565
	Degree	-14.69758292	-14.7°
c) $\sin^{-1}0.20345$	Radian	.2048803359	0.2049
	Degree	11.73877855	11.7°
d) $\cos^{-1}1.318$	Radian	ERR:DOMAIN	
	Degree	ERR:DOMAIN	

The value 1.318 is not in $[-1, 1]$, the domain of the arccosine function.

e) The cosecant function is the reciprocal of the sine function:

$\csc^{-1}8.205 =$

$\sin^{-1}(1/8.205)$	Radian	.1221806653	0.1222
	Degree	7.000436462	7.0°

> *Now Try Exercises 21, 25, and 27.*

❯ Composition of Trigonometric Functions and Their Inverses

Various compositions of trigonometric functions and their inverses often occur in practice. For example, we might want to try to simplify an expression such as

$$\sin\left(\sin^{-1}x\right) \quad \text{or} \quad \sin\left(\cot^{-1}\frac{x}{2}\right).$$

COMPOSITION OF FUNCTIONS

REVIEW SECTION 2.3.

In the expression on the left, we are finding "the sine of a number whose sine is x." Recall from Section 5.1 that if a function f has an inverse that is also a function, then

$$f(f^{-1}(x)) = x, \quad \text{for all } x \text{ in the } domain \text{ of } f^{-1},$$

and

$$f^{-1}(f(x)) = x, \quad \text{for all } x \text{ in the } domain \text{ of } f.$$

Thus, if $f(x) = \sin x$ and $f^{-1}(x) = \sin^{-1}x$, then

$$\sin(\sin^{-1}x) = x, \textbf{ for all } x \textbf{ in the } \textit{domain} \textbf{ of } \sin^{-1},$$

which is any number on the interval $[-1, 1]$. Similar results hold for the other trigonometric functions.

COMPOSITION OF TRIGONOMETRIC FUNCTIONS

$\sin\left(\sin^{-1}x\right) = x, \quad$ for all x in the domain of $\sin^{-1}$.

$\cos\left(\cos^{-1}x\right) = x, \quad$ for all x in the domain of $\cos^{-1}$.

$\tan\left(\tan^{-1}x\right) = x, \quad$ for all x in the domain of $\tan^{-1}$.

EXAMPLE 3 Simplify each of the following.

a) $\cos\left(\cos^{-1}\dfrac{\sqrt{3}}{2}\right)$ **b)** $\sin\left(\sin^{-1}1.8\right)$

Solution

a) Since $\sqrt{3}/2$ is in $[-1, 1]$, the domain of $\cos^{-1}$, it follows that

$$\cos\left(\cos^{-1}\frac{\sqrt{3}}{2}\right) = \frac{\sqrt{3}}{2}.$$

b) Since 1.8 is not in $[-1, 1]$, the domain of $\sin^{-1}$, we cannot evaluate this expression. We know that there is no number with a sine of 1.8. Because we cannot find $\sin^{-1}1.8$, we state that $\sin\left(\sin^{-1}1.8\right)$ does not exist.

> *Now Try Exercise 37.*

Now let's consider an expression like $\sin^{-1}(\sin x)$. We might also suspect that this is equal to x for any x in the domain of $\sin x$, but this is not true unless x is in the range of the $\sin^{-1}$ function. Note that in order to define $\sin^{-1}$, we had to restrict the domain of the sine function. In doing so, we restricted the range of the inverse sine function. Thus,

$$\sin^{-1}(\sin x) = x, \quad \text{for all } x \text{ in the } \textit{range} \text{ of } \sin^{-1}.$$

Similar results hold for the other trigonometric functions.

SPECIAL CASES

$\sin^{-1}(\sin x) = x, \quad$ for all x in the range of $\sin^{-1}$.
$\cos^{-1}(\cos x) = x, \quad$ for all x in the range of $\cos^{-1}$.
$\tan^{-1}(\tan x) = x, \quad$ for all x in the range of $\tan^{-1}$.

EXAMPLE 4 Simplify each of the following.

a) $\tan^{-1}\left(\tan\dfrac{\pi}{6}\right)$ **b)** $\sin^{-1}\left(\sin\dfrac{3\pi}{4}\right)$

Solution

a) Since $\pi/6$ is in $(-\pi/2, \pi/2)$, the range of the $\tan^{-1}$ function, we can use $\tan^{-1}(\tan x) = x$. Thus,

$$\tan^{-1}\left(\tan\frac{\pi}{6}\right) = \frac{\pi}{6}.$$

b) Note that $3\pi/4$ is not in $[-\pi/2, \pi/2]$, the range of the $\sin^{-1}$ function. Thus we *cannot* apply $\sin^{-1}(\sin x) = x$. Instead we first find $\sin(3\pi/4)$, which is $\sqrt{2}/2$, and substitute:

$$\sin^{-1}\left(\sin\frac{3\pi}{4}\right) = \sin^{-1}\left(\frac{\sqrt{2}}{2}\right) = \frac{\pi}{4}.$$

> *Now Try Exercises 41 and 43.*

Now we find some additional function compositions.

EXAMPLE 5 Simplify each of the following.

a) $\sin\left[\tan^{-1}(-1)\right]$ **b)** $\cos^{-1}\left(\sin\dfrac{\pi}{2}\right)$

Solution

a) We know that $\tan^{-1}(-1)$ is the number (or angle) θ in $(-\pi/2, \pi/2)$ whose tangent is -1. That is, $\tan\theta = -1$. Thus, $\theta = -\pi/4$ and

$$\sin\left[\tan^{-1}(-1)\right] = \sin\left[-\frac{\pi}{4}\right] = -\frac{\sqrt{2}}{2}.$$

b) $\cos^{-1}\left(\sin\dfrac{\pi}{2}\right) = \cos^{-1}(1) = 0 \qquad \sin\dfrac{\pi}{2} = 1$

> *Now Try Exercises 47 and 49.*

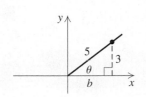

Next, let's consider

$$\cos\left(\sin^{-1}\tfrac{3}{5}\right).$$

Without using a calculator, we cannot find $\sin^{-1}\tfrac{3}{5}$. However, we can still evaluate the entire expression by sketching a reference triangle. We are looking for angle θ such that $\sin^{-1}\tfrac{3}{5} = \theta$, or $\sin\theta = \tfrac{3}{5}$. Since $\sin^{-1}$ is defined in $[-\pi/2, \pi/2]$ and $\tfrac{3}{5} > 0$, we know that θ is in quadrant I. We sketch a reference right triangle, as shown at left. The angle θ in this triangle is an angle whose sine is $\tfrac{3}{5}$. We wish to find the cosine of this angle. Since the triangle is a right triangle, we can find the length of the base, b. It is 4. Thus we know that $\cos\theta = b/5$, or $\tfrac{4}{5}$. Therefore,

$$\cos\left(\sin^{-1}\tfrac{3}{5}\right) = \tfrac{4}{5}.$$

EXAMPLE 6 Find $\sin\left(\cot^{-1}\dfrac{x}{2}\right)$.

Solution Since $\cot^{-1}$ is defined in $(0, \pi)$, we consider quadrants I and II. We draw right triangles, as shown at left, whose legs have lengths x and 2, so that $\cot\theta = x/2$.

In each, we find the length of the hypotenuse and then read off the sine ratio. We get

$$\sin\left(\cot^{-1}\frac{x}{2}\right) = \frac{2}{\sqrt{x^2 + 4}}.$$

> *Now Try Exercise 55.*

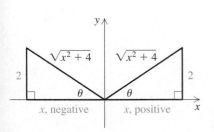

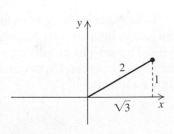

In the following example, we use a sum identity to evaluate an expression.

EXAMPLE 7 Evaluate:

$$\sin\left(\sin^{-1}\tfrac{1}{2} + \cos^{-1}\tfrac{5}{13}\right).$$

Solution Since $\sin^{-1}\tfrac{1}{2}$ and $\cos^{-1}\tfrac{5}{13}$ are both angles, the expression is the sine of a sum of two angles, so we use the identity

$$\sin(u + v) = \sin u \cos v + \cos u \sin v.$$

Thus,

$$\sin\left(\sin^{-1}\tfrac{1}{2} + \cos^{-1}\tfrac{5}{13}\right)$$
$$= \sin\left(\sin^{-1}\tfrac{1}{2}\right)\cdot\cos\left(\cos^{-1}\tfrac{5}{13}\right) + \cos\left(\sin^{-1}\tfrac{1}{2}\right)\cdot\sin\left(\cos^{-1}\tfrac{5}{13}\right)$$
$$= \tfrac{1}{2}\cdot\tfrac{5}{13} + \cos\left(\sin^{-1}\tfrac{1}{2}\right)\cdot\sin\left(\cos^{-1}\tfrac{5}{13}\right). \qquad \textbf{Using composition identities}$$

Now since $\sin^{-1}\tfrac{1}{2} = \pi/6$, $\cos\left(\sin^{-1}\tfrac{1}{2}\right)$ simplifies to $\cos\pi/6$, or $\sqrt{3}/2$. We can illustrate this with a reference triangle in quadrant I. (See Fig. 1.)

FIGURE 1.

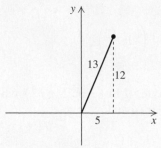

FIGURE 2.

To find $\sin\left(\cos^{-1}\frac{5}{13}\right)$, we use a reference triangle in quadrant I and determine that the sine of the angle whose cosine is $\frac{5}{13}$ is $\frac{12}{13}$. (See Fig. 2.)

Our expression now simplifies to

$$\frac{1}{2}\cdot\frac{5}{13}+\frac{\sqrt{3}}{2}\cdot\frac{12}{13}, \quad \text{or} \quad \frac{5+12\sqrt{3}}{26}.$$

Thus,

$$\sin\left(\sin^{-1}\frac{1}{2}+\cos^{-1}\frac{5}{13}\right)=\frac{5+12\sqrt{3}}{26}.$$

> **Now Try Exercise 63.**

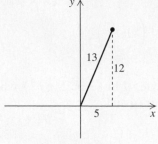

7.4   Exercise Set

Find each of the following exactly in radians and degrees.

1. $\sin^{-1}\left(-\dfrac{\sqrt{3}}{2}\right)$ **2.** $\cos^{-1}\dfrac{1}{2}$

3. $\tan^{-1} 1$ **4.** $\sin^{-1} 0$

5. $\cos^{-1}\dfrac{\sqrt{2}}{2}$ **6.** $\sec^{-1}\sqrt{2}$

7. $\tan^{-1} 0$ **8.** $\tan^{-1}\dfrac{\sqrt{3}}{3}$

9. $\cos^{-1}\dfrac{\sqrt{3}}{2}$ **10.** $\cot^{-1}\left(-\dfrac{\sqrt{3}}{3}\right)$

11. $\csc^{-1} 2$ **12.** $\sin^{-1}\dfrac{1}{2}$

13. $\cot^{-1}(-\sqrt{3})$ **14.** $\tan^{-1}(-1)$

15. $\sin^{-1}\left(-\dfrac{1}{2}\right)$ **16.** $\cos^{-1}\left(-\dfrac{\sqrt{2}}{2}\right)$

17. $\cos^{-1} 0$ **18.** $\sin^{-1}\dfrac{\sqrt{3}}{2}$

19. $\sec^{-1} 2$ **20.** $\csc^{-1}(-1)$

Use a calculator to find each of the following in radians, rounded to four decimal places, and in degrees, rounded to the nearest tenth of a degree.

21. $\tan^{-1} 0.3673$

22. $\cos^{-1}(-0.2935)$

23. $\sin^{-1} 0.9613$

24. $\sin^{-1}(-0.6199)$

25. $\cos^{-1}(-0.9810)$

26. $\tan^{-1} 158$

27. $\csc^{-1}(-6.2774)$

28. $\sec^{-1} 1.1677$

29. $\tan^{-1}(1.091)$

30. $\cot^{-1} 1.265$

31. $\sin^{-1}(-0.8192)$

32. $\cos^{-1}(-0.2716)$

33. State the domains of the inverse sine, inverse cosine, and inverse tangent functions.

34. State the ranges of the inverse sine, inverse cosine, and inverse tangent functions.

35. *Angle of Depression.* An airplane is flying at an altitude of 2000 ft toward Logan International Airport, in Boston, Massachusetts. The straight-line distance from the airplane to the airport is d feet. Express θ, the angle of depression, as a function of d.

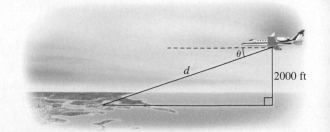

36. *Angle of Inclination.* A guy wire is attached to the top of a 50-ft pole and stretched to a point that is d feet from the bottom of the pole. Express β, the angle of inclination, as a function of d.

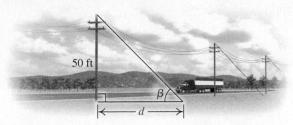

50 ft

β

d

Evaluate.

37. $\sin\left(\sin^{-1} 0.3\right)$

38. $\tan\left[\tan^{-1}\left(-4.2\right)\right]$

39. $\cos^{-1}\left[\cos\left(-\dfrac{\pi}{4}\right)\right]$

40. $\sin^{-1}\left(\sin\dfrac{2\pi}{3}\right)$

41. $\sin^{-1}\left(\sin\dfrac{\pi}{5}\right)$

42. $\cot^{-1}\left(\cot\dfrac{2\pi}{3}\right)$

43. $\tan^{-1}\left(\tan\dfrac{2\pi}{3}\right)$

44. $\cos^{-1}\left(\cos\dfrac{\pi}{7}\right)$

45. $\sin\left(\tan^{-1}\dfrac{\sqrt{3}}{3}\right)$

46. $\cos\left(\sin^{-1}\dfrac{\sqrt{3}}{2}\right)$

47. $\tan\left(\cos^{-1}\dfrac{\sqrt{2}}{2}\right)$

48. $\cos^{-1}\left(\sin\pi\right)$

49. $\sin^{-1}\left(\cos\dfrac{\pi}{6}\right)$

50. $\sin^{-1}\left[\tan\left(-\dfrac{\pi}{4}\right)\right]$

51. $\tan\left(\sin^{-1} 0.1\right)$

52. $\cos\left(\tan^{-1}\dfrac{\sqrt{3}}{4}\right)$

53. $\sin^{-1}\left(\sin\dfrac{7\pi}{6}\right)$

54. $\tan^{-1}\left[\tan\left(-\dfrac{3\pi}{4}\right)\right]$

Find each of the following.

55. $\sin\left(\tan^{-1}\dfrac{a}{3}\right)$

56. $\tan\left(\cos^{-1}\dfrac{3}{x}\right)$

57. $\cot\left(\sin^{-1}\dfrac{p}{q}\right)$

58. $\sin\left(\cos^{-1} x\right)$

59. $\tan\left(\sin^{-1}\dfrac{p}{\sqrt{p^2+9}}\right)$

60. $\tan\left(\dfrac{1}{2}\sin^{-1}\dfrac{1}{2}\right)$

61. $\cos\left(\dfrac{1}{2}\sin^{-1}\dfrac{\sqrt{3}}{2}\right)$

62. $\sin\left(2\cos^{-1}\dfrac{3}{5}\right)$

Evaluate.

63. $\cos\left(\sin^{-1}\dfrac{\sqrt{2}}{2}+\cos^{-1}\dfrac{3}{5}\right)$

64. $\sin\left(\sin^{-1}\dfrac{1}{2}+\cos^{-1}\dfrac{3}{5}\right)$

65. $\sin\left(\sin^{-1} x+\cos^{-1} y\right)$

66. $\cos\left(\sin^{-1} x-\cos^{-1} y\right)$

67. $\sin\left(\sin^{-1} 0.6032+\cos^{-1} 0.4621\right)$

68. $\cos\left(\sin^{-1} 0.7325-\cos^{-1} 0.4838\right)$

❯ Skill Maintenance

Vocabulary Reinforcement

In each of Exercises 69–76, fill in the blank with the correct term. Some of the given choices will not be used.

linear speed	congruent
angular speed	circular
angle of elevation	periodic
angle of depression	period
complementary	amplitude
supplementary	quadrantal
similar	radian measure

69. A function f is said to be _____ if there exists a positive constant p such that $f(s+p)=f(s)$ for all s in the domain of f. **[6.5]**

70. The _____ of a rotation is the ratio of the distance s traveled by a point at a radius r from the center of rotation to the length of the radius r. **[6.4]**

71. Triangles are _____ if their corresponding angles have the same measure. **[6.1]**

72. The angle between the horizontal and a line of sight below the horizontal is called a(n) _____. **[6.2]**

73. _____ is the amount of rotation per unit of time. **[6.4]**

74. Two positive angles are _____ if their sum is 180°. **[6.3]**

75. The _____ of a periodic function is one half of the distance between its maximum and minimum function values. **[6.5]**

76. Trigonometric functions with domains composed of real numbers are called _____ functions. **[6.5]**

❯ Synthesis

Prove the identity.

77. $\sin^{-1} x+\cos^{-1} x=\dfrac{\pi}{2}$

78. $\tan^{-1} x=\sin^{-1}\dfrac{x}{\sqrt{x^2+1}}$

79. $\sin^{-1} x=\cos^{-1}\sqrt{1-x^2}$, for $x\ge 0$

80. $\cos^{-1} x=\tan^{-1}\dfrac{\sqrt{1-x^2}}{x}$, for $x>0$

81. *Height of a Mural.*   An art student's eye is at a point A, looking at a mural of height h, with the bottom of the mural y feet above the eye (see the figure). The eye is x feet from the wall. Write an expression for θ in terms of x, y, and h. Then evaluate the expression when $x = 20$ ft, $y = 7$ ft, and $h = 25$ ft.

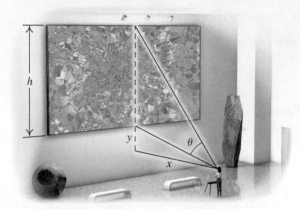

82. Use a calculator to approximate the following expression:

$$16 \tan^{-1} \tfrac{1}{5} - 4 \tan^{-1} \tfrac{1}{239}.$$

What number does this expression seem to approximate?

7.5 Solving Trigonometric Equations

❯ Solve trigonometric equations.

Just in Time 17

When an equation contains a trigonometric expression with a variable, such as $\cos x$, it is called a *trigonometric equation*. Some trigonometric equations are identities, such as $\sin^2 x + \cos^2 x = 1$. Now we consider equations, such as $2 \cos x = -1$, that are usually not identities. As we have done for other types of equations, we will solve such equations by finding all values for x that make the equation true.

EXAMPLE 1   Solve: $2 \cos x = -1$.

Solution   We first solve for $\cos x$:

$$2 \cos x = -1$$
$$\cos x = -\tfrac{1}{2}.$$

The solutions are numbers whose cosine is $-\tfrac{1}{2}$. To find them, we use the unit circle (see Section 6.5).

There are just two points on the unit circle for which the cosine is $-\tfrac{1}{2}$, as shown in the figure at left. They are the points corresponding to $2\pi/3$ and $4\pi/3$. These numbers, plus any multiple of 2π, are the solutions:

$$\frac{2\pi}{3} + 2k\pi \quad \text{and} \quad \frac{4\pi}{3} + 2k\pi,$$

where k is any integer. In degrees, the solutions are

$$120° + k \cdot 360° \quad \text{and} \quad 240° + k \cdot 360°,$$

where k is any integer.

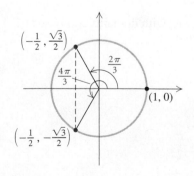

To check the solution to $2\cos x = -1$, we can graph $y_1 = 2\cos x$ and $y_2 = -1$ on the same set of axes and find the *first* coordinates of the points of intersection. Using $\pi/3$ as the Xscl facilitates our reading of the solutions. First, let's graph these equations on the interval from 0 to 2π, as shown in the figure on the left below. The only solutions in $[0, 2\pi)$ are $2\pi/3$ and $4\pi/3$.

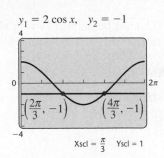 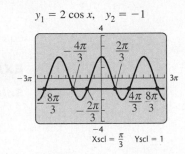

Next, let's change the viewing window to $[-3\pi, 3\pi, -4, 4]$ and graph again. Since the cosine function is periodic, there is an infinite number of solutions. A few of these appear in the graph on the right above. From the graph, we see that the solutions are $2\pi/3 + 2k\pi$ and $4\pi/3 + 2k\pi$, where k is any integer.

> *Now Try Exercise 1.*

EXAMPLE 2 Solve: $4\sin^2 x = 1$.

Solution We begin by solving for $\sin x$:

$$4\sin^2 x = 1$$
$$\sin^2 x = \tfrac{1}{4}$$
$$\sin x = \pm\tfrac{1}{2}.$$

Again, we use the unit circle to find those numbers whose sine is $\tfrac{1}{2}$ or $-\tfrac{1}{2}$. The solutions are

$$\frac{\pi}{6} + 2k\pi, \qquad \frac{5\pi}{6} + 2k\pi, \qquad \frac{7\pi}{6} + 2k\pi, \quad \text{and} \quad \frac{11\pi}{6} + 2k\pi,$$

where k is any integer. In degrees, the solutions are

$$30° + k \cdot 360°, \qquad 150° + k \cdot 360°,$$
$$210° + k \cdot 360°, \quad \text{and} \quad 330° + k \cdot 360°,$$

where k is any integer.

The general solutions listed above could be condensed using odd as well as even multiples of π:

$$\frac{\pi}{6} + k\pi \quad \text{and} \quad \frac{5\pi}{6} + k\pi,$$

or, in degrees,

$$30° + k \cdot 180° \quad \text{and} \quad 150° + k \cdot 180°,$$

where k is any integer.

Let's do a partial check using a graphing calculator, checking only the solutions in $[0, 2\pi)$. We graph $y_1 = 4\sin^2 x$ and $y_2 = 1$ and note that the solutions in $[0, 2\pi)$ are $\pi/6$, $5\pi/6$, $7\pi/6$, and $11\pi/6$.

> *Now Try Exercise 13.*

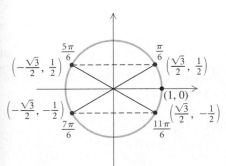

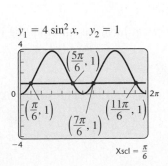

In most applications, it is sufficient to find just the solutions from 0 to 2π or from 0° to 360°. We then remember that any multiple of 2π, or 360°, can be added to obtain the rest of the solutions.

We must be careful to find all solutions in $[0, 2\pi)$ when solving trigonometric equations involving double angles.

EXAMPLE 3   Solve $3 \tan 2x = -3$ on the interval $[0, 2\pi)$.

Solution   We first solve for $\tan 2x$:

$$3 \tan 2x = -3$$
$$\tan 2x = -1.$$

We are looking for solutions x to the equation for which

$$0 \le x < 2\pi.$$

Multiplying by 2, we get

$$0 \le 2x < 4\pi,$$

which is the interval that we use when solving $\tan 2x = -1$.

Using the unit circle, we find points $2x$ in $[0, 4\pi)$ for which $\tan 2x = -1$. These values of $2x$ are as follows:

$$2x = \frac{3\pi}{4}, \quad \frac{7\pi}{4}, \quad \frac{11\pi}{4}, \quad \text{and} \quad \frac{15\pi}{4}.$$

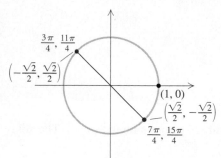

Thus the desired values of x in $[0, 2\pi)$ are each of these values divided by 2— hence,

$$x = \frac{3\pi}{8}, \quad \frac{7\pi}{8}, \quad \frac{11\pi}{8}, \quad \text{and} \quad \frac{15\pi}{8}.$$

Calculators are needed to solve some trigonometric equations. Answers can be found in radians or degrees, depending on the mode setting.

EXAMPLE 4   Solve $\frac{1}{2} \cos \phi + 1 = 1.2108$ in $[0, 360°)$.

Solution   We have

$$\frac{1}{2} \cos \phi + 1 = 1.2108$$
$$\frac{1}{2} \cos \phi = 0.2108$$
$$\cos \phi = 0.4216.$$

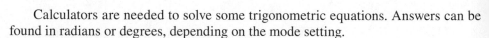

Using a calculator set in DEGREE mode (see the window at left), we find that the reference angle, $\cos^{-1} 0.4216$, is

$$\phi \approx 65.06°.$$

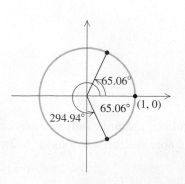

Since $\cos \phi$ is positive, the solutions are in quadrants I and IV. The solutions in $[0, 360°)$ are

$$65.06° \quad \text{and} \quad 360° - 65.06° = 294.94°.$$

Now Try Exercise 9.

EXAMPLE 5 Solve $2\cos^2 u = 1 - \cos u$ in $[0°, 360°)$.

Algebraic Solution

We use the principle of zero products:

$$2\cos^2 u = 1 - \cos u$$
$$2\cos^2 u + \cos u - 1 = 0$$
$$(2\cos u - 1)(\cos u + 1) = 0$$
$$2\cos u - 1 = 0 \quad or \quad \cos u + 1 = 0$$
$$2\cos u = 1 \quad or \quad \cos u = -1$$
$$\cos u = \tfrac{1}{2} \quad or \quad \cos u = -1.$$

Thus,
$$u = 60°, 300° \quad or \quad u = 180°.$$

The solutions in $[0°, 360°)$ are $60°$, $180°$, and $300°$.

Graphical Solution

We can use either the Intersect method or the Zero method to solve trigonometric equations. Here we illustrate by solving the equation using both methods. We set the calculator in DEGREE mode. With the Intersect method, we graph the equations $y_1 = 2\cos^2 x$ and $y_2 = 1 - \cos x$. With the Zero method, we write the equation in the form $2\cos^2 u + \cos u - 1 = 0$. Then we graph $y = 2\cos^2 x + \cos x - 1$.

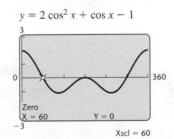

The leftmost solution is $60°$. Using the INTERSECT feature or the ZERO feature two more times, we find the other solutions: $180°$ and $300°$. The solutions in $[0°, 360°)$ are $60°$, $180°$, and $300°$.

> *Now Try Exercise 15.*

EXAMPLE 6 Solve $\sin^2\beta - \sin\beta = 0$ in $[0, 2\pi)$.

Solution We factor and use the principle of zero products:

$$\sin^2\beta - \sin\beta = 0$$
$$\sin\beta(\sin\beta - 1) = 0 \qquad \textbf{Factoring}$$
$$\sin\beta = 0 \quad or \quad \sin\beta - 1 = 0$$
$$\sin\beta = 0 \quad or \quad \sin\beta = 1$$
$$\beta = 0, \pi \quad or \quad \beta = \pi/2.$$

The solutions in $[0, 2\pi)$ are 0, $\pi/2$, and π.

> *Now Try Exercise 17.*

$y = \sin^2 x - \sin x$

$\text{Xscl} = \dfrac{\pi}{4}$

QUADRATIC FORMULA
REVIEW SECTION 3.2.

If a trigonometric equation is quadratic but difficult or impossible to factor, we use the quadratic formula.

EXAMPLE 7 Solve $10 \sin^2 x - 12 \sin x - 7 = 0$ in $[0°, 360°)$.

Solution This equation is quadratic in $\sin x$ with $a = 10$, $b = -12$, and $c = -7$. Substituting into the quadratic formula, we get

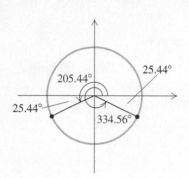

$$\sin x = \frac{-b \pm \sqrt{b^2 - 4ac}}{2a} \quad \textbf{Using the quadratic formula}$$

$$= \frac{-(-12) \pm \sqrt{(-12)^2 - 4(10)(-7)}}{2 \cdot 10} \quad \textbf{Substituting}$$

$$= \frac{12 \pm \sqrt{144 + 280}}{20}$$

$$= \frac{12 \pm \sqrt{424}}{20}$$

$$\approx \frac{12 \pm 20.5913}{20}$$

$$\sin x \approx 1.6296 \quad or \quad \sin x \approx -0.4296.$$

Since sine values are never greater than 1, the first equation has no solution. Using the other equation, we find the reference angle to be $25.44°$. Since $\sin x$ is negative, the solutions are in quadrants III and IV.

Thus the solutions in $[0°, 360°)$ are

$$180° + 25.44° = 205.44° \quad \text{and} \quad 360° - 25.44° = 334.56°.$$

> *Now Try Exercise 23.*

Trigonometric equations can involve more than one function.

EXAMPLE 8 Solve $2 \cos^2 x \tan x = \tan x$ in $[0, 2\pi)$.

Solution Using a graphing calculator, we can determine that there are six solutions. If we let Xscl $= \pi/4$, the solutions are easily read from the graph. In the figures at left, we show the Intersect method and the Zero method of solving graphically. Each illustrates that the solutions in $[0, 2\pi)$ are

$$0, \quad \frac{\pi}{4}, \quad \frac{3\pi}{4}, \quad \pi, \quad \frac{5\pi}{4}, \quad \text{and} \quad \frac{7\pi}{4}.$$

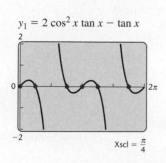

We can verify these solutions algebraically, as follows:

$$2 \cos^2 x \tan x = \tan x$$

$$2 \cos^2 x \tan x - \tan x = 0$$

$$\tan x (2 \cos^2 x - 1) = 0$$

$$\tan x = 0 \quad or \quad 2 \cos^2 x - 1 = 0$$

$$\tan x = 0 \quad or \quad \cos^2 x = \frac{1}{2}$$

$$\tan x = 0 \quad or \quad \cos x = \pm \frac{\sqrt{2}}{2}$$

$$x = 0, \pi \quad or \quad x = \frac{\pi}{4}, \frac{3\pi}{4}, \frac{5\pi}{4}, \frac{7\pi}{4}.$$

Thus, $x = 0$, $\pi/4$, $3\pi/4$, π, $5\pi/4$, and $7\pi/4$.

> *Now Try Exercise 29.*

When a trigonometric equation involves more than one function, it is sometimes helpful to use identities to rewrite the equation in terms of a single function.

EXAMPLE 9   Solve $\sin x + \cos x = 1$ in $[0, 2\pi)$.

Algebraic Solution

We have

$$\sin x + \cos x = 1$$
$$(\sin x + \cos x)^2 = 1^2 \quad \text{Squaring both sides}$$
$$\sin^2 x + 2 \sin x \cos x + \cos^2 x = 1$$
$$2 \sin x \cos x + 1 = 1 \quad \text{Using } \sin^2 x + \cos^2 x = 1$$
$$2 \sin x \cos x = 0$$
$$\sin x = 0 \quad or \quad \cos x = 0$$
$$x = 0, \pi \quad or \quad x = \pi/2, 3\pi/2.$$

Now we check these in the original equation $\sin x + \cos x = 1$:

$$\sin 0 + \cos 0 = 0 + 1 = 1,$$
$$\sin \frac{\pi}{2} + \cos \frac{\pi}{2} = 1 + 0 = 1,$$
$$\sin \pi + \cos \pi = 0 + (-1) = -1,$$
$$\sin \frac{3\pi}{2} + \cos \frac{3\pi}{2} = (-1) + 0 = -1.$$

We find that π and $3\pi/2$ do not check, but the other values do. Thus the solutions in $[0, 2\pi)$ are

$$0 \quad \text{and} \quad \frac{\pi}{2}.$$

When the solution process involves squaring both sides, values are sometimes obtained that are not solutions of the original equation. As we saw in this example, it is necessary to check the possible solutions.

Graphical Solution

We can graph the left side and then the right side of the equation as seen in the first window below. Then we look for points of intersection. We could also rewrite the equation as $\sin x + \cos x - 1 = 0$, graph the left side, and look for the zeros of the function, as illustrated in the second window below. In each window, we see that the solutions in $[0, 2\pi)$ are 0 and $\pi/2$.

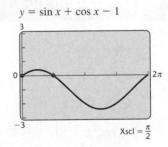

$y_1 = \sin x + \cos x, \quad y_2 = 1$

$y = \sin x + \cos x - 1$

This example illustrates a valuable advantage of the calculator—that is, with a graphing calculator, extraneous solutions do not appear.

> *Now Try Exercise 39.*

EXAMPLE 10 Solve $\cos 2x + \sin x = 1$ in $[0, 2\pi)$.

| **Algebraic Solution** | **Graphical Solution** |

Algebraic Solution

We have

$$\cos 2x + \sin x = 1$$

$$1 - 2\sin^2 x + \sin x = 1 \qquad \text{**Using the identity** } \cos 2x = 1 - 2\sin^2 x$$

$$-2\sin^2 x + \sin x = 0$$

$$\sin x (-2\sin x + 1) = 0 \qquad \text{**Factoring**}$$

$$\sin x = 0 \quad or \quad -2\sin x + 1 = 0 \qquad \text{**Principle of zero products**}$$

$$\sin x = 0 \quad or \quad \sin x = \frac{1}{2}$$

$$x = 0, \pi \quad or \quad x = \frac{\pi}{6}, \frac{5\pi}{6}.$$

All four values check. The solutions in $[0, 2\pi)$ are 0, $\pi/6$, $5\pi/6$, and π.

Graphical Solution

We graph $y_1 = \cos 2x + \sin x - 1$ and look for the zeros of the function.

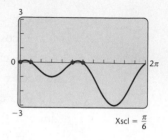

The solutions in $[0, 2\pi)$ are 0, $\pi/6$, $5\pi/6$, and π.

> **Now Try Exercise 27.**

Just in Time
14

EXAMPLE 11 Solve $\tan^2 x + \sec x - 1 = 0$ in $[0, 2\pi)$.

Algebraic Solution

We have

$$\tan^2 x + \sec x - 1 = 0$$

$$\sec^2 x - 1 + \sec x - 1 = 0 \qquad \text{**Using the identity** } 1 + \tan^2 x = \sec^2 x, \text{ or } \tan^2 x = \sec^2 x - 1$$

$$\sec^2 x + \sec x - 2 = 0$$

$$(\sec x + 2)(\sec x - 1) = 0 \qquad \text{**Factoring**}$$

$$\sec x = -2 \quad or \quad \sec x = 1 \qquad \text{**Principle of zero products**}$$

$$\cos x = -\frac{1}{2} \quad or \quad \cos x = 1 \qquad \text{**Using the identity** } \cos x = 1/\sec x$$

$$x = \frac{2\pi}{3}, \frac{4\pi}{3} \quad or \quad x = 0.$$

All these values check. The solutions in $[0, 2\pi)$ are 0, $2\pi/3$, and $4\pi/3$.

Graphical Solution

We graph $y = \tan^2 x + \sec x - 1$, but we enter this equation in the form

$$y_1 = \tan^2 x + \frac{1}{\cos x} - 1.$$

We use the ZERO feature to find zeros of the function.

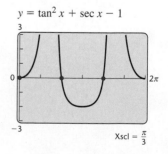

The solutions in $[0, 2\pi)$ are 0, $2\pi/3$, and $4\pi/3$.

> **Now Try Exercise 37.**

Sometimes we cannot find solutions algebraically, but we can approximate them with a graphing calculator.

EXAMPLE 12 Solve $\sin x - \cos x = \cot x$ in $[0, 2\pi)$.

Solution In the screen on the left below, we graph $y_1 = \sin x - \cos x$ and $y_2 = \cot x$ and determine the points of intersection. In the screen on the right, we graph the function $y_1 = \sin x - \cos x - \cot x$ and determine the zeros.

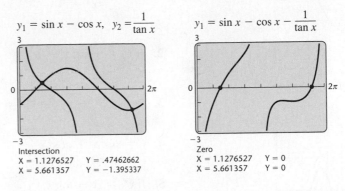

Each method leads to the approximate solutions 1.13 and 5.66 in $[0, 2\pi)$.

> **Now Try Exercise 47.**

Visualizing the Graph

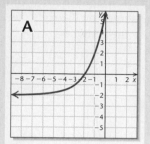

A

Match the equation with its graph.

1. $f(x) = \dfrac{4}{x^2 - 9}$

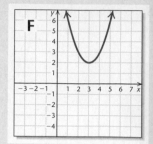

F

2. $f(x) = \dfrac{1}{2}\sin x - 1$

3. $(x - 2)^2 + (y + 3)^2 = 4$

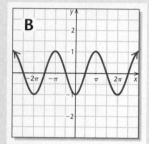

B

4. $y = \sin^2 x + \cos^2 x$

5. $f(x) = 3 - \log x$

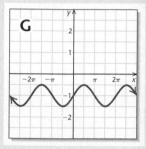

G

6. $f(x) = 2^{x+3} - 2$

7. $y = 2\cos\left(x - \dfrac{\pi}{2}\right)$

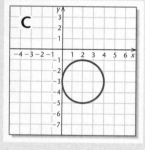

C

8. $y = -x^3 + 3x^2$

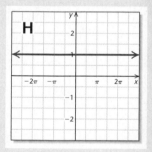

H

9. $f(x) = (x - 3)^2 + 2$

10. $f(x) = -\cos x$

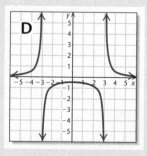

D

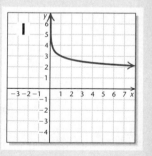

I

Answers on page A-45

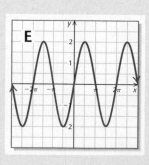

E

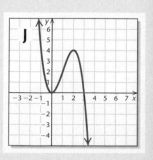

J

7.5   Exercise Set

Solve, finding all solutions. Express the solutions in both radians and degrees.

1. $\cos x = \dfrac{\sqrt{3}}{2}$

2. $\sin x = -\dfrac{\sqrt{2}}{2}$

3. $\tan x = -\sqrt{3}$

4. $\cos x = -\dfrac{1}{2}$

5. $\sin x = \dfrac{1}{2}$

6. $\tan x = -1$

7. $\cos x = -\dfrac{\sqrt{2}}{2}$

8. $\sin x = \dfrac{\sqrt{3}}{2}$

Solve, finding all solutions in $[0, 2\pi)$ or $[0°, 360°)$. Verify your answer using a graphing calculator.

9. $2\cos x - 1 = -1.2814$

10. $\sin x + 3 = 2.0816$

11. $2\sin x + \sqrt{3} = 0$

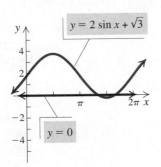

12. $2\tan x - 4 = 1$

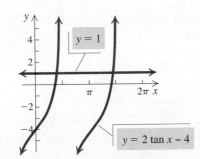

13. $2\cos^2 x = 1$

14. $\csc^2 x - 4 = 0$

15. $2\sin^2 x + \sin x = 1$

16. $\cos^2 x + 2\cos x = 3$

17. $2\cos^2 x - \sqrt{3}\cos x = 0$

18. $2\sin^2 \theta + 7\sin \theta = 4$

19. $6\cos^2 \phi + 5\cos \phi + 1 = 0$

20. $2\sin t \cos t + 2\sin t - \cos t - 1 = 0$

21. $\sin 2x \cos x - \sin x = 0$

22. $5\sin^2 x - 8\sin x = 3$

23. $\cos^2 x + 6\cos x + 4 = 0$

24. $2\tan^2 x = 3\tan x + 7$

25. $7 = \cot^2 x + 4\cot x$

26. $3\sin^2 x = 3\sin x + 2$

Solve, finding all solutions in $[0, 2\pi)$.

27. $\cos 2x - \sin x = 1$

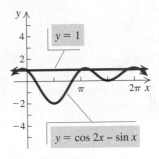

28. $2\sin x \cos x + \sin x = 0$

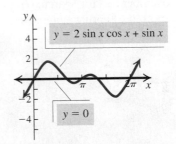

29. $\tan x \sin x - \tan x = 0$

30. $\sin 4x - 2\sin 2x = 0$

31. $\sin 2x \cos x + \sin x = 0$

32. $\cos 2x \sin x + \sin x = 0$

33. $2\sec x \tan x + 2\sec x + \tan x + 1 = 0$

34. $\sin 2x \sin x - \cos 2x \cos x = -\cos x$

35. $\sin 2x + \sin x + 2\cos x + 1 = 0$

36. $\tan^2 x + 4 = 2\sec^2 x + \tan x$

37. $\sec^2 x - 2\tan^2 x = 0$

38. $\cot x = \tan (2x - 3\pi)$

39. $2\cos x + 2\sin x = \sqrt{6}$

40. $\sqrt{3}\cos x - \sin x = 1$

41. $\sec^2 x + 2\tan x = 6$

42. $5 \cos 2x + \sin x = 4$

43. $\cos(\pi - x) + \sin\left(x - \dfrac{\pi}{2}\right) = 1$

44. $\dfrac{\sin^2 x - 1}{\cos\left(\dfrac{\pi}{2} - x\right) + 1} = \dfrac{\sqrt{2}}{2} - 1$

Solve using a calculator, finding all solutions in $[0, 2\pi)$.

45. $x \sin x = 1$

46. $x^2 + 2 = \sin x$

47. $2 \cos^2 x = x + 1$

48. $x \cos x - 2 = 0$

49. $\cos x - 2 = x^2 - 3x$

50. $\sin x = \tan \dfrac{x}{2}$

51. *Sales of Fishing Boats.* Sales of fishing boats fluctuate in cycles. The following cosine function can be used to estimate the total amount of sales of fishing boats, y, in thousands of dollars, in month x, for a business:

$$y = 15.328 \cos(0.475x - 1.728) + 87.223.$$

Approximate the total amount of sales to the nearest dollar for November and for March. (*Hint*: 1 represents January, 2 represents February, and so on.)

52. *Sales of Skis.* Sales of certain products fluctuate in cycles. The following sine function can be used to estimate the total amount of sales of skis, y, in thousands of dollars, in month x, for a business in a northern climate:

$$y = 9.584 \sin(0.436x + 2.097) + 10.558.$$

Approximate the total amount of sales to the nearest dollar for December and for July. (*Hint*: 1 represents January, 2 represents February, and so on.)

53. *Average High Temperature in Chicago.* The data in the following table give the average high temperature in Chicago for certain months.

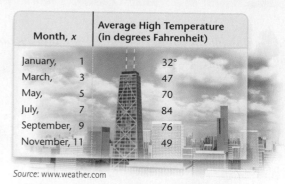

Month, x		Average High Temperature (in degrees Fahrenheit)
January,	1	32°
March,	3	47
May,	5	70
July,	7	84
September,	9	76
November,	11	49

Source: www.weather.com

a) Using the SINE REGRESSION feature on a graphing calculator, fit a sine function of the form $y = A \sin(BX - C) + D$ to this set of data.

b) Approximate the average high temperature in Chicago in April and in December.

c) Determine in which months the average high temperature is about 63°.

54. *Daylight Hours.* The data in the following table give the number of daylight hours for certain days in Kajaani, Finland.

Day, x		Number of Daylight Hours, y
January 10,	10	5.0
February 19,	50	9.1
March 3,	62	10.4
April 28,	118	16.4
May 14,	134	18.2
June 11,	162	20.7
July 17,	198	19.5
August 22,	234	15.7
September 19,	262	12.7
October 1,	274	11.4
November 14,	318	6.7
December 28,	362	4.3

Source: The Astronomical Almanac, 1995, Washington: U.S. Government Printing Office

a) Using the SINE REGRESSION feature on a graphing calculator, model these data with an equation of the form $y = A \sin(Bx - C) + D$.

b) Approximate the number of daylight hours in Kajaani for April 22 ($x = 112$), July 4 ($x = 185$), and December 15 ($x = 349$).

c) Determine on which days of the year there will be about 12 hr of daylight.

❯ Skill Maintenance

Solve the right triangle. [6.2]

55.

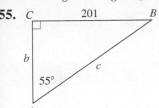

56.

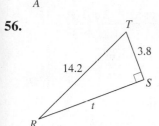

Solve. [1.5]

57. $\dfrac{x}{27} = \dfrac{4}{3}$

58. $\dfrac{0.01}{0.7} = \dfrac{0.2}{h}$

❯ Synthesis

Solve in $[0, 2\pi)$.

59. $|\sin x| = \dfrac{\sqrt{3}}{2}$

60. $|\cos x| = \dfrac{1}{2}$

61. $\sqrt{\tan x} = \sqrt[4]{3}$

62. $e^{\sin x} = 1$

63. $\ln (\cos x) = 0$

64. $e^{\ln (\sin x)} = 1$

65. $\sin (\ln x) = -1$

66. $12 \sin x - 7\sqrt{\sin x} + 1 = 0$

67. *Temperature During an Illness.*   The temperature T, in degrees Fahrenheit, of a patient t days into a 12-day illness is given by

$$T(t) = 101.6° + 3° \sin \left(\dfrac{\pi}{8}t \right).$$

Find the times t during the illness at which the patient's temperature was 103°.

68. *Satellite Location.*   A satellite circles the earth in such a manner that it is y miles from the equator (north or south, height from the surface not considered) t minutes after its launch, where

$$Y = 5000\left[\cos \dfrac{\pi}{45}(t - 10) \right].$$

At what times t on the interval $[0, 240]$, the first 4 hr, is the satellite 3000 mi north of the equator?

69. *Nautical Mile.*   (See Exercise 60 in Exercise Set 7.2.) In Great Britain, the *nautical mile* is defined as the length of a minute of arc of the earth's radius. Since the earth is flattened at the poles, a British nautical mile varies with latitude. In fact, it is given, in feet, by the function

$$N(\phi) = 6066 - 31 \cos 2\phi,$$

where ϕ is the latitude in degrees. At what latitude north is the length of a British nautical mile found to be 6040 ft?

70. *Acceleration Due to Gravity.*   (See Exercise 59 in Exercise Set 7.2.) The acceleration due to gravity is often denoted by g in a formula such as $S = \frac{1}{2}gt^2$, where S is the distance that an object falls in t seconds. The number g is generally considered constant, but in fact it varies slightly with latitude. If ϕ stands for latitude, in degrees, an excellent approximation of g is given by the formula

$$g = 9.78049(1 + 0.005288 \sin^2\phi - 0.000006 \sin^2 2\phi),$$

where g is measured in meters per second per second at sea level. At what latitude north does $g = 9.8$?

Solve.

71. $\cos^{-1}x = \cos^{-1}\frac{3}{5} - \sin^{-1}\frac{4}{5}$

72. $\sin^{-1}x = \tan^{-1}\frac{1}{3} + \tan^{-1}\frac{1}{2}$

73. Suppose that $\sin x = 5 \cos x$. Find $\sin x \cos x$.

STUDY GUIDE

KEY TERMS AND CONCEPTS	EXAMPLES

SECTION 7.1: IDENTITIES: PYTHAGOREAN AND SUM AND DIFFERENCE

Identity

An **identity** is an equation that is true for all possible replacements of the variables.

Basic Identities

$$\sin x = \frac{1}{\csc x}, \qquad \tan x = \frac{\sin x}{\cos x},$$

$$\cos x = \frac{1}{\sec x}, \qquad \cot x = \frac{\cos x}{\sin x},$$

$$\tan x = \frac{1}{\cot x},$$

$$\sin(-x) = -\sin x,$$

$$\cos(-x) = \cos x,$$

$$\tan(-x) = -\tan x$$

Pythagorean Identities

$$\sin^2 x + \cos^2 x = 1,$$

$$1 + \cot^2 x = \csc^2 x,$$

$$1 + \tan^2 x = \sec^2 x$$

Simplify.

a) $\dfrac{\tan^2 x \sec x}{\sec^2 x \tan x}$

b) $\dfrac{\cos^2 \alpha - 2\cos\alpha - 15}{\cos\alpha + 3}$

c) $\dfrac{\cos\theta}{\sin(-\theta)} + \dfrac{1}{\tan\theta}$

d) $\sqrt{\dfrac{1 + \sin x}{1 - \sin x}}$

a) $\dfrac{\tan^2 x \sec x}{\sec^2 x \tan x} = \dfrac{\tan x}{\sec x} = \dfrac{\dfrac{\sin x}{\cos x}}{\dfrac{1}{\cos x}} = \dfrac{\sin x}{\cos x} \cdot \dfrac{\cos x}{1} = \sin x$

b) $\dfrac{\cos^2 \alpha - 2\cos\alpha - 15}{\cos\alpha + 3} = \dfrac{(\cos\alpha - 5)(\cos\alpha + 3)}{\cos\alpha + 3}$

$$= \cos\alpha - 5$$

c) $\dfrac{\cos\theta}{\sin(-\theta)} + \dfrac{1}{\tan\theta} = -\dfrac{\cos\theta}{\sin\theta} + \dfrac{\cos\theta}{\sin\theta} = 0$

d) $\sqrt{\dfrac{1 + \sin x}{1 - \sin x}} = \sqrt{\dfrac{1 + \sin x}{1 - \sin x}} \cdot \sqrt{\dfrac{1 - \sin x}{1 - \sin x}}$

$$= \dfrac{\sqrt{1 - \sin^2 x}}{1 - \sin x}$$

$$= \dfrac{\sqrt{\cos^2 x}}{1 - \sin x} = \dfrac{\cos x}{1 - \sin x}$$

Sum and Difference Identities

$$\sin(u \pm v) = \sin u \cos v \pm \cos u \sin v,$$

$$\cos(u \pm v) = \cos u \cos v \mp \sin u \sin v,$$

$$\tan(u \pm v) = \frac{\tan u \pm \tan v}{1 \mp \tan u \tan v}$$

Evaluate $\cos\dfrac{7\pi}{12}$ exactly.

$$\cos\frac{7\pi}{12} = \cos\left(\frac{3\pi}{4} - \frac{\pi}{6}\right) = \cos\frac{3\pi}{4}\cos\frac{\pi}{6} + \sin\frac{3\pi}{4}\sin\frac{\pi}{6}$$

$$= \left(-\frac{\sqrt{2}}{2}\right)\left(\frac{\sqrt{3}}{2}\right) + \left(\frac{\sqrt{2}}{2}\right)\left(\frac{1}{2}\right) = \frac{\sqrt{2} - \sqrt{6}}{4}$$

Write $\sin 51° \cos 24° - \cos 51° \sin 24°$ as a trigonometric function of a single angle. Then evaluate.

We use the identity

$$\sin(u - v) = \sin u \cos v - \cos u \sin v.$$

Then

$$\sin 51° \cos 24° - \cos 51° \sin 24° = \sin(51° - 24°)$$

$$= \sin 27° \approx 0.4540.$$

Assume that $\sin \alpha = \frac{7}{8}$ and $\sin \beta = \frac{9}{10}$ and that α and β are between 0 and $\pi/2$. Then evaluate $\sin (\alpha + \beta)$.

We use the identity

$$\sin (\alpha + \beta) = \sin \alpha \cos \beta + \cos \alpha \sin \beta.$$

Substituting, we have

$$\sin (\alpha + \beta) = \tfrac{7}{8} \cos \beta + \tfrac{9}{10} \cos \alpha.$$

To determine $\cos \alpha$ and $\cos \beta$, we use reference triangles and the Pythagorean theorem.

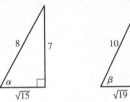

$$\cos \alpha = \frac{\sqrt{15}}{8} \quad \text{and} \quad \cos \beta = \frac{\sqrt{19}}{10}$$

Then

$$\sin (\alpha + \beta) = \frac{7}{8} \cdot \frac{\sqrt{19}}{10} + \frac{9}{10} \cdot \frac{\sqrt{15}}{8} = \frac{7\sqrt{19} + 9\sqrt{15}}{80}.$$

SECTION 7.2: IDENTITIES: COFUNCTION, DOUBLE-ANGLE, AND HALF-ANGLE

Cofunction Identities

$$\sin\left(\frac{\pi}{2} - x\right) = \cos x,$$

$$\cos\left(\frac{\pi}{2} - x\right) = \sin x,$$

$$\tan\left(\frac{\pi}{2} - x\right) = \cot x,$$

$$\cot\left(\frac{\pi}{2} - x\right) = \tan x,$$

$$\sec\left(\frac{\pi}{2} - x\right) = \csc x,$$

$$\csc\left(\frac{\pi}{2} - x\right) = \sec x,$$

$$\sin\left(x \pm \frac{\pi}{2}\right) = \pm\cos x,$$

$$\cos\left(x \pm \frac{\pi}{2}\right) = \mp\sin x$$

Given that $\cos \theta = \frac{3}{5}$ and that the terminal side of θ is in quadrant IV, find the exact function value for $\cos\left(\frac{\pi}{2} - \theta\right)$ and $\cos\left(\theta + \frac{\pi}{2}\right)$.

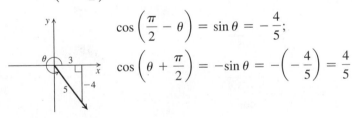

$$\cos\left(\frac{\pi}{2} - \theta\right) = \sin \theta = -\frac{4}{5};$$

$$\cos\left(\theta + \frac{\pi}{2}\right) = -\sin \theta = -\left(-\frac{4}{5}\right) = \frac{4}{5}$$

Double-Angle Identities

$\sin 2x = 2 \sin x \cos x,$

$\cos 2x = \cos^2 x - \sin^2 x$

$\quad\quad = 1 - 2 \sin^2 x$

$\quad\quad = 2 \cos^2 x - 1,$

$\tan 2x = \dfrac{2 \tan x}{1 - \tan^2 x}$

Given that $\tan \theta = -3$ and that θ is in quadrant II, find $\sin 2\theta$, $\cos 2\theta$, and the quadrant in which 2θ lies.

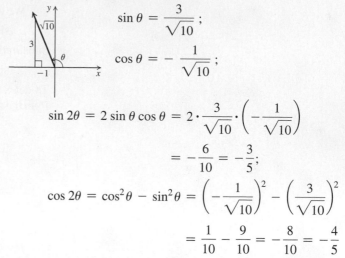

$$\sin \theta = \frac{3}{\sqrt{10}};$$

$$\cos \theta = -\frac{1}{\sqrt{10}};$$

$$\sin 2\theta = 2 \sin \theta \cos \theta = 2 \cdot \frac{3}{\sqrt{10}} \cdot \left(-\frac{1}{\sqrt{10}}\right)$$

$$= -\frac{6}{10} = -\frac{3}{5};$$

$$\cos 2\theta = \cos^2 \theta - \sin^2 \theta = \left(-\frac{1}{\sqrt{10}}\right)^2 - \left(\frac{3}{\sqrt{10}}\right)^2$$

$$= \frac{1}{10} - \frac{9}{10} = -\frac{8}{10} = -\frac{4}{5}$$

Since both $\sin 2\theta$ and $\cos 2\theta$ are negative, we have 2θ in quadrant III.

Simplify: $\dfrac{\frac{1}{2} \sin 2x}{1 - \cos^2 x}$.

$$\frac{\frac{1}{2} \sin 2x}{1 - \cos^2 x} = \frac{\frac{1}{2} \cdot 2 \sin x \cos x}{\sin^2 x} = \frac{\cos x}{\sin x} = \cot x$$

Half-Angle Identities

$\sin \dfrac{x}{2} = \pm \sqrt{\dfrac{1 - \cos x}{2}},$

$\cos \dfrac{x}{2} = \pm \sqrt{\dfrac{1 + \cos x}{2}},$

$\tan \dfrac{x}{2} = \pm \sqrt{\dfrac{1 - \cos x}{1 + \cos x}}$

$\quad\quad = \dfrac{\sin x}{1 + \cos x}$

$\quad\quad = \dfrac{1 - \cos x}{\sin x}$

Evaluate $\sin \dfrac{3\pi}{8}$ exactly.

Note that $\dfrac{3\pi}{8}$ is in quadrant I. Thus $\sin \dfrac{3\pi}{8}$ is positive.

$$\sin \frac{3\pi}{8} = \frac{\frac{3\pi}{4}}{2} = \sqrt{\frac{1 - \cos \frac{3\pi}{4}}{2}} = \sqrt{\frac{1 - \left(-\frac{\sqrt{2}}{2}\right)}{2}}$$

$$= \sqrt{\frac{2 + \sqrt{2}}{4}} = \frac{\sqrt{2 + \sqrt{2}}}{2}$$

Simplify: $\sin^2 \dfrac{x}{2} + \cos x$.

$$\sin^2 \frac{x}{2} + \cos x = \frac{1 - \cos x}{2} + \frac{2 \cos x}{2} = \frac{1 + \cos x}{2}$$

SECTION 7.3: PROVING TRIGONOMETRIC IDENTITIES

Proving Identities

Method 1: Start with either the left side or the right side of the equation and obtain the other side.

Method 2: Work with each side separately until you obtain the same expression.

Using method 1, prove $\dfrac{\tan x \sin^2 x}{1 + \cos x} = \tan x - \sin x$.

We begin with the left side and obtain the right side:

$$\frac{\tan x \sin^2 x}{1 + \cos x} = \frac{\tan x\,(1 - \cos^2 x)}{1 + \cos x}$$

$$= \frac{\tan x\,(1 - \cos x)\,(1 + \cos x)}{(1 + \cos x)}$$

$$= \tan x\,(1 - \cos x)$$

$$= \tan x - \tan x \cos x$$

$$= \tan x - \sin x.$$

Using method 2, prove $\dfrac{\sec \alpha}{1 + \cot^2 \alpha} = \dfrac{1 - \cos^2 \alpha}{\cos \alpha}$.

We begin with the left side:

$$\frac{\sec \alpha}{1 + \cot^2 \alpha} = \frac{\dfrac{1}{\cos \alpha}}{\csc^2 \alpha} = \frac{1}{\cos \alpha} \cdot \frac{\sin^2 \alpha}{1}$$

$$= \sin \alpha \cdot \frac{\sin \alpha}{\cos \alpha} = \sin \alpha \cdot \tan \alpha.$$

Next, we work with the right side:

$$\frac{1 - \cos^2 \alpha}{\cos \alpha} = \frac{\sin^2 \alpha}{\cos \alpha} = \sin \alpha \cdot \frac{\sin \alpha}{\cos \alpha} = \sin \alpha \cdot \tan \alpha.$$

We have obtained the same expression from each side, so the proof is complete.

Product-to-Sum Identities

$$\sin x \cdot \sin y = \frac{1}{2}[\cos\,(x - y) - \cos\,(x + y)],$$

$$\cos x \cdot \cos y = \frac{1}{2}[\cos\,(x - y) + \cos\,(x + y)],$$

$$\sin x \cdot \cos y = \frac{1}{2}[\sin\,(x + y) + \sin\,(x - y)],$$

$$\cos x \cdot \sin y = \frac{1}{2}[\sin\,(x + y) - \sin\,(x - y)]$$

Sum-to-Product Identities

$$\sin x + \sin y = 2 \sin \frac{x + y}{2} \cos \frac{x - y}{2},$$

$$\sin x - \sin y = 2 \cos \frac{x + y}{2} \sin \frac{x - y}{2},$$

$$\cos y + \cos x = 2 \cos \frac{x + y}{2} \cos \frac{x - y}{2},$$

$$\cos y - \cos x = 2 \sin \frac{x + y}{2} \sin \frac{x - y}{2}$$

Use the product-to-sum or the sum-to-product identities to find an identity for $3 \sin 2\beta \sin 5\beta$ and for $\cos 3\alpha - \cos \alpha$.

Using $\sin x \cdot \sin y = \dfrac{1}{2}[\cos\,(x - y) - \cos\,(x + y)]$, we have

$$3 \sin 2\beta \sin 5\beta = 3 \cdot \frac{1}{2}[\cos\,(-3\beta) - \cos 7\beta]$$

$$= \frac{3}{2}[\cos 3\beta - \cos 7\beta].$$

Using $\cos y - \cos x = 2 \sin \dfrac{x + y}{2} \sin \dfrac{x - y}{2}$, we have

$$\cos 3\alpha - \cos \alpha = 2 \sin \frac{4\alpha}{2} \sin \frac{(-2\alpha)}{2}$$

$$= 2 \sin 2\alpha \sin\,(-\alpha)$$

$$= -2 \sin 2\alpha \sin \alpha$$

$$= -2\,(2 \sin \alpha \cos \alpha)\,(\sin \alpha)$$

$$= -4 \sin^2 \alpha \cos \alpha.$$

SECTION 7.4: INVERSES OF THE TRIGONOMETRIC FUNCTIONS

Inverse Trigonometric Functions

FUNCTION	DOMAIN	RANGE
$y = \sin^{-1}x$	$[-1, 1]$	$\left[-\dfrac{\pi}{2}, \dfrac{\pi}{2}\right]$
$y = \cos^{-1}x$	$[-1, 1]$	$[0, \pi]$
$y = \tan^{-1}x$	$(-\infty, \infty)$	$\left(-\dfrac{\pi}{2}, \dfrac{\pi}{2}\right)$

The notation $y = \sin^{-1}x$ is equivalent to $y = \arcsin x$. The notation can be read:

- the inverse sine of x,
- the arcsine of x, or
- the number, or the angle, whose sine is x.

Find each of the following function values.

a) $\sin^{-1}\left(-\dfrac{1}{2}\right)$ **b)** $\cos^{-1}\left(-\dfrac{\sqrt{2}}{2}\right)$ **c)** $\tan^{-1}\sqrt{3}$

a)

In $[-\pi/2, \pi/2]$, the only number whose sine is $-\dfrac{1}{2}$ is $-\pi/6$. Thus,

$$\sin^{-1}\left(-\dfrac{1}{2}\right) = -\pi/6, \text{ or } -30°.$$

b)

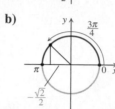

In $[0, \pi]$, the only number whose cosine is $-\dfrac{\sqrt{2}}{2}$ is $3\pi/4$. Thus,

$$\cos^{-1}\left(-\dfrac{\sqrt{2}}{2}\right) = 3\pi/4, \text{ or } 135°.$$

c)

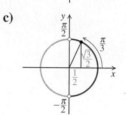

In $(-\pi/2, \pi/2)$, the only number whose tangent is $\sqrt{3}$ is $\pi/3$. Thus,

$$\tan^{-1}\sqrt{3} = \pi/3, \text{ or } 60°.$$

Approximate each of the following function values in both radians and degrees. Round radian measure to four decimal places and degree measure to the nearest tenth of a degree.

a) $\cos^{-1}0.3281 \approx 1.2365$, or $70.8°$

b) $\tan^{-1}(-7.1154) \approx -1.4312$, or $-82.0°$

c) $\sin^{-1}(-0.5492) \approx -0.5814$, or $-33.3°$

Composition of Trigonometric Functions

The following are true for any x in the domain of the inverse function:

$$\sin(\sin^{-1}x) = x,$$
$$\cos(\cos^{-1}x) = x,$$
$$\tan(\tan^{-1}x) = x.$$

The following are true for any x in the range of the inverse function:

$$\sin^{-1}(\sin x) = x,$$
$$\cos^{-1}(\cos x) = x,$$
$$\tan^{-1}(\tan x) = x.$$

Simplify each of the following.

a) $\tan\left(\tan^{-1}\sqrt{3}\right)$

Since $\sqrt{3}$ is in $(-\infty, \infty)$, the domain of $\tan^{-1}$,
$\tan\left(\tan^{-1}\sqrt{3}\right) = \sqrt{3}$.

b) $\cos\left[\cos^{-1}\left(-\dfrac{1}{2}\right)\right]$

Since $-\dfrac{1}{2}$ is in $[-1, 1]$, the domain of $\cos^{-1}$,

$$\cos\left[\cos^{-1}\left(-\dfrac{1}{2}\right)\right] = -\dfrac{1}{2}.$$

c) $\sin^{-1}[\sin(\pi/6)]$

Since $\pi/6$ is in $[-\pi/2, \pi/2]$, the range of $\sin^{-1}$,
$\sin^{-1}[\sin(\pi/6)] = \pi/6$.

d) $\cos^{-1}[\cos(3\pi/2)]$

Since $3\pi/2$ is not in $[0, \pi]$, the range of $\cos^{-1}$, we cannot apply $\cos^{-1}(\cos x) = x$. Instead, we find $\cos(3\pi/2)$, which is 0, and substitute to get $\cos^{-1} 0 = \pi/2$. Thus, $\cos^{-1}[\cos(3\pi/2)] = \pi/2$.

e) $\sin^{-1}[\cos(-\pi/6)] = \sin^{-1}\dfrac{\sqrt{3}}{2}$ $\cos(-\pi/6) = \sqrt{3}/2$

$$= \pi/3$$

f) $\tan\left[\cos^{-1}\left(-\dfrac{1}{2}\right)\right] = \tan\dfrac{2\pi}{3}$ $\cos^{-1}\left(-\dfrac{1}{2}\right) = 2\pi/3$

$$= -\sqrt{3}$$

Find: $\sin\left(\cos^{-1}\dfrac{x}{5}\right)$.

Let θ be the angle whose cosine is $x/5$: $\cos\theta = x/5$. Considering all values of x, we draw right triangles, in which the length of the hypotenuse is 5 and the length of one leg is x. The other leg in each triangle is $\sqrt{25 - x^2}$.

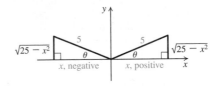

Thus, $\sin\left(\cos^{-1}\dfrac{x}{5}\right) = \dfrac{\sqrt{25 - x^2}}{5}$.

SECTION 7.5: SOLVING TRIGONOMETRIC EQUATIONS

Trigonometric Equations

When an equation contains a trigonometric expression with a variable, it is called a trigonometric equation. To solve such equations, we find all values for the variable that make the equation true.

In most applications, it is sufficient to find just the solutions from 0 to 2π, or from 0° to 360°. We then remember that any multiple of 2π, or 360°, can be added to obtain the rest of the solutions.

Solve $3\tan x = \sqrt{3}$. Find *all* solutions.

$$3\tan x = \sqrt{3}$$
$$\tan x = \sqrt{3}/3$$

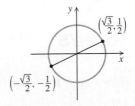

The solutions are numbers whose tangent is $\sqrt{3}/3$. There are only two points on the unit circle for which the tangent is $\sqrt{3}/3$. They are the points corresponding to $\pi/6$ and $7\pi/6$. The solutions are $\pi/6 + k\pi$, where k is any integer.

Solve: $\sin x = 1 - 2 \sin^2 x$ in $[0°, 360°)$.

$$\sin x = 1 - 2 \sin^2 x$$
$$2 \sin^2 x + \sin x - 1 = 0$$
$$(2 \sin x - 1)(\sin x + 1) = 0$$
$$2 \sin x - 1 = 0 \quad or \quad \sin x + 1 = 0$$
$$\sin x = \tfrac{1}{2} \quad or \quad \sin x = -1$$

Thus, $x = 30°, 150°$ *or* $x = 270°$. All values check. The solutions in $[0°, 360°)$ are $30°$, $150°$, and $270°$.

Solve: $\sin 2x \cos x + \cos x = 0$ in $[0, 2\pi)$.

$$\sin 2x \cos x + \cos x = 0$$
$$\cos x (\sin 2x + 1) = 0$$
$$\cos x = 0 \qquad or \quad \sin 2x + 1 = 0$$
$$\cos x = 0 \qquad or \qquad \sin 2x = -1$$
$$x = \pi/2, 3\pi/2 \quad or \qquad 2x = 3\pi/2, 7\pi/2$$
$$x = 3\pi/4, 7\pi/4$$

All values check. The solutions in $[0, 2\pi)$ are $\pi/2, 3\pi/4$, $3\pi/2$, and $7\pi/4$.

Solve $5 \cos^2 x = 2 \cos x + 6$ in $[0°, 360°)$.

$$5 \cos^2 x - 2 \cos x - 6 = 0$$

$$\cos x = \frac{2 \pm \sqrt{124}}{10} \qquad \textbf{Using the quadratic formula}$$

$$\cos x \approx 1.3136 \quad or \quad \cos x \approx -0.9136$$

Since cosine values are never greater than 1, $\cos x \approx 1.3136$ has no solution. Using $\cos x \approx -0.9136$, we find that $\cos^{-1}(-0.9136) \approx 156.01°$. Thus the solutions in $[0°, 360°)$ are $156.01°$ and $360° - 156.01°$, or $203.99°$.

REVIEW EXERCISES

Determine whether the statement is true or false.

1. $\sin^2 s \neq \sin s^2$. **[7.1]**

2. Given $0 < \alpha < \pi/2$ and $0 < \beta < \pi/2$ and that $\sin(\alpha + \beta) = 1$ and $\sin(\alpha - \beta) = 0$, then $\alpha = \pi/4$. **[7.1]**

3. If the terminal side of θ is in quadrant IV, then $\tan \theta < \cos \theta$. **[7.1]**

4. $\cos 5\pi/12 = \cos 7\pi/12$. **[7.2]**

5. Given that $\sin \theta = -\tfrac{2}{5}$, $\tan \theta < \cos \theta$. **[7.1]**

Complete the Pythagorean identity. **[7.1]**

6. $1 + \cot^2 x =$

7. $\sin^2 x + \cos^2 x =$

Multiply and simplify. Check using a graphing calculator. **[7.1]**

8. $(\tan y - \cot y)(\tan y + \cot y)$

9. $(\cos x + \sec x)^2$

Factor and simplify. Check using a graphing calculator. **[7.1]**

10. $\sec x \csc x - \csc^2 x$

11. $3 \sin^2 y - 7 \sin y - 20$

12. $1000 - \cos^3 u$

Simplify and check using a graphing calculator. **[7.1]**

13. $\dfrac{\sec^4 x - \tan^4 x}{\sec^2 x + \tan^2 x}$

14. $\dfrac{2 \sin^2 x}{\cos^3 x} \cdot \left(\dfrac{\cos x}{2 \sin x} \right)^2$

15. $\dfrac{3 \sin x}{\cos^2 x} \cdot \dfrac{\cos^2 x + \cos x \sin x}{\sin^2 x - \cos^2 x}$

16. $\dfrac{3}{\cos y - \sin y} - \dfrac{2}{\sin^2 y - \cos^2 y}$

17. $\left(\dfrac{\cot x}{\csc x} \right)^2 + \dfrac{1}{\csc^2 x}$

18. $\dfrac{4 \sin x \cos^2 x}{16 \sin^2 x \cos x}$

In Exercises 19–21, assume that all radicands are nonnegative.

19. Simplify:

$$\sqrt{\sin^2 x + 2 \cos x \sin x + \cos^2 x}. \text{ [7.1]}$$

20. Rationalize the denominator: $\sqrt{\dfrac{1 + \sin x}{1 - \sin x}}.$ **[7.1]**

21. Rationalize the numerator: $\sqrt{\dfrac{\cos x}{\tan x}}.$ **[7.1]**

22. Given that $x = 3 \tan \theta$, express $\sqrt{9 + x^2}$ as a trigonometric function without radicals. Assume that $0 < \theta < \pi/2$. **[7.1]**

Use the sum and difference formulas to write equivalent expressions. You need not simplify. **[7.1]**

23. $\cos \left(x + \dfrac{3\pi}{2} \right)$ **24.** $\tan (45° - 30°)$

25. Simplify: $\cos 27° \cos 16° + \sin 27° \sin 16°$. **[7.1]**

26. Find $\cos 165°$ exactly. **[7.1]**

27. Given that $\tan \alpha = \sqrt{3}$ and $\sin \beta = \sqrt{2}/2$ and that α and β are between 0 and $\pi/2$, evaluate $\tan (\alpha - \beta)$ exactly. **[7.1]**

28. Assume that $\sin \theta = 0.5812$ and $\cos \phi = 0.2341$ and that both θ and ϕ are first-quadrant angles. Evaluate $\cos (\theta + \phi)$. **[7.1]**

Complete the cofunction identity. **[7.2]**

29. $\cos \left(x + \dfrac{\pi}{2} \right) =$

30. $\cos \left(\dfrac{\pi}{2} - x \right) =$

31. $\sin \left(x - \dfrac{\pi}{2} \right) =$

32. Given that $\cos \alpha = -\dfrac{3}{5}$ and that the terminal side is in quadrant III:

  a) Find the other function values for α. **[7.2]**
  b) Find the six function values for $\pi/2 - \alpha$. **[7.2]**
  c) Find the six function values for $\alpha + \pi/2$. **[7.2]**

33. Find an equivalent expression for $\csc \left(x - \dfrac{\pi}{2} \right)$. **[7.2]**

34. Find $\tan 2\theta$, $\cos 2\theta$, and $\sin 2\theta$ and the quadrant in which 2θ lies, where $\cos \theta = -\dfrac{4}{5}$ and θ is in quadrant III. **[7.2]**

35. Find $\sin \dfrac{\pi}{8}$ exactly. **[7.2]**

36. Given that $\sin \beta = 0.2183$ and β is in quadrant I, find $\sin 2\beta$, $\cos \dfrac{\beta}{2}$, and $\cos 4\beta$. **[7.2]**

Simplify and check using a graphing calculator. **[7.2]**

37. $1 - 2 \sin^2 \dfrac{x}{2}$

38. $(\sin x + \cos x)^2 - \sin 2x$

39. $2 \sin x \cos^3 x + 2 \sin^3 x \cos x$

40. $\dfrac{2 \cot x}{\cot^2 x - 1}$

Prove the identity. **[7.3]**

41. $\dfrac{1 - \sin x}{\cos x} = \dfrac{\cos x}{1 + \sin x}$

42. $\dfrac{1 + \cos 2\theta}{\sin 2\theta} = \cot \theta$

43. $\dfrac{\tan y + \sin y}{2 \tan y} = \cos^2 \dfrac{y}{2}$

44. $\dfrac{\sin x - \cos x}{\cos^2 x} = \dfrac{\tan^2 x - 1}{\sin x + \cos x}$

Use the product-to-sum identities and the sum-to-product identities to find identities for each of the following. **[7.3]**

45. $3 \cos 2\theta \sin \theta$

46. $\sin \theta - \sin 4\theta$

In Exercises 47–50, use a graphing calculator to determine which expression (A)–(D) on the right can be used to complete the identity. Then prove the identity algebraically. **[7.3]**

47. $\csc x - \cos x \cot x$

48. $\dfrac{1}{\sin x \cos x} - \dfrac{\cos x}{\sin x}$

49. $\dfrac{\cot x - 1}{1 - \tan x}$

50. $\dfrac{\cos x + 1}{\sin x} + \dfrac{\sin x}{\cos x + 1}$

A. $\dfrac{\csc x}{\sec x}$

B. $\sin x$

C. $\dfrac{2}{\sin x}$

D. $\dfrac{\sin x \cos x}{1 - \sin^2 x}$

Find each of the following exactly in both radians and degrees. **[7.4]**

51. $\sin^{-1}\left(-\dfrac{1}{2}\right)$

52. $\cos^{-1}\dfrac{\sqrt{3}}{2}$

53. $\tan^{-1}1$

54. $\sin^{-1}0$

Use a calculator to find each of the following in radians, rounded to four decimal places, and in degrees, rounded to the nearest tenth of a degree. **[7.4]**

55. $\cos^{-1}(-0.2194)$

56. $\cot^{-1}2.381$

Evaluate. **[7.4]**

57. $\cos\left(\cos^{-1}\dfrac{1}{2}\right)$

58. $\tan^{-1}\left(\tan\dfrac{\sqrt{3}}{3}\right)$

59. $\sin^{-1}\left(\sin\dfrac{\pi}{7}\right)$

60. $\cos\left(\sin^{-1}\dfrac{\sqrt{2}}{2}\right)$

Find each of the following. **[7.4]**

61. $\cos\left(\tan^{-1}\dfrac{b}{3}\right)$

62. $\cos\left(2\sin^{-1}\dfrac{4}{5}\right)$

Solve, finding all solutions. Express the solutions in both radians and degrees. **[7.5]**

63. $\cos x = -\dfrac{\sqrt{2}}{2}$

64. $\tan x = \sqrt{3}$

Solve, finding all solutions in $[0, 2\pi)$. **[7.5]**

65. $4\sin^2 x = 1$

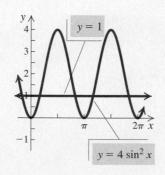

66. $\sin 2x \, \sin x - \cos x = 0$

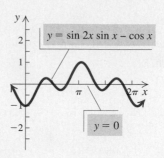

67. $2\cos^2 x + 3\cos x = -1$

68. $\sin^2 x - 7\sin x = 0$

69. $\csc^2 x - 2\cot^2 x = 0$

70. $\sin 4x + 2\sin 2x = 0$

71. $2\cos x + 2\sin x = \sqrt{2}$

72. $6\tan^2 x = 5\tan x + \sec^2 x$

Solve using a graphing calculator, finding all solutions in $[0, 2\pi)$. **[7.5]**

73. $x\cos x = 1$

74. $2\sin^2 x = x + 1$

75. Which of the following is the domain of the function $\cos^{-1}x$? **[7.4]**

A. $(0, \pi)$ **B.** $[-1, 1]$
C. $[-\pi/2, \pi/2]$ **D.** $(-\infty, \infty)$

76. Simplify: $\sin^{-1}\left(\sin\dfrac{7\pi}{6}\right)$. **[7.4]**

A. $-\pi/6$ **B.** $7\pi/6$
C. $-1/2$ **D.** $11\pi/6$

77. The graph of $f(x) = \sin^{-1}x$ is which of the following? **[7.4]**

A.
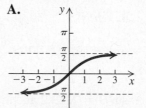

B.
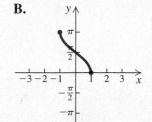

C.
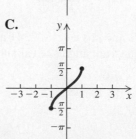

D.
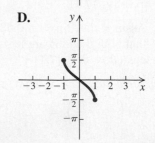

❯ Synthesis

78. Find the measure of the angle from l_1 to l_2:

l_1: $x + y = 3$; l_2: $2x - y = 5$. **[7.1]**

79. Find an identity for $\cos(u + v)$ involving only cosines. **[7.1], [7.2]**

80. Simplify: $\cos\left(\dfrac{\pi}{2} - x\right)[\csc x - \sin x]$. **[7.2]**

81. Find $\sin\theta$, $\cos\theta$, and $\tan\theta$ under the given conditions:

$\sin 2\theta = \dfrac{1}{5}$, $\dfrac{\pi}{2} \le 2\theta < \pi$. **[7.2]**

82. Show that

$$\tan^{-1}x = \frac{\sin^{-1}x}{\cos^{-1}x}$$

is *not* an identity. **[7.4]**

83. Solve $e^{\cos x} = 1$ in $[0, 2\pi)$. **[7.5]**

❯ Collaborative Discussion and Writing

84. Why are the ranges of the inverse trigonometric functions restricted? **[7.4]**

85. Miles lists his answer to a problem as $\pi/6 + k\pi$, for any integer k, while Jaylen lists his answer as $\pi/6 + 2k\pi$ and $7\pi/6 + 2k\pi$, for any integer k. Are their answers equivalent? Why or why not? **[7.5]**

86. How does the graph of $y = \sin^{-1}x$ differ from the graph of $y = \sin x$? **[7.4]**

87. What is the difference between a trigonometric equation that is an identity and a trigonometric equation that is not an identity? Give an example of each. **[7.1], [7.5]**

88. Why is it that

$$\sin\frac{5\pi}{6} = \frac{1}{2}, \quad \text{but} \quad \sin^{-1}\left(\frac{1}{2}\right) \ne \frac{5\pi}{6}? \text{ **[7.4]**}$$

---

7 | **Chapter Test**

Simplify.

1. $\dfrac{2\cos^2 x - \cos x - 1}{\cos x - 1}$

2. $\left(\dfrac{\sec x}{\tan x}\right)^2 - \dfrac{1}{\tan^2 x}$

3. Rationalize the denominator:

$$\sqrt{\frac{1 - \sin\theta}{1 + \sin\theta}}.$$

Assume that the radicand is nonnegative.

4. Given that $x = 2\sin\theta$, express $\sqrt{4 - x^2}$ as a trigonometric function without radicals. Assume that $0 < \theta < \pi/2$.

Use the sum or difference identities to evaluate exactly.

5. $\sin 75°$

6. $\tan\dfrac{\pi}{12}$

7. Assuming that $\cos u = \frac{5}{13}$ and $\cos v = \frac{12}{13}$ and that u and v are between 0 and $\pi/2$, evaluate $\cos(u - v)$ exactly.

8. Given that $\cos\theta = -\frac{2}{3}$ and that the terminal side is in quadrant II, find $\cos(\pi/2 - \theta)$.

9. Given that $\sin\theta = -\frac{4}{5}$ and θ is in quadrant III, find $\sin 2\theta$ and the quadrant in which 2θ lies.

10. Use a half-angle identity to evaluate $\cos\dfrac{\pi}{12}$ exactly.

11. Given that $\sin\theta = 0.6820$ and that θ is in quadrant I, find $\cos(\theta/2)$.

12. Simplify: $(\sin x + \cos x)^2 - 1 + 2\sin 2x$.

Prove each of the following identities.

13. $\csc x - \cos x \cot x = \sin x$

14. $(\sin x + \cos x)^2 = 1 + \sin 2x$

15. $(\csc\beta + \cot\beta)^2 = \dfrac{1 + \cos\beta}{1 - \cos\beta}$

16. $\dfrac{1 + \sin\alpha}{1 + \csc\alpha} = \dfrac{\tan\alpha}{\sec\alpha}$

Use the product-to-sum identities and the sum-to-product identities to find identities for each of the following.

17. $\cos 8\alpha - \cos \alpha$

18. $4 \sin \beta \cos 3\beta$

19. Find $\sin^{-1}\left(-\dfrac{\sqrt{2}}{2}\right)$ exactly in degrees.

20. Find $\tan^{-1}\sqrt{3}$ exactly in radians.

21. Use a calculator to find $\cos^{-1}(-0.6716)$ in radians, rounded to four decimal places.

22. Evaluate $\cos\left(\sin^{-1}\dfrac{1}{2}\right)$.

23. Find $\tan\left(\sin^{-1}\dfrac{5}{x}\right)$.

24. Evaluate $\cos\left(\sin^{-1}\frac{1}{2} + \cos^{-1}\frac{1}{2}\right)$.

Solve, finding all solutions in $[0, 2\pi)$.

25. $4 \cos^2 x = 3$

26. $2 \sin^2 x = \sqrt{2} \sin x$

27. $\sqrt{3} \cos x + \sin x = 1$

28. The graph of $f(x) = \cos^{-1}x$ is which of the following?

A.
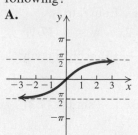

B.
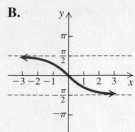

C.
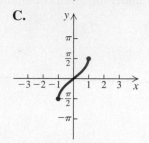

D.
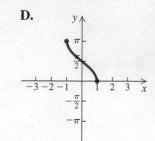

❯ Synthesis

29. Find $\cos \theta$, given that

$$\cos 2\theta = \frac{5}{6}, \quad \frac{3\pi}{2} < \theta < 2\pi.$$

Applications of Trigonometry

8

8.1 The Law of Sines

8.2 The Law of Cosines

8.3 Complex Numbers:
Trigonometric Notation

Mid-Chapter Mixed Review

8.4 Polar Coordinates
and Graphs

Visualizing the Graph

8.5 Vectors and Applications

8.6 Vector Operations

Study Guide

Review Exercises

Test

**This problem appears as
Exercise 25 in Section 8.2.**

A teacup ride for children at an amusement park consists of five teacups equally spaced around a circle. Each cup holds 6 passengers and is at the end of an arm 20 ft long. Find the linear distance between a pair of adjacent cups. Round the length to the nearest tenth of a foot.

8.1 ▷ The Law of Sines

> ❯ Use the law of sines to solve triangles.
>
> ❯ Find the area of any triangle given the lengths of two sides and the measure of the included angle.

To **solve a triangle** means to find the lengths of all its sides and the measures of all its angles. We solved right triangles in Section 6.2. For review, let's solve the right triangle shown below. We begin by listing the known measures.

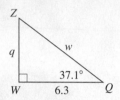

$$Q = 37.1°, \qquad q = ?,$$
$$W = 90°, \qquad w = ?,$$
$$Z = ?, \qquad z = 6.3$$

Since the sum of the three angle measures of any triangle is 180°, we can immediately find the measure of the third angle:

$$Z = 180° - (90° + 37.1°)$$
$$= 52.9°.$$

Then using the tangent and cosine ratios, respectively, we can find q and w:

$$\tan 37.1° = \frac{q}{6.3}, \quad \text{or}$$
$$q = 6.3 \tan 37.1° \approx 4.8,$$

and $\quad \cos 37.1° = \dfrac{6.3}{w}, \quad$ or

$$w = \frac{6.3}{\cos 37.1°} \approx 7.9.$$

Now all six measures are known and we have solved triangle QWZ.

$$Q = 37.1°, \qquad q \approx 4.8,$$
$$W = 90°, \qquad w \approx 7.9,$$
$$Z = 52.9°, \qquad z = 6.3$$

❯ Solving Oblique Triangles

The trigonometric functions can also be used to solve triangles that are not right triangles. Such triangles are called **oblique.** Any triangle, right or oblique, can be solved *if at least one side and any other two measures are known.* The five possible situations are illustrated on the next page.

1. **AAS:** Two angles of a triangle and a side opposite one of them are known.

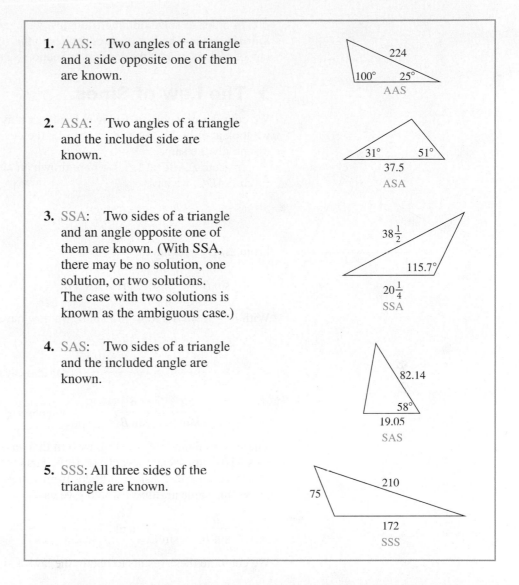

2. **ASA:** Two angles of a triangle and the included side are known.

3. **SSA:** Two sides of a triangle and an angle opposite one of them are known. (With SSA, there may be no solution, one solution, or two solutions. The case with two solutions is known as the ambiguous case.)

4. **SAS:** Two sides of a triangle and the included angle are known.

5. **SSS:** All three sides of the triangle are known.

The list above does not include the situation in which only the three angle measures are given. The reason for this lies in the fact that the angle measures determine *only the shape* of the triangle and *not the size*, as shown with the following triangles. Thus we cannot solve a triangle when only the three angle measures are given.

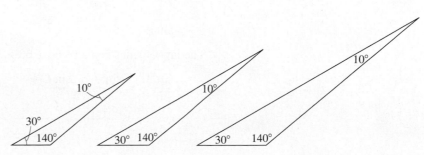

In order to solve oblique triangles, we need to derive the *law of sines* and the *law of cosines*. The law of sines applies to the first three situations listed above. The law of cosines, which we develop in Section 8.2, applies to the last two situations.

❯ The Law of Sines

We consider any oblique triangle. It may or may not have an obtuse angle. Although we look at only the acute-triangle case, the derivation of the obtuse-triangle case is essentially the same.

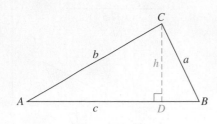

In acute $\triangle ABC$ at left, we have drawn an altitude from vertex C. It has length h. From $\triangle ADC$, we have

$$\sin A = \frac{h}{b}, \quad \text{or} \quad h = b \sin A.$$

From $\triangle BDC$, we have

$$\sin B = \frac{h}{a}, \quad \text{or} \quad h = a \sin B.$$

With $h = b \sin A$ and $h = a \sin B$, we now have

$$a \sin B = b \sin A$$

$$\frac{a \sin B}{\sin A \sin B} = \frac{b \sin A}{\sin A \sin B} \qquad \textbf{Dividing by } \sin A \sin B$$

$$\frac{a}{\sin A} = \frac{b}{\sin B}. \qquad \textbf{Simplifying}$$

There is no danger of dividing by 0 in this case because we are dealing with triangles whose angles are never 0° or 180°. Thus the sine value will never be 0.

If we were to consider altitudes from vertex A and vertex B in the triangle shown above, the same argument would give us

$$\frac{b}{\sin B} = \frac{c}{\sin C} \quad \text{and} \quad \frac{a}{\sin A} = \frac{c}{\sin C}.$$

We combine these results to obtain the law of sines.

THE LAW OF SINES

In any triangle ABC,

$$\frac{a}{\sin A} = \frac{b}{\sin B} = \frac{c}{\sin C}.$$

The law of sines can also be expressed as

$$\frac{\sin A}{a} = \frac{\sin B}{b} = \frac{\sin C}{c}.$$

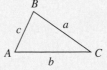

❯ Solving Triangles (AAS and ASA)

When two angles and a side of any triangle are known, the law of sines can be used to solve the triangle.

EXAMPLE 1 In $\triangle EFG$, $e = 4.56$, $E = 43°$, and $G = 57°$. Solve the triangle.

Solution We first make a drawing. We know three of the six measures.

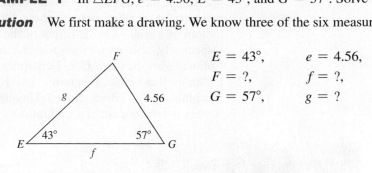

$$E = 43°, \quad e = 4.56,$$
$$F = ?, \quad f = ?,$$
$$G = 57°, \quad g = ?$$

From the figure, we see that we have the AAS situation. We begin by finding F:

$$F = 180° - (43° + 57°) = 80°.$$

We can now find the other two sides, using the law of sines:

$$\frac{f}{\sin F} = \frac{e}{\sin E}$$

$$\frac{f}{\sin 80°} = \frac{4.56}{\sin 43°} \qquad \textbf{Substituting}$$

$$f = \frac{4.56 \sin 80°}{\sin 43°} \qquad \textbf{Solving for } f$$

$$f \approx 6.58;$$

$$\frac{g}{\sin G} = \frac{e}{\sin E}$$

$$\frac{g}{\sin 57°} = \frac{4.56}{\sin 43°} \qquad \textbf{Substituting}$$

$$g = \frac{4.56 \sin 57°}{\sin 43°} \qquad \textbf{Solving for } g$$

$$g \approx 5.61.$$

Thus we have solved the triangle:

$$E = 43°, \qquad e = 4.56,$$
$$F = 80°, \qquad f \approx 6.58,$$
$$G = 57°, \qquad g \approx 5.61.$$

❯ *Now Try Exercise 1.*

The law of sines is frequently used in determining distances.

EXAMPLE 2 *Vietnam Veterans Memorial.* Designed by Maya Lin, the Vietnam Veterans Memorial, in Washington, D.C., consists of two congruent black granite walls on which 58,307 (as of May 2015) names are inscribed in chronological order of the date of the casualty. Each wall closely approximates a triangle. The height of the memorial at its tallest point is about 120.5 in. The angles formed by the top and the bottom of a wall with the height of the memorial are about 89.4056° and 88.2625°, respectively. (*Sources*: www.tourofdc.org; Maya Lin, Designer, New York, NY; National Park Service, U.S. Department of the Interior; Jennifer Talken-Spaulding, Cultural Resources Program manager, National Mall and Memorial Parks, Washington, D.C.; Bryan Swank, Unique Products, Columbus, IN). Find the lengths of the top and the bottom of a wall rounded to the nearest tenth of an inch.

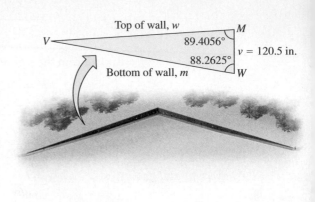

Solution We first find angle V:

$$180° - (89.4056° + 88.2625°) = 2.3319°.$$

Because the application involves the ASA situation, we use the law of sines to determine m and w.

$$\frac{m}{\sin M} = \frac{v}{\sin V}$$

$$\frac{m}{\sin 89.4056°} = \frac{120.5 \text{ in.}}{\sin 2.3319°} \qquad \textbf{Substituting}$$

$$m = \frac{120.5 \text{ in.} \, (\sin 89.4056°)}{\sin 2.3319°} \qquad \textbf{Solving for } m$$

$$m \approx 2961.4 \text{ in.}$$

$$\frac{w}{\sin W} = \frac{v}{\sin V}$$

$$\frac{w}{\sin 88.2625°} = \frac{120.5 \text{ in.}}{\sin 2.3319°} \qquad \textbf{Substituting}$$

$$w = \frac{120.5 \text{ in.} \, (\sin 88.2625°)}{\sin 2.3319°} \qquad \textbf{Solving for } w$$

$$w \approx 2960.2 \text{ in.}$$

Thus, $m \approx 2961.4$ in. and $w \approx 2960.2$ in.

Now Try Exercise 23.

❯ Solving Triangles (SSA)

When two sides of a triangle and an angle opposite one of them are known, the law of sines can be used to solve the triangle.

Suppose for $\triangle ABC$ that b, c, and B are given. The various possibilities are as shown in the eight cases below: five cases when B is acute and three cases when B is obtuse. Note that $b < c$ in cases 1, 2, 3, and 6; $b = c$ in cases 4 and 7; and $b > c$ in cases 5 and 8.

ANGLE *B* IS ACUTE

Case 1: No solution
$b < c$; side b is too short to reach the base. No triangle is formed.

Case 2: One solution
$b < c$; side b just reaches the base and is perpendicular to it.

Case 3: Two solutions
$b < c$; an arc of radius b meets the base at two points. (This case is called the **ambiguous case**.)

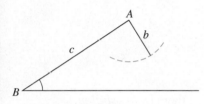

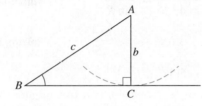

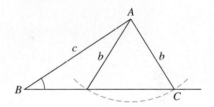

Case 4: One solution
$b = c$; an arc of radius b meets the base at just one point other than B.

Case 5: One solution
$b > c$; an arc of radius b meets the base at just one point.

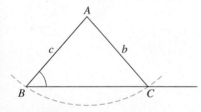

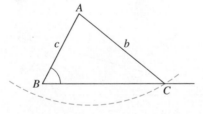

ANGLE *B* IS OBTUSE

Case 6: No solution
$b < c$; side b is too short to reach the base. No triangle is formed.

Case 7: No solution
$b = c$; an arc of radius b meets the base only at point B. No triangle is formed.

Case 8: No solution
$b > c$; an arc of radius b meets the base at just one point.

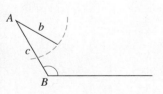

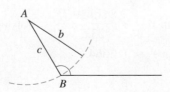

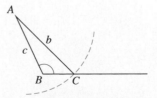

The eight cases illustrated above lead us to three possibilities in the SSA situation: *no* solution, *one* solution, or *two* solutions. Let's investigate these possibilities further, looking for ways to recognize the number of solutions.

EXAMPLE 3 *No solution.*   In $\triangle QRS$, $q = 15$, $r = 28$, and $Q = 43.6°$. Solve the triangle.

Solution   We make a drawing and list the known measures.

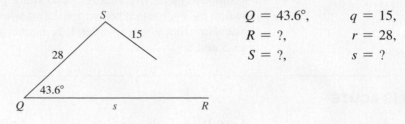

$$Q = 43.6°, \qquad q = 15,$$
$$R = ?, \qquad r = 28,$$
$$S = ?, \qquad s = ?$$

We note the SSA situation and use the law of sines to find R:

$$\frac{q}{\sin Q} = \frac{r}{\sin R}$$

$$\frac{15}{\sin 43.6°} = \frac{28}{\sin R} \qquad \textbf{Substituting}$$

$$\sin R = \frac{28 \sin 43.6°}{15} \qquad \textbf{Solving for sin } R$$

$$\sin R \approx 1.2873.$$

Since there is no angle with a sine greater than 1, there is *no solution*.

> **Now Try Exercise 13.**

EXAMPLE 4 *One solution.*   In $\triangle XYZ$, $x = 23.5$, $y = 9.8$, and $X = 39.7°$. Solve the triangle.

Solution   We make a drawing and organize the given information.

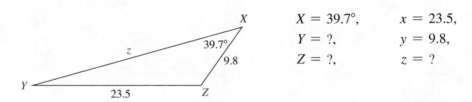

$$X = 39.7°, \qquad x = 23.5,$$
$$Y = ?, \qquad y = 9.8,$$
$$Z = ?, \qquad z = ?$$

We see the SSA situation and begin by finding Y with the law of sines:

$$\frac{x}{\sin X} = \frac{y}{\sin Y}$$

$$\frac{23.5}{\sin 39.7°} = \frac{9.8}{\sin Y} \qquad \textbf{Substituting}$$

$$\sin Y = \frac{9.8 \sin 39.7°}{23.5} \qquad \textbf{Solving for sin } Y$$

$$\sin Y \approx 0.2664.$$

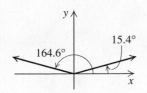

There are two angles less than $180°$ with a sine of 0.2664. They are $15.4°$ and $164.6°$, to the nearest tenth of a degree. An angle of $164.6°$ cannot be an angle of this triangle because it already has an angle of $39.7°$ and these two angles would total more than $180°$. Thus, $15.4°$ is the only possibility for Y. Therefore,

$$Z \approx 180° - (39.7° + 15.4°) \approx 124.9°.$$

We now find z:

$$\frac{z}{\sin Z} = \frac{x}{\sin X}$$

$$\frac{z}{\sin 124.9°} = \frac{23.5}{\sin 39.7°} \qquad \textbf{Substituting}$$

$$z = \frac{23.5 \sin 124.9°}{\sin 39.7°} \qquad \textbf{Solving for } z$$

$$z \approx 30.2.$$

We have now solved the triangle:

$$X = 39.7°, \qquad x = 23.5,$$
$$Y \approx 15.4°, \qquad y = 9.8,$$
$$Z \approx 124.9°, \qquad z \approx 30.2.$$

> **Now Try Exercise 5.**

The next example illustrates the ambiguous case in which there are two possible solutions.

EXAMPLE 5 *Two solutions.* In $\triangle ABC$, $b = 15$, $c = 20$, and $B = 29°$. Solve the triangle.

Solution We make a drawing, list the known measures, and see that we again have the SSA situation.

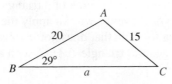

$$A = ?, \qquad a = ?,$$
$$B = 29°, \qquad b = 15,$$
$$C = ?, \qquad c = 20$$

We first find C:

$$\frac{b}{\sin B} = \frac{c}{\sin C}$$

$$\frac{15}{\sin 29°} = \frac{20}{\sin C} \qquad \textbf{Substituting}$$

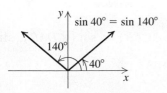

$$\sin C = \frac{20 \sin 29°}{15} \approx 0.6464. \qquad \textbf{Solving for } \sin C$$

There are two angles less than $180°$ with a sine of 0.6464. They are $40°$ and $140°$, to the nearest degree. This gives us two possible solutions.

Possible Solution I.

If $C = 40°$, then

$$A = 180° - (29° + 40°) = 111°.$$

Then we find a:

$$\frac{a}{\sin A} = \frac{b}{\sin B}$$

$$\frac{a}{\sin 111°} = \frac{15}{\sin 29°}$$

$$a = \frac{15 \sin 111°}{\sin 29°} \approx 29.$$

These measures make a triangle as shown below; thus we have a solution.

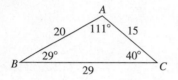

Possible Solution II.

If $C = 140°$, then

$$A = 180° - (29° + 140°) = 11°.$$

Then we find a:

$$\frac{a}{\sin A} = \frac{b}{\sin B}$$

$$\frac{a}{\sin 11°} = \frac{15}{\sin 29°}$$

$$a = \frac{15 \sin 11°}{\sin 29°} \approx 6.$$

These measures make a triangle as shown below; thus we have a second solution.

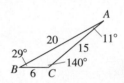

> **Now Try Exercise 3.**

Examples 3–5 illustrate the SSA situation. Note that we need not memorize the relationship between each of the eight cases and the outcome it yields—no solution, one solution, or two solutions. When we are using the law of sines, the sine value leads us directly to the correct solution or solutions.

❯ The Area of a Triangle

The familiar formula for the area of a triangle, $A = \frac{1}{2}bh$, can be used only when the height h is known. However, we can apply the method used to derive the law of sines to derive an area formula that does not involve the height.

Consider a general triangle $\triangle ABC$, with area K, as shown below.

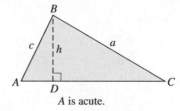

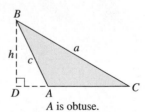

Note that in the triangle on the right, $\sin(\angle CAB) = \sin(\angle DAB)$, since $\sin A = \sin(180° - A)$. Then in each $\triangle ADB$,

$$\sin A = \frac{h}{c}, \quad \text{or} \quad h = c \sin A.$$

Substituting into the formula $K = \frac{1}{2}bh$, we get

$$K = \tfrac{1}{2}bc \sin A.$$

Any pair of sides and the included angle could have been used. Thus we also have

$$K = \tfrac{1}{2}ab \sin C \quad \text{and} \quad K = \tfrac{1}{2}ac \sin B.$$

THE AREA OF A TRIANGLE

The area K of any $\triangle ABC$ is one-half of the product of the lengths of two sides and the sine of the included angle:

$$K = \tfrac{1}{2}bc \sin A = \tfrac{1}{2}ab \sin C = \tfrac{1}{2}ac \sin B.$$

EXAMPLE 6 *Area of a Triangular Garden.* A university landscaping architecture department is designing a garden for a triangular area in a dormitory complex. Two sides of the garden, formed by the sidewalks in front of buildings A and B, measure 172 ft and 186 ft, respectively, and together form a 53° angle. The third side of the garden, formed by the sidewalk along Crossroads Avenue, measures 160 ft. What is the area of the garden to the nearest square foot?

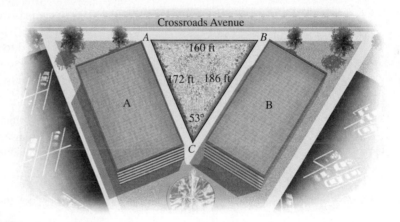

Solution Since we do not know a height of the triangle, we use the area formula:

$$K = \tfrac{1}{2}ab \sin C$$
$$K = \tfrac{1}{2} \cdot 186 \text{ ft} \cdot 172 \text{ ft} \cdot \sin 53°$$
$$K \approx 12{,}775 \text{ ft}^2.$$

The area of the garden is approximately 12,775 ft².

> **Now Try Exercise 25.**

8.1 Exercise Set

Solve the triangle, if possible.

1.
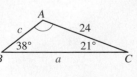

2.
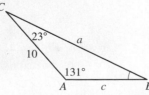

3.
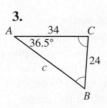

4.
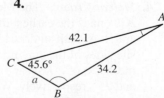

5. $C = 61°10'$, $c = 30.3$, $b = 24.2$

6. $A = 126.5°$, $a = 17.2$, $c = 13.5$

7. $c = 3$ mi, $B = 37.48°$, $C = 32.16°$

8. $a = 2345$ mi, $b = 2345$ mi, $A = 124.67°$

9. $b = 56.78$ yd, $c = 56.78$ yd, $C = 83.78°$

10. $A = 129°32'$, $C = 18°28'$, $b = 1204$ in.

11. $a = 20.01$ cm, $b = 10.07$ cm, $A = 30.3°$

12. $b = 4.157$ km, $c = 3.446$ km, $C = 51°48'$

13. $A = 89°$, $a = 15.6$ in., $b = 18.4$ in.

14. $C = 46°32'$, $a = 56.2$ m, $c = 22.1$ m

15. $a = 200$ m, $A = 32.76°$, $C = 21.97°$

16. $B = 115°$, $c = 45.6$ yd, $b = 23.8$ yd

Find the area of the triangle.

17. $B = 42°$, $a = 7.2$ ft, $c = 3.4$ ft

18. $A = 17°12'$, $b = 10$ in., $c = 13$ in.

19. $C = 82°54'$, $a = 4$ yd, $b = 6$ yd

20. $C = 75.16°$, $a = 1.5$ m, $b = 2.1$ m

21. $B = 135.2°$, $a = 46.12$ ft, $c = 36.74$ ft

22. $A = 113°$, $b = 18.2$ cm, $c = 23.7$ cm

23. *Lawn Irrigation.* Sanchez Irrigation is installing a lawn irrigation system with three heads. They determine that the best locations for the heads A, B, and C are such that $\angle CAB$ is $40°$ and $\angle ACB$ is $45°$ and that the distance from A to B is 34 ft. Find the distance from B to C.

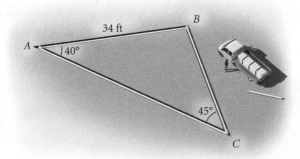

24. *Meteor Crater.* The Meteor Crater in northern Arizona is the earliest discovered meteorite impact crater. A math student locates points R and S on opposite sides of the crater and point T 700 ft from S. The measures of $\angle RST$ and $\angle RTS$ are estimated to be $95°$ and $75°$, respectively. What is the width of the crater from point R to point S?

25. *Area of a Back Yard.* A new homeowner has a triangular-shaped back yard. Two of the three sides measure 53 ft and 42 ft and form an included angle of $135°$. To determine the amount of fertilizer and grass seed to be purchased, the owner must know, or at least approximate, the area of the yard. Find the area of the yard to the nearest square foot.

26. *Elephant Zoo Exhibit.* A zoo is expanding its African elephant exhibit and needs to fence an outdoor triangular area, with a barn forming one side of the triangle. The maintenance department has only 168 ft of fencing in stock. If one side of the outdoor area is 92 ft and the angle that it makes with the barn is $78°$, does the zoo need to order more fencing?

27. *Length of a Pole.*   A pole leans away from the sun at an angle of 7° to the vertical. When the angle of elevation of the sun is 51°, the pole casts a shadow 47 ft long on level ground. How long is the pole?

In Exercises 28–31, keep in mind the two types of bearing considered in Sections 6.2 and 6.3.

28. *Reconnaissance Airplane.*   A reconnaissance airplane leaves its airport on the east coast of the United States and flies in a direction of 85°. Because of bad weather, it returns to another airport 230 km due north of its home base. To get to the new airport, it flies in a direction of 283°. What is the total distance that the airplane flew?

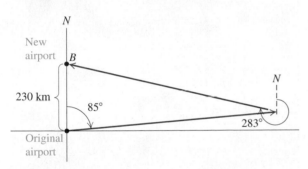

29. *Fire Tower.*   A ranger in fire tower *A* spots a fire at a direction of 295°. A ranger in fire tower *B*, located 45 mi at a direction of 45° from tower *A*, spots the same fire at a direction of 255°. How far from tower *A* is the fire? from tower *B*?

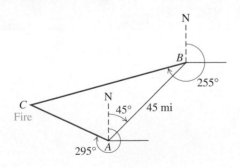

30. *Lighthouse.*   A boat leaves lighthouse *A* and sails 5.1 km. At this time it is sighted from lighthouse *B*, 7.2 km west of *A*. The bearing of the boat from *B* is N65°10′E. How far is the boat from *B*?

31. *Distance to Nassau.*   Miami, Florida, is located 178 mi N73°10′W of Nassau. Because of an approaching hurricane, a cruise ship sailing in the region needs to know how far it is from Nassau. The ship's position is N77°43′E of Miami and N19°35′E of Nassau. How far is the ship from Nassau?

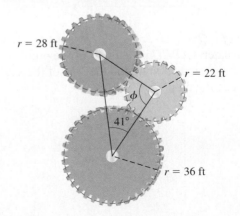

32. *Gears.*   Three gears are arranged as shown in the following figure. Find the angle ϕ.

❯ Skill Maintenance

Find the acute angle A, in both radians and degrees, for the given function value. **[6.1]**

33. $\cos A = 0.2213$

34. $\cos A = 1.5612$

Convert to decimal degree notation. Round to the nearest hundredth. **[6.1]**

35. $18°14'20''$

36. $125°3'42''$

37. Given that $f(x) = \sqrt{x - 6}$, find $f(31)$. **[1.2]**

Find the value. **[6.3]**

38. $\cos \dfrac{\pi}{6}$

39. $\sin 45°$

40. $\sin 300°$

41. $\cos \left(-\dfrac{2\pi}{3} \right)$

42. Multiply: $(1 - i)(1 + i)$. **[3.1]**

❯ Synthesis

43. Prove the following area formulas for a general triangle *ABC* with area represented by *K*.

$$K = \frac{a^2 \sin B \sin C}{2 \sin A}, \qquad K = \frac{c^2 \sin A \sin B}{2 \sin C},$$

$$K = \frac{b^2 \sin C \sin A}{2 \sin B}$$

44. *Area of a Parallelogram.* Prove that the area of a parallelogram is the product of two adjacent sides and the sine of the included angle.

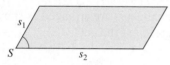

45. *Area of a Quadrilateral.* Prove that the area of a quadrilateral *ABCD* is one-half of the product of the lengths of its diagonals and the sine of the angle θ between the diagonals.

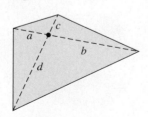

46. Find *d*.

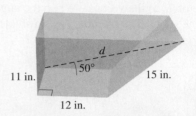

47. *Recording Studio.* A musician is constructing an octagonal recording studio in his home. The studio with dimensions as shown below is to be built within a rectangular $31'9''$ by $29'9''$ room (*Source*: Tony Medeiros, Indianapolis, IN). Point *D* is 9'' from wall 2, and points *C* and *B* are each 9'' from wall 1. Using the law of sines and right triangles, determine to the nearest tenth of an inch how far point *A* is from wall 1 and from wall 4. (For more information on this studio, see Example 2 in Section 8.2.)

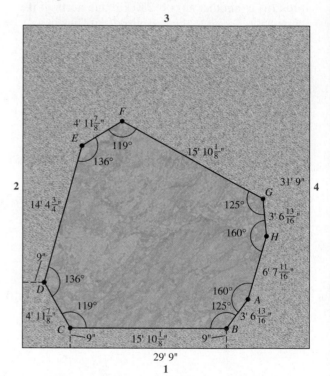

8.2 The Law of Cosines

> ❯ Use the law of cosines to solve triangles.
>
> ❯ Determine whether the law of sines or the law of cosines should be applied to solve a triangle.

The law of sines is used to solve triangles given a side and two angles (AAS and ASA) or given two sides and an angle opposite one of them (SSA). A second law, called the *law of cosines*, is needed to solve triangles given two sides and the included angle (SAS) or given three sides (SSS).

❯ The Law of Cosines

To derive this property, we consider any $\triangle ABC$ placed on a coordinate system. We position the origin at one of the vertices—say, C—and the positive half of the x-axis along one of the sides—say, CB. Let (x, y) be the coordinates of vertex A. Point B has coordinates $(a, 0)$ and point C has coordinates $(0, 0)$.

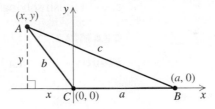

Then $\quad \cos C = \dfrac{x}{b}, \quad$ so $\quad x = b \cos C$

and $\quad \sin C = \dfrac{y}{b}, \quad$ so $\quad y = b \sin C.$

Thus point A has coordinates $(b \cos C, b \sin C)$.

Next, we use the distance formula to determine c^2:

$$c^2 = (x - a)^2 + (y - 0)^2,$$

or $\quad c^2 = (b \cos C - a)^2 + (b \sin C - 0)^2.$

Now we multiply and simplify:

$$
\begin{aligned}
c^2 &= b^2 \cos^2 C - 2ab \cos C + a^2 + b^2 \sin^2 C \\
&= a^2 + b^2(\sin^2 C + \cos^2 C) - 2ab \cos C \\
&= a^2 + b^2 - 2ab \cos C.
\end{aligned}
$$

Using the identity $\sin^2 x + \cos^2 x = 1$

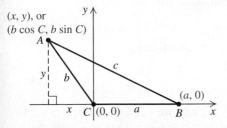
(x, y), or
(b cos C, b sin C)

Had we placed the origin at one of the other vertices, we would have obtained

$$a^2 = b^2 + c^2 - 2bc \cos A$$

or $\quad b^2 = a^2 + c^2 - 2ac \cos B.$

THE LAW OF COSINES

In any triangle ABC,

$$a^2 = b^2 + c^2 - 2bc \cos A,$$
$$b^2 = a^2 + c^2 - 2ac \cos B,$$
and $\quad c^2 = a^2 + b^2 - 2ab \cos C.$

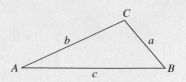

Thus, in any triangle, the square of a side is the sum of the squares of the other two sides, minus twice the product of those sides and the cosine of the included angle. When the included angle is 90°, the law of cosines reduces to the Pythagorean theorem.

The law of cosines can also be expressed as

$$\cos A = \frac{b^2 + c^2 - a^2}{2bc}, \qquad \cos B = \frac{a^2 + c^2 - b^2}{2ac}, \quad \text{and} \quad \cos C = \frac{a^2 + b^2 - c^2}{2ab}$$

❯ Solving Triangles (SAS)

When two sides of a triangle and the included angle are known, we can use the law of cosines to find the third side. The law of cosines or the law of sines can then be used to finish solving the triangle.

EXAMPLE 1   Solve $\triangle ABC$ if $a = 32$, $b = 71$, and $C = 32.8°$.

Solution   We first label a triangle with the known and unknown measures:

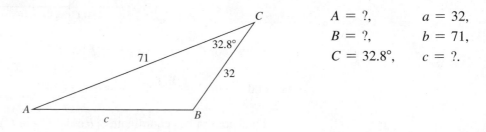

$$
\begin{aligned}
A &= ?, & a &= 32, \\
B &= ?, & b &= 71, \\
C &= 32.8°, & c &= ?.
\end{aligned}
$$

We can find the third side using the law of cosines, as follows:

$$
\begin{aligned}
c^2 &= a^2 + b^2 - 2ab \cos C \\
c^2 &= 32^2 + 71^2 - 2 \cdot 32 \cdot 71 \cos 32.8° \qquad \textbf{Substituting} \\
c^2 &\approx 2245.5 \\
c &\approx 47.
\end{aligned}
$$

We now have $a = 32$, $b = 71$, and $c \approx 47$, and we need to find the other two angle measures. At this point, we can find them in two ways. One way uses the law of sines. The ambiguous case may arise, however, and we would need to be alert to this possibility. The advantage of using the law of cosines again is that if we solve for the cosine and find that its value is *negative*, then we know that the angle is obtuse. If the value of the cosine is *positive*, then the angle is acute. Thus we use the law of cosines to find a second angle.

Let's find angle A. We select the formula from the law of cosines that contains $\cos A$ and substitute:

$$a^2 = b^2 + c^2 - 2bc \cos A$$
$$32^2 \approx 71^2 + 47^2 - 2 \cdot 71 \cdot 47 \cos A \qquad \text{Substituting}$$
$$1024 \approx 5041 + 2209 - 6674 \cos A$$
$$-6226 \approx -6674 \cos A$$
$$\cos A \approx 0.9328738$$
$$A \approx 21.1°.$$

The third angle is now easy to find:

$$B \approx 180° - (32.8° + 21.1°)$$
$$\approx 126.1°.$$

Thus,

$$A \approx 21.1°, \qquad a = 32,$$
$$B \approx 126.1°, \qquad b = 71,$$
$$C = 32.8°, \qquad c \approx 47.$$

> **Now Try Exercise 1.**

Due to differences created by rounding, answers may vary depending on the order in which they are found. Had we found the measure of angle B first in Example 1, the angle measures would have been $B \approx 126.9°$ and $A \approx 20.3°$. Variances in rounding also change the answers. Had we used 47.4 for c in Example 1, the angle measures would have been $A \approx 21.5°$ and $B \approx 125.7°$.

Suppose that we used the law of sines at the outset in Example 1 to find c. We were given only three measures: $a = 32$, $b = 71$, and $C = 32.8°$. When substituting these measures into the proportions, we see that there is not enough information to use the law of sines:

$$\frac{a}{\sin A} = \frac{b}{\sin B} \rightarrow \frac{32}{\sin A} = \frac{71}{\sin B},$$

$$\frac{b}{\sin B} = \frac{c}{\sin C} \rightarrow \frac{71}{\sin B} = \frac{c}{\sin 32.8°},$$

$$\frac{a}{\sin A} = \frac{c}{\sin C} \rightarrow \frac{32}{\sin A} = \frac{c}{\sin 32.8°}.$$

In all three situations, the resulting equation, after the substitutions, still has two unknowns. Thus we cannot use the law of sines to find b.

EXAMPLE 2   *Recording Studio.*   A musician is constructing an octagonal recording studio in his home and needs to determine two distances for the electrician. The dimensions for the most acoustically perfect studio are shown in the figure on the following page (*Source*: Tony Medeiros, Indianapolis, IN). Determine the distances from D to F and from D to B to the nearest tenth of an inch.

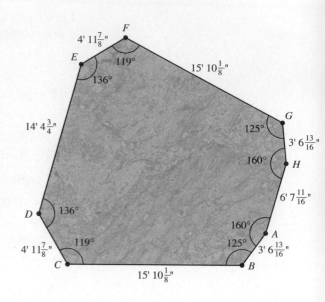

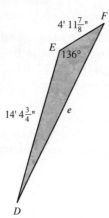

Solution We begin by connecting points D and F and labeling the known measures of $\triangle DEF$. Converting the linear measures to decimal notation in inches, we have

$$d = 4'11\tfrac{7}{8}'' = 59.875 \text{ in.,}$$
$$f = 14'4\tfrac{3}{4}'' = 172.75 \text{ in.,}$$
$$E = 136°.$$

We can find the measure of the third side, e, using the law of cosines:

$$e^2 = d^2 + f^2 - 2 \cdot d \cdot f \cdot \cos E \qquad \textbf{Using the law of cosines}$$
$$e^2 = (59.875 \text{ in.})^2 + (172.75 \text{ in.})^2$$
$$\qquad - 2(59.875 \text{ in.})(172.75 \text{ in.}) \cos 136° \qquad \textbf{Substituting}$$
$$e^2 \approx 48{,}308.4257 \text{ in}^2$$
$$e \approx 219.8 \text{ in.}$$

Thus it is approximately 219.8 in. from D to F.

We continue by connecting points D and B and labeling the known measures of $\triangle DCB$:

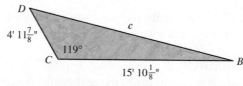

$$b = 4'11\tfrac{7}{8}'' = 59.875 \text{ in.,}$$
$$d = 15'10\tfrac{1}{8}'' = 190.125 \text{ in.,}$$
$$C = 119°.$$

Using the law of cosines, we can determine c, the length of the third side:

$$c^2 = b^2 + d^2 - 2 \cdot b \cdot d \cdot \cos C \qquad \textbf{Using the law of cosines}$$
$$c^2 = (59.875 \text{ in.})^2 + (190.125 \text{ in.})^2$$
$$\qquad - 2(59.875 \text{ in.})(190.125 \text{ in.}) \cos 119° \qquad \textbf{Substituting}$$
$$c^2 \approx 50{,}770.4191 \text{ in}^2$$
$$c \approx 225.3 \text{ in.}$$

The distance from D to B is approximately 225.3 in.

Now Try Exercise 25.

> Solving Triangles (SSS)

When all three sides of a triangle are known, the law of cosines can be used to solve the triangle.

EXAMPLE 3   Solve $\triangle RST$ if $r = 3.5$, $s = 4.7$, and $t = 2.8$.

Solution   We sketch a triangle and label it with the given measures.

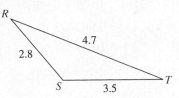

$$
\begin{array}{ll}
R = ?, & r = 3.5, \\
S = ?, & s = 4.7, \\
T = ?, & t = 2.8
\end{array}
$$

Since we do not know any of the angle measures, we cannot use the law of sines. We begin instead by finding an angle with the law of cosines. We choose to find S first and select the formula that contains $\cos S$:

$$s^2 = r^2 + t^2 - 2rt \cos S$$
$$(4.7)^2 = (3.5)^2 + (2.8)^2 - 2(3.5)(2.8) \cos S \qquad \textbf{Substituting}$$
$$\cos S = \frac{(3.5)^2 + (2.8)^2 - (4.7)^2}{2(3.5)(2.8)}$$
$$\cos S \approx -0.1020408$$
$$S \approx 95.86°.$$

In a similar manner, we find angle R:

$$r^2 = s^2 + t^2 - 2st \cos R$$
$$(3.5)^2 = (4.7)^2 + (2.8)^2 - 2(4.7)(2.8) \cos R$$
$$\cos R = \frac{(4.7)^2 + (2.8)^2 - (3.5)^2}{2(4.7)(2.8)}$$
$$\cos R \approx 0.6717325$$
$$R \approx 47.80°.$$

Then

$$T \approx 180° - (95.86° + 47.80°) \approx 36.34°.$$

Thus,

$$
\begin{array}{ll}
R \approx 47.80°, & r = 3.5, \\
S \approx 95.86°, & s = 4.7, \\
T \approx 36.34°, & t = 2.8.
\end{array}
$$

> *Now Try Exercise 3.*

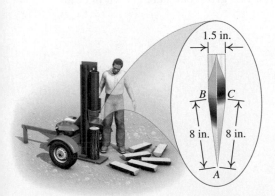

EXAMPLE 4   *Wedge Bevel.*   The *bevel* of the wedge (the angle formed at the cutting edge of the wedge) of a cutting tool determines the cutting characteristics of the tool. A small bevel like that of a straight razor makes for a keen edge, but is impractical for heavy-duty cutting because the edge dulls quickly and is prone to chipping. A large bevel is suitable for heavy-duty work like chopping wood. The diagram at left illustrates the wedge of a Huskee log splitter (*Source*: Huskee). What is its bevel?

Solution   Since we know three sides of a triangle, we can use the law of cosines to find the bevel, angle A:

$$a^2 = b^2 + c^2 - 2bc \cos A$$
$$(1.5)^2 = 8^2 + 8^2 - 2 \cdot 8 \cdot 8 \cdot \cos A$$
$$2.25 = 64 + 64 - 128 \cos A$$
$$\cos A = \frac{64 + 64 - 2.25}{128}$$
$$\cos A \approx 0.982422$$
$$A \approx 10.76°.$$

Thus the bevel is approximately $10.76°$.

> *Now Try Exercise 31.*

CONNECTING THE CONCEPTS

Choosing the Appropriate Law

The following summarizes the situations in which to use the law of sines and the law of cosines.

To solve an oblique triangle:

Use the *law of sines* for:	Use the *law of cosines* for:
AAS	SAS
ASA	SSS
SSA	

The law of cosines can also be used for the SSA situation, but since the process involves solving a quadratic equation, we do not include that option here.

EXAMPLE 5   In $\triangle ABC$, three measures are given. Determine which law to use when solving the triangle. You need not solve the triangle.

a) $a = 14$, $b = 23$, $c = 10$

b) $a = 207$, $B = 43.8°$, $C = 57.6°$

c) $A = 112°$, $C = 37°$, $a = 84.7$

d) $B = 101°$, $a = 960$, $c = 1042$

e) $b = 17.26$, $a = 27.29$, $A = 39°$

f) $A = 61°$, $B = 39°$, $C = 80°$

Solution   It is helpful to make a drawing of a triangle with the given information. The triangle need not be drawn to scale. The given parts are shown in color.

FIGURE		SITUATION	LAW TO USE
a)	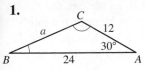	SSS	Law of Cosines
b)		ASA	Law of Sines
c)		AAS	Law of Sines
d)		SAS	Law of Cosines
e)		SSA	Law of Sines
f)		AAA	Cannot be solved

> *Now Try Exercises 17 and 19.*

8.2   Exercise Set

Solve the triangle, if possible.

1.

2.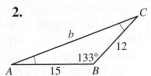

3.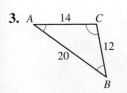

4.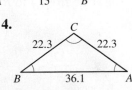

5. $B = 72°40'$, $c = 16$ m, $a = 78$ m

6. $C = 22.28°$, $a = 25.4$ cm, $b = 73.8$ cm

7. $a = 16$ m, $b = 20$ m, $c = 32$ m

8. $B = 72.66°$, $a = 23.78$ km, $c = 25.74$ km

9. $a = 2$ ft, $b = 3$ ft, $c = 8$ ft

10. $A = 96°13'$, $b = 15.8$ yd, $c = 18.4$ yd

11. $a = 26.12$ km, $b = 21.34$ km, $c = 19.25$ km

12. $C = 28°43'$, $a = 6$ mm, $b = 9$ mm

13. $a = 60.12$ mi, $b = 40.23$ mi, $C = 48.7°$

14. $a = 11.2$ cm, $b = 5.4$ cm, $c = 7$ cm

15. $b = 10.2$ in., $c = 17.3$ in., $A = 53.456°$

16. $a = 17$ yd, $b = 15.4$ yd, $c = 1.5$ yd

Determine which law applies. Then solve the triangle.

17. $A = 70°$, $B = 12°$, $b = 21.4$

18. $a = 15$, $c = 7$, $B = 62°$

19. $a = 3.3$, $b = 2.7$, $c = 2.8$

20. $a = 1.5$, $b = 2.5$, $A = 58°$

21. $A = 40.2°$, $B = 39.8°$, $C = 100°$

22. $a = 60$, $b = 40$, $C = 47°$

23. $a = 3.6$, $b = 6.2$, $c = 4.1$

24. $B = 110°30'$, $C = 8°10'$, $c = 0.912$

25. *Amusement Park Ride.* A teacup ride for children at an amusement park consists of five teacups equally spaced around a circle. Each cup holds 6 passengers and is at the end of an arm 20 ft long. Find the linear distance between a pair of adjacent cups. Round the length to the nearest tenth of a foot.

26. *Shark Pool.* Feeding and observing sharks has become a popular attraction at water parks. The floor of a newly constructed shark pool in the shape of a parallelogram has sides that measure 38 ft and 57 ft. To meet the minimum required length for the shortest diagonal, the angles must be 80° and 100°. Find the lengths of the diagonals of the pool. Round the lengths to the nearest foot.

27. *A-Frame Architecture.* William O'Brien, Jr., Assistant Professor of Architecture at the MIT School of Architecture and Planning, designed Allandale House, the asymmetrical A-frame house pictured below (*Source*: William O'Brien, Jr., LLC, 188 Prospect Street, Unit 5, Cambridge, MA 02139). The included angle between sides *AB* and *CB*, which measure 32.67 ft and 19.25 ft, respectively, measures 67°. Find the length of side *AC*.

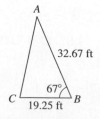

28. *Fish Attractor.* Each year at Cedar Resort, discarded Christmas trees are collected and sunk in the lake to form a fish attractor. Visitors are told that it is 253 ft from the pier to the fish attractor and 415 ft to another pier across the lake. Using a compass, a fisherman finds that the attractor's azimuth (the direction measured as an angle from north) is 340° and that of the other pier is 35°. What is the distance between the fish attractor and the pier across the lake?

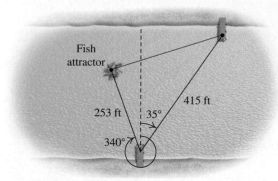

29. *In-line Skater.* An in-line skater skates on a fitness trail along the Pacific Ocean from point *A* to point *B*. As shown below, two streets intersecting at point *C* also intersect the trail at *A* and *B*. In her car, the skater found the lengths of *AC* and *BC* to be approximately 0.5 mi and 1.3 mi, respectively. From a map, she estimates the included angle at *C* to be 110°. How far did she skate from *A* to *B*?

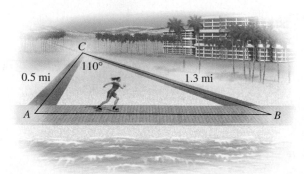

30. *Baseball Bunt.*   A batter in a baseball game drops a bunt down the first-base line. It rolls 34 ft at an angle of 25° with the base path. The pitcher's mound is 60.5 ft from home plate. How far must the pitcher travel to pick up the ball? (*Hint*: A baseball diamond is a square.)

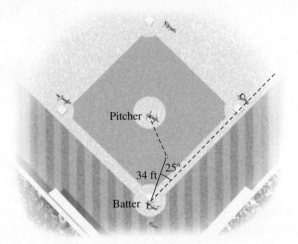

31. *Survival Trip.*   A group of college students is learning to navigate for an upcoming survival trip. On a map, they have been given three points at which they are to check in. The map also shows the distances between the points. However, in order to navigate they need to know the angle measurements. Calculate the angles for them.

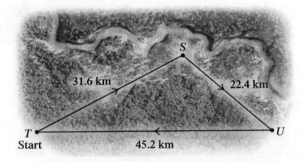

32. *Ships.*   Two ships leave harbor at the same time. The first sails N15°W at 25 knots. (A knot is one nautical mile per hour.) The second sails N32°E at 20 knots. After 2 hr, how far apart are the ships?

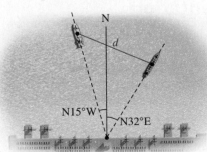

33. *Airplanes.*   Two airplanes leave an airport at the same time. The first flies 150 km/h in a direction of 320°. The second flies 200 km/h in a direction of 200°. After 3 hr, how far apart are the planes?

34. *Slow-Pitch Softball.*   A slow-pitch softball diamond is a square 65 ft on a side. The pitcher's mound is 46 ft from home plate. How far is it from the pitcher's mound to first base?

35. *Isosceles Trapezoid.*   The longer base of an isosceles trapezoid measures 14 ft. The nonparallel sides measure 10 ft, and the base angles measure 80°.

   a) Find the length of a diagonal.
   b) Find the area.

36. *Dimensions of Sail.*   A sail that is in the shape of an isosceles triangle has a vertex angle of 38°. The angle is included by two sides, each measuring 20 ft. Find the length of the other side of the sail.

37. Three circles are arranged as shown in the following figure. Find the length *PQ*.

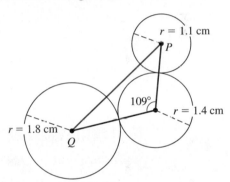

38. *Swimming Pool.*   A triangular swimming pool measures 44 ft on one side and 32.8 ft on another side. These sides form an angle that measures 40.8°. How long is the other side?

> **Skill Maintenance**

Classify the function as linear, quadratic, cubic, quartic, rational, exponential, logarithmic, or trigonometric.

39. $f(x) = -\frac{3}{4}x^4$ **[4.1]**

40. $y - 3 = 17x$ **[1.3]**

41. $y = \sin^2 x - 3 \sin x$ **[6.5]**

42. $f(x) = 2^{x-1/2}$ **[5.2]**

43. $f(x) = \dfrac{x^2 - 2x + 3}{x - 1}$ **[4.5]**

44. $f(x) = 27 - x^3$ **[4.1]**

45. $y = e^x + e^{-x} - 4$ **[5.2]**

46. $y = \log_2(x - 2) - \log_2(x + 3)$ **[5.3]**

47. $f(x) = -\cos(\pi x - 3)$ **[6.5]**

48. $y = \frac{1}{2}x^2 - 2x + 2$ **[3.2]**

❯ Synthesis

49. *Canyon Depth.*   A bridge is being built across a canyon. The length of the bridge is 5045 ft. From the deepest point in the canyon, the angles of elevation of the ends of the bridge are 78° and 72°. How deep is the canyon?

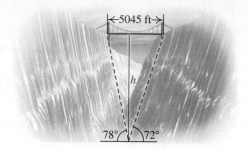

50. *Reconnaissance Plane.*   A reconnaissance plane patrolling at 5000 ft sights a submarine at bearing 35° and at an angle of depression of 25°. A carrier is at bearing 105° and at an angle of depression of 60°. How far is the submarine from the carrier?

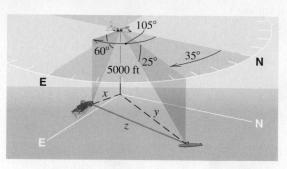

51. *Area of Isosceles Triangle.*   Find a formula for the area of an isosceles triangle in terms of the congruent sides and their included angle. Under what conditions will the area of a triangle with fixed congruent sides be maximum?

8.3 ❯ Complex Numbers: Trigonometric Notation

❯ Graph complex numbers.

❯ Given a complex number in standard form, find trigonometric, or polar, notation; and given a complex number in trigonometric form, find standard notation.

❯ Use trigonometric notation to multiply and divide complex numbers.

❯ Use DeMoivre's theorem to raise complex numbers to powers.

❯ Find the *n*th roots of a complex number.

Just in Time
1

❯ Graphical Representation

Just as real numbers can be graphed on a line, complex numbers can be graphed on a plane. We graph a complex number $a + bi$ in the same way that we graph an ordered pair of real numbers (a, b). However, in place of an *x*-axis, we have a real axis, and in place of a *y*-axis, we have an imaginary axis. Horizontal distances correspond to the real part of a complex number. Vertical distances correspond to the imaginary part. Recall that $i = \sqrt{-1}$.

COMPLEX NUMBERS

REVIEW SECTION 3.1.

EXAMPLE 1   Graph each of the following complex numbers.

a) $3 + 2i$

b) $-4 - 5i$

c) $-3i$

d) $-1 + 3i$

e) 2

Solution

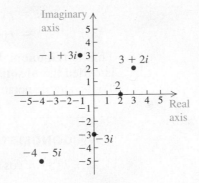

We recall that the absolute value of a real number is its distance from 0 on the number line. The absolute value of a complex number is its distance from the origin in the complex plane. For a point $a + bi$, using the distance formula, we have

$$|a + bi| = \sqrt{(a - 0)^2 + (b - 0)^2} = \sqrt{a^2 + b^2}.$$

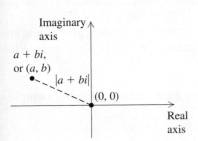

ABSOLUTE VALUE OF A COMPLEX NUMBER

The **absolute value of a complex number** $a + bi$ is

$$|a + bi| = \sqrt{a^2 + b^2}.$$

EXAMPLE 2 Find the absolute value of each of the following.

a) $3 + 4i$ **b)** $-2 - i$ **c)** $\frac{4}{5}i$

Solution

a) $|3 + 4i| = \sqrt{3^2 + 4^2} = \sqrt{9 + 16} = \sqrt{25} = 5$

b) $|-2 - i| = \sqrt{(-2)^2 + (-1)^2} = \sqrt{5}$

c) $\left|\frac{4}{5}i\right| = \left|0 + \frac{4}{5}i\right| = \sqrt{0^2 + \left(\frac{4}{5}\right)^2} = \frac{4}{5}$

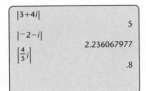

We can check these results using a graphing calculator in complex mode as shown at left. Note that $\sqrt{5} \approx 2.236067977$ and $\frac{4}{5} = 0.8$.

> *Now Try Exercises 3 and 5.*

❯ Trigonometric Notation for Complex Numbers

Now let's consider a nonzero complex number $a + bi$. Suppose that its absolute value is r. If we let θ be an angle in standard position whose terminal side passes through the point (a, b), as shown in the figure, then

$$\cos \theta = \frac{a}{r}, \quad \text{or} \quad a = r \cos \theta$$

and

$$\sin \theta = \frac{b}{r}, \quad \text{or} \quad b = r \sin \theta.$$

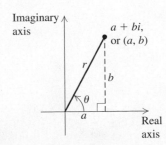

Substituting these values for a and b into the $(a + bi)$ notation, we get

$$a + bi = r\cos\theta + (r\sin\theta)i$$
$$= r(\cos\theta + i\sin\theta).$$

This is **trigonometric notation** for a complex number $a + bi$. The number r is called the **absolute value** of $a + bi$, and θ is called the **argument** of $a + bi$. Trigonometric notation for a complex number is also called **polar notation**.

TRIGONOMETRIC NOTATION FOR COMPLEX NUMBERS

$$a + bi = r(\cos\theta + i\sin\theta)$$

In order to find trigonometric notation for a complex number given in **standard notation**, $a + bi$, we must find r and determine the angle θ for which $\sin\theta = b/r$ and $\cos\theta = a/r$.

EXAMPLE 3 Find trigonometric notation for each of the following complex numbers.

a) $1 + i$

b) $\sqrt{3} - i$

Solution

a) We note that $a = 1$ and $b = 1$. Then

$$r = \sqrt{a^2 + b^2} = \sqrt{1^2 + 1^2} = \sqrt{2},$$

$$\sin\theta = \frac{b}{r} = \frac{1}{\sqrt{2}}, \quad \text{or} \quad \frac{\sqrt{2}}{2},$$

and

$$\cos\theta = \frac{a}{r} = \frac{1}{\sqrt{2}}, \quad \text{or} \quad \frac{\sqrt{2}}{2}.$$

Since θ is in quadrant I, $\theta = \pi/4$, or $45°$, and we have

$$1 + i = \sqrt{2}\left(\cos\frac{\pi}{4} + i\sin\frac{\pi}{4}\right),$$

or

$$1 + i = \sqrt{2}(\cos 45° + i\sin 45°).$$

b) We see that $a = \sqrt{3}$ and $b = -1$. Then

$$r = \sqrt{(\sqrt{3})^2 + (-1)^2} = 2,$$

$$\sin\theta = \frac{-1}{2} = -\frac{1}{2},$$

and

$$\cos\theta = \frac{\sqrt{3}}{2}.$$

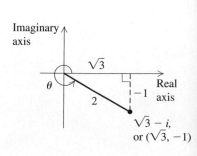

Since θ is in quadrant IV, $\theta = 11\pi/6$, or $330°$, and we have

$$\sqrt{3} - i = 2\left(\cos\frac{11\pi}{6} + i\sin\frac{11\pi}{6}\right),$$

or

$$\sqrt{3} - i = 2(\cos 330° + i\sin 330°).$$

angle(1+i)
45
angle(√3−i)
−30

As shown at left, a graphing calculator (in DEGREE mode) can be used to determine angle values in degrees.

> *Now Try Exercise 13.*

In changing to trigonometric notation, note that there are many angles satisfying the given conditions. We ordinarily choose the *smallest positive* angle.

To change from trigonometric notation to standard notation, $a + bi$, we recall that $a = r \cos \theta$ and $b = r \sin \theta$.

EXAMPLE 4 Find standard notation, $a + bi$, for each of the following complex numbers.

a) $2(\cos 120° + i \sin 120°)$

b) $\sqrt{8}\left(\cos \frac{7\pi}{4} + i \sin \frac{7\pi}{4}\right)$

Solution

a) Rewriting, we have

$$2(\cos 120° + i \sin 120°) = 2 \cos 120° + (2 \sin 120°)i.$$

Thus,

$$a = 2 \cos 120° = 2 \cdot \left(-\frac{1}{2}\right) = -1$$

and

$$b = 2 \sin 120° = 2 \cdot \frac{\sqrt{3}}{2} = \sqrt{3},$$

Degree Mode

2(cos(120)+*i*sin(120))
−1+1.732050808*i*

so

$$2(\cos 120° + i \sin 120°) = -1 + \sqrt{3}i.$$

b) Rewriting, we have

$$\sqrt{8}\left(\cos \frac{7\pi}{4} + i \sin \frac{7\pi}{4}\right) = \sqrt{8} \cos \frac{7\pi}{4} + \left(\sqrt{8} \sin \frac{7\pi}{4}\right)i.$$

Thus,

$$a = \sqrt{8} \cos \frac{7\pi}{4} = \sqrt{8} \cdot \frac{\sqrt{2}}{2} = \frac{\sqrt{16}}{2} = \frac{4}{2} = 2$$

and

$$b = \sqrt{8} \sin \frac{7\pi}{4} = \sqrt{8} \cdot \left(-\frac{\sqrt{2}}{2}\right) = -\frac{\sqrt{16}}{2} = -\frac{4}{2} = -2,$$

Radian Mode

√8(cos($\frac{7\pi}{4}$)+*i*sin($\frac{7\pi}{4}$))
2−2*i*

so

$$\sqrt{8}\left(\cos \frac{7\pi}{4} + i \sin \frac{7\pi}{4}\right) = 2 - 2i.$$

> *Now Try Exercises 23 and 27.*

> ## Multiplication and Division with Trigonometric Notation

Multiplication of complex numbers is easier to manage with trigonometric notation than with standard notation. We simply multiply the absolute values and add the arguments. Let's state this in a more formal manner.

COMPLEX NUMBERS: MULTIPLICATION

For any complex numbers $r_1(\cos \theta_1 + i \sin \theta_1)$ and $r_2(\cos \theta_2 + i \sin \theta_2)$,

$$r_1(\cos \theta_1 + i \sin \theta_1) \cdot r_2(\cos \theta_2 + i \sin \theta_2)$$
$$= r_1 r_2 [\cos (\theta_1 + \theta_2) + i \sin (\theta_1 + \theta_2)].$$

Proof. We have

$$r_1(\cos \theta_1 + i \sin \theta_1) \cdot r_2(\cos \theta_2 + i \sin \theta_2)$$
$$= r_1 r_2(\cos \theta_1 \cos \theta_2 - \sin \theta_1 \sin \theta_2) + r_1 r_2(\sin \theta_1 \cos \theta_2 + \cos \theta_1 \sin \theta_2)i.$$

Now, using identities for sums of angles, we simplify, obtaining

$$r_1 r_2 \cos (\theta_1 + \theta_2) + r_1 r_2 \sin (\theta_1 + \theta_2)i,$$

or

$$r_1 r_2 [\cos (\theta_1 + \theta_2) + i \sin (\theta_1 + \theta_2)],$$

which was to be shown.

EXAMPLE 5 Multiply and express the answer to each of the following in standard notation.

a) $3(\cos 40° + i \sin 40°)$ and $4(\cos 20° + i \sin 20°)$

b) $2(\cos \pi + i \sin \pi)$ and $3\left[\cos \left(-\dfrac{\pi}{2}\right) + i \sin \left(-\dfrac{\pi}{2}\right)\right]$

Solution

a) $3(\cos 40° + i \sin 40°) \cdot 4(\cos 20° + i \sin 20°)$
$$= 3 \cdot 4 \cdot [\cos (40° + 20°) + i \sin (40° + 20°)]$$
$$= 12(\cos 60° + i \sin 60°)$$
$$= 12\left(\dfrac{1}{2} + \dfrac{\sqrt{3}}{2}i\right)$$
$$= 6 + 6\sqrt{3}i$$

Degree Mode

> 3(cos(40)+*i*sin(40))*4(cos(20)+
> *i*sin(20))
>
>   6+10.39230485*i*

b) $2(\cos \pi + i \sin \pi) \cdot 3\left[\cos \left(-\dfrac{\pi}{2}\right) + i \sin \left(-\dfrac{\pi}{2}\right)\right]$
$$= 2 \cdot 3 \cdot \left[\cos \left(\pi + \left(-\dfrac{\pi}{2}\right)\right) + i \sin \left(\pi + \left(-\dfrac{\pi}{2}\right)\right)\right]$$
$$= 6\left(\cos \dfrac{\pi}{2} + i \sin \dfrac{\pi}{2}\right)$$
$$= 6(0 + i \cdot 1)$$
$$= 6i$$

Radian Mode

> 2(cos(π)+*i*sin(π))*3(cos(−π/2)+
> *i*sin(−π/2))
>
>   6*i*

> **Now Try Exercise 35.**

EXAMPLE 6 Convert to trigonometric notation and multiply:
$$(1 + i)(\sqrt{3} - i).$$

Solution We first find trigonometric notation:

$$1 + i = \sqrt{2}(\cos 45° + i \sin 45°), \qquad \text{See Example 3(a).}$$
$$\sqrt{3} - i = 2(\cos 330° + i \sin 330°). \qquad \text{See Example 3(b).}$$

Then we multiply:

$$\sqrt{2}(\cos 45° + i \sin 45°) \cdot 2(\cos 330° + i \sin 330°)$$
$$= 2\sqrt{2}[\cos(45° + 330°) + i \sin(45° + 330°)]$$
$$= 2\sqrt{2}(\cos 375° + i \sin 375°)$$
$$= 2\sqrt{2}(\cos 15° + i \sin 15°). \quad \textbf{375° has the same}$$
$$\textbf{terminal side as 15°.}$$

Now Try Exercise 37.

To divide complex numbers, we divide the absolute values and subtract the arguments. We state this fact below, but omit the proof.

COMPLEX NUMBERS: DIVISION

For any complex numbers $r_1(\cos \theta_1 + i \sin \theta_1)$ and
$r_2(\cos \theta_2 + i \sin \theta_2)$, $r_2 \neq 0$,

$$\frac{r_1(\cos \theta_1 + i \sin \theta_1)}{r_2(\cos \theta_2 + i \sin \theta_2)} = \frac{r_1}{r_2}[\cos(\theta_1 - \theta_2) + i \sin(\theta_1 - \theta_2)].$$

EXAMPLE 7   Divide

$$2\left(\cos \frac{3\pi}{2} + i \sin \frac{3\pi}{2}\right) \quad \text{by} \quad 4\left(\cos \frac{\pi}{2} + i \sin \frac{\pi}{2}\right)$$

and express the solution in standard notation.

Solution   We have

$$\frac{2\left(\cos \dfrac{3\pi}{2} + i \sin \dfrac{3\pi}{2}\right)}{4\left(\cos \dfrac{\pi}{2} + i \sin \dfrac{\pi}{2}\right)} = \frac{2}{4}\left[\cos\left(\frac{3\pi}{2} - \frac{\pi}{2}\right) + i \sin\left(\frac{3\pi}{2} - \frac{\pi}{2}\right)\right]$$

$$= \frac{1}{2}(\cos \pi + i \sin \pi)$$

$$= \frac{1}{2}(-1 + i \cdot 0)$$

$$= -\frac{1}{2}.$$

EXAMPLE 8   Convert to trigonometric notation and divide:

$$\frac{1+i}{1-i}.$$

Solution   We first convert to trigonometric notation:

$$1 + i = \sqrt{2}(\cos 45° + i \sin 45°), \quad \textbf{See Example 3(a).}$$
$$1 - i = \sqrt{2}(\cos 315° + i \sin 315°).$$

We now divide:

$$\frac{\sqrt{2}(\cos 45° + i \sin 45°)}{\sqrt{2}(\cos 315° + i \sin 315°)}$$

$$= 1[\cos (45° - 315°) + i \sin (45° - 315°)]$$

$$= \cos (-270°) + i \sin (-270°)$$

$$= 0 + i \cdot 1$$

$$= i.$$

Now Try Exercise 39.

$\frac{1+i}{1-i}$

i

❯ Powers of Complex Numbers

An important theorem about powers and roots of complex numbers is named for the French mathematician Abraham DeMoivre (1667–1754). Let's consider the square of a complex number $r(\cos \theta + i \sin \theta)$:

$$[r(\cos \theta + i \sin \theta)]^2 = [r(\cos \theta + i \sin \theta)] \cdot [r(\cos \theta + i \sin \theta)]$$

$$= r \cdot r \cdot [\cos (\theta + \theta) + i \sin (\theta + \theta)]$$

$$= r^2(\cos 2\theta + i \sin 2\theta).$$

Similarly, we see that

$$[r(\cos \theta + i \sin \theta)]^3$$

$$= r \cdot r \cdot r \cdot [\cos (\theta + \theta + \theta) + i \sin (\theta + \theta + \theta)]$$

$$= r^3(\cos 3\theta + i \sin 3\theta).$$

DeMoivre's theorem is the generalization of these results.

DEMOIVRE'S THEOREM

For any complex number $r(\cos \theta + i \sin \theta)$ and any natural number n,

$$[r(\cos \theta + i \sin \theta)]^n = r^n(\cos n\theta + i \sin n\theta).$$

EXAMPLE 9 Find each of the following.

a) $(1 + i)^9$ **b)** $(\sqrt{3} - i)^{10}$

Solution

a) We first find trigonometric notation:

$$1 + i = \sqrt{2}(\cos 45° + i \sin 45°).$$ **See Example 3(a).**

Then

$$(1 + i)^9 = [\sqrt{2}(\cos 45° + i \sin 45°)]^9$$

$$= (\sqrt{2})^9[\cos (9 \cdot 45°) + i \sin (9 \cdot 45°)] \quad \text{DeMoivre's theorem}$$

$$= 2^{9/2}(\cos 405° + i \sin 405°)$$

$$= 16\sqrt{2}(\cos 45° + i \sin 45°) \quad \text{405° has the same terminal side as 45°.}$$

$$= 16\sqrt{2}\left(\frac{\sqrt{2}}{2} + i \frac{\sqrt{2}}{2}\right)$$

$$= 16 + 16i.$$

b) We first convert to trigonometric notation:

$$\sqrt{3} - i = 2(\cos 330° + i \sin 330°). \quad \textbf{See Example 3(b).}$$

Then

$$
\begin{aligned}
\left(\sqrt{3} - i\right)^{10} &= \left[2(\cos 330° + i \sin 330°)\right]^{10}\\
&= 2^{10}(\cos 3300° + i \sin 3300°)\\
&= 1024(\cos 60° + i \sin 60°) \qquad \text{\small 3300° has the same}\\
&\qquad\qquad\qquad\qquad\qquad\qquad\qquad \text{\small terminal side as 60°.}\\
&= 1024\left(\frac{1}{2} + i\frac{\sqrt{3}}{2}\right)\\
&= 512 + 512\sqrt{3}i.
\end{aligned}
$$

> **Now Try Exercise 47.**

❯ Roots of Complex Numbers

As we will see, every nonzero complex number has two square roots. A nonzero complex number has three cube roots, four fourth roots, and so on. In general, a nonzero complex number has n different nth roots. They can be found using the formula that we now state but do not prove.

Just in Time

27

> ### ROOTS OF COMPLEX NUMBERS
>
> The nth roots of a complex number $r(\cos \theta + i \sin \theta)$, $r \neq 0$, are given by
>
> $$r^{1/n}\left[\cos\left(\frac{\theta}{n} + k\cdot\frac{360°}{n}\right) + i \sin\left(\frac{\theta}{n} + k\cdot\frac{360°}{n}\right)\right],$$
>
> where $k = 0, 1, 2, \ldots, n - 1$.

EXAMPLE 10   Find the square roots of $2 + 2\sqrt{3}i$.

Solution   We first find trigonometric notation:

$$2 + 2\sqrt{3}i = 4(\cos 60° + i \sin 60°).$$

Then $n = 2, 1/n = 1/2$, and $k = 0, 1$. We have

$$
\begin{aligned}
&\left[4(\cos 60° + i \sin 60°)\right]^{1/2}\\
&\quad = 4^{1/2}\left[\cos\left(\frac{60°}{2} + k\cdot\frac{360°}{2}\right) + i \sin\left(\frac{60°}{2} + k\cdot\frac{360°}{2}\right)\right], \quad k = 0, 1\\
&\quad = 2\left[\cos\left(30° + k\cdot 180°\right) + i \sin\left(30° + k\cdot 180°\right)\right], \quad k = 0, 1.
\end{aligned}
$$

Thus the roots are

$$2(\cos 30° + i \sin 30°) \text{ for } k = 0$$

and $\quad 2(\cos 210° + i \sin 210°) \text{ for } k = 1,$

or $\qquad \sqrt{3} + i \quad$ and $\quad -\sqrt{3} - i.$

> **Now Try Exercise 57.**

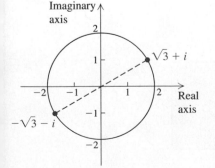

In Example 10, we see that the two square roots of the number are opposites of each other. We can illustrate this graphically. We also note that the roots are equally spaced about a circle of radius r—in this case, $r = 2$. The roots are $360°/2$, or $180°$ apart.

EXAMPLE 11 Find the cube roots of 1. Then locate them on a graph.

Solution We begin by finding trigonometric notation:

$$1 = 1(\cos 0° + i \sin 0°).$$

Then $n = 3$, $1/n = 1/3$, and $k = 0, 1, 2$. We have

$$[1(\cos 0° + i \sin 0°)]^{1/3}$$

$$= 1^{1/3}\left[\cos\left(\frac{0°}{3} + k \cdot \frac{360°}{3}\right) + i \sin\left(\frac{0°}{3} + k \cdot \frac{360°}{3}\right)\right], \quad k = 0, 1, 2.$$

The roots are

$$1(\cos 0° + i \sin 0°), \quad 1(\cos 120° + i \sin 120°),$$

and $\quad 1(\cos 240° + i \sin 240°),$

or $\quad 1, \quad -\frac{1}{2} + \frac{\sqrt{3}}{2}i, \quad$ and $\quad -\frac{1}{2} - \frac{\sqrt{3}}{2}i.$

The graphs of the cube roots lie equally spaced about a circle of radius 1. The roots are $360°/3$, or $120°$ apart.

> **Now Try Exercise 59.**

The *n*th roots of 1 are often referred to as the **nth roots of unity.** In Example 11, we found the cube roots of unity.

EXPLORING WITH TECHNOLOGY

We can approximate the *n*th roots of a number p with a graphing calculator set in PARAMETRIC mode.

WINDOW

Tmin $= 0$

Tmax $= 360$, or 2π

Tstep $= 360/n$, or $2\pi/n$

To find the fifth roots of 8, enter

$X_{IT} = (8^\wedge(1/5)) \cos T$

and

$Y_{IT} = (8^\wedge(1/5)) \sin T.$

Here we use DEGREE mode. Use the TRACE feature to locate the fifth roots. The T, X, and Y values appear on the screen.

$X_{1T} = 8^{1/5} \cos T$
$Y_{1T} = 8^{1/5} \sin T$

$T = 72$
$X = .46838218, \ Y = 1.4415321$

One of the fifth roots of 8 is approximately 0.4684. Find the other four.

EXAMPLE 12 Solve: $x^5 + i = 0$.

Solution To find all complex solutions of $x^5 + i = 0$, or $x^5 = -i$, we find the fifth roots of $-i$. We begin by converting $-i$ to trigonometric notation:

$$-i = 1(\cos 270° + i \sin 270°).$$

Then $n = 5$, $1/n = 1/5$, and $k = 0, 1, 2, 3, 4$. We have

$$[1(\cos 270° + i \sin 270°)]^{1/5}$$
$$= 1^{1/5}\left[\cos\left(\frac{270°}{5} + k \cdot \frac{360°}{5}\right) + i \sin\left(\frac{270°}{5} + k \cdot \frac{360°}{5}\right)\right], \quad k = 0, 1, 2, 3, 4.$$

The solutions are

$$\cos 54° + i \sin 54°, \qquad \cos 126° + i \sin 126°, \qquad \cos 198° + i \sin 198°,$$
$$\cos 270° + i \sin 270° \text{ (or } -i), \quad \text{and} \quad \cos 342° + i \sin 342°.$$

> **Now Try Exercise 73.**

8.3 Exercise Set

Graph the complex number and find its absolute value.

1. $4 + 3i$

2. $-2 - 3i$

3. i

4. $-5 - 2i$

5. $4 - i$

6. $6 + 3i$

7. 3

8. $-2i$

Express the indicated number in both standard notation and trigonometric notation.

9.

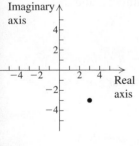

10.

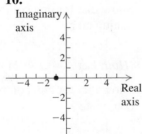

11.

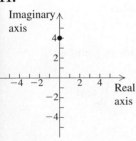

12.

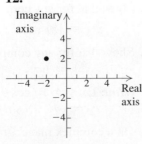

Find trigonometric notation.

13. $1 - i$

14. $-10\sqrt{3} + 10i$

15. $-3i$

16. $-5 + 5i$

17. $\sqrt{3} + i$

18. 4

19. $\dfrac{2}{5}$

20. $7.5i$

21. $-3\sqrt{2} - 3\sqrt{2}i$

22. $-\dfrac{9}{2} - \dfrac{9\sqrt{3}}{2}i$

Find standard notation, $a + bi$.

23. $3(\cos 30° + i \sin 30°)$

24. $6(\cos 120° + i \sin 120°)$

25. $10(\cos 270° + i \sin 270°)$

26. $3(\cos 0° + i \sin 0°)$

27. $\sqrt{8}\left(\cos\dfrac{\pi}{4} + i \sin\dfrac{\pi}{4}\right)$

28. $5\left(\cos\dfrac{\pi}{3} + i \sin\dfrac{\pi}{3}\right)$

29. $2\left(\cos\dfrac{\pi}{2} + i \sin\dfrac{\pi}{2}\right)$

30. $3\left[\cos\left(-\dfrac{3\pi}{4}\right) + i \sin\left(-\dfrac{3\pi}{4}\right)\right]$

31. $\sqrt{2}[\cos(-60°) + i \sin(-60°)]$

32. $4(\cos 135° + i \sin 135°)$

Multiply or divide and leave the answer in trigonometric notation.

33. $\dfrac{12(\cos 48° + i \sin 48°)}{3(\cos 6° + i \sin 6°)}$

34. $5\left(\cos\dfrac{\pi}{3} + i\sin\dfrac{\pi}{3}\right) \cdot 2\left(\cos\dfrac{\pi}{4} + i\sin\dfrac{\pi}{4}\right)$

35. $2.5(\cos 35° + i\sin 35°) \cdot 4.5(\cos 21° + i\sin 21°)$

36. $\dfrac{\dfrac{1}{2}\left(\cos\dfrac{2\pi}{3} + i\sin\dfrac{2\pi}{3}\right)}{\dfrac{3}{8}\left(\cos\dfrac{\pi}{6} + i\sin\dfrac{\pi}{6}\right)}$

Convert to trigonometric notation and then multiply or divide.

37. $(1 - i)(2 + 2i)$

38. $(1 + i\sqrt{3})(1 + i)$

39. $\dfrac{1 - i}{1 + i}$

40. $\dfrac{1 - i}{\sqrt{3} - i}$

41. $(3\sqrt{3} - 3i)(2i)$

42. $(2\sqrt{3} + 2i)(2i)$

43. $\dfrac{2\sqrt{3} - 2i}{1 + \sqrt{3}i}$

44. $\dfrac{3 - 3\sqrt{3}i}{\sqrt{3} - i}$

Raise the number to the given power and write trigonometric notation for the answer.

45. $\left[2\left(\cos\dfrac{\pi}{3} + i\sin\dfrac{\pi}{3}\right)\right]^3$

46. $[2(\cos 120° + i\sin 120°)]^4$

47. $(1 + i)^6$

48. $(-\sqrt{3} + i)^5$

Raise the number to the given power and write standard notation for the answer.

49. $[3(\cos 20° + i\sin 20°)]^3$

50. $[2(\cos 10° + i\sin 10°)]^9$

51. $(1 - i)^5$

52. $(2 + 2i)^4$

53. $\left(\dfrac{1}{\sqrt{2}} - \dfrac{1}{\sqrt{2}}i\right)^{12}$

54. $\left(\dfrac{\sqrt{3}}{2} + \dfrac{1}{2}i\right)^{10}$

Find the square roots of the number.

55. $-i$

56. $1 + i$

57. $2\sqrt{2} - 2\sqrt{2}i$

58. $-\sqrt{3} - i$

Find the cube roots of the number.

59. i

60. $-64i$

61. $2\sqrt{3} - 2i$

62. $1 - \sqrt{3}i$

63. Find and graph the fourth roots of 16.

64. Find and graph the fourth roots of i.

65. Find and graph the fifth roots of -1.

66. Find and graph the sixth roots of 1.

67. Find the tenth roots of 8.

68. Find the ninth roots of -4.

69. Find the sixth roots of -1.

70. Find the fourth roots of 12.

Solve.

71. $x^3 = 1$

72. $x^5 - 1 = 0$

73. $x^4 + i = 0$

74. $x^4 + 81 = 0$

75. $x^6 + 64 = 0$

76. $x^5 + \sqrt{3} + i = 0$

❯ Skill Maintenance

Convert to degree measure. **[6.4]**

77. $\dfrac{\pi}{12}$

78. 3π

Convert to radian measure. **[6.4]**

79. $330°$

80. $-225°$

Find the function value using coordinates of points on the unit circle. **[6.5]**

81. $\sin\dfrac{2\pi}{3}$

82. $\cos\dfrac{\pi}{6}$

83. $\cos\dfrac{\pi}{4}$

84. $\sin\dfrac{5\pi}{6}$

❯ Synthesis

85. Find polar notation for $(\cos\theta + i\sin\theta)^{-1}$.

86. Show that for any complex number z,
$$|z| = |-z|.$$

87. Show that for any complex number z and its conjugate $\bar{z}$,
$$|z| = |\bar{z}|.$$
(*Hint*: Let $z = a + bi$ and $\bar{z} = a - bi$.)

88. Show that for any complex number z and its conjugate $\bar{z}$,
$$|z\bar{z}| = |z^2|.$$
(*Hint*: Let $z = a + bi$ and $\bar{z} = a - bi$.)

89. Show that for any complex number z,
$$|z^2| = |z|^2.$$

90. Show that for any complex numbers z and w,
$$|z \cdot w| = |z| \cdot |w|.$$
(*Hint*: Let $z = r_1(\cos\theta_1 + i\sin\theta_1)$ and $w = r_2(\cos\theta_2 + i\sin\theta_2)$.)

91. On a complex plane, graph $z + \bar{z} = 3$.

92. On a complex plane, graph $|z| = 1$.

Determine whether the statement is true or false.

1. Any triangle, right or oblique, can be solved if at least one side and any other two measures are known. **[8.1]**

2. The absolute value of $-i$ is 1. **[8.3]**

3. The law of cosines cannot be used to solve a triangle when all three sides are known. **[8.2]**

4. Since angle measures determine only the shape of a triangle and not the size, we cannot solve a triangle when only the three angle measures are given. **[8.1]**

Solve △ABC, if possible. **[8.1], [8.2]**

5. $a = 8.3$ in., $A = 52°$, $C = 65°$

6. $A = 27.2°$, $c = 33$ m, $a = 14$ m

7. $a = 17.8$ yd, $b = 13.1$ yd, $c = 25.6$ yd

8. $a = 29.4$ cm, $b = 40.8$ cm, $A = 42.7°$

9. $A = 148°$, $b = 200$ yd, $c = 185$ yd

10. $b = 18$ ft, $c = 27$ ft, $B = 28°$

11. Find the area of the triangle with $C = 54°$, $a = 38$ in., and $b = 29$ in. **[8.2]**

Graph the complex number and find its absolute value. **[8.3]**

12. $-5 + 3i$

13. $-i$

14. 4

15. $1 - 5i$

Find trigonometric notation. **[8.3]**

16. $\dfrac{\sqrt{2}}{2} + \dfrac{\sqrt{6}}{2}i$

17. $1 - \sqrt{3}i$

18. $5i$

19. $-2 - 2i$

Find standard notation. **[8.3]**

20. $2\left(\cos\dfrac{7\pi}{4} + i\sin\dfrac{7\pi}{4}\right)$

21. $12(\cos 30° + i\sin 30°)$

22. $\sqrt{5}(\cos 0° + i\sin 0°)$

23. $4\left[\cos\left(-\dfrac{3\pi}{2}\right) + i\sin\left(-\dfrac{3\pi}{2}\right)\right]$

Multiply or divide and leave the answer in trigonometric notation. **[8.3]**

24. $8(\cos 20° + i\sin 20°) \cdot 2(\cos 25° + i\sin 25°)$

25. $3\left(\cos\dfrac{\pi}{3} + i\sin\dfrac{\pi}{3}\right) \div \left[\dfrac{1}{3}\left(\cos\dfrac{\pi}{4} + i\sin\dfrac{\pi}{4}\right)\right]$

Convert to trigonometric notation and then multiply or divide. **[8.3]**

26. $(1 - i)(\sqrt{3} - i)$

27. $\dfrac{1 - \sqrt{3}i}{1 + i}$

28. Find $(1 - i)^7$ and write trigonometric notation for the answer. **[8.3]**

29. Find $[2(\cos 15° + i\sin 15°)]^4$ and write standard notation for the answer. **[8.3]**

30. Find the square roots of $-2 - 2\sqrt{3}i$. **[8.3]**

31. Find the cube roots of -1. **[8.3]**

COLLABORATIVE DISCUSSION AND WRITING

32. Try to solve this triangle using the law of cosines. Then explain why it is easier to solve it using the law of sines. **[8.2]**

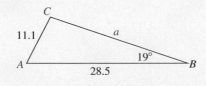

33. Explain why the following statements are not contradictory. **[8.3]**

The number 1 has one real cube root.

The number 1 has three complex cube roots.

34. Explain why we cannot solve a triangle given SAS with the law of sines. **[8.2]**

35. Explain why the law of sines cannot be used to find the first angle when solving a triangle given three sides. **[8.1]**

36. Explain why trigonometric notation for a complex number is not unique, but rectangular, or standard, notation is unique. **[8.3]**

37. Explain why $x^6 - 2x^3 + 1 = 0$ has three distinct solutions, $x^6 - 2x^3 = 0$ has four distinct solutions, and $x^6 - 2x = 0$ has six distinct solutions. **[8.3]**

8.4 ▷ Polar Coordinates and Graphs

> ❯ Graph points given their polar coordinates.

> ❯ Convert from rectangular coordinates to polar coordinates and from polar coordinates to rectangular coordinates.

> ❯ Convert from rectangular equations to polar equations and from polar equations to rectangular equations.

> ❯ Graph polar equations.

❯ Polar Coordinates

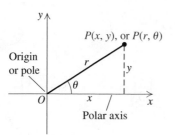

All graphing throughout this text has been done with rectangular coordinates, (x, y), in the Cartesian coordinate system. We now introduce the polar coordinate system. As shown in the diagram at left, any point P has rectangular coordinates (x, y) and polar coordinates (r, θ). On a polar graph, the origin is called the **pole** and the positive half of the x-axis is called the **polar axis.** The point P can be plotted given the directed angle θ from the polar axis to the ray OP and the directed distance r from the pole to the point. The angle θ can be expressed in degrees or radians.

To plot points on a polar graph:

1. Locate the directed angle θ.
2. Move a directed distance r from the pole. If $r > 0$, move along ray OP. If $r < 0$, move in the opposite direction of ray OP.

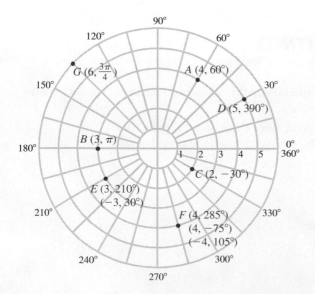

Polar graph paper, shown at left, facilitates plotting. Points B and G illustrate that θ may be in radians. Points E and F illustrate that the polar coordinates of a point are not unique.

EXAMPLE 1   Graph each of the following points.

a) $A(3, 60°)$

b) $B(0, 10°)$

c) $C(-5, 120°)$

d) $D(1, -60°)$

e) $E\left(2, \dfrac{3\pi}{2}\right)$

f) $F\left(-4, \dfrac{\pi}{3}\right)$

g) $G\left(-5, -\dfrac{\pi}{4}\right)$

h) $H\left(5, \dfrac{3\pi}{4}\right)$

Solution

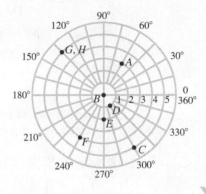

> *Now Try Exercises 3 and 7.*

To convert from rectangular coordinates to polar coordinates and from polar coordinates to rectangular coordinates, we need to recall the following relationships.

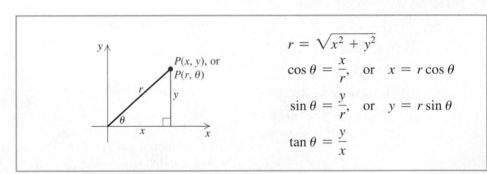

$$r = \sqrt{x^2 + y^2}$$

$$\cos \theta = \frac{x}{r}, \quad \text{or} \quad x = r \cos \theta$$

$$\sin \theta = \frac{y}{r}, \quad \text{or} \quad y = r \sin \theta$$

$$\tan \theta = \frac{y}{x}$$

EXAMPLE 2   Convert each of the following to polar coordinates.

a) $(3, 3)$

b) $(2\sqrt{3}, -2)$

Solution

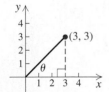

a) We first find r:

$$r = \sqrt{3^2 + 3^2} = \sqrt{18} = 3\sqrt{2}.$$

Then we determine θ:

$$\tan \theta = \frac{3}{3} = 1; \quad \text{therefore,} \quad \theta = 45°, \text{ or } \frac{\pi}{4}.$$

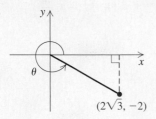

$(2\sqrt{3}, -2)$

We know that for $r = 3\sqrt{2}, \theta = \pi/4$ and not $5\pi/4$ since $(3, 3)$ is in quadrant I. Thus, $(r, \theta) = (3\sqrt{2}, 45°)$, or $(3\sqrt{2}, \pi/4)$. Other possibilities for polar coordinates include $(3\sqrt{2}, -315°)$ and $(-3\sqrt{2}, 5\pi/4)$.

b) We first find r:

$$r = \sqrt{(2\sqrt{3})^2 + (-2)^2} = \sqrt{12 + 4} = \sqrt{16} = 4.$$

Then we determine θ:

$$\tan \theta = \frac{-2}{2\sqrt{3}} = -\frac{1}{\sqrt{3}}; \quad \text{therefore,} \quad \theta = 330°, \text{ or } \frac{11\pi}{6}.$$

Thus, $(r, \theta) = (4, 330°)$, or $(4, 11\pi/6)$. Other possibilities for polar coordinates for this point include $(4, -\pi/6)$ and $(-4, 150°)$.

> *Now Try Exercise 19.*

It is easier to convert from polar coordinates to rectangular coordinates than from rectangular coordinates to polar coordinates.

EXAMPLE 3　Convert each of the following to rectangular coordinates.

a) $\left(10, \dfrac{\pi}{3}\right)$　　　　　　　　　　　**b)** $(-5, 135°)$

Solution

a) The ordered pair $(10, \pi/3)$ gives us $r = 10$ and $\theta = \pi/3$. We now find x and y:

$$x = r \cos \theta = 10 \cos \frac{\pi}{3} = 10 \cdot \frac{1}{2} = 5$$

and

$$y = r \sin \theta = 10 \sin \frac{\pi}{3} = 10 \cdot \frac{\sqrt{3}}{2} = 5\sqrt{3}.$$

Thus, $(x, y) = (5, 5\sqrt{3})$.

b) From the ordered pair $(-5, 135°)$, we know that $r = -5$ and $\theta = 135°$. We now find x and y:

$$x = -5 \cos 135° = -5 \cdot \left(-\frac{\sqrt{2}}{2}\right) = \frac{5\sqrt{2}}{2}$$

and

$$y = -5 \sin 135° = -5 \cdot \left(\frac{\sqrt{2}}{2}\right) = -\frac{5\sqrt{2}}{2}.$$

Thus, $(x, y) = \left(\dfrac{5\sqrt{2}}{2}, -\dfrac{5\sqrt{2}}{2}\right)$.

> *Now Try Exercises 31 and 37.*

The conversions above can be easily made with some graphing calculators.

❯ Polar Equations and Rectangular Equations

Some curves have simpler equations in polar coordinates than in rectangular coordinates. For others, the reverse is true.

EXAMPLE 4 Convert each of the following to a polar equation.

a) $x^2 + y^2 = 25$ **b)** $2x - y = 5$

Solution

a) We have

$$x^2 + y^2 = 25$$
$$(r \cos \theta)^2 + (r \sin \theta)^2 = 25 \qquad \text{Substituting for } x \text{ and } y$$
$$r^2 \cos^2 \theta + r^2 \sin^2 \theta = 25$$
$$r^2(\cos^2 \theta + \sin^2 \theta) = 25$$
$$r^2 = 25 \qquad \cos^2 \theta + \sin^2 \theta = 1$$
$$r = 5.$$

This example illustrates that the polar equation of a circle centered at the origin is much simpler than the rectangular equation.

b) We have

$$2x - y = 5$$
$$2(r \cos \theta) - (r \sin \theta) = 5$$
$$r(2 \cos \theta - \sin \theta) = 5.$$

In this example, we see that the rectangular equation is simpler than the polar equation.

❯ *Now Try Exercises 47 and 51.*

EXAMPLE 5 Convert each of the following to a rectangular equation.

a) $r = 4$
b) $r \cos \theta = 6$
c) $r = 2 \cos \theta + 3 \sin \theta$

Solution

a) We have

$$r = 4$$
$$\sqrt{x^2 + y^2} = 4 \qquad \text{Substituting for } r$$
$$x^2 + y^2 = 16. \qquad \text{Squaring}$$

In squaring, we must be careful not to introduce solutions of the equation that are not already present. In this case, we did not, because the graph of either equation is a circle of radius 4 centered at the origin.

b) We have

$$r \cos \theta = 6$$
$$x = 6. \qquad x = r \cos \theta$$

The graph of $r \cos \theta = 6$, or $x = 6$, is a vertical line.

c) We have

$$r = 2 \cos \theta + 3 \sin \theta$$
$$r^2 = 2r \cos \theta + 3r \sin \theta \qquad \text{Multiplying by } r \text{ on both sides}$$
$$x^2 + y^2 = 2x + 3y. \qquad \text{Substituting } x^2 + y^2 \text{ for } r^2,$$
$$\text{x for } r \cos \theta, \text{ and } y \text{ for } r \sin \theta$$

> *Now Try Exercises 59 and 63.*

〉 Graphing Polar Equations

To graph a polar equation, we can make a table of values, choosing values of θ and calculating corresponding values of r. We plot the points and complete the graph, as we do when graphing a rectangular equation. A difference occurs in the case of a polar equation, however, because as θ increases sufficiently, points may begin to repeat and the curve will be traced again and again. When this happens, the curve is complete.

EXAMPLE 6   Graph: $r = 1 - \sin \theta$.

Solution   We first make a table of values. The TABLE feature on a graphing calculator is the most efficient way to create this list. Note that the points begin to repeat at $\theta = 360°$. We plot these points and draw the curve, as shown below.

θ	r
0°	1
15°	0.7412
30°	0.5
45°	0.2929
60°	0.1340
75°	0.0341
90°	0
105°	0.0341
120°	0.1340
135°	0.2929
150°	0.5
165°	0.7412
180°	1

θ	r
195°	1.2588
210°	1.5
225°	1.7071
240°	1.8660
255°	1.9659
270°	2
285°	1.9659
300°	1.8660
315°	1.7071
330°	1.5
345°	1.2588
360°	1
375°	0.7412
390°	0.5

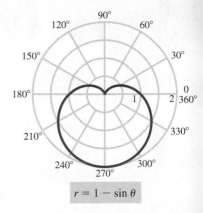

$r = 1 - \sin \theta$

Because of its heart shape, this curve is called a *cardioid*.

> *Now Try Exercise 77.*

We plotted points in Example 6 because we feel that it is important to understand how these curves are developed. We can also graph polar equations using a graphing calculator. The equation usually must be written first in the form $r = f(\theta)$. It is necessary to decide on not only the best window dimensions but also the range of values for θ. Typically, we begin with a range of 0 to 2π for θ in radians and 0° to 360° for θ in degrees. Because most polar graphs are curved, it is important to *square* the window to minimize distortion.

EXPLORING WITH TECHNOLOGY

Graph $r = 4 \sin 3\theta$. Begin by setting the calculator in POLAR mode, and use either of the following windows. The calculator allows us to view the curve as it is formed.

WINDOW (Radians)	WINDOW (Degrees)
$\theta\text{min} = 0$	$\theta\text{min} = 0$
$\theta\text{max} = 2\pi$	$\theta\text{max} = 360$
$\theta\text{step} = \pi/24$	$\theta\text{step} = 1$

$r = 4 \sin 3\theta$

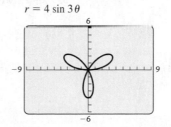

Now graph each of the following equations and note the effect of changing the coefficient of $\sin 3\theta$ and changing the coefficient of θ:

$r = 2 \sin 3\theta,$

$r = 6 \sin 3\theta,$

$r = 4 \sin \theta,$ and

$r = 4 \sin 5\theta.$

Polar equations of the form $r = a \cos n\theta$ and $r = a \sin n\theta$ have rose-shaped curves. The number a determines the length of the petals, and the number n determines the number of petals. If n is odd, there are n petals. If n is even, there are $2n$ petals.

EXAMPLE 7 Graph each of the following polar equations. Try to visualize the shape of the curve before graphing it.

a) $r = 3$

b) $r = 5 \sin \theta$

c) $r = 2 \csc \theta$

Solution For each graph, we begin with a table of values. Then we plot points and complete the graph.

a) $r = 3$

For all values of θ, r is 3. Thus the graph of $r = 3$ is a circle of radius 3 centered at the origin.

θ	r
0°	3
60°	3
135°	3
210°	3
300°	3
360°	3

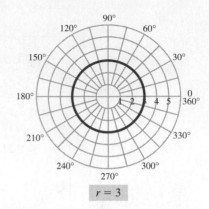

$r = 3$

We can verify our graph by converting to the equivalent rectangular equation. For $r = 3$, we substitute $\sqrt{x^2 + y^2}$ for r and square. The resulting equation,

$$x^2 + y^2 = 3^2,$$

is the equation of a circle with radius 3 centered at the origin.

b) $r = 5 \sin \theta$

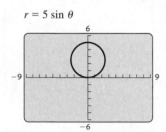

$r = 5 \sin \theta$

θ	r
0°	0
15°	1.2941
30°	2.5
45°	3.5355
60°	4.3301
75°	4.8296
90°	5
105°	4.8296
120°	4.3301
135°	3.5355
150°	2.5
165°	1.2941
180°	0

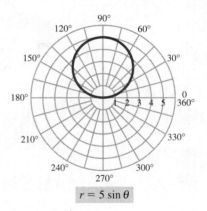

$r = 5 \sin \theta$

c) $r = 2 \csc \theta$

We can rewrite $r = 2 \csc \theta$ as $r = 2/\sin \theta$.

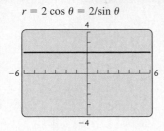

$r = 2 \cos \theta = 2/\sin \theta$

θ	r
0°	Not defined
15°	7.7274
30°	4
45°	2.8284
60°	2.3094
75°	2.0706
90°	2
105°	2.0706
120°	2.3094
135°	2.8284
150°	4
165°	7.7274
180°	Not defined

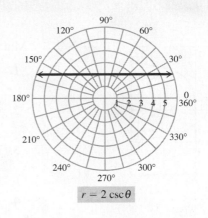

$r = 2 \csc \theta$

> *Now Try Exercise 71.*

We can check our graph in Example 7(c) by converting the polar equation to the equivalent rectangular equation:

$$r = 2 \csc \theta$$
$$r = \frac{2}{\sin \theta}$$
$$r \sin \theta = 2$$
$$y = 2. \qquad \textbf{Substituting } y \textbf{ for } r \sin \theta$$

The graph of $y = 2$ is a horizontal line passing through $(0, 2)$ on a rectangular grid, as shown at left.

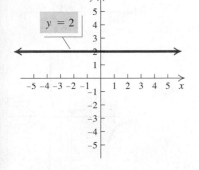

$y = 2$

EXAMPLE 8 Graph the equation $r + 1 = 2 \cos 2\theta$ with a graphing calculator.

Solution We first solve for r:

$$r = 2 \cos 2\theta - 1.$$

We then obtain the graph shown at left.

> *Now Try Exercise 91.*

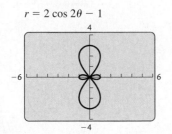

$r = 2 \cos 2\theta - 1$

Visualizing the Graph

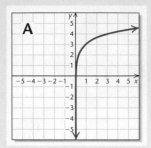

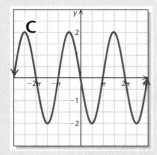

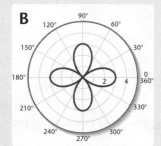

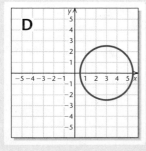

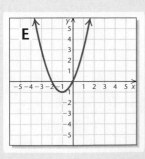

Match the equation with its graph.

1. $f(x) = 2^{(1/2)x}$

2. $y = -2 \sin x$

3. $y = (x + 1)^2 - 1$

4. $f(x) = \dfrac{x - 3}{x^2 + x - 6}$

5. $r = 1 + \sin \theta$

6. $f(x) = 2 \log x + 3$

7. $(x - 3)^2 + y^2 = \dfrac{25}{4}$

8. $y = -\cos\left(x - \dfrac{\pi}{2}\right)$

9. $r = 3 \cos 2\theta$

10. $f(x) = x^4 - x^3 + x^2 - x$

Answers on page A-49

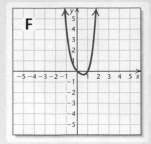

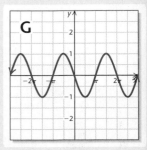

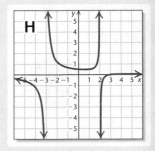

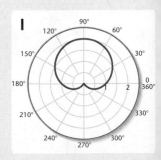

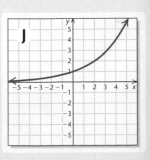

8.4 Exercise Set

Graph the point on a polar grid.

1. $\left(2, \dfrac{\pi}{4}\right)$ **2.** $(4, \pi)$

3. $(3.5, 210°)$ **4.** $(-3, 135°)$

5. $\left(1, \dfrac{\pi}{6}\right)$ **6.** $\left(2.75, \dfrac{5\pi}{6}\right)$

7. $\left(-5, \dfrac{\pi}{2}\right)$ **8.** $(0, 15°)$

9. $(3, -315°)$ **10.** $\left(1.2, -\dfrac{2\pi}{3}\right)$

11. $(4.3, -60°)$ **12.** $(3, 405°)$

Find polar coordinates of points A, B, C, and D. Give three answers for each point.

13.

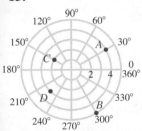

14.

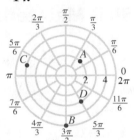

Find the polar coordinates of the point. Express the angle in degrees and then in radians, using the smallest possible positive angle.

15. $(0, -3)$ **16.** $(-4, 4)$

17. $\left(3, -3\sqrt{3}\right)$ **18.** $\left(-\sqrt{3}, 1\right)$

19. $\left(4\sqrt{3}, -4\right)$ **20.** $\left(2\sqrt{3}, 2\right)$

21. $\left(-\sqrt{2}, -\sqrt{2}\right)$ **22.** $\left(-3, 3\sqrt{3}\right)$

23. $\left(1, \sqrt{3}\right)$ **24.** $(0, -1)$

25. $\left(\dfrac{5\sqrt{2}}{2}, -\dfrac{5\sqrt{2}}{2}\right)$ **26.** $\left(-\dfrac{3}{2}, -\dfrac{3\sqrt{3}}{2}\right)$

Use a graphing calculator to convert from rectangular coordinates to polar coordinates. Express the answer in both degrees and radians, using the smallest possible positive angle.

27. $(3, 7)$ **28.** $\left(-2, -\sqrt{5}\right)$

29. $\left(-\sqrt{10}, 3.4\right)$ **30.** $(0.9, -6)$

Find the rectangular coordinates of the point.

31. $(5, 60°)$ **32.** $(0, -23°)$

33. $(-3, 45°)$ **34.** $(6, 30°)$

35. $(3, -120°)$ **36.** $\left(7, \dfrac{\pi}{6}\right)$

37. $\left(-2, \dfrac{5\pi}{3}\right)$ **38.** $(1.4, 225°)$

39. $(2, 210°)$ **40.** $\left(1, \dfrac{7\pi}{4}\right)$

41. $\left(-6, \dfrac{5\pi}{6}\right)$ **42.** $(4, 180°)$

Use a graphing calculator to convert from polar coordinates to rectangular coordinates. Round the coordinates to the nearest hundredth.

43. $(3, -43°)$ **44.** $\left(-5, \dfrac{\pi}{7}\right)$

45. $\left(-4.2, \dfrac{3\pi}{5}\right)$ **46.** $(2.8, 166°)$

Convert to a polar equation.

47. $3x + 4y = 5$ **48.** $5x + 3y = 4$

49. $x = 5$ **50.** $y = 4$

51. $x^2 + y^2 = 36$ **52.** $x^2 - 4y^2 = 4$

53. $x^2 = 25y$ **54.** $2x - 9y + 3 = 0$

55. $y^2 - 5x - 25 = 0$ **56.** $x^2 + y^2 = 8y$

57. $x^2 - 2x + y^2 = 0$ **58.** $3x^2y = 81$

Convert to a rectangular equation.

59. $r = 5$ **60.** $\theta = \dfrac{3\pi}{4}$

61. $r \sin \theta = 2$ **62.** $r = -3 \sin \theta$

63. $r + r \cos \theta = 3$ **64.** $r = \dfrac{2}{1 - \sin \theta}$

65. $r - 9 \cos \theta = 7 \sin \theta$

66. $r + 5 \sin \theta = 7 \cos \theta$

67. $r = 5 \sec \theta$ **68.** $r = 3 \cos \theta$

69. $\theta = \dfrac{5\pi}{3}$ **70.** $r = \cos \theta - \sin \theta$

Graph the equation by plotting points. Then check your work using a graphing calculator.

71. $r = \sin \theta$

72. $r = 1 - \cos \theta$

73. $r = 4 \cos 2\theta$

74. $r = 1 - 2 \sin \theta$

75. $r = \cos \theta$

76. $r = 2 \sec \theta$

77. $r = 2 - \cos 3\theta$

78. $r = \dfrac{1}{1 + \cos \theta}$

In Exercises 79–90, use a graphing calculator to match the equation with one of figures (a)–(l) that follow. Try matching the graphs mentally before using a calculator.

a)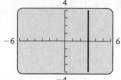

b)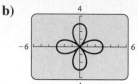

c)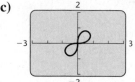

d)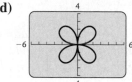

e)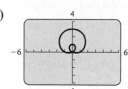

f)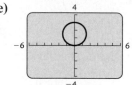

g)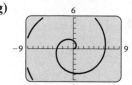

h)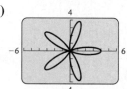

i)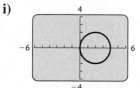

j)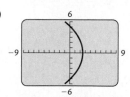

k)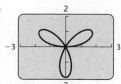

l)

79. $r = 3 \sin 2\theta$

80. $r = 4 \cos \theta$

81. $r = \theta$

82. $r^2 = \sin 2\theta$

83. $r = \dfrac{5}{1 + \cos \theta}$

84. $r = 1 + 2 \sin \theta$

85. $r = 3 \cos 2\theta$

86. $r = 3 \sec \theta$

87. $r = 3 \sin \theta$

88. $r = 4 \cos 5\theta$

89. $r = 2 \sin 3\theta$

90. $r \sin \theta = 6$

Graph the equation using a graphing calculator.

91. $r = \sin \theta \tan \theta$ (Cissoid)

92. $r = 3\theta$ (Spiral of Archimedes)

93. $r = e^{\theta/10}$ (Logarithmic spiral)

94. $r = 10^{2\theta}$ (Logarithmic spiral)

95. $r = \cos 2\theta \sec \theta$ (Strophoid)

96. $r = \cos 2\theta - 2$ (Peanut)

97. $r = \frac{1}{4} \tan^2 \theta \sec \theta$ (Semicubical parabola)

98. $r = \sin 2\theta + \cos \theta$ (Twisted sister)

❯ Skill Maintenance

Solve. [1.5]

99. $2x - 4 = x + 8$

100. $4 - 5y = 3$

Graph. [1.3]

101. $y = 2x - 5$

102. $4x - y = 6$

103. $x = -3$

104. $y = 0$

❯ Synthesis

105. Convert to a rectangular equation:
$$r = \sec^2 \frac{\theta}{2}.$$

106. The center of a regular hexagon is at the origin, and one vertex is the point $(4, 0°)$. Find the coordinates of the other vertices.

8.5 > Vectors and Applications

> Determine whether two vectors are equivalent.

> Find the sum, or resultant, of two vectors.

> Resolve a vector into its horizontal component and its vertical component.

> Solve applied problems involving vectors.

We measure some quantities using only their magnitudes. For example, we describe time, length, and mass using units like seconds, feet, and kilograms, respectively. However, to measure quantities like **displacement**, **velocity**, or **force**, we need to describe a *magnitude* and a *direction*. Together magnitude and direction describe a **vector.** The following are some examples.

Displacement.   An object moves a certain distance in a certain direction.

A surveyor steps 20 yd to the northeast.

A hiker follows a trail 5 mi to the west.

A batter hits a ball 100 m along the left-field line.

Velocity.   An object travels at a certain speed in a certain direction.

A breeze is blowing 15 mph from the northwest.

An airplane is traveling 450 km/h in a direction of 243°.

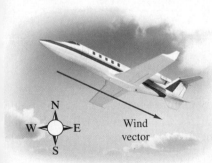

Force.   A push or pull is exerted on an object in a certain direction.

A force of 200 lb is required to pull a cart up a 30° incline.

A 25-lb force is required to lift a box upward.

A force of 15 newtons is exerted downward on the handle of a jack. (A newton, abbreviated N, is a unit of force used in physics; 1 N ≈ 0.22 lb.)

> Vectors

Vectors can be graphically represented by directed line segments. The length is chosen, according to some scale, to represent the **magnitude of the vector**, and the direction of the directed line segment represents the **direction of the vector**. For example, if we let 1 cm represent 5 km/h, then a 15-km/h wind from the northwest would be represented by a directed line segment 3 cm long, as shown in the figure at left.

Wind vector

VECTOR

A **vector** in the plane is a directed line segment. Two vectors are **equivalent** if they have the same *magnitude* and the same *direction*.

Consider a vector drawn from point *A* to point *B*. Point *A* is called the **initial point** of the vector, and point *B* is called the **terminal point.** Symbolic notation for this vector is $\overrightarrow{AB}$ (read "vector *AB*"). Vectors are also denoted by boldface letters such as **u**, **v**, and **w**. The four vectors in the figure at left have the *same* length and the *same* direction. Thus they represent **equivalent** vectors; that is,

$$\overrightarrow{AB} = \overrightarrow{CD} = \overrightarrow{OP} = \mathbf{v}.$$

In the context of vectors, we use = to mean equivalent.

The length, or **magnitude**, of $\overrightarrow{AB}$ is expressed as $|\overrightarrow{AB}|$. In order to determine whether vectors are equivalent, we find their magnitudes and directions.

EXAMPLE 1 The vectors **u**, $\overrightarrow{OR}$, and **w** are shown in the following figure. Show that $\mathbf{u} = \overrightarrow{OR} = \mathbf{w}$.

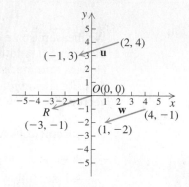

Solution We first find the length of each vector using the distance formula:

$$|\mathbf{u}| = \sqrt{[2 - (-1)]^2 + (4 - 3)^2} = \sqrt{9 + 1} = \sqrt{10},$$
$$|\overrightarrow{OR}| = \sqrt{[0 - (-3)]^2 + [0 - (-1)]^2} = \sqrt{9 + 1} = \sqrt{10},$$
$$|\mathbf{w}| = \sqrt{(4 - 1)^2 + [-1 - (-2)]^2} = \sqrt{9 + 1} = \sqrt{10}.$$

Thus,

$$|\mathbf{u}| = |\overrightarrow{OR}| = |\mathbf{w}|.$$

The vectors **u**, $\overrightarrow{OR}$, and **w** appear to go in the same direction so we check their slopes. If the lines that they are on all have the same slope, the vectors have the same direction. We calculate the slopes:

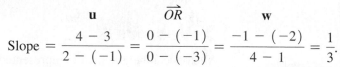

$$\text{Slope} = \frac{4 - 3}{2 - (-1)} = \frac{0 - (-1)}{0 - (-3)} = \frac{-1 - (-2)}{4 - 1} = \frac{1}{3}.$$

Since **u**, $\overrightarrow{OR}$, and **w** have the *same* magnitude and the *same* direction,

$$\mathbf{u} = \overrightarrow{OR} = \mathbf{w}.$$

> **Now Try Exercise 1.**

Keep in mind that the equivalence of vectors requires only the same magnitude and the same direction—not the same location. In the illustrations at left, each of the first three pairs of vectors are not equivalent. The fourth set of vectors is an example of equivalence.

u ≠ **v** (not equivalent)
Different magnitudes;
different directions

u ≠ **v**
Same magnitude;
different directions

u ≠ **v**
Different magnitudes;
same direction

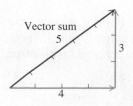

u = **v**
Same magnitude;
same direction

› Vector Addition

Suppose that a person takes 4 steps east and then 3 steps north. He or she will then be 5 steps from the starting point in the direction shown at left. A vector 4 units long and pointing to the right represents 4 steps east and a vector 3 units long and pointing up represents 3 steps north. The **sum** of the two vectors is the vector 5 steps in magnitude and in the direction shown. The sum is also called the **resultant** of the two vectors.

In general, two nonzero vectors **u** and **v** can be added geometrically by placing the initial point of **v** at the terminal point of **u** and then finding the vector that has the same initial point as **u** and the same terminal point as **v**, as shown in the following figure.

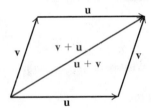

The sum **u** + **v** is the vector represented by the directed line segment from the initial point A of **u** to the terminal point C of **v**. That is, if

$$\mathbf{u} = \overrightarrow{AB} \quad \text{and} \quad \mathbf{v} = \overrightarrow{BC},$$

then

$$\mathbf{u} + \mathbf{v} = \overrightarrow{AB} + \overrightarrow{BC} = \overrightarrow{AC}.$$

We can also describe vector addition by placing the initial points of the vectors together, completing a parallelogram, and finding the diagonal of the parallelogram. (See the figure on the left below.) This description of addition is sometimes called the **parallelogram law** of vector addition. Vector addition is **commutative.** As shown in the figure on the right below, both **u** + **v** and **v** + **u** are represented by the same directed line segment.

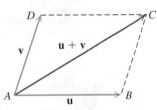
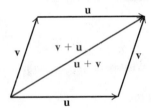

❯ Applications

If two forces F_1 and F_2 act on an object, the *combined* effect is the sum, or the resultant, $F_1 + F_2$ of the separate forces.

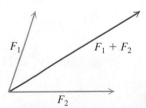

EXAMPLE 2   Forces of 15 newtons and 25 newtons act on an object at right angles to each other. Find their sum, or resultant, giving the magnitude of the resultant and the angle that it makes with the larger force.

Solution   We make a drawing—this time, a rectangle—using **v** or $\overrightarrow{OB}$ to represent the resultant. To find the magnitude, we use the Pythagorean equation:

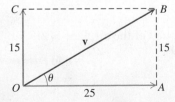

$$|\mathbf{v}|^2 = 15^2 + 25^2 \qquad |\mathbf{v}| \text{ denotes the length, or magnitude, of } \mathbf{v}.$$
$$|\mathbf{v}| = \sqrt{15^2 + 25^2}$$
$$\approx 29.2.$$

To find the direction, we note that since *OAB* is a right triangle,

$$\tan \theta = \tfrac{15}{25} = 0.6.$$

Using a calculator, we find θ, the angle that the resultant makes with the larger force:

$$\theta = \tan^{-1}(0.6) \approx 31°.$$

The resultant $\overrightarrow{OB}$ has a magnitude of 29.2 newtons and makes an angle of 31° with the larger force.

> *Now Try Exercise 13.*

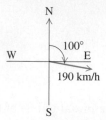

Pilots must adjust the direction of their flight when there is a crosswind. Both the wind and the aircraft velocities can be described by vectors.

EXAMPLE 3 *Airplane Speed and Direction.* An airplane travels on a bearing of 100° at an airspeed of 190 km/h while a wind is blowing 48 km/h from 220°. Find the ground speed of the airplane and the direction of its track, or course, over the ground.

Solution We first make a drawing. The wind is represented by $\overrightarrow{OC}$ and the velocity vector of the airplane by $\overrightarrow{OA}$. The resultant velocity vector is **v**, the sum of the two vectors. The angle θ between **v** and $\overrightarrow{OA}$ is called a **drift angle.**

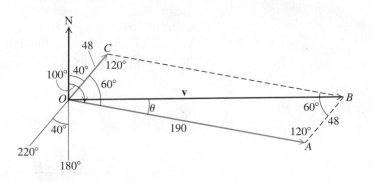

Note that the measure of $\angle COA = 100° - 40° = 60°$. Thus the measure of $\angle CBA$ is also 60° (opposite angles of a parallelogram are equal). Since the sum of all the angles of the parallelogram is 360° and $\angle OCB$ and $\angle OAB$ have the same measure, each must be 120°. By the *law of cosines* in $\triangle OAB$, we have

$$|\mathbf{v}|^2 = 48^2 + 190^2 - 2 \cdot 48 \cdot 190 \cos 120°$$
$$|\mathbf{v}|^2 = 47{,}524$$
$$|\mathbf{v}| = 218.$$

Thus, $|\mathbf{v}|$ is 218 km/h. By the *law of sines* in the same triangle,

$$\frac{48}{\sin \theta} = \frac{218}{\sin 120°},$$

or

$$\sin \theta = \frac{48 \sin 120°}{218} \approx 0.1907$$
$$\theta \approx 11°.$$

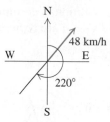

Thus, $\theta = 11°$, to the nearest degree. The ground speed of the airplane is 218 km/h, and its track is in the direction of $100° - 11°$, or 89°.

> *Now Try Exercise 27.*

› Components

Given a vector **w**, we may want to find two other vectors **u** and **v** whose sum is **w**. The vectors **u** and **v** are called **components** of **w** and the process of finding them is called **resolving**, or **representing**, a vector into its vector components.

When we resolve a vector, we generally look for perpendicular components. Most often, one component will be parallel to the *x*-axis and the other will be parallel to the *y*-axis. For this reason, they are often called the **horizontal** component and the **vertical** component of a vector. In the figure at left, the vector $\mathbf{w} = \overrightarrow{AC}$ is resolved as the sum of $\mathbf{u} = \overrightarrow{AB}$ and $\mathbf{v} = \overrightarrow{BC}$. The horizontal component of **w** is **u** and the vertical component is **v**.

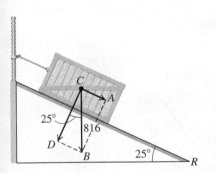

EXAMPLE 4   A vector **w** has a magnitude of 130 and is inclined 40° with the horizontal. Resolve the vector into its horizontal component and its vertical component.

Solution   We first make a drawing showing a horizontal vector **u** and a vertical vector **v** whose sum is **w**.

From $\triangle ABC$, we find $|\mathbf{u}|$ and $|\mathbf{v}|$ using the definitions of the cosine and sine functions:

$$\cos 40° = \frac{|\mathbf{u}|}{130}, \quad \text{or} \quad |\mathbf{u}| = 130 \cos 40° \approx 100,$$

$$\sin 40° = \frac{|\mathbf{v}|}{130}, \quad \text{or} \quad |\mathbf{v}| = 130 \sin 40° \approx 84.$$

Thus the horizontal component of **w** is 100 right, and the vertical component of **w** is 84 up.

› *Now Try Exercise 31.*

EXAMPLE 5   *Shipping Crate.*   A wooden shipping crate that weighs 816 lb is placed on a loading ramp that makes an angle of 25° with the horizontal. To keep the crate from sliding, a chain is hooked to the crate and to a pole at the top of the ramp. Find the magnitude of the components of the crate's weight (disregarding friction) perpendicular to and parallel to the incline.

Solution   We first make a drawing illustrating the forces with a rectangle. We let

$|\overrightarrow{CB}|$ = the weight of the crate = 816 lb (force of gravity),

$|\overrightarrow{CD}|$ = the magnitude of the component of the crate's weight perpendicular to the incline (force against the ramp), and

$|\overrightarrow{CA}|$ = the magnitude of the component of the crate's weight parallel to the incline (force that pulls the crate down the ramp).

The angle at *R* is given to be 25° and $\angle BCD = \angle R = 25°$ because the sides of these angles are, respectively, perpendicular. Using the cosine and sine functions, we find that

$$\cos 25° = \frac{|\overrightarrow{CD}|}{816}, \quad \text{or} \quad |\overrightarrow{CD}| = 816 \cos 25° \approx 740 \text{ lb}, \quad \text{and}$$

$$\sin 25° = \frac{|\overrightarrow{DB}|}{816} = \frac{|\overrightarrow{CA}|}{816}, \quad \text{or} \quad |\overrightarrow{CA}| = 816 \sin 25° \approx 345 \text{ lb}.$$

› *Now Try Exercise 39.*

8.5 Exercise Set

Sketch the pair of vectors and determine whether they are equivalent. Use the following ordered pairs for the initial point and the terminal point.

$A(-2, 2)$	$E(-4, 1)$	$I(-6, -3)$
$B(3, 4)$	$F(2, 1)$	$J(3, 1)$
$C(-2, 5)$	$G(-4, 4)$	$K(-3, -3)$
$D(-1, -1)$	$H(1, 2)$	$O(0, 0)$

1. $\overrightarrow{GE}$, $\overrightarrow{BJ}$
2. $\overrightarrow{DJ}$, $\overrightarrow{OF}$
3. $\overrightarrow{DJ}$, $\overrightarrow{AB}$
4. $\overrightarrow{CG}$, $\overrightarrow{FO}$
5. $\overrightarrow{DK}$, $\overrightarrow{BH}$
6. $\overrightarrow{BA}$, $\overrightarrow{DI}$
7. $\overrightarrow{EG}$, $\overrightarrow{BJ}$
8. $\overrightarrow{GC}$, $\overrightarrow{FO}$
9. $\overrightarrow{GA}$, $\overrightarrow{BH}$
10. $\overrightarrow{JD}$, $\overrightarrow{CG}$
11. $\overrightarrow{AB}$, $\overrightarrow{ID}$
12. $\overrightarrow{OF}$, $\overrightarrow{HB}$

13. Two forces of 32 N (newtons) and 45 N act on an object at right angles. Find the magnitude of the resultant and the angle that it makes with the smaller force.

14. Two forces of 50 N and 60 N act on an object at right angles. Find the magnitude of the resultant and the angle that it makes with the larger force.

15. Two forces of 410 N and 600 N act on an object. The angle between the forces is 47°. Find the magnitude of the resultant and the angle that it makes with the larger force.

16. Two forces of 255 N and 325 N act on an object. The angle between the forces is 64°. Find the magnitude of the resultant and the angle that it makes with the smaller force.

*In Exercises 17–24, the magnitudes of vectors **u** and **v** and the angle θ between the vectors are given. Find the sum of **u** + **v**. Give the magnitude to the nearest tenth and give the direction by specifying to the nearest degree the angle that the resultant makes with **u**.*

17. $|\mathbf{u}| = 45$, $|\mathbf{v}| = 35$, $\theta = 90°$
18. $|\mathbf{u}| = 54$, $|\mathbf{v}| = 43$, $\theta = 150°$
19. $|\mathbf{u}| = 10$, $|\mathbf{v}| = 12$, $\theta = 67°$
20. $|\mathbf{u}| = 25$, $|\mathbf{v}| = 30$, $\theta = 75°$
21. $|\mathbf{u}| = 20$, $|\mathbf{v}| = 20$, $\theta = 117°$
22. $|\mathbf{u}| = 30$, $|\mathbf{v}| = 30$, $\theta = 123°$
23. $|\mathbf{u}| = 23$, $|\mathbf{v}| = 47$, $\theta = 27°$
24. $|\mathbf{u}| = 32$, $|\mathbf{v}| = 74$, $\theta = 72°$

25. *Hot-Air Balloon.* A hot-air balloon is rising vertically 10 ft/sec while the wind is blowing horizontally 5 ft/sec. Find the speed **v** of the balloon and the angle θ that it makes with the horizontal.

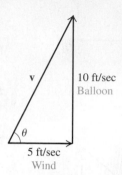

26. *Ship.* A ship first sails N80°E for 120 nautical mi, and then S20°W for 200 nautical mi. How far is the ship, then, from the starting point and in what direction?

27. *Boat.* A boat heads 35°, propelled by a force of 750 lb. A wind from 320° exerts a force of 150 lb on the boat. How large is the resultant force **F**, and in what direction is the boat moving?

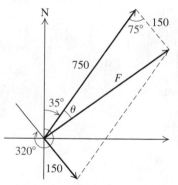

28. *Airplane.* An airplane flies 32° for 210 km, and then 280° for 170 km. How far is the airplane, then, from the starting point and in what direction?

29. *Airplane.* An airplane has an airspeed of 150 km/h. It is to make a flight in a direction of 70° while there is a 25-km/h wind from 340°. What will the airplane's actual heading be?

30. *Wind.* A wind has an easterly component (*from* the east) of 10 km/h and a southerly component (*from* the south) of 16 km/h. Find the magnitude and the direction of the wind.

31. A vector **w** has magnitude 100 and points southeast. Resolve the vector into an easterly component and a southerly component.

32. A vector **u** with a magnitude of 150 lb is inclined to the right and upward 52° from the horizontal. Resolve the vector into components.

33. *Airplane.*   An airplane takes off at a speed **S** of 225 mph at an angle of 17° with the horizontal. Resolve the vector **S** into components.

34. *Wheelbarrow.*   A wheelbarrow is pushed by applying a 97-lb force **F** that makes a 38° angle with the horizontal. Resolve **F** into its horizontal component and its vertical component. (The horizontal component is the effective force in the direction of motion and the vertical component adds weight to the wheelbarrow.)

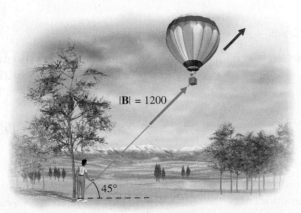

|F| = 97

38°

35. *Luggage Wagon.*   A luggage wagon is being pulled with vector force **V**, which has a magnitude of 780 lb at an angle of elevation of 60°. Resolve the vector **V** into components.

|V| = 780

60°

36. *Hot-air Balloon.*   A hot-air balloon exerts a 1200-lb pull on a tether line at a 45° angle with the horizontal. Resolve the vector **B** into components.

|B| = 1200

45°

37. *Airplane.*   An airplane is flying at 200 km/h in a direction of 305°. Find the westerly component and the northerly component of its velocity.

38. *Baseball.*   A baseball player throws a baseball with a speed **S** of 72 mph at an angle of 45° with the horizontal. Resolve the vector **S** into components.

39. A block weighing 100 lb rests on a 25° incline. Find the magnitude of the components of the block's weight perpendicular to and parallel to the incline.

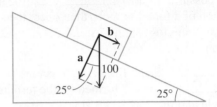

b

a

100

25°   25°

40. A shipping crate that weighs 450 kg is placed on a loading ramp that makes an angle of 30° with the horizontal. Find the magnitude of the components of the crate's weight perpendicular to and parallel to the incline.

41. An 80-lb block of ice rests on a 37° incline. What force parallel to the incline is necessary in order to keep the ice from sliding down?

42. What force is necessary to pull a 3500-lb truck up a 9° incline?

❯ Skill Maintenance

Vocabulary Reinforcement

In each of Exercises 43–52, fill in the blank with the correct term. Some of the given choices will not be used.

angular speed	cosine
linear speed	common
acute	natural
obtuse	horizontal line
secant of θ	vertical line
cotangent of θ	double-angle
identity	half-angle
inverse	coterminal
absolute value	reference angle
sines	

43. Logarithms, base *e*, are called _____ logarithms. **[5.3]**

44. _____ identities give trigonometric function values of $x/2$ in terms of function values of *x*. **[7.2]**

45. _____ is distance traveled per unit of time. **[6.4]**

46. The sine of an angle is also the _____ of the angle's complement. **[6.1]**

47. A(n) _____ is an equation that is true for all possible replacements of the variables. **[7.1]**

48. The _____ is the length of the side adjacent to θ divided by the length of the side opposite θ. **[6.1]**

49. If two or more angles have the same terminal side, the angles are said to be _____. **[6.3]**

50. In any triangle, the sides are proportional to the _____ of the opposite angles. **[8.1]**

51. If it is possible for a(n) _____ to intersect the graph of a function more than once, then the function is not one-to-one and its _____ is not a function. **[5.1]**

52. The _____ for an angle is the _____ angle formed by the terminal side of the angle and the x-axis. **[6.3]**

❯ Synthesis

53. *Eagle's Flight.* An eagle flies from its nest 7 mi in the direction northeast, where it stops to rest on a cliff.

It then flies 8 mi in the direction S30°W to land on top of a tree. Place an xy-coordinate system so that the origin is the bird's nest, the x-axis points east, and the y-axis points north.

a) At what point is the cliff located?
b) At what point is the tree located?

8.6 ❯ Vector Operations

❯ Perform calculations with vectors in component form.

❯ Express a vector as a linear combination of unit vectors.

❯ Express a vector in terms of its magnitude and its direction.

❯ Find the angle between two vectors using the dot product.

❯ Solve applied problems involving forces in equilibrium.

❯ Position Vectors

Let's consider a vector **v** whose initial point is the *origin* in an xy-coordinate system and whose terminal point is (a, b). We say that the vector is in **standard position** and refer to it as a **position vector**. Note that the ordered pair (a, b) defines the vector uniquely. Thus we can use (a, b) to denote the vector. To emphasize that we are thinking of a vector and to avoid the confusion of notation with ordered-pair notation and interval notation, we generally write

$$\mathbf{v} = \langle a, b \rangle.$$

The coordinate a is the *scalar* **horizontal component** of the vector, and the coordinate b is the *scalar* **vertical component** of the vector. By **scalar**, we mean a *numerical* quantity rather than a *vector* quantity. Thus, $\langle a, b \rangle$ is considered to be the *component form* of **v**. Note that a and b are *not* vectors and should not be confused with the vector component definition given in Section 8.5.

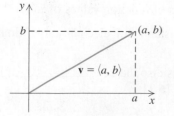

Now consider $\overrightarrow{AC}$ with $A = (x_1, y_1)$ and $C = (x_2, y_2)$. Let's see how to find the position vector equivalent to $\overrightarrow{AC}$. As you can see in the following figure, the initial point A is relocated to the origin $(0, 0)$. The coordinates of P are found by subtracting the coordinates of A from the coordinates of C. Thus, $P = (x_2 - x_1, y_2 - y_1)$ and the position vector is $\overrightarrow{OP}$.

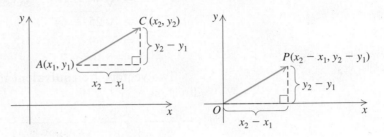

It can be shown that $\overrightarrow{OP}$ and $\overrightarrow{AC}$ have the same magnitude and the same direction and are therefore equivalent. Thus, $\overrightarrow{AC} = \overrightarrow{OP} = \langle x_2 - x_1, y_2 - y_1 \rangle$.

COMPONENT FORM OF A VECTOR

The **component form** of $\overrightarrow{AC}$ with $A = (x_1, y_1)$ and $C = (x_2, y_2)$ is

$$\overrightarrow{AC} = \langle x_2 - x_1, y_2 - y_1 \rangle.$$

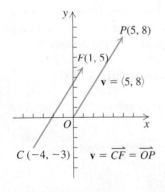

EXAMPLE 1 Find the component form of $\overrightarrow{CF}$ if $C = (-4, -3)$ and $F = (1, 5)$.

Solution We have

$$\overrightarrow{CF} = \langle 1 - (-4), 5 - (-3) \rangle = \langle 5, 8 \rangle.$$

Note that vector $\overrightarrow{CF}$ is equivalent to *position vector* $\overrightarrow{OP}$ with $P = (5, 8)$ as shown in the figure at left.

Now that we know how to write vectors in component form, let's restate some definitions that we first considered in Section 8.5.

The length of a vector $\mathbf{v}$ is easy to determine when the components of the vector are known. For $\mathbf{v} = \langle v_1, v_2 \rangle$, we have

$$|\mathbf{v}|^2 = v_1^2 + v_2^2 \quad \text{Using the Pythagorean equation}$$
$$|\mathbf{v}| = \sqrt{v_1^2 + v_2^2}.$$

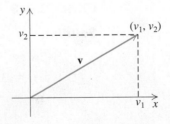

LENGTH OF A VECTOR

The **length**, or **magnitude**, of a vector $\mathbf{v} = \langle v_1, v_2 \rangle$ is given by

$$|\mathbf{v}| = \sqrt{v_1^2 + v_2^2}.$$

EXAMPLE 2   Find the length, or magnitude, of vector $\mathbf{v} = \langle 5, 8 \rangle$, illustrated in Example 1.

Solution

$$
\begin{aligned}
|\mathbf{v}| &= \sqrt{v_1^2 + v_2^2} & \text{\textbf{Length of vector} } \mathbf{v} = \langle v_1, v_2 \rangle \\
&= \sqrt{5^2 + 8^2} & \text{\textbf{Substituting 5 for} } v_1 \text{ \textbf{and 8 for} } v_2 \\
&= \sqrt{25 + 64} \\
&= \sqrt{89}
\end{aligned}
$$

> *Now Try Exercises 1 and 7.*

Two vectors are **equivalent** if they have the *same* magnitude and the *same* direction.

EQUIVALENT VECTORS

Let $\mathbf{u} = \langle u_1, u_2 \rangle$ and $\mathbf{v} = \langle v_1, v_2 \rangle$. Then

$$\langle u_1, u_2 \rangle = \langle v_1, v_2 \rangle \quad \text{if and only if} \quad u_1 = v_1 \quad \text{and} \quad u_2 = v_2.$$

❯ Operations on Vectors

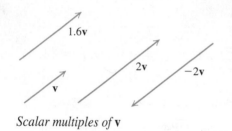

Scalar multiples of **v**

To multiply a vector $\mathbf{v}$ by a positive real number, we multiply its length by the number. Its direction stays the same. When a vector $\mathbf{v}$ is multiplied by 2, for instance, its length is doubled and its direction is not changed. When a vector is multiplied by 1.6, its length is increased by 60% and its direction stays the same. To multiply a vector $\mathbf{v}$ by a negative real number, we multiply its length by the number and reverse its direction. When a vector is multiplied by -2, its length is doubled and its direction is reversed. Since real numbers work like scaling factors in vector multiplication, we call them **scalars** and the products $k\mathbf{v}$ are called **scalar multiples** of $\mathbf{v}$.

SCALAR MULTIPLICATION

For a real number k and a vector $\mathbf{v} = \langle v_1, v_2 \rangle$, the **scalar product** of k and $\mathbf{v}$ is

$$k\mathbf{v} = k\langle v_1, v_2 \rangle = \langle kv_1, kv_2 \rangle.$$

The vector $k\mathbf{v}$ is a **scalar multiple** of the vector $\mathbf{v}$.

EXAMPLE 3   Let $\mathbf{u} = \langle -5, 4 \rangle$ and $\mathbf{w} = \langle 1, -1 \rangle$. Find $-7\mathbf{w}$, $3\mathbf{u}$, and $-1\mathbf{w}$.

Solution

$$
\begin{aligned}
-7\mathbf{w} &= -7\langle 1, -1 \rangle = \langle -7, 7 \rangle, \\
3\mathbf{u} &= 3\langle -5, 4 \rangle = \langle -15, 12 \rangle, \\
-1\mathbf{w} &= -1\langle 1, -1 \rangle = \langle -1, 1 \rangle
\end{aligned}
$$

In Section 8.5, we used the parallelogram law to add two vectors, but now we can add two vectors using components. To add two vectors given in component form, we add the corresponding components. Let $\mathbf{u} = \langle u_1, u_2 \rangle$ and $\mathbf{v} = \langle v_1, v_2 \rangle$. Then

$$\mathbf{u} + \mathbf{v} = \langle u_1 + v_1, u_2 + v_2 \rangle.$$

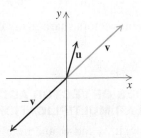

For example, if $\mathbf{v} = \langle -3, 2 \rangle$ and $\mathbf{w} = \langle 5, -9 \rangle$, then

$$\mathbf{v} + \mathbf{w} = \langle -3 + 5, 2 + (-9) \rangle = \langle 2, -7 \rangle.$$

VECTOR ADDITION

If $\mathbf{u} = \langle u_1, u_2 \rangle$ and $\mathbf{v} = \langle v_1, v_2 \rangle$, then

$$\mathbf{u} + \mathbf{v} = \langle u_1 + v_1, u_2 + v_2 \rangle.$$

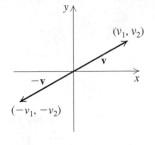

Before we define vector subtraction, we must define $-\mathbf{v}$. The opposite of $\mathbf{v} = \langle v_1, v_2 \rangle$, shown at left, is

$$-\mathbf{v} = (-1)\mathbf{v} = (-1)\langle v_1, v_2 \rangle = \langle -v_1, -v_2 \rangle.$$

Vector subtraction such as $\mathbf{u} - \mathbf{v}$ involves subtracting corresponding components. We show this by rewriting $\mathbf{u} - \mathbf{v}$ as $\mathbf{u} + (-\mathbf{v})$. If $\mathbf{u} = \langle u_1, u_2 \rangle$ and $\mathbf{v} = \langle v_1, v_2 \rangle$, then

$$\begin{aligned}
\mathbf{u} - \mathbf{v} = \mathbf{u} + (-\mathbf{v}) &= \langle u_1, u_2 \rangle + \langle -v_1, -v_2 \rangle \\
&= \langle u_1 + (-v_1), u_2 + (-v_2) \rangle \\
&= \langle u_1 - v_1, u_2 - v_2 \rangle.
\end{aligned}$$

We can illustrate vector subtraction with parallelograms, just as we did vector addition.

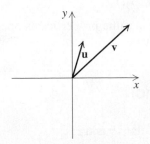

Sketch $\mathbf{u}$ and $\mathbf{v}$.

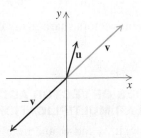

Sketch $-\mathbf{v}$.

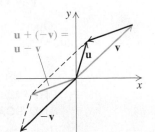

Sketch $\mathbf{u} + (-\mathbf{v})$, or $\mathbf{u} - \mathbf{v}$, using the parallelogram law.

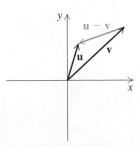

$\mathbf{u} - \mathbf{v}$ is the vector from the terminal point of $\mathbf{v}$ to the terminal point of $\mathbf{u}$.

VECTOR SUBTRACTION

If $\mathbf{u} = \langle u_1, u_2 \rangle$ and $\mathbf{v} = \langle v_1, v_2 \rangle$, then

$$\mathbf{u} - \mathbf{v} = \langle u_1 - v_1, u_2 - v_2 \rangle.$$

It is interesting to compare the sum of two vectors with the difference of the same two vectors in the same parallelogram. The vectors **u** + **v** and **u** − **v** are the diagonals of the parallelogram.

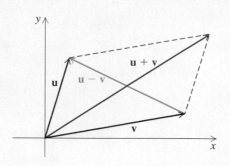

EXAMPLE 4 Do the following calculations, where $\mathbf{u} = \langle 7, 2 \rangle$ and $\mathbf{v} = \langle -3, 5 \rangle$.

a) **u** + **v**   **b)** **u** − 6**v**

c) 3**u** + 4**v**   **d)** $|5\mathbf{v} - 2\mathbf{u}|$

Solution

a) $\mathbf{u} + \mathbf{v} = \langle 7, 2 \rangle + \langle -3, 5 \rangle = \langle 7 + (-3), 2 + 5 \rangle = \langle 4, 7 \rangle$

b) $\mathbf{u} - 6\mathbf{v} = \langle 7, 2 \rangle - 6\langle -3, 5 \rangle = \langle 7, 2 \rangle - \langle -18, 30 \rangle = \langle 25, -28 \rangle$

c) $3\mathbf{u} + 4\mathbf{v} = 3\langle 7, 2 \rangle + 4\langle -3, 5 \rangle = \langle 21, 6 \rangle + \langle -12, 20 \rangle = \langle 9, 26 \rangle$

d) $\begin{aligned}|5\mathbf{v} - 2\mathbf{u}| &= |5\langle -3, 5 \rangle - 2\langle 7, 2 \rangle| = |\langle -15, 25 \rangle - \langle 14, 4 \rangle| \\ &= |\langle -29, 21 \rangle| \\ &= \sqrt{(-29)^2 + 21^2} \\ &= \sqrt{1282} \\ &\approx 35.8 \end{aligned}$

> **Now Try Exercises 9 and 11.**

Before we state the properties of vector addition and scalar multiplication, we must define another special vector—the zero vector. The vector whose initial and terminal points are both $(0, 0)$ is the **zero vector**, denoted by **O**, or $\langle 0, 0 \rangle$. Its magnitude is 0. In vector addition, the zero vector is the additive identity vector:

$$\mathbf{v} + \mathbf{O} = \mathbf{v}. \qquad \langle v_1, v_2 \rangle + \langle 0, 0 \rangle = \langle v_1, v_2 \rangle$$

Operations on vectors share many of the same properties as operations on real numbers.

**PROPERTIES OF VECTOR ADDITION
AND SCALAR MULTIPLICATION**

For all vectors **u**, **v**, and **w**, and for all scalars b and c:

1. **u** + **v** = **v** + **u**.

2. **u** + (**v** + **w**) = (**u** + **v**) + **w**.

3. **v** + **O** = **v**.

4. $1\mathbf{v} = \mathbf{v}$; $0\mathbf{v} = \mathbf{O}$.

5. **v** + (−**v**) = **O**.

6. $b(c\mathbf{v}) = (bc)\mathbf{v}$.

7. $(b + c)\mathbf{v} = b\mathbf{v} + c\mathbf{v}$.

8. $b(\mathbf{u} + \mathbf{v}) = b\mathbf{u} + b\mathbf{v}$.

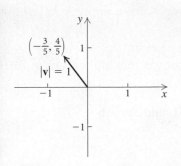

❯ Unit Vectors

A vector of magnitude, or length, 1 is called a **unit vector**. The vector $\mathbf{v} = \left\langle -\frac{3}{5}, \frac{4}{5} \right\rangle$ is a unit vector because

$$\begin{aligned}
|\mathbf{v}| &= \left| \left\langle -\tfrac{3}{5}, \tfrac{4}{5} \right\rangle \right| = \sqrt{\left(-\tfrac{3}{5}\right)^2 + \left(\tfrac{4}{5}\right)^2} \\
&= \sqrt{\tfrac{9}{25} + \tfrac{16}{25}} \\
&= \sqrt{\tfrac{25}{25}} \\
&= \sqrt{1} = 1.
\end{aligned}$$

EXAMPLE 5 Find a unit vector that has the same direction as the vector $\mathbf{w} = \langle -3, 5 \rangle$.

Solution We first find the length of $\mathbf{w}$:

$$|\mathbf{w}| = \sqrt{(-3)^2 + 5^2} = \sqrt{34}.$$

Thus we want a vector whose length is $1/\sqrt{34}$ of $\mathbf{w}$ and whose direction is the same as vector $\mathbf{w}$. That vector is

$$\mathbf{u} = \frac{1}{\sqrt{34}}\mathbf{w} = \frac{1}{\sqrt{34}}\langle -3, 5 \rangle = \left\langle \frac{-3}{\sqrt{34}}, \frac{5}{\sqrt{34}} \right\rangle.$$

The vector $\mathbf{u}$ is a *unit vector* because

$$\begin{aligned}
|\mathbf{u}| &= \left| \frac{1}{\sqrt{34}}\mathbf{w} \right| = \sqrt{\left(\frac{-3}{\sqrt{34}}\right)^2 + \left(\frac{5}{\sqrt{34}}\right)^2} = \sqrt{\frac{9}{34} + \frac{25}{34}} \\
&= \sqrt{\frac{34}{34}} = \sqrt{1} = 1.
\end{aligned}$$

❯ *Now Try Exercise 33.*

UNIT VECTOR

If $\mathbf{v}$ is a vector and $\mathbf{v} \neq \mathbf{O}$, then

$$\frac{1}{|\mathbf{v}|} \cdot \mathbf{v}, \quad \text{or} \quad \frac{\mathbf{v}}{|\mathbf{v}|},$$

is a **unit vector** in the direction of $\mathbf{v}$.

Although unit vectors can have any direction, the unit vectors parallel to the x- and y-axes are particularly useful. They are defined as

$$\mathbf{i} = \langle 1, 0 \rangle \quad \text{and} \quad \mathbf{j} = \langle 0, 1 \rangle.$$

Any vector can be expressed as a **linear combination** of unit vectors $\mathbf{i}$ and $\mathbf{j}$. For example, let $\mathbf{v} = \langle v_1, v_2 \rangle$. Then

$$\begin{aligned}
\mathbf{v} = \langle v_1, v_2 \rangle &= \langle v_1, 0 \rangle + \langle 0, v_2 \rangle \\
&= v_1 \langle 1, 0 \rangle + v_2 \langle 0, 1 \rangle = v_1\mathbf{i} + v_2\mathbf{j}.
\end{aligned}$$

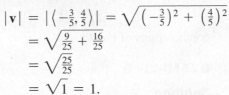

EXAMPLE 6 Express the vector $\mathbf{r} = \langle 2, -6 \rangle$ as a linear combination of $\mathbf{i}$ and $\mathbf{j}$.

Solution We have

$$\mathbf{r} = \langle 2, -6 \rangle = 2\mathbf{i} + (-6)\mathbf{j} = 2\mathbf{i} - 6\mathbf{j}.$$

❯ *Now Try Exercise 39.*

EXAMPLE 7   Write the vector $\mathbf{q} = -\mathbf{i} + 7\mathbf{j}$ in component form.

Solution   We have

$$\mathbf{q} = -\mathbf{i} + 7\mathbf{j} = -1\mathbf{i} + 7\mathbf{j} = \langle -1, 7 \rangle.$$

Vector operations can also be performed when vectors are written as linear combinations of $\mathbf{i}$ and $\mathbf{j}$.

EXAMPLE 8   If $\mathbf{a} = 5\mathbf{i} - 2\mathbf{j}$ and $\mathbf{b} = -\mathbf{i} + 8\mathbf{j}$, find $3\mathbf{a} - \mathbf{b}$.

Solution   We have

$$\begin{aligned}
3\mathbf{a} - \mathbf{b} &= 3(5\mathbf{i} - 2\mathbf{j}) - (-\mathbf{i} + 8\mathbf{j}) \\
&= 15\mathbf{i} - 6\mathbf{j} + \mathbf{i} - 8\mathbf{j} \\
&= 16\mathbf{i} - 14\mathbf{j}.
\end{aligned}$$

> *Now Try Exercise 45.*

❯ Direction Angles

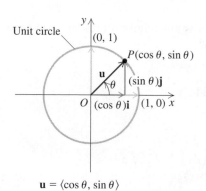

$$\mathbf{u} = \langle \cos \theta, \sin \theta \rangle$$
$$= (\cos \theta)\mathbf{i} + (\sin \theta)\mathbf{j}$$

UNIT CIRCLE

REVIEW SECTION 6.5.

The terminal point P of a unit vector in standard position is a point on the unit circle denoted by $(\cos \theta, \sin \theta)$. Thus the unit vector can be expressed in component form,

$$\mathbf{u} = \langle \cos \theta, \sin \theta \rangle,$$

or as a linear combination of the unit vectors $\mathbf{i}$ and $\mathbf{j}$,

$$\mathbf{u} = (\cos \theta)\mathbf{i} + (\sin \theta)\mathbf{j},$$

where the components of $\mathbf{u}$ are functions of the **direction angle** θ measured counterclockwise from the x-axis to the vector. As θ varies from 0 to 2π, the point P traces the circle $x^2 + y^2 = 1$. This takes in all possible directions for unit vectors so the equation $\mathbf{u} = (\cos \theta)\mathbf{i} + (\sin \theta)\mathbf{j}$ describes every possible unit vector in the plane.

EXAMPLE 9   Calculate and sketch the unit vector

$$\mathbf{u} = (\cos \theta)\mathbf{i} + (\sin \theta)\mathbf{j}$$

for $\theta = 2\pi/3$. Include the unit circle in your sketch.

Solution   We have

$$\begin{aligned}
\mathbf{u} &= \left(\cos \frac{2\pi}{3} \right)\mathbf{i} + \left(\sin \frac{2\pi}{3} \right)\mathbf{j} \\
&= \left(-\frac{1}{2} \right)\mathbf{i} + \left(\frac{\sqrt{3}}{2} \right)\mathbf{j}.
\end{aligned}$$

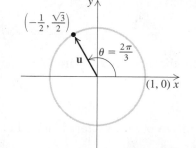

> *Now Try Exercise 49.*

Let $\mathbf{v} = \langle v_1, v_2 \rangle$ with direction angle θ. Using the definition of the tangent function, we can determine the direction angle from the components of $\mathbf{v}$:

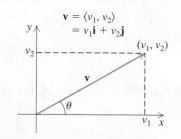

$$\tan \theta = \frac{v_2}{v_1}$$

$$\theta = \tan^{-1} \frac{v_2}{v_1}.$$

EXAMPLE 10 Determine the direction angle θ of the vector $\mathbf{w} = -4\mathbf{i} - 3\mathbf{j}$.

Solution We know that

$$\mathbf{w} = -4\mathbf{i} - 3\mathbf{j} = \langle -4, -3 \rangle.$$

Thus we have

$$\tan \theta = \frac{-3}{-4} = \frac{3}{4} \quad \text{and} \quad \theta = \tan^{-1}\frac{3}{4}.$$

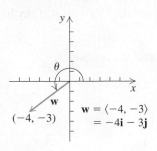

$\mathbf{w} = \langle -4, -3 \rangle$
$= -4\mathbf{i} - 3\mathbf{j}$

Since $\mathbf{w}$ is in the third quadrant, we know that θ is a third-quadrant angle. The reference angle is

$$\tan^{-1}\frac{3}{4} \approx 37°, \quad \text{and} \quad \theta \approx 180° + 37°, \text{or } 217°.$$

> **Now Try Exercise 55.**

It is convenient for work with applied problems and in subsequent courses, such as calculus, to have a way to express a vector so that both its magnitude and its direction can be determined, or read, easily. Let $\mathbf{v}$ be a vector. Then $\mathbf{v}/|\mathbf{v}|$ is a unit vector in the same direction as $\mathbf{v}$. Thus we have

$$\frac{\mathbf{v}}{|\mathbf{v}|} = (\cos \theta)\mathbf{i} + (\sin \theta)\mathbf{j}$$

$$\mathbf{v} = |\mathbf{v}|[(\cos \theta)\mathbf{i} + (\sin \theta)\mathbf{j}] \qquad \textbf{Multiplying by } |\mathbf{v}|$$

$$= |\mathbf{v}|(\cos \theta)\mathbf{i} + |\mathbf{v}|(\sin \theta)\mathbf{j}.$$

Let's revisit the applied problem in Example 3 of Section 8.5 and use this new notation.

EXAMPLE 11 *Airplane Speed and Direction.* An airplane travels on a bearing of 100° at an airspeed of 190 km/h while a wind is blowing 48 km/h from 220°. Find the ground speed of the airplane and the direction of its track, or course, over the ground.

Solution We first make a drawing. The wind is represented by $\overrightarrow{OC}$ and the velocity vector of the airplane by $\overrightarrow{OA}$. The resultant velocity vector is $\mathbf{v}$, the sum of the two vectors:

$$\mathbf{v} = \overrightarrow{OC} + \overrightarrow{OA}.$$

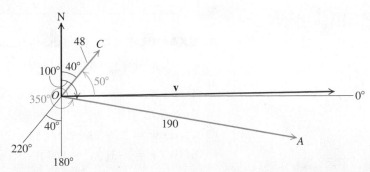

The bearing (measured from north) of the airspeed vector $\overrightarrow{OA}$ is 100°. Its *direction angle* (measured counterclockwise from the positive x-axis) is 350°. The bearing (measured from north) of the wind vector $\overrightarrow{OC}$ is 40°. Its direction angle (measured counterclockwise from the positive x-axis) is 50°. The magnitudes of $\overrightarrow{OA}$ and $\overrightarrow{OC}$ are 190 and 48, respectively. We have

$$\overrightarrow{OA} = 190(\cos 350°)\mathbf{i} + 190(\sin 350°)\mathbf{j}, \quad \text{and}$$

$$\overrightarrow{OC} = 48(\cos 50°)\mathbf{i} + 48(\sin 50°)\mathbf{j}.$$

Thus,

$$\mathbf{v} = \overrightarrow{OA} + \overrightarrow{OC}$$
$$= [190(\cos 350°)\mathbf{i} + 190(\sin 350°)\mathbf{j}] + [48(\cos 50°)\mathbf{i} + 48(\sin 50°)\mathbf{j}]$$
$$= [190(\cos 350°) + 48(\cos 50°)]\mathbf{i} + [190(\sin 350°) + 48(\sin 50°)]\mathbf{j}$$
$$\approx 217.97\mathbf{i} + 3.78\mathbf{j}.$$

From this form, we can determine the ground speed and the course:

$$\text{Ground speed} \approx \sqrt{(217.97)^2 + (3.78)^2}$$
$$\approx 218 \text{ km/h}.$$

We let α be the direction angle of $\mathbf{v}$. Then

$$\tan \alpha = \frac{3.78}{217.97}$$

$$\alpha = \tan^{-1}\frac{3.78}{217.97} \approx 1°.$$

Thus the course of the airplane (the direction from north) is $90° - 1°$, or $89°$.

> *Now Try Exercise 79.*

❯ Angle Between Vectors

When a vector is multiplied by a scalar, the result is a vector. When two vectors are added, the result is also a vector. Thus we might expect the product of two vectors to be a vector as well, but it is not. The *dot product* of two vectors is a real number, or scalar. This product is useful in finding the angle between two vectors and in determining whether two vectors are perpendicular.

DOT PRODUCT

The **dot product** of two vectors $\mathbf{u} = \langle u_1, u_2 \rangle$ and $\mathbf{v} = \langle v_1, v_2 \rangle$ is

$$\mathbf{u} \cdot \mathbf{v} = u_1v_1 + u_2v_2.$$

(Note that $u_1v_1 + u_2v_2$ is a *scalar*, not a vector.)

EXAMPLE 12   Find the indicated dot product when

$$\mathbf{u} = \langle 2, -5 \rangle, \quad \mathbf{v} = \langle 0, 4 \rangle, \quad \text{and} \quad \mathbf{w} = \langle -3, 1 \rangle.$$

a) $\mathbf{u} \cdot \mathbf{w}$ **b)** $\mathbf{w} \cdot \mathbf{v}$

Solution

a) $\mathbf{u} \cdot \mathbf{w} = 2(-3) + (-5)1 = -6 - 5 = -11$
b) $\mathbf{w} \cdot \mathbf{v} = -3(0) + 1(4) = 0 + 4 = 4$

The dot product can be used to find the angle between two vectors. The angle *between* two vectors is the smallest positive angle formed by the two directed line segments. Thus the angle θ between $\mathbf{u}$ and $\mathbf{v}$ is the same angle as that between $\mathbf{v}$ and $\mathbf{u}$, and $0 \leq \theta \leq \pi$.

ANGLE BETWEEN TWO VECTORS

If θ is the angle between two *nonzero* vectors **u** and **v**, then

$$\cos \theta = \frac{\mathbf{u} \cdot \mathbf{v}}{|\mathbf{u}||\mathbf{v}|}.$$

EXAMPLE 13   Find the angle between $\mathbf{u} = \langle 3, 7 \rangle$ and $\mathbf{v} = \langle -4, 2 \rangle$.

Solution   We begin by finding $\mathbf{u} \cdot \mathbf{v}, |\mathbf{u}|$, and $|\mathbf{v}|$:

$$\mathbf{u} \cdot \mathbf{v} = 3(-4) + 7(2) = 2,$$
$$|\mathbf{u}| = \sqrt{3^2 + 7^2} = \sqrt{58}, \quad \text{and}$$
$$|\mathbf{v}| = \sqrt{(-4)^2 + 2^2} = \sqrt{20}.$$

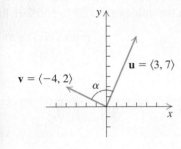

Then

$$\cos \alpha = \frac{\mathbf{u} \cdot \mathbf{v}}{|\mathbf{u}||\mathbf{v}|} = \frac{2}{\sqrt{58}\sqrt{20}}$$

$$\alpha = \cos^{-1} \frac{2}{\sqrt{58}\sqrt{20}}$$

$$\approx 86.6°.$$

Now Try Exercise 63.

❯ Forces in Equilibrium

When several forces act through the same point on an object, their vector sum must be **O** in order for a balance to occur. When a balance occurs, then the object is either stationary or moving in a straight line without acceleration. The fact that the vector sum must be **O** for a balance, and vice versa, allows us to solve many applied problems involving forces.

EXAMPLE 14   *Suspended Block.*   A 350-lb block is suspended by two cables, as shown at left. At point *A*, there are three forces acting: **W**, the block pulling down, and **R** and **S**, the two cables pulling upward and outward, respectively. Find the tension in each cable.

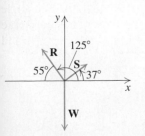

Solution   We draw a force diagram with the initial points of each vector at the origin. In order for there to be a balance, the vector sum must be the vector **O**:

$$\mathbf{R} + \mathbf{S} + \mathbf{W} = \mathbf{O}.$$

We can express each vector in terms of its magnitude and its direction angle:

$$\mathbf{R} = |\mathbf{R}|[(\cos 125°)\mathbf{i} + (\sin 125°)\mathbf{j}],$$
$$\mathbf{S} = |\mathbf{S}|[(\cos 37°)\mathbf{i} + (\sin 37°)\mathbf{j}], \quad \text{and}$$
$$\mathbf{W} = |\mathbf{W}|[(\cos 270°)\mathbf{i} + (\sin 270°)\mathbf{j}]$$
$$= 350(\cos 270°)\mathbf{i} + 350(\sin 270°)\mathbf{j}$$
$$= -350\mathbf{j}. \quad \cos 270° = 0; \sin 270° = -1$$

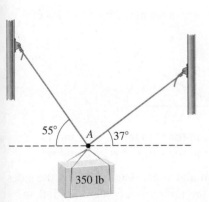

Substituting for **R**, **S**, and **W** in $\mathbf{R} + \mathbf{S} + \mathbf{W} = \mathbf{O}$, we have

$$[|\mathbf{R}|(\cos 125°) + |\mathbf{S}|(\cos 37°)]\mathbf{i} + [|\mathbf{R}|(\sin 125°) + |\mathbf{S}|(\sin 37°) - 350]\mathbf{j} = 0\mathbf{i} + 0\mathbf{j}.$$

This gives us two equations:

$$|\mathbf{R}|(\cos 125°) + |\mathbf{S}|(\cos 37°) = 0 \quad \text{and} \qquad \textbf{(1)}$$
$$|\mathbf{R}|(\sin 125°) + |\mathbf{S}|(\sin 37°) - 350 = 0. \qquad \textbf{(2)}$$

Solving equation (1) for $|\mathbf{R}|$, we get

$$|\mathbf{R}| = -\frac{|\mathbf{S}|(\cos 37°)}{\cos 125°}. \quad \textbf{(3)}$$

Substituting this expression for $|\mathbf{R}|$ in equation (2) gives us

$$-\frac{|\mathbf{S}|(\cos 37°)}{\cos 125°}(\sin 125°) + |\mathbf{S}|(\sin 37°) - 350 = 0.$$

Then solving this equation for $|\mathbf{S}|$, we get $|\mathbf{S}| \approx 201$, and substituting 201 for $|\mathbf{S}|$ in equation (3), we get $|\mathbf{R}| \approx 280$. The tensions in the cables are 280 lb and 201 lb.

> *Now Try Exercise 83.*

8.6 Exercise Set

Find the component form of the vector given the initial point and the terminal point. Then find the length of the vector.

1. $\overrightarrow{MN}$; $M(6, -7)$, $N(-3, -2)$

2. $\overrightarrow{CD}$; $C(1, 5)$, $D(5, 7)$

3. $\overrightarrow{FE}$; $E(8, 4)$, $F(11, -2)$

4. $\overrightarrow{BA}$; $A(9, 0)$, $B(9, 7)$

5. $\overrightarrow{KL}$; $K(4, -3)$, $L(8, -3)$

6. $\overrightarrow{GH}$; $G(-6, 10)$, $H(-3, 2)$

7. Find the magnitude of vector $\mathbf{u}$ if $\mathbf{u} = \langle -1, 6 \rangle$.

8. Find the magnitude of vector $\overrightarrow{ST}$ if $\overrightarrow{ST} = \langle -12, 5 \rangle$.

Do the indicated calculations in Exercises 9–26 for the vectors

$$\mathbf{u} = \langle 5, -2 \rangle, \quad \mathbf{v} = \langle -4, 7 \rangle, \quad \text{and} \quad \mathbf{w} = \langle -1, -3 \rangle.$$

9. $\mathbf{u} + \mathbf{w}$
10. $\mathbf{w} + \mathbf{u}$
11. $|3\mathbf{w} - \mathbf{v}|$
12. $6\mathbf{v} + 5\mathbf{u}$
13. $\mathbf{v} - \mathbf{u}$
14. $|2\mathbf{w}|$
15. $5\mathbf{u} - 4\mathbf{v}$
16. $-5\mathbf{v}$
17. $|3\mathbf{u}| - |\mathbf{v}|$
18. $|\mathbf{v}| + |\mathbf{u}|$
19. $\mathbf{v} + \mathbf{u} + 2\mathbf{w}$
20. $\mathbf{w} - (\mathbf{u} + 4\mathbf{v})$
21. $2\mathbf{v} + \mathbf{O}$
22. $10|7\mathbf{w} - 3\mathbf{u}|$
23. $\mathbf{u} \cdot \mathbf{w}$
24. $\mathbf{w} \cdot \mathbf{u}$
25. $\mathbf{u} \cdot \mathbf{v}$
26. $\mathbf{v} \cdot \mathbf{w}$

The vectors $\mathbf{u}$, $\mathbf{v}$, and $\mathbf{w}$ are drawn below. Copy them on a sheet of paper. Then sketch each of the vectors in Exercises 27–30.

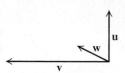

27. $\mathbf{u} + \mathbf{v}$
28. $\mathbf{u} - 2\mathbf{v}$
29. $\mathbf{u} + \mathbf{v} + \mathbf{w}$
30. $\frac{1}{2}\mathbf{u} - \mathbf{w}$

31. Vectors $\mathbf{u}$, $\mathbf{v}$, and $\mathbf{w}$ are determined by the sides of $\triangle ABC$ below.

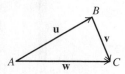

a) Find an expression for $\mathbf{w}$ in terms of $\mathbf{u}$ and $\mathbf{v}$.
b) Find an expression for $\mathbf{v}$ in terms of $\mathbf{u}$ and $\mathbf{w}$.

32. In $\triangle ABC$, vectors $\mathbf{u}$ and $\mathbf{w}$ are determined by the sides shown, where P is the midpoint of side BC. Find an expression for $\mathbf{v}$ in terms of $\mathbf{u}$ and $\mathbf{w}$.

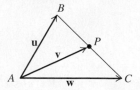

Find a unit vector that has the same direction as the given vector.

33. $\mathbf{v} = \langle -5, 12 \rangle$ **34.** $\mathbf{u} = \langle 3, 4 \rangle$

35. $\mathbf{w} = \langle 1, -10 \rangle$ **36.** $\mathbf{a} = \langle 6, -7 \rangle$

37. $\mathbf{r} = \langle -2, -8 \rangle$ **38.** $\mathbf{t} = \langle -3, -3 \rangle$

Express the vector as a linear combination of the unit vectors $\mathbf{i}$ and $\mathbf{j}$.

39. $\mathbf{w} = \langle -4, 6 \rangle$ **40.** $\mathbf{r} = \langle -15, 9 \rangle$

41. $\mathbf{s} = \langle 2, 5 \rangle$ **42.** $\mathbf{u} = \langle 2, -1 \rangle$

Express the vector as a linear combination of $\mathbf{i}$ and $\mathbf{j}$.

43.   **44.**

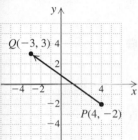   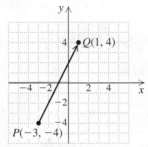

For Exercises 45–48, use the vectors

$\mathbf{u} = 2\mathbf{i} + \mathbf{j}$, $\mathbf{v} = -3\mathbf{i} - 10\mathbf{j}$,   and $\mathbf{w} = \mathbf{i} - 5\mathbf{j}$.

Perform the indicated vector operations and state the answer in two forms: (a) as a linear combination of $\mathbf{i}$ and $\mathbf{j}$ and (b) in component form.

45. $4\mathbf{u} - 5\mathbf{w}$ **46.** $\mathbf{v} + 3\mathbf{w}$

47. $\mathbf{u} - (\mathbf{v} + \mathbf{w})$ **48.** $(\mathbf{u} - \mathbf{v}) + \mathbf{w}$

Sketch (include the unit circle) and calculate the unit vector $\mathbf{u} = (\cos \theta)\mathbf{i} + (\sin \theta)\mathbf{j}$ for the given direction angle.

49. $\theta = \dfrac{\pi}{2}$ **50.** $\theta = \dfrac{\pi}{3}$

51. $\theta = \dfrac{4\pi}{3}$ **52.** $\theta = \dfrac{3\pi}{2}$

Determine the direction angle θ of the vector, to the nearest degree.

53. $\mathbf{u} = \langle -2, -5 \rangle$ **54.** $\mathbf{w} = \langle 4, -3 \rangle$

55. $\mathbf{q} = \mathbf{i} + 2\mathbf{j}$ **56.** $\mathbf{w} = 5\mathbf{i} - \mathbf{j}$

57. $\mathbf{t} = \langle 5, 6 \rangle$ **58.** $\mathbf{b} = \langle -8, -4 \rangle$

Find the magnitude and the direction angle θ of the vector.

59. $\mathbf{u} = 3[(\cos 45°)\mathbf{i} + (\sin 45°)\mathbf{j}]$

60. $\mathbf{w} = 6[(\cos 150°)\mathbf{i} + (\sin 150°)\mathbf{j}]$

61. $\mathbf{v} = \left\langle -\dfrac{1}{2}, \dfrac{\sqrt{3}}{2} \right\rangle$

62. $\mathbf{u} = -\mathbf{i} - \mathbf{j}$

Find the angle between the given vectors, to the nearest tenth of a degree.

63. $\mathbf{u} = \langle 2, -5 \rangle$, $\mathbf{v} = \langle 1, 4 \rangle$

64. $\mathbf{a} = \langle -3, -3 \rangle$, $\mathbf{b} = \langle -5, 2 \rangle$

65. $\mathbf{w} = \langle 3, 5 \rangle$, $\mathbf{r} = \langle 5, 5 \rangle$

66. $\mathbf{v} = \langle -4, 2 \rangle$, $\mathbf{t} = \langle 1, -4 \rangle$

67. $\mathbf{a} = \mathbf{i} + \mathbf{j}$, $\mathbf{b} = 2\mathbf{i} - 3\mathbf{j}$

68. $\mathbf{u} = 3\mathbf{i} + 2\mathbf{j}$, $\mathbf{v} = -\mathbf{i} + 4\mathbf{j}$

Express each vector in Exercises 69–72 in the form $a\mathbf{i} + b\mathbf{j}$ and sketch each in the coordinate plane.

69. The unit vectors $\mathbf{u} = (\cos \theta)\mathbf{i} + (\sin \theta)\mathbf{j}$ for $\theta = \pi/6$ and $\theta = 3\pi/4$. Include the unit circle $x^2 + y^2 = 1$ in your sketch.

70. The unit vectors $\mathbf{u} = (\cos \theta)\mathbf{i} + (\sin \theta)\mathbf{j}$ for $\theta = -\pi/4$ and $\theta = -3\pi/4$. Include the unit circle $x^2 + y^2 = 1$ in your sketch.

71. The unit vector obtained by rotating $\mathbf{j}$ counterclockwise $3\pi/4$ radians about the origin

72. The unit vector obtained by rotating $\mathbf{j}$ clockwise $2\pi/3$ radians about the origin

For the vectors in Exercises 73 and 74, find the unit vectors $\mathbf{u} = (\cos \theta)\mathbf{i} + (\sin \theta)\mathbf{j}$ in the same direction.

73. $-\mathbf{i} + 3\mathbf{j}$ **74.** $6\mathbf{i} - 8\mathbf{j}$

For the vectors in Exercises 75 and 76, express each vector in terms of its magnitude and its direction.

75. $2\mathbf{i} - 3\mathbf{j}$ **76.** $5\mathbf{i} + 12\mathbf{j}$

77. Use a sketch to show that

   $\mathbf{v} = 3\mathbf{i} - 6\mathbf{j}$ and $\mathbf{u} = -\mathbf{i} + 2\mathbf{j}$

   have opposite directions.

78. Use a sketch to show that

   $\mathbf{v} = 3\mathbf{i} - 6\mathbf{j}$ and $\mathbf{u} = \frac{1}{2}\mathbf{i} - \mathbf{j}$

   have the same direction.

Exercises 79–82 appeared first in Exercise Set 8.5, where we used the law of cosines and the law of sines to solve the applied problems. For this exercise set, solve the problem using the vector form

   $$\mathbf{v} = |\mathbf{v}|[(\cos \theta)\mathbf{i} + (\sin \theta)\mathbf{j}].$$

79. *Ship.* A ship first sails N80°E for 120 nautical mi, and then S20°W for 200 nautical mi. How far is the ship, then, from the starting point, and in what direction is the ship moving?

80. *Boat.* A boat heads 35°, propelled by a force of 750 lb. A wind from 320° exerts a force of 150 lb on the boat. How large is the resultant force, and in what direction is the boat moving?

81. *Airplane.* An airplane has an airspeed of 150 km/h. It is to make a flight in a direction of 70° while there is a 25-km/h wind from 340°. What will the airplane's actual heading be?

82. *Airplane.* An airplane flies 032° for 210 mi, and then 280° for 170 mi. How far is the airplane, then, from the starting point, and in what direction is the plane moving?

83. Two cables support a 1000-lb weight, as shown. Find the tension in each cable.

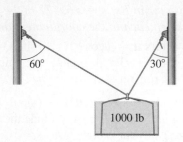

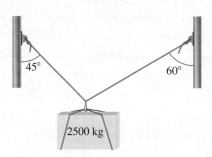

84. A 2500-kg block is suspended by two ropes, as shown. Find the tension in each rope.

85. A 150-lb sign is hanging from the end of a hinged boom, supported by a cable inclined 42° with the horizontal. Find the tension in the cable and the compression in the boom.

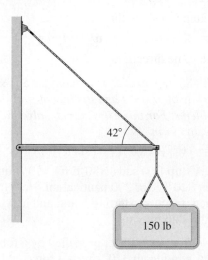

86. A weight of 200 lb is supported by a frame made of two rods and hinged at points A, B, and C. Find the forces exerted by the two rods.

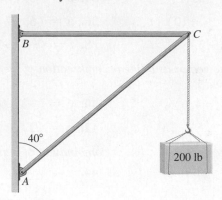

❯ Skill Maintenance

Find the zeros of the function.

87. $x^3 - 4x^2 = 0$ **[4.1]**

88. $6x^2 + 7x = 55$ **[3.2]**

❯ Synthesis

89. If the dot product of two nonzero vectors $\mathbf{u}$ and $\mathbf{v}$ is 0, then the vectors are perpendicular (**orthogonal**). Let $\mathbf{u} = \langle u_1, u_2 \rangle$ and $\mathbf{v} = \langle v_1, v_2 \rangle$. Prove that if $\mathbf{u} \cdot \mathbf{v} = 0$, then $\mathbf{u}$ and $\mathbf{v}$ are perpendicular.

90. If $\overrightarrow{PQ}$ is any vector, what is $\overrightarrow{PQ} + \overrightarrow{QP}$?

91. Find all the unit vectors that are parallel to the vector $\langle 3, -4 \rangle$.

92. Find vector $\mathbf{v}$ from point A to the origin, where $\overrightarrow{AB} = 4\mathbf{i} - 2\mathbf{j}$ and B is the point $(-2, 5)$.

93. Given the vector $\overrightarrow{AB} = 3\mathbf{i} - \mathbf{j}$ and A is the point $(2, 9)$, find the point B.

STUDY GUIDE

KEY TERMS AND CONCEPTS	EXAMPLES

SECTION 8.1: THE LAW OF SINES

Solving a Triangle

To **solve a triangle** means to find the lengths of all its sides and the measures of all its angles.

Any triangle, right or oblique, can be solved *if at least one side and any other two measures are known*. We cannot solve a triangle when only the three angle measures are given.

The Law of Sines

In any $\triangle ABC$,

$$\frac{a}{\sin A} = \frac{b}{\sin B} = \frac{c}{\sin C}.$$

The law of sines is used to solve triangles given a side and two angles (AAS and ASA) or given two sides and an angle opposite one of them (SSA). In the SSA situation, there are three possibilities: *no* solution, *one* solution, or *two* solutions.

Solve $\triangle RST$, if $r = 47.6$, $S = 123.5°$, and $T = 31.4°$.

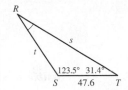

$$\begin{aligned} R &= ?, & r &= 47.6, \\ S &= 123.5°, & s &= ?, \\ T &= 31.4°, & t &= ? \end{aligned}$$

We have the ASA situation. We first find R:

$$R = 180° - (123.5° + 31.4°) = 25.1°.$$

We then find the other two sides using the law of sines. We have

$$\frac{47.6}{\sin 25.1°} = \frac{s}{\sin 123.5°}. \quad \text{Using } \frac{r}{\sin R} = \frac{s}{\sin S}$$

Solving for s, we get $s \approx 93.6$. We also have

$$\frac{47.6}{\sin 25.1°} = \frac{t}{\sin 31.4°}. \quad \text{Using } \frac{r}{\sin R} = \frac{t}{\sin T}$$

Solving for t, we get $t \approx 58.5$.

We have solved the triangle:

$$\begin{aligned} R &= 25.1°, & r &= 47.6, \\ S &= 123.5°, & s &\approx 93.6, \\ T &= 31.4°, & t &\approx 58.5. \end{aligned}$$

Solve $\triangle DEF$, if $d = 35.6$, $f = 48.1$, and $D = 32.2°$.

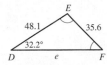

$$\begin{aligned} D &= 32.2°, & d &= 35.6, \\ E &= ?, & e &= ?, \\ F &= ?, & f &= 48.1 \end{aligned}$$

We have the SSA situation. We first find F:

$$\frac{48.1}{\sin F} = \frac{35.6}{\sin 32.2°} \quad \text{Using } \frac{f}{\sin F} = \frac{d}{\sin D}$$

$$\sin F = \frac{48.1 \sin 32.2°}{35.6} \approx 0.7200.$$

There are two angles less than $180°$ with a sine of 0.7200. They are $46.1°$ and $133.9°$. This gives us two possible solutions.

(continued)

Possible solution I:

If $F \approx 46.1°$, then $E = 180° - (32.2° + 46.1°) = 101.7°$. Then we find e using the law of sines:

$$\frac{e}{\sin 101.7°} = \frac{35.6}{\sin 32.2°}. \qquad \text{Using } \frac{e}{\sin E} = \frac{d}{\sin D}$$

Solving for e, we get $e \approx 65.4$. The solution is

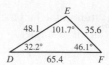

Possible solution II:

If $F \approx 133.9°$, then $E = 180° - (32.2° + 133.9°) = 13.9°$. Then we find e using the law of sines:

$$\frac{e}{\sin 13.9°} = \frac{35.6}{\sin 32.2°}. \qquad \text{Using } \frac{e}{\sin E} = \frac{d}{\sin D}$$

Solving for e, we get $e \approx 16.0$. The solution is

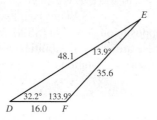

The Area of a Triangle

The area K of any $\triangle ABC$ is one-half of the product of the lengths of two sides and the sine of the included angle:

$$K = \frac{1}{2}bc \sin A = \frac{1}{2}ab \sin C = \frac{1}{2}ac \sin B.$$

Find the area of $\triangle ABC$ if $C = 115°$, $a = 10$ m, and $b = 13$ m.

$$K = \frac{1}{2}ab \sin C$$

$$= \frac{1}{2} \cdot 10 \text{ m} \cdot 13 \text{ m} \cdot \sin 115°$$

$$\approx 59 \text{ m}^2$$

SECTION 8.2: THE LAW OF COSINES

The Law of Cosines

In any $\triangle ABC$,

$$a^2 = b^2 + c^2 - 2bc \cos A,$$
$$b^2 = a^2 + c^2 - 2ac \cos B,$$
$$c^2 = a^2 + b^2 - 2ab \cos C.$$

The law of cosines is used to solve triangles given two sides and the included angle (SAS) or given three sides (SSS).

Solve $\triangle PQR$, if $p = 27$, $r = 39$, and $Q = 110.8°$.

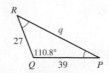

$$P = ?, \qquad p = 27,$$
$$Q = 110.8°, \qquad q = ?,$$
$$R = ?, \qquad r = 39$$

We can find the third side using the law of cosines:

$$q^2 = p^2 + r^2 - 2pr \cos Q$$
$$q^2 = 27^2 + 39^2 - 2 \cdot 27 \cdot 39 \cdot \cos 110.8°$$
$$q^2 \approx 2997.9$$
$$q \approx 55.$$

(continued)

We then find angle R using the law of cosines:

$$r^2 = p^2 + q^2 - 2pq \cos R$$
$$39^2 = 27^2 + 55^2 - 2 \cdot 27 \cdot 55 \cdot \cos R$$
$$\cos R = \frac{27^2 + 55^2 - 39^2}{2 \cdot 27 \cdot 55} \approx 0.7519.$$

Since $\cos R$ is positive, R is acute:

$$R \approx 41.2°.$$

Thus angle $P \approx 180° - (41.2° + 110.8°) \approx 28.0°$. We have solved the triangle:

$$P \approx 28.0°, \qquad p = 27,$$
$$Q = 110.8°, \qquad q \approx 55,$$
$$R \approx 41.2°, \qquad r = 39.$$

SECTION 8.3: COMPLEX NUMBERS: TRIGONOMETRIC NOTATION

Complex numbers can be graphed on a plane. We graph $a + bi$ in the same way that we graph an ordered pair of numbers (a, b).

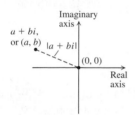

Absolute Value of a Complex Number

$$|a + bi| = \sqrt{a^2 + b^2}$$

Graph the complex numbers $2 - 4i$ and $\frac{7}{3}i$ and find the absolute value of each.

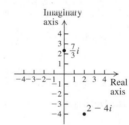

$$|2 - 4i| = \sqrt{2^2 + (-4)^2} = \sqrt{4 + 16} = \sqrt{20} = 2\sqrt{5};$$
$$\left|\frac{7}{3}i\right| = \sqrt{0^2 + \left(\frac{7}{3}\right)^2} = \sqrt{\left(\frac{7}{3}\right)^2} = \frac{7}{3}$$

Trigonometric Notation for Complex Numbers

$$a + bi = r(\cos \theta + i \sin \theta)$$

To find trigonometric notation for a complex number given in *standard notation*, $a + bi$, we find r and determine the angle θ for which $\sin \theta = b/r$ and $\cos \theta = a/r$.

Find trigonometric notation for $-1 + i$.

We have

$$a = -1 \text{ and } b = 1,$$
$$r = \sqrt{a^2 + b^2} = \sqrt{(-1)^2 + 1^2} = \sqrt{1 + 1} = \sqrt{2},$$
$$\sin \theta = \frac{b}{r} = \frac{1}{\sqrt{2}} = \frac{\sqrt{2}}{2} \quad \text{and} \quad \cos \theta = \frac{a}{r} = \frac{-1}{\sqrt{2}} = -\frac{\sqrt{2}}{2}.$$

Since $\sin \theta$ is positive and $\cos \theta$ is negative, θ is in quadrant II. We have $\theta = \frac{3\pi}{4}$, or $135°$, and

$$-1 + i = \sqrt{2}\left(\cos \frac{3\pi}{4} + i \sin \frac{3\pi}{4}\right), \quad \text{or}$$
$$-1 + i = \sqrt{2}(\cos 135° + i \sin 135°).$$

Find standard notation, $a + bi$, for $2(\cos 210° + i \sin 210°)$.

We have

$$2(\cos 210° + i \sin 210°) = 2 \cos 210° + (2 \sin 210°)i.$$

Thus,

$$a = 2 \cos 210° = 2 \cdot \left(-\frac{\sqrt{3}}{2}\right) = -\sqrt{3},$$

and

$$b = 2 \sin 210° = 2 \cdot \left(-\frac{1}{2}\right) = -1,$$

so

$$2(\cos 210° + i \sin 210°) = -\sqrt{3} - i.$$

Multiplication of Complex Numbers

$r_1(\cos \theta_1 + i \sin \theta_1) \cdot r_2(\cos \theta_2 + i \sin \theta_2)$
$\quad = r_1 r_2 [\cos (\theta_1 + \theta_2) + i \sin (\theta_1 + \theta_2)]$

Multiply $5(\cos \pi + i \sin \pi)$ and $2\left(\cos \dfrac{\pi}{6} + i \sin \dfrac{\pi}{6}\right)$ and express the answer in standard notation.

$$5(\cos \pi + i \sin \pi) \cdot 2\left(\cos \frac{\pi}{6} + i \sin \frac{\pi}{6}\right)$$
$$= 5 \cdot 2\left[\cos \left(\pi + \frac{\pi}{6}\right) + i \sin \left(\pi + \frac{\pi}{6}\right)\right]$$
$$= 10\left(\cos \frac{7\pi}{6} + i \sin \frac{7\pi}{6}\right)$$
$$= 10\left(-\frac{\sqrt{3}}{2} - \frac{1}{2}i\right) = -5\sqrt{3} - 5i$$

Division of Complex Numbers

$\dfrac{r_1(\cos \theta_1 + i \sin \theta_1)}{r_2(\cos \theta_2 + i \sin \theta_2)}$

$\quad = \dfrac{r_1}{r_2}[\cos (\theta_1 - \theta_2) + i \sin (\theta_1 - \theta_2)],$

$$r_2 \neq 0$$

Divide $3(\cos 315° + i \sin 315°)$ by $6(\cos 135° + i \sin 135°)$ and express the answer in standard notation.

$$\frac{3(\cos 315° + i \sin 315°)}{6(\cos 135° + i \sin 135°)}$$
$$= \frac{1}{2}[\cos (315° - 135°) + i \sin (315° - 135°)]$$
$$= \frac{1}{2}(\cos 180° + i \sin 180°)$$
$$= \frac{1}{2}(-1 + i \cdot 0) = -\frac{1}{2}$$

DeMoivre's Theorem

$[r(\cos \theta + i \sin \theta)]^n$
$\quad = r^n(\cos n\theta + i \sin n\theta)$

Find $\left(-\sqrt{3} - i\right)^5$.

We first find trigonometric notation:

$$-\sqrt{3} - i = 2(\cos 210° + i \sin 210°).$$

Then

$$\left(-\sqrt{3} - i\right)^5 = [2(\cos 210° + i \sin 210°)]^5$$
$$= 2^5(\cos 1050° + i \sin 1050°)$$
$$= 32(\cos 330° + i \sin 330°)$$
$$= 32\left[\frac{\sqrt{3}}{2} - \frac{1}{2}i\right]$$
$$= 16\sqrt{3} - 16i.$$

Roots of Complex Numbers

The *n*th roots of $r(\cos \theta + i \sin \theta)$ are

$$r^{1/n}\left[\cos\left(\frac{\theta}{n} + k \cdot \frac{360°}{n}\right) + i \sin\left(\frac{\theta}{n} + k \cdot \frac{360°}{n}\right)\right],$$

$r \neq 0, k = 0, 1, 2, \ldots, n - 1.$

Find the cube roots of -8. Then locate them on a graph.

We first find trigonometric notation:

$$-8 = 8(\cos 180° + i \sin 180°).$$

Then $n = 3$, $1/n = 1/3$, and $k = 0, 1, 2.$ We have

$$[8(\cos 180° + i \sin 180°)]^{1/3}$$
$$= 8^{1/3}\left[\cos\left(\frac{180°}{3} + k \cdot \frac{360°}{3}\right) + i \sin\left(\frac{180°}{3} + k \cdot \frac{360°}{3}\right)\right],$$

$k = 0, 1, 2.$

The roots are

$$2(\cos 60° + i \sin 60°),$$
$$2(\cos 180° + i \sin 180°), \quad \text{and}$$
$$2(\cos 300° + i \sin 300°), \quad \text{or}$$
$$1 + \sqrt{3}i, \quad -2, \quad \text{and} \quad 1 - \sqrt{3}i.$$

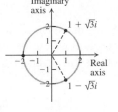

SECTION 8.4: POLAR COORDINATES AND GRAPHS

Plotting Points on a Polar Graph

Any point P has rectangular coordinates (x, y) and polar coordinates (r, θ). To plot points on a polar graph:

1. Locate the direction angle θ.
2. Move a directed distance r from the pole. If $r > 0$, move along ray OP. If $r < 0$, move in the opposite direction of ray OP.

Graph each of the following points:

$$A(4, 240°),$$
$$B\left(-2, \frac{2\pi}{3}\right),$$
$$C(3, -45°),$$
$$D(0, \pi),$$
$$E\left(-5, \frac{3\pi}{2}\right).$$

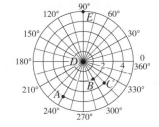

To convert from rectangular coordinates to polar coordinates and from polar coordinates to rectangular coordinates, recall the following relationships:

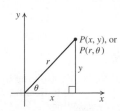

$$r = \sqrt{x^2 + y^2},$$
$$\cos \theta = \frac{x}{r}, \quad \text{or} \quad x = r \cos \theta,$$
$$\sin \theta = \frac{y}{r}, \quad \text{or} \quad y = r \sin \theta,$$
$$\tan \theta = \frac{y}{x}.$$

Convert $(-5, 5\sqrt{3})$ to polar coordinates.

We first find r:

$$r = \sqrt{(-5)^2 + (5\sqrt{3})^2}$$
$$= \sqrt{25 + 75} = \sqrt{100} = 10.$$

Then we determine θ:

$$\tan \theta = \frac{5\sqrt{3}}{-5} = -\sqrt{3};$$

therefore, $\theta = 120°$, or $\frac{2\pi}{3}$.

Thus, $(r, \theta) = (10, 120°)$, or $(10, 2\pi/3)$.

Other possibilities for polar coordinates for this point include $(10, -4\pi/3)$ and $(-10, 300°)$.

Convert $(4, 210°)$ to rectangular coordinates.

The ordered pair $(4, 210°)$ gives us $r = 4$ and $\theta = 210°$. We now find x and y:

$$x = r \cos \theta = 4 \cos 210° = 4 \cdot \left(-\frac{\sqrt{3}}{2}\right) = -2\sqrt{3};$$

$$y = r \sin \theta = 4 \sin 210° = 4\left(-\frac{1}{2}\right) = -2.$$

Thus, $(x, y) = \left(-2\sqrt{3}, -2\right)$.

Some curves have simpler equations in polar coordinates than in rectangular coordinates. For others, the reverse is true.

Convert each of the following rectangular equations to a polar equation.

a) $x^2 + y^2 = 100$

b) $y - 3x = 11$

a)
$$x^2 + y^2 = 100$$
$$(r \cos \theta)^2 + (r \sin \theta)^2 = 100$$
$$r^2 \cos^2 \theta + r^2 \sin^2 \theta = 100$$
$$r^2(\cos^2 \theta + \sin^2 \theta) = 100$$
$$r^2 = 100 \qquad \cos^2 \theta + \sin^2 \theta = 1$$
$$r = 10$$

b)
$$y - 3x = 11$$
$$(r \sin \theta) - 3(r \cos \theta) = 11$$
$$r(\sin \theta - 3 \cos \theta) = 11$$

Convert each of the following polar equations to a rectangular equation.

a) $r = 7$

b) $r = 3 \sin \theta - 5 \cos \theta$

a)
$$r = 7$$
$$\sqrt{x^2 + y^2} = 7$$
$$x^2 + y^2 = 49$$

b)
$$r = 3 \sin \theta - 5 \cos \theta$$
$$r^2 = 3r \sin \theta - 5r \cos \theta \qquad \textbf{Multiplying by } r \textbf{ on both sides}$$
$$x^2 + y^2 = 3y - 5x$$

To graph a polar equation:

1. Make a table of values, choosing values of θ and calculating corresponding values of r. As θ increases sufficiently, points may begin to repeat.

2. Plot the points and complete the graph.

Graph: $r = 2 - 3 \cos \theta$.

θ	r	θ	r
0°	−1	195°	4.8978
15°	−0.8978	210°	4.5981
30°	−0.5981	225°	4.1213
45°	−0.1213	240°	3.5
60°	0.5	255°	2.7765
75°	1.2235	270°	2
90°	2	285°	1.2235
105°	2.7765	300°	0.5
120°	3.5	315°	−0.1213
135°	4.1213	330°	−0.5981
150°	4.5981	345°	−0.8978
165°	4.8978	360°	−1
180°	5		

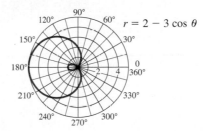

SECTION 8.5: VECTORS AND APPLICATIONS

Vector

A vector in the plane is a directed line segment. Two vectors are equivalent if they have the same *magnitude* and the same *direction*.

The vectors $\mathbf{v}$ and $\overrightarrow{AB}$ are shown in the following figure. Show that $\mathbf{v}$ and $\overrightarrow{AB}$ are equivalent.

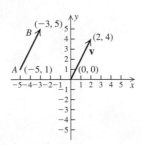

We first find the lengths:

$$|\mathbf{v}| = \sqrt{(2 - 0)^2 + (4 - 0)^2}$$
$$= \sqrt{2^2 + 4^2}$$
$$= \sqrt{20};$$
$$|\overrightarrow{AB}| = \sqrt{[-3 - (-5)]^2 + (5 - 1)^2}$$
$$= \sqrt{2^2 + 4^2}$$
$$= \sqrt{20}.$$

Thus, $|\mathbf{v}| = |\overrightarrow{AB}|$.

(*continued*)

The slopes of the lines that the vectors are on are:

$$\text{Slope of } \mathbf{v} = \frac{4 - 0}{2 - 0}$$

$$= \frac{4}{2} = 2;$$

$$\text{Slope of } \overrightarrow{AB} = \frac{5 - 1}{-3 - (-5)}$$

$$= \frac{4}{2} = 2.$$

Since the slopes are the same, vectors $\mathbf{v}$ and $\overrightarrow{AB}$ have the same direction.

Since $\mathbf{v}$ and $\overrightarrow{AB}$ have the same magnitude and the same direction, $\mathbf{v} = \overrightarrow{AB}$.

If two forces F_1 and F_2 act on an object, the combined effect is the sum, or resultant, $F_1 + F_2$ of the separate forces.

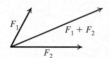

Two forces of 85 N and 120 N act on an object. The angle between the forces is 62°. Find the magnitude of the resultant and the angle that it makes with the larger force.

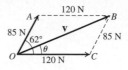

We have

$$\angle A = 180° - 62° = 118°.$$

We use the law of cosines to find the magnitude of the resultant $\mathbf{v}$:

$$|\mathbf{v}|^2 = 120^2 + 85^2 - 2 \cdot 120 \cdot 85 \cdot \cos 118°$$
$$|\mathbf{v}| \approx \sqrt{31{,}202}$$
$$\approx 177 \text{ N}.$$

We use the law of sines to find θ:

$$\frac{177}{\sin 118°} = \frac{85}{\sin \theta}$$

$$\sin \theta = \frac{85 \sin 118°}{177} \approx 0.4240$$

$$\theta \approx 25°.$$

The magnitude of $\mathbf{v}$ is approximately 177 N, and it makes a 25° angle with the larger force, 120 N.

Horizontal Components and Vertical Components of a Vector

The components of vector $\mathbf{w}$ are vectors $\mathbf{u}$ and $\mathbf{v}$ such that $\mathbf{w} = \mathbf{u} + \mathbf{v}$. We generally look for perpendicular components, with one component parallel to the x-axis (horizontal component) and the other parallel to the y-axis (vertical component).

A vector $\mathbf{w}$ has a magnitude of 200 and is inclined 52° with the horizontal. Resolve the vector into its horizontal component and its vertical component.

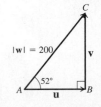

(continued)

From $\triangle ABC$, we find $|\mathbf{u}|$ and $|\mathbf{v}|$:

$$\cos 52° = \frac{|\mathbf{u}|}{200}, \quad \text{or} \quad |\mathbf{u}| = 200 \cos 52° \approx 123;$$

$$\sin 52° = \frac{|\mathbf{v}|}{200}, \quad \text{or} \quad |\mathbf{v}| = 200 \sin 52° \approx 158.$$

Thus the horizontal component of $\mathbf{w}$ is 123 right, and the vertical component of $\mathbf{w}$ is 158 up.

Given $|\mathbf{u}| = 16$ and $|\mathbf{v}| = 19$, and that the angle between the vectors is 130°, find the sum $\mathbf{u} + \mathbf{v}$. Give the magnitude to the nearest tenth and give the direction by specifying to the nearest degree the angle that the resultant makes with $\mathbf{u}$.

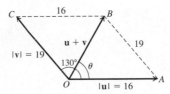

We first find A:

$$A = 180° - 130° = 50°.$$

We use the law of cosines to find $|\mathbf{u} + \mathbf{v}|$:

$$|\mathbf{u} + \mathbf{v}|^2 = 19^2 + 16^2 - 2 \cdot 19 \cdot 16 \cos 50°$$
$$|\mathbf{u} + \mathbf{v}| \approx \sqrt{226.19} \approx 15.0.$$

Next, we use the law of cosines to find θ:

$$19^2 \approx 15^2 + 16^2 - 2 \cdot 15 \cdot 16 \cos \theta$$
$$\cos \theta \approx 0.25$$
$$\theta \approx 76°.$$

The magnitude of $\mathbf{u} + \mathbf{v} \approx 15$, and the angle that $\mathbf{u} + \mathbf{v}$ makes with $\mathbf{u}$ is about 76°.

SECTION 8.6: VECTOR OPERATIONS

Component Form of a Vector
$$\mathbf{v} = \langle a, b \rangle$$

The coordinate a is the scalar **horizontal component** of the vector, and the coordinate b is the scalar **vertical component**.

The **component form** of $\overrightarrow{AC}$ with $A = (x_1, y_1)$ and $C = (x_2, y_2)$ is

$$\overrightarrow{AC} = \langle x_2 - x_1, y_2 - y_1 \rangle.$$

Find the component form of $\overrightarrow{OR}$ given $O(3, -9)$ and $R(-5, 4)$. Then find the length, or magnitude, of $\overrightarrow{OR}$.

$$\overrightarrow{OR} = \langle -5 - 3, 4 - (-9) \rangle = \langle -8, 13 \rangle;$$
$$|\overrightarrow{OR}| = \sqrt{(-8)^2 + 13^2} = \sqrt{64 + 169} = \sqrt{233} \approx 15.3$$

Vectors

If $\mathbf{u} = \langle u_1, u_2 \rangle$ and $\mathbf{v} = \langle v_1, v_2 \rangle$ and k is a scalar, then:

Length: $\quad |\mathbf{v}| = \sqrt{v_1^2 + v_2^2}$;

Addition: $\quad \mathbf{u} + \mathbf{v} = \langle u_1 + v_1, u_2 + v_2 \rangle$;

Subtraction: $\mathbf{u} - \mathbf{v} = \langle u_1 - v_1, u_2 - v_2 \rangle$;

Scalar Multiplication: $k\mathbf{v} = \langle kv_1, kv_2 \rangle$;

Dot Product: $\mathbf{u} \cdot \mathbf{v} = u_1 v_1 + u_2 v_2$.

Equivalent Vectors

Let $\mathbf{u} = \langle u_1, u_2 \rangle$ and $\mathbf{v} = \langle v_1, v_2 \rangle$. Then $\langle u_1, u_2 \rangle = \langle v_1, v_2 \rangle$ if and only if $u_1 = v_1$ and $u_2 = v_2$.

Do the indicated calculations, where $\mathbf{u} = \langle -6, 2 \rangle$ and $\mathbf{v} = \langle 8, 3 \rangle$.

a) $\mathbf{v} - 2\mathbf{u}$ **b)** $5\mathbf{u} + 3\mathbf{v}$

c) $|\mathbf{u} - \mathbf{v}|$ **d)** $\mathbf{u} \cdot \mathbf{v}$

a) $\mathbf{v} - 2\mathbf{u} = \langle 8, 3 \rangle - 2\langle -6, 2 \rangle$
$$= \langle 8, 3 \rangle - \langle -12, 4 \rangle$$
$$= \langle 20, -1 \rangle$$

b) $5\mathbf{u} + 3\mathbf{v} = 5\langle -6, 2 \rangle + 3\langle 8, 3 \rangle$
$$= \langle -30, 10 \rangle + \langle 24, 9 \rangle$$
$$= \langle -6, 19 \rangle$$

c) $|\mathbf{u} - \mathbf{v}| = |\langle -6, 2 \rangle - \langle 8, 3 \rangle|$
$$= |\langle -14, -1 \rangle|$$
$$= \sqrt{(-14)^2 + (-1)^2}$$
$$= \sqrt{197}$$
$$\approx 14.0$$

d) $\mathbf{u} \cdot \mathbf{v} = -6 \cdot 8 + 2 \cdot 3$
$$= -42$$

Zero Vector

The vector whose initial and terminal points are both $(0, 0)$ is the **zero vector**, denoted by $\mathbf{O}$, or $\langle 0, 0 \rangle$. Its magnitude is 0. In vector addition, the zero vector is the additive identity vector:

$$\mathbf{v} + \mathbf{O} = \mathbf{v}.$$

Unit Vector

A vector of magnitude, or length, 1 is called a **unit vector**.

If $\mathbf{v}$ is a vector and $\mathbf{v} \neq \mathbf{O}$, then

$$\frac{1}{|\mathbf{v}|} \cdot \mathbf{v}, \quad \text{or} \quad \frac{\mathbf{v}}{|\mathbf{v}|},$$

is a unit vector in the direction of $\mathbf{v}$.

Unit vectors parallel to the x- and y-axes are defined as

$$\mathbf{i} = \langle 1, 0 \rangle \quad \text{and} \quad \mathbf{j} = \langle 0, 1 \rangle.$$

Any vector can be expressed as a linear combination of unit vectors $\mathbf{i}$ and $\mathbf{j}$:

$$\mathbf{v} = \langle v_1, v_2 \rangle = v_1 \mathbf{i} + v_2 \mathbf{j}.$$

Find a unit vector that has the same direction as the vector $\mathbf{w} = \langle -9, 4 \rangle$.

We first find the length of $\mathbf{w}$:

$$|\mathbf{w}| = \sqrt{(-9)^2 + 4^2} = \sqrt{81 + 16} = \sqrt{97}.$$

We are looking for a vector $\mathbf{u}$ whose length is $1/\sqrt{97}$ of $\mathbf{w}$ and whose direction is the same as vector $\mathbf{w}$:

$$\mathbf{u} = \frac{1}{\sqrt{97}}\mathbf{w} = \frac{1}{\sqrt{97}}\langle -9, 4 \rangle = \left\langle -\frac{9}{\sqrt{97}}, \frac{4}{\sqrt{97}} \right\rangle.$$

Express the vector $\mathbf{q} = \langle 18, -7 \rangle$ as a linear combination of $\mathbf{i}$ and $\mathbf{j}$.

$$\mathbf{q} = \langle 18, -7 \rangle = 18\mathbf{i} + (-7)\mathbf{j} = 18\mathbf{i} - 7\mathbf{j}$$

Write the vector $\mathbf{r} = -4\mathbf{i} + \mathbf{j}$ in component form.

$$\mathbf{r} = -4\mathbf{i} + \mathbf{j} = -4\mathbf{i} + 1\mathbf{j} = \langle -4, 1 \rangle$$

Direction Angles

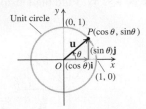

The unit vector can be expressed in component form,

$$\mathbf{u} = \langle \cos \theta, \sin \theta \rangle,$$

or

$$\mathbf{u} = (\cos \theta)\mathbf{i} + (\sin \theta)\mathbf{j},$$

where the components of **u** are functions of the direction angle θ measured counterclockwise from the x-axis to the vector.

Calculate and sketch the unit vector $\mathbf{u} = (\cos \theta)\mathbf{i} + (\sin \theta)\mathbf{j}$ for $\theta = 5\pi/4$.

$$\mathbf{u} = \left(\cos \frac{5\pi}{4} \right)\mathbf{i} + \left(\sin \frac{5\pi}{4} \right)\mathbf{j}$$

$$\mathbf{u} = \left(-\frac{\sqrt{2}}{2} \right)\mathbf{i} + \left(-\frac{\sqrt{2}}{2} \right)\mathbf{j}$$

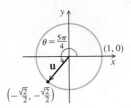

Determine the direction angle θ of the vector to the nearest degree.

a) $\mathbf{t} = \langle -2, 9 \rangle$ **b)** $\mathbf{b} = 5\mathbf{i} + 3\mathbf{j}$

a) $\mathbf{t} = \langle -2, 9 \rangle$

$$\tan \theta = \frac{9}{-2} = -\frac{9}{2} \quad \text{and} \quad \theta = \tan^{-1}\left(-\frac{9}{2} \right) \approx -77°$$

Since **t** is in quadrant II, we know that θ is a second-quadrant angle. The reference angle is 77°. Thus,

$$\theta \approx 180° - 77°, \quad \text{or} \quad 103°.$$

b) $\mathbf{b} = 5\mathbf{i} + 3\mathbf{j} = \langle 5, 3 \rangle$

$$\tan \theta = \frac{3}{5} \quad \text{and} \quad \theta = \tan^{-1} \frac{3}{5} \approx 31°$$

Since **b** is in quadrant I, $\theta \approx 31°$.

Angle Between Two Vectors

If θ is the angle between two nonzero vectors **u** and **v**, then

$$\cos \theta = \frac{\mathbf{u} \cdot \mathbf{v}}{|\mathbf{u}||\mathbf{v}|}.$$

Find the angle between $\mathbf{u} = \langle -2, 1 \rangle$ and $\mathbf{v} = \langle 3, -4 \rangle$.

We first find $\mathbf{u} \cdot \mathbf{v}$, $|\mathbf{u}|$, and $|\mathbf{v}|$:

$$\mathbf{u} \cdot \mathbf{v} = (-2) \cdot 3 + 1 \cdot (-4) = -10,$$
$$|\mathbf{u}| = \sqrt{(-2)^2 + 1^2} = \sqrt{5}, \quad \text{and}$$
$$|\mathbf{v}| = \sqrt{3^2 + (-4)^2} = \sqrt{25} = 5.$$

Then

$$\cos \theta = \frac{\mathbf{u} \cdot \mathbf{v}}{|\mathbf{u}||\mathbf{v}|} = \frac{-10}{\sqrt{5} \cdot 5}$$

$$\theta = \cos^{-1}\left(\frac{-10}{\sqrt{5} \cdot 5} \right) \approx 153.4°.$$

Forces in Equilibrium

When several forces act through the same point on an object, their vector sum must be **O** in order for a balance to occur.

A 600-lb block is suspended by two cables as shown. At point *B*, there are three forces acting: **R**, **S**, and **W**. Find the tension in each cable.

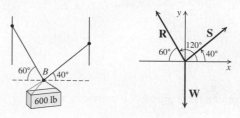

We have

$$\mathbf{R} = |\mathbf{R}|[(\cos 120°)\mathbf{i} + (\sin 120°)\mathbf{j}],$$
$$\mathbf{S} = |\mathbf{S}|[(\cos 40°)\mathbf{i} + (\sin 40°)\mathbf{j}], \quad \text{and}$$
$$\mathbf{W} = 600[(\cos 270°)\mathbf{i} + (\sin 270°)\mathbf{j}] = -600\mathbf{j}.$$

For a balance, the vector sum must be the vector **O**:

$$\mathbf{R} + \mathbf{S} + \mathbf{W} = \mathbf{O}.$$
$$|\mathbf{R}|[(\cos 120°)\mathbf{i} + (\sin 120°)\mathbf{j}]$$
$$+ |\mathbf{S}|[(\cos 40°)\mathbf{i} + (\sin 40°)\mathbf{j}] - 600\mathbf{j} = 0\mathbf{i} + 0\mathbf{j}$$

This gives us two equations:

$$|\mathbf{R}|(\cos 120°) + |\mathbf{S}|(\cos 40°) = 0,$$
$$|\mathbf{R}|(\sin 120°) + |\mathbf{S}|(\sin 40°) - 600 = 0.$$

Solving this system of equations for $|\mathbf{R}|$ and $|\mathbf{S}|$, we get $|\mathbf{R}| \approx 467$ and $|\mathbf{S}| \approx 305$. The tensions in the cables are 467 lb and 305 lb.

REVIEW EXERCISES

Determine whether the statement is true or false.

1. For any point (x, y) on the unit circle, $\langle x, y \rangle$ is a unit vector. **[8.6]**

2. The law of sines can be used to solve a triangle when all three sides are known. **[8.1]**

3. Two vectors are equivalent if they have the same magnitude and the lines that they are on have the same slope. **[8.5]**

4. Vectors $\langle 8, -2 \rangle$ and $\langle -8, 2 \rangle$ are equivalent. **[8.6]**

5. Any triangle, right or oblique, can be solved if at least one angle and any other two measures are known. **[8.1]**

6. When two angles and an included side of a triangle are known, the triangle cannot be solved using the law of cosines. **[8.2]**

Solve △ABC, if possible. **[8.1], [8.2]**

7. $a = 23.4$ ft, $b = 15.7$ ft, $c = 8.3$ ft

8. $B = 27°$, $C = 35°$, $b = 19$ in.

9. $A = 133°28'$, $C = 31°42'$, $b = 890$ m

10. $B = 37°$, $b = 4$ yd, $c = 8$ yd

11. Find the area of △ABC if $b = 9.8$ m, $c = 7.3$ m, and $A = 67.3°$. **[8.1]**

12. A parallelogram has sides of lengths 3.21 ft and 7.85 ft. One of its angles measures 147°. Find the area of the parallelogram. **[8.1]**

13. *Flower Garden.* A triangular flower garden has sides of lengths 11 m, 9 m, and 6 m. Find the angles of the garden to the nearest degree. **[8.2]**

14. In an isosceles triangle, the base angles each measure 52.3° and the base is 513 ft long. Find the lengths of the other two sides to the nearest foot. **[8.1]**

15. *Airplanes.* Two airplanes leave an airport at the same time. The first flies 175 km/h in a direction of 305.6°. The second flies 220 km/h in a direction of 195.5°. After 2 hr, how far apart are the planes? **[8.2]**

16. *Sandbox.* A child-care center has a triangular-shaped sandbox. Two of the three sides measure 15 ft and 12.5 ft and form an included angle of 42°. To determine the amount of sand that is needed to fill the box, the director must determine the area of the floor of the box. Find the area of the floor of the box to the nearest square foot. **[8.1]**

Graph the complex number and find its absolute value. **[8.3]**

17. $2 - 5i$

18. 4

19. $2i$

20. $-3 + i$

Find trigonometric notation. **[8.3]**

21. $1 + i$

22. $-4i$

23. $-5\sqrt{3} + 5i$

24. $\frac{3}{4}$

Find standard notation, a + bi. **[8.3]**

25. $4(\cos 60° + i \sin 60°)$

26. $7(\cos 0° + i \sin 0°)$

27. $5\left(\cos \dfrac{2\pi}{3} + i \sin \dfrac{2\pi}{3}\right)$

28. $2\left[\cos \left(-\dfrac{\pi}{6}\right) + i \sin \left(-\dfrac{\pi}{6}\right)\right]$

Convert to trigonometric notation and then multiply or divide, expressing the answer in standard notation. **[8.3]**

29. $(1 + i\sqrt{3})(1 - i)$

30. $\dfrac{2 - 2i}{2 + 2i}$

31. $\dfrac{2 + 2\sqrt{3}i}{\sqrt{3} - i}$

32. $i(3 - 3\sqrt{3}i)$

Raise the number to the given power and write trigonometric notation for the answer. **[8.3]**

33. $[2(\cos 60° + i \sin 60°)]^3$

34. $(1 - i)^4$

Raise the number to the given power and write standard notation for the answer. **[8.3]**

35. $(1 + i)^6$

36. $\left(\dfrac{1}{2} + \dfrac{\sqrt{3}}{2}i\right)^{10}$

37. Find the square roots of $-1 + i$. **[8.3]**

38. Find the cube roots of $3\sqrt{3} - 3i$. **[8.3]**

39. Find and graph the fourth roots of 81. **[8.3]**

40. Find and graph the fifth roots of 1. **[8.3]**

Find all the complex solutions of the equation. **[8.3]**

41. $x^4 - i = 0$

42. $x^3 + 1 = 0$

43. Find the polar coordinates of each of these points. Give three answers for each point. **[8.4]**

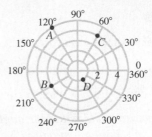

Find the polar coordinates of the point. Express the answer in degrees and then in radians. **[8.4]**

44. $(-4\sqrt{2}, 4\sqrt{2})$

45. $(0, -5)$

Use a graphing calculator to convert from rectangular coordinates to polar coordinates. Express the answer in degrees and then in radians. **[8.4]**

46. $(-2, 5)$

47. $(-4.2, \sqrt{7})$

Find the rectangular coordinates of the point. **[8.4]**

48. $\left(3, \dfrac{\pi}{4}\right)$

49. $(-6, -120°)$

Use a graphing calculator to convert from polar coordinates to rectangular coordinates. Round the coordinates to the nearest hundredth. **[8.4]**

50. $(2, -15°)$

51. $\left(-2.3, \dfrac{\pi}{5}\right)$

Convert to a polar equation. **[8.4]**

52. $5x - 2y = 6$

53. $y = 3$

54. $x^2 + y^2 = 9$

55. $y^2 - 4x - 16 = 0$

Convert to a rectangular equation. **[8.4]**

56. $r = 6$

57. $r + r \sin \theta = 1$

58. $r = \dfrac{3}{1 - \cos \theta}$

59. $r - 2 \cos \theta = 3 \sin \theta$

In Exercises 60–63, match the equation with one of figures (a)–(d) that follow. **[8.4]**

a)

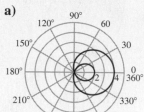

b)

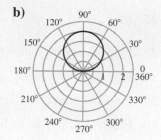

c)

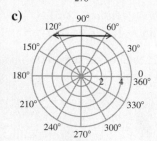

d)

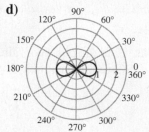

60. $r = 2 \sin \theta$

61. $r^2 = \cos 2\theta$

62. $r = 1 + 3 \cos \theta$

63. $r \sin \theta = 4$

Magnitudes of vectors **u** *and* **v** *and the angle* θ *between the vectors are given. Find the magnitude of the sum,* **u** + **v**, *to the nearest tenth and give the direction by specifying to the nearest degree the angle that it makes with the vector* **u**. **[8.5]**

64. $|\mathbf{u}| = 12$, $|\mathbf{v}| = 15$, $\theta = 120°$

65. $|\mathbf{u}| = 41$, $|\mathbf{v}| = 60$, $\theta = 25°$

The vectors **u**, **v**, *and* **w** *are drawn below. Copy them on a sheet of paper. Then sketch each of the vectors in Exercises 66 and 67.* **[8.5]**

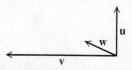

66. $\mathbf{u} - \mathbf{v}$

67. $\mathbf{u} + \frac{1}{2}\mathbf{w}$

68. Forces of 230 N and 500 N act on an object. The angle between the forces is 52°. Find the resultant, giving the angle that it makes with the smaller force. **[8.5]**

69. *Wind.* A wind has an easterly component of 15 km/h and a southerly component of 25 km/h. Find the magnitude and the direction of the wind. **[8.5]**

70. *Ship.* A ship first sails N75°E for 90 nautical mi, and then S10°W for 100 nautical mi. How far is the ship, then, from the starting point, and in what direction? **[8.5]**

Find the component form of the vector given the initial point and the terminal point. **[8.6]**

71. $\overrightarrow{AB}$; $A(2, -8)$, $B(-2, -5)$

72. $\overrightarrow{TR}$; $R(0, 7)$, $T(-2, 13)$

73. Find the magnitude of vector **u** if $\mathbf{u} = \langle 5, -6 \rangle$. **[8.6]**

Do the calculations in Exercises 74–77 for the vectors $\mathbf{u} = \langle 3, -4 \rangle$, $\mathbf{v} = \langle -3, 9 \rangle$, and $\mathbf{w} = \langle -2, -5 \rangle$. **[8.6]**

74. $4\mathbf{u} + \mathbf{w}$

75. $2\mathbf{w} - 6\mathbf{v}$

76. $|\mathbf{u}| + |2\mathbf{w}|$

77. $\mathbf{u} \cdot \mathbf{w}$

78. Find a unit vector that has the same direction as $\mathbf{v} = \langle -6, -2 \rangle$. **[8.6]**

79. Express the vector $\mathbf{t} = \langle -9, 4 \rangle$ as a linear combination of the unit vectors **i** and **j**. **[8.6]**

80. Determine the direction angle θ of the vector $\mathbf{w} = \langle -4, -1 \rangle$ to the nearest degree. **[8.6]**

81. Find the magnitude and the direction angle θ of $\mathbf{u} = -5\mathbf{i} - 3\mathbf{j}$. **[8.6]**

82. Find the angle between $\mathbf{u} = \langle 3, -7 \rangle$ and $\mathbf{v} = \langle 2, 2 \rangle$ to the nearest tenth of a degree. **[8.6]**

83. *Airplane.* An airplane has an airspeed of 160 mph. It is to make a flight in a direction of 80° while there is a 20-mph wind from 310°. What will the airplane's actual heading be? **[8.6]**

Do the calculations in Exercises 84–87 for the vectors $\mathbf{u} = 2\mathbf{i} + 5\mathbf{j}$, $\mathbf{v} = -3\mathbf{i} + 10\mathbf{j}$, and $\mathbf{w} = 4\mathbf{i} + 7\mathbf{j}$. **[8.6]**

84. $5\mathbf{u} - 8\mathbf{v}$

85. $\mathbf{u} - (\mathbf{v} + \mathbf{w})$

86. $|\mathbf{u} - \mathbf{v}|$

87. $3|\mathbf{w}| + |\mathbf{v}|$

88. Express the vector $\overrightarrow{PQ}$ in the form $a\mathbf{i} + b\mathbf{j}$, if P is the point $(1, -3)$ and Q is the point $(-4, 2)$. **[8.6]**

Express each vector in Exercises 89 and 90 in the form $a\mathbf{i} + b\mathbf{j}$ *and sketch each in the coordinate plane.* **[8.6]**

89. The unit vectors $\mathbf{u} = (\cos \theta)\mathbf{i} + (\sin \theta)\mathbf{j}$ for $\theta = \pi/4$ and $\theta = 5\pi/4$. Include the unit circle $x^2 + y^2 = 1$ in your sketch. **[8.6]**

90. The unit vector obtained by rotating **j** counterclockwise $2\pi/3$ radians about the origin.

91. Express the vector $3\mathbf{i} - \mathbf{j}$ as a product of its magnitude and its direction. **[8.6]**

92. Which of the following is the trigonometric notation for $1 - i$? **[8.3]**

A. $\sqrt{2}\left(\cos\dfrac{5\pi}{4} + i\sin\dfrac{5\pi}{4}\right)$

B. $\sqrt{2}\left(\cos\dfrac{7\pi}{4} - \sin\dfrac{7\pi}{4}\right)$

C. $\cos\dfrac{7\pi}{4} + i\sin\dfrac{7\pi}{4}$

D. $\sqrt{2}\left(\cos\dfrac{7\pi}{4} + i\sin\dfrac{7\pi}{4}\right)$

93. Convert the polar equation $r = 100$ to a rectangular equation. **[8.4]**

A. $x^2 + y^2 = 10{,}000$

B. $x^2 + y^2 = 100$

C. $\sqrt{x^2 + y^2} = 10$

D. $\sqrt{x^2 + y^2} = 1000$

94. The graph of $r = 1 - 2\cos\theta$ is which of the following? **[8.4]**

A.
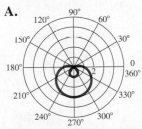

B.
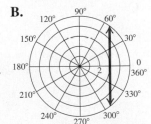

C.
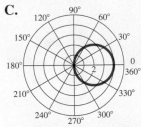

D.
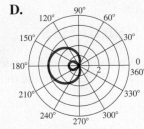

> ## Synthesis

95. Let $\mathbf{u} = 12\mathbf{i} + 5\mathbf{j}$. Find a vector that has the same direction as **u** but has length 3. **[8.6]**

96. A parallelogram has sides of lengths 3.42 and 6.97. Its area is 18.4. Find the sizes of its angles. **[8.1]**

> ## Collaborative Discussion and Writing

97. Summarize how you can determine algebraically when solving triangles whether there is no solution, one solution, or two solutions. **[8.1]**, **[8.2]**

98. Give an example of an equation that is easier to graph in polar notation than in rectangular notation and explain why. **[8.4]**

99. Explain why the rectangular coordinates of a point are unique and the polar coordinates of a point are not unique. **[8.4]**

100. Explain why vectors $\overrightarrow{QR}$ and $\overrightarrow{RQ}$ are not equivalent. **[8.5]**

101. Explain how unit vectors are related to the unit circle. **[8.6]**

102. Write a vector sum problem for a classmate for which the answer is $\mathbf{v} = 5\mathbf{i} - 8\mathbf{j}$. **[8.6]**

8 | Chapter Test

Solve △ABC, if possible.

1. $a = 18$ ft, $B = 54°$, $C = 43°$

2. $b = 8$ m, $c = 5$ m, $C = 36°$

3. $a = 16.1$ in., $b = 9.8$ in., $c = 11.2$ in.

4. Find the area of △ABC if $C = 106.4°$, $a = 7$ cm, and $b = 13$ cm.

5. *Distance Across a Lake.* Points A and B are on opposite sides of a lake. Point C is 52 m from A. The measure of $\angle BAC$ is determined to be 108°, and the measure of $\angle ACB$ is determined to be 44°. What is the distance from A to B?

6. *Location of Airplanes.* Two airplanes leave an airport at the same time. The first flies 210 km/h in a direction of 290°. The second flies 180 km/h in a direction of 185°. After 3 hr, how far apart are the planes?

7. Graph: $-4 + i$.

8. Find the absolute value of $2 - 3i$.

9. Find trigonometric notation for $3 - 3i$.

10. Divide and express the result in standard notation $a + bi$:

$$\frac{2\left(\cos \dfrac{2\pi}{3} + i \sin \dfrac{2\pi}{3}\right)}{8\left(\cos \dfrac{\pi}{6} + i \sin \dfrac{\pi}{6}\right)}.$$

11. Find $(1 - i)^8$ and write standard notation for the answer.

12. Find the polar coordinates of $(-1, \sqrt{3})$. Express the angle in degrees using the smallest possible positive angle.

13. Convert $\left(-1, \dfrac{2\pi}{3}\right)$ to rectangular coordinates.

14. Convert to a polar equation: $x^2 + y^2 = 10$.

15. Graph: $r = 1 - \cos \theta$.

16. For vectors **u** and **v**, $|\mathbf{u}| = 8$, $|\mathbf{v}| = 5$, and the angle between the vectors is 63°. Find $\mathbf{u} + \mathbf{v}$. Give the magnitude to the nearest tenth, and give the direction by specifying the angle that the resultant makes with **u**, to the nearest degree.

17. For $\mathbf{u} = 2\mathbf{i} - 7\mathbf{j}$ and $\mathbf{v} = 5\mathbf{i} + \mathbf{j}$, find $2\mathbf{u} - 3\mathbf{v}$.

18. Find a unit vector in the same direction as $-4\mathbf{i} + 3\mathbf{j}$.

19. Which of the following is the graph of $r = 3 \cos \theta$?

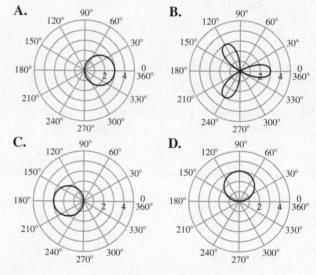

❯ Synthesis

20. A parallelogram has sides of length 15.4 and 9.8. Its area is 72.9. Find the measures of the angles.

Systems of Equations and Matrices

CHAPTER 9

APPLICATION

This problem appears as Exercise 55 in Section 9.1.

About 791,000 Chinese–Americans live in New York City and Boston. The number of Chinese–Americans in New York City is 95,000 more than five times the number in Boston. (*Source*: U.S. Census Bureau) How many Chinese–Americans live in each city?

637

9.1 > Systems of Equations in Two Variables

> Solve a system of two linear equations in two variables by graphing.

> Solve a system of two linear equations in two variables using the substitution method and the elimination method.

> Use systems of two linear equations to solve applied problems.

A **system of equations** is composed of two or more equations considered simultaneously. For example,

$$x - y = 5,$$
$$2x + y = 1$$

is a **system of two linear equations in two variables**. The solution set of this system consists of all ordered pairs that make *both* equations true. The ordered pair $(2, -3)$ is a solution of the system of equations above. We can verify this by substituting 2 for x and -3 for y in *each* equation.

$x - y = 5$			$2x + y = 1$	
$2 - (-3)\ ?\ 5$			$2 \cdot 2 + (-3)\ ?\ 1$	
$2 + 3$			$4 - 3$	
5	5	TRUE	1	1 TRUE

GRAPHS OF EQUATIONS

REVIEW SECTION 1.1.

> Solving Systems of Equations Graphically

Recall that the graph of a linear equation is a line that contains all the ordered pairs in the solution set of the equation. When we graph a system of linear equations, each point at which the graphs intersect is a solution of *both* equations and therefore a **solution of the system of equations**.

EXAMPLE 1   Solve the following system of equations graphically:

$$x - y = 5,$$
$$2x + y = 1.$$

Solution   We graph the equations on the same set of axes, as shown below.

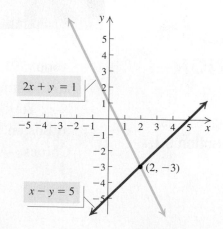

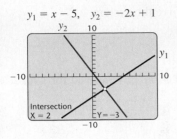

$y_1 = x - 5, \quad y_2 = -2x + 1$

We see that the graphs intersect at a single point, $(2, -3)$, so $(2, -3)$ is the solution of the system of equations. To check this solution, we substitute 2 for x and -3 for y in both equations as we did on the preceding page.

To use a graphing calculator to solve this system of equations, it might be necessary to write each equation in "Y $= \cdots$" form. If so, we would graph $y_1 = x - 5$ and $y_2 = -2x + 1$ and then use the INTERSECT feature. We see in the window at left that the solution is $(2, -3)$.

> **Now Try Exercise 7.**

The graphs of most of the systems of linear equations that we use to model applications intersect at a single point, like the system above. However, it is possible that the graphs will have no points in common or infinitely many points in common. Each of these possibilities is illustrated below.

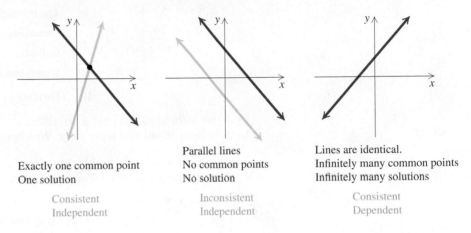

Exactly one common point   Parallel lines   Lines are identical.
One solution   No common points   Infinitely many common points
                                  No solution   Infinitely many solutions

Consistent   Inconsistent   Consistent
Independent   Independent   Dependent

If a system of equations has at least one solution, it is **consistent**. If the system has no solutions, it is **inconsistent**. In addition, for a system of two linear equations in two variables, if one equation can be obtained by multiplying by a constant on both sides of the other equation, the equations are **dependent**. Otherwise, they are **independent**. A system of two dependent linear equations in two variables has an infinite number of solutions.

❯ The Substitution Method

Solving a system of equations graphically is not always accurate when the solutions are not integers. A solution like $\left(\frac{43}{27}, -\frac{19}{27}\right)$, for instance, will be difficult to determine from a hand-drawn graph.

Algebraic methods for solving systems of equations, when used correctly, always give accurate results. One such technique is the **substitution method**. It is used most often when a variable is alone on one side of an equation or when it is easy to solve for a variable. To apply the substitution method, we begin by using one of the equations to express one variable in terms of the other. Then we substitute that expression in the other equation of the system.

EXAMPLE 2   Use the substitution method to solve the system

$$x - y = 5, \quad \textbf{(1)}$$
$$2x + y = 1. \quad \textbf{(2)}$$

Just in Time

17

Solution   First, we solve equation (1) for x. (We could have solved for y instead.) We have

$$x - y = 5 \quad \textbf{(1)}$$
$$x = y + 5. \quad \text{Solving for } x$$

Then we substitute $y + 5$ for x in equation (2). This gives an equation in one variable, which we know how to solve:

$$2x + y = 1 \quad \textbf{(2)}$$
$$2(y + 5) + y = 1 \quad \text{The parentheses are necessary.}$$
$$2y + 10 + y = 1 \quad \text{Removing parentheses}$$
$$3y + 10 = 1 \quad \text{Collecting like terms on the left}$$
$$3y = -9 \quad \text{Subtracting 10 on both sides}$$
$$y = -3. \quad \text{Dividing by 3 on both sides}$$

Now we substitute -3 for y in either of the original equations (this is called **back-substitution**) and solve for x. We choose equation (1):

$$x - y = 5 \quad \textbf{(1)}$$
$$x - (-3) = 5 \quad \text{Substituting } -3 \text{ for } y$$
$$x + 3 = 5$$
$$x = 2. \quad \text{Subtracting 3 on both sides}$$

We have previously checked the pair $(2, -3)$ in both equations. The solution of the system of equations is $(2, -3)$. Since there is exactly one solution, the system of equations is consistent, and the equations are independent.

> *Now Try Exercise 17.*

❭ The Elimination Method

Another algebraic technique for solving systems of equations is the **elimination method**. With this method, we eliminate a variable by adding two equations. If the coefficients of a particular variable are opposites, we can eliminate that variable simply by adding the original equations. For example, if the x-coefficient is -3 in one equation and is 3 in the other equation, then the sum of the x-terms will be 0 and thus the variable x will be eliminated when we add the equations.

EXAMPLE 3   Use the elimination method to solve the system

$$2x + y = 2, \quad \textbf{(1)}$$
$$x - y = 7. \quad \textbf{(2)}$$

Algebraic Solution	**Graphical Solution**

Since the y-coefficients, 1 and -1, are opposites, we can eliminate y by adding the equations:

$$\begin{array}{ll} 2x + y = 2 & \textbf{(1)} \\ \underline{x - y = 7} & \textbf{(2)} \\ 3x \quad\;\; = 9 & \text{Adding} \\ x = 3. \end{array}$$

We then back-substitute 3 for x in either equation and solve for y. We choose equation (1):

$$\begin{array}{ll} 2x + y = 2 & \textbf{(1)} \\ 2\cdot 3 + y = 2 & \text{Substituting 3 for } x \\ 6 + y = 2 \\ y = -4. \end{array}$$

We check the solution by substituting the pair $(3, -4)$ in both equations.

$$\begin{array}{c|c} 2x + y = 2 & x - y = 7 \\ \hline 2\cdot 3 + (-4) \stackrel{?}{=} 2 & 3 - (-4) \stackrel{?}{=} 7 \\ 6 - 4 & 3 + 4 \\ 2 \;\big|\; 2 \quad \text{TRUE} & 7 \;\big|\; 7 \quad \text{TRUE} \end{array}$$

The solution is $(3, -4)$. Since there is exactly one solution, the system of equations is consistent, and the equations are independent.

We solve each equation for y, getting

$$y = -2x + 2 \quad \text{and} \quad y = x - 7.$$

Next, we graph these equations and find the point of intersection of the graphs using the INTERSECT feature.

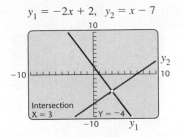

The graphs intersect at the point $(3, -4)$, so the solution of the system of equations is $(3, -4)$.

Now Try Exercise 31.

Before we add, it might be necessary to multiply one or both equations by suitable constants in order to find two equations in which the coefficients of a variable are opposites.

EXAMPLE 4 Use the elimination method to solve the system

$$\begin{array}{ll} 4x + 3y = 11, & \textbf{(1)} \\ -5x + 2y = 15. & \textbf{(2)} \end{array}$$

Solution We can obtain x-coefficients that are opposites by multiplying the first equation by 5 and the second equation by 4:

$$\begin{array}{ll} 20x + 15y = 55 & \text{Multiplying equation (1) by 5} \\ \underline{-20x + \;\; 8y = 60} & \text{Multiplying equation (2) by 4} \\ 23y = 115 & \text{Adding} \\ y = 5. \end{array}$$

We then back-substitute 5 for y in either equation (1) or (2) and solve for x. We choose equation (1):

$$\begin{array}{ll} 4x + 3y = 11 & \textbf{(1)} \\ 4x + 3\cdot 5 = 11 & \text{Substituting 5 for } y \\ 4x + 15 = 11 \\ 4x = -4 \\ x = -1. \end{array}$$

We can check the pair $(-1, 5)$ by substituting in both equations or by graphing. The solution is $(-1, 5)$. The system of equations is consistent and the equations are independent.

> *Now Try Exercise 33.*

In Example 4, the two systems

$$\begin{array}{cc} 4x + 3y = 11, & 20x + 15y = 55, \\ -5x + 2y = 15 & \text{and} \quad -20x + 8y = 60 \end{array}$$

are **equivalent** because they have exactly the same solutions. When we use the elimination method, we often multiply one or both equations by constants to find equivalent equations that allow us to eliminate a variable by adding.

EQUATION-SOLVING SPECIAL CASES

REVIEW SECTION 1.5.

EXAMPLE 5 Solve each of the following systems using the elimination method.

a) $x - 3y = 1,$ **(1)**   **b)** $2x + 3y = 6,$ **(1)**
      $-2x + 6y = 5$ **(2)** $4x + 6y = 12$ **(2)**

Solution

a) We multiply equation (1) by 2 and add:

$$\begin{array}{ll} 2x - 6y = 2 & \textbf{Multiplying equation (1) by 2} \\ \underline{-2x + 6y = 5} & \textbf{(2)} \\ \quad\quad 0 = 7. & \textbf{Adding} \end{array}$$

There are no values of x and y for which $0 = 7$ is true, so the system has *no solution*. The solution set is $\varnothing$. The system of equations is inconsistent, and the equations are independent. The graphs of the equations are parallel lines, as shown in Fig. 1.

b) We multiply equation (1) by -2 and add:

$$\begin{array}{ll} -4x - 6y = -12 & \textbf{Multiplying equation (1) by } -\textbf{2} \\ \underline{4x + 6y = 12} & \textbf{(2)} \\ \quad\quad 0 = 0. & \textbf{Adding} \end{array}$$

We obtain the equation $0 = 0$, which is true for all values of x and y. This tells us that the equations are dependent, so there are *infinitely many solutions*. That is, any solution of one equation of the system is also a solution of the other. The system of equations is consistent. The graphs of the equations are identical, as shown in Fig. 2.

Solving either equation for y, we have $y = -\frac{2}{3}x + 2$, so we can write the solutions of the system as ordered pairs (x, y), where y is expressed as $-\frac{2}{3}x + 2$. Thus the solutions can be written in the form $\left(x, -\frac{2}{3}x + 2\right)$. Any real value that we choose for x then gives us a value for y and thus an ordered pair in the solution set. For example,

$$\begin{array}{lll} \text{if } x = -3, & \text{then } -\frac{2}{3}x + 2 = -\frac{2}{3}(-3) + 2 = 4, \\ \text{if } x = 0, & \text{then } -\frac{2}{3}x + 2 = -\frac{2}{3}\cdot 0 + 2 = 2, & \text{and} \\ \text{if } x = 6, & \text{then } -\frac{2}{3}x + 2 = -\frac{2}{3}\cdot 6 + 2 = -2. \end{array}$$

Thus some of the solutions are $(-3, 4)$, $(0, 2)$, and $(6, -2)$.

Similarly, solving either equation for x, we have $x = -\frac{3}{2}y + 3$, so the solutions (x, y) can also be written, expressing x as $-\frac{3}{2}y + 3$, in the form $\left(-\frac{3}{2}y + 3, y\right)$.

Since the two forms of the solutions are equivalent, they yield the same solution set, as illustrated in the table at left. Note, for example, that when $y = 4$, we have the solution $(-3, 4)$; when $y = 2$, we have $(0, 2)$; and when $y = -2$, we have $(6, -2)$.

> *Now Try Exercises 35 and 37.*

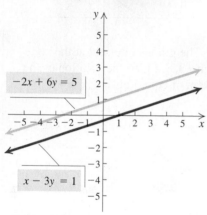

FIGURE 1.

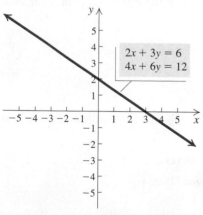

FIGURE 2.

x	$-\frac{2}{3}x + 2$
$-\frac{3}{2}y + 3$	y
-3	4
0	2
6	-2

❯ Applications

Frequently the most challenging and time-consuming step in the problem-solving process is translating a situation to mathematical language. However, in many cases, this task is made easier if we translate to more than one equation in more than one variable.

EXAMPLE 6 *Snack Mixtures.* At Max's Munchies, caramel corn worth $2.50 per pound is mixed with honey roasted mixed nuts worth $7.50 per pound in order to get 20 lb of a mixture worth $4.50 per pound. How much of each snack is used?

Solution We use the five-step problem-solving process.

1. **Familiarize.** Let's begin by making a guess. Suppose that 16 lb of caramel corn and 4 lb of nuts are used. Then the total weight of the mixture would be 16 lb + 4 lb, or 20 lb, the desired weight. The total values of these amounts of ingredients are found by multiplying the price per pound by the number of pounds used:

Caramel corn:	$2.50(16) =	$40
Nuts:	$7.50(4) =	$30
Total value:		$70.

   The desired value of the mixture is $4.50 per pound, so the value of 20 lb would be $4.50(20), or $90. Thus we see that our guess, which led to a total of $70, is incorrect. Nevertheless, these calculations will help us to translate.

2. **Translate.** We organize the information in a table. We let $x =$ the number of pounds of caramel corn in the mixture and $y =$ the number of pounds of nuts.

	Caramel Corn	Nuts	Mixture	
Price per Pound	$2.50	$7.50	$4.50	
Number of Pounds	x	y	20	$\longrightarrow x + y = 20$
Value of Mixture	$2.50x$	$7.50y$	4.50(20), or 90	$\longrightarrow 2.50x + 7.50y = 90$

From the second row of the table, we get one equation:

$$x + y = 20.$$

The last row of the table yields a second equation:

$$2.50x + 7.50y = 90, \quad \text{or} \quad 2.5x + 7.5y = 90.$$

We can multiply by 10 on both sides of the second equation to clear the decimals. This gives us the following system of equations:

$$x + y = 20, \qquad \textbf{(1)}$$
$$25x + 75y = 900. \qquad \textbf{(2)}$$

3. **Carry out.** We carry out the solution as follows.

Algebraic Solution	**Graphical Solution**

Algebraic Solution

Using the elimination method, we multiply equation (1) by -25 and add it to equation (2):

$$-25x - 25y = -500$$
$$\underline{25x + 75y = 900}$$
$$50y = 400$$
$$y = 8.$$

Then we back-substitute to find x:

$$x + y = 20 \qquad \textbf{(1)}$$
$$x + 8 = 20 \qquad \textbf{Substituting 8 for } y$$
$$x = 12.$$

Graphical Solution

We solve each equation for y, getting

$$y = 20 - x \quad \text{and} \quad y = \frac{900 - 25x}{75}.$$

Next, we graph these equations and find the point of intersection of the graphs. The graphs intersect at the point $(12, 8)$, so the possible solution is $(12, 8)$.

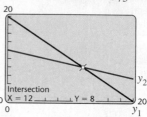

4. **Check.** If 12 lb of caramel corn and 8 lb of nuts are used, the mixture weighs $12 + 8$, or 20 lb. The value of the mixture is $\$2.50(12) + \$7.50(8)$, or $\$30 + \$60$, or $\$90$. Since the possible solution yields the desired weight and value of the mixture, our result checks.

5. **State.** The mixture should consist of 12 lb of caramel corn and 8 lb of honey roasted mixed nuts.

> **Now Try Exercise 61.**

EXAMPLE 7 *Airplane Travel.* An airplane flies the 3000-mi distance from Los Angeles to New York, with a tailwind, in 5 hr. The return trip, against the wind, takes 6 hr. Find the speed of the airplane in still air and the speed of the wind.

Solution

1. **Familiarize.** We first make a drawing, letting $p =$ the speed of the plane in still air, in miles per hour, and $w =$ the speed of the wind, also in miles per hour. When the plane is traveling with a tailwind, the wind increases the speed of the plane, so the speed with the tailwind is $p + w$. On the other hand, the headwind slows the plane down, so the speed with the headwind is $p - w$.

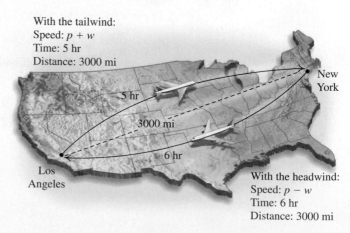

With the tailwind:
Speed: $p + w$
Time: 5 hr
Distance: 3000 mi

With the headwind:
Speed: $p - w$
Time: 6 hr
Distance: 3000 mi

2. Translate. We organize the information in a table. Using the formula *Distance = Rate* (or *Speed*) · *Time*, we find that each row of the table yields an equation.

	Distance	Rate	Time
With Tailwind	3000	$p + w$	5
With Headwind	3000	$p - w$	6

$\longrightarrow 3000 = (p + w)5$

$\longrightarrow 3000 = (p - w)6$

We now have a system of equations:

$$3000 = (p + w)5, \quad \text{or} \quad 600 = p + w, \quad \textbf{(1)} \quad \textbf{Dividing by 5}$$
$$3000 = (p - w)6, \qquad \qquad 500 = p - w. \quad \textbf{(2)} \quad \textbf{Dividing by 6}$$

3. Carry out. We use the elimination method:

$$
\begin{aligned}
600 &= p + w &\textbf{(1)} \\
\underline{500} &= \underline{p - w} &\textbf{(2)} \\
1100 &= 2p &\textbf{Adding} \\
550 &= p. &\textbf{Dividing by 2 on both sides}
\end{aligned}
$$

Now we substitute in one of the equations to find w:

$$
\begin{aligned}
600 &= p + w &\textbf{(1)} \\
600 &= 550 + w &\textbf{Substituting 550 for } p \\
50 &= w. &\textbf{Subtracting 550 on both sides}
\end{aligned}
$$

4. Check. If $p = 550$ and $w = 50$, then the speed of the plane with the tailwind is $550 + 50$, or 600 mph, and the speed with the headwind is $550 - 50$, or 500 mph. At 600 mph, the time it takes to travel 3000 mi is $3000/600$, or 5 hr. At 500 mph, the time it takes to travel 3000 mi is $3000/500$, or 6 hr. The times check, so the answer is correct.

5. State. The speed of the plane in still air is 550 mph, and the speed of the wind is 50 mph.

> *Now Try Exercise 67.*

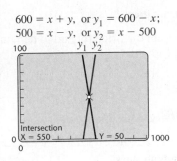

$600 = x + y$, or $y_1 = 600 - x$;
$500 = x - y$, or $y_2 = x - 500$

EXAMPLE 8 *Supply and Demand.*   Suppose that the price and the supply of the Star Station satellite radio are related by the equation

$$y = 90 + 30x,$$

where y is the price, in dollars, at which the seller is willing to supply x thousand units. Also suppose that the price and the demand for the same model of satellite radio are related by the equation

$$y = 200 - 25x,$$

where y is the price, in dollars, at which the consumer is willing to buy x thousand units.

The **equilibrium point** for this radio is the pair (x, y) that is a solution of both equations. The **equilibrium price** is the price at which the amount of the product that the seller is willing to supply is the same as the amount demanded by the consumer. Find the equilibrium point for this radio.

Solution

1., 2. Familiarize and Translate. We are given a system of equations in the statement of the problem, so no further translation is necessary.

$$y = 90 + 30x, \qquad \textbf{(1)}$$
$$y = 200 - 25x \qquad \textbf{(2)}$$

We substitute some values for x in each equation to get an idea of the corresponding prices. When $x = 1$,

$$y = 90 + 30 \cdot 1 = 120, \qquad \textbf{Substituting in equation (1)}$$
$$y = 200 - 25 \cdot 1 = 175. \qquad \textbf{Substituting in equation (2)}$$

This indicates that the price when 1 thousand units are supplied is lower than the price when 1 thousand units are demanded.

When $x = 4$,

$$y = 90 + 30 \cdot 4 = 210, \qquad \textbf{Substituting in equation (1)}$$
$$y = 200 - 25 \cdot 4 = 100. \qquad \textbf{Substituting in equation (2)}$$

In this case, the price related to supply is higher than the price related to demand. It would appear that the x-value we are looking for is between 1 and 4.

3. Carry out. We use the substitution method:

$$y = 90 + 30x \qquad \textbf{Equation (1)}$$
$$200 - 25x = 90 + 30x \qquad \textbf{Substituting } 200 - 25x \textbf{ for } y \textbf{ in equation (1)}$$
$$110 = 55x \qquad \textbf{Adding 25x and subtracting 90 on both sides}$$
$$2 = x. \qquad \textbf{Dividing by 55 on both sides}$$

We now back-substitute 2 for x in either equation and find y:

$$y = 200 - 25x \qquad \textbf{(2)}$$
$$= 200 - 25 \cdot 2 \qquad \textbf{Substituting 2 for } x$$
$$= 200 - 50$$
$$= 150.$$

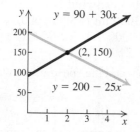

We can visualize the solution as the coordinates of the point of intersection of the graphs of the equations $y = 90 + 30x$ and $y = 200 - 25x$, as shown at left.

4. Check. We can check by substituting 2 for x and 150 for y in both equations. Also note that 2 is between 1 and 4 as expected from the *Familiarize* and *Translate* steps.

5. State. The equilibrium point is $(2, \$150)$. That is, the equilibrium quantity is 2 thousand units and the equilibrium price is $150.

> **Now Try Exercise 69.**

Visualizing the Graph

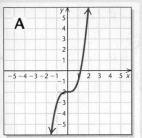

A

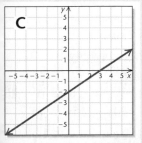

B

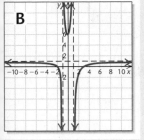

C

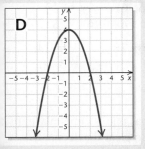

D

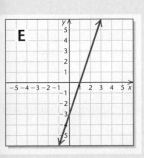

E

Match the equation or system of equations with its graph.

1. $2x - 3y = 6$

2. $f(x) = x^2 - 2x - 3$

3. $f(x) = -x^2 + 4$

4. $(x - 2)^2 + (y + 3)^2 = 9$

5. $f(x) = x^3 - 2$

6. $f(x) = -(x - 1)^2(x + 1)^2$

7. $f(x) = \dfrac{x - 1}{x^2 - 4}$

8. $f(x) = \dfrac{x^2 - x - 6}{x^2 - 1}$

9. $x - y = -1,$
$2x - y = 2$

10. $3x - y = 3,$
$2y = 6x - 6$

Answers on page A-52

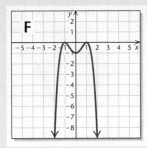

F

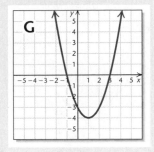

G

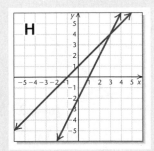

H

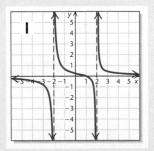

I

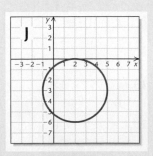

J

9.1 Exercise Set

In Exercises 1–6, match the system of equations with one of the graphs (a)–(f) that follow.

a)

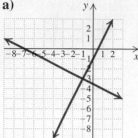

b)

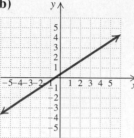

c)

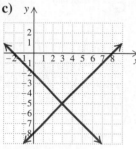

d)

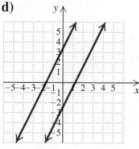

e)

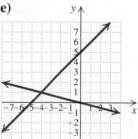

f)

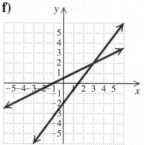

1. $x + y = -2,$
$y = x - 8$

2. $x - y = -5,$
$x = -4y$

3. $x - 2y = -1,$
$4x - 3y = 6$

4. $2x - y = 1,$
$x + 2y = -7$

5. $2x - 3y = -1,$
$-4x + 6y = 2$

6. $4x - 2y = 5,$
$6x - 3y = -10$

Solve graphically.

7. $x + y = 2,$
$3x + y = 0$

8. $x + y = 1,$
$3x + y = 7$

9. $x + 2y = 1,$
$x + 4y = 3$

10. $3x + 4y = 5,$
$x - 2y = 5$

11. $y + 1 = 2x,$
$y - 1 = 2x$

12. $2x - y = 1,$
$3y = 6x - 3$

13. $x - y = -6,$
$y = -2x$

14. $2x + y = 5,$
$x = -3y$

15. $2y = x - 1,$
$3x = 6y + 3$

16. $y = 3x + 2,$
$3x - y = -3$

Solve using the substitution method. Use a graphing calculator to check your answer.

17. $x + y = 9,$
$2x - 3y = -2$

18. $3x - y = 5,$
$x + y = \frac{1}{2}$

19. $x - 2y = 7,$
$x = y + 4$

20. $x + 4y = 6,$
$x = -3y + 3$

21. $y = 2x - 6,$
$5x - 3y = 16$

22. $3x + 5y = 2,$
$2x - y = -3$

23. $x + y = 3,$
$y = 4 - x$

24. $x - 2y = 3,$
$2x = 4y + 6$

25. $x - 5y = 4,$
$y = 7 - 2x$

26. $5x + 3y = -1,$
$x + y = 1$

27. $x + 2y = 2,$
$4x + 4y = 5$

28. $2x - y = 2,$
$4x + y = 3$

29. $3x - y = 5,$
$3y = 9x - 15$

30. $2x - y = 7,$
$y = 2x - 5$

Solve using the elimination method. Also determine whether the system is consistent or inconsistent and whether the equations are dependent or independent. Use a graphing calculator to check your answer.

31. $x + 2y = 7,$
$x - 2y = -5$

32. $3x + 4y = -2,$
$-3x - 5y = 1$

33. $x - 3y = 2,$
$6x + 5y = -34$

34. $x + 3y = 0,$
$20x - 15y = 75$

35. $3x - 12y = 6,$
$2x - 8y = 4$

36. $2x + 6y = 7,$
$3x + 9y = 10$

37. $4x - 2y = 3,$
$2x - y = 4$

38. $6x + 9y = 12,$
$4x + 6y = 8$

39. $2x = 5 - 3y,$
$4x = 11 - 7y$

40. $7(x - y) = 14,$
$2x = y + 5$

41. $0.3x - 0.2y = -0.9,$
$0.2x - 0.3y = -0.6$
(*Hint:* Since each coefficient has one decimal place, first multiply each equation by 10 to clear the decimals.)

42. $0.2x - 0.3y = 0.3,$
$0.4x + 0.6y = -0.2$
(*Hint:* Since each coefficient has one decimal place, first multiply each equation by 10 to clear the decimals.)

43. $\frac{1}{5}x + \frac{1}{2}y = 6,$
$\frac{3}{5}x - \frac{1}{2}y = 2$
(*Hint:* First multiply by the least common denominator to clear fractions.)

44. $\frac{2}{3}x + \frac{3}{5}y = -17$,

$\frac{1}{2}x - \frac{1}{3}y = -1$

(*Hint*: First multiply by the least common denominator to clear fractions.)

Determine whether the statement is true or false.

45. If the graph of a system of equations is a pair of parallel lines, then the system of equations is inconsistent.

46. If we obtain the equation $0 = 0$ when using the elimination method to solve a system of equations, then the system has no solution.

47. If a system of two linear equations in two variables is consistent, then it has exactly one solution.

48. If a system of two linear equations in two variables is dependent, then it has infinitely many solutions.

49. It is possible for a system of two linear equations in two variables to be consistent and dependent.

50. It is possible for a system of two linear equations in two variables to be inconsistent and dependent.

51. *Cosmetic Surgery.* Liposuction and nose reshaping are two of the five most popular cosmetic surgeries in the United States. Together, these two procedures accounted for 427,676 surgeries in 2014. The number of liposuction surgeries was 6572 fewer than the number of nose-shaping surgeries. (*Source*: American Society of Plastic Surgeons) Find the number of each type of surgery.

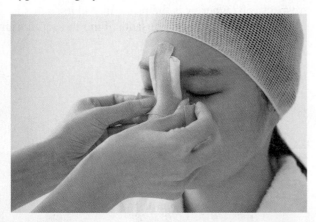

52. *Military Spending.* In 2014, the United States spent $452 billion more on its military than China spent on its military. Together, China and the United States spent $710 billion. (*Source*: International Institute for Strategic Studies) Find the amount spent on the military in China and in the United States.

53. *Baggage Fees.* In 2013 and 2014, U.S. airlines collected a total of $6.85 billion in baggage fees, with the baggage fees in 2014 exceeding those in 2013 by $0.15 billion (*Source*: U.S. Department of Transportation). Find the amount collected in baggage fees in 2013 and in 2014.

54. *Apartment Rent.* The average apartment rent in the United States is $1230 per month (*Source*: realtor.com). Jacob has an apartment in Boston and one in San Francisco. The total monthly rent for the two apartments is $4904. The rent in Boston is $1142 less than the rent in San Francisco. Find the rent for each apartment.

55. *Chinese–Americans.* About 791,000 Chinese–Americans live in New York City and Boston. The number of Chinese–Americans in New York City is 95,000 more than five times the number in Boston. (*Source*: U.S. Census Bureau) How many Chinese–Americans live in each city?

56. *Calories in Pie.* Using the calorie count of one-eighth of a 9-in. pie, we find that the number of calories in a piece of pecan pie is 221 less than twice the number of calories in a piece of lemon meringue pie. If one eats a piece of each, a total of 865 calories is consumed. (*Source*: *Good Housekeeping*, Good Health, p. 41, November 2007) How many calories are there in each piece of pie?

57. *Mail-Order Business.* A mail-order gardening equipment business shipped 120 packages one day. Customers are charged $6.50 for each standard-delivery package and $10.00 for each express-delivery package. Total shipping charges for the day were $934. How many of each kind of package were shipped?

58. *Concert Ticket Prices.* One evening 1500 concert tickets were sold for the Fairmont Summer Jazz Festival. Tickets cost $25 for a covered pavilion seat and $15 for a lawn seat. Total receipts were $28,500. How many of each type of ticket were sold?

59. *Investment.* Charles inherited $15,000 and invested it in two municipal bonds that pay 4% and 5% simple interest. The annual interest is $690. Find the amount invested at each rate.

60. *Sales of Scarves.* During the holiday season, Brianna sold scarves at a kiosk in a shopping mall. Embroidered floral scarves cost $24 each, and sheer chevron scarves cost $18 each. One day she sold 39 scarves. Total receipts for the day were $798. How many of each type of scarf did she sell?

61. *Coffee Mixtures.* The owner of The Daily Grind coffee shop mixes French roast coffee worth $12.00 per pound with Colombian coffee worth $9.50 per pound in order to get 20 lb of a mixture worth $10.50 per pound. How much of each type of coffee was used?

62. *Commissions.* Dyer's Office Solutions offers its sales representatives a choice between being paid a commission of 8% of sales or being paid a monthly salary of $1500 plus a commission of 1% of sales. For what amount of monthly sales do the two plans pay the same amount?

63. *Nutrition.* A one-cup serving of spaghetti with meatballs contains 260 Cal (calories) and 32 g of carbohydrates. A one-cup serving of chopped iceberg lettuce contains 5 Cal and 1 g of carbohydrates. (*Source*: U.S. Department of Agriculture) How many servings of each would be required in order to obtain 400 Cal and 50 g of carbohydrates?

64. *Nutrition.* One serving of tomato soup contains 100 Cal and 18 g of carbohydrates. One slice of whole wheat bread contains 70 Cal and 13 g of carbohydrates. (*Source*: U.S. Department of Agriculture) How many servings of each would be required in order to obtain 230 Cal and 42 g of carbohydrates?

65. *Motion.* A Leisure Time Cruises riverboat travels 46 km downstream in 2 hr. It travels 51 km upstream in 3 hr. Find the speed of the boat and the speed of the stream.

66. *Motion.* A DC10 airplane travels 3000 km with a tailwind in 3 hr. It travels 3000 km with a headwind in 4 hr. Find the speed of the plane and the speed of the wind.

67. *Motion.* Two private airplanes travel toward each other from cities that are 780 km apart at speeds of 190 km/h and 200 km/h. They leave at the same time. In how many hours will they meet?

68. *Motion.* Aaron's boat travels 45 mi downstream in 3 hr. The return trip upstream takes 5 hr. Find the speed of the boat in still water and the speed of the current.

69. *Supply and Demand.* The supply and demand for an all-terrain skateboard are related to price by the equations

$$y = 140 + 4x,$$
$$y = 275 - 5x,$$

respectively, where y is the price, in dollars, and x is the number of units, in thousands. Find the equilibrium point for this product.

70. *Supply and Demand.* The supply and demand for a particular model of treadmill are related to price by the equations

$$y = 240 + 40x,$$
$$y = 500 - 25x,$$

respectively, where y is the price, in dollars, and x is the number of units, in thousands. Find the equilibrium point for this product.

The point at which a company's costs equal its revenues is the **break-even point**. In Exercises 71–74, C represents the production cost, in dollars, of x units of a product and R represents the revenue, in dollars, from the sale of x units. Find the number of units that must be produced and sold in order to break even. That is, find the value of x for which C = R.

71. $C = 14x + 350$,
$R = 16.5x$

72. $C = 8.5x + 75$,
$R = 10x$

73. $C = 15x + 12{,}000$,
$R = 18x - 6000$

74. $C = 3x + 400$,
$R = 7x - 600$

75. *Red Meat and Poultry Consumption.*   The amount of red meat consumed in the United States has decreased about 7% in recent years while the amount of poultry consumed has increased about 3% during those years, as shown by the data in the following table.

Year	U.S. Red Meat Consumption (in billions of pounds)	U.S. Poultry Consumption (in billions of pounds)
2005	47.385	34.947
2010	45.931	35.201
2014	43.887	36.107

Source: U.S. Department of Agriculture, Economic Research Service

a) Find linear regression functions $r(x)$ and $p(x)$ that represent red meat consumption and poultry consumption, respectively, in billions of pounds x years after 2005.

b) Use the functions found in part (a) to estimate when poultry consumption will equal red meat consumption.

❯ Skill Maintenance

Solve. [1.5]

76. *Paramedic Pay.*   In 2012, the annual mean wage of a paramedic in the state of Washington was $50,980. This wage was $8600 less than twice the annual mean wage of a paramedic in Kentucky. (*Source*: U.S. Bureau of Labor Statistics) Find the annual median wage of a paramedic in Kentucky.

77. *Registered Snowmobiles.*   There were 251,986 registered snowmobiles in Minnesota in 2013. This was 21,952 more than twice the number of registered snowmobiles in New York. (*Sources*: International Snowmobile Manufacturers Association; Maine Snowmobile Association) Find the number of registered snowmobiles in New York.

78. *International Adoptions.*   In 2014, the number of international adoptions in the United States was at its lowest level since 2004. The number of international adoptions in 2014 totaled 6441, a decrease of 71.9% from 2004 (*Source*: U.S. State Department). Find the number of international adoptions in 2004. Round to the nearest ten.

Consider the function
$$f(x) = x^2 - 4x + 3$$
in Exercises 79–82.

79. What are the inputs if the output is 15? **[3.2]**

80. Given an output of 8, find the corresponding inputs. **[3.2]**

81. What is the output if the input is -9? **[1.2]**

82. Find the zeros of the function. **[3.2]**

❯ Synthesis

83. *Motion.*   Nancy jogs and walks to campus each day. She averages 4 km/h walking and 8 km/h jogging. The distance from home to the campus is 6 km, and she makes the trip in 1 hr. How far does she jog on each trip?

84. *e-Commerce.*   Shirts.com advertises a limited-time sale, offering 1 turtleneck for $15 and 2 turtlenecks for $25. A total of 1250 turtlenecks are sold and $16,750 is taken in. How many customers ordered 2 turtlenecks?

85. *Motion.*   A train leaves Union Station for Central Station, 216 km away, at 9 A.M. One hour later, a train leaves Central Station for Union Station. They meet at noon. If the second train had started at 9 A.M. and the first train at 10:30 A.M., they would still have met at noon. Find the speed of each train.

86. *Antifreeze Mixtures.*   An automobile radiator contains 16 L of antifreeze and water. This mixture is 30% antifreeze. How much of this mixture should be drained and replaced with pure antifreeze so that the final mixture will be 50% antifreeze?

87. Two solutions of the equation $Ax + By = 1$ are $(3, -1)$ and $(-4, -2)$. Find A and B.

88. *Ticket Line.*   You are in line at a ticket window. There are 2 more people ahead of you in line than there are behind you. In the entire line, there are three times as many people as there are behind you. How many people are ahead of you?

89. *Gas Mileage.*   The Jeep Renegade Sport 4 × 4 vehicle gets 23 miles per gallon (mpg) in city driving and 32 mpg in highway driving (*Source: Car and Driver,* May 2015, p. 114). The car is driven 403 mi on 14 gal of gasoline. How many miles were driven in the city and how many were driven on the highway?

9.2 ▷ Systems of Equations in Three Variables

> ❯ Solve systems of linear equations in three variables.

> ❯ Use systems of three equations to solve applied problems.

> ❯ Model a situation using a quadratic function.

A **linear equation in three variables** is an equation equivalent to one of the form $Ax + By + Cz = D$, where A, B, C, and D are real numbers and none of A, B, and C is 0. A **solution of a system of three equations in three variables** is an ordered triple that makes all three equations true. For example, the triple $(2, -1, 0)$ is a solution of the system of equations

$$4x + 2y + 5z = 6,$$
$$2x - y + z = 5,$$
$$3x + 2y - z = 4.$$

We can verify this by substituting 2 for x, -1 for y, and 0 for z in each equation.

❯ Solving Systems of Equations in Three Variables

We will solve systems of equations in three variables using an algebraic method called **Gaussian elimination**, named for the German mathematician Karl Friedrich Gauss (1777–1855). Our goal is to transform the original system to an equivalent system (one with the same solution set) of the form

$$Ax + By + Cz = D,$$
$$Ey + Fz = G,$$
$$Hz = K.$$

Then we solve the third equation for z and back-substitute to find y and then x.

Each of the following operations can be used to transform the original system to an equivalent system in the desired form.

1. Interchange any two equations.
2. Multiply by a nonzero constant on both sides of one of the equations.
3. Add a nonzero multiple of one equation to another equation.

EXAMPLE 1   Solve the following system:

$$x - 2y + 3z = 11, \quad \textbf{(1)}$$
$$4x + 2y - 3z = 4, \quad \textbf{(2)}$$
$$3x + 3y - z = 4. \quad \textbf{(3)}$$

Solution First, we eliminate x from two pairs of equations. We multiply equation (1) by -4 and add it to equation (2). We also multiply equation (1) by -3 and add it to equation (3).

$$\begin{array}{rl} -4x + 8y - 12z = -44 & \textbf{Multiplying (1) by} -\textbf{4} \\ \underline{4x + 2y - 3z = 4} & \textbf{(2)} \\ 10y - 15z = -40; & \textbf{(4)} \end{array}$$

$$\begin{array}{rl} -3x + 6y - 9z = -33 & \textbf{Multiplying (1) by} -\textbf{3} \\ \underline{3x + 3y - z = 4} & \textbf{(3)} \\ 9y - 10z = -29. & \textbf{(5)} \end{array}$$

Now we have

$$\begin{array}{rl} x - 2y + 3z = 11, & \textbf{(1)} \\ 10y - 15z = -40, & \textbf{(4)} \\ 9y - 10z = -29. & \textbf{(5)} \end{array}$$

Next, we multiply equation (5) by 10 to make the y-coefficient a multiple of the y-coefficient in the equation above it:

$$\begin{array}{rl} x - 2y + 3z = 11, & \textbf{(1)} \\ 10y - 15z = -40, & \textbf{(4)} \\ 90y - 100z = -290. & \textbf{(6)} \end{array}$$

Next, we multiply equation (4) by -9 and add it to equation (6):

$$\begin{array}{rl} -90y + 135z = 360 & \textbf{Multiplying (4) by} -\textbf{9} \\ \underline{90y - 100z = -290} & \textbf{(6)} \\ 35z = 70. & \textbf{(7)} \end{array}$$

We now have the system of equations

$$\begin{array}{rl} x - 2y + 3z = 11, & \textbf{(1)} \\ 10y - 15z = -40, & \textbf{(4)} \\ 35z = 70. & \textbf{(7)} \end{array}$$

Now we solve equation (7) for z:

$$35z = 70$$
$$z = 2.$$

Then we back-substitute 2 for z in equation (4) and solve for y:

$$10y - 15 \cdot 2 = -40$$
$$10y - 30 = -40$$
$$10y = -10$$
$$y = -1.$$

Finally, we back-substitute -1 for y and 2 for z in equation (1) and solve for x:

$$x - 2(-1) + 3 \cdot 2 = 11$$
$$x + 2 + 6 = 11$$
$$x = 3.$$

We can check the triple $(3, -1, 2)$ in each of the three original equations. Since it makes all three equations true, the solution is $(3, -1, 2)$.

> **Now Try Exercise 1.**

EXAMPLE 2 Solve the following system:

$$x + y + z = 7, \qquad \textbf{(1)}$$
$$3x - 2y + z = 3, \qquad \textbf{(2)}$$
$$x + 6y + 3z = 25. \qquad \textbf{(3)}$$

Solution We multiply equation (1) by -3 and add it to equation (2), getting equation (4). We also multiply equation (1) by -1 and add it to equation (3), getting equation (5).

$$x + y + z = 7, \qquad \textbf{(1)}$$
$$-5y - 2z = -18, \qquad \textbf{(4)}$$
$$5y + 2z = 18 \qquad \textbf{(5)}$$

Next, we add equation (4) to equation (5), getting equation (6).

$$x + y + z = 7, \qquad \textbf{(1)}$$
$$-5y - 2z = -18, \qquad \textbf{(4)}$$
$$0 = 0. \qquad \textbf{(6)}$$

The equation $0 = 0$ tells us that equations (1), (2), and (3) are dependent. This means that the original system of three equations is equivalent to a system of two equations. One way to see this is to note that four times equation (1) minus equation (2) is equation (3). Thus removing equation (3) from the system does not affect the solution of the system. We can say that the original system is equivalent to

$$x + y + z = 7, \qquad \textbf{(1)}$$
$$3x - 2y + z = 3. \qquad \textbf{(2)}$$

In this particular case, the original system has infinitely many solutions. (In some cases, a system containing dependent equations is inconsistent.) To find an expression for these solutions, we first solve equation (4) for either y or z. We choose to solve for y:

$$-5y - 2z = -18 \qquad \textbf{(4)}$$
$$-5y = 2z - 18$$
$$y = -\tfrac{2}{5}z + \tfrac{18}{5}.$$

Then we back-substitute in equation (1) to find an expression for x in terms of z:

$$x - \tfrac{2}{5}z + \tfrac{18}{5} + z = 7 \qquad \textbf{Substituting } -\tfrac{2}{5}z + \tfrac{18}{5} \textbf{ for } y$$
$$x + \tfrac{3}{5}z + \tfrac{18}{5} = 7$$
$$x + \tfrac{3}{5}z = \tfrac{17}{5}$$
$$x = -\tfrac{3}{5}z + \tfrac{17}{5}.$$

The solutions of the system of equations are ordered triples of the form $\left(-\tfrac{3}{5}z + \tfrac{17}{5}, -\tfrac{2}{5}z + \tfrac{18}{5}, z\right)$, where z can be any real number. Any real number that we use for z then gives us values for x and y and thus an ordered triple in the solution set. For example, if we choose $z = 0$, we have the solution $\left(\tfrac{17}{5}, \tfrac{18}{5}, 0\right)$. If we choose $z = -1$, we have $(4, 4, -1)$.

> **Now Try Exercise 9.**

If we get a false equation, such as $0 = -5$, at some stage of the elimination process, we conclude that the original system is *inconsistent*; that is, it has no solutions.

Although systems of three linear equations in three variables do not lend themselves well to graphical solutions, it is of interest to picture some possible solutions. The graph of a linear equation in three variables is a plane. Thus the solution set of such a system is the intersection of three planes. Some possibilities are shown below.

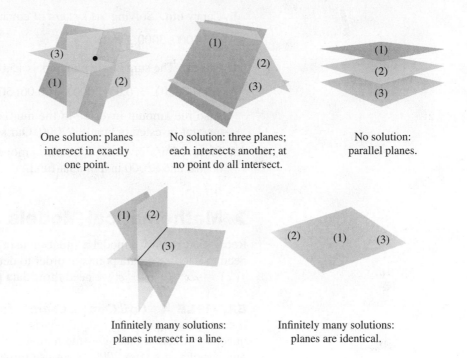

One solution: planes intersect in exactly one point.

No solution: three planes; each intersects another; at no point do all intersect.

No solution: parallel planes.

Infinitely many solutions: planes intersect in a line.

Infinitely many solutions: planes are identical.

❯ Applications

Systems of equations in three or more variables allow us to solve many problems in fields such as business, the social and natural sciences, and engineering.

EXAMPLE 3 *Investment.* Moira inherited $15,000 and invested part of it in a money market account, part in municipal bonds, and part in a mutual fund. After 1 year, she received a total of $730 in simple interest from the three investments. The money market account paid 4% annually, the bonds paid 5% annually, and the mutual fund paid 6% annually. There was $2000 more invested in the mutual fund than in bonds. Find the amount that Moira invested in each category.

Solution

1. **Familiarize.** We let x, y, and z represent the amounts invested in the money market account, the bonds, and the mutual fund, respectively. Then the amounts of income produced annually by each investment are given by 4%x, 5%y, and 6%z, or $0.04x$, $0.05y$, and $0.06z$.

2. **Translate.** The fact that a total of $15,000 is invested gives us one equation:

   $$x + y + z = 15,000.$$

   Since the total amount of interest is $730, we have a second equation:

   $$0.04x + 0.05y + 0.06z = 730.$$

   Another statement in the problem gives us a third equation.

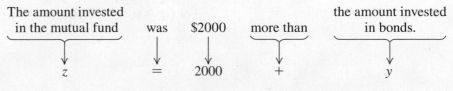

The amount invested in the mutual fund	was	$2000	more than	the amount invested in bonds.
z	$=$	2000	$+$	y

We now have a system of three equations:

$$x + y + z = 15,000, \qquad\qquad x + y + z = 15,000,$$
$$0.04x + 0.05y + 0.06z = 730, \quad \text{or} \quad 4x + 5y + 6z = 73,000,$$
$$z = 2000 + y; \qquad\qquad\qquad -y + z = 2000.$$

3. **Carry out.** Solving the system of equations, we get

(7000, 3000, 5000).

4. **Check.** The sum of the numbers is 15,000. The income produced is

$$0.04(7000) + 0.05(3000) + 0.06(5000) = 280 + 150 + 300, \quad \text{or} \quad \$730$$

Also the amount invested in the mutual fund, $5000, is $2000 more than the amount invested in bonds, $3000. Our solution checks in the original problem.

5. **State.** Moira invested $7000 in a money market account, $3000 in municipal bonds, and $5000 in a mutual fund.

> *Now Try Exercise 29.*

❯ Mathematical Models and Applications

Recall that when we model a situation using a linear function $f(x) = mx + b$, we need to know two data points in order to determine m and b. For a quadratic model $f(x) = ax^2 + bx + c$, we need three data points in order to determine a, b, and c.

EXAMPLE 4 *Civil Cases in Court.* The following table lists the number of civil cases pending in U.S. federal courts in three recent years. Use the data to find a quadratic function that gives the number of pending civil cases as a function of the number of years after 2009. Then use this function to estimate the number of civil cases pending in U.S. federal courts in 2014.

Year, x	Number of Civil Cases Pending in U.S. Federal Courts
2009, 0	311,353
2011, 2	267,495
2013, 4	300,469

Source: Administrative Office of the U.S. Courts

Solution We let $x = $ the number of years after 2009 and $n(x) = $ the number of pending civil cases. Then $x = 0$ corresponds to 2009, $x = 2$ corresponds to 2011, and $x = 4$ corresponds to 2013. We use three data points $(0, 311,353)$, $(2, 267,495)$, and $(4, 300,469)$ to find a, b, and c in the function $n(x) = ax^2 + bx + c$. First, we substitute:

$$n(x) = ax^2 + bx + c.$$

For $(0, 311{,}353)$: $\quad 311{,}353 = a \cdot 0^2 + b \cdot 0 + c,$
For $(2, 267{,}495)$: $\quad 267{,}495 = a \cdot 2^2 + b \cdot 2 + c,$
For $(4, 300{,}469)$: $\quad 300{,}469 = a \cdot 4^2 + b \cdot 4 + c.$

We now have a system of equations in the variables a, b, and c:

$$c = 311{,}353,$$
$$4a + 2b + c = 267{,}495,$$
$$16a + 4b + c = 300{,}469.$$

Solving this system, we get $(9604, -41{,}137, 311{,}353)$. Thus,

$$n(x) = 9604x^2 - 41{,}137x + 311{,}353.$$

To estimate the number of civil cases pending in the U.S. federal courts in 2014, we find $n(5)$, since 2014 is 5 years after 2009:

$$n(5) = 9604 \cdot 5^2 - 41{,}137 \cdot 5 + 311{,}353$$
$$= 345{,}768.$$

We estimate that there were 345,768 civil cases pending in U.S. federal courts in 2014.

> **Now Try Exercise 37.**

The function in Example 4 can also be found using the QUADRATIC REGRESSION feature on a graphing calculator. Note that the method of Example 4 works when we have exactly three data points, whereas the QUADRATIC REGRESSION feature on a graphing calculator can be used for *three or more* points.

QuadReg
$y = ax^2 + bx + c$
$a = 9604$
$b = -41137$
$c = 311353$
$R^2 = 1$

$Y_1(5)$
345768

9.2 Exercise Set

Solve the system of equations.

1. $x + y + z = 2,$
$6x - 4y + 5z = 31,$
$5x + 2y + 2z = 13$

2. $x + 6y + 3z = 4,$
$2x + y + 2z = 3,$
$3x - 2y + z = 0$

3. $x - y + 2z = -3,$
$x + 2y + 3z = 4,$
$2x + y + z = -3$

4. $x + y + z = 6,$
$2x - y - z = -3,$
$x - 2y + 3z = 6$

5. $x + 2y - z = 5,$
$2x - 4y + z = 0,$
$3x + 2y + 2z = 3$

6. $2x + 3y - z = 1,$
$x + 2y + 5z = 4,$
$3x - y - 8z = -7$

7. $x + 2y - z = -8,$
$2x - y + z = 4,$
$8x + y + z = 2$

8. $x + 2y - z = 4,$
$4x - 3y + z = 8,$
$5x - y = 12$

9. $2x + y - 3z = 1,$
$x - 4y + z = 6,$
$4x - 7y - z = 13$

10. $x + 3y + 4z = 1,$
$3x + 4y + 5z = 3,$
$x + 8y + 11z = 2$

11. $4a + 9b = 8,$
$8a + 6c = -1,$
$6b + 6c = -1$

12. $3p + 2r = 11,$
$q - 7r = 4,$
$p - 6q = 1$

13. $2x + z = 1,$
$3y - 2z = 6,$
$x - 2y = -9$

14. $3x + 4z = -11,$
$x - 2y = 5,$
$4y - z = -10$

15. $w + x + y + z = 2,$
$w + 2x + 2y + 4z = 1,$
$-w + x - y - z = -6,$
$-w + 3x + y - z = -2$

16. $w + x - y + z = 0,$
$-w + 2x + 2y + z = 5,$
$-w + 3x + y - z = -4,$
$-2w + x + y - 3z = -7$

17. *Paralympic Medals.* At the 2014 Paralympic Games in Sochi, Russia, the top three countries—the Russian Federation, Ukraine, and the United States—won a total of 123 medals. Ukraine won 7 more medals than the United States. The Russian Federation won 37 more medals than the total amount won by Ukraine and the United States. (*Source:* International

Paralympic Committee) How many medals did each of the top three countries win?

18. *Restaurant Meals.* The total number of restaurant-purchased meals that the average person will eat in a restaurant, in a car, or at home in a year is 170. The total number of these meals eaten in a car or at home exceeds the number eaten in a restaurant by 14. Twenty more restaurant-purchased meals will be eaten in a restaurant than at home. (*Source*: The NPD Group) Find the number of restaurant-purchased meals eaten in a restaurant, the number eaten in a car, and the number eaten at home.

19. *Top Corn Producers.* The top three corn producers in the world—the United States, China, and Brazil—grew a total of about 652 million metric tons (MT) of corn in 2014. The United States produced 70 million MT more than the combined production of China and Brazil. China produced 141 million MT more than Brazil. (*Source*: worldcornproduction.com) Find the number of metric tons of corn produced by each country.

20. *Acres Planted in Crops.* The top three crops in the United States in 2013, in terms of the number of acres planted, were corn, soybeans, and wheat. The total number of acres planted in these crops was 227.9 million. The number of acres planted in wheat was 39.2 million fewer than the number of acres planted in corn. The number of acres planted in soybeans was 35.7 million fewer than twice the number of acres planted in wheat. (*Source*: U.S. Department of Agriculture) Find the number of acres planted in each crop.

21. *Top Art Auction Sales.* The top three prices for works of art sold at auction in 2013 totaled $306.2 million. These works included Francis Bacon's 1969 *Three Studies of Lucian Freud*, Andy Warhol's *Silver Car Crash* (*Double Disaster*), and Jeff Koons's *Balloon Dog* (*Orange*). The selling price of the

Warhol art was $47 million more than that of the Koons art. Together, the Warhol art and the Koons art sold for $21.4 million more than the Bacon art. (*Source*: Bloomberg.com, Katya Kazakina and Scott Reyburn, December 24, 2013) What was the selling price of each work?

22. *Jolts of Caffeine.* One 8-oz serving each of brewed coffee, Red Bull energy drink, and Mountain Dew soda contains a total of 197 mg of caffeine. One serving of brewed coffee has 6 mg more caffeine than two servings of Mountain Dew. One serving of Red Bull contains 37 mg less caffeine than one serving each of brewed coffee and Mountain Dew. (*Source*: Australian Institute of Sport) Find the amount of caffeine in one serving of each beverage.

23. *Pet Ownership.* In a recent year, Americans owned a total of about 152.3 million dogs, cats, and birds. The number of dogs owned was 12.5 million less than the total number of cats and birds owned, and 4.2 million more cats were owned than dogs. (*Source*: American Veterinary Medical Association) How many of each type of pet did Americans own?

24. *Mail-Order Business.* Natural Fibers Clothing charges $4 for shipping orders of $25 or less, $8 for orders from $25.01 to $75, and $10 for orders over $75. One week, shipping charges for 600 orders totaled $4280. Eighty more orders for $25 or less were shipped than orders for more than $75. Find the number of orders shipped at each rate.

25. *Foreign Aid.* In 2013, the top three donors of foreign aid were the United States, Great Britain, and Germany. The total amount of their aid was $63.5 billion. Together, Great Britain and Germany donated $0.5 billion more than the United States did. The United States donated $17.4 billion more than Germany did. (*Source*: Organization for Economic Cooperation and Development) How much in foreign aid did each country donate?

26. *Spring Cleaning.* In a group of 100 adults, 70 say they are most likely to do spring housecleaning in March, April, or May. Of these 70, the number who clean in April is 14 more than the total number who clean in March and May. The total number who clean in April and May is 2 more than three times the number who clean in March. (*Source*: Zoomerang online survey) Find the number who clean in each month.

27. *Nutrition.* A hospital dietician must plan a lunch menu that provides 485 Cal, 41.5 g of carbohydrates, and 35 mg of calcium. A 3-oz serving of broiled ground beef contains 245 Cal, 0 g of carbohydrates, and 9 mg of calcium. One baked potato contains 145 Cal, 34 g of carbohydrates, and 8 mg of calcium. A one-cup serving of strawberries contains 45 Cal, 10 g of carbohydrates, and 21 mg of calcium. (*Source*: U.S. Department of Agriculture) How many servings of each are required in order to provide the desired nutritional values?

28. *Nutrition.* A diabetic patient wishes to prepare a meal consisting of roasted chicken breast, mashed potatoes, and peas. A 3-oz serving of roasted skinless chicken breast contains 140 Cal, 27 g of protein, and 64 mg of sodium. A one-cup serving of mashed potatoes contains 160 Cal, 4 g of protein, and 636 mg of sodium, and a one-cup serving of peas contains 125 Cal, 8 g of protein, and 139 mg of sodium. (*Source*: U.S. Department of Agriculture) How many servings of each should be used if the meal is to contain 415 Cal, 50.5 g of protein, and 553 mg of sodium?

29. *Investment.* Santiago receives $126 per year in simple interest from three investments. Part is invested at 2%, part at 3%, and part at 4%. There is $500 more invested at 3% than at 2%. The amount invested at 4% is three times the amount invested at 3%. Find the amount invested at each rate.

30. *Investment.* Walter earns a year-end bonus of $5000 and puts it in 3 one-year investments that pay $243 in simple interest. Part is invested at 3%, part at 4%, and part at 6%. There is $1500 more invested at 6% than at 3%. Find the amount invested at each rate.

31. *Price Increases.* In San Francisco, orange juice, a raisin bagel, and a cup of coffee from Katie's Koffee Kart cost a total of $8.15. Katie posts a notice announcing that, effective the following week, the price of orange juice will increase 25% and the price of bagels will increase 20%. After the increase, the same purchase will cost a total of $9.30, and the raisin bagel

will cost 30¢ more than coffee. Find the price of each item before the increase.

32. *Cost of Snack Food.* Chad and Brittany pool their loose change to buy snacks on their coffee break. One day, they spent $6.75 on 1 carton of milk, 2 donuts, and 1 cup of coffee. The next day, they spent $8.50 on 3 donuts and 2 cups of coffee. The third day, they bought 1 carton of milk, 1 donut, and 2 cups of coffee and spent $7.25. On the fourth day, they have a total of $6.45 left. Is this enough to buy 2 cartons of milk and 2 donuts?

33. *Hours Spent Studying.* The following table lists the percent of college freshmen who responded that they spent 6 or more hours per week studying during their high school senior year, which is represented in terms of the number of years after 1992.

Year, x	Percent Who Say They Studied 6 or More Hours per Week
1992, 0	43
2002, 10	33
2012, 20	38

Source: 2014 Brown Center Report on Education, compiled from "The American Freshman," UCLA Higher Education Research Institute

a) Use a system of equations to fit a quadratic function $f(x) = ax^2 + bx + c$ to the data.

b) Use the function to estimate the percent of college freshmen who say they spent 6 or more hours per week studying during their high school senior year in 2007 and in 2014.

34. *Existing Home Sales.* The following table lists the number of U.S. existing home sales, in millions, represented in terms of years after 2006.

Year, x	Existing Home Sales (in millions)
2006, 0	6.5
2009, 3	4.3
2012, 6	4.7

Source: National Association of Realtors from Haver Analytics

a) Use a system of equations to fit a quadratic function $f(x) = ax^2 + bx + c$ to the data.

b) Use the function to estimate the number of existing home sales in 2015.

35. *Deportations.* The following table lists the number of foreigners deported, in thousands, represented in terms of the number of years after 2007.

Year, x	U.S. Deportations (in thousands)
2007, 0	291
2010, 3	393
2013, 6	369

Source: U.S. Immigration and Customs Enforcement

a) Use a system of equations to fit a quadratic function $f(x) = ax^2 + bx + c$ to the data.
b) Use the function to estimate the number of deportations in 2014.

36. *Strength of Army.* The following table lists the number of active-duty personnel in the U.S. army, in thousands, represented in terms of the number of years after 2009.

Year, x	Active-Duty Army Personnel (in thousands)
2009, 0	553
2011, 2	565
2013, 4	541

Source: Department of the Army, U.S. Department of Defense

a) Use a system of equations to fit a quadratic function $f(x) = ax^2 + bx + c$ to the data.
b) Use the function to estimate the number of active-duty Army personnel in 2010 and in 2014.

37. *Gasoline Prices.* The following table lists the average U.S. retail price per gallon of gasoline on December 12 for years from 2009 to 2013.

Year, x	U.S. Average Retail Price Per Gallon of Gasoline on December 12
2009, 0	$2.61
2010, 1	2.97
2011, 2	3.27
2012, 3	3.32
2013, 4	3.25

Source: American Automobile Association

a) Use a graphing calculator to fit a quadratic function $f(x) = ax^2 + bx + c$ to the data, where x is the number of years after 2009.
b) Use the function found in part (a) to estimate the average retail price per gallon of gasoline on December 12 in 2014.

38. *Unemployment Rate.* The following table lists the U.S. unemployment rate in October for selected years from 2002 to 2010.

Year	U.S. Unemployment Rate in October
2002, 0	5.7%
2004, 2	5.5
2005, 3	5.0
2008, 6	6.6
2010, 8	9.7

Source: U.S. Bureau of Labor Statistics

a) Use a graphing calculator to fit a quadratic function $f(x) = ax^2 + bx + c$ to the data, where x is the number of years after 2002.
b) Use the function found in part (a) to estimate the unemployment rate in 2003, in 2007, and in 2009.

› Skill Maintenance

Vocabulary Reinforcement

In each of Exercises 39–46, fill in the blank with the correct term. Some of the given choices will not be used.

Descartes' rule of signs	constant function
the leading-term test	horizontal asymptote
the intermediate value theorem	vertical asymptote
	oblique asymptote
the fundamental theorem of algebra	direct variation
	inverse variation
polynomial function	horizontal line
rational function	vertical line
one-to-one function	parallel
	perpendicular

39. Two lines with slopes m_1 and m_2 are _____ if and only if the product of their slopes is -1. **[1.4]**

40. We can use _____ to determine the behavior of the graph of a polynomial function as $x \to \infty$ or as $x \to -\infty$. **[4.1]**

41. If it is possible for a(n) _____ to cross a graph more than once, then the graph is not the graph of a function. **[1.2]**

42. A function is a(n) _____ if different inputs have different outputs. **[5.1]**

43. A(n) _____ is a function that is a quotient of two polynomials. **[4.5]**

44. If a situation gives rise to a function $f(x) = k/x$, or $y = k/x$, where k is a positive constant, we say that we have _____. **[2.5]**

45. A(n) _____ of a rational function $p(x)/q(x)$, where $p(x)$ and $q(x)$ have no common factors other than constants, occurs at an x-value that makes the denominator 0. **[4.5]**

46. When the numerator and the denominator of a rational function have the same degree, the graph of the function has a(n) _____. **[4.5]**

❯ Synthesis

47. Let u represent $1/x$, v represent $1/y$, and w represent $1/z$. Solve first for u, v, and w. Then solve the following system of equations:

$$\frac{2}{x} + \frac{2}{y} - \frac{3}{z} = 3,$$
$$\frac{1}{x} - \frac{2}{y} - \frac{3}{z} = 9,$$
$$\frac{7}{x} - \frac{2}{y} + \frac{9}{z} = -39.$$

48. *Transcontinental Railroad.* Use the following facts to find the year in which the first U.S. transcontinental railroad was completed: The sum of the digits in the year is 24. The units digit is 1 more than the hundreds digit. Both the tens and the units digits are multiples of three.

In Exercises 49 and 50, three solutions of an equation are given. Use a system of three equations in three variables to find the constants and write the equation.

49. $Ax + By + Cz = 12$;
$\left(1, \frac{3}{4}, 3\right)$, $\left(\frac{4}{3}, 1, 2\right)$, and $(2, 1, 1)$

50. $y = B - Mx - Nz$;
$(1, 1, 2)$, $(3, 2, -6)$, and $\left(\frac{3}{2}, 1, 1\right)$

In Exercises 51 and 52, four solutions of the equation $y = ax^3 + bx^2 + cx + d$ are given. Use a system of four equations in four variables to find the constants a, b, c, and d and write the equation.

51. $(-2, 59)$, $(-1, 13)$, $(1, -1)$, and $(2, -17)$

52. $(-2, -39)$, $(-1, -12)$, $(1, -6)$, and $(3, 16)$

9.3 ❯ Matrices and Systems of Equations

> ❯ Solve systems of equations using matrices.

❯ Matrices and Row-Equivalent Operations

In this section, we consider additional techniques for solving systems of equations. You have probably noted that when we solve a system of equations, we perform computations with the coefficients and the constants and continually rewrite the variables. We can streamline the solution process by omitting the variables until a solution is found. For example, the system

$$2x - 3y = 7,$$
$$x + 4y = -2$$

can be written more simply as

$$\begin{bmatrix} 2 & -3 & 7 \\ 1 & 4 & -2 \end{bmatrix}.$$

The vertical line replaces the equals signs.

A rectangular array of numbers like the one above is called a **matrix** (pl., **matrices**). The matrix above is called an **augmented matrix** for the given system of equations, because it contains not only the coefficients but also the constant terms. The matrix

$$\begin{bmatrix} 2 & -3 \\ 1 & 4 \end{bmatrix}$$

is called the **coefficient matrix** of the system.

The **rows** of a matrix are horizontal, and the **columns** are vertical. The augmented matrix above has 2 rows and 3 columns, and the coefficient matrix has 2 rows and 2 columns. A matrix with m rows and n columns is said to be of **order** $m \times n$. Thus the order of the augmented matrix above is 2×3, and the order of the coefficient matrix is 2×2. When $m = n$, a matrix is said to be **square**. The coefficient matrix above is a square matrix. The numbers 2 and 4 lie on the **main diagonal** of the coefficient matrix. The numbers in a matrix are called **entries** or **elements**.

❭ Gaussian Elimination with Matrices

In Section 9.2, we described a series of operations that can be used to transform a system of equations to an equivalent system. Each of these operations corresponds to one that can be used to produce *row-equivalent matrices*.

ROW-EQUIVALENT OPERATIONS

1. Interchange any two rows.
2. Multiply each entry in a row by the same nonzero constant.
3. Add a nonzero multiple of one row to another row.

We can use these operations on the augmented matrix of a system of equations to solve the system.

EXAMPLE 1 Solve the following system:

$$2x - y + 4z = -3,$$
$$x - 2y - 10z = -6,$$
$$3x \qquad + 4z = 7.$$

Solution First, we write the augmented matrix, writing 0 for the missing y-term in the last equation:

$$\begin{bmatrix} 2 & -1 & 4 & | & -3 \\ 1 & -2 & -10 & | & -6 \\ 3 & 0 & 4 & | & 7 \end{bmatrix}.$$

Our goal is to find a row-equivalent matrix of the form shown below.

$$\begin{bmatrix} 1 & a & b & | & c \\ 0 & 1 & d & | & e \\ 0 & 0 & 1 & | & f \end{bmatrix}$$

The variables can then be reinserted to form equations from which we can complete the solution. This is done by working from the bottom equation to the top and using back-substitution.

The first step is to multiply and/or interchange rows so that each number in the first column below the first number is a multiple of that number. In this case, we interchange the first and second rows to obtain a 1 in the upper left-hand corner.

$$\begin{bmatrix} 1 & -2 & -10 & | & -6 \\ 2 & -1 & 4 & | & -3 \\ 3 & 0 & 4 & | & 7 \end{bmatrix} \qquad \begin{aligned} &\text{New row 1} = \text{row 2} \\ &\text{New row 2} = \text{row 1} \end{aligned}$$

Next, we multiply the first row by -2 and add it to the second row. We also multiply the first row by -3 and add it to the third row.

$$\begin{bmatrix} 1 & -2 & -10 & | & -6 \\ 0 & 3 & 24 & | & 9 \\ 0 & 6 & 34 & | & 25 \end{bmatrix}$$ **Row 1 is unchanged.**
New row 2 $= -2(\text{row } 1) + \text{row } 2$
New row 3 $= -3(\text{row } 1) + \text{row } 3$

Now we multiply the second row by $\frac{1}{3}$ to get a 1 in the second row, second column.

$$\begin{bmatrix} 1 & -2 & -10 & | & -6 \\ 0 & 1 & 8 & | & 3 \\ 0 & 6 & 34 & | & 25 \end{bmatrix}$$ **New row 2 $= \frac{1}{3}(\text{row } 2)$**

Then we multiply the second row by -6 and add it to the third row.

$$\begin{bmatrix} 1 & -2 & -10 & | & -6 \\ 0 & 1 & 8 & | & 3 \\ 0 & 0 & -14 & | & 7 \end{bmatrix}$$ **New row 3 $= -6(\text{row } 2) + \text{row } 3$**

Finally, we multiply the third row by $-\frac{1}{14}$ to get a 1 in the third row, third column.

$$\begin{bmatrix} 1 & -2 & -10 & | & -6 \\ 0 & 1 & 8 & | & 3 \\ 0 & 0 & 1 & | & -\frac{1}{2} \end{bmatrix}$$ **New row 3 $= -\frac{1}{14}(\text{row } 3)$**

Now we can write the system of equations that corresponds to the last matrix above:

$$\begin{aligned} x - 2y - 10z &= -6, & \textbf{(1)} \\ y + 8z &= 3, & \textbf{(2)} \\ z &= -\tfrac{1}{2}. & \textbf{(3)} \end{aligned}$$

We back-substitute $-\frac{1}{2}$ for z in equation (2) and solve for y:

$$y + 8\left(-\tfrac{1}{2}\right) = 3$$
$$y - 4 = 3$$
$$y = 7.$$

Next, we back-substitute 7 for y and $-\frac{1}{2}$ for z in equation (1) and solve for x:

$$x - 2 \cdot 7 - 10\left(-\tfrac{1}{2}\right) = -6$$
$$x - 14 + 5 = -6$$
$$x - 9 = -6$$
$$x = 3.$$

The triple $\left(3, 7, -\frac{1}{2}\right)$ checks in the original system of equations, so it is the solution.

▶ *Now Try Exercise 27.*

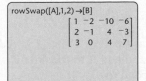

rowSwap([A],1,2) →[B]
$$\begin{bmatrix} 1 & -2 & -10 & -6 \\ 2 & -1 & 4 & -3 \\ 3 & 0 & 4 & 7 \end{bmatrix}$$

Row-equivalent operations can be performed on a graphing calculator. For example, to interchange the first and second rows of the augmented matrix, as we did in the first step in Example 1, we enter the matrix as matrix **A** and select "row-Swap" from the MATRIX MATH menu. Some graphing calculators will not automatically store the matrix produced using a row-equivalent operation, so when several operations are to be performed in succession, it is helpful to store the result of each operation as it is produced. In the window at left, we see both the matrix produced by the rowSwap operation and the indication that this matrix is stored as matrix **B**.

The procedure followed in Example 1 is called **Gaussian elimination with matrices**. The last matrix in Example 1 is in **row-echelon form**. To be in this form, a matrix must have the following properties.

ROW-ECHELON FORM

1. If a row does not consist entirely of 0's, then the first nonzero element in the row is a 1 (called a **leading 1**).
2. For any two successive nonzero rows, the leading 1 in the lower row is farther to the right than the leading 1 in the higher row.
3. All the rows consisting entirely of 0's are at the bottom of the matrix.

If a fourth property is also satisfied, a matrix is said to be in **reduced row-echelon form**:

4. Each column that contains a leading 1 has 0's everywhere else.

EXAMPLE 2   Which of the following matrices are in row-echelon form? Which, if any, are in reduced row-echelon form?

a) $\begin{bmatrix} 1 & -3 & 5 & | & -2 \\ 0 & 1 & -4 & | & 3 \\ 0 & 0 & 1 & | & 10 \end{bmatrix}$ b) $\begin{bmatrix} 0 & -1 & | & 2 \\ 0 & 1 & | & 5 \end{bmatrix}$ c) $\begin{bmatrix} 1 & -2 & -6 & 4 & | & 7 \\ 0 & 3 & 5 & -8 & | & -1 \\ 0 & 0 & 1 & 9 & | & 2 \end{bmatrix}$

d) $\begin{bmatrix} 1 & 0 & 0 & | & -2.4 \\ 0 & 1 & 0 & | & 0.8 \\ 0 & 0 & 1 & | & 5.6 \end{bmatrix}$ e) $\begin{bmatrix} 1 & 0 & 0 & 0 & | & \frac{2}{3} \\ 0 & 1 & 0 & 0 & | & -\frac{1}{4} \\ 0 & 0 & 1 & 0 & | & \frac{6}{7} \\ 0 & 0 & 0 & 0 & | & 0 \end{bmatrix}$ f) $\begin{bmatrix} 1 & -4 & 2 & | & 5 \\ 0 & 0 & 0 & | & 0 \\ 0 & 1 & -3 & | & -8 \end{bmatrix}$

Solution   The matrices in (a), (d), and (e) satisfy the row-echelon criteria and, thus, are in row-echelon form. In (b) and (c), the first nonzero elements of the first and second rows, respectively, are not 1. In (f), the row consisting entirely of 0's is not at the bottom of the matrix. Thus the matrices in (b), (c), and (f) are not in row-echelon form. In (d) and (e), not only are the row-echelon criteria met but each column that contains a leading 1 also has 0's elsewhere, so these matrices are in reduced row-echelon form.

❯ Gauss–Jordan Elimination

We have seen that with Gaussian elimination we perform row-equivalent operations on a matrix to obtain a row-equivalent matrix in row-echelon form. When we continue to apply these operations until we have a matrix in *reduced* row-echelon form, we are using **Gauss–Jordan elimination**. This method is named for Karl Friedrich Gauss and Wilhelm Jordan (1842–1899).

EXAMPLE 3   Use Gauss–Jordan elimination to solve the system of equations in Example 1.

Solution   Using Gaussian elimination in Example 1, we obtained the matrix

$$\begin{bmatrix} 1 & -2 & -10 & | & -6 \\ 0 & 1 & 8 & | & 3 \\ 0 & 0 & 1 & | & -\frac{1}{2} \end{bmatrix}.$$

We continue to perform row-equivalent operations until we have a matrix in reduced row-echelon form. We multiply the third row by 10 and add it to the first row. We also multiply the third row by -8 and add it to the second row.

$$\begin{bmatrix} 1 & -2 & 0 & | & -11 \\ 0 & 1 & 0 & | & 7 \\ 0 & 0 & 1 & | & -\frac{1}{2} \end{bmatrix} \qquad \begin{array}{l} \textbf{New row 1} = \textbf{10(row 3)} + \textbf{row 1} \\ \textbf{New row 2} = \textbf{\textminus 8(row 3)} + \textbf{row 2} \end{array}$$

Next, we multiply the second row by 2 and add it to the first row.

$$\begin{bmatrix} 1 & 0 & 0 & | & 3 \\ 0 & 1 & 0 & | & 7 \\ 0 & 0 & 1 & | & -\frac{1}{2} \end{bmatrix}$$ New row 1 = 2(row 2) + row 1

Writing the system of equations that corresponds to this matrix, we have

$$x \qquad\qquad = 3,$$
$$y \qquad = 7,$$
$$z = -\tfrac{1}{2}.$$

We can actually read the solution, $\left(3, 7, -\frac{1}{2}\right)$, directly from the last column of the reduced row-echelon matrix.

We can also use a graphing calculator to solve this system of equations. After the augmented matrix is entered, reduced row-echelon form can be found directly using the "rref" operation from the MATRIX MATH menu.

> *Now Try Exercise 27.*

rref([A])►Frac
$$\begin{bmatrix} 1 & 0 & 0 & 3 \\ 0 & 1 & 0 & 7 \\ 0 & 0 & 1 & -\frac{1}{2} \end{bmatrix}$$

EXAMPLE 4   Solve the following system:

$$3x - 4y - \;\;z = 6,$$
$$2x - \;\;y + \;\;z = -1,$$
$$4x - 7y - 3z = 13.$$

Solution   We write the augmented matrix and use Gauss–Jordan elimination.

$$\begin{bmatrix} 3 & -4 & -1 & | & 6 \\ 2 & -1 & 1 & | & -1 \\ 4 & -7 & -3 & | & 13 \end{bmatrix}$$

We begin by multiplying the second and third rows by 3 so that each number in the first column below the first number, 3, is a multiple of that number.

$$\begin{bmatrix} 3 & -4 & -1 & | & 6 \\ 6 & -3 & 3 & | & -3 \\ 12 & -21 & -9 & | & 39 \end{bmatrix}$$ New row 2 = 3(row 2)
New row 3 = 3(row 3)

Next, we multiply the first row by -2 and add it to the second row. We also multiply the first row by -4 and add it to the third row.

$$\begin{bmatrix} 3 & -4 & -1 & | & 6 \\ 0 & 5 & 5 & | & -15 \\ 0 & -5 & -5 & | & 15 \end{bmatrix}$$ New row 2 = -2(row 1) + row 2
New row 3 = -4(row 1) + row 3

Now we add the second row to the third row.

$$\begin{bmatrix} 3 & -4 & -1 & | & 6 \\ 0 & 5 & 5 & | & -15 \\ 0 & 0 & 0 & | & 0 \end{bmatrix}$$ New row 3 = row 2 + row 3

We can stop at this stage because we have a row consisting entirely of 0's. The last row of the matrix corresponds to the equation $0 = 0$, which is true for all values of x, y, and z. Therefore, the equations are dependent and the system is equivalent to

$$3x - 4y - \;\;z = 6,$$
$$5y + 5z = -15.$$

This particular system has infinitely many solutions. (A system containing dependent equations could be inconsistent.)

Solving the second equation for y gives us

$$y = -z - 3.$$

Substituting $-z - 3$ for y in the first equation and solving for x, we get

$$3x - 4(-z - 3) - z = 6$$
$$x = -z - 2.$$

Then the solutions of this system are of the form

$$(-z - 2, -z - 3, z),$$

where z can be any real number.

> **Now Try Exercise 33.**

Similarly, if we obtain a row whose only nonzero entry occurs in the last column, we have an inconsistent system of equations. For example, in the matrix

$$\begin{bmatrix} 1 & 0 & 3 & | & -2 \\ 0 & 1 & 5 & | & 4 \\ 0 & 0 & 0 & | & 6 \end{bmatrix},$$

the last row corresponds to the false equation $0 = 6$, so we know the original system of equations has no solution.

9.3 Exercise Set

Determine the order of the matrix.

1. $\begin{bmatrix} 1 & -6 \\ -3 & 2 \\ 0 & 5 \end{bmatrix}$

2. $\begin{bmatrix} 7 \\ -5 \\ -1 \\ 3 \end{bmatrix}$

3. $\begin{bmatrix} 2 & -4 & 0 & 9 \end{bmatrix}$

4. $\begin{bmatrix} -8 \end{bmatrix}$

5. $\begin{bmatrix} 1 & -5 & -8 \\ 6 & 4 & -2 \\ -3 & 0 & 7 \end{bmatrix}$

6. $\begin{bmatrix} 13 & 2 & -6 & 4 \\ -1 & 18 & 5 & -12 \end{bmatrix}$

Write the augmented matrix for the system of equations.

7. $2x - y = 7,$
$x + 4y = -5$

8. $3x + 2y = 8,$
$2x - 3y = 15$

9. $x - 2y + 3z = 12,$
$2x \qquad - 4z = 8,$
$3y + z = 7$

10. $x + y - z = 7,$
$3y + 2z = 1,$
$-2x - 5y \qquad = 6$

Write the system of equations that corresponds to the augmented matrix.

11. $\begin{bmatrix} 3 & -5 & | & 1 \\ 1 & 4 & | & -2 \end{bmatrix}$

12. $\begin{bmatrix} 1 & 2 & | & -6 \\ 4 & 1 & | & -3 \end{bmatrix}$

13. $\begin{bmatrix} 2 & 1 & -4 & | & 12 \\ 3 & 0 & 5 & | & -1 \\ 1 & -1 & 1 & | & 2 \end{bmatrix}$

14. $\begin{bmatrix} -1 & -2 & 3 & | & 6 \\ 0 & 4 & 1 & | & 2 \\ 2 & -1 & 0 & | & 9 \end{bmatrix}$

Solve the system of equations using Gaussian elimination or Gauss–Jordan elimination. Use a graphing calculator to check your answer.

15. $4x + 2y = 11,$
$3x - y = 2$

16. $2x + y = 1,$
$3x + 2y = -2$

17. $5x - 2y = -3,$
$2x + 5y = -24$

18. $2x + y = 1,$
$3x - 6y = 4$

19. $3x + 4y = 7,$
$-5x + 2y = 10$

20. $5x - 3y = -2,$
$4x + 2y = 5$

21. $3x + 2y = 6,$
$2x - 3y = -9$

22. $x - 4y = 9,$
$2x + 5y = 5$

23. $x - 3y = 8,$
$-2x + 6y = 3$

24. $4x - 8y = 12,$
$-x + 2y = -3$

25. $-2x + 6y = 4,$
$3x - 9y = -6$

26. $6x + 2y = -10,$
$-3x - y = 6$

27. $x + 2y - 3z = 9,$
$2x - y + 2z = -8,$
$3x - y - 4z = 3$

28. $x - y + 2z = 0,$
$x - 2y + 3z = -1,$
$2x - 2y + z = -3$

29. $4x - y - 3z = 1,$
$8x + y - z = 5,$
$2x + y + 2z = 5$

30. $3x + 2y + 2z = 3,$
$x + 2y - z = 5,$
$2x - 4y + z = 0$

31. $x - 2y + 3z = -4,$
$3x + y - z = 0,$
$2x + 3y - 5z = 1$

32. $2x - 3y + 2z = 2,$
$x + 4y - z = 9,$
$-3x + y - 5z = 5$

33. $2x - 4y - 3z = 3,$
$x + 3y + z = -1,$
$5x + y - 2z = 2$

34. $x + y - 3z = 4,$
$4x + 5y + z = 1,$
$2x + 3y + 7z = -7$

35. $p + q + r = 1,$
$p + 2q + 3r = 4,$
$4p + 5q + 6r = 7$

36. $m + n + t = 9,$
$m - n - t = -15,$
$3m + n + t = 2$

37. $a + b - c = 7,$
$a - b + c = 5,$
$3a + b - c = -1$

38. $a - b + c = 3,$
$2a + b - 3c = 5,$
$4a + b - c = 11$

39. $-2w + 2x + 2y - 2z = -10,$
$w + x + y + z = -5,$
$3w + x - y + 4z = -2,$
$w + 3x - 2y + 2z = -6$

40. $-w + 2x - 3y + z = -8,$
$-w + x + y - z = -4,$
$w + x + y + z = 22,$
$-w + x - y - z = -14$

Use Gaussian elimination or Gauss–Jordan elimination in Exercises 41–44.

41. *Borrowing.*   Greenfield Manufacturing borrowed $30,000 to buy a new piece of equipment. Part of the money was borrowed at 8%, part at 10%, and part at 12%. The annual interest was $3040, and the total amount borrowed at 8% and at 10% was twice the amount borrowed at 12%. How much was borrowed at each rate?

42. *Summer Sports Camps.*   One year, a family spent a total of $2500 on summer sports camps for their three sons. The amount spent on camps for the oldest son was $200 more than four times what was spent for the youngest son. The total spent for the two younger sons was $1100. How much was spent on summer sports camps for each son?

43. *Stamp Purchase.*   For her business, Olivia spent $86.80 on both 49¢ and 21¢ stamps. She bought a total of 200 stamps. How many of each type did she buy?

44. *Time of Return.*   The Patels pay their babysitter $11 per hour before 11 P.M. and $14.50 after 11 P.M. One evening, they went out for 6 hr and paid the sitter $73. What time did they return?

❯ Skill Maintenance

Classify the function as linear, quadratic, cubic, quartic, rational, exponential, or logarithmic.

45. $f(x) = 3^{x-1}$ **[5.2]**

46. $f(x) = 3x - 1$ **[1.3]**

47. $f(x) = \dfrac{3x - 1}{x^2 + 4}$ **[4.5]**

48. $f(x) = -\frac{3}{4}x^4 + \frac{9}{2}x^3 + 2x^2 - 4$ **[4.1]**

49. $f(x) = \ln(3x - 1)$ **[5.3]**

50. $f(x) = \frac{3}{4}x^3 - x$ **[4.1]**

51. $f(x) = 3$ **[1.3]**

52. $f(x) = 2 - x - x^2$ **[3.2]**

❯ Synthesis

In Exercises 53 and 54, three solutions of the equation $y = ax^2 + bx + c$ are given. Use a system of three equations in three variables and Gaussian elimination or Gauss–Jordan elimination to find the constants a, b, and c and write the equation.

53. $(-3, 12), (-1, -7),$ and $(1, -2)$

54. $(-1, 0), (1, -3),$ and $(3, -22)$

55. Find two different row-echelon forms of
$$\begin{bmatrix} 1 & 5 \\ 3 & 2 \end{bmatrix}.$$

56. Consider the system of equations
$$x - y + 3z = -8,$$
$$2x + 3y - z = 5,$$
$$3x + 2y + 2kz = -3k.$$

For what value(s) of k, if any, will the system have

a) no solution?
b) exactly one solution?
c) infinitely many solutions?

Solve using matrices.

57. $y = x + z,$
$3y + 5z = 4,$
$x + 4 = y + 3z$

58. $x + y = 2z,$
$2x - 5z = 4,$
$x - z = y + 8$

59. $x - 4y + 2z = 7,$
$3x + y + 3z = -5$

60. $x - y - 3z = 3,$
$-x + 3y + z = -7$

61. $4x + 5y = 3,$
$-2x + y = 9,$
$3x - 2y = -15$

62. $2x - 3y = -1,$
$-x + 2y = -2,$
$3x - 5y = 1$

9.4 › Matrix Operations

> ❯ Add, subtract, and multiply matrices when possible.
> ❯ Write a matrix equation equivalent to a system of equations.

In addition to being used to solve systems of equations, matrices are useful in many other types of applications. In this section, we study matrices and some of their properties.

An uppercase letter is generally used to name a matrix, and lower-case letters with double subscripts generally denote its entries. For example, a_{47}, read "*a* sub four seven," indicates the entry in the fourth row and the seventh column. A general term is represented by a_{ij}. The notation a_{ij} indicates the entry in row i and column j. In general, we can write a matrix as

$$\mathbf{A} = [a_{ij}] = \begin{bmatrix} a_{11} & a_{12} & a_{13} & \cdots & a_{1n} \\ a_{21} & a_{22} & a_{23} & \cdots & a_{2n} \\ a_{31} & a_{32} & a_{33} & \cdots & a_{3n} \\ \vdots & \vdots & \vdots & & \vdots \\ a_{m1} & a_{m2} & a_{m3} & \cdots & a_{mn} \end{bmatrix}.$$

The matrix above has m rows and n columns; that is, its order is $m \times n$.

Two matrices are **equal** if they have the same order and corresponding entries are equal.

❯ Matrix Addition and Subtraction

To add or subtract matrices, we add or subtract their corresponding entries. For this to be possible, the matrices must have the same order.

ADDITION AND SUBTRACTION OF MATRICES

Given two $m \times n$ matrices $\mathbf{A} = [a_{ij}]$ and $\mathbf{B} = [b_{ij}]$, their sum is

$$\mathbf{A} + \mathbf{B} = [a_{ij} + b_{ij}]$$

and their difference is

$$\mathbf{A} - \mathbf{B} = [a_{ij} - b_{ij}].$$

Addition of matrices is both commutative and associative.

EXAMPLE 1 Find $\mathbf{A} + \mathbf{B}$ for each of the following.

a) $\mathbf{A} = \begin{bmatrix} -5 & 0 \\ 4 & \frac{1}{2} \end{bmatrix}$, $\mathbf{B} = \begin{bmatrix} 6 & -3 \\ 2 & 3 \end{bmatrix}$

b) $\mathbf{A} = \begin{bmatrix} 1 & 3 \\ -1 & 5 \\ 6 & 0 \end{bmatrix}$, $\mathbf{B} = \begin{bmatrix} -1 & -2 \\ 1 & -2 \\ -3 & 1 \end{bmatrix}$

Solution We have a pair of 2×2 matrices in part (a) and a pair of 3×2 matrices in part (b). Since the matrices within each pair of matrices have the same order, we can add the corresponding entries.

[A]+[B]
$$\begin{bmatrix} 1 & -3 \\ 6 & 3.5 \end{bmatrix}$$

a) $\mathbf{A} + \mathbf{B} = \begin{bmatrix} -5 & 0 \\ 4 & \frac{1}{2} \end{bmatrix} + \begin{bmatrix} 6 & -3 \\ 2 & 3 \end{bmatrix}$

$$= \begin{bmatrix} -5 + 6 & 0 + (-3) \\ 4 + 2 & \frac{1}{2} + 3 \end{bmatrix} = \begin{bmatrix} 1 & -3 \\ 6 & 3\frac{1}{2} \end{bmatrix}$$

We can also enter $\mathbf{A}$ and $\mathbf{B}$ in a graphing calculator and then find $\mathbf{A} + \mathbf{B}$.

b) $\mathbf{A} + \mathbf{B} = \begin{bmatrix} 1 & 3 \\ -1 & 5 \\ 6 & 0 \end{bmatrix} + \begin{bmatrix} -1 & -2 \\ 1 & -2 \\ -3 & 1 \end{bmatrix}$

[A]+[B]
$$\begin{bmatrix} 0 & 1 \\ 0 & 3 \\ 3 & 1 \end{bmatrix}$$

$$= \begin{bmatrix} 1 + (-1) & 3 + (-2) \\ -1 + 1 & 5 + (-2) \\ 6 + (-3) & 0 + 1 \end{bmatrix} = \begin{bmatrix} 0 & 1 \\ 0 & 3 \\ 3 & 1 \end{bmatrix}$$

This sum can also be found on a graphing calculator after $\mathbf{A}$ and $\mathbf{B}$ have been entered.

> **Now Try Exercise 5.**

EXAMPLE 2 Find $\mathbf{C} - \mathbf{D}$ for each of the following.

a) $\mathbf{C} = \begin{bmatrix} 1 & 2 \\ -2 & 0 \\ -3 & -1 \end{bmatrix}$, $\mathbf{D} = \begin{bmatrix} 1 & -1 \\ 1 & 3 \\ 2 & 3 \end{bmatrix}$

b) $\mathbf{C} = \begin{bmatrix} 5 & -6 \\ -3 & 4 \end{bmatrix}$, $\mathbf{D} = \begin{bmatrix} -4 \\ 1 \end{bmatrix}$

Solution

a) Since the order of each matrix is 3×2, we can subtract corresponding entries:

$$\mathbf{C} - \mathbf{D} = \begin{bmatrix} 1 & 2 \\ -2 & 0 \\ -3 & -1 \end{bmatrix} - \begin{bmatrix} 1 & -1 \\ 1 & 3 \\ 2 & 3 \end{bmatrix}$$

[C]−[D]
$$\begin{bmatrix} 0 & 3 \\ -3 & -3 \\ -5 & -4 \end{bmatrix}$$

$$= \begin{bmatrix} 1 - 1 & 2 - (-1) \\ -2 - 1 & 0 - 3 \\ -3 - 2 & -1 - 3 \end{bmatrix} = \begin{bmatrix} 0 & 3 \\ -3 & -3 \\ -5 & -4 \end{bmatrix}.$$

This subtraction can also be done using a graphing calculator.

b) $\mathbf{C}$ is a 2×2 matrix and $\mathbf{D}$ is a 2×1 matrix. Since the matrices do not have the same order, we cannot subtract. If we try to do this subtraction on a graphing calculator, we get an ERROR message indicating that the dimensions are mismatched.

ERR:DIM MISMATCH
1: Quit
2: Goto

> **Now Try Exercise 13.**

The **opposite**, or **additive inverse**, of a matrix is obtained by replacing each entry with its opposite.

EXAMPLE 3 Find $-\mathbf{A}$ and $\mathbf{A} + (-\mathbf{A})$ for

$$\mathbf{A} = \begin{bmatrix} 1 & 0 & 2 \\ 3 & -1 & 5 \end{bmatrix}.$$

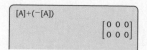

Solution   To find $-\mathbf{A}$, we replace each entry of $\mathbf{A}$ with its opposite.

$$-\mathbf{A} = \begin{bmatrix} -1 & 0 & -2 \\ -3 & 1 & -5 \end{bmatrix},$$

$$\mathbf{A} + (-\mathbf{A}) = \begin{bmatrix} 1 & 0 & 2 \\ 3 & -1 & 5 \end{bmatrix} + \begin{bmatrix} -1 & 0 & -2 \\ -3 & 1 & -5 \end{bmatrix}$$

$$= \begin{bmatrix} 0 & 0 & 0 \\ 0 & 0 & 0 \end{bmatrix}.$$

A matrix having 0's for all its entries is called a **zero matrix**. When a zero matrix is added to a second matrix of the same order, the second matrix is unchanged. Thus a zero matrix is an **additive identity**. For example,

$$\begin{bmatrix} 2 & 3 & -4 \\ 0 & 6 & 5 \end{bmatrix} + \begin{bmatrix} 0 & 0 & 0 \\ 0 & 0 & 0 \end{bmatrix} = \begin{bmatrix} 2 & 3 & -4 \\ 0 & 6 & 5 \end{bmatrix}.$$

The matrix

$$\begin{bmatrix} 0 & 0 & 0 \\ 0 & 0 & 0 \end{bmatrix}$$

is the additive identity for any 2×3 matrix.

❯ Scalar Multiplication

When we find the product of a number and a matrix, we obtain a **scalar product**.

SCALAR PRODUCT

The **scalar product** of a number k and a matrix $\mathbf{A}$ is the matrix denoted $k\mathbf{A}$, obtained by multiplying each entry of $\mathbf{A}$ by the number k. The number k is called a **scalar**.

EXAMPLE 4   Find $3\mathbf{A}$ and $(-1)\mathbf{A}$ for

$$\mathbf{A} = \begin{bmatrix} -3 & 0 \\ 4 & 5 \end{bmatrix}.$$

Solution   We have

$$3\mathbf{A} = 3\begin{bmatrix} -3 & 0 \\ 4 & 5 \end{bmatrix} = \begin{bmatrix} 3(-3) & 3 \cdot 0 \\ 3 \cdot 4 & 3 \cdot 5 \end{bmatrix} = \begin{bmatrix} -9 & 0 \\ 12 & 15 \end{bmatrix},$$

$$(-1)\mathbf{A} = -1\begin{bmatrix} -3 & 0 \\ 4 & 5 \end{bmatrix} = \begin{bmatrix} -1(-3) & -1 \cdot 0 \\ -1 \cdot 4 & -1 \cdot 5 \end{bmatrix} = \begin{bmatrix} 3 & 0 \\ -4 & -5 \end{bmatrix}.$$

These scalar products can also be found on a graphing calculator after $\mathbf{A}$ has been entered.

❯ **Now Try Exercise 9.**

The properties of matrix addition and scalar multiplication are similar to the properties of addition and multiplication of real numbers.

> **PROPERTIES OF MATRIX ADDITION
> AND SCALAR MULTIPLICATION**
>
> For any $m \times n$ matrices **A**, **B**, and **C** and any scalars k and l:
>
> $\quad$ **A** + **B** = **B** + **A**. $\qquad\qquad$ Commutative property of addition
>
> $\quad$ **A** + (**B** + **C**) = (**A** + **B**) + **C**. $\quad$ Associative property of addition
>
> $\quad$ (kl)**A** = $k(l$**A**$)$. $\qquad\qquad$ Associative property of scalar
> $\qquad\qquad\qquad\qquad\qquad\qquad\qquad$ multiplication
>
> $\quad$ $k($**A** + **B**$)$ = k**A** + k**B**. $\qquad$ Distributive property
>
> $\quad$ $(k + l)$**A** = k**A** + l**A**. $\qquad$ Distributive property
>
> There exists a unique matrix **0** such that:
>
> $\quad$ **A** + **0** = **0** + **A** = **A**. $\qquad$ Additive identity property
>
> There exists a unique matrix $-$**A** such that:
>
> $\quad$ **A** + $(-$**A**$)$ = $-$**A** + **A** = **0**. $\qquad$ Additive inverse property

EXAMPLE 5 *Production.* Waterworks, Inc., manufactures three types of kayaks in its two plants. The following table lists the number of each style produced at each plant in April.

	Whitewater Kayak	Ocean Kayak	Crossover Kayak
Madison Plant	150	120	100
Greensburg Plant	180	90	130

a) Write a 2×3 matrix **A** that represents the information in the table.

b) The manufacturer increased production by 20% in May. Find a matrix **M** that represents the increased production figures.

c) Find the matrix **A** + **M** and tell what it represents.

Solution

a) We write the entries in the table in a 2×3 matrix **A**.

$$\mathbf{A} = \begin{bmatrix} 150 & 120 & 100 \\ 180 & 90 & 130 \end{bmatrix}$$

b) The production in May will be represented by **A** + 20%**A**, or **A** + 0.2**A**, or 1.2**A**. Thus,

$$\mathbf{M} = (1.2)\begin{bmatrix} 150 & 120 & 100 \\ 180 & 90 & 130 \end{bmatrix} = \begin{bmatrix} 180 & 144 & 120 \\ 216 & 108 & 156 \end{bmatrix}.$$

c) $\mathbf{A} + \mathbf{M} = \begin{bmatrix} 150 & 120 & 100 \\ 180 & 90 & 130 \end{bmatrix} + \begin{bmatrix} 180 & 144 & 120 \\ 216 & 108 & 156 \end{bmatrix}$

$\qquad\qquad = \begin{bmatrix} 330 & 264 & 220 \\ 396 & 198 & 286 \end{bmatrix}$

The matrix **A** + **M** represents the total production of each of the three types of kayaks at each plant in April and May.

> **Now Try Exercise 29.**

❯ Products of Matrices

Matrix multiplication is defined in such a way that it can be used in solving systems of equations and in many applications.

MATRIX MULTIPLICATION

For an $m \times n$ matrix $\mathbf{A} = [a_{ij}]$ and an $n \times p$ matrix $\mathbf{B} = [b_{ij}]$, the **product $\mathbf{AB} = [c_{ij}]$** is an $m \times p$ matrix, where

$$c_{ij} = a_{i1} \cdot b_{1j} + a_{i2} \cdot b_{2j} + a_{i3} \cdot b_{3j} + \cdots + a_{in} \cdot b_{nj}.$$

In other words, the entry c_{ij} in $\mathbf{AB}$ is obtained by multiplying the entries in row i of $\mathbf{A}$ by the corresponding entries in column j of $\mathbf{B}$ and adding the results.

Note that we can multiply two matrices only when the number of columns in the first matrix is equal to the number of rows in the second matrix.

EXAMPLE 6   For

$$\mathbf{A} = \begin{bmatrix} 3 & 1 & -1 \\ 2 & 0 & 3 \end{bmatrix}, \quad \mathbf{B} = \begin{bmatrix} 1 & 6 \\ 3 & -5 \\ -2 & 4 \end{bmatrix}, \quad \text{and} \quad \mathbf{C} = \begin{bmatrix} 4 & -6 \\ 1 & 2 \end{bmatrix},$$

find each of the following.

a) **AB**
b) **BA**
c) **BC**
d) **AC**

Solution

a) $\mathbf{A}$ is a 2×3 matrix and $\mathbf{B}$ is a 3×2 matrix, so $\mathbf{AB}$ will be a 2×2 matrix.

$$\mathbf{AB} = \begin{bmatrix} 3 & 1 & -1 \\ 2 & 0 & 3 \end{bmatrix} \begin{bmatrix} 1 & 6 \\ 3 & -5 \\ -2 & 4 \end{bmatrix}$$

$$= \begin{bmatrix} 3 \cdot 1 + 1 \cdot 3 + (-1)(-2) & 3 \cdot 6 + 1(-5) + (-1)(4) \\ 2 \cdot 1 + 0 \cdot 3 + 3(-2) & 2 \cdot 6 + 0(-5) + 3 \cdot 4 \end{bmatrix} = \begin{bmatrix} 8 & 9 \\ -4 & 24 \end{bmatrix}$$

b) $\mathbf{B}$ is a 3×2 matrix and $\mathbf{A}$ is a 2×3 matrix, so $\mathbf{BA}$ will be a 3×3 matrix.

$$\mathbf{BA} = \begin{bmatrix} 1 & 6 \\ 3 & -5 \\ -2 & 4 \end{bmatrix} \begin{bmatrix} 3 & 1 & -1 \\ 2 & 0 & 3 \end{bmatrix}$$

$$= \begin{bmatrix} 1 \cdot 3 + 6 \cdot 2 & 1 \cdot 1 + 6 \cdot 0 & 1(-1) + 6 \cdot 3 \\ 3 \cdot 3 + (-5)(2) & 3 \cdot 1 + (-5)(0) & 3(-1) + (-5)(3) \\ -2 \cdot 3 + 4 \cdot 2 & -2 \cdot 1 + 4 \cdot 0 & -2(-1) + 4 \cdot 3 \end{bmatrix} = \begin{bmatrix} 15 & 1 & 17 \\ -1 & 3 & -18 \\ 2 & -2 & 14 \end{bmatrix}$$

Note in parts (a) and (b) that $\mathbf{AB} \neq \mathbf{BA}$. Multiplication of matrices is generally not commutative.

Matrix multiplication can be performed on a graphing calculator. The products in parts (a) and (b) are shown at left.

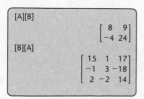

[A][B]
$$\begin{bmatrix} 8 & 9 \\ -4 & 24 \end{bmatrix}$$

[B][A]
$$\begin{bmatrix} 15 & 1 & 17 \\ -1 & 3 & -18 \\ 2 & -2 & 14 \end{bmatrix}$$

c) **B** is a 3 × 2 matrix and **C** is a 2 × 2 matrix, so **BC** will be a 3 × 2 matrix.

$$BC = \begin{bmatrix} 1 & 6 \\ 3 & -5 \\ -2 & 4 \end{bmatrix} \begin{bmatrix} 4 & -6 \\ 1 & 2 \end{bmatrix}$$

$$= \begin{bmatrix} 1 \cdot 4 + 6 \cdot 1 & 1(-6) + 6 \cdot 2 \\ 3 \cdot 4 + (-5)(1) & 3(-6) + (-5)(2) \\ -2 \cdot 4 + 4 \cdot 1 & -2(-6) + 4 \cdot 2 \end{bmatrix} = \begin{bmatrix} 10 & 6 \\ 7 & -28 \\ -4 & 20 \end{bmatrix}$$

d) The product **AC** is not defined because the number of columns of **A**, 3, is not equal to the number of rows of **C**, 2.

When the product **AC** is entered on a graphing calculator, an ERROR message is returned, indicating that the dimensions of the matrices are mismatched.

> *Now Try Exercises 23 and 25.*

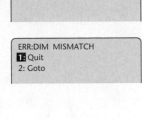

EXAMPLE 7 *Bakery Profit.*   Two of the items sold at Sweet Treats Bakery are gluten-free bagels and gluten-free doughnuts. The following table lists the number of dozens of each product that were sold at the bakery's three stores one week.

	Main Street Store	Avon Road Store	Dalton Avenue Store
Bagels (in dozens)	25	30	20
Doughnuts (in dozens)	40	35	15

The bakery's profit on one dozen bagels is $5, and its profit on one dozen doughnuts is $6. Use matrices to find the total profit on these items at each store for the given week.

Solution   We can write the table showing the sales of the products as a 2 × 3 matrix:

$$S = \begin{bmatrix} 25 & 30 & 20 \\ 40 & 35 & 15 \end{bmatrix}.$$

The profit per dozen for each product can also be written as a matrix:

$$P = \begin{bmatrix} 5 & 6 \end{bmatrix}.$$

Then the total profit at each store is given by the matrix product **PS**:

$$PS = \begin{bmatrix} 5 & 6 \end{bmatrix} \begin{bmatrix} 25 & 30 & 20 \\ 40 & 35 & 15 \end{bmatrix}$$

$$= \begin{bmatrix} 5 \cdot 25 + 6 \cdot 40 & 5 \cdot 30 + 6 \cdot 35 & 5 \cdot 20 + 6 \cdot 15 \end{bmatrix}$$

$$= \begin{bmatrix} 365 & 360 & 190 \end{bmatrix}.$$

The total profit on gluten-free bagels and gluten-free doughnuts for the given week was $365 at the Main Street store, $360 at the Avon Road store, and $190 at the Dalton Avenue store.

> *Now Try Exercise 33.*

A matrix that consists of a single row, like **P** in Example 7, is called a **row matrix**. Similarly, a matrix that consists of a single column, like

$$\begin{bmatrix} 8 \\ -3 \\ 5 \end{bmatrix},$$

is called a **column matrix**.

We have already seen that matrix multiplication is generally not commutative. Nevertheless, matrix multiplication does have some properties that are similar to those for multiplication of real numbers.

PROPERTIES OF MATRIX MULTIPLICATION

For matrices $\mathbf{A}$, $\mathbf{B}$, and $\mathbf{C}$, assuming that the indicated operations are possible:

$\mathbf{A}(\mathbf{BC}) = (\mathbf{AB})\mathbf{C}$.	Associative property of multiplication
$\mathbf{A}(\mathbf{B} + \mathbf{C}) = \mathbf{AB} + \mathbf{AC}$.	Distributive property
$(\mathbf{B} + \mathbf{C})\mathbf{A} = \mathbf{BA} + \mathbf{CA}$.	Distributive property

---

❭ Matrix Equations

We can write a matrix equation equivalent to a system of equations.

EXAMPLE 8 Write a matrix equation equivalent to the following system of equations:

$$\begin{aligned} 4x + 2y - z &= 3, \\ 9x \quad\quad + z &= 5, \\ 4x + 5y - 2z &= 1. \end{aligned}$$

Solution We write the coefficients on the left in a matrix. We then write the product of that matrix and the column matrix containing the variables and set the result equal to the column matrix containing the constants on the right:

$$\begin{bmatrix} 4 & 2 & -1 \\ 9 & 0 & 1 \\ 4 & 5 & -2 \end{bmatrix} \begin{bmatrix} x \\ y \\ z \end{bmatrix} = \begin{bmatrix} 3 \\ 5 \\ 1 \end{bmatrix}.$$

If we let

$$\mathbf{A} = \begin{bmatrix} 4 & 2 & -1 \\ 9 & 0 & 1 \\ 4 & 5 & -2 \end{bmatrix}, \quad \mathbf{X} = \begin{bmatrix} x \\ y \\ z \end{bmatrix}, \quad \text{and} \quad \mathbf{B} = \begin{bmatrix} 3 \\ 5 \\ 1 \end{bmatrix},$$

we can write this matrix equation as $\mathbf{AX} = \mathbf{B}$. ❭ *Now Try Exercise 41.*

9.4   Exercise Set

Find x and y.

1. $\begin{bmatrix} 5 & x \end{bmatrix} = \begin{bmatrix} y & -3 \end{bmatrix}$

2. $\begin{bmatrix} 6x \\ 25 \end{bmatrix} = \begin{bmatrix} -9 \\ 5y \end{bmatrix}$

3. $\begin{bmatrix} 3 & 2x \\ y & -8 \end{bmatrix} = \begin{bmatrix} 3 & -2 \\ 1 & -8 \end{bmatrix}$

4. $\begin{bmatrix} x-1 & 4 \\ y+3 & -7 \end{bmatrix} = \begin{bmatrix} 0 & 4 \\ -2 & -7 \end{bmatrix}$

For Exercises 5–20, let

$$\mathbf{A} = \begin{bmatrix} 1 & 2 \\ 4 & 3 \end{bmatrix}, \quad \mathbf{B} = \begin{bmatrix} -3 & 5 \\ 2 & -1 \end{bmatrix},$$

$$\mathbf{C} = \begin{bmatrix} 1 & -1 \\ -1 & 1 \end{bmatrix}, \quad \mathbf{D} = \begin{bmatrix} 1 & 1 \\ 1 & 1 \end{bmatrix},$$

$$\mathbf{E} = \begin{bmatrix} 1 & 3 \\ 2 & 6 \end{bmatrix}, \quad \mathbf{F} = \begin{bmatrix} 3 & 3 \\ -1 & -1 \end{bmatrix},$$

$$\mathbf{0} = \begin{bmatrix} 0 & 0 \\ 0 & 0 \end{bmatrix}, \quad \mathbf{I} = \begin{bmatrix} 1 & 0 \\ 0 & 1 \end{bmatrix}.$$

Find each of the following.

5. A + B   **6. B + A**   **7. E + 0**

8. 2A   **9. 3F**   **10. (−1)D**

11. 3F + 2A   **12. A − B**   **13. B − A**

14. AB   **15. BA**   **16. 0F**

17. CD   **18. EF**   **19. AI**

20. IA

Find the product, if possible.

21. $\begin{bmatrix} -1 & 0 & 7 \\ 3 & -5 & 2 \end{bmatrix} \begin{bmatrix} 6 \\ -4 \\ 1 \end{bmatrix}$

22. $\begin{bmatrix} 6 & -1 & 2 \end{bmatrix} \begin{bmatrix} 1 & 4 \\ -2 & 0 \\ 5 & -3 \end{bmatrix}$

23. $\begin{bmatrix} -2 & 4 \\ 5 & 1 \\ -1 & -3 \end{bmatrix} \begin{bmatrix} 3 & -6 \\ -1 & 4 \end{bmatrix}$

24. $\begin{bmatrix} 2 & -1 & 0 \\ 0 & 5 & 4 \end{bmatrix} \begin{bmatrix} -3 & 1 & 0 \\ 0 & 2 & -1 \\ 5 & 0 & 4 \end{bmatrix}$

25. $\begin{bmatrix} 1 \\ -5 \\ 3 \end{bmatrix} \begin{bmatrix} -6 & 5 & 8 \\ 0 & 4 & -1 \end{bmatrix}$

26. $\begin{bmatrix} 2 & 0 & 0 \\ 0 & -1 & 0 \\ 0 & 0 & 3 \end{bmatrix} \begin{bmatrix} 0 & -4 & 3 \\ 2 & 1 & 0 \\ -1 & 0 & 6 \end{bmatrix}$

27. $\begin{bmatrix} 1 & -4 & 3 \\ 0 & 8 & 0 \\ -2 & -1 & 5 \end{bmatrix} \begin{bmatrix} 3 & 0 & 0 \\ 0 & -4 & 0 \\ 0 & 0 & 1 \end{bmatrix}$

28. $\begin{bmatrix} 4 \\ -5 \end{bmatrix} \begin{bmatrix} 2 & 0 \\ 6 & -7 \\ 0 & -3 \end{bmatrix}$

29. *Produce.*   The produce manager at Stan's Market orders 40 lb of tomatoes, 20 lb of zucchini, and 30 lb of onions from a local farmer one week.

a) Write a 1 × 3 matrix **A** that represents the amount of each item ordered.

b) The following week the produce manager increases his order by 10%. Find a matrix **B** that represents this order.

c) Find **A + B** and tell what the entries represent.

30. *Budget.*   For the month of June, Lexi's budgets $400 for food, $160 for clothes, and $60 for entertainment.

a) Write a 1 × 3 matrix **B** that represents the amounts budgeted for these items.

b) After receiving a raise, Lexi increases the amount budgeted for each item in July by 5%. Find a matrix **R** that represents the new amounts.

c) Find **B + R** and tell what the entries represent.

31. *Nutrition.*   A 3-oz serving of roasted, skinless chicken breast contains 140 Cal, 27 g of protein, 3 g of fat, 13 mg of calcium, and 64 mg of sodium. One-half cup of potato salad contains 180 Cal, 4 g of protein, 11 g of fat, 24 mg of calcium, and 662 mg of sodium. One broccoli spear contains 50 Cal, 5 g of protein, 1 g of fat, 82 mg of calcium, and 20 mg of sodium. (*Source*: U.S. Department of Agriculture)

a) Write 1 × 5 matrices **C**, **P**, and **B** that represent the nutritional values of each food.

b) Find **C + 2P + 3B** and tell what the entries represent.

32. *Nutrition.*   One slice of cheese pizza contains 290 Cal, 15 g of protein, 9 g of fat, and 39 g of carbohydrates. One-half cup of gelatin dessert contains 70 Cal, 2 g of protein, 0 g of fat, and 17 g of carbohydrates. One cup of whole milk contains 150 Cal, 8 g of protein, 8 g of fat, and 11 g of carbohydrates. (*Source*: U.S. Department of Agriculture)

a) Write 1 × 4 matrices **P**, **G**, and **M** that represent the nutritional values of each food.

b) Find **3P + 2G + 2M** and tell what the entries represent.

33. *Food Service Management.*   The food service manager at a large hospital is concerned about maintaining reasonable food costs. The following table lists the cost per serving, in dollars, for items on four menus.

Menu	Meat	Potato	Vegetable	Salad	Dessert
1	1.50	0.30	0.36	0.45	0.64
2	1.55	0.28	0.48	0.57	0.75
3	1.62	0.52	0.65	0.38	0.53
4	1.70	0.43	0.40	0.42	0.68

On a particular day, a dietician orders 65 meals from menu 1, 48 from menu 2, 93 from menu 3, and 57 from menu 4.

a) Write the information in the table as a 4 × 5 matrix **M**.

b) Write a row matrix **N** that represents the number of each menu ordered.

c) Find the product **NM**.

d) State what the entries of **NM** represent.

34. *Food Service Management.* A college food service manager uses a table like the one below to list the number of units of ingredients, by weight, required for various menu items.

	White Cake	Bread	Coffee Cake	Sugar Cookies
Flour	1	2.5	0.75	0.5
Milk	0	0.5	0.25	0
Eggs	0.75	0.25	0.5	0.5
Butter	0.5	0	0.5	1

The cost per unit of each ingredient is 25 cents for flour, 34 cents for milk, 54 cents for eggs, and 83 cents for butter.

a) Write the information in the table as a 4 × 4 matrix **M**.

b) Write a row matrix **C** that represents the cost per unit of each ingredient.

c) Find the product **CM**.

d) State what the entries of **CM** represent.

35. *Production Cost.* Anja supplies two small campus coffee shops with homemade chocolate chip cookies, oatmeal cookies, and peanut butter cookies. The following table shows the number of each type of cookie, in dozens, that Anja sold in one week.

	Skip's Coffee Shop	The Coffee Club
Chocolate Chip	8	15
Oatmeal	6	10
Peanut Butter	4	3

Anja spends $4 for the ingredients for one dozen chocolate chip cookies, $2.50 for the ingredients for one dozen oatmeal cookies, and $3 for the ingredients for one dozen peanut butter cookies.

a) Write the information in the table as a 3 × 2 matrix **S**.

b) Write a row matrix **C** that represents the cost, per dozen, of the ingredients for each type of cookie.

c) Find the product **CS**.

d) State what the entries of **CS** represent.

36. *Profit.* A manufacturer produces exterior plywood, interior plywood, and fiberboard, which are shipped to two distributors. The following table lists the number of units of each type of product that are shipped to each warehouse.

	Distributor 1	Distributor 2
Exterior Plywood	900	500
Interior Plywood	450	1000
Fiberboard	600	700

The profits from each unit of exterior plywood, interior plywood, and fiberboard are $8, $10, and $7, respectively.

a) Write the information in the table as a 3 × 2 matrix **M**.

b) Write a row matrix **P** that represents the profit per unit of each type of product.

c) Find the product **PM**.

d) State what the entries of **PM** represent.

37. *Profit.* In Exercise 35, suppose that Anja's profits on one dozen chocolate chip, oatmeal, and peanut butter cookies are $7.50, $4.80, and $6.25, respectively.

a) Write a row matrix **P** that represents this information.

b) Use the matrices **S** and **P** to find Karin's total profit from each coffee shop.

38. *Production Cost.* In Exercise 36, suppose that the manufacturer's production costs for each unit of exterior plywood, interior plywood, and fiberboard are $20, $25, and $15, respectively.

a) Write a row matrix **C** that represents this information.

b) Use the matrices **M** and **C** to find the total production cost for the products shipped to each distributor.

Write a matrix equation equivalent to the system of equations.

39. $2x - 3y = 7,$
$\quad x + 5y = -6$

40. $-x + y = 3,$
$\quad 5x - 4y = 16$

41. $x + y - 2z = 6,$
$\quad 3x - y + z = 7,$
$\quad 2x + 5y - 3z = 8$

42. $3x - y + z = 1,$
$\quad x + 2y - z = 3,$
$\quad 4x + 3y - 2z = 11$

43. $3x - 2y + 4z = 17,$
$\quad 2x + y - 5z = 13$

44. $3x + 2y + 5z = 9,$
$\quad 4x - 3y + 2z = 10$

45. $-4w + x - y + 2z = 12,$
$\quad w + 2x - y - z = 0,$
$\quad -w + x + 4y - 3z = 1,$
$\quad 2w + 3x + 5y - 7z = 9$

46. $12w + 2x + 4y - 5z = 2,$
$\quad -w + 4x - y + 12z = 5,$
$\quad 2w - x + 4y = 13,$
$\quad 2x + 10y + z = 5$

> ## Skill Maintenance

In Exercises 47–50:

a) *Find the vertex.* **[3.3]**
b) *Find the axis of symmetry.* **[3.3]**
c) *Determine whether there is a maximum or a minimum value and find that value.* **[3.3]**
d) *Graph the function.* **[3.3]**

47. $f(x) = x^2 - x - 6$

48. $f(x) = 2x^2 - 5x - 3$

49. $f(x) = -x^2 - 3x + 2$

50. $f(x) = -3x^2 + 4x + 4$

> ## Synthesis

For Exercises 51–54, let

$$A = \begin{bmatrix} -1 & 0 \\ 2 & 1 \end{bmatrix} \quad \text{and} \quad B = \begin{bmatrix} 1 & -1 \\ 0 & 2 \end{bmatrix}.$$

51. Show that
$$(A + B)(A - B) \neq A^2 - B^2,$$
where
$$A^2 = AA \quad \text{and} \quad B^2 = BB.$$

52. Show that
$$(A + B)(A + B) \neq A^2 + 2AB + B^2.$$

53. Show that
$$(A + B)(A - B) = A^2 + BA - AB - B^2.$$

54. Show that
$$(A + B)(A + B) = A^2 + BA + AB + B^2.$$

Mid-Chapter Mixed Review

Determine whether the statement is true or false.

1. For a system of two linear equations in two variables, if the graphs of the equations are parallel lines, then the system of equations has infinitely many solutions. **[9.1]**

2. One of the properties of a matrix written in row-echelon form is that all the rows consisting entirely of 0's are at the bottom of the matrix. **[9.3]**

3. We can multiply two matrices only when the number of columns in the first matrix is equal to the number of rows in the second matrix. **[9.4]**

4. Addition of matrices is not commutative. **[9.4]**

Solve. **[9.1], [9.2]**

5. $2x + y = -4,$
$\quad x = y - 5$

6. $x + y = 4,$
$\quad y = 2 - x$

7. $2x - 3y = 8,$
$\quad 3x + 2y = -1$

8. $x - 3y = 1,$
$\quad 6y = 2x - 2$

9. $x + 2y + 3z = 4,$
$\quad x - 2y + \ z = 2,$
$\quad 2x - 6y + 4z = 7$

10. *e-Commerce.* computerwarehouse.com charges $8 to ship orders up to 10 lb, $12 for orders from 10 lb up to 15 lb, and $15 for orders of 15 lb or more. One day, shipping charges for 150 orders totaled $1620. The number of orders under 10 lb was three times the number of orders weighing 15 lb or more. Find the number of packages shipped at each rate. **[9.2]**

Solve the system of equations using Gaussian elimination or Gauss–Jordan elimination. **[9.3]**

11. $2x + \ y = 5,$
$\quad 3x + 2y = 6$

12. $3x + 2y - 3z = -2,$
$\quad 2x + 3y + 2z = -2,$
$\quad x + 4y + 4z = 1$

For Exercises 13–20, let

$$A = \begin{bmatrix} 3 & -1 \\ 5 & 4 \end{bmatrix}, \quad B = \begin{bmatrix} -2 & 6 \\ 1 & -3 \end{bmatrix}, \quad C = \begin{bmatrix} -4 & 1 & -1 \\ 2 & 3 & -2 \end{bmatrix}, \quad and \quad D = \begin{bmatrix} -2 & 3 & 0 \\ 1 & -1 & 2 \\ -3 & 4 & 1 \end{bmatrix}.$$

Find each of the following. **[9.4]**

13. $A + B$

14. $B - A$

15. $4D$

16. $2A + 3B$

17. AB

18. BA

19. BC

20. DC

21. Write a matrix equation equivalent to the following system of equations: **[9.4]**

$\quad 2x - \ y + 3z = 7,$
$\quad \ x + 2y - \ z = 3,$
$\quad 3x - 4y + 2z = 5.$

COLLABORATIVE DISCUSSION AND WRITING

22. Explain in your own words when using the elimination method for solving a system of equations is preferable to using the substitution method. **[9.1]**

23. Given two linear equations in three variables,

$\quad Ax + By + Cz = D \quad$ and $\quad Ex + Fy + Gz = H$

explain how you could find a third equation such that the system contains dependent equations. **[9.2]**

24. Explain in your own words why the augmented matrix below represents a system of dependent equations. **[9.3]**

$$\begin{bmatrix} 1 & -3 & 2 & | & -5 \\ 0 & 1 & -4 & | & 8 \\ 0 & 0 & 0 & | & 0 \end{bmatrix}$$

25. Is it true that if $AB = 0$, for matrices A and B, then $A = 0$ or $B = 0$? Why or why not? **[9.4]**

9.5 Inverses of Matrices

> Find the inverse of a square matrix, if it exists.
> Use inverses of matrices to solve systems of equations.

In this section, we continue our study of matrix algebra, finding the **multiplicative inverse**, or simply **inverse**, of a square matrix, if it exists. Then we use such inverses to solve systems of equations.

> The Identity Matrix

Recall that, for real numbers, $a \cdot 1 = 1 \cdot a = a$; 1 is the multiplicative identity. A multiplicative identity matrix is very similar to the number 1.

IDENTITY MATRIX

For any positive integer n, the $n \times n$ **identity matrix** is an $n \times n$ matrix with 1's on the main diagonal and 0's elsewhere and is denoted by

$$\mathbf{I} = \begin{bmatrix} 1 & 0 & 0 & \cdots & 0 \\ 0 & 1 & 0 & \cdots & 0 \\ 0 & 0 & 1 & \cdots & 0 \\ \vdots & \vdots & \vdots & & \vdots \\ 0 & 0 & 0 & \cdots & 1 \end{bmatrix}.$$

Then $\mathbf{AI} = \mathbf{IA} = \mathbf{A}$, for any $n \times n$ matrix $\mathbf{A}$.

EXAMPLE 1 For

$$\mathbf{A} = \begin{bmatrix} 4 & -7 \\ -3 & 2 \end{bmatrix} \quad \text{and} \quad \mathbf{I} = \begin{bmatrix} 1 & 0 \\ 0 & 1 \end{bmatrix},$$

find each of the following.

a) AI   **b) IA**

Solution

a) $\mathbf{AI} = \begin{bmatrix} 4 & -7 \\ -3 & 2 \end{bmatrix}\begin{bmatrix} 1 & 0 \\ 0 & 1 \end{bmatrix}$

$= \begin{bmatrix} 4 \cdot 1 - 7 \cdot 0 & 4 \cdot 0 - 7 \cdot 1 \\ -3 \cdot 1 + 2 \cdot 0 & -3 \cdot 0 + 2 \cdot 1 \end{bmatrix} = \begin{bmatrix} 4 & -7 \\ -3 & 2 \end{bmatrix} = \mathbf{A}$

b) $\mathbf{IA} = \begin{bmatrix} 1 & 0 \\ 0 & 1 \end{bmatrix}\begin{bmatrix} 4 & -7 \\ -3 & 2 \end{bmatrix}$

$= \begin{bmatrix} 1 \cdot 4 + 0(-3) & 1(-7) + 0 \cdot 2 \\ 0 \cdot 4 + 1(-3) & 0(-7) + 1 \cdot 2 \end{bmatrix} = \begin{bmatrix} 4 & -7 \\ -3 & 2 \end{bmatrix} = \mathbf{A}$

These products can also be found using a graphing calculator after $\mathbf{A}$ and $\mathbf{I}$ have been entered.

[A] [I]

$\begin{bmatrix} 4 & -7 \\ -3 & 2 \end{bmatrix}$

[I] [A]

$\begin{bmatrix} 4 & -7 \\ -3 & 2 \end{bmatrix}$

❯ The Inverse of a Matrix

Recall that for every nonzero real number a, there is a multiplicative inverse $1/a$, o a^{-1}, such that $a \cdot a^{-1} = a^{-1} \cdot a = 1$. The multiplicative inverse of a matrix behave in a similar manner.

> **INVERSE OF A MATRIX**
>
> For an $n \times n$ matrix $\mathbf{A}$, if there is a matrix $\mathbf{A}^{-1}$ for which
> $\mathbf{A}^{-1} \cdot \mathbf{A} = \mathbf{I} = \mathbf{A} \cdot \mathbf{A}^{-1}$, then $\mathbf{A}^{-1}$ is the **inverse** of $\mathbf{A}$.

We read $\mathbf{A}^{-1}$ as "$\mathbf{A}$ inverse." Note that not every matrix has an inverse.

EXAMPLE 2 Verify that

$$\mathbf{B} = \begin{bmatrix} 4 & -3 \\ 3 & -2 \end{bmatrix} \text{ is the inverse of } \mathbf{A} = \begin{bmatrix} -2 & 3 \\ -3 & 4 \end{bmatrix}.$$

Solution We show that $\mathbf{BA} = \mathbf{I} = \mathbf{AB}$.

$$\mathbf{BA} = \begin{bmatrix} 4 & -3 \\ 3 & -2 \end{bmatrix} \begin{bmatrix} -2 & 3 \\ -3 & 4 \end{bmatrix} = \begin{bmatrix} 1 & 0 \\ 0 & 1 \end{bmatrix} = \mathbf{I}$$

$$\mathbf{AB} = \begin{bmatrix} -2 & 3 \\ -3 & 4 \end{bmatrix} \begin{bmatrix} 4 & -3 \\ 3 & -2 \end{bmatrix} = \begin{bmatrix} 1 & 0 \\ 0 & 1 \end{bmatrix} = \mathbf{I}$$

❯ *Now Try Exercise 1.*

We can find the inverse of a square matrix, if it exists, by using row-equivalent operations as in the Gauss–Jordan elimination method. For example, consider the matrix

$$\mathbf{A} = \begin{bmatrix} -2 & 3 \\ -3 & 4 \end{bmatrix}.$$

To find its inverse, we first form an **augmented matrix** consisting of $\mathbf{A}$ on the lef side and the 2×2 identity matrix on the right side:

$$\begin{bmatrix} -2 & 3 & | & 1 & 0 \\ -3 & 4 & | & 0 & 1 \end{bmatrix}.$$

The 2×2 The 2×2
matrix $\mathbf{A}$ identity matrix

Then we attempt to transform the augmented matrix to one of the form

$$\begin{bmatrix} 1 & 0 & | & a & b \\ 0 & 1 & | & c & d \end{bmatrix}.$$

The 2×2 The matrix $\mathbf{A}^{-1}$
identity matrix

If we can do this, the matrix on the right, $\begin{bmatrix} a & b \\ c & d \end{bmatrix}$, is $\mathbf{A}^{-1}$.

EXAMPLE 3 Find $\mathbf{A}^{-1}$, where

$$\mathbf{A} = \begin{bmatrix} -2 & 3 \\ -3 & 4 \end{bmatrix}.$$

Solution First, we write the augmented matrix. Then we transform it to the desired form.

$$\begin{bmatrix} -2 & 3 & | & 1 & 0 \\ -3 & 4 & | & 0 & 1 \end{bmatrix}$$

$$\begin{bmatrix} 1 & -\frac{3}{2} & | & -\frac{1}{2} & 0 \\ -3 & 4 & | & 0 & 1 \end{bmatrix} \qquad \text{New row 1} = -\tfrac{1}{2}(\text{row 1})$$

$$\begin{bmatrix} 1 & -\frac{3}{2} & | & -\frac{1}{2} & 0 \\ 0 & -\frac{1}{2} & | & -\frac{3}{2} & 1 \end{bmatrix} \qquad \text{New row 2} = 3(\text{row 1}) + \text{row 2}$$

$$\begin{bmatrix} 1 & -\frac{3}{2} & | & -\frac{1}{2} & 0 \\ 0 & 1 & | & 3 & -2 \end{bmatrix} \qquad \text{New row 2} = -2(\text{row 2})$$

$$\begin{bmatrix} 1 & 0 & | & 4 & -3 \\ 0 & 1 & | & 3 & -2 \end{bmatrix} \qquad \text{New row 1} = \tfrac{3}{2}(\text{row 2}) + \text{row 1}$$

Thus,

$$\mathbf{A}^{-1} = \begin{bmatrix} 4 & -3 \\ 3 & -2 \end{bmatrix},$$

which we verified in Example 2.

The $\boxed{x^{-1}}$ key on a graphing calculator can also be used to find the inverse of a matrix.

> *Now Try Exercise 5.*

[A]$^{-1}$

$$\begin{bmatrix} 4 & -3 \\ 3 & -2 \end{bmatrix}$$

EXAMPLE 4 Find $\mathbf{A}^{-1}$, where

$$\mathbf{A} = \begin{bmatrix} 1 & 2 & -1 \\ 3 & 5 & 3 \\ 2 & 4 & 3 \end{bmatrix}.$$

Solution First, we write the augmented matrix. Then we transform it to the desired form.

$$\begin{bmatrix} 1 & 2 & -1 & | & 1 & 0 & 0 \\ 3 & 5 & 3 & | & 0 & 1 & 0 \\ 2 & 4 & 3 & | & 0 & 0 & 1 \end{bmatrix}$$

$$\begin{bmatrix} 1 & 2 & -1 & | & 1 & 0 & 0 \\ 0 & -1 & 6 & | & -3 & 1 & 0 \\ 0 & 0 & 5 & | & -2 & 0 & 1 \end{bmatrix} \qquad \begin{matrix} \text{New row 2} = -3(\text{row 1}) + \text{row 2} \\ \text{New row 3} = -2(\text{row 1}) + \text{row 3} \end{matrix}$$

$$\begin{bmatrix} 1 & 2 & -1 & | & 1 & 0 & 0 \\ 0 & -1 & 6 & | & -3 & 1 & 0 \\ 0 & 0 & 1 & | & -\frac{2}{5} & 0 & \frac{1}{5} \end{bmatrix} \qquad \text{New row 3} = \tfrac{1}{5}(\text{row 3})$$

$$\begin{bmatrix} 1 & 2 & 0 & | & \frac{3}{5} & 0 & \frac{1}{5} \\ 0 & -1 & 0 & | & -\frac{3}{5} & 1 & -\frac{6}{5} \\ 0 & 0 & 1 & | & -\frac{2}{5} & 0 & \frac{1}{5} \end{bmatrix} \qquad \begin{matrix} \text{New row 1} = \text{row 3} + \text{row 1} \\ \text{New row 2} = -6(\text{row 3}) + \text{row 2} \end{matrix}$$

$$\begin{bmatrix} 1 & 0 & 0 & | & -\frac{3}{5} & 2 & -\frac{11}{5} \\ 0 & -1 & 0 & | & -\frac{3}{5} & 1 & -\frac{6}{5} \\ 0 & 0 & 1 & | & -\frac{2}{5} & 0 & \frac{1}{5} \end{bmatrix} \qquad \text{New row 1} = 2(\text{row 2}) + \text{row 1}$$

$$\begin{bmatrix} 1 & 0 & 0 & | & -\frac{3}{5} & 2 & -\frac{11}{5} \\ 0 & 1 & 0 & | & \frac{3}{5} & -1 & \frac{6}{5} \\ 0 & 0 & 1 & | & -\frac{2}{5} & 0 & \frac{1}{5} \end{bmatrix} \qquad \text{New row 2} = -1(\text{row 2})$$

Thus,

$$\mathbf{A}^{-1} = \begin{bmatrix} -\frac{3}{5} & 2 & -\frac{11}{5} \\ \frac{3}{5} & -1 & \frac{6}{5} \\ -\frac{2}{5} & 0 & \frac{1}{5} \end{bmatrix}.$$

> **Now Try Exercise 9.**

If a matrix has an inverse, we say that it is **invertible**, or **nonsingular**. When we cannot obtain the identity matrix on the left using the Gauss–Jordan method, then no inverse exists. This occurs when we obtain a row consisting entirely of 0's in either of the two matrices in the augmented matrix. In this case, we say that **A** is a **singular matrix**.

When we try to find the inverse of a noninvertible, or singular, matrix using a graphing calculator, the calculator returns an error message similar to ERR: SINGULAR MATRIX.

> ## Solving Systems of Equations

MATRIX EQUATIONS

REVIEW SECTION 9.4.

We can write a system of n linear equations in n variables as a matrix equation $\mathbf{AX} = \mathbf{B}$. If **A** has an inverse, then the system of equations has a unique solution that can be found by solving for **X**, as follows:

$$\mathbf{AX} = \mathbf{B}$$

$$\mathbf{A}^{-1}(\mathbf{AX}) = \mathbf{A}^{-1}\mathbf{B} \qquad \text{Multiplying by } \mathbf{A}^{-1} \text{ on the left on both sides}$$

$$(\mathbf{A}^{-1}\mathbf{A})\mathbf{X} = \mathbf{A}^{-1}\mathbf{B} \qquad \text{Using the associative property of matrix multiplication}$$

$$\mathbf{IX} = \mathbf{A}^{-1}\mathbf{B} \qquad \mathbf{A}^{-1}\mathbf{A} = \mathbf{I}$$

$$\mathbf{X} = \mathbf{A}^{-1}\mathbf{B}. \qquad \mathbf{IX} = \mathbf{X}$$

MATRIX SOLUTIONS OF SYSTEMS OF EQUATIONS

For a system of n linear equations in n variables, $\mathbf{AX} = \mathbf{B}$, if **A** is an invertible matrix, then the unique solution of the system is given by

$$\mathbf{X} = \mathbf{A}^{-1}\mathbf{B}.$$

Since matrix multiplication is not commutative in general, care must be taken to multiply *on the left* by $\mathbf{A}^{-1}$.

EXAMPLE 5   Use an inverse matrix to solve the following system of equations:

$$-2x + 3y = 4,$$
$$-3x + 4y = 5.$$

Solution   We write an equivalent matrix equation, $\mathbf{AX} = \mathbf{B}$:

$$\begin{bmatrix} -2 & 3 \\ -3 & 4 \end{bmatrix} \cdot \begin{bmatrix} x \\ y \end{bmatrix} = \begin{bmatrix} 4 \\ 5 \end{bmatrix}$$
$$\mathbf{A} \quad \cdot \quad \mathbf{X} \ = \ \mathbf{B}$$

In Example 3, we found that

$$\mathbf{A}^{-1} = \begin{bmatrix} 4 & -3 \\ 3 & -2 \end{bmatrix}.$$

We also verified this in Example 2. Now we have

$$\mathbf{X} = \mathbf{A}^{-1}\mathbf{B}$$

$$\begin{bmatrix} x \\ y \end{bmatrix} = \begin{bmatrix} 4 & -3 \\ 3 & -2 \end{bmatrix} \begin{bmatrix} 4 \\ 5 \end{bmatrix} = \begin{bmatrix} 1 \\ 2 \end{bmatrix}.$$

The solution of the system of equations is $(1, 2)$.

To use a graphing calculator to solve this system of equations, we enter **A** and **B** and then enter the notation $\mathbf{A}^{-1}\mathbf{B}$ on the home screen.

> *Now Try Exercises 25 and 35.*

9.5 Exercise Set

Determine whether **B** *is the inverse of* **A**.

1. $\mathbf{A} = \begin{bmatrix} 1 & -3 \\ -2 & 7 \end{bmatrix},\quad \mathbf{B} = \begin{bmatrix} 7 & 3 \\ 2 & 1 \end{bmatrix}$

2. $\mathbf{A} = \begin{bmatrix} 3 & 2 \\ 4 & 3 \end{bmatrix},\quad \mathbf{B} = \begin{bmatrix} 3 & -2 \\ -4 & 3 \end{bmatrix}$

3. $\mathbf{A} = \begin{bmatrix} -1 & -1 & 6 \\ 1 & 0 & -2 \\ 1 & 0 & -3 \end{bmatrix},\quad \mathbf{B} = \begin{bmatrix} 2 & 3 & 2 \\ 3 & 3 & 4 \\ 1 & 1 & 1 \end{bmatrix}$

4. $\mathbf{A} = \begin{bmatrix} -2 & 0 & -3 \\ 5 & 1 & 7 \\ -3 & 0 & 4 \end{bmatrix},\quad \mathbf{B} = \begin{bmatrix} 4 & 0 & -3 \\ 1 & 1 & 1 \\ -3 & 0 & 2 \end{bmatrix}$

Use the Gauss–Jordan method to find $\mathbf{A}^{-1}$, *if it exists.*
Check your answers by using a graphing calculator to find
$\mathbf{A}^{-1}\mathbf{A}$ *and* $\mathbf{A}\mathbf{A}^{-1}$.

5. $\mathbf{A} = \begin{bmatrix} 3 & 2 \\ 5 & 3 \end{bmatrix}$

6. $\mathbf{A} = \begin{bmatrix} 3 & 5 \\ 1 & 2 \end{bmatrix}$

7. $\mathbf{A} = \begin{bmatrix} 6 & 9 \\ 4 & 6 \end{bmatrix}$

8. $\mathbf{A} = \begin{bmatrix} -4 & -6 \\ 2 & 3 \end{bmatrix}$

9. $\mathbf{A} = \begin{bmatrix} 3 & 1 & 0 \\ 1 & 1 & 1 \\ 1 & -1 & 2 \end{bmatrix}$

10. $\mathbf{A} = \begin{bmatrix} 1 & 0 & 1 \\ 2 & 1 & 0 \\ 1 & -1 & 1 \end{bmatrix}$

11. $\mathbf{A} = \begin{bmatrix} 1 & -4 & 8 \\ 1 & -3 & 2 \\ 2 & -7 & 10 \end{bmatrix}$

12. $\mathbf{A} = \begin{bmatrix} -2 & 5 & 3 \\ 4 & -1 & 3 \\ 7 & -2 & 5 \end{bmatrix}$

Use a graphing calculator to find $\mathbf{A}^{-1}$, *if it exists.*

13. $\mathbf{A} = \begin{bmatrix} 4 & -3 \\ 1 & -2 \end{bmatrix}$

14. $\mathbf{A} = \begin{bmatrix} 0 & -1 \\ 1 & 0 \end{bmatrix}$

15. $\mathbf{A} = \begin{bmatrix} 2 & 3 & 2 \\ 3 & 3 & 4 \\ -1 & -1 & -1 \end{bmatrix}$

16. $\mathbf{A} = \begin{bmatrix} 1 & 2 & 3 \\ 2 & -1 & -2 \\ -1 & 3 & 3 \end{bmatrix}$

17. $\mathbf{A} = \begin{bmatrix} 1 & 2 & -1 \\ -2 & 0 & 1 \\ 1 & -1 & 0 \end{bmatrix}$

18. $\mathbf{A} = \begin{bmatrix} 7 & -1 & -9 \\ 2 & 0 & -4 \\ -4 & 0 & 6 \end{bmatrix}$

19. $\mathbf{A} = \begin{bmatrix} 1 & 3 & -1 \\ 0 & 2 & -1 \\ 1 & 1 & 0 \end{bmatrix}$ **20.** $\mathbf{A} = \begin{bmatrix} -1 & 0 & -1 \\ -1 & 1 & 0 \\ 0 & 1 & 1 \end{bmatrix}$

21. $\mathbf{A} = \begin{bmatrix} 1 & 2 & 3 & 4 \\ 0 & 1 & 3 & -5 \\ 0 & 0 & 1 & -2 \\ 0 & 0 & 0 & -1 \end{bmatrix}$

22. $\mathbf{A} = \begin{bmatrix} -2 & -3 & 4 & 1 \\ 0 & 1 & 1 & 0 \\ 0 & 4 & -6 & 1 \\ -2 & -2 & 5 & 1 \end{bmatrix}$

23. $\mathbf{A} = \begin{bmatrix} 1 & -14 & 7 & 38 \\ -1 & 2 & 1 & -2 \\ 1 & 2 & -1 & -6 \\ 1 & -2 & 3 & 6 \end{bmatrix}$

24. $\mathbf{A} = \begin{bmatrix} 10 & 20 & -30 & 15 \\ 3 & -7 & 14 & -8 \\ -7 & -2 & -1 & 2 \\ 4 & 4 & -3 & 1 \end{bmatrix}$

In Exercises 25–28, a system of equations is given, together with the inverse of the coefficient matrix. Use the inverse of the coefficient matrix to solve the system of equations.

25. $11x + 3y = -4,$
$\quad 7x + 2y = 5;$ $\quad \mathbf{A}^{-1} = \begin{bmatrix} 2 & -3 \\ -7 & 11 \end{bmatrix}$

26. $8x + 5y = -6,$
$\quad 5x + 3y = 2;$ $\quad \mathbf{A}^{-1} = \begin{bmatrix} -3 & 5 \\ 5 & -8 \end{bmatrix}$

27. $3x + y \quad\;\; = 2,$
$\quad 2x - y + 2z = -5,$ $\quad \mathbf{A}^{-1} = \dfrac{1}{9}\begin{bmatrix} 3 & 1 & -2 \\ 0 & -3 & 6 \\ -3 & 2 & 5 \end{bmatrix}$
$\quad\; x + y + \;\; z = 5;$

28. $\quad\;\; y - \;\; z = -4,$
$\quad 4x + y \quad\;\; = -3,$ $\quad \mathbf{A}^{-1} = \dfrac{1}{5}\begin{bmatrix} -3 & 2 & -1 \\ 12 & -3 & 4 \\ 7 & -3 & 4 \end{bmatrix}$
$\quad 3x - y + 3z = 1;$

Solve the system of equations using the inverse of the coefficient matrix of the equivalent matrix equation.

29. $4x + 3y = 2,$
$\quad\; x - 2y = 6$

30. $2x - 3y = 7,$
$\quad 4x + \;\; y = -7$

31. $5x + \;\; y = 2,$
$\quad 3x - 2y = -4$

32. $\quad\; x - 6y = 5,$
$\quad -x + 4y = -5$

33. $x \quad\quad\; + z = 1,$
$\quad 2x + y \quad\;\; = 3,$
$\quad\; x - y + z = 4$

34. $\quad\quad\; x + 2y + 3z = -1,$
$\quad 2x - 3y + 4z = 2,$
$\quad -3x + 5y - 6z = 4$

35. $2x + 3y + 4z = 2,$
$\quad\; x - 4y + 3z = 2,$
$\quad 5x + \;\; y + \;\; z = -4$

36. $x + \;\; y \quad\quad\; = 2,$
$\quad 3x \quad\quad\; + 2z = 5,$
$\quad 2x + 3y - 3z = 9$

37. $2w - 3x + 4y - 5z = 0,$
$\quad 3w - 2x + 7y - 3z = 2,$
$\quad\; w + \;\; x - \;\; y + \;\; z = 1,$
$\quad -w - 3x - 6y + 4z = 6$

38. $5w - 4x + 3y - 2z = -6,$
$\quad\; w + 4x - 2y + 3z = -5,$
$\quad 2w - 3x + 6y - 9z = 14,$
$\quad 3w - 5x + 2y - 4z = -3$

39. *Cranberry Production.* In 2012, a total of 660 million lb of cranberries were produced in Wisconsin and Massachusetts. Wisconsin grew 240 million lb more than Massachusetts grew. (*Source*: U.S. Department of Agriculture) How many pounds of cranberries were grown in each state?

40. *Lunch Cost.* Coworkers Jan and Richard purchase lunch from a food truck. Jan buys 1 beef taco and 2 fruit cups for $5.25. Richard buys 3 beef tacos and 1 fruit cup for $8.25. Find the price of each item.

41. *Cost of Landscape Materials.* Cheryl's Landscaping bought 15 cubic yards of topsoil, 30 cubic yards of mulch, and 11 cubic yards of decorative rock for $1693. The next week, the business bought 8 cubic yards of topsoil, 19 cubic yards of mulch, and 7 cubic yards of decorative rock for $1030. Decorative rock costs $10 more per cubic yard than mulch. Find the price per cubic yard of each material.

42. *Investment.* Trevor receives $230 per year in simple interest from three investments totaling $8500. Part is invested at 2.2%, part at 2.65%, and the rest at 3.05%. There is $1500 more invested at 3.05% than at 2.2%. Find the amount invested at each rate.

❯ Skill Maintenance

Use synthetic division to find the function values. **[4.3]**

43. $f(x) = x^3 - 6x^2 + 4x - 8;$ find $f(-2)$

44. $f(x) = 2x^4 - x^3 + 5x^2 + 6x - 4;$ find $f(3)$

Solve.

45. $2x^2 + x = 7$ **[3.2]**

46. $\dfrac{1}{x+1} - \dfrac{6}{x-1} = 1$ **[3.4]**

47. $\sqrt{2x+1} - 1 = \sqrt{2x-4}$ **[3.2], [3.4]**

48. $x - \sqrt{x} - 6 = 0$ **[3.2], [3.4]**

Factor the polynomial $f(x)$. **[4.3]**

49. $f(x) = x^3 - 3x^2 - 6x + 8$

50. $f(x) = x^4 + 2x^3 - 16x^2 - 2x + 15$

❯ Synthesis

State the conditions under which $\mathbf{A}^{-1}$ exists. Then find a formula for $\mathbf{A}^{-1}$.

51. $\mathbf{A} = \begin{bmatrix} x \end{bmatrix}$

52. $\mathbf{A} = \begin{bmatrix} x & 0 \\ 0 & y \end{bmatrix}$

53. $\mathbf{A} = \begin{bmatrix} 0 & 0 & x \\ 0 & y & 0 \\ z & 0 & 0 \end{bmatrix}$

54. $\mathbf{A} = \begin{bmatrix} x & 1 & 1 & 1 \\ 0 & y & 0 & 0 \\ 0 & 0 & z & 0 \\ 0 & 0 & 0 & w \end{bmatrix}$

9.6

Determinants and Cramer's Rule

❯ Evaluate determinants of square matrices.

❯ Use Cramer's rule to solve systems of equations.

❯ Determinants of Square Matrices

With every square matrix, we associate a number called its *determinant*.

DETERMINANT OF A 2 × 2 MATRIX

The **determinant** of the matrix $\begin{bmatrix} a & c \\ b & d \end{bmatrix}$ is denoted $\begin{vmatrix} a & c \\ b & d \end{vmatrix}$ and is defined as

$$\begin{vmatrix} a & c \\ b & d \end{vmatrix} = ad - bc.$$

EXAMPLE 1   Evaluate: $\begin{vmatrix} \sqrt{2} & -3 \\ -4 & -\sqrt{2} \end{vmatrix}$.

Solution

$$\begin{vmatrix} \sqrt{2} & -3 \\ -4 & \sqrt{2} \end{vmatrix} \qquad \text{The arrows indicate the products involved.}$$

$$= \sqrt{2}(-\sqrt{2}) - (-4)(-3)$$
$$= -2 - 12$$
$$= -14$$

❯ **Now Try Exercise 5.**

We now consider a way to evaluate determinants of square matrices of order 3 × 3 or higher.

❯ Evaluating Determinants Using Cofactors

Often we first find minors and cofactors of matrices in order to evaluate determinants.

MINOR

For a square matrix $\mathbf{A} = [a_{ij}]$, the **minor** M_{ij} of an entry a_{ij} is the determinant of the matrix formed by deleting the ith row and the jth column of $\mathbf{A}$.

EXAMPLE 2   For the matrix

$$\mathbf{A} = [a_{ij}] = \begin{bmatrix} -8 & 0 & 6 \\ 4 & -6 & 7 \\ -1 & -3 & 5 \end{bmatrix},$$

find each of the following.

a) M_{11}

b) M_{23}

Solution

a) For M_{11}, we delete the first row and the first column and find the determinant of the 2×2 matrix formed by the remaining entries.

$$\begin{bmatrix} -8 & 0 & 6 \\ 4 & -6 & 7 \\ -1 & -3 & 5 \end{bmatrix} \qquad M_{11} = \begin{vmatrix} -6 & 7 \\ -3 & 5 \end{vmatrix}$$

$$= (-6) \cdot 5 - (-3) \cdot 7$$
$$= -30 - (-21)$$
$$= -30 + 21$$
$$= -9$$

b) For M_{23}, we delete the second row and the third column and find the determinant of the 2×2 matrix formed by the remaining entries.

$$\begin{bmatrix} -8 & 0 & 6 \\ 4 & -6 & 7 \\ -1 & -3 & 5 \end{bmatrix} \qquad M_{23} = \begin{vmatrix} -8 & 0 \\ -1 & -3 \end{vmatrix}$$

$$= -8(-3) - (-1)0$$
$$= 24$$

> **Now Try Exercise 9.**

COFACTOR

For a square matrix $\mathbf{A} = [a_{ij}]$, the **cofactor** A_{ij} of an entry a_{ij} is given by

$$A_{ij} = (-1)^{i+j} M_{ij},$$

where M_{ij} is the minor of a_{ij}.

EXAMPLE 3 For the matrix given in Example 2, find each of the following.

a) A_{11} **b)** A_{23}

Solution

a) In Example 2, we found that $M_{11} = -9$. Then

$$A_{11} = (-1)^{1+1} M_{11} = (1)(-9) = -9.$$

b) In Example 2, we found that $M_{23} = 24$. Then

$$A_{23} = (-1)^{2+3} M_{24} = (-1)(24) = -24.$$

> **Now Try Exercise 11.**

Note that minors and cofactors are *numbers*. They are *not matrices*.

Consider the matrix $\mathbf{A}$ given by

$$\mathbf{A} = \begin{bmatrix} a_{11} & a_{12} & a_{13} \\ a_{21} & a_{22} & a_{23} \\ a_{31} & a_{32} & a_{33} \end{bmatrix}.$$

The determinant of the matrix, denoted $|\mathbf{A}|$, can be found by multiplying each element of the first column by its cofactor and adding:

$$|\mathbf{A}| = a_{11}A_{11} + a_{21}A_{21} + a_{31}A_{31}.$$

Because

$$A_{11} = (-1)^{1+1}M_{11} = M_{11},$$
$$A_{21} = (-1)^{2+1}M_{21} = -M_{21},$$

and $A_{31} = (-1)^{3+1}M_{31} = M_{31},$

we can write

$$|\mathbf{A}| = a_{11} \cdot \begin{vmatrix} a_{22} & a_{23} \\ a_{32} & a_{33} \end{vmatrix} - a_{21} \cdot \begin{vmatrix} a_{12} & a_{13} \\ a_{32} & a_{33} \end{vmatrix} + a_{31} \cdot \begin{vmatrix} a_{12} & a_{13} \\ a_{22} & a_{23} \end{vmatrix}.$$

It can be shown that we can find $|\mathbf{A}|$ by choosing *any* row or column, multiplying each element in that row or column by its cofactor, and adding. This is called *expanding* across a row or down a column. We just expanded down the first column. We now define the determinant of a square matrix of any order.

DETERMINANT OF ANY SQUARE MATRIX

For any square matrix $\mathbf{A}$ of order $n \times n$ $(n > 1)$, we define the **determinant** of $\mathbf{A}$, denoted $|\mathbf{A}|$, as follows. Choose any row or column. Multiply each element in that row or column by its cofactor and add the results. The determinant of a 1×1 matrix is simply the element of the matrix. The value of a determinant will be the same no matter which row or column is chosen.

EXAMPLE 4   Evaluate $|\mathbf{A}|$ by expanding across the third row:

$$\mathbf{A} = \begin{bmatrix} -8 & 0 & 6 \\ 4 & -6 & 7 \\ -1 & -3 & 5 \end{bmatrix}.$$

Solution   We have

$$|\mathbf{A}| = (-1)A_{31} + (-3)A_{32} + 5A_{33}$$

$$= (-1)(-1)^{3+1} \cdot \begin{vmatrix} 0 & 6 \\ -6 & 7 \end{vmatrix} + (-3)(-1)^{3+2} \cdot \begin{vmatrix} -8 & 6 \\ 4 & 7 \end{vmatrix}$$

$$+ 5(-1)^{3+3} \cdot \begin{vmatrix} -8 & 0 \\ 4 & -6 \end{vmatrix}$$

$$= (-1) \cdot 1 \cdot [0 \cdot 7 - (-6)6] + (-3)(-1)[-8 \cdot 7 - 4 \cdot 6]$$

$$+ 5 \cdot 1 \cdot [-8(-6) - 4 \cdot 0]$$

$$= -[36] + 3[-80] + 5[48]$$

$$= -36 - 240 + 240 = -36.$$

The value of this determinant is -36 no matter which row or column we expand on.

> **Now Try Exercise 13.**

Determinants can also be evaluated using a graphing calculator.

EXAMPLE 5   Use a graphing calculator to evaluate $|\mathbf{A}|$:

$$\mathbf{A} = \begin{bmatrix} 1 & 6 & -1 \\ -3 & -5 & 3 \\ 0 & 4 & 2 \end{bmatrix}.$$

det ([A])	
	26

Solution   First, we enter $\mathbf{A}$. Then we select the determinant operation, det, from the MATRIX MATH menu and enter the name of the matrix, $\mathbf{A}$. The calculator will return the value of the determinant of the matrix, 26.

> **Now Try Exercise 17.**

❯ Cramer's Rule

Determinants can be used to solve systems of linear equations. Consider a system of two linear equations:

$$a_1x + b_1y = c_1,$$
$$a_2x + b_2y = c_2.$$

Solving this system using the elimination method, we obtain

$$x = \frac{c_1b_2 - c_2b_1}{a_1b_2 - a_2b_1} \quad \text{and} \quad y = \frac{a_1c_2 - a_2c_1}{a_1b_2 - a_2b_1}.$$

The numerators and the denominators of these expressions can be written as determinants:

$$x = \frac{\begin{vmatrix} c_1 & b_1 \\ c_2 & b_2 \end{vmatrix}}{\begin{vmatrix} a_1 & b_1 \\ a_2 & b_2 \end{vmatrix}} \quad \text{and} \quad y = \frac{\begin{vmatrix} a_1 & c_1 \\ a_2 & c_2 \end{vmatrix}}{\begin{vmatrix} a_1 & b_1 \\ a_2 & b_2 \end{vmatrix}}.$$

If we let

$$D = \begin{vmatrix} a_1 & b_1 \\ a_2 & b_2 \end{vmatrix}, \quad D_x = \begin{vmatrix} c_1 & b_1 \\ c_2 & b_2 \end{vmatrix}, \quad \text{and} \quad D_y = \begin{vmatrix} a_1 & c_1 \\ a_2 & c_2 \end{vmatrix},$$

we have

$$x = \frac{D_x}{D} \quad \text{and} \quad y = \frac{D_y}{D}.$$

This procedure for solving systems of equations is known as *Cramer's rule*.

CRAMER'S RULE FOR 2 × 2 SYSTEMS

The solution of the system of equations

$$a_1x + b_1y = c_1,$$
$$a_2x + b_2y = c_2$$

is given by

$$x = \frac{D_x}{D}, \quad y = \frac{D_y}{D},$$

where

$$D = \begin{vmatrix} a_1 & b_1 \\ a_2 & b_2 \end{vmatrix}, \quad D_x = \begin{vmatrix} c_1 & b_1 \\ c_2 & b_2 \end{vmatrix},$$

$$D_y = \begin{vmatrix} a_1 & c_1 \\ a_2 & c_2 \end{vmatrix}, \quad \text{and} \quad D \neq 0.$$

Note that the denominator D contains the coefficients of x and y, in the same position as in the original equations. For x, the numerator is obtained by replacing the x-coefficients in D (the a's) with the c's. For y, the numerator is obtained by replacing the y-coefficients in D (the b's) with the c's.

EXAMPLE 6 Solve using Cramer's rule:

$$2x + 5y = 7,$$
$$5x - 2y = -3.$$

Algebraic Solution

We have

$$x = \frac{\begin{vmatrix} 7 & 5 \\ -3 & -2 \end{vmatrix}}{\begin{vmatrix} 2 & 5 \\ 5 & -2 \end{vmatrix}} = \frac{7(-2) - (-3)5}{2(-2) - 5 \cdot 5}$$

$$= \frac{-14 + 15}{-4 - 25}$$

$$= \frac{1}{-29} = -\frac{1}{29};$$

$$y = \frac{\begin{vmatrix} 2 & 7 \\ 5 & -3 \end{vmatrix}}{\begin{vmatrix} 2 & 5 \\ 5 & -2 \end{vmatrix}} = \frac{2(-3) - 5 \cdot 7}{-29}$$

$$= \frac{-6 - 35}{-29} = \frac{-41}{-29} = \frac{41}{29}.$$

The solution is $\left(-\frac{1}{29}, \frac{41}{29} \right)$.

Graphical Solution

To use Cramer's rule to solve this system of equations on a graphing calculator, we first enter the matrices corresponding to D, D_x, and D_y. We enter

$$\mathbf{A} = \begin{bmatrix} 2 & 5 \\ 5 & -2 \end{bmatrix}, \quad \mathbf{B} = \begin{bmatrix} 7 & 5 \\ -3 & -2 \end{bmatrix},$$

and $\quad \mathbf{C} = \begin{bmatrix} 2 & 7 \\ 5 & -3 \end{bmatrix}.$

Then

$$x = \frac{\det(\mathbf{B})}{\det(\mathbf{A})} \quad \text{and} \quad y = \frac{\det(\mathbf{C})}{\det(\mathbf{A})}.$$

```
det ([B])/det([A]) ▶ Frac
                                      -1/29
det ([C])/det([A]) ▶ Frac
                                      41/29
```

The solution is $\left(-\frac{1}{29}, \frac{41}{29} \right)$.

> **Now Try Exercise 31.**

Cramer's rule works only when a system of equations has a unique solution. This occurs when $D \neq 0$. If $D = 0$ and D_x and D_y are also 0, then the equations are dependent. If $D = 0$ and D_x and/or D_y is not 0, then the system is inconsistent.

Cramer's rule can be extended to a system of n linear equations in n variables. We consider a 3 × 3 system.

CRAMER'S RULE FOR 3 × 3 SYSTEMS

The solution of the system of equations

$$a_1 x + b_1 y + c_1 z = d_1,$$
$$a_2 x + b_2 y + c_2 z = d_2,$$
$$a_3 x + b_3 y + c_3 z = d_3$$

is given by

$$x = \frac{D_x}{D}, \quad y = \frac{D_y}{D}, \quad z = \frac{D_z}{D},$$

where

$$D = \begin{vmatrix} a_1 & b_1 & c_1 \\ a_2 & b_2 & c_2 \\ a_3 & b_3 & c_3 \end{vmatrix}, \quad D_x = \begin{vmatrix} d_1 & b_1 & c_1 \\ d_2 & b_2 & c_2 \\ d_3 & b_3 & c_3 \end{vmatrix},$$

$$D_y = \begin{vmatrix} a_1 & d_1 & c_1 \\ a_2 & d_2 & c_2 \\ a_3 & d_3 & c_3 \end{vmatrix}, \quad D_z = \begin{vmatrix} a_1 & b_1 & d_1 \\ a_2 & b_2 & d_2 \\ a_3 & b_3 & d_3 \end{vmatrix}, \quad \text{and} \quad D \neq 0.$$

Note that the determinant D_x is obtained from D by replacing the x-coefficients with d_1, d_2, and d_3. D_y and D_z are obtained in a similar manner. As with a system of two equations, Cramer's rule cannot be used if $D = 0$. If $D = 0$ and D_x, D_y, and D_z are 0, then the equations are dependent. If $D = 0$ and at least one of D_x, D_y, or D_z is not 0, then the system is inconsistent.

EXAMPLE 7   Solve using Cramer's rule:

$$x - 3y + 7z = 13,$$
$$x + y + z = 1,$$
$$x - 2y + 3z = 4.$$

Solution   We have

$$D = \begin{vmatrix} 1 & -3 & 7 \\ 1 & 1 & 1 \\ 1 & -2 & 3 \end{vmatrix} = -10, \qquad D_x = \begin{vmatrix} 13 & -3 & 7 \\ 1 & 1 & 1 \\ 4 & -2 & 3 \end{vmatrix} = 20,$$

$$D_y = \begin{vmatrix} 1 & 13 & 7 \\ 1 & 1 & 1 \\ 1 & 4 & 3 \end{vmatrix} = -6, \qquad D_z = \begin{vmatrix} 1 & -3 & 13 \\ 1 & 1 & 1 \\ 1 & -2 & 4 \end{vmatrix} = -24.$$

Then

$$x = \frac{D_x}{D} = \frac{20}{-10} = -2, \qquad y = \frac{D_y}{D} = \frac{-6}{-10} = \frac{3}{5}, \qquad z = \frac{D_z}{D} = \frac{-24}{-10} = \frac{12}{5}.$$

The solution is $\left(-2, \frac{3}{5}, \frac{12}{5}\right)$.

In practice, it is not necessary to evaluate D_z. When we have found values for x and y, we can substitute them into one of the equations to find z.

> **Now Try Exercise 39.**

9.6   Exercise Set

Evaluate the determinant.

1. $\begin{vmatrix} 5 & 3 \\ -2 & -4 \end{vmatrix}$

2. $\begin{vmatrix} -8 & 6 \\ -1 & 2 \end{vmatrix}$

3. $\begin{vmatrix} 4 & -7 \\ -2 & 3 \end{vmatrix}$

4. $\begin{vmatrix} -9 & -6 \\ 5 & 4 \end{vmatrix}$

5. $\begin{vmatrix} -2 & -\sqrt{5} \\ -\sqrt{5} & 3 \end{vmatrix}$

6. $\begin{vmatrix} \sqrt{5} & -3 \\ 4 & 2 \end{vmatrix}$

7. $\begin{vmatrix} x & 4 \\ x & x^2 \end{vmatrix}$

8. $\begin{vmatrix} y^2 & -2 \\ y & 3 \end{vmatrix}$

Use the following matrix for Exercises 9–17:

$$\mathbf{A} = \begin{bmatrix} 7 & -4 & -6 \\ 2 & 0 & -3 \\ 1 & 2 & -5 \end{bmatrix}.$$

9. Find M_{11}, M_{32}, and M_{22}.

10. Find M_{13}, M_{31}, and M_{23}.

11. Find A_{11}, A_{32}, and A_{22}.

12. Find A_{13}, A_{31}, and A_{23}.

13. Evaluate $|\mathbf{A}|$ by expanding across the second row.

14. Evaluate $|\mathbf{A}|$ by expanding down the second column.

15. Evaluate $|\mathbf{A}|$ by expanding down the third column.

16. Evaluate $|\mathbf{A}|$ by expanding across the first row.

17. Use a graphing calculator to evaluate $|\mathbf{A}|$.

Use the following matrix for Exercises 18–24:

$$\mathbf{A} = \begin{bmatrix} 1 & 0 & 0 & -2 \\ 4 & 1 & 0 & 0 \\ 5 & 6 & 7 & 8 \\ -2 & -3 & -1 & 0 \end{bmatrix}.$$

18. Find M_{12} and M_{44}.

19. Find M_{41} and M_{33}.

20. Find A_{22} and A_{34}.

21. Find A_{24} and A_{43}.

22. Evaluate $|\mathbf{A}|$ by expanding down the third column.

23. Evaluate $|\mathbf{A}|$ by expanding across the first row.

24. Use a graphing calculator to evaluate $|\mathbf{A}|$.

Evaluate the determinant.

25. $\begin{vmatrix} 3 & 1 & 2 \\ -2 & 3 & 1 \\ 3 & 4 & -6 \end{vmatrix}$

26. $\begin{vmatrix} 3 & -2 & 1 \\ 2 & 4 & 3 \\ -1 & 5 & 1 \end{vmatrix}$

27. $\begin{vmatrix} x & 0 & -1 \\ 2 & x & x^2 \\ -3 & x & 1 \end{vmatrix}$

28. $\begin{vmatrix} x & 1 & -1 \\ x^2 & x & x \\ 0 & x & 1 \end{vmatrix}$

Solve using Cramer's rule.

29. $\begin{aligned} -2x + 4y &= 3, \\ 3x - 7y &= 1 \end{aligned}$

30. $\begin{aligned} 5x - 4y &= -3, \\ 7x + 2y &= 6 \end{aligned}$

31. $\begin{aligned} 2x - y &= 5, \\ x - 2y &= 1 \end{aligned}$

32. $\begin{aligned} 3x + 4y &= -2, \\ 5x - 7y &= 1 \end{aligned}$

33. $\begin{aligned} 2x + 9y &= -2, \\ 4x - 3y &= 3 \end{aligned}$

34. $\begin{aligned} 2x + 3y &= -1, \\ 3x + 6y &= -0.5 \end{aligned}$

35. $\begin{aligned} 2x + 5y &= 7, \\ 3x - 2y &= 1 \end{aligned}$

36. $\begin{aligned} 3x + 2y &= 7, \\ 2x + 3y &= -2 \end{aligned}$

37. $\begin{aligned} 3x + 2y - z &= 4, \\ 3x - 2y + z &= 5, \\ 4x - 5y - z &= -1 \end{aligned}$

38. $\begin{aligned} 3x - y + 2z &= 1, \\ x - y + 2z &= 3, \\ -2x + 3y + z &= 1 \end{aligned}$

39. $\begin{aligned} 3x + 5y - z &= -2, \\ x - 4y + 2z &= 13, \\ 2x + 4y + 3z &= 1 \end{aligned}$

40. $\begin{aligned} 3x + 2y + 2z &= 1, \\ 5x - y - 6z &= 3, \\ 2x + 3y + 3z &= 4 \end{aligned}$

41. $\begin{aligned} x - 3y - 7z &= 6, \\ 2x + 3y + z &= 9, \\ 4x + y &= 7 \end{aligned}$

42. $\begin{aligned} x - 2y - 3z &= 4, \\ 3x - 2z &= 8, \\ 2x + y + 4z &= 13 \end{aligned}$

43. $\begin{aligned} 6y + 6z &= -1, \\ 8x + 6z &= -1, \\ 4x + 9y &= 8 \end{aligned}$

44. $\begin{aligned} 3x + 5y &= 2, \\ 2x - 3z &= 7, \\ 4y + 2z &= -1 \end{aligned}$

❯ Skill Maintenance

Determine whether the function is one-to-one, and if it is, find a formula for $f^{-1}(x)$. **[5.1]**

45. $f(x) = 3x + 2$

46. $f(x) = x^2 - 4$

47. $f(x) = |x| + 3$

48. $f(x) = \sqrt[3]{x} + 1$

Simplify. Write answers in the form $a + bi$, where a and b are real numbers. **[3.1]**

49. $(3 - 4i) - (-2 - i)$

50. $(5 + 2i) + (1 - 4i)$

51. $(1 - 2i)(6 + 2i)$

52. $\dfrac{3 + i}{4 - 3i}$

❯ Synthesis

Solve.

53. $\begin{vmatrix} y & 2 \\ 3 & y \end{vmatrix} = y$

54. $\begin{vmatrix} x & -3 \\ -1 & x \end{vmatrix} \geq 0$

55. $\begin{vmatrix} 2 & x & 1 \\ 1 & 2 & -1 \\ 3 & 4 & -2 \end{vmatrix} = -6$

56. $\begin{vmatrix} m + 2 & -3 \\ m + 5 & -4 \end{vmatrix} = 3m - 5$

Rewrite the expression using a determinant. Answers may vary.

57. $a^2 + b^2$

58. $\frac{1}{2}h(a + b)$

59. $2\pi r^2 + 2\pi rh$

60. $x^2y^2 - Q^2$

Systems of Inequalities and Linear Programming

> Graph linear inequalities.

> Graph systems of linear inequalities.

> Solve linear programming problems.

A graph of an inequality is a drawing that represents its solutions. We have already seen that an inequality in one variable can be graphed on the number line. An inequality in two variables can be graphed on a coordinate plane.

SOLVE LINEAR INEQUALITIES

REVIEW SECTION 1.6.

❯ Graphs of Linear Inequalities

A statement like $5x - 4y < 20$ is a linear inequality in two variables.

LINEAR INEQUALITY IN TWO VARIABLES

A **linear inequality in two variables** is an inequality that can be written in the form

$$Ax + By < C,$$

where A, B, and C are real numbers and A and B are not both zero. The symbol $<$ may be replaced with $\leq$, $>$, or $\geq$.

A solution of a linear inequality in two variables is an ordered pair (x, y) for which the inequality is true. For example, $(1, 3)$ is a solution of $5x - 4y < 20$ because $5 \cdot 1 - 4 \cdot 3 < 20$, or $-7 < 20$, is true. On the other hand, $(2, -6)$ is not a solution of $5x - 4y < 20$ because $5 \cdot 2 - 4 \cdot (-6) \not< 20$, or $34 \not< 20$.

The **solution set** of an inequality is the set of all ordered pairs that make it true. The **graph of an inequality** represents its solution set.

EXAMPLE 1 Graph: $y < x + 3$.

Solution We begin by graphing the **related equation** $y = x + 3$. We use a dashed line because the inequality symbol is $<$. This indicates that the line itself is not in the solution set of the inequality.

Note that the line divides the coordinate plane into two regions called **half-planes**. One of these half-planes satisfies the inequality. Either *all* points in a half-plane are in the solution set of the inequality or *none* is.

To determine which half-plane satisfies the inequality, we try a test point in either region. The point $(0, 0)$ is usually a convenient choice so long as it does not lie on the line.

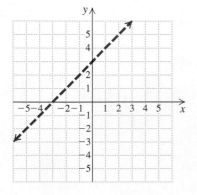

$$\begin{array}{c|c} y < x + 3 \\ \hline 0 \ ? \ 0 + 3 \\ 0 \ | \ 3 \end{array} \qquad \text{TRUE} \qquad 0 < 3 \text{ is true.}$$

Since $(0, 0)$ satisfies the inequality, so do all points in the half-plane that contains $(0, 0)$. We shade this region to show the solution set of the inequality.

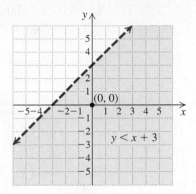

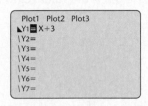

$y = x + 3$

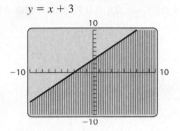

There are several ways to graph this inequality on a graphing calculator. One method is to first enter the related equation, $y = x + 3$, as shown at left. Then, after using a test point to determine which half-plane to shade as described above, we select the "shade below" graph style. Note that we must keep in mind that the line $y = x + 3$ is not included in the solution set. (Even if DOT mode or the dot graph style is selected, the line appears to be solid rather than dashed.)

We could also use the Shade option from the DRAW menu to graph this inequality. Some calculators have a pre-loaded application that can be used to graph an inequality. This application, Inequalz, is found on the APPS menu. Note that when this application is used, the inequality $y < x + 3$ is entered directly and the graph of the related equation appears as a dashed line.

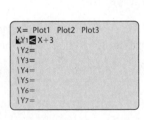

$y < x + 3$

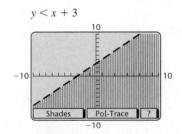

> **Now Try Exercise 13.**

In general, we use the following procedure to graph linear inequalities in two variables by hand.

To graph a linear inequality in two variables:

1. Replace the inequality symbol with an equals sign and graph this related equation. If the inequality symbol is $<$ or $>$, draw the line dashed. If the inequality symbol is $\leq$ or $\geq$, draw the line solid.

2. The graph consists of a half-plane on one side of the line and, if the line is solid, the line as well. To determine which half-plane to shade, test a point not on the line in the original inequality. If that point is a solution, shade the half-plane containing that point. If not, shade the opposite half-plane.

EXAMPLE 2 Graph: $3x + 4y \geq 12$.

Solution

1. First, we graph the related equation $3x + 4y = 12$. We use a solid line because the inequality symbol is $\geq$. This indicates that the line is included in the solution set.

2. To determine which half-plane to shade, we test a point in either region. We choose $(0, 0)$.

$$\frac{3x + 4y \geq 12}{3 \cdot 0 + 4 \cdot 0 \ ? \ 12}$$
$$0 \ | \ 12 \quad \text{FALSE} \qquad 0 \geq 12 \text{ is false.}$$

Because $(0, 0)$ is *not* a solution, all the points in the half-plane that does *not* contain $(0, 0)$ are solutions. We shade that region, as shown in the following figure.

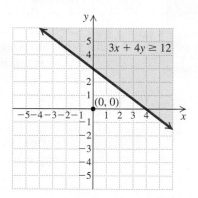

$$y = \frac{-3x + 12}{4}$$

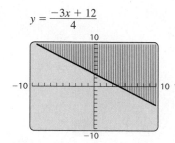

To graph this inequality on a graphing calculator using either the "shade above" graph style or the Shade option from the DRAW menu, we must first solve the related equation for y and enter $y = \dfrac{-3x + 12}{4}$. Similarly, to use the Inequalz application, we solve the inequality for y and enter $y \geq \dfrac{-3x + 12}{4}$.

 *Now Try Exercise 17.*

EXAMPLE 3 Graph $x > -3$ on a plane.

Solution

1. First, we graph the related equation $x = -3$. We use a dashed line because the inequality symbol is $>$. This indicates that the line is not included in the solution set.

2. The inequality tells us that all points (x, y) for which $x > -3$ are solutions. These are the points to the right of the line. We can also use a test point to determine the solutions. We choose $(5, 1)$.

$$\frac{x > -3}{5 \ ? \ -3} \quad \text{TRUE} \qquad 5 > -3 \text{ is true.}$$

Because $(5, 1)$ is a solution, we shade the region containing that point—that is, the region to the right of the dashed line.

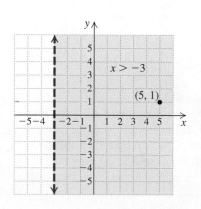

We can also graph this inequality on a graphing calculator that has the Inequalz application on the APPS menu.

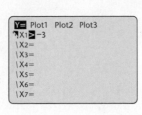

$x > -3$

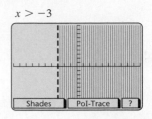

> *Now Try Exercise 23.*

EXAMPLE 4   Graph $y \le 4$ on a plane.

Solution

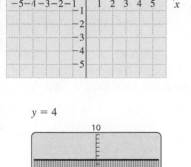

1. First, we graph the related equation $y = 4$. We use a solid line because the inequality symbol is $\le$.

2. The inequality tells us that all points (x, y) for which $y \le 4$ are solutions of the inequality. These are the points on or below the line. We can also use a test point to determine the solutions. We choose $(-2, 5)$.

$$\frac{y \le 4}{5 \,\,?\,\, 4}$$ **FALSE**   **$5 \le 4$ is false.**

Because $(-2, 5)$ is not a solution, we shade the half-plane that does not contain that point.

We can graph the inequality $y \le 4$ using Inequalz or by graphing $y = 4$ and then using the "shade below" graph style, as shown at left.

> *Now Try Exercise 25.*

❯ Systems of Linear Inequalities

A system of inequalities consists of two or more inequalities considered simultaneously. For example,

$$x + y \le 4,$$
$$x - y \ge 2$$

is a system of *two linear inequalities in two variables*.

A solution of a system of inequalities is an ordered pair that is a solution of each inequality in the system. To graph a system of linear inequalities, we graph each inequality and determine the region that is common to *all* the solution sets.

EXAMPLE 5   Graph the solution set of the system

$$x + y \le 4,$$
$$x - y \ge 2.$$

Solution   We graph $x + y \le 4$ by first graphing the equation $x + y = 4$ using a solid line. Next, we choose $(0, 0)$ as a test point and find that it is a solution of $x + y \le 4$, so we shade the half-plane containing $(0, 0)$ using red. Next, we graph $x - y = 2$ using a solid line. We find that $(0, 0)$ is not a solution of $x - y \ge 2$, so we shade the half-plane that does not contain $(0, 0)$ using green. The arrows near the ends of each line help to indicate the half-plane that contains each solution set.

The solution set of the system of equations is the region shaded both red and green, or brown, including parts of the lines $x + y = 4$ and $x - y = 2$.

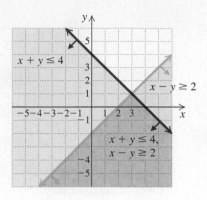

We can use different shading patterns on a graphing calculator to graph this system of inequalities. The solution set is the region shaded using both patterns. See Fig. 1 at left.

We can also use the Inequalz application to graph this system of inequalities. If we choose the Ineq Intersection option from the Shades menu, only the solution set is shaded, as shown in Fig. 2.

A system of inequalities may have a graph that consists of a polygon and its interior. As we will see later in this section, in many applications we will need to know the vertices of such a polygon.

EXAMPLE 6 Graph the following system of inequalities and find the coordinates of any vertices formed:

$$3x - y \leq 6, \quad \textbf{(1)}$$
$$y - 3 \leq 0, \quad \textbf{(2)}$$
$$x + y \geq 0. \quad \textbf{(3)}$$

Solution We graph the related equations $3x - y = 6$, $y - 3 = 0$, and $x + y = 0$ using solid lines. The half-plane containing the solution set for each inequality is indicated by the arrows near the ends of each line. We shade the region common to all three solution sets.

To find the vertices, we solve three systems of equations. The system of equations from inequalities (1) and (2) is

$$3x - y = 6,$$
$$y - 3 = 0.$$

Solving, we obtain the vertex $(3, 3)$.

The system of equations from inequalities (1) and (3) is

$$3x - y = 6,$$
$$x + y = 0.$$

Solving, we obtain the vertex $\left(\frac{3}{2}, -\frac{3}{2}\right)$.

The system of equations from inequalities (2) and (3) is

$$y - 3 = 0,$$
$$x + y = 0.$$

Solving, we obtain the vertex $(-3, 3)$.

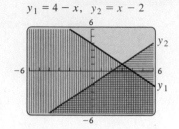

$y_1 = 4 - x, \quad y_2 = x - 2$

FIGURE 1.

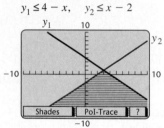

$y_1 \leq 4 - x, \quad y_2 \leq x - 2$

FIGURE 2.

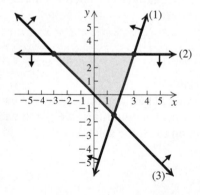

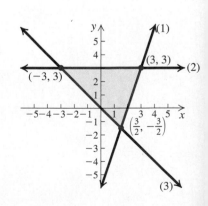

If a system of inequalities is graphed on a graphing calculator using a shading option, the coordinates of the vertices can be found using the INTERSECT feature. If the Inequalz application from the APPS menu is used to graph a system of inequalities, the PoI-Trace feature can be used to find the coordinates of the vertices.

▶ *Now Try Exercise 55.*

❯ Applications: Linear Programming

In many applications, we want to find a maximum value or a minimum value. In business, for example, we might want to both maximize profit and minimize cost. **Linear programming** can tell us how to do this.

In our study of linear programming, we will consider linear functions of two variables that are to be maximized or minimized subject to several conditions, or **constraints**. These constraints are expressed as inequalities. The solution set of the system of inequalities made up of the constraints contains all the **feasible solutions** of a linear programming problem. The function that we want to maximize or minimize is called the **objective function**.

It can be shown that the maximum and the minimum values of the objective function occur at a vertex of the region of feasible solutions. Thus we have the following procedure.

LINEAR PROGRAMMING PROCEDURE

To find the maximum or the minimum value of a linear objective function subject to a set of constraints:

1. Graph the region of feasible solutions.
2. Determine the coordinates of the vertices of the region.
3. Evaluate the objective function at each vertex. The largest and the smallest of those values are the maximum and the minimum values of the function, respectively.

EXAMPLE 7 *Maximizing Profit.* Aspen Carpentry makes bookcases and desks. Each bookcase requires 5 hr of woodworking and 4 hr of finishing. Each desk requires 10 hr of woodworking and 3 hr of finishing. Each month the shop has 600 hr of labor available for woodworking and 240 hr for finishing. The profit on each bookcase is $75 and on each desk is $140. Assume that all that are made are sold. How many of each product should be made each month in order to maximize profit? What is the maximum profit?

Solution We let x = the number of bookcases to be produced and y = the number of desks. Then the profit P is given by the function

$P = 75x + 140y.$ **To emphasize that P is a function of two variables, we sometimes write $P(x, y) = 75x + 140y.$**

We know that x bookcases require $5x$ hr of woodworking and that y desks require $10y$ hr of woodworking. Since there is no more than 600 hr of labor available for woodworking, we have one constraint:

$5x + 10y \leq 600.$

Similarly, the bookcases and desks require $4x$ hr and $3y$ hr of finishing, respectively. There is no more than 240 hr of labor available for finishing, so we have a second constraint:

$4x + 3y \leq 240.$

We also know that $x \geq 0$ and $y \geq 0$ because the carpentry shop cannot make a negative number of either product.

Thus we want to maximize the objective function $P = 75x + 140y$ subject to the constraints

$$5x + 10y \leq 600,$$
$$4x + 3y \leq 240,$$
$$x \geq 0,$$
$$y \geq 0.$$

We graph the system of inequalities and determine the vertices. Then we evaluate the objective function P at each vertex.

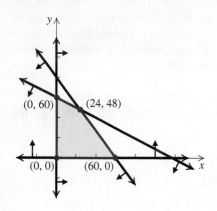

Vertices (x, y)	Profit $P = 75x + 140y$
$(0, 0)$	$P = 75 \cdot 0 + 140 \cdot 0 = 0$
$(60, 0)$	$P = 75 \cdot 60 + 140 \cdot 0 = 4500$
$(24, 48)$	$P = 75 \cdot 24 + 140 \cdot 48 = 8520$ ← Maximum
$(0, 60)$	$P = 75 \cdot 0 + 140 \cdot 60 = 8400$

The carpentry shop will make a maximum profit of $8520 when 24 bookcases and 48 desks are produced and sold.

We can create a table in which an objective function is evaluated at each vertex of a system of inequalities if the system has been graphed using the Inequalz application in the APPS menu.

> *Now Try Exercise 65.*

9.7 Exercise Set

In Exercises 1–8, match the inequality with one of the graphs (a)–(h) that follow.

a)
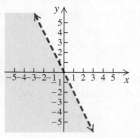

b)
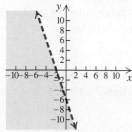

c)
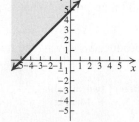

d)
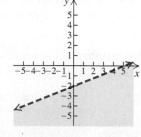

e)
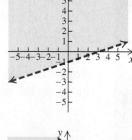

f)
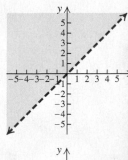

g)
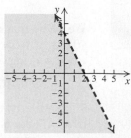

h)
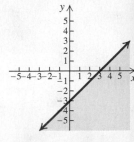

1. $y > x$ **2.** $y < -2x$

3. $y \leq x - 3$ **4.** $y \geq x + 5$

5. $2x + y < 4$ **6.** $3x + y < -6$

7. $2x - 5y > 10$ **8.** $3x - 9y < 9$

Graph.

9. $y > 2x$ **10.** $2y < x$

11. $y + x \geq 0$ **12.** $y - x < 0$

13. $y > x - 3$ **14.** $y \leq x + 4$

15. $x + y < 4$ **16.** $x - y \geq 5$

17. $3x - 2y \leq 6$ **18.** $2x - 5y < 10$

19. $3y + 2x \geq 6$ **20.** $2y + x \leq 4$

21. $3x - 2 \leq 5x + y$ **22.** $2x - 6y \geq 8 + 2y$

23. $x < -4$ **24.** $y > -3$

25. $y \geq 5$ **26.** $x \leq 5$

27. $-4 < y < -1$
  (*Hint*: Think of this as $-4 < y$ *and* $y < -1$.)

28. $-3 \leq x \leq 3$
  (*Hint*: Think of this as $-3 \leq x$ *and* $x \leq 3$.)

29. $y \geq |x|$ **30.** $y \leq |x + 2|$

In Exercises 31–36, match the system of inequalities with one of the graphs (a)–(f) that follow.

a)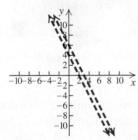

b)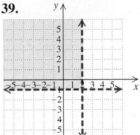

c)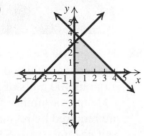

d)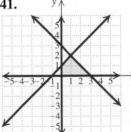

e)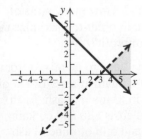

f)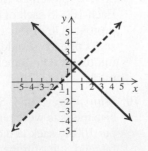

31. $y > x + 1,$
  $y \leq 2 - x$

32. $y < x - 3,$
  $y \geq 4 - x$

33. $2x + y < 4,$
  $4x + 2y > 12$

34. $x \leq 5,$
  $y \geq 1$

35. $x + y \leq 4,$
  $x - y \geq -3,$
  $x \geq 0,$
  $y \geq 0$

36. $x - y \geq -2,$
  $x + y \leq 6,$
  $x \geq 0,$
  $y \geq 0$

Find a system of inequalities with the given graph. Answers may vary.

37.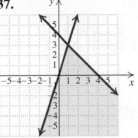

38.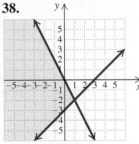

39.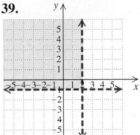

40.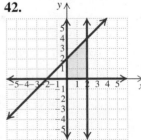

41.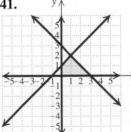

42.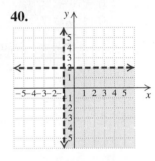

Graph the system of inequalities. Then find the coordinates of the vertices.

43. $y \leq x,$
  $y \geq 3 - x$

44. $y \leq x,$
  $y \geq 5 - x$

45. $y \geq x,$
  $y \leq 4 - x$

46. $y \geq x,$
  $y \leq 2 - x$

47. $y \geq -3,$
  $x \geq 1$

48. $y \leq -2,$
  $x \geq 2$

49. $x \leq 3,$
  $y \geq 2 - 3x$

50. $x \geq -2,$
  $y \leq 3 - 2x$

51. $x + y \leq 1,$
  $x - y \leq 2$

52. $y + 3x \geq 0,$
  $y + 3x \leq 2$

53. $2y - x \leq 2,$
$y + 3x \geq -1$

54. $y \leq 2x + 1,$
$y \geq -2x + 1,$
$x - 2 \leq 0$

55. $x - y \leq 2,$
$x + 2y \geq 8,$
$y - 4 \leq 0$

56. $x + 2y \leq 12,$
$2x + y \leq 12,$
$x \geq 0,$
$y \geq 0$

57. $4y - 3x \geq -12,$
$4y + 3x \geq -36,$
$y \leq 0,$
$x \leq 0$

58. $8x + 5y \leq 40,$
$x + 2y \leq 8,$
$x \geq 0,$
$y \geq 0$

59. $3x + 4y \geq 12,$
$5x + 6y \leq 30,$
$1 \leq x \leq 3$

60. $y - x \geq 1,$
$y - x \leq 3,$
$2 \leq x \leq 5$

Find the maximum value and the minimum value of the function and the values of x and y for which they occur.

61. $P = 17x - 3y + 60,$ subject to

$6x + 8y \leq 48,$
$0 \leq y \leq 4,$
$0 \leq x \leq 7.$

62. $Q = 28x - 4y + 72,$ subject to

$5x + 4y \geq 20,$
$0 \leq y \leq 4,$
$0 \leq x \leq 3.$

63. $F = 5x + 36y,$ subject to

$5x + 3y \leq 34,$
$3x + 5y \leq 30,$
$x \geq 0,$
$y \geq 0.$

64. $G = 16x + 14y,$ subject to

$3x + 2y \leq 12,$
$7x + 5y \leq 29,$
$x \geq 0,$
$y \geq 0.$

65. *Maximizing Mileage.*   Jazmin owns a pickup truck and a moped. He can afford 12 gal of gasoline to be split between the truck and the moped. Jazmin's truck gets 20 mpg and, with the fuel currently in the tank, can hold at most an additional 10 gal of gas. His moped gets 100 mpg and can hold at most 3 gal of gas. How many gallons of gasoline should each vehicle use if Jazmin wants to travel as far as possible on the 12 gal of gas? What is the maximum number of miles that he can travel?

20 mpg
088 123
100 mpg

66. *Maximizing Income.*   Golden Harvest Foods makes jumbo biscuits and regular biscuits. The oven can cook at most 200 biscuits per day. Each jumbo biscuit requires 2 oz of flour, each regular biscuit requires 1 oz of flour, and there is 300 oz of flour available. The income from each jumbo biscuit is $0.80 and from each regular biscuit is $0.50. Assume that all that are made are sold. How many of each size biscuit should be made in order to maximize income? What is the maximum income?

67. *Maximizing Profit.*   Waterbrook Farm includes 240 acres of cropland. The farm owner wishes to plant this acreage in both corn and soybeans. The profit per acre in corn production is $325 and in soybeans is $180. A total of 320 hr of labor is available. Each acre of corn requires 2 hr of labor, whereas each acre of soybeans requires 1 hr of labor. How should the land be divided between corn and soybeans in order to yield the maximum profit? What is the maximum profit?

68. *Maximizing Profit.*   Norris Mill can convert logs into lumber and plywood. In a given week, the mill can turn out 400 units of production, of which at least 100 units of lumber and at least 150 units of plywood are required by regular customers. The profit is $25 per unit of lumber and $38 per unit of plywood. Assume that all units produced are sold. How many units of each should the mill produce in order to maximize the profit? What is the maximum profit?

69. *Minimizing Cost.*   An animal feed to be mixed from soybean meal and oats must contain at least 120 lb of protein, 24 lb of fat, and 10 lb of mineral ash. Each 100-lb sack of soybean meal costs $20 and contains 50 lb of protein, 8 lb of fat, and 5 lb of mineral ash. Each 100-lb sack of oats costs $8 and contains 15 lb of protein, 5 lb of fat, and 1 lb of mineral ash. How many sacks of each should be used in order to satisfy the minimum requirements at minimum cost? What is the minimum cost?

70. *Minimizing Cost.*   Suppose that in the preceding exercise the oats were replaced by alfalfa, which costs $10 per 100-lb sack and contains 20 lb of protein, 6 lb of fat, and 8 lb of mineral ash. How much of each would now be required in order to minimize the cost? What is the minimum cost?

71. *Maximizing Income.* Francisco is planning to invest up to $40,000 in corporate and municipal bonds. The least he is allowed to invest in corporate bonds is $6000, and he does not want to invest more than $22,000 in corporate bonds. He also does not want to invest more than $30,000 in municipal bonds. The interest is 3% on corporate bonds and $4\frac{1}{4}\%$ on municipal bonds. This is simple interest for one year. How much should he invest in each type of bond in order to maximize his income? What is the maximum income?

72. *Maximizing Income.* Mila is planning to invest up to $22,000 in certificates of deposit at City Bank and People's Bank. She wants to invest at least $2000 but no more than $14,000 at City Bank. People's Bank does not insure more than a $15,000 investment, so she will invest no more than that in People's Bank. The interest is $2\frac{1}{2}\%$ at City Bank and $1\frac{3}{4}\%$ at People's Bank. This is simple interest for one year. How much should she invest in each bank in order to maximize her income? What is the maximum income?

73. *Minimizing Transportation Cost.* An airline with two types of airplanes, P_1 and P_2, has contracted with a tour group to provide transportation for a minimum of 2000 first-class, 1500 tourist-class, and 2400 economy-class passengers. For a certain trip, airplane P_1 costs $12 thousand to operate and can accommodate 40 first-class, 40 tourist-class, and 120 economy-class passengers, whereas airplane P_2 costs $10 thousand to operate and can accommodate 80 first-class, 30 tourist-class, and 40 economy-class passengers. How many of each type of airplane should be used in order to minimize the operating cost? What is the minimum operating cost?

74. *Minimizing Transportation Cost.* Suppose that in the preceding problem a new airplane P_3 becomes available, having an operating cost for the same trip of $15 thousand and accommodating 40 first-class, 40 tourist-class, and 80 economy-class passengers. If airplane P_1 were replaced by airplane P_3, how many of P_2 and P_3 should be used in order to minimize the operating cost? What is the minimum operating cost?

75. *Maximizing Profit.* It takes Fena Tailoring 3 hr of cutting and 6 hr of sewing to make a tiered silk organza bridal dress. It takes 6 hr of cutting and 3 hr of sewing to make a lace sheath bridal dress. The shop has at most 27 hr per week available for cutting and at most 36 hr per week for sewing. The profit is

$320 on an organza dress and $305 on a lace dress. Assume that all that are made are sold. How many of each kind of bridal dress should be made each week in order to maximize profit? What is the maximum profit?

76. *Maximizing Profit.* Cambridge Metal Works manufactures two sizes of gears. The smaller gear requires 4 hr of machining and 1 hr of polishing and yields a profit of $45. The larger gear requires 1 hr of machining and 1 hr of polishing and yields a profit of $30. The firm has available at most 24 hr per day for machining and 9 hr per day for polishing. Assume that all that are made are sold. How many of each type of gear should be produced each day in order to maximize profit? What is the maximum profit?

77. *Minimizing Nutrition Cost.* Suppose that it takes 12 units of carbohydrates and 6 units of protein to satisfy Jacob's minimum weekly requirements. A particular type of meat contains 2 units of carbohydrates and 2 units of protein per pound. A particular cheese contains 3 units of carbohydrates and 1 unit of protein per pound. The meat costs $3.50 per pound and the cheese costs $4.60 per pound. How many pounds of each are needed in order to minimize the cost and still meet the minimum requirements?

78. *Minimizing Salary Cost.* The Spring Hill school board is analyzing education costs for Hill Top School. It wants to hire teachers and teacher's aides to make up a faculty that satisfies its needs at minimum cost. The average annual salary for a teacher is $53,000 and for a teacher's aide is $23,600. The school building can accommodate a faculty of no more than 50 but needs at least 20 faculty members to function properly. The school must have at least 12 aides, but the number of teachers must be at least twice the number of aides in order to accommodate the expectations of the community. How many teachers and teacher's aides should be hired in order to minimize salary costs? What is the minimum salary cost?

79. *Maximizing Animal Support in a Forest.*   A certain area of forest is populated by two species of animal, which scientists refer to as A and B for simplicity. The forest supplies two kinds of food, referred to as F_1 and F_2. For one year, each member of species A requires 1 unit of F_1 and 0.5 unit of F_2. Each member of species B requires 0.2 unit of F_1 and 1 unit of F_2. The forest can normally supply at most 600 units of F_1 and 525 units of F_2 per year. What is the maximum total number of these animals that the forest can support?

80. *Maximizing Animal Support in a Forest.*   Refer to Exercise 79. If there is a wet spring, then supplies of food increase to 1080 units of F_1 and 810 units of F_2. In this case, what is the maximum total number of these animals that the forest can support?

❯ Skill Maintenance

Solve.

81. $-5 \leq x + 2 < 4$ **[1.6]**

82. $|x - 3| \geq 2$ **[3.5]**

83. $x^2 - 2x \leq 3$ **[4.6]**

84. $\dfrac{x - 1}{x + 2} > 4$ **[4.6]**

❯ Synthesis

Graph the system of inequalities.

85. $y \geq x^2 - 2$,
     $y \leq 2 - x^2$

86. $y < x + 1$,
     $y \geq x^2$

Graph the inequality.

87. $|x + y| \leq 1$

88. $|x| + |y| \leq 1$

89. $|x| > |y|$

90. $|x - y| > 0$

91. *Allocation of Resources.*   Comfort-by-Design Furniture produces chairs and sofas. Each chair requires 20 ft of wood, 1 lb of foam rubber, and 2 yd^2 of fabric. Each sofa requires 100 ft of wood, 50 lb of foam rubber, and 20 yd^2 of fabric. The manufacturer has in stock 1900 ft of wood, 500 lb of foam rubber, and 240 yd^2 of fabric. The chairs can be sold for $200 each and the sofas for $750 each. Assume that all that are made are sold. How many of each should be produced in order to maximize income? What is the maximum income?

9.8   Partial Fractions

> ❯ Decompose rational expressions into partial fractions.

There are situations in calculus in which it is useful to write a rational expression as a sum of two or more simpler rational expressions. In the equation

$$\frac{4x - 13}{2x^2 + x - 6} = \frac{3}{x + 2} + \frac{-2}{2x - 3},$$

each fraction on the right side is called a **partial fraction**. The expression on the right side is the **partial fraction decomposition** of the rational expression on the left side. In this section, we learn how such decompositions are created.

❯ Partial Fraction Decompositions

The procedure for finding the partial fraction decomposition of a rational expression involves factoring its denominator into linear factors and quadratic factors.

PROCEDURE FOR DECOMPOSING A RATIONAL EXPRESSION INTO PARTIAL FRACTIONS

Consider any rational expression $P(x)/Q(x)$ such that $P(x)$ and $Q(x)$ have no common factor other than 1 or -1.

1. If the degree of $P(x)$ is greater than or equal to the degree of $Q(x)$, divide to express $P(x)/Q(x)$ as a quotient $+$ remainder$/Q(x)$ and follow steps (2)–(5) to decompose the resulting rational expression.

2. If the degree of $P(x)$ is less than the degree of $Q(x)$, factor $Q(x)$ into linear factors of the form $(px + q)^n$ and/or quadratic factors of the form $(ax^2 + bx + c)^m$. Any quadratic factor $ax^2 + bx + c$ must be *irreducible*, meaning that it cannot be factored into linear factors with rational coefficients.

3. Assign to each linear factor $(px + q)^n$ the sum of n partial fractions:

$$\frac{A_1}{px + q} + \frac{A_2}{(px + q)^2} + \cdots + \frac{A_n}{(px + q)^n}.$$

4. Assign to each quadratic factor $(ax^2 + bx + c)^m$ the sum of m partial fractions:

$$\frac{B_1 x + C_1}{ax^2 + bx + c} + \frac{B_2 x + C_2}{(ax^2 + bx + c)^2} + \cdots + \frac{B_m x + C_m}{(ax^2 + bx + c)^m}.$$

5. Apply algebraic methods, as illustrated in the following examples, to find the constants in the numerators of the partial fractions.

Just in Time

23

EXAMPLE 1 Decompose into partial fractions:

$$\frac{4x - 13}{2x^2 + x - 6}.$$

Solution The degree of the numerator is less than the degree of the denominator. We begin by factoring the denominator: $(x + 2)(2x - 3)$. We find constants A and B such that

$$\frac{4x - 13}{(x + 2)(2x - 3)} = \frac{A}{x + 2} + \frac{B}{2x - 3}.$$

To determine A and B, we add the expressions on the right:

$$\frac{4x - 13}{(x + 2)(2x - 3)} = \frac{A(2x - 3) + B(x + 2)}{(x + 2)(2x - 3)}.$$

Next, we equate the numerators:

$$4x - 13 = A(2x - 3) + B(x + 2).$$

Since the last equation containing A and B is true for all x, we can substitute any value of x and still have a true equation. If we choose $x = \frac{3}{2}$, then $2x - 3 = 0$ and A will be eliminated when we make the substitution. This gives us

$$4\left(\tfrac{3}{2}\right) - 13 = A\left(2 \cdot \tfrac{3}{2} - 3\right) + B\left(\tfrac{3}{2} + 2\right)$$
$$-7 = 0 + \tfrac{7}{2}B.$$

Solving, we obtain $B = -2$.

In order to have $x + 2 = 0$, we let $x = -2$. Then B will be eliminated when we make the substitution. This gives us

$$4(-2) - 13 = A[2(-2) - 3] + B(-2 + 2)$$
$$-21 = -7A + 0.$$

Solving, we obtain $A = 3$.

The decomposition is as follows:

$$\frac{4x - 13}{2x^2 + x - 6} = \frac{3}{x + 2} + \frac{-2}{2x - 3}, \quad \text{or} \quad \frac{3}{x + 2} - \frac{2}{2x - 3}.$$

To check, we can add to see if we get the expression on the left. We can also use the TABLE feature on a graphing calculator, comparing values of

$$y_1 = \frac{4x - 13}{2x^2 + x - 6} \quad \text{and} \quad y_2 = \frac{3}{x + 2} - \frac{2}{2x - 3}$$

X	Y₁	Y₂
-1	3.4	3.4
0	2.1667	2.1667
1	3	3
2	-1.25	-1.25
3	-.0667	-.0667
4	.1	.1
5	.14286	.14286

X = -1

for the same values of x. Since $y_1 = y_2$ for the given values of x as we scroll through the table, the decomposition appears to be correct.

> **Now Try Exercise 3.**

EXAMPLE 2 Decompose into partial fractions:

$$\frac{7x^2 - 29x + 24}{(2x - 1)(x - 2)^2}.$$

Solution The degree of the numerator is 2 and the degree of the denominator is 3, so the degree of the numerator is less than the degree of the denominator. The denominator is given in factored form. The decomposition has the following form:

$$\frac{7x^2 - 29x + 24}{(2x - 1)(x - 2)^2} = \frac{A}{2x - 1} + \frac{B}{x - 2} + \frac{C}{(x - 2)^2}.$$

As in Example 1, we add the expressions on the right:

$$\frac{7x^2 - 29x + 24}{(2x - 1)(x - 2)^2} = \frac{A(x - 2)^2 + B(2x - 1)(x - 2) + C(2x - 1)}{(2x - 1)(x - 2)^2}.$$

Then we equate the numerators. This gives us

$$7x^2 - 29x + 24 = A(x - 2)^2 + B(2x - 1)(x - 2) + C(2x - 1).$$

Since the equation containing A, B, and C is true for all x, we can substitute any value of x and still have a true equation. In order to have $2x - 1 = 0$, we let $x = \frac{1}{2}$. This gives us

$$7\left(\tfrac{1}{2}\right)^2 - 29 \cdot \tfrac{1}{2} + 24 = A\left(\tfrac{1}{2} - 2\right)^2 + 0 + 0$$
$$\tfrac{45}{4} = \tfrac{9}{4}A.$$

Solving, we obtain $A = 5$.

In order to have $x - 2 = 0$, we let $x = 2$. Substituting gives us

$$7(2)^2 - 29(2) + 24 = 0 + 0 + C(2 \cdot 2 - 1)$$
$$-6 = 3C.$$

Solving, we obtain $C = -2$.

To find B, we choose any value for x except $\frac{1}{2}$ or 2 and replace A with 5 and C with -2. We let $x = 1$:

$$7 \cdot 1^2 - 29 \cdot 1 + 24 = 5(1 - 2)^2 + B(2 \cdot 1 - 1)(1 - 2)$$
$$+ (-2)(2 \cdot 1 - 1)$$
$$2 = 5 - B - 2$$
$$B = 1.$$

The decomposition is as follows:

$$\frac{7x^2 - 29x + 24}{(2x - 1)(x - 2)^2} = \frac{5}{2x - 1} + \frac{1}{x - 2} - \frac{2}{(x - 2)^2}.$$

X	Y₁	Y₂
-5	-.6382	-.6382
-4	-.7778	-.7778
-3	-.9943	-.9943
-2	-1.375	-1.375
-1	-2.222	-2.222
0	-6	-6
1	2	2

X = -5

POLYNOMIAL DIVISION

REVIEW SECTION 4.3.

We can check the result using a table of values. We let

$$y_1 = \frac{7x^2 - 29x + 24}{(2x - 1)(x - 2)^2} \quad \text{and} \quad y_2 = \frac{5}{2x - 1} + \frac{1}{x - 2} - \frac{2}{(x - 2)^2}.$$

Since $y_1 = y_2$ for given values of x as we scroll through the table, the decomposition appears to be correct.

⟩ *Now Try Exercise 7.*

EXAMPLE 3   Decompose into partial fractions:

$$\frac{6x^3 + 5x^2 - 7}{3x^2 - 2x - 1}.$$

Solution   The degree of the numerator is greater than that of the denominator. Therefore, we divide and find an equivalent expression:

$$
\begin{array}{r}
2x + 3 \\
3x^2 - 2x - 1\overline{)6x^3 + 5x^2 \quad\quad - 7} \\
\underline{6x^3 - 4x^2 - 2x} \\
9x^2 + 2x - 7 \\
\underline{9x^2 - 6x - 3} \\
8x - 4.
\end{array}
$$

The original expression is thus equivalent to

$$2x + 3 + \frac{8x - 4}{3x^2 - 2x - 1}.$$

We decompose the fraction to get

$$\frac{8x - 4}{(3x + 1)(x - 1)} = \frac{5}{3x + 1} + \frac{1}{x - 1}.$$

The final result is

$$2x + 3 + \frac{5}{3x + 1} + \frac{1}{x - 1}.$$

⟩ *Now Try Exercise 17.*

Systems of equations can be used to decompose rational expressions. Let's reconsider Example 2.

EXAMPLE 4   Decompose into partial fractions:

$$\frac{7x^2 - 29x + 24}{(2x - 1)(x - 2)^2}.$$

Solution   The decomposition has the following form:

$$\frac{A}{2x - 1} + \frac{B}{x - 2} + \frac{C}{(x - 2)^2}.$$

We first add as in Example 2:

$$\frac{7x^2 - 29x + 24}{(2x - 1)(x - 2)^2} = \frac{A}{2x - 1} + \frac{B}{x - 2} + \frac{C}{(x - 2)^2}$$

$$= \frac{A(x - 2)^2 + B(2x - 1)(x - 2) + C(2x - 1)}{(2x - 1)(x - 2)^2}.$$

Then we equate numerators:

$$7x^2 - 29x + 24$$
$$= A(x - 2)^2 + B(2x - 1)(x - 2) + C(2x - 1)$$
$$= A(x^2 - 4x + 4) + B(2x^2 - 5x + 2) + C(2x - 1)$$
$$= Ax^2 - 4Ax + 4A + 2Bx^2 - 5Bx + 2B + 2Cx - C,$$

or, combining like terms,

$$7x^2 - 29x + 24$$
$$= (A + 2B)x^2 + (-4A - 5B + 2C)x + (4A + 2B - C).$$

Next, we equate corresponding coefficients:

$$7 = A + 2B, \qquad \text{The coefficients of the } x^2\text{-terms must be the same.}$$
$$-29 = -4A - 5B + 2C, \qquad \text{The coefficients of the } x\text{-terms must be the same.}$$
$$24 = 4A + 2B - C. \qquad \text{The constant terms must be the same.}$$

SYSTEMS OF EQUATIONS IN THREE VARIABLES

REVIEW SECTION 9.2, 9.5, OR 9.6.

We now have a system of three equations. You should confirm that the solution of the system is

$$A = 5, \qquad B = 1, \quad \text{and} \quad C = -2.$$

The decomposition is as follows:

$$\frac{7x^2 - 29x + 24}{(2x - 1)(x - 2)^2} = \frac{5}{2x - 1} + \frac{1}{x - 2} - \frac{2}{(x - 2)^2}.$$

> **Now Try Exercise 15.**

EXAMPLE 5   Decompose into partial fractions:

$$\frac{11x^2 - 8x - 7}{(2x^2 - 1)(x - 3)}.$$

Solution   The decomposition has the following form:

$$\frac{11x^2 - 8x - 7}{(2x^2 - 1)(x - 3)} = \frac{Ax + B}{2x^2 - 1} + \frac{C}{x - 3}.$$

Adding and equating numerators, we get

$$11x^2 - 8x - 7 = (Ax + B)(x - 3) + C(2x^2 - 1)$$
$$= Ax^2 - 3Ax + Bx - 3B + 2Cx^2 - C,$$

or $$11x^2 - 8x - 7 = (A + 2C)x^2 + (-3A + B)x + (-3B - C).$$

We then equate corresponding coefficients:

$$11 = A + 2C, \qquad \text{The coefficients of the } x^2\text{-terms}$$
$$-8 = -3A + B, \qquad \text{The coefficients of the } x\text{-terms}$$
$$-7 = -3B - C. \qquad \text{The constant terms}$$

We solve this system of three equations and obtain

$$A = 3, \qquad B = 1, \quad \text{and} \quad C = 4.$$

The decomposition is as follows:

$$\frac{11x^2 - 8x - 7}{(2x^2 - 1)(x - 3)} = \frac{3x + 1}{2x^2 - 1} + \frac{4}{x - 3}.$$

> **Now Try Exercise 13.**

9.8 | Exercise Set

Decompose into partial fractions. Check your answers using a graphing calculator.

1. $\dfrac{x + 7}{(x - 3)(x + 2)}$

2. $\dfrac{2x}{(x + 1)(x - 1)}$

3. $\dfrac{7x - 1}{6x^2 - 5x + 1}$

4. $\dfrac{13x + 46}{12x^2 - 11x - 15}$

5. $\dfrac{3x^2 - 11x - 26}{(x^2 - 4)(x + 1)}$

6. $\dfrac{5x^2 + 9x - 56}{(x - 4)(x - 2)(x + 1)}$

7. $\dfrac{9}{(x + 2)^2(x - 1)}$

8. $\dfrac{x^2 - x - 4}{(x - 2)^3}$

9. $\dfrac{2x^2 + 3x + 1}{(x^2 - 1)(2x - 1)}$

10. $\dfrac{x^2 - 10x + 13}{(x^2 - 5x + 6)(x - 1)}$

11. $\dfrac{x^4 - 3x^3 - 3x^2 + 10}{(x + 1)^2(x - 3)}$

12. $\dfrac{10x^3 - 15x^2 - 35x}{x^2 - x - 6}$

13. $\dfrac{-x^2 + 2x - 13}{(x^2 + 2)(x - 1)}$

14. $\dfrac{26x^2 + 208x}{(x^2 + 1)(x + 5)}$

15. $\dfrac{6 + 26x - x^2}{(2x - 1)(x + 2)^2}$

16. $\dfrac{5x^3 + 6x^2 + 5x}{(x^2 - 1)(x + 1)^3}$

17. $\dfrac{6x^3 + 5x^2 + 6x - 2}{2x^2 + x - 1}$

18. $\dfrac{2x^3 + 3x^2 - 11x - 10}{x^2 + 2x - 3}$

19. $\dfrac{2x^2 - 11x + 5}{(x - 3)(x^2 + 2x - 5)}$

20. $\dfrac{3x^2 - 3x - 8}{(x - 5)(x^2 + x - 4)}$

21. $\dfrac{-4x^2 - 2x + 10}{(3x + 5)(x + 1)^2}$

22. $\dfrac{26x^2 - 36x + 22}{(x - 4)(2x - 1)^2}$

23. $\dfrac{36x + 1}{12x^2 - 7x - 10}$

24. $\dfrac{-17x + 61}{6x^2 + 39x - 21}$

25. $\dfrac{-4x^2 - 9x + 8}{(3x^2 + 1)(x - 2)}$

26. $\dfrac{11x^2 - 39x + 16}{(x^2 + 4)(x - 8)}$

❯ Skill Maintenance

Find the zeros of the polynomial function.

27. $f(x) = x^3 + x^2 + 9x + 9$ **[4.1], [4.3], [4.4]**

28. $f(x) = x^3 - 3x^2 + x - 3$ **[4.1], [4.3], [4.4]**

29. $f(x) = x^3 + x^2 - 3x - 2$ **[4.4]**

30. $f(x) = x^4 - x^3 - 5x^2 - x - 6$ **[4.3]**

31. $f(x) = x^3 + 5x^2 + 5x - 3$ **[4.1], [4.3], [4.4]**

❯ Synthesis

Decompose into partial fractions.

32. $\dfrac{9x^3 - 24x^2 + 48x}{(x - 2)^4(x + 1)}$

[*Hint*: Let the expression equal

$$\frac{A}{x + 1} + \frac{P(x)}{(x - 2)^4}$$

and find $P(x)$].

33. $\dfrac{x}{x^4 - a^4}$

34. $\dfrac{1}{e^{-x} + 3 + 2e^x}$

35. $\dfrac{1 + \ln x^2}{(\ln x + 2)(\ln x - 3)^2}$

STUDY GUIDE

KEY TERMS AND CONCEPTS   **EXAMPLES**

SECTION 9.1: SYSTEMS OF EQUATIONS IN TWO VARIABLES

A **system of two linear equations in two variables** is composed of two linear equations that are considered simultaneously.

  The **solutions** of the system of equations are all ordered pairs that make *both* equations true.

  A system of equations is **consistent** if it has at least one solution. A system of equations that has no solution is **inconsistent**.

  The equations are **dependent** if one equation can be obtained by multiplying on both sides of the other equation by a constant. Otherwise, the equations are **independent**.

  Systems of two equations in two variables can be solved graphically.

Solve: $x + y = 2,$
$\quad\quad\quad y = x - 4.$

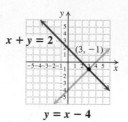

The solution is the point of intersection, $(3, -1)$. The system is consistent. The equations are independent.

Solve: $x + y = 2,$
$\quad\quad\quad x + y = -2.$

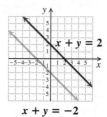

The graphs do not intersect, so there is no solution. The system is inconsistent. The equations are independent.

Solve: $x + y = 2,$
$\quad\quad\quad 3x + 3y = 6.$

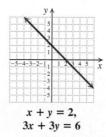

$$x + y = 2,$$
$$3x + 3y = 6$$

The graphs are the same. There are infinitely many common points, so there are infinitely many solutions. The solutions are of the form $(x, 2 - x)$ or $(2 - y, y)$. The system is consistent. The equations are dependent.

Systems of two equations in two variables can be solved using substitution.	Solve: $x = y - 5,$ $\quad\ \ 2x + 3y = 5.$

Substitute and solve for y:

$2(y - 5) + 3y = 5$

$2y - 10 + 3y = 5$

$\quad\quad 5y - 10 = 5$

$\quad\quad\quad\ \ 5y = 15$

$\quad\quad\quad\ \ \ y = 3.$

Back-substitute and solve for x:

$x = y - 5$

$x = 3 - 5$

$x = -2.$

The solution is $(-2, 3)$.

Systems of two equations in two variables can be solved using elimination.

Solve: $3x + \ y = -1,$

$\quad\quad\ \ x - 3y = 8.$

Eliminate y and solve for x:

$\quad 9x + 3y = -3$

$\quad \underline{\ x - 3y = \ \ \ 8}$

$10x \quad\quad = \ \ \ 5$

$\quad\ x = \ \ \ \frac{1}{2}$

Back-substitute and solve for y:

$3x + y = -1$

$3 \cdot \frac{1}{2} + y = -1$

$\quad \frac{3}{2} + y = -1$

$\quad\quad\ y = -\frac{5}{2}.$

The solution is $\left(\frac{1}{2}, -\frac{5}{2}\right)$.

Some applied problems can be solved by translating to a system of two equations in two variables.

See Examples 6–8 on pp. 643–646.

SECTION 9.2: SYSTEMS OF EQUATIONS IN THREE VARIABLES

A **solution** of a system of equations in three variables is an ordered triple that makes *all* three equations true.

We can use **Gaussian elimination** to solve a system of three equations in three variables by using the operations listed on p. 652 to transform the original system to one of the form

$$Ax + By + Cz = D,$$
$$Ey + Fz = G,$$
$$Hz = K.$$

Then we solve the third equation for z and back-substitute to find y and x.

As we see in Example 1 on p. 652, Gaussian elimination can be used to transform the system of equations

$$x - 2y + 3z = 11,$$
$$4x + 2y - 3z = 4,$$
$$3x + 3y - \ z = 4$$

to the equivalent form

$$x - 2y + \ 3z = 11,$$
$$10y - 15z = -40,$$
$$35z = 70.$$

Solve for z: $\quad 35z = 70$

$\quad\quad\quad\quad\quad z = 2.$

Back-substitute to find y and x:

$10y - 15 \cdot 2 = -40 \quad\quad x - 2(-1) + 3 \cdot 2 = 11$

$\quad 10y - 30 = -40 \quad\quad\quad\quad\ x + 2 + 6 = 11$

$\quad\quad\quad 10y = -10 \quad\quad\quad\quad\quad\quad\ x + 8 = 11$

$\quad\quad\quad\quad\ y = -1. \quad\quad\quad\quad\quad\quad\quad\quad x = 3.$

The solution is $(3, -1, 2)$.

Some applied problems can be solved by translating to a system of three equations in three variables.	See Example 3 on p. 655.
We can use a system of three equations to model a situation with a quadratic function.	Find a quadratic function that fits the data points $(0, -5)$, $(1, -4)$, and $(2, 1)$. We substitute in the function $f(x) = ax^2 + bx + c$: For $(0, -5)$: $\quad -5 = a \cdot 0^2 + b \cdot 0 + c$, For $(1, -4)$: $\quad -4 = a \cdot 1^2 + b \cdot 1 + c$, For $(2, 1)$: $\quad\quad 1 = a \cdot 2^2 + b \cdot 2 + c$. We now have a system of equations: $$c = -5,$$ $$a + b + c = -4,$$ $$4a + 2b + c = 1.$$ Solving this system of equations gives $(2, -1, -5)$. Thus, $$f(x) = 2x^2 - x - 5.$$

SECTION 9.3: MATRICES AND SYSTEMS OF EQUATIONS

A **matrix** (pl., **matrices**) is a rectangular array of numbers called **entries**, or **elements**, of the matrix.	Row 1 $\rightarrow$ $\begin{bmatrix} 3 & -2 & 5 \\ -1 & 4 & -3 \end{bmatrix}$ $\leftarrow$ Row 2 Column 1 Column 2 Column 3 This matrix has 2 rows and 3 columns. Its **order** is 2×3.
We can apply the **row-equivalent operations** on p. 662 to use **Gaussian elimination** with matrices to solve systems of equations.	Solve: $\quad x - 2y = 8$, $\quad\quad\quad 2x + y = 1$. We write the augmented matrix and transform it to **row-echelon form** or **reduced row-echelon form**: $$\begin{bmatrix} 1 & -2 & \vert & 8 \\ 2 & 1 & \vert & 1 \end{bmatrix} \rightarrow \begin{bmatrix} 1 & -2 & \vert & 8 \\ 0 & 1 & \vert & -3 \end{bmatrix} \rightarrow \begin{bmatrix} 1 & 0 & \vert & 2 \\ 0 & 1 & \vert & -3 \end{bmatrix}$$ Row-echelon form Reduced row-echelon form Thus we have $x = 2$, $y = -3$. The solution is $(2, -3)$.

SECTION 9.4: MATRIX OPERATIONS

Matrices of the same order can be added or subtracted by adding or subtracting their corresponding entries.	Find each of the following. $$\begin{bmatrix} 3 & -4 \\ -1 & 2 \end{bmatrix} + \begin{bmatrix} -5 & -1 \\ 3 & 0 \end{bmatrix} = \begin{bmatrix} 3 + (-5) & -4 + (-1) \\ -1 + 3 & 2 + 0 \end{bmatrix}$$ $$= \begin{bmatrix} -2 & -5 \\ 2 & 2 \end{bmatrix}$$ $$\begin{bmatrix} 3 & -4 \\ -1 & 2 \end{bmatrix} - \begin{bmatrix} -5 & -1 \\ 3 & 0 \end{bmatrix} = \begin{bmatrix} 3 - (-5) & -4 - (-1) \\ -1 - 3 & 2 - 0 \end{bmatrix}$$ $$= \begin{bmatrix} 8 & -3 \\ -4 & 2 \end{bmatrix}$$

The **scalar product** of a number k and a matrix $\mathbf{A}$ is the matrix $k\mathbf{A}$ obtained by multiplying each entry of $\mathbf{A}$ by k. The number k is called a **scalar**.

The properties of matrix addition and scalar multiplication are given on p. 671.

For $\mathbf{A} = \begin{bmatrix} 2 & 3 & -1 \\ -4 & -2 & 5 \end{bmatrix}$, find $2\mathbf{A}$.

$$2\mathbf{A} = 2\begin{bmatrix} 2 & 3 & -1 \\ -4 & -2 & 5 \end{bmatrix} = \begin{bmatrix} 2 \cdot 2 & 2 \cdot 3 & 2 \cdot (-1) \\ 2 \cdot (-4) & 2 \cdot (-2) & 2 \cdot 5 \end{bmatrix}$$

$$= \begin{bmatrix} 4 & 6 & -2 \\ -8 & -4 & 10 \end{bmatrix}$$

For an $m \times n$ matrix $\mathbf{A} = [a_{ij}]$ and an $n \times p$ matrix $\mathbf{B} = [b_{ij}]$, the **product** $\mathbf{AB} = [c_{ij}]$ is an $m \times p$ matrix, where

$$c_{ij} = a_{i1} \cdot b_{1j} + a_{i2} \cdot b_{2j}$$
$$+ a_{i3} \cdot b_{3j} + \cdots + a_{in} \cdot b_{nj}.$$

The properties of matrix multiplication are given on p. 674.

For $\mathbf{A} = \begin{bmatrix} 4 & -1 & 3 \\ 0 & -2 & 1 \end{bmatrix}$ and $\mathbf{B} = \begin{bmatrix} -3 & 1 \\ 3 & 4 \\ 2 & -1 \end{bmatrix}$, find $\mathbf{AB}$.

$$\mathbf{AB} = \begin{bmatrix} 4 & -1 & 3 \\ 0 & -2 & 1 \end{bmatrix}\begin{bmatrix} -3 & 1 \\ 3 & 4 \\ 2 & -1 \end{bmatrix}$$

$$= \begin{bmatrix} 4 \cdot (-3) + (-1) \cdot 3 + 3 \cdot 2 & 4 \cdot 1 + (-1) \cdot 4 + 3 \cdot (-1) \\ 0 \cdot (-3) + (-2) \cdot 3 + 1 \cdot 2 & 0 \cdot 1 + (-2) \cdot 4 + 1 \cdot (-1) \end{bmatrix}$$

$$= \begin{bmatrix} -9 & -3 \\ -4 & -9 \end{bmatrix}$$

We can write a matrix equation equivalent to a system of equations.

Write a matrix equation equivalent to the system of equations:

$$2x - 3y = 6,$$
$$x - 4y = 1.$$

This system of equations can be written as

$$\begin{bmatrix} 2 & -3 \\ 1 & -4 \end{bmatrix}\begin{bmatrix} x \\ y \end{bmatrix} = \begin{bmatrix} 6 \\ 1 \end{bmatrix}.$$

SECTION 9.5: INVERSES OF MATRICES

The $n \times n$ **identity matrix I** is an $n \times n$ matrix with 1's on the main diagonal and 0's elsewhere.

For any $n \times n$ matrix $\mathbf{A}$,

$$\mathbf{AI} = \mathbf{IA} = \mathbf{A}.$$

For an $n \times n$ matrix $\mathbf{A}$, if there is a matrix $\mathbf{A}^{-1}$ for which $\mathbf{A}^{-1} \cdot \mathbf{A} = \mathbf{I} = \mathbf{A} \cdot \mathbf{A}^{-1}$, then $\mathbf{A}^{-1}$ is the **inverse** of $\mathbf{A}$.

For a system of n linear equations in n variables, $\mathbf{AX} = \mathbf{B}$, if $\mathbf{A}$ has an inverse, then the solution of the system of equations is given by

$$\mathbf{X} = \mathbf{A}^{-1}\mathbf{B}.$$

Since matrix multiplication is not commutative, in general, $\mathbf{B}$ *must* be multiplied *on the left* by $\mathbf{A}^{-1}$.

The inverse of an $n \times n$ matrix $\mathbf{A}$ can be found by first writing an augmented matrix consisting of $\mathbf{A}$ on the left side and the $n \times n$ identity matrix on the right side. Then row-equivalent operations are used to transform the augmented matrix to a matrix with the $n \times n$ identity matrix on the left side and the inverse on the right side.

See Examples 3 and 4 on pp. 680 and 681.

Use an inverse matrix to solve the following system of equations:

$$x - y = 1,$$
$$x - 2y = -1.$$

First, we write an equivalent matrix equation:

$$\underset{\mathbf{A}}{\begin{bmatrix} 1 & -1 \\ 1 & -2 \end{bmatrix}} \cdot \underset{\mathbf{X}}{\begin{bmatrix} x \\ y \end{bmatrix}} = \underset{\mathbf{B}}{\begin{bmatrix} 1 \\ -1 \end{bmatrix}}.$$

(continued)

Then we find $\mathbf{A}^{-1}$ and multiply *on the left* by $\mathbf{A}^{-1}$:

$$\mathbf{X} = \mathbf{A}^{-1} \cdot \mathbf{B}$$

$$\begin{bmatrix} x \\ y \end{bmatrix} = \begin{bmatrix} 2 & -1 \\ 1 & -1 \end{bmatrix}\begin{bmatrix} 1 \\ -1 \end{bmatrix} = \begin{bmatrix} 3 \\ 2 \end{bmatrix}.$$

The solution is $(3, 2)$.

SECTION 9.6: DETERMINANTS AND CRAMER'S RULE

Determinant of a 2 × 2 Matrix

The determinant of the matrix $\begin{bmatrix} a & c \\ b & d \end{bmatrix}$ is denoted by $\begin{vmatrix} a & c \\ b & d \end{vmatrix}$ and is defined as

$$\begin{vmatrix} a & c \\ b & d \end{vmatrix} = ad - bc.$$

Evaluate: $\begin{vmatrix} 3 & -4 \\ 2 & 1 \end{vmatrix}$.

$$\begin{vmatrix} 3 & -4 \\ 2 & 1 \end{vmatrix} = 3 \cdot 1 - 2(-4) = 3 + 8 = 11$$

The **determinant** of any **square matrix** can be found by *expanding across a row* or *down a column*. See p. 687.

See Example 4 on p. 687.

We can use determinants to solve systems of linear equations.

Cramer's rule for a 2 × 2 system is given on p. 688. Cramer's rule for a 3 × 3 system is given on p. 689.

Solve: $2x - 3y = 2,$
    $6x + 6y = 1.$

$$x = \frac{\begin{vmatrix} 2 & -3 \\ 1 & 6 \end{vmatrix}}{\begin{vmatrix} 2 & -3 \\ 6 & 6 \end{vmatrix}}, \qquad y = \frac{\begin{vmatrix} 2 & 2 \\ 6 & 1 \end{vmatrix}}{\begin{vmatrix} 2 & -3 \\ 6 & 6 \end{vmatrix}}$$

$$x = \frac{15}{30} = \frac{1}{2}, \qquad y = \frac{-10}{30} = -\frac{1}{3}.$$

The solution is $\left(\frac{1}{2}, -\frac{1}{3}\right)$.

SECTION 9.7: SYSTEMS OF INEQUALITIES AND LINEAR PROGRAMMING

To graph a linear inequality in two variables:

1. Graph the related equation. Draw a dashed line if the inequality symbol is $<$ or $>$. Draw a solid line if the inequality symbol is $\leq$ or $\geq$.

2. Use a test point to determine which half-plane to shade.

Graph: $x + y > 2$.

1. Graph $x + y = 2$ using a dashed line.

2. Test a point not on the line. We use $(0, 0)$.

$$\frac{x + y > 2}{0 + 0 \;?\; 2}$$
$$0 \quad | \quad \text{FALSE}$$

Since $0 > 2$ is false, we shade the half-plane that does not contain $(0, 0)$.

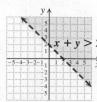

To graph a system of inequalities, graph each inequality and determine the region that is common to all the solution sets.	Graph the solution set of the system $x - y \le 3,$ $2x + y \ge 4.$ 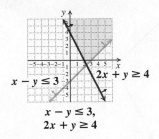 $x - y \le 3,$ $2x + y \ge 4$
The maximum or the minimum value of an **objective function** over a region of **feasible solutions** is the maximum or the minimum value of the function at a vertex of that region.	Maximize $G = 8x - 5y$ subject to $x + y \le 3,$ $x \ge 0,$ $y \ge 1.$ 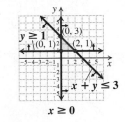

Vertex	$G = 8x - 5y$	
$(0, 1)$	$G = 8 \cdot 0 - 5 \cdot 1 = -5$	
$(0, 3)$	$G = 8 \cdot 0 - 5 \cdot 3 = -15$	
$(2, 1)$	$G = 8 \cdot 2 - 5 \cdot 1 = 11$	← Maximum

SECTION 9.8: PARTIAL FRACTIONS

The procedure for **decomposing a rational expression into partial fractions** is given on p. 703.	See Examples 1–5 on pp. 703–706.

REVIEW EXERCISES

Determine whether the statement is true or false.

1. A system of equations with exactly one solution is consistent and has independent equations. **[9.1]**

2. A system of two linear equations in two variables can have exactly two solutions. **[9.1]**

3. For any $m \times n$ matrices **A** and **B**,

   $\mathbf{A} + \mathbf{B} = \mathbf{B} + \mathbf{A}.$ **[9.4]**

4. In general, matrix multiplication is commutative. **[9.4]**

In Exercises 5–12, match the equations or inequalities with one of the graphs (a)–(h) that follow.

a)

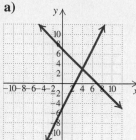

b)

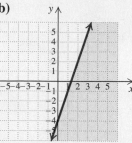

c)

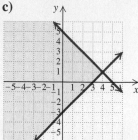

d)

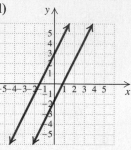

e)

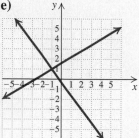

f)

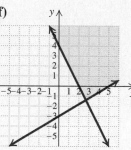

g)

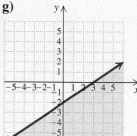

h)

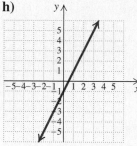

5. $x + y = 7,$
$2x - y = 5$ **[9.1]**

6. $3x - 5y = -8,$
$4x + 3y = -1$ **[9.1]**

7. $y = 2x - 1,$
$4x - 2y = 2$ **[9.1]**

8. $6x - 3y = 5,$
$y = 2x + 3$ **[9.1]**

9. $y \leq 3x - 4$ **[9.7]**

10. $2x - 3y \geq 6$ **[9.7]**

11. $x - y \leq 3,$
$x + y \leq 5$ **[9.7]**

12. $2x + y \geq 4,$
$3x - 5y \leq 15$ **[9.7]**

Solve.

13. $5x - 3y = -4,$
$3x - y = -4$ **[9.1]**

14. $2x + 3y = 2,$
$5x - y = -29$ **[9.1]**

15. $x + 5y = 12,$
$5x + 25y = 12$ **[9.1]**

16. $x + y = -2,$
$-3x - 3y = 6$ **[9.1]**

17. $x + 5y - 3z = 4,$
$3x - 2y + 4z = 3,$
$2x + 3y - z = 5$ **[9.2]**

18. $2x - 4y + 3z = -3,$
$-5x + 2y - z = 7,$
$3x + 2y - 2z = 4$ **[9.2]**

19. $x - y = 5,$
$y - z = 6,$
$-w + z = 7,$
$w + x = 8$ **[9.2]**

20. Classify each of the systems in Exercises 13–19 as consistent or inconsistent. **[9.1]**, **[9.2]**

21. Classify each of the systems in Exercises 13–19 as having dependent or independent equations. **[9.1]**, **[9.2]**

Solve the system of equations using Gaussian elimination or Gauss–Jordan elimination. **[9.3]**

22. $x + 2y = 5,$
$2x - 5y = -8$

23. $3x + 4y + 2z = 3,$
$5x - 2y - 13z = 3,$
$4x + 3y - 3z = 6$

24. $3x + 5y + z = 0,$
$2x - 4y - 3z = 0,$
$x + 3y + z = 0$

25. $w + x + y + z = -2,$
$-3w - 2x + 3y + 2z = 10,$
$2w + 3x + 2y - z = -12,$
$2w + 4x - y + z = 1$

26. *Coins.* The value of 75 coins, consisting of only nickels and dimes, is $5.95. How many of each kind of coin are there? **[9.1]**

27. *Investment.* The Davidson family invested $5000, part at 3% and the remainder at 3.5%. The annual income from both investments is $167. What is the amount invested at each rate? **[9.1]**

28. *Nutrition.* A dietician must plan a breakfast menu that provides 460 Cal, 9 g of fat, and 55 mg of calcium. One plain bagel contains 200 Cal, 2 g of fat, and 29 mg of calcium. A one-tablespoon serving of cream cheese contains 100 Cal, 10 g of fat, and 24 mg of calcium. One banana contains 105 Cal, 1 g of fat, and 7 g of calcium. (*Source*: U.S. Department of Agriculture) How many servings of each are required to provide the desired nutritional values? **[9.2]**

29. *Test Scores.* A student has a total of 226 points on three tests. The sum of the scores on the first and second tests exceeds the score on the third test by 62. The first score exceeds the second by 6. Find the three scores. **[9.2]**

60. *Employed Civilians.* The following table lists the number of persons, ages 16 and older, employed in the United States, represented in terms of the number of years after 2008. **[9.2]**

Year, x	Persons Ages 16 and Older Employed (in millions)
2008, 0	145
2010, 2	139
2012, 4	142

Source: Bureau of Labor Statistics, U.S. Department of Labor

a) Use a system of equations to fit a quadratic function $f(x) = ax^2 + bx + c$ to the data.
b) Use the function to estimate the number of persons employed in 2014.

For Exercises 31–38, let

$$\mathbf{A} = \begin{bmatrix} 1 & -1 & 0 \\ 2 & 3 & -2 \\ -2 & 0 & 1 \end{bmatrix}, \quad \mathbf{B} = \begin{bmatrix} -1 & 0 & 6 \\ 1 & -2 & 0 \\ 0 & 1 & -3 \end{bmatrix},$$

and

$$\mathbf{C} = \begin{bmatrix} -2 & 0 \\ 1 & 3 \end{bmatrix}.$$

Find each of the following, if possible. **[9.4]**

31. $\mathbf{A} + \mathbf{B}$ **32.** $-3\mathbf{A}$

33. $-\mathbf{A}$ **34.** $\mathbf{AB}$

35. $\mathbf{B} + \mathbf{C}$ **36.** $\mathbf{A} - \mathbf{B}$

37. $\mathbf{BA}$ **38.** $\mathbf{A} + 3\mathbf{B}$

39. *Food Service Management.* The following table lists the cost per serving, in dollars, for items on four menus that are served at an NFL training camp.

Menu	Meat	Potato	Vegetable	Salad	Dessert
1	2.25	0.38	0.55	0.33	0.85
2	3.09	0.42	0.46	0.48	0.51
3	2.40	0.31	0.59	0.36	0.64
4	1.80	0.29	0.34	0.55	0.52

On a particular day, a dietician orders 41 meals from menu 1, 18 from menu 2, 39 from menu 3, and 36 from menu 4.

a) Write the information in the table as a 4×5 matrix **M**. **[9.4]**
b) Write a row matrix **N** that represents the number of each menu ordered. **[9.4]**
c) Find the product **NM**. **[9.4]**
d) State what the entries of **NM** represent. **[9.4]**

Find $\mathbf{A}^{-1}$, if it exists. **[9.5]**

40. $\mathbf{A} = \begin{bmatrix} -2 & 0 \\ 1 & 3 \end{bmatrix}$

41. $\mathbf{A} = \begin{bmatrix} 0 & 0 & 3 \\ 0 & -2 & 0 \\ 4 & 0 & 0 \end{bmatrix}$

42. $\mathbf{A} = \begin{bmatrix} 1 & 0 & 0 & 0 \\ 0 & 4 & -5 & 0 \\ 0 & 2 & 2 & 0 \\ 0 & 0 & 0 & 1 \end{bmatrix}$

43. Write a matrix equation equivalent to this system of equations:

$$3x - 2y + 4z = 13,$$
$$x + 5y - 3z = 7,$$
$$2x - 3y + 7z = -8. \quad \textbf{[9.4]}$$

Solve the system of equations using the inverse of the coefficient matrix of the equivalent matrix equation. **[9.5]**

44. $2x + 3y = 5,$
$3x + 5y = 11$

45. $5x - y + 2z = 17,$
$3x + 2y - 3z = -16,$
$4x - 3y - z = 5$

46. $w - x - y + z = -1,$
$2w + 3x - 2y - z = 2,$
$-w + 5x + 4y - 2z = 3,$
$3w - 2x + 5y + 3z = 4$

Evaluate the determinant. **[9.6]**

47. $\begin{vmatrix} 1 & -2 \\ 3 & 4 \end{vmatrix}$ **48.** $\begin{vmatrix} \sqrt{3} & -5 \\ -3 & -\sqrt{3} \end{vmatrix}$

49. $\begin{vmatrix} 2 & -1 & 1 \\ 1 & 2 & -1 \\ 3 & 4 & -3 \end{vmatrix}$ **50.** $\begin{vmatrix} 1 & -1 & 2 \\ -1 & 2 & 0 \\ -1 & 3 & 1 \end{vmatrix}$

Solve using Cramer's rule. **[9.6]**

51. $5x - 2y = 19,$
$7x + 3y = 15$

52. $x + y = 4,$
$4x + 3y = 11$

53. $3x - 2y + z = 5,$
$4x - 5y - z = -1,$
$3x + 2y - z = 4$

54. $2x - y - z = 2,$
$3x + 2y + 2z = 10,$
$x - 5y - 3z = -2$

Graph. **[9.7]**

55. $y \le 3x + 6$ **56.** $4x - 3y \ge 12$

57. Graph this system of inequalities and find the coordinates of any vertices formed. **[9.7]**

$2x + y \ge 9,$
$4x + 3y \ge 23,$
$x + 3y \ge 8,$
$x \ge 0,$
$y \ge 0$

58. Find the maximum value and the minimum value of $T = 6x + 10y$ subject to

$x + y \le 10,$
$5x + 10y \ge 50,$
$x \ge 2,$
$y \ge 0.$ **[9.7]**

59. *Maximizing a Test Score.* Jackson is taking a test that contains questions in group A worth 7 points each and questions in group B worth 12 points each. The total number of questions answered must be at least 8. If Jackson knows that group A questions take 8 min each and group B questions take 10 min each and the maximum time for the test is 80 min, how many questions from each group must he answer correctly in order to maximize his score? What is the maximum score? **[9.7]**

Decompose into partial fractions. **[9.8]**

60. $\dfrac{5}{(x + 2)^2(x + 1)}$

61. $\dfrac{-8x + 23}{2x^2 + 5x - 12}$

62. Solve: $2x + y = 7,$
$x - 2y = 6.$ **[9.1]**

A. x and y are both positive numbers.
B. x and y are both negative numbers.
C. x is positive and y is negative.
D. x is negative and y is positive.

63. Which of the following is *not* a row-equivalent operation on a matrix? **[9.3]**

A. Interchange any two columns.
B. Interchange any two rows.
C. Add two rows.
D. Multiply each entry in a row by -3.

64. The graph of the given system of inequalities is which of the following? **[9.7]**

$x + y \le 3,$
$x - y \le 4$

A. 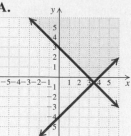   **B.**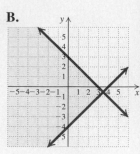

C. 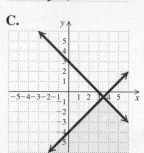   **D.**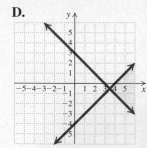

❯ Synthesis

65. One year, Lucia invested a total of $40,000, part at 4%, part at 5%, and the rest at $5\frac{1}{2}$%. The total amount of interest received on the investments was $1990. The interest received on the $5\frac{1}{2}$% investment was $590 more than the interest received on the 4% investment. How much was invested at each rate? **[9.2]**

Solve.

66. $\dfrac{2}{3x} + \dfrac{4}{5y} = 8,$

$\dfrac{5}{4x} - \dfrac{3}{2y} = -6$ **[9.1]**

67. $\dfrac{3}{x} - \dfrac{4}{y} + \dfrac{1}{z} = -2,$

$\dfrac{5}{x} + \dfrac{1}{y} - \dfrac{2}{z} = 1,$

$\dfrac{7}{x} + \dfrac{3}{y} + \dfrac{2}{z} = 19$ **[9.2]**

Graph. **[9.7]**

68. $|x| - |y| \le 1$ **69.** $|xy| > 1$

❯ Collaborative Discussion and Writing

70. Dylon solves the equation $2x + 5 = 3x - 7$ by finding the point of intersection of the graphs of $y_1 = 2x + 5$ and $y_2 = 3x - 7$. She finds the same point when she solves the system of equations

$$y = 2x + 5,$$
$$y = 3x - 7.$$

Explain the difference between the solution of the equation and the solution of the system of equations. **[9.1]**

71. For square matrices **A** and **B**, is it true, in general, that $(\mathbf{AB})^2 = \mathbf{A}^2\mathbf{B}^2$? Explain. **[9.4]**

72. Given the system of equations

$$a_1x + b_1y = c_1,$$
$$a_2x + b_2y = c_2,$$

explain why the equations are dependent or the system is inconsistent when

$$\begin{vmatrix} a_1 & b_1 \\ a_2 & b_2 \end{vmatrix} = 0. \quad \textbf{[9.6]}$$

73. If the lines $a_1x + b_1y = c_1$ and $a_2x + b_2y = c_2$ are parallel, what can you say about the values of

$$\begin{vmatrix} a_1 & b_1 \\ a_2 & b_2 \end{vmatrix}, \quad \begin{vmatrix} c_1 & b_1 \\ c_2 & b_2 \end{vmatrix}, \quad \text{and} \quad \begin{vmatrix} a_1 & c_1 \\ a_2 & c_2 \end{vmatrix}? \quad \textbf{[9.6]}$$

74. Describe how the graph of a linear inequality differs from the graph of a linear equation. **[9.7]**

75. What would you say to a classmate who tells you that the partial fraction decomposition of

$$\frac{3x^2 - 8x + 9}{(x + 3)(x^2 - 5x + 6)}$$

is

$$\frac{2}{x + 3} + \frac{x - 1}{x^2 - 5x + 6}?$$

Explain. **[9.8]**

---

9 Chapter Test

Solve. Use any method. Also determine whether the system is consistent or inconsistent and whether the equations are dependent or independent.

1. $3x + 2y = 1,$
$\quad 2x - y = -11$

2. $2x - y = 3,$
$\quad 2y = 4x - 6$

3. $x - y = 4,$
$\quad 3y = 3x - 8$

4. $2x - 3y = 8,$
$\quad 5x - 2y = 9$

Solve.

5. $4x + 2y + z = 4,$
$\quad 3x - y + 5z = 4,$
$\quad 5x + 3y - 3z = -2$

6. *Ticket Sales.* One evening, 620 tickets were sold for Clearview Community College's talent show. Tickets cost $8 each for students and $12 each for nonstudents. Total receipts were $5592. How many of each type of ticket were sold?

7. Hui, Ashlyn, and Sheriann can process 352 online orders per day. Hui and Ashlyn together can process 224 orders per day while Hui and Sheriann together can process 248 orders per day. How many orders can each of them process alone?

For Exercises 8–13, let

$$\mathbf{A} = \begin{bmatrix} 1 & -1 & 3 \\ -2 & 5 & 2 \end{bmatrix}, \quad \mathbf{B} = \begin{bmatrix} -5 & 1 \\ -2 & 4 \end{bmatrix},$$

and

$$\mathbf{C} = \begin{bmatrix} 3 & -4 \\ -1 & 0 \end{bmatrix}.$$

Find each of the following, if possible.

8. $\mathbf{B} + \mathbf{C}$

9. $\mathbf{A} - \mathbf{C}$

10. $\mathbf{CB}$

11. $\mathbf{AB}$

12. $2\mathbf{A}$

13. $\mathbf{C}^{-1}$

14. *Food Service Management.* The following table lists the cost per serving, in dollars, for items on three lunch menus served at a senior citizens' center.

Menu	Main Dish	Side Dish	Dessert
1	1.55	1.00	0.99
2	1.70	0.95	1.01
3	1.65	0.99	0.96

On a particular day, 26 Menu 1 meals, 18 Menu 2 meals, and 23 Menu 3 meals are served.

a) Write the information in the table as a 3×3 matrix **M**.
b) Write a row matrix **N** that represents the number of each menu served.
c) Find the product **NM**.
d) State what the entries of **NM** represent.

15. Write a matrix equation equivalent to the system of equations

$$3x - 4y + 2z = -8,$$
$$2x + 3y + z = 7,$$
$$x - 5y - 3z = 3.$$

16. Solve the system of equations using the inverse of the coefficient matrix of the equivalent matrix equation.

$$3x + 2y + 6z = 2,$$
$$x + y + 2z = 1,$$
$$2x + 2y + 5z = 3$$

Evaluate the determinant.

17. $\begin{vmatrix} 3 & -5 \\ 8 & 7 \end{vmatrix}$

18. $\begin{vmatrix} 2 & -1 & 4 \\ -3 & 1 & -2 \\ 5 & 3 & -1 \end{vmatrix}$

19. Solve using Cramer's rule. Show your work.

$$5x + 2y = -1,$$
$$7x + 6y = 1$$

20. Graph: $3x + 4y \le -12$.

21. Find the maximum value and the minimum value of $Q = 2x + 3y$ subject to

$$x + y \le 6,$$
$$2x - 3y \ge -3,$$
$$x \ge 1,$$
$$y \ge 0.$$

22. *Maximizing Profit.* Jane's Cakes prepares pound cakes and carrot cakes. In a given week, at most 100 cakes can be prepared, of which 25 pound cakes and 15 carrot cakes are required by regular customers. The profit from each pound cake is $6, and the profit from each carrot cake is $8. How many of each type of cake should be prepared in order to maximize the profit? What is the maximum profit?

23. Decompose into partial fractions:

$$\frac{3x - 11}{x^2 + 2x - 3}.$$

24. The graph of the given system of inequalities is which of the following?

$$x + 2y \ge 4,$$
$$x - y \le 2$$

A.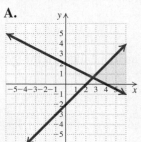

B.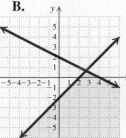

C.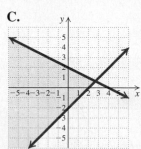

D.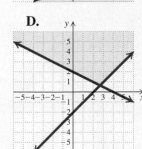

❯ Synthesis

25. Three solutions of the equation $Ax - By = Cz - 8$ are $(2, -2, 2)$, $(-3, -1, 1)$, and $(4, 2, 9)$. Find A, B, and C.

Analytic Geometry Topics

APPLICATION

This problem appears as Exercise 33 in Section 10.1.

An engineer designs a satellite dish with a parabolic cross section. The dish is 15 ft wide at the opening, and the focus is placed 4 ft from the vertex. Find the depth of the satellite dish at the vertex.

10.1 The Parabola

> Given an equation of a parabola, complete the square, if necessary, and then find the vertex, the focus, and the directrix and graph the parabola.

A **conic section** is formed when a right circular cone with two parts, called *nappes* is intersected by a plane. One of four types of curves can be formed: a parabola, circle, an ellipse, or a hyperbola.

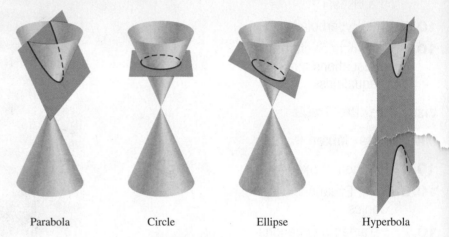

| Parabola | Circle | Ellipse | Hyperbola |

Conic Sections

Conic sections can be defined algebraically using second-degree equations o the form $Ax^2 + Bxy + Cy^2 + Dx + Ey + F = 0$. In addition, they can be define geometrically as a set of points that satisfy certain conditions.

❯ Parabolas

The graph of the quadratic function $f(x) = ax^2 + bx + c$, $a \neq 0$, is a parabola. A parabola can be defined geometrically.

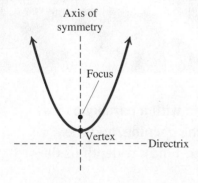

> **PARABOLA**
>
> A **parabola** is the set of all points in a plane equidistant from a fixed line (the **directrix**) and a fixed point not on the line (the **focus**).

The line that is perpendicular to the directrix and contains the focus is the **axi of symmetry**. The **vertex** is the midpoint of the segment between the focus and th directrix. (See the figure at left.)

Let's derive the standard equation of a parabola with vertex $(0, 0)$ and direc trix $y = -p$, where $p > 0$. We place the coordinate axes as shown in Fig. 1 on th following page. The y-axis is the axis of symmetry and contains the focus F. Th distance from the focus to the vertex is the same as the distance from the vertex t the directrix. Thus the coordinates of F are $(0, p)$.

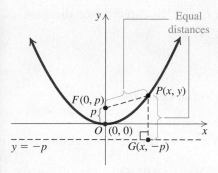

FIGURE 1.

Let $P(x, y)$ be any point on the parabola and consider $\overline{PG}$ perpendicular to the line $y = -p$. The coordinates of G are $(x, -p)$. By the definition of a parabola,

$$PF = PG.$$ **The distance from P to the focus is the same as the distance from P to the directrix.**

Then using the distance formula, we have

$$\sqrt{(x - 0)^2 + (y - p)^2} = \sqrt{(x - x)^2 + [y - (-p)]^2}$$
$$x^2 + y^2 - 2py + p^2 = y^2 + 2py + p^2 \qquad \text{**Squaring both sides and squaring the binomials**}$$
$$x^2 = 4py.$$

We have shown that if $P(x, y)$ is on the parabola shown in Fig. 1 above, then its coordinates satisfy this equation. The converse is also true, but we will not prove it here.

Note that if $p > 0$, as above, the graph opens up. If $p < 0$, the graph opens down.

The equation of a parabola with vertex $(0, 0)$ and directrix $x = -p$ is derived in a similar manner. Such a parabola opens either to the right $(p > 0)$, as shown in Fig. 2 below, or to the left $(p < 0)$.

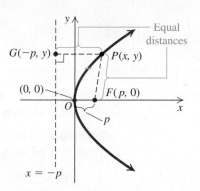

FIGURE 2.

STANDARD EQUATION OF A PARABOLA WITH VERTEX (0, 0) AND VERTICAL AXIS OF SYMMETRY

The standard equation of a parabola with vertex $(0, 0)$ and directrix $y = -p$ is

$$x^2 = 4py.$$

The focus is $(0, p)$ and the y-axis is the axis of symmetry. When $p > 0$, the parabola opens up; when $p < 0$, the parabola opens down.

STANDARD EQUATION OF A PARABOLA WITH VERTEX (0, 0) AND HORIZONTAL AXIS OF SYMMETRY

The standard equation of a parabola with vertex $(0, 0)$ and directrix $x = -p$ is

$$y^2 = 4px.$$

The focus is $(p, 0)$ and the x-axis is the axis of symmetry. When $p > 0$, the parabola opens to the right; when $p < 0$, the parabola opens to the left.

EXAMPLE 1 Find the focus, the vertex, and the directrix of the parabola $y = -\frac{1}{12}x^2$. Then graph the parabola.

Solution We write $y = -\frac{1}{12}x^2$ in the form $x^2 = 4py$:

$$-\frac{1}{12}x^2 = y \qquad \text{**Given equation**}$$
$$x^2 = -12y \qquad \text{**Multiplying by -12 on both sides**}$$
$$x^2 = 4(-3)y. \qquad \text{**Standard form**}$$

Since the equation can be written in the form $x^2 = 4py$, we know that the vertex is $(0, 0)$.

We have $p = -3$, so the focus is $(0, p)$, or $(0, -3)$. The directrix is $y = -p = -(-3) = 3$.

x	y
0	0
± 1	$-\frac{1}{12}$
± 2	$-\frac{1}{3}$
± 3	$-\frac{3}{4}$
± 4	$-\frac{4}{3}$

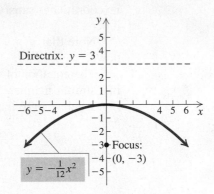

Now Try Exercise 7.

EXAMPLE 2 Find an equation of the parabola with vertex $(0, 0)$ and focus $(5, 0)$. Then graph the parabola.

Solution The focus is on the x-axis so the line of symmetry is the x-axis. The equation is of the type

$$y^2 = 4px.$$

Since the focus $(5, 0)$ is 5 units to the right of the vertex, $p = 5$ and the equation is

$$y^2 = 4(5)x, \quad \text{or} \quad y^2 = 20x.$$

x	y^2	y	(x, y)
0	0	0	$(0, 0)$
1	20	$\pm\sqrt{20}$	$(1, 4.47)$
			$(1, -4.47)$
2	40	$\pm\sqrt{40}$	$(2, 6.32)$
			$(2, -6.32)$
3	60	$\pm\sqrt{60}$	$(3, 7.75)$
			$(3, -7.75)$

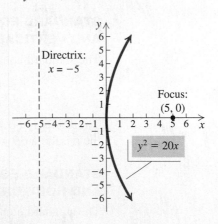

We can also use a graphing calculator to graph parabolas. It might be necessary to solve the equation for y before entering it in the calculator:

$$y^2 = 20x$$
$$y = \pm\sqrt{20x}.$$

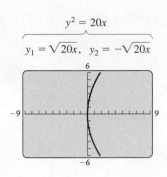

$$y^2 = 20x$$

$$y_1 = \sqrt{20x}, \quad y_2 = -\sqrt{20x}$$

We now graph

$$y_1 = \sqrt{20x} \quad \text{and} \quad y_2 = -\sqrt{20x}, \quad \text{or} \quad y_1 = \sqrt{20x} \quad \text{and} \quad y_2 = -y_1,$$

in a squared viewing window.

On some graphing calculators, the Conics application from the APPS menu can be used to graph parabolas. This method will be discussed in Example 4.

> *Now Try Exercise 15.*

❯ Finding Standard Form by Completing the Square

If a parabola with vertex at the origin is translated horizontally $|h|$ units and vertically $|k|$ units, it has an equation as follows.

STANDARD EQUATION OF A PARABOLA WITH VERTEX (h, k) AND VERTICAL AXIS OF SYMMETRY

The standard equation of a parabola with vertex (h, k) and vertical axis of symmetry is

$$(x - h)^2 = 4p(y - k),$$

where the vertex is (h, k), the focus is $(h, k + p)$, and the directrix is $y = k - p$.

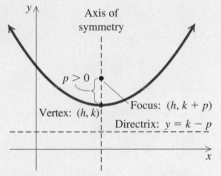

(When $p < 0$, the parabola opens down.)

STANDARD EQUATION OF A PARABOLA WITH VERTEX (h, k) AND HORIZONTAL AXIS OF SYMMETRY

The standard equation of a parabola with vertex (h, k) and horizontal axis of symmetry is

$$(y - k)^2 = 4p(x - h),$$

where the vertex is (h, k), the focus is $(h + p, k)$, and the directrix is $x = h - p$.

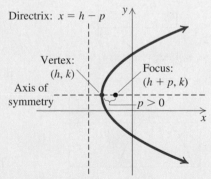

(When $p < 0$, the parabola opens to the left.)

COMPLETING THE SQUARE

REVIEW SECTION 3.2.

Just in Time

16

We can complete the square on equations of the form $y = ax^2 + bx + c$ or $x = ay^2 + by + c$ in order to write them in standard form.

EXAMPLE 3 For the parabola

$$x^2 + 6x + 4y + 5 = 0,$$

find the vertex, the focus, and the directrix. Then draw the graph.

Solution We first complete the square:

$$x^2 + 6x + 4y + 5 = 0$$

$$x^2 + 6x \qquad\quad = -4y - 5 \qquad\qquad \text{Subtracting } 4y \text{ and } 5 \text{ on both sides}$$

$$x^2 + 6x + 9 = -4y - 5 + 9 \qquad \text{Adding 9 on both sides to complete the square on the left side}$$

$$x^2 + 6x + 9 = -4y + 4$$

$$(x + 3)^2 = -4(y - 1) \qquad\qquad \text{Factoring}$$

$$[x - (-3)]^2 = 4(-1)(y - 1). \qquad \text{Writing standard form:}$$
$$\quad (x - h)^2 = 4p(y - k)$$

We see that $h = -3$, $k = 1$, and $p = -1$, so we have the following:

Vertex (h, k): $\qquad\qquad (-3, 1)$;

Focus $(h, k + p)$: $\qquad (-3, 1 + (-1))$, or $(-3, 0)$;

Directrix $y = k - p$: $\quad y = 1 - (-1)$, or $y = 2$.

x	y
-4	$\frac{3}{4}$
-2	$\frac{3}{4}$
-5	0
-1	0
-6	$-\frac{5}{4}$
0	$-\frac{5}{4}$

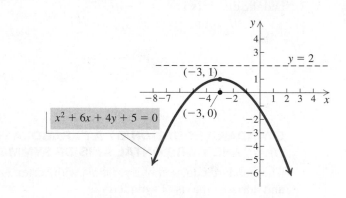

$x^2 + 6x + 4y + 5 = 0$

$(-3, 1)$

$(-3, 0)$

$y = 2$

$y = \frac{1}{4}(-x^2 - 6x - 5)$

We can check the graph on a graphing calculator using a squared viewing window, as shown at left. It might be necessary to solve for y first:

$$x^2 + 6x + 4y + 5 = 0$$

$$4y = -x^2 - 6x - 5$$

$$y = \frac{1}{4}(-x^2 - 6x - 5).$$

The hand-drawn graph appears to be correct.

> *Now Try Exercise 25.*

EXAMPLE 4 For the parabola

$$y^2 - 2y - 8x - 31 = 0,$$

find the vertex, the focus, and the directrix. Then draw the graph.

Solution We first complete the square:

$$y^2 - 2y - 8x - 31 = 0$$

$$y^2 - 2y \qquad\quad = 8x + 31 \qquad\qquad \text{Adding } 8x \text{ and } 31 \text{ on both sides}$$

$$y^2 - 2y + 1 = 8x + 31 + 1. \qquad \text{Adding 1 on both sides to complete the square on the left side}$$

Then

$$y^2 - 2y + 1 = 8x + 32$$

$$(y - 1)^2 = 8(x + 4) \qquad \text{Factoring}$$

$$(y - 1)^2 = 4(2)[x - (-4)]. \qquad \begin{array}{l}\textbf{Writing standard form:} \\ (y - k)^2 = 4p(x - h)\end{array}$$

We see that $h = -4$, $k = 1$, and $p = 2$, so we have the following:

Vertex (h, k): $(-4, 1)$;

Focus $(h + p, k)$: $(-4 + 2, 1)$, or $(-2, 1)$;

Directrix $x = h - p$: $x = -4 - 2$, or $x = -6$.

x	y
$-\frac{31}{8}$	2
$-\frac{31}{8}$	0
$-\frac{7}{2}$	3
$-\frac{7}{2}$	-1
$-\frac{23}{8}$	4
$-\frac{23}{8}$	-2

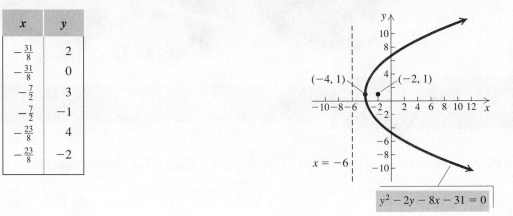

When the equation of a parabola is written in standard form, we can use the Conics PARABOLA APP to graph it, as shown on the left below.

$$(y - 1)^2 = 4(2)[x - (-4)]$$

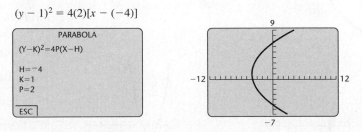

We can also draw the graph or check the hand-drawn graph on a graphing calculator by first solving the original equation for y using the quadratic formula.

> **Now Try Exercise 31.**

❯ Applications

Parabolas have many applications. For example, cross sections of car headlights, flashlights, and searchlights are parabolas. The bulb is located at the focus and light from that point is reflected outward parallel to the axis of symmetry. Satellite dishes and field microphones used at sporting events often have parabolic cross sections. Incoming radio waves or sound waves parallel to the axis are reflected into the focus.

Similarly, in solar cooking, a parabolic mirror is mounted on a rack with a cooking pot hung in the focal area. Incoming sun rays parallel to the axis are reflected into the focus, producing a temperature high enough for cooking.

10.1   Exercise Set

In Exercises 1–6, match the equation with one of the graphs (a)–(f) that follow.

a)

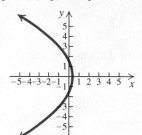

b)

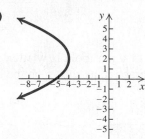

c)

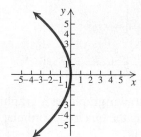

d)

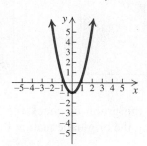

e)

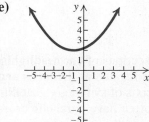

f)

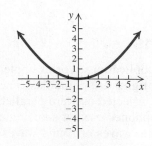

1. $x^2 = 8y$

2. $y^2 = -10x$

3. $(y - 2)^2 = -3(x + 4)$

4. $(x + 1)^2 = 5(y - 2)$

5. $13x^2 - 8y - 9 = 0$

6. $41x + 6y^2 = 12$

Find the vertex, the focus, and the directrix. Then draw the graph.

7. $x^2 = 20y$

8. $x^2 = 16y$

9. $y^2 = -6x$

10. $y^2 = -2x$

11. $x^2 - 4y = 0$

12. $y^2 + 4x = 0$

13. $x = 2y^2$

14. $y = \frac{1}{2}x^2$

Find an equation of a parabola satisfying the given conditions.

15. Vertex $(0, 0)$, focus $(-3, 0)$

16. Vertex $(0, 0)$, focus $(0, 10)$

17. Focus $(7, 0)$, directrix $x = -7$

18. Focus $\left(0, \frac{1}{4}\right)$, directrix $y = -\frac{1}{4}$

19. Focus $(0, -\pi)$, directrix $y = \pi$

20. Focus $(-\sqrt{2}, 0)$, directrix $x = \sqrt{2}$

21. Focus $(3, 2)$, directrix $x = -4$

22. Focus $(-2, 3)$, directrix $y = -3$

Find the vertex, the focus, and the directrix. Then draw the graph.

23. $(x + 2)^2 = -6(y - 1)$

24. $(y - 3)^2 = -20(x + 2)$

25. $x^2 + 2x + 2y + 7 = 0$

26. $y^2 + 6y - x + 16 = 0$

27. $x^2 - y - 2 = 0$

28. $x^2 - 4x - 2y = 0$

29. $y = x^2 + 4x + 3$

30. $y = x^2 + 6x + 10$

31. $y^2 - y - x + 6 = 0$

32. $y^2 + y - x - 4 = 0$

33. *Satellite Dish.* An engineer designs a satellite dish with a parabolic cross section. The dish is 15 ft wide at the opening, and the focus is placed 4 ft from the vertex.

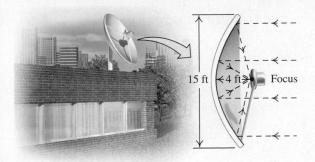

15 ft   4 ft   Focus

a) Position a coordinate system with the origin at the vertex and the *x*-axis on the parabola's axis of symmetry and find an equation of the parabola.
b) Find the depth of the satellite dish at the vertex.

34. *Flashlight Mirror.* A heavy-duty flashlight mirror has a parabolic cross section with diameter 6 in. and depth 1 in.

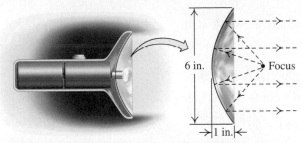

6 in.   Focus

1 in.

a) Position a coordinate system with the origin at the vertex and the *x*-axis on the parabola's axis of symmetry and find an equation of the parabola.
b) How far from the vertex should the bulb be positioned if it is to be placed at the focus?

35. *Ultrasound Receiver.* Information Unlimited designed and sells the Ultrasonic Receiver, which detects sounds unable to be heard by the human ear. The HT90P can detect mechanical and electrical sounds such as leaking gases, air, corona, and motor friction noises. It can also be used to hear bats, insects, and even beading water. The receiver has a parabolic cross section and is 2.625 in. deep. The focus is 3.287 in. from the vertex. (*Source*: Information Unlimited, Amherst, NH, Robert Iannini, President) Find the diameter of the outside edge of the receiver.

36. *Spotlight.* A spotlight has a parabolic cross section that is 4 ft wide at the opening and 1.5 ft deep at the vertex. How far from the vertex is the focus?

❯ Skill Maintenance

Consider the following linear equations. Without graphing them, answer the questions below.

a) $y = 2x$ **b)** $y = \frac{1}{3}x + 5$
c) $y = -3x - 2$ **d)** $y = -0.9x + 7$
e) $y = -5x + 3$ **f)** $y = x + 4$
g) $8x - 4y = 7$ **h)** $3x + 6y = 2$

37. Which has/have *x*-intercept $\left(\frac{2}{3}, 0\right)$? **[1.1]**

38. Which has/have *y*-intercept $(0, 7)$? **[1.1], [1.4]**

39. Which slant up from left to right? **[1.3]**

40. Which has/have the least steep slant? **[1.3]**

41. Which has/have slope $\frac{1}{3}$? **[1.4]**

42. Which, if any, contain the point $(3, 7)$? **[1.1]**

43. Which, if any, are parallel? **[1.4]**

44. Which, if any, are perpendicular? **[1.4]**

❯ Synthesis

45. Find an equation of the parabola with a vertical axis of symmetry and vertex $(-1, 2)$ and containing the point $(-3, 1)$.

46. Find an equation of a parabola with a horizontal axis of symmetry and vertex $(-2, 1)$ and containing the point $(-3, 5)$.

Use a graphing calculator to find the vertex, the focus, and the directrix of each of the following.

47. $4.5x^2 - 7.8x + 9.7y = 0$

48. $134.1y^2 + 43.4x - 316.6y - 122.4 = 0$

49. *Suspension Bridge.* The cables of a 200-ft portion of the roadbed of a suspension bridge are positioned as shown below. Vertical cables are to be spaced every 20 ft along this portion of the roadbed. Calculate the lengths of these vertical cables.

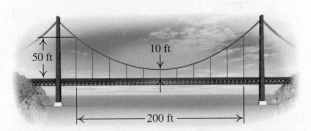

50 ft   10 ft

200 ft

10.2 ▷ The Circle and the Ellipse

> ❯ Given an equation of a circle, complete the square, if necessary, and then find the center and the radius and graph the circle.
>
> ❯ Given an equation of an ellipse, complete the square, if necessary, and then find the center, the vertices, and the foci and graph the ellipse.

❯ Circles

CIRCLES
REVIEW SECTION 1.1.

We can define a circle geometrically.

CIRCLE

A **circle** is the set of all points in a plane that are at a fixed distance from a fixed point (the **center**) in the plane.

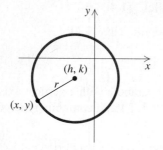

Recall the standard equation of a circle with center (h, k) and radius r.

STANDARD EQUATION OF A CIRCLE

The standard equation of a circle with center (h, k) and radius r is
$$(x - h)^2 + (y - k)^2 = r^2.$$

EXAMPLE 1   For the circle
$$x^2 + y^2 - 16x + 14y + 32 = 0,$$
find the center and the radius. Then graph the circle.

Solution   First, we complete the square twice:

$$x^2 + y^2 - 16x + 14y + 32 = 0$$
$$x^2 - 16x \qquad + y^2 + 14y \qquad = -32$$
$$x^2 - 16x + 64 + y^2 + 14y + 49 = -32 + 64 + 49$$

$[\frac{1}{2}(-16)]^2 = (-8)^2 = 64$ and $(\frac{1}{2} \cdot 14)^2 = 7^2 = 49$; **adding 64 and 49 on both sides to complete the square twice on the left side**

$$(x - 8)^2 + (y + 7)^2 = 81$$
$$(x - 8)^2 + [y - (-7)]^2 = 9^2. \qquad \textbf{Writing standard form}$$

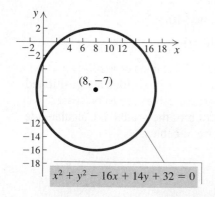

The center is $(8, -7)$ and the radius is 9. We graph the circle as shown at left.

To use a graphing calculator to graph the circle, it might be necessary to solve for y first. The original equation can be solved using the quadratic formula, or the standard form of the equation can be solved using the principle of square roots.

When we use the Conics CIRCLE APP on a graphing calculator to graph a circle, it is not necessary to write the equation in standard form or to solve it for y first. We enter the coefficients of x^2, y^2, x, and y and also the constant term when the equation is written in the form $ax^2 + ay^2 + bx + cy + d = 0$.

$$x^2 + y^2 - 16x + 14y + 32 = 0$$

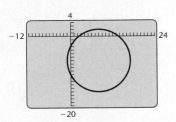

Some graphing calculators have a DRAW feature that provides a quick way to graph a circle when the center and the radius are known. This feature is described on p. 11.

> **Now Try Exercise 7.**

❯ Ellipses

We have studied two conic sections, the parabola and the circle. Now we turn our attention to a third, the *ellipse*.

ELLIPSE

An **ellipse** is the set of all points in a plane, the sum of whose distances from two fixed points (the **foci**) is constant. The **center** of an ellipse is the mid-point of the segment between the foci.

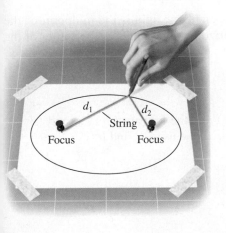

We can draw an ellipse by first placing two thumbtacks in a piece of cardboard, as shown at left. These are the foci (singular, *focus*). We then attach a piece of string to the tacks. Its length is the constant sum of the distances $d_1 + d_2$ from the foci to any point on the ellipse. Next, we trace a curve with a pencil held tight against the string. The figure traced is an ellipse.

Let's first consider the ellipse shown below with center at the origin. The points F_1 and F_2 are the foci. The segment $\overline{A'A}$ is the **major axis**, and the points A' and A are the **vertices**. The segment $\overline{B'B}$ is the **minor axis**, and the points B' and B are the **y-intercepts**. Note that the major axis of an ellipse is longer than the minor axis.

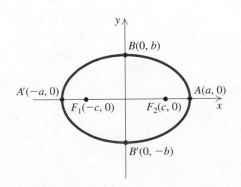

STANDARD EQUATION OF AN ELLIPSE WITH CENTER AT THE ORIGIN

Major Axis Horizontal

$$\frac{x^2}{a^2} + \frac{y^2}{b^2} = 1, \quad a > b > 0$$

Vertices: $(-a, 0), (a, 0)$
y-intercepts: $(0, -b), (0, b)$
Foci: $(-c, 0), (c, 0)$, where $c^2 = a^2 - b^2$

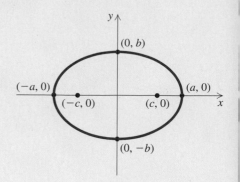

Major Axis Vertical

$$\frac{x^2}{b^2} + \frac{y^2}{a^2} = 1, \quad a > b > 0$$

Vertices: $(0, -a), (0, a)$
x-intercepts: $(-b, 0), (b, 0)$
Foci: $(0, -c), (0, c)$, where $c^2 = a^2 - b^2$

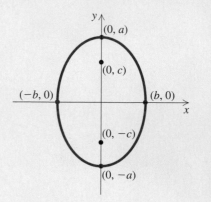

EXAMPLE 2   Find the standard equation of the ellipse with vertices $(-5, 0)$ and $(5, 0)$ and foci $(-3, 0)$ and $(3, 0)$. Then graph the ellipse.

Solution   Since the foci are on the x-axis and the origin is the midpoint of the segment between them, the major axis is horizontal and $(0, 0)$ is the center of the ellipse. Thus the equation is of the form

$$\frac{x^2}{a^2} + \frac{y^2}{b^2} = 1.$$

Since the vertices are $(-5, 0)$ and $(5, 0)$ and the foci are $(-3, 0)$ and $(3, 0)$, we know that $a = 5$ and $c = 3$. These values can be used to find b^2:

$$c^2 = a^2 - b^2$$
$$3^2 = 5^2 - b^2$$
$$9 = 25 - b^2$$
$$b^2 = 16.$$

Thus the equation of the ellipse is

$$\frac{x^2}{25} + \frac{y^2}{16} = 1, \quad \text{or} \quad \frac{x^2}{5^2} + \frac{y^2}{4^2} = 1.$$

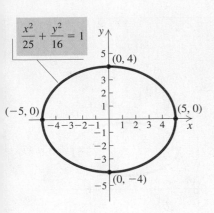

$$\frac{x^2}{25} + \frac{y^2}{16} = 1$$

To graph the ellipse, we plot the vertices $(-5, 0)$ and $(5, 0)$. Since $b^2 = 16$, we know that $b = 4$ and the y-intercepts are $(0, -4)$ and $(0, 4)$. We plot these points as well and connect the four points that we have plotted with a smooth curve.

When the equation of an ellipse is written in standard form, we can use the Conics ELLIPSE APP on a graphing calculator to graph it. Note that the center is $(0, 0)$, so $H = 0$ and $K = 0$.

$$\frac{x^2}{5^2} + \frac{y^2}{4^2} = 1$$

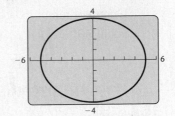

We can also draw the graph or check the hand-drawn graph on a graphing calculator by first solving for y.

> *Now Try Exercise 31.*

EXAMPLE 3 For the ellipse

$$9x^2 + 4y^2 = 36,$$

find the vertices and the foci. Then draw the graph.

Solution We first find standard form:

$$9x^2 + 4y^2 = 36$$

$$\frac{9x^2}{36} + \frac{4y^2}{36} = \frac{36}{36} \qquad \text{**Dividing by 36 on both sides to get 1 on the right side**}$$

$$\frac{x^2}{4} + \frac{y^2}{9} = 1$$

$$\frac{x^2}{2^2} + \frac{y^2}{3^2} = 1. \qquad \text{**Writing standard form**}$$

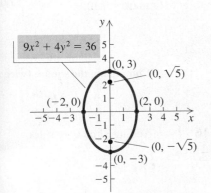

$$9x^2 + 4y^2 = 36$$

Thus, $a = 3$ and $b = 2$. The major axis is vertical, so the vertices are $(0, -3)$ and $(0, 3)$. Since we know that $c^2 = a^2 - b^2$, we have $c^2 = 3^2 - 2^2 = 9 - 4 = 5$, so $c = \sqrt{5}$ and the foci are $(0, -\sqrt{5})$ and $(0, \sqrt{5})$.

To graph the ellipse, we plot the vertices. Note also that since $b = 2$, the x-intercepts are $(-2, 0)$ and $(2, 0)$. We plot these points as well and connect the four points we have plotted with a smooth curve.

> *Now Try Exercise 25.*

If the center of an ellipse is not at the origin but at some point (h, k), then we can think of an ellipse with center at the origin being translated horizontally $|h|$ units and vertically $|k|$ units.

STANDARD EQUATION OF AN ELLIPSE WITH CENTER AT (h, k)

Major Axis Horizontal

$$\frac{(x-h)^2}{a^2} + \frac{(y-k)^2}{b^2} = 1, \quad a > b > 0$$

Vertices: $(h-a, k)$, $(h+a, k)$

Length of minor axis: $2b$

Foci: $(h-c, k)$, $(h+c, k)$, where $c^2 = a^2 - b^2$

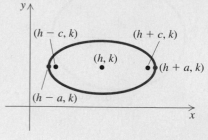

Major Axis Vertical

$$\frac{(x-h)^2}{b^2} + \frac{(y-k)^2}{a^2} = 1, \quad a > b > 0$$

Vertices: $(h, k-a)$, $(h, k+a)$

Length of minor axis: $2b$

Foci: $(h, k-c)$, $(h, k+c)$, where $c^2 = a^2 - b^2$

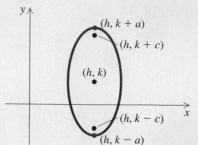

EXAMPLE 4   For the ellipse

$$4x^2 + y^2 + 24x - 2y + 21 = 0,$$

find the center, the vertices, and the foci. Then draw the graph.

Solution   First, we complete the square twice to get standard form:

$$4x^2 + y^2 + 24x - 2y + 21 = 0$$
$$4(x^2 + 6x \quad\quad) + (y^2 - 2y \quad\quad) = -21$$
$$4(x^2 + 6x + 9 - 9) + (y^2 - 2y + 1 - 1) = -21 \qquad \textbf{Completing the square twice}$$
$$4(x^2 + 6x + 9) + 4(-9) + (y^2 - 2y + 1) + (-1) = -21$$
$$4(x+3)^2 - 36 + (y-1)^2 - 1 = -21$$
$$4(x+3)^2 + (y-1)^2 = 16 \qquad \textbf{Adding 37 on both sides}$$
$$\tfrac{1}{16}[4(x+3)^2 + (y-1)^2] = \tfrac{1}{16}\cdot 16$$
$$\frac{(x+3)^2}{4} + \frac{(y-1)^2}{16} = 1$$
$$\frac{[x-(-3)]^2}{2^2} + \frac{(y-1)^2}{4^2} = 1. \qquad \textbf{Writing standard form}$$

The center is $(-3, 1)$. Note that $a = 4$ and $b = 2$. The major axis is vertical, so the vertices are 4 units above and below the center:

$$(-3, 1+4) \text{ and } (-3, 1-4), \quad \text{or} \quad (-3, 5) \text{ and } (-3, -3).$$

We know that $c^2 = a^2 - b^2$, so $c^2 = 4^2 - 2^2 = 16 - 4 = 12$ and $c = \sqrt{12}$, or $2\sqrt{3}$. Then the foci are $2\sqrt{3}$ units above and below the center:

$$(-3, 1 + 2\sqrt{3}) \quad \text{and} \quad (-3, 1 - 2\sqrt{3}).$$

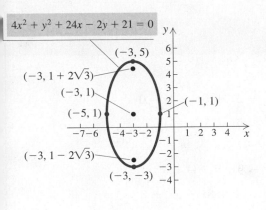

$$4x^2 + y^2 + 24x - 2y + 21 = 0$$

To graph the ellipse, we plot the vertices. Note also that since $b = 2$, two other points on the graph are the endpoints of the minor axis, 2 units right and left of the center:

$$(-3 + 2, 1) \quad \text{and} \quad (-3 - 2, 1),$$

or

$$(-1, 1) \quad \text{and} \quad (-5, 1).$$

We plot these points as well and connect the four points with a smooth curve, as shown at left.

When the equation of an ellipse is written in standard form, we can use the Conics ELLIPSE APP to graph it.

$$\frac{[x - (-3)]^2}{2^2} + \frac{(y - 1)^2}{4^2} = 1$$

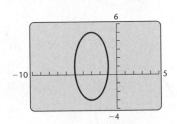

> **Now Try Exercise 43.**

❯ Applications

An exciting medical application of an ellipse is a device called a *lithotripter*. One type of this device uses electromagnetic technology to generate a shock wave to pulverize kidney stones. The wave originates at one focus of an ellipse and is reflected to the kidney stone, which is positioned at the other focus. Recovery time following the use of this technique is much shorter than with conventional surgery.

Ellipses have many other applications. Planets travel around the sun in elliptical orbits with the sun at one focus, for example, and satellites travel around the earth in elliptical orbits as well.

A room with an ellipsoidal ceiling is known as a *whispering gallery*. In such a room, a word whispered at one focus can be clearly heard at the other. Whispering galleries are found in the rotunda of the Capitol Building in Washington, D.C., and in St. Paul's Cathedral in London.

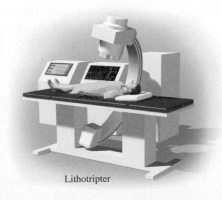

Lithotripter

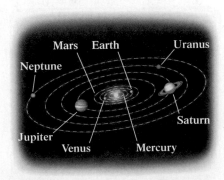

10.2 Exercise Set

In Exercises 1–6, match the equation with one of the graphs (a)–(f) that follow.

a)
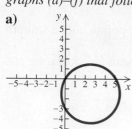

b)
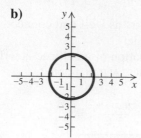

c)
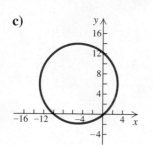

d)
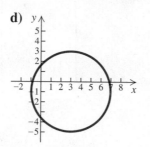

e)
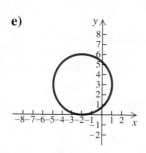

f)
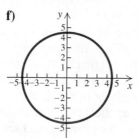

1. $x^2 + y^2 = 5$

2. $y^2 = 20 - x^2$

3. $x^2 + y^2 - 6x + 2y = 6$

4. $x^2 + y^2 + 10x - 12y = 3$

5. $x^2 + y^2 - 5x + 3y = 0$

6. $x^2 + 4x - 2 = 6y - y^2 - 6$

Find the center and the radius of the circle with the given equation. Then draw the graph.

7. $x^2 + y^2 - 14x + 4y = 11$

8. $x^2 + y^2 + 2x - 6y = -6$

9. $x^2 + y^2 + 6x - 2y = 6$

10. $x^2 + y^2 - 4x + 2y = 4$

11. $x^2 + y^2 + 4x - 6y - 12 = 0$

12. $x^2 + y^2 - 8x - 2y - 19 = 0$

13. $x^2 + y^2 - 6x - 8y + 16 = 0$

14. $x^2 + y^2 - 2x + 6y + 1 = 0$

15. $x^2 + y^2 + 6x - 10y = 0$

16. $x^2 + y^2 - 7x - 2y = 0$

17. $x^2 + y^2 - 9x = 7 - 4y$

18. $y^2 - 6y - 1 = 8x - x^2 + 3$

In Exercises 19–22, match the equation with one of the graphs (a)–(d) that follow.

a)
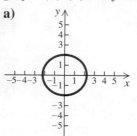

b)
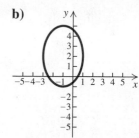

c)
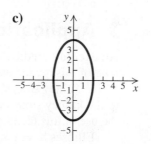

d)
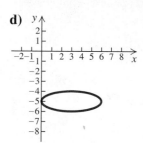

19. $16x^2 + 4y^2 = 64$

20. $4x^2 + 5y^2 = 20$

21. $x^2 + 9y^2 - 6x + 90y = -225$

22. $9x^2 + 4y^2 + 18x - 16y = 11$

Find the vertices and the foci of the ellipse with the given equation. Then draw the graph.

23. $\dfrac{x^2}{4} + \dfrac{y^2}{1} = 1$

24. $\dfrac{x^2}{25} + \dfrac{y^2}{36} = 1$

25. $16x^2 + 9y^2 = 144$

26. $9x^2 + 4y^2 = 36$

27. $2x^2 + 3y^2 = 6$

28. $5x^2 + 7y^2 = 35$

29. $4x^2 + 9y^2 = 1$

30. $25x^2 + 16y^2 = 1$

Find an equation of an ellipse satisfying the given conditions.

31. Vertices: $(-7, 0)$ and $(7, 0)$;
foci: $(-3, 0)$ and $(3, 0)$

32. Vertices: $(0, -6)$ and $(0, 6)$;
foci: $(0, -4)$ and $(0, 4)$

33. Vertices: $(0, -8)$ and $(0, 8)$;
length of minor axis: 10

34. Vertices: $(-5, 0)$ and $(5, 0)$;
length of minor axis: 6

35. Foci: $(-2, 0)$ and $(2, 0)$;
length of major axis: 6

36. Foci: $(0, -3)$ and $(0, 3)$;
length of major axis: 10

Find the center, the vertices, and the foci of the ellipse.
Then draw the graph.

37. $\dfrac{(x-1)^2}{9} + \dfrac{(y-2)^2}{4} = 1$

38. $\dfrac{(x-1)^2}{1} + \dfrac{(y-2)^2}{4} = 1$

39. $\dfrac{(x+3)^2}{25} + \dfrac{(y-5)^2}{36} = 1$

40. $\dfrac{(x-2)^2}{16} + \dfrac{(y+3)^2}{25} = 1$

41. $3(x+2)^2 + 4(y-1)^2 = 192$

42. $4(x-5)^2 + 3(y-4)^2 = 48$

43. $4x^2 + 9y^2 - 16x + 18y - 11 = 0$

44. $x^2 + 2y^2 - 10x + 8y + 29 = 0$

45. $4x^2 + y^2 - 8x - 2y + 1 = 0$

46. $9x^2 + 4y^2 + 54x - 8y + 49 = 0$

*The **eccentricity** of an ellipse is defined as $e = c/a$. For*
an ellipse, $0 < c < a$, so $0 < e < 1$. When e is close
to 0, an ellipse appears to be nearly circular. When e is
close to 1, an ellipse is very flat.

47. Observe the shapes of the ellipses in Examples 2
and 4. Which ellipse has the smaller eccentricity?
Confirm your answer by computing the
eccentricity of each ellipse.

48. Which ellipse has the smaller eccentricity? (Assume
that the coordinate systems have the same scale.)

a)   **b)**

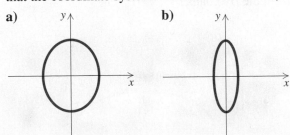

49. Find an equation of an ellipse with vertices
$(0, -4)$ and $(0, 4)$ and $e = \frac{1}{4}$.

50. Find an equation of an ellipse with vertices
$(-3, 0)$ and $(3, 0)$ and $e = \frac{7}{10}$.

51. *The Ellipse.*   The lighting of the National Christmas
Tree located on the Ellipse, a large grassy area south
of the White House, marks the beginning of the holiday season in Washington, D.C. This area of the lawn
is actually an ellipse with major axis of length 1048 ft
and minor axis of length 898 ft. Assuming that a
coordinate system is superimposed on the area in such
a way that the center is at the origin and the major and
minor axes are on the x- and y-axes of the coordinate
system, respectively, find an equation of the ellipse.

52. *Bridge Supports.*   The bridge support shown in the
following figure is the top half of an ellipse. Assuming that a coordinate system is super-imposed on the
drawing in such a way that point Q, the center of the
ellipse, is at the origin, find an equation of the ellipse.

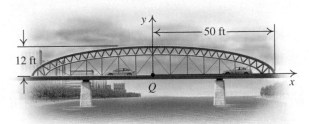

53. *Whispering Gallery.*   A whispering gallery, often
elliptical in shape, has acoustic properties such that a
whisper made at one point can be heard at other dis-
tant points. A science museum is designing a new exhibit hall that will illustrate a whispering gallery. The
hall will be 90 ft in length with the ceiling 30 ft high at
the center. How far are the foci from the center of the
ellipse? Round to the nearest tenth of a foot.

54. *Whispering Gallery.*   An art museum is adding a new
exhibit room in the shape of an ellipse. The director
wants to mark the foci so that a tour guide can stand at
one focus and without speaking loudly can be clearly
heard by a group touring the museum. If the room is
64 ft long and each focus is 5 ft from the outside wall
along the major axis, how high is the ceiling? Round
to the nearest tenth of a foot.

55. *Carpentry.* A carpenter is cutting a 3-ft by 4-ft elliptical sign from a 3-ft by 4-ft piece of plywood. The ellipse will be drawn using a string attached to the board at the foci of the ellipse.

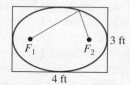

a) How far from the ends of the board should the string be attached?
b) How long should the string be?

56. *The Earth's Orbit.* The maximum distance of the earth from the sun is 9.3×10^7 mi. The minimum distance is 9.1×10^7 mi. The sun is at one focus of the elliptical orbit. Find the distance from the sun to the other focus.

❯ Skill Maintenance

Vocabulary Reinforcement

In each of Exercises 57–64, fill in the blank with the correct term. Some of the given choices will not be used.

piecewise function	ellipse
linear equation	midpoint
factor	distance
remainder	one real-number
solution	solution
zero	two different real-number
x-intercept	solutions
y-intercept	two different imaginary-
parabola	number solutions
circle	

57. The _____ between two points (x_1, y_1) and (x_2, y_2) is given by $\left(\dfrac{x_1 + x_2}{2}, \dfrac{y_1 + y_2}{2} \right)$. **[1.1]**

58. An input c of a function f is a(n) _____ of the function if $f(c) = 0$. **[1.5]**

59. A(n) _____ of the graph of an equation is a point $(0, b)$. **[1.1]**

60. For a quadratic equation $ax^2 + bx + c = 0$, if $b^2 - 4ac > 0$, the equation has _____. **[3.2]**

61. Given a polynomial $f(x)$, then $f(c)$ is the _____ that would be obtained by dividing $f(x)$ by $x - c$. **[4.3]**

62. A(n) _____ is the set of all points in a plane the sum of whose distances from two fixed points is constant. **[10.2]**

63. A(n) _____ is the set of all points in a plane equidistant from a fixed line and a fixed point not on the line. **[10.1]**

64. A(n) _____ is the set of all points in a plane that are at a fixed distance from a fixed point in the plane. **[10.2]**

❯ Synthesis

Find an equation of an ellipse satisfying the given conditions.

65. Vertices: $(3, -4)$, $(3, 6)$;
endpoints of minor axis: $(1, 1)$, $(5, 1)$

66. Vertices: $(-1, -1)$, $(-1, 5)$;
endpoints of minor axis: $(-3, 2)$, $(1, 2)$

67. Vertices: $(-3, 0)$ and $(3, 0)$;
passing through $\left(2, \frac{22}{3} \right)$

68. Center: $(-2, 3)$; major axis vertical;
length of major axis: 4;
length of minor axis: 1

Use a graphing calculator to find the center and the vertices of each of the following.

69. $4x^2 + 9y^2 - 16.025x + 18.0927y - 11.346 = 0$

70. $9x^2 + 4y^2 + 54.063x - 8.016y + 49.872 = 0$

71. *Bridge Arch.* A bridge with a semielliptical arch spans a river as shown here. What is the clearance 6 ft from the riverbank?

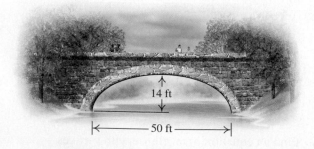

10.3 › The Hyperbola

> › Given an equation of a hyperbola, complete the square, if necessary, and then find the center, the vertices, and the foci and graph the hyperbola.

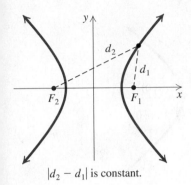

The last type of conic section that we will study is the *hyperbola*.

HYPERBOLA

A **hyperbola** is the set of all points in a plane for which the absolute value of the difference of the distances from two fixed points (the **foci**) is constant. The midpoint of the segment between the foci is the **center** of the hyperbola.

$|d_2 - d_1|$ is constant.

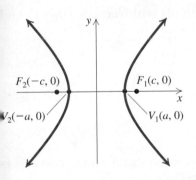

› Standard Equations of Hyperbolas

We first consider the equation of a hyperbola with center at the origin. In the figure at left, F_1 and F_2 are the foci. The segment $\overline{V_2 V_1}$ is the **transverse axis**, and the points V_2 and V_1 are the **vertices**.

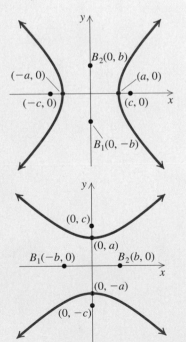

STANDARD EQUATION OF A HYPERBOLA WITH CENTER AT THE ORIGIN

Transverse Axis Horizontal

$$\frac{x^2}{a^2} - \frac{y^2}{b^2} = 1$$

Vertices: $(-a, 0)$, $(a, 0)$

Foci: $(-c, 0)$, $(c, 0)$,
    where $c^2 = a^2 + b^2$

Transverse Axis Vertical

$$\frac{y^2}{a^2} - \frac{x^2}{b^2} = 1$$

Vertices: $(0, -a)$, $(0, a)$

Foci: $(0, -c)$, $(0, c)$,
    where $c^2 = a^2 + b^2$

The segment $\overline{B_1B_2}$ is the **conjugate axis** of the hyperbola.

To graph a hyperbola with a horizontal transverse axis, it is helpful to begin by graphing the lines $y = -(b/a)x$ and $y = (b/a)x$. These are the **asymptotes** of the hyperbola. For a hyperbola with a vertical transverse axis, the asymptotes are $y = -(a/b)x$ and $y = (a/b)x$. As $|x|$ gets larger and larger, the graph of the hyperbola gets closer and closer to the asymptotes.

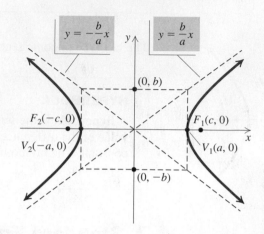

EXAMPLE 1   Find an equation of the hyperbola with vertices $(0, -4)$ and $(0, 4)$ and foci $(0, -6)$ and $(0, 6)$.

Solution   We know that $a = 4$ and $c = 6$. We find b^2:

$$c^2 = a^2 + b^2$$
$$6^2 = 4^2 + b^2$$
$$36 = 16 + b^2$$
$$20 = b^2.$$

Since the vertices and the foci are on the y-axis, we know that the transverse axis is vertical. We can now write the equation of the hyperbola:

$$\frac{y^2}{16} - \frac{x^2}{20} = 1. \qquad \frac{y^2}{a^2} - \frac{x^2}{b^2} = 1$$

> **Now Try Exercise 7.**

EXAMPLE 2   For the hyperbola given by

$$9x^2 - 16y^2 = 144,$$

find the vertices, the foci, and the asymptotes. Then graph the hyperbola.

Solution   First, we find standard form:

$$9x^2 - 16y^2 = 144$$

$$\frac{1}{144}(9x^2 - 16y^2) = \frac{1}{144} \cdot 144 \qquad \textbf{Multiplying by } \tfrac{1}{144} \textbf{ to get 1 on the right side}$$

$$\frac{x^2}{16} - \frac{y^2}{9} = 1$$

$$\frac{x^2}{4^2} - \frac{y^2}{3^2} = 1. \qquad \textbf{Writing standard form}$$

The hyperbola has a horizontal transverse axis, so the vertices are $(-a, 0)$ and $(a, 0)$, or $(-4, 0)$ and $(4, 0)$. From the standard form of the equation, we know that $a^2 = 4^2$, or 16, and $b^2 = 3^2$, or 9. We find the foci:

$$c^2 = a^2 + b^2$$
$$c^2 = 16 + 9$$
$$c^2 = 25$$
$$c = 5.$$

Thus the foci are $(-5, 0)$ and $(5, 0)$.

Next, we find the asymptotes:

$$y = -\frac{b}{a}x = -\frac{3}{4}x \quad \text{and} \quad y = \frac{b}{a}x = \frac{3}{4}x.$$

To draw the graph, we sketch the asymptotes first. This is easily done by drawing the rectangle with horizontal sides passing through $(0, 3)$ and $(0, -3)$ and vertical sides through $(4, 0)$ and $(-4, 0)$. Then we draw and extend the diagonals of this rectangle. The two extended diagonals are the asymptotes of the hyperbola. Next, we plot the vertices and draw the branches of the hyperbola outward from the vertices toward the asymptotes.

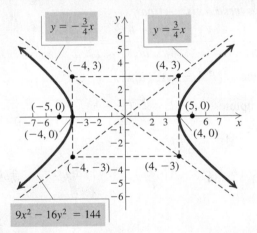

To graph this hyperbola on a graphing calculator, it might be necessary to solve for y first and then graph the top and bottom halves of the hyperbola in the same squared viewing window.

On some graphing calculators, the Conics HYPERBOLA APP can be used to graph hyperbolas, as shown below. Note that the center is $(0, 0)$, so $H = 0$ and $K = 0$.

$$\frac{x^2}{4^2} - \frac{y^2}{3^2} = 1 \qquad\qquad 9x^2 - 16y^2 = 144$$

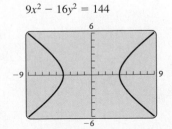

> *Now Try Exercise 17.*

If a hyperbola with center at the origin is translated horizontally $|h|$ units and vertically $|k|$ units, the center is at the point (h, k).

STANDARD EQUATION OF A HYPERBOLA WITH CENTER AT (h, k)

Transverse Axis Horizontal

$$\frac{(x - h)^2}{a^2} - \frac{(y - k)^2}{b^2} = 1$$

Vertices: $(h - a, k), (h + a, k)$

Asymptotes: $y - k = \dfrac{b}{a}(x - h)$,

$$y - k = -\frac{b}{a}(x - h)$$

Foci: $(h - c, k), (h + c, k)$, where $c^2 = a^2 + b^2$

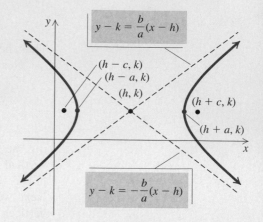

Transverse Axis Vertical

$$\frac{(y - k)^2}{a^2} - \frac{(x - h)^2}{b^2} = 1$$

Vertices: $(h, k - a), (h, k + a)$

Asymptotes: $y - k = \dfrac{a}{b}(x - h)$,

$$y - k = -\frac{a}{b}(x - h)$$

Foci: $(h, k - c), (h, k + c)$, where $c^2 = a^2 + b^2$

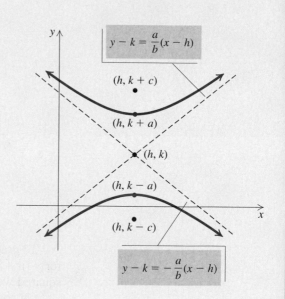

EXAMPLE 3   For the hyperbola given by

$$4y^2 - x^2 + 24y + 4x + 28 = 0,$$

find the center, the vertices, the foci, and the asymptotes. Then draw the graph.

Solution First, we complete the square to get standard form:

$$4y^2 - x^2 + 24y + 4x + 28 = 0$$

$$4(y^2 + 6y \qquad\quad) - (x^2 - 4x \qquad\quad) = -28$$

$$4(y^2 + 6y + 9 - 9) - (x^2 - 4x + 4 - 4) = -28$$

$$4(y^2 + 6y + 9) + 4(-9) - (x^2 - 4x + 4) - (-4) = -28$$

$$4(y^2 + 6y + 9) - 36 - (x^2 - 4x + 4) + 4 = -28$$

$$4(y^2 + 6y + 9) - (x^2 - 4x + 4) = -28 + 36 - 4$$

$$4(y + 3)^2 - (x - 2)^2 = 4$$

$$\frac{(y + 3)^2}{1} - \frac{(x - 2)^2}{4} = 1 \qquad \text{Dividing by 4}$$

$$\frac{[y - (-3)]^2}{1^2} - \frac{(x - 2)^2}{2^2} = 1. \qquad \text{Standard form}$$

The center is $(2, -3)$. Note that $a = 1$ and $b = 2$. The transverse axis is vertical, so the vertices are 1 unit below and above the center:

$$(2, -3 - 1) \text{ and } (2, -3 + 1), \quad \text{or} \quad (2, -4) \text{ and } (2, -2).$$

We know that $c^2 = a^2 + b^2$, so $c^2 = 1^2 + 2^2 = 1 + 4 = 5$ and $c = \sqrt{5}$. Thus the foci are $\sqrt{5}$ units below and above the center:

$$(2, -3 - \sqrt{5}) \quad \text{and} \quad (2, -3 + \sqrt{5}).$$

The asymptotes are

$$y - (-3) = \tfrac{1}{2}(x - 2) \quad \text{and} \quad y - (-3) = -\tfrac{1}{2}(x - 2),$$

or $\quad y + 3 = \tfrac{1}{2}(x - 2) \quad \text{and} \quad y + 3 = -\tfrac{1}{2}(x - 2).$

We sketch the asymptotes, plot the vertices, and draw the graph.

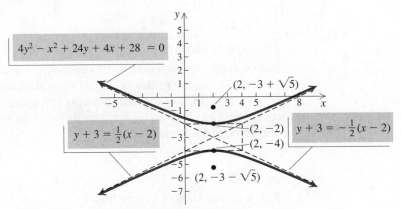

When the equation of a hyperbola is written in standard form, we can use the Conics HYPERBOLA APP to graph it, as shown in the following windows.

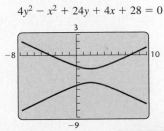

> ***Now Try Exercise 29.***

CONNECTING THE CONCEPTS   **Classifying Equations of Conic Sections**

EQUATION	TYPE OF CONIC SECTION	GRAPH
$x - 4 + 4y = y^2$	Only one variable is squared, so this cannot be a circle, an ellipse, or a hyperbola. Find an equivalent equation: $$x = (y - 2)^2.$$ This is an equation of a parabola.	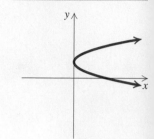
$3x^2 + 3y^2 = 75$	Both variables are squared, so this cannot be a parabola. The squared terms are added, so this cannot be a hyperbola. Divide by 3 on both sides to find an equivalent equation: $$x^2 + y^2 = 25.$$ This is an equation of a circle.	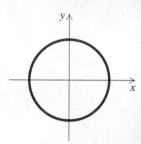
$y^2 = 16 - 4x^2$	Both variables are squared, so this cannot be a parabola. Add $4x^2$ on both sides to find an equivalent equation: $4x^2 + y^2 = 16$. The squared terms are added, so this cannot be a hyperbola. The coefficients of x^2 and y^2 are not the same, so this is not a circle. Divide by 16 on both sides to find an equivalent equation: $$\frac{x^2}{4} + \frac{y^2}{16} = 1.$$ This is an equation of an ellipse.	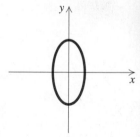
$x^2 = 4y^2 + 36$	Both variables are squared, so this cannot be a parabola. Subtract $4y^2$ on both sides to find an equivalent equation: $x^2 - 4y^2 = 36$. The squared terms are not added, so this cannot be a circle or an ellipse. Divide by 36 on both sides to find an equivalent equation: $$\frac{x^2}{36} - \frac{y^2}{9} = 1.$$ This is an equation of a hyperbola.	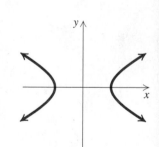

❯ Applications

Some comets travel in hyperbolic paths with the sun at one focus. Such comets pass by the sun only one time, unlike those with elliptical orbits, which reappear at intervals. We also see hyperbolas in architecture, such as in a cross section of a planetarium, an amphitheater, or a cooling tower for a steam or nuclear power plant.

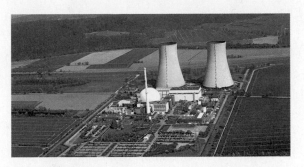

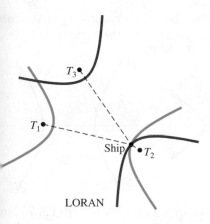

LORAN

Another application of hyperbolas is in the long-range navigation system LORAN. This system uses transmitting stations in three locations to send out simultaneous signals to a ship or an aircraft. The difference in the arrival times of the signals from one pair of transmitters is recorded on the ship or aircraft. This difference is also recorded for signals from another pair of transmitters. For each pair, a computation is performed to determine the difference in the distances from each member of the pair to the ship or aircraft. If each pair of differences is kept constant, two hyperbolas can be drawn. Each has one of the pairs of transmitters as foci, and the ship or aircraft lies on the intersection of two of their branches.

10.3 | Exercise Set

In Exercises 1–6, match the equation with one of the graphs (a)–(f) that follow.

a)
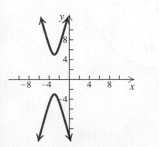

b)
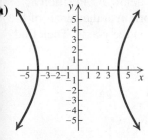

c)
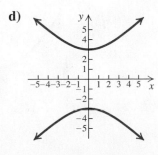

d)
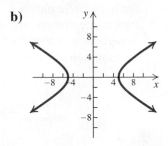

e)
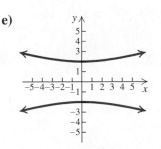

f)
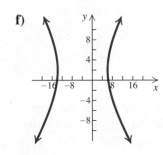

1. $\dfrac{x^2}{25} - \dfrac{y^2}{9} = 1$

2. $\dfrac{y^2}{4} - \dfrac{x^2}{36} = 1$

3. $\dfrac{(y-1)^2}{16} - \dfrac{(x+3)^2}{1} = 1$

4. $\dfrac{(x+4)^2}{100} - \dfrac{(y-2)^2}{81} = 1$

5. $25x^2 - 16y^2 = 400$

6. $y^2 - x^2 = 9$

Find an equation of a hyperbola satisfying the given conditions.

7. Vertices: $(0, 3)$ and $(0, -3)$;
   foci: $(0, 5)$ and $(0, -5)$

8. Vertices: $(1, 0)$ and $(-1, 0)$;
   foci: $(2, 0)$ and $(-2, 0)$

9. Asymptotes: $y = \frac{3}{2}x$, $y = -\frac{3}{2}x$;
   one vertex: $(2, 0)$

10. Asymptotes: $y = \frac{5}{4}x$, $y = -\frac{5}{4}x$;
   one vertex: $(0, 3)$

Find the center, the vertices, the foci, and the asymptotes. Then draw the graph.

11. $\dfrac{x^2}{4} - \dfrac{y^2}{4} = 1$

12. $\dfrac{x^2}{1} - \dfrac{y^2}{9} = 1$

13. $\dfrac{(x - 2)^2}{9} - \dfrac{(y + 5)^2}{1} = 1$

14. $\dfrac{(x - 5)^2}{16} - \dfrac{(y + 2)^2}{9} = 1$

15. $\dfrac{(y + 3)^2}{4} - \dfrac{(x + 1)^2}{16} = 1$

16. $\dfrac{(y + 4)^2}{25} - \dfrac{(x + 2)^2}{16} = 1$

17. $x^2 - 4y^2 = 4$ **18.** $4x^2 - y^2 = 16$

19. $9y^2 - x^2 = 81$ **20.** $y^2 - 4x^2 = 4$

21. $x^2 - y^2 = 2$ **22.** $x^2 - y^2 = 3$

23. $y^2 - x^2 = \frac{1}{4}$ **24.** $y^2 - x^2 = \frac{1}{9}$

Find the center, the vertices, the foci, and the asymptotes of the hyperbola. Then draw the graph.

25. $x^2 - y^2 - 2x - 4y - 4 = 0$

26. $4x^2 - y^2 + 8x - 4y - 4 = 0$

27. $36x^2 - y^2 - 24x + 6y - 41 = 0$

28. $9x^2 - 4y^2 + 54x + 8y + 41 = 0$

29. $9y^2 - 4x^2 - 18y + 24x - 63 = 0$

30. $x^2 - 25y^2 + 6x - 50y = 41$

31. $x^2 - y^2 - 2x - 4y = 4$

32. $9y^2 - 4x^2 - 54y - 8x + 41 = 0$

33. $y^2 - x^2 - 6x - 8y - 29 = 0$

34. $x^2 - y^2 = 8x - 2y - 13$

*The **eccentricity** of a hyperbola is defined as $e = c/a$. For a hyperbola, $c > a > 0$, so $e > 1$. When e is close to 1, a hyperbola appears to be very narrow. As the eccentricity increases, the hyperbola becomes "wider."*

35. Observe the shapes of the hyperbolas in Examples 2 and 3. Which hyperbola has the larger eccentricity? Confirm your answer by computing the eccentricity of each hyperbola.

36. Which hyperbola has the larger eccentricity? (Assume that the coordinate systems have the same scale.)

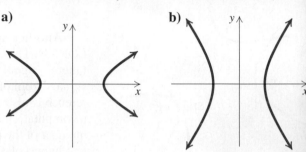

a)   **b)**

37. Find an equation of a hyperbola with vertices $(3, 7)$ and $(-3, 7)$ and $e = \frac{5}{3}$.

38. Find an equation of a hyperbola with vertices $(-1, 3)$ and $(-1, 7)$ and $e = 4$.

39. *Hyperbolic Mirror.*   Certain telescopes contain both a parabolic mirror and a hyperbolic mirror. In the telescope shown in the figure, the parabola and the hyperbola share focus F_1, which is 14 m above the vertex of the parabola. The hyperbola's second focus F_2 is 2 m above the parabola's vertex. The vertex of the hyperbolic mirror is 1 m below F_1. Position a coordinate system with the origin at the center of the hyperbola and with the foci on the y-axis. Then find the equation of the hyperbola.

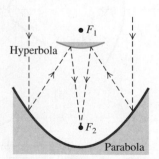

40. *Nuclear Cooling Tower.* A cross section of a nuclear cooling tower is a hyperbola with equation

$$\frac{x^2}{90^2} - \frac{y^2}{130^2} = 1.$$

The tower is 450 ft tall, and the distance from the top of the tower to the center of the hyperbola is half the distance from the base of the tower to the center of the hyperbola. Find the diameter of the top and the base of the tower.

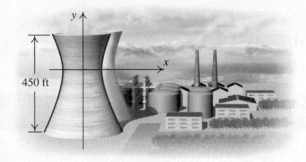

❯ Skill Maintenance

In Exercises 41–44, given the function:

a) *Determine whether it is one-to-one.* **[5.1]**

b) *If it is one-to-one, find a formula for the inverse.* **[5.1]**

41. $f(x) = 2x - 3$

42. $f(x) = x^3 + 2$

43. $f(x) = \dfrac{5}{x - 1}$

44. $f(x) = \sqrt{x + 4}$

Solve. **[9.1], [9.3], [9.5], [9.6]**

45. $x + y = 5,$
$x - y = 7$

46. $3x - 2y = 5,$
$5x + 2y = 3$

47. $2x - 3y = 7,$
$3x + 5y = 1$

48. $3x + 2y = -1,$
$2x + 3y = 6$

❯ Synthesis

Find an equation of a hyperbola satisfying the given conditions.

49. Vertices: $(3, -8)$ and $(3, -2)$;
asymptotes: $y = 3x - 14,\ y = -3x + 4$

50. Vertices: $(-9, 4)$ and $(-5, 4)$;
asymptotes: $y = 3x + 25,\ y = -3x - 17$

Use a graphing calculator to find the center, the vertices, and the asymptotes.

51. $5x^2 - 3.5y^2 + 14.6x - 6.7y + 3.4 = 0$

52. $x^2 - y^2 - 2.046x - 4.088y - 4.228 = 0$

53. *Navigation.* Two radio transmitters positioned 300 mi apart along the shore send simultaneous signals to a ship that is 200 mi offshore, sailing parallel to the shoreline. The signal from transmitter S reaches the ship 200 microseconds later than the signal from transmitter T. The signals travel at a speed of 186,000 miles per second, or 0.186 mile per microsecond. Find the equation of the hyperbola with foci S and T on which the ship is located. (*Hint*: For any point on the hyperbola, the absolute value of the difference of its distances from the foci is $2a$.)

10.4 ❯ Nonlinear Systems of Equations and Inequalities

> ❯ Solve a nonlinear system of equations.
> ❯ Use nonlinear systems of equations to solve applied problems.
> ❯ Graph nonlinear systems of inequalities.

The systems of equations that we have studied so far have been composed of linear equations. Now we consider systems of two equations in two variables in which at least one equation is not linear.

❯ Nonlinear Systems of Equations

The graphs of the equations in a nonlinear system of equations can have no point of intersection or one or more points of intersection. The coordinates of each point of intersection represent a solution of the system of equations. When no point of intersection exists, the system of equations has no real-number solution.

Solutions of nonlinear systems of equations can be found using the substitution method or the elimination method. The substitution method is preferable for a system consisting of one linear equation and one nonlinear equation. The elimination method is preferable in most, but not all, cases when both equations are nonlinear.

EXAMPLE 1   Solve the following system of equations:

$$x^2 + y^2 = 25, \qquad \textbf{(1)} \qquad \textbf{The graph is a circle.}$$
$$3x - 4y = 0. \qquad \textbf{(2)} \qquad \textbf{The graph is a line.}$$

Algebraic Solution

We use the substitution method. First, we solve equation (2) for x:

$$x = \tfrac{4}{3}y. \qquad \textbf{(3)} \qquad \textbf{We could have solved for } y \textbf{ instead.}$$

Next, we substitute $\frac{4}{3}y$ for x in equation (1) and solve for y:

$$\left(\tfrac{4}{3}y\right)^2 + y^2 = 25$$
$$\tfrac{16}{9}y^2 + y^2 = 25$$
$$\tfrac{25}{9}y^2 = 25$$
$$y^2 = 9 \qquad \textbf{Multiplying by } \tfrac{9}{25}$$
$$y = \pm 3.$$

Now we substitute these numbers for y in equation (3) and solve for x:

$$x = \tfrac{4}{3}(3) = 4, \qquad \textbf{(4, 3) appears to be a solution.}$$
$$x = \tfrac{4}{3}(-3) = -4. \qquad \textbf{(−4, −3) appears to be a solution.}$$

Check: For $(4, 3)$:

$$
\begin{array}{c|c}
x^2 + y^2 = 25 & 3x - 4y = 0 \\
\hline
4^2 + 3^2 \;?\; 25 & 3(4) - 4(3) \;?\; 0 \\
16 + 9 & 12 - 12 \\
\quad 25 \;\big|\; 25 \quad \text{TRUE} & \quad 0 \;\big|\; 0 \quad \text{TRUE}
\end{array}
$$

For $(-4, -3)$:

$$
\begin{array}{c|c}
x^2 + y^2 = 25 & 3x - 4y = 0 \\
\hline
(-4)^2 + (-3)^2 \;?\; 25 & 3(-4) - 4(-3) \;?\; 0 \\
16 + 9 & -12 + 12 \\
\quad 25 \;\big|\; 25 \quad \text{TRUE} & \quad 0 \;\big|\; 0 \quad \text{TRUE}
\end{array}
$$

The pairs $(4, 3)$ and $(-4, -3)$ check, so they are the solutions.

Graphical Solution

We graph both equations in the same viewing window. Note that there are two points of intersection. We can find their coordinates using the INTERSECT feature.

$x^2 + y^2 = 25$

$y_1 = \sqrt{25 - x^2}, \quad y_2 = -\sqrt{25 - x^2},$
$y_3 = \frac{3}{4}x$

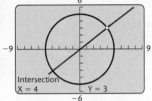

$x^2 + y^2 = 25$

$y_1 = \sqrt{25 - x^2}, \quad y_2 = -\sqrt{25 - x^2},$
$y_3 = \frac{3}{4}x$

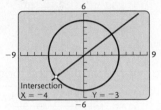

The solutions are $(4, 3)$ and $(-4, -3)$.

> **Now Try Exercise 7.**

In the algebraic solution in Example 1, suppose that to find x we had substituted 3 and -3 in equation (1) rather than equation (3). If $y = 3$, $y^2 = 9$, and if $y = -3$, $y^2 = 9$, so both substitutions can be performed at the same time:

$$x^2 + y^2 = 25 \quad \textbf{(1)}$$
$$x^2 + (\pm 3)^2 = 25$$
$$x^2 + 9 = 25$$
$$x^2 = 16$$
$$x = \pm 4.$$

Each y-value produces two values for x. Thus, if $y = 3$, $x = 4$ or $x = -4$, and if $y = -3$, $x = 4$ or $x = -4$. The possible solutions are $(4, 3)$, $(-4, 3)$, $(4, -3)$, and $(-4, -3)$. A check reveals that $(4, -3)$ and $(-4, 3)$ are not solutions of equation (2). Since a circle and a line can intersect in at most two points, it is clear that there can be at most two real-number solutions.

EXAMPLE 2 Solve the following system of equations:

$$x + y = 5, \qquad \textbf{(1)} \qquad \textbf{The graph is a line.}$$
$$y = 3 - x^2. \qquad \textbf{(2)} \qquad \textbf{The graph is a parabola.}$$

Algebraic Solution

We use the substitution method, substituting $3 - x^2$ for y in equation (1):

$$x + 3 - x^2 = 5$$
$$-x^2 + x - 2 = 0 \qquad \textbf{Subtracting 5 and rearranging}$$
$$x^2 - x + 2 = 0. \qquad \textbf{Multiplying by } -1$$

Next, we use the quadratic formula:

$$x = \frac{-b \pm \sqrt{b^2 - 4ac}}{2a}$$
$$= \frac{-(-1) \pm \sqrt{(-1)^2 - 4(1)(2)}}{2(1)}$$
$$= \frac{1 \pm \sqrt{1 - 8}}{2}$$
$$= \frac{1 \pm \sqrt{-7}}{2}$$
$$= \frac{1 \pm i\sqrt{7}}{2}$$
$$= \frac{1}{2} \pm \frac{\sqrt{7}}{2}i.$$

Now, we substitute these values for x in equation (1) and solve for y:

$$\frac{1}{2} + \frac{\sqrt{7}}{2}i + y = 5$$
$$y = 5 - \frac{1}{2} - \frac{\sqrt{7}}{2}i = \frac{9}{2} - \frac{\sqrt{7}}{2}i$$

and $\dfrac{1}{2} - \dfrac{\sqrt{7}}{2}i + y = 5$

$$y = 5 - \frac{1}{2} + \frac{\sqrt{7}}{2}i = \frac{9}{2} + \frac{\sqrt{7}}{2}i.$$

The solutions are

$$\left(\frac{1}{2} + \frac{\sqrt{7}}{2}i, \frac{9}{2} - \frac{\sqrt{7}}{2}i \right) \quad \text{and} \quad \left(\frac{1}{2} - \frac{\sqrt{7}}{2}i, \frac{9}{2} + \frac{\sqrt{7}}{2}i \right).$$

There are no real-number solutions.

Graphical Solution

We graph both equations in the same viewing window.

$$y_1 = 5 - x, \quad y_2 = 3 - x^2$$

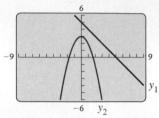

Note that there are no points of intersection. This indicates that there are no real-number solutions. Algebra must be used, as at left, to find the imaginary-number solutions.

Now Try Exercise 17.

EXAMPLE 3   Solve the following system of equations:

$$2x^2 + 5y^2 = 39, \qquad \textbf{(1)} \qquad \text{The graph is an ellipse.}$$
$$3x^2 - y^2 = -1. \qquad \textbf{(2)} \qquad \text{The graph is a hyperbola.}$$

Algebraic Solution

We use the elimination method. First, we multiply equation (2) by 5 and add to eliminate the y^2-term:

$$
\begin{array}{ll}
2x^2 + 5y^2 = 39 & \textbf{(1)} \\
\underline{15x^2 - 5y^2 = -5} & \textbf{Multiplying} \\
& \textbf{(2) by 5} \\
17x^2 \qquad\quad = 34 & \textbf{Adding} \\
\qquad x^2 = 2 & \\
\qquad\quad x = \pm\sqrt{2}. &
\end{array}
$$

If $x = \sqrt{2}$, $x^2 = 2$, and if $x = -\sqrt{2}$, $x^2 = 2$. Thus substituting $\sqrt{2}$ or $-\sqrt{2}$ for x in equation (2) gives us

$$
\begin{aligned}
3(\pm\sqrt{2})^2 - y^2 &= -1 \\
3 \cdot 2 - y^2 &= -1 \\
6 - y^2 &= -1 \\
-y^2 &= -7 \\
y^2 &= 7 \\
y &= \pm\sqrt{7}.
\end{aligned}
$$

Each x-value produces two values for y. Thus, for $x = \sqrt{2}$, we have $y = \sqrt{7}$ or $y = -\sqrt{7}$, and for $x = -\sqrt{2}$, we have $y = \sqrt{7}$ or $y = -\sqrt{7}$. The possible solutions are $\left(\sqrt{2}, \sqrt{7}\right)$, $\left(\sqrt{2}, -\sqrt{7}\right)$, $\left(-\sqrt{2}, \sqrt{7}\right)$, and $\left(-\sqrt{2}, -\sqrt{7}\right)$. All four pairs check, so they are the solutions.

Graphical Solution

We graph both equations in the same viewing window. There are four points of intersection. We can use the INTERSECT feature to find their coordinates.

$y_1 = \sqrt{(39 - 2x^2)/5}, \quad y_2 = -\sqrt{(39 - 2x^2)/5},$
$y_3 = \sqrt{3x^2 + 1}, \quad y_4 = -\sqrt{3x^2 + 1}$

$(-1.414, 2.646)$ $(1.414, 2.646)$
$(-1.414, -2.646)$ $(1.414, -2.646)$

Note that the algebraic solution yields exact solutions, whereas the graphical solution yields decimal approximations of the solutions on most graphing calculators.

The solutions are approximately $(1.414, 2.646)$, $(1.414, -2.646)$, $(-1.414, 2.646)$, and $(-1.414, -2.646)$.

Now Try Exercise 27.

EXAMPLE 4 Solve the following system of equations:

$$x^2 - 3y^2 = 6, \qquad \textbf{(1)}$$
$$xy = 3. \qquad \textbf{(2)}$$

Algebraic Solution

We use the substitution method. First, we solve equation (2) for y:

$$xy = 3 \qquad \textbf{(2)}$$

$$y = \frac{3}{x}. \qquad \textbf{(3)} \qquad \textbf{Dividing by } x$$

Next, we substitute $3/x$ for y in equation (1) and solve for x:

$$x^2 - 3\left(\frac{3}{x}\right)^2 = 6$$

$$x^2 - 3 \cdot \frac{9}{x^2} = 6$$

$$x^2 - \frac{27}{x^2} = 6$$

$$x^4 - 27 = 6x^2 \qquad \textbf{Multiplying by } x^2$$

$$x^4 - 6x^2 - 27 = 0$$

$$u^2 - 6u - 27 = 0 \qquad \textbf{Letting } u = x^2$$

$$(u - 9)(u + 3) = 0 \qquad \textbf{Factoring}$$

$$u = 9 \quad or \quad u = -3 \qquad \textbf{Principle of zero products}$$

$$x^2 = 9 \quad or \quad x^2 = -3 \qquad \textbf{Substituting } x^2 \textbf{ for } u$$

$$x = \pm 3 \quad or \quad x = \pm \sqrt{3}i.$$

Since $y = 3/x$,

when $x = 3$, $\qquad y = \dfrac{3}{3} = 1$;

when $x = -3$, $\qquad y = \dfrac{3}{-3} = -1$;

when $x = \sqrt{3}i$, $\qquad y = \dfrac{3}{\sqrt{3}i} = \dfrac{3}{\sqrt{3}i} \cdot \dfrac{-\sqrt{3}i}{-\sqrt{3}i} = -\sqrt{3}i$;

when $x = -\sqrt{3}i$, $\quad y = \dfrac{3}{-\sqrt{3}i} = \dfrac{3}{-\sqrt{3}i} \cdot \dfrac{\sqrt{3}i}{\sqrt{3}i} = \sqrt{3}i$.

The pairs $(3, 1)$, $(-3, -1)$, $(\sqrt{3}i, -\sqrt{3}i)$, and $(-\sqrt{3}i, \sqrt{3}i)$ check, so they are the solutions.

Graphical Solution

We graph both equations in the same viewing window and find the coordinates of their points of intersection.

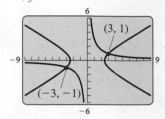

$$x^2 - 3y^2 = 6$$
$$y_1 = \sqrt{(x^2 - 6)/3}, \quad y_2 = -\sqrt{(x^2 - 6)/3},$$
$$y_3 = 3/x$$

Note that the graphical method yields only the real-number solutions of the system of equations. The algebraic method must be used in order to find *all* the solutions.

> **Now Try Exercise 19.**

❯ Modeling and Problem Solving

EXAMPLE 5 *Dimensions of a Piece of Land.* For a student recreation building at Southport Community College, an architect wants to lay out a rectangular piece of land that has a perimeter of 204 m and an area of 2565 m². Find the dimensions of the piece of land.

Solution

1. **Familiarize.** We make a drawing and label it, letting l = the length of the piece of land, in meters, and w = the width, in meters.

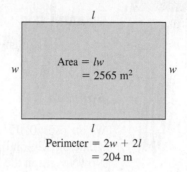

l

w Area = lw
     = 2565 m² w

l

Perimeter = $2w + 2l$
          = 204 m

2. **Translate.** We now have the following:

   Perimeter: $2w + 2l = 204,$ **(1)**
   Area: $lw = 2565.$ **(2)**

3. **Carry out.** We solve the system of equations both algebraically and graphically.

Algebraic Solution	Graphical Solution

Algebraic Solution

We solve the system of equations

$$2w + 2l = 204, \quad \textbf{(1)}$$
$$lw = 2565. \quad \textbf{(2)}$$

Solving the second equation for l gives us $l = 2565/w$. We then substitute $2565/w$ for l in equation (1) and solve for w:

$$2w + 2\left(\frac{2565}{w}\right) = 204$$

$$2w^2 + 2(2565) = 204w \qquad \textbf{Multiplying by } w$$
$$2w^2 - 204w + 2(2565) = 0$$
$$w^2 - 102w + 2565 = 0 \qquad \textbf{Multiplying by } \tfrac{1}{2}$$
$$(w - 57)(w - 45) = 0$$
$$w = 57 \quad or \quad w = 45. \qquad \textbf{Principle of zero products}$$

If $w = 57$, then $l = 2565/w = 2565/57 = 45$. If $w = 45$, then $l = 2565/w = 2565/45 = 57$. Since length is generally considered to be longer than width, we have the solution $l = 57$ and $w = 45$, or $(57, 45)$.

Graphical Solution

We replace l with x and w with y, graph $y_1 = (204 - 2x)/2$ and $y_2 = 2565/x$, and find the point(s) of intersection of the graphs.

$y_1 = (204 - 2x)/2, \ y_2 = 2565/x$

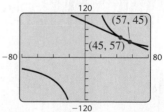

As in the algebraic solution, we have two possible solutions: $(45, 57)$ and $(57, 45)$. Since length, x, is generally considered to be longer than width, y, we have the solution $(57, 45)$.

4. Check. If $l = 57$ and $w = 45$, the perimeter is $2 \cdot 45 + 2 \cdot 57$, or 204. The area is $57 \cdot 45$, or 2565. The numbers check.

5. State. The length of the piece of land is 57 m and the width is 45 m.

Now Try Exercise 61.

❭ Nonlinear Systems of Inequalities

SYSTEMS OF INEQUALITIES

REVIEW SECTION 9.7.

Recall that a solution of a system of inequalities is an ordered pair that is a solution of each inequality in the system. Now we graph a system of nonlinear inequalities.

EXAMPLE 6   Graph the solution set of the system

$$x^2 + y^2 \le 25,$$
$$3x - 4y > 0.$$

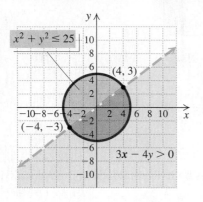

Solution   We graph $x^2 + y^2 \le 25$ by first graphing the related equation of the circle $x^2 + y^2 = 25$. We use a solid curve since the inequality symbol is $\le$. Next we choose $(0, 0)$ as a test point and find that it is a solution of $x^2 + y^2 \le 25$, so we shade the region that contains $(0, 0)$ using red. This is the region inside the circle. Now we graph the line $3x - 4y = 0$ using a dashed line since the inequality symbol is $>$. The point $(0, 0)$ is on the line, so we choose another test point, say, $(0, 2)$. We find that this point is not a solution of $3x - 4y > 0$, so we shade the half-plane that does not contain $(0, 2)$ using green. The solution set of the system of inequalities is the region shaded both red and green, or brown, including part of the circle $x^2 + y^2 = 25$.

To find the points of intersection of the graphs of the related equations, we solve the system composed of those equations:

$$x^2 + y^2 = 25,$$
$$3x - 4y = 0.$$

In Example 1, we found that these points are $(4, 3)$ and $(-4, -3)$.

Now Try Exercise 75.

EXAMPLE 7   Use a graphing calculator to graph the system

$$y \le 4 - x^2,$$
$$x + y \ge 2.$$

Solution   We graph $y_1 = 4 - x^2$ and $y_2 = 2 - x$. Using the test point $(0, 0)$ for each inequality, we find that we should shade below y_1 and above y_2. We can find the points of intersection of the graphs of the related equations, $(-1, 3)$ and $(2, 0)$, using the INTERSECT feature.

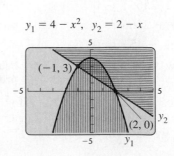

Now Try Exercise 81.

Visualizing the Graph

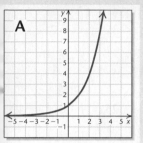

A

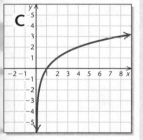

B

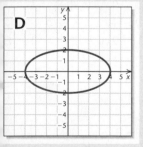

C

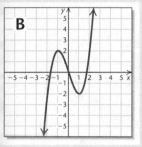

D

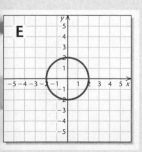

E

Match the equation or system of equations with its graph.

1. $y = x^3 - 3x$

2. $y = x^2 + 2x - 3$

3. $y = \dfrac{x - 1}{x^2 - x - 2}$

4. $y = -3x + 2$

5. $x + y = 3,$
   $2x + 5y = 3$

6. $9x^2 - 4y^2 = 36,$
   $x^2 + y^2 = 9$

7. $5x^2 + 5y^2 = 20$

8. $4x^2 + 16y^2 = 64$

9. $y = \log_2 x$

10. $y = 2^x$

Answers on page A-61

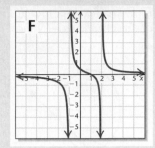

F

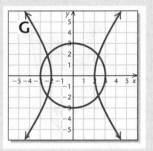

G

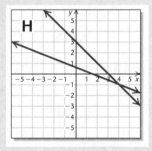

H

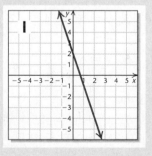

I

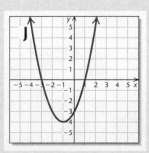

J

10.4 | Exercise Set

In Exercises 1–6, match the system of equations with one of the graphs (a)–(f) that follow.

a)
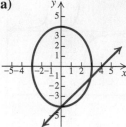

b)
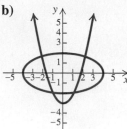

c)
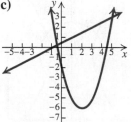

d)
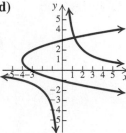

e)
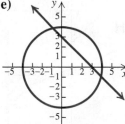

f)
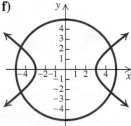

1. $x^2 + y^2 = 16$,
$x + y = 3$

2. $16x^2 + 9y^2 = 144$,
$x - y = 4$

3. $y = x^2 - 4x - 2$,
$2y - x = 1$

4. $4x^2 - 9y^2 = 36$,
$x^2 + y^2 = 25$

5. $y = x^2 - 3$,
$x^2 + 4y^2 = 16$

6. $y^2 - 2y = x + 3$,
$xy = 4$

Solve.

7. $x^2 + y^2 = 25$,
$y - x = 1$

8. $x^2 + y^2 = 100$,
$y - x = 2$

9. $4x^2 + 9y^2 = 36$,
$3y + 2x = 6$

10. $9x^2 + 4y^2 = 36$,
$3x + 2y = 6$

11. $x^2 + y^2 = 25$,
$y^2 = x + 5$

12. $y = x^2$,
$x = y^2$

13. $x^2 + y^2 = 9$,
$x^2 - y^2 = 9$

14. $y^2 - 4x^2 = 4$,
$4x^2 + y^2 = 4$

15. $y^2 - x^2 = 9$,
$2x - 3 = y$

16. $x + y = -6$,
$xy = -7$

17. $y^2 = x + 3$,
$2y = x + 4$

18. $y = x^2$,
$3x = y + 2$

19. $x^2 + y^2 = 25$,
$xy = 12$

20. $x^2 - y^2 = 16$,
$x + y^2 = 4$

21. $x^2 + y^2 = 4$,
$16x^2 + 9y^2 = 144$

22. $x^2 + y^2 = 25$,
$25x^2 + 16y^2 = 400$

23. $x^2 + 4y^2 = 25$,
$x + 2y = 7$

24. $y^2 - x^2 = 16$,
$2x - y = 1$

25. $x^2 - xy + 3y^2 = 27$,
$x - y = 2$

26. $2y^2 + xy + x^2 = 7$,
$x - 2y = 5$

27. $x^2 + y^2 = 16$,
$y^2 - 2x^2 = 10$

28. $x^2 + y^2 = 14$,
$x^2 - y^2 = 4$

29. $x^2 + y^2 = 5$,
$xy = 2$

30. $x^2 + y^2 = 20$,
$xy = 8$

31. $3x + y = 7$,
$4x^2 + 5y = 56$

32. $2y^2 + xy = 5$,
$4y + x = 7$

33. $a + b = 7$,
$ab = 4$

34. $p + q = -4$,
$pq = -5$

35. $x^2 + y^2 = 13$,
$xy = 6$

36. $x^2 + 4y^2 = 20$,
$xy = 4$

37. $x^2 + y^2 + 6y + 5 = 0$,
$x^2 + y^2 - 2x - 8 = 0$

38. $2xy + 3y^2 = 7$,
$3xy - 2y^2 = 4$

39. $2a + b = 1$,
$b = 4 - a^2$

40. $4x^2 + 9y^2 = 36$,
$x + 3y = 3$

41. $a^2 + b^2 = 89$,
$a - b = 3$

42. $xy = 4$,
$x + y = 5$

43. $xy - y^2 = 2$,
$2xy - 3y^2 = 0$

44. $4a^2 - 25b^2 = 0$,
$2a^2 - 10b^2 = 3b + 4$

45. $m^2 - 3mn + n^2 + 1 = 0$,
$3m^2 - mn + 3n^2 = 13$

46. $ab - b^2 = -4$,
$ab - 2b^2 = -6$

47. $x^2 + y^2 = 5$,
$x - y = 8$

48. $4x^2 + 9y^2 = 36$,
$y - x = 8$

49. $a^2 + b^2 = 14$,
$ab = 3\sqrt{5}$

50. $x^2 + xy = 5$,
$2x^2 + xy = 2$

51. $x^2 + y^2 = 25,$
$9x^2 + 4y^2 = 36$

52. $x^2 + y^2 = 1,$
$9x^2 - 16y^2 = 144$

53. $5y^2 - x^2 = 1,$
$xy = 2$

54. $x^2 - 7y^2 = 6,$
$xy = 1$

In Exercises 55–58, determine whether the statement is true or false.

55. A nonlinear system of equations can have both real-number solutions and imaginary-number solutions.

56. If the graph of a nonlinear system of equations consists of a line and a parabola, then the system has two real-number solutions.

57. If the graph of a nonlinear system of equations consists of a line and a circle, then the system has at most two real-number solutions.

58. If the graph of a nonlinear system of equations consists of a line and an ellipse, then it is possible for the system to have exactly one real-number solution.

59. *Photo Dimensions.*   Hailey's Frame Shop has been commissioned to frame 5 black-and-white photos for an island resort. Each photo has a perimeter of 68 in. and a diagonal of 26 in. Find the dimensions of the photos.

60. *Sign Dimensions.*   Alison's Advertising is building a rectangular sign with an area of 2 yd^2 and a perimeter of 6 yd. Find the dimensions of the sign.

61. *Graphic Design.*   Marcia Graham, owner of Graham's Graphics, is designing an advertising brochure for the Art League's spring show. Each page of the brochure is rectangular with an area of 20 in^2 and a perimeter of 18 in. Find the dimensions of the brochure.

62. *Landscaping.*   Green Leaf Landscaping is planting a rectangular wildflower garden with a perimeter of 6 m and a diagonal of $\sqrt{5}$ m. Find the dimensions of the garden.

63. *Fencing.*   Clark's Country Pet Resort is fencing a new play area for dogs. The manager has purchased 210 yd of fence to enclose a rectangular pen. The area of the pen must be 2250 yd^2. What are the dimensions of the pen?

64. *Carpentry.*   Ted Hansen of Hansen Woodworking Designs has been commissioned to make a rectangular tabletop with an area of $\sqrt{2}$ m^2 and a diagonal of $\sqrt{3}$ m for the Decorators' Show House. Find the dimensions of the tabletop.

65. *Banner Design.*   A rectangular banner with an area of $\sqrt{3}$ m^2 is being designed to advertise an exhibit at the Davis Gallery. The length of a diagonal is 2 m. Find the dimensions of the banner.

66. *Investment.*   Jenna made an investment for 1 year that earned $7.50 simple interest. If the principal had been $25 more and the interest rate 1% less, then the interest would have been the same. Find the principal and the interest rate.

67. *Office Dimensions.*   The diagonal of the floor of a rectangular office cubicle is 1 ft longer than the length of the cubicle and 3 ft longer than twice the width. Find the dimensions of the cubicle.

68. *Seed Test Plots.* The Burton Seed Company has two square test plots. The sum of their areas is 832 ft^2, and the difference of their areas is 320 ft^2. Find the length of a side of each plot.

In Exercises 69–74, match the system of inequalities with one of the graphs (a)–(f) that follow.

a)

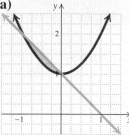

b)

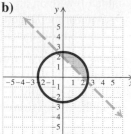

c)

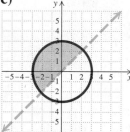

d)

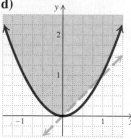

e)

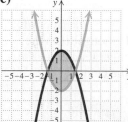

f)

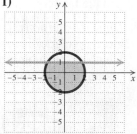

69. $x^2 + y^2 \le 5,$
$x + y > 2$

70. $y \le 2 - x^2,$
$y \ge x^2 - 2$

71. $y \ge x^2,$
$y > x$

72. $x^2 + y^2 \le 4,$
$y \le 1$

73. $y \ge x^2 + 1,$
$x + y \le 1$

74. $x^2 + y^2 \le 9,$
$y > x$

Graph the system of inequalities. Then find the coordinates of the points of intersection of the graphs of the related equations.

75. $x^2 + y^2 \le 16,$
$y < x$

76. $x^2 + y^2 \le 10,$
$y > x$

77. $x^2 \le y,$
$x + y \ge 2$

78. $x \ge y^2,$
$x - y \le 2$

79. $x^2 + y^2 \le 25,$
$x - y > 5$

80. $x^2 + y^2 \ge 9,$
$x - y > 3$

81. $y \ge x^2 - 3,$
$y \le 2x$

82. $y \le 3 - x^2,$
$y \ge x + 1$

83. $y \ge x^2,$
$y < x + 2$

84. $y \le 1 - x^2,$
$y > x - 1$

❭ Skill Maintenance

Solve. [5.5]

85. $2^{3x} = 64$

86. $5^x = 27$

87. $\log_3 x = 4$

88. $\log (x - 3) + \log x = 1$

❭ Synthesis

89. Find an equation of the circle that passes through the points $(2, 4)$ and $(3, 3)$ and whose center is on the line $3x - y = 3$.

90. Find an equation of the circle that passes through the points $(2, 3)$, $(4, 5)$, and $(0, -3)$.

91. Find an equation of an ellipse centered at the origin that passes through the points $(1, \sqrt{3}/2)$ and $(\sqrt{3}, 1/2)$.

92. Find an equation of a hyperbola of the type

$$\frac{x^2}{b^2} - \frac{y^2}{a^2} = 1$$

that passes through the points $(-3, -3\sqrt{5}/2)$ and $(-3/2, 0)$.

93. Show that a hyperbola does not intersect its asymptotes. That is, solve the system of equations

$$\frac{x^2}{a^2} - \frac{y^2}{b^2} = 1,$$

$$y = \frac{b}{a}x \left(\text{or } y = -\frac{b}{a}x \right).$$

94. *Numerical Relationship.* Find two numbers whose product is 2 and the sum of whose reciprocals is $\frac{33}{8}$.

95. *Numerical Relationship.* The sum of two numbers is 1, and their product is 1. Find the sum of their cubes. There is a method to solve this problem that is easier than solving a nonlinear system of equations. Can you discover it?

96. *Box Dimensions.* Four squares with sides 5 in. long are cut from the corners of a rectangular metal sheet that has an area of 340 in². The edges are bent up to form an open box with a volume of 350 in³. Find the dimensions of the box.

Solve.

97. $x^3 + y^3 = 72,$
$x + y = 6$

98. $a + b = \dfrac{5}{6},$
$\dfrac{a}{b} + \dfrac{b}{a} = \dfrac{13}{6}$

99. $p^2 + q^2 = 13,$
$\dfrac{1}{pq} = -\dfrac{1}{6}$

100. $e^x - e^{x+y} = 0,$
$e^y - e^{x-y} = 0$

Solve using a graphing calculator. Find all real solutions.

101. $y - \ln x = 2,$
$y = x^2$

102. $y = \ln(x + 4),$
$x^2 + y^2 = 6$

103. $y = e^x,$
$x - y = -2$

104. $y - e^{-x} = 1,$
$y = 2x + 5$

105. $14.5x^2 - 13.5y^2 - 64.5 = 0,$
$5.5x - 6.3y - 12.3 = 0$

106. $2x + 2y = 1660,$
$xy = 35{,}325$

107. $0.319x^2 + 2688.7y^2 = 56{,}548,$
$0.306x^2 - 2688.7y^2 = 43{,}452$

108. $13.5xy + 15.6 = 0,$
$5.6x - 6.7y - 42.3 = 0$

Mid-Chapter Mixed Review

Determine whether the statement is true or false.

1. The graph of $(x + 3)^2 = 8(y - 2)$ is a parabola with vertex $(-3, 2)$. **[10.1]**

2. The graph of $(x - 4)^2 + (y + 1)^2 = 9$ is a circle with radius 9. **[10.2]**

3. The hyperbola $\dfrac{x^2}{5} - \dfrac{y^2}{10} = 1$ has a horizontal transverse axis. **[10.3]**

4. Every nonlinear system of equations has at least one real-number solution. **[10.4]**

In Exercises 5–12, match the equation with one of the graphs (a)–(h) that follow. **[10.1], [10.2], [10.3]**

a)
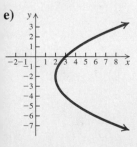

b)
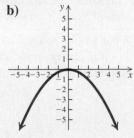

c)
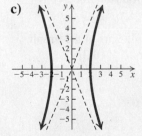

d)
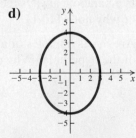

e)
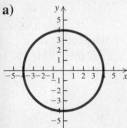

f)
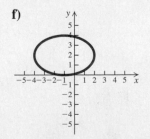

g)
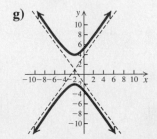

h)
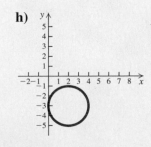

5. $x^2 = -4y$

6. $(y + 2)^2 = 4(x - 2)$

7. $16x^2 + 9y^2 = 144$

8. $x^2 + y^2 = 16$

9. $4(y - 1)^2 - 9(x + 2)^2 = 36$

10. $4(x + 1)^2 + 9(y - 2)^2 = 36$

11. $(x - 2)^2 + (y + 3)^2 = 4$

12. $25x^2 - 4y^2 = 100$

Find the vertex, the focus, and the directrix of the parabola. Then draw the graph. **[10.1]**

13. $y^2 = 12x$

14. $x^2 - 6x - 4y = -17$

Find the center and the radius of the circle. Then draw the graph. **[10.2]**

15. $x^2 + y^2 + 4x - 8y = 5$

16. $x^2 + y^2 - 6x + 2y - 6 = 0$

Find the vertices and the foci of the ellipse. Then draw the graph. **[10.2]**

17. $\dfrac{x^2}{1} + \dfrac{y^2}{9} = 1$

18. $25x^2 + 4y^2 - 50x + 8y = 71$

Find the center, the vertices, the foci, and the asymptotes of the hyperbola. Then draw the graph. **[10.3]**

19. $9y^2 - 16x^2 = 144$

20. $\dfrac{(x + 3)^2}{1} - \dfrac{(y - 2)^2}{4} = 1$

Solve. **[10.4]**

21. $x^2 + y^2 = 29,$
$\quad x - y = 3$

22. $x^2 + y^2 = 8,$
$\quad xy = 4$

23. $x^2 + 2y^2 = 20,$
$\quad y^2 - x^2 = 28$

24. $2x - y = -4,$
$\quad 3x^2 + 2y = 7$

25. The sum of two numbers is 1 and the sum of their squares is 13. Find the numbers. **[10.4]**

Graph the system of inequalities. Then find the coordinates of the points of intersection of the graphs of the related equations. **[10.4]**

26. $x^2 + y^2 \leq 8,$
$\quad x > y$

27. $y \geq x^2 - 1,$
$\quad y \leq x + 1$

COLLABORATIVE DISCUSSION AND WRITING

28. Is a parabola always the graph of a function? Why or why not? **[10.1]**

29. Is the center of an ellipse part of the graph of the ellipse? Why or why not? **[10.2]**

30. Are the asymptotes of a hyperbola part of the graph of the hyperbola? Why or why not? **[10.3]**

31. What would you say to a classmate who tells you that it is always possible to visualize all the solutions of a nonlinear system of equations? **[10.4]**

10.5 ▷ Rotation of Axes

> ❯ Use rotation of axes to graph conic sections.
>
> ❯ Use the discriminant to determine the type of conic represented by a given equation.

CONIC SECTIONS

REVIEW SECTIONS 10.1–10.3.

In Section 10.1, we saw that conic sections can be defined algebraically using a second-degree equation of the form $Ax^2 + Bxy + Cy^2 + Dx + Ey + F = 0$. Up to this point, we have considered only equations of this form for which $B = 0$. Now we turn our attention to equations of conics that contain an xy-term.

❯ Rotation of Axes

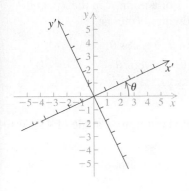

When B is nonzero, the graph of $Ax^2 + Bxy + Cy^2 + Dx + Ey + F = 0$ is a conic section with an axis that is parallel to neither the x-axis nor the y-axis. We use a technique called **rotation of axes** when we graph such an equation. The goal is to rotate the x- and y-axes through a positive angle θ to yield an $x'y'$-coordinate system, as shown at left. For the appropriate choice of θ, the graph of any conic section with an xy-term will have its axis parallel to the x'-axis or the y'-axis.

Algebraically we want to rewrite an equation

$$Ax^2 + Bxy + Cy^2 + Dx + Ey + F = 0$$

in the xy-coordinate system in the form

$$A'(x')^2 + C'(y')^2 + D'x' + E'y' + F' = 0$$

in the $x'y'$-coordinate system. Equations of this second type were graphed in Sections 10.1–10.3.

To achieve our goal, we find formulas relating the xy-coordinates of a point and the $x'y'$-coordinates of the same point. We begin by letting P be a point with coordinates (x, y) in the xy-coordinate system and (x', y') in the $x'y'$-coordinate system, as shown at left.

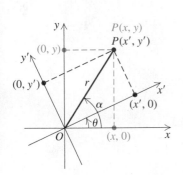

We let r represent the distance OP, and we let α represent the angle from the x-axis to OP. Then

$$\cos \alpha = \frac{x}{r} \quad \text{and} \quad \sin \alpha = \frac{y}{r},$$

so

$$x = r \cos \alpha \quad \text{and} \quad y = r \sin \alpha.$$

We also see from the figure above that

$$\cos (\alpha - \theta) = \frac{x'}{r} \quad \text{and} \quad \sin (\alpha - \theta) = \frac{y'}{r},$$

so

$$x' = r \cos (\alpha - \theta) \quad \text{and} \quad y' = r \sin (\alpha - \theta).$$

Then

$$x' = r \cos \alpha \cos \theta + r \sin \alpha \sin \theta$$

and

$$y' = r \sin \alpha \cos \theta - r \cos \alpha \sin \theta.$$

Substituting x for $r\cos\alpha$ and y for $r\sin\alpha$ gives us

$$x' = x\cos\theta + y\sin\theta \tag{1}$$

and

$$y' = y\cos\theta - x\sin\theta. \tag{2}$$

We can use these formulas to find the $x'y'$-coordinates of any point given that point's xy-coordinates and an angle of rotation θ. To express xy-coordinates in terms of $x'y'$-coordinates and an angle of rotation θ, we solve the system composed of equations (1) and (2) above for x and y. (See Exercise 43.) We get

$$x = x'\cos\theta - y'\sin\theta \quad \text{and} \quad y = x'\sin\theta + y'\cos\theta.$$

ROTATION OF AXES FORMULAS

If the x- and y-axes are rotated about the origin through a positive acute angle θ, then the coordinates (x, y) and (x', y') of a point P in the xy- and $x'y'$-coordinate systems are related by the following formulas:

$$x' = x\cos\theta + y\sin\theta, \qquad y' = -x\sin\theta + y\cos\theta;$$
$$x = x'\cos\theta - y'\sin\theta, \qquad y = x'\sin\theta + y'\cos\theta.$$

EXAMPLE 1   Suppose that the xy-axes are rotated through an angle of $45°$. Write the equation $xy = 1$ in the $x'y'$-coordinate system.

Solution   We substitute $45°$ for θ in the rotation of axes formulas for x and y:

$$x = x'\cos 45° - y'\sin 45°,$$
$$y = x'\sin 45° + y'\cos 45°.$$

We know that

$$\sin 45° = \frac{\sqrt{2}}{2} \quad \text{and} \quad \cos 45° = \frac{\sqrt{2}}{2},$$

so we have

$$x = x'\left(\frac{\sqrt{2}}{2}\right) - y'\left(\frac{\sqrt{2}}{2}\right) = \frac{\sqrt{2}}{2}(x' - y')$$

and

$$y = x'\left(\frac{\sqrt{2}}{2}\right) + y'\left(\frac{\sqrt{2}}{2}\right) = \frac{\sqrt{2}}{2}(x' + y').$$

Next, we substitute these expressions for x and y in the equation $xy = 1$:

$$\frac{\sqrt{2}}{2}(x' - y') \cdot \frac{\sqrt{2}}{2}(x' + y') = 1$$

$$\frac{1}{2}\left[(x')^2 - (y')^2\right] = 1$$

$$\frac{(x')^2}{2} - \frac{(y')^2}{2} = 1, \quad \text{or} \quad \frac{(x')^2}{(\sqrt{2})^2} - \frac{(y')^2}{(\sqrt{2})^2} = 1.$$

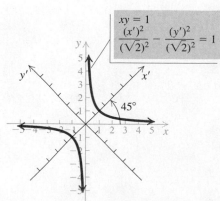

We have the equation of a hyperbola in the $x'y'$-coordinate system with its transverse axis on the x'-axis and with vertices $(-\sqrt{2}, 0)$ and $(\sqrt{2}, 0)$. Its asymptotes are $y' = -x'$ and $y' = x'$. These correspond to the axes of the xy-coordinate system. We now sketch the graph as shown at left.

Now let's substitute the rotation of axes formulas for x and y in the equation

$$Ax^2 + Bxy + Cy^2 + Dx + Ey + F = 0.$$

We have

$$A(x' \cos \theta - y' \sin \theta)^2 + B(x' \cos \theta - y' \sin \theta)(x' \sin \theta + y' \cos \theta)$$
$$+ C(x' \sin \theta + y' \cos \theta)^2 + D(x' \cos \theta - y' \sin \theta)$$
$$+ E(x' \sin \theta + y' \cos \theta) + F = 0.$$

Performing the operations indicated and collecting like terms yields the equation

$$A'(x')^2 + B'x'y' + C'(y')^2 + D'x' + E'y' + F' = 0, \qquad \text{(3)}$$

where

$$A' = A \cos^2 \theta + B \sin \theta \cos \theta + C \sin^2 \theta,$$
$$B' = 2(C - A) \sin \theta \cos \theta + B(\cos^2 \theta - \sin^2 \theta),$$
$$C' = A \sin^2 \theta - B \sin \theta \cos \theta + C \cos^2 \theta,$$
$$D' = D \cos \theta + E \sin \theta,$$
$$E' = -D \sin \theta + E \cos \theta, \quad \text{and}$$
$$F' = F.$$

Recall that our goal is to produce an equation without an $x'y'$-term, or with $B' = 0$. Then we must have

$$2(C - A) \sin \theta \cos \theta + B(\cos^2 \theta - \sin^2 \theta) = 0$$

$$(C - A) \sin 2\theta + B \cos 2\theta = 0 \qquad \text{\textbf{Using double-angle formulas}}$$

$$B \cos 2\theta = (A - C) \sin 2\theta$$

$$\frac{\cos 2\theta}{\sin 2\theta} = \frac{A - C}{B}$$

$$\cot 2\theta = \frac{A - C}{B}.$$

Thus, when θ is chosen so that

$$\cot 2\theta = \frac{A - C}{B},$$

equation (3) will have no $x'y'$-term. Although we will not do so here, it can be shown that we can always find θ such that $0° < 2\theta < 180°$, or $0° < \theta < 90°$.

ELIMINATING THE *xy*-TERM

To eliminate the xy-term from the equation

$$Ax^2 + Bxy + Cy^2 + Dx + Ey + F = 0, \quad B \neq 0,$$

select an angle θ such that

$$\cot 2\theta = \frac{A - C}{B}, \quad 0° < 2\theta < 180°,$$

and use the rotation of axes formulas.

EXAMPLE 2   Graph the equation

$$3x^2 - 2\sqrt{3}xy + y^2 + 2x + 2\sqrt{3}y = 0.$$

Solution   We have

$$A = 3, \quad B = -2\sqrt{3}, \quad C = 1, \quad D = 2, \quad E = 2\sqrt{3}, \quad \text{and} \quad F = 0.$$

To select the angle of rotation θ, we must have

$$\cot 2\theta = \frac{A - C}{B} = \frac{3 - 1}{-2\sqrt{3}} = \frac{2}{-2\sqrt{3}} = -\frac{1}{\sqrt{3}}.$$

Thus, $2\theta = 120°$, and $\theta = 60°$. We substitute this value for θ in the rotation of axes formulas for x and y:

$$x = x' \cos 60° - y' \sin 60°,$$
$$y = x' \sin 60° + y' \cos 60°.$$

This gives us

$$x = x' \cdot \frac{1}{2} - y' \cdot \frac{\sqrt{3}}{2} = \frac{x'}{2} - \frac{y'\sqrt{3}}{2}$$

and

$$y = x' \cdot \frac{\sqrt{3}}{2} + y' \cdot \frac{1}{2} = \frac{x'\sqrt{3}}{2} + \frac{y'}{2}.$$

Now we substitute these expressions for x and y in the given equation:

$$3\left(\frac{x'}{2} - \frac{y'\sqrt{3}}{2}\right)^2 - 2\sqrt{3}\left(\frac{x'}{2} - \frac{y'\sqrt{3}}{2}\right)\left(\frac{x'\sqrt{3}}{2} + \frac{y'}{2}\right) +$$
$$\left(\frac{x'\sqrt{3}}{2} + \frac{y'}{2}\right)^2 + 2\left(\frac{x'}{2} - \frac{y'\sqrt{3}}{2}\right) + 2\sqrt{3}\left(\frac{x'\sqrt{3}}{2} + \frac{y'}{2}\right) = 0.$$

After simplifying, we get

$$4(y')^2 + 4x' = 0, \quad \text{or}$$
$$(y')^2 = -x'.$$

This is the equation of a parabola with its vertex at $(0, 0)$ of the $x'y'$-coordinate system and axis of symmetry $y' = 0$. We sketch the graph.

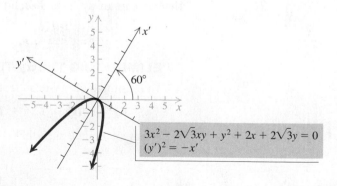

$$3x^2 - 2\sqrt{3}xy + y^2 + 2x + 2\sqrt{3}y = 0$$
$$(y')^2 = -x'$$

> **Now Try Exercise 23.**

❯ The Discriminant

It is possible to determine the type of conic represented by the equation $Ax^2 + Bxy + Cy^2 + Dx + Ey + F = 0$ before rotating the axes. Using the expressions for A', B', and C' in terms of A, B, C, and θ developed earlier, it can be shown that

$$(B')^2 - 4A'C' = B^2 - 4AC.$$

Now when θ is chosen so that

$$\cot 2\theta = \frac{A - C}{B},$$

rotation of axes gives us an equation

$$A'(x')^2 + C'(y')^2 + D'x' + E'y' + F' = 0.$$

If A' and C' have the same sign, or $A'C' > 0$, then the graph of this equation is an ellipse or a circle. If A' and C' have different signs, or $A'C' < 0$, then the graph is a hyperbola. And, if either $A' = 0$ or $C' = 0$, or $A'C' = 0$, the graph is a parabola.

Because $B' = 0$ and $(B')^2 - 4A'C' = B^2 - 4AC$, it follows that $B^2 - 4AC = -4A'C'$. Then the graph is an ellipse or a circle if $B^2 - 4AC < 0$, a hyperbola if $B^2 - 4AC > 0$, or a parabola if $B^2 - 4AC = 0$. (There are certain special cases, called *degenerate conics*, where these statements do not hold, but we will not concern ourselves with these here.) The expression $B^2 - 4AC$ is the **discriminant** of the equation $Ax^2 + Bxy + Cy^2 + Dx + Ey + F = 0$.

The graph of the equation

$$Ax^2 + Bxy + Cy^2 + Dx + Ey + F = 0$$

is, except in degenerate cases,

1. an ellipse or a circle if $B^2 - 4AC < 0$,
2. a hyperbola if $B^2 - 4AC > 0$, and
3. a parabola if $B^2 - 4AC = 0$.

EXAMPLE 3   Graph the equation $3x^2 + 2xy + 3y^2 = 16$.

Solution   We have

$$A = 3, \quad B = 2, \quad \text{and} \quad C = 3,$$

so $B^2 - 4AC = 2^2 - 4 \cdot 3 \cdot 3 = 4 - 36 = -32.$

Since the discriminant is negative, the graph is an ellipse or a circle. Now, to rotate the axes, we begin by determining θ:

$$\cot 2\theta = \frac{A - C}{B} = \frac{3 - 3}{2} = \frac{0}{2} = 0.$$

Then $2\theta = 90°$ and $\theta = 45°$, so

$$\sin \theta = \frac{\sqrt{2}}{2} \quad \text{and} \quad \cos \theta = \frac{\sqrt{2}}{2}.$$

As we saw in Example 1, substituting these values for $\sin \theta$ and $\cos \theta$ in the rotation of axes formulas gives

$$x = \frac{\sqrt{2}}{2}(x' - y') \quad \text{and} \quad y = \frac{\sqrt{2}}{2}(x' + y').$$

$2^2 - 4\cdot3\cdot3$

-32

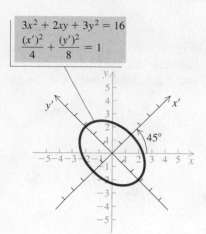

$3x^2 + 2xy + 3y^2 = 16$

$\dfrac{(x')^2}{4} + \dfrac{(y')^2}{8} = 1$

Now we substitute for x and y in the given equation:

$$3\left[\frac{\sqrt{2}}{2}(x' - y')\right]^2 + 2\left[\frac{\sqrt{2}}{2}(x' - y')\right]\left[\frac{\sqrt{2}}{2}(x' + y')\right] +$$

$$3\left[\frac{\sqrt{2}}{2}(x' + y')\right]^2 = 16.$$

After simplifying, we have

$$4(x')^2 + 2(y')^2 = 16, \quad \text{or}$$

$$\frac{(x')^2}{4} + \frac{(y')^2}{8} = 1.$$

This is the equation of an ellipse with vertices $(0, -\sqrt{8})$ and $(0, \sqrt{8})$, or $(0, -2\sqrt{2})$ and $(0, 2\sqrt{2})$, on the y'-axis. The x'-intercepts are $(-2, 0)$ and $(2, 0)$. We sketch the graph, shown at left.

> **Now Try Exercise 19.**

EXAMPLE 4 Graph the equation $4x^2 - 24xy - 3y^2 - 156 = 0$.

Solution We have

$$A = 4, \quad B = -24, \quad \text{and} \quad C = -3,$$

so $\quad B^2 - 4AC = (-24)^2 - 4 \cdot 4(-3) = 576 + 48 = 624.$

Since the discriminant is positive, the graph is a hyperbola. To rotate the axes, we begin by determining θ:

$$\cot 2\theta = \frac{A - C}{B} = \frac{4 - (-3)}{-24} = -\frac{7}{24}.$$

Since $\cot 2\theta < 0$, we have $90° < 2\theta < 180°$. From the triangle at left, we see that $\cos 2\theta = -\frac{7}{25}$.

Using half-angle formulas, we have

$$\sin \theta = \sqrt{\frac{1 - \cos 2\theta}{2}} = \sqrt{\frac{1 - \left(-\frac{7}{25}\right)}{2}} = \frac{4}{5}$$

and

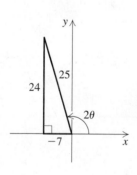

$$\cos \theta = \sqrt{\frac{1 + \cos 2\theta}{2}} = \sqrt{\frac{1 + \left(-\frac{7}{25}\right)}{2}} = \frac{3}{5}.$$

Substituting in the rotation of axes formulas gives us

$$x = x' \cos \theta - y' \sin \theta = \tfrac{3}{5}x' - \tfrac{4}{5}y'$$

and

$$y = x' \sin \theta + y' \cos \theta = \tfrac{4}{5}x' + \tfrac{3}{5}y'.$$

Now we substitute for x and y in the given equation:

$$4\left(\tfrac{3}{5}x' - \tfrac{4}{5}y'\right)^2 - 24\left(\tfrac{3}{5}x' - \tfrac{4}{5}y'\right)\left(\tfrac{4}{5}x' + \tfrac{3}{5}y'\right) - 3\left(\tfrac{4}{5}x' + \tfrac{3}{5}y'\right)^2 - 156 = 0.$$

After simplifying, we have

$$13(y')^2 - 12(x')^2 - 156 = 0$$

$$13(y')^2 - 12(x')^2 = 156$$

$$\frac{(y')^2}{12} - \frac{(x')^2}{13} = 1.$$

The graph of this equation is a hyperbola with vertices $(0, -\sqrt{12})$ and $(0, \sqrt{12})$, or $(0, -2\sqrt{3})$ and $(0, 2\sqrt{3})$, on the y'-axis. Since we know that $\sin\theta = \frac{4}{5}$ and $0° < \theta < 90°$, we can use a calculator to find that $\theta \approx 53.1°$. Thus the xy-axes are rotated through an angle of about $53.1°$ in order to obtain the $x'y'$-axes. We sketch the graph.

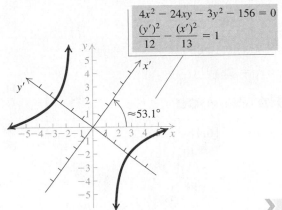

$$4x^2 - 24xy - 3y^2 - 156 = 0$$
$$\frac{(y')^2}{12} - \frac{(x')^2}{13} = 1$$

> **Now Try Exercise 35.**

10.5   Exercise Set

For the given angle of rotation and coordinates of a point in the xy-coordinate system, find the coordinates of the point in the x'y'-coordinate system.

1. $\theta = 45°$, $(\sqrt{2}, -\sqrt{2})$

2. $\theta = 45°$, $(-1, 3)$

3. $\theta = 30°$, $(0, 2)$

4. $\theta = 60°$, $(0, \sqrt{3})$

For the given angle of rotation and coordinates of a point in the x'y'-coordinate system, find the coordinates of the point in the xy-coordinate system.

5. $\theta = 45°$, $(1, -1)$

6. $\theta = 45°$, $(-3\sqrt{2}, \sqrt{2})$

7. $\theta = 30°$, $(2, 0)$

8. $\theta = 60°$, $(-1, -\sqrt{3})$

Use the discriminant to determine whether the graph of the equation is an ellipse (or a circle), a hyperbola, or a parabola.

9. $3x^2 - 5xy + 3y^2 - 2x + 7y = 0$

10. $5x^2 + 6xy - 4y^2 + x - 3y + 4 = 0$

11. $x^2 - 3xy - 2y^2 + 12 = 0$

12. $4x^2 + 7xy + 2y^2 - 3x + y = 0$

13. $4x^2 - 12xy + 9y^2 - 3x + y = 0$

14. $6x^2 + 5xy + 6y^2 + 15 = 0$

15. $2x^2 - 8xy + 7y^2 + x - 2y + 1 = 0$

16. $x^2 + 6xy + 9y^2 - 3x + 4y = 0$

17. $8x^2 - 7xy + 5y^2 - 17 = 0$

18. $x^2 + xy - y^2 - 4x + 3y - 2 = 0$

Graph the equation.

19. $4x^2 + 2xy + 4y^2 = 15$

20. $3x^2 + 10xy + 3y^2 + 8 = 0$

21. $x^2 - 10xy + y^2 + 36 = 0$

22. $x^2 + 2xy + y^2 + 4\sqrt{2}x - 4\sqrt{2}y = 0$

23. $x^2 - 2\sqrt{3}xy + 3y^2 - 12\sqrt{3}x - 12y = 0$

24. $13x^2 + 6\sqrt{3}xy + 7y^2 - 16 = 0$

25. $7x^2 + 6\sqrt{3}xy + 13y^2 - 32 = 0$

26. $x^2 + 4xy + y^2 - 9 = 0$

27. $11x^2 + 10\sqrt{3}xy + y^2 = 32$

28. $5x^2 - 8xy + 5y^2 = 81$

29. $\sqrt{2}x^2 + 2\sqrt{2}xy + \sqrt{2}y^2 - 8x + 8y = 0$

30. $x^2 + 2\sqrt{3}xy + 3y^2 - 8x + 8\sqrt{3}y = 0$

31. $x^2 + 6\sqrt{3}xy - 5y^2 + 8x - 8\sqrt{3}y - 48 = 0$

32. $3x^2 - 2xy + 3y^2 - 6\sqrt{2}x + 2\sqrt{2}y - 26 = 0$

33. $x^2 + xy + y^2 = 24$

34. $4x^2 + 3\sqrt{3}xy + y^2 = 55$

35. $4x^2 - 4xy + y^2 - 8\sqrt{5}x - 16\sqrt{5}y = 0$

36. $9x^2 - 24xy + 16y^2 - 400x - 300y = 0$

37. $11x^2 + 7xy - 13y^2 = 621$

38. $3x^2 + 4xy + 6y^2 = 28$

❯ Skill Maintenance

Convert to radian measure. **[6.4]**

39. $120°$

40. $-315°$

Convert to degree measure.

41. $\dfrac{\pi}{3}$

42. $\dfrac{3\pi}{4}$

❯ Synthesis

43. Solve this system of equations for x and y:
$$x' = x\cos\theta + y\sin\theta,$$
$$y' = y\cos\theta - x\sin\theta.$$
Show your work.

44. Show that for any angle θ, the equation $x^2 + y^2 = r^2$ becomes $(x')^2 + (y')^2 = r^2$ when the rotation of axes formulas are applied.

45. Show that $A + C = A' + C'$.

10.6 ❯ Polar Equations of Conics

❯ Graph polar equations of conics.

❯ Convert from polar equations of conics to rectangular equations of conics.

❯ Find polar equations of conics.

In Sections 10.1–10.3, we saw that the parabola, the ellipse, and the hyperbola have different definitions in rectangular coordinates. When polar coordinates are used, we can give a single definition that applies to all three conics.

CONIC SECTIONS

REVIEW SECTIONS 10.1–10.3.

AN ALTERNATIVE DEFINITION OF CONICS

Let L be a fixed line (the **directrix**), let F be a fixed point (the **focus**) not on L, and let e be a positive constant (the **eccentricity**). A **conic** is the set of all points P in the plane such that

$$\frac{PF}{PL} = e,$$

where PF is the distance from P to F and PL is the distance from P to L. The conic is a parabola if $e = 1$, an ellipse if $0 < e < 1$, and a hyperbola if $e > 1$.

Note that if $e = 1$, then $PF = PL$ and the alternative definition of a parabola is identical to the definition presented in Section 10.1.

〉 **Polar Equations of Conics**

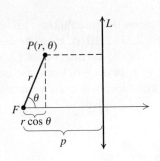

To derive equations for the conics in polar coordinates, we position the focus F at the pole and position the directrix L either perpendicular to the polar axis or parallel to it. In the figure at left, we place L perpendicular to the polar axis and p units to the right of the focus, or pole.

Note that $PL = p - r\cos\theta$. Then if P is any point on the conic, we have

$$\frac{PF}{PL} = e$$

$$\frac{r}{p - r\cos\theta} = e$$

$$r = ep - er\cos\theta$$

$$r + er\cos\theta = ep$$

$$r(1 + e\cos\theta) = ep$$

$$r = \frac{ep}{1 + e\cos\theta}.$$

Thus we see that the polar equation of a conic with focus at the pole and directrix perpendicular to the polar axis and p units to the right of the pole is

$$r = \frac{ep}{1 + e\cos\theta},$$

where e is the eccentricity of the conic.

For an ellipse and a hyperbola, we can make the following statement regarding eccentricity.

ECCENTRICITY

For an ellipse and a hyperbola, the **eccentricity e** is given by

$$e = \frac{c}{a},$$

where c is the distance from the center to a focus and a is the distance from the center to a vertex.

EXAMPLE 1   Describe and graph the conic $r = \dfrac{18}{6 + 3\cos\theta}$.

Solution   We begin by dividing the numerator and the denominator by 6 to obtain a constant term of 1 in the denominator:

$$r = \frac{3}{1 + 0.5\cos\theta}.$$

This equation is in the form

$$r = \frac{ep}{1 + e\cos\theta}$$

with $e = 0.5$. Since $e < 1$, the graph is an ellipse. Also, since $e = 0.5$ and $ep = 0.5p = 3$, we have $p = 6$. Thus the ellipse has a vertical directrix that

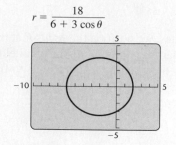

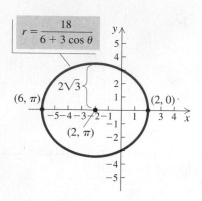

lies 6 units to the right of the pole. We graph the equation in the square window $[-10, 5, -5, 5]$, as shown at left.

It follows that the major axis is horizontal and lies on the polar axis. The vertices are found by letting $\theta = 0$ and $\theta = \pi$. They are $(2, 0)$ and $(6, \pi)$. The center of the ellipse is at the midpoint of the segment connecting the vertices, or at $(2, \pi)$.

The length of the major axis is 8, so we have $2a = 8$, or $a = 4$. From the equation of the conic, we know that $e = 0.5$. Using the equation $e = c/a$, we can find that $c = 2$. Finally, using $a = 4$ and $c = 2$ in $b^2 = a^2 - c^2$ gives us

$$b^2 = 4^2 - 2^2 = 16 - 4 = 12$$
$$b = \sqrt{12}, \text{ or } 2\sqrt{3},$$

so the length of the minor axis is $2\sqrt{12}$, or $4\sqrt{3}$. This is useful to know when sketching a hand-drawn graph of the conic.

▷ **Now Try Exercise 15.**

Other derivations similar to the one on p. 767 lead to the following result.

POLAR EQUATIONS OF CONICS

A polar equation of any of the four forms

$$r = \frac{ep}{1 \pm e \cos \theta}, \qquad r = \frac{ep}{1 \pm e \sin \theta}$$

is a conic section. The conic is a parabola if $e = 1$, an ellipse if $0 < e < 1$, and a hyperbola if $e > 1$.

The following table describes the polar equations of conics with a focus at the pole and the directrix either perpendicular to or parallel to the polar axis.

Equation	Description
$r = \dfrac{ep}{1 + e \cos \theta}$	Vertical directrix p units to the right of the pole (or focus)
$r = \dfrac{ep}{1 - e \cos \theta}$	Vertical directrix p units to the left of the pole (or focus)
$r = \dfrac{ep}{1 + e \sin \theta}$	Horizontal directrix p units above the pole (or focus)
$r = \dfrac{ep}{1 - e \sin \theta}$	Horizontal directrix p units below the pole (or focus)

EXAMPLE 2 Describe and graph the conic $r = \dfrac{10}{5 - 5 \sin \theta}$.

Solution We first divide the numerator and the denominator by 5:

$$r = \frac{2}{1 - \sin \theta}.$$

This equation is in the form

$$r = \frac{ep}{1 - e \sin \theta}$$

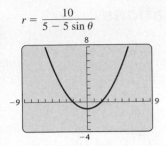

$$r = \frac{10}{5 - 5\sin\theta}$$

with $e = 1$, so the graph is a parabola. Since $e = 1$ and $ep = 1 \cdot p = 2$, we have $p = 2$. Thus the parabola has a horizontal directrix 2 units below the pole. We graph the equation in the square window $[-9, 9, -4, 8]$, as shown at left.

It follows that the parabola has a vertical axis of symmetry. Since the directrix lies below the focus, or pole, the parabola opens up. The vertex is the midpoint of the segment of the axis of symmetry from the focus to the directrix. We find it by letting $\theta = 3\pi/2$. It is $(1, 3\pi/2)$.

> **Now Try Exercise 17.**

EXAMPLE 3   Describe and graph the conic $r = \dfrac{4}{2 + 6\sin\theta}$.

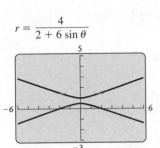

$$r = \frac{4}{2 + 6\sin\theta}$$

Solution   We first divide the numerator and the denominator by 2:

$$r = \frac{2}{1 + 3\sin\theta}.$$

This equation is in the form

$$r = \frac{ep}{1 + e\sin\theta}$$

with $e = 3$. Since $e > 1$, the graph is a hyperbola. We have $e = 3$ and $ep = 3p = 2$, so $p = \frac{2}{3}$. Thus the hyperbola has a horizontal directrix that lies $\frac{2}{3}$ unit above the pole. We graph the equation as shown at left.

It follows that the transverse axis is vertical. To find the vertices, we let $\theta = \pi/2$ and $\theta = 3\pi/2$. The vertices are $(1/2, \pi/2)$ and $(-1, 3\pi/2)$. The center of the hyperbola is the midpoint of the segment connecting the vertices, or $(3/4, \pi/2)$. Thus the distance c from the center to a focus is $3/4$. Using $c = 3/4$, $e = 3$, and $e = c/a$, we have $a = 1/4$. Then since $c^2 = a^2 + b^2$, we have

$$b^2 = \left(\frac{3}{4}\right)^2 - \left(\frac{1}{4}\right)^2 = \frac{9}{16} - \frac{1}{16} = \frac{1}{2}$$

$$b = \frac{1}{\sqrt{2}}, \text{ or } \frac{\sqrt{2}}{2}.$$

Knowing the values of a and b allows us to sketch the asymptotes if we are graphing the hyperbola by hand. We can also easily plot the points $(2, 0)$ and $(2, \pi)$ on the polar axis.

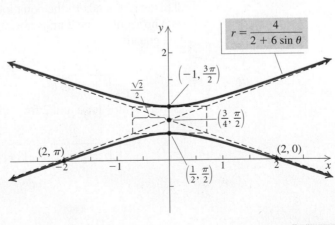

> **Now Try Exercise 13.**

❯ Converting from Polar Equations to Rectangular Equations

We can use the relationships between polar coordinates and rectangular coordinates that were developed in Section 8.4 to convert polar equations of conics to rectangular equations.

EXAMPLE 4 Convert to a rectangular equation: $r = \dfrac{2}{1 - \sin \theta}$.

Solution We have

$$r = \frac{2}{1 - \sin \theta}$$

$$r - r \sin \theta = 2 \qquad \textbf{Multiplying by } 1 - \sin \theta$$

$$r = r \sin \theta + 2$$

$$\sqrt{x^2 + y^2} = y + 2 \qquad \textbf{Substituting } \sqrt{x^2 + y^2} \textbf{ for } r \textbf{ and } y \textbf{ for } r \sin \theta$$

$$x^2 + y^2 = y^2 + 4y + 4 \qquad \textbf{Squaring both sides}$$

$$x^2 = 4y + 4, \quad \text{or}$$

$$x^2 - 4y - 4 = 0.$$

This is the equation of a parabola, as we should have anticipated, since $e = 1$.

> **Now Try Exercise 23.**

❯ Finding Polar Equations of Conics

We can find the polar equation of a conic with a focus at the pole if we know its eccentricity and the equation of the directrix.

EXAMPLE 5 Find a polar equation of the conic with a focus at the pole, eccentricity $\frac{1}{3}$, and directrix $r = 2 \csc \theta$.

Solution The equation of the directrix can be written

$$r = \frac{2}{\sin \theta}, \quad \text{or} \quad r \sin \theta = 2.$$

This corresponds to the equation $y = 2$ in rectangular coordinates, so the directrix is a horizontal line 2 units above the polar axis. Using the table on p. 768, we see that the equation is of the form

$$r = \frac{ep}{1 + e \sin \theta}.$$

Substituting $\frac{1}{3}$ for e and 2 for p gives us

$$r = \frac{\frac{1}{3} \cdot 2}{1 + \frac{1}{3} \sin \theta} = \frac{\frac{2}{3}}{1 + \frac{1}{3} \sin \theta} = \frac{2}{3 + \sin \theta}.$$

> **Now Try Exercise 39.**

10.6 | Exercise Set

In Exercises 1–6, match the equation with one of the graphs (a)–(f) that follow.

a)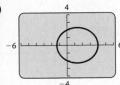

b)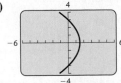

c)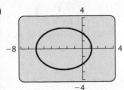

d)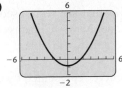

e)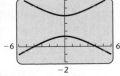

f)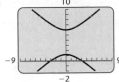

1. $r = \dfrac{3}{1 + \cos \theta}$

2. $r = \dfrac{4}{1 + 2 \sin \theta}$

3. $r = \dfrac{8}{4 - 2 \cos \theta}$

4. $r = \dfrac{12}{4 + 6 \sin \theta}$

5. $r = \dfrac{5}{3 - 3 \sin \theta}$

6. $r = \dfrac{6}{3 + 2 \cos \theta}$

For each equation:

a) *Tell whether the equation describes a parabola, an ellipse, or a hyperbola.*

b) *State whether the directrix is vertical or horizontal and give its location in relation to the pole.*

c) *Find the vertex or vertices.*

d) *Graph the equation.*

7. $r = \dfrac{1}{1 + \cos \theta}$

8. $r = \dfrac{4}{2 + \cos \theta}$

9. $r = \dfrac{15}{5 - 10 \sin \theta}$

10. $r = \dfrac{12}{4 + 8 \sin \theta}$

11. $r = \dfrac{8}{6 - 3 \cos \theta}$

12. $r = \dfrac{6}{2 + 2 \sin \theta}$

13. $r = \dfrac{20}{10 + 15 \sin \theta}$

14. $r = \dfrac{10}{8 - 2 \cos \theta}$

15. $r = \dfrac{9}{6 + 3 \cos \theta}$

16. $r = \dfrac{4}{3 - 9 \sin \theta}$

17. $r = \dfrac{3}{2 - 2 \sin \theta}$

18. $r = \dfrac{12}{3 + 9 \cos \theta}$

19. $r = \dfrac{4}{2 - \cos \theta}$

20. $r = \dfrac{5}{1 - \sin \theta}$

21. $r = \dfrac{7}{2 + 10 \sin \theta}$

22. $r = \dfrac{3}{8 - 4 \cos \theta}$

23.–38. Convert the equations in Exercises 7–22 to rectangular equations.

Find a polar equation of the conic with a focus at the pole and the given eccentricity and directrix.

39. $e = 2$, $r = 3 \csc \theta$

40. $e = \frac{2}{3}$, $r = -\sec \theta$

41. $e = 1$, $r = 4 \sec \theta$

42. $e = 3$, $r = 2 \csc \theta$

43. $e = \frac{1}{2}$, $r = -2 \sec \theta$

44. $e = 1$, $r = 4 \csc \theta$

45. $e = \frac{3}{4}$, $r = 5 \csc \theta$

46. $e = \frac{4}{5}$, $r = 2 \sec \theta$

47. $e = 4$, $r = -2 \csc \theta$

48. $e = 3$, $r = 3 \csc \theta$

❯ Skill Maintenance

For $f(x) = (x - 3)^2 + 4$, find each of the following.
[1.2]

49. $f(t)$

50. $f(2t)$

51. $f(t - 1)$

52. $f(t + 2)$

❯ Synthesis

Parabolic Orbit. *Suppose that a comet travels in a parabolic orbit with the sun as its focus. Position a polar coordinate system with the pole at the sun and the axis of the orbit perpendicular to the polar axis. When the comet is the given distance from the sun, the segment from the comet to the sun makes the given angle with the polar axis. Find a polar equation of the orbit, assuming that the directrix lies above the pole.*

53. 100 million miles, $\dfrac{\pi}{6}$

54. 120 million miles, $\dfrac{\pi}{4}$

10.7 ▶ Parametric Equations

> ❯ Graph parametric equations.

> ❯ Determine an equivalent rectangular equation for parametric equations.

> ❯ Determine parametric equations for a rectangular equation.

> ❯ Solve applied problems involving projectile motion.

❯ Graphing Parametric Equations

We have graphed *plane curves* that are composed of sets of ordered pairs (x, y) in the rectangular coordinate plane. Now we discuss a way to represent plane curves in which x and y are functions of a third variable, t.

EXAMPLE 1   Graph the curve represented by the equations

$$x = \tfrac{1}{2}t, \qquad y = t^2 - 3; \quad -3 \le t \le 3.$$

Solution   We can choose values for t between -3 and 3 and find the corresponding values of x and y. When $t = -3$, we have

$$x = \tfrac{1}{2}(-3) = -\tfrac{3}{2}, \qquad y = (-3)^2 - 3 = 6.$$

The following table lists other ordered pairs. We plot these points and then draw the curve.

t	x	y	(x, y)
-3	$-\tfrac{3}{2}$	6	$\left(-\tfrac{3}{2}, 6\right)$
-2	-1	1	$(-1, 1)$
-1	$-\tfrac{1}{2}$	-2	$\left(-\tfrac{1}{2}, -2\right)$
0	0	-3	$(0, -3)$
1	$\tfrac{1}{2}$	-2	$\left(\tfrac{1}{2}, -2\right)$
2	1	1	$(1, 1)$
3	$\tfrac{3}{2}$	6	$\left(\tfrac{3}{2}, 6\right)$

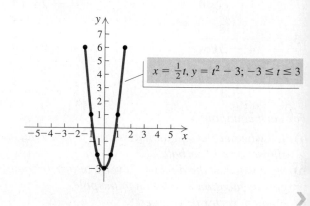

$x = \tfrac{1}{2}t, \, y = t^2 - 3; \, -3 \le t \le 3$

The curve above appears to be part of a parabola. Let's verify this by finding the equivalent rectangular equation. Solving $x = \tfrac{1}{2}t$ for t, we get $t = 2x$. Substituting $2x$ for t in $y = t^2 - 3$, we have

$$y = (2x)^2 - 3 = 4x^2 - 3.$$

This is a quadratic equation. Hence its graph is a parabola. The curve is part of the parabola $y = 4x^2 - 3$. Since $-3 \le t \le 3$ and $x = \tfrac{1}{2}t$, we must include the restriction $-\tfrac{3}{2} \le x \le \tfrac{3}{2}$ when we write the equivalent rectangular equation:

$$y = 4x^2 - 3, \quad -\tfrac{3}{2} \le x \le \tfrac{3}{2}.$$

The equations $x = \tfrac{1}{2}t$ and $y = t^2 - 3$ are **parametric equations** for the curve. The variable t is the **parameter.** When we write the corresponding rectangular equation, we say that we **eliminate the parameter.**

> PARAMETRIC EQUATIONS
>
> If f and g are continuous functions of t on an interval I, then the set of ordered pairs (x, y) such that $x = f(t)$ and $y = g(t)$ is a **plane curve.** The equations $x = f(t)$ and $y = g(t)$ are **parametric equations** for the curve. The variable t is the **parameter**.

❯ Determining a Rectangular Equation for Given Parametric Equations

EXAMPLE 2 Using a graphing calculator, graph each of the following plane curves given their respective parametric equations and the restriction on the parameter. Then find the equivalent rectangular equation.

a) $x = t^2$, $y = t - 1$; $-1 \le t \le 4$

b) $x = \sqrt{t}$, $y = 2t + 3$; $0 \le t \le 3$

Solution

$x = t^2$, $y = t - 1$; $-1 \le t \le 4$

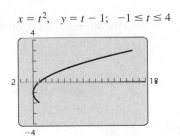

a) To graph the curve (see the window at left), we set the graphing calculator in PARAMETRIC mode, enter the equations, and select minimum and maximum values for x, y, and t.

WINDOW

Tmin $= -1$	Xmin $= -2$	Ymin $= -4$
Tmax $= 4$	Xmax $= 18$	Ymax $= 4$
Tstep $= .1$	Xscl $= 1$	Yscl $= 1$

To find an equivalent rectangular equation, we first solve either equation for t. We choose the equation $y = t - 1$:

$$y = t - 1$$
$$y + 1 = t.$$

We then substitute $y + 1$ for t in $x = t^2$:

$$x = t^2$$
$$x = (y + 1)^2. \qquad \text{Substituting}$$

This is an equation of a parabola that opens to the right. Given that $-1 \le t \le 4$, we have the corresponding restrictions on x and y: $0 \le x \le 16$ and $-2 \le y \le 3$. Thus the equivalent rectangular equation is

$$x = (y + 1)^2, \quad -2 \le y \le 3.$$

$x = \sqrt{t}$, $y = 2t + 3$;
$0 \le t \le 3$

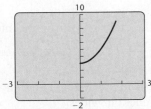

b) To graph the curve (see the window at left), we use PARAMETRIC mode and enter the equations and window settings.

WINDOW

Tmin $= 0$	Xmin $= -3$	Ymin $= -2$
Tmax $= 3$	Xmax $= 3$	Ymax $= 10$
Tstep $= .1$	Xscl $= 1$	Yscl $= 1$

To find an equivalent rectangular equation, we first solve $x = \sqrt{t}$ for t:

$$x = \sqrt{t}$$
$$x^2 = t.$$

Then we substitute x^2 for t in $y = 2t + 3$:

$$y = 2t + 3$$
$$y = 2x^2 + 3. \quad \textbf{Substituting}$$

When $0 \leq t \leq 3$, we have $0 \leq x \leq \sqrt{3}$. The equivalent rectangular equation is

$$y = 2x^2 + 3, \quad 0 \leq x \leq \sqrt{3}.$$

> *Now Try Exercise 5.*

We first graphed in parametric mode in Section 6.5. There we used an angle measure as the parameter as we do in the next example.

EXAMPLE 3   Graph the plane curve represented by $x = \cos t$ and $y = \sin t$ with t in $[0, 2\pi]$. Then determine an equivalent rectangular equation.

$x = \cos t, \ y = \sin t; \ 0 \leq t \leq 2\pi$

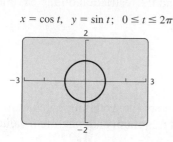

Solution   Using a squared window and a Tstep of $\pi/48$, we obtain the graph at left. It appears to be the unit circle.

The equivalent rectangular equation can be obtained by squaring both sides of each parametric equation:

$$x^2 = \cos^2 t \quad \text{and} \quad y^2 = \sin^2 t.$$

This allows us to use the trigonometric identity $\sin^2 \theta + \cos^2 \theta = 1$. Substituting, we get

$$x^2 + y^2 = 1.$$

As expected, this is an equation of the unit circle.

> *Now Try Exercise 13.*

EXAMPLE 4   Graph the plane curve represented by

$$x = 5 \cos t \quad \text{and} \quad y = 3 \sin t; \quad 0 \leq t \leq 2\pi.$$

Then eliminate the parameter to find the rectangular equation.

$x = 5 \cos t, \ y = 3 \sin t; \ 0 \leq t \leq 2\pi$

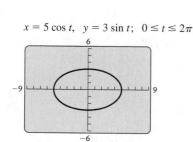

Solution   The graph at left appears to be the graph of an ellipse. To find the rectangular equation, we first solve for $\cos t$ and $\sin t$ in the parametric equations:

$$x = 5 \cos t \qquad y = 3 \sin t$$
$$\frac{x}{5} = \cos t, \qquad \frac{y}{3} = \sin t.$$

Using the identity $\sin^2 \theta + \cos^2 \theta = 1$, we can substitute to eliminate the parameter:

$$\sin^2 t + \cos^2 t = 1$$
$$\left(\frac{y}{3}\right)^2 + \left(\frac{x}{5}\right)^2 = 1 \quad \textbf{Substituting}$$
$$\frac{x^2}{25} + \frac{y^2}{9} = 1. \quad \textbf{Ellipse}$$

The rectangular form of the equation confirms that the graph is an ellipse centered at the origin with vertices at $(5, 0)$ and $(-5, 0)$.

> *Now Try Exercise 15.*

One advantage of graphing the unit circle parametrically, as we did in Example 3, is that it provides a method of finding trigonometric function values.

EXAMPLE 5   Using the VALUE feature from the CALC menu and the parametric graph of the unit circle, find each of the following function values.

a) $\cos \dfrac{7\pi}{6}$

b) $\sin 4.13$

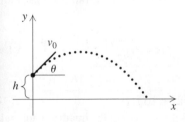

T = 3.6651914
X = −.8660254   Y = −.5

Solution

a) With the calculator set in RADIAN mode and the VALUE feature from the CALC menu, we enter $7\pi/6$ for t. The value of x, which is $\cos t$, appears, as shown at left. Thus,

$$\cos\frac{7\pi}{6} \approx -0.8660.$$

b) With the calculator set in RADIAN mode, we enter 4.13 for t. The value of y, which is $\sin t$, will appear on the screen. The calculator will show that $\sin 4.13 \approx -0.8352$.

> *Now Try Exercise 21.*

❯ Determining Parametric Equations for a Given Rectangular Equation

Many sets of parametric equations can represent the same plane curve. In fact, there are infinitely many such equations.

EXAMPLE 6   Find three sets of parametric equations for the parabola

$$y = 4 - (x + 3)^2.$$

Solution

If $x = t$, then $y = 4 - (t + 3)^2$, or $-t^2 - 6t - 5$.

If $x = t - 3$, then $y = 4 - (t - 3 + 3)^2$, or $4 - t^2$.

If $x = \dfrac{t}{3}$, then $y = 4 - \left(\dfrac{t}{3} + 3\right)^2$, or $-\dfrac{t^2}{9} - 2t - 5$.

> *Now Try Exercise 29.*

❯ Applications

The motion of an object that is propelled upward can be described with parametric equations. Such motion is called **projectile motion.** It can be shown using more advanced mathematics that, neglecting air resistance, the following equations describe the path of a projectile propelled upward at an angle θ with the horizontal from a height h, in feet, at an initial speed v_0, in feet per second:

$$x = (v_0 \cos\theta)t, \qquad y = h + (v_0 \sin\theta)t - 16t^2.$$

We can use these equations to determine the location of the object at time t, in seconds.

EXAMPLE 7 *Projectile Motion.*   A baseball is thrown from a height of 6 ft with an initial speed of 100 ft/sec at an angle of 45° with the horizontal.

a) Find parametric equations that give the position of the ball at time t, in seconds.

b) Graph the plane curve represented by the equations found in part (a).

c) Find the height of the ball after 1 sec, after 2 sec, and after 3 sec.

d) Determine how long the ball is in the air.

e) Determine the horizontal distance that the ball travels.

f) Find the maximum height of the ball.

Solution

a) We substitute 6 for h, 100 for v_0, and 45° for θ in the equations above:

$$x = (v_0 \cos \theta)t$$
$$= (100 \cos 45°)t$$
$$= \left(100 \cdot \frac{\sqrt{2}}{2}\right)t = 50\sqrt{2}t;$$

$$y = h + (v_0 \sin \theta)t - 16t^2$$
$$= 6 + (100 \sin 45°)t - 16t^2$$
$$= 6 + \left(100 \cdot \frac{\sqrt{2}}{2}\right)t - 16t^2$$
$$= 6 + 50\sqrt{2}t - 16t^2.$$

b) $x = 50\sqrt{2}t, \quad y = 6 + 50\sqrt{2}t - 16t^2$

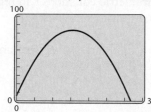

c) The height of the ball at time t is represented by y. We can use a table set in ASK mode to find the desired values of y as shown at left, or we can substitute in the equation for y as shown below.

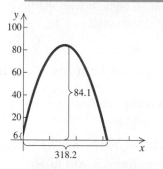

When $t = 1$, $y = 6 + 50\sqrt{2}(1) - 16(1)^2 \approx 60.7$ ft.
When $t = 2$, $y = 6 + 50\sqrt{2}(2) - 16(2)^2 \approx 83.4$ ft.
When $t = 3$, $y = 6 + 50\sqrt{2}(3) - 16(3)^2 \approx 74.1$ ft.

d) The ball hits the ground when $y = 0$. Thus, in order to determine how long the ball is in the air, we solve the equation $y = 0$:

$$6 + 50\sqrt{2}t - 16t^2 = 0$$
$$-16t^2 + 50\sqrt{2}t + 6 = 0 \qquad \textbf{Standard form}$$
$$t = \frac{-50\sqrt{2} \pm \sqrt{(50\sqrt{2})^2 - 4(-16)(6)}}{2(-16)}$$

Using the quadratic formula

$$t \approx -0.1 \quad \text{or} \quad t \approx 4.5.$$

The negative value for t has no meaning in this application. Thus we determine that the ball is in the air for about 4.5 sec.

e) Since the ball is in the air for about 4.5 sec, the horizontal distance that it travels is given by

$$x = 50\sqrt{2}(4.5) \approx 318.2 \text{ ft.}$$

f) To find the maximum height of the ball, we find the maximum value of y. This occurs at the vertex of the quadratic function represented by y. At the vertex, we have

$$t = -\frac{b}{2a} = -\frac{50\sqrt{2}}{2(-16)} \approx 2.2.$$

When $t = 2.2$,

$$y = 6 + 50\sqrt{2}(2.2) - 16(2.2)^2 \approx 84.1 \text{ ft.}$$

> *Now Try Exercise 33.*

QUADRATIC FORMULA

REVIEW SECTION 3.2.

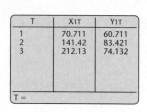

The path of a fixed point on the circumference of a circle as it rolls along a line is called a **cycloid.** For example, a point on the rim of a bicycle wheel traces a cycloid curve.

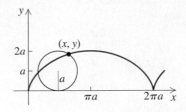

The parametric equations of a cycloid are

$$x = a(t - \sin t), \qquad y = a(1 - \cos t),$$

where a is the radius of the circle that traces the curve and t is in radian measure.

EXAMPLE 8 Graph the cycloid described by the parametric equations

$$x = 3(t - \sin t), \qquad y = 3(1 - \cos t); \quad 0 \le t \le 6\pi.$$

Solution

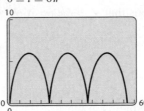

> **Now Try Exercise 35.**

10.7 | Exercise Set

Graph the plane curve given by the parametric equations. Then find an equivalent rectangular equation.

1. $x = \frac{1}{2}t, \ y = 6t - 7; \ -1 \le t \le 6$

2. $x = t, \ y = 5 - t; \ -2 \le t \le 3$

3. $x = 4t^2, \ y = 2t; \ -1 \le t \le 1$

4. $x = \sqrt{t}, \ y = 2t + 3; \ 0 \le t \le 8$

5. $x = t^2, \ y = \sqrt{t}; \ 0 \le t \le 4$

6. $x = t^3 + 1, \ y = t; \ -3 \le t \le 3$

7. $x = t + 3, \ y = \dfrac{1}{t + 3}; \ -2 \le t \le 2$

8. $x = 2t^3 + 1, \ y = 2t^3 - 1; \ -4 \le t \le 4$

9. $x = 2t - 1, \ y = t^2; \ -3 \le t \le 3$

10. $x = \frac{1}{3}t, \ y = t; \ -5 \le t \le 5$

11. $x = e^{-t}, \ y = e^t; \ -\infty < t < \infty$

12. $x = 2 \ln t, \ y = t^2; \ 0 < t < \infty$

13. $x = 3 \cos t, \ y = 3 \sin t; \ 0 \le t \le 2\pi$

14. $x = 2 \cos t, \ y = 4 \sin t; \ 0 \le t \le 2\pi$

15. $x = \cos t, \ y = 2 \sin t; \ 0 \le t \le 2\pi$

16. $x = 2 \cos t, \ y = 2 \sin t; \ 0 \le t \le 2\pi$

17. $x = \sec t, \ y = \cos t; \ -\dfrac{\pi}{2} < t < \dfrac{\pi}{2}$

18. $x = \sin t, \ y = \csc t; \ 0 < t < \pi$

19. $x = 1 + 2 \cos t, \ y = 2 + 2 \sin t; \ 0 \le t \le 2\pi$

20. $x = 2 + \sec t, \ y = 1 + 3 \tan t; \ 0 < t < \dfrac{\pi}{2}$

Using a parametric graph of the unit circle, find the function value.

21. $\sin \dfrac{\pi}{4}$

22. $\cos \dfrac{2\pi}{3}$

23. $\cos \dfrac{17\pi}{12}$

24. $\sin \dfrac{4\pi}{5}$

25. $\tan \dfrac{\pi}{5}$

26. $\tan \dfrac{2\pi}{7}$

27. $\cos 5.29$

28. $\sin 1.83$

Find two sets of parametric equations for the rectangular equation.

29. $y = 4x - 3$

30. $y = x^2 - 1$

31. $y = (x - 2)^2 - 6x$

32. $y = x^3 + 3$

33. *Projectile Motion.* A ball is thrown from a height of 7 ft with an initial speed of 80 ft/sec at an angle of 30° with the horizontal.

  a) Find parametric equations that give the position of the ball at time t, in seconds.

  b) Graph the plane curve represented by the equations found in part (a).

  c) Find the height of the ball after 1 sec and after 2 sec.

  d) Determine how long the ball is in the air.

  e) Determine the horizontal distance that the ball travels.

  f) Find the maximum height of the ball.

34. *Projectile Motion.* A projectile is launched from the ground with an initial speed of 200 ft/sec at an angle of 60° with the horizontal.

  a) Find parametric equations that give the position of the projectile at time t, in seconds.

  b) Graph the plane curve represented by the equations found in part (a).

  c) Find the height of the projectile after 4 sec and after 8 sec.

  d) Determine how long the projectile is in the air.

  e) Determine the horizontal distance that the projectile travels.

  f) Find the maximum height of the projectile.

Graph the cycloid.

35. $x = 2(t - \sin t)$, $y = 2(1 - \cos t)$; $0 \le t \le 4\pi$

36. $x = 4t - 4\sin t$, $y = 4 - 4\cos t$; $0 \le t \le 6\pi$

37. $x = t - \sin t$, $y = 1 - \cos t$; $-2\pi \le t \le 2\pi$

38. $x = 5(t - \sin t)$, $y = 5(1 - \cos t)$; $-4\pi \le t \le 4\pi$

❯ Skill Maintenance

Graph.

39. $y = x^3$ **[1.1]**

40. $x = y^3$ **[1.1]**

41. $f(x) = \sqrt{x - 2}$ **[1.2]**

42. $f(x) = \dfrac{3}{x^2 - 1}$ **[4.5]**

❯ Synthesis

43. Graph the curve described by

$$x = 3\cos t, \qquad y = 3\sin t; \quad 0 \le t \le 2\pi.$$

As t increases, the path of the curve is generated in the counterclockwise direction. How can this set of equations be changed so that the curve is generated in the clockwise direction?

44. Graph the plane curve described by

$$x = \cos^3 t, \qquad y = \sin^3 t; \quad 0 \le t \le 2\pi.$$

Then find the equivalent rectangular equation.

STUDY GUIDE

KEY TERMS AND CONCEPTS	EXAMPLES

SECTION 10.1: THE PARABOLA

Standard Equation of a Parabola with Vertex $(0, 0)$ and Vertical Axis of Symmetry

The standard equation of a parabola with vertex $(0, 0)$ and directrix $y = -p$ is

$$x^2 = 4py.$$

The focus is $(0, p)$ and the y-axis is the axis of symmetry.

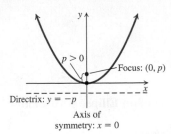

(When $p < 0$, the parabola opens down.)

See also Example 1 on p. 722.

Standard Equation of a Parabola with Vertex $(0, 0)$ and Horizontal Axis of Symmetry

The standard equation of a parabola with vertex $(0, 0)$ and directrix $x = -p$ is

$$y^2 = 4px.$$

The focus is $(p, 0)$ and the x-axis is the axis of symmetry.

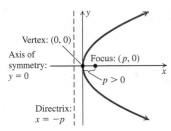

(When $p < 0$, the parabola opens to the left.)

See also Example 2 on p. 722.

Standard Equation of a Parabola with Vertex (h, k) and Vertical Axis of Symmetry

The standard equation of a parabola with vertex (h, k) and vertical axis of symmetry is

$$(x - h)^2 = 4p(y - k),$$

where the vertex is (h, k), the focus is $(h, k + p)$, and the directrix is $y = k - p$.

The parabola opens up if $p > 0$. It opens down if $p < 0$.

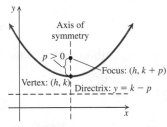

(When $p < 0$, the parabola opens down.)

See also Example 3 on p. 724.

Standard Equation of a Parabola with Vertex (h, k) and Horizontal Axis of Symmetry

The standard equation of a parabola with vertex (h, k) and horizontal axis of symmetry is

$$(y - k)^2 = 4p(x - h),$$

where the vertex is (h, k), the focus is $(h + p, k)$, and the directrix is $x = h - p$.

The parabola opens to the right if $p > 0$. It opens to the left if $p < 0$.

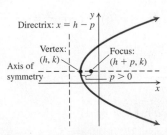

(When $p < 0$, the parabola opens to the left.)

See also Example 4 on p. 724.

SECTION 10.2: THE CIRCLE AND THE ELLIPSE

Standard Equation of a Circle

The standard equation of a circle with center (h, k) and radius r is

$$(x - h)^2 + (y - k)^2 = r^2.$$

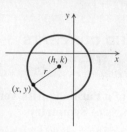

See also Example 1 on p. 728.

Standard Equation of an Ellipse with Center at the Origin

Major Axis Horizontal

$$\frac{x^2}{a^2} + \frac{y^2}{b^2} = 1, \quad a > b > 0$$

Vertices: $(-a, 0), (a, 0)$
y-intercepts: $(0, -b), (0, b)$
Foci: $(-c, 0), (c, 0)$, where $c^2 = a^2 - b^2$

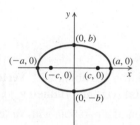

Major Axis Vertical

$$\frac{x^2}{b^2} + \frac{y^2}{a^2} = 1, \quad a > b > 0$$

Vertices: $(0, -a), (0, a)$
x-intercepts: $(-b, 0), (b, 0)$
Foci: $(0, -c), (0, c)$, where $c^2 = a^2 - b^2$

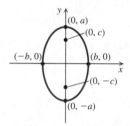

See also Examples 2 and 3 on pp. 730 and 731.

Standard Equation of an Ellipse with Center at (h, k)

Major Axis Horizontal

$$\frac{(x - h)^2}{a^2} + \frac{(y - k)^2}{b^2} = 1, \quad a > b > 0$$

Vertices: $(h - a, k), (h + a, k)$
Length of minor axis: $2b$
Foci: $(h - c, k), (h + c, k)$, where $c^2 = a^2 - b^2$

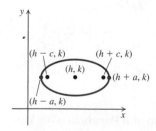

Major Axis Vertical

$$\frac{(x - h)^2}{b^2} + \frac{(y - k)^2}{a^2} = 1, \quad a > b > 0$$

Vertices: $(h, k - a), (h, k + a)$
Length of minor axis: $2b$
Foci: $(h, k - c), (h, k + c)$, where $c^2 = a^2 - b^2$

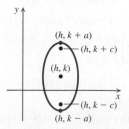

See also Example 4 on p. 732.

SECTION 10.3: THE HYPERBOLA

Standard Equation of a Hyperbola with Center at the Origin

Transverse Axis Horizontal

$$\frac{x^2}{a^2} - \frac{y^2}{b^2} = 1$$

Vertices: $(-a, 0)$, $(a, 0)$

Asymptotes: $y = -\frac{b}{a}x$, $y = \frac{b}{a}x$

Foci: $(-c, 0)$, $(c, 0)$, where $c^2 = a^2 + b^2$

Transverse Axis Vertical

$$\frac{y^2}{a^2} - \frac{x^2}{b^2} = 1$$

Vertices: $(0, -a)$, $(0, a)$

Asymptotes: $y = -\frac{a}{b}x$, $y = \frac{a}{b}x$

Foci: $(0, -c)$, $(0, c)$, where $c^2 = a^2 + b^2$

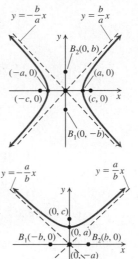

See also Examples 1 and 2 on p. 738.

Standard Equation of a Hyperbola with Center at (h, k)

Transverse Axis Horizontal

$$\frac{(x - h)^2}{a^2} - \frac{(y - k)^2}{b^2} = 1$$

Vertices: $(h - a, k)$, $(h + a, k)$

Asymptotes: $y - k = \frac{b}{a}(x - h)$,

$$y - k = -\frac{b}{a}(x - h)$$

Foci: $(h - c, k)$, $(h + c, k)$, where $c^2 = a^2 + b^2$

Transverse Axis Vertical

$$\frac{(y - k)^2}{a^2} - \frac{(x - h)^2}{b^2} = 1$$

Vertices: $(h, k - a)$, $(h, k + a)$

Asymptotes: $y - k = \frac{a}{b}(x - h)$,

$$y - k = -\frac{a}{b}(x - h)$$

Foci: $(h, k - c)$, $(h, k + c)$, where $c^2 = a^2 + b^2$

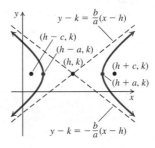

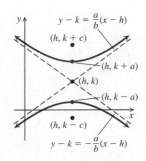

See also Example 3 on p. 740.

SECTION 10.4: NONLINEAR SYSTEMS OF EQUATIONS AND INEQUALITIES

Substitution or elimination can be used to solve **systems of equations containing at least one nonlinear equation**.	Solve: $x^2 - y = 2$, **(1)** **The graph is a parabola.** $x - y = -4$. **(2)** **The graph is a line.**

$$x = y - 4 \qquad \text{Solving equation (2) for } x$$
$$(y - 4)^2 - y = 2 \qquad \text{Substituting for } x \text{ in equation (1)}$$
$$y^2 - 8y + 16 - y = 2$$
$$y^2 - 9y + 14 = 0$$
$$(y - 2)(y - 7) = 0$$
$$y - 2 = 0 \quad or \quad y - 7 = 0$$
$$y = 2 \quad or \qquad y = 7$$

If $y = 2$, then $x = 2 - 4 = -2$.

If $y = 7$, then $x = 7 - 4 = 3$.

The pairs $(-2, 2)$ and $(3, 7)$ check, so they are the solutions.

Some applied problems translate to a nonlinear system of equations.	See Example 5 on p. 751.

To graph a **nonlinear system of inequalities**, graph each inequality in the system and then shade the region where their solution sets overlap. To find the point(s) of intersection of the graphs of the related equations, solve the system of equations composed of those equations.	Graph: $x^2 - y \leq 2$, $x - y > -4$. 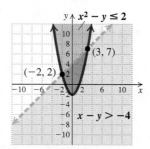

To find the points of intersection of the graphs of the related equations, we solve the system of equations

$$x^2 - y = 2,$$
$$x - y = -4.$$

We saw in the example above that these points are $(-2, 2)$ and $(3, 7)$.

SECTION 10.5: ROTATION OF AXES

To eliminate the xy-term from the equation

$Ax^2 + Bxy + Cy^2 + Dx + Ey + F = 0$,
$B \neq 0$,

select an angle θ such that

$$\cot 2\theta = \frac{A - C}{B}, \quad 0° < 2\theta < 180°,$$

and use the rotation of axes formulas below.

If the x- and y-axes are rotated about the origin through a positive acute angle θ, then the coordinates (x, y) and (x', y') of a point P in the xy- and $x'y'$-coordinate systems are related by the following formulas:

$x' = x \cos \theta + y \sin \theta,$
$y' = -x \sin \theta + y \cos \theta;$
$x = x' \cos \theta - y' \sin \theta,$
$y = x' \sin \theta + y' \cos \theta.$

See Example 2 on p. 762.

The expression $B^2 - 4AC$ is the **discriminant** of the equation

$Ax^2 + Bxy + Cy^2 + Dx + Ey + F = 0.$

The graph of the equation

$Ax^2 + Bxy + Cy^2 + Dx + Ey + F = 0$

is, except in degenerate cases,

1. an ellipse or a circle if $B^2 - 4AC < 0$,
2. a hyperbola if $B^2 - 4AC > 0$, and
3. a parabola if $B^2 - 4AC = 0$.

Use the discriminant to determine whether the equation is an ellipse or a circle, a hyperbola, or a parabola.

a) $x^2 - 2xy - 3y^2 + 5x + 1 = 0$
b) $3x^2 + 6xy + 3y^2 + 4x - 7y + 2 = 0$
c) $5x^2 + 3xy + y^2 - 8 = 0$

a) $A = 1, B = -2$, and $C = -3$, so

$B^2 - 4AC = (-2)^2 - 4 \cdot 1 \cdot (-3) = 4 + 12 = 16 > 0.$

The graph is a hyperbola.

b) $A = 3, B = 6$, and $C = 3$, so

$B^2 - 4AC = 6^2 - 4 \cdot 3 \cdot 3 = 36 - 36 = 0.$

The graph is a parabola.

c) $A = 5, B = 3$, and $C = 1$, so

$B^2 - 4AC = 3^2 - 4 \cdot 5 \cdot 1 = 9 - 20 = -11 < 0.$

The graph is an ellipse or a circle.

SECTION 10.6: POLAR EQUATIONS OF CONICS

A polar equation of any of the four forms

$$r = \frac{ep}{1 \pm e \cos \theta}, \qquad r = \frac{ep}{1 \pm e \sin \theta}$$

is a conic section. The conic is a parabola if $e = 1$, an ellipse if $0 < e < 1$, and a hyperbola if $e > 1$.

Equation	Description
$r = \dfrac{ep}{1 + e \cos \theta}$	Vertical directrix p units to the right of the pole (or focus)
$r = \dfrac{ep}{1 - e \cos \theta}$	Vertical directrix p units to the left of the pole (or focus)
$r = \dfrac{ep}{1 + e \sin \theta}$	Horizontal directrix p units above the pole (or focus)
$r = \dfrac{ep}{1 - e \sin \theta}$	Horizontal directrix p units below the pole (or focus)

Describe and graph the conic

$$r = \frac{3}{3 + \cos \theta}.$$

We first divide the numerator and the denominator by 3:

$$r = \frac{1}{1 + \frac{1}{3} \cos \theta}.$$

The equation is of the form

$$r = \frac{ep}{1 + e \cos \theta},$$

with $e = \frac{1}{3}$. Since $0 < \frac{1}{3} < 1$, the graph is an ellipse.

Since $e = \frac{1}{3}$ and $ep = \frac{1}{3} \cdot p = 1$, we have $p = 3$. Thus the ellipse has a vertical directrix 3 units to the right of the pole. We find the vertices by letting $\theta = 0$ and $\theta = \pi$. When $\theta = 0$,

$$r = \frac{3}{3 + \cos 0} = \frac{3}{3 + 1} = \frac{3}{4}.$$

When $\theta = \pi$,

$$r = \frac{3}{3 + \cos \pi} = \frac{3}{3 - 1} = \frac{3}{2}.$$

The vertices are $\left(\frac{3}{4}, 0 \right)$ and $\left(\frac{3}{2}, \pi \right)$.

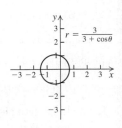

We can use the relationships between polar coordinates and rectangular coordinates to **convert polar equations of conics to rectangular equations.**

Convert to a rectangular equation:

$$r = \frac{4}{3 - \cos \theta}.$$

We have

$$r = \frac{4}{3 - \cos \theta}$$

$$3r - r \cos \theta = 4 \quad \textbf{Multiplying by } \mathbf{3 - \cos \theta}$$

$$3r = r \cos \theta + 4$$

$$3\sqrt{x^2 + y^2} = x + 4 \quad \textbf{Substituting}$$

$$9x^2 + 9y^2 = x^2 + 8x + 16 \quad \textbf{Squaring both sides}$$

$$8x^2 + 9y^2 - 8x - 16 = 0.$$

SECTION 10.7: PARAMETRIC EQUATIONS

Parametric Equations

If f and g are continuous functions of t on an interval I, then the set of ordered pairs (x, y) such that $x = f(t)$ and $y = g(t)$ is a **plane curve.** The equations $x = f(t)$ and $y = g(t)$ are **parametric equations** for the curve. The variable t is the **parameter.**

Graph the plane curve given by the following pair of parametric equations. Then find an equivalent rectangular equation.

$$x = \frac{3}{2}t, \qquad y = 2t - 1; \quad -2 \le t \le 3$$

We can choose values of t between -2 and 3 and find the corresponding values of x and y.

t	x	y	(x, y)
-2	-3	-5	$(-3, -5)$
-1	$-\frac{3}{2}$	-3	$\left(-\frac{3}{2}, -3\right)$
0	0	-1	$(0, -1)$
1	$\frac{3}{2}$	1	$\left(\frac{3}{2}, 1\right)$
2	3	3	$(3, 3)$
3	$\frac{9}{2}$	5	$\left(\frac{9}{2}, 5\right)$

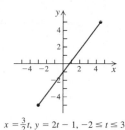

$x = \frac{3}{2}t, \ y = 2t - 1, \ -2 \le t \le 3$

Solving $x = \frac{3}{2}t$ for t, we have $\frac{2}{3}x = t$. Then we have

$$y = 2t - 1 = 2 \cdot \tfrac{2}{3}x - 1 = \tfrac{4}{3}x - 1.$$

Since $-2 \le t \le 3$ and $x = \frac{3}{2}t$, we have $-3 \le x \le \frac{9}{2}$. Thus we have the equivalent rectangular equation

$$y = \tfrac{4}{3}x - 1, \quad -3 \le x \le \tfrac{9}{2}.$$

REVIEW EXERCISES

Determine whether the statement is true or false.

1. The graph of $x + y^2 = 1$ is a parabola that opens to the left. **[10.1]**

2. The graph of $\dfrac{(x - 2)^2}{4} + \dfrac{(y + 3)^2}{9} = 1$ is an ellipse with center $(-2, 3)$. **[10.2]**

3. A parabola must open up or down. **[10.1]**

4. The major axis of the ellipse $\dfrac{x^2}{4} + \dfrac{y^2}{16} = 1$ is vertical. **[10.2]**

5. The graph of a nonlinear system of equations shows all the solutions of the system of equations. **[10.4]**

In Exercises 6–13, match the equation with one of the graphs (a)–(h) that follow.

a)

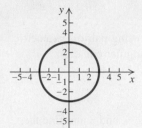

b)

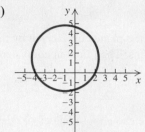

c)

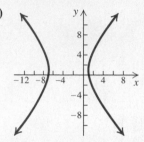

d)

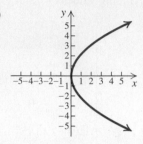

e)

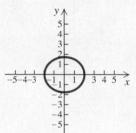

f)

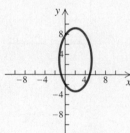

g)

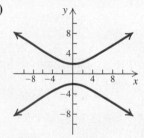

h)

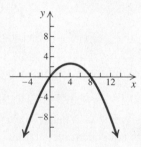

6. $y^2 = 5x$ **[10.1]**

7. $y^2 = 9 - x^2$ **[10.2]**

8. $3x^2 + 4y^2 = 12$ **[10.2]**

9. $9y^2 - 4x^2 = 36$ **[10.3]**

10. $x^2 + y^2 + 2x - 3y = 8$ **[10.2]**

11. $4x^2 + y^2 - 16x - 6y = 15$ **[10.2]**

12. $x^2 - 8x + 6y = 0$ **[10.1]**

13. $\dfrac{(x+3)^2}{16} - \dfrac{(y-1)^2}{25} = 1$ **[10.3]**

14. Find an equation of the parabola with directrix $y = \frac{3}{2}$ and focus $\left(0, -\frac{3}{2}\right)$. **[10.1]**

15. Find the focus, the vertex, and the directrix of the parabola given by $y^2 = -12x$. **[10.1]**

16. Find the vertex, the focus, and the directrix of the parabola given by

$$x^2 + 10x + 2y + 9 = 0. \ \textbf{[10.1]}$$

17. Find the center, the vertices, and the foci of the ellipse given by

$$16x^2 + 25y^2 - 64x + 50y - 311 = 0.$$

Then draw the graph. **[10.2]**

18. Find an equation of the ellipse having vertices $(0, -4)$ and $(0, 4)$ with minor axis of length 6. **[10.2]**

19. Find the center, the vertices, the foci, and the asymptotes of the hyperbola given by

$$x^2 - 2y^2 + 4x + y - \tfrac{1}{8} = 0. \ \textbf{[10.3]}$$

20. *Spotlight.* A spotlight has a parabolic cross section that is 2 ft wide at the opening and 1.5 ft deep at the vertex. How far from the vertex is the focus? **[10.1]**

Solve. **[10.4]**

21. $x^2 - 16y = 0,$
$x^2 - y^2 = 64$

22. $4x^2 + 4y^2 = 65,$
$6x^2 - 4y^2 = 25$

23. $x^2 - y^2 = 33,$
$x + y = 11$

24. $x^2 - 2x + 2y^2 = 8,$
$2x + y = 6$

25. $x^2 - y = 3,$
$2x - y = 3$

26. $x^2 + y^2 = 25,$
$x^2 - y^2 = 7$

27. $x^2 - y^2 = 3,$
$y = x^2 - 3$

28. $x^2 + y^2 = 18,$
$2x + y = 3$

29. $x^2 + y^2 = 100,$
$2x^2 - 3y^2 = -120$

30. $x^2 + 2y^2 = 12,$
$xy = 4$

31. *Numerical Relationship.* The sum of two numbers is 11, and the sum of their squares is 65. Find the numbers. **[10.4]**

32. *Dimensions of a Rectangle.* A rectangle has a perimeter of 38 m and an area of 84 m². What are the dimensions of the rectangle? **[10.4]**

33. *Numerical Relationship.* Find two positive integers whose sum is 12 and the sum of whose reciprocals is $\frac{3}{8}$. **[10.4]**

34. *Perimeter.* The perimeter of a square is 12 cm more than the perimeter of another square. The area of the first square exceeds the area of the other by 39 cm². Find the perimeter of each square. **[10.4]**

35. *Radius of a Circle.* The sum of the areas of two circles is 130π ft². The difference of the areas is 112π ft². Find the radius of each circle. **[10.4]**

Graph the system of inequalities. Then find the coordinates of the points of intersection of the graphs of the related equations. **[10.4]**

36. $y \le 4 - x^2,$
$x - y \le 2$

37. $x^2 + y^2 \le 16,$
$x + y < 4$

38. $y \ge x^2 - 1,$
$y < 1$

39. $x^2 + y^2 \le 9,$
$x \le -1$

Graph the equation. **[10.5]**

40. $5x^2 - 2xy + 5y^2 - 24 = 0$

41. $x^2 - 10xy + y^2 + 12 = 0$

42. $5x^2 + 6\sqrt{3}xy - y^2 = 16$

43. $x^2 + 2xy + y^2 - \sqrt{2}x + \sqrt{2}y = 0$

Graph the equation. State whether the directrix is vertical or horizontal, describe its location in relation to the pole, and find the vertex or vertices. **[10.6]**

44. $r = \dfrac{6}{3 - 3\sin\theta}$

45. $r = \dfrac{8}{2 + 4\cos\theta}$

46. $r = \dfrac{4}{2 - \cos\theta}$

47. $r = \dfrac{18}{9 + 6\sin\theta}$

48.–51. Convert the equations in Exercises 44–47 to rectangular equations. **[10.6]**

Find a polar equation of the conic with a focus at the pole and the given eccentricity and directrix. **[10.6]**

52. $e = \frac{1}{2}, \ r = 2\sec\theta$

53. $e = 3, \ r = -6\csc\theta$

54. $e = 1, \ r = -4\sec\theta$

55. $e = 2, \ r = 3\csc\theta$

Graph the plane curve given by the set of parametric equations and the restrictions for the parameter. Then find the equivalent rectangular equation. **[10.7]**

56. $x = t, \ y = 2 + t; \ -3 \le t \le 3$

57. $x = \sqrt{t}, \ y = t - 1; \ 0 \le t \le 9$

58. $x = 2\cos t, \ y = 2\sin t; \ 0 \le t \le 2\pi$

59. $x = 3\sin t, \ y = \cos t; \ 0 \le t \le 2\pi$

Find two sets of parametric equations for the given rectangular equation. **[10.7]**

60. $y = 2x - 3$

61. $y = x^2 + 4$

62. *Projectile Motion.* A projectile is launched from the ground with an initial speed of 150 ft/sec at an angle of 45° with the horizontal. **[10.7]**

   a) Find parametric equations that give the position of the projectile at time t, in seconds.
   b) Find the height of the projectile after 3 sec and after 6 sec.
   c) Determine how long the projectile is in the air.
   d) Determine the horizontal distance that the projectile travels.
   e) Find the maximum height of the projectile.

63. The vertex of the parabola $y^2 - 4y - 12x - 8 = 0$ is which of the following? **[10.1]**

   A. $(1, -2)$ **B.** $(-1, 2)$
   C. $(2, -1)$ **D.** $(-2, 1)$

64. Which of the following cannot be a number of solutions possible for a system of equations representing an ellipse and a straight line? **[10.4]**

   A. 0   **B.** 1
   C. 2   **D.** 4

65. The graph of $x^2 + 4y^2 = 4$ is which of the following? **[10.2], [10.3]**

A.

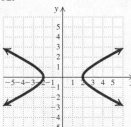

B.

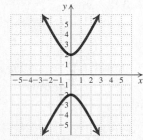

C.

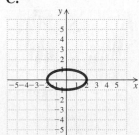

D.

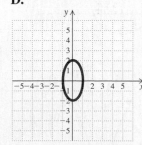

❯ Synthesis

66. Find two numbers whose product is 4 and the sum of whose reciprocals is $\frac{65}{56}$. **[10.4]**

67. Find an equation of the circle that passes through the points $(10, 7)$, $(-6, 7)$, and $(-8, 1)$. **[10.2], [10.4]**

68. Find an equation of the ellipse containing the point $(-1/2, 3\sqrt{3}/2)$ and with vertices $(0, -3)$ and $(0, 3)$. **[10.2]**

69. *Navigation.* Two radio transmitters positioned 400 mi apart along the shore send simultaneous signals to a ship that is 250 mi offshore, sailing parallel to the shoreline.

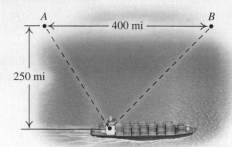

The signal from transmitter A reaches the ship 300 microseconds before the signal from transmitter B. The signals travel at a speed of 186,000 miles per second, or 0.186 mile per microsecond. Find the equation of the hyperbola with foci A and B on which the ship is located. (*Hint*: For any point on the hyperbola, the absolute value of the difference of its distances from the foci is $2a$.) **[10.3]**

❯ Collaborative Discussion and Writing

70. Explain how the distance formula is used to find the standard equation of a parabola. **[10.1]**

71. Explain why function notation is not used in Section 10.2. **[10.2]**

72. Explain how the procedure you would follow for graphing an equation of the form $Ax^2 + Bxy + Cy^2 + Dx + Ey + F = 0$ when $B \neq 0$ differs from the procedure you would follow when $B = 0$. **[10.5]**

73. Consider the graphs of

$$r = \frac{e}{1 - e \sin \theta}$$

for $e = 0.2, 0.4, 0.6,$ and 0.8. Explain the effect of the value of e on the graph. **[10.6]**

10 | Chapter Test

In Exercises 1–4, match the equation with one of the graphs (a)–(d) that follow.

a)
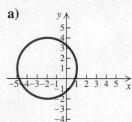

b)
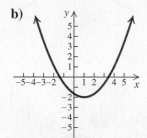

c)
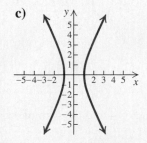

d)
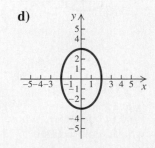

1. $4x^2 - y^2 = 4$

2. $x^2 - 2x - 3y = 5$

3. $x^2 + 4x + y^2 - 2y - 4 = 0$

4. $9x^2 + 4y^2 = 36$

Find the vertex, the focus, and the directrix of the parabola. Then draw the graph.

5. $x^2 = 12y$

6. $y^2 + 2y - 8x - 7 = 0$

7. Find an equation of the parabola with focus $(0, 2)$ and directrix $y = -2$.

8. Find the center and the radius of the circle given by $x^2 + y^2 + 2x - 6y - 15 = 0$. Then draw the graph.

Find the center, the vertices, and the foci of the ellipse. Then draw the graph.

9. $9x^2 + 16y^2 = 144$

10. $\dfrac{(x+1)^2}{4} + \dfrac{(y-2)^2}{9} = 1$

11. Find an equation of the ellipse having vertices $(0, -5)$ and $(0, 5)$ and with minor axis of length 4.

Find the center, the vertices, the foci, and the asymptotes of the hyperbola. Then draw the graph.

12. $4x^2 - y^2 = 4$

13. $\dfrac{(y-2)^2}{4} - \dfrac{(x+1)^2}{9} = 1$

14. Find the asymptotes of the hyperbola given by $2y^2 - x^2 = 18$.

15. *Satellite Dish.* A satellite dish has a parabolic cross section that is 18 in. wide at the opening and 6 in. deep at the vertex. How far from the vertex is the focus?

Solve.

16. $2x^2 - 3y^2 = -10$,
$x^2 + 2y^2 = 9$

17. $x^2 + y^2 = 13$,
$x + y = 1$

18. $x + y = 5$,
$xy = 6$

19. *Landscaping.* Leisurescape is planting a rectangular flower garden with a perimeter of 18 ft and a diagonal of $\sqrt{41}$ ft. Find the dimensions of the garden.

20. *Fencing.* It will take 210 ft of fencing to enclose a rectangular playground with an area of 2700 ft². Find the dimensions of the playground.

21. Graph the system of inequalities. Then find the coordinates of the points of intersection of the graphs of the related equations.
$$y \geq x^2 - 4,$$
$$y < 2x - 1$$

22. Graph: $5x^2 - 8xy + 5y^2 = 9$.

23. Graph $r = \dfrac{2}{1 - \sin \theta}$. State whether the directrix is vertical or horizontal, describe its location in relation to the pole, and find the vertex or vertices.

24. Find a polar equation of the conic with a focus at the pole, eccentricity 2, and directrix $r = 3 \sec \theta$.

25. Graph the plane curve given by the parametric equations $x = \sqrt{t}, y = t + 2; 0 \leq t \leq 16$.

26. Find a rectangular equation equivalent to $x = 3 \cos \theta$, $y = 3 \sin \theta; 0 \leq \theta \leq 2\pi$.

27. Find two sets of parametric equations for the rectangular equation $y = x - 5$.

28. *Projectile Motion.* A projectile is launched from a height of 10 ft with an initial speed of 250 ft/sec at an angle of 30° with the horizontal.

 a) Find parametric equations that give the position of the projectile at time t, in seconds.
 b) Find the height of the projectile after 1 sec and after 3 sec.
 c) Determine how long the projectile is in the air.
 d) Determine the horizontal distance that the projectile travels.
 e) Find the maximum height of the projectile.

29. The graph of $(y - 1)^2 = 4(x + 1)$ is which of the following?

A.

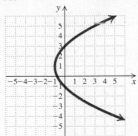

B.

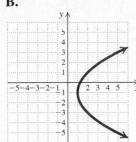

C.

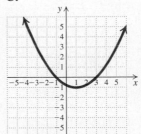

D.

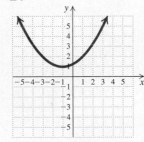

❯ Synthesis

30. Find an equation of the circle for which the endpoints of a diameter are $(1, 1)$ and $(5, -3)$.

CHAPTER

11

Sequences, Series, and Combinatorics

APPLICATION

This problem appears as Exercise 21 in Section 11.8.

An American roulette wheel contains 38 slots numbered 00, 0, 1, 2, 3, . . . , 35, 36. Eighteen of the slots numbered 1–36 are colored red and 18 are colored black. The 00 and 0 slots are considered to be uncolored. The wheel is spun, and a ball is rolled around the rim until it falls into a slot. What is the probability that the ball falls into a black slot?

11.1 › Sequences and Series

> ❯ Find terms of sequences given the *n*th term.
>
> ❯ Look for a pattern in a sequence and try to determine a general term.
>
> ❯ Convert between sigma notation and other notation for a series.
>
> ❯ Construct the terms of a recursively defined sequence.

In this section, we discuss sets or lists of numbers, considered in order, and their sums.

Suppose that $1000 is invested at 4%, compounded annually. The amounts to which the account will grow after 1 year, 2 years, 3 years, 4 years, and so on, form the following sequence of numbers:

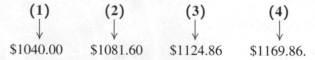

We can think of this as a function that pairs 1 with $1040.00, 2 with $1081.60, 3 with $1124.86, and so on. A **sequence** is thus a *function*, where the domain is a set of consecutive positive integers beginning with 1.

If we continue to compute the amounts of money in the account forever, we obtain an **infinite sequence** with function values

$1040.00, $1081.60, $1124.86, $1169.86, $1216.65, $1265.32, ...

The dots "..." at the end indicate that the sequence goes on without stopping. If we stop after a certain number of years, we obtain a **finite sequence**:

$1040.00, $1081.60, $1124.86, $1169.86.

SEQUENCES

An **infinite sequence** is a function having for its domain the set of positive integers, $\{1, 2, 3, 4, 5, \ldots\}$.

A **finite sequence** is a function having for its domain a set of positive integers, $\{1, 2, 3, 4, 5, \ldots, n\}$, for some positive integer n.

Consider the sequence given by the formula

$$a(n) = 2^n, \quad \text{or} \quad a_n = 2^n.$$

Some of the function values, also known as the **terms** of the sequence, are as follows:

$$a_1 = 2^1 = 2,$$
$$a_2 = 2^2 = 4,$$
$$a_3 = 2^3 = 8,$$
$$a_4 = 2^4 = 16,$$
$$a_5 = 2^5 = 32.$$

The first term of the sequence is denoted as a_1, the fifth term as a_5, and the *n*th term, or **general term,** as a_n. This sequence can also be denoted as

$$2, 4, 8, \ldots, \quad \text{or as} \quad 2, 4, 8, \ldots, 2^n, \ldots.$$

EXAMPLE 1 Find the first 4 terms and the 23rd term of the sequence whose general term is given by $a_n = (-1)^n n^2$.

Solution We have $a_n = (-1)^n n^2$, so

$$a_1 = (-1)^1 \cdot 1^2 = -1,$$
$$a_2 = (-1)^2 \cdot 2^2 = 4,$$
$$a_3 = (-1)^3 \cdot 3^2 = -9,$$
$$a_4 = (-1)^4 \cdot 4^2 = 16,$$
$$a_{23} = (-1)^{23} \cdot 23^2 = -529.$$

We can also use a graphing calculator to find the desired terms of this sequence. We enter $y_1 = (-1)^x x^2$. We then set up a table in ASK mode and enter 1, 2, 3, 4, and 23 as values for x.

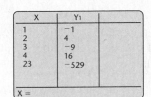

X	Y₁
1	−1
2	4
3	−9
4	16
23	−529

X =

> *Now Try Exercise 1.*

Note in Example 1 that the power $(-1)^n$ causes the signs of the terms to alternate between positive and negative, depending on whether n is even or odd. This kind of sequence is called an **alternating sequence.**

EXAMPLE 2 Use a graphing calculator to find the first 5 terms of the sequence whose general term is given by $a_n = n/(n+1)$.

Solution We can use a table or the SEQ feature, as shown here. We select SEQ from the LIST OPS menu and enter the general term, the variable, and the numbers of the first and last terms desired. The calculator will write the terms horizontally as a list. The list can also be written in fraction notation. The first 5 terms of the sequence are

$$\frac{1}{2}, \quad \frac{2}{3}, \quad \frac{3}{4}, \quad \frac{4}{5}, \quad \text{and} \quad \frac{5}{6}.$$

seq(X/(X+1),X,1,5)▶Frac
$\{\frac{1}{2} \ \frac{2}{3} \ \frac{3}{4} \ \frac{4}{5} \ \frac{5}{6}\}$

We can graph a sequence just as we graph other functions. Consider the function given by $f(x) = x + 1$ and the sequence whose general term is given by $a_n = n + 1$. The graph of $f(x) = x + 1$ is shown on the left below. Since the domain of a sequence is a set of positive integers, the graph of a sequence is a set of points that are not connected. Thus if we use only positive integers for inputs of $f(x) = x + 1$, we have the graph of the sequence $a_n = n + 1$, as shown on the right below.

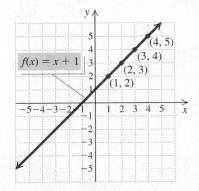

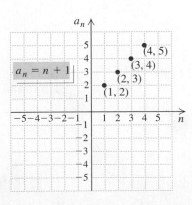

We can also use a graphing calculator to graph a sequence. Since we are graphing a set of unconnected points, we use DOT mode. We also select SEQUENCE mode. In this mode, the variable is n and functions are named $u(n)$, $v(n)$, and $w(n)$ rather than y_1, y_2, and y_3. On many graphing calculators, **X,T,Θ,n** is used to enter n in SEQUENCE mode.

EXAMPLE 3   Use a graphing calculator to graph the sequence whose general term is given by $a_n = n/(n + 1)$.

Solution   With the calculator set in DOT mode and SEQUENCE mode, we enter $u(n) = n/(n + 1)$. All the function values will be positive numbers that are less than 1, so we choose the window $[0, 10, 0, 1]$ and we also choose $n\text{Min} = 1$, $n\text{Max} = 10$, PlotStart $= 1$, and PlotStep $= 1$.

> **Now Try Exercise 19.**

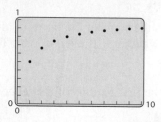

❯ Finding the General Term

When only the first few terms of a sequence are known, we do not know for sure what the general term is, but we might be able to make a prediction by looking for a pattern.

EXAMPLE 4   For each of the following sequences, predict the general term.

a) $1, \sqrt{2}, \sqrt{3}, 2, \ldots$ **b)** $-1, 3, -9, 27, -81, \ldots$

c) $2, 4, 8, \ldots$

Solution

a) These are square roots of consecutive integers, so the general term might be $\sqrt{n}$.

b) These are powers of 3 with alternating signs, so the general term might be $(-1)^n 3^{n-1}$.

c) If we see the pattern of powers of 2, we will see 16 as the next term and guess 2^n for the general term. Then the sequence could be written with more terms as

   $2, 4, 8, 16, 32, 64, 128, \ldots$.

If we see that we can get the second term by adding 2, the third term by adding 4, and the next term by adding 6, and so on, we will see 14 as the next term. A general term for the sequence is $n^2 - n + 2$, and the sequence can be written with more terms as

   $2, 4, 8, 14, 22, 32, 44, 58, \ldots$.

> **Now Try Exercise 23.**

Example 4(c) illustrates that, in fact, you can never be certain about the general term when only a few terms are given. The fewer the given terms, the greater the uncertainty.

❯ Sums and Series

> **SERIES**
>
> Given the infinite sequence
>
> $$a_1, a_2, a_3, a_4, \ldots, a_n, \ldots,$$
>
> the sum of the terms
>
> $$a_1 + a_2 + a_3 + \cdots + a_n + \cdots$$
>
> is called an **infinite series.** A **partial sum** is the sum of the first n terms:
>
> $$a_1 + a_2 + a_3 + \cdots + a_n.$$
>
> A partial sum is also called a **finite series,** or ***n*th partial sum,** and is denoted S_n.

EXAMPLE 5 For the sequence $-2, 4, -6, 8, -10, 12, -14, \ldots$, find each of the following.

a) S_1 **b)** S_4 **c)** S_5

Solution

a) $S_1 = -2$

b) $S_4 = -2 + 4 + (-6) + 8 = 4$

c) $S_5 = -2 + 4 + (-6) + 8 + (-10) = -6$ ❯ *Now Try Exercise 33.*

We can also use a graphing calculator to find partial sums of a sequence when a formula for the general term is known.

EXAMPLE 6 Use a graphing calculator to find S_1, S_2, S_3, and S_4 for the sequence whose general term is given by $a_n = n^2 - 3$.

Solution We can use the CUMSUM feature from the LIST OPS menu. The calculator will write the partial sums as a list. (Note that the calculator can be set in either FUNCTION mode or SEQUENCE mode. Here we show SEQUENCE mode.)

We have $S_1 = -2$, $S_2 = -1$, $S_3 = 5$, and $S_4 = 18$. ❯

```
cumSum(seq(n² − 3,n,1,4))
            {−2 −1 5 18}
```

❯ Sigma Notation

The Greek letter Σ (sigma) can be used to denote a sum when the general term of a sequence is a formula. For example, the sum of the first four terms of the sequence $3, 5, 7, 9, \ldots, 2k + 1, \ldots$ can be named as follows, using what is called **sigma notation,** or **summation notation:**

$$\sum_{k=1}^{4} (2k + 1).$$

This is read "the sum as k goes from 1 to 4 of $2k + 1$." The letter k is called the **index of summation.** The index of summation might start at a number other than 1, and letters other than k can be used.

EXAMPLE 7 Find and evaluate each of the following sums.

a) $\displaystyle\sum_{k=1}^{5} k^3$ b) $\displaystyle\sum_{k=0}^{4} (-1)^k 5^k$ c) $\displaystyle\sum_{i=8}^{11} \left(2 + \frac{1}{i}\right)$

Solution

a) We replace k with 1, 2, 3, 4, and 5. Then we add the results.

$$\sum_{k=1}^{5} k^3 = 1^3 + 2^3 + 3^3 + 4^3 + 5^3$$

$$= 1 + 8 + 27 + 64 + 125 = 225$$

We can also find this sum on a graphing calculator using the SUM feature from the LIST MATH menu and the SEQ feature from the LIST OPS menu.

```
sum(seq(n^3,n,1,5))
                 225
```

b) $\displaystyle\sum_{k=0}^{4} (-1)^k 5^k = (-1)^0 5^0 + (-1)^1 5^1 + (-1)^2 5^2 + (-1)^3 5^3 + (-1)^4 5^4$

$$= 1 - 5 + 25 - 125 + 625 = 521$$

c) $\displaystyle\sum_{i=8}^{11} \left(2 + \frac{1}{i}\right) = \left(2 + \frac{1}{8}\right) + \left(2 + \frac{1}{9}\right) + \left(2 + \frac{1}{10}\right) + \left(2 + \frac{1}{11}\right)$

$$= 8\frac{1691}{3960}$$

> *Now Try Exercise 37.*

EXAMPLE 8 Write sigma notation for each sum.

a) $1 + 2 + 4 + 8 + 16 + 32 + 64$

b) $-2 + 4 - 6 + 8 - 10$

c) $x + \dfrac{x^2}{2} + \dfrac{x^3}{3} + \dfrac{x^4}{4} + \cdots$

Solution

a) $1 + 2 + 4 + 8 + 16 + 32 + 64$

This is the sum of powers of 2, beginning with 2^0, or 1, and ending with 2^6, or 64. Sigma notation is $\Sigma_{k=0}^{6} 2^k$.

b) $-2 + 4 - 6 + 8 - 10$

Disregarding the alternating signs, we see that this is the sum of the first 5 even integers. Note that $2k$ is a formula for the kth positive even integer, and $(-1)^k = -1$ when k is odd and $(-1)^k = 1$ when k is even. Thus the general term is $(-1)^k(2k)$. The sum begins with $k = 1$ and ends with $k = 5$, so sigma notation is $\Sigma_{k=1}^{5}(-1)^k(2k)$.

c) $x + \dfrac{x^2}{2} + \dfrac{x^3}{3} + \dfrac{x^4}{4} + \cdots$

The general term is x^k/k, beginning with $k = 1$. This is also an infinite series. We use the symbol ∞ for infinity and write the series using sigma notation: $\Sigma_{k=1}^{\infty}(x^k/k)$.

> *Now Try Exercise 55.*

❯ Recursive Definitions

A sequence may be defined **recursively**, or by using a **recursion formula**. Such a definition lists the first term, or the first few terms, and then describes how to determine the remaining terms from the given terms.

EXAMPLE 9 Find the first 5 terms of the sequence defined by

$$a_1 = 5, \qquad a_{n+1} = 2a_n - 3, \quad \text{for } n \geq 1.$$

Solution We have

$$a_1 = 5,$$

$$a_2 = 2a_1 - 3 = 2 \cdot 5 - 3 = 7,$$

$$a_3 = 2a_2 - 3 = 2 \cdot 7 - 3 = 11,$$

$$a_4 = 2a_3 - 3 = 2 \cdot 11 - 3 = 19,$$

$$a_5 = 2a_4 - 3 = 2 \cdot 19 - 3 = 35.$$

Many graphing calculators have the capability to work with recursively defined sequences when they are set in SEQUENCE mode. For this sequence, for instance, the function could be entered as $u(n) = 2 * u(n-1) - 3$ with $u(n\text{Min}) = 5$. We can read the terms of the sequence from a table.

> *Now Try Exercise 65.*

```
Plot1  Plot2  Plot3
 nMin=1
\u(n)■2*u(n−1)−3
 u(nMin)■{5}
\v(n)=
 v(nMin)=
\w(n)=
 w(nMin)=
```

n	u(n)
1	5
2	7
3	11
4	19
5	35
6	67
7	131

n = 1

11.1   Exercise Set

In each of the following, the nth term of a sequence is given. Find the first 4 terms, a_{10}, and a_{15}.

1. $a_n = 4n - 1$

2. $a_n = (n-1)(n-2)(n-3)$

3. $a_n = \dfrac{n}{n-1}, n \geq 2$

4. $a_n = n^2 - 1, n \geq 3$

5. $a_n = \dfrac{n^2 - 1}{n^2 + 1}$

6. $a_n = \left(-\dfrac{1}{2}\right)^{n-1}$

7. $a_n = (-1)^n n^2$

8. $a_n = (-1)^{n-1}(3n - 5)$

9. $a_n = 5 + \dfrac{(-2)^{n+1}}{2^n}$

10. $a_n = \dfrac{2n - 1}{n^2 + 2n}$

Find the indicated term of the given sequence.

11. $a_n = 5n - 6; \ a_8$

12. $a_n = (3n - 4)(2n + 5); \ a_7$

13. $a_n = (2n + 3)^2; \ a_6$

14. $a_n = (-1)^{n-1}(4.6n - 18.3); \ a_{12}$

15. $a_n = 5n^2(4n - 100); \ a_{11}$

16. $a_n - \left(1 + \dfrac{1}{n}\right)^2; \ a_{80}$

17. $a_n = \ln e^n; \ a_{67}$

18. $a_n = 2 - \dfrac{1000}{n}; \ a_{100}$

Use a graphing calculator to construct a table of values and a graph for the first 10 terms of the sequence.

19. $a_n = \left(1 + \dfrac{1}{n}\right)^n$

20. $a_n = \sqrt{n + 1} - \sqrt{n}$

21. $a_1 = 2, \ a_{n+1} = \sqrt{1 + \sqrt{a_n}}$

22. $a_1 = 2, \ a_{n+1} = \dfrac{1}{2}\left(a_n + \dfrac{2}{a_n}\right)$

Predict the general term, or nth term, a_n, of the sequence. Answers may vary.

23. $2, 4, 6, 8, 10, \ldots$

24. $3, 9, 27, 81, 243, \ldots$

25. $-2, 6, -18, 54, \ldots$

26. $-2, 3, 8, 13, 18, \ldots$

27. $\dfrac{2}{3}, \dfrac{3}{4}, \dfrac{4}{5}, \dfrac{5}{6}, \dfrac{6}{7}, \ldots$

28. $\sqrt{2}, 2, \sqrt{6}, 2\sqrt{2}, \sqrt{10}, \ldots$

29. $1 \cdot 2, 2 \cdot 3, 3 \cdot 4, 4 \cdot 5, \ldots$

30. $-1, -4, -7, -10, -13, \ldots$

31. $0, \log 10, \log 100, \log 1000, \ldots$

32. $\ln e^2, \ln e^3, \ln e^4, \ln e^5, \ldots$

Find the indicated partial sums for the sequence.

33. $1, 2, 3, 4, 5, 6, 7, \ldots$; S_3 and S_7

34. $1, -3, 5, -7, 9, -11, \ldots$; S_2 and S_5

35. $2, 4, 6, 8, \ldots$; S_4 and S_5

36. $1, \frac{1}{4}, \frac{1}{9}, \frac{1}{16}, \frac{1}{25}, \ldots$; S_1 and S_5

Find and evaluate the sum.

37. $\displaystyle\sum_{k=1}^{5} \frac{1}{2k}$

38. $\displaystyle\sum_{i=1}^{6} \frac{1}{2i+1}$

39. $\displaystyle\sum_{i=0}^{6} 2^i$

40. $\displaystyle\sum_{k=4}^{7} \sqrt{2k-1}$

41. $\displaystyle\sum_{k=7}^{10} \ln k$

42. $\displaystyle\sum_{k=1}^{4} \pi k$

43. $\displaystyle\sum_{k=1}^{8} \frac{k}{k+1}$

44. $\displaystyle\sum_{i=1}^{5} \frac{i-1}{i+3}$

45. $\displaystyle\sum_{i=1}^{5} (-1)^i$

46. $\displaystyle\sum_{k=0}^{5} (-1)^{k+1}$

47. $\displaystyle\sum_{k=1}^{8} (-1)^{k+1} 3k$

48. $\displaystyle\sum_{k=0}^{7} (-1)^k 4^{k+1}$

49. $\displaystyle\sum_{k=0}^{6} \frac{2}{k^2+1}$

50. $\displaystyle\sum_{i=1}^{10} i(i+1)$

51. $\displaystyle\sum_{k=0}^{5} (k^2 - 2k + 3)$

52. $\displaystyle\sum_{k=1}^{10} \frac{1}{k(k+1)}$

53. $\displaystyle\sum_{i=0}^{10} \frac{2^i}{2^i+1}$

54. $\displaystyle\sum_{k=0}^{3} (-2)^{2k}$

Write sigma notation. Answers may vary.

55. $5 + 10 + 15 + 20 + 25 + \cdots$

56. $7 + 14 + 21 + 28 + 35 + \cdots$

57. $2 - 4 + 8 - 16 + 32 - 64$

58. $3 + 6 + 9 + 12 + 15$

59. $-\frac{1}{2} + \frac{2}{3} - \frac{3}{4} + \frac{4}{5} - \frac{5}{6} + \frac{6}{7}$

60. $\frac{1}{1^2} + \frac{1}{2^2} + \frac{1}{3^2} + \frac{1}{4^2} + \frac{1}{5^2}$

61. $4 - 9 + 16 - 25 + \cdots + (-1)^n n^2$

62. $9 - 16 + 25 + \cdots + (-1)^{n+1} n^2$

63. $\dfrac{1}{1 \cdot 2} + \dfrac{1}{2 \cdot 3} + \dfrac{1}{3 \cdot 4} + \dfrac{1}{4 \cdot 5} + \cdots$

64. $\dfrac{1}{1 \cdot 2^2} + \dfrac{1}{2 \cdot 3^2} + \dfrac{1}{3 \cdot 4^2} + \dfrac{1}{4 \cdot 5^2} + \cdots$

Find the first 4 terms of the recursively defined sequence.

65. $a_1 = 4, \ a_{n+1} = 1 + \dfrac{1}{a_n}$

66. $a_1 = 256, \ a_{n+1} = \sqrt{a_n}$

67. $a_1 = 6561, \ a_{n+1} = (-1)^n \sqrt{a_n}$

68. $a_1 = e^Q, \ a_{n+1} = \ln a_n$

69. $a_1 = 2, \ a_2 = 3, \ a_{n+1} = a_n + a_{n-1}$

70. $a_1 = -10, \ a_2 = 8, \ a_{n+1} = a_n - a_{n-1}$

71. *Compound Interest.* Suppose that $4000 is invested at 3.75%, compounded annually. The value of the investment after n years is given by the sequence

$$a_n = \$4000(1.0375)^n, \quad n = 1, 2, 3, \ldots.$$

a) Find the first 10 terms of the sequence.

b) Find the value of the investment after 20 years.

72. *Salvage Value.* The value of a post-hole digger is $5200. Its salvage value each year is 75% of its value the year before. Give a sequence that lists the salvage value of the post-hole digger for each year of a 10-year period.

73. *Wage Sequence.* Xavier is paid $16.20 per hour for working at Red Freight Limited. Each year he receives a $2.25 hourly raise. Give a sequence that lists Xavier's hourly wage over a 10-year period.

74. *Bacteria Growth.* Suppose that a single cell of bacteria divides into two every 15 min. Suppose that the same rate of division is maintained for 4 hr. Give a sequence that lists the number of cells after successive 15-min periods.

75. *Fibonacci Sequence: Rabbit Population Growth.* One of the most famous recursively defined sequences is the **Fibonacci sequence**. In 1202, the Italian mathematician Leonardo da Pisa, also called Fibonacci, proposed the following model for rabbit population growth. Suppose that every month each mature pair of rabbits in the population produces a new pair that begins reproducing after two months, and also suppose that no rabbits die. Beginning with one pair of newborn rabbits, the population can be modeled by the following recursively defined sequence:

$$a_1 = 1, \ a_2 = 1, \ a_n = a_{n-1} + a_{n-2}, \text{ for } n \geq 3,$$

where a_n is the total number of pairs of rabbits in month n. Find the first 7 terms of the Fibonacci sequence.

76. *Weekly Earnings.* The following table lists the average weekly earnings of U.S. production workers in recent years.

Year, n	Average Weekly Earnings of U.S. Production Workers
2008, 0	$607.53
2009, 1	616.01
2010, 2	636.25
2011, 3	653.19
2012, 4	665.82
2013, 5	677.67

Sources: U.S. Bureau of Labor Statistics, U.S. Department of Labor

a) Use a graphing calculator to fit a linear sequence regression function

$$a_n = an + b$$

to the data, where n is the number of years since 2008.

b) Estimate the average weekly earnings in 2014, in 2016, and in 2017.

77. *Price of Silver.* The following table lists the price of silver per troy ounce in recent years.

Year, n	Price of Silver per Troy Ounce
2009, 0	$14.69
2010, 1	20.20
2011, 2	35.26
2012, 3	31.21
2013, 4	23.80

Source: Mineral Commodity Summaries 2014, U.S. Geological Survey, U.S. Department of the Interior

a) Use a graphing calculator to fit a quadratic sequence regression function

$$a_n = an^2 + bn + c$$

to the data, where n is the number of years since 2009.

b) Estimate the price of silver per troy ounce in 2014.

❯ Skill Maintenance

Solve. **[9.1], [9.3], [9.5], [9.6]**

78. $3x - 2y = 3,$
$2x + 3y = -11$

79. *Harvesting Pumpkins.* A total of 23,400 acres of pumpkins were harvested in Illinois and Ohio in 2012. The number of acres of pumpkins harvested in Ohio was 9000 fewer than the number of acres of pumpkins harvested in Illinois. (*Source*: U. S. Department of Agriculture) Find the number of acres of pumpkins harvested in Illinois and in Ohio in 2012.

Find the center and the radius of the circle with the given equation. **[10.2]**

80. $x^2 + y^2 - 6x + 4y = 3$

81. $x^2 + y^2 + 5x - 8y = 2$

❯ Synthesis

Find the first 5 terms of the sequence, and then find S_5.

82. $a_n = \dfrac{1}{2^n} \log 1000^n$

83. $a_n = i^n, i = \sqrt{-1}$

84. $a_n = \ln (1 \cdot 2 \cdot 3 \cdot \cdots \cdot n)$

For each sequence, find a formula for S_n.

85. $a_n = \ln n$

86. $a_n = \dfrac{1}{n} - \dfrac{1}{n + 1}$

11.2 Arithmetic Sequences and Series

> ❯ For any arithmetic sequence, find the nth term when n is given and n when the nth term is given, and given two terms, find the common difference and construct the sequence.

> ❯ Find the sum of the first n terms of an arithmetic sequence.

A sequence in which each term after the first term is found by adding the same number to the preceding term is an **arithmetic sequence.**

❯ Arithmetic Sequences

The sequence 2, 5, 8, 11, 14, 17, ... is arithmetic because adding 3 to any term produces the next term. In other words, the difference between any term and the preceding one is 3. Arithmetic sequences are also called *arithmetic progressions*.

ARITHMETIC SEQUENCE

A sequence is **arithmetic** if there exists a number d, called the **common difference,** such that $a_{n+1} = a_n + d$ for any integer $n \geq 1$.

EXAMPLE 1   For each of the following arithmetic sequences, identify the first term, a_1, and the common difference, d.

a) 4, 9, 14, 19, 24, ...

b) 34, 27, 20, 13, 6, −1, −8, ...

c) 2, $2\frac{1}{2}$, 3, $3\frac{1}{2}$, 4, $4\frac{1}{2}$, ...

Solution   The first term, a_1, is the first term listed. To find the common difference, d, we choose any term beyond the first and subtract the preceding term from it.

SEQUENCE	FIRST TERM, a_1	COMMON DIFFERENCE, d
a) 4, 9, 14, 19, 24, ...	4	5 $(9 - 4 = 5)$
b) 34, 27, 20, 13, 6, −1, −8, ...	34	−7 $(27 - 34 = -7)$
c) 2, $2\frac{1}{2}$, 3, $3\frac{1}{2}$, 4, $4\frac{1}{2}$, ...	2	$\frac{1}{2}$ $\left(2\frac{1}{2} - 2 = \frac{1}{2}\right)$

We obtained the common difference by subtracting a_1 from a_2. Had we subtracted a_2 from a_3 or a_3 from a_4, we would have obtained the same values for d. Thus we can check by adding d to each term in a sequence to see if we progress correctly to the next term.

Check:

a) $4 + 5 = 9$, $9 + 5 = 14$, $14 + 5 = 19$, $19 + 5 = 24$

b) $34 + (-7) = 27$, $27 + (-7) = 20$, $20 + (-7) = 13$,
    $13 + (-7) = 6$, $6 + (-7) = -1$, $-1 + (-7) = -8$

c) $2 + \frac{1}{2} = 2\frac{1}{2}$, $2\frac{1}{2} + \frac{1}{2} = 3$, $3 + \frac{1}{2} = 3\frac{1}{2}$, $3\frac{1}{2} + \frac{1}{2} = 4$,
    $4 + \frac{1}{2} = 4\frac{1}{2}$

> ❯ **Now Try Exercise 1.**

To find a formula for the general, or *n*th, term of any arithmetic sequence, we denote the common difference by d, write out the first few terms, and look for a pattern:

$$a_1,$$
$$a_2 = a_1 + d,$$
$$a_3 = a_2 + d = (a_1 + d) + d = a_1 + 2d, \qquad \text{Substituting for } a_2$$
$$a_4 = a_3 + d = (a_1 + 2d) + d = a_1 + 3d. \qquad \text{Substituting for } a_3$$

Note that the coefficient of d in each case is 1 less than the subscript.

Generalizing, we obtain the following formula.

---

*n*TH TERM OF AN ARITHMETIC SEQUENCE

The ***n*th term** of an arithmetic sequence is given by $a_n = a_1 + (n-1)d$, for any integer $n \geq 1$.

EXAMPLE 2 Find the 14th term of the arithmetic sequence

$$4, 7, 10, 13, \ldots.$$

Solution We first note that $a_1 = 4$, $d = 7 - 4$, or 3, and $n = 14$. Then using the formula for the *n*th term, we obtain

$$a_n = a_1 + (n-1)d$$
$$a_{14} = 4 + (14-1) \cdot 3 \qquad \text{Substituting}$$
$$= 4 + 13 \cdot 3 = 4 + 39$$
$$= 43.$$

The 14th term is 43.

> *Now Try Exercise 9.*

EXAMPLE 3 In the sequence of Example 2, which term is 301? That is, find n if $a_n = 301$.

Solution We substitute 301 for a_n, 4 for a_1, and 3 for d in the formula for the *n*th term and solve for n:

$$a_n = a_1 + (n-1)d$$
$$301 = 4 + (n-1) \cdot 3 \qquad \text{Substituting}$$
$$301 = 4 + 3n - 3$$
$$301 = 3n + 1$$
$$300 = 3n$$
$$100 = n.$$

Solving for n

The term 301 is the 100th term of the sequence.

> *Now Try Exercise 15.*

Given two terms and their places in an arithmetic sequence, we can construct the sequence.

EXAMPLE 4 The 3rd term of an arithmetic sequence is 8, and the 16th term is 47. Find a_1 and d and construct the sequence.

Solution We know that $a_3 = 8$ and $a_{16} = 47$. Thus we would need to add d 13 times to get from 8 to 47. That is,

$$8 + 13d = 47. \qquad a_3 \text{ and } a_{16} \text{ are } 16 - 3, \text{ or } 13, \text{ terms apart.}$$

Solving $8 + 13d = 47$, we obtain

$$13d = 39$$

$$d = 3.$$

Since $a_3 = 8$, we subtract d twice to get a_1. Thus,

$$a_1 = 8 - 2 \cdot 3 = 2. \qquad \text{a_1 and a_3 are $3 - 1$, or 2, terms apart.}$$

The sequence is 2, 5, 8, 11, ... Note that we could also subtract d 15 times from a_1 in order to find a_1.

> **Now Try Exercise 23.**

In general, d should be subtracted $n - 1$ times from a_n in order to find a_1.

❯ Sum of the First *n* Terms of an Arithmetic Sequence

Consider the arithmetic sequence

$$3, 5, 7, 9, \ldots.$$

When we add the first 4 terms of the sequence, we get S_4, which is

$$3 + 5 + 7 + 9, \quad \text{or} \quad 24.$$

This sum is called an **arithmetic series**. To find a formula for the sum of the first *n* terms, S_n, of an arithmetic sequence, we first denote an arithmetic sequence, as follows:

> **This term is two terms back from the last. If you add d to this term, the result is the next-to-last term, $a_n - d$.**

$$a_1, \quad (a_1 + d), \quad (a_1 + 2d), \ldots, \quad \overbrace{(a_n - 2d)}, \quad \underbrace{(a_n - d)}, \quad a_n.$$

> **This is the next-to-last term. If you add d to this term, the result is a_n.**

Then S_n is given by

$$S_n = a_1 + (a_1 + d) + (a_1 + 2d) + \cdots + (a_n - 2d)$$
$$+ (a_n - d) + a_n. \tag{1}$$

Reversing the order of the addition gives us

$$S_n = a_n + (a_n - d) + (a_n - 2d) + \cdots + (a_1 + 2d)$$
$$+ (a_1 + d) + a_1. \tag{2}$$

If we add corresponding terms of each side of equations (1) and (2), we get

$$2S_n = [a_1 + a_n] + [(a_1 + d) + (a_n - d)] + [(a_1 + 2d) + (a_n - 2d)]$$
$$+ \cdots + [(a_n - 2d) + (a_1 + 2d)]$$
$$+ [(a_n - d) + (a_1 + d)] + [a_n + a_1].$$

In the expression for $2S_n$, there are *n* expressions in square brackets. Each of these expressions is equivalent to $a_1 + a_n$. Thus the expression for $2S_n$ can be written in simplified form as

$$2S_n = [a_1 + a_n] + [a_1 + a_n] + [a_1 + a_n] + \cdots + [a_n + a_1]$$
$$+ [a_n + a_1] + [a_n + a_1].$$

Since $a_1 + a_n$ is being added n times, it follows that

$$2S_n = n(a_1 + a_n),$$

from which we get the following formula.

SUM OF THE FIRST n TERMS
The sum of the first n terms of an arithmetic sequence is given by

$$S_n = \frac{n}{2}(a_1 + a_n).$$

EXAMPLE 5   Find the sum of the first 100 natural numbers.

Solution   The sum is

$$1 + 2 + 3 + \cdots + 99 + 100.$$

This is the sum of the first 100 terms of the arithmetic sequence for which

$$a_1 = 1, \quad a_n = 100, \quad \text{and} \quad n = 100.$$

Thus substituting into the formula

$$S_n = \frac{n}{2}(a_1 + a_n),$$

we get

$$S_{100} = \frac{100}{2}(1 + 100) = 50(101) = 5050.$$

The sum of the first 100 natural numbers is 5050.   ▶ *Now Try Exercise 27.*

EXAMPLE 6   Find the sum of the first 15 terms of the arithmetic sequence

$$4, 7, 10, 13, \ldots.$$

Solution   Note that $a_1 = 4$, $d = 3$, and $n = 15$. Before using the formula

$$S_n = \frac{n}{2}(a_1 + a_n),$$

we find the last term, a_{15}:

$$a_{15} = 4 + (15 - 1)3 \qquad \textbf{Substituting into the formula } a_n = a_1 + (n - 1)d$$
$$= 4 + 14 \cdot 3 = 46.$$

Thus,

$$S_{15} = \frac{15}{2}(4 + 46) = \frac{15}{2}(50) = 375.$$

The sum of the first 15 terms is 375.   ▶ *Now Try Exercise 25.*

EXAMPLE 7   Find the sum: $\displaystyle\sum_{k=1}^{130}(4k + 5)$.

Solution   It is helpful to first write out a few terms:

$$9 + 13 + 17 + \cdots.$$

It appears that this is an arithmetic series coming from an arithmetic sequence with $a_1 = 9$, $d = 4$, and $n = 130$. Before using the formula

$$S_n = \frac{n}{2}(a_1 + a_n),$$

we find the last term, a_{130}:

$$a_{130} = 4 \cdot 130 + 5 \qquad \text{The } k\text{th term is } 4k + 5.$$
$$= 520 + 5$$
$$= 525.$$

Thus,

$$S_{130} = \frac{130}{2}(9 + 525) \qquad \text{Substituting into } S_n = \frac{n}{2}(a_1 + a_n)$$
$$= 34{,}710.$$

```
sum(seq(4X+5,X,1,130))
                  34710
```

This sum can also be found on a graphing calculator. It is not necessary to have the calculator set in SEQUENCE mode in order to do this.

> *Now Try Exercise 33.*

> Applications

The translation of some applications and problem-solving situations may involve arithmetic sequences or series. We consider some examples.

EXAMPLE 8 *Hourly Wages.* Kendall accepts a job, starting with an hourly wage of $14.25, and is promised a raise of 15¢ per hour every 2 months for 5 years. At the end of 5 years, what will Kendall's hourly wage be?

Solution It helps to first write down the hourly wage for several 2-month time periods:

Beginning: $14.25,
After 2 months: $14.40,
After 4 months: $14.55,
and so on.

What appears is a sequence of numbers: 14.25, 14.40, 14.55, ... This sequence is arithmetic, because adding 0.15 each time gives us the next term.

We want to find the last term of an arithmetic sequence, so we use the formula $a_n = a_1 + (n - 1)d$. We know that $a_1 = 14.25$ and $d = 0.15$, but what is n? That is, how many terms are in the sequence? Each year there are $12/2$, or 6 raises, since Kendall gets a raise every 2 months. There are 5 years, so the total number of raises will be $5 \cdot 6$, or 30. Thus there will be 31 terms: the original wage and 30 increased rates.

Substituting in the formula $a_n = a_1 + (n - 1)d$ gives us

$$a_{31} = 14.25 + (31 - 1) \cdot 0.15 = 18.75.$$

Thus, at the end of 5 years, Kendall's hourly wage will be $18.75.

> *Now Try Exercise 43.*

The calculations in Example 8 could be done in a number of ways. There is often a variety of ways in which a problem can be solved. In this chapter, we concentrate on the use of sequences and series and their related formulas.

EXAMPLE 9 *Total in a Stack.* A stack of electrical poles has 30 poles in the bottom row. There are 29 poles in the second row, 28 in the next row, and so on. How many poles are in the stack if there are 5 poles in the top row?

Solution A drawing will help in this case. The following figure shows the ends of the poles and the way in which they stack.

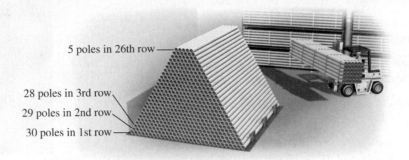

5 poles in 26th row

28 poles in 3rd row
29 poles in 2nd row
30 poles in 1st row

Since the number of poles goes from 30 in a row up to 5 in the top row, there must be 26 rows. We want the sum

$$30 + 29 + 28 + \cdots + 5.$$

Thus we have an arithmetic series. We use the formula

$$S_n = \frac{n}{2}(a_1 + a_n),$$

with $n = 26$, $a_1 = 30$, and $a_{26} = 5$.
Substituting, we get

$$S_{26} = \frac{26}{2}(30 + 5) = 455.$$

There are 455 poles in the stack.

> *Now Try Exercise 39.*

11.2 | Exercise Set

Find the first term and the common difference.

1. 3, 8, 13, 18, . . .

2. $1.08, $1.16, $1.24, $1.32, . . .

3. 9, 5, 1, −3, . . .

4. −8, −5, −2, 1, 4, . . .

5. $\frac{3}{2}, \frac{9}{4}, 3, \frac{15}{4}, \ldots$

6. $\frac{3}{5}, \frac{1}{10}, -\frac{2}{5}, \ldots$

7. $316, $313, $310, $307, . . .

8. Find the 11th term of the arithmetic sequence 0.07, 0.12, 0.17, ...

9. Find the 12th term of the arithmetic sequence 2, 6, 10, ...

10. Find the 17th term of the arithmetic sequence 7, 4, 1, ...

11. Find the 14th term of the arithmetic sequence $3, \frac{7}{3}, \frac{5}{3}, \ldots$.

12. Find the 13th term of the arithmetic sequence $1200, $964.32, $728.64, ...

13. Find the 10th term of the arithmetic sequence $2345.78, $2967.54, $3589.30, ...

14. In the sequence of Exercise 8, what term is the number 1.67?

15. In the sequence of Exercise 9, what term is the number 106?

16. In the sequence of Exercise 10, what term is −296?

17. In the sequence of Exercise 11, what term is -27?

18. Find a_{20} when $a_1 = 14$ and $d = -3$.

19. Find a_1 when $d = 4$ and $a_8 = 33$.

20. Find d when $a_1 = 8$ and $a_{11} = 26$.

21. Find n when $a_1 = 25$, $d = -14$, and $a_n = -507$.

22. In an arithmetic sequence, $a_{17} = -40$ and $a_{28} = -73$. Find a_1 and d. Write the first 5 terms of the sequence.

23. In an arithmetic sequence, $a_{17} = \frac{25}{3}$ and $a_{32} = \frac{95}{6}$. Find a_1 and d. Write the first 5 terms of the sequence.

24. Find the sum of the first 14 terms of the series $11 + 7 + 3 + \cdots$.

25. Find the sum of the first 20 terms of the series $5 + 8 + 11 + 14 + \cdots$.

26. Find the sum of the first 300 natural numbers.

27. Find the sum of the first 400 even natural numbers.

28. Find the sum of the odd numbers 1 to 199, inclusive.

29. Find the sum of the multiples of 7 from 7 to 98, inclusive.

30. Find the sum of all multiples of 4 that are between 14 and 523.

31. If an arithmetic series has $a_1 = 2$, $d = 5$, and $n = 20$, what is S_n?

32. If an arithmetic series has $a_1 = 7$, $d = -3$, and $n = 32$, what is S_n?

Find the sum.

33. $\displaystyle\sum_{k=1}^{40} (2k + 3)$

34. $\displaystyle\sum_{k=5}^{20} 8k$

35. $\displaystyle\sum_{k=0}^{19} \frac{k - 3}{4}$

36. $\displaystyle\sum_{k=2}^{50} (2000 - 3k)$

37. $\displaystyle\sum_{k=12}^{57} \frac{7 - 4k}{13}$

38. $\displaystyle\sum_{k=101}^{200} (1.14k - 2.8) - \sum_{k=1}^{5} \left(\frac{k + 4}{10}\right)$

39. *Total Savings.* If 10¢ is saved on October 1, 20¢ is saved on October 2, 30¢ on October 3, and so on, how much is saved during the 31 days of October?

40. *Stacking Poles.* How many poles will be in a stack of telephone poles if there are 50 in the first layer, 49 in the second, and so on, with 6 in the top layer?

41. *Auditorium Seating.* Auditoriums are often built with more seats per row as the rows move toward the back. Suppose that the first balcony of a theater has 28 seats in the first row, 32 in the second, 36 in the third, and so on, for 20 rows. How many seats are in the first balcony altogether?

42. *Investment Return.* Max, an investment counselor, sets up an investment situation for a client that will return $5000 the first year, $6125 the second year, $7250 the third year, and so on, for 25 years. How much is received from the investment altogether?

43. *Parachutist Free Fall.* When a parachutist jumps from an airplane, the distances, in feet, that the parachutist falls in each successive second before pulling the ripcord to release the parachute are as follows:

16, 48, 80, 112, 144, ...

Is this sequence arithmetic? What is the common difference? What is the total distance fallen in 10 sec?

44. *Lightning Distance.* The following table lists the distance in miles from lightning d_n when thunder is heard n seconds after lightning is seen. Is this sequence arithmetic? What is the common difference?

n (in seconds)	d_n (in miles)
5	1
6	1.2
7	1.4
8	1.6
9	1.8
10	2

45. *Garden Plantings.* A gardener is making a planting in the shape of a trapezoid. It will have 35 plants in the first row, 31 in the second row, 27 in the third row, and so on. If the pattern is consistent, how many plants will there be in the last row? How many plants are there altogether?

6. *Band Formation.* A formation of a marching band has 10 marchers in the first row, 12 in the second row, 14 in the third row, and so on, for 8 rows. How many marchers are in the last row? How many marchers are there altogether?

7. *Raw Material Production.* In a manufacturing process, it took 3 units of raw materials to produce 1 unit of a product. The raw material needs thus formed the sequence 3, 6, 9, . . . , $3n$, ... Is this sequence arithmetic? What is the common difference?

Skill Maintenance

Solve.

48. $7x - 2y = 4,$
$x + 3y = 17$
[9.1], [9.3], [9.5], [9.6]

49. $2x + y + 3z = 12,$
$x - 3y + 2z = 11,$
$5x + 2y - 4z = -4$
[9.2], [9.3], [9.5], [9.6]

50. Find the vertices and the foci of the ellipse with equation $9x^2 + 16y^2 = 144$. **[10.2]**

51. Find an equation of the ellipse with vertices $(0, -5)$ and $(0, 5)$ and minor axis of length 4. **[10.2]**

Synthesis

52. *Straight-Line Depreciation.* An architectural firm buys a large-format copier for $10,300 on January 1 of a given year. The machine is expected to last for 8 years, at the end of which time its **trade-in value**, or **salvage value**, will be $2100. If the company's accountant figures the decline in value to be the same each year, then its **book values**, or **salvage values**, after t years, $0 \le t \le 8$, form an arithmetic sequence given by

$$a_t = C - t\left(\frac{C - S}{N}\right),$$

where C is the original cost of the item ($10,300), N is the number of years of expected life (8), and S is the salvage value ($2100).

a) Find the formula for a_t for the straight-line depreciation of the copier.

b) Find the salvage value after 0 year, 1 year, 2 years, 3 years, 4 years, 7 years, and 8 years.

53. Find a formula for the sum of the first n odd natural numbers:

$$1 + 3 + 5 + \cdots + (2n - 1).$$

54. Find three numbers in an arithmetic sequence such that the sum of the first and third is 10 and the product of the first and second is 15.

55. Find the first term and the common difference for the arithmetic sequence for which

$$a_2 = 40 - 3q \quad \text{and} \quad a_4 = 10p + q.$$

If $p, m,$ and q form an arithmetic sequence, it can be shown that $m = (p + q)/2$. The number m is the **arithmetic mean,** *or* **average,** *of p and q. Given two numbers p and q, if we find k other numbers m_1, $m_2, \ldots, m_k$ such that*

$$p, m_1, m_2, \ldots, m_k, q$$

forms an arithmetic sequence, we say that we have "inserted k arithmetic means between p and q."

56. Insert three arithmetic means between -3 and 5.

57. Insert four arithmetic means between 4 and 13.

11.3 Geometric Sequences and Series

> Identify the common ratio of a geometric sequence, and find a given term and the sum of the first n terms.

> Find the sum of an infinite geometric series, if it exists.

A sequence in which each term after the first term is found by multiplying the preceding term by the same number is a **geometric sequence**.

❯ Geometric Sequences

Consider the sequence:

$$2, \ 6, \ 18, \ 54, \ 162, \ldots.$$

Note that multiplying each term by 3 produces the next term. We call the number 3 the **common ratio** because it can be found by dividing any term by the preceding term. A geometric sequence is also called a *geometric progression*.

GEOMETRIC SEQUENCE

A sequence is **geometric** if there is a number r, called the **common ratio**, such that

$$\frac{a_{n+1}}{a_n} = r, \quad \text{or} \quad a_{n+1} = a_n r, \quad \text{for any integer } n \geq 1.$$

EXAMPLE 1 For each of the following geometric sequences, identify the common ratio.

a) $3, 6, 12, 24, 48, \ldots$ **b)** $1, -\dfrac{1}{2}, \dfrac{1}{4}, -\dfrac{1}{8}, \ldots$

c) $\$5200, \$3900, \$2925, \$2193.75, \ldots$ **d)** $\$1000, \$1060, \$1123.60, \ldots$

Solution

SEQUENCE	COMMON RATIO
a) $3, \ 6, \ 12, \ 24, \ 48, \ldots$	2 $\left(\frac{6}{3} = 2, \frac{12}{6} = 2, \text{ and so on}\right)$
b) $1, \ -\dfrac{1}{2}, \dfrac{1}{4}, -\dfrac{1}{8}, \ldots$	$-\dfrac{1}{2}$ $\left(\frac{-\frac{1}{2}}{1} = -\frac{1}{2}, \frac{\frac{1}{4}}{-\frac{1}{2}} = -\frac{1}{2}, \text{ and so on}\right)$
c) $\$5200, \$3900, \$2925, \$2193.75, \ldots$	0.75 $\left(\dfrac{\$3900}{\$5200} = 0.75, \dfrac{\$2925}{\$3900} = 0.75, \text{ and so on}\right)$
d) $\$1000, \$1060, \$1123.60, \ldots$	1.06 $\left(\dfrac{\$1060}{\$1000} = 1.06, \dfrac{\$1123.60}{\$1060} = 1.06, \text{ and so on}\right)$

❯ **Now Try Exercise 1.**

We now find a formula for the general, or *n*th, term of a geometric sequence. If we let a_1 be the first term and r the common ratio, then the first few terms are as follows:

$$a_1,$$
$$a_2 = a_1 r,$$
$$a_3 = a_2 r = (a_1 r)r = a_1 r^2, \quad \text{Substituting } a_1 r \text{ for } a_2$$
$$a_4 = a_3 r = (a_1 r^2)r = a_1 r^3. \quad \text{Substituting } a_1 r^2 \text{ for } a_3$$

Note that the exponent is 1 less than the subscript.

Generalizing, we obtain the following.

_n_TH TERM OF A GEOMETRIC SEQUENCE

The **_n_th term** of a geometric sequence is given by

$$a_n = a_1 r^{n-1}, \quad \text{for any integer } n \geq 1.$$

EXAMPLE 2   Find the 7th term of the geometric sequence 4, 20, 100, ...

Solution   We first note that

$$a_1 = 4 \quad \text{and} \quad n = 7.$$

To find the common ratio, we can divide any term (other than the first) by the preceding term. Since the second term is 20 and the first is 4, we get

$$r = \frac{20}{4}, \quad \text{or} \quad 5.$$

Then using the formula $a_n = a_1 r^{n-1}$, we have

$$a_7 = 4 \cdot 5^{7-1} = 4 \cdot 5^6 = 4 \cdot 15{,}625 = 62{,}500.$$

Thus the 7th term is 62,500. ❭ *Now Try Exercise 11.*

EXAMPLE 3   Find the 10th term of the geometric sequence 64, -32, 16, -8, ...

Solution   We first note that

$$a_1 = 64, \qquad n = 10, \quad \text{and} \quad r = \frac{-32}{64}, \text{or} -\frac{1}{2}.$$

Then using the formula $a_n = a_1 r^{n-1}$, we have

$$a_{10} = 64 \cdot \left(-\frac{1}{2}\right)^{10-1} = 64 \cdot \left(-\frac{1}{2}\right)^9 = 2^6 \cdot \left(-\frac{1}{2^9}\right) = -\frac{1}{2^3} = -\frac{1}{8}.$$

Thus the 10th term is $-\frac{1}{8}$. ❭ *Now Try Exercise 15.*

❭ Sum of the First _n_ Terms of a Geometric Sequence

Next, we develop a formula for the sum S_n of the first n terms of a geometric sequence:

$$a_1, \ a_1 r, \ a_1 r^2, \ a_1 r^3, \ \ldots, \ a_1 r^{n-1}, \ldots .$$

The associated **geometric series** is given by

$$S_n = a_1 + a_1 r + a_1 r^2 + a_1 r^3 + \cdots + a_1 r^{n-1}. \tag{1}$$

We want to find a formula for this sum. If we multiply by r on both sides of equation (1), we have

$$r S_n = a_1 r + a_1 r^2 + a_1 r^3 + a_1 r^4 + \cdots + a_1 r^n. \tag{2}$$

Subtracting equation (2) from equation (1), we see that the differences of the terms shown in red are 0, leaving

$$S_n - rS_n = a_1 - a_1r^n,$$

or

$$S_n(1 - r) = a_1(1 - r^n). \quad \textbf{Factoring}$$

Dividing by $1 - r$ on both sides gives us the following formula.

SUM OF THE FIRST *n* TERMS

The sum of the first *n* terms of a geometric sequence is given by

$$S_n = \frac{a_1(1 - r^n)}{1 - r}, \quad \text{for any } r \neq 1.$$

EXAMPLE 4 Find the sum of the first 7 terms of the geometric sequence 3, 15, 75, 375, ...

Solution We first note that

$$a_1 = 3, \quad n = 7, \quad \text{and} \quad r = \tfrac{15}{3}, \text{ or } 5.$$

Then using the formula

$$S_n = \frac{a_1(1 - r^n)}{1 - r},$$

we have

$$S_7 = \frac{3(1 - 5^7)}{1 - 5} = \frac{3(1 - 78,125)}{-4} = 58,593.$$

Thus the sum of the first 7 terms is 58,593.

> *Now Try Exercise 23.*

EXAMPLE 5 Find the sum: $\displaystyle\sum_{k=1}^{11}(0.3)^k$.

Solution This is a geometric series with $a_1 = 0.3$, $r = 0.3$, and $n = 11$. Thus,

$$S_{11} = \frac{0.3(1 - 0.3^{11})}{1 - 0.3} \approx 0.42857.$$

sum(seq(.3^X,X,1,11))
.4285706694

We can also find this sum using a graphing calculator set in either FUNCTION mode or SEQUENCE mode.

> *Now Try Exercise 41.*

❯ Infinite Geometric Series

The sum of the terms of an infinite geometric sequence is an **infinite geometric series.** For some geometric sequences, S_n gets close to a specific number as *n* gets large. For example, consider the infinite series

$$\frac{1}{2} + \frac{1}{4} + \frac{1}{8} + \frac{1}{16} + \cdots + \frac{1}{2^n} + \cdots.$$

We can visualize S_n by considering the area of a square. For S_1, we shade half the square. For S_2, we shade half the square plus half the remaining half, or $\frac{1}{4}$. For S_3, we shade the parts shaded in S_2 plus half the remaining part. We see that the values of S_n will continue to get close to 1 (shading the complete square).

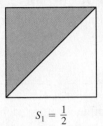

$$S_1 = \frac{1}{2}$$

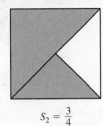

$$S_2 = \frac{3}{4}$$

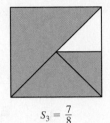

$$S_3 = \frac{7}{8}$$

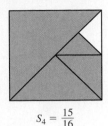

$$S_4 = \frac{15}{16}$$

We examine some partial sums. Note that each of the partial sums is less than 1, but S_n gets very close to 1 as n gets large.

n	S_n
1	0.5
5	0.96875
10	0.9990234375
20	0.9999990463
30	0.9999999991

```
sum(seq(1/2^X,X,1,20))
                .9999990463
sum(seq(1/2^X,X,1,30))
                .9999999991
```

We say that 1 is the **limit** of S_n and also that 1 is the **sum of the infinite geometric sequence**. The sum of an infinite geometric sequence is denoted S_∞. In this case, $S_\infty = 1$.

Some infinite sequences do not have sums. Consider the infinite geometric series

$$2 + 4 + 8 + 16 + \cdots + 2^n + \cdots.$$

We again examine some partial sums. Note that as n gets large, S_n gets large without bound. This sequence does not have a sum.

n	S_n
1	2
5	62
10	2,046
20	2,097,150
30	2,147,483,646

```
sum(seq(2^X,X,1,20))
                2097150
sum(seq(2^X,X,1,30))
             2147483646
```

It can be shown (but we will not do so here) that the sum of an infinite geometric series exists if and only if $|r| < 1$ (that is, the absolute value of the common ratio is less than 1).

To find a formula for the sum of an infinite geometric series, we first consider the sum of the first n terms:

$$S_n = \frac{a_1(1 - r^n)}{1 - r} = \frac{a_1 - a_1 r^n}{1 - r}. \qquad \textbf{Using the distributive law}$$

For $|r| < 1$, values of r^n get close to 0 as n gets large. As r^n gets close to 0, so does $a_1 r^n$. Thus, S_n gets close to $a_1/(1 - r)$.

LIMIT OR SUM OF AN INFINITE GEOMETRIC SERIES

When $|r| < 1$, the limit or sum of an infinite geometric series is given by

$$S_\infty = \frac{a_1}{1 - r}.$$

EXAMPLE 6   Determine whether each of the following infinite geometric series has a limit. If a limit exists, find it.

a) $1 + 3 + 9 + 27 + \cdots$ \qquad\qquad **b)** $-2 + 1 - \frac{1}{2} + \frac{1}{4} - \frac{1}{8} + \cdots$

Solution

a) Here $r = 3$, so $|r| = |3| = 3$. Since $|r| > 1$, the series *does not* have a limit.

b) Here $r = -\frac{1}{2}$, so $|r| = \left|-\frac{1}{2}\right| = \frac{1}{2}$. Since $|r| < 1$, the series *does* have a limit. We find the limit:

$$S_\infty = \frac{a_1}{1 - r} = \frac{-2}{1 - \left(-\frac{1}{2}\right)} = \frac{-2}{\frac{3}{2}} = -\frac{4}{3}.$$

> *Now Try Exercises 33 and 37.*

EXAMPLE 7   Find fraction notation for $0.78787878\ldots$, or $0.\overline{78}$.

Solution   We can express this as

$$0.78 + 0.0078 + 0.000078 + \cdots.$$

Then we see that this is an infinite geometric series, where $a_1 = 0.78$ and $r = 0.01$. Since $|r| < 1$, this series has a limit:

$$S_\infty = \frac{a_1}{1 - r} = \frac{0.78}{1 - 0.01} = \frac{0.78}{0.99} = \frac{78}{99}, \quad \text{or} \quad \frac{26}{33}.$$

Thus fraction notation for $0.78787878\ldots$ is $\frac{26}{33}$. You can check this on your calculator.

> *Now Try Exercise 51.*

❯ Applications

The translation of some applications and problem-solving situations may involve geometric sequences or series. Examples 9 and 10, in particular, show applications in business and economics.

EXAMPLE 8   *A Daily Doubling Salary.*   Suppose that someone offers you a job for the month of September (30 days) under the following conditions. You will be paid \$0.01 for the first day, \$0.02 for the second, \$0.04 for the third, and so on, doubling your previous day's salary each day. How much would you earn? (Would you take the job? Make a conjecture before reading further.)

Solution   You earn \$0.01 the first day, \$0.01(2) the second day, \$0.01(2)(2) the third day, and so on. The amount earned is the geometric series

$$\$0.01 + \$0.01(2) + \$0.01(2^2) + \$0.01(2^3) + \cdots + \$0.01(2^{29}),$$

where $a_1 = \$0.01$, $r = 2$, and $n = 30$. Using the formula

$$S_n = \frac{a_1(1 - r^n)}{1 - r},$$

we have

$$S_{30} = \frac{\$0.01(1 - 2^{30})}{1 - 2} = \$10,737,418.23.$$

The pay exceeds $10.7 million for the month.

> *Now Try Exercise 57.*

EXAMPLE 9   *The Amount of an Annuity.*   An **annuity** is a sequence of equal payments, made at equal time intervals, that earn interest. Fixed deposits in a savings account are an example of an annuity. Suppose that to save money to buy a car, Janelle deposits $2000 at the *end* of each of 5 years in an account that pays 3% interest, compounded annually. The total amount in the account at the end of 5 years is called the **amount of the annuity**. Find that amount.

Solution   The following time diagram can help visualize the problem. Note that no deposit is made until the end of the first year.

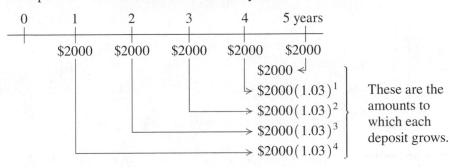

The amount of the annuity is the geometric series

$$\$2000 + \$2000(1.03)^1 + \$2000(1.03)^2 + \$2000(1.03)^3 + \$2000(1.03)^4,$$

where $a_1 = \$2000$, $n = 5$, and $r = 1.03$. Using the formula

$$S_n = \frac{a_1(1 - r^n)}{1 - r},$$

we have

$$S_5 = \frac{\$2000(1 - 1.03^5)}{1 - 1.03} \approx \$10,618.27.$$

The amount of the annuity is $10,618.27.

> *Now Try Exercise 61.*

EXAMPLE 10   *The Economic Multiplier.*   Large sporting events have a significant impact on the economy of the host city. Super Bowl XLVII, hosted by New Orleans, generated a $480-million net impact for the region (*Source*: NewOrleansSaints.com, posted April 18, 2013, Marius M. Mihai, Research Analyst of the Division of Business and Economic Research at the University of New Orleans (DBER)). Assume that 60% of that amount is spent again in the area, and then 60% of that amount is spent again, and so on. This is known as the *economic multiplier effect*. Find the total effect on the economy.

Solution   The total economic effect is given by the infinite series

$$\$480,000,000 + \$480,000,000(0.6) + \$480,000,000(0.6)^2 + \cdots.$$

Since $|r| = |0.6| = 0.6 < 1$, the series has a sum. Using the formula for the sum of an infinite geometric series, we have

$$S_\infty = \frac{a_1}{1 - r} = \frac{\$480,000,000}{1 - 0.6} = \$1,200,000,000.$$

The total effect of the spending on the economy is $1,200,000,000.

> *Now Try Exercise 67.*

Visualizing the Graph

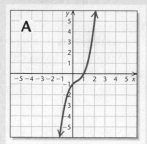

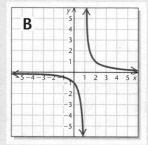

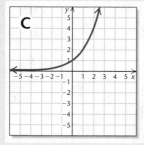

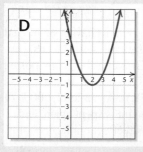

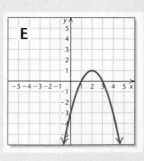

Match the equation with its graph.

1. $(x - 1)^2 + (y + 2)^2 = 9$

2. $y = x^3 - x^2 + x - 1$

3. $f(x) = 2^x$

4. $f(x) = x$

5. $a_n = n$

6. $y = \log (x + 3)$

7. $f(x) = -(x - 2)^2 + 1$

8. $f(x) = (x - 2)^2 - 1$

9. $y = \dfrac{1}{x - 1}$

10. $y = -3x + 4$

Answers on page A-69

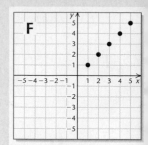

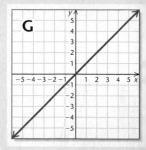

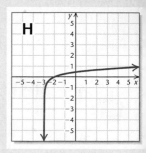

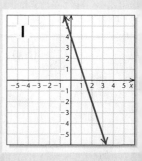

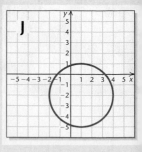

11.3 | Exercise Set

Find the common ratio.

1. 2, 4, 8, 16, . . .

2. 18, −6, 2, −$\frac{2}{3}$, . . .

3. −1, 1, −1, 1, . . .

4. −8, −0.8, −0.08, −0.008, . . .

5. $\frac{2}{3}$, −$\frac{4}{3}$, $\frac{8}{3}$, −$\frac{16}{3}$, . . .

6. 75, 15, 3, $\frac{3}{5}$, . . .

7. 6.275, 0.6275, 0.06275, . . .

8. $\dfrac{1}{x}, \dfrac{1}{x^2}, \dfrac{1}{x^3}, \cdots$

9. 5, $\dfrac{5a}{2}, \dfrac{5a^2}{4}, \dfrac{5a^3}{8}, \cdots$

10. $780, $858, $943.80, $1038.18, . . .

Find the indicated term of the geometric sequence.

11. 2, 4, 8, 16, . . . ; the 7th term

12. 2, −10, 50, −250, . . . ; the 9th term

13. 2, 2$\sqrt{3}$, 6, . . . ; the 9th term

14. 1, −1, 1, −1, . . . ; the 57th term

15. $\frac{7}{625}$, −$\frac{7}{25}$, . . . ; the 23rd term

16. $1000, $1060, $1123.60, . . . ; the 5th term

Find the nth, or general, term.

17. 1, 3, 9, . . .   **18.** 25, 5, 1, . . .

19. 1, −1, 1, −1, . . .   **20.** −2, 4, −8, . . .

21. $\dfrac{1}{x}, \dfrac{1}{x^2}, \dfrac{1}{x^3}, \cdots$ **22.** 5, $\dfrac{5a}{2}, \dfrac{5a^2}{4}, \dfrac{5a^3}{8}, \cdots$

23. Find the sum of the first 7 terms of the geometric series

$$6 + 12 + 24 + \cdots.$$

24. Find the sum of the first 10 terms of the geometric series

$$16 - 8 + 4 - \cdots.$$

25. Find the sum of the first 9 terms of the geometric series

$$\tfrac{1}{18} - \tfrac{1}{6} + \tfrac{1}{2} - \cdots.$$

26. Find the sum of the geometric series

$$-8 + 4 + (-2) + \cdots + \left(-\tfrac{1}{32}\right).$$

Determine whether the statement is true or false.

27. The sequence 2, −2$\sqrt{2}$, 4, −4$\sqrt{2}$, 8, . . . is geometric.

28. The sequence with general term $3n$ is geometric.

29. The sequence with general term 2^n is geometric.

30. Multiplying a term of a geometric sequence by the common ratio produces the next term of the sequence.

31. An infinite geometric series with common ratio −0.75 has a sum.

32. Every infinite geometric series has a limit.

Find the sum, if it exists.

33. $4 + 2 + 1 + \cdots$

34. $7 + 3 + \frac{9}{7} + \cdots$

35. $25 + 20 + 16 + \cdots$

36. $100 - 10 + 1 - \frac{1}{10} + \cdots$

37. $8 + 40 + 200 + \cdots$

38. $-6 + 3 - \frac{3}{2} + \frac{3}{4} - \cdots$

39. $0.6 + 0.06 + 0.006 + \cdots$

40. $\displaystyle\sum_{k=0}^{10} 3^k$ **41.** $\displaystyle\sum_{k=1}^{11} 15\left(\frac{2}{3}\right)^k$

42. $\displaystyle\sum_{k=0}^{50} 200(1.08)^k$ **43.** $\displaystyle\sum_{k=1}^{\infty} \left(\frac{1}{2}\right)^{k-1}$

44. $\displaystyle\sum_{k=1}^{\infty} 2^k$ **45.** $\displaystyle\sum_{k=1}^{\infty} 12.5^k$

46. $\displaystyle\sum_{k=1}^{\infty} 400(1.0625)^k$ **47.** $\displaystyle\sum_{k=1}^{\infty} \$500(1.11)^{-k}$

48. $\displaystyle\sum_{k=1}^{\infty} \$1000(1.06)^{-k}$ **49.** $\displaystyle\sum_{k=1}^{\infty} 16(0.1)^{k-1}$

50. $\displaystyle\sum_{k=1}^{\infty} \frac{8}{3}\left(\frac{1}{2}\right)^{k-1}$

Find fraction notation.

51. 0.131313. . . , or 0.$\overline{13}$ **52.** 0.2222. . . , or 0.$\overline{2}$

53. 8.999$\overline{9}$ **54.** 6.1616$\overline{16}$

55. 3.4125$\overline{125}$ **56.** 12.7809$\overline{809}$

57. *Daily Doubling Salary.* Suppose that someone offers you a job for the month of February (28 days) under the following conditions. You will be paid $0.01 the 1st day, $0.02 the 2nd, $0.04 the 3rd, and so on, doubling your previous day's salary each day. How much would you earn altogether?

58. *Bouncing Ping-Pong Ball.* A ping-pong ball is dropped from a height of 16 ft and always rebounds $\frac{1}{4}$ of the distance fallen.

   a) How high does it rebound the 6th time?

   b) Find the total sum of the rebound heights of the ball.

59. *Bungee Jumping.* A bungee jumper always rebounds 60% of the distance fallen. A bungee jump is made using a cord that stretches to 200 ft.

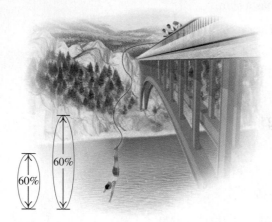

   a) After jumping and then rebounding 9 times, how far has a bungee jumper traveled upward (the total rebound distance)?

   b) About how far will a jumper have traveled upward (bounced) before coming to rest?

60. *Population Growth.* A coastal town has a present population of 32,100, and the population is increasing by 3% each year.

   a) What will the population be in 15 years?

   b) How long will it take for the population to double?

61. *Amount of an Annuity.* To save for the down payment on a house, the Ramirez family makes a sequence of 10 yearly deposits of $3200 each in a savings account on which interest is compounded annually at 4.6%. Find the amount of the annuity.

62. *Amount of an Annuity.* To create a college fund, a parent makes a sequence of 18 yearly deposits of $1000 each in a savings account on which interest is compounded annually at 3.2%. Find the amount of the annuity.

63. *Doubling the Thickness of Paper.* A piece of paper is 0.01 in. thick. It is cut and stacked repeatedly in such a way that its thickness is doubled each time for 20 times. How thick is the result?

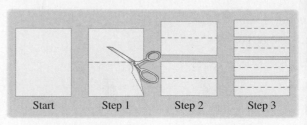

Start   Step 1   Step 2   Step 3

64. *Amount of an Annuity.* A sequence of yearly payments of P dollars is invested at the end of each of N years at interest rate i, compounded annually. The total amount in the account, or the amount of the annuity, is V.

   a) Show that

   $$V = \frac{P\left[(1 + i)^N - 1\right]}{i}.$$

   b) Suppose that interest is compounded n times per year and deposits are made every compounding period. Show that the formula for V is then given by

   $$V = \frac{P\left[\left(1 + \dfrac{i}{n}\right)^{nN} - 1\right]}{i/n}.$$

65. *Amount of an Annuity.* A sequence of payments of $300 is invested over 12 years at the end of each quarter at 5.1%, compounded quarterly. Find the amount of the annuity. Use the formula in Exercise 64(b).

66. *Amount of an Annuity.* A sequence of yearly payments of $750 is invested at the end of each of 10 years at 4.75%, compounded annually. Find the amount of the annuity. Use the formula in Exercise 64(a).

67. *The Economic Multiplier.* Suppose that the government is making a $13,000,000,000 expenditure for educational improvement. If 85% of this is spent again, and so on, what is the total effect on the economy?

68. *Advertising Effect.* Gigi's Cupcake Truck is about to open for business in a city of 3,000,000 people, traveling to several curbside locations in the city each day to sell cupcakes. The owners plan an advertising campaign that they think will induce 30% of the people to buy their cupcakes. They estimate that if those people like the product, they will induce

$30\% \cdot 30\% \cdot 3{,}000{,}000$ more to buy the product, and those will induce $30\% \cdot 30\% \cdot 30\% \cdot 3{,}000{,}000$ and so on. In all, how many people will buy Gigi's cupcakes as a result of the advertising campaign? What percentage of the population is this?

❯ Skill Maintenance

For each pair of functions, find $(f \circ g)(x)$ and $(g \circ f)(x)$.
[2.3]

69. $f(x) = x^2, g(x) = 4x + 5$

70. $f(x) = x - 1, g(x) = x^2 + x + 3$

Solve. **[5.5]**

71. $5^x = 35$ **72.** $\log_2 x = -4$

❯ Synthesis

73. Prove that
$$\sqrt{3} - \sqrt{2}, \quad 4 - \sqrt{6}, \quad \text{and} \quad 6\sqrt{3} - 2\sqrt{2}$$
form a geometric sequence.

74. Assume that $a_1, a_2, a_3, \ldots$ is a geometric sequence. Prove that $\ln a_1, \ln a_2, \ln a_3, \ldots$ is an arithmetic sequence.

75. Consider the sequence
$$x + 3, \quad x + 7, \quad 4x - 2, \ldots.$$

   a) If the sequence is arithmetic, find x and then determine each of the 3 terms and the 4th term.
   b) If the sequence is geometric, find x and then determine each of the 3 terms and the 4th term.

76. Find the sum of the first n terms of
$$1 + x + x^2 + \cdots.$$

77. Find the sum of the first n terms of
$$x^2 - x^3 + x^4 - x^5 + \cdots.$$

78. The sides of a square are 16 cm long. A second square is inscribed by joining the midpoints of the sides, successively. In the second square, we repeat the process, inscribing a third square. If this process is continued indefinitely, what is the sum of all the areas of all the squares? (*Hint*: Use an infinite geometric series.)

11.4 ❯ **Mathematical Induction**

❯ Prove infinite sequences of statements using mathematical induction.

In this section, we learn to prove a sequence of mathematical statements using a procedure called *mathematical induction*.

❯ Proving Infinite Sequences of Statements

Infinite sequences of statements occur often in mathematics. In an infinite sequence of statements, there is a statement for each natural number. For example, consider the sequence of statements represented by the following:

"The sum of the first n positive odd integers is n^2,"   or
$$1 + 3 + 5 + \cdots + (2n - 1) = n^2.$$

Let's think of this as $S(n)$, or S_n. Substituting natural numbers for n gives a sequence of statements. We list the first four:

S_1: $1 = 1^2$;
S_2: $1 + 3 = 4 = 2^2$;
S_3: $1 + 3 + 5 = 9 = 3^2$;
S_4: $1 + 3 + 5 + 7 = 16 = 4^2$.

The fact that the statement is true for $n = 1, 2, 3,$ and 4 might tempt us to conclude that the statement is true for any natural number n, but we cannot be sure that this is the case. We can, however, use the principle of mathematical induction to prove that the statement is true for all natural numbers.

THE PRINCIPLE OF MATHEMATICAL INDUCTION

We can prove an infinite sequence of statements S_n by showing the following.

1. *Basis step:* S_1 is true.
2. *Induction step:* For all natural numbers k, $S_k \rightarrow S_{k+1}$.

Mathematical induction is analogous to lining up a sequence of dominoes. The induction step tells us that if any one domino is knocked over, then the one next to it will be hit and knocked over. The basis step tells us that the first domino can indeed be knocked over. Note that in order for all dominoes to fall, *both* conditions must be satisfied.

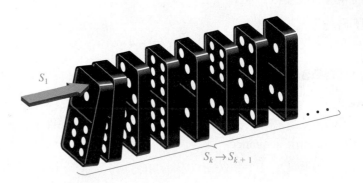

When you are learning to do proofs by mathematical induction, it is helpful to first write out S_n, S_1, S_k, and S_{k+1}. This helps to identify what is to be assumed and what is to be deduced.

EXAMPLE 1 Prove: For every natural number n,

$$1 + 3 + 5 + \cdots + (2n - 1) = n^2.$$

Proof.   We first list S_n, S_1, S_k, and S_{k+1}.

S_n: $1 + 3 + 5 + \cdots + (2n - 1) = n^2$
S_1: $1 = 1^2$
S_k: $1 + 3 + 5 + \cdots + (2k - 1) = k^2$
S_{k+1}: $1 + 3 + 5 + \cdots + (2k - 1) + [2(k + 1) - 1] = (k + 1)^2$

1. *Basis step*: S_1, as listed, is true since $1 = 1^2$, or $1 = 1$.

2. *Induction step*: We let k be any natural number. We assume S_k to be true and try to show that it implies that S_{k+1} is true. Now S_k is

   $$1 + 3 + 5 + \cdots + (2k - 1) = k^2.$$

   Starting with the left side of S_{k+1} and substituting k^2 for $1 + 3 + 5 + \cdots + (2k - 1)$, we have

   $$\underbrace{1 + 3 + \cdots + (2k - 1)} + [2(k + 1) - 1]$$

   $$\begin{aligned}
   &= k^2 + [2(k + 1) - 1] \qquad \textbf{We assume } S_k \textbf{ is true.} \\
   &= k^2 + 2k + 2 - 1 \\
   &= k^2 + 2k + 1 \\
   &= (k + 1)^2.
   \end{aligned}$$

   We have shown that for all natural numbers k, $S_k \rightarrow S_{k+1}$. This completes the induction step. It and the basis step tell us that the proof is complete.

   > *Now Try Exercise 5.*

EXAMPLE 2   Prove: For every natural number n,

$$\frac{1}{2} + \frac{1}{4} + \frac{1}{8} + \cdots + \frac{1}{2^n} = \frac{2^n - 1}{2^n}.$$

Proof.   We first list S_n, S_1, S_k, and S_{k+1}.

$$S_n: \qquad \frac{1}{2} + \frac{1}{4} + \frac{1}{8} + \cdots + \frac{1}{2^n} = \frac{2^n - 1}{2^n}$$

$$S_1: \qquad \frac{1}{2^1} = \frac{2^1 - 1}{2^1}$$

$$S_k: \qquad \frac{1}{2} + \frac{1}{4} + \frac{1}{8} + \cdots + \frac{1}{2^k} = \frac{2^k - 1}{2^k}$$

$$S_{k+1}: \quad \frac{1}{2} + \frac{1}{4} + \frac{1}{8} + \cdots + \frac{1}{2^k} + \frac{1}{2^{k+1}} = \frac{2^{k+1} - 1}{2^{k+1}}$$

1. *Basis step*: We show S_1 to be true as follows:

   $$\frac{2^1 - 1}{2^1} = \frac{2 - 1}{2} = \frac{1}{2}.$$

2. *Induction step*: We let k be any natural number. We assume S_k to be true and try to show that it implies that S_{k+1} is true. Now S_k is

   $$\frac{1}{2} + \frac{1}{4} + \frac{1}{8} + \cdots + \frac{1}{2^k} = \frac{2^k - 1}{2^k}.$$

   We start with the left side of S_{k+1}. Since we assume S_k is true, we can substitute

   $$\frac{2^k - 1}{2^k} \quad \text{for} \quad \frac{1}{2} + \frac{1}{4} + \cdots + \frac{1}{2^k}.$$

We have

$$\underbrace{\frac{1}{2} + \frac{1}{4} + \frac{1}{8} + \cdots + \frac{1}{2^k}} + \frac{1}{2^{k+1}}$$

$$= \frac{2^k - 1}{2^k} + \frac{1}{2^{k+1}} = \frac{2^k - 1}{2^k} \cdot \frac{2}{2} + \frac{1}{2^{k+1}}$$

$$= \frac{(2^k - 1) \cdot 2 + 1}{2^{k+1}}$$

$$= \frac{2^{k+1} - 2 + 1}{2^{k+1}}$$

$$= \frac{2^{k+1} - 1}{2^{k+1}}.$$

We have shown that for all natural numbers k, $S_k \rightarrow S_{k+1}$. This completes the induction step. It and the basis step tell us that the proof is complete.

> *Now Try Exercise 15.*

EXAMPLE 3 Prove: For every natural number n, $n < 2^n$.

Proof. We first list S_n, S_1, S_k, and S_{k+1}.

S_n: $n < 2^n$
S_1: $1 < 2^1$
S_k: $k < 2^k$
S_{k+1}: $k + 1 < 2^{k+1}$

1. *Basis step*: S_1, as listed, is true since $2^1 = 2$ and $1 < 2$.
2. *Induction step*: We let k be any natural number. We assume S_k to be true and try to show that it implies that S_{k+1} is true. Now

$$k < 2^k \qquad \text{This is } S_k.$$
$$2k < 2 \cdot 2^k \qquad \text{Multiplying by 2 on both sides}$$
$$2k < 2^{k+1} \qquad \text{Adding exponents on the right}$$
$$k + k < 2^{k+1}. \qquad \text{Rewriting } 2k \text{ as } k + k$$

Since k is any natural number, we know that $1 \leq k$. Thus,

$$k + 1 \leq k + k. \qquad \text{Adding } k \text{ on both sides of } 1 \leq k$$

Putting the results $k + 1 \leq k + k$ and $k + k < 2^{k+1}$ together gives us

$$k + 1 < 2^{k+1}. \qquad \text{This is } S_{k+1}.$$

We have shown that for all natural numbers k, $S_k \rightarrow S_{k+1}$. This completes the induction step. It and the basis step tell us that the proof is complete.

> *Now Try Exercise 11.*

11.4 | Exercise Set

List the first five statements in the sequence that can be obtained from each of the following. Determine whether each of the statements is true or false.

1. $n^2 < n^3$

2. $n^2 - n + 41$ is prime. Find a value for n for which the statement is false.

3. A polygon of n sides has $[n(n-3)]/2$ diagonals.

4. The sum of the angles of a polygon of n sides is $(n-2) \cdot 180°$.

Use mathematical induction to prove each of the following.

5. $2 + 4 + 6 + \cdots + 2n = n(n+1)$

6. $4 + 8 + 12 + \cdots + 4n = 2n(n+1)$

7. $1 + 5 + 9 + \cdots + (4n-3) = n(2n-1)$

8. $3 + 6 + 9 + \cdots + 3n = \dfrac{3n(n+1)}{2}$

9. $2 + 4 + 8 + \cdots + 2^n = 2(2^n - 1)$

10. $2 \le 2^n$ **11.** $n < n + 1$

12. $3^n < 3^{n+1}$ **13.** $2n \le 2^n$

14. $\dfrac{1}{1 \cdot 2} + \dfrac{1}{2 \cdot 3} + \cdots + \dfrac{1}{n(n+1)} = \dfrac{n}{n+1}$

15. $\dfrac{1}{1 \cdot 2 \cdot 3} + \dfrac{1}{2 \cdot 3 \cdot 4} + \dfrac{1}{3 \cdot 4 \cdot 5} + \cdots$

$\quad + \dfrac{1}{n(n+1)(n+2)} = \dfrac{n(n+3)}{4(n+1)(n+2)}$

16. If x is any real number greater than 1, then for any natural number n, $x \le x^n$.

The following formulas can be used to find sums of powers of natural numbers. Use mathematical induction to prove each formula.

17. $1 + 2 + 3 + \cdots + n = \dfrac{n(n+1)}{2}$

18. $1^2 + 2^2 + 3^2 + \cdots + n^2 = \dfrac{n(n+1)(2n+1)}{6}$

19. $1^3 + 2^3 + 3^3 + \cdots + n^3 = \dfrac{n^2(n+1)^2}{4}$

20. $1^4 + 2^4 + 3^4 + \cdots + n^4$

$\quad = \dfrac{n(n+1)(2n+1)(3n^2 + 3n - 1)}{30}$

Use mathematical induction to prove each of the following.

21. $\displaystyle\sum_{i=1}^{n} i(i+1) = \dfrac{n(n+1)(n+2)}{3}$

22. $\left(1 + \dfrac{1}{1}\right)\left(1 + \dfrac{1}{2}\right)\left(1 + \dfrac{1}{3}\right) \cdots \left(1 + \dfrac{1}{n}\right)$

$\quad = n + 1$

23. The sum of n terms of an arithmetic sequence:

$a_1 + (a_1 + d) + (a_1 + 2d) + \cdots + [a_1 + (n-1)d]$

$\quad = \dfrac{n}{2}[2a_1 + (n-1)d]$

❯ Skill Maintenance

Solve.

24. $2x - 3y = 1,$
$3x - 4y = 3$ **[9.1], [9.3], [9.5], [9.6]**

25. *Investment.* Huiliang received \$150 in simple interest one year from three investments. Part is invested at 1.5%, part at 2%, and part at 3%. The amount invested at 2% is twice the amount invested at 1.5%. There is \$400 more invested at 3% than at 2%. Find the amount invested at each rate. **[9.2], [9.3], [9.5], [9.6]**

❯ Synthesis

Use mathematical induction to prove each of the following.

26. The sum of n terms of a geometric sequence:

$a_1 + a_1 r + a_1 r^2 + \cdots + a_1 r^{n-1} = \dfrac{a_1 - a_1 r^n}{1 - r}$

27. $x + y$ is a factor of $x^{2n} - y^{2n}$.

Prove each of the following using mathematical induction. Do the basis step for $n = 2$.

28. For every natural number $n \ge 2$,

$\quad 2n + 1 < 3^n$.

29. For every natural number $n \ge 2$,

$\quad \log_a (b_1 b_2 \cdots b_n)$

$\quad\quad = \log_a b_1 + \log_a b_2 + \cdots + \log_a b_n$.

Prove each of the following for any complex numbers $z_1, z_2, \ldots, z_n$, where $i^2 = -1$ and $\bar{z}$ is the conjugate of z. (See Section 3.1.)

30. $\overline{z^n} = \bar{z}^n$

31. $\overline{z_1 + z_2 + \cdots + z_n} = \bar{z_1} + \bar{z_2} + \cdots + \bar{z_n}$

32. *The Tower of Hanoi Problem.* There are three pegs on a board. On one peg are *n* disks, each smaller than the one on which it rests. The problem is to move this pile of disks to another peg. The final order must be the same, but you can move only one disk at a time and can never place a larger disk on a smaller one.

a) What is the *least* number of moves needed to move 3 disks? 4 disks? 2 disks? 1 disk?

b) Conjecture a formula for the *least* number of moves needed to move *n* disks. Prove it by mathematical induction.

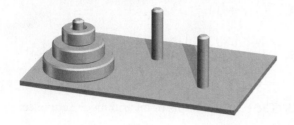

Mid-Chapter Mixed Review

Determine whether the statement is true or false.

1. The general term of the sequence $1, -2, 3, -4, \ldots$ can be expressed as $a_n = n$. **[11.1]**

2. To find the common difference of an arithmetic sequence, choose any term except the first and then subtract the preceding term from it. **[11.2]**

3. The sequence $7, 3, -1, -5, \ldots$ is geometric. **[11.2], [11.3]**

4. If we can show that $S_k \rightarrow S_{k+1}$ for some natural number k, then we know that S_n is true for all natural numbers n. **[11.4]**

In each of the following, the nth term of a sequence is given. Find the first 4 terms, a_9, and a_{14}.

5. $a_n = 3n + 5$ **[11.1]**

6. $a_n = (-1)^{n+1}(n - 1)$ **[11.1]**

Predict the general term, or nth term, a_n, of the sequence. Answers may vary.

7. $3, 6, 9, 12, 15, \ldots$ **[11.1]**

8. $-1, 4, -9, 16, -25, \ldots$ **[11.1]**

9. Find the partial sum S_4 for the sequence $1, \frac{1}{2}, \frac{1}{4}, \frac{1}{8}, \frac{1}{16} \ldots$ **[11.1]**

10. Find and evaluate the sum $\displaystyle\sum_{k=1}^{5} k(k + 1)$. **[11.1]**

11. Write sigma notation for the sum $-4 + 8 - 12 + 16 - 20 + \cdots$. **[11.1]**

12. Find the first 4 terms of the sequence defined by $a_1 = 2, a_{n+1} = 4a_n - 2$. **[11.1]**

13. Find the common difference of the arithmetic sequence $12, 7, 2, -3, \ldots$. **[11.2]**

14. Find the 10th term of the arithmetic sequence $4, 6, 8, 10, \ldots$. **[11.2]**

15. In the sequence in Exercise 14, what term is the number 44? **[11.2]**

16. Find the sum of the first 16 terms of the arithmetic series $6 + 11 + 16 + 21 + \cdots$. **[11.2]**

17. Find the common ratio of the geometric sequence $16, -8, 4, -2, 1, \ldots$. **[11.3]**

18. Find **(a)** the 8th term and **(b)** the sum of the first 10 terms of the geometric sequence $\frac{1}{16}, \frac{1}{8}, \frac{1}{4}, \frac{1}{2}, 1, \ldots$. **[11.3]**

Find the sum, if it exists.

19. $-8 + 4 - 2 + 1 - \cdots$ **[11.3]**

20. $\displaystyle\sum_{k=0}^{\infty} 5^k$ **[11.3]**

21. *Landscaping.*   A landscaper is planting a triangular flower bed with 36 plants in the first row, 30 plants in the second row, 24 in the third row, and so on, for a total of 6 rows. How many plants will be planted in all? [11.2]

22. *Amount of an Annuity.*   To save money for adding a bedroom to their home, at the end of each of 4 years the Jacobsons deposit $3200 in an account that pays 4% interest, compounded annually. Find the total amount of the annuity. [11.3]

23. Prove: For every natural number n,

$$1 + 4 + 7 + \cdots + (3n - 2) = \tfrac{1}{2}n(3n - 1).$$

[11.4]

Collaborative Discussion and Writing

24. The sum of the first n terms of an arithmetic sequence can be given by

$$S_n = \frac{n}{2}[2a_1 + (n - 1)d].$$

Compare this formula to

$$S_n = \frac{n}{2}(a_1 + a_n).$$

Discuss the reasons for the use of one formula over the other. [11.2]

25. It is said that as a young child, the mathematician Karl F. Gauss (1777–1855) was able to compute the sum $1 + 2 + 3 + \cdots + 100$ very quickly in his head to the amazement of a teacher. Explain how Gauss might have done this had he possessed some knowledge of arithmetic sequences and series. Then give a formula for the sum of the first n natural numbers. [11.2]

26. Write a problem for a classmate to solve. Devise the problem so that a geometric series is involved and the solution is "The total amount in the bank is $900(1.08)^{40}$, or about $19,552." [11.3]

27. Write an explanation of the idea behind mathematical induction for a fellow student. [11.4]

11.5 Combinatorics: Permutations

> ❯ Evaluate factorial notation and permutation notation and solve related applied problems.

In order to study probability, it is first necessary to learn about **combinatorics**, the theory of counting.

❯ Permutations

In this section, we will consider the part of combinatorics called *permutations*.

> The study of permutations involves *order* and *arrangements*.

EXAMPLE 1 How many 3-letter code symbols can be formed with the letters A, B, C *without* repetition (that is, using each letter only once)?

Solution Consider placing the letters in these boxes.

We can select any of the 3 letters for the first letter in the symbol. Once this letter has been selected, the second must be selected from the 2 remaining letters. After this, the third letter is already determined, since only 1 possibility is left. That is, we can place any of the 3 letters in the first box, either of the remaining 2 letters in the second box, and the only remaining letter in the third box. The possibilities can be determined using a **tree diagram,** as shown below.

TREE DIAGRAM   **OUTCOMES**

$$
\begin{array}{llll}
 & & B \text{——} C & \text{ABC} \\
 & A < & & \\
 & & C \text{——} B & \text{ACB} \\
 & & A \text{——} C & \text{BAC} \\
< & B < & & \\
 & & C \text{——} A & \text{BCA} \\
 & & A \text{——} B & \text{CAB} \\
 & C < & & \\
 & & B \text{——} A & \text{CBA}
\end{array}
$$

Each outcome represents one permutation of the letters A, B, C.

We see that there are 6 possibilities. The set of all the possibilities is

{ABC, ACB, BAC, BCA, CAB, CBA}.

This is the set of all *permutations* of the letters A, B, C.

Suppose that we perform an experiment such as selecting letters (as in the preceding example), flipping a coin, or drawing a card. The results are called **outcomes.** An **event** is a set of outcomes. The following principle enables us to count actions that are combined to form an event.

THE FUNDAMENTAL COUNTING PRINCIPLE

Given a combined action, or *event*, in which the first action can be performed in n_1 ways, the second action can be performed in n_2 ways, and so on, the total number of ways in which the combined action can be performed is the product

$$n_1 \cdot n_2 \cdot n_3 \cdots \cdots n_k.$$

Thus, in Example 1, there are 3 choices for the first letter, 2 for the second letter, and 1 for the third letter, making a total of $3 \cdot 2 \cdot 1$, or 6 possibilities.

EXAMPLE 2 How many 3-letter code symbols can be formed with the letters A, B, C, D, E *with* repetition (that is, allowing letters to be repeated)?

Solution Since repetition is allowed, there are 5 choices for the first letter, 5 choices for the second, and 5 for the third. Thus, by the fundamental counting principle, there are $5 \cdot 5 \cdot 5$, or 125 code symbols.

> **PERMUTATION**
>
> A **permutation** of a set of n objects is an ordered arrangement of all n objects.

We can use the fundamental counting principle to count the number of permutations of the objects in a set. Consider, for example, a set of 4 objects

$$\{A, B, C, D\}.$$

To find the number of ordered arrangements of the set, we select a first letter: There are 4 choices. Then we select a second letter: There are 3 choices. Then we select a third letter: There are 2 choices. Finally, there is 1 choice for the last selection. Thus, by the fundamental counting principle, there are $4 \cdot 3 \cdot 2 \cdot 1$, or 24, permutations of a set of 4 objects.

We can find a formula for the total number of permutations of all objects in a set of n objects. We have n choices for the first selection, $n - 1$ choices for the second, $n - 2$ for the third, and so on. For the nth selection, there is only 1 choice.

> **THE TOTAL NUMBER OF PERMUTATIONS OF**
> **n OBJECTS**
>
> The total number of permutations of n objects, denoted $_nP_n$, is given by
>
> $$_nP_n = n(n - 1)(n - 2) \cdots 3 \cdot 2 \cdot 1.$$

EXAMPLE 3   Find each of the following.

a) $_4P_4$ **b)** $_7P_7$

Solution

Start with 4.

a) $_4P_4 = \underbrace{4 \cdot 3 \cdot 2 \cdot 1}_{} = 24$

                                    4 factors

b) $_7P_7 = 7 \cdot 6 \cdot 5 \cdot 4 \cdot 3 \cdot 2 \cdot 1 = 5040$

We can also find the total number of permutations of n objects using the $_nP_r$ operation from the MATH PRB (probability) menu on a graphing calculator.

> **Now Try Exercise 1.**

EXAMPLE 4   In how many ways can 9 packages be placed in 9 mailboxes, one package in a box?

Solution   We have

$$_9P_9 = 9 \cdot 8 \cdot 7 \cdot 6 \cdot 5 \cdot 4 \cdot 3 \cdot 2 \cdot 1 = 362{,}880.$$

> **Now Try Exercise 23.**

〉 Factorial Notation

We will use products such as $7 \cdot 6 \cdot 5 \cdot 4 \cdot 3 \cdot 2 \cdot 1$ so often that it is convenient to adopt a notation for them. For the product

$$7 \cdot 6 \cdot 5 \cdot 4 \cdot 3 \cdot 2 \cdot 1,$$

we write 7!, read "7 factorial."

We now define factorial notation for natural numbers and for 0.

FACTORIAL NOTATION

For any natural number n,

$$n! = n(n-1)(n-2) \cdots 3 \cdot 2 \cdot 1.$$

For the number 0,

$$0! = 1.$$

We define 0! as 1 so that certain formulas can be stated concisely and with a consistent pattern.

Here are some examples of factorial notation.

$$7! = 7 \cdot 6 \cdot 5 \cdot 4 \cdot 3 \cdot 2 \cdot 1 = 5040$$
$$6! = 6 \cdot 5 \cdot 4 \cdot 3 \cdot 2 \cdot 1 = 720$$
$$5! = 5 \cdot 4 \cdot 3 \cdot 2 \cdot 1 = 120$$
$$4! = 4 \cdot 3 \cdot 2 \cdot 1 = 24$$
$$3! = 3 \cdot 2 \cdot 1 = 6$$
$$2! = 2 \cdot 1 = 2$$
$$1! = 1 = 1$$
$$0! = 1 = 1$$

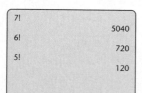

Factorial notation can also be evaluated using the **!** operation from the MATH PRB (probability) menu, as shown at left.

We now see that the following statement is true.

$$_nP_n = n!$$

We will often need to manipulate factorial notation. For example, note that

$$8! = 8 \cdot 7 \cdot 6 \cdot 5 \cdot 4 \cdot 3 \cdot 2 \cdot 1$$
$$= 8 \cdot (7 \cdot 6 \cdot 5 \cdot 4 \cdot 3 \cdot 2 \cdot 1) = 8 \cdot 7!.$$

Generalizing, we get the following.

For any natural number n, $n! = n(n-1)!$.

By using this result repeatedly, we can further manipulate factorial notation.

EXAMPLE 5   Rewrite 7! with a factor of 5!.

Solution   We have

$$7! = 7 \cdot 6! = 7 \cdot 6 \cdot 5!.$$

In general, we have the following.

> For any natural numbers k and n, with $k < n$,
>
> $$n! = \underbrace{n(n-1)(n-2) \cdots [n-(k-1)]}_{k \text{ factors}} \cdot \underbrace{(n-k)!}_{n-k \text{ factors}}$$

❯ Permutations of *n* Objects Taken *k* at a Time

Consider a set of 5 objects

$$\{A, B, C, D, E\}.$$

How many ordered arrangements can be formed using 3 objects from the set without repetition? Examples of such an arrangement are EBA, CAB, and BCD. There are 5 choices for the first object, 4 choices for the second, and 3 choices for the third. By the fundamental counting principle, there are

$$5 \cdot 4 \cdot 3,$$

or   60 *permutations* of a set of 5 objects taken 3 at a time.

Note that

$$5 \cdot 4 \cdot 3 = \frac{5 \cdot 4 \cdot 3 \cdot 2 \cdot 1}{2 \cdot 1}, \quad \text{or} \quad \frac{5!}{2!}.$$

> **PERMUTATION OF *n* OBJECTS TAKEN *k* AT A TIME**
>
> A **permutation** of a set of n objects taken k at a time is an ordered arrangement of k objects taken from the set.

Consider a set of n objects and the selection of an ordered arrangement of k of them. There would be n choices for the first object. Then there would remain $n - 1$ choices for the second, $n - 2$ choices for the third, and so on. We make k choices in all, so there are k factors in the product. By the fundamental counting principle, the total number of permutations is

$$\underbrace{n(n-1)(n-2) \cdots [n-(k-1)]}_{k \text{ factors}}.$$

We can express this in another way by multiplying by 1, as follows:

$$n(n-1)(n-2) \cdots [n-(k-1)] \cdot \frac{(n-k)!}{(n-k)!}$$

$$= \frac{n(n-1)(n-2) \cdots [n-(k-1)](n-k)!}{(n-k)!}$$

$$= \frac{n!}{(n-k)!}.$$

This gives us the following.

**THE NUMBER OF PERMUTATIONS OF *n* OBJECTS
TAKEN *k* AT A TIME**

The number of permutations of a set of *n* objects taken *k* at a time, denoted
$_nP_k$, is given by

$$_nP_k = \underbrace{n(n-1)(n-2) \cdots [n-(k-1)]}_{k \text{ factors}} \qquad (1)$$

$$= \frac{n!}{(n-k)!}. \qquad (2)$$

EXAMPLE 6   Compute $_8P_4$ using both forms of the formula.

Solution   Using form (1), we have

The 8 tells where to start.

$$_8P_4 = \underbrace{8 \cdot 7 \cdot 6 \cdot 5}_{} = 1680.$$

The 4 tells how many factors.

Using form (2), we have

$$_8P_4 = \frac{8!}{(8-4)!}$$

$$= \frac{8!}{4!}$$

$$= \frac{8 \cdot 7 \cdot 6 \cdot 5 \cdot 4!}{4!} = \frac{8 \cdot 7 \cdot 6 \cdot 5 \cdot 4\!\!\!/\,!}{4\!\!\!/\,!}$$

$$= 8 \cdot 7 \cdot 6 \cdot 5 = 1680.$$

```
8 nPr 4
                1680
```

We can also evaluate $_8P_4$ using the $_nP_r$ operation from the MATH PRB menu on a
graphing calculator.

> *Now Try Exercise 3.*

EXAMPLE 7   *Flags of Nations.*   The flags of many nations consist of three
vertical stripes. For example, the flag of Ireland, shown here, has its first stripe green,
second white, and third orange.

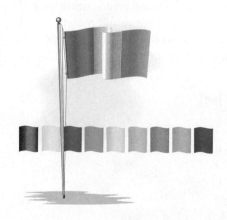

Suppose that the following 9 colors are available:

{black, yellow, red, blue, white, gold, orange, pink, purple}.

How many different flags of 3 colors can be made without repetition of colors in a
flag? This assumes that the order in which the stripes appear is considered.

Solution   We are determining the number of permutations of 9 objects taken 3 at a
time. There is no repetition of colors. Using form (1), we get

$$_9P_3 = 9 \cdot 8 \cdot 7 = 504.$$

> *Now Try Exercise 37(a).*

EXAMPLE 8   *Batting Orders.*   A baseball manager arranges the batting order as follows: The 4 infielders will bat first. Then the 3 outfielders, the catcher, and the pitcher will follow, not necessarily in that order. How many different batting orders are possible?

Solution   The infielders can bat in $_4P_4$ different ways, the rest in $_5P_5$ different ways. Then by the fundamental counting principle, we have

$$_4P_4 \cdot {_5P_5} = 4! \cdot 5!, \quad \text{or} \quad 2880 \text{ possible batting orders.}$$ ▶ *Now Try Exercise 31.*

If we allow repetition, a situation like the following can occur.

EXAMPLE 9   How many 5-letter code symbols can be formed with the letters A, B, C, and D if we allow a letter to occur more than once?

Solution   We can select each of the 5 letters of the code in 4 ways. That is, we can select the first letter in 4 ways, the second in 4 ways, and so on. Thus there are 4^5, or 1024 arrangements.

▶ *Now Try Exercise 37(b).*

The number of distinct arrangements of n objects taken k at a time, allowing repetition, is n^k.

❯ Permutations of Sets with Nondistinguishable Objects

Consider a set of 7 marbles, 4 of which are blue and 3 of which are red. When they are lined up, one red marble will look just like any other red marble. In this sense, we say that the red marbles are nondistinguishable and, similarly, the blue marbles are nondistinguishable.

We know that there are 7! permutations of this set. Many of them will look alike, however. We develop a formula for finding the number of distinguishable permutations.

Consider a set of n objects in which n_1 are of one kind, n_2 are of a second kind, ..., and n_k are of a kth kind. The total number of permutations of the set is $n!$, but this includes many that are nondistinguishable. Let N be the total number of distinguishable permutations. For each of these N permutations, there are $n_1!$ actual, nondistinguishable permutations, obtained by permuting the objects of the first kind. For each of these $N \cdot n_1!$ permutations, there are $n_2!$ nondistinguishable permutations, obtained by permuting the objects of the second kind, and so on. By the fundamental counting principle, the total number of permutations, including those that are nondistinguishable, is

$$N \cdot n_1! \cdot n_2! \cdot \cdots \cdot n_k!.$$

Then we have $N \cdot n_1! \cdot n_2! \cdot \cdots \cdot n_k! = n!$. Solving for N, we obtain

$$N = \frac{n!}{n_1! \cdot n_2! \cdot \cdots \cdot n_k!}.$$

Now, to finish our discussion of the marbles, we have

$$N = \frac{7!}{4! \, 3!}$$

$$= \frac{7 \cdot 6 \cdot 5 \cdot 4!}{4! \cdot 3 \cdot 2 \cdot 1} = \frac{7 \cdot 3 \cdot 2 \cdot 5 \cdot 4!}{4! \cdot 3 \cdot 2 \cdot 1}$$

$$= \frac{7 \cdot 5}{1}, \quad \text{or} \quad 35$$

distinguishable permutations of the marbles.

In general, we have the following.

For a set of n objects in which n_1 are of one kind, n_2 are of another kind, $\ldots$, and n_k are of a kth kind, the number of distinguishable permutations is

$$\frac{n!}{n_1! \cdot n_2! \cdot \cdots \cdot n_k!}.$$

EXAMPLE 10 In how many distinguishable ways can the letters of the word CINCINNATI be arranged?

Solution There are 2 C's, 3 I's, 3 N's, 1 A, and 1 T for a total of 10 letters. Thus,

$$N = \frac{10!}{2! \cdot 3! \cdot 3! \cdot 1! \cdot 1!}, \quad \text{or} \quad 50,400.$$

The letters of the word CINCINNATI can be arranged in 50,400 distinguishable ways.

> **Now Try Exercise 35.**

11.5 Exercise Set

Evaluate.

1. $_6P_6$

2. $_4P_3$

3. $_{10}P_7$

4. $_{10}P_3$

5. $5!$

6. $7!$

7. $0!$

8. $1!$

9. $\dfrac{9!}{5!}$

10. $\dfrac{9!}{4!}$

11. $(8 - 3)!$

12. $(8 - 5)!$

13. $\dfrac{10!}{7!3!}$

14. $\dfrac{7!}{(7 - 2)!}$

15. $_8P_0$

16. $_{13}P_1$

17. $_{52}P_4$

18. $_{52}P_5$

19. $_nP_3$

20. $_nP_2$

21. $_nP_1$

22. $_nP_0$

In each of Exercises 23–41, give your answer using permutation notation, factorial notation, or other operations. Then evaluate.

How many permutations are there of the letters in each of the following words, if all the letters are used without repetition?

23. CREDIT

24. FRUIT

25. EDUCATION

26. TOURISM

27. How many permutations are there of the letters of the word EDUCATION if the letters are taken 4 at a time?

28. How many permutations are there of the letters of the word TOURISM if the letters are taken 5 at a time?

29. How many 5-digit numbers can be formed using the digits 2, 4, 6, 8, and 9 without repetition? with repetition?

30. In how many ways can 7 athletes be arranged in a straight line?

31. *Program Planning.*   A program is planned to have 5 musical numbers and 4 speeches. In how many ways can this be done if a musical number and a speech are to alternate and a musical number is to come first?

32. A professor is going to grade her 24 students on a curve. She will give 3 A's, 5 B's, 9 C's, 4 D's, and 3 F's. In how many ways can she do this?

33. *Phone Numbers.*   How many 7-digit phone numbers can be formed with the digits 0, 1, 2, 3, 4, 5, 6, 7, 8, and 9, assuming that the first number cannot be 0 or 1? Accordingly, how many telephone numbers can there be within a given area code, before the area needs to be split with a new area code?

34. How many distinguishable code symbols can be formed from the letters of the word BUSINESS? BIOLOGY? MATHEMATICS?

35. Suppose that the expression $a^2b^3c^4$ is rewritten without exponents. In how many distinguishable ways can this be done?

36. *Coin Arrangements.*   A penny, a nickel, a dime, and a quarter are arranged in a straight line.

a) Considering just the coins, in how many ways can they be lined up?

b) Considering the coins and heads and tails, in how many ways can they be lined up?

37. How many code symbols can be formed using 5 of the 6 letters A, B, C, D, E, F if the letters:

a) are not repeated?

b) can be repeated?

c) are not repeated but must begin with D?

d) are not repeated but must begin with DE?

38. *License Plates.*   A state forms its license plates by first listing a number that corresponds to the county in which the owner of the car resides. (The names of the counties are alphabetized and the number is its location in that order.) Then the plate lists a letter of the alphabet, and this is followed by a number from 1 to 9999. How many such plates are possible if there are 80 counties?

39. *Zip Codes.*   A U.S. postal zip code is a five-digit number.

a) How many zip codes are possible if any of the digits 0 to 9 can be used?

b) If each post office has its own zip code, how many possible post offices can there be?

40. *Zip-Plus-4 Codes.*   A zip-plus-4 postal code uses a 9-digit number like 75247-5456. How many 9-digit zip-plus-4 postal codes are possible?

41. *Social Security Numbers.*   A social security number is a 9-digit number like 243-47-0825.

a) How many different social security numbers can there be?

b) There are about 310 million people in the United States. Can each person have a unique social security number?

❯ Skill Maintenance

Find the zero(s) of the function.

42. $f(x) = 4x - 9$ **[1.5]**

43. $f(x) = x^2 + x - 6$ **[3.2]**

44. $f(x) = 2x^2 - 3x - 1$ **[3.2]**

45. $f(x) = x^3 - 4x^2 - 7x + 10$ **[4.4]**

❯ Synthesis

Solve for n.

46. $_nP_5 = 7 \cdot {}_nP_4$ **47.** $_nP_4 = 8 \cdot {}_{n-1}P_3$

48. $_nP_5 = 9 \cdot {}_{n-1}P_4$ **49.** $_nP_4 = 8 \cdot {}_nP_3$

50. Show that $n! = n(n-1)(n-2)(n-3)!$.

51. *Single-Elimination Tournaments.*   In a single-elimination sports tournament consisting of *n* teams, a team is eliminated when it loses one game. How many games are required to complete the tournament?

52. *Double-Elimination Tournaments.*   In a double-elimination softball tournament consisting of *n* teams, a team is eliminated when it loses two games. At most, how many games are required to complete the tournament?

11.6   Combinatorics: Combinations

> ❯ Evaluate combination notation and solve related applied problems.

We now consider counting techniques in which order is not considered.

❯ Combinations

We sometimes make a selection from a set *without regard to order.* Such a selection is called a *combination.* If you play cards, for example, you know that in most situations the *order* in which you hold cards is not important. That is,

The hand

is "equivalent" to these hands.

Each hand contains the same combination of three cards.

EXAMPLE 1   Find all the combinations of 3 letters taken from the set of 5 letters $\{A, B, C, D, E\}$.

Solution   The combinations are

$$\{A, B, C\}, \quad \{A, B, D\},$$
$$\{A, B, E\}, \quad \{A, C, D\},$$
$$\{A, C, E\}, \quad \{A, D, E\},$$
$$\{B, C, D\}, \quad \{B, C, E\},$$
$$\{B, D, E\}, \quad \{C, D, E\}.$$

There are 10 combinations of the 5 letters taken 3 at a time.

When we find all the combinations from a set of 5 objects taken 3 at a time, we are finding all the 3-element subsets. When a set is named, the order of the elements is *not* considered. Thus,

$$\{A, C, B\} \quad \text{names the same set as} \quad \{A, B, C\}.$$

COMBINATION; COMBINATION NOTATION

A **combination** containing k objects chosen from a set of n objects, $k \leq n$, is denoted using **combination notation** $_nC_k$.

We want to derive a general formula for $_nC_k$ for any $k \leq n$. First, it is true that $_nC_n = 1$, because a set with n objects has only 1 subset with n objects, the set itself. Second, $_nC_1 = n$, because a set with n objects has n subsets with 1 object each. Finally, $_nC_0 = 1$, because a set with n objects has only one subset with 0 objects, namely, the empty set $\varnothing$. To consider other possibilities, let's return to Example 1 and compare the number of combinations with the number of permutations.

COMBINATIONS　　　　　　**PERMUTATIONS**

$_5C_3$ of these $\left\{\begin{array}{l}\{A, B, C\} \longrightarrow \text{ABC} \quad \text{BCA} \quad \text{CAB} \quad \text{CBA} \quad \text{BAC} \quad \text{ACB} \\ \{A, B, D\} \longrightarrow \text{ABD} \quad \text{BDA} \quad \text{DAB} \quad \text{DBA} \quad \text{BAD} \quad \text{ADB} \\ \{A, B, E\} \longrightarrow \text{ABE} \quad \text{BEA} \quad \text{EAB} \quad \text{EBA} \quad \text{BAE} \quad \text{AEB} \\ \{A, C, D\} \longrightarrow \text{ACD} \quad \text{CDA} \quad \text{DAC} \quad \text{DCA} \quad \text{CAD} \quad \text{ADC} \\ \{A, C, E\} \longrightarrow \text{ACE} \quad \text{CEA} \quad \text{EAC} \quad \text{ECA} \quad \text{CAE} \quad \text{AEC} \\ \{A, D, E\} \longrightarrow \text{ADE} \quad \text{DEA} \quad \text{EAD} \quad \text{EDA} \quad \text{DAE} \quad \text{AED} \\ \{B, C, D\} \longrightarrow \text{BCD} \quad \text{CDB} \quad \text{DBC} \quad \text{DCB} \quad \text{CBD} \quad \text{BDC} \\ \{B, C, E\} \longrightarrow \text{BCE} \quad \text{CEB} \quad \text{EBC} \quad \text{ECB} \quad \text{CBE} \quad \text{BEC} \\ \{B, D, E\} \longrightarrow \text{BDE} \quad \text{DEB} \quad \text{EBD} \quad \text{EDB} \quad \text{DBE} \quad \text{BED} \\ \{C, D, E\} \longrightarrow \text{CDE} \quad \text{DEC} \quad \text{ECD} \quad \text{EDC} \quad \text{DCE} \quad \text{CED}\end{array}\right\}$ $3! \cdot {}_5C_3$ of these

Note that each combination of 3 objects yields 6, or 3!, permutations.

$$3! \cdot {}_5C_3 = 60 = {}_5P_3 = 5 \cdot 4 \cdot 3,$$

so

$$_5C_3 = \frac{_5P_3}{3!} = \frac{5 \cdot 4 \cdot 3}{3 \cdot 2 \cdot 1} = 10.$$

In general, the number of combinations of n objects taken k at a time, $_nC_k$, times the number of permutations of these objects, $k!$, must equal the number of permutations of n objects taken k at a time:

$$k! \cdot {}_nC_k = {}_nP_k$$

$$_nC_k = \frac{_nP_k}{k!} = \frac{1}{k!} \cdot {}_nP_k$$

$$= \frac{1}{k!} \cdot \frac{n!}{(n-k)!} = \frac{n!}{k!(n-k)!}.$$

COMBINATIONS OF *n* OBJECTS TAKEN *k* AT A TIME

The total number of combinations of n objects taken k at a time, denoted $_nC_k$, is given by

$$_nC_k = \frac{n!}{k!\,(n-k)!},\tag{1}$$

or

$$_nC_k = \frac{_nP_k}{k!} = \frac{n(n-1)(n-2)\cdots[n-(k-1)]}{k!}.\tag{2}$$

Another kind of notation for $_nC_k$ is **binomial coefficient notation.** The reaso[n] for such terminology will be seen later.

BINOMIAL COEFFICIENT NOTATION

$$\binom{n}{k} = {}_nC_k$$

You should be able to use either notation and either form of the formula.

EXAMPLE 2   Evaluate $\binom{7}{5}$, using forms (1) and (2).

Solution

a) By form (1),

$$\binom{7}{5} = \frac{7!}{5!(7-5)!} = \frac{7!}{5!\,2!}$$

$$= \frac{7 \cdot 6 \cdot 5!}{5! \cdot 2!} = \frac{7 \cdot 6 \cdot \cancel{5}!}{\cancel{5}! \cdot 2!} = \frac{7 \cdot 6}{2 \cdot 1} = 21.$$

b) By form (2),

The 7 tells where to start.

$$\binom{7}{5} = \frac{7 \cdot 6 \cdot 5 \cdot 4 \cdot 3}{5 \cdot 4 \cdot 3 \cdot 2 \cdot 1} = \frac{7 \cdot 6}{2 \cdot 1} = 21$$

The 5 tells how many factors there are in both the numerator and the denominator and where to start the denominator.

We can also find combinations using the $_nC_r$ operation from the MATH PR[OB] (probability) menu on a graphing calculator.

> **Now Try Exercise 11.**

```
7 nCr 5
                    21
```

Be sure to keep in mind that $\binom{n}{k}$ does not mean $n \div k$, or n/k.

EXAMPLE 3   Evaluate $\binom{n}{0}$ and $\binom{n}{2}$.

Solution   We use form (1) for the first expression and form (2) for the second. Then

$$\binom{n}{0} = \frac{n!}{0!(n-0)!} = \frac{n!}{1 \cdot n!} = 1,$$

using form (1), and

$$\binom{n}{2} = \frac{n(n-1)}{2!} = \frac{n(n-1)}{2}, \quad \text{or} \quad \frac{n^2 - n}{2},$$

using form (2).

Now Try Exercise 19.

Note that

$$\binom{7}{2} = \frac{7 \cdot 6}{2 \cdot 1} = 21.$$

Using the result of Example 2 gives us

$$\binom{7}{5} = \binom{7}{2}.$$

This says that the number of 5-element subsets of a set of 7 objects is the same as the number of 2-element subsets of a set of 7 objects. When 5 elements are chosen from a set, one also chooses *not* to include 2 elements. To see this, consider the set $\{A, B, C, D, E, F, G\}$:

$$\{B, G\}$$

$$\{A, B, C, D, E, F, G\}$$

Each time we form a subset with 5 elements, we leave behind a subset with 2 elements, and vice versa.

$$\{A, C, D, E, F\}$$

In general, we have the following. This result provides an alternative way to compute combinations.

SUBSETS OF SIZE k AND OF SIZE $n - k$

$$\binom{n}{k} = \binom{n}{n-k} \quad \text{and} \quad {}_nC_k = {}_nC_{n-k}$$

The number of subsets of size k of a set with n objects is the same as the number of subsets of size $n - k$. The number of combinations of n objects taken k at a time is the same as the number of combinations of n objects taken $n - k$ at a time.

We now solve problems involving combinations.

EXAMPLE 4 *Indiana Lottery.* Run by the state of Indiana, Hoosier Lotto is a twice-weekly lottery game with jackpots starting at $1 million. For a wager of $1, a player can choose 6 numbers from 1 through 48. If the numbers match those drawn by the state, the player wins the jackpot. (*Source*: www.hoosierlottery.com)

a) How many 6-number combinations are there?

b) Suppose that it takes you 10 min to pick your numbers and buy a game ticket. How many tickets can you buy in 4 days?

c) How many people would you have to hire for 4 days to buy tickets with all the possible combinations and ensure that you win?

Solution

a) No order is implied here. You pick any 6 different numbers from 1 through 48. Thus the number of combinations is

$$_{48}C_6 = \binom{48}{6} = \frac{48!}{6!(48-6)!} = \frac{48!}{6!\,42!}$$

$$= \frac{48 \cdot 47 \cdot 46 \cdot 45 \cdot 44 \cdot 43 \cdot 42!}{6 \cdot 5 \cdot 4 \cdot 3 \cdot 2 \cdot 1 \cdot 42!}$$

$$= \frac{48 \cdot 47 \cdot 46 \cdot 45 \cdot 44 \cdot 43}{6 \cdot 5 \cdot 4 \cdot 3 \cdot 2 \cdot 1} = 12{,}271{,}512.$$

b) First, we find the number of minutes in 4 days:

$$4 \text{ days} = 4 \text{ days} \cdot \frac{24 \text{ hr}}{1 \text{ day}} \cdot \frac{60 \text{ min}}{1 \text{ hr}} = 5760 \text{ min}.$$

Thus you could buy 5760/10, or 576 tickets in 4 days.

c) You would need to hire 12,271,512/576, or about 21,305 people for four days, to buy tickets with all the possible combinations and ensure a win. (This presumes lottery tickets can be bought 24 hours a day.)

> *Now Try Exercise 23.*

EXAMPLE 5 How many committees can be formed from a group of 5 governors and 7 senators if each committee consists of 3 governors and 4 senators?

Solution The 3 governors can be selected in $_5C_3$ ways and the 4 senators can be selected in $_7C_4$ ways. If we use the fundamental counting principle, it follows that the number of possible committees is

$$_5C_3 \cdot {_7C_4} = \frac{5!}{3!\,2!} \cdot \frac{7!}{4!\,3!} = \frac{5 \cdot 4 \cdot 3!}{3! \cdot 2 \cdot 1} \cdot \frac{7 \cdot 6 \cdot 5 \cdot 4!}{4! \cdot 3 \cdot 2 \cdot 1}$$

$$= \frac{5 \cdot 2 \cdot 2 \cdot 3!}{3! \cdot 2 \cdot 1} \cdot \frac{7 \cdot 3 \cdot 2 \cdot 5 \cdot 4!}{4! \cdot 3 \cdot 2 \cdot 1}$$

$$= 10 \cdot 35$$

$$= 350.$$

> *Now Try Exercise 27.*

```
5 nCr 3*7 nCr 4
                        350
```

CONNECTING THE CONCEPTS

Permutations and Combinations

PERMUTATIONS

Permutations involve order and arrangements of objects.

Given 5 books, we can arrange 3 of them on a shelf in $_5P_3$, or 60 ways.

Placing the books in different orders produces different arrangements.

COMBINATIONS

Combinations do not involve order or arrangements of objects.

Given 5 books, we can select 3 of them in $_5C_3$, or 10 ways.

The order in which the books are chosen does not matter.

11.6   Exercise Set

Evaluate.

1. $_{13}C_2$

2. $_9C_6$

3. $\dbinom{13}{11}$

4. $\dbinom{9}{3}$

5. $\dbinom{7}{1}$

6. $\dbinom{8}{8}$

7. $\dfrac{_5P_3}{3!}$

8. $\dfrac{_{10}P_5}{5!}$

9. $\dbinom{6}{0}$

10. $\dbinom{6}{1}$

11. $\dbinom{6}{2}$

12. $\dbinom{6}{3}$

13. $\dbinom{7}{0} + \dbinom{7}{1} + \dbinom{7}{2} + \dbinom{7}{3} + \dbinom{7}{4} + \dbinom{7}{5}$
$+ \dbinom{7}{6} + \dbinom{7}{7}$

14. $\dbinom{6}{0} + \dbinom{6}{1} + \dbinom{6}{2} + \dbinom{6}{3} + \dbinom{6}{4}$
$+ \dbinom{6}{5} + \dbinom{6}{6}$

15. $_{52}C_4$

16. $_{52}C_5$

17. $\dbinom{27}{11}$

18. $\dbinom{37}{8}$

19. $\dbinom{n}{1}$

20. $\dbinom{n}{3}$

21. $\dbinom{m}{m}$

22. $\dbinom{t}{4}$

In each of the following exercises, give an expression for the answer using permutation notation, combination notation, factorial notation, or other operations. Then evaluate.

23. *Key Club Officers.*   There are 36 students in a high school Key Club, a service organization for teens. How many sets of 4 officers can be selected?

24. *League Games.*   How many games can be played in a 9-team sports league if each team plays all other teams once? twice?

25. *Test Options.*   On a test, a student is to select 10 out of 13 questions. In how many ways can this be done?

26. *Senate Committees.*   Suppose the Senate of the United States consists of 58 Republicans and 42 Democrats. How many committees can be formed consisting of 6 Republicans and 4 Democrats?

27. *Test Options.*   Of the first 10 questions on a test, a student must answer 7. Of the second 5 questions, the student must answer 3. In how many ways can this be done?

28. *Lines and Triangles from Points.*   How many lines are determined by 8 points, no 3 of which are collinear? How many triangles are determined by the same points?

29. *Poker Hands.*   How many 5-card poker hands are possible with a 52-card deck?

30. *Bridge Hands.*   How many 13-card bridge hands are possible with a 52-card deck?

31. *Baskin-Robbins Ice Cream.* Burt Baskin and Irv Robbins began making ice cream in 1945. Initially they developed 31 flavors—one for each day of the month. (*Source*: Baskin-Robbins)

a) How many 2-dip cones are possible using the 31 original flavors if order of flavors is to be considered and no flavor is repeated?

b) How many 2-dip cones are possible if order is to be considered and a flavor can be repeated?

c) How many 2-dip cones are possible if order is not considered and no flavor is repeated?

32. *Powerball®.* Powerball® is a biweekly lottery game in which 5 white balls are drawn from a drum of 59 balls numbered 1–59 and 1 red ball is drawn from a drum of 35 balls numbered 1–35. To win the jackpot, a player must select numbers to match in any order the 5 white balls and the 1 red ball. (*Source*: www.powerball.com) How many 6-number combinations are there?

> ## Skill Maintenance

Solve.

33. $3x - 7 = 5x + 10$ **[1.5]**

34. $2x^2 - x = 3$ **[3.2]**

35. $x^2 + 5x + 1 = 0$ **[3.2]**

36. $x^3 + 3x^2 - 10x = 24$ **[4.4]**

> ## Synthesis

37. *Flush.* A flush in poker consists of a 5-card hand with all cards of the same suit. How many 5-card hands (flushes) are there that consist of all diamonds?

38. *Full House.* A full house in poker consists of three of a kind and a pair (two of a kind). How many full houses are there that consist of 3 aces and 2 queens? (See Section 11.8 for a description of a 52-card deck.)

39. *League Games.* How many games are played in a league with n teams if each team plays each other team once? twice?

40. There are n points on a circle. How many quadrilaterals can be inscribed with these points as vertices?

Solve for n.

41. $\begin{pmatrix} n \\ n - 2 \end{pmatrix} = 6$

42. $\begin{pmatrix} n + 1 \\ 3 \end{pmatrix} = 2 \cdot \begin{pmatrix} n \\ 2 \end{pmatrix}$

43. $\begin{pmatrix} n + 2 \\ 4 \end{pmatrix} = 6 \cdot \begin{pmatrix} n \\ 2 \end{pmatrix}$

44. $\begin{pmatrix} n \\ 3 \end{pmatrix} = 2 \cdot \begin{pmatrix} n - 1 \\ 2 \end{pmatrix}$

45. Prove that

$$\begin{pmatrix} n \\ k - 1 \end{pmatrix} + \begin{pmatrix} n \\ k \end{pmatrix} = \begin{pmatrix} n + 1 \\ k \end{pmatrix}$$

for any natural numbers n and k, $k \leq n$.

11.7 The Binomial Theorem

> Expand a power of a binomial using Pascal's triangle or factorial notation.

> Find a specific term of a binomial expansion.

> Find the total number of subsets of a set of n objects.

In this section, we consider ways of expanding a binomial $(a + b)^n$.

❯ Binomial Expansion Using Pascal's Triangle

Consider the following expanded powers of $(a + b)^n$, where $a + b$ is any binomial and n is a whole number. Look for patterns.

$$(a + b)^0 = \quad\quad\quad\quad 1$$
$$(a + b)^1 = \quad\quad\quad\quad a + b$$
$$(a + b)^2 = \quad\quad\quad a^2 + 2ab + b^2$$
$$(a + b)^3 = \quad\quad a^3 + 3a^2b + 3ab^2 + b^3$$
$$(a + b)^4 = \quad a^4 + 4a^3b + 6a^2b^2 + 4ab^3 + b^4$$
$$(a + b)^5 = a^5 + 5a^4b + 10a^3b^2 + 10a^2b^3 + 5ab^4 + b^5$$

Each expansion is a polynomial. There are some patterns to be noted.

1. There is one more term than the power of the exponent, n. That is, there are $n + 1$ terms in the expansion of $(a + b)^n$.

2. In each term, the sum of the exponents is n, the power to which the binomial is raised.

3. The exponents of a start with n, the power of the binomial, and decrease to 0. The last term has no factor of a. The first term has no factor of b, so powers of b start with 0 and increase to n.

4. The coefficients start at 1 and increase through certain values about "halfway" and then decrease through these same values back to 1.

Let's explore the coefficients further. Suppose that we want to find an expansion of $(a + b)^6$. The patterns we just noted indicate that there are 7 terms in the expansion:

$$a^6 + c_1a^5b + c_2a^4b^2 + c_3a^3b^3 + c_4a^2b^4 + c_5ab^5 + b^6.$$

How can we determine the value of each coefficient, c_i? We can do so in two ways. The first method involves writing the coefficients in a triangular array, as follows. This is known as **Pascal's triangle**:

$$(a + b)^0: \quad\quad\quad\quad\quad\quad 1$$
$$(a + b)^1: \quad\quad\quad\quad\quad 1 \quad\quad 1$$
$$(a + b)^2: \quad\quad\quad\quad 1 \quad\quad 2 \quad\quad 1$$
$$(a + b)^3: \quad\quad\quad 1 \quad\quad 3 \quad\quad 3 \quad\quad 1$$
$$(a + b)^4: \quad\quad 1 \quad\quad 4 \quad\quad 6 \quad\quad 4 \quad\quad 1$$
$$(a + b)^5: \quad 1 \quad\quad 5 \quad\quad 10 \quad\quad 10 \quad\quad 5 \quad\quad 1$$

There are many patterns in the triangle. Find as many as you can.

Perhaps you discovered a way to write the next row of numbers, given the numbers in the row above it. There are always 1's on the outside. Each remaining number is the sum of the two numbers above it. Let's try to find an expansion for $(a + b)^6$ by adding another row using the patterns we have discovered:

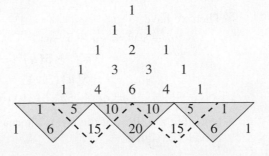

We see that in the last row

the 1st and last numbers are **1**;

the 2nd number is $1 + 5$, or **6**;

the 3rd number is $5 + 10$, or **15**;

the 4th number is $10 + 10$, or **20**;

the 5th number is $10 + 5$, or **15**; and

the 6th number is $5 + 1$, or **6**.

Thus the expansion for $(a + b)^6$ is

$$(a + b)^6 = 1a^6 + 6a^5b + 15a^4b^2 + 20a^3b^3 + 15a^2b^4 + 6ab^5 + 1b^6.$$

To find an expansion for $(a + b)^8$, we complete two more rows of Pascal's triangl

$$
\begin{array}{ccccccccccccccccc}
 & & & & & & & & 1 & & & & & & & & \\
 & & & & & & & 1 & & 1 & & & & & & & \\
 & & & & & & 1 & & 2 & & 1 & & & & & & \\
 & & & & & 1 & & 3 & & 3 & & 1 & & & & & \\
 & & & & 1 & & 4 & & 6 & & 4 & & 1 & & & & \\
 & & & 1 & & 5 & & 10 & & 10 & & 5 & & 1 & & & \\
 & & 1 & & 6 & & 15 & & 20 & & 15 & & 6 & & 1 & & \\
 & 1 & & 7 & & 21 & & 35 & & 35 & & 21 & & 7 & & 1 & \\
1 & & 8 & & 28 & & 56 & & 70 & & 56 & & 28 & & 8 & & 1 \\
\end{array}
$$

Thus the expansion of $(a + b)^8$ is

$$
\begin{aligned}
(a + b)^8 = {}& a^8 + 8a^7b + 28a^6b^2 + 56a^5b^3 + 70a^4b^4 + 56a^3b^5 \\
& + 28a^2b^6 + 8ab^7 + b^8.
\end{aligned}
$$

We can generalize our results as follows.

THE BINOMIAL THEOREM USING PASCAL'S TRIANGLE

For any binomial $a + b$ and any natural number n,

$$
\begin{aligned}
(a + b)^n = {}& c_0a^nb^0 + c_1a^{n-1}b^1 + c_2a^{n-2}b^2 + \cdots \\
& + c_{n-1}a^1b^{n-1} + c_na^0b^n,
\end{aligned}
$$

where the numbers $c_0, c_1, c_2, \ldots, c_{n-1}, c_n$ are from the $(n + 1)$st row of Pascal's triangle.

EXAMPLE 1 Expand: $(u - v)^5$.

Solution We have $(a + b)^n$, where $a = u$, $b = -v$, and $n = 5$. We use the 6t
row of Pascal's triangle:

$$\mathbf{1} \quad 5 \quad 10 \quad 10 \quad 5 \quad 1$$

Then we have

$$
\begin{aligned}
(u - v)^5 &= [u + (-v)]^5 \\
&= 1(u)^5 + 5(u)^4(-v)^1 + 10(u)^3(-v)^2 + 10(u)^2(-v)^3 \\
&\quad + 5(u)(-v)^4 + 1(-v)^5 \\
&= u^5 - 5u^4v + 10u^3v^2 - 10u^2v^3 + 5uv^4 - v^5.
\end{aligned}
$$

Note that the signs of the terms alternate between $+$ and $-$. When the power of $-$
is odd, the sign is $-$.

Now Try Exercise 5.

EXAMPLE 2 Expand: $\left(2t + \dfrac{3}{t}\right)^4$.

Solution We have $(a + b)^n$, where $a = 2t$, $b = 3/t$, and $n = 4$. We use the 5th row of Pascal's triangle:

$$1 \quad 4 \quad 6 \quad 4 \quad 1$$

Then we have

$$\left(2t + \frac{3}{t}\right)^4 = 1(2t)^4 + 4(2t)^3\left(\frac{3}{t}\right)^1 + 6(2t)^2\left(\frac{3}{t}\right)^2 + 4(2t)^1\left(\frac{3}{t}\right)^3 + 1\left(\frac{3}{t}\right)^4$$

$$= 1(16t^4) + 4(8t^3)\left(\frac{3}{t}\right) + 6(4t^2)\left(\frac{9}{t^2}\right) + 4(2t)\left(\frac{27}{t^3}\right) + 1\left(\frac{81}{t^4}\right)$$

$$= 16t^4 + 96t^2 + 216 + 216t^{-2} + 81t^{-4}.$$

> **Now Try Exercise 9.**

❯ Binomial Expansion Using Combination Notation

Suppose that we want to find the expansion of $(a + b)^{11}$. The disadvantage in using Pascal's triangle is that we must compute all the preceding rows of the triangle to obtain the row needed for the expansion. The following method avoids this. It also enables us to find a specific term— say, the 8th term—without computing all the other terms of the expansion. This method is useful in such courses as finite mathematics, calculus, and statistics, and it uses the *binomial coefficient notation* $\dbinom{n}{k}$ developed in Section 11.6.

We can restate the binomial theorem as follows.

THE BINOMIAL THEOREM USING COMBINATION NOTATION

For any binomial $a + b$ and any natural number n,

$$(a + b)^n = \binom{n}{0}a^n b^0 + \binom{n}{1}a^{n-1}b^1 + \binom{n}{2}a^{n-2}b^2 + \cdots$$

$$+ \binom{n}{n-1}a^1 b^{n-1} + \binom{n}{n}a^0 b^n$$

$$= \sum_{k=0}^{n} \binom{n}{k}a^{n-k}b^k.$$

The binomial theorem can be proved by mathematical induction. (See Exercise 57.) This form shows why $\dbinom{n}{k}$ is called a *binomial coefficient*.

EXAMPLE 3 Expand: $(x^2 - 2y)^5$.

Solution We have $(a + b)^n$, where $a = x^2$, $b = -2y$, and $n = 5$. Then using the binomial theorem, we have

$$(x^2 - 2y)^5 = \binom{5}{0}(x^2)^5 + \binom{5}{1}(x^2)^4(-2y) + \binom{5}{2}(x^2)^3(-2y)^2$$
$$+ \binom{5}{3}(x^2)^2(-2y)^3 + \binom{5}{4}x^2(-2y)^4 + \binom{5}{5}(-2y)^5$$
$$= \frac{5!}{0!\,5!}x^{10} + \frac{5!}{1!\,4!}x^8(-2y) + \frac{5!}{2!\,3!}x^6(4y^2) + \frac{5!}{3!\,2!}x^4(-8y^3)$$
$$+ \frac{5!}{4!\,1!}x^2(16y^4) + \frac{5!}{5!\,0!}(-32y^5)$$
$$= 1 \cdot x^{10} + 5x^8(-2y) + 10x^6(4y^2) + 10x^4(-8y^3)$$
$$+ 5x^2(16y^4) + 1 \cdot (-32y^5)$$
$$= x^{10} - 10x^8y + 40x^6y^2 - 80x^4y^3 + 80x^2y^4 - 32y^5.$$

> *Now Try Exercise 11.*

EXAMPLE 4 Expand: $\left(\dfrac{2}{x} + 3\sqrt{x}\right)^4$.

Solution We have $(a + b)^n$, where $a = 2/x$, $b = 3\sqrt{x}$, and $n = 4$. Then using the binomial theorem, we have

$$\left(\frac{2}{x} + 3\sqrt{x}\right)^4 = \binom{4}{0}\left(\frac{2}{x}\right)^4 + \binom{4}{1}\left(\frac{2}{x}\right)^3(3\sqrt{x}) + \binom{4}{2}\left(\frac{2}{x}\right)^2(3\sqrt{x})^2$$
$$+ \binom{4}{3}\left(\frac{2}{x}\right)(3\sqrt{x})^3 + \binom{4}{4}(3\sqrt{x})^4$$
$$= \frac{4!}{0!\,4!}\left(\frac{16}{x^4}\right) + \frac{4!}{1!\,3!}\left(\frac{8}{x^3}\right)(3x^{1/2})$$
$$+ \frac{4!}{2!\,2!}\left(\frac{4}{x^2}\right)(9x) + \frac{4!}{3!\,1!}\left(\frac{2}{x}\right)(27x^{3/2}) + \frac{4!}{4!\,0!}(81x^2)$$
$$= \frac{16}{x^4} + \frac{96}{x^{5/2}} + \frac{216}{x} + 216x^{1/2} + 81x^2.$$

> *Now Try Exercise 13.*

❯ Finding a Specific Term

Suppose that we want to determine only a particular term of an expansion. The method we have developed will allow us to find such a term without computing all the rows of Pascal's triangle or all the preceding coefficients.

Note that in the binomial theorem, $\binom{n}{0}a^n b^0$ gives us the 1st term, $\binom{n}{1}a^{n-1}b$ gives us the 2nd term, $\binom{n}{2}a^{n-2}b^2$ gives us the 3rd term, and so on. This can be generalized as follows.

FINDING THE $(k + 1)$st TERM

The $(k + 1)$st term of $(a + b)^n$ is $\binom{n}{k} a^{n-k} b^k$.

EXAMPLE 5   Find the 5th term in the expansion of $(2x - 5y)^6$.

Solution   First, we note that $5 = 4 + 1$. Thus, $k = 4$, $a = 2x$, $b = -5y$, and $n = 6$. Then the 5th term of the expansion is

$$\binom{6}{4} (2x)^{6-4} (-5y)^4, \quad \text{or} \quad \frac{6!}{4! \, 2!} (2x)^2 (-5y)^4, \quad \text{or} \quad 37{,}500 x^2 y^4.$$

> *Now Try Exercise 21.*

EXAMPLE 6   Find the 8th term in the expansion of $(3x - 2)^{10}$.

Solution   First, we note that $8 = 7 + 1$. Thus, $k = 7$, $a = 3x$, $b = -2$, and $n = 10$. Then the 8th term of the expansion is

$$\binom{10}{7} (3x)^{10-7} (-2)^7, \quad \text{or} \quad \frac{10!}{7! \, 3!} (3x)^3 (-2)^7, \quad \text{or} \quad -414{,}720 x^3.$$

> *Now Try Exercise 27.*

❯ Total Number of Subsets

Suppose that a set has n objects. The number of subsets containing k elements is $\binom{n}{k}$ by a result of Section 11.6. The total number of subsets of a set is the number of subsets with 0 elements, plus the number of subsets with 1 element, plus the number of subsets with 2 elements, and so on. The total number of subsets of a set with n elements is

$$\binom{n}{0} + \binom{n}{1} + \binom{n}{2} + \cdots + \binom{n}{n}.$$

Now consider the expansion of $(1 + 1)^n$:

$$(1 + 1)^n = \binom{n}{0} \cdot 1^n + \binom{n}{1} \cdot 1^{n-1} \cdot 1^1 + \binom{n}{2} \cdot 1^{n-2} \cdot 1^2$$

$$+ \cdots + \binom{n}{n} \cdot 1^n$$

$$= \binom{n}{0} + \binom{n}{1} + \binom{n}{2} + \cdots + \binom{n}{n}.$$

Thus the total number of subsets is $(1 + 1)^n$, or 2^n. We have proved the following.

TOTAL NUMBER OF SUBSETS

The total number of subsets of a set with n elements is 2^n.

EXAMPLE 7 The set $\{A, B, C, D, E\}$ has how many subsets?

Solution The set has 5 elements, so the number of subsets is 2^5, or 32.

Now Try Exercise 31.

EXAMPLE 8 Wendy's, a national restaurant chain, offers the following topping for its hamburgers:

$$\{\text{catsup, mustard, mayonnaise, tomato, lettuce, onions, pickle}\}.$$

How many different kinds of hamburgers can Wendy's serve, excluding size of han burger or number of patties?

Solution The toppings on each hamburger are the elements of a subset of the s of all possible toppings, the empty set being a plain hamburger. The total number possible hamburgers is

$$\binom{7}{0} + \binom{7}{1} + \binom{7}{2} + \cdots + \binom{7}{7} = 2^7 = 128.$$

Thus Wendy's serves hamburgers in 128 different ways.

Now Try Exercise 33.

11.7 Exercise Set

Expand.

1. $(x + 5)^4$

2. $(x - 1)^4$

3. $(x - 3)^5$

4. $(x + 2)^9$

5. $(x - y)^5$

6. $(x + y)^8$

7. $(5x + 4y)^6$

8. $(2x - 3y)^5$

9. $\left(2t + \dfrac{1}{t}\right)^7$

10. $\left(3y - \dfrac{1}{y}\right)^4$

11. $(x^2 - 1)^5$

12. $(1 + 2q^3)^8$

13. $\left(\sqrt{5} + t\right)^6$

14. $(x - \sqrt{2})^6$

15. $\left(a - \dfrac{2}{a}\right)^9$

16. $(1 + 3)^n$

17. $\left(\sqrt{2} + 1\right)^6 - \left(\sqrt{2} - 1\right)^6$

18. $(1 - \sqrt{2})^4 + (1 + \sqrt{2})^4$

19. $(x^{-2} + x^2)^4$

20. $\left(\dfrac{1}{\sqrt{x}} - \sqrt{x}\right)^6$

Find the indicated term of the binomial expansion.

21. 3rd; $(a + b)^7$

22. 6th; $(x + y)^8$

23. 6th; $(x - y)^{10}$

24. 5th; $(p - 2q)^9$

25. 12th; $(a - 2)^{14}$

26. 11th; $(x - 3)^{12}$

27. 5th; $(2x^3 - \sqrt{y})^8$

28. 4th; $\left(\dfrac{1}{b^2} + \dfrac{b}{3}\right)^7$

29. Middle; $(2u - 3v^2)^{10}$

30. Middle two; $\left(\sqrt{x} + \sqrt{3}\right)^5$

Determine the number of subsets of each of the following.

31. A set of 7 elements

32. A set of 6 members

33. The set of letters of the Greek alphabet, which contains 24 letters

34. The set of letters of the English alphabet, which contains 26 letters

35. What is the degree of $(x^5 + 3)^4$?

36. What is the degree of $(2 - 5x^3)^7$?

Expand each of the following, where $i^2 = -1$.

37. $(3 + i)^5$

38. $(1 + i)^6$

39. $\left(\sqrt{2} - i\right)^4$

40. $\left(\dfrac{\sqrt{3}}{2} - \dfrac{1}{2}i\right)^{11}$

41. Find a formula for $(a - b)^n$. Use sigma notation.

42. Expand and simplify:

$$\dfrac{(x + h)^{13} - x^{13}}{h}.$$

43. Expand and simplify:

$$\frac{(x + h)^n - x^n}{h}.$$

Use sigma notation.

> Skill Maintenance

Given that $f(x) = x^2 + 1$ *and* $g(x) = 2x - 3$, *find each of the following.*

44. $(f + g)(x)$ **[2.2]**   **45.** $(fg)(x)$ **[2.2]**

46. $(f \circ g)(x)$ **[2.3]**   **47.** $(g \circ f)(x)$ **[2.3]**

> Synthesis

Solve for x.

48. $\displaystyle\sum_{k=0}^{8} \binom{8}{k} x^{8-k} 3^k = 0$

49. $\displaystyle\sum_{k=0}^{4} \binom{4}{k} (-1)^k x^{4-k} 6^k = 81$

50. Find the ratio of the 4th term of

$$\left(p^2 - \frac{1}{2} \sqrt[3]{q} \right)^5$$

to the 3rd term.

51. Find the term of

$$\left(\sqrt[3]{x} - \frac{1}{\sqrt{x}} \right)^7$$

containing $1/x^{1/6}$.

52. *Money Combinations.* A money clip contains one each of the following bills: \$1, \$2, \$5, \$10, \$20, \$50, and \$100. How many different sums of money can be formed using the bills?

Find the sum.

53. $_{100}C_0 + {}_{100}C_1 + \cdots + {}_{100}C_{100}$

54. $_nC_0 + {}_nC_1 + \cdots + {}_nC_n$

Simplify.

55. $\displaystyle\sum_{k=0}^{23} \binom{23}{k} (\log_a x)^{23-k} (\log_a t)^k$

56. $\displaystyle\sum_{k=0}^{15} \binom{15}{k} i^{30-2k}$

57. Use mathematical induction and the property

$$\binom{n}{r-1} + \binom{n}{r} = \binom{n+1}{r}$$

to prove the binomial theorem.

11.8 > Probability

> Compute the probability of a simple event.

When a coin is tossed, we can reason that the chance, or likelihood, that it will fall heads is 1 out of 2, or the **probability** that it will fall heads is $\frac{1}{2}$. Of course, this does not mean that if a coin is tossed 10 times it will necessarily fall heads 5 times. If the coin is a "fair coin" and it is tossed a great many times, however, it will fall heads very nearly half of the time. Here we give an introduction to two kinds of probability, **experimental** and **theoretical**.

> Experimental Probability and Theoretical Probability

If we toss a coin a great number of times—say, 1000—and count the number of times it falls heads, we can determine the probability that it will fall heads. If it falls heads 503 times, we would calculate the probability of its falling heads to be

$$\frac{503}{1000}, \quad \text{or} \quad 0.503.$$

This is an **experimental** determination of probability. Such a determination of probability is discovered by the observation and study of data and is quite common and very useful. Here, for example, are some probabilities that have been determined *experimentally*:

1. 60% of all freshmen entering four-year colleges graduate in 6 years (*Source*: www.satprepct.com, College Planning Partnership's Blog, February 24, 2011; Sam Rosensohn).

2. The probability that a woman will be diagnosed with breast cancer in her lifetime is $\frac{1}{8}$ (*Source*: National Cancer Institute).

3. Anyone who reaches the age of 65 has a 0.4 probability of entering a nursing home during the remaining years of life (*Source*: "Facing the Future," Russ Banham, *Wall Street Journal*).

If we consider a coin and reason that it is just as likely to fall heads as tails, we would calculate the probability that it will fall heads to be $\frac{1}{2}$. This is a **theoretical** determination of probability. Here are some other probabilities that have been determined *theoretically*, using mathematics:

1. If there are 30 people in a room, the probability that two of them have the same birthday (excluding year) is 0.706.

2. While on a trip, you meet someone and, after a period of conversation, discover that you have a common acquaintance. The typical reaction, "It's a small world!" is actually not appropriate, because the probability of such an occurrence is quite high—just over 22%.

In summary, experimental probabilities are determined by making observations and gathering data. Theoretical probabilities are determined by reasoning mathematically. Examples of experimental probability and theoretical probability like those above, especially those we do not expect, lead us to see the value of a study of probability. You might ask, "What is the *true* probability?" In fact, there is none. Experimentally, we can determine probabilities within certain limits. These may or may not agree with the probabilities that we obtain theoretically. There are situations in which it is much easier to determine one of these types of probabilities than the other. For example, it would be quite difficult to arrive at the probability of catching a cold using theoretical probability.

❯ Computing Experimental Probabilities

We first consider experimental determination of probability. The basic principle we use in computing such probabilities is as follows.

PRINCIPLE *P* (EXPERIMENTAL)

Given an experiment in which n observations are made, if a situation, or event, E occurs m times out of n observations, then we say that the **experimental probability** of the event, $P(E)$, is given by

$$P(E) = \frac{m}{n}.$$

EXAMPLE 1   *Television Ratings.*   There are an estimated 114,200,000 households in the United States that have at least one television. Each week, viewing information is collected and reported. One week, 28,510,000 households tuned in to the 2013 Grammy Awards ceremony on CBS, and 14,204,000 households tuned in to the action series "NCIS" on CBS (*Source*: Nielsen Media Research). What is the probability that a television household tuned in to the Grammy Awards ceremony during the given week? to "NCIS"?

Solution   The probability that a television household was tuned in to the Grammy Awards ceremony is P, where

$$P = \frac{28,510,000}{114,200,000} \approx 0.2496 \approx 24.96\%.$$

The probability that a television household was tuned in to "NCIS" is P, where

$$P = \frac{14,204,000}{114,200,000} \approx 0.1244 \approx 12.44\%.$$

> **Now Try Exercise 1.**

EXAMPLE 2   *Sociological Survey.*   The authors of this text conducted a survey to determine the number of people who are left-handed, right-handed, or both. The results are shown in the graph at left.

a) Determine the probability that a person is right-handed.

b) Determine the probability that a person is left-handed.

c) Determine the probability that a person is ambidextrous (uses both hands with equal ability).

d) There are 130 students signed up for tennis lessons in a summer program offered by a community school corporation. On the basis of the data in this experiment, how many of the students would you expect to be left-handed?

Solution

a) The number of people who are right-handed is 82, the number who are left-handed is 17, and the number who are ambidextrous is 1. The total number of observations is $82 + 17 + 1$, or 100. Thus the probability that a person is right-handed is P, where

$$P = \frac{82}{100}, \quad \text{or} \quad 0.82, \quad \text{or} \quad 82\%.$$

Tennis Player Handedness

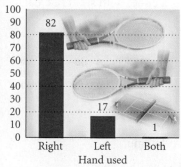

b) The probability that a person is left-handed is P, where

$$P = \frac{17}{100}, \quad \text{or} \quad 0.17, \quad \text{or} \quad 17\%.$$

c) The probability that a person is ambidextrous is P, where

$$P = \frac{1}{100}, \quad \text{or} \quad 0.01, \quad \text{or} \quad 1\%.$$

d) There are 130 students, and from part (b) we can expect 17% to be left-handed. Since

$$17\% \text{ of } 130 = 0.17 \cdot 130 = 22.1,$$

we can expect that about 22 of the students will be left-handed.

> *Now Try Exercise 3.*

❯ Theoretical Probability

Suppose that we perform an experiment such as flipping a coin, throwing a dart, drawing a card from a deck, or checking an item off an assembly line for quality. Each possible result of such an experiment is called an **outcome**. The set of all possible outcomes is called the **sample space**. An **event** is a set of outcomes, that is, a subset of the sample space.

EXAMPLE 3 *Dart Throwing.* Consider this dartboard. Assume that the experiment is "throwing a dart" and that the dart hits the board. Find each of the following.

a) The outcomes **b)** The sample space

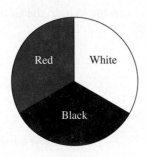

Solution

a) The outcomes are *hitting black* (B), *hitting red* (R), and *hitting white* (W).

b) The sample space is {*hitting black, hitting red, hitting white*}, which can be simply stated as {B, R, W}.

EXAMPLE 4 *Die Rolling.* A die (pl., dice) is a cube, with six faces, each containing a number of dots from 1 to 6.

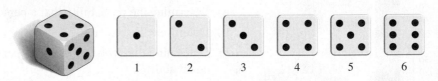

Suppose that a die is rolled. Find each of the following.

a) The outcomes **b)** The sample space

Solution

a) The outcomes are 1, 2, 3, 4, 5, 6.

b) The sample space is {1, 2, 3, 4, 5, 6}.

We denote the probability that an event E occurs as $P(E)$. For example, "a coin falling heads" may be denoted H. Then $P(H)$ represents the probability of the coin falling heads. When all the outcomes of an experiment have the same probability of occurring, we say that they are *equally likely*. To see the distinction between events that are equally likely and those that are not, consider the dartboards shown below.

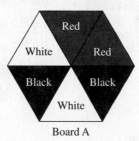

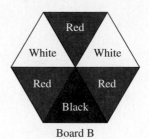

Board A   Board B

For board A, the events *hitting black*, *hitting red*, and *hitting white* are equally likely, because the black, red, and white areas are the same. However, for board B the areas are not the same so these events are not equally likely.

PRINCIPLE *P* (THEORETICAL)

If an event E can occur m ways out of n possible equally likely outcomes of a sample space S, then the **theoretical probability** of the event, $P(E)$, is given by

$$P(E) = \frac{m}{n}.$$

EXAMPLE 5   Suppose that we select, without looking, one marble from a bag containing 3 red marbles and 4 green marbles. What is the probability of selecting a red marble?

Solution   There are 7 equally likely ways of selecting any marble, and since the number of ways of getting a red marble is 3, we have

$$P(\text{selecting a red marble}) = \frac{3}{7}.$$

> *Now Try Exercise 5(a).*

EXAMPLE 6   What is the probability of rolling an even number on a die?

Solution   The event is rolling an *even* number. It can occur 3 ways (rolling 2, 4, or 6). The number of equally likely outcomes is 6. By Principle P, we have

$$P(\text{even}) = \frac{3}{6}, \quad \text{or} \quad \frac{1}{2}.$$

> *Now Try Exercise 7.*

We will use a number of examples related to a standard bridge deck of 52 cards. Such a deck is made up as shown in the following figure.

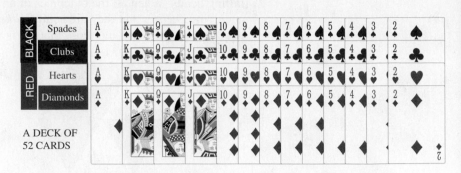

A DECK OF
52 CARDS

EXAMPLE 7 What is the probability of drawing an ace from a well-shuffled deck of cards?

Solution There are 52 outcomes (the number of cards in the deck), they are equally likely (from a well-shuffled deck), and there are 4 ways to obtain an ace, so by Principle P, we have

$$P(\text{drawing an ace}) = \frac{4}{52}, \quad \text{or} \quad \frac{1}{13}.$$

> *Now Try Exercise 9(a).*

The following are some results that follow from Principle P.

PROBABILITY PROPERTIES

a) If an event E cannot occur, then $P(E) = 0$.
b) If an event E is certain to occur, then $P(E) = 1$.
c) The probability that an event E will occur is a number from 0 to 1:
$0 \le P(E) \le 1$.

For example, in coin tossing, the event that a coin will land on its edge has probability 0. The event that a coin falls either heads or tails has probability 1.

In the following examples, we use the combinatorics that we studied in Sections 11.5 and 11.6 to calculate theoretical probabilities.

EXAMPLE 8 Suppose that 2 cards are drawn from a well-shuffled deck of 52 cards. What is the probability that both of them are spades?

Solution The number of ways n of drawing 2 cards from a well-shuffled deck of 52 cards is $_{52}C_2$. Since 13 of the 52 cards are spades, the number of ways m of drawing 2 spades is $_{13}C_2$. Thus,

13 nCr 2/52 nCr 2►Frac
$\frac{1}{17}$

$$P(\text{drawing 2 spades}) = \frac{m}{n} = \frac{_{13}C_2}{_{52}C_2} = \frac{78}{1326} = \frac{1}{17}.$$

> *Now Try Exercise 11.*

EXAMPLE 9 Suppose that 3 people are selected at random from a group that consists of 6 men and 4 women. What is the probability that 1 man and 2 women are selected?

Solution   The number of ways of selecting 3 people from a group of 10 is $_{10}C_3$. One man can be selected in $_6C_1$ ways, and 2 women can be selected in $_4C_2$ ways. By the fundamental counting principle, the number of ways of selecting 1 man and 2 women is $_6C_1 \cdot _4C_2$. Thus the probability that 1 man and 2 women are selected is

$$P = \frac{_6C_1 \cdot _4C_2}{_{10}C_3} = \frac{3}{10}.$$

```
6 nCr 1*4 nCr 2/10 nCr
3▶Frac
                    3
                    ──
                    10
```

> *Now Try Exercise 13.*

EXAMPLE 10   *Rolling Two Dice.*   What is the probability of getting a total of 8 on a roll of a pair of dice?

Solution   On each die, there are 6 possible outcomes. The outcomes are paired so there are $6 \cdot 6$, or 36, possible ways in which the two can fall. (Assuming that the dice are different—say, one red and one blue—can help in visualizing this.)

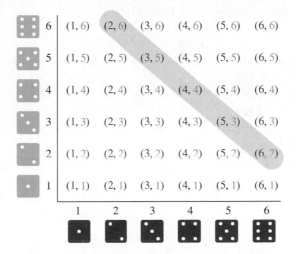

The pairs that total 8 are as shown in the figure above. There are 5 possible ways of getting a total of 8, so the probability is $\frac{5}{36}$.

> *Now Try Exercise 19.*

11.8   Exercise Set

1. *Select a Number.*   In a survey conducted by the authors, 100 people were polled and asked to select a number from 1 to 5. The results are shown in the following table.

Number Chosen	1	2	3	4	5
Number Who Chose That Number	18	24	23	23	12

a) What is the probability that the number chosen is 1? 2? 3? 4? 5?

b) What general conclusion might be made from the results of the experiment?

2. *Mason Dots*®.   Made by the Tootsie Industries of Chicago, Illinois, Mason Dots® is a gumdrop candy. A box was opened by the authors and was found to contain the following number of gumdrops:

Orange   9
Lemon   8
Strawberry   7
Grape   6
Lime   5
Cherry   4

If we take one gumdrop out of the box, what is the probability of getting lemon? lime? orange? grape? strawberry? licorice?

3. *Marketing via E-mail.* In the second quarter of 2013, the probability that a marketing e-mail would be opened was 28.5% (*Source*: Q2 2013 Email Trends and Benchmarks, Epsilon). A business sent a marketing e-mail to 18,200 subscribers. How many of these e-mails can the business expect will be opened?

4. *Linguistics.* An experiment was conducted by the authors to determine the relative occurrence of various letters of the English alphabet. The front page of a newspaper was considered. In all, there were 9136 letters. The number of occurrences of each letter of the alphabet is listed in the following table.

Letter	Number of Occurrences	Probability
A	853	$853/9136 \approx 9.3\%$
B	136	
C	273	
D	286	
E	1229	
F	173	
G	190	
H	399	
I	539	
J	21	
K	57	
L	417	
M	231	
N	597	
O	705	
P	238	
Q	4	
R	609	
S	745	
T	789	
U	240	
V	113	
W	127	
X	20	
Y	124	
Z	21	$21/9136 \approx 0.2\%$

a) Complete the table of probabilities with the percentage, to the nearest tenth of a percent, of the occurrence of each letter.
b) What is the probability of a vowel occurring?
c) What is the probability of a consonant occurring?

5. *Marbles.* Suppose that we select, without looking, one marble from a bag containing 4 red marbles and 10 green marbles. What is the probability of selecting each of the following?
a) A red marble
b) A green marble
c) A purple marble
d) A red marble or a green marble

6. *Selecting Coins.* Suppose that we select, without looking, one coin from a bag containing 5 pennies, 3 dimes, and 7 quarters. What is the probability of selecting each of the following?
a) A dime
b) A quarter
c) A nickel
d) A penny, a dime, or a quarter

7. *Rolling a Die.* What is the probability of rolling a number less than 4 on a die?

8. *Rolling a Die.* What is the probability of rolling either a 1 or a 6 on a die?

9. *Drawing a Card.* Suppose that a card is drawn from a well-shuffled deck of 52 cards. What is the probability of drawing each of the following?
a) A queen
b) An ace or a 10
c) A heart
d) A black 6

10. *Drawing a Card.* Suppose that a card is drawn from a well-shuffled deck of 52 cards. What is the probability of drawing each of the following?
a) A 7
b) A jack or a king
c) A black ace
d) A red card

11. *Drawing Cards.* Suppose that 3 cards are drawn from a well-shuffled deck of 52 cards. What is the probability that they are all aces?

12. *Drawing Cards.* Suppose that 4 cards are drawn from a well-shuffled deck of 52 cards. What is the probability that they are all red?

13. *Production Unit.* The sales force of a business consists of 10 men and 10 women. A production unit of 4 people is set up at random. What is the probability that 2 men and 2 women are chosen?

14. *Coin Drawing.* A sack contains 7 dimes, 5 nickels, and 10 quarters. Eight coins are drawn at random. What is the probability of getting 4 dimes, 3 nickels, and 1 quarter?

Five-Card Poker Hands.   Suppose that 5 cards are drawn from a deck of 52 cards. What is the probability of drawing each of the following?

15. 3 sevens and 2 kings

16. 5 aces

17. 5 spades

18. 4 aces and 1 five

19. *Tossing Three Coins.*   Three coins are flipped. An outcome might be HTH.

   a) Find the sample space.

   What is the probability of getting each of the following?

   b) Exactly one head
   c) At most two tails
   d) At least one head
   e) Exactly two tails

Roulette.   An American roulette wheel contains 38 slots numbered 00, 0, 1, 2, 3, . . . , 35, 36. Eighteen of the slots numbered 1–36 are colored red and 18 are colored black. The 00 and 0 slots are considered to be uncolored. The wheel is spun, and a ball is rolled around the rim until it falls into a slot. What is the probability that the ball falls in each of the following?

20. A red slot

21. A black slot

22. The 00 slot

23. The 0 slot

24. Either the 00 or the 0 slot

25. A red slot or a black slot

26. The number 24

27. An odd-numbered slot

28. *Dartboard.*   The following figure shows a dartboard. A dart is thrown and hits the board. Find the probabilities

   $P(\text{red}), P(\text{green}), P(\text{blue}), P(\text{yellow}).$

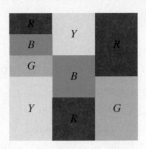

❯ Skill Maintenance

Vocabulary Reinforcement

In each of Exercises 29–36, fill in the blank with the correct term. Some of the given choices will be used more than once. Others will not be used.

range	zero
domain	*y*-intercept
function	one-to-one
inverse function	rational
composite function	permutation
direct variation	combination
inverse variation	arithmetic sequence
factor	geometric sequence
solution	

29. A(n) _____ of a function is an input for which the output is 0. **[1.5]**

30. A function is _____ if different inputs have different outputs. **[5.1]**

31. A(n) _____ is a correspondence between a first set, called the _____ , and a second set, called the _____ , such that each member of the _____ corresponds to exactly one member of the _____ . **[1.2]**

32. The first coordinate of an *x*-intercept of a function is a(n) _____ of the function. **[1.5]**

33. A selection made from a set without regard to order is a(n) _____ . **[11.6]**

34. If we have a function $f(x) = k/x$, where k is a positive constant, we have _____ . **[2.6]**

35. For a polynomial function $f(x)$, if $f(c) = 0$, then $x - c$ is a(n) _____ of the polynomial. **[4.3]**

36. We have $\dfrac{a_{n+1}}{a_n} = r$, for any integer $n \geq 1$, in a(n) _____ . **[11.3]**

❯ Synthesis

Five-Card Poker Hands.   Suppose that 5 cards are drawn from a deck of 52 cards. For the following exercises, give both a reasoned expression and an answer.

37. *Two Pairs.* A hand with *two pairs* is a hand like Q-Q-3-3-A.

   a) How many are there?

   b) What is the probability of getting two pairs?

38. *Full House.* A *full house* consists of 3 of a kind and a pair such as Q-Q-Q-4-4.

   a) How many full houses are there?

   b) What is the probability of getting a full house?

39. *Three of a Kind.* A *three-of-a-kind* is a 5-card hand in which exactly 3 of the cards are of the same denomination and the other 2 are not a pair, such as Q-Q-Q-10-7.

   a) How many three-of-a-kind hands are there?

   b) What is the probability of getting three of a kind?

40. *Four of a Kind.* A *four-of-a-kind* is a 5-card hand in which 4 of the cards are of the same denomination, such as J-J-J-J-6, 7-7-7-7-A, or 2-2-2-2-5.

   a) How many four-of-a-kind hands are there?

   b) What is the probability of getting four of a kind?

Chapter 11 Summary and Review

STUDY GUIDE

KEY TERMS AND CONCEPTS	EXAMPLES
SECTION 11.1: SEQUENCES AND SERIES	
An **infinite sequence** is a function having for its domain the set of positive integers $\{1, 2, 3, 4, 5, \ldots\}$. A **finite sequence** is a function having for its domain a set of positive integers $\{1, 2, 3, 4, 5, \ldots, n\}$ for some positive integer n.	The first four terms of the sequence whose general term is given by $a_n = 3n + 2$ are $$a_1 = 3 \cdot 1 + 2 = 5,$$ $$a_2 = 3 \cdot 2 + 2 = 8,$$ $$a_3 = 3 \cdot 3 + 2 = 11, \quad \text{and}$$ $$a_4 = 3 \cdot 4 + 2 = 14.$$
The sum of the terms of an infinite sequence is an **infinite series**. A **partial sum** is the sum of the first n terms. It is also called a **finite series** or the ***n*th partial sum** and is denoted S_n.	For the sequence above, $S_4 = 5 + 8 + 11 + 14$, or 38. We can denote this sum using **sigma notation** as $$\sum_{k=1}^{4} (3k + 2).$$

A sequence can be defined **recursively** by listing the first term, or the first few terms, and then using a **recursion formula** to determine the remaining terms from the given term.

The first four terms of the recursively defined sequence

$$a_1 = 3, \ a_{n+1} = (a_n - 1)^2$$

are

$$a_1 = 3,$$
$$a_2 = (a_1 - 1)^2 = (3 - 1)^2 = 4,$$
$$a_3 = (a_2 - 1)^2 = (4 - 1)^2 = 9, \quad \text{and}$$
$$a_4 = (a_3 - 1)^2 = (9 - 1)^2 = 64.$$

SECTION 11.2: ARITHMETIC SEQUENCES AND SERIES

For an arithmetic sequence:

$a_{n+1} = a_n + d;$ **d is the common difference.**

$a_n = a_1 + (n - 1)d;$ **The nth term**

$S_n = \dfrac{n}{2}(a_1 + a_n).$ **The sum of the first n terms**

For the arithmetic sequence 5, 8, 11, 14, . . . :

$$a_1 = 5;$$
$$d = 3 \quad (8 - 5 = 3, 11 - 8 = 3, \text{ and so on});$$
$$a_6 = 5 + (6 - 1)3 = 5 + 15 = 20;$$
$$S_6 = \frac{6}{2}(5 + 20) = 3(25) = 75.$$

SECTION 11.3: GEOMETRIC SEQUENCES AND SERIES

For a geometric sequence:

$a_{n+1} = a_n r;$ **r is the common ratio.**

$a_n = a_1 r^{n-1};$ **The nth term**

$S_n = \dfrac{a_1(1 - r^n)}{1 - r};$ **The sum of the first n terms**

$S_\infty = \dfrac{a_1}{1 - r}, \quad |r| < 1.$ **The limit, or sum, of an infinite geometric series**

For the geometric sequence $12, -6, 3, -\frac{3}{2}, \ldots$:

$$a_1 = 12;$$
$$r = -\frac{1}{2} \quad \left(\frac{-6}{12} = -\frac{1}{2}, \frac{3}{-6} = -\frac{1}{2}, \text{and so on} \right);$$
$$a_6 = 12\left(-\frac{1}{2} \right)^{6-1} = 12\left(-\frac{1}{2^5} \right) = -\frac{3}{8};$$
$$S_6 = \frac{12\left[1 - \left(-\frac{1}{2} \right)^6 \right]}{1 - \left(-\frac{1}{2} \right)} = \frac{12\left(1 - \frac{1}{64} \right)}{\frac{3}{2}} = \frac{63}{8};$$
$$|r| = \left| -\tfrac{1}{2} \right| = \tfrac{1}{2} < 1, \text{ so we have}$$
$$S_\infty = \frac{12}{1 - \left(-\frac{1}{2} \right)} = \frac{12}{\frac{3}{2}} = 8.$$

SECTION 11.4: MATHEMATICAL INDUCTION

The Principle of Mathematical Induction

We can prove an infinite sequence of statements S_n by showing the following.

(1) *Basis step.* S_1 is true.

(2) *Induction step.* For all natural numbers k,
$$S_k \rightarrow S_{k+1}.$$

See Examples 1–3 on pp. 818–820.

SECTION 11.5: COMBINATORICS: PERMUTATIONS

The Fundamental Counting Principle

Given a combined action, or *event*, in which the first action can be performed in n_1 ways, the second action can be performed in n_2 ways, and so on, the total number of ways in which the combined action can be performed is the product

$$n_1 \cdot n_2 \cdot n_3 \cdots \cdots n_k.$$

The product $n(n-1)(n-2)\cdots 3 \cdot 2 \cdot 1$, for any natural number n, can also be written in **factorial notation** as $n!$. For the number 0, $0! = 1$.

The total number of permutations, or ordered arrangements, of n objects, denoted $_nP_n$, is given by

$$_nP_n = n(n-1)(n-2)\cdots 3 \cdot 2 \cdot 1, \quad \text{or} \quad n!.$$

In how many ways can 7 books be arranged in a straight line?

We have

$$_7P_7 = 7! = 7 \cdot 6 \cdot 5 \cdot 4 \cdot 3 \cdot 2 \cdot 1 = 5040.$$

The Number of Permutations of n Objects Taken k at a Time

$$_nP_k = n(n-1)(n-2)\cdots[n-(k-1)] \quad \text{(1)}$$

$$= \frac{n!}{(n-k)!} \quad \text{(2)}$$

Compute $_7P_4$.

Using form (1), we have

$$_7P_4 = 7 \cdot 6 \cdot 5 \cdot 4 = 840.$$

Using form (2), we have

$$_7P_4 = \frac{7!}{(7-4)!} = \frac{7 \cdot 6 \cdot 5 \cdot 4 \cdot 3!}{3!}$$

$$= \frac{7 \cdot 6 \cdot 5 \cdot 4 \cdot 3!}{3!} = 840.$$

The number of distinct arrangements of n objects taken k at a time, allowing repetition, is n^k.

The number of 4-number code symbols that can be formed with the numbers 5, 6, 7, 8, and 9, if we allow a number to occur more than once, is 5^4, or 625.

For a set of n objects in which n_1 are of one kind, n_2 are of another kind, ..., and n_k are of a kth kind, the number of distinguishable permutations is

$$\frac{n!}{n_1! \cdot n_2! \cdot \cdots \cdot n_k!}.$$

Find the number of distinguishable code symbols that can be formed using the letters in the word MISSISSIPPI.

There are 1 M, 4 I's, 4 S's, and 2 P's, for a total of 11 letters, so we have

$$\frac{11!}{1! \, 4! \, 4! \, 2!}, \quad \text{or} \quad 34{,}650.$$

SECTION 11.6: COMBINATORICS: COMBINATIONS

The Number of Combinations of
n Objects Taken k at a Time

$$_nC_k = \frac{n!}{k!(n-k)!} \qquad \textbf{(1)}$$

$$= \frac{_nP_k}{k!}$$

$$= \frac{n(n-1)(n-2) \cdots [n-(k-1)]}{k!}. \qquad \textbf{(2)}$$

We can also use **binomial coefficient notation**:

$$\binom{n}{k} = {_nC_k}.$$

Compute: $_6C_4$, or $\binom{6}{4}$.

Using form (1), we have

$$\binom{6}{4} = \frac{6!}{4!(6-4)!} = \frac{6!}{4!\,2!}$$

$$= \frac{6 \cdot 5 \cdot 4!}{4!\,2!} = \frac{6 \cdot 5 \cdot \cancel{4!}}{\cancel{4!} \cdot 2 \cdot 1} = 15.$$

Using form (2), we have

$$\binom{6}{4} = \frac{_6P_4}{4!} = \frac{6 \cdot 5 \cdot 4 \cdot 3}{4 \cdot 3 \cdot 2 \cdot 1} = 15.$$

SECTION 11.7: THE BINOMIAL THEOREM

The Binomial Theorem Using
Pascal's Triangle

For any binomial $a + b$ and any natural number n,

$$(a+b)^n = c_0 a^n b^0 + c_1 a^{n-1} b^1 + c_2 a^{n-2} b^2$$
$$+ \cdots + c_{n-1} a^1 b^{n-1} + c_n a^0 b^n,$$

where the numbers $c_0, c_1, c_2, \ldots, c_{n-1}, c_n$ are from the $(n+1)$st row of Pascal's triangle. (See Pascal's triangle on p. 839.)

Expand: $(x-2)^4$.

We have $a = x$, $b = -2$, and $n = 4$. We use the fifth row of Pascal's triangle.

$$(x-2)^4 = 1 \cdot x^4 + 4 \cdot x^3(-2)^1$$
$$+ 6 \cdot x^2(-2)^2 + 4 \cdot x^1(-2)^3 + 1(-2)^4$$
$$= x^4 + 4x^3(-2) + 6x^2 \cdot 4 + 4x(-8) + 16$$
$$= x^4 - 8x^3 + 24x^2 - 32x + 16$$

The Binomial Theorem Using
Combination Notation

For any binomial $a + b$ and any natural number n,

$$(a+b)^n = \binom{n}{0} a^n b^0 + \binom{n}{1} a^{n-1} b^1$$

$$+ \binom{n}{2} a^{n-2} b^2 + \cdots$$

$$+ \binom{n}{n-1} a^1 b^{n-1} + \binom{n}{n} a^0 b^n$$

$$= \sum_{k=0}^{n} \binom{n}{k} a^{n-k} b^k.$$

Expand: $(x^2 + 3)^3$.

We have $a = x^2$, $b = 3$, and $n = 3$.

$$(x^2+3)^3 = \binom{3}{0}(x^2)^3 + \binom{3}{1}(x^2)^2(3)$$

$$+ \binom{3}{2}(x^2)3^2 + \binom{3}{3}3^3$$

$$= \frac{3!}{0!\,3!}x^6 + \frac{3!}{1!\,2!}(x^4)(3) + \frac{3!}{2!\,1!}(x^2)(9)$$

$$+ \frac{3!}{3!\,0!}(27)$$

$$= 1 \cdot x^6 + 3 \cdot 3x^4 + 3 \cdot 9x^2 + 1 \cdot 27$$

$$= x^6 + 9x^4 + 27x^2 + 27$$

The $(k+1)$st term of $(a+b)^n$ is

$$\binom{n}{k} a^{n-k} b^k.$$

The third term of $(x^2 + 3)^3$ is

$$\binom{3}{2}(x^2)^{3-2} \cdot 3^2 = 3 \cdot x^2 \cdot 9 = 27x^2. \qquad (k = 2)$$

The **total number of subsets** of a set with n elements is 2^n.

How many subsets does the set $\{W, X, Y, Z\}$ have?

The set has 4 elements, so we have

$$2^4, \quad \text{or} \quad 16.$$

Principle P (Experimental)

Given an experiment in which n observations are made, if a situation, or event, E occurs m times out of n observations, then we say that the **experimental probability** of the event, $P(E)$, is given by

$$P(E) = \frac{m}{n}.$$

From a batch of 1000 gears, 35 were found to be defective. The probability that a defective gear is produced is

$$\frac{35}{1000} = 0.035, \quad \text{or} \quad 3.5\%.$$

Principle P (Theoretical)

If an event E can occur m ways out of n possible equally likely outcomes of a sample space S, then the **theoretical probability** of the event, $P(E)$, is given by

$$P(E) = \frac{m}{n}.$$

What is the probability of drawing 2 red marbles and 1 green marble from a bag containing 5 red marbles, 6 green marbles, and 4 white marbles?

Number of ways of drawing 3 marbles from a bag of 15: $_{15}C_3$

Number of ways of drawing 2 red marbles from 5 red marbles: $_5C_2$

Number of ways of drawing 1 green marble from 6 green marbles: $_6C_1$

Probability that 2 red marbles and 1 green marble are drawn:

$$\frac{_5C_2 \cdot {_6C_1}}{_{15}C_3} = \frac{10 \cdot 6}{455} = \frac{12}{91}$$

REVIEW EXERCISES

Determine whether the statement is true or false.

1. A sequence is a function. **[11.1]**

2. An infinite geometric series with $r = -1$ has a limit. **[11.3]**

3. Permutations involve order and arrangements of objects. **[11.5]**

4. The total number of subsets of a set with n elements is n^2. **[11.7]**

5. Find the first 4 terms, a_{11}, and a_{23}:

$$a_n = (-1)^n \left(\frac{n^2}{n^4 + 1} \right). \ \textbf{[11.1]}$$

6. Predict the general, or nth, term. Answers may vary.
   $2, -5, 10, -17, 26, \ldots$ **[11.1]**

7. Find and evaluate:

$$\sum_{k=1}^{4} \frac{(-1)^{k+1} 3^k}{3^k - 1}. \ \textbf{[11.1]}$$

8. Use a graphing calculator to construct a table of values and a graph for the first 10 terms of this sequence.
   $a_1 = 0.3, \quad a_{k+1} = 5a_k + 1$ **[11.1]**

9. Write sigma notation. Answers may vary.
   $0 + 3 + 8 + 15 + 24 + 35 + 48$ **[11.1]**

10. Find the 10th term of the arithmetic sequence
    $\frac{3}{4}, \frac{13}{12}, \frac{17}{12}, \ldots$ **[11.2]**

11. Find the 6th term of the arithmetic sequence
    $a - b, a, a + b, \ldots$ **[11.2]**

12. Find the sum of the first 18 terms of the arithmetic sequence
    $4, 7, 10, \ldots$ **[11.2]**

13. Find the sum of the first 200 natural numbers. **[11.2]**

14. The 1st term in an arithmetic sequence is 5, and the 17th term is 53. Find the 3rd term. **[11.2]**

15. The common difference in an arithmetic sequence is 3. The 10th term is 23. Find the first term. **[11.2]**

16. For a geometric sequence, $a_1 = -2$, $r = 2$, and $a_n = -64$. Find n and S_n. **[11.3]**

17. For a geometric sequence, $r = \frac{1}{2}$ and $S_5 = \frac{31}{2}$. Find a_1 and a_5. **[11.3]**

Find the sum of each infinite geometric series, if it exists. **[11.3]**

18. $25 + 27.5 + 30.25 + 33.275 + \cdots$

19. $0.27 + 0.0027 + 0.000027 + \cdots$

20. $\frac{1}{2} - \frac{1}{6} + \frac{1}{18} - \cdots$

21. Find fraction notation for $2.\overline{43}$. **[11.3]**

22. Insert four arithmetic means between 5 and 9. **[11.2]**

23. *Bouncing Golfball.* A golfball is dropped from a height of 30 ft to the pavement. It always rebounds three-fourths of the distance that it drops. How far (up and down) will the ball have traveled when it hits the pavement for the 6th time? **[11.3]**

24. *Amount of an Annuity.* To create a college fund, a parent makes a sequence of 18 yearly deposits of $2000 each in a savings account on which interest is compounded annually at 2.8%. Find the amount of the annuity. **[11.3]**

25. *Total Gift.* Suppose that you receive 10¢ on the first day of the year, 12¢ on the 2nd day, 14¢ on the 3rd day, and so on.

a) How much will you receive on the 365th day? **[11.2]**

b) What is the sum of these 365 gifts? **[11.2]**

26. *The Economic Multiplier.* Suppose that the government is making a $24,000,000,000 expenditure for travel to Mars. If 73% of this amount is spent again, and so on, what is the total effect on the economy? **[11.3]**

Use mathematical induction to prove each of the following. **[11.4]**

27. For every natural number n,
$$1 + 4 + 7 + \cdots + (3n - 2) = \frac{n(3n - 1)}{2}.$$

28. For every natural number n,
$$1 + 3 + 3^2 + \cdots + 3^{n-1} = \frac{3^n - 1}{2}.$$

29. For every natural number $n \geq 2$,
$$\left(1 - \frac{1}{2}\right)\left(1 - \frac{1}{3}\right) \cdots \left(1 - \frac{1}{n}\right) = \frac{1}{n}.$$

30. *Book Arrangements.* In how many ways can 6 books be arranged on a shelf? **[11.5]**

31. *Flag Displays.* If 9 different signal flags are available, how many different displays are possible using 4 flags in a row? **[11.5]**

32. *Prize Choices.* The winner of a contest can choose any 8 of 15 prizes. How many different sets of prizes can be chosen? **[11.6]**

33. *Fraternity–Sorority Names.* The Greek alphabet contains 24 letters. How many fraternity or sorority names can be formed using 3 different letters? **[11.5]**

34. *Letter Arrangements.* In how many distinguishable ways can the letters of the word TENNESSEE be arranged? **[11.5]**

35. *Floor Plans.* A manufacturer of houses has 1 floor plan but achieves variety by having 3 different roofs, 4 different ways of attaching the garage, and 3 different types of entrances. Find the number of different houses that can be produced. **[11.5]**

36. *Code Symbols.* How many code symbols can be formed using 5 out of 6 of the letters of G, H, I, J, K, L if the letters:

a) cannot be repeated? **[11.5]**

b) can be repeated? **[11.5]**

c) cannot be repeated but must begin with K? **[11.5]**

d) cannot be repeated but must end with IGH? **[11.5]**

37. Determine the number of subsets of a set containing 8 members. **[11.7]**

Expand. **[11.7]**

38. $(m + n)^7$

39. $(x - \sqrt{2})^5$

40. $(x^2 - 3y)^4$

41. $\left(a + \frac{1}{a}\right)^8$

42. $(1 + 5i)^6$, where $i^2 = -1$

43. Find the 4th term of $(a + x)^{12}$. **[11.7]**

44. Find the 12th term of $(2a - b)^{18}$. Do not multiply out the factorials. **[11.7]**

45. *Rolling Dice.* What is the probability of getting a 10 on a roll of a pair of dice? on a roll of 1 die? **[11.8]**

46. *Drawing a Card.* From a deck of 52 cards, 1 card is drawn at random. What is the probability that it is a club? **[11.8]**

47. *Drawing Three Cards.* From a deck of 52 cards, 3 are drawn at random without replacement. What is the probability that 2 are aces and 1 is a king? **[11.8]**

48. *Election Poll.* Three people were running for mayor in an election campaign. A poll was conducted to see which candidate was favored. During the polling, 86 favored candidate A, 97 favored B, and 23 favored C. Assuming that the poll is a valid indicator of the election results, what is the probability that the election will be won by A? B? C? **[11.8]**

49. *Consumption of American Cheese.* The following table lists the number of pounds of American cheese consumed per capita for selected years.

Year, n	American Cheese Consumed per Capita (in pounds)
1930, 0	3.2
1950, 20	5.5
1970, 40	7.0
1990, 60	11.1
2000, 70	12.7
2010, 80	13.3

Sources: Economic Research Service; U.S. Department of Agriculture

a) Find a linear sequence function $a_n = an + b$ that models the data. Let n represent the number of years since 1930. **[11.1]**

b) Use the sequence found in part (a) to estimate the number of pounds of American cheese consumed per capita in 2016. **[11.1]**

50. Which of the following is the 25th term of the arithmetic sequence $12, 10, 8, 6, \ldots$? **[11.2]**

A. -38 B. -36

C. 32 D. 60

51. What is the probability of getting a total of 4 on a roll of a pair of dice? **[11.8]**

A. $\frac{1}{12}$ B. $\frac{1}{9}$

C. $\frac{1}{6}$ D. $\frac{5}{36}$

52. The graph of the sequence whose general term is $a_n = n - 1$ is which of the following? **[11.1]**

A.

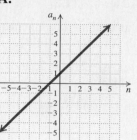

B.

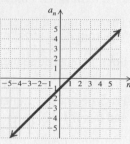

C.

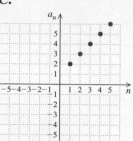

D.

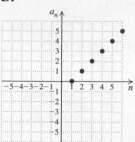

❯ Synthesis

53. Suppose that $a_1, a_2, \ldots, a_n$ is an arithmetic sequence. Is $b_1, b_2, \ldots, b_n$ an arithmetic sequence if:

a) $b_n = |a_n|$? **[11.2]**   **b)** $b_n = a_n + 8$? **[11.2]**

c) $b_n = 7a_n$? **[11.2]**   **d)** $b_n = \dfrac{1}{a_n}$? **[11.2]**

e) $b_n = \log a_n$? **[11.2]**   **f)** $b_n = a_n^3$? **[11.2]**

54. Suppose that $a_1, a_2, \ldots, a_n$ and $b_1, b_2, \ldots, b_n$ are geometric sequences. Prove that $c_1, c_2, \ldots, c_n$ is a geometric sequence, where $c_n = a_n b_n$. **[11.3]**

55. Write the first 3 terms of the infinite geometric series with $r = -\frac{1}{3}$ and $S_\infty = \frac{3}{8}$. **[11.3]**

56. The zeros of this polynomial function form an arithmetic sequence. Find them. **[11.2]**

$$f(x) = x^4 - 4x^3 - 4x^2 + 16x$$

57. Simplify:

$$\sum_{k=0}^{10} (-1)^k \binom{10}{k} (\log x)^{10-k} (\log y)^k. \quad \textbf{[11.6]}$$

Solve for n. **[11.6]**

58. $\binom{n}{6} = 3 \cdot \binom{n-1}{5}$ **59.** $\binom{n}{n-1} = 36$

60. Solve for a:

$$\sum_{k=0}^{5} \binom{5}{k} 9^{5-k} a^k = 0. \quad \textbf{[11.7]}$$

> # Collaborative Discussion and Writing

61. *Circular Arrangements.* In how many ways can the numbers on a clock face be arranged? See if you can derive a formula for the number of distinct circular arrangements of *n* objects. Explain your reasoning. [11.5]

62. How "long" is 15!? Suppose that you own 15 books and decide to make up all the possible arrangements of the books on a shelf. About how long, in years, would it take you if you were to make one arrangement per second? Write out the reasoning you used for this problem in the form of a paragraph. [11.5]

63. Explain why a "combination" lock should really be called a "permutation" lock. [11.6]

64. Give an explanation that you might use with a fellow student to explain that

$$\binom{n}{k} = \binom{n}{n-k}. \quad [11.6]$$

11 | Chapter Test

1. For the sequence whose *n*th term is $a_n = (-1)^n(2n + 1)$, find a_{21}.

2. Find the first 5 terms of the sequence with general term
$$a_n = \frac{n + 1}{n + 2}.$$

3. Find and evaluate:
$$\sum_{k=1}^{4} (k^2 + 1).$$

4. Use a graphing calculator to construct a table of values and a graph for the first 10 terms of the sequence with general term
$$a_n = \frac{n + 1}{n + 2}.$$

Write sigma notation. Answers may vary.

5. $4 + 8 + 12 + 16 + 20 + 24$

6. $2 + 4 + 8 + 16 + 32 + \cdots$

7. Find the first 4 terms of the recursively defined sequence
$$a_1 = 3, \quad a_{n+1} = 2 + \frac{1}{a_n}.$$

8. Find the 15th term of the arithmetic sequence $2, 5, 8, \ldots$.

9. The 1st term of an arithmetic sequence is 8 and the 21st term is 108. Find the 7th term.

10. Find the sum of the first 20 terms of the series $17 + 13 + 9 + \cdots$.

11. Find the sum: $\sum_{k=1}^{25} (2k + 1)$.

12. Find the 11th term of the geometric sequence $10, -5, \frac{5}{2}, -\frac{5}{4}, \ldots$.

13. For a geometric sequence, $r = 0.2$ and $S_4 = 1248$. Find a_1.

Find the sum, if it exists.

14. $\sum_{k=1}^{8} 2^k$

15. $18 + 6 + 2 + \cdots$

16. Find fraction notation for $0.\overline{56}$.

17. *Salvage Value.* The value of an office machine is $10,000. Its salvage value each year is 80% of its value the year before. Give a sequence that lists the salvage value of the machine for each year of a 6-year period.

18. *Hourly Wage.* William accepts a job, starting with an hourly wage of $13.40, and is promised a raise of 30¢ per hour every three months for 4 years. What will William's hourly wage be at the end of the 4-year period?

19. *Amount of an Annuity.* To create a college fund, a parent makes a sequence of 18 equal yearly deposits of $2500 in a savings account on which interest is compounded annually at 5.6%. Find the amount of the annuity.

20. Use mathematical induction to prove that, for every natural number *n*,
$$2 + 5 + 8 + \cdots + (3n - 1) = \frac{n(3n + 1)}{2}.$$

Evaluate.

21. $_{15}P_6$

22. $_{21}C_{10}$

23. $\binom{n}{4}$

24. How many 4-digit numbers can be formed using the digits 1, 3, 5, 6, 7, and 9 without repetition?

25. How many code symbols can be formed using 4 of the 6 letters A, B, C, X, Y, Z if the letters:

   a) can be repeated?

   b) are not repeated and must begin with Z?

26. *Scuba Club Officers.* The Bay Woods Scuba Club has 28 members. How many sets of 4 officers can be selected from this group?

27. *Test Options.*   On a test with 20 questions, a student must answer 8 of the first 12 questions and 4 of the last 8. In how many ways can this be done?

28. Expand: $(x + 1)^5$.

29. Find the 5th term of the binomial expansion $(x - y)^7$.

30. Determine the number of subsets of a set containing 9 members.

31. *Marbles.*   Suppose that we select, without looking, one marble from a bag containing 6 red marbles and 8 blue marbles. What is the probability of selecting a blue marble?

32. *Drawing Coins.*   Ethan has 6 pennies, 5 dimes, and 4 quarters in his pocket. Six coins are drawn at random. What is the probability of getting 1 penny, 2 dimes, and 3 quarters?

33. The graph of the sequence whose general term is $a_n = 2n - 2$ is which of the following?

A.

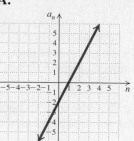

B.

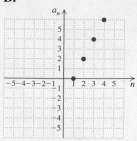

C.

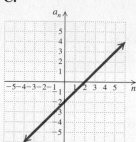

D.

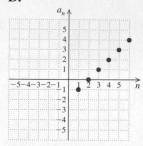

❯ Synthesis

34. Solve for n: $_nP_7 = 9 \cdot {_nP_6}$.

Answers

Chapter 1

Visualizing the Graph

1. H **2.** B **3.** D **4.** A **5.** G **6.** I **7.** C **8.** J
9. F **10.** E

Exercise Set 1.1

1. *A:* $(-5, 4)$; *B:* $(2, -2)$; *C:* $(0, -5)$;
D: $(3, 5)$; *E:* $(-5, -4)$; *F:* $(3, 0)$
3. 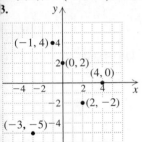 **5.**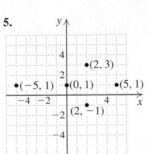

7. $(1971, 3), (1981, 15), (1991, 32), (2001, 59), (2011, 72),$
$(2014, 96)$ **9.** Yes; no **11.** Yes; no **13.** No; yes
15. No; yes
17. *x*-intercept: $(-3, 0)$; **19.** *x*-intercept: $(2, 0)$;
y-intercept: $(0, 5)$; *y*-intercept: $(0, 4)$;
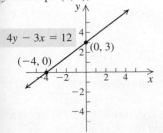 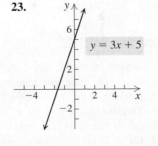

21. *x*-intercept: $(-4, 0)$; **23.**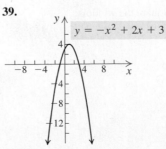
y-intercept: $(0, 3)$;

25. 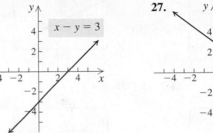 **27.**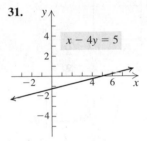

29. 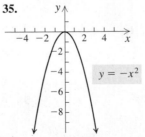 **31.**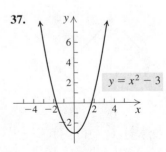

33.

35. **37.**

39.

A-1

41. (b)   **43.** (a)

45. $y = 2x + 1$

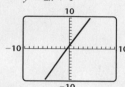

47. $4x + y = 7$

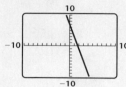

49. $y = \frac{1}{3}x + 2$

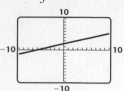

51. $2x + 3y = -5$

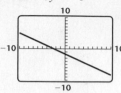

53. $y = x^2 + 6$

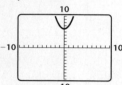

55. $y = 2 - x^2$

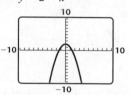

57. $y = x^2 + 4x - 2$

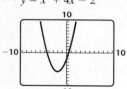

59. Standard window

61. $[-1, 1, -0.3, 0.3]$ **63.** $\sqrt{10}$, 3.162   **65.** 13
67. $\sqrt{45}$, 6.708   **69.** 16   **71.** $\frac{14}{3}$ **73.** $\sqrt{128.05}$, 11.316
75. $\sqrt{a^2 + b^2}$ **77.** 6.5   **79.** Yes   **81.** No
83. $(-4, -6)$ **85.** $\left(-\frac{1}{5}, \frac{1}{4}\right)$ **87.** $(4.95, -4.95)$
89. $\left(-6, \frac{13}{2}\right)$ **91.** $\left(-\frac{5}{12}, \frac{13}{40}\right)$
93.

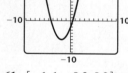

$\left(-\frac{1}{2}, \frac{3}{2}\right), \left(\frac{7}{2}, \frac{1}{2}\right), \left(\frac{5}{2}, \frac{9}{2}\right), \left(-\frac{3}{2}, \frac{11}{2}\right)$; no

95. $\left(\dfrac{\sqrt{7} + \sqrt{2}}{2}, -\dfrac{1}{2}\right)$ **97.** Square the window;

for example, use $[-12, 9, -4, 10]$.
99. $(x - 2)^2 + (y - 3)^2 = \frac{25}{9}$
101. $(x + 1)^2 + (y - 4)^2 = 25$
103. $(x - 2)^2 + (y - 1)^2 = 169$
105. $(x + 2)^2 + (y - 3)^2 = 4$

107. $(0, 0)$; 2;

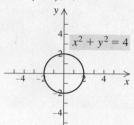

$x^2 + y^2 = 4$

109. $(0, 3)$; 4;

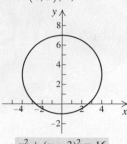

$x^2 + (y - 3)^2 = 16$

111. $(1, 5)$; 6;

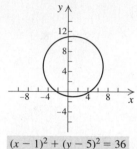

$(x - 1)^2 + (y - 5)^2 = 36$

113. $(-4, -5)$; 3;

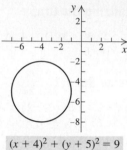

$(x + 4)^2 + (y + 5)^2 = 9$

115. $(x + 2)^2 + (y - 1)^2 = 3^2$
117. $(x - 5)^2 + (y + 5)^2 = 15^2$ **119.** Third
121. $\sqrt{h^2 + h + 2a - 2\sqrt{a^2 + ah}}$,
$\left(\dfrac{2a + h}{2}, \dfrac{\sqrt{a} + \sqrt{a + h}}{2}\right)$
123. $(x - 2)^2 + (y + 7)^2 = 36$ **125.** $(0, 4)$
127. (a) $(0, -3)$; (b) 5 ft   **129.** Yes   **131.** Yes
133. Let $P_1 = (x_1, y_1)$, $P_2 = (x_2, y_2)$, and
$M = \left(\dfrac{x_1 + x_2}{2}, \dfrac{y_1 + y_2}{2}\right)$. Let $d(AB)$ denote the distance from
point A to point B.

$$d(P_1M) = \sqrt{\left(\dfrac{x_1 + x_2}{2} - x_1\right)^2 + \left(\dfrac{y_1 + y_2}{2} - y_1\right)^2}$$
$$= \frac{1}{2}\sqrt{(x_2 - x_1)^2 + (y_2 - y_1)^2};$$
$$d(P_2M) = \sqrt{\left(\dfrac{x_1 + x_2}{2} - x_2\right)^2 + \left(\dfrac{y_1 + y_2}{2} - y_2\right)^2}$$
$$= \frac{1}{2}\sqrt{(x_1 - x_2)^2 + (y_1 - y_2)^2}$$
$$= \frac{1}{2}\sqrt{(x_2 - x_1)^2 + (y_2 - y_1)^2} = d(P_1M).$$

Exercise Set 1.2

1. Yes   **3.** Yes   **5.** No   **7.** Yes   **9.** Yes   **11.** Yes
13. No   **15.** Function; domain: $\{2, 3, 4\}$; range: $\{10, 15, 20\}$
17. Not a function; domain: $\{-7, -2, 0\}$; range: $\{3, 1, 4, 7\}$
19. Function; domain: $\{-2, 0, 2, 4, -3\}$; range: $\{1\}$
21. (a) 1; (b) 6; (c) 22; (d) $3x^2 + 2x + 1$; (e) $3t^2 - 4t + 2$
23. (a) 8; (b) -8; (c) $-x^3$; (d) $27y^3$; (e) $8 + 12h + 6h^2 + h^3$
25. (a) $\frac{1}{8}$; (b) 0; (c) does not exist; (d) $\frac{81}{53}$, or approximately
1.5283; (e) $\dfrac{x + h - 4}{x + h + 3}$ **27.** 0; does not exist; does not exist as a

real number; $\dfrac{1}{\sqrt{3}}$, or $\dfrac{\sqrt{3}}{3}$ **29.** $g(-2.1) \approx -21.8$;
$(5.08) \approx -130.4$; $g(10.003) \approx -468.3$

1.

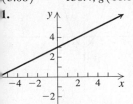

$f(x) = \frac{1}{2}x + 3$

33.

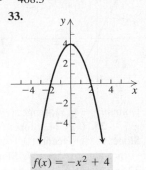

$f(x) = -x^2 + 4$

5.

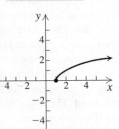

$f(x) - \sqrt{x - 1}$

7. $h(1) = -2$; $h(3) = 2$; $h(4) = 1$
9. $s(-4) = 3$; $s(-2) = 0$; $s(0) = -3$
1. $f(-1) = 2$; $f(0) = 0$; $f(1) = -2$
3. No **45.** Yes **47.** Yes **49.** No
1. All real numbers, or $(-\infty, \infty)$
3. All real numbers, or $(-\infty, \infty)$
5. $\{x \mid x \neq 0\}$, or $(-\infty, 0) \cup (0, \infty)$
7. $\{x \mid x \neq 2\}$, or $(-\infty, 2) \cup (2, \infty)$
9. $\{x \mid x \neq -1 \text{ and } x \neq 5\}$, or $(-\infty, -1) \cup (-1, 5) \cup (5, \infty)$
1. All real numbers, or $(-\infty, \infty)$
3. $\{x \mid x \neq 0 \text{ and } x \neq 7\}$, or $(-\infty, 0) \cup (0, 7) \cup (7, \infty)$
5. All real numbers, or $(-\infty, \infty)$
7. Domain: $[0, 5]$; range: $[0, 3]$
9. Domain: $[-2\pi, 2\pi]$; range: $[-1, 1]$
1. Domain: $(-\infty, \infty)$; range: $\{-3\}$
3. Domain: $[-5, 3]$; range: $[-2, 2]$
5. Domain: $(-\infty, \infty)$; range: $[0, \infty)$
7. Domain: $(-\infty, \infty)$; range: $(-\infty, \infty)$
9. Domain: $(-\infty, 3) \cup (3, \infty)$; range: $(-\infty, 0) \cup (0, \infty)$
1. Domain: $(-\infty, \infty)$; range: $(-\infty, \infty)$
3. Domain: $(-\infty, 7]$; range: $[0, \infty)$
5. Domain: $(-\infty, \infty)$; range: $(-\infty, 3]$
7. 645 m; 0 m   **89. (a)** 2018: $25.21; 2025: $28.23;
(b) about 49 years after 1985, or in 2034   **91.** $(-3, -2)$, yes;
$(2, -3)$, no **92.** $\left(\frac{4}{5}, -2\right)$, yes; $\left(\frac{11}{5}, \frac{1}{10}\right)$, yes
93.

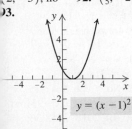

$y = (x - 1)^2$

94.

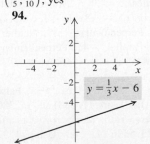

$y = \frac{1}{3}x - 6$

95.

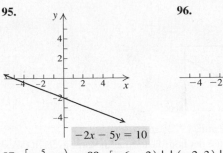

$-2x - 5y = 10$

96.

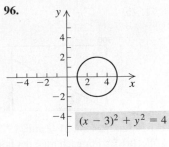

$(x - 3)^2 + y^2 = 4$

97. $\left[-\frac{5}{2}, \infty\right)$ **99.** $[-6, -2) \cup (-2, 3) \cup (3, \infty)$
101. $f(x) = x$, $g(x) = x + 1$ **103.** -7

Visualizing the Graph

1. E   **2.** D   **3.** A   **4.** J   **5.** C   **6.** F   **7.** H   **8.** G
9. B   **10.** I

Exercise Set 1.3

1. (a) Yes; **(b)** yes; **(c)** yes   **3. (a)** Yes; **(b)** no; **(c)** no
5. $\frac{6}{5}$ **7.** $-\frac{3}{5}$ **9.** 0   **11.** $\frac{1}{5}$ **13.** Not defined   **15.** 0.3
17. 0   **19.** $-\frac{6}{5}$ **21.** $-\frac{1}{3}$ **23.** Not defined   **25.** -2
27. 5   **29.** 0   **31.** 1.3   **33.** Not defined   **35.** $-\frac{1}{2}$ **37.** -1
39. 0   **41.** The average rate of change in the lowest price of a
World Series ticket from 1946 to 2012 was an increase of about
$1.65 per year.   **43.** The average rate of change in the popula-
tion of Cleveland, Ohio, over the 13-year period was a decrease
of about 6792 people per year.   **45.** The average rate of change
in per-capita consumption of whole milk from 1970 to 2011 was
a decrease of about 0.5 gal per year.   **47.** The average rate
of change in the number of acres used for growing almonds in
California from 2003 to 2012 was an increase of about 28,889 acres
per year.   **49.** $\frac{3}{5}$; $(0, -7)$ **51.** Slope is not defined; there is no
y-intercept.   **53.** $-\frac{1}{2}$; $(0, 5)$ **55.** $-\frac{3}{2}$; $(0, 5)$
57. 0; $(0, -6)$ **59.** $\frac{4}{5}$; $\left(0, \frac{8}{5}\right)$ **61.** $\frac{1}{4}$; $\left(0, -\frac{1}{2}\right)$
63.

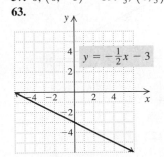

$y = -\frac{1}{2}x - 3$

65.

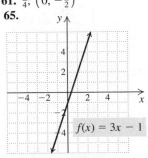

$f(x) = 3x - 1$

67.

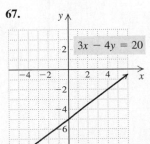

$3x - 4y = 20$

69.

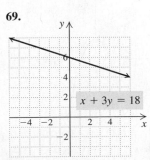

$x + 3y = 18$

71. (a)

$$y = \frac{1}{33}x + 1$$

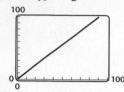

(b) 1 atm, 2 atm, $31\frac{10}{33}$ atm, $152\frac{17}{33}$ atm, $213\frac{4}{33}$ atm
73. (a) $\frac{11}{10}$. For each mile per hour faster that the car travels, it takes $\frac{11}{10}$ ft longer to stop;

(b)

$$y = \frac{11}{10}x + \frac{1}{2}$$

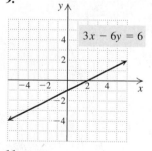

(c) 6 ft, 11.5 ft, 22.5 ft, 55.5 ft, 72 ft; **(d)** $\{r | r > 0\}$, or $(0, \infty)$. If r is allowed to be 0, the function says that a stopped car has a reaction distance of $\frac{1}{2}$ ft. **75.** $C(t) = 2250 + 3380t$; $C(20) = \$69,850$ **77.** $C(x) = 750 + 15x$; $C(32) = \$1230$
79. $-\frac{5}{4}$ **80.** 10 **81.** 40 **82.** $a^2 + 3a$
83. $a^2 + 2ah + h^2 - 3a - 3h$ **85.** $2a + h$ **87.** False
89. $f(x) = x + b$

Mid-Chapter Mixed Review: Chapter 1

1. False **2.** True **3.** False **4.** x-intercept: $(5, 0)$;
y-intercept: $(0, -8)$ **5.** $\sqrt{605} = 11\sqrt{5} \approx 24.6$; $\left(-\frac{5}{2}, -4\right)$
6. $\sqrt{2} \approx 1.4$; $\left(-\frac{1}{4}, -\frac{3}{10}\right)$ **7.** $(x + 5)^2 + (y - 2)^2 = 169$
8. Center: $(3, -1)$; radius: 2
9.

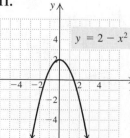

10.

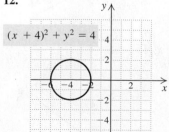

11.

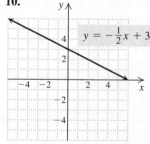

12.

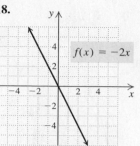

13. $f(-4) = -36; f(0) = 0; f(1) = -1$
14. $g(-6) = 0; g(0) = -2; g(3)$ is not defined
15. All real numbers, or $(-\infty, \infty)$
16. $\{x | x \neq -5\}$, or $(-\infty, -5) \cup (-5, \infty)$
17. $\{x | x \neq -3 \text{ and } x \neq 1\}$, or $(-\infty, -3) \cup (-3, 1) \cup (1, \infty)$

18.

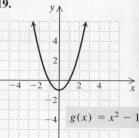

19.

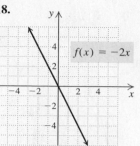

20. Domain: $[-4, 3)$; range: $[-4, 5)$ **21.** Not defined
22. $-\frac{1}{4}$ **23.** 0 **24.** Slope: $-\frac{1}{9}$; y-intercept: $(0, 12)$
25. Slope: 0; y-intercept: $(0, -6)$ **26.** Slope is not defined; there is no y-intercept **27.** Slope: $\frac{3}{16}$; y-intercept: $\left(0, \frac{1}{16}\right)$
28. The sign of the slope indicates the slant of a line. A line that slants up from left to right has positive slope, because corresponding changes in x and y have the same sign. A line that slants down from left to right has negative slope, because corresponding changes in x and y have opposite signs. A horizontal line has zero slope, because there is no change in y for a given change in x. The slope of a vertical line is not defined, because there is no change in x for a given change in y and division by 0 is not defined. The larger the absolute value of slope, the steeper the line. This is because a larger absolute value corresponds to a greater change in y, compared to the change in x, than a smaller absolute value. **29.** A vertical line $(x = a)$ crosses the graph more than once. Thus, $x = a$ fails the vertical-line test. **30.** The domain of a function is the set of all inputs of the function. The range is the set of all outputs. The range depends on the domain. **31.** Let $A = (a, b)$ and $B = (c, d)$. The coordinates of a point C one-half of the way from A to B are $\left(\frac{a + c}{2}, \frac{b + d}{2}\right)$. A point D that is one-half of the way from C to B is $\frac{1}{2} + \frac{1}{2} \cdot \frac{1}{2}$, or $\frac{3}{4}$, of the way from A to B. Its coordinates are $\left(\dfrac{\frac{a + c}{2} + c}{2}, \dfrac{\frac{b + d}{2} + d}{2}\right)$, or $\left(\frac{a + 3c}{4}, \frac{b + 3d}{4}\right)$. Then a point E that is one-half of the way from D to B is $\frac{3}{4} + \frac{1}{2} \cdot \frac{1}{4}$, or $\frac{7}{8}$, of the way from A to B. Its coordinates are $\left(\dfrac{\frac{a + 3c}{4} + c}{2}, \dfrac{\frac{b + 3d}{4} + d}{2}\right)$, or $\left(\frac{a + 7c}{8}, \frac{b + 7d}{8}\right)$.

Exercise Set 1.4

1. 4; $(0, -2)$; $y = 4x - 2$ **3.** -1; $(0, 0)$; $y = -x$
5. 0; $(0, -3)$; $y = -3$ **7.** $y = \frac{2}{9}x + 4$ **9.** $y = -4x - 7$
11. $y = -4.2x + \frac{3}{4}$ **13.** $y = \frac{2}{9}x + \frac{19}{3}$ **15.** $y = 8$
17. $y = -\frac{3}{5}x - \frac{17}{5}$ **19.** $y = -3x + 2$ **21.** $y = -\frac{1}{2}x + \frac{7}{2}$
23. $y = \frac{2}{3}x - 6$ **25.** $y = 7.3$ **27.** Horizontal: $y = -3$; vertical: $x = 0$ **29.** Horizontal: $y = -1$; vertical: $x = \frac{2}{11}$
31. $h(x) = -3x + 7$; 1 **33.** $f(x) = \frac{2}{5}x - 1$; -1
35. Perpendicular **37.** Neither parallel nor perpendicular
39. Parallel **41.** Perpendicular **43.** $y = \frac{2}{7}x + \frac{29}{7}$; $y = -\frac{7}{2}x + \frac{31}{2}$ **45.** $y = -0.3x - 2.1$; $y = \frac{10}{3}x + \frac{70}{3}$
47. $y = -\frac{3}{4}x + \frac{1}{4}$; $y = \frac{4}{3}x - 6$ **49.** $x = 3$; $y = -3$
51. True **53.** True **55.** False **57.** No **59.** Yes
61. (a) Using $(1, 333)$ and $(4, 380)$ gives us $y = 15.67x + 317.33$, where x is the number of years after 2009; **(b)** 2018: $\$458.36$; 2023: $\$536.71$ **63.** Using $(1, 32.5)$ and $(4, 49.9)$ gives us

= 5.8x + 26.7, where x is the number of years after 2010 and is in billions of dollars; 2019: $78.9 billion   **65.** Using (1, 28.3) nd (3, 30.8) gives us $y = 1.25x + 27.05$, where x is the number of ears after 2009 and y is in gallons; 2017: about 37.1 gal
7. (a) $y = 20.05714286x + 301.8571429$, where x is the number of f years after 2009; **(b)** $482.37; this value is $24.01 more than the alue found in Exercise 61; **(c)** $r \approx 0.9851$; the line fits the data airly well.   **69. (a)** $y = 6.47x + 23.8$, where x is the number of ears after 2010 and y is in billions; **(b)** 2019: $82.03 billion; this alue is $3.13 billion more than the value found in Exercise 63; **c)** $r \approx 0.9915$; the line fits the data well.
1. (a) $M = 0.2H + 156$; **(b)** 164, 169, 171, 173; **(c)** $r = 1$; he regression line fits the data perfectly and should be a ood predictor.   **73.** -1 **74.** Not defined
5. $(x + 7)^2 + (y + 1)^2 = \frac{81}{25}$ **76.** $x^2 + (y - 3)^2 = 6.25$
7. -7.75 **79.** $y = -\frac{1}{2}x + 7$

Exercise Set 1.5

. 4   **3.** All real numbers, or $(-\infty, \infty)$ **5.** $-\frac{3}{4}$ **7.** -9
. 6   **11.** No solution   **13.** $\frac{11}{5}$ **15.** $\frac{35}{6}$ **17.** 8   **19.** -4
1. 6   **23.** -1 **25.** $\frac{4}{5}$ **27.** $-\frac{3}{2}$ **29.** $-\frac{2}{3}$ **31.** $\frac{1}{2}$
3. About 51,075 words   **35.** $1300   **37.** 26°, 130°, 24°
9. $3.609 billion   **41.** 3 hr   **43.** 2.1 million tweets
5. $5000   **47.** About 287,000 students   **49.** Length: 100 yd; width: 65 yd   **51.** Length: 93 m; width: 68 m   **53.** 2.5 hr
5. $2400 at 3%; $2600 at 4%   **57.** IBM: 6809 patents; Samsung: 4676 patents   **59.** Michael Jackson: $140 million; Elvis Presley: $55 million   **61.** 74.25 lb   **63.** 4.5 hr
5. 12 mi   **67.** Italy: 30%; Spain: 20%; United States: 8%
9. -5 **71.** $\frac{11}{2}$ **73.** 16   **75.** -12 **77.** 6   **79.** 20
1. 25   **83.** 15   **85. (a)** $(4, 0)$; **(b)** 4   **87. (a)** $(-2, 0)$;
b) -2 **89. (a)** $(-4, 0)$; **(b)** -4 **91.** $y = -\frac{3}{4}x + \frac{13}{4}$
2. $y = -\frac{3}{4}x + \frac{1}{4}$ **93.** 13   **94.** $(-1, \frac{1}{2})$ **95.** $f(-3) = \frac{1}{2}$;
$f(0) = 0$; $f(3)$ does not exist.   **96.** $m = 7$; y-intercept: $(0, -\frac{1}{2})$ **97.** Yes   **99.** No   **101.** $-\frac{2}{3}$ **103.** No; the 6-oz cup costs about 6.4% more per ounce.   **105.** 11.25 mi

Exercise Set 1.6

1. $\{x | x > 5\}$, or $(5, \infty)$;
3. $\{x | x > 3\}$, or $(3, \infty)$;
5. $\{x | x \geq -3\}$, or $[-3, \infty)$;
7. $\{y | y \geq \frac{22}{13}\}$, or $[\frac{22}{13}, \infty)$;
9. $\{x | x > 6\}$, or $(6, \infty)$;
11. $\{x | x \geq -\frac{5}{12}\}$, or $[-\frac{5}{12}, \infty)$;
13. $\{x | x \leq \frac{15}{34}\}$, or $(-\infty, \frac{15}{34}]$;
15. $\{x | x < 1\}$, or $(-\infty, 1)$;

17. $\{x | x \geq 7\}$, or $[7, \infty)$ **19.** $\{x | x \leq \frac{1}{5}\}$, or $(-\infty, \frac{1}{5}]$
21. $\{x | x > -4\}$, or $(-4, \infty)$
23. $[-3, 3)$;
25. $[8, 10]$;
27. $[-7, -1]$;
29. $(-\frac{3}{2}, 2)$;
31. $(1, 5]$;
33. $(-\frac{11}{3}, \frac{13}{3})$;
35. $(-\infty, -2] \cup (1, \infty)$;
37. $(-\infty, -\frac{7}{2}] \cup [\frac{1}{2}, \infty)$;
39. $(-\infty, 9.6) \cup (10.4, \infty)$;
41. $(-\infty, -\frac{57}{4}] \cup [-\frac{55}{4}, \infty)$;
43. More than 45 years after 1980   **45.** Less than 10 hr
47. $5000   **49.** $300,000   **51.** Sales greater than $18,000
53. Function; domain; range; domain; exactly one; range
54. Midpoint formula   **55.** x-intercept   **56.** Constant; identity
57. $(-\frac{1}{4}, \frac{5}{9})$ **59.** $(-\frac{1}{8}, \frac{1}{2})$

Review Exercises: Chapter 1

1. True   **2.** True   **3.** False   **4.** False   **5.** True   **6.** False
7. Yes; no   **8.** Yes; no
9. x-intercept: (3, 0); y-intercept: (0, −2);
10. x-intercept: (2, 0); y-intercept: (0, 5);

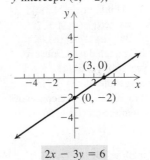

$2x - 3y = 6$

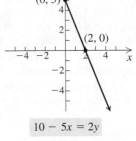

$10 - 5x = 2y$

11.

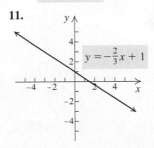

$y = -\frac{2}{3}x + 1$

12.

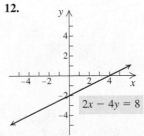

$2x - 4y = 8$

13.

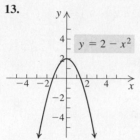

$y = 2 - x^2$

14. $\sqrt{34} \approx 5.831$ **15.** $\left(\frac{1}{2}, \frac{11}{2}\right)$
16. Center: $(-1, 3)$; radius: 3;

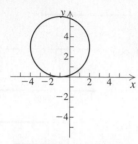

$(x + 1)^2 + (y - 3)^2 = 9$

17. $x^2 + (y + 4)^2 = \frac{9}{4}$ **18.** $(x + 2)^2 + (y - 6)^2 = 13$
19. $(x - 2)^2 + (y - 4)^2 = 26$ **20.** No   **21.** Yes
22. Not a function; domain: $\{3, 5, 7\}$; range: $\{1, 3, 5, 7\}$
23. Function; domain: $\{-2, 0, 1, 2, 7\}$; range: $\{-7, -4, -2, 2, 7\}$
24. (a) -3; **(b)** 9; **(c)** $a^2 - 3a - 1$; **(d)** $x^2 + x - 3$
25. (a) 0; **(b)** $\dfrac{x - 6}{x + 6}$; **(c)** does not exist; **(d)** $-\frac{5}{3}$
26. $f(2) = -1$; $f(-4) = -3$; $f(0) = -1$ **27.** No   **28.** Yes
29. No   **30.** Yes   **31.** All real numbers, or $(-\infty, \infty)$
32. $\{x | x \neq 0\}$, or $(-\infty, 0) \cup (0, \infty)$
33. $\{x | x \neq 5 \text{ and } x \neq 1\}$, or $(-\infty, 1) \cup (1, 5) \cup (5, \infty)$
34. $\{x | x \neq -4 \text{ and } x \neq 4\}$, or $(-\infty, -4) \cup (-4, 4) \cup (4, \infty)$
35. Domain: $[-4, 4]$; range: $[0, 4]$ **36.** Domain: $(-\infty, \infty)$;
range: $[0, \infty)$ **37.** Domain: $(-\infty, \infty)$; range: $(-\infty, \infty)$
38. Domain: $(-\infty, \infty)$; range: $[0, \infty)$ **39. (a)** Yes; **(b)** no;
(c) no, strictly speaking, but data might be modeled by a linear regression function.   **40. (a)** Yes; **(b)** yes; **(c)** yes   **41.** $\frac{5}{3}$
42. 0   **43.** Not defined   **44.** The average rate of change in per-capita coffee consumption from 1990 to 2012 was a decrease of about 0.1 gal per year.   **45.** $m = -\frac{7}{11}$; y-intercept: $(0, -6)$
46. $m = -2$; y-intercept: $(0, -7)$
47.

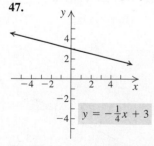

$y = -\frac{1}{4}x + 3$

48. $C(t) = 110 + 85t$; $1130   **49. (a)** 70°C, 220°C, 10,020°C;
(b) $[0, 5600]$ **50.** $y = -\frac{2}{3}x - 4$ **51.** $y = 3x + 5$
52. $y = \frac{1}{3}x - \frac{1}{3}$ **53.** Horizontal: $y = \frac{2}{5}$; vertical: $x = -4$
54. $h(x) = 2x - 5$; -5 **55.** Parallel   **56.** Neither
57. Perpendicular   **58.** $y = -\frac{2}{3}x - \frac{1}{3}$ **59.** $y = \frac{3}{2}x - \frac{5}{2}$
60. (a) Using $(2, 7969)$ and $(8, 8576)$ gives us
$W(x) = 101.17x + 7766.64$; 2013: 8475 female graduates;

(b) $W(x) = 98.9x + 7747.8$, where x is the number of years after 2006; 2013: 8440 female graduates; $r \approx 0.9942$; the line fits the data well.   **61.** $\frac{3}{2}$ **62.** -6 **63.** -1 **64.** -21
65. $\frac{95}{24}$ **66.** No solution   **67.** All real numbers, or $(-\infty, \infty)$
68. 568 million quarters   **69.** $2300   **70.** 3.4 hr
71. 3   **72.** 4   **73.** 0.2, or $\frac{1}{5}$ **74.** 4
75. $(-\infty, 12)$;

76. $(-\infty, -4]$;

77. $\left[-\frac{4}{3}, \frac{4}{3}\right]$;

78. $\left(\frac{2}{5}, 2\right]$;

79. $\left(-\infty, -\frac{1}{2}\right) \cup (3, \infty)$;

80. $\left(-\infty, -\frac{5}{3}\right] \cup [1, \infty)$;

81. Years after 2019   **82.** Fahrenheit temperatures less than 113°
83. B   **84.** B   **85.** C   **86.** $\left(\frac{5}{2}, 0\right)$
87. $\{x | x < 0\}$, or $(-\infty, 0)$ **88.** $\{x | x \neq -3 \text{ and }$
$x \neq 0 \text{ and } x \neq 3\}$, or $(-\infty, -3) \cup (-3, 0) \cup$
$(0, 3) \cup (3, \infty)$ **89.** Think of the slopes as $\dfrac{-3/5}{1}$ and $\dfrac{1/2}{1}$.
The graph of $f(x)$ changes $\frac{3}{5}$ unit vertically for each unit of horizontal change, whereas the graph of $g(x)$ changes $\frac{1}{2}$ unit vertically for each unit of horizontal change. Since $\frac{3}{5} > \frac{1}{2}$, the graph of $f(x) = -\frac{3}{5}x + 4$ is steeper than the graph of $g(x) = \frac{1}{2}x - 6$.
90. If an equation contains no fractions, using the addition principle before using the multiplication principle eliminates the need to add or subtract fractions.   **91.** The solution set of a disjunction is a union of sets, so it is only possible for a disjunction to have no solution when the solution set of each inequality is the empty set.
92. The graph of $f(x) = mx + b, m \neq 0$, is a straight line that is not horizontal. The graph of such a line intersects the x-axis exactly once. Thus the function has exactly one zero.   **93.** By definition, the notation $3 < x < 4$ indicates that $3 < x \text{ and } x < 4$. The disjunction $x < 3 \text{ or } x > 4$ cannot be written $3 > x > 4$, or $4 < x < 3$, because it is not possible for x to be greater than 4 *and* less than 3.   **94.** A function is a correspondence between two sets in which each member of the first set corresponds to exactly one member of the second set.

Test: Chapter 1

1. [1.1] Yes   **2.** [1.1] x-intercept: $(-2, 0)$; y-intercept: $(0, 5)$;

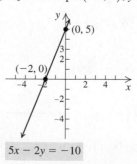

$5x - 2y = -10$

3. [1.1] $\sqrt{45} \approx 6.708$ **4.** [1.1] $\left(-3, \frac{9}{2}\right)$ **5.** [1.1] Center: $(-4, 5)$; radius: 6   **6.** [1.1] $(x + 1)^2 + (y - 2)^2 = 5$
7. [1.2] **(a)** Yes; **(b)** $\{-4, 3, 1, 0\}$; **(c)** $\{7, 0, 5\}$

[1.2] **(a)** 8; **(b)** $2a^2 + 7a + 11$ **9.** [1.2] **(a)** Does not exist; **(b)** 0   **10.** [1.2] 0   **11.** [1.2] **(a)** No; **(b)** yes

12. [1.2] $\{x \mid x \ne 4\}$, or $(-\infty, 4) \cup (4, \infty)$ **13.** [1.2] All real numbers, or $(-\infty, \infty)$ **14.** [1.2] $\{x \mid -5 \le x \le 5\}$, or $[-5, 5]$

15. [1.2] **(a)**

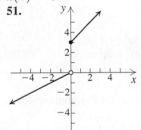

(b) $(-\infty, \infty)$;
(c) $[3, \infty)$

$f(x) = |x - 2| + 3$

16. [1.3] Not defined   **17.** [1.3] $-\frac{11}{6}$ **18.** [1.3] 0

19. [1.3] The average rate of change in the percent of adults who are married for the years from 1960 to 2012 was a decrease of about 0.4% per year.   **20.** [1.3] Slope: $\frac{3}{2}$; y-intercept: $\left(0, \frac{5}{2}\right)$

21. [1.3] $C(t) = 65 + 48t$; \$173   **22.** [1.4] $y = -\frac{5}{8}x - 5$

23. [1.4] $y - 4 = -\frac{3}{4}(x - (-5))$, or $y - (-2) = -\frac{3}{4}(x - 3)$, or $y = -\frac{3}{4}x + \frac{1}{4}$ **24.** [1.4] $x = -\frac{3}{8}$ **25.** [1.4] Perpendicular

26. [1.4] $y - 3 = -\frac{1}{2}(x + 1)$, or $y = -\frac{1}{2}x + \frac{5}{2}$ **27.** [1.4] $y - 3 = 2(x + 1)$, or $y = 2x + 5$ **28.** [1.4] **(a)** Using (2, 544.05) and (8, 653.19) gives us $y = 18.19x + 507.67$, where x is the number of years after 2003; 2017: \$762.33; **(b)** $y = 16.47728571x + 517.2819048$; 2017: \$747.96; $r \approx 0.9971$ **29.** [1.5] -1 **30.** [1.5] All real numbers, or $(-\infty, \infty)$ **31.** [1.5] -60 **32.** [1.5] $\frac{21}{11}$ **33.** [1.5] length: 60 m; width: 45 m   **34.** [1.5] \$1.80   **35.** [1.5] -3

36. [1.6] $(-\infty, -3]$;   (number line with bracket at −3)

37. [1.6] $(-5, 3)$;   (number line with parentheses at −5 and 3)

38. [1.6] $(-\infty, 2] \cup [4, \infty)$;   (number line with brackets at 2 and 4)

39. [1.6] More than 6 hr   **40.** [1.3] B   **41.** [1.2] -2

Chapter 2

Exercise Set 2.1

1. (a) $(-5, 1)$; **(b)** $(3, 5)$; **(c)** $(1, 3)$ **3. (a)** $(-3, -1)$, $(3, 5)$; **(b)** $(1, 3)$; **(c)** $(-5, -3)$ **5. (a)** $(-\infty, -8)$, $(-3, -2)$; **(b)** $(-8, -6)$; **(c)** $(-6, -3)$, $(-2, \infty)$

7. Domain: $[-5, 5]$; range: $[-3, 3]$ **9.** Domain: $[-5, -1] \cup [1, 5]$; range: $[-4, 6]$ **11.** Domain: $(-\infty, \infty)$; range: $(-\infty, 3]$ **13.** Relative maximum: 3.25 at $x = 2.5$; increasing: $(-\infty, 2.5)$; decreasing: $(2.5, \infty)$ **15.** Relative maximum: 2.370 at $x = -0.667$; relative minimum: 0 at $x = 2$; increasing: $(-\infty, -0.667)$, $(2, \infty)$; decreasing: $(-0.667, 2)$

17. Increasing: $(0, \infty)$; decreasing: $(-\infty, 0)$; relative minimum: 0 at $x = 0$ **19.** Increasing: $(-\infty, 0)$; decreasing: $(0, \infty)$; relative maximum: 5 at $x = 0$ **21.** Increasing: $(3, \infty)$; decreasing: $(-\infty, 3)$; relative minimum: 1 at $x = 3$ **23.** Increasing: $(1, 3)$; decreasing: $(-\infty, 1)$, $(3, \infty)$; relative maximum: -4 at $x = 3$; relative minimum: -8 at $x = 1$ **25.** Increasing: $(-1.552, 0)$, $(1.552, \infty)$; decreasing: $(-\infty, -1.552)$, $(0, 1.552)$; relative maximum: 4.07 at $x = 0$; relative minima: -2.314 at $x = -1.552$, -2.314 at $x = 1.552$

27. (a) $y = -x^2 + 300x + 6$; **(b)** 22,506 fruit trees

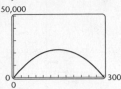

(c) When \$150 thousand is spent on advertising, 22,506 fruit trees will be sold.   **29.** Increasing: $(-1, 1)$; decreasing: $(-\infty, -1)$, $(1, \infty)$ **31.** Increasing: $(-1.414, 1.414)$; decreasing: $(-2, -1.414)$, $(1.414, 2)$ **33.** $A(x) = x(240 - x)$, or $240x - x^2$ **35.** $h(d) = \sqrt{d^2 - 3500^2}$

37. $A(w) = 10w - \dfrac{w^2}{2}$ **39.** $d(s) = \dfrac{14}{s}$

41. (a) $A(x) = x(240 - 4x)$, or $240x - 4x^2$; **(b)** $\{x \mid 0 < x < 60\}$; **(c)** 120 ft by 30 ft

43. (a) $V(x) = x(12 - 2x)(12 - 2x)$, or $4x(6 - x)^2$; **(b)** $\{x \mid 0 < x < 6\}$; **(c)** $y = 4x(6 - x)^2$;

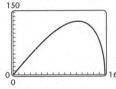

(d) 8 cm by 8 cm by 2 cm   **45. (a)** $A(x) = x\sqrt{256 - x^2}$; **(b)** $\{x \mid 0 < x < 16\}$; **(c)** $y = x\sqrt{256 - x^2}$

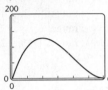

(d) 11.314 ft by 11.314 ft   **47.** $g(-4) = 0$; $g(0) = 4$; $g(1) = 5$; $g(3) = 5$ **49.** $h(-6) = 0$; $h(0) = 1$; $h(1) = 3$; $h(4) = 6$

51.

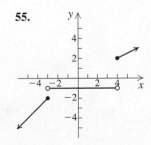

53.

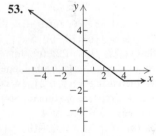

55.

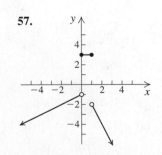

57.

59.
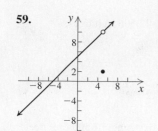

61.
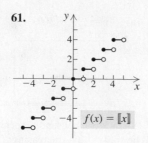
$f(x) = [\![x]\!]$

63.
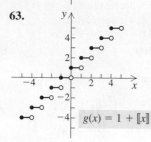
$g(x) = 1 + [\![x]\!]$

65. Domain: $(-\infty, \infty)$; range: $(-\infty, 0] \cup [3, \infty)$
67. Domain: $(-\infty, \infty)$; range: $[-1, \infty)$
69. Domain: $(-\infty, \infty)$; range: $(-\infty, -2] \cup \{-1\} \cup [2, \infty)$
71. Domain: $(-\infty, \infty)$; range: $\{-5, -2, 4\}$;
$$f(x) = \begin{cases} -2, & \text{for } x < 2, \\ -5, & \text{for } x = 2, \\ 4, & \text{for } x > 2 \end{cases}$$
73. Domain: $(-\infty, \infty)$; range: $(-\infty, -1] \cup [2, \infty)$;
$$g(x) = \begin{cases} x, & \text{for } x \le -1, \\ 2, & \text{for } -1 < x < 2, \\ x, & \text{for } x \ge 2 \end{cases}$$
or
$$g(x) = \begin{cases} x, & \text{for } x \le -1, \\ 2, & \text{for } -1 < x \le 2, \\ x, & \text{for } x > 2 \end{cases}$$
75. Domain: $[-5, 3]$; range: $(-3, 5)$;
$$h(x) = \begin{cases} x + 8, & \text{for } -5 \le x < -3, \\ 3, & \text{for } -3 \le x \le 1, \\ 3x - 6, & \text{for } 1 < x \le 3 \end{cases}$$
77. (a) 38; (b) 38; (c) $5a^2 - 7$; (d) $5a^2 - 7$
78. (a) 22; (b) -22; (c) $4a^3 - 5a$; (d) $-4a^3 + 5a$
79. $y = -\frac{1}{8}x + \frac{7}{8}$ **80.** Slope is $\frac{2}{9}$; y-intercept is $\left(0, \frac{1}{9}\right)$.
81. Increasing: $(-5, -2)$, $(4, \infty)$; decreasing: $(-\infty, -5)$, $(-2, 4)$; relative maximum: 560 at $x = -2$; relative minima: 425 at $x = -5$, -304 at $x = 4$
83. (a)

C↑
$12
9
6
3
2   4 t

(b) $C(t) = 3([\![t]\!] + 1), t > 0$
85. $\{x \mid -5 \le x < -4 \text{ or } 5 \le x < 6\}$
87. (a) $h(r) = \dfrac{30 - 5r}{3}$; (b) $V(r) = \pi r^2 \left(\dfrac{30 - 5r}{3}\right)$;
(c) $V(h) = \pi h \left(\dfrac{30 - 3h}{5}\right)^2$

Exercise Set 2.2

1. 33   **3.** -1 **5.** Does not exist   **7.** 0   **9.** 1
11. Does not exist   **13.** 0   **15.** 5   **17.** (a) Domain of f, g, $f + g$, $f - g$, fg, and ff: $(-\infty, \infty)$; domain of f/g: $\left(-\infty, \frac{3}{5}\right) \cup \left(\frac{3}{5}, \infty\right)$; domain of g/f: $\left(-\infty, -\frac{3}{2}\right) \cup \left(-\frac{3}{2}, \infty\right)$;
(b) $(f + g)(x) = -3x + 6$; $(f - g)(x) = 7x$;
$(fg)(x) = -10x^2 - 9x + 9$; $(ff)(x) = 4x^2 + 12x + 9$;
$(f/g)(x) = \dfrac{2x + 3}{3 - 5x}$; $(g/f)(x) = \dfrac{3 - 5x}{2x + 3}$
19. (a) Domain of f: $(-\infty, \infty)$; domain of g: $[-4, \infty)$; domain of $f + g$, $f - g$, and fg: $[-4, \infty)$; domain of ff: $(-\infty, \infty)$; domain of f/g: $(-4, \infty)$; domain of g/f: $[-4, 3) \cup (3, \infty)$; (b) $(f + g)(x) = x - 3 + \sqrt{x + 4}$;
$(f - g)(x) = x - 3 - \sqrt{x + 4}$; $(fg)(x) = (x - 3)\sqrt{x + 4}$;
$(ff)(x) = x^2 - 6x + 9$; $(f/g)(x) = \dfrac{x - 3}{\sqrt{x + 4}}$;
$(g/f)(x) = \dfrac{\sqrt{x + 4}}{x - 3}$ **21.** (a) Domain of f, g, $f + g$, $f - g$, fg, and ff: $(-\infty, \infty)$; domain of f/g: $(-\infty, 0) \cup (0, \infty)$; domain of g/f: $\left(-\infty, \frac{1}{2}\right) \cup \left(\frac{1}{2}, \infty\right)$ (b) $(f + g)(x) = -2x^2 + 2x - 1$;
$(f - g)(x) = 2x^2 + 2x - 1$; $(fg)(x) = -4x^3 + 2x^2$;
$(ff)(x) = 4x^2 - 4x + 1$; $(f/g)(x) = \dfrac{2x - 1}{-2x^2}$;
$(g/f)(x) = \dfrac{-2x^2}{2x - 1}$ **23.** (a) Domain of f: $[3, \infty)$; domain of g: $[-3, \infty)$; domain of $f + g$, $f - g$, fg, and ff: $[3, \infty)$; domain of f/g: $[3, \infty)$; domain of g/f: $(3, \infty)$;
(b) $(f + g)(x) = \sqrt{x - 3} + \sqrt{x + 3}$;
$(f - g)(x) = \sqrt{x - 3} - \sqrt{x + 3}$; $(fg)(x) = \sqrt{x^2 - 9}$;
$(ff)(x) = |x - 3|$; $(f/g)(x) = \dfrac{\sqrt{x - 3}}{\sqrt{x + 3}}$;
$(g/f)(x) = \dfrac{\sqrt{x + 3}}{\sqrt{x - 3}}$ **25.** (a) Domain of f, g, $f + g$, $f - g$, fg, and ff: $(-\infty, \infty)$; domain of f/g: $(-\infty, 0) \cup (0, \infty)$; domain of g/f: $(-\infty, -1) \cup (-1, \infty)$; (b) $(f + g)(x) = x + 1 + |x|$;
$(f - g)(x) = x + 1 - |x|$; $(fg)(x) = (x + 1)|x|$;
$(ff)(x) = x^2 + 2x + 1$; $(f/g)(x) = \dfrac{x + 1}{|x|}$; $(g/f)(x) = \dfrac{|x|}{x + 1}$
27. (a) Domain of f, g, $f + g$, $f - g$, fg, and ff: $(-\infty, \infty)$; domain of f/g: $(-\infty, -3) \cup \left(-3, \frac{1}{2}\right) \cup \left(\frac{1}{2}, \infty\right)$; domain of g/f: $(-\infty, 0) \cup (0, \infty)$; (b) $(f + g)(x) = x^3 + 2x^2 + 5x - 3$;
$(f - g)(x) = x^3 - 2x^2 - 5x + 3$; $(fg)(x) = 2x^5 + 5x^4 - 3x^3$;
$(ff)(x) = x^6$; $(f/g)(x) = \dfrac{x^3}{2x^2 + 5x - 3}$;
$(g/f)(x) = \dfrac{2x^2 + 5x - 3}{x^3}$ **29.** (a) Domain of f: $(-\infty, -1) \cup (-1, \infty)$; domain of g: $(-\infty, 6) \cup (6, \infty)$; domain of $f + g$, $f - g$, and fg: $(-\infty, -1) \cup (-1, 6) \cup (6, \infty)$; domain of ff: $(-\infty, -1) \cup (-1, \infty)$; domain of f/g and g/f: $(-\infty, -1) \cup (-1, 6) \cup (6, \infty)$;
(b) $(f + g)(x) = \dfrac{4}{x + 1} + \dfrac{1}{6 - x}$;

$ - g)(x) = \dfrac{4}{x+1} - \dfrac{1}{6-x}; (fg)(x) = \dfrac{4}{(x+1)(6-x)};$

$f)(x) = \dfrac{16}{(x+1)^2}; (f/g)(x) = \dfrac{4(6-x)}{x+1};$

$g/f)(x) = \dfrac{x+1}{4(6-x)}$

1. **(a)** Domain of f: $(-\infty,0)\cup(0,\infty)$; domain of g: $(-\infty,\infty)$;
omain of $f+g$, $f-g$, fg, and ff: $(-\infty,0)\cup(0,\infty)$;
omain of f/g: $(-\infty,0)\cup(0,3)\cup(3,\infty)$; domain of g/f:
$-\infty,0)\cup(0,\infty)$; **(b)** $(f+g)(x)=\dfrac{1}{x}+x-3$;

$ - g)(x) = \dfrac{1}{x} - x + 3; (fg)(x) = 1 - \dfrac{3}{x}; (ff)(x) = \dfrac{1}{x^2};$

$/g)(x) = \dfrac{1}{x(x-3)}; (g/f)(x) = x(x-3)$

3. **(a)** Domain of f: $(-\infty,2)\cup(2,\infty)$; domain of g: $[1,\infty)$;
omain of $f+g$, $f-g$, and fg: $[1,2)\cup(2,\infty)$; domain of ff:
$-\infty,2)\cup(2,\infty)$; domain of f/g: $(1,2)\cup(2,\infty)$; domain of
f: $[1,2)\cup(2,\infty)$; **(b)** $(f+g)(x)=\dfrac{3}{x-2}+\sqrt{x-1};$

$ - g)(x) = \dfrac{3}{x-2} - \sqrt{x-1};$

$g)(x) = \dfrac{3\sqrt{x-1}}{x-2}; (ff)(x) = \dfrac{9}{(x-2)^2};$

$/g)(x) = \dfrac{3}{(x-2)\sqrt{x-1}}; (g/f)(x) = \dfrac{(x-2)\sqrt{x-1}}{3}$

5. Domain of F: $[2,11]$; domain of G: $[1,9]$; domain of $F+G$:
$2,9]$ **37.** $[2,3)\cup(3,9]$

9.
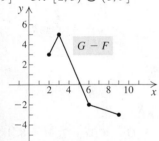

1. Domain of F: $[0,9]$; domain of G: $[3,10]$; domain of $F+G$:
$3,9]$ **43.** $[3,6)\cup(6,8)\cup(8,9]$

5.
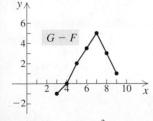

7. **(a)** $P(x) = -0.4x^2 + 57x - 13$; **(b)** $R(100) = 2000$;
$(100) = 313$; $P(100) = 1687$; **(c)** Left to the student

9. 3   **51.** 6   **53.** $\frac{1}{3}$ **55.** $\dfrac{-1}{3x(x+h)}$, or $-\dfrac{1}{3x(x+h)}$

7. $\dfrac{1}{4x(x+h)}$ **59.** $2x+h$ **61.** $-2x-h$

3. $6x+3h-2$ **65.** $\dfrac{5|x+h|-5|x|}{h}$

7. $3x^2+3xh+h^2$ **69.** $\dfrac{7}{(x+h+3)(x+3)}$

71.
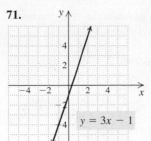

72.
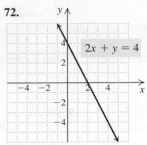

73.
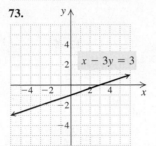

74.
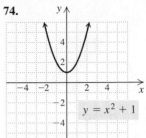

75. $f(x) = \dfrac{1}{x+7}, g(x) = \dfrac{1}{x-3}$; answers may vary

77. $(-\infty,-1)\cup(-1,1)\cup\left(1,\frac{7}{3}\right)\cup\left(\frac{7}{3},3\right)\cup(3,\infty)$

Exercise Set 2.3

1. -8 **3.** 64   **5.** 218   **7.** -80 **9.** -6 **11.** 512
13. -32 **15.** x^9 **17.** $(f\circ g)(x) = (g\circ f)(x) = x$;
domain of $f\circ g$ and $g\circ f$: $(-\infty,\infty)$ **19.** $(f\circ g)(x) = 3x^2 - 2x$;
$(g\circ f)(x) = 3x^2 + 4x$; domain of $f\circ g$ and $g\circ f$: $(-\infty,\infty)$
21. $(f\circ g)(x) = 16x^2 - 24x + 6$; $(g\circ f)(x) = 4x^2 - 15$;
domain of $f\circ g$ and $g\circ f$: $(-\infty,\infty)$
23. $(f\circ g)(x) = \dfrac{4x}{x-5}$; $(g\circ f)(x) = \dfrac{1-5x}{4}$; domain of $f\circ g$:
$(-\infty,0)\cup(0,5)\cup(5,\infty)$; domain of $g\circ f$: $\left(-\infty,\frac{1}{5}\right)\cup\left(\frac{1}{5},\infty\right)$
25. $(f\circ g)(x) = (g\circ f)(x) = x$; domain of $f\circ g$ and $g\circ f$:
$(-\infty,\infty)$ **27.** $(f\circ g)(x) = 2\sqrt{x} + 1$; $(g\circ f)(x) = \sqrt{2x+1}$;
domain of $f\circ g$: $[0,\infty)$; domain of $g\circ f$: $\left[-\frac{1}{2},\infty\right)$
29. $(f\circ g)(x) = 20$; $(g\circ f)(x) = 0.05$; domain of $f\circ g$ and $g\circ f$:
$(-\infty,\infty)$ **31.** $(f\circ g)(x) = |x|$; $(g\circ f)(x) = x$; domain of $f\circ g$:
$(-\infty,\infty)$; domain of $g\circ f$: $[-5,\infty)$ **33.** $(f\circ g)(x) = 5-x$;
$(g\circ f)(x) = \sqrt{1-x^2}$; domain of $f\circ g$: $(-\infty,3]$; domain of $g\circ f$:
$[-1,1]$ **35.** $(f\circ g)(x) = (g\circ f)(x) = x$; domain of $f\circ g$:
$(-\infty,-1)\cup(-1,\infty)$; domain of $g\circ f$: $(-\infty,0)\cup(0,\infty)$
37. $(f\circ g)(x) = x^3 - 2x^2 - 4x + 6$;
$(g\circ f)(x) = x^3 - 5x^2 + 3x + 8$; domain of $f\circ g$ and $g\circ f$:
$(-\infty,\infty)$ **39.** $f(x) = x^5$; $g(x) = 4 + 3x$

41. $f(x) = \dfrac{1}{x}; g(x) = (x-2)^4$ **43.** $f(x) = \dfrac{x-1}{x+1};$

$g(x) = x^3$ **45.** $f(x) = x^6; g(x) = \dfrac{2+x^3}{2-x^3}$

47. $f(x) = \sqrt{x}; g(x) = \dfrac{x-5}{x+2}$

49. $f(x) = x^3 - 5x^2 + 3x - 1; g(x) = x + 2$
51. **(a)** $r(t) = 3t$; **(b)** $A(r) = \pi r^2$; **(c)** $(A\circ r)(t) = 9\pi t^2$; the
function gives the area of the ripple in terms of time t.
53. $f(x) = x + 1$ **55.** (c)   **56.** None   **57.** (b), (d), (f), and
(h)   **58.** (b)   **59.** (a)   **60.** (c) and (g)   **61.** (c) and (g)

62. (a) and (f)   **63.** Only $(c \circ p)(a)$ makes sense. It represents the cost of the grass seed required to seed a lawn with area a.

Mid-Chapter Mixed Review: Chapter 2

1. True   **2.** False   **3.** True   **4. (a)** $(2, 4)$;
(b) $(-5, -3)$, $(4, 5)$; **(c)** $(-3, -1)$ **5.** Relative maximum:
6.30 at $x = -1.29$; relative minimum: -2.30 at $x = 1.29$;
increasing: $(-\infty, -1.29)$, $(1.29, \infty)$; decreasing: $(-1.29, 1.29)$
6. Domain: $[-5, -1] \cup [2, 5]$; range: $[-3, 5]$
7. $A(h) = \dfrac{h^2}{2} + h$ **8.** -10; -8; 1; 3

9.

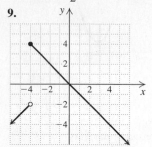

10. 1   **11.** -4 **12.** 5
13. Does not exist

14. (a) Domain of f, g, $f + g$, $f - g$, fg, and ff: $(-\infty, \infty)$;
domain of f/g: $(-\infty, -4) \cup (-4, \infty)$; domain of
g/f: $\left(-\infty, -\frac{5}{2}\right) \cup \left(-\frac{5}{2}, \infty\right)$; **(b)** $(f + g)(x) = x + 1$;
$(f - g)(x) = 3x + 9$; $(fg)(x) = -2x^2 - 13x - 20$;
$(ff)(x) = 4x^2 + 20x + 25$; $(f/g)(x) = \dfrac{2x + 5}{-x - 4}$;
$(g/f)(x) = \dfrac{-x - 4}{2x + 5}$ **15. (a)** Domain of f: $(-\infty, \infty)$;
domain of g, $f + g$, $f - g$, and fg: $[-2, \infty)$; domain
of ff: $(-\infty, \infty)$; domain of f/g: $(-2, \infty)$; domain of
g/f: $[-2, 1) \cup (1, \infty)$; **(b)** $(f + g)(x) = x - 1 + \sqrt{x + 2}$;
$(f - g)(x) = x - 1 - \sqrt{x + 2}$; $(fg)(x) = (x - 1)\sqrt{x + 2}$;
$(ff)(x) = x^2 - 2x + 1$; $(f/g)(x) = \dfrac{x - 1}{\sqrt{x + 2}}$;
$(g/f)(x) = \dfrac{\sqrt{x + 2}}{x - 1}$ **16.** 4   **17.** $-2x - h$ **18.** 6

19. 28   **20.** -24 **21.** 102   **22.** $(f \circ g)(x) = 3x + 2$;
$(g \circ f)(x) = 3x + 4$; domain of $f \circ g$ and $g \circ f$: $(-\infty, \infty)$
23. $(f \circ g)(x) = 3\sqrt{x} + 2$; $(g \circ f)(x) = \sqrt{3x + 2}$; domain
of $f \circ g$: $[0, \infty)$; domain of $g \circ f$: $\left[-\frac{2}{3}, \infty\right)$ **24.** The graph of
$y = (h - g)(x)$ will be the same as the graph of $y = h(x)$ shifted
down b units.   **25.** Under the given conditions, $(f + g)(x)$ and
$(f/g)(x)$ have different domains if $g(x) = 0$ for one or more
real numbers x.   **26.** If f and g are linear functions, then any
real number can be an input for each function. Thus the domain of
$f \circ g = $ the domain of $g \circ f = (-\infty, \infty)$.   **27.** This approach is
not valid. Consider Exercise 23 in Section 2.3, for example. Since
$(f \circ g)(x) = \dfrac{4x}{x - 5}$, an examination of only this composed function
would lead to the incorrect conclusion that the domain of $f \circ g$ is
$(-\infty, 5) \cup (5, \infty)$. However, we must also exclude from the
domain of $f \circ g$ those values of x that are not in the domain of g.
Thus the domain of $f \circ g$ is $(-\infty, 0) \cup (0, 5) \cup (5, \infty)$.

Exercise Set 2.4

1. x-axis, no; y-axis, yes; origin, no   **3.** x-axis, yes; y-axis, no;
origin, no   **5.** x-axis, no; y-axis, no; origin, yes   **7.** x-axis, no;
y-axis, yes; origin, no   **9.** x-axis, no; y-axis, no; origin, no
11. x-axis, no; y-axis, yes; origin, no   **13.** x-axis, no; y-axis, no;
origin, yes   **15.** x-axis, no; y-axis, no; origin, yes   **17.** x-axis,
yes; y-axis, yes; origin, yes   **19.** x-axis, no; y-axis, yes; origin, no
21. x-axis, yes; y-axis, yes; origin, yes   **23.** x-axis, no; y-axis, no;
origin, no   **25.** x-axis, no; y-axis, no; origin, yes   **27.** x-axis,
$(-5, -6)$; y-axis, $(5, 6)$; origin, $(5, -6)$ **29.** x-axis, $(-10, 7)$;
y-axis, $(10, -7)$; origin, $(10, 7)$ **31.** x-axis, $(0, 4)$; y-axis,
$(0, -4)$; origin, $(0, 4)$ **33.** Even   **35.** Odd   **37.** Neither
39. Odd   **41.** Even   **43.** Odd   **45.** Neither   **47.** Even
49.

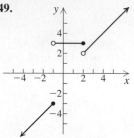

50. University of California–Berkeley: 3576 volunteers; University
of Wisconsin–Madison: 3112 volunteers   **51.** Odd
53. x-axis, yes; y-axis, no; origin, no

55. $E(-x) = \dfrac{f(-x) + f(-(-x))}{2} = \dfrac{f(-x) + f(x)}{2} = E(x)$

57. (a) $E(x) + O(x) = \dfrac{f(x) + f(-x)}{2} + \dfrac{f(x) - f(-x)}{2} = \dfrac{2f(x)}{2} = f(x)$; **(b)** $f(x) = \dfrac{-22x^2 + \sqrt{x} + \sqrt{-x} - 20}{2} + \dfrac{8x^3 + \sqrt{x} - \sqrt{-x}}{2}$ **59.** True

Visualizing the Graph

1. C   **2.** B   **3.** A   **4.** E   **5.** G   **6.** D   **7.** H   **8.** I
9. F

Exercise Set 2.5

1. Start with the graph of
$y = x^2$. Shift it right 3 units.

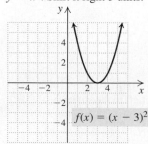

$f(x) = (x - 3)^2$

3. Start with the graph of
$y = x$. Shift it down 3 units.

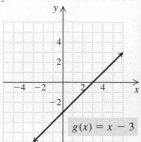

$g(x) = x - 3$

. Start with the graph of
$= \sqrt{x}$. Reflect it across the
-axis.

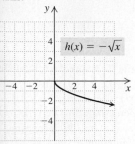

. Start with the graph of
$= x$. Stretch it vertically by
multiplying each y-coordinate
y 3. Then reflect it across the
-axis and shift it up 3 units.

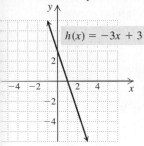

3. Start with the graph of
$= x^3$. Shift it right 2 units.
hen reflect it across the x-axis.

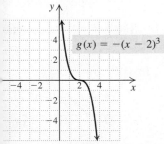

7. Start with the graph of
$= x^3$. Shrink it vertically by
multiplying each y-coordinate
y $\frac{1}{3}$. Then shift it up 2 units.

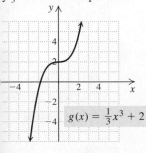

7. Start with the graph of
$y = \frac{1}{x}$. Shift it up 4 units.

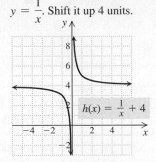

11. Start with the graph of
$y = |x|$. Shrink it vertically by
multiplying each y-coordinate
by $\frac{1}{2}$. Then shift it down 2 units.

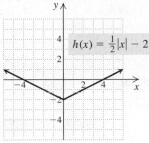

15. Start with the graph of
$y = x^2$. Shift it left 1 unit. Then
shift it down 1 unit.

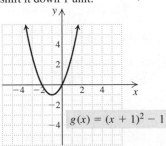

19. Start with the graph of
$y = \sqrt{x}$. Shift it left 2 units.

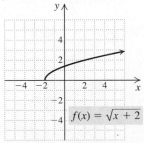

21. Start with the graph of $y = \sqrt[3]{x}$. Shift it down 2 units.

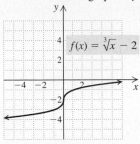

23. Start with the graph of $y = |x|$. Shrink it horizontally by multiplying each x-coordinate by $\frac{1}{3}$ (or dividing each x-coordinate by 3).

25. Start with the graph of $y = \frac{1}{x}$. Stretch it vertically by multiplying each y-coordinate by 2.   **27.** Start with the graph of $y = \sqrt{x}$. Stretch it vertically by multiplying each y-coordinate by 3. Then shift it down 5 units.   **29.** Start with the graph of $y = |x|$. Stretch it horizontally by multiplying each x-coordinate by 3. Then shift it down 4 units.   **31.** Start with the graph of $y = x^2$. Shift it right 5 units, shrink it vertically by multiplying each y-coordinate by $\frac{1}{4}$, and then reflect it across the x-axis.   **33.** Start with the graph of $y = \frac{1}{x}$. Shift it left 3 units, then up 2 units.   **35.** Start with the graph of $y = x^2$. Shift it right 3 units. Then reflect it across the x-axis and shift it up 5 units.   **37.** $(-12, 2)$ **39.** $(12, 4)$
41. $(-12, 2)$ **43.** $(-12, 16)$ **45.** B   **47.** A
49. $f(x) = -(x - 8)^2$ **51.** $f(x) = |x + 7| + 2$
53. $f(x) = \frac{1}{2x} - 3$ **55.** $f(x) = -(x - 3)^2 + 4$
57. $f(x) = \sqrt{-(x + 2)} - 1$
59.

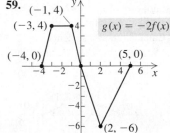

61.

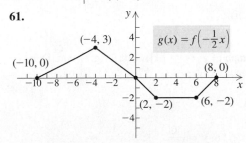

63.

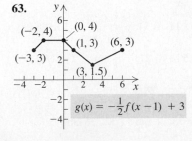

65.

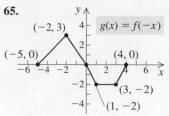

67.

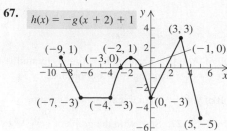

69.

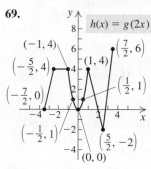

71. (f)   **73.** (f)   **75.** (d)   **77.** (c)
79. $f(-x) = 2(-x)^4 - 35(-x)^3 + 3(-x) - 5 =$
$2x^4 + 35x^3 - 3x - 5 = g(x)$ **81.** $g(x) = x^3 - 3x^2 + 2$
83. $k(x) = (x + 1)^3 - 3(x + 1)^2$ **85.** *x*-axis, no; *y*-axis, yes;
origin, no   **86.** *x*-axis, yes; *y*-axis, no; origin, no   **87.** *x*-axis, no;
y-axis, no; origin, yes   **88.** \$44,070   **89.** \$7500
90. Saudi Arabia: 53,919 students; Canada: 28,304 students
91.   **93.**

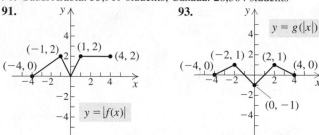

95. Start with the graph of $g(x) = [\![x]\!]$. Shift it right $\frac{1}{2}$ unit. Domain:
all real numbers; range: all integers.   **97.** $(3, 8)$; $(3, 6)$; $(\frac{3}{2}, 4)$

Exercise Set 2.6

1. 4.5; $y = 4.5x$ **3.** 36; $y = \dfrac{36}{x}$ **5.** 4; $y = 4x$

7. 4; $y = \dfrac{4}{x}$ **9.** $\dfrac{3}{8}$; $y = \dfrac{3}{8}x$ **11.** 0.54; $y = \dfrac{0.54}{x}$ **13.** \$8.25

15. $5\frac{5}{7}$ hr   **17.** 90 g   **19.** 3.5 hr   **21.** $66\frac{2}{3}$ cm   **23.** 1.92 ft

25. $y = \dfrac{0.0015}{x^2}$ **27.** $y = 15x^2$ **29.** $y = xz$ **31.** $y = \frac{3}{10}xz^2$

33. $y = \dfrac{1}{5} \cdot \dfrac{xz}{wp}$, or $y = \dfrac{xz}{5wp}$ **35.** 2.5 m   **37.** 36 mph

39. About 43 earned runs   **41.** Parallel   **42.** Zero
43. Relative minimum   **44.** Odd function   **45.** Inverse variation
47. \$3.56; \$3.53   **49.** $\dfrac{\pi}{4}$

Review Exercises: Chapter 2

1. True   **2.** False   **3.** True   **4.** True   **5.** (a) $(-4, -2)$;
(b) $(2, 5)$; **(c)** $(-2, 2)$ **6.** (a) $(-1, 0)$, $(2, \infty)$; **(b)** $(0, 2)$;
(c) $(-\infty, -1)$ **7.** Increasing: $(0, \infty)$; decreasing: $(-\infty, 0)$;
relative minimum: -1 at $x = 0$ **8.** Increasing: $(-\infty, 0)$; decreasing: $(0, \infty)$; relative maximum: 2 at $x = 0$ **9.** Increasing:
$(2, \infty)$; decreasing: $(-\infty, 2)$; relative minimum: -1 at $x = 2$
10. Increasing: $(-\infty, 0.5)$; decreasing: $(0.5, \infty)$;
relative maximum: 6.25 at $x = 0.5$ **11.** Increasing:
$(-\infty, -1.155)$, $(1.155, \infty)$; decreasing: $(-1.155, 1.155)$;
relative maximum: 3.079 at $x = -1.155$; relative minimum:
-3.079 at $x = 1.155$ **12.** Increasing: $(-1.155, 1.155)$;
decreasing: $(-\infty, -1.155)$, $(1.155, \infty)$; relative maximum:
1.540 at $x = 1.155$; relative minimum: -1.540 at $x = -1.155$
13. $A(x) = x(48 - 2x)$, or $48x - 2x^2$ **14.** $A(x) = 2x\sqrt{4 - x}$

15. (a) $A(x) = x\left(33 - \dfrac{x}{2}\right)$, or $33x - \dfrac{x^2}{2}$;

(b) $\{x \mid 0 < x < 66\}$;
(c) $y_1 = x\left(33 - \dfrac{x}{2}\right)$ **(d)** 33 ft by 16.5 ft

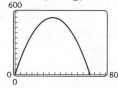
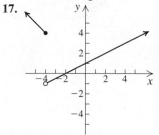

16. (a) $A(x) = x^2 + \dfrac{432}{x}$; **(b)** $(0, \infty)$;

(c) $x = 6$ in., height $= 3$ in.
17.   **18.**

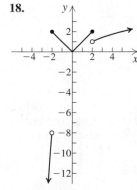

19.   **20.**

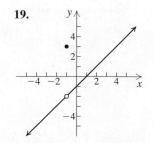
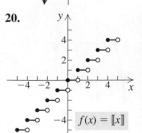

1.

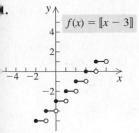

$f(x) = [\![x - 3]\!]$

22. $f(-1) = 1; f(5) = 2; f(-2) = 2; f(-3) = -27$
23. $f(-2) = -3; f(-1) = 3; f(0) = -1; f(4) = 3$
24. -33 **25.** 0 **26.** Does not exist
27. (a) Domain of f: $(-\infty, 0) \cup (0, \infty)$; domain of g: $(-\infty, \infty)$;
domain of $f + g$, $f - g$, and fg: $(-\infty, 0) \cup (0, \infty)$; domain of f/g:
$(-\infty, 0) \cup \left(0, \frac{3}{2}\right) \cup \left(\frac{3}{2}, \infty\right)$ (b) $(f + g)(x) = \frac{4}{x^2} + 3 - 2x$;

$(f - g)(x) = \frac{4}{x^2} - 3 + 2x; (fg)(x) = \frac{12}{x^2} - \frac{8}{x}$;

$(f/g)(x) = \frac{4}{x^2(3 - 2x)}$ **28.** (a) Domain of f, g, $f + g$, $f - g$,

and fg: $(-\infty, \infty)$; domain of f/g: $\left(-\infty, \frac{1}{2}\right) \cup \left(\frac{1}{2}, \infty\right)$;
(b) $(f + g)(x) = 3x^2 + 6x - 1; (f - g)(x) = 3x^2 + 2x + 1$;

$(fg)(x) = 6x^3 + 5x^2 - 4x; (f/g)(x) = \frac{3x^2 + 4x}{2x - 1}$

29. $P(x) = -0.5x^2 + 105x - 6$ **30.** 2 **31.** $-2x - h$

32. $\frac{-4}{x(x + h)}$, or $-\frac{4}{x(x + h)}$ **33.** 9 **34.** 5 **35.** 128

36. 580 **37.** 7 **38.** -509 **39.** $4x - 3$

40. $-24 + 27x^3 - 9x^6 + x^9$ **41.** (a) $(f \circ g)(x) = \frac{4}{(3 - 2x)^2}$;

$(g \circ f)(x) = 3 - \frac{8}{x^2}$; (b) domain of $f \circ g$: $\left(-\infty, \frac{3}{2}\right) \cup \left(\frac{3}{2}, \infty\right)$;

domain of $g \circ f$: $(-\infty, 0) \cup (0, \infty)$
42. (a) $(f \circ g)(x) = 12x^2 - 4x - 1$;
$(g \circ f)(x) = 6x^2 + 8x - 1$; (b) domain of $f \circ g$ and
$g \circ f$: $(-\infty, \infty)$ **43.** $f(x) = \sqrt{x}, g(x) = 5x + 2$; answers
may vary. **44.** $f(x) = 4x^2 + 9, g(x) = 5x - 1$; answers may
vary. **45.** x-axis, yes; y-axis, yes; origin, yes **46.** x-axis, yes;
y-axis, yes; origin, yes **47.** x-axis, no; y-axis, no; origin, no
48. x-axis, no; y-axis, yes; origin, no **49.** x-axis, no; y-axis, no;
origin, yes **50.** x-axis, no; y-axis, yes; origin, no **51.** Even
52. Even **53.** Odd **54.** Even **55.** Even **56.** Neither
57. Odd **58.** Even **59.** Even **60.** Odd
61. $f(x) = (x + 3)^2$ **62.** $f(x) = -\sqrt{x - 3} + 4$
63. $f(x) = 2|x - 3|$
64.

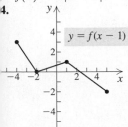

$y = f(x - 1)$

65.

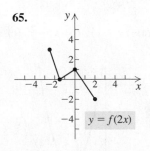

$y = f(2x)$

66.

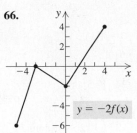

$y = -2f(x)$

67.

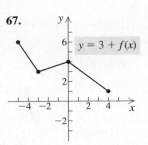

$y = 3 + f(x)$

68. $y = 4x$ **69.** $y = \frac{2}{3}x$ **70.** $y = \frac{2500}{x}$ **71.** $y = \frac{54}{x}$

72. $y = \frac{48}{x^2}$ **73.** $y = \frac{1}{10} \cdot \frac{xz^2}{w}$ **74.** 20 min **75.** 75

76. 500 watts **77.** A **78.** C **79.** B **80.** Let $f(x)$ and
$g(x)$ be odd functions. Then by definition, $f(-x) = -f(x)$, or
$f(x) = -f(-x)$, and $g(-x) = -g(x)$, or $g(x) = -g(-x)$.
Thus, $(f + g)(x) = f(x) + g(x) = -f(-x) + [-g(-x)] = -[f(-x) + g(-x)] = -(f + g)(-x)$ and $f + g$ is odd.
81. Reflect the graph of $y = f(x)$ across the x-axis and then across
the y-axis. **82.** (a) $4x^3 - 2x + 9$; (b) $4x^3 + 24x^2 + 46x + 35$;
(c) $4x^3 - 2x + 42$. (a) Adds 2 to each function value; (b) adds 2 to
each input before finding a function value; (c) adds the output for 2
to the output for x **83.** In the graph of $y = f(cx)$, the constant c
stretches or shrinks the graph of $y = f(x)$ horizontally. The constant
c in $y = c f(x)$ stretches or shrinks the graph of $y = f(x)$ vertically.
For $y = f(cx)$, the x-coordinates of $y = f(x)$ are divided by c; for
$y = c f(x)$, the y-coordinates of $y = f(x)$ are multiplied by c.
84. The graph of $f(x) = 0$ is symmetric with respect to the x-axis,
the y-axis, and the origin. This function is both even and odd.
85. If all the exponents are even numbers, then $f(x)$ is an even
function. If $a_0 = 0$ and all the exponents are odd numbers,
then $f(x)$ is an odd function. **86.** Let $y(x) = kx^2$. Then
$y(2x) = k(2x)^2 = k \cdot 4x^2 = 4 \cdot kx^2 = 4 \cdot y(x)$. Thus doubling x

causes y to be quadrupled. **87.** Let $y = k_1 x$ and $x = \frac{k_2}{z}$. Then

$y = k_1 \cdot \frac{k_2}{z}$, or $y = \frac{k_1 k_2}{z}$, so y varies inversely as z.

Test: Chapter 2

1. [2.1] (a) $(-5, -2)$; (b) $(2, 5)$; (c) $(-2, 2)$
2. [2.1] Increasing: $(-\infty, 0)$; decreasing: $(0, \infty)$; relative maximum: 2 at $x = 0$ **3.** [2.1] Increasing: $(-\infty, -2.667), (0, \infty)$;
decreasing: $(-2.667, 0)$; relative maximum: 9.481 at
$x = -2.667$; relative minimum: 0 at $x = 0$
4. [2.1] $A(b) = \frac{1}{2}b(4b - 6)$, or $2b^2 - 3b$
5. [2.1]

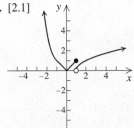

6. [2.1] $f\left(-\frac{7}{8}\right) = \frac{7}{8}; f(5) = 2; f(-4) = 16$ **7.** [2.2] 66
8. [2.2] 6 **9.** [2.2] -1 **10.** [2.2] 0 **11.** [2.1] $(-\infty, \infty)$
12. [2.1] $[3, \infty)$ **13.** [2.2] $[3, \infty)$ **14.** [2.2] $[3, \infty)$
15. [2.2] $[3, \infty)$ **16.** [2.2] $(3, \infty)$
17. [2.2] $(f + g)(x) = x^2 + \sqrt{x - 3}$

18. [2.2] $(f - g)(x) = x^2 - \sqrt{x - 3}$
19. [2.2] $(fg)(x) = x^2\sqrt{x - 3}$
20. [2.2] $(f/g)(x) = \dfrac{x^2}{\sqrt{x - 3}}$ **21.** [2.2] $\frac{1}{2}$
22. [2.2] $4x + 2h - 1$ **23.** [2.3] 83 **24.** [2.3] 0
25. [2.3] 4 **26.** [2.3] $16x + 15$
27. [2.3] $(f \circ g)(x) = \sqrt{x^2 - 4}$; $(g \circ f)(x) = x - 4$
28. [2.3] Domain of $(f \circ g)(x) = (-\infty, -2] \cup [2, \infty)$; domain of $(g \circ f)(x) = [5, \infty)$ **29.** [2.3] $f(x) = x^4$; $g(x) = 2x - 7$; answers may vary **30.** [2.4] x-axis: no; y-axis: yes; origin: no
31. [2.4] Odd **32.** [2.5] $f(x) = (x - 2)^2 - 1$
33. [2.5] $f(x) = -(x + 2)^2 + 3$
34. [2.5]

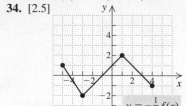

35. [2.6] $y = \dfrac{30}{x}$ **36.** [2.6] $y = 5x$ **37.** [2.6] $y = \dfrac{50xz^2}{w}$
38. [2.6] 50 ft **39.** [2.5] C **40.** [2.5] $(-1, 1)$

Chapter 3

Exercise Set 3.1

1. $\sqrt{3}i$ **3.** $5i$ **5.** $-\sqrt{33}i$ **7.** $-9i$ **9.** $7\sqrt{2}i$
11. $2 + 11i$ **13.** $5 - 12i$ **15.** $4 + 8i$ **17.** $-4 - 2i$
19. $5 + 9i$ **21.** $5 + 4i$ **23.** $5 + 7i$ **25.** $11 - 5i$
27. $-1 + 5i$ **29.** $2 - 12i$ **31.** -12 **33.** -45
35. $35 + 14i$ **37.** $6 + 16i$ **39.** $13 - i$ **41.** $-11 + 16i$
43. $-10 + 11i$ **45.** $-31 - 34i$ **47.** $-14 + 23i$ **49.** 41
51. 13 **53.** 74 **55.** $12 + 16i$ **57.** $-45 - 28i$
59. $-8 - 6i$ **61.** $2i$ **63.** $-7 + 24i$ **65.** $\frac{15}{146} + \frac{33}{146}i$
67. $\frac{10}{13} - \frac{15}{13}i$ **69.** $-\frac{14}{13} + \frac{5}{13}i$ **71.** $\frac{11}{25} - \frac{27}{25}i$
73. $\dfrac{-4\sqrt{3} + 10}{41} + \dfrac{5\sqrt{3} + 8}{41}i$ **75.** $-\frac{1}{2} + \frac{1}{2}i$ **77.** $-\frac{1}{2} - \frac{13}{2}i$
79. $-i$ **81.** $-i$ **83.** 1 **85.** i **87.** 625
89. $y = -2x + 1$ **90.** All real numbers, or $(-\infty, \infty)$
91. $\left(-\infty, -\frac{5}{3}\right) \cup \left(-\frac{5}{3}, \infty\right)$ **92.** $x^2 - 3x - 1$ **93.** $\frac{8}{11}$
94. $2x + h - 3$ **95.** True **97.** True **99.** $a^2 + b^2$
101. $x^2 - 6x + 25$

Exercise Set 3.2

1. $\frac{2}{3}, \frac{3}{2}$ **3.** $-2, 10$ **5.** $-1, \frac{2}{3}$ **7.** $-\sqrt{3}, \sqrt{3}$
9. $-\sqrt{7}, \sqrt{7}$ **11.** $-\sqrt{2}i, \sqrt{2}i$ **13.** $-4i, 4i$ **15.** $0, 3$
17. $-\frac{1}{3}, 0, 2$ **19.** $-1, -\frac{1}{7}, 1$ **21. (a)** $(-4, 0), (2, 0)$;
(b) $-4, 2$ **23. (a)** $(-1, 0), (3, 0)$; **(b)** $-1, 3$
25. (a) $(-2, 0), (2, 0)$; **(b)** $-2, 2$ **27. (a)** $(1, 0)$; **(b)** 1
29. $-7, 1$ **31.** $4 \pm \sqrt{7}$ **33.** $-4 \pm 3i$ **35.** $-2, \frac{1}{3}$
37. $-3, 5$ **39.** $-1, \frac{2}{5}$ **41.** $\dfrac{5 \pm \sqrt{7}}{3}$ **43.** $-\frac{1}{2} \pm \dfrac{\sqrt{7}}{2}i$
45. $\dfrac{4 \pm \sqrt{31}}{5}$ **47.** $\frac{5}{6} \pm \dfrac{\sqrt{23}}{6}i$ **49.** $4 \pm \sqrt{11}$

51. $\dfrac{-1 \pm \sqrt{61}}{6}$ **53.** $\dfrac{5 \pm \sqrt{17}}{4}$ **55.** $-\frac{1}{5} \pm \frac{3}{5}i$
57. 144; two real **59.** -7; two imaginary **61.** 49; two real
63. 2, 6 **65.** 0.143, 6 **67.** $-0.151, 1.651$ **69.** $-0.637, 3.13$
71. $-5, -1$ **73.** $\dfrac{3 \pm \sqrt{21}}{2}$ **75.** $\dfrac{5 \pm \sqrt{21}}{2}$
77. $-1 \pm \sqrt{6}$ **79.** $\frac{1}{4} \pm \dfrac{\sqrt{31}}{4}i$ **81.** $\dfrac{1 \pm \sqrt{13}}{6}$
83. $\dfrac{1 \pm \sqrt{6}}{5}$ **85.** $\dfrac{-3 \pm \sqrt{57}}{8}$ **87.** $-1.535, 0.869$
89. $-0.347, 1.181$ **91.** $\pm 1, \pm\sqrt{2}$ **93.** $\pm\sqrt{2}, \pm\sqrt{5}i$
95. $\pm 1, \pm\sqrt{5}i$ **97.** 16 **99.** $-8, 64$ **101.** 1, 16
103. $\frac{5}{2}, 3$ **105.** $-\frac{3}{2}, -1, \frac{1}{2}, 1$ **107.** 1995 **109.** 2011
111. About 10.216 sec **113.** Length: 4 ft; width: 3 ft
115. 4 and 9; -9 and -4 **117.** 2 cm **119.** Length: 8 ft; width:
6 ft **121.** Linear **123.** Quadratic **125.** Linear
127. About $3.95 million **128.** About 12 years after 2004, or in
2016 **129.** x-axis: yes; y-axis: yes; origin: yes **130.** x-axis: no;
y-axis: yes; origin: no **131.** Odd **132.** Neither **133. (a)** 2;
(b) $\frac{11}{2}$ **135. (a)** 2; **(b)** $1 - i$ **137.** 1
139. $-\sqrt{7}, -\frac{3}{2}, 0, \frac{1}{3}, \sqrt{7}$ **141.** $\dfrac{-1 \pm \sqrt{1 + 4\sqrt{2}}}{2}$
143. $3 \pm \sqrt{5}$ **145.** 19 **147.** $-2 \pm \sqrt{2}, \frac{1}{2} \pm \dfrac{\sqrt{7}}{2}i$

Visualizing the Graph

1. C **2.** B **3.** A **4.** J **5.** F **6.** D **7.** I **8.** G
9. H **10.** E

Exercise Set 3.3

1. (a) $\left(-\frac{1}{2}, -\frac{9}{4}\right)$; **(b)** $x = -\frac{1}{2}$; **(c)** minimum: $-\frac{9}{4}$
3. (a) $(4, -4)$; **(b)** $x = 4$; **(c)** minimum: -4;
(d)

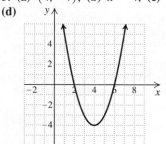

$f(x) = x^2 - 8x + 12$

5. (a) $\left(\frac{7}{2}, -\frac{1}{4}\right)$; **(b)** $x = \frac{7}{2}$; **(c)** minimum: $-\frac{1}{4}$;
(d)

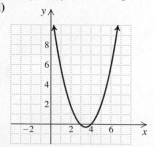

$f(x) = x^2 - 7x + 12$

7. (a) $(-2, 1)$; **(b)** $x = -2$; **(c)** minimum: 1;
(d)

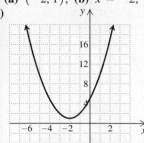

$$f(x) = x^2 + 4x + 5$$

9. (a) $(-4, -2)$; **(b)** $x = -4$; **(c)** minimum: -2;
(d)

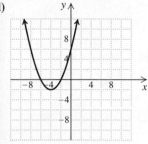

$$g(x) = \frac{x^2}{2} + 4x + 6$$

11. (a) $\left(-\frac{3}{2}, \frac{7}{2}\right)$; **(b)** $x = -\frac{3}{2}$; **(c)** minimum: $\frac{7}{2}$;
(d)

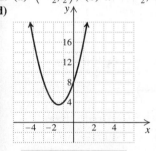

$$g(x) = 2x^2 + 6x + 8$$

13. (a) $(-3, 12)$; **(b)** $x = -3$; **(c)** maximum: 12;
(d)

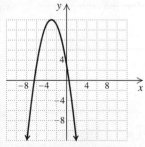

$$f(x) = -x^2 - 6x + 3$$

15. (a) $\left(\frac{1}{2}, \frac{3}{2}\right)$; **(b)** $x = \frac{1}{2}$; **(c)** maximum: $\frac{3}{2}$;
(d)

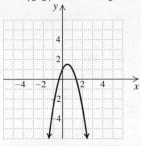

$$g(x) = -2x^2 + 2x + 1$$

17. (f)   **19.** (b)   **21.** (h)   **23.** (c)   **25.** True   **27.** False
29. True   **31. (a)** $(3, -4)$; **(b)** minimum: -4; **(c)** $[-4, \infty)$;
(d) increasing: $(3, \infty)$; decreasing: $(-\infty, 3)$
33. (a) $(-1, -18)$; **(b)** minimum: -18; **(c)** $[-18, \infty)$;
(d) increasing: $(-1, \infty)$; decreasing: $(-\infty, -1)$
35. (a) $\left(5, \frac{9}{2}\right)$; **(b)** maximum: $\frac{9}{2}$; **(c)** $\left(-\infty, \frac{9}{2}\right]$; **(d)** increasing:
$(-\infty, 5)$; decreasing: $(5, \infty)$ **37. (a)** $(-1, 2)$; **(b)** minimum:
2; **(c)** $[2, \infty)$; **(d)** increasing: $(-1, \infty)$; decreasing: $(-\infty, -1)$
39. (a) $\left(-\frac{3}{2}, 18\right)$; **(b)** maximum: 18; **(c)** $(-\infty, 18]$;
(d) increasing: $\left(-\infty, -\frac{3}{2}\right)$; decreasing: $\left(-\frac{3}{2}, \infty\right)$
41. 0.625 sec; 12.25 ft   **43.** 3.75 sec; 305 ft   **45.** 4.5 in.
47. Base: 10 cm; height: 10 cm   **49.** 2100 portable doghouses
51. \$797; 40 units   **53.** 4800 yd^2 **55.** About 60.5 ft   **57.** 3
58. $4x + 2h - 1$
59.

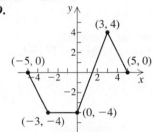

$$g(x) = -2f(x)$$

60.

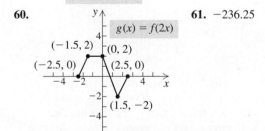

$$g(x) = f(2x)$$

61. -236.25

63.

$$f(x) = (|x| - 5)^2 - 3$$

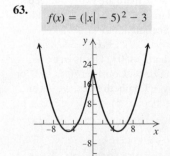

65. Pieces should be
$\dfrac{24\pi}{4 + \pi}$ in. and $\dfrac{96}{4 + \pi}$ in.

Mid-Chapter Mixed Review: Chapter 3

1. True   **2.** False   **3.** True   **4.** False   **5.** $6i$ **6.** $\sqrt{5}i$
7. $-4i$ **8.** $4\sqrt{2}i$ **9.** $-1+i$ **10.** $-7+5i$ **11.** $23+2i$
12. $-\frac{1}{29}-\frac{17}{29}i$ **13.** i **14.** 1 **15.** $-i$ **16.** -64
17. $-4, 1$ **18.** $-2, -\frac{3}{2}$ **19.** $\pm\sqrt{6}$ **20.** $\pm 10i$
21. $4x^2 - 8x - 3 = 0;\ 4x^2 - 8x = 3;$
$x^2 - 2x = \frac{3}{4};\ x^2 - 2x + 1 = \frac{3}{4} + 1;\ (x-1)^2 = \frac{7}{4};$
$x - 1 = \pm\dfrac{\sqrt{7}}{2};\ x = 1 \pm \dfrac{\sqrt{7}}{2} = \dfrac{2 \pm \sqrt{7}}{2}$
22. (a) 29; two real; (b) $\dfrac{3 \pm \sqrt{29}}{2}$; $-1.193, 4.193$
23. (a) 0; one real; (b) $\frac{3}{2}$ **24.** (a) -8; two nonreal;
(b) $-\dfrac{1}{3} \pm \dfrac{\sqrt{2}}{3}i$ **25.** $\pm 1, \pm\sqrt{6}i$ **26.** $\frac{1}{4}, 4$
27. 5 and 7; -7 and -5 **28.** (a) $(3, -2)$; (b) $x = 3$;
(c) minimum: -2; (d) $[-2, \infty)$;
(e) increasing: $(3, \infty)$; decreasing: $(-\infty, 3)$;
(f)

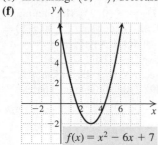

$f(x) = x^2 - 6x + 7$

29. (a) $(-1, -3)$; (b) $x = -1$; (c) maximum: -3;
(d) $(-\infty, -3]$; (e) increasing: $(-\infty, -1)$; decreasing: $(-1, \infty)$;
(f)

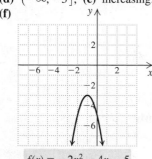

$f(x) = -2x^2 - 4x - 5$

30. Base: 8 in., height: 8 in.   **31.** The sum of two imaginary numbers is not always an imaginary number. For example, $(2 + i) + (3 - i) = 5$, a real number.   **32.** Use the discriminant. If $b^2 - 4ac < 0$, there are no x-intercepts. If $b^2 - 4ac = 0$, there is one x-intercept. If $b^2 - 4ac > 0$, there are two x-intercepts.
33. Completing the square was used in Section 3.2 to solve quadratic equations. It was used again in Section 3.3 to write quadratic functions in the form $f(x) = a(x - h)^2 + k$.   **34.** The x-intercepts of $g(x)$ are also $(x_1, 0)$ and $(x_2, 0)$. This is true because $f(x)$ and $g(x)$ have the same zeros. Consider $g(x) = 0$, or $-ax^2 - bx - c = 0$. Multiplying by -1 on both sides, we get an equivalent equation $ax^2 + bx + c = 0$, or $f(x) = 0$.

Exercise Set 3.4

1. $\frac{20}{9}$ **3.** 286   **5.** 6   **7.** 6   **9.** 4   **11.** 2, 3   **13.** $-1, 6$
15. $\frac{1}{2}, 5$ **17.** 7   **19.** No solution   **21.** $-\frac{69}{14}$ **23.** $-\frac{37}{18}$

25. 2   **27.** No solution   **29.** $\{x \mid x \text{ is a real number } and \ x \neq 0$
$and \ x \neq 6\}$ **31.** $\frac{5}{3}$ **33.** $\frac{9}{2}$ **35.** 3   **37.** -4 **39.** -5
41. $\pm\sqrt{2}$ **43.** No solution   **45.** 6   **47.** -1 **49.** $\frac{35}{2}$
51. -98 **53.** -6 **55.** 5   **57.** 7   **59.** 2   **61.** $-1, 2$
63. 7   **65.** 7   **67.** No solution   **69.** 1   **71.** 3, 7
73. 5   **75.** -1 **77.** -8 **79.** 81   **81.** $T_1 = \dfrac{P_1 V_1 T_2}{P_2 V_2}$

83. $C = \dfrac{1}{LW^2}$ **85.** $R_2 = \dfrac{RR_1}{R_1 - R}$

87. $P = \dfrac{A}{I^2 + 2I + 1}$, or $\dfrac{A}{(I + 1)^2}$ **89.** $p = \dfrac{Fm}{m - F}$

91. $\frac{15}{2}$, or 7.5   **92.** 3   **93.** China: 53,800,000 metric tons;
United States: 10,508,000 metric tons   **94.** About $18,609
95. $3 \pm 2\sqrt{2}$ **97.** -1 **99.** 0, 1

Exercise Set 3.5

1. $-7, 7$ **3.** 0   **5.** $-\frac{5}{6}, \frac{5}{6}$ **7.** No solution   **9.** $-\frac{1}{3}, \frac{1}{3}$
11. $-3, 3$ **13.** $-3, 5$ **15.** $-8, 4$ **17.** $-1, -\frac{1}{3}$
19. $-24, 44$ **21.** $-2, 4$ **23.** $-13, 7$ **25.** $-\frac{4}{3}, \frac{2}{3}$
27. $-\frac{3}{4}, \frac{9}{4}$ **29.** $-13, 1$ **31.** 0, 1
33. $(-7, 7)$;

35. $[-2, 2]$;

37. $(-\infty, -4.5] \cup [4.5, \infty)$;

39. $(-\infty, -3) \cup (3, \infty)$;

41. $\left(-\frac{1}{3}, \frac{1}{3}\right)$;

43. $(-\infty, -3] \cup [3, \infty)$;

45. $(-17, 1)$;

47. $(-\infty, -17] \cup [1, \infty)$;

49. $\left(-\frac{1}{4}, \frac{3}{4}\right)$;

51. $[-6, 3]$;

53. $(-\infty, 4.9) \cup (5.1, \infty)$;

55. $\left(-\infty, -\frac{1}{2}\right] \cup \left[\frac{7}{2}, \infty\right)$;

57. $\left[-\frac{7}{3}, 1\right]$;

59. $(-\infty, -8) \cup (7, \infty)$;

61. No solution   **63.** $(-\infty, \infty)$ **65.** y-intercept
66. Distance formula   **67.** Relation   **68.** Function
69. Horizontal lines   **70.** Parallel   **71.** Decreasing
72. Symmetric with respect to the y-axis   **73.** $\left(-\infty, \frac{1}{2}\right)$
75. No solution   **77.** $\left(-\infty, -\frac{8}{3}\right) \cup (-2, \infty)$

Review Exercises: Chapter 3

1. True   **2.** True   **3.** False   **4.** False   **5.** $-\frac{5}{2}, \frac{1}{3}$ **6.** $-5, 1$
7. $-2, \frac{4}{3}$ **8.** $-\sqrt{3}, \sqrt{3}$ **9.** $-\sqrt{10}i, \sqrt{10}i$

10. 1 **11.** −5, 3 **12.** $\dfrac{1 \pm \sqrt{41}}{4}$ **13.** $-\dfrac{1}{3} \pm \dfrac{2\sqrt{2}}{3}i$

14. $\dfrac{27}{7}$ **15.** $-\dfrac{1}{2}, \dfrac{9}{4}$ **16.** 0, 3 **17.** 5 **18.** 1, 7 **19.** −8, 1

20. $(-\infty, -3] \cup [3, \infty)$;

21. $\left(-\dfrac{14}{3}, 2\right)$;

22. $\left(-\dfrac{2}{3}, 1\right)$;

23. $(-\infty, -6] \cup [-2, \infty)$;

24. $P = \dfrac{MN}{M + N}$ **25.** $-2\sqrt{10}i$ **26.** $-4\sqrt{15}$ **27.** $-\dfrac{7}{8}$

28. $2 - i$ **29.** $1 - 4i$ **30.** $-18 - 26i$ **31.** $\dfrac{11}{10} + \dfrac{3}{10}i$

32. $-i$ **33.** $x^2 - 3x + \dfrac{9}{4} = 18 + \dfrac{9}{4}; \left(x - \dfrac{3}{2}\right)^2 = \dfrac{81}{4};$
$x - \dfrac{3}{2} = \pm\dfrac{9}{2}; x = \dfrac{3}{2} \pm \dfrac{9}{2}; -3, 6$ **34.** $3x^2 - 12x = 6;$
$x^2 - 4x = 2; x^2 - 4x + 4 = 2 + 4; (x - 2)^2 = 6;$
$x - 2 = \pm\sqrt{6}; x = 2 \pm \sqrt{6}; 2 - \sqrt{6}, 2 + \sqrt{6}$

35. $-4, \dfrac{2}{3}$ **36.** $1 - 3i, 1 + 3i$ **37.** $-2, 5$ **38.** 1

39. $\pm\sqrt{\dfrac{3 \pm \sqrt{5}}{2}}$ **40.** $-\sqrt{3}, 0, \sqrt{3}$ **41.** $-2, -\dfrac{2}{3}, 3$

42. −5, −2, 2 **43. (a)** $\left(\dfrac{3}{8}, -\dfrac{7}{16}\right)$; **(b)** $x = \dfrac{3}{8}$; **(c)** maximum: $-\dfrac{7}{16}$; **(d)** $\left(-\infty, -\dfrac{7}{16}\right]$;
(e)

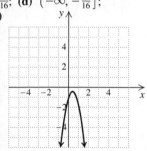

$f(x) = -4x^2 + 3x - 1$

44. (a) $(1, -2)$; **(b)** $x = 1$; **(c)** minimum: −2; **(d)** $[-2, \infty)$;
(e)

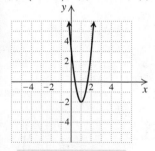

$f(x) = 5x^2 - 10x + 3$

45. (d) **46.** (c) **47.** (b) **48.** (a) **49.** 30 ft, 40 ft
50. Rebecca: 15 km/h; Harry: 8 km/h **51.** $35 - 5\sqrt{33}$ ft, or
about 6.3 ft **52.** 6 ft by 6 ft **53.** $\dfrac{15 - \sqrt{115}}{2}$ cm, or
about 2.1 cm **54.** B **55.** B **56.** A **57.** 256
58. $4 \pm 3\sqrt[4]{3}$, or 0.052, 7.948 **59.** −7, 9 **60.** $-\dfrac{1}{4}, 2$
61. −1 **62.** 9% **63.** ± 6 **64.** The product of two imaginary numbers is not always an imaginary number. For example,
$i \cdot i = i^2 = -1$, a real number. **65.** No; consider the quadratic

formula $x = \dfrac{-b \pm \sqrt{b^2 - 4ac}}{2a}$. If $b^2 - 4ac = 0$, then $x = \dfrac{-b}{2a}$,
so there is one real zero. If $b^2 - 4ac > 0$, then $\sqrt{b^2 - 4ac}$ is a
real number and there are two real zeros. If $b^2 - 4ac < 0$, then
$\sqrt{b^2 - 4ac}$ is an imaginary number and there are two imaginary
zeros. Thus a quadratic function cannot have one real zero and one
imaginary zero. **66.** You can conclude that $|a_1| = |a_2|$ since
these constants determine how wide the parabolas are. Nothing
can be concluded about the h's and the k's. **67.** When both sides
of an equation are multiplied by the LCD, the resulting equation
might not be equivalent to the original equation. One or more of
the possible solutions of the resulting equation might make a
denominator of the original equation 0. **68.** When both sides of
an equation are raised to an even power, the resulting equation might
not be equivalent to the original equation. For example, the solution
set of $x = -2$ is $\{-2\}$, but the solution set of $x^2 = (-2)^2$, or
$x^2 = 4$, is $\{-2, 2\}$. **69.** Absolute value is nonnegative.
70. $|x| \geq 0 > p$ for any real number x.

Test: Chapter 3

1. [3.2] $\dfrac{1}{2}, -5$ **2.** [3.2] $-\sqrt{6}, \sqrt{6}$ **3.** [3.2] $-2i, 2i$

4. [3.2] $-1, 3$ **5.** [3.2] $\dfrac{5 \pm \sqrt{13}}{2}$ **6.** [3.2] $\dfrac{3}{4} \pm \dfrac{\sqrt{23}}{4}i$

7. [3.2] 16 **8.** [3.4] $-1, \dfrac{13}{6}$ **9.** [3.4] 5 **10.** [3.4] 5

11. [3.5] $-11, 3$ **12.** [3.5] $-\dfrac{1}{2}, 2$

13. [3.5] $[-7, 1]$;

14. [3.5] $(-2, 3)$;

15. [3.5] $(-\infty, -7) \cup (-3, \infty)$;

16. [3.5] $\left(-\infty, -\dfrac{2}{3}\right] \cup [4, \infty)$;

17. [3.4] $B = \dfrac{AC}{A - C}$ **18.** [3.4] $n = \dfrac{R^2}{3p}$

19. [3.2] $x^2 + 4x = 1; x^2 + 4x + 4 = 1 + 4; (x + 2)^2 = 5;$
$x + 2 = \pm\sqrt{5}; x = -2 \pm \sqrt{5}; -2 - \sqrt{5}, -2 + \sqrt{5}$
20. [3.2] About 11.4 sec **21.** [3.1] $\sqrt{43}i$ **22.** [3.1] $-5i$
23. [3.1] $3 - 5i$ **24.** [3.1] $10 + 5i$ **25.** [3.1] $\dfrac{1}{10} - \dfrac{1}{5}i$

26. [3.1] i **27.** [3.2] $-\dfrac{1}{4}, 3$ **28.** [3.2] $\dfrac{1 \pm \sqrt{57}}{4}$

29. [3.3] **(a)** $(1, 9)$; **(b)** $x = 1$; **(c)** maximum: 9; **(d)** $(-\infty, 9]$;
(e)

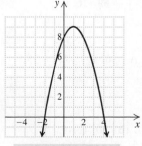

$f(x) = -x^2 + 2x + 8$

30. [3.3] 20 ft by 40 ft **31.** [3.3] C **32.** [3.3], [3.4] $-\dfrac{4}{9}$

Chapter 4

Exercise Set 4.1

1. $\frac{1}{2}x^3$; $\frac{1}{2}$; 3; cubic **3.** $0.9x$; 0.9; 1; linear **5.** $305x^4$; 305;
4; quartic **7.** x^4; 1; 4; quartic **9.** $4x^3$; 4; 3; cubic **11.** (d)
13. (b) **15.** (c) **17.** (a) **19.** (c) **21.** (d)
23. Yes; no; no **25.** No; yes; yes **27.** -3, multiplicity 2;
1, multiplicity 1 **29.** 4, multiplicity 3; -6, multiplicity 1
31. ± 3, each has multiplicity 3 **33.** 0, multiplicity 3; 1,
multiplicity 2; -4, multiplicity 1 **35.** 3, multiplicity 2; -4,
multiplicity 3; 0, multiplicity 4 **37.** $\pm \sqrt{3}$, ± 1, each has
multiplicity 1 **39.** -3, ± 1, each has multiplicity 1
41. ± 2, $\frac{1}{2}$, each has multiplicity 1 **43.** -1.532, -0.347, 1.879
45. -1.414, 0, 1.414 **47.** -1, 0, 1 **49.** -10.153, -1.871,
-0.821, -0.303, 0.098, 0.535, 1.219, 3.297 **51.** -1.386;
relative maximum: 1.506 at $x = -0.632$, relative minimum: 0.494
at $x = 0.632$; $(-\infty, \infty)$ **53.** -1.249, 1.249; relative minimum:
-3.8 at $x = 0$, no relative maxima; $[-3.8, \infty)$ **55.** -1.697, 0,
1.856; relative maximum: 11.012 at $x = 1.258$, relative
minimum: -8.183 at $x = -1.116$; $(-\infty, \infty)$ **57.** False
59. True **61.** 2008: 1.7 million albums; 2012: 4.9 million
albums; 2016: 15.6 million albums **63.** 2009: about 771 million
print books; 2014: about 635 million print books; 2015: about
795 million print books **65.** 5 sec **67.** 2003: 684,025
admissions; 2006: 739,119 admissions; 2011: 665,806 admissions
69. 6.3% **71.** (b) **73.** (c) **75.** (a) **77.** (a) Quadratic:
$y = -83.83100233x^2 - 930.0808858x + 23,360.76224$,
$R^2 \approx 0.9444$; cubic: $y = 52.94133644x^3 - 877.951049x^2 +$
$2098.163559x + 21,454.87413$, $R^2 \approx 0.9911$; quartic:
$y = -2.758741259x^4 + 108.1161616x^3 - 1222.793706x^2 +$
$2787.848873x + 21,256.24476$, $R^2 \approx 0.9919$; the R^2-value
for the quartic function, 0.9919, is the closest to 1; thus
the quartic function is the best fit; (b) 2014: 7476
foreign adoptions **79.** (a) Cubic: $y = -0.0018497401x^3 +$
$0.0678652219x^2 + 0.0636134735x + 2.244194051$, $R^2 \approx 0.8658$;
quartic: $y = 0.000024423807x^4 - 0.0037660904x^3 +$
$0.11514113x^2 - 0.3046875884x + 2.594806309$, $R^2 \approx 0.8686$;
the R^2-value for the quartic function, 0.8686, is the closest to 1; thus
the quartic function is the better fit. (b) 1988: \$10.516 billion; 2002:
\$17.158 billion **81.** 5 **82.** $6\sqrt{2}$ **83.** Center: $(3, -5)$;
radius: 7 **84.** Center: $(-4, 3)$; radius: $2\sqrt{2}$ **85.** $\{y | y \geq 3\}$,
or $[3, \infty)$ **86.** $\{x | x > \frac{5}{3}\}$, or $(\frac{5}{3}, \infty)$
87. $\{x | x \leq -13 \text{ or } x \geq 1\}$, or $(-\infty, -13] \cup [1, \infty)$
88. $\{x | -\frac{11}{12} \leq x \leq \frac{5}{12}\}$, or $[-\frac{11}{12}, \frac{5}{12}]$ **89.** 16; x^{16}

Visualizing the Graph

1. H **2.** D **3.** J **4.** B **5.** A **6.** C **7.** I
8. E **9.** G **10.** F

Exercise Set 4.2

1. (a) 5; (b) 5; (c) 4 **3.** (a) 10; (b) 10; (c) 9
5. (a) 3; (b) 3; (c) 2 **7.** (d) **9.** (f) **11.** (b)

13.
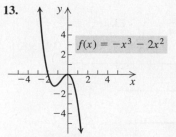
$f(x) = -x^3 - 2x^2$

15.
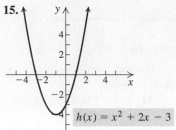
$h(x) = x^2 + 2x - 3$

17.
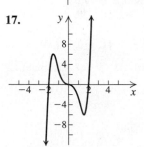
$h(x) = x^5 - 4x^3$

19.
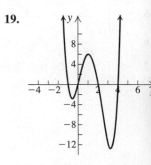
$h(x) = x(x - 4)(x + 1)(x - 2$

21.
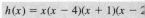
$g(x) = -\frac{1}{4}x^3 - \frac{3}{4}x^2$

23.
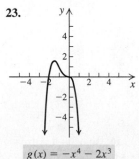
$g(x) = -x^4 - 2x^3$

25.
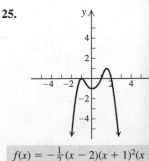
$f(x) = -\frac{1}{2}(x - 2)(x + 1)^2(x$

27.

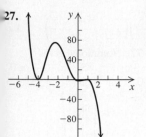

$g(x) = -x(x - 1)^2(x + 4)^2$

29.

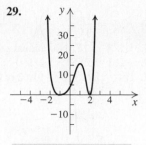

$f(x) = (x - 2)^2(x + 1)^4$

31.

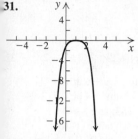

$g(x) = -(x - 1)^4$

33.

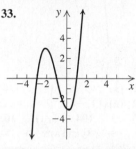

$h(x) = x^3 + 3x^2 - x - 3$

35.

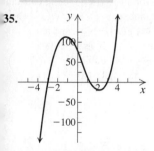

$f(x) = 6x^3 - 8x^2 - 54x + 72$

37.

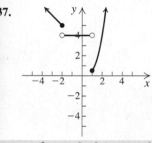

$$g(x) = \begin{cases} -x + 3, & \text{for } x \le -2, \\ 4, & \text{for } -2 < x < 1, \\ \frac{1}{2}x^3, & \text{for } x \ge 1 \end{cases}$$

39. $f(-5) = -18$ and $f(-4) = 7$. By the intermediate value theorem, since $f(-5)$ and $f(-4)$ have opposite signs, then $f(x)$ has a zero between -5 and -4.   **41.** $f(-3) = 22$ and $f(-2) = 5$. Both $f(-3)$ and $f(-2)$ are positive. We cannot use the intermediate value theorem to determine if there is a zero between -3 and -2.   **43.** $f(2) = 2$ and $f(3) = 57$. Both $f(2)$ and $f(3)$ are positive. We cannot use the intermediate value theorem to determine if there is a zero between 2 and 3.   **45.** $f(4) = -12$ and $f(5) = 4$. By the intermediate value theorem, since $f(4)$ and $f(5)$ have opposite signs, then $f(x)$ has a zero between 4 and 5.   **47.** (d)   **48.** (f)
49. (e)   **50.** (a)   **51.** (b)   **52.** (c)   **53.** $\frac{9}{10}$ **54.** $-3, 0, 4$
55. $-\frac{5}{3}, \frac{11}{2}$ **56.** $\frac{196}{25}$

Exercise Set 4.3

1. (a) No; (b) yes; (c) no   **3.** (a) Yes; (b) no; (c) yes
5. $P(x) = (x + 2)(x^2 - 2x + 4) - 16$
7. $P(x) = (x + 9)(x^2 - 3x + 2) + 0$
9. $P(x) = (x + 2)(x^3 - 2x^2 + 2x - 4) + 11$
11. $Q(x) = 2x^3 + x^2 - 3x + 10, R(x) = -42$
13. $Q(x) = x^2 - 4x + 8, R(x) = -24$
15. $Q(x) = 3x^2 - 4x + 8, R(x) = -18$
17. $Q(x) = x^4 + 3x^3 + 10x^2 + 30x + 89, R(x) = 267$
19. $Q(x) = x^3 + x^2 + x + 1, R(x) = 0$
21. $Q(x) = 2x^3 + x^2 + \frac{7}{2}x + \frac{7}{4}, R(x) = -\frac{1}{8}$ **23.** $0; -60; 0$
25. $10; 80; 998$ **27.** $5,935,988; -772$ **29.** $0; 0; 65;$
$1 - 12\sqrt{2}$ **31.** Yes; no   **33.** Yes; yes   **35.** No; yes
37. No; no   **39.** $f(x) = (x - 1)(x + 2)(x + 3); 1, -2, -3$
41. $f(x) = (x - 2)(x - 5)(x + 1); 2, 5, -1$
43. $f(x) = (x - 2)(x - 3)(x + 4); 2, 3, -4$
45. $f(x) = (x - 3)^3(x + 2); 3, -2$
47. $f(x) = (x - 1)(x - 2)(x - 3)(x + 5); 1, 2, 3, -5$

49.

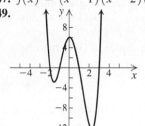

$f(x) = x^4 - x^3 - 7x^2 + x + 6$

51.

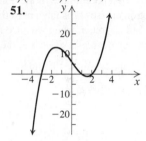

$f(x) = x^3 - 7x + 6$

53.

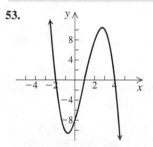

$f(x) = -x^3 + 3x^2 + 6x - 8$

55. $\frac{5}{4} \pm \frac{\sqrt{71}}{4}i$ **56.** $-1, \frac{3}{7}$ **57.** $-5, 0$ **58.** 10
59. $-3, -2$ **60.** $f(x) = 1313x + 339; 2011: 5591$ cases, 2017:
13,469 cases   **61.** $b = 15$ in., $h = 15$ in.
63. (a) $x + 4, x + 3, x - 2, x - 5;$
(b) $P(x) = (x + 4)(x + 3)(x - 2)(x - 5);$ (c) yes; two
examples are $f(x) = c \cdot P(x)$ for any nonzero constant c and
$g(x) = (x - a)P(x);$ (d) no
65. $\frac{14}{3}$ **67.** $0, -6$ **69.** Answers may vary. One possibility is
$P(x) = x^{15} - x^{14}$.   **71.** $x^2 + 2ix + (2 - 4i), R -6 - 2i$
73. $x - 3 + i, R 6 - 3i$

Mid-Chapter Mixed Review: Chapter 4

1. False   **2.** True   **3.** True   **4.** False   **5.** 5; multiplicity 6
6. $-5, -\frac{1}{2}, 5;$ each has multiplicity 1   **7.** $\pm 1, \pm\sqrt{2};$ each has
multiplicity 1   **8.** 3, multiplicity 2; -4, multiplicity 1
9. (d)   **10.** (a)   **11.** (b)   **12.** (c)   **13.** $f(-2) = -13$ and
$f(0) = 3$. By the intermediate value theorem, since $f(-2)$ and
$f(0)$ have opposite signs, then $f(x)$ has a zero between -2 and 0.

14. $f\left(-\frac{1}{2}\right) = \frac{19}{8}$ and $f(1) = 2$. Both $f\left(-\frac{1}{2}\right)$ and $f(1)$ are positive. We cannot use the intermediate value theorem to determine if there is a zero between $-\frac{1}{2}$ and 1.

15. $P(x) = (x - 1)(x^3 - 5x^2 - 5x - 4) - 6$

16. $Q(x) = 3x^3 + 5x^2 + 12x + 18, R(x) = 42$

17. $Q(x) = x^4 - x^3 + x^2 - x + 1, R(x) = -6$

18. -380 **19.** -15 **20.** $20 - \sqrt{2}$ **21.** Yes; no

22. Yes; yes **23.** $f(x) = (x - 1)(x - 8)(x + 7); 1, 8, -7$

24. $f(x) = (x + 1)(x - 2)(x - 4)(x + 3); -1, 2, 4, -3$

25. The range of a polynomial function with an odd degree is $(-\infty, \infty)$. The range of a polynomial function with an even degree is $[s, \infty)$ for some real number s if $a_n > 0$ and is $(-\infty, s]$ for some real number s if $a_n < 0$. **26.** Since we can find $f(0)$ for any polynomial function $f(x)$, it is not possible for the graph of a polynomial function to have no y-intercept. It is possible for a polynomial function to have no x-intercepts. For instance, a function of the form $f(x) = x^2 + a, a > 0$, has no x-intercepts. There are other examples as well. **27.** The zeros of a polynomial function are the first coordinates of the points at which the graph of the function crosses or is tangent to the x-axis. **28.** For a polynomial $P(x)$ of degree n, when we have $P(x) = d(x) \cdot Q(x) + R(x)$, where the degree of $d(x)$ is 1, then the degree of $Q(x)$ must be $n - 1$.

Exercise Set 4.4

1. $f(x) = x^3 - 6x^2 - x + 30$

3. $f(x) = x^3 + 3x^2 + 4x + 12$

5. $f(x) = x^3 - 3x^2 - 2x + 6$ **7.** $f(x) = x^3 - 6x - 4$

9. $f(x) = x^3 + 2x^2 + 29x + 148$

11. $f(x) = x^3 - \frac{5}{3}x^2 - \frac{2}{3}x$

13. $f(x) = x^5 + 2x^4 - 2x^2 - x$

15. $f(x) = x^4 + 3x^3 + 3x^2 + x$ **17.** $-\sqrt{3}$ **19.** $i, 2 + \sqrt{5}$

21. $-3i$ **23.** $-4 + 3i, 2 + \sqrt{3}$ **25.** $-\sqrt{5}, 4i$

27. $2 + i$ **29.** $-3 - 4i, 4 + \sqrt{5}$ **31.** $4 + i$

33. $f(x) = x^3 - 4x^2 + 6x - 4$ **35.** $f(x) = x^2 + 16$

37. $f(x) = x^3 - 5x^2 + 16x - 80$

39. $f(x) = x^4 - 2x^3 - 3x^2 + 10x - 10$

41. $f(x) = x^4 + 4x^2 - 45$ **43.** $-\sqrt{2}, \sqrt{2}$ **45.** $i, 2, 3$

47. $1 + 2i, 1 - 2i$ **49.** ± 1 **51.** $\pm 1, \pm\frac{1}{2}, \pm 2, \pm 4, \pm 8$

53. $\pm 1, \pm 2, \pm\frac{1}{3}, \pm\frac{1}{5}, \pm\frac{2}{3}, \pm\frac{2}{5}, \pm\frac{1}{15}, \pm\frac{2}{15}$ **55. (a)** Rational: -3; other: $\pm\sqrt{2}$; **(b)** $f(x) = (x + 3)(x + \sqrt{2})(x - \sqrt{2})$

57. (a) Rational: $\frac{1}{3}$; other: $\pm\sqrt{5}$;

(b) $f(x) = 3\left(x - \frac{1}{3}\right)(x + \sqrt{5})(x - \sqrt{5})$, or $(3x - 1)(x + \sqrt{5})(x - \sqrt{5})$ **59. (a)** Rational: $-2, 1$; other: none; **(b)** $f(x) = (x + 2)(x - 1)^2$ **61. (a)** Rational: $-\frac{3}{2}$; other: $\pm 3i$; **(b)** $f(x) = 2\left(x + \frac{3}{2}\right)(x + 3i)(x - 3i)$, or $(2x + 3)(x + 3i)(x - 3i)$ **63. (a)** Rational: $-\frac{1}{5}, 1$; other: $\pm 2i$; **(b)** $f(x) = 5\left(x + \frac{1}{5}\right)(x - 1)(x + 2i)(x - 2i)$

65. (a) Rational: $-2, -1$; other: $3 \pm \sqrt{13}$;

(b) $f(x) = (x + 2)(x + 1)(x - 3 - \sqrt{13})(x - 3 + \sqrt{13})$

67. (a) Rational: 2; other: $1 \pm \sqrt{3}$;

(b) $f(x) = (x - 2)(x - 1 - \sqrt{3})(x - 1 + \sqrt{3})$

69. (a) Rational: -2; other: $1 \pm \sqrt{3}i$;

(b) $f(x) = (x + 2)(x - 1 - \sqrt{3}i)(x - 1 + \sqrt{3}i)$

71. (a) Rational: $\dfrac{1}{2}$; other: $\dfrac{1 \pm \sqrt{5}}{2}$;

(b) $f(x) = \dfrac{1}{3}\left(x - \dfrac{1}{2}\right)\left(x - \dfrac{1 + \sqrt{5}}{2}\right)\left(x - \dfrac{1 - \sqrt{5}}{2}\right)$

73. $1, -3$ **75.** No rational zeros **77.** No rational zeros

79. $-2, 1, 2$ **81.** 3 or 1; 0 **83.** 0; 3 or 1 **85.** 2 or 0; 2 or 0

87. 1; 1 **89.** 1; 0 **91.** 2 or 0; 2 or 0 **93.** 3 or 1; 1 **95.** 1; 1

97.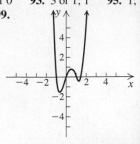

$f(x) = 4x^3 + x^2 - 8x - 2$

99.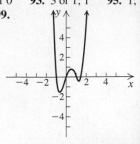

$f(x) = 2x^4 - 3x^3 - 2x^2 + 3x$

101. (a) $(4, -6)$; **(b)** $x = 4$; **(c)** minimum: -6 at $x = 4$

102. (a) $(1, -4)$; **(b)** $x = 1$; **(c)** minimum: -4 at $x = 1$

103. 10 **104.** $-3, 11$ **105.** $-x^3$; -1; 3; as $x \to \infty$, $g(x) \to -\infty$, and as $x \to -\infty, g(x) \to \infty$; cubic **106.** $-x^2$; -1; 2; as $x \to \infty, f(x) \to -\infty$, and as $x \to -\infty, f(x) \to -\infty$; quadratic **107.** $-\frac{4}{9}$; $-\frac{4}{9}$; 0 degree; for all x, $f(x) = -\frac{4}{9}$; constant **108.** x; 1; 1; as $x \to \infty, h(x) \to \infty$, and as $x \to -\infty, h(x) \to -\infty$; linear **109.** x^4; 1; 4; as $x \to \infty, g(x) \to \infty$, and as $x \to -\infty, g(x) \to \infty$; quartic **110.** x^3; 1; 3; as $x \to \infty$, $h(x) \to \infty$, and as $x \to -\infty, h(x) \to -\infty$; cubic

111. (a) $-1, \frac{1}{2}, 3$; **(b)** $0, \frac{3}{2}, 4$; **(c)** $-3, -\frac{3}{2}, 1$; **(d)** $-\frac{1}{2}, \frac{1}{4}, \frac{3}{2}$

113. $-8, -\frac{3}{2}, 4, 7, 15$

Visualizing the Graph

1. A **2.** C **3.** D **4.** H **5.** G **6.** F **7.** B

8. I **9.** J **10.** E

Exercise Set 4.5

1. $\{x | x \neq 2\}$, or $(-\infty, 2) \cup (2, \infty)$ **3.** $\{x | x \neq 1 \text{ and } x \neq 5\}$, or $(-\infty, 1) \cup (1, 5) \cup (5, \infty)$ **5.** $\{x | x \neq -5\}$, or $(-\infty, -5) \cup (-5, \infty)$ **7.** (d); $x = 2, x = -2$, $y = 0$ **9.** (e); $x = 2, x = -2, y = 0$ **11.** (c); $x = 2$, $x = -2, y = 8x$ **13.** $x = 0$ **15.** $x = 2$ **17.** $x = 4$, $x = -6$ **19.** $x = \frac{3}{2}, x = -1$ **21.** $y = \frac{3}{4}$ **23.** $y = 0$

25. No horizontal asymptote **27.** $y = x + 1$

29. $y = x$ **31.** $y = x - 3$

33. Domain: $(-\infty, 0) \cup (0, \infty)$; no x-intercepts, no y-intercept;

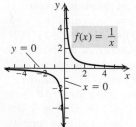

35. Domain: $(-\infty, 0) \cup (0, \infty)$; no x-intercepts, no y-intercept;

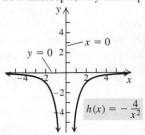

37. Domain:
$(-\infty, -1) \cup (-1, \infty)$;
x-intercepts: $(1, 0)$ and
$(3, 0)$, *y*-intercept: $(0, 3)$;

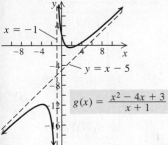

$$g(x) = \frac{x^2 - 4x + 3}{x + 1}$$

39. Domain:
$(-\infty, 5) \cup (5, \infty)$;
no *x*-intercepts,
y-intercept: $\left(0, \frac{2}{5}\right)$;

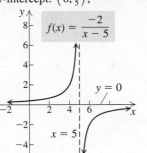

$$f(x) = \frac{-2}{x - 5}$$

51. Domain:
$(-\infty, \infty)$; no *x*-intercepts,
y-intercept: $\left(0, \frac{1}{3}\right)$;

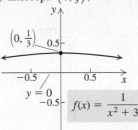

$$f(x) = \frac{1}{x^2 + 3}$$

53. Domain:
$(-\infty, 2) \cup (2, \infty)$;
x-intercept: $(-2, 0)$,
y-intercept: $(0, 2)$;

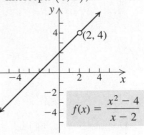

$$f(x) = \frac{x^2 - 4}{x - 2}$$

41. Domain:
$(-\infty, 0) \cup (0, \infty)$;
x-intercept: $\left(-\frac{1}{2}, 0\right)$,
no *y*-intercept;

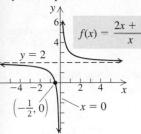

$$f(x) = \frac{2x + 1}{x}$$

43. Domain: $(-\infty, -3) \cup$
$(-3, 3) \cup (3, \infty)$;
no *x*-intercepts,
y-intercept: $\left(0, -\frac{1}{3}\right)$;

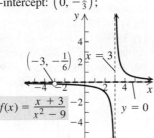

$$f(x) = \frac{x + 3}{x^2 - 9}$$

55. Domain:
$(-\infty, -2) \cup (-2, \infty)$;
x-intercept: $(1, 0)$,
y-intercept: $\left(0, -\frac{1}{2}\right)$;

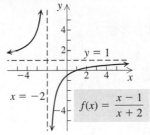

$$f(x) = \frac{x - 1}{x + 2}$$

57. Domain:
$\left(-\infty, -\frac{1}{2}\right) \cup \left(-\frac{1}{2}, 0\right) \cup$
$(0, 3) \cup (3, \infty)$; *x*-intercept:
$(-3, 0)$, no *y*-intercept;

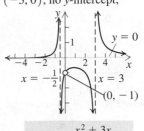

$$f(x) = \frac{x^2 + 3x}{2x^3 - 5x^2 - 3x}$$

45. Domain:
$(-\infty, -3) \cup (-3, 0) \cup (0, \infty)$;
no *x*-intercepts, no *y*-intercept;

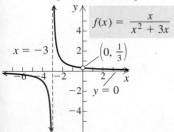

$$f(x) = \frac{x}{x^2 + 3x}$$

59. Domain:
$(-\infty, -1) \cup (-1, \infty)$;
x-intercepts: $(-3, 0)$ and
$(3, 0)$, *y*-intercept: $(0, -9)$;

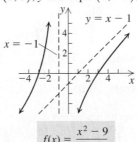

$$f(x) = \frac{x^2 - 9}{x + 1}$$

61. Domain:
$(-\infty, \infty)$; *x*-intercepts:
$(-2, 0)$ and $(1, 0)$,
y-intercept: $(0, -2)$;

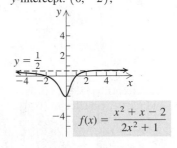

$$f(x) = \frac{x^2 + x - 2}{2x^2 + 1}$$

47. Domain:
$(-\infty, 2) \cup (2, \infty)$;
no *x*-intercepts,
y-intercept: $\left(0, \frac{1}{4}\right)$;

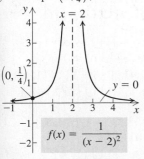

$$f(x) = \frac{1}{(x - 2)^2}$$

49. Domain:
$(-\infty, -3) \cup (-3, -1) \cup$
$(-1, \infty)$; *x*-intercept: $(1, 0)$,
y-intercept: $(0, -1)$;

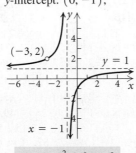

$$f(x) = \frac{x^2 + 2x - 3}{x^2 + 4x + 3}$$

63. Domain: $(-\infty, 1) \cup (1, \infty)$;
x-intercept: $\left(-\frac{2}{3}, 0\right)$,
y-intercept: $(0, 2)$;

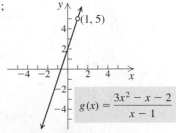

$$g(x) = \frac{3x^2 - x - 2}{x - 1}$$

65. Domain: $(-\infty, -1) \cup$ $(-1, 3) \cup (3, \infty)$; x-intercept: $(1, 0)$, y-intercept: $\left(0, \frac{1}{3}\right)$;

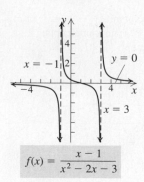

$$f(x) = \frac{x - 1}{x^2 - 2x - 3}$$

67. Domain: $(-\infty, -4) \cup$ $(-4, 2) \cup (2, \infty)$; x-intercept: $\left(\frac{1}{3}, 0\right)$, y-intercept: $\left(0, \frac{1}{2}\right)$;

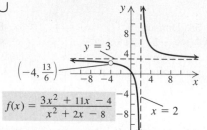

$$f(x) = \frac{3x^2 + 11x - 4}{x^2 + 2x - 8}$$

69. Domain: $(-\infty, -1) \cup (-1, \infty)$; x-intercept: $(3, 0)$, y-intercept: $(0, -3)$;

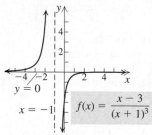

$$f(x) = \frac{x - 3}{(x + 1)^3}$$

71. Domain: $(-\infty, 0) \cup (0, \infty)$; x-intercept: $(-1, 0)$, no y-intercept;

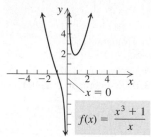

$$f(x) = \frac{x^3 + 1}{x}$$

73. Domain: $(-\infty, -2) \cup$ $(-2, 7) \cup (7, \infty)$; x-intercepts: $(-5, 0)$, $(0, 0)$, and $(3, 0)$, y-intercept: $(0, 0)$;

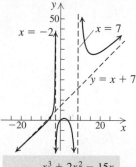

$$f(x) = \frac{x^3 + 2x^2 - 15x}{x^2 - 5x - 14}$$

75. Domain: $(-\infty, \infty)$; x-intercept: $(0, 0)$, y-intercept: $(0, 0)$;

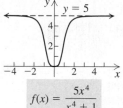

$$f(x) = \frac{5x^4}{x^4 + 1}$$

77. Domain: $(-\infty, -1) \cup$ $(-1, 2) \cup (2, \infty)$; x-intercept: $(0, 0)$, y-intercept: $(0, 0)$;

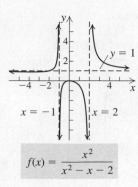

$$f(x) = \frac{x^2}{x^2 - x - 2}$$

79. $f(x) = \dfrac{1}{x^2 - x - 20}$ **81.** $f(x) = \dfrac{3x^2 + 6x}{2x^2 - 2x - 40}$, or $f(x) = \dfrac{3x^2 + 12x + 12}{2x^2 - 2x - 40}$ **83. (a)** $N(t) \rightarrow 0.16$ as $t \rightarrow \infty$;

(b) The medication never completely disappears from the body; a trace amount remains. **85. (a)** $P(0) = 0$; $P(1) = 45{,}455$; $P(3) = 55{,}556$; $P(8) = 29{,}197$; **(b)** $P(t) \rightarrow 0$ as $t \rightarrow \infty$; **(c)** In time, no one lives in this community. **87.** Domain, range, domain, range **88.** Slope **89.** Slope–intercept equation **90.** Point–slope equation **91.** x-intercept **92.** $f(-x) = -f(x)$ **93.** Vertical lines **94.** Midpoint formula **95.** y-intercept **97.** $y = x^3 + 4$ **99.**

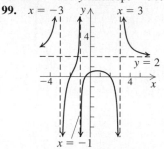

$$f(x) = \frac{2x^3 + x^2 - 8x - 4}{x^3 + x^2 - 9x - 9}$$

Exercise Set 4.6

1. $-5, 3$ **3.** $[-5, 3]$ **5.** $(-\infty, -5] \cup [3, \infty)$
7. $(-\infty, -4) \cup (2, \infty)$ **9.** $(-\infty, -4) \cup [2, \infty)$
11. 0 **13.** $(-5, 0] \cup (1, \infty)$ **15.** $(-\infty, -5) \cup (0, 1)$
17. $(-\infty, -3) \cup (0, 3)$ **19.** $(-3, 0) \cup (3, \infty)$
21. $(-\infty, -5) \cup (-3, 2)$ **23.** $(-2, 0] \cup (2, \infty)$
25. $(-4, 1)$ **27.** $(-\infty, -2) \cup (1, \infty)$
29. $(-\infty, -1] \cup [3, \infty)$ **31.** $(-\infty, -5) \cup (5, \infty)$
33. $(-\infty, -2] \cup [2, \infty)$ **35.** $(-\infty, 3) \cup (3, \infty)$
37. $\varnothing$ **39.** $\left(-\infty, -\frac{5}{4}\right] \cup [0, 3]$ **41.** $[-3, -1] \cup [1, \infty)$
43. $(-\infty, -2) \cup (1, 3)$ **45.** $\left[-\sqrt{2}, -1\right] \cup \left[\sqrt{2}, \infty\right)$
47. $(-\infty, -1] \cup \left[\frac{3}{2}, 2\right]$ **49.** $(-\infty, 5]$
51. $(-\infty, -1.680) \cup (2.154, 5.526)$ **53.** $-4; (-4, \infty)$
55. $-\frac{5}{2}; \left(-\frac{5}{2}, \infty\right)$ **57.** $0, 4; (-\infty, 0] \cup (4, \infty)$
59. $2, \frac{7}{2}; \left(2, \frac{7}{2}\right]$ **61.** $-3, -\frac{1}{5}, 1; \left(-3, -\frac{1}{5}\right] \cup (1, \infty)$
63. $2, \frac{46}{11}, 5; \left(2, \frac{46}{11}\right) \cup (5, \infty)$
65. $1 - \sqrt{2}, 0, 1 + \sqrt{2}; \left(1 - \sqrt{2}, 0\right) \cup \left(1 + \sqrt{2}, \infty\right)$
67. $-3, 1, 3, \frac{11}{3}; (-\infty, -3) \cup (1, 3) \cup \left[\frac{11}{3}, \infty\right)$

59. 0; $(-\infty, \infty)$ **71.** $-3, \dfrac{1 - \sqrt{61}}{6}, -\dfrac{1}{2}, 0, \dfrac{1 + \sqrt{61}}{6}$;

$\left(-3, \dfrac{1 - \sqrt{61}}{6}\right) \cup \left(-\dfrac{1}{2}, 0\right) \cup \left(\dfrac{1 + \sqrt{61}}{6}, \infty\right)$

73. $-1, 0, \dfrac{2}{7}, \dfrac{7}{2}$; $(-1, 0) \cup \left(\dfrac{2}{7}, \dfrac{7}{2}\right)$

75. $-6 - \sqrt{33}, -5, -6 + \sqrt{33}, 1, 5$;

$\left[-6 - \sqrt{33}, -5\right) \cup \left[-6 + \sqrt{33}, 1\right) \cup (5, \infty)$

77. $(0.408, 2.449)$ **79. (a)** $(10, 200)$;

(b) $(0, 10) \cup (200, \infty)$ **81.** $\{n \mid 9 \le n \le 23\}$

83. $(x + 2)^2 + (y - 4)^2 = 9$ **84.** $x^2 + (y + 3)^2 = \dfrac{49}{16}$

85. (a) $\left(\dfrac{3}{4}, -\dfrac{55}{8}\right)$; **(b)** maximum: $-\dfrac{55}{8}$ when $x = \dfrac{3}{4}$;

(c) $\left(-\infty, -\dfrac{55}{8}\right]$ **86. (a)** $(5, -23)$; **(b)** minimum: -23 when

$x = 5$; **(c)** $[-23, \infty)$ **87.** $\left[-\sqrt{5}, \sqrt{5}\right]$ **89.** $\left[-\dfrac{3}{2}, \dfrac{3}{2}\right]$

91. $\left(-\infty, -\dfrac{1}{4}\right) \cup \left(\dfrac{1}{2}, \infty\right)$ **93.** $x^2 + x - 12 < 0$; answers may

vary   **95.** $(-\infty, -3) \cup (7, \infty)$

Review Exercises: Chapter 4

1. True   **2.** True   **3.** False   **4.** False   **5.** False

6. (a) $-2.637, 1.137$; **(b)** relative maximum: 7.125 at $x = -0.75$;

(c) none; **(d)** domain: all real numbers; range: $(-\infty, 7.125]$

7. (a) $-3, -1.414, 1.414$; **(b)** relative maximum: 2.303 at

$x = -2.291$; **(c)** relative minimum: -6.303 at $x = 0.291$;

(d) domain: all real numbers; range: all real numbers

8. (a) $0, 1, 2$; **(b)** relative maximum: 0.202 at $x = 0.610$;

(c) relative minima: 0 at $x = 0$, -0.620 at $x = 1.640$;

(d) domain: all real numbers; range: $[-0.620, \infty)$

9. $0.45x^4$; 0.45; 4; quartic   **10.** -25; -25; 0; constant

11. $-0.5x$; -0.5; 1; linear   **12.** $\dfrac{1}{3}x^3$; $\dfrac{1}{3}$; 3; cubic

13. As $x \to \infty$, $f(x) \to -\infty$, and as $x \to -\infty$, $f(x) \to -\infty$.

14. As $x \to \infty$, $f(x) \to \infty$, and as $x \to -\infty$, $f(x) \to -\infty$.

15. $\dfrac{2}{3}$, multiplicity 1; -2, multiplicity 3; 5, multiplicity 2

16. ± 1, ± 5, each has multiplicity 1   **17.** ± 3, -4, each has

multiplicity 1   **18. (a)** 4%; **(b)** 5%   **19. (a)** Linear:

$f(x) = 0.5408695652x - 30.30434783$; quadratic:

$f(x) = 0.0030322581x^2 - 0.5764516129x + 57.53225806$; cubic:

$f(x) = 0.0000247619x^3 - 0.0112857143x^2 + 2.002380952x - 82.14285714$; **(b)** the cubic function; **(c)** $298, 498$

20.

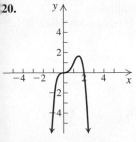

$f(x) = -x^4 + 2x^3$

21.

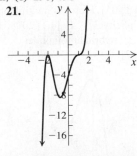

$g(x) = (x - 1)^3(x + 2)^2$

22.

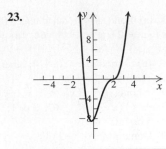

$h(x) = x^3 + 3x^2 - x - 3$

23.

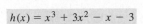

$f(x) = x^4 - 5x^3 + 6x^2 + 4x - 8$

24.

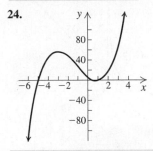

$g(x) = 2x^3 + 7x^2 - 14x + 5$

25. $f(1) = -4$ and $f(2) = 3$. Since $f(1)$ and $f(2)$ have opposite signs, $f(x)$ has a zero between 1 and 2.   **26.** $f(-1) = -3.5$ and $f(1) = -0.5$. Since $f(-1)$ and $f(1)$ have the same sign, the intermediate value theorem does not allow us to determine whether there is a zero between -1 and 1.   **27.** $Q(x) = 6x^2 + 16x + 52$, $R(x) = 155$; $P(x) = (x - 3)(6x^2 + 16x + 52) + 155$

28. $Q(x) = x^3 - 3x^2 + 3x - 2$, $R(x) = 7$; $P(x) = (x + 1)(x^3 - 3x^2 + 3x - 2) + 7$

29. $x^2 + 7x + 22$, R 120   **30.** $x^3 + x^2 + x + 1$, R 0

31. $x^4 - x^3 + x^2 - x - 1$, R 1   **32.** 36   **33.** 0

34. $-141{,}220$ **35.** Yes, no   **36.** No, yes   **37.** Yes, no

38. No, yes   **39.** $f(x) = (x - 1)^2(x + 4)$; $-4, 1$

40. $f(x) = (x - 2)(x + 3)^2$; $-3, 2$

41. $f(x) = (x - 2)^2(x - 5)(x + 5)$; $-5, 2, 5$

42. $f(x) = (x - 1)(x + 1)(x - \sqrt{2})(x + \sqrt{2})$; $-\sqrt{2}$, $-1, 1, \sqrt{2}$ **43.** $f(x) = x^3 + 3x^2 - 6x - 8$

44. $f(x) = x^3 + x^2 - 4x + 6$

45. $f(x) = x^3 - \dfrac{5}{2}x^2 + \dfrac{1}{2}$, or $2x^3 - 5x^2 + 1$

46. $f(x) = x^4 + \dfrac{29}{2}x^3 + \dfrac{135}{2}x^2 + \dfrac{175}{2}x - \dfrac{125}{2}$, or $2x^4 + 29x^3 + 135x^2 + 175x - 125$

47. $f(x) = x^5 + 4x^4 - 3x^3 - 18x^2$ **48.** $-\sqrt{5}, 4 - i$

49. $1 - \sqrt{3}, \sqrt{3}$ **50.** $\sqrt{2}$ **51.** $f(x) = x^2 - 11$

52. $f(x) = x^3 - 6x^2 + x - 6$

53. $f(x) = x^4 - 5x^3 + 4x^2 + 2x - 8$

54. $f(x) = x^4 - x^2 - 20$ **55.** $f(x) = x^3 + \dfrac{8}{3}x^2 - x$,

or $3x^3 + 8x^2 - 3x$ **56.** $\pm\dfrac{1}{4}, \pm\dfrac{1}{2}, \pm\dfrac{3}{4}, \pm 1, \pm\dfrac{3}{2}, \pm 2, \pm 3, \pm 4$, $\pm 6, \pm 12$ **57.** $\pm\dfrac{1}{3}, \pm 1$ **58.** $\pm 1, \pm 2, \pm 3, \pm 4$, $\pm 6, \pm 8, \pm 12, \pm 24$ **59. (a)** Rational: $0, -2, \dfrac{1}{3}, 3$;

other: none; **(b)** $f(x) = 3x\left(x - \dfrac{1}{3}\right)(x + 2)^2(x - 3)$, or

$x(3x - 1)(x + 2)^2(x - 3)$ **60. (a)** Rational: 2; other: $\pm\sqrt{3}$;

(b) $f(x) = (x - 2)(x + \sqrt{3})(x - \sqrt{3})$

61. (a) Rational: $-1, 1$; other: $3 \pm i$;

(b) $f(x) = (x + 1)(x - 1)(x - 3 - i)(x - 3 + i)$

62. (a) Rational: -5; other: $1 \pm \sqrt{2}$;

(b) $f(x) = (x + 5)(x - 1 - \sqrt{2})(x - 1 + \sqrt{2})$

63. (a) Rational: $\frac{2}{3}$, 1; other: none; **(b)** $f(x) = 3\left(x - \frac{2}{3}\right)(x - 1)^2$, or $(3x - 2)(x - 1)^2$ **64. (a)** Rational: 2; other: $1 \pm \sqrt{5}$; **(b)** $f(x) = (x - 2)^3(x - 1 + \sqrt{5})(x - 1 - \sqrt{5})$
65. (a) Rational: -4, 0, 3, 4; other: none; **(b)** $f(x) = x^2(x + 4)^2(x - 3)(x - 4)$ **66. (a)** Rational: $\frac{5}{2}$, 1; other: none; **(b)** $f(x) = 2\left(x - \frac{5}{2}\right)(x - 1)^4$, or $(2x - 5)(x - 1)^4$ **67.** 3 or 1; 0 **68.** 4 or 2 or 0; 2 or 0 **69.** 3 or 1; 0
70. Domain: $(-\infty, -2) \cup (-2, \infty)$; x-intercepts: $\left(-\sqrt{5}, 0\right)$ and $\left(\sqrt{5}, 0\right)$, y-intercept: $\left(0, -\frac{5}{2}\right)$

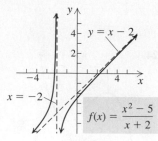

$y = x - 2$
$x = -2$
$f(x) = \dfrac{x^2 - 5}{x + 2}$

71. Domain: $(-\infty, 2) \cup (2, \infty)$; x-intercepts: none, y-intercept: $\left(0, \frac{5}{4}\right)$

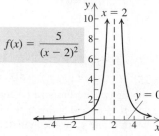

$x = 2$
$f(x) = \dfrac{5}{(x - 2)^2}$
$y = 0$

72. Domain: $(-\infty, -4) \cup (-4, 5) \cup (5, \infty)$; x-intercepts: $(-3, 0)$ and $(2, 0)$, y-intercept: $\left(0, \frac{3}{10}\right)$

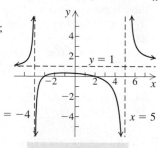

$y = 1$
$x = -4$
$x = 5$
$f(x) = \dfrac{x^2 + x - 6}{x^2 - x - 20}$

73. Domain: $(-\infty, -3) \cup (-3, 5) \cup (5, \infty)$; x-intercept: $(2, 0)$, y-intercept: $\left(0, \frac{2}{15}\right)$

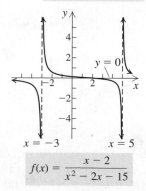

$y = 0$
$x = -3$
$x = 5$
$f(x) = \dfrac{x - 2}{x^2 - 2x - 15}$

74. $f(x) = \dfrac{1}{x^2 - x - 6}$ **75.** $f(x) = \dfrac{4x^2 + 12x}{x^2 - x - 6}$
76. (a) $N(t) \to 0.0875$ as $t \to \infty$; **(b)** The medication never completely disappears from the body; a trace amount remains.
77. $(-3, 3)$ **78.** $\left(-\infty, -\frac{1}{2}\right) \cup (2, \infty)$
79. $[-4, 1] \cup [2, \infty)$ **80.** $\left(-\infty, -\frac{14}{3}\right) \cup (-3, \infty)$

81. (a) 7 sec after launch; **(b)** $(2, 3)$
82. $\left[\dfrac{5 - \sqrt{15}}{2}, \dfrac{5 + \sqrt{15}}{2}\right]$ **83.** A **84.** C **85.** B
86. $(-\infty, -1 - \sqrt{6}] \cup [-1 + \sqrt{6}, \infty)$
87. $\left(-\infty, -\frac{1}{2}\right) \cup \left(\frac{1}{2}, \infty\right)$ **88.** $1 + i, 1 - i, i, -i$
89. $(-\infty, 2)$ **90.** $(x - 1)\left(x + \dfrac{1}{2} - \dfrac{\sqrt{3}}{2}i\right)\left(x + \dfrac{1}{2} + \dfrac{\sqrt{3}}{2}i\right)$
91. 7 **92.** -4 **93.** $(-\infty, -5] \cup [2, \infty)$
94. $(-\infty, 1.1] \cup [2, \infty)$ **95.** $\left(-1, \frac{3}{7}\right)$
96. A polynomial function is a function that can be defined by a polynomial expression. A rational function is a function that can be defined as a quotient of two polynomials. **97.** No; since imaginary zeros of polynomials with rational coefficients occur in conjugate pairs, a third-degree polynomial with rational coefficients can have at most two imaginary zeros. Thus there must be at least one real zero. **98.** Vertical asymptotes occur at any x-values that make the denominator zero and do not also make the numerator zero. The graph of a rational function does not cross any vertical asymptotes. Horizontal asymptotes occur when the degree of the numerator is less than or equal to the degree of the denominator. Oblique asymptotes occur when the degree of the numerator is 1 greater than the degree of the denominator. Graphs of rational functions may cross horizontal or oblique asymptotes.
99. If $P(-x)$ is an even function, then $P(-x) = P(x)$ and thus $P(-x)$ has the same number of sign changes as $P(x)$. Hence, $P(x)$ has one negative real zero also. **100.** A horizontal asymptote occurs when the degree of the numerator of a rational function is less than or equal to the degree of the denominator. An oblique asymptote occurs when the degree of the numerator is 1 greater than the degree of the denominator. Thus a rational function cannot have both a horizontal asymptote and an oblique asymptote. **101.** A quadratic inequality $ax^2 + bx + c \le 0, a > 0$, or $ax^2 + bx + c \ge 0, a < 0$, has a solution set that is a closed interval.

Test: Chapter 4

1. [4.1] $-x^4$; -1; 4; quartic **2.** [4.1] $-4.7x$; -4.7; 1; linear
3. [4.1] 0, $\frac{5}{3}$, each has multiplicity 1; 3, multiplicity 2; -1, multiplicity 3 **4.** [4.1] 2008: 312,848 hybrid automobiles; 2012: 372,750 hybrid automobiles
5. [4.2]

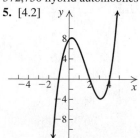

$f(x) = x^3 - 5x^2 + 2x + 8$

6. [4.2]

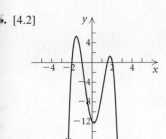

$f(x) = -2x^4 + x^3 + 11x^2 - 4x - 12$

7. [4.2] $f(0) = 3$ and $f(2) = -17$. Since $f(0)$ and $f(2)$ have opposite signs, $f(x)$ has a zero between 0 and 2.

8. [4.2] $g(-2) = 5$ and $g(-1) = 1$. Both $g(-2)$ and $g(-1)$ are positive. We cannot use the intermediate value theorem to determine if there is a zero between -2 and -1.

9. [4.3] $Q(x) = x^3 + 4x^2 + 4x + 6$, $R(x) = 1$;
$P(x) = (x - 1)(x^3 + 4x^2 + 4x + 6) + 1$

10. [4.3] $3x^2 + 15x + 63$, R 322 **11.** [4.3] -115

12. [4.3] Yes **13.** [4.4] $f(x) = x^4 - 27x^2 - 54x$

14. [4.4] $-\sqrt{3}, 2 + i$ **15.** [4.4] $f(x) = x^3 + 10x^2 + 9x + 90$

16. [4.4] $f(x) = x^5 - 2x^4 - x^3 + 6x^2 - 6x$

17. [4.4] $\pm 1, \pm 2, \pm 3, \pm 4, \pm 6, \pm 12, \pm \frac{1}{2}, \pm \frac{3}{2}$

18. [4.4] $\pm \frac{1}{10}, \pm \frac{1}{5}, \pm \frac{1}{2}, \pm 1, \pm \frac{5}{2}, \pm 5$

19. [4.4] **(a)** Rational: -1; other: $\pm \sqrt{5}$;
(b) $f(x) = (x + 1)(x - \sqrt{5})(x + \sqrt{5})$ **20.** [4.4]
(a) Rational: $-\frac{1}{2}, 1, 2, 3$; other: none;
(b) $f(x) = 2(x + \frac{1}{2})(x - 1)(x - 2)(x - 3)$, or
$(2x + 1)(x - 1)(x - 2)(x - 3)$

21. [4.4] **(a)** Rational: -4; other: $\pm 2i$;
(b) $f(x) = (x - 2i)(x + 2i)(x + 4)$

22. [4.4] **(a)** Rational: $\frac{2}{3}, 1$; other: none;
(b) $f(x) = 3(x - \frac{2}{3})(x - 1)^3$, or $(3x - 2)(x - 1)^3$

23. [4.4] 2 or 0; 2 or 0

24. [4.5] Domain:
$(-\infty, 3) \cup (3, \infty)$;
x-intercepts: none,
y-intercept: $(0, \frac{2}{9})$;

$f(x) = \dfrac{2}{(x - 3)^2}$

$y = 0$ $x = 3$

25. [4.5] Domain:
$(-\infty, -1) \cup$
$(-1, 4) \cup (4, \infty)$;
x-intercept: $(-3, 0)$,
y-intercept: $(0, -\frac{3}{4})$;

$y = 0$ $x = -1$ $x = 4$

$f(x) = \dfrac{x + 3}{x^2 - 3x - 4}$

26. [4.5] Answers may vary; $f(x) = \dfrac{x + 4}{x^2 - x - 2}$

27. [4.6] $\left(-\infty, -\frac{1}{2}\right) \cup (3, \infty)$

28. [4.6] $(-\infty, 4) \cup \left[\frac{13}{2}, \infty\right)$ **29.** **(a)** [4.1] 6 sec;

(b) [4.1], [4.6] $(1, 3)$ **30.** [4.2] D

31. [4.1], [4.6] $(-\infty, -4] \cup [3, \infty)$

Chapter 5

Exercise Set 5.1

1. $\{(8, 7), (8, -2), (-4, 3), (-8, 8)\}$ **3.** $\{(-1, -1), (4, -3)\}$

5. $x = 4y - 5$ **7.** $y^3x = -5$ **9.** $y = x^2 - 2x$

11.

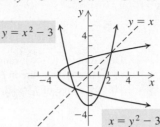

$y = x^2 - 3$ $y = x$ $x = y^2 - 3$

13.

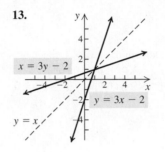

$x = 3y - 2$ $y = 3x - 2$ $y = x$

15.

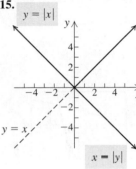

$y = |x|$ $y = x$ $x = |y|$

17. Assume that $f(a) = f(b)$ for any numbers a and b in the domain of f. Since $f(a) = \frac{1}{3}a - 6$ and $f(b) = \frac{1}{3}b - 6$, we have

$\quad \frac{1}{3}a - 6 = \frac{1}{3}b - 6$
$\quad \frac{1}{3}a - \frac{1}{3}b \qquad$ Adding 6
$\quad a = b. \qquad$ Multiplying by 3

Thus, if $f(a) = f(b)$, then $a = b$ and f is one-to-one.

19. Assume that $f(a) = f(b)$ for any numbers a and b in the domain of f. Since $f(a) = a^3 + \frac{1}{2}$ and $f(b) = b^3 + \frac{1}{2}$, we have

$\quad a^3 + \frac{1}{2} = b^3 + \frac{1}{2}$
$\quad a^3 = b^3 \qquad$ Subtracting $\frac{1}{2}$
$\quad a = b. \qquad$ Taking the cube root

Thus, if $f(a) = f(b)$, then $a = b$ and f is one-to-one.

21. Find two numbers a and b for which $a \neq b$ and $g(a) = g(b)$. Two such numbers are -2 and 2, because $g(-2) = g(2) = -3$. Thus, g is not one-to-one.

23. Find two numbers a and b for which $a \neq b$ and $g(a) = g(b)$. Two such numbers are -1 and 1, because $g(-1) = g(1) = 0$. Thus, g is not one-to-one.

25. Yes **27.** No **29.** No **31.** Yes **33.** Yes

35. No **37.** No **39.** Yes **41.** No **43.** No

45. $y_1 = 0.8x + 1.7$,   Domain and range of both f
$y_2 = \dfrac{x - 1.7}{0.8}$ and f^{-1}: $(-\infty, \infty)$

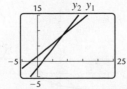

47.
$y_1 = \frac{1}{2}x - 4,$
$y_2 = 2x + 8$

Domain and range of both f and f^{-1}: $(-\infty, \infty)$

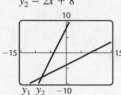

49.
$y_1 = \sqrt{x - 3},$
$y_2 = x^2 + 3, x \geq 0$

Domain of f: $[3, \infty)$, range of f: $[0, \infty)$; domain of f^{-1}: $[0, \infty)$, range of f^{-1}: $[3, \infty)$

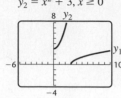

51.
$y_1 = x^2 - 4, x \geq 0;$
$y_2 = \sqrt{4 + x}$

Domain of f: $[0, \infty)$, range of f: $[-4, \infty)$; domain of f^{-1}: $[-4, \infty)$, range of f^{-1}: $[0, \infty)$

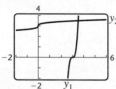
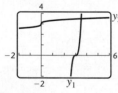

53.
$y_1 = (3x - 9)^3,$
$y_2 = \dfrac{\sqrt[3]{x} + 9}{3}$

Domain and range of both f and f^{-1}: $(-\infty, \infty)$

55. (a) Yes; **(b)** $f^{-1}(x) = x - 4$

57. (a) Yes; **(b)** $f^{-1}(x) = \dfrac{x + 1}{2}$

59. (a) Yes; **(b)** $f^{-1}(x) = \dfrac{4}{x} - 7$

61. (a) Yes; **(b)** $f^{-1}(x) = \dfrac{3x + 4}{x - 1}$

63. (a) Yes; **(b)** $f^{-1}(x) = \sqrt[3]{x} + 1$
65. (a) No

67. (a) Yes; **(b)** $f^{-1}(x) = \sqrt{\dfrac{x + 2}{5}}$

69. (a) Yes; **(b)** $f^{-1}(x) = x^2 - 1, x \geq 0$
71. $\frac{1}{3}x$ **73.** $-x$ **75.** $x^3 + 5$
77.

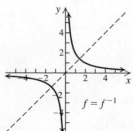

79.

81.

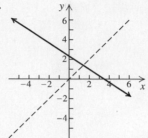

83. $f^{-1}(f(x)) = f^{-1}\left(\frac{7}{8}x\right) = \frac{8}{7} \cdot \frac{7}{8}x = x;$
$f(f^{-1}(x)) = f\left(\frac{8}{7}x\right) = \frac{7}{8} \cdot \frac{8}{7}x = x$

85. $f^{-1}(f(x)) = f^{-1}\left(\dfrac{1 - x}{x}\right) = \dfrac{1}{\dfrac{1 - x}{x} + 1} =$

$\dfrac{1}{\dfrac{1 - x + x}{x}} = \dfrac{1}{\dfrac{1}{x}} = 1 \cdot \dfrac{x}{1} = x; f(f^{-1}(x)) = f\left(\dfrac{1}{x + 1}\right) =$

$\dfrac{1 - \dfrac{1}{x + 1}}{\dfrac{1}{x + 1}} = \dfrac{\dfrac{x + 1 - 1}{x + 1}}{\dfrac{1}{x + 1}} = \dfrac{x}{x + 1} \cdot \dfrac{x + 1}{1} = x$

87. $f^{-1}(f(x)) = f^{-1}\left(\dfrac{2}{5}x + 1\right) = \dfrac{5\left(\dfrac{2}{5}x + 1\right) - 5}{2} =$

$\dfrac{2x + 5 - 5}{2} = \dfrac{2x}{2} = x; f(f^{-1}(x)) = f\left(\dfrac{5x - 5}{2}\right) =$

$\dfrac{2}{5}\left(\dfrac{5x - 5}{2}\right) + 1 = x - 1 + 1 = x$

89. $f^{-1}(x) = \frac{1}{5}x + \frac{3}{5}$; domain and range of both f and f^{-1}: $(-\infty, \infty)$

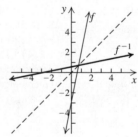

91. $f^{-1}(x) = \dfrac{2}{x}$; domain and range of both f and f^{-1}: $(-\infty, 0) \cup (0, \infty)$

3. $f^{-1}(x) = \sqrt[3]{3x + 6}$; domain of f and f^{-1}: $(-\infty, \infty)$; range of f and f^{-1}: $(-\infty, \infty)$

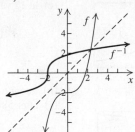

5. $f^{-1}(x) = \dfrac{3x + 1}{x - 1}$; domain of f: $(-\infty, 3) \cup (3, \infty)$;

range of f: $(-\infty, 1) \cup (1, \infty)$;

domain of f^{-1}: $(-\infty, 1) \cup (1, \infty)$;

range of f^{-1}: $(-\infty, 3) \cup (3, \infty)$;

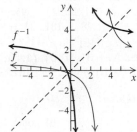

7. 5; a **99. (a)** \$38, \$16.40, \$11; **(b)** $C^{-1}(x) = \dfrac{72}{x - 2}$;

$C^{-1}(x)$ represents the number of players in the group lesson, where x is the cost per player, in dollars; **(c)** 1 player, 4 players, players **101. (a)** 2012: \$42.572 billion; 2016: \$61.764 billion;

(b) $H^{-1}(x) = \dfrac{x - 28.1778}{4.798}$; $H^{-1}(x)$ represents the number of years after 2009, where x is the amount of e-commerce holiday season sales, in billions of dollars

103. (b), (d), (f), (h) **104.** (a), (c), (e), (g) **105.** (a)
106. (d) **107.** (f) **108.** (a), (b), (c), (d) **109.** Yes
111. $f(x) = x^2 - 3, x \geq 0$; $f^{-1}(x) = \sqrt{x + 3}, x \geq -3$
113. Answers may vary. $f(x) = 3/x$, $f(x) = 1 - x$, $f(x) = x$

Exercise Set 5.2

1. 54.5982 **3.** 0.0856 **5.** (f) **7.** (e) **9.** (a)

11.

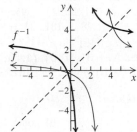

$f(x) = 3^x$

13.

$f(x) = 6^x$

15.

$f(x) = \left(\frac{1}{4}\right)^x$

17.

$y = -2^x$

19.

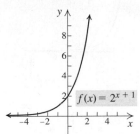

$f(x) = -0.25^x + 4$

21.

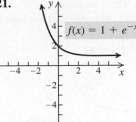

$f(x) = 1 + e^{-x}$

23.

$y = \frac{1}{4}e^x$

25.

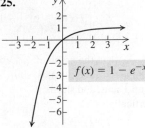

$f(x) = 1 - e^{-x}$

27. Shift the graph of $y = 2^x$ left 1 unit.

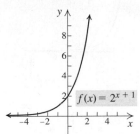

$f(x) = 2^{x+1}$

29. Shift the graph of $y = 2^x$ down 3 units.

$f(x) = 2^x - 3$

31. Shift the graph of $y = 2^x$ left 1 unit, reflect it across the y-axis, and shift it up 2 units.

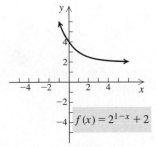

$f(x) = 2^{1-x} + 2$

33. Reflect the graph of $y = 3^x$ across the y-axis and then across the x-axis and then shift it up 4 units.

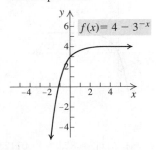

$f(x) = 4 - 3^{-x}$

35. Shift the graph of $y = \left(\frac{3}{2}\right)^x$ right 1 unit.

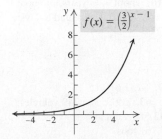

$f(x) = \left(\frac{3}{2}\right)^{x-1}$

37. Shift the graph of $y = 2^x$ left 3 units and then down 5 units.

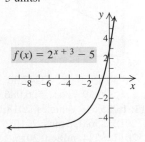

$f(x) = 2^{x+3} - 5$

39. Shift the graph of $y = 2^x$ right 1 unit, stretch it vertically, and shift it up 1 unit.

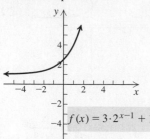

$f(x) = 3 \cdot 2^{x-1} + 1$

41. Shrink the graph of $y = e^x$ horizontally.

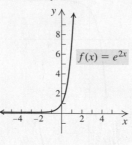

$f(x) = e^{2x}$

43. Reflect the graph of $y = e^x$ across the x-axis, shift it up 1 unit, and shrink it vertically.

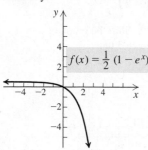

$f(x) = \frac{1}{2}(1 - e^x)$

45. Shift the graph of $y = e^x$ left 1 unit and then reflect it across the y-axis.

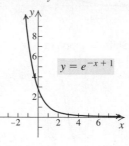

$y = e^{-x+1}$

47. Reflect the graph of $y = e^x$ across the y-axis, then across the x-axis, then shift it up 1 unit, and then stretch it vertically.

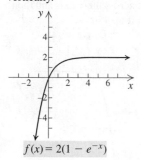

$f(x) = 2(1 - e^{-x})$

49.

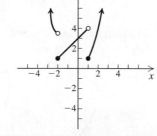

$$f(x) = \begin{cases} e^{-x} - 4, & \text{for } x < -2, \\ x + 3, & \text{for } -2 \le x < 1, \\ x^2, & \text{for } x \ge 1 \end{cases}$$

51. (a) $A(t) = 82,000(1.01125)^{4t}$;

(b) $y = 82,000(1.01125)^{4x}$;

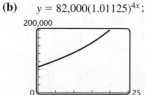

(c) $82,000; $89,677.22; $102,561.54; $128,278.90;

(d) about 4.43 years, or about 4 years, 5 months, and 5 days

53. $4930.86   **55.** $3247.30   **57.** $153,610.15

59. $76,305.59   **61.** $26,086.69   **63.** 2000: 385,249 vehicles; 2013: 1,225,519 vehicles; 2018: 1,912,580 vehicles

65. 2012: 356,022 tons; 2016: 5,605,761 tons

67. 2011: $234 million; 2015: $5844 million, or $5.844 billion

69. 2020: 101,234 centenarians; 2050: 414,387 centenarians

71. 1982: $48 billion; 1995: $109 billion; 2010: $284 billion; 2017: $445 billion

73. (a) $y = 6982(0.85)^x$;

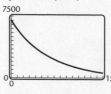

(b) $6982; $5935; $5044; $3098; $1903;

(c) in 12 years

75. (a) $y = 100(1 - e^{-0.04x})$

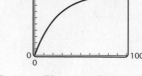

(b) about 63%;

(c) after 58 days

77. (c)   **79.** (a)   **81.** (l)   **83.** (g)   **85.** (i)

87. (k)   **89.** (m)   **91.** $(1.481, 4.090)$

93. $(-0.402, -1.662), (1.051, 2.722)$ **95.** 4.448

97. $(0, \infty)$ **99.** 2.294, 3.228   **101.** $31 - 22i$

102. $\frac{1}{2} - \frac{1}{2}i$ **103.** $\left(-\frac{1}{2}, 0\right), (7, 0); -\frac{1}{2}, 7$

104. $(1, 0); 1$ **105.** $(-1, 0), (0, 0), (1, 0); -1, 0, 1$

106. $(-4, 0), (0, 0), (3, 0); -4, 0, 3$ **107.** $-8, 0, 2$

108. $\dfrac{5 \pm \sqrt{97}}{6}$ **109.** $\pi^7; 70^{80}$

111. (a)

$y = e^{-x^2}$

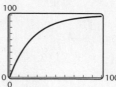

(b) none; **(c)** relative maximum: 1 at $x = 0$

113. $f(x) = \dfrac{2e^x(e^h - 1)}{h}$

Visualizing the Graph

1. J   **2.** F   **3.** H   **4.** B   **5.** E   **6.** A   **7.** C
8. I   **9.** D   **10.** G

Exercise Set 5.3

1.

$x = 3^y$

3.

$x = \left(\frac{1}{2}\right)^y$

5.

$y = \log_3 x$

7.

$f(x) = \log x$

. 4 **11.** 3 **13.** −3 **15.** −2 **17.** 0 **19.** 1
1. 4 **23.** $\frac{1}{4}$ **25.** −7 **27.** $\frac{1}{2}$ **29.** $\frac{3}{4}$ **31.** 0 **33.** $\frac{1}{2}$
5. $\log_{10} 1000 = 3$, or $\log 1000 = 3$ **37.** $\log_8 2 = \frac{1}{3}$
9. $\log_e t = 3$, or $\ln t = 3$
1. $\log_e 7.3891 = 2$, or $\ln 7.3891 = 2$ **43.** $\log_p 3 = k$
5. $5^1 = 5$ **47.** $10^{-2} = 0.01$ **49.** $e^{3.4012} = 30$
1. $a^{-x} = M$ **53.** $a^x = T^3$ **55.** 0.4771 **57.** 2.7259
9. −0.2441 **61.** Does not exist **63.** 0.6931 **65.** 6.6962
7. Does not exist **69.** 3.3219 **71.** −0.2614
3. 0.7384 **75.** 2.2619 **77.** 0.5880

9. $y_1 = 3^x$,

$y_2 = \dfrac{\log x}{\log 3}$

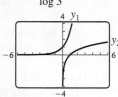

81. $y_1 = \log x$,
$y_2 = 10^x$

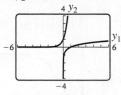

3. Shift the graph of $= \log_2 x$ left 3 units. Domain: $(-3, \infty)$; vertical asymptote: $x = -3$;
$$y = \frac{\log (x + 3)}{\log 2}$$

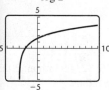

85. Shift the graph of $y = \log_3 x$ down 1 unit. Domain: $(0, \infty)$; vertical asymptote: $x = 0$;
$$y = \frac{\log x}{\log 3} - 1$$

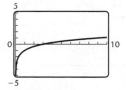

7. Stretch the graph of $= \ln x$ vertically. Domain: $(0, \infty)$; vertical asymptote: $x = 0$;
$y = 4 \ln x$

89. Reflect the graph of $y = \ln x$ across the x-axis and shift it up 2 units. Domain: $(0, \infty)$; vertical asymptote: $x = 0$;
$y = 2 - \ln x$

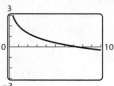

1. Shift the graph of $y = \log x$ right 1 unit, shrink it vertically, and shift it down 2 units. Domain: $(1, \infty)$; vertical asymptote: $x = 1$.

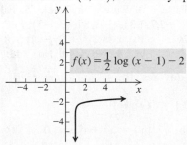

$f(x) = \frac{1}{2} \log (x - 1) - 2$

93.

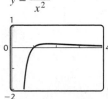

$$g(x) = \begin{cases} 5, & \text{for } x \le 0, \\ \log x + 1, & \text{for } x > 0 \end{cases}$$

95. **(a)** 2.5 ft/sec;
(b) 2.8 ft/sec; **(c)** 2.0 ft/sec;
(d) 2.4 ft/sec; **(e)** 2.2 ft/sec;
(f) 2.5 ft/sec; **(g)** 2.3 ft/sec;
(h) 3.1 ft/sec
97. **(a)** 7.7; **(b)** 9.5; **(c)** 6.6;
(d) 7.6; **(e)** 8.2; **(f)** 7.9;
(g) 5.1; **(h)** 9.3
99. **(a)** 10^{-7}; **(b)** 4.0×10^{-6};
(c) 6.3×10^{-4}; **(d)** 1.6×10^{-5}

101. **(a)** 140 decibels; **(b)** 115 decibels; **(c)** 40 decibels;
(d) 65 decibels; **(e)** 120 decibels; **(f)** 194 decibels
102. $m = \frac{3}{10}$; y-intercept: $\left(0, -\frac{7}{5}\right)$ **103.** $m = 0$; y-intercept:
$(0, 6)$ **104.** Slope is not defined; no y-intercept
105. −280 **106.** −4 **107.** $f(x) = x^3 - 7x$
108. $f(x) = x^3 - x^2 + 16x - 16$ **109.** 3 **111.** $(0, \infty)$
113. $(-\infty, 0) \cup (0, \infty)$ **115.** $\left(-\frac{5}{2}, -2\right)$ **117.** (d) **119.** (b)
121. **(a)** $y = x \ln x$

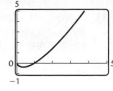

(b) 1; **(c)** relative minimum: -0.368 at $x = 0.368$

123. **(a)** $y = \dfrac{\ln x}{x^2}$

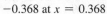

(b) 1; **(c)** relative maximum: 0.184 at $x = 1.649$

125. $(1.250, 0.891)$

Mid-Chapter Mixed Review: Chapter 5

1. False **2.** True **3.** False **4.** Yes; $f^{-1}(x) = -\dfrac{2}{x}$
5. No **6.** Yes; $f^{-1}(x) = \dfrac{5}{x} + 2$
7. $(f^{-1} \circ f)(x) = f^{-1}(\sqrt{x - 5}) = (\sqrt{x - 5})^2 + 5 = x - 5 + 5 = x$; $(f \circ f^{-1})(x) = f(x^2 + 5) = \sqrt{(x^2 + 5) - 5} = \sqrt{x^2} = x$ **8.** $f^{-1}(x) = \sqrt[3]{x} - 2$; domain and range of both f and f^{-1}: $(-\infty, \infty)$ **9.** (d)
10. (h) **11.** (c) **12.** (g) **13.** (b) **14.** (f) **15.** (e)
16. (a) **17.** \$4185.57 **18.** 0 **19.** $-\frac{4}{5}$ **20.** −2
21. 2 **22.** 0 **23.** −4 **24.** 0 **25.** 3 **26.** $\frac{1}{4}$ **27.** 1
28. $\log_e 0.0025 = -6$, or $\ln 0.0025 = -6$
29. $10^r = T$ **30.** 2.7268 **31.** 2.0115 **32.** For an even function $f, f(x) = f(-x)$, so we have $f(x) = f(-x)$ but $x \ne -x$ (for $x \ne 0$). Thus, f is not one-to-one and hence it does not have an inverse. **33.** The most interest will be earned the eighth year, because the principal is greatest during that year.
34. In $f(x) = x^3$, the variable x is the base. The range of f is $(-\infty, \infty)$. In $g(x) = 3^x$, the variable x is the exponent. The range of g is $(0, \infty)$. The graph of f does not have an asymptote. The graph of g has an asymptote $y = 0$.
35. If $\log b < 0$, then $0 < b < 1$.

Exercise Set 5.4

1. $\log_3 81 + \log_3 27 = 4 + 3 = 7$
3. $\log_5 5 + \log_5 125 = 1 + 3 = 4$
5. $\log_t 8 + \log_t Y$ **7.** $\ln x + \ln y$ **9.** $3 \log_b t$
11. $8 \log y$ **13.** $-6 \log_c K$ **15.** $\frac{1}{3} \ln 4$
17. $\log_t M - \log_t 8$ **19.** $\log x - \log y$ **21.** $\ln r - \ln s$
23. $\log_a 6 + \log_a x + 5 \log_a y + 4 \log_a z$
25. $2 \log_b p + 5 \log_b q - 4 \log_b m - 9$
27. $\ln 2 - \ln 3 - 3 \ln x - \ln y$
29. $\frac{3}{2} \log r + \frac{1}{2} \log t$ **31.** $3 \log_a x - \frac{5}{2} \log_a p - 4 \log_a q$
33. $2 \log_a m + 3 \log_a n - \frac{3}{4} - \frac{5}{4} \log_a b$ **35.** $\log_a 150$
37. $\log 100 = 2$ **39.** $\log m^3 \sqrt{n}$
41. $\log_a x^{-5/2} y^4$, or $\log_a \dfrac{y^4}{x^{5/2}}$ **43.** $\ln x$ **45.** $\ln(x-2)$
47. $\log \dfrac{x-7}{x-2}$ **49.** $\ln \dfrac{x}{(x^2-25)^3}$ **51.** $\ln \dfrac{2^{11/5} x^9}{y^8}$
53. -0.74 **55.** 1.991 **57.** 0.356 **59.** 4.827
61. -1.792 **63.** 0.099 **65.** 3 **67.** $|x-4|$ **69.** $4x$
71. w **73.** $8t$ **75.** $\frac{1}{2}$ **77.** Quartic
78. Exponential   **79.** Linear (constant)   **80.** Exponential
81. Rational   **82.** Logarithmic   **83.** Cubic   **84.** Rational
85. Linear   **86.** Quadratic   **87.** 4 **89.** $\log_a(x^3 - y^3)$
91. $\frac{1}{2} \log_a(x-y) - \frac{1}{2} \log_a(x+y)$ **93.** 7 **95.** True
97. True   **99.** True   **101.** -2 **103.** 3
105. $e^{-xy} = \dfrac{a}{b}$
107. $\log_a \left(\dfrac{x + \sqrt{x^2 - 5}}{5} \cdot \dfrac{x - \sqrt{x^2 - 5}}{x - \sqrt{x^2 - 5}} \right)$

$\qquad = \log_a \dfrac{5}{5\left(x - \sqrt{x^2 - 5}\right)} = \log_a \dfrac{1}{x - \sqrt{x^2 - 5}}$

$\qquad = \log_a \left(x - \sqrt{x^2 - 5}\right)^{-1} = -\log_a \left(x - \sqrt{x^2 - 5}\right)$

Exercise Set 5.5

1. 4 **3.** $\frac{3}{2}$ **5.** 5.044 **7.** $\frac{5}{2}$ **9.** $-3, \frac{1}{2}$ **11.** 0.959
13. 0 **15.** 0 **17.** 6.908 **19.** 84.191 **21.** -1.710
23. 2.844 **25.** $-1.567, 1.567$ **27.** 1.869
29. $-1.518, 0.825$ **31.** 625 **33.** 0.0001 **35.** e
37. $-\frac{1}{3}$ **39.** $\frac{22}{3}$ **41.** 10 **43.** 4 **45.** $\frac{1}{63}$ **47.** 2
49. $\frac{2}{5}$ **51.** 5 **53.** $\frac{21}{8}$ **55.** $\frac{8}{7}$ **57.** 6 **59.** 6.192
61. 0 **63.** $-1.911, 4.222$ **65.** 0.621 **67.** $-10, 0.366$
69. 7.062 **71.** 2.444 **73.** $(4.093, 0.786)$
75. $(7.586, 6.684)$ **77.** (a) $(0, -6)$; (b) $x = 0$;
(c) minimum: -6 when $x = 0$ **78.** (a) $(3, 1)$; (b) $x = 3$;
(c) maximum: 1 when $x = 3$ **79.** (a) $(-1, -5)$; (b) $x = -1$;
(c) maximum: -5 when $x = -1$ **80.** (a) $(2, 4)$; (b) $x = 2$;
(c) minimum: 4 when $x = 2$ **81.** $\dfrac{\ln 2}{2}$, or 0.347 **83.** $1, e^4$ or

$1, 54.598$ **85.** $\frac{1}{3}, 27$ **87.** $1, e^2$ or $1, 7.389$ **89.** $0, \dfrac{\ln 2}{\ln 5}$, or

$0, 0.431$ **91.** e^{-2}, e^2 or $0.135, 7.389$ **93.** $\frac{7}{4}$ **95.** $a = \frac{2}{3} b$

Exercise Set 5.6

1. (a) $P(t) = 438{,}000 \, e^{0.034t}$, where t is the number of years after
2014; (b) $485{,}034$; (c) about 4 years after 2014; (d) about
20 years

3. (a) 0.70%; (b) 1.63%; (c) 20.9 years; (d) 62.4 years;
(e) 0.18%; (f) 29.9 years; (g) 54.2 years; (h) 0.46%; (i) 2.64%
(j) 177.7 years   **5.** About 750 years after 2015
7. (a) $P(t) = 10{,}000 e^{0.054t}$; (b) \$10,554.85; \$11,140.48;
\$13,099.64; \$17,160.07; (c) about 12.8 years
9. About 12,320 years   **11.** (a) 22.4% per minute; (b) 3.1%
per year; (c) 60.3 days; (d) 10.7 years; (e) 2.4% per year;
(f) 1.0% per year; (g) 0.0029% per year   **13.** (a) $k \approx 0.0437$;
$V(t) = 8000 e^{0.0437t}$; (b) \$12,938; (c) about 15.9 years;
(d) about 26 years after 2000, or in 2026   **15.** (a) $k \approx 0.0681$;
$S(t) = 935{,}000 e^{0.0681t}$; (b) \$3,650,234; (c) about 10 years;
(d) about 37 years after 1980, or in 2017
17. (a)
$$y = \frac{3500}{1 + 19.9 e^{-0.6x}}$$
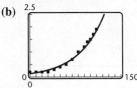
(b) 167; (c) 500; 1758; 3007; 3449; 3495; (d) as $t \to \infty$, $N(t) \to 3500$; the number approaches 3500 but never actually reaches it.
19. $46.7°F$ **21.** $59.6°F$
23. (d)   **25.** (a)   **27.** (e)

29. (a) $y = 0.1377082721(1.023820625)^x$; $r \approx 0.9824$, the function is a good fit;
(b)
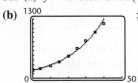
(c) 2007: 1.7%; 2015: 2.1%; 2020: 2.3%

31. (a) $y = 173.8943128(1.051418179)^x$;
(b)
(c) 2016: \$234.9 billion; 2023: \$333.71 billion

33. (a) $y = 1454.850619(0.9265336199)^x$;
(b) 2008: 1348 architects; 2012: 993 architects;
(c) about 6 years after 2007, or in 2013
35. Multiplication principle for inequalities   **36.** Product rule
37. Principle of zero products   **38.** Principle of square roots
39. Power rule   **40.** Multiplication principle for equations
41. \$166.16   **43.** \$19,609.67   **45.** (a) 24.7%; 1.5%; 0.09%;
(3.98×10^{-29}); (b) 0.00008% **47.** Linear

Review Exercises: Chapter 5

1. True   **2.** False   **3.** False   **4.** True   **5.** False
6. True   **7.** $\{(-2.7, 1.3), (-3, 8), (3, -5), (-3, 6), (-5, 7)\}$
8. (a) $x = -2y + 3$; (b) $x = 3y^2 + 2y - 1$;
(c) $0.8y^3 - 5.4x^2 = 3y$ **9.** No   **10.** No   **11.** Yes
12. Yes   **13.** (a) Yes; (b) $f^{-1}(x) = \dfrac{-x + 2}{3}$

14. (a) Yes; (b) $f^{-1}(x) = \dfrac{x + 2}{x - 1}$

15. (a) Yes; (b) $f^{-1}(x) = x^2 + 6$, $x \geq 0$
16. (a) Yes; (b) $f^{-1}(x) = \sqrt[3]{x + 8}$ **17.** (a) No
18. (a) Yes; (b) $f^{-1}(x) = \ln x$

19. $(f^{-1} \circ f)(x) =$
$f^{-1}(f(x)) = f^{-1}(6x - 5) = \dfrac{6x - 5 + 5}{6} = \dfrac{6x}{6} = x;$
$(f \circ f^{-1})(x) =$
$f(f^{-1}(x)) = f\left(\dfrac{x + 5}{6}\right) = 6\left(\dfrac{x + 5}{6}\right) - 5 = x + 5 - 5 = x$

20. $(f^{-1} \circ f)(x) =$
$f^{-1}(f(x)) = f^{-1}\left(\dfrac{x + 1}{x}\right) = \dfrac{1}{\dfrac{x + 1}{x} - 1} =$
$\dfrac{1}{\dfrac{x + 1 - x}{x}} = \dfrac{1}{\dfrac{1}{x}} = x;\ (f \circ f^{-1})(x) =$
$f(f^{-1}(x)) = f\left(\dfrac{1}{x - 1}\right) =$
$\dfrac{\dfrac{1}{x - 1} + 1}{\dfrac{1}{x - 1}} = \dfrac{\dfrac{1 + x - 1}{x - 1}}{\dfrac{1}{x - 1}} = \dfrac{x}{x - 1} \cdot \dfrac{x - 1}{1} = x$

21. $f^{-1}(x) = \dfrac{2 - x}{5}$; domain and range of both f and f^{-1}:
$(-\infty, \infty)$;

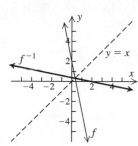

22. $f^{-1}(x) = \dfrac{-2x - 3}{x - 1}$;
domain of f: $(-\infty, -2) \cup (-2, \infty)$;
range of f: $(-\infty, 1) \cup (1, \infty)$;
domain of f^{-1}: $(-\infty, 1) \cup (1, \infty)$;
range of f^{-1}: $(-\infty, -2) \cup (-2, \infty)$

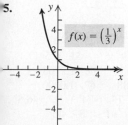

23. 657 **24.** a
25.

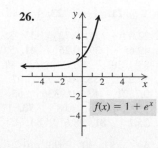

26.

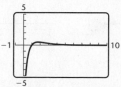

27.

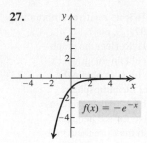

28.

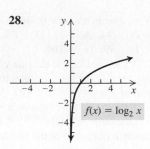

29.

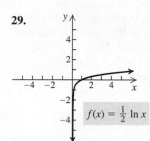

30.

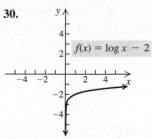

31. (c) **32.** (a) **33.** (b) **34.** (f) **35.** (e)
36. (d) **37.** 3 **38.** 5 **39.** 1 **40.** 0 **41.** $\frac{1}{4}$
42. $\frac{1}{2}$ **43.** 0 **44.** 1 **45.** $\frac{1}{3}$ **46.** -2 **47.** $4^2 = x$
48. $a^k = Q$ **49.** $\log_4 \frac{1}{64} = -3$ **50.** $\ln 80 = x$, or
$\log_e 80 = x$ **51.** 1.0414 **52.** -0.6308 **53.** 1.0986
54. -3.6119 **55.** Does not exist **56.** Does not exist
57. 1.9746 **58.** 0.5283 **59.** $\log_b \dfrac{x^3 \sqrt{z}}{y^4}$, or $\log_b \dfrac{x^3 z^{1/2}}{y^4}$
60. $\ln(x^2 - 4)$ **61.** $\frac{1}{4} \ln w + \frac{1}{2} \ln r$ **62.** $\frac{2}{3} \log M - \frac{1}{3} \log N$
63. 0.477 **64.** 1.699 **65.** -0.699 **66.** 0.233
67. $-5k$ **68.** $-6t$ **69.** 16 **70.** $\frac{1}{5}$ **71.** 4.382 **72.** 2
73. $\frac{1}{2}$ **74.** 5 **75.** 4 **76.** 9 **77.** 1 **78.** 3.912
79. (a) $A(t) = 30{,}000(1.0105)^{4t}$; (b) \$30,000; \$38,547.20;
\$49,529.56; \$63,640.87 **80.** 2014: 3466 breweries; 2018: 6905
breweries **81.** 15.4 years **82.** 2.7% **83.** About 2623 years
84. 5.6 **85.** 7.0 **86.** 30 decibels
87. (a) 2.2 ft/sec; (b) 8,553,143
88. (a) $k \approx 0.1392$; (b) $S(t) = 0.035e^{0.1392t}$, where t is the
number of years after 1940 and S is in billions of dollars; (c) 1970:
\$2.279 billion; 2000: \$148.353 billion; 2015: \$1197.023 billion, or
about \$1.197 trillion; (d) in 2019 **89.** (a) $P(t) = 15.2e^{0.0167t}$,
where t is the number of years after 2013 and P is in millions;
(b) 2017: 16.3 million; 2020: 17.1 million; (c) about 10 years
after 2013; (d) 41.5 years
90. (a) $y = 99.16310468(1.02186993)^x$;
(b)

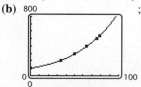

; (c) 1970: 221 billion
metric tons;
2000: 423 billion metric tons;
2015: 585 billion metric tons;
(d) about 107 years after 1933,
or in 2040

91. No **92.** (a) $y = 5e^{-x} \ln x$

(b) relative maximum: 0.486 at $x = 1.763$; no relative minimum
93. D　**94.** A　**95.** D　**96.** B　**97.** $\frac{1}{64}$, 64
98. 1　**99.** 16　**100.** $(1, \infty)$　**101.** Reflect the graph
of $f(x) = \ln x$ across the line $y = x$ to obtain the graph of
$h(x) = e^x$. Then shift this graph right 2 units to obtain the
graph of $g(x) = e^{x-2}$.　**102.** Measure the atmospheric
pressure P at the top of the building. Substitute that value in
the equation $P = 14.7e^{-0.00005a}$, and solve for the height, or
altitude, a of the top of the building. Also measure the atmospheric pressure at the base of the building and solve for the
altitude of the base. Then subtract to find the height of the building.　**103.** $\log_a ab^3 \neq (\log_a a)(\log_a b^3)$. If the first step had been
correct, then the second step would be as well. The correct procedure follows: $\log_a ab^3 = \log_a a + \log_a b^3 = 1 + 3\log_a b$.
104. The inverse of a function $f(x)$ is written $f^{-1}(x)$, whereas
$[f(x)]^{-1}$ means $\dfrac{1}{f(x)}$.

Test: Chapter 5

1. [5.1] $\{(5, -2), (3, 4), (-1, 0), (-3, -6)\}$　**2.** [5.1] No
3. [5.1] Yes　**4.** [5.1] **(a)** Yes; **(b)** $f^{-1}(x) = \sqrt[3]{x - 1}$
5. [5.1] **(a)** Yes; **(b)** $f^{-1}(x) = 1 - x$
6. [5.1] **(a)** Yes; **(b)** $f^{-1}(x) = \dfrac{2x}{1 + x}$　**7.** [5.1] **(a)** No
8. [5.1] $(f^{-1} \circ f)(x) =$
$f^{-1}(f(x)) = f^{-1}(-4x + 3) = \dfrac{3 - (-4x + 3)}{4} =$
$\dfrac{4x}{4} = x;\ (f \circ f^{-1})(x) =$
$f(f^{-1}(x)) = f\!\left(\dfrac{3 - x}{4}\right) = -4\!\left(\dfrac{3 - x}{4}\right) + 3 =$
$-3 + x + 3 = x$
9. [5.1] $f^{-1}(x) = \dfrac{4x + 1}{x}$; domain of f: $(-\infty, 4) \cup (4, \infty)$;
range of f: $(-\infty, 0) \cup (0, \infty)$;
domain of f^{-1}: $(-\infty, 0) \cup (0, \infty)$;
range of f^{-1}: $(-\infty, 4) \cup (4, \infty)$;

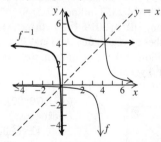

10. [5.2]　**11.** [5.3]

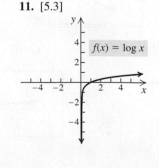

12. [5.2]　**13.** [5.3]

[graph: $f(x) = e^x - 3$]　[graph: $f(x) = \ln(x + 2)$]

14. [5.3] -5　**15.** [5.3] 1　**16.** [5.3] 0　**17.** [5.3] $\frac{1}{5}$
18. [5.3] $x = e^4$　**19.** [5.3] $x = \log_3 5.4$　**20.** [5.3] 2.7726
21. [5.3] -0.5331　**22.** [5.3] 1.2851　**23.** [5.4] $\log_a \dfrac{x^2\sqrt{z}}{y}$,
or $\log_a \dfrac{x^2 z^{1/2}}{y}$　**24.** [5.4] $\frac{2}{5}\ln x + \frac{1}{5}\ln y$　**25.** [5.4] 2.322
26. [5.4] $-4t$　**27.** [5.5] $\frac{1}{2}$　**28.** [5.5] 1　**29.** [5.5] 1
30. [5.5] 4.174　**31.** [5.3] 9.0　**32.** [5.6] 0.0154
33. [5.6] **(a)** 4.5%; **(b)** $P(t) = 1000e^{0.045t}$; **(c)** \$1433.33;
(d) 15.4 years　**34.** [5.2] C　**35.** [5.5] $\frac{27}{8}$

Chapter 6

Exercise Set 6.1

1. $\sin \phi = \frac{15}{17}$, $\cos \phi = \frac{8}{17}$, $\tan \phi = \frac{15}{8}$, $\csc \phi = \frac{17}{15}$, $\sec \phi = \frac{17}{8}$,
$\cot \phi = \frac{8}{15}$　**3.** $\sin \alpha = \dfrac{\sqrt{3}}{2}$, $\cos \alpha = \dfrac{1}{2}$, $\tan \alpha = \sqrt{3}$,
$\csc \alpha = \dfrac{2\sqrt{3}}{3}$, $\sec \alpha = 2$, $\cot \alpha = \dfrac{\sqrt{3}}{3}$
5. $\sin \phi = \dfrac{27}{5\sqrt{37}}$, or $\dfrac{27\sqrt{37}}{185}$; $\cos \phi = \dfrac{14}{5\sqrt{37}}$, or $\dfrac{14\sqrt{37}}{185}$;
$\tan \phi = \dfrac{27}{14}$; $\csc \phi = \dfrac{5\sqrt{37}}{27}$; $\sec \phi = \dfrac{5\sqrt{37}}{14}$; $\cot \phi = \dfrac{14}{27}$
7. $\csc \alpha = \dfrac{3}{\sqrt{5}}$, or $\dfrac{3\sqrt{5}}{5}$; $\sec \alpha = \dfrac{3}{2}$; $\cot \alpha = \dfrac{2}{\sqrt{5}}$, or $\dfrac{2\sqrt{5}}{5}$
9. $\cos \theta = \frac{7}{25}$, $\tan \theta = \frac{24}{7}$, $\csc \theta = \frac{25}{24}$, $\sec \theta = \frac{25}{7}$, $\cot \theta = \frac{7}{24}$
11. $\sin \phi = \dfrac{2\sqrt{5}}{5}$, $\cos \phi = \dfrac{\sqrt{5}}{5}$, $\csc \phi = \dfrac{\sqrt{5}}{2}$, $\sec \phi = \sqrt{5}$,
$\cot \phi = \dfrac{1}{2}$　**13.** $\sin \theta = \dfrac{2}{3}$, $\cos \theta = \dfrac{\sqrt{5}}{3}$, $\tan \theta = \dfrac{2\sqrt{5}}{5}$,
$\sec \theta = \dfrac{3\sqrt{5}}{5}$, $\cot \theta = \dfrac{\sqrt{5}}{2}$　**15.** $\sin \beta = \dfrac{2\sqrt{5}}{5}$, $\tan \beta = 2$,
$\csc \beta = \dfrac{\sqrt{5}}{2}$, $\sec \beta = \sqrt{5}$, $\cot \beta = \dfrac{1}{2}$　**17.** $\dfrac{\sqrt{2}}{2}$
19. 2　**21.** $\dfrac{\sqrt{3}}{3}$　**23.** $\frac{1}{2}$　**25.** 1　**27.** 2
29. 22.6 ft　**31.** 9.72°　**33.** 35.01°　**35.** 3.03°
37. 49.65°　**39.** 0.25°　**41.** 5.01°　**43.** 17°36′
45. 83°1′30″　**47.** 11°45′　**49.** 47°49′36″　**51.** 0°54′
53. 39°27′　**55.** 0.6293　**57.** 0.0737　**59.** 1.2765
61. 0.7621　**63.** 0.9336　**65.** 12.4288　**67.** 1.0000
69. 1.7032　**71.** 30.8°　**73.** 12.5°　**75.** 64.4°　**77.** 46.5°
79. 25.2°　**81.** 38.6°　**83.** 45°　**85.** 60°　**87.** 45°
89. 60°　**91.** 30°　**93.** $\cos 20° = \sin 70° = \dfrac{1}{\sec 20°}$
95. $\tan 52° = \cot 38° = \dfrac{1}{\cot 52°}$　**97.** $\sin 25° \approx 0.4226$,

os 25° ≈ 0.9063, tan 25° ≈ 0.4663, csc 25° ≈ 2.3662,
ec 25° ≈ 1.1034, cot 25° ≈ 2.1445
9. sin 18°49′55″ ≈ 0.3228, cos 18°49′55″ ≈ 0.9465,
an 18°49′55″ ≈ 0.3411, csc 18°49′55″ ≈ 3.0979,
ec 18°49′55″ ≈ 1.0565, cot 18°49′55″ ≈ 2.9321
01. $\sin 8° = q$, $\cos 8° = p$, $\tan 8° = \frac{1}{r}$, $\csc 8° = \frac{1}{q}$,

$\sec 8° = \frac{1}{p}$, $\cot 8° = r$

02. 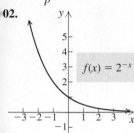   **103.**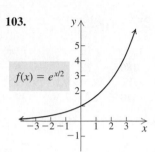

04. 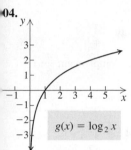   **105.**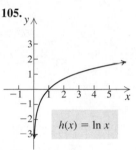

06. About 9.21   **107.** 4   **108.** $\frac{101}{97}$ **109.** 343
11. 0.6534   **113.** Let $h =$ the height of the triangle. Then

Area $= \frac{1}{2}bh$, where $\sin \theta = \frac{h}{a}$, or $h = a \sin \theta$, so Area $= \frac{1}{2} ab \sin \theta$.

Exercise Set 6.2

. $F = 60°$, $d = 3$, $f \approx 5.2$ **3.** $A = 22.7°$, $a \approx 52.7$,
≈ 136.6 **5.** $P = 47°38′$, $n \approx 34.4$, $p \approx 25.4$
. $B = 2°17′$, $b \approx 0.39$, $c \approx 9.74$ **9.** $A \approx 77.2°$, $B \approx 12.8°$,
≈ 439 **11.** $B = 42.42°$, $a \approx 35.7$, $b \approx 32.6$
3. $B = 55°$, $a \approx 28.0$, $c \approx 48.8$ **15.** $A \approx 62.4°$, $B \approx 27.6°$,
≈ 3.56 **17.** Approximately 34°   **19.** About 13.9°
1. 154 in., or 12 ft 10 in.   **23.** 10.4°   **25.** About 424 ft
7. About 92.9 cm   **29.** About 45 ft   **31.** Radius: 9.15 in.;
ength: 73.20 in.; width: 54.90 in.   **33.** 17.9 ft
5. About 8 km   **37.** About 19.5 mi   **39.** About 24 km
0. $3\sqrt{10}$, or about 9.487   **41.** $10\sqrt{2}$, or about 14.142
2. $\ln t = 4$ **43.** $10^{-3} = 0.001$ **45.** 3.3

Exercise Set 6.3

. III   **3.** III   **5.** I   **7.** III   **9.** II   **11.** II   **13.** 434°,
94°, −286°, −646°   **15.** 475.3°, 835.3°, −244.7°, −604.7°
7. 180°, 540°, −540°, −900°   **19.** 72.89°, 162.89°
1. 77°56′46″, 167°56′46″   **23.** 44.8°, 134.8°
5. $\sin \beta = \frac{5}{13}$, $\cos \beta = -\frac{12}{13}$, $\tan \beta = -\frac{5}{12}$, $\csc \beta = \frac{13}{5}$,
$\sec \beta = -\frac{13}{12}$, $\cot \beta = -\frac{12}{5}$
7. $\sin \alpha = -\frac{4\sqrt{97}}{97}$; $\cos \alpha = \frac{9\sqrt{97}}{97}$; $\tan \alpha = -\frac{4}{9}$;

$\sec \alpha = -\frac{\sqrt{97}}{4}$; $\sec \alpha = \frac{\sqrt{97}}{9}$; $\cot \alpha = -\frac{9}{4}$

29. $\sin \phi = -\frac{2\sqrt{7}}{7}$, $\cos \phi = -\frac{\sqrt{21}}{7}$, $\tan \phi = \frac{2\sqrt{3}}{3}$,

$\csc \phi = -\frac{\sqrt{7}}{2}$, $\sec \phi = -\frac{\sqrt{21}}{3}$, $\cot \phi = \frac{\sqrt{3}}{2}$

31. $\sin \theta = -\frac{2\sqrt{13}}{13}$, $\cos \theta = \frac{3\sqrt{13}}{13}$, $\tan \theta = -\frac{2}{3}$

33. $\sin \theta = \frac{5\sqrt{41}}{41}$, $\cos \theta = \frac{4\sqrt{41}}{41}$, $\tan \theta = \frac{5}{4}$

35. $\cos \theta = -\frac{2\sqrt{2}}{3}$, $\tan \theta = \frac{\sqrt{2}}{4}$, $\csc \theta = -3$, $\sec \theta = -\frac{3\sqrt{2}}{4}$,

$\cot \theta = 2\sqrt{2}$ **37.** $\sin \theta = -\frac{\sqrt{5}}{5}$, $\cos \theta = \frac{2\sqrt{5}}{5}$, $\tan \theta = -\frac{1}{2}$,

$\csc \theta = -\sqrt{5}$, $\sec \theta = \frac{\sqrt{5}}{2}$ **39.** $\sin \phi = -\frac{4}{5}$, $\tan \phi = -\frac{4}{3}$,

$\csc \phi = -\frac{5}{4}$, $\sec \phi = \frac{5}{3}$, $\cot \phi = -\frac{3}{4}$ **41.** $30°$; $-\frac{\sqrt{3}}{2}$

43. $45°$; 1 **45.** 0 **47.** $45°$; $-\frac{\sqrt{2}}{2}$ **49.** $30°$; 2

51. $30°$; $\sqrt{3}$ **53.** $30°$; $-\frac{\sqrt{3}}{3}$ **55.** Not defined   **57.** -1

59. $60°$; $\sqrt{3}$ **61.** $45°$; $\frac{\sqrt{2}}{2}$ **63.** $45°$; $-\sqrt{2}$ **65.** 1

67. 0   **69.** 0   **71.** 0   **73.** Positive: cos, sec; negative: sin,
csc, tan, cot   **75.** Positive: tan, cot; negative: sin, csc, cos, sec
77. Positive: sin, csc; negative: cos, sec, tan, cot   **79.** Positive: all
81. sin 319° = −0.6561, cos 319° = 0.7547, tan 319° = −0.8693,
csc 319° ≈ −1.5242, sec 319° ≈ 1.3250, cot 319° ≈ −1.1504
83. sin 115° = 0.9063, cos 115° = −0.4226, tan 115° = −2.1445,
csc 115° ≈ 1.1034, sec 115° ≈ −2.3663, cot 115° ≈ −0.4663
85. −1.1585   **87.** −1.4910   **89.** 0.8771   **91.** 0.4352
93. 0.9563   **95.** 2.9238   **97.** 275.4°   **99.** 200.1°
101. 288.1°   **103.** 72.6°   **105.** East: about 130 km; south: 75 km
107. About 223 km
109. 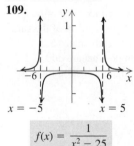   **110.**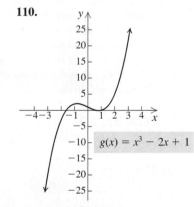

111. Domain: $\{x | x \neq -2\}$; range: $\{x | x \neq 1\}$
112. Domain: $\{x | x \neq -\frac{3}{2} \text{ and } x \neq 5\}$; range: all real numbers
113. 12   **114.** −2, 3   **115.** $(12, 0)$ **116.** $(-2, 0)$, $(3, 0)$
117. About 70 ft

Mid-Chapter Mixed Review: Chapter 6

1. True   **2.** True   **3.** True   **4.** $S = 47.5°$, $s \approx 59.9$,
$q \approx 54.9$ **5.** $A \approx 27.8°$, $B \approx 62.2°$, $b \approx 27.2$
6. 285°, 645°; −435°, −1155°   **7.** 574°30′, 1294°30′;
−145°30′, −505°30′   **8.** 71.8°; 161.8°   **9.** 2°44′50″;
92°44′50″   **10.** sin 155° = 0.4226, cos 155° = −0.9063,

tan $155° = -0.4663$, csc $155° \approx 2.3663$, sec $155° \approx -1.1034$, cot $155° \approx -2.1445$ **11.** sin $\alpha = -\frac{5}{13}$, cos $\alpha = -\frac{12}{13}$, tan $\alpha = \frac{5}{12}$, csc $\alpha = -\frac{13}{5}$, sec $\alpha = -\frac{13}{12}$, cot $\alpha = \frac{12}{5}$

12. sin $\theta = -\frac{1}{\sqrt{5}}$, or $-\frac{\sqrt{5}}{5}$; cos $\theta = -\frac{2}{\sqrt{5}}$, or $-\frac{2\sqrt{5}}{5}$; tan $\theta = \frac{1}{2}$; csc $\theta = -\sqrt{5}$; sec $\theta = -\frac{\sqrt{5}}{2}$

13. sin $\alpha = \frac{\sqrt{77}}{9}$; tan $\alpha = \frac{\sqrt{77}}{2}$; csc $\alpha = \frac{9}{\sqrt{77}}$, or $\frac{9\sqrt{77}}{77}$; sec $\alpha = \frac{9}{2}$; cot $\alpha = \frac{2}{\sqrt{77}}$, or $\frac{2\sqrt{77}}{77}$

14. $42.1472°$ **15.** $51°10'48''$ **16.** sin $81° \approx 0.9877$, cos $81° \approx 0.1564$, tan $81° \approx 6.3131$, csc $81° \approx 1.0125$, sec $81° \approx 6.3939$, cot $81° \approx 0.1584$ **17.** $67.5°$

18. About 290 mi   **19.** $\frac{\sqrt{3}}{3}$ **20.** $\frac{\sqrt{2}}{2}$ **21.** $\sqrt{3}$

22. $-\sqrt{2}$ **23.** $\frac{\sqrt{2}}{2}$ **24.** -2 **25.** 1 **26.** 0

27. $\frac{\sqrt{3}}{2}$ **28.** -1 **29.** $-\sqrt{3}$ **30.** 1 **31.** $\frac{1}{2}$

32. $\sqrt{2}$ **33.** $-\frac{\sqrt{3}}{2}$ **34.** Not defined   **35.** 2

36. Not defined   **37.** 1 **38.** $-\frac{\sqrt{2}}{2}$ **39.** 0.7683

40. 1.5557 **41.** 0.4245 **42.** 0.1817 **43.** -1.0403
44. -1.3127 **45.** -0.6441 **46.** 0.0480

47. Given points P and Q on the terminal side of an angle θ, the reference triangles determined by them are similar. Thus corresponding sides are proportional and the trigonometric ratios are the same. See the specific example on p. 423.

48. If f and g are reciprocal functions, then $f(\theta) = \frac{1}{g(\theta)}$. If f and g are cofunctions, then $f(\theta) = g(90° - \theta)$.   **49.** Sine: $(0, 1)$; cosine: $(0, 1)$; tangent: $(0, \infty)$ **50.** Since sin $\theta = y/r$ and cos $\theta = x/r$ and $r > 0$ for all angles θ, the domain of the sine function and of the cosine function is the set of all angles θ. However, tan $\theta = y/x$ and $x = 0$ for all angles that are odd multiples of $90°$. Thus the domain of the tangent function must be restricted to avoid division by 0.

Exercise Set 6.4

1.

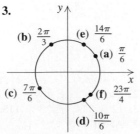

3.

(a) $\frac{\pi}{6}$; (b) $\frac{2\pi}{3}$; (c) $\frac{7\pi}{6}$; (d) $\frac{10\pi}{6}$; (e) $\frac{14\pi}{6}$; (f) $\frac{23\pi}{4}$

5. $M: \frac{2\pi}{3}, -\frac{4\pi}{3}$; $N: \frac{3\pi}{2}, -\frac{\pi}{2}$; $P: \frac{5\pi}{4}, -\frac{3\pi}{4}$; $Q: \frac{11\pi}{6}, -\frac{\pi}{6}$

7.

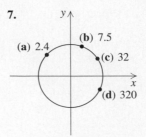

(a) 2.4 (b) 7.5 (c) 32 (d) 320

9. $\frac{5\pi}{12}$ **11.** $\frac{10\pi}{9}$ **13.** $-\frac{214.6\pi}{180}$, or $-\frac{1073\pi}{900}$ **15.** $-\pi$

17. $\frac{12.5\pi}{180}$, or $\frac{5\pi}{72}$ **19.** $-\frac{17\pi}{9}$ **21.** 4.19 **23.** -1.05

25. 2.06 **27.** 0.02 **29.** 6.02 **31.** 1.66 **33.** $-135°$

35. $1440°$ **37.** $57.30°$ **39.** $134.47°$ **41.** $225°$

43. $-5156.62°$ **45.** $51.43°$ **47.** $0° = 0$ radians, $30° = \frac{\pi}{6}$, $45° = \frac{\pi}{4}$, $60° = \frac{\pi}{3}$, $90° = \frac{\pi}{2}$, $135° = \frac{3\pi}{4}$, $180° = \pi$, $225° = \frac{5\pi}{4}$, $270° = \frac{3\pi}{2}$, $315° = \frac{7\pi}{4}$, $360° = 2\pi$ **49.** $\frac{9\pi}{4}, -\frac{7\pi}{4}$

51. $\frac{19\pi}{6}, -\frac{5\pi}{6}$ **53.** $\frac{4\pi}{3}, -\frac{8\pi}{3}$ **55.** Complement: $\frac{\pi}{6}$; supplement: $\frac{2\pi}{3}$ **57.** Complement: $\frac{\pi}{8}$; supplement: $\frac{5\pi}{8}$

59. Complement: $\frac{5\pi}{12}$; supplement: $\frac{11\pi}{12}$ **61.** 2.29

63. 5.50 in.   **65.** 1.1; $63°$ **67.** 3.2 yd   **69.** 3489

71. $3150 \frac{\text{cm}}{\text{min}}$ **73.** 0.92 mph   **75.** 1047 mph

77. 10 mph   **79.** About 202 **81.** 1.676 radians/sec
82. One-to-one   **83.** Cosine of θ **84.** Exponential function   **85.** Horizontal asymptote   **86.** Odd function   **87.** Natural   **88.** Horizontal line; inverse
89. Logarithm   **91.** 111.7 km; 69.8 mi   **93.** (a) $5°37'30''$;
(b) $19°41'15''$ **95.** 1.46 nautical miles

Exercise Set 6.5

1. (a) $\left(-\frac{3}{4}, -\frac{\sqrt{7}}{4}\right)$; (b) $\left(\frac{3}{4}, \frac{\sqrt{7}}{4}\right)$; (c) $\left(\frac{3}{4}, -\frac{\sqrt{7}}{4}\right)$

3. (a) $\left(\frac{2}{5}, \frac{\sqrt{21}}{5}\right)$; (b) $\left(-\frac{2}{5}, -\frac{\sqrt{21}}{5}\right)$; (c) $\left(-\frac{2}{5}, \frac{\sqrt{21}}{5}\right)$

5. $\left(\frac{\sqrt{2}}{2}, -\frac{\sqrt{2}}{2}\right)$ **7.** 0 **9.** $\sqrt{3}$ **11.** 0 **13.** $-\frac{\sqrt{3}}{2}$

15. Not defined   **17.** $\frac{\sqrt{3}}{2}$ **19.** $-\frac{\sqrt{2}}{2}$ **21.** 0 **23.** 0

25. 0.4816 **27.** 1.3065 **29.** -2.1599 **31.** 1
33. -1.1747 **35.** -1 **37.** -0.7071 **39.** 0 **41.** 0.8391

43. (a)

$y = \sin x$

(b)

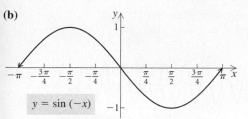

$$y = \sin(-x)$$

(c) same as (b); **(d)** the same
45. (a)

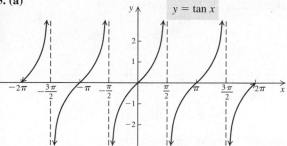

$$y = \tan x$$

(b)

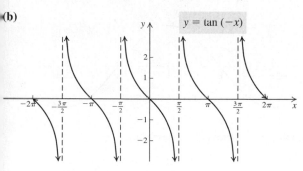

$$y = \tan(-x)$$

(c) same as (b); **(d)** the same
47. (a) See Exercise 43(a);
(b)

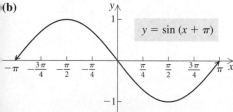

$$y = \sin(x + \pi)$$

(c) same as (b); **(d)** the same
49. (a)

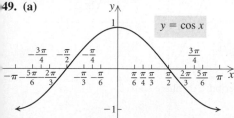

$$y = \cos x$$

(b)

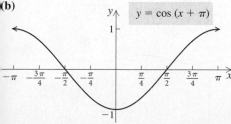

$$y = \cos(x + \pi)$$

(c) same as (b); **(d)** the same   **51.** Even: cosine, secant; odd:
sine, tangent, cosecant, cotangent   **53.** Positive: I, III;
negative: II, IV   **55.** Positive: I, IV; negative: II, III

57.

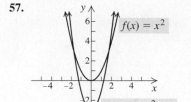

$$f(x) = x^2$$
$$g(x) = 2x^2 - 3$$

Stretch the graph of f vertically, then shift it down 3 units.
58.

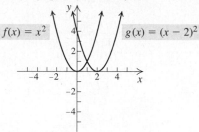

$$f(x) = x^2 \qquad g(x) = (x - 2)^2$$

Shift the graph of f right 2 units.
59.

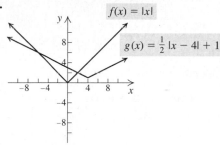

$$f(x) = |x|$$
$$g(x) = \tfrac{1}{2}|x - 4| + 1$$

Shift the graph of f right 4 units, shrink it vertically, then shift
it up 1 unit.
60.

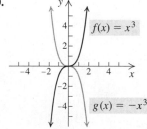

$$f(x) = x^3$$
$$g(x) = -x^3$$

Reflect the graph of f across the x-axis.

61. $y = -(x - 2)^3 - 1$ **62.** $y = \dfrac{1}{4x} + 3$ **63.** $\sin x$

65. $\sin x$ **67.** $-\cos x$ **69.** $-\sin x$ **71. (a)** $\dfrac{\pi}{2} + 2k\pi$, k an

integer; **(b)** $\pi + 2k\pi$, k an integer; **(c)** $k\pi$, k an integer
73. Domain: $(-\infty, \infty)$; range: $[0, 1]$; period: π; amplitude: $\tfrac{1}{2}$
75. $\left[-\dfrac{\pi}{2} + 2k\pi, \dfrac{\pi}{2} + 2k\pi\right]$, k an integer

77. $\left\{x \mid x \neq \dfrac{\pi}{2} + k\pi, k \text{ an integer}\right\}$
79.

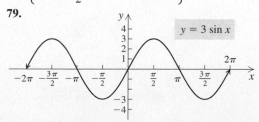

$$y = 3 \sin x$$

81.

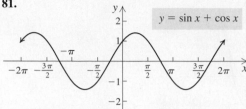

$y = \sin x + \cos x$

83. (a) $\triangle OPA \sim \triangle ODB$;

Thus, $\dfrac{AP}{OA} = \dfrac{BD}{OB}$

$\dfrac{\sin \theta}{\cos \theta} = \dfrac{BD}{1}$

$\tan \theta = BD$

(b) $\triangle OPA \sim \triangle ODB$;

$\dfrac{OD}{OP} = \dfrac{OB}{OA}$

$\dfrac{OD}{1} = \dfrac{1}{\cos \theta}$

$OD = \sec \theta$

(c) $\triangle OAP \sim \triangle ECO$;

$\dfrac{OE}{PO} = \dfrac{CO}{AP}$

$\dfrac{OE}{1} = \dfrac{1}{\sin \theta}$

$OE = \csc \theta$

(d) $\triangle OAP \sim \triangle ECO$;

$\dfrac{CE}{AO} = \dfrac{CO}{AP}$

$\dfrac{CE}{\cos \theta} = \dfrac{1}{\sin \theta}$

$CE = \dfrac{\cos \theta}{\sin \theta}$

$CE = \cot \theta$

85. 1

Visualizing the Graph

1. J **2.** H **3.** E **4.** F **5.** B **6.** D **7.** G
8. A **9.** C **10.** I

Exercise Set 6.6

1. Amplitude: 1; period: 2π; phase shift: 0

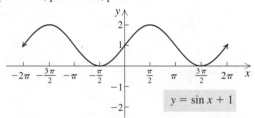

$y = \sin x + 1$

3. Amplitude: 3; period: 2π; phase shift: 0

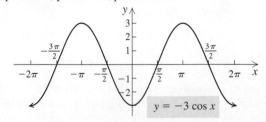

$y = -3 \cos x$

5. Amplitude: $\frac{1}{2}$; period: 2π; phase shift: 0

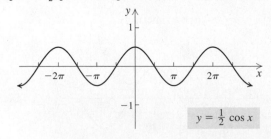

$y = \frac{1}{2} \cos x$

7. Amplitude: 1; period: π; phase shift: 0

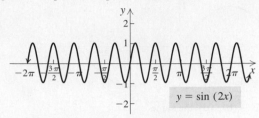

$y = \sin (2x)$

9. Amplitude: 2; period: 4π; phase shift: 0

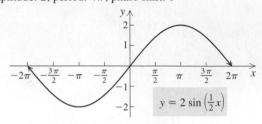

$y = 2 \sin \left(\frac{1}{2}x\right)$

11. Amplitude: 1; period: 4π; phase shift: 0

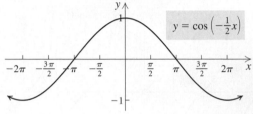

$y = \cos \left(-\frac{1}{2}x\right)$

13. Amplitude: $\dfrac{1}{2}$; period: 2π; phase shift: $-\dfrac{\pi}{2}$

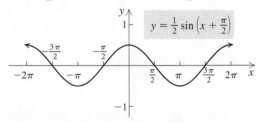

$y = \frac{1}{2} \sin \left(x + \frac{\pi}{2}\right)$

15. Amplitude: 3; period: 2π; phase shift: π

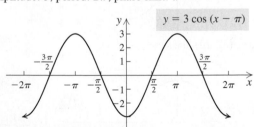

$y = 3 \cos (x - \pi)$

17. Amplitude: $\frac{1}{3}$; period: 2π; phase shift: 0

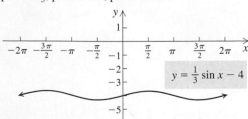

$y = \frac{1}{3} \sin x - 4$

19. Amplitude: 1; period: 2π; phase shift: 0

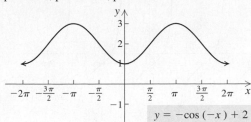

$$y = -\cos(-x) + 2$$

21. Amplitude: 2; period: 4π; phase shift: π

23. Amplitude: $\dfrac{1}{2}$; period: π; phase shift: $-\dfrac{\pi}{4}$

25. Amplitude: 3; period: 2; phase shift: $\dfrac{3}{\pi}$

27. Amplitude: $\frac{1}{2}$; period: 1; phase shift: 0

29. Amplitude: 1; period: 4π; phase shift: π

31. Amplitude: 1; period: 1; phase shift: 0

33. Amplitude: $\dfrac{1}{4}$; period: 2; phase shift: $\dfrac{4}{\pi}$

35. (b)   **37.** (h)   **39.** (a)   **41.** (f)

43. $y = \frac{1}{2}\cos x + 1$ **45.** $y = \cos\left(x + \dfrac{\pi}{2}\right) - 2$

47.

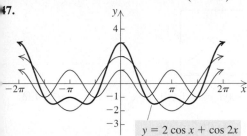

$$y = 2\cos x + \cos 2x$$

49.

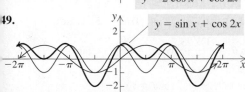

$$y = \sin x + \cos 2x$$

51.

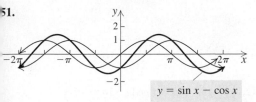

$$y = \sin x - \cos x$$

53.

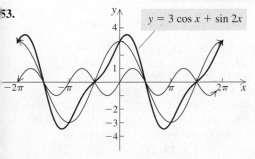

$$y = 3\cos x + \sin 2x$$

55. $y = x + \sin x$

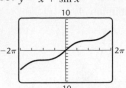

57. $y = \cos x - x$

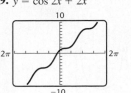

59. $y = \cos 2x + 2x$

61. $y = 4\cos 2x - 2\sin x$

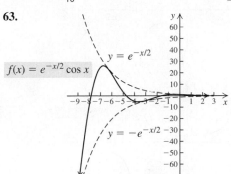

63.

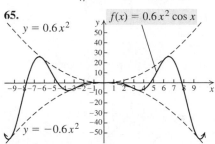

$y = e^{-x/2}$

$f(x) = e^{-x/2}\cos x$

$y = -e^{-x/2}$

65.

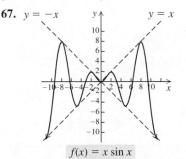

$f(x) = 0.6x^2\cos x$

$y = 0.6x^2$

$y = -0.6x^2$

67. $y = -x$ $y = x$

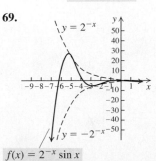

$f(x) = x\sin x$

69.

$y = 2^{-x}$

$y = -2^{-x}$

$f(x) = 2^{-x}\sin x$

71. Rational   **72.** Logarithmic   **73.** Quartic
74. Linear   **75.** Trigonometric   **76.** Exponential

77. Linear **78.** Trigonometric **79.** Cubic
80. Exponential **81.** Maximum: 8; minimum: 4
83.

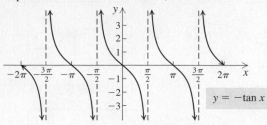

$$y = -\tan x$$

85.

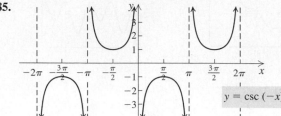

$$y = \csc(-x)$$

87.

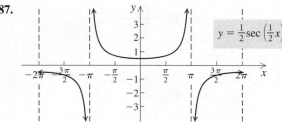

$$y = \frac{1}{2}\sec\left(\frac{1}{2}x\right)$$

89.

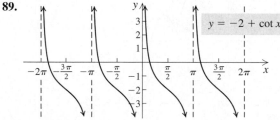

$$y = -2 + \cot x$$

91.

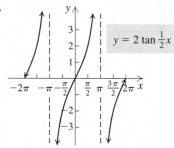

$$y = 2\tan\frac{1}{2}x$$

93.

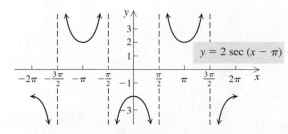

$$y = 2\sec(x - \pi)$$

95.

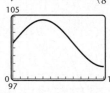

$$y = \cot\left(x + \frac{\pi}{2}\right) - 1$$

97.

$$y = 2\csc\left(\frac{1}{2}x - \frac{3\pi}{4}\right)$$

99. $-9.42, -6.28, -3.14, 3.14, 6.28, 9.42$ **101.** $-3.14, 0, 3.14$
103. (a) $y = 101.6 + 3\sin\left(\frac{\pi}{8}x\right)$ **(b)** $104.6°, 98.6°$

105. Amplitude: 3000; period: 90; phase shift: 10 **107.** 4 in.

Review Exercises: Chapter 6

1. False **2.** False **3.** True **4.** True **5.** False

6. False **7.** $\sin\theta = \dfrac{3\sqrt{73}}{73}$, $\cos\theta = \dfrac{8\sqrt{73}}{73}$, $\tan\theta = \dfrac{3}{8}$,

$\csc\theta = \dfrac{\sqrt{73}}{3}$, $\sec\theta = \dfrac{\sqrt{73}}{8}$, $\cot\theta = \dfrac{8}{3}$

8. $\cos\beta = \dfrac{3}{10}$, $\tan\beta = \dfrac{\sqrt{91}}{3}$, $\csc\beta = \dfrac{10\sqrt{91}}{91}$,

$\sec\beta = \dfrac{10}{3}$, $\cot\beta = \dfrac{3\sqrt{91}}{91}$ **9.** $\dfrac{\sqrt{2}}{2}$ **10.** $\dfrac{\sqrt{3}}{3}$

11. $-\dfrac{\sqrt{2}}{2}$ **12.** $\dfrac{1}{2}$ **13.** Not defined **14.** $-\sqrt{3}$

15. $\dfrac{2\sqrt{3}}{3}$ **16.** -1 **17.** $22°16'12''$ **18.** $47.56°$

19. 0.4452 **20.** 1.1315 **21.** 0.9498 **22.** -0.9092
23. -1.5282 **24.** -0.2778 **25.** $205.3°$ **26.** $47.2°$
27. $60°$ **28.** $60°$ **29.** $45°$ **30.** $30°$
31. $\sin 30.9° \approx 0.5135$, $\cos 30.9° \approx 0.8581$,
$\tan 30.9° \approx 0.5985$, $\csc 30.9° \approx 1.9474$, $\sec 30.9° \approx 1.1654$,
$\cot 30.9° \approx 1.6709$ **32.** $b \approx 4.5$, $A \approx 58.1°$, $B \approx 31.9°$
33. $A = 38.83°$, $b \approx 37.9$, $c \approx 48.6$ **34.** 1748 m
35. 14 ft **36.** II **37.** I **38.** IV **39.** $425°, -295°$
40. $\dfrac{\pi}{3}, -\dfrac{5\pi}{3}$ **41.** Complement: $76.6°$; supplement: $166.6°$

42. Complement: $\dfrac{\pi}{3}$; supplement: $\dfrac{5\pi}{6}$ **43.** $\sin\theta = \dfrac{3\sqrt{13}}{13}$,

$\cos\theta = \dfrac{-2\sqrt{13}}{13}$, $\tan\theta = -\dfrac{3}{2}$, $\csc\theta = \dfrac{\sqrt{13}}{3}$, $\sec\theta = -\dfrac{\sqrt{13}}{2}$,

$\cot\theta = -\dfrac{2}{3}$ **44.** $\sin\theta = -\dfrac{2}{3}$, $\cos\theta = -\dfrac{\sqrt{5}}{3}$, $\cot\theta = \dfrac{\sqrt{5}}{2}$,

$\sec\theta = -\dfrac{3\sqrt{5}}{5}$, $\csc\theta = -\dfrac{3}{2}$ **45.** About 1743 mi

46.

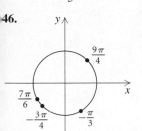

47. $\dfrac{121}{150}\pi$, 2.53   **48.** $-\dfrac{\pi}{6}$, -0.52 **49.** $270°$ **50.** $171.89°$

51. $-257.83°$ **52.** $1980°$ **53.** $\dfrac{7\pi}{4}$, or 5.5 cm

54. 2.25, $129°$ **55.** About 37.7 ft/min

56. 497,829 radians/hr   **57.** $\left(\tfrac{3}{5},\tfrac{4}{5}\right)$, $\left(-\tfrac{3}{5},-\tfrac{4}{5}\right)$, $\left(-\tfrac{3}{5},\tfrac{4}{5}\right)$

58. -1 **59.** 1   **60.** $-\dfrac{\sqrt{3}}{2}$ **61.** $\tfrac{1}{2}$ **62.** $\dfrac{\sqrt{3}}{3}$ **63.** -1

64. -0.9056 **65.** 0.9218   **66.** Not defined   **67.** 4.3813
68. -6.1685 **69.** 0.8090   **70.** $y = \sin x$: see p. 453;
$y = \cos x$: see p. 453; $y = \tan x$: see p. 457; $y = \cot x$: see p. 458;
$y = \sec x$: see p. 458; $y = \csc x$: see p. 458   **71.** Period of sin,
cos, sec, csc: 2π; period of tan, cot: π

72.

Function	Domain	Range
Sine	$(-\infty,\infty)$	$[-1,1]$
Cosine	$(-\infty,\infty)$	$[-1,1]$
Tangent	All real numbers except $(\pi/2) + k\pi$, where k is an integer	$(-\infty,\infty)$

73.

Function	I	II	III	IV
Sine	$+$	$+$	$-$	$-$
Cosine	$+$	$-$	$-$	$+$
Tangent	$+$	$-$	$+$	$-$

74. Amplitude: 1; period: 2π; phase shift: $-\dfrac{\pi}{2}$

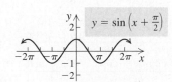

75. Amplitude: $\dfrac{1}{2}$; period: π; phase shift: $\dfrac{\pi}{4}$

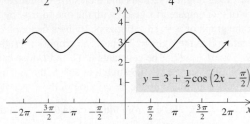

76. (d)   **77.** (a)   **78.** (c)   **79.** (b)
80.   **81.**

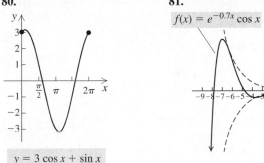

82. C   **83.** B   **84.** B
85. Domain: $(-\infty,\infty)$; range: $[-3,3]$; period 4π

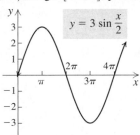

86. $y_2 = 2\sin\left(x + \dfrac{\pi}{2}\right) - 2$ **87.** The domain consists of the

intervals $\left(-\dfrac{\pi}{2} + 2k\pi, \dfrac{\pi}{2} + 2k\pi\right)$, k an integer.

88. Both degrees and radians are units of angle measure. Degree
notation has been in use since Babylonian times. A degree is defined
to be $\frac{1}{360}$ of one complete positive revolution. Radians are defined
in terms of intercepted arc length on a circle, with one radian being
the measure of the angle for which the arc length equals the radius.
There are 2π radians in one complete revolution.   **89.** For a
point at a distance r from the center of rotation with a fixed angular

speed k, the linear speed is given by $v = r \cdot k$, or $r = \dfrac{1}{k}v$. Thus the

length of the radius is directly proportional to the linear speed.
90. The numbers for which the value of the cosine function is 0 are
not in the domain of the tangent function.   **91.** The denominator
B in the phase shift C/B serves to shrink or stretch the translation of
C units by the same factor as the horizontal shrinking or stretching
of the period. Thus the translation must be done after the horizontal
shrinking or stretching. For example, consider $y = \sin(2x - \pi)$.

The phase shift of this function is $\pi/2$. First translate the graph of $y = \sin x$ to the right $\pi/2$ units and then shrink it horizontally by a factor of 2. Compare this graph with the one formed by first shrinking the graph of $y = \sin x$ horizontally by a factor of 2 and then translating it to the right $\pi/2$ units. The graphs differ; the second one is correct. **92.** The constants B, C, and D translate the graphs, and the constants A and B stretch or shrink the graphs. See the chart on p. 470 for a complete description of the effect of each constant. **93.** We see from the formula $\theta = \dfrac{s}{r}$ that the tire with the 15-in. diameter will rotate through a larger angle than the tire with the 16-in. diameter. Thus the car with the 15-in. tires will probably need new tires first.

Test: Chapter 6

1. [6.1] $\sin \theta = \dfrac{4}{\sqrt{65}}$, or $\dfrac{4\sqrt{65}}{65}$; $\cos \theta = \dfrac{7}{\sqrt{65}}$, or $\dfrac{7\sqrt{65}}{65}$; $\tan \theta = \dfrac{4}{7}$; $\csc \theta = \dfrac{\sqrt{65}}{4}$; $\sec \theta = \dfrac{\sqrt{65}}{7}$; $\cot \theta = \dfrac{7}{4}$

2. [6.3] $\dfrac{\sqrt{3}}{2}$ **3.** [6.3] -1 **4.** [6.4] -1 **5.** [6.4] $-\sqrt{2}$

6. [6.1] $38.47°$ **7.** [6.3] -0.2419 **8.** [6.3] -0.2079
9. [6.4] -5.7588 **10.** [6.4] 0.7827 **11.** [6.1] $30°$
12. [6.1] $\sin 61.6° \approx 0.8796$; $\cos 61.6° \approx 0.4756$;
$\tan 61.6° \approx 1.8495$; $\csc 61.6° \approx 1.1369$; $\sec 61.6° \approx 2.1026$;
$\cot 61.6° \approx 0.5407$ **13.** [6.2] $B = 54.1°$, $a \approx 32.6$, $c \approx 55.7$

14. [6.3] Answers may vary; $472°, -248°$ **15.** [6.4] $\dfrac{\pi}{6}$

16. [6.3] $\cos \theta = \dfrac{5}{\sqrt{41}}$, or $\dfrac{5\sqrt{41}}{41}$; $\tan \theta = -\dfrac{4}{5}$; $\csc \theta = -\dfrac{\sqrt{41}}{4}$;
$\sec \theta = \dfrac{\sqrt{41}}{5}$; $\cot \theta = -\dfrac{5}{4}$ **17.** [6.4] $\dfrac{7\pi}{6}$ **18.** [6.4] $135°$

19. [6.4] $\dfrac{16\pi}{3} \approx 16.755$ cm   **20.** [6.6] 1 **21.** [6.6] 2π

22. [6.6] $\dfrac{\pi}{2}$ **23.** [6.6] (c)   **24.** [6.2] 401 ft

25. [6.2] About 272 mi   **26.** [6.4] $18\pi \approx 56.55$ m/min

27. [6.6]

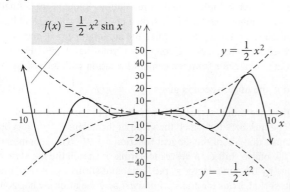

28. [6.6] C
29. [6.5] $\left\{ x \left| -\dfrac{\pi}{2} + 2k\pi < x < \dfrac{\pi}{2} + 2k\pi, k \text{ an integer} \right. \right\}$

Chapter 7

Exercise Set 7.1

1. $\sin^2 x - \cos^2 x$ **3.** $\sin y + \cos y$ **5.** $1 - 2\sin \phi \cos \phi$
7. $\sin^3 x + \csc^3 x$ **9.** $\cos x (\sin x + \cos x)$
11. $(\sin x + \cos x)(\sin x - \cos x)$
13. $(2\cos x + 3)(\cos x - 1)$
15. $(\sin x + 3)(\sin^2 x - 3\sin x + 9)$ **17.** $\tan x$
19. $\sin x + 1$ **21.** $\dfrac{2\tan t + 1}{3\tan t + 1}$ **23.** 1 **25.** $\dfrac{5\cot \phi}{\sin \phi + \cos \phi}$
27. $\dfrac{1 + 2\sin s + 2\cos s}{\sin^2 s - \cos^2 s}$ **29.** $\dfrac{5(\sin \theta - 3)}{3}$ **31.** $\sin x \cos x$
33. $\sqrt{\cos \alpha}\,(\sin \alpha - \cos \alpha)$ **35.** $1 - \sin y$
37. $\dfrac{\sqrt{\sin x \cos x}}{\cos x}$ **39.** $\dfrac{\sqrt{2}\cot y}{2}$ **41.** $\dfrac{\cos x}{\sqrt{\sin x \cos x}}$
43. $\dfrac{1 + \sin y}{\cos y}$ **45.** $\cos \theta = \dfrac{\sqrt{a^2 - x^2}}{a}$, $\tan \theta = \dfrac{x}{\sqrt{a^2 - x^2}}$
47. $\sin \theta = \dfrac{\sqrt{x^2 - 9}}{x}$, $\cos \theta = \dfrac{3}{x}$ **49.** $\sin \theta \tan \theta$
51. $\dfrac{\sqrt{6} - \sqrt{2}}{4}$ **53.** $\dfrac{\sqrt{3} + 1}{1 - \sqrt{3}}$, or $-2 - \sqrt{3}$
55. $\dfrac{\sqrt{6} + \sqrt{2}}{4}$ **57.** $\sin 59° \approx 0.8572$
59. $\cos 24° \approx 0.9135$ **61.** $\tan 52° \approx 1.2799$
63. $\tan (\mu + \nu) = \dfrac{\sin (\mu + \nu)}{\cos (\mu + \nu)}$
$= \dfrac{\sin \mu \cos \nu + \cos \mu \sin \nu}{\cos \mu \cos \nu - \sin \mu \sin \nu}$
$= \dfrac{\sin \mu \cos \nu + \cos \mu \sin \nu}{\cos \mu \cos \nu - \sin \mu \sin \nu} \cdot \dfrac{\dfrac{1}{\cos \mu \cos \nu}}{\dfrac{1}{\cos \mu \cos \nu}}$
$= \dfrac{\dfrac{\sin \mu}{\cos \mu} + \dfrac{\sin \nu}{\cos \nu}}{1 - \dfrac{\sin \mu \sin \nu}{\cos \mu \cos \nu}}$
$= \dfrac{\tan \mu + \tan \nu}{1 - \tan \mu \tan \nu}$

65. 0 **67.** $-\frac{7}{25}$ **69.** $\dfrac{-24 + 2\sqrt{170}}{63}$ **71.** -1.5789

73. 0.7071 **75.** $2\sin \alpha \cos \beta$ **77.** $\cos u$ **79.** All real numbers   **80.** No solution   **81.** 1.9417 **82.** 1.6645

83. $0°$; the lines are parallel   **85.** $\dfrac{3\pi}{4}$, or $135°$ **87.** $22.83°$

89. $\dfrac{\cos (x + h) - \cos x}{h}$
$= \dfrac{\cos x \cos h - \sin x \sin h - \cos x}{h}$
$= \dfrac{\cos x \cos h - \cos x}{h} - \dfrac{\sin x \sin h}{h}$
$= \cos x \left(\dfrac{\cos h - 1}{h} \right) - \sin x \left(\dfrac{\sin h}{h} \right)$

91. Let $x = \dfrac{\pi}{5}$. Then $\dfrac{\sin 5x}{x} = \dfrac{\sin \pi}{\pi/5} = 0 \neq \sin 5$. Answers may vary.

93. Let $\alpha = \dfrac{\pi}{4}$. Then $\cos (2\alpha) = \cos \dfrac{\pi}{2} = 0$, but $2\cos \alpha = 2\cos \dfrac{\pi}{4} = \sqrt{2}$. Answers may vary.

95. Let $x = \dfrac{\pi}{6}$. Then $\dfrac{\cos 6x}{\cos x} = \dfrac{\cos \pi}{\cos \dfrac{\pi}{6}} = \dfrac{-1}{\sqrt{3}/2} \neq 6.$

Answers may vary.   **97.** $\dfrac{6 - 3\sqrt{3}}{9 + 2\sqrt{3}} \approx 0.0645$

99. $168.7°$ **101.** $\cos 2\theta = \cos^2 \theta - \sin^2 \theta$, or $1 - 2 \sin^2 \theta$, or

$2\cos^2 \theta - 1$ **103.** $\tan\left(x + \dfrac{\pi}{4}\right) = \dfrac{\tan x + \tan \dfrac{\pi}{4}}{1 - \tan x \tan \dfrac{\pi}{4}} = \dfrac{1 + \tan x}{1 - \tan x}$

105. $\sin(\alpha + \beta) + \sin(\alpha - \beta) = \sin \alpha \cos \beta + \cos \alpha \sin \beta + \sin \alpha \cos \beta - \cos \alpha \sin \beta = 2 \sin \alpha \cos \beta$

Exercise Set 7.2

1. (a) $\tan \dfrac{3\pi}{10} \approx 1.3763$, $\csc \dfrac{3\pi}{10} \approx 1.2361$, $\sec \dfrac{3\pi}{10} \approx 1.7013$,

$\cot \dfrac{3\pi}{10} \approx 0.7266$; **(b)** $\sin \dfrac{\pi}{5} \approx 0.5878$, $\cos \dfrac{\pi}{5} \approx 0.8090$,

$\tan \dfrac{\pi}{5} \approx 0.7266$, $\csc \dfrac{\pi}{5} \approx 1.7013$, $\sec \dfrac{\pi}{5} \approx 1.2361$,

$\cot \dfrac{\pi}{5} \approx 1.3763$ **3. (a)** $\cos \theta = -\dfrac{2\sqrt{2}}{3}$, $\tan \theta = -\dfrac{\sqrt{2}}{4}$,

$\csc \theta = 3$, $\sec \theta = -\dfrac{3\sqrt{2}}{4}$, $\cot \theta = -2\sqrt{2}$;

(b) $\sin\left(\dfrac{\pi}{2} - \theta\right) = -\dfrac{2\sqrt{2}}{3}$, $\cos\left(\dfrac{\pi}{2} - \theta\right) = \dfrac{1}{3}$,

$\tan\left(\dfrac{\pi}{2} - \theta\right) = -2\sqrt{2}$, $\csc\left(\dfrac{\pi}{2} - \theta\right) = -\dfrac{3\sqrt{2}}{4}$,

$\sec\left(\dfrac{\pi}{2} - \theta\right) = 3$, $\cot\left(\dfrac{\pi}{2} - \theta\right) = -\dfrac{\sqrt{2}}{4}$;

(c) $\sin\left(\theta - \dfrac{\pi}{2}\right) = \dfrac{2\sqrt{2}}{3}$, $\cos\left(\theta - \dfrac{\pi}{2}\right) = \dfrac{1}{3}$,

$\tan\left(\theta - \dfrac{\pi}{2}\right) = 2\sqrt{2}$, $\csc\left(\theta - \dfrac{\pi}{2}\right) = \dfrac{3\sqrt{2}}{4}$,

$\sec\left(\theta - \dfrac{\pi}{2}\right) = 3$, $\cot\left(\theta - \dfrac{\pi}{2}\right) = \dfrac{\sqrt{2}}{4}$

5. $\sec\left(x + \dfrac{\pi}{2}\right) = -\csc x$ **7.** $\tan\left(x - \dfrac{\pi}{2}\right) = -\cot x$

9. $\sin 2\theta = \dfrac{24}{25}$, $\cos 2\theta = -\dfrac{7}{25}$, $\tan 2\theta = -\dfrac{24}{7}$; II

11. $\sin 2\theta = \dfrac{24}{25}$, $\cos 2\theta = -\dfrac{7}{25}$, $\tan 2\theta = -\dfrac{24}{7}$; II

13. $\sin 2\theta = -\dfrac{120}{169}$, $\cos 2\theta = \dfrac{119}{169}$, $\tan 2\theta = -\dfrac{120}{119}$; IV

15. $\cos 4x = 1 - 8 \sin^2 x \cos^2 x$, or $\cos^4 x - 6 \sin^2 x \cos^2 x + \sin^4 x$, or $8 \cos^4 x - 8 \cos^2 x + 1$

17. $\dfrac{\sqrt{2 + \sqrt{3}}}{2}$ **19.** $\dfrac{\sqrt{2 + \sqrt{2}}}{2}$ **21.** $2 + \sqrt{3}$

23. 0.6421 **25.** 0.1735

27. (d); $\dfrac{\cos 2x}{\cos x - \sin x} = \dfrac{\cos^2 x - \sin^2 x}{\cos x - \sin x}$

$= \dfrac{(\cos x + \sin x)(\cos x - \sin x)}{\cos x - \sin x}$

$= \cos x + \sin x$

$= \dfrac{\sin x}{\sin x}(\cos x + \sin x)$

$= \sin x\left(\dfrac{\cos x}{\sin x} + \dfrac{\sin x}{\sin x}\right)$

$= \sin x (\cot x + 1)$

29. (d); $\dfrac{\sin 2x}{2\cos x} = \dfrac{2 \sin x \cos x}{2 \cos x} = \sin x$ **31.** $\cos x$

33. $\cos 2x$ **35.** 1 **37.** 8 **39.** $\sin^2 x$ **40.** 1
41. $-\cos^2 x$ **42.** $\csc^2 x$ **43.** 1 **44.** $\sec^2 x$ **45.** $\cos^2 x$
46. $\tan^2 x$ **47.** (a), (e)   **48.** (b), (c), (f)   **49.** (d)
50. (e)   **51.** $\sin 141° \approx 0.6293$, $\cos 141° \approx -0.7772$,
$\tan 141° \approx -0.8097$, $\csc 141° \approx 1.5891$, $\sec 141° \approx -1.2867$,
$\cot 141° \approx -1.2350$ **53.** $-\cos x (1 + \cot x)$ **55.** $\cot^2 y$
57. $\sin \theta = -\dfrac{15}{17}$, $\cos \theta = -\dfrac{8}{17}$, $\tan \theta = \dfrac{15}{8}$
59. (a) 9.80359 m/sec²; **(b)** 9.80180 m/sec²;
(c) $g = 9.78049(1 + 0.005264 \sin^2 \phi + 0.000024 \sin^4 \phi)$

Exercise Set 7.3

1.

$\sec x - \sin x \tan x$	$\cos x$
$\dfrac{1}{\cos x} - \sin x \cdot \dfrac{\sin x}{\cos x}$	
$\dfrac{1 - \sin^2 x}{\cos x}$	
$\dfrac{\cos^2 x}{\cos x}$	
$\cos x$	

3.

$1 - \cos x$	$\sin x$
$\sin x$	$1 + \cos x$
	$\dfrac{\sin x}{1 + \cos x} \cdot \dfrac{1 - \cos x}{1 - \cos x}$
	$\dfrac{\sin x (1 - \cos x)}{1 - \cos^2 x}$
	$\dfrac{\sin x (1 - \cos x)}{\sin^2 x}$
	$\dfrac{1 - \cos x}{\sin x}$

5.

$\dfrac{1 + \tan \theta}{1 - \tan \theta} + \dfrac{1 + \cot \theta}{1 - \cot \theta}$	0
$\dfrac{1 + \dfrac{\sin \theta}{\cos \theta}}{1 - \dfrac{\sin \theta}{\cos \theta}} + \dfrac{1 + \dfrac{\cos \theta}{\sin \theta}}{1 - \dfrac{\cos \theta}{\sin \theta}}$	
$\dfrac{\dfrac{\cos \theta + \sin \theta}{\cos \theta}}{\dfrac{\cos \theta - \sin \theta}{\cos \theta}} + \dfrac{\dfrac{\sin \theta + \cos \theta}{\sin \theta}}{\dfrac{\sin \theta - \cos \theta}{\sin \theta}}$	
$\dfrac{\cos \theta + \sin \theta}{\cos \theta} \cdot \dfrac{\cos \theta}{\cos \theta - \sin \theta} +$	
$\dfrac{\sin \theta + \cos \theta}{\sin \theta} \cdot \dfrac{\sin \theta}{\sin \theta - \cos \theta}$	
$\dfrac{\cos \theta + \sin \theta}{\cos \theta - \sin \theta} + \dfrac{\sin \theta + \cos \theta}{\sin \theta - \cos \theta}$	
$\dfrac{\cos \theta + \sin \theta}{\cos \theta - \sin \theta} - \dfrac{\cos \theta + \sin \theta}{\cos \theta - \sin \theta}$	
0	

7.

$$
\begin{array}{c|c}
\dfrac{\cos^2 \alpha + \cot \alpha}{\cos^2 \alpha - \cot \alpha} & \dfrac{\cos^2 \alpha \tan \alpha + 1}{\cos^2 \alpha \tan \alpha - 1} \\[2ex]
\dfrac{\cos^2 \alpha + \dfrac{\cos \alpha}{\sin \alpha}}{\cos^2 \alpha - \dfrac{\cos \alpha}{\sin \alpha}} & \dfrac{\cos^2 \alpha \dfrac{\sin \alpha}{\cos \alpha} + 1}{\cos^2 \alpha \dfrac{\sin \alpha}{\cos \alpha} - 1} \\[3ex]
\dfrac{\cos \alpha \left(\cos \alpha + \dfrac{1}{\sin \alpha} \right)}{\cos \alpha \left(\cos \alpha - \dfrac{1}{\sin \alpha} \right)} & \dfrac{\sin \alpha \cos \alpha + 1}{\sin \alpha \cos \alpha - 1} \\[3ex]
\dfrac{\cos \alpha + \dfrac{1}{\sin \alpha}}{\cos \alpha - \dfrac{1}{\sin \alpha}} & \\[3ex]
\dfrac{\dfrac{\sin \alpha \cos \alpha + 1}{\sin \alpha}}{\dfrac{\sin \alpha \cos \alpha - 1}{\sin \alpha}} & \\[3ex]
\dfrac{\sin \alpha \cos \alpha + 1}{\sin \alpha \cos \alpha - 1} &
\end{array}
$$

9.

$$
\begin{array}{c|c}
\dfrac{2 \tan \theta}{1 + \tan^2 \theta} & \sin 2\theta \\[2ex]
\dfrac{2 \tan \theta}{\sec^2 \theta} & 2 \sin \theta \cos \theta \\[2ex]
\dfrac{2 \sin \theta}{\cos \theta} \cdot \dfrac{\cos^2 \theta}{1} & \\[2ex]
2 \sin \theta \cos \theta &
\end{array}
$$

11.

$$
\begin{array}{c|c}
1 - \cos 5\theta \cos 3\theta - \sin 5\theta \sin 3\theta & 2 \sin^2 \theta \\
1 - [\cos 5\theta \cos 3\theta + \sin 5\theta \sin 3\theta] & 1 - \cos 2\theta \\
1 - \cos (5\theta - 3\theta) & \\
1 - \cos 2\theta &
\end{array}
$$

13.

$$
\begin{array}{c|c}
2 \sin \theta \cos^3 \theta + 2 \sin^3 \theta \cos \theta & \sin 2\theta \\
2 \sin \theta \cos \theta (\cos^2 \theta + \sin^2 \theta) & 2 \sin \theta \cos \theta \\
2 \sin \theta \cos \theta &
\end{array}
$$

15.

$$
\begin{array}{c|c}
\dfrac{\tan x - \sin x}{2 \tan x} & \sin^2 \dfrac{x}{2} \\[2ex]
\dfrac{1}{2} \left[\dfrac{\dfrac{\sin x}{\cos x} - \sin x}{\dfrac{\sin x}{\cos x}} \right] & \dfrac{1 - \cos x}{2} \\[3ex]
\dfrac{1}{2} \dfrac{\sin x - \sin x \cos x}{\cos x} \cdot \dfrac{\cos x}{\sin x} & \\[2ex]
\dfrac{1 - \cos x}{2} &
\end{array}
$$

17.

$$
\begin{array}{c|c}
\sin (\alpha + \beta) \sin (\alpha - \beta) & \sin^2 \alpha - \sin^2 \beta \\
\left(\begin{array}{c} \sin \alpha \cos \beta + \\ \cos \alpha \sin \beta \end{array} \right) \left(\begin{array}{c} \sin \alpha \cos \beta - \\ \cos \alpha \sin \beta \end{array} \right) & \begin{array}{c} 1 - \cos^2 \alpha - \\ (1 - \cos^2 \beta) \end{array} \\
\sin^2 \alpha \cos^2 \beta - \cos^2 \alpha \sin^2 \beta & \cos^2 \beta - \cos^2 \alpha \\
\begin{array}{c} \cos^2 \beta (1 - \cos^2 \alpha) - \\ \cos^2 \alpha (1 - \cos^2 \beta) \end{array} & \\
\begin{array}{c} \cos^2 \beta - \cos^2 \alpha \cos^2 \beta - \\ \cos^2 \alpha + \cos^2 \alpha \cos^2 \beta \end{array} & \\
\cos^2 \beta - \cos^2 \alpha &
\end{array}
$$

19.

$$
\begin{array}{c|c}
\tan \theta (\tan \theta + \cot \theta) & \sec^2 \theta \\
\tan^2 \theta + \tan \theta \cot \theta & \\
\tan^2 \theta + 1 & \\
\sec^2 \theta &
\end{array}
$$

21.

$$
\begin{array}{c|c}
\dfrac{1 + \cos^2 x}{\sin^2 x} & 2 \csc^2 x - 1 \\[2ex]
\dfrac{1}{\sin^2 x} + \dfrac{\cos^2 x}{\sin^2 x} & \\[2ex]
\csc^2 x + \cot^2 x & \\
\csc^2 x + \csc^2 x - 1 & \\
2 \csc^2 x - 1 &
\end{array}
$$

23.

$$
\begin{array}{c|c}
\dfrac{1 + \sin x}{1 - \sin x} + \dfrac{\sin x - 1}{1 + \sin x} & 4 \sec x \tan x \\[2ex]
\dfrac{(1 + \sin x)^2 - (1 - \sin x)^2}{1 - \sin^2 x} & 4 \cdot \dfrac{1}{\cos x} \cdot \dfrac{\sin x}{\cos x} \\[2ex]
\dfrac{(1 + 2 \sin x + \sin^2 x) - (1 - 2 \sin x + \sin^2 x)}{\cos^2 x} & \dfrac{4 \sin x}{\cos^2 x} \\[2ex]
\dfrac{4 \sin x}{\cos^2 x} &
\end{array}
$$

25.

$$
\begin{array}{c|c}
\cos^2 \alpha \cot^2 \alpha & \cot^2 \alpha - \cos^2 \alpha \\
(1 - \sin^2 \alpha) \cot^2 \alpha & \\
\cot^2 \alpha - \sin^2 \alpha \cdot \dfrac{\cos^2 \alpha}{\sin^2 \alpha} & \\
\cot^2 \alpha - \cos^2 \alpha &
\end{array}
$$

27.

$$
\begin{array}{c|c}
2 \sin^2 \theta \cos^2 \theta + \cos^4 \theta & 1 - \sin^4 \theta \\
\cos^2 \theta (2 \sin^2 \theta + \cos^2 \theta) & (1 + \sin^2 \theta)(1 - \sin^2 \theta) \\
\cos^2 \theta (\sin^2 \theta + \sin^2 \theta + \cos^2 \theta) & (1 + \sin^2 \theta)(\cos^2 \theta) \\
\cos^2 \theta (\sin^2 \theta + 1) &
\end{array}
$$

29.

$$
\begin{array}{c|c}
\dfrac{1 + \sin x}{1 - \sin x} & (\sec x + \tan x)^2 \\[2ex]
\dfrac{1 + \sin x}{1 - \sin x} \cdot \dfrac{1 + \sin x}{1 + \sin x} & \left(\dfrac{1}{\cos x} + \dfrac{\sin x}{\cos x} \right)^2 \\[2ex]
\dfrac{(1 + \sin x)^2}{1 - \sin^2 x} & \dfrac{(1 + \sin x)^2}{\cos^2 x} \\[2ex]
\dfrac{(1 + \sin x)^2}{\cos^2 x} &
\end{array}
$$

31. Sine sum and difference identities:
$$\sin(x+y) = \sin x \cos y + \cos x \sin y,$$
$$\sin(x-y) = \sin x \cos y - \cos x \sin y.$$
Add the sum and difference identities:
$$\sin(x+y) + \sin(x-y) = 2\sin x \cos y$$
$$\tfrac{1}{2}[\sin(x+y) + \sin(x-y)] = \sin x \cos y. \qquad (3)$$
Subtract the difference identity from the sum identity:
$$\sin(x+y) - \sin(x-y) = 2\cos x \sin y$$
$$\tfrac{1}{2}[\sin(x+y) - \sin(x-y)] = \cos x \sin y. \qquad (4)$$

33. $\sin 3\theta - \sin 5\theta = 2\cos\dfrac{8\theta}{2}\sin\dfrac{-2\theta}{2} = -2\cos 4\theta \sin\theta$

35. $\sin 8\theta + \sin 5\theta = 2\sin\dfrac{13\theta}{2}\cos\dfrac{3\theta}{2}$

37. $\sin 7u \sin 5u = \tfrac{1}{2}(\cos 2u - \cos 12u)$

39. $7\cos\theta\sin 7\theta = \dfrac{7}{2}[\sin 8\theta - \sin(-6\theta)]$
$$= \dfrac{7}{2}(\sin 8\theta + \sin 6\theta)$$

41. $\cos 55° \sin 25° = \tfrac{1}{2}(\sin 80° - \sin 30°) = \tfrac{1}{2}\sin 80° - \tfrac{1}{4}$

43.

$\sin 4\theta + \sin 6\theta$	$\cot\theta\,(\cos 4\theta - \cos 6\theta)$
$2\sin\dfrac{10\theta}{2}\cos\dfrac{-2\theta}{2}$	$\dfrac{\cos\theta}{\sin\theta}\left(2\sin\dfrac{10\theta}{2}\sin\dfrac{2\theta}{2}\right)$
$2\sin 5\theta\cos(-\theta)$	$\dfrac{\cos\theta}{\sin\theta}(2\sin 5\theta\sin\theta)$
$2\sin 5\theta\cos\theta$	$2\sin 5\theta\cos\theta$

45.

$\cot 4x\,(\sin x + \sin 4x + \sin 7x)$	$\cos x + \cos 4x + \cos 7x$
$\dfrac{\cos 4x}{\sin 4x}\left(\sin 4x + 2\sin\dfrac{8x}{2}\cos\dfrac{-6x}{2}\right)$	$\cos 4x + 2\cos\dfrac{8x}{2}\cdot\cos\dfrac{6x}{2}$
$\dfrac{\cos 4x}{\sin 4x}(\sin 4x + 2\sin 4x\cos 3x)$	$\cos 4x + 2\cos 4x\cdot\cos 3x$
$\cos 4x\,(1 + 2\cos 3x)$	$\cos 4x\,(1 + 2\cos 3x)$

47.

$\cot\dfrac{x+y}{2}$	$\dfrac{\sin y - \sin x}{\cos x - \cos y}$
$\dfrac{\cos\dfrac{x+y}{2}}{\sin\dfrac{x+y}{2}}$	$\dfrac{2\cos\dfrac{x+y}{2}\sin\dfrac{y-x}{2}}{2\sin\dfrac{x+y}{2}\sin\dfrac{y-x}{2}}$
	$\dfrac{\cos\dfrac{x+y}{2}}{\sin\dfrac{x+y}{2}}$

49.

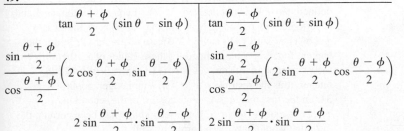

51. B;

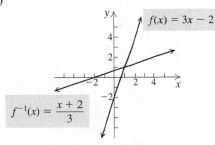

53. A;

$\sin x \cos x + 1$	$\dfrac{\sin^3 x - \cos^3 x}{\sin x - \cos x}$
	$\dfrac{(\sin x - \cos x)(\sin^2 x + \sin x \cos x + \cos^2 x)}{\sin x - \cos x}$
	$\sin^2 x + \sin x \cos x + \cos^2 x$
	$\sin x \cos x + 1$

55. C;

$\dfrac{1}{\cot x \sin^2 x}$	$\tan x + \cot x$
$\dfrac{1}{\dfrac{\cos x}{\sin x}\cdot\sin^2 x}$	$\dfrac{\sin x}{\cos x} + \dfrac{\cos x}{\sin x}$
$\dfrac{1}{\cos x \sin x}$	$\dfrac{\sin^2 x + \cos^2 x}{\cos x \sin x}$
	$\dfrac{1}{\cos x \sin x}$

57. (a), (d)

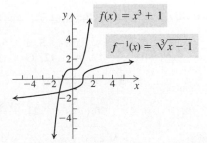

(b) yes; **(c)** $f^{-1}(x) = \dfrac{x+2}{3}$

58. (a), (d)

(b) yes; **(c)** $f^{-1}(x) = \sqrt[3]{x-1}$

59. (a), (d)

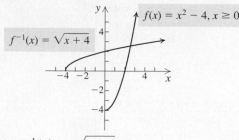

(b) yes; **(c)** $f^{-1}(x) = \sqrt{x+4}$

60. (a), (d)

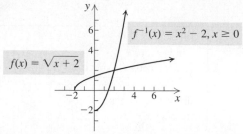

(b) yes; **(c)** $f^{-1}(x) = x^2 - 2,\ x \geq 0$ **61.** $0, \frac{5}{2}$ **62.** $-4, \frac{7}{3}$

63. $\pm 2, \pm 3i$ **64.** $5 \pm 2\sqrt{6}$ **65.** 27   **66.** 9

67.

$\ln |\tan x|$	$-\ln |\cot x|$
$\ln \left| \dfrac{1}{\cot x} \right|$	
$\ln |1| - \ln |\cot x|$	
$0 - \ln |\cot x|$	
$-\ln |\cot x|$	

69. $\log (\cos x - \sin x) + \log (\cos x + \sin x)$
$\quad = \log \left[(\cos x - \sin x)(\cos x + \sin x) \right]$
$\quad = \log (\cos^2 x - \sin^2 x) = \log \cos 2x$

71.

$$\frac{1}{\omega C(\tan \theta + \tan \phi)} = \frac{1}{\omega C\left(\dfrac{\sin \theta}{\cos \theta} + \dfrac{\sin \phi}{\cos \phi} \right)}$$

$$= \frac{1}{\omega C\left(\dfrac{\sin \theta \cos \phi + \sin \phi \cos \theta}{\cos \theta \cos \phi} \right)}$$

$$= \frac{\cos \theta \cos \phi}{\omega C \sin (\theta + \phi)}$$

Mid-Chapter Mixed Review: Chapter 7

1. True   **2.** False   **3.** True   **4.** True   **5.** J   **6.** H
7. D   **8.** I   **9.** C   **10.** E   **11.** A   **12.** G   **13.** F
14. B   **15.** $\dfrac{\sqrt{\cos x}}{\sin x}$ **16.** 1   **17.** $2 \cos x + 1$
18. $-\cos x$ **19.** $1 - \sin 2x$ **20.** $\cos x$ **21.** $\dfrac{\sqrt{\sec x + 1}}{\sin x}$
22. $\cos 12° \approx 0.9781$ **23.** $\dfrac{\sqrt{2 - \sqrt{2}}}{2}$ **24.** $\dfrac{\sqrt{6} + \sqrt{2}}{4}$
25. $-\frac{119}{120}$ **26.** $-\frac{24}{25}$; quadrant IV

27.

$\cos^2 \dfrac{x}{2}$	$\dfrac{\tan x + \sin x}{2 \tan x}$
$\dfrac{1 + \cos x}{2}$	$\dfrac{1}{2}\left[\dfrac{\dfrac{\sin x}{\cos x} + \sin x}{\dfrac{\sin x}{\cos x}} \right]$
	$\dfrac{1}{2} \dfrac{\sin x + \sin x \cos x}{\cos x} \cdot \dfrac{\cos x}{\sin x}$
	$\dfrac{1}{2} \dfrac{\sin x (1 + \cos x)}{\sin x}$
	$\dfrac{1 + \cos x}{2}$

28.

$\dfrac{1 - \sin x}{\cos x}$	$\dfrac{\cos x}{1 + \sin x}$
$\dfrac{1 - \sin x}{\cos x} \cdot \dfrac{1 + \sin x}{1 + \sin x}$	
$\dfrac{1 - \sin^2 x}{\cos x (1 + \sin x)}$	
$\dfrac{\cos^2 x}{\cos x (1 + \sin x)}$	
$\dfrac{\cos x}{1 + \sin x}$	

29.

$\dfrac{\sin^3 x - \cos^3 x}{\sin x - \cos x}$	$\dfrac{2 + \sin 2x}{2}$
$\dfrac{(\sin x - \cos x)(\sin^2 x + \sin x \cos x + \cos^2 x)}{\sin x - \cos x}$	$\dfrac{2 + 2 \sin x \cos x}{2}$
$\sin^2 x + \cos^2 x + \sin x \cos x$	$\dfrac{2 (1 + \sin x \cos x)}{2}$
$1 + \sin x \cos x$	$1 + \sin x \cos x$

30.

$\sin 6\theta - \sin 2\theta$	$\tan 2\theta \,(\cos 2\theta + \cos 6\theta)$
$2 \cos \dfrac{8\theta}{2} \sin \dfrac{4\theta}{2}$	$\dfrac{\sin 2\theta}{\cos 2\theta}\left(2 \cos \dfrac{8\theta}{2} \cos \dfrac{4\theta}{2} \right)$
$2 \cos 4\theta \sin 2\theta$	$\dfrac{\sin 2\theta}{\cos 2\theta} (2 \cos 4\theta \cos 2\theta)$
	$2 \cos 4\theta \sin 2\theta$

31. The expression $\tan (x + 450°)$ can be simplified using the sine and cosine sum formulas but cannot be simplified using the tangent sum formula because although $\sin 450°$ and $\cos 450°$ are both defined, $\tan 450°$ is not defined.   **32.** Each has amplitude 1 and is periodic. The period of $y_1 = \sin x$ is 2π, of $y_2 = \sin 2x$ is π, and of $y_3 = \sin (x/2)$ is 4π.   **33. (a)** $x \neq k\pi/2$, k odd; the tangent function is not defined for these values of x; **(b)** $\sin x = 0$ for $x = k\pi$, k an integer; $\cos x = -1$ for $x = k\pi$, k an odd integer; thus the restriction $x \neq k\pi$, k an integer, applies. **34.** In the first line, $\cos 4x \neq 2 \cos 2x$. In the second line, $\cos 2x \neq \cos^2 x + \sin^2 x$. If the second line had been correct, the third line would have been correct as well.

Exercise Set 7.4

1. $-\dfrac{\pi}{3}, -60°$ **3.** $\dfrac{\pi}{4}, 45°$ **5.** $\dfrac{\pi}{4}, 45°$ **7.** $0, 0°$

9. $\dfrac{\pi}{6}, 30°$ **11.** $\dfrac{\pi}{6}, 30°$ **13.** $\dfrac{5\pi}{6}, 150°$ **15.** $-\dfrac{\pi}{6}, -30°$

17. $\frac{\pi}{2}, 90°$ **19.** $\frac{\pi}{3}, 60°$ **21.** $0.3520, 20.2°$

23. $1.2917, 74.0°$ **25.** $2.9463, 168.8°$ **27.** $-0.1600, -9.2°$

29. $0.8289, 47.5°$ **31.** $-0.9600, -55.0°$ **33.** $\sin^{-1}: [-1, 1];$

$\cos^{-1}: [-1, 1]; \tan^{-1}: (-\infty, \infty)$ **35.** $\theta = \sin^{-1}\left(\frac{2000}{d}\right)$

37. 0.3 **39.** $\frac{\pi}{4}$ **41.** $\frac{\pi}{5}$ **43.** $-\frac{\pi}{3}$ **45.** $\frac{1}{2}$ **47.** 1

49. $\frac{\pi}{3}$ **51.** $\frac{\sqrt{11}}{33}$ **53.** $-\frac{\pi}{6}$ **55.** $\frac{a}{\sqrt{a^2+9}}$

57. $\frac{\sqrt{q^2-p^2}}{p}$ **59.** $\frac{p}{3}$ **61.** $\frac{\sqrt{3}}{2}$ **63.** $-\frac{\sqrt{2}}{10}$

65. $xy + \sqrt{(1-x^2)(1-y^2)}$ **67.** 0.9861 **69.** Periodic
70. Radian measure   **71.** Similar   **72.** Angle of depression
73. Angular speed   **74.** Supplementary   **75.** Amplitude
76. Circular
77.

	$\frac{\pi}{2}$
$\sin^{-1} x + \cos^{-1} x$	
$\sin(\sin^{-1} x + \cos^{-1} x)$	$\sin\frac{\pi}{2}$
$[\sin(\sin^{-1} x)][\cos(\cos^{-1} x)] +$ $[\cos(\sin^{-1} x)][\sin(\cos^{-1} x)]$	1
$x \cdot x + \sqrt{1-x^2} \cdot \sqrt{1-x^2}$	
$x^2 + 1 - x^2$	
1	

79.

$\sin^{-1} x$	$\cos^{-1}\sqrt{1-x^2}$
$\sin(\sin^{-1} x)$	$\sin(\cos^{-1}\sqrt{1-x^2})$
x	x

81. $\theta = \tan^{-1}\frac{y+h}{x} - \tan^{-1}\frac{y}{x}; 38.7°$

Visualizing the Graph

1. D   **2.** G   **3.** C   **4.** H   **5.** I   **6.** A   **7.** E
8. J   **9.** F   **10.** B

Exercise Set 7.5

1. $\frac{\pi}{6} + 2k\pi, \frac{11\pi}{6} + 2k\pi,$ or $30° + k \cdot 360°, 330° + k \cdot 360°$

3. $\frac{2\pi}{3} + k\pi,$ or $120° + k \cdot 180°$

5. $\frac{\pi}{6} + 2k\pi, \frac{5\pi}{6} + 2k\pi,$ or $30° + k \cdot 360°, 150° + k \cdot 360°$

7. $\frac{3\pi}{4} + 2k\pi, \frac{5\pi}{4} + 2k\pi,$ or $135° + k \cdot 360°, 225° + k \cdot 360°$

9. $1.7120, 4.5712,$ or $98.09°, 261.91°$ **11.** $\frac{4\pi}{3}, \frac{5\pi}{3},$ or

$240°, 300°$ **13.** $\frac{\pi}{4}, \frac{3\pi}{4}, \frac{5\pi}{4}, \frac{7\pi}{4},$ or $45°, 135°, 225°, 315°$

15. $\frac{\pi}{6}, \frac{5\pi}{6}, \frac{3\pi}{2},$ or $30°, 150°, 270°$ **17.** $\frac{\pi}{6}, \frac{\pi}{2}, \frac{3\pi}{2}, \frac{11\pi}{6},$ or

$30°, 90°, 270°, 330°$ **19.** $1.9106, \frac{2\pi}{3}, \frac{4\pi}{3}, 4.3726,$ or

$109.47°, 120°, 240°, 250.53°$ **21.** $0, \frac{\pi}{4}, \frac{3\pi}{4}, \pi, \frac{5\pi}{4}, \frac{7\pi}{4},$ or $0°,$

$45°, 135°, 180°, 225°, 315°$ **23.** $2.4402, 3.8430,$ or $139.81°,$
$220.19°$ **25.** $0.6496, 2.9557, 3.7912, 6.0973,$ or $37.22°,$

$169.35°, 217.22°, 349.35°$ **27.** $0, \pi, \frac{7\pi}{6}, \frac{11\pi}{6}$

29. $0, \pi$ **31.** $0, \pi$ **33.** $\frac{3\pi}{4}, \frac{7\pi}{4}$

35. $\frac{2\pi}{3}, \frac{4\pi}{3}, \frac{3\pi}{2}$ **37.** $\frac{\pi}{4}, \frac{3\pi}{4}, \frac{5\pi}{4}, \frac{7\pi}{4}$ **39.** $\frac{\pi}{12}, \frac{5\pi}{12}$

41. $0.967, 1.853, 4.108, 4.994$ **43.** $\frac{2\pi}{3}, \frac{4\pi}{3}$ **45.** $1.114,$

2.773 **47.** 0.515 **49.** $0.422, 1.756$
51. November: \$72,853; March: \$101,853
53. (a) $y = 26.3148 \sin(0.4927x - 1.9612) + 58.2283;$
(b) April: 58°; December: 39°; (c) April and October
55. $B = 35°, b \approx 140.7, c \approx 245.4$
56. $R \approx 15.5°, T \approx 74.5°, t \approx 13.7$ **57.** 36   **58.** 14

59. $\frac{\pi}{3}, \frac{2\pi}{3}, \frac{4\pi}{3}, \frac{5\pi}{3}$ **61.** $\frac{\pi}{3}, \frac{4\pi}{3}$ **63.** 0

65. $e^{3\pi/2 + 2k\pi},$ where k (an integer) ≤ -1
67. 1.24 days, 6.76 days   **69.** $16.5°$N
71. 1   **73.** $\frac{5}{26},$ or about 0.1923

Review Exercises: Chapter 7

1. True   **2.** True   **3.** True   **4.** False   **5.** False
6. $\csc^2 x$ **7.** 1   **8.** $\tan^2 y - \cot^2 y$
9. $\frac{(\cos^2 x + 1)^2}{\cos^2 x}$ **10.** $\csc x(\sec x - \csc x)$
11. $(3 \sin y + 5)(\sin y - 4)$
12. $(10 - \cos u)(100 + 10 \cos u + \cos^2 u)$
13. 1   **14.** $\frac{1}{2} \sec x$ **15.** $\frac{3 \tan x}{\sin x - \cos x}$
16. $\frac{3 \cos y + 3 \sin y + 2}{\cos^2 y - \sin^2 y}$ **17.** 1   **18.** $\frac{1}{4} \cot x$
19. $\sin x + \cos x$ **20.** $\frac{\cos x}{1 - \sin x}$ **21.** $\frac{\cos x}{\sqrt{\sin x}}$
22. $3 \sec \theta$ **23.** $\cos x \cos \frac{3\pi}{2} - \sin x \sin \frac{3\pi}{2}$
24. $\frac{\tan 45° - \tan 30°}{1 + \tan 45° \tan 30°}$ **25.** $\cos(27° - 16°),$ or $\cos 11°$
26. $\frac{-\sqrt{6} - \sqrt{2}}{4}$ **27.** $2 - \sqrt{3}$ **28.** -0.3745
29. $-\sin x$ **30.** $\sin x$ **31.** $-\cos x$
32. (a) $\sin \alpha = -\frac{4}{5}, \tan \alpha = \frac{4}{3}, \cot \alpha = \frac{3}{4}, \sec \alpha = -\frac{5}{3},$
$\csc \alpha = -\frac{5}{4};$ (b) $\sin\left(\frac{\pi}{2} - \alpha\right) = -\frac{3}{5}, \cos\left(\frac{\pi}{2} - \alpha\right) = -\frac{4}{5},$
$\tan\left(\frac{\pi}{2} - \alpha\right) = \frac{3}{4}, \cot\left(\frac{\pi}{2} - \alpha\right) = \frac{4}{3}, \sec\left(\frac{\pi}{2} - \alpha\right) = -\frac{5}{4},$
$\csc\left(\frac{\pi}{2} - \alpha\right) = -\frac{5}{3};$ (c) $\sin\left(\alpha + \frac{\pi}{2}\right) = -\frac{3}{5},$
$\cos\left(\alpha + \frac{\pi}{2}\right) = \frac{4}{5}, \tan\left(\alpha + \frac{\pi}{2}\right) = -\frac{3}{4}, \cot\left(\alpha + \frac{\pi}{2}\right) = -\frac{4}{3},$
$\sec\left(\alpha + \frac{\pi}{2}\right) = \frac{5}{4}, \csc\left(\alpha + \frac{\pi}{2}\right) = -\frac{5}{3}$ **33.** $-\sec x$
34. $\tan 2\theta = \frac{24}{7}, \cos 2\theta = \frac{7}{25}, \sin 2\theta = \frac{24}{25};$ I

35. $\dfrac{\sqrt{2-\sqrt{2}}}{2}$ **36.** $\sin 2\beta = 0.4261$, $\cos \dfrac{\beta}{2} = 0.9940$,

$\cos 4\beta = 0.6369$ **37.** $\cos x$ **38.** 1   **39.** $\sin 2x$ **40.** $\tan 2x$

41.

$\dfrac{1-\sin x}{\cos x}$	$\dfrac{\cos x}{1+\sin x}$
$\dfrac{1-\sin x}{\cos x} \cdot \dfrac{\cos x}{\cos x}$	$\dfrac{\cos x}{1+\sin x} \cdot \dfrac{1-\sin x}{1-\sin x}$
$\dfrac{\cos x - \sin x \cos x}{\cos^2 x}$	$\dfrac{\cos x - \sin x \cos x}{1-\sin^2 x}$
	$\dfrac{\cos x - \sin x \cos x}{\cos^2 x}$

42.

$\dfrac{1+\cos 2\theta}{\sin 2\theta}$	$\cot \theta$
$\dfrac{1+2\cos^2 \theta - 1}{2\sin \theta \cos \theta}$	$\dfrac{\cos \theta}{\sin \theta}$
$\dfrac{\cos \theta}{\sin \theta}$	

43.

$\dfrac{\tan y + \sin y}{2\tan y}$	$\cos^2 \dfrac{y}{2}$
$\dfrac{1}{2}\left[\dfrac{\frac{\sin y + \sin y \cos y}{\cos y}}{\frac{\sin y}{\cos y}}\right]$	$\dfrac{1+\cos y}{2}$
$\dfrac{1}{2}\left[\dfrac{\sin y\,(1+\cos y)}{\cos y} \cdot \dfrac{\cos y}{\sin y}\right]$	
$\dfrac{1+\cos y}{2}$	

44.

$\dfrac{\sin x - \cos x}{\cos^2 x}$	$\dfrac{\tan^2 x - 1}{\sin x + \cos x}$
	$\dfrac{\frac{\sin^2 x}{\cos^2 x} - 1}{\sin x + \cos x}$
	$\dfrac{\sin^2 x - \cos^2 x}{\cos^2 x} \cdot \dfrac{1}{\sin x + \cos x}$
	$\dfrac{\sin x - \cos x}{\cos^2 x}$

45. $3\cos 2\theta \sin \theta = \frac{3}{2}(\sin 3\theta - \sin \theta)$

46. $\sin \theta - \sin 4\theta = -2\cos \dfrac{5\theta}{2} \sin \dfrac{3\theta}{2}$

47. B;

$\csc x - \cos x \cot x$	$\sin x$
$\dfrac{1}{\sin x} - \cos x \dfrac{\cos x}{\sin x}$	
$\dfrac{1-\cos^2 x}{\sin x}$	
$\dfrac{\sin^2 x}{\sin x}$	
$\sin x$	

48. D;

$\dfrac{1}{\sin x \cos x} - \dfrac{\cos x}{\sin x}$	$\dfrac{\sin x \cos x}{1-\sin^2 x}$
$\dfrac{1}{\sin x \cos x} - \dfrac{\cos^2 x}{\sin x \cos x}$	$\dfrac{\sin x \cos x}{\cos^2 x}$
$\dfrac{1-\cos^2 x}{\sin x \cos x}$	$\dfrac{\sin x}{\cos x}$
$\dfrac{\sin^2 x}{\sin x \cos x}$	
$\dfrac{\sin x}{\cos x}$	

49. A;

$\dfrac{\cot x - 1}{1-\tan x}$	$\dfrac{\csc x}{\sec x}$
$\dfrac{\frac{\cos x}{\sin x} - \frac{\sin x}{\sin x}}{\frac{\sin x}{\sin x} - \frac{\sin x}{\sin x}}$	$\dfrac{\frac{1}{\sin x}}{\frac{1}{\cos x}}$
$\dfrac{\frac{\cos x}{\cos x} - \frac{\sin x}{\cos x}}{}$	$\dfrac{\frac{1}{\sin x}}{\frac{1}{\cos x}}$
$\dfrac{\cos x - \sin x}{\sin x} \cdot \dfrac{\cos x}{\cos x - \sin x}$	$\dfrac{1}{\sin x} \cdot \dfrac{\cos x}{1}$
$\dfrac{\cos x}{\sin x}$	$\dfrac{\cos x}{\sin x}$

50. C;

$\dfrac{\cos x + 1}{\sin x} + \dfrac{\sin x}{\cos x + 1}$	$\dfrac{2}{\sin x}$
$\dfrac{(\cos x + 1)^2 + \sin^2 x}{\sin x\,(\cos x + 1)}$	
$\dfrac{\cos^2 x + 2\cos x + 1 + \sin^2 x}{\sin x\,(\cos x + 1)}$	
$\dfrac{2\cos x + 2}{\sin x\,(\cos x + 1)}$	
$\dfrac{2(\cos x + 1)}{\sin x\,(\cos x + 1)}$	
$\dfrac{2}{\sin x}$	

51. $-\dfrac{\pi}{6}$, $-30°$ **52.** $\dfrac{\pi}{6}$, $30°$ **53.** $\dfrac{\pi}{4}$, $45°$ **54.** 0, $0°$

55. 1.7920, 102.7°   **56.** 0.3976, 22.8°   **57.** $\frac{1}{2}$ **58.** $\dfrac{\sqrt{3}}{3}$

59. $\dfrac{\pi}{7}$ **60.** $\dfrac{\sqrt{2}}{2}$ **61.** $\dfrac{3}{\sqrt{b^2+9}}$ **62.** $-\frac{7}{25}$

63. $\dfrac{3\pi}{4} + 2k\pi$, $\dfrac{5\pi}{4} + 2k\pi$, or $135° + k \cdot 360°$, $225° + k \cdot 360°$

64. $\dfrac{\pi}{3} + k\pi$, or $60° + k \cdot 180°$ **65.** $\dfrac{\pi}{6}, \dfrac{5\pi}{6}, \dfrac{7\pi}{6}, \dfrac{11\pi}{6}$

66. $\dfrac{\pi}{4}, \dfrac{\pi}{2}, \dfrac{3\pi}{4}, \dfrac{5\pi}{4}, \dfrac{3\pi}{2}, \dfrac{7\pi}{4}$ **67.** $\dfrac{2\pi}{3}, \pi, \dfrac{4\pi}{3}$ **68.** $0, \pi$

69. $\dfrac{\pi}{4}, \dfrac{3\pi}{4}, \dfrac{5\pi}{4}, \dfrac{7\pi}{4}$ **70.** $0, \dfrac{\pi}{2}, \pi, \dfrac{3\pi}{2}$ **71.** $\dfrac{7\pi}{12}, \dfrac{23\pi}{12}$

72. 0.864, 2.972, 4.006, 6.114   **73.** 4.917   **74.** No solution in $[0, 2\pi)$ **75.** B   **76.** A   **77.** C   **78.** 108.4°

79. $\cos(u+v) = \cos u \cos v - \sin u \sin v$

$= \cos u \cos v - \cos\left(\dfrac{\pi}{2} - u\right)\cos\left(\dfrac{\pi}{2} - v\right)$

80. $\cos^2 x$

81. $\sin\theta = \sqrt{\dfrac{1}{2} + \dfrac{\sqrt{6}}{5}}$; $\cos\theta = \sqrt{\dfrac{1}{2} - \dfrac{\sqrt{6}}{5}}$;

$\tan\theta = \sqrt{\dfrac{5 + 2\sqrt{6}}{5 - 2\sqrt{6}}}$

82. Let $x = \dfrac{\sqrt{2}}{2}$. Then $\tan^{-1}\dfrac{\sqrt{2}}{2} \approx 0.6155$ and

$\dfrac{\sin^{-1}\dfrac{\sqrt{2}}{2}}{\cos^{-1}\dfrac{\sqrt{2}}{2}} = \dfrac{\dfrac{\pi}{4}}{\dfrac{\pi}{4}} = 1.$ **83.** $\dfrac{\pi}{2}, \dfrac{3\pi}{2}$

84. The ranges of the inverse trigonometric functions are restricted in order that they might be functions.   **85.** Yes; first note that $7\pi/6 = \pi/6 + \pi$. Since $\pi/6 + k\pi$ includes both odd and even multiples of π, it is equivalent to $\pi/6 + 2k\pi$ and $7\pi/6 + 2k\pi$.   **86.** The graphs have different domains and ranges. The graph of $y = \sin^{-1} x$ is the reflection of the portion of the graph of $y = \sin x$ for $-\pi/2 \le x \le \pi/2$, across the line $y = x$.   **87.** A trigonometric equation that is an identity is true for all possible replacements of the variables. A trigonometric equation that is not true for all possible replacements is not an identity. The equation $\sin^2 x + \cos^2 x = 1$ is an identity whereas $\sin^2 x = 1$ is not.   **88.** The range of the arcsine function does not include $5\pi/6$. It is $[-\pi/2, \pi/2]$.

Test: Chapter 7

1. [7.1] $2\cos x + 1$ **2.** [7.1] 1 **3.** [7.1] $\dfrac{\cos\theta}{1 + \sin\theta}$

4. [7.1] $2\cos\theta$ **5.** [7.1] $\dfrac{\sqrt{2} + \sqrt{6}}{4}$ **6.** [7.1] $2 - \sqrt{3}$

7. [7.1] $\dfrac{120}{169}$ **8.** [7.2] $\dfrac{\sqrt{5}}{3}$ **9.** [7.2] $\dfrac{24}{25}$, II

10. [7.2] $\dfrac{\sqrt{2 + \sqrt{3}}}{2}$ **11.** [7.2] 0.9304 **12.** [7.2] $3\sin 2x$

13. [7.3]

$\csc x - \cos x \cot x$	$\sin x$
$\dfrac{1}{\sin x} - \cos x \cdot \dfrac{\cos x}{\sin x}$	
$\dfrac{1 - \cos^2 x}{\sin x}$	
$\dfrac{\sin^2 x}{\sin x}$	
$\sin x$	

14. [7.3]

$(\sin x + \cos x)^2$	$1 + \sin 2x$
$\sin^2 x + 2\sin x \cos x + \cos^2 x$	
$1 + 2\sin x \cos x$	
$1 + \sin 2x$	

15. [7.3]

$(\csc\beta + \cot\beta)^2$	$\dfrac{1 + \cos\beta}{1 - \cos\beta}$
$\left(\dfrac{1}{\sin\beta} + \dfrac{\cos\beta}{\sin\beta}\right)^2$	$\dfrac{1 + \cos\beta}{1 - \cos\beta} \cdot \dfrac{1 + \cos\beta}{1 + \cos\beta}$
$\left(\dfrac{1 + \cos\beta}{\sin\beta}\right)^2$	$\dfrac{(1 + \cos\beta)^2}{1 - \cos^2\beta}$
$\dfrac{(1 + \cos\beta)^2}{\sin^2\beta}$	$\dfrac{(1 + \cos\beta)^2}{\sin^2\beta}$

16. [7.3]

$\dfrac{1 + \sin\alpha}{1 + \csc\alpha}$	$\tan\alpha$
	$\sec\alpha$
$\dfrac{1 + \sin\alpha}{1 + \dfrac{1}{\sin\alpha}}$	$\dfrac{\dfrac{\sin\alpha}{\cos\alpha}}{\dfrac{1}{\cos\alpha}}$
$\dfrac{1 + \sin\alpha}{\dfrac{\sin\alpha + 1}{\sin\alpha}}$	$\sin\alpha$
$\sin\alpha$	

17. [7.4] $\cos 8\alpha - \cos\alpha = -2\sin\dfrac{9\alpha}{2}\sin\dfrac{7\alpha}{2}$

18. [7.4] $4\sin\beta\cos 3\beta = 2(\sin 4\beta - \sin 2\beta)$

19. [7.4] $-45°$ **20.** [7.4] $\dfrac{\pi}{3}$ **21.** [7.4] 2.3072

22. [7.4] $\dfrac{\sqrt{3}}{2}$ **23.** [7.4] $\dfrac{5}{\sqrt{x^2 - 25}}$ **24.** [7.4] 0

25. [7.5] $\dfrac{\pi}{6}, \dfrac{5\pi}{6}, \dfrac{7\pi}{6}, \dfrac{11\pi}{6}$ **26.** [7.5] $0, \dfrac{\pi}{4}, \dfrac{3\pi}{4}, \pi$

27. [7.5] $\dfrac{\pi}{2}, \dfrac{11\pi}{6}$ **28.** [7.4] D   **29.** [7.2] $\sqrt{\dfrac{11}{12}}$

Chapter 8

Exercise Set 8.1

1. $A = 121°, a \approx 33, c \approx 14$ **3.** $B \approx 57.4°, C \approx 86.1°, c \approx 40$ or $B \approx 122.6°, C \approx 20.9°, c \approx 14$
5. $B \approx 44°24', A \approx 74°26', a \approx 33.3$ **7.** $A = 110.36°, a \approx 5$ mi, $b \approx 3.4$ mi   **9.** $B \approx 83.78°, A \approx 12.44°, a \approx 12.30$ yd   **11.** $B \approx 14.7°, C \approx 135.0°, c \approx 28.04$ cm
13. No solution   **15.** $B = 125.27°, b \approx 302$ m, $c \approx 138$ m
17. 8.2 ft^2 **19.** 12 yd^2 **21.** 596.98 ft^2 **23.** About 31 ft
25. 787 ft^2 **27.** About 51 ft   **29.** From A: about 35 mi; from B: about 66 mi   **31.** About 102 mi
33. $1.348, 77.2°$ **34.** No angle
35. $18.24°$ **36.** $125.06°$ **37.** 5
38. $\dfrac{\sqrt{3}}{2}$ **39.** $\dfrac{\sqrt{2}}{2}$ **40.** $-\dfrac{\sqrt{3}}{2}$

41. $-\dfrac{1}{2}$ **42.** 2

43. Use the formula for the area of a triangle and the law of sines.

$$K = \frac{1}{2}ab\sin C \quad\text{and}\quad b = \frac{a\sin B}{\sin A},$$

$$\text{so } K = \frac{a^2\sin B\sin C}{2\sin A}.$$

$$K = \frac{1}{2}bc\sin A \quad\text{and}\quad b = \frac{c\sin B}{\sin C},$$

$$\text{so } K = \frac{c^2\sin A\sin B}{2\sin C}.$$

$$K = \frac{1}{2}bc\sin A \quad\text{and}\quad c = \frac{b\sin C}{\sin B},$$

$$\text{so } K = \frac{b^2\sin A\sin C}{2\sin B}.$$

45. For the quadrilateral $ABCD$, we have

$$\text{Area} = \tfrac{1}{2}bd \sin \theta + \tfrac{1}{2}ac \sin \theta$$
$$+ \tfrac{1}{2}ad(\sin 180° - \theta) + \tfrac{1}{2}bc \sin (180° - \theta)$$

***Note*: $\sin \theta = \sin (180° - \theta)$.**

$$= \tfrac{1}{2}(bd + ac + ad + bc)\sin \theta$$
$$= \tfrac{1}{2}(a + b)(c + d)\sin \theta$$
$$= \tfrac{1}{2}d_1 d_2 \sin \theta,$$

where $d_1 = a + b$ and $d_2 = c + d$.   **45.** $d \approx 18.8$ in.
47. 44.1″ from wall 1 and 104.3″ from wall 4

Exercise Set 8.2

1. $a \approx 15, B \approx 24°, C \approx 126°$ **3.** $A \approx 36.18°,$
$B \approx 43.53°, C \approx 100.29°$ **5.** $b \approx 75$ m, $A \approx 94°51',$
$C \approx 12°29'$ **7.** $A \approx 24.15°, B \approx 30.75°, C \approx 125.10°$
9. No solution   **11.** $A \approx 79.93°, B \approx 53.55°, C \approx 46.52°$
13. $c \approx 45.17$ mi, $A \approx 89.3°, B \approx 42.0°$ **15.** $a \approx 13.9$ in.,
$B \approx 36.127°, C \approx 90.417°$ **17.** Law of sines; $C = 98°,$
$a \approx 96.7, c \approx 101.9$ **19.** Law of cosines; $A \approx 73.71°,$
$B \approx 51.75°, C \approx 54.54°$ **21.** Cannot be solved
23. Law of cosines; $A \approx 33.71°, B \approx 107.08°, C \approx 39.21°$
25. 23.5 ft   **27.** 30.76 ft   **29.** About 1.5 mi
31. $S \approx 112.5°, T \approx 27.2°, U \approx 40.3°$ **33.** About 912 km
35. (a) About 16 ft; **(b)** about 122 ft²   **37.** About 4.7 cm
39. Quartic   **40.** Linear   **41.** Trigonometric
42. Exponential   **43.** Rational   **44.** Cubic
45. Exponential   **46.** Logarithmic   **47.** Trigonometric
48. Quadratic   **49.** About 9386 ft
51. $A = \tfrac{1}{2}a^2 \sin \theta$; when $\theta = 90°$

Exercise Set 8.3

1. 5;

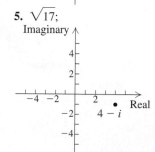

3. 1;

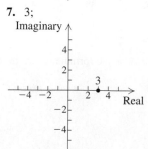

5. $\sqrt{17}$;

7. 3;

9. $3 - 3i$; $3\sqrt{2}\left(\cos \dfrac{7\pi}{4} + i \sin \dfrac{7\pi}{4}\right)$, or
$3\sqrt{2}(\cos 315° + i \sin 315°)$

11. $4i$; $4\left(\cos \dfrac{\pi}{2} + i \sin \dfrac{\pi}{2}\right)$, or $4(\cos 90° + i \sin 90°)$

13. $\sqrt{2}\left(\cos \dfrac{7\pi}{4} + i \sin \dfrac{7\pi}{4}\right)$, or $\sqrt{2}(\cos 315° + i \sin 315°)$

15. $3\left(\cos \dfrac{3\pi}{2} + i \sin \dfrac{3\pi}{2}\right)$, or $3(\cos 270° + i \sin 270°)$

17. $2\left(\cos \dfrac{\pi}{6} + i \sin \dfrac{\pi}{6}\right)$, or $2(\cos 30° + i \sin 30°)$

19. $\dfrac{2}{5}(\cos 0 + i \sin 0)$, or $\dfrac{2}{5}(\cos 0° + i \sin 0°)$

21. $6\left(\cos \dfrac{5\pi}{4} + i \sin \dfrac{5\pi}{4}\right)$, or $6(\cos 225° + i \sin 225°)$

23. $\dfrac{3\sqrt{3}}{2} + \dfrac{3}{2}i$ **25.** $-10i$ **27.** $2 + 2i$ **29.** $2i$

31. $\dfrac{\sqrt{2}}{2} - \dfrac{\sqrt{6}}{2}i$ **33.** $4(\cos 42° + i \sin 42°)$

35. $11.25(\cos 56° + i \sin 56°)$ **37.** 4   **39.** $-i$
41. $6 + 6\sqrt{3}i$ **43.** $-2i$ **45.** $8(\cos \pi + i \sin \pi)$

47. $8\left(\cos \dfrac{3\pi}{2} + i \sin \dfrac{3\pi}{2}\right)$ **49.** $\dfrac{27}{2} + \dfrac{27\sqrt{3}}{2}i$

51. $-4 + 4i$ **53.** -1 **55.** $-\dfrac{\sqrt{2}}{2} + \dfrac{\sqrt{2}}{2}i, \dfrac{\sqrt{2}}{2} - \dfrac{\sqrt{2}}{2}i$

57. $2(\cos 157.5° + i \sin 157.5°), 2(\cos 337.5° + i \sin 337.5°)$

59. $\dfrac{\sqrt{3}}{2} + \dfrac{1}{2}i, -\dfrac{\sqrt{3}}{2} + \dfrac{1}{2}i, -i$

61. $\sqrt[3]{4}(\cos 110° + i \sin 110°), \sqrt[3]{4}(\cos 230° + i \sin 230°),$
$\sqrt[3]{4}(\cos 350° + i \sin 350°)$

63. $2, 2i, -2, -2i;$

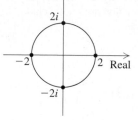

65. $\cos 36° + i \sin 36°,$
$\cos 108° + i \sin 108°, -1,$
$\cos 252° + i \sin 252°,$
$\cos 324° + i \sin 324°;$

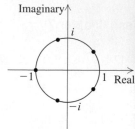

67. $\sqrt[10]{8}, \sqrt[10]{8}(\cos 36° + i \sin 36°),$
$\sqrt[10]{8}(\cos 72° + i \sin 72°), \sqrt[10]{8}(\cos 108° + i \sin 108°),$
$\sqrt[10]{8}(\cos 144° + i \sin 144°), -\sqrt[10]{8},$
$\sqrt[10]{8}(\cos 216° + i \sin 216°), \sqrt[10]{8}(\cos 252° + i \sin 252°),$
$\sqrt[10]{8}(\cos 288° + i \sin 288°), \sqrt[10]{8}(\cos 324° + i \sin 324°)$

69. $\dfrac{\sqrt{3}}{2} + \dfrac{1}{2}i, i, -\dfrac{\sqrt{3}}{2} + \dfrac{1}{2}i, -\dfrac{\sqrt{3}}{2} - \dfrac{1}{2}i, -i, \dfrac{\sqrt{3}}{2} - \dfrac{1}{2}i$

71. $1, -\dfrac{1}{2} + \dfrac{\sqrt{3}}{2}i, -\dfrac{1}{2} - \dfrac{\sqrt{3}}{2}i$

73. $\cos 67.5° + i \sin 67.5°, \cos 157.5° + i \sin 157.5°,$
$\cos 247.5° + i \sin 247.5°, \cos 337.5° + i \sin 337.5°$
75. $\sqrt{3} + i, 2i, -\sqrt{3} + i, -\sqrt{3} - i, -2i, \sqrt{3} - i$

77. $15°$ **78.** $540°$ **79.** $\dfrac{11\pi}{6}$ **80.** $-\dfrac{5\pi}{4}$ **81.** $\dfrac{\sqrt{3}}{2}$

82. $\dfrac{\sqrt{3}}{2}$ **83.** $\dfrac{\sqrt{2}}{2}$ **84.** $\tfrac{1}{2}$ **85.** $\cos \theta - i \sin \theta$

87. $z = a + bi, |z| = \sqrt{a^2 + b^2}; \bar{z} = a - bi,$
$|\bar{z}| = \sqrt{a^2 + (-b)^2} = \sqrt{a^2 + b^2}, \therefore |z| = |\bar{z}|$
89. $|(a + bi)^2| = |a^2 - b^2 + 2abi|$
$$= \sqrt{(a^2 - b^2)^2 + 4a^2b^2}$$
$$= \sqrt{a^4 + 2a^2b^2 + b^4} = a^2 + b^2,$$
$$|a + bi|^2 = (\sqrt{a^2 + b^2})^2 = a^2 + b^2$$

1.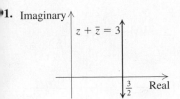

Mid-Chapter Mixed Review: Chapter 8

1. True **2.** True **3.** False **4.** True **5.** $B = 63°$, $b \approx 9.4$ in., $c \approx 9.5$ in. **6.** No solution **7.** $A \approx 40.5°$, $B \approx 28.5°$, $C \approx 111.0°$ **8.** $B \approx 70.2°$, $C \approx 67.1°$, $c \approx 39.9$ cm or $B \approx 109.8°$, $C \approx 27.5°$, $c \approx 20.0$ cm **9.** $a \approx 370$ yd, $B \approx 16.6°$, $C \approx 15.4°$ **10.** $C \approx 45°$, $A \approx 107°$, $a \approx 37$ ft, or $C \approx 135°$, $A \approx 17°$, $a \approx 11$ ft **11.** About 446 in^2

12. 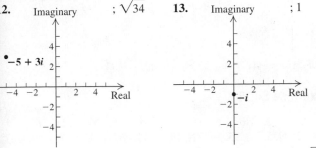 ; $\sqrt{34}$ **13.** ; 1

14. ; 4 **15.** 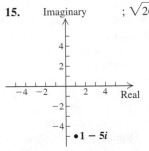 ; $\sqrt{26}$

16. $\sqrt{2}\left(\cos\dfrac{\pi}{3} + i\sin\dfrac{\pi}{3}\right)$, or $\sqrt{2}\,(\cos 60° + i\sin 60°)$

17. $2\left(\cos\dfrac{5\pi}{3} + i\sin\dfrac{5\pi}{3}\right)$, or $2(\cos 300° + i\sin 300°)$

18. $5\left(\cos\dfrac{\pi}{2} + i\sin\dfrac{\pi}{2}\right)$, or $5(\cos 90° + i\sin 90°)$

19. $2\sqrt{2}\left(\cos\dfrac{5\pi}{4} + i\sin\dfrac{5\pi}{4}\right)$, or $2\sqrt{2}\,(\cos 225° + i\sin 225°)$ **20.** $\sqrt{2} - \sqrt{2}i$ **21.** $6\sqrt{3} + 6i$ **22.** $\sqrt{5}$ **23.** $4i$

24. $16(\cos 45° + i\sin 45°)$

25. $9\left[\cos\left(\dfrac{\pi}{12}\right) + i\sin\left(\dfrac{\pi}{12}\right)\right]$

26. $2\sqrt{2}\,(\cos 285° + i\sin 285°)$ **27.** $\sqrt{2}\,(\cos 255° + i\sin 255°)$ **28.** $8\sqrt{2}\,(\cos 45° + i\sin 45°)$ **29.** $8 + 8\sqrt{3}i$ **30.** $2(\cos 120° + i\sin 120°)$ and $2(\cos 300° + i\sin 300°)$, or $-1 + \sqrt{3}i$ and $1 - \sqrt{3}i$ **31.** $1(\cos 60° + i\sin 60°)$, $1(\cos 180° + i\sin 180°)$, and $1(\cos 300° + i\sin 300°)$, or $\dfrac{1}{2} + \dfrac{\sqrt{3}}{2}i$, -1, and $\dfrac{1}{2} - \dfrac{\sqrt{3}}{2}i$ **32.** Using the law of

cosines, it is necessary to solve the quadratic equation:
$(11.1)^2 = a^2 + (28.5)^2 - 2a(28.5)\cos 19°$, or
$0 = a^2 - [2(28.5)\cos 19°]a + [(28.5)^2 - (11.1)^2]$.
The law of sines requires less complicated computations.
33. A nonzero complex number has n different complex nth roots. Thus, 1 has three different complex cube roots, one of which is the real number 1. The other two roots are complex conjugates. Since the set of real numbers is a subset of the set of complex numbers, the real cube root of 1 is also a complex root of 1. **34.** The law of sines involves two angles of a triangle and the sides opposite them. Three of these four values must be known in order to find the fourth. Given SAS, only two of these four values are known. **35.** The law of sines involves two angles of a triangle and the sides opposite them. Three of these four values must be known in order to find the fourth. Thus we must know the measure of one angle in order to use the law of sines. **36.** Trigonometric notation is not unique because there are infinitely many angles coterminal with a given angle. Standard notation is unique because any point has a unique ordered pair (a, b) associated with it.
37. $x^6 - 2x^3 + 1 = 0$
$(x^3 - 1)^2 = 0$
$x^3 - 1 = 0$
$x^3 = 1$
$x = 1^{1/3}$

This equation has three distinct solutions because there are three distinct cube roots of 1.

$x^6 - 2x^3 = 0$
$x^3(x^3 - 2) = 0$
$x^3 = 0 \quad or \quad x^3 - 2 = 0$
$x^3 = 0 \quad or \quad x^3 = 2$
$x = 0 \quad or \quad x = 2^{1/3}$

This equation has four distinct solutions because 0 is one solution and the three distinct cube roots of 2 provide an additional three solutions.

$x^6 - 2x = 0$
$x(x^5 - 2) = 0$
$x = 0 \quad or \quad x^5 - 2 = 0$
$x = 0 \quad or \quad x^5 = 2$
$x = 0 \quad or \quad x = 2^{1/5}$

This equation has six distinct solutions because 0 is one solution and the five fifth roots of 2 provide an additional five solutions.

Visualizing the Graph

1. J **2.** C **3.** E **4.** H **5.** I **6.** A **7.** D **8.** G **9.** B **10.** F

Exercise Set 8.4

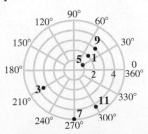

13. A: $(4, 30°)$, $(4, 390°)$, $(-4, 210°)$; B: $(5, 300°)$, $(5, -60°)$, $(-5, 120°)$; C: $(2, 150°)$, $(2, 510°)$, $(-2, 330°)$; D: $(3, 225°)$, $(3, -135°)$, $(-3, 45°)$; answers may vary

15. $(3, 270°), \left(3, \dfrac{3\pi}{2}\right)$ **17.** $(6, 300°), \left(6, \dfrac{5\pi}{3}\right)$

19. $(8, 330°), \left(8, \dfrac{11\pi}{6}\right)$ **21.** $(2, 225°), \left(2, \dfrac{5\pi}{4}\right)$

23. $(2, 60°), \left(2, \dfrac{\pi}{3}\right)$ **25.** $(5, 315°), \left(5, \dfrac{7\pi}{4}\right)$

27. $(7.616, 66.8°), (7.616, 1.166)$
29. $(4.643, 132.9°), (4.643, 2.320)$
31. $\left(\dfrac{5}{2}, \dfrac{5\sqrt{3}}{2}\right)$ **33.** $\left(-\dfrac{3\sqrt{2}}{2}, -\dfrac{3\sqrt{2}}{2}\right)$

35. $\left(-\dfrac{3}{2}, -\dfrac{3\sqrt{3}}{2}\right)$ **37.** $(-1, \sqrt{3})$ **39.** $(-\sqrt{3}, -1)$

41. $(3\sqrt{3}, -3)$ **43.** $(2.19, -2.05)$ **45.** $(1.30, -3.99)$
47. $r(3\cos\theta + 4\sin\theta) = 5$ **49.** $r\cos\theta = 5$
51. $r = 6$ **53.** $r^2\cos^2\theta = 25r\sin\theta$
55. $r^2\sin^2\theta - 5r\cos\theta - 25 = 0$ **57.** $r^2 = 2r\cos\theta$
59. $x^2 + y^2 = 25$ **61.** $y = 2$ **63.** $y^2 = -6x + 9$
65. $x^2 - 9x + y^2 - 7y = 0$ **67.** $x = 5$ **69.** $y = -\sqrt{3}x$

71.
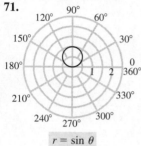
$r = \sin\theta$

73.
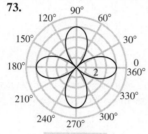
$r = 4\cos 2\theta$

75.
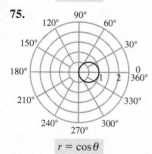
$r = \cos\theta$

77.
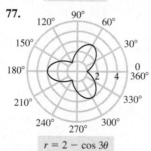
$r = 2 - \cos 3\theta$

79. (d) **81.** (g) **83.** (j) **85.** (b) **87.** (e) **89.** (k)
91. $r = \sin\theta\,\tan\theta$ **93.** $r = e^{\theta/10}$

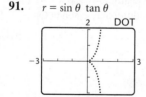

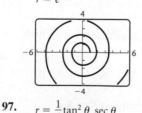

95. $r = \cos 2\theta\,\sec\theta$ **97.** $r = \dfrac{1}{4}\tan^2\theta\,\sec\theta$

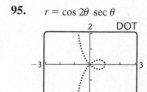

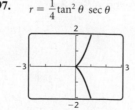

99. 12 **100.** $\dfrac{1}{5}$
101.

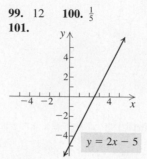

$y = 2x - 5$

102.

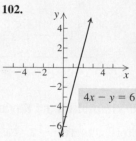

$4x - y = 6$

103.

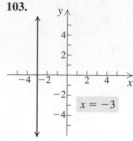

$x = -3$

104.

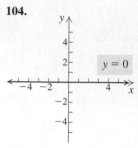

$y = 0$

105. $y^2 = -4x + 4$

Exercise Set 8.5

1. Yes **3.** No **5.** Yes **7.** No **9.** No **11.** Yes
13. 55 N, 55° **15.** 929 N, 19° **17.** 57.0, 38° **19.** 18.4, 37°
21. 20.9, 58° **23.** 68.3, 18° **25.** 11 ft/sec, 63°
27. 726 lb, 47° **29.** 60° **31.** 70.7 east; 70.7 south
33. Horizontal: 215.17 mph forward; vertical: 65.78 mph up
35. Horizontal: 390 lb forward; vertical: 675.5 lb up
37. Northerly: 115 km/h; westerly: 164 km/h
39. Perpendicular: 90.6 lb; parallel: 42.3 lb **41.** 48.1 lb
43. Natural **44.** Half-angle **45.** Linear speed
46. Cosine **47.** Identity **48.** Cotangent of θ
49. Coterminal **50.** Sines **51.** Horizontal line; inverse
52. Reference angle; acute
53. (a) $(4.950, 4.950)$; (b) $(0.950, -1.978)$

Exercise Set 8.6

1. $\langle -9, 5\rangle; \sqrt{106}$ **3.** $\langle -3, 6\rangle; 3\sqrt{5}$ **5.** $\langle 4, 0\rangle; 4$
7. $\sqrt{37}$ **9.** $\langle 4, -5\rangle$ **11.** $\sqrt{257}$ **13.** $\langle -9, 9\rangle$
15. $\langle 41, -38\rangle$ **17.** $\sqrt{261} - \sqrt{65}$ **19.** $\langle -1, -1\rangle$
21. $\langle -8, 14\rangle$ **23.** 1 **25.** -34
27.

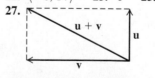

29.

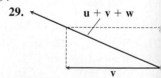

31. (a) $\mathbf{w} = \mathbf{u} + \mathbf{v}$; (b) $\mathbf{v} = \mathbf{w} - \mathbf{u}$ **33.** $\left\langle -\dfrac{5}{13}, \dfrac{12}{13}\right\rangle$
35. $\left\langle \dfrac{1}{\sqrt{101}}, -\dfrac{10}{\sqrt{101}}\right\rangle$ **37.** $\left\langle -\dfrac{1}{\sqrt{17}}, -\dfrac{4}{\sqrt{17}}\right\rangle$
39. $\mathbf{w} = -4\mathbf{i} + 6\mathbf{j}$ **41.** $\mathbf{s} = 2\mathbf{i} + 5\mathbf{j}$ **43.** $-7\mathbf{i} + 5\mathbf{j}$
45. (a) $3\mathbf{i} + 29\mathbf{j}$; (b) $\langle 3, 29\rangle$ **47.** (a) $4\mathbf{i} + 16\mathbf{j}$; (b) $\langle 4, 16\rangle$
49. $\mathbf{j}$, or $\langle 0, 1\rangle$ **51.** $-\dfrac{1}{2}\mathbf{i} - \dfrac{\sqrt{3}}{2}\mathbf{j}$, or $\left\langle -\dfrac{1}{2}, -\dfrac{\sqrt{3}}{2}\right\rangle$
53. 248° **55.** 63° **57.** 50° **59.** $|\mathbf{u}| = 3; \theta = 45°$
61. 1; 120° **63.** 144.2° **65.** 14.0° **67.** 101.3°

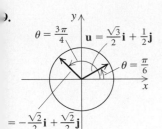

$\theta = \frac{3\pi}{4}$ $\mathbf{u} = \frac{\sqrt{3}}{2}\mathbf{i} + \frac{1}{2}\mathbf{j}$ $\theta = \frac{\pi}{6}$

71.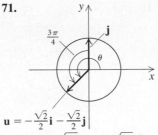

$= -\frac{\sqrt{2}}{2}\mathbf{i} + \frac{\sqrt{2}}{2}\mathbf{j}$ $\mathbf{u} = -\frac{\sqrt{2}}{2}\mathbf{i} - \frac{\sqrt{2}}{2}\mathbf{j}$

3. $\mathbf{u} = -\frac{\sqrt{10}}{10}\mathbf{i} + \frac{3\sqrt{10}}{10}\mathbf{j}$ **75.** $\sqrt{13}\left(\frac{2\sqrt{13}}{13}\mathbf{i} - \frac{3\sqrt{13}}{13}\mathbf{j}\right)$

7.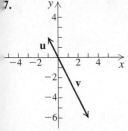

9. 174 nautical mi, S17°E **81.** 60° **83.** 500 lb on left, 66 lb on right **85.** Cable: 224-lb tension; boom: 167-lb compression **87.** 0, 4 **88.** $-\frac{11}{3}, \frac{5}{2}$

9. $\cos\theta = \dfrac{\mathbf{u}\cdot\mathbf{v}}{|\mathbf{u}||\mathbf{v}|} = \dfrac{0}{|\mathbf{u}||\mathbf{v}|}, \therefore \cos\theta = 0$ and $\theta = 90°$.

1. $\frac{3}{5}\mathbf{i} - \frac{4}{5}\mathbf{j}, -\frac{3}{5}\mathbf{i} + \frac{4}{5}\mathbf{j}$ **93.** $(5, 8)$

Review Exercises: Chapter 8

. True **2.** False **3.** False **4.** False **5.** False
. True **7.** $A \approx 153°, B \approx 18°, C \approx 9°$ **8.** $A = 118°$,
≈ 37 in., $c \approx 24$ in. **9.** $B = 14°50', a \approx 2523$ m,
≈ 1827 m **10.** No solution **11.** 33 m² **12.** 13.72 ft²
3. 63 ft² **14.** 92°, 33°, 55° **15.** 419 ft **16.** About 650 km
7. $\sqrt{29}$; **18.** 4;

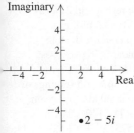 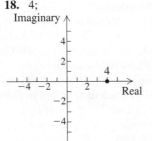

9. 2; **20.** $\sqrt{10}$;

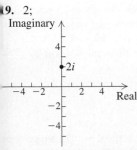 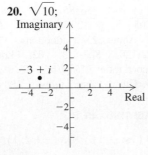

21. $\sqrt{2}\left(\cos\frac{\pi}{4} + i\sin\frac{\pi}{4}\right)$, or $\sqrt{2}(\cos 45° + i\sin 45°)$

22. $4\left(\cos\frac{3\pi}{2} + i\sin\frac{3\pi}{2}\right)$, or $4(\cos 270° + i\sin 270°)$

23. $10\left(\cos\frac{5\pi}{6} + i\sin\frac{5\pi}{6}\right)$, or $10(\cos 150° + i\sin 150°)$
24. $\frac{3}{4}(\cos 0 + i\sin 0)$, or $\frac{3}{4}(\cos 0° + i\sin 0°)$ **25.** $2 + 2\sqrt{3}i$
26. 7 **27.** $-\frac{5}{2} + \frac{5\sqrt{3}}{2}i$ **28.** $\sqrt{3} - i$
29. $1 + \sqrt{3} + (-1 + \sqrt{3})i$
30. $-i$ **31.** $2i$ **32.** $3\sqrt{3} + 3i$
33. $8(\cos 180° + i\sin 180°)$ **34.** $4(\cos 7\pi + i\sin 7\pi)$
35. $-8i$ **36.** $-\frac{1}{2} - \frac{\sqrt{3}}{2}i$
37. $\sqrt[4]{2}\left(\cos\frac{3\pi}{8} + i\sin\frac{3\pi}{8}\right), \sqrt[4]{2}\left(\cos\frac{11\pi}{8} + i\sin\frac{11\pi}{8}\right)$
38. $\sqrt[3]{6}(\cos 110° + i\sin 110°), \sqrt[3]{6}(\cos 230° + i\sin 230°),$
$\sqrt[3]{6}(\cos 350° + i\sin 350°)$
39. $3, 3i, -3, -3i$ **40.** $1, \cos 72° + i\sin 72°,$
$\cos 144° + i\sin 144°,$
$\cos 216° + i\sin 216°,$
$\cos 288° + i\sin 288°$

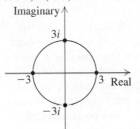 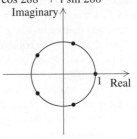

41. $\cos 22.5° + i\sin 22.5°, \cos 112.5° + i\sin 112.5°,$
$\cos 202.5° + i\sin 202.5°, \cos 292.5° + i\sin 292.5°$
42. $\frac{1}{2} + \frac{\sqrt{3}}{2}i, -1, \frac{1}{2} - \frac{\sqrt{3}}{2}i$
43. $A: (5, 120°), (5, 480°), (-5, 300°);$
$B: (3, 210°), (-3, 30°), (-3, 390°); C: (4, 60°),$
$(4, 420°), (-4, 240°); D: (1, 300°), (1, -60°), (-1, 120°);$
answers may vary **44.** $(8, 135°), \left(8, \frac{3\pi}{4}\right)$
45. $(5, 270°), \left(5, \frac{3\pi}{2}\right)$ **46.** $(5.385, 111.8°), (5.385, 1.951)$
47. $(4.964, 147.8°), (4.964, 2.579)$ **48.** $\left(\frac{3\sqrt{2}}{2}, \frac{3\sqrt{2}}{2}\right)$
49. $(3, 3\sqrt{3})$ **50.** $(1.93, -0.52)$ **51.** $(-1.86, -1.35)$
52. $r(5\cos\theta - 2\sin\theta) = 6$ **53.** $r\sin\theta = 3$ **54.** $r = 3$
55. $r^2\sin^2\theta - 4r\cos\theta - 16 = 0$ **56.** $x^2 + y^2 = 36$
57. $x^2 + 2y = 1$ **58.** $y^2 - 6x = 9$
59. $x^2 - 2x + y^2 - 3y = 0$ **60.** (b) **61.** (d)
62. (a) **63.** (c) **64.** 13.7, 71° **65.** 98.7, 15°
66. 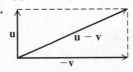 **67.**

68. 666.7 N, 36° **69.** 29 km/h, 149° **70.** 102.4 nautical mi,
S43°E **71.** $\langle -4, 3\rangle$ **72.** $\langle 2, -6\rangle$ **73.** $\sqrt{61}$
74. $\langle 10, -21\rangle$ **75.** $\langle 14, -64\rangle$ **76.** $5 + \sqrt{116}$
77. 14 **78.** $\left\langle -\frac{3}{\sqrt{10}}, -\frac{1}{\sqrt{10}}\right\rangle$ **79.** $-9\mathbf{i} + 4\mathbf{j}$
80. 194.0° **81.** $\sqrt{34}; \theta = 211.0°$ **82.** 111.8° **83.** 85.1°

84. $34\mathbf{i} - 55\mathbf{j}$ **85.** $\mathbf{i} - 12\mathbf{j}$
86. $5\sqrt{2}$ **87.** $3\sqrt{65} + \sqrt{109}$ **88.** $-5\mathbf{i} + 5\mathbf{j}$
89.

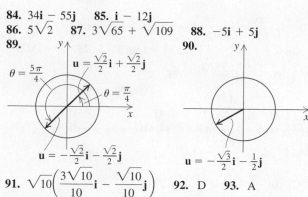

91. $\sqrt{10}\left(\dfrac{3\sqrt{10}}{10}\mathbf{i} - \dfrac{\sqrt{10}}{10}\mathbf{j}\right)$ **92.** D **93.** A
94. D **95.** $\frac{36}{13}\mathbf{i} + \frac{15}{13}\mathbf{j}$ **96.** $50.52°, 129.48°$
97. A triangle has no solution when a sine value or a cosine value found is less than -1 or greater than 1. A triangle also has no solution if the sum of the angle measures calculated is greater than $180°$. A triangle has only one solution if only one possible answer is found, or if one of the possible answers has an angle sum greater than $180°$. A triangle has two solutions when two possible answers are found and neither results in an angle sum greater than $180°$.
98. One example is the equation of a circle not centered at the origin. Often, in rectangular coordinates, we must complete the square in order to graph the circle. **99.** Rectangular coordinates are unique because any point has a unique ordered pair (x, y) associated with it. Polar coordinates are not unique because there are infinitely many angles coterminal with a given angle and also because r can be positive or negative depending on the angle used.
100. Vectors $\overrightarrow{QR}$ and $\overrightarrow{RQ}$ have opposite directions, so they are not equivalent. **101.** The terminal point of a unit vector in standard position is a point on the unit circle. **102.** Answers may vary. For $\mathbf{u} = 3\mathbf{i} - 4\mathbf{j}$ and $\mathbf{w} = 2\mathbf{i} - 4\mathbf{j}$, find $\mathbf{v}$, where $\mathbf{v} = \mathbf{u} + \mathbf{w}$.

Test: Chapter 8

1. [8.1] $A = 83°, b \approx 14.7$ ft, $c \approx 12.4$ ft
2. [8.1] $A \approx 73.9°, B \approx 70.1°, a \approx 8.2$ m, or $A \approx 34.1°$,
$B \approx 109.9°, a \approx 4.8$ m **3.** [8.2] $A \approx 99.9°, B \approx 36.8°$,
$C \approx 43.3°$ **4.** [8.1] About 43.6 cm² **5.** [8.1] About 77 m
6. [8.5] About 930 km
7. [8.3]

```
              Imaginary
                 5
                 4
                 3
       -4 + i    2
         •       1
       -5-4-3-2-1  1 2 3 4 5  → Real
                -1
                -2
                -3
                -4
                -5
```

8. [8.3] $\sqrt{13}$ **9.** [8.3] $3\sqrt{2}(\cos 315° + i\sin 315°)$, or
$3\sqrt{2}\left(\cos \dfrac{7\pi}{4} + i\sin \dfrac{7\pi}{4}\right)$ **10.** [8.3] $\frac{1}{4}i$
11. [8.3] 16 **12.** [8.4] $2(\cos 120° + i\sin 120°)$
13. [8.4] $\left(\dfrac{1}{2}, -\dfrac{\sqrt{3}}{2}\right)$ **14.** [8.4] $r = \sqrt{10}$

15. [8.4]

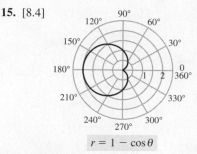

$r = 1 - \cos\theta$

16. [8.5] Magnitude: 11.2; direction: $23.4°$
17. [8.6] $-11\mathbf{i} - 17\mathbf{j}$ **18.** [8.6] $-\frac{4}{5}\mathbf{i} + \frac{3}{5}\mathbf{j}$
19. [8.4] A **20.** [8.1] $28.9°, 151.1°$

Chapter 9

Visualizing the Graph

1. C **2.** G **3.** D **4.** J **5.** A **6.** F **7.** I
8. B **9.** H **10.** E

Exercise Set 9.1

1. (c) **3.** (f) **5.** (b) **7.** $(-1, 3)$ **9.** $(-1, 1)$
11. No solution **13.** $(-2, 4)$ **15.** Infinitely many solutions;
$\left(x, \dfrac{x - 1}{2}\right)$, or $(2y + 1, y)$ **17.** $(5, 4)$ **19.** $(1, -3)$
21. $(2, -2)$ **23.** No solution **25.** $\left(\frac{39}{11}, -\frac{1}{11}\right)$ **27.** $\left(\frac{1}{2}, \frac{3}{4}\right)$
29. Infinitely many solutions; $(x, 3x - 5)$ or $\left(\frac{1}{3}y + \frac{5}{3}, y\right)$
31. $(1, 3)$; consistent, independent **33.** $(-4, -2)$; consistent, independent **35.** Infinitely many solutions; $(4y + 2, y)$ or $\left(x, \frac{1}{4}x - \frac{1}{2}\right)$; consistent, dependent **37.** No solution; inconsistent, independent **39.** $(1, 1)$; consistent, independent
41. $(-3, 0)$; consistent, independent **43.** $(10, 8)$; consistent, independent **45.** True **47.** False **49.** True
51. Liposuction: 210,552 surgeries; nose reshaping: 217,124 surgeries **53.** 2013: \$3.35 billion; 2014: \$3.5 billion
55. New York City: 675,000 Chinese–Americans; Boston: 116,000 Chinese–Americans **57.** Standard: 76 packages; express: 44 packages **59.** 4%: \$6000; 5%: \$9000
61. French roast: 8 lb; Colombian: 12 lb **63.** 1.5 servings of spaghetti, 2 servings of lettuce **65.** Boat: 20 km/h; stream: 3 km/h **67.** 2 hr **69.** (15, \$200) **71.** 140
73. 6000 **75.** (a) $r(x) = -0.3846557377x + 47.52939344$;
$p(x) = 0.1256885246x + 34.83178689$; (b) about 25 years after 2005 **76.** \$29,790 **77.** 115,017 registered snowmobiles
78. About 22,920 adoptions **79.** $-2, 6$ **80.** $-1, 5$
81. 120 **82.** 1, 3 **83.** 4 km **85.** First train: 36 km/h; second train: 54 km/h **87.** $A = \frac{1}{10}, B = -\frac{7}{10}$
89. City: 115 mi; highway: 288 mi

Exercise Set 9.2

1. $(3, -2, 1)$ **3.** $(-3, 2, 1)$ **5.** $\left(2, \frac{1}{2}, -2\right)$ **7.** No solution
9. Infinitely many solutions; $\left(\dfrac{11y + 19}{5}, y, \dfrac{9y + 11}{5}\right)$
11. $\left(\frac{1}{2}, \frac{2}{3}, -\frac{5}{6}\right)$ **13.** $(-1, 4, 3)$ **15.** $(1, -2, 4, -1)$

7. Russian Federation: 80 medals; Ukraine: 25 medals; United States: 18 medals   **19.** United States: 361 million MT; China: 416 million MT; Brazil: 75 million MT   **21.** Bacon: $142.4 million; Warhol: $105.4 million; Koons: $58.4 million   **23.** Dogs: 89.9 million; cats: 74.1 million; birds: 8.3 million   **25.** United States: $31.5 billion; Great Britain: $17.9 billion; Germany: $14.1 billion   **27.** $1\frac{1}{4}$ servings of beef, 1 baked potato, $\frac{1}{2}$ serving of strawberries   **29.** 2%: $300; 3%: $800; 4%: $2400   **31.** Orange juice: $2.40; bagel: $2.75; coffee: $3.00
33. (a) $f(x) = \frac{3}{40}x^2 - \frac{7}{4}x + 43$; **(b)** 2007: 33.625%; 2014: 40.8%   **35. (a)** $f(x) = -7x^2 + 55x + 291$; **(b)** 333,000 deportations   **37. (a)** $f(x) = -0.0792857143x^2 + 3.4801428571x + 2.599428571$; **(b)** $3.02   **39.** Perpendicular   **40.** The leading-term test   **41.** A vertical line   **42.** A one-to-one function   **43.** A rational function   **44.** Inverse variation   **45.** A vertical asymptote   **46.** A horizontal asymptote   **47.** $\left(-\frac{1}{2}, -1, -\frac{1}{3}\right)$ **49.** $3x + 4y + 2z = 12$ **51.** $y = -4x^3 + 5x^2 - 3x + 1$

Exercise Set 9.3

1. 3×2 **3.** 1×4 **5.** 3×3 **7.** $\begin{bmatrix} 2 & -1 & 7 \\ 1 & 4 & -5 \end{bmatrix}$
9. $\begin{bmatrix} 1 & -2 & 3 & 12 \\ 2 & 0 & -4 & 8 \\ 0 & 3 & 1 & 7 \end{bmatrix}$
11. $3x - 5y = 1,$
$\quad x + 4y = -2$
13. $2x + y - 4z = 12,$
$\quad 3x \quad + 5z = -1,$
$\quad x - y + z = 2$
15. $\left(\frac{3}{2}, \frac{5}{2}\right)$ **17.** $\left(-\frac{63}{29}, -\frac{114}{29}\right)$ **19.** $\left(-1, \frac{5}{2}\right)$ **21.** $(0, 3)$
23. No solution   **25.** Infinitely many solutions; $(3y - 2, y)$
27. $(-1, 2, -2)$ **29.** $\left(\frac{3}{2}, -4, 3\right)$ **31.** $(-1, 6, 3)$
33. Infinitely many solutions; $\left(\frac{1}{2}z + \frac{1}{2}, -\frac{1}{2}z - \frac{1}{2}, z\right)$
35. Infinitely many solutions; $(r - 2, -2r + 3, r)$
37. No solution   **39.** $(1, -3, -2, -1)$ **41.** 8%: $8000; 10%: $12,000; 12%: $10,000   **43.** 49¢: 160 stamps; 21¢: 40 stamps
45. Exponential   **46.** Linear   **47.** Rational   **48.** Quartic
49. Logarithmic   **50.** Cubic   **51.** Linear   **52.** Quadratic
53. $y = 3x^2 + \frac{5}{2}x - \frac{15}{2}$ **55.** $\begin{bmatrix} 1 & 5 \\ 0 & 1 \end{bmatrix}, \begin{bmatrix} 1 & 0 \\ 0 & 1 \end{bmatrix}$
57. $\left(-\frac{4}{3}, -\frac{1}{3}, 1\right)$ **59.** Infinitely many solutions; $\left(-\frac{14}{13}z - 1, \frac{3}{13}z - 2, z\right)$ **61.** $(-3, 3)$

Exercise Set 9.4

1. $x = -3, y = 5$ **3.** $x = -1, y = 1$
5. $\begin{bmatrix} -2 & 7 \\ 6 & 2 \end{bmatrix}$ **7.** $\begin{bmatrix} 1 & 3 \\ 2 & 6 \end{bmatrix}$ **9.** $\begin{bmatrix} 9 & 9 \\ -3 & -3 \end{bmatrix}$
11. $\begin{bmatrix} 11 & 13 \\ 5 & 3 \end{bmatrix}$ **13.** $\begin{bmatrix} -4 & 3 \\ -2 & -4 \end{bmatrix}$ **15.** $\begin{bmatrix} 17 & 9 \\ -2 & 1 \end{bmatrix}$
17. $\begin{bmatrix} 0 & 0 \\ 0 & 0 \end{bmatrix}$ **19.** $\begin{bmatrix} 1 & 2 \\ 4 & 3 \end{bmatrix}$ **21.** $\begin{bmatrix} 1 \\ 40 \end{bmatrix}$

23. $\begin{bmatrix} -10 & 28 \\ 14 & -26 \\ 0 & -6 \end{bmatrix}$ **25.** Not defined   **27.** $\begin{bmatrix} 3 & 16 & 3 \\ 0 & -32 & 0 \\ -6 & 4 & 5 \end{bmatrix}$
29. (a) $\begin{bmatrix} 40 & 20 & 30 \end{bmatrix}$; **(b)** $\begin{bmatrix} 44 & 22 & 33 \end{bmatrix}$; **(c)** $\begin{bmatrix} 84 & 42 & 63 \end{bmatrix}$; the total amount of each type of produce ordered for both weeks
31. (a) $C = \begin{bmatrix} 140 & 27 & 3 & 13 & 64 \end{bmatrix}$,
$P = \begin{bmatrix} 180 & 4 & 11 & 24 & 662 \end{bmatrix}$,
$B = \begin{bmatrix} 50 & 5 & 1 & 82 & 20 \end{bmatrix}$; **(b)** $\begin{bmatrix} 650 & 50 & 28 & 307 & 1448 \end{bmatrix}$,
the total nutritional value of a meal of 1 3-oz serving of chicken, 1 cup of potato salad, and 3 broccoli spears
33. (a) $\begin{bmatrix} 1.50 & 0.30 & 0.36 & 0.45 & 0.64 \\ 1.55 & 0.28 & 0.48 & 0.57 & 0.75 \\ 1.62 & 0.52 & 0.65 & 0.38 & 0.53 \\ 1.70 & 0.43 & 0.40 & 0.42 & 0.68 \end{bmatrix}$;
(b) $\begin{bmatrix} 65 & 48 & 93 & 57 \end{bmatrix}$;
(c) $\begin{bmatrix} 419.46 & 105.81 & 129.69 & 115.89 & 165.65 \end{bmatrix}$;
(d) the total cost, in dollars, for each item for the day's meals
35. (a) $\begin{bmatrix} 8 & 15 \\ 6 & 10 \\ 4 & 3 \end{bmatrix}$; **(b)** $\begin{bmatrix} 4 & 2.50 & 3 \end{bmatrix}$; **(c)** $\begin{bmatrix} 59 & 94 \end{bmatrix}$;
(d) the total cost, in dollars, of ingredients for each coffee shop
37. (a) $\begin{bmatrix} 7.50 & 4.80 & 6.25 \end{bmatrix}$; **(b)** $PS = \begin{bmatrix} 113.80 & 179.25 \end{bmatrix}$
39. $\begin{bmatrix} 2 & -3 \\ 1 & 5 \end{bmatrix} \begin{bmatrix} x \\ y \end{bmatrix} = \begin{bmatrix} 7 \\ -6 \end{bmatrix}$
41. $\begin{bmatrix} 1 & 1 & -2 \\ 3 & -1 & 1 \\ 2 & 5 & -3 \end{bmatrix} \begin{bmatrix} x \\ y \\ z \end{bmatrix} = \begin{bmatrix} 6 \\ 7 \\ 8 \end{bmatrix}$
43. $\begin{bmatrix} 3 & -2 & 4 \\ 2 & 1 & -5 \end{bmatrix} \begin{bmatrix} x \\ y \\ z \end{bmatrix} = \begin{bmatrix} 17 \\ 13 \end{bmatrix}$
45. $\begin{bmatrix} -4 & 1 & -1 & 2 \\ 1 & 2 & -1 & -1 \\ -1 & 1 & 4 & -3 \\ 2 & 3 & 5 & -7 \end{bmatrix} \begin{bmatrix} w \\ x \\ y \\ z \end{bmatrix} = \begin{bmatrix} 12 \\ 0 \\ 1 \\ 9 \end{bmatrix}$
47. (a) $\left(\frac{1}{2}, -\frac{25}{4}\right)$;
(b) $x = \frac{1}{2}$;
(c) minimum: $-\frac{25}{4}$;
(d)
48. (a) $\left(\frac{5}{4}, -\frac{49}{8}\right)$;
(b) $x = \frac{5}{4}$;
(c) minimum: $-\frac{49}{8}$;
(d)

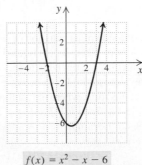

$f(x) = x^2 - x - 6$ $f(x) = 2x^2 - 5x - 3$

49. (a) $\left(-\frac{3}{2}, \frac{17}{4}\right)$;
(b) $x = -\frac{3}{2}$;
(c) maximum: $\frac{17}{4}$;
(d)

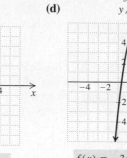

$f(x) = -x^2 - 3x + 2$

50. (a) $\left(\frac{2}{3}, \frac{16}{3}\right)$;
(b) $x = \frac{2}{3}$;
(c) maximum: $\frac{16}{3}$;
(d)

$f(x) = -3x^2 + 4x + 4$

51. $(\mathbf{A} + \mathbf{B})(\mathbf{A} - \mathbf{B}) = \begin{bmatrix} -2 & 1 \\ 2 & -1 \end{bmatrix}$; $\mathbf{A}^2 - \mathbf{B}^2 = \begin{bmatrix} 0 & 3 \\ 0 & -3 \end{bmatrix}$

53. $(\mathbf{A} + \mathbf{B})(\mathbf{A} - \mathbf{B}) = \begin{bmatrix} -2 & 1 \\ 2 & -1 \end{bmatrix}$
$= \mathbf{A}^2 + \mathbf{BA} - \mathbf{AB} - \mathbf{B}^2$

Mid-Chapter Mixed Review: Chapter 9

1. False **2.** True **3.** True **4.** False **5.** $(-3, 2)$
6. No solution **7.** $(1, -2)$ **8.** Infinitely many solutions;
$\left(x, \dfrac{x-1}{3}\right)$ or $(3y + 1, y)$ **9.** $\left(\frac{1}{3}, -\frac{1}{6}, \frac{4}{3}\right)$
10. Under 10 lb: 60 packages; 10 lb up to 15 lb: 70 packages; 15 lb
or more: 20 packages **11.** $(4, -3)$ **12.** $(-3, 2, -1)$

13. $\begin{bmatrix} 1 & 5 \\ 6 & 1 \end{bmatrix}$ **14.** $\begin{bmatrix} -5 & 7 \\ -4 & -7 \end{bmatrix}$ **15.** $\begin{bmatrix} -8 & 12 & 0 \\ 4 & -4 & 8 \\ -12 & 16 & 4 \end{bmatrix}$

16. $\begin{bmatrix} 0 & 16 \\ 13 & -1 \end{bmatrix}$ **17.** $\begin{bmatrix} -7 & 21 \\ -6 & 18 \end{bmatrix}$ **18.** $\begin{bmatrix} 24 & 26 \\ -12 & -13 \end{bmatrix}$

19. $\begin{bmatrix} 20 & 16 & -10 \\ -10 & -8 & 5 \end{bmatrix}$ **20.** Not defined

21. $\begin{bmatrix} 2 & -1 & 3 \\ 1 & 2 & -1 \\ 3 & -4 & 2 \end{bmatrix} \begin{bmatrix} x \\ y \\ z \end{bmatrix} = \begin{bmatrix} 7 \\ 3 \\ 5 \end{bmatrix}$

22. When a variable is not alone on one side of an equation or
when solving for a variable is difficult or produces an expression
containing fractions, the elimination method is preferable to the
substitution method. **23.** Add a nonzero multiple of one equation to a nonzero multiple of the other equation, where the multiples
are not opposites. **24.** See Example 4 in Section 9.3.
25. No; see Exercise 17 in Section 9.4, for example.

Exercise Set 9.5

1. Yes **3.** No **5.** $\begin{bmatrix} -3 & 2 \\ 5 & -3 \end{bmatrix}$ **7.** Does not exist

9. $\begin{bmatrix} \frac{3}{8} & -\frac{1}{4} & \frac{1}{8} \\ -\frac{1}{8} & \frac{3}{4} & -\frac{3}{8} \\ -\frac{1}{4} & \frac{1}{2} & \frac{1}{4} \end{bmatrix}$ **11.** Does not exist **13.** $\begin{bmatrix} 0.4 & -0.6 \\ 0.2 & -0.8 \end{bmatrix}$

15. $\begin{bmatrix} -1 & -1 & -6 \\ 1 & 0 & 2 \\ 0 & 1 & 3 \end{bmatrix}$ **17.** $\begin{bmatrix} 1 & 1 & 2 \\ 1 & 1 & 1 \\ 2 & 3 & 4 \end{bmatrix}$ **19.** Does not exist

21. $\begin{bmatrix} 1 & -2 & 3 & 8 \\ 0 & 1 & -3 & 1 \\ 0 & 0 & 1 & -2 \\ 0 & 0 & 0 & -1 \end{bmatrix}$

23. $\begin{bmatrix} 0.25 & 0.25 & 1.25 & -0.25 \\ 0.5 & 1.25 & 1.75 & -1 \\ -0.25 & -0.25 & -0.75 & 0.75 \\ 0.25 & 0.5 & 0.75 & -0.5 \end{bmatrix}$

25. $(-23, 83)$ **27.** $(-1, 5, 1)$ **29.** $(2, -2)$ **31.** $(0, 2)$
33. $(3, -3, -2)$ **35.** $(-1, 0, 1)$ **37.** $(1, -1, 0, 1)$
39. Wisconsin: 450 million lb; Massachusetts: 210 million lb
41. Topsoil: $29 per cubic yard; mulch: $28 per cubic yard; decora
tive rock: $38 per cubic yard **43.** -48 **44.** 194
45. $\dfrac{-1 \pm \sqrt{57}}{4}$ **46.** $-3, -2$ **47.** 4 **48.** 9
49. $(x + 2)(x - 1)(x - 4)$
50. $(x + 5)(x + 1)(x - 1)(x - 3)$

51. $\mathbf{A}^{-1}$ exists if and only if $x \neq 0$. $\mathbf{A}^{-1} = \begin{bmatrix} \frac{1}{x} \end{bmatrix}$

53. $\mathbf{A}^{-1}$ exists if and only if $xyz \neq 0$. $\mathbf{A}^{-1} = \begin{bmatrix} 0 & 0 & \frac{1}{z} \\ 0 & \frac{1}{y} & 0 \\ \frac{1}{x} & 0 & 0 \end{bmatrix}$

Exercise Set 9.6

1. -14 **3.** -2 **5.** -11 **7.** $x^3 - 4x$ **9.** $M_{11} = 6$,
$M_{32} = -9$, $M_{22} = -29$ **11.** $A_{11} = 6$, $A_{32} = 9$, $A_{22} = -29$
13. -10 **15.** -10 **17.** -10 **19.** $M_{41} = -14$, $M_{33} = 20$
21. $A_{24} = 15$, $A_{43} = 30$ **23.** 110 **25.** -109
27. $-x^4 + x^2 - 5x$ **29.** $\left(-\frac{25}{2}, -\frac{11}{2}\right)$ **31.** $(3, 1)$
33. $\left(\frac{1}{2}, -\frac{1}{3}\right)$ **35.** $(1, 1)$ **37.** $\left(\frac{3}{2}, \frac{13}{14}, \frac{33}{14}\right)$ **39.** $(3, -2, 1)$
41. $(1, 3, -2)$ **43.** $\left(\frac{1}{2}, \frac{2}{3}, -\frac{5}{6}\right)$ **45.** $f^{-1}(x) = \dfrac{x - 2}{3}$
46. Not one-to-one **47.** Not one-to-one
48. $f^{-1}(x) = (x - 1)^3$ **49.** $5 - 3i$ **50.** $6 - 2i$
51. $10 - 10i$ **52.** $\frac{9}{25} + \frac{13}{25}i$ **53.** $3, -2$ **55.** 4

57. Answers may vary. $\begin{vmatrix} a & b \\ -b & a \end{vmatrix}$

59. Answers may vary. $\begin{vmatrix} 2\pi r & 2\pi r \\ -h & r \end{vmatrix}$

Exercise Set 9.7

1. (f) **3.** (h) **5.** (g) **7.** (b)
9.

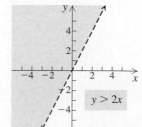

$y > 2x$

11.

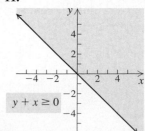

$y + x \geq 0$

3.

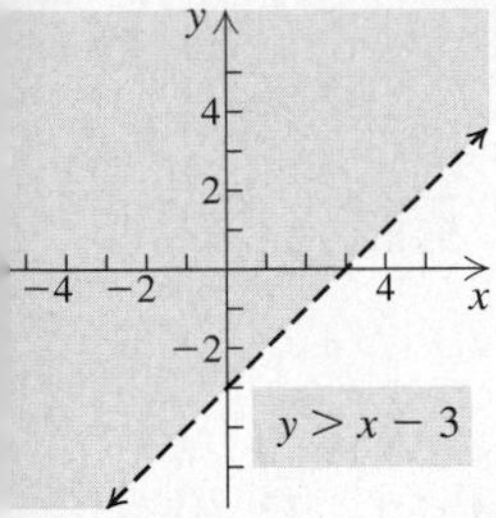

15.

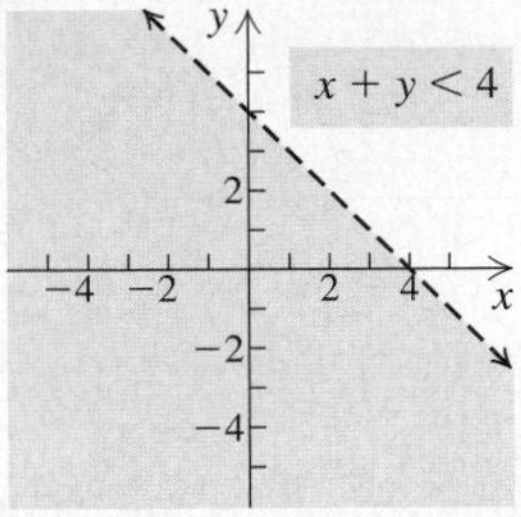

7.

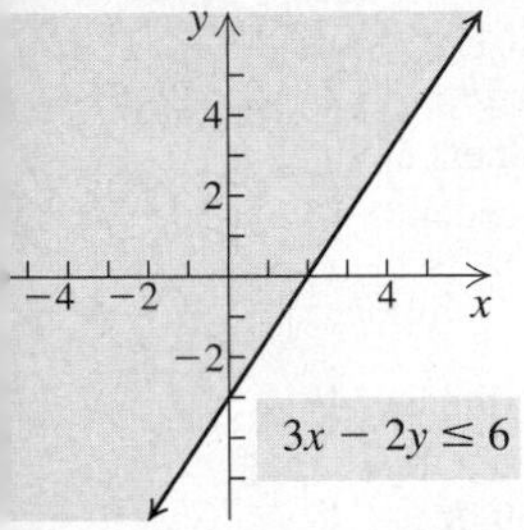

19.

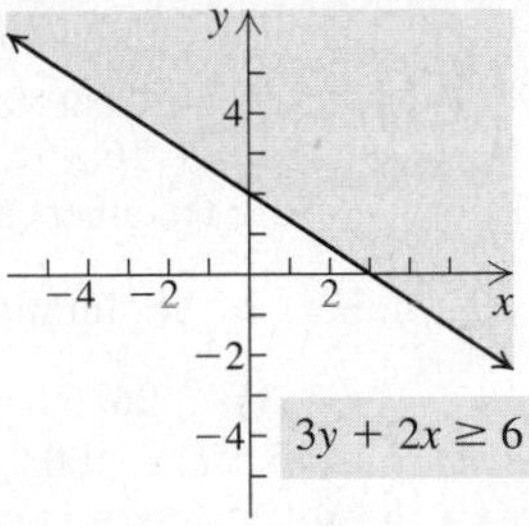

1.

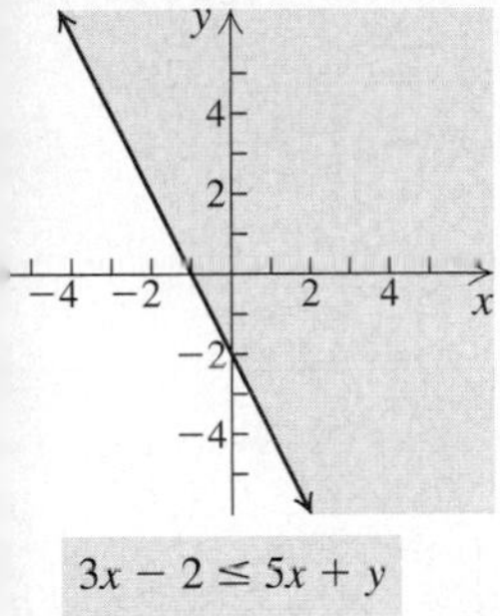

23.

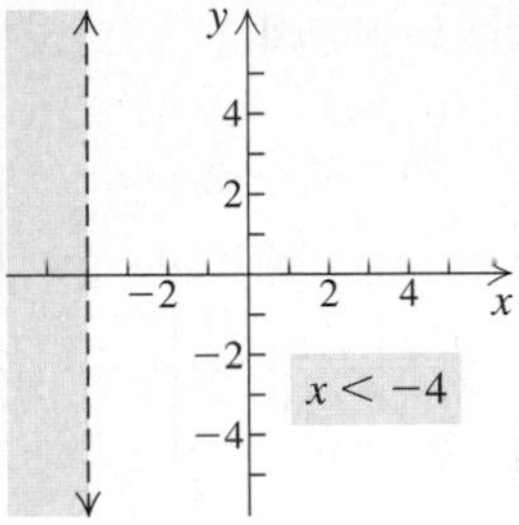

5.

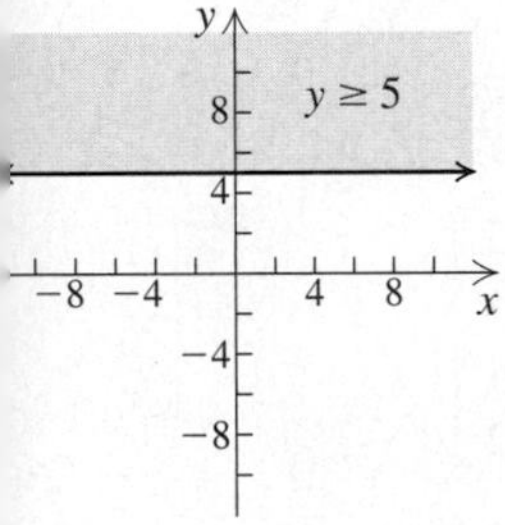

27.

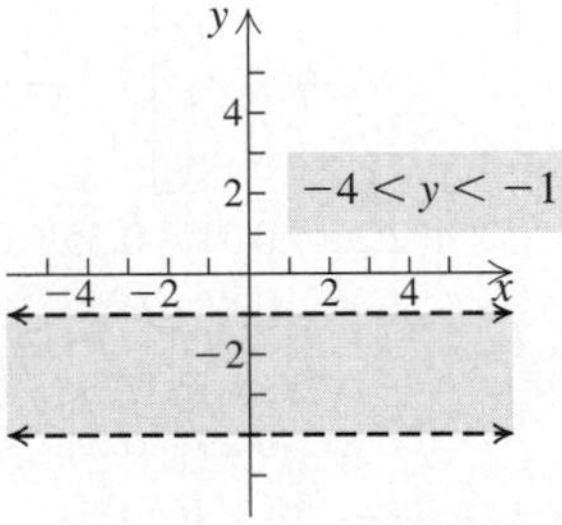

31. (f) **33.** (a) **35.** (b)

9.

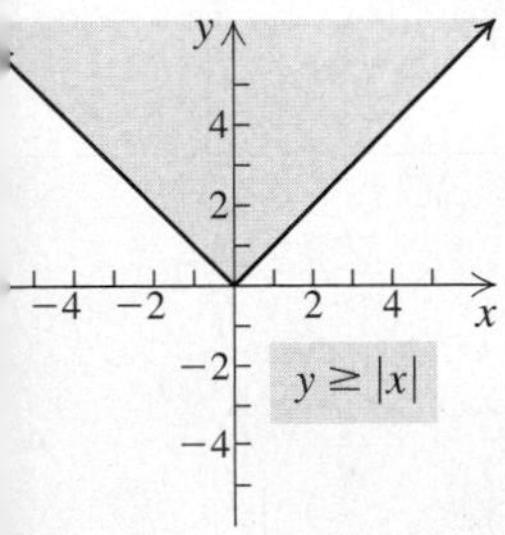

7. $y \le -x + 4$, $y \le 3x$ **39.** $x < 2$, $y > -1$ **41.** $y \le -x + 3$, $y \le x + 1$, $x \ge 0$, $y \ge 0$

43.

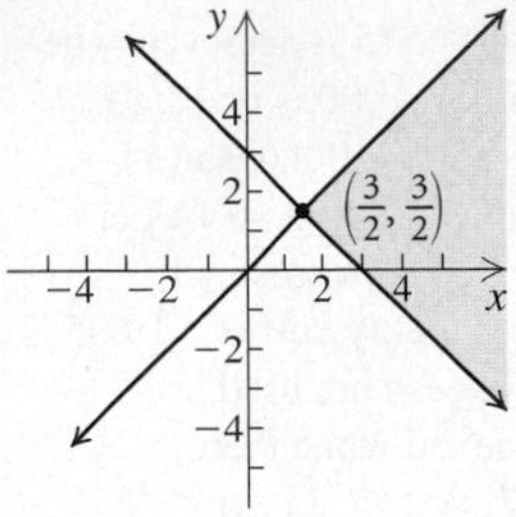

45.

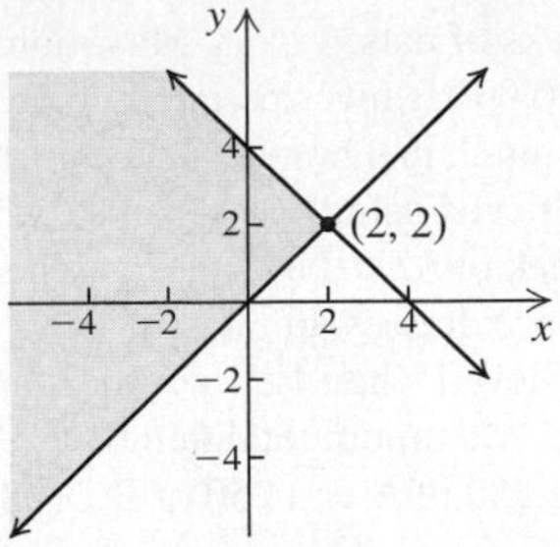

47.

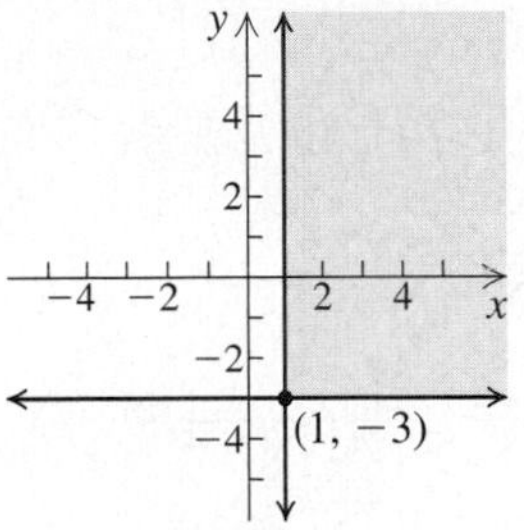

49.

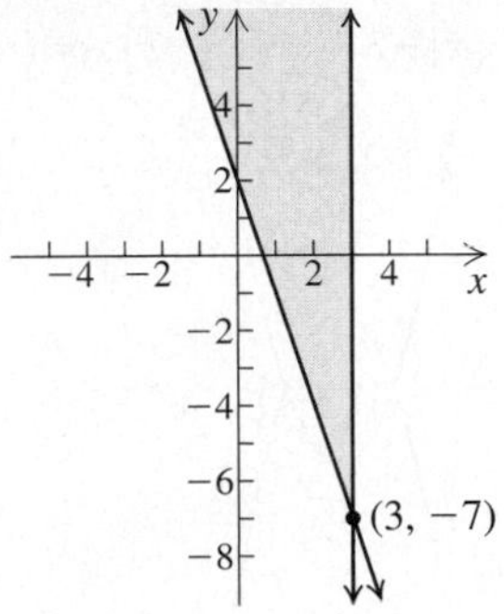

51.

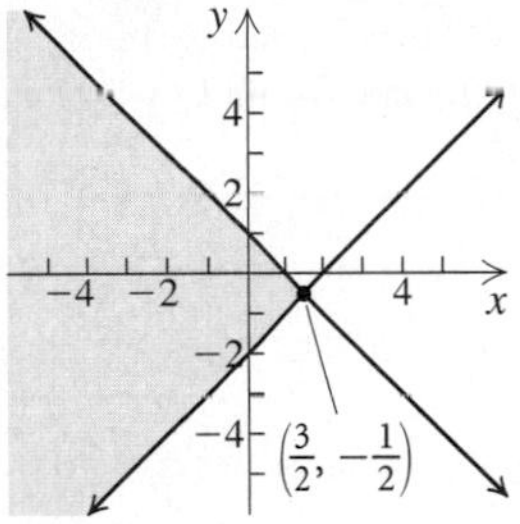

53.

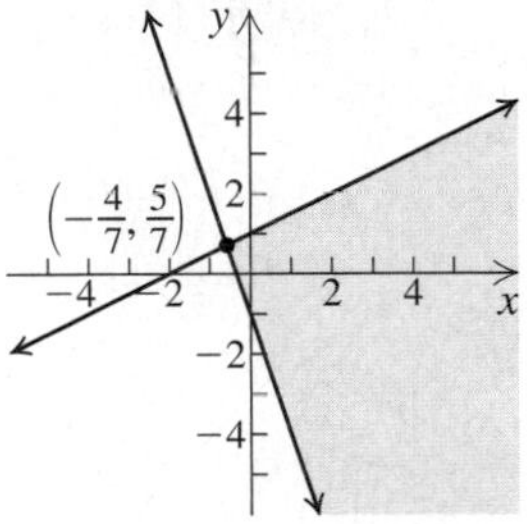

55.

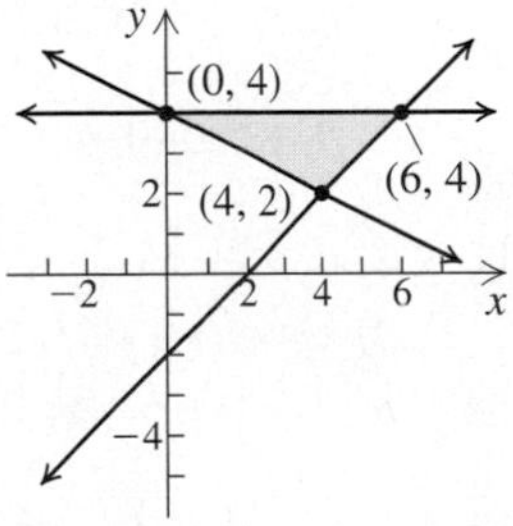

57.

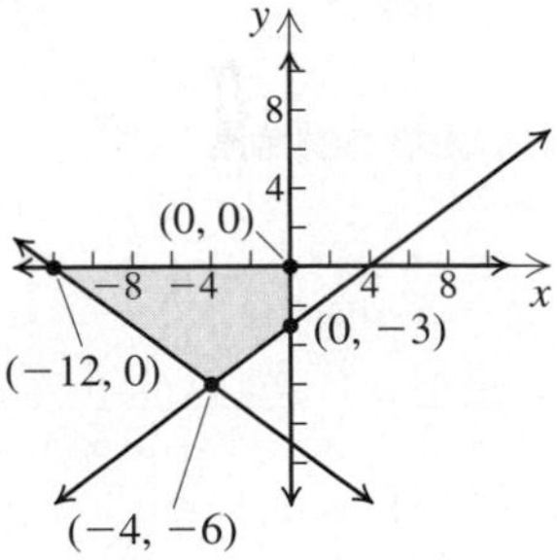

59.

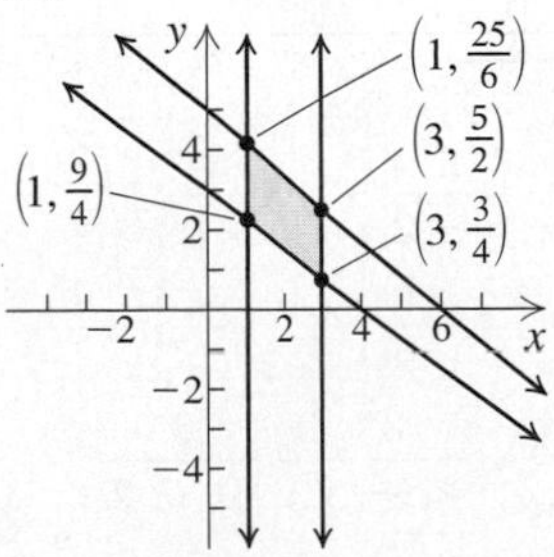

61. Maximum: 179 when $x = 7$ and $y = 0$; minimum: 48 when $x = 0$ and $y = 4$ **63.** Maximum: 216 when $x = 0$ and $y = 6$; minimum: 0 when $x = 0$ and $y = 0$ **65.** Maximum number of miles is 480 when the truck uses 9 gal and the moped uses 3 gal. **67.** Maximum profit of $54,800 is achieved when 80 acres of corn and 160 acres of soybeans are planted. **69.** Minimum cost

of $51\frac{9}{13}$ is achieved by using $1\frac{11}{13}$ sacks of soybean meal and $1\frac{11}{13}$ sacks of oats.   **71.** Maximum income of $1575 is achieved when $10,000 is invested in corporate bonds and $30,000 is invested in municipal bonds.   **73.** Minimum cost of $460 thousand is achieved using 30 P_1's and 10 P_2's.   **75.** Maximum profit per week of $2210 is achieved when 5 silk organza dresses and 2 lace dresses are made.   **77.** Minimum weekly cost of $19.05 is achieved when 1.5 lb of meat and 3 lb of cheese are used.
79. Maximum total number of 800 is achieved when there are 550 of A and 250 of B.   **81.** $\{x \mid -7 \le x < 2\}$, or $[-7, 2)$ **82.** $\{x \mid x \le 1 \ or \ x \ge 5\}$, or $(-\infty, 1] \cup [5, \infty)$
83. $\{x \mid -1 \le x \le 3\}$, or $[-1, 3]$
84. $\{x \mid -3 < x < -2\}$, or $(-3, -2)$
85.

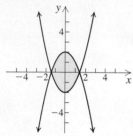

87.

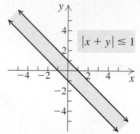

$|x + y| \le 1$

89.

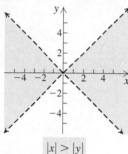

$|x| > |y|$

91. Maximum income of $19,000 is achieved by making 95 chairs and 0 sofas.

Exercise Set 9.8

1. $\dfrac{2}{x - 3} - \dfrac{1}{x + 2}$ **3.** $\dfrac{5}{2x - 1} - \dfrac{4}{3x - 1}$

5. $\dfrac{2}{x + 2} - \dfrac{3}{x - 2} + \dfrac{4}{x + 1}$

7. $-\dfrac{3}{(x + 2)^2} - \dfrac{1}{x + 2} + \dfrac{1}{x - 1}$ **9.** $\dfrac{3}{x - 1} - \dfrac{4}{2x - 1}$

11. $x - 2 + \dfrac{\frac{17}{16}}{x + 1} - \dfrac{\frac{11}{4}}{(x + 1)^2} - \dfrac{\frac{17}{16}}{x - 3}$

13. $\dfrac{3x + 5}{x^2 + 2} - \dfrac{4}{x - 1}$ **15.** $\dfrac{3}{2x - 1} - \dfrac{2}{x + 2} + \dfrac{10}{(x + 2)^2}$

17. $3x + 1 + \dfrac{2}{2x - 1} + \dfrac{3}{x + 1}$ **19.** $-\dfrac{1}{x - 3} + \dfrac{3x}{x^2 + 2x - 5}$

21. $\dfrac{5}{3x + 5} - \dfrac{3}{x + 1} + \dfrac{4}{(x + 1)^2}$ **23.** $\dfrac{8}{4x - 5} + \dfrac{3}{3x + 2}$

25. $\dfrac{2x - 5}{3x^2 + 1} - \dfrac{2}{x - 2}$ **27.** $-1, \pm 3i$ **28.** $3, \pm i$

29. $-2, \dfrac{1 \pm \sqrt{5}}{2}$ **30.** $-2, 3, \pm i$ **31.** $-3, -1 \pm \sqrt{2}$

33. $-\dfrac{\frac{1}{2a^2}x}{x^2 + a^2} + \dfrac{\frac{1}{4a^2}}{x - a} + \dfrac{\frac{1}{4a^2}}{x + a}$

35. $-\dfrac{3}{25(\ln x + 2)} + \dfrac{3}{25(\ln x - 3)} + \dfrac{7}{5(\ln x - 3)^2}$

Review Exercises: Chapter 9

1. True   **2.** False   **3.** True   **4.** False   **5.** (a)   **6.** (e)
7. (h)   **8.** (d)   **9.** (b)   **10.** (g)   **11.** (c)   **12.** (f)
13. $(-2, -2)$ **14.** $(-5, 4)$ **15.** No solution
16. Infinitely many solutions; $(-y - 2, y)$, or $(x, -x - 2)$
17. $(3, -1, -2)$ **18.** No solution   **19.** $(-5, 13, 8, 2)$
20. Consistent: 13, 14, 16, 17, 19; the others are inconsistent.
21. Dependent: 16; the others are independent.   **22.** $(1, 2)$

23. $(-3, 4, -2)$ **24.** Infinitely many solutions; $\left(\dfrac{z}{2}, -\dfrac{z}{2}, z\right)$

25. $(-4, 1, -2, 3)$ **26.** Nickels: 31; dimes: 44
27. 3%: $1600; 3.5%: $3400   **28.** 1 bagel, $\frac{1}{2}$ serving of cream cheese, 2 bananas   **29.** 75, 69, 82
30. (a) $f(x) = 1.125x^2 - 5.25x + 145$; (b) 154 million persons employed

31. $\begin{bmatrix} 0 & -1 & 6 \\ 3 & 1 & -2 \\ -2 & 1 & -2 \end{bmatrix}$ **32.** $\begin{bmatrix} -3 & 3 & 0 \\ -6 & -9 & 6 \\ 6 & 0 & -3 \end{bmatrix}$

33. $\begin{bmatrix} -1 & 1 & 0 \\ -2 & -3 & 2 \\ 2 & 0 & -1 \end{bmatrix}$ **34.** $\begin{bmatrix} -2 & 2 & 6 \\ 1 & -8 & 18 \\ 2 & 1 & -15 \end{bmatrix}$

35. Not defined   **36.** $\begin{bmatrix} 2 & -1 & -6 \\ 1 & 5 & -2 \\ -2 & -1 & 4 \end{bmatrix}$

37. $\begin{bmatrix} -13 & 1 & 6 \\ -3 & -7 & 4 \\ 8 & 3 & -5 \end{bmatrix}$ **38.** $\begin{bmatrix} -2 & -1 & 18 \\ 5 & -3 & -2 \\ -2 & 3 & -8 \end{bmatrix}$

39. (a) $\begin{bmatrix} 2.25 & 0.38 & 0.55 & 0.33 & 0.85 \\ 3.09 & 0.42 & 0.46 & 0.48 & 0.51 \\ 2.40 & 0.31 & 0.59 & 0.36 & 0.64 \\ 1.80 & 0.29 & 0.34 & 0.55 & 0.52 \end{bmatrix}$;

(b) $\begin{bmatrix} 41 & 18 & 39 & 36 \end{bmatrix}$;
(c) $\begin{bmatrix} 306.27 & 45.67 & 66.08 & 56.01 & 87.71 \end{bmatrix}$;
(d) the total cost, in dollars, for each item for the day's meals

40. $\begin{bmatrix} -\frac{1}{2} & 0 \\ \frac{1}{6} & \frac{1}{3} \end{bmatrix}$ **41.** $\begin{bmatrix} 0 & 0 & \frac{1}{4} \\ 0 & -\frac{1}{2} & 0 \\ \frac{1}{3} & 0 & 0 \end{bmatrix}$

42. $\begin{bmatrix} 1 & 0 & 0 & 0 \\ 0 & \frac{1}{9} & \frac{5}{18} & 0 \\ 0 & -\frac{1}{9} & \frac{2}{9} & 0 \\ 0 & 0 & 0 & 1 \end{bmatrix}$

43. $\begin{bmatrix} 3 & -2 & 4 \\ 1 & 5 & -3 \\ 2 & -3 & 7 \end{bmatrix} \begin{bmatrix} x \\ y \\ z \end{bmatrix} = \begin{bmatrix} 13 \\ 7 \\ -8 \end{bmatrix}$

44. $(-8, 7)$ **45.** $(1, -2, 5)$ **46.** $(2, -1, 1, -3)$ **47.** 10
48. -18 **49.** -6 **50.** -1 **51.** $(3, -2)$ **52.** $(-1, 5)$
53. $\left(\frac{3}{2}, \frac{13}{14}, \frac{33}{14}\right)$ **54.** $(2, -1, 3)$

55.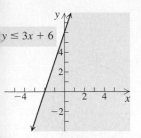

56.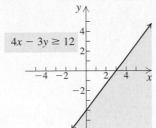

57.

58. Minimum: 52 when $x = 2$ and $y = 4$; maximum: 92 when $x = 2$ and $y = 8$ **59.** Maximum score of 96 is achieved when 4 group A questions and 8 group B questions are answered.
60. $\dfrac{5}{x + 1} - \dfrac{5}{x + 2} - \dfrac{5}{(x + 2)^2}$ **61.** $\dfrac{2}{2x - 3} - \dfrac{5}{x + 4}$
62. C   **63.** A   **64.** B   **65.** 4%: \$10,000; 5%: \$12,000; $5\frac{1}{2}$%: \$8,000   **66.** $\left(\frac{5}{18}, \frac{1}{7}\right)$ **67.** $\left(1, \frac{1}{2}, \frac{1}{3}\right)$
68.

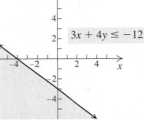

$|x| - |y| \le 1$

69.

$|xy| > 1$

70. The solution of the equation $2x + 5 = 3x - 7$ is the first coordinate of the point of intersection of the graphs of $y_1 = 2x + 5$ and $y_2 = 3x - 7$. The solution of the system of equations $y = 2x + 5$, $y = 3x - 7$ is the ordered pair that is the point of intersection of the two lines.   **71.** In general, $(\mathbf{AB})^2 \ne \mathbf{A}^2\mathbf{B}^2$. $(\mathbf{AB})^2 = \mathbf{ABAB}$ and $\mathbf{A}^2\mathbf{B}^2 = \mathbf{AABB}$. Since matrix multiplication is not commutative, $\mathbf{BA} \ne \mathbf{AB}$, so $(\mathbf{AB})^2 \ne \mathbf{A}^2\mathbf{B}^2$.   **72.** If $\begin{vmatrix} a_1 & b_1 \\ a_2 & b_2 \end{vmatrix} = 0$, then $a_1 = ka_2$ and $b_1 = kb_2$ for some number k. This means that the equations $a_1x + b_1y = c_1$ and $a_2x + b_2y = c_2$ are dependent if $c_1 = kc_2$, or the system is inconsistent if $c_1 \ne kc_2$.
73. If $a_1x + b_1y = c_1$ and $a_2x + b_2y = c_2$ are parallel lines, then $a_1 = ka_2$, $b_1 = kb_2$, and $c_1 \ne kc_2$, for some number k. Then $\begin{vmatrix} a_1 & b_1 \\ a_2 & b_2 \end{vmatrix} = 0$, $\begin{vmatrix} c_1 & b_1 \\ c_2 & b_2 \end{vmatrix} \ne 0$, and $\begin{vmatrix} a_1 & c_1 \\ a_2 & c_2 \end{vmatrix} \ne 0$.
74. The graph of a linear equation consists of a set of points on a line. The graph of a linear inequality consists of the set of points in a half-plane and might also include the points on the line that is the boundary of the half-plane.   **75.** The denominator of the second fraction, $x^2 - 5x + 6$, can be factored into linear factors with

rational coefficients: $(x - 3)(x - 2)$. Thus the given expression is not a partial fraction decomposition.

Test: Chapter 9

1. [9.1] $(-3, 5)$; consistent, independent   **2.** [9.1] Infinitely many solutions; $(x, 2x - 3)$ or $\left(\dfrac{y + 3}{2}, y\right)$; consistent, dependent
3. [9.1] No solution; inconsistent, independent
4. [9.1] $(1, -2)$; consistent, independent   **5.** [9.2] $(-1, 3, 2)$
6. [9.1] Student: 462 tickets; nonstudent: 158 tickets
7. [9.2] Hui: 120 orders; Ashlyn: 104 orders; Sheriann: 128 orders
8. [9.4] $\begin{bmatrix} -2 & -3 \\ -3 & 4 \end{bmatrix}$ **9.** [9.4] Not defined
10. [9.4] $\begin{bmatrix} -7 & -13 \\ 5 & -1 \end{bmatrix}$ **11.** [9.4] Not defined
12. [9.4] $\begin{bmatrix} 2 & -2 & 6 \\ -4 & 10 & 4 \end{bmatrix}$ **13.** [9.5] $\begin{bmatrix} 0 & -1 \\ -\frac{1}{4} & -\frac{3}{4} \end{bmatrix}$
14. [9.4] **(a)** $\begin{bmatrix} 1.55 & 1.00 & 0.99 \\ 1.70 & 0.95 & 1.01 \\ 1.65 & 0.99 & 0.96 \end{bmatrix}$; **(b)** $\begin{bmatrix} 26 & 18 & 23 \end{bmatrix}$;
(c) $\begin{bmatrix} 108.85 & 65.87 & 66.00 \end{bmatrix}$; **(d)** the total cost, in dollars, for each type of menu item served on the given day
15. [9.4] $\begin{bmatrix} 3 & -4 & 2 \\ 2 & 3 & 1 \\ 1 & -5 & -3 \end{bmatrix}\begin{bmatrix} x \\ y \\ z \end{bmatrix} = \begin{bmatrix} -8 \\ 7 \\ 3 \end{bmatrix}$ **16.** [9.5] $(-2, 1, 1)$
17. [9.6] 61   **18.** [9.6] -33 **19.** [9.6] $\left(-\frac{1}{2}, \frac{3}{4}\right)$
20. [9.7]

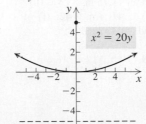

$3x + 4y \le -12$

21. [9.7] Maximum: 15 when $x = 3$ and $y = 3$; minimum: 2 when $x = 1$ and $y = 0$ **22.** [9.7] Maximum profit of \$750 occurs when 25 pound cakes and 75 carrot cakes are prepared.
23. [9.8] $-\dfrac{2}{x - 1} + \dfrac{5}{x + 3}$ **24.** [9.7] D   **25.** [9.2] $A = 1$, $B = -3$, $C = 2$

Chapter 10

Exercise Set 10.1

1. (f)   **3.** (b)   **5.** (d)
7. $V: (0, 0)$; $F: (0, 5)$; $D: y = -5$
9. $V: (0, 0)$; $F: \left(-\frac{3}{2}, 0\right)$; $D: x = \frac{3}{2}$

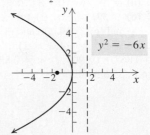

$x^2 = 20y$

$y^2 = -6x$

11. $V: (0,0)$; $F: (0,1)$;
$D: y = -1$

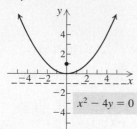

$x^2 - 4y = 0$

13. $V: (0,0)$; $F: \left(\frac{1}{8}, 0\right)$;
$D: x = -\frac{1}{8}$

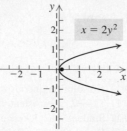

$x = 2y^2$

15. $y^2 = -12x$ **17.** $y^2 = 28x$ **19.** $x^2 = -4\pi y$
21. $(y - 2)^2 = 14\left(x + \frac{1}{2}\right)$
23. $V: (-2,1)$; $F: \left(-2, -\frac{1}{2}\right)$;
$D: y = \frac{5}{2}$

25. $V: (-1,-3)$;
$F: \left(-1, -\frac{7}{2}\right)$; $D: y = -\frac{5}{2}$

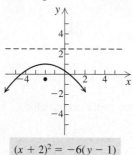

$(x + 2)^2 = -6(y - 1)$

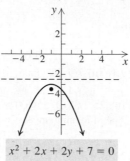

$x^2 + 2x + 2y + 7 = 0$

27. $V: (0,-2)$; $F: \left(0, -1\frac{3}{4}\right)$;
$D: y = -2\frac{1}{4}$

29. $V: (-2,-1)$;
$F: \left(-2, -\frac{3}{4}\right)$; $D: y = -1\frac{1}{4}$

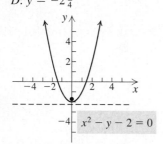

$x^2 - y - 2 = 0$

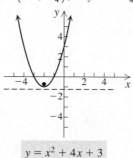

$y = x^2 + 4x + 3$

31. $V: \left(5\frac{3}{4}, \frac{1}{2}\right)$; $F: \left(6, \frac{1}{2}\right)$; $D: x = 5\frac{1}{2}$

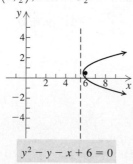

$y^2 - y - x + 6 = 0$

33. (a) $y^2 = 16x$; (b) $3\frac{33}{64}$ ft   **35.** About 11.75 in.   **37.** (h)
38. (d)   **39.** (a), (b), (f), (g)   **40.** (b)   **41.** (b)   **42.** (f)
43. (a) and (g)   **44.** (a) and (h); (g) and (h); (b) and (c)
45. $(x + 1)^2 = -4(y - 2)$ **47.** $V: (0.867, 0.348)$;
$F: (0.867, -0.190)$; $D: y = 0.887$ **49.** 10 ft, 11.6 ft, 16.4 ft,
24.4 ft, 35.6 ft, 50 ft

Exercise Set 10.2

1. (b)   **3.** (d)   **5.** (a)
7. $(7, -2)$; 8

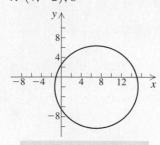

$x^2 + y^2 - 14x + 4y = 11$

9. $(-3, 1)$; 4

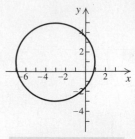

$x^2 + y^2 + 6x - 2y = 6$

11. $(-2, 3)$; 5

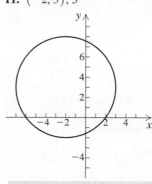

$x^2 + y^2 + 4x - 6y - 12 = 0$

13. $(3, 4)$; 3

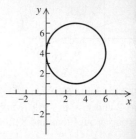

$x^2 + y^2 - 6x - 8y + 16 = 0$

15. $(-3, 5)$; $\sqrt{34}$

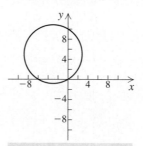

$x^2 + y^2 + 6x - 10y = 0$

17. $\left(\frac{9}{2}, -2\right)$; $\frac{5\sqrt{5}}{2}$

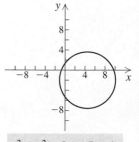

$x^2 + y^2 - 9x = 7 - 4y$

19. (c)   **21.** (d)
23. $V: (2, 0), (-2, 0)$;
$F: (\sqrt{3}, 0), (-\sqrt{3}, 0)$

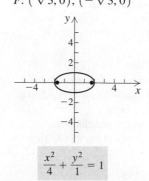

$\frac{x^2}{4} + \frac{y^2}{1} = 1$

25. $V: (0, 4), (0, -4)$;
$F: (0, \sqrt{7}), (0, -\sqrt{7})$

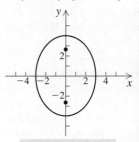

$16x^2 + 9y^2 = 144$

7. V: $(-\sqrt{3}, 0)$, $(\sqrt{3}, 0)$;
: $(-1, 0)$, $(1, 0)$

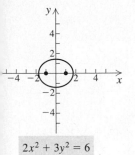

$$2x^2 + 3y^2 = 6$$

29. V: $\left(-\dfrac{1}{2}, 0\right)$, $\left(\dfrac{1}{2}, 0\right)$;

F: $\left(-\dfrac{\sqrt{5}}{6}, 0\right)$, $\left(\dfrac{\sqrt{5}}{6}, 0\right)$

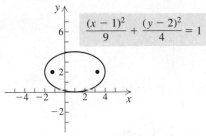

$$4x^2 + 9y^2 = 1$$

1. $\dfrac{x^2}{49} + \dfrac{y^2}{40} = 1$ **33.** $\dfrac{x^2}{25} + \dfrac{y^2}{64} = 1$ **35.** $\dfrac{x^2}{9} + \dfrac{y^2}{5} = 1$

7. C: $(1, 2)$; V: $(4, 2)$, $(-2, 2)$; F: $(1 + \sqrt{5}, 2)$, $(1 - \sqrt{5}, 2)$

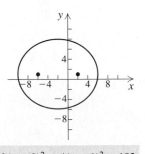

$$\dfrac{(x - 1)^2}{9} + \dfrac{(y - 2)^2}{4} = 1$$

9. C: $(-3, 5)$; V: $(-3, 11)$,
$-3, -1)$; F: $(-3, 5 + \sqrt{11})$,
$-3, 5 - \sqrt{11})$

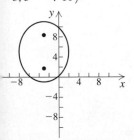

$$\dfrac{(x + 3)^2}{25} + \dfrac{(y - 5)^2}{36} = 1$$

41. C: $(-2, 1)$; V: $(-10, 1)$,
$(6, 1)$; F: $(-6, 1)$, $(2, 1)$

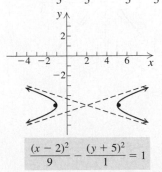

$$3(x + 2)^2 + 4(y - 1)^2 = 192$$

3. C: $(2, -1)$; V: $(-1, -1)$, $(5, -1)$; F: $(2 + \sqrt{5}, -1)$,
$2 - \sqrt{5}, -1)$

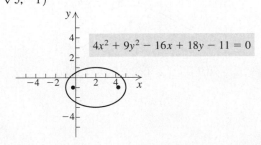

$$4x^2 + 9y^2 - 16x + 18y - 11 = 0$$

45. C: $(1, 1)$; V: $(1, 3)$, $(1, -1)$; F: $(1, 1 + \sqrt{3})$,
$(1, 1 - \sqrt{3})$

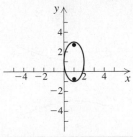

$$4x^2 + y^2 - 8x - 2y + 1 = 0$$

47. Example 2; $\dfrac{3}{5} < \dfrac{\sqrt{12}}{4}$ **49.** $\dfrac{x^2}{15} + \dfrac{y^2}{16} = 1$

51. $\dfrac{x^2}{524^2} + \dfrac{y^2}{449^2} = 1$, or $\dfrac{x^2}{274{,}576} + \dfrac{y^2}{201{,}601} = 1$

53. 33.5 ft **55.** (a) 0.7 ft; (b) 4 ft **57.** Midpoint **58.** Zero
59. y-intercept **60.** Two different real-number solutions
61. Remainder **62.** Ellipse **63.** Parabola **64.** Circle

65. $\dfrac{(x - 3)^2}{4} + \dfrac{(y - 1)^2}{25} = 1$ **67.** $\dfrac{x^2}{9} + \dfrac{y^2}{484/5} - 1$

69. C: $(2.003, -1.005)$; V: $(-1.017, -1.005)$, $(5.023, -1.005)$
71. About 9.1 ft

Exercise Set 10.3

1. (b) **3.** (c) **5.** (a) **7.** $\dfrac{y^2}{9} - \dfrac{x^2}{16} = 1$ **9.** $\dfrac{x^2}{4} - \dfrac{y^2}{9} = 1$

11. C: $(0, 0)$; V: $(2, 0)$, $(-2, 0)$; F: $(2\sqrt{2}, 0)$, $(-2\sqrt{2}, 0)$;
A: $y = x$, $y = -x$

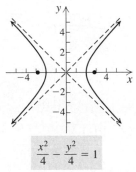

$$\dfrac{x^2}{4} - \dfrac{y^2}{4} = 1$$

13. C: $(2, -5)$; V: $(-1, -5)$, $(5, -5)$; F: $(2 - \sqrt{10}, -5)$,
$(2 + \sqrt{10}, -5)$; A: $y = -\dfrac{x}{3} - \dfrac{13}{3}$, $y = \dfrac{x}{3} - \dfrac{17}{3}$

$$\dfrac{(x - 2)^2}{9} - \dfrac{(y + 5)^2}{1} = 1$$

15. $C: (-1, -3); V: (-1, -1), (-1, -5); F: (-1, -3 + 2\sqrt{5}),$
$(-1, -3 - 2\sqrt{5}); A: y = \frac{1}{2}x - \frac{5}{2}, y = -\frac{1}{2}x - \frac{7}{2}$

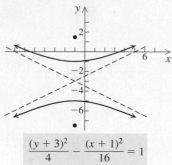

$$\frac{(y+3)^2}{4} - \frac{(x+1)^2}{16} = 1$$

17. $C: (0, 0); V: (-2, 0), (2, 0); F: (-\sqrt{5}, 0), (\sqrt{5}, 0);$
$A: y = -\frac{1}{2}x, y = \frac{1}{2}x$

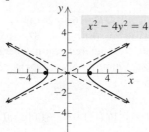

$x^2 - 4y^2 = 4$

19. $C: (0, 0); V: (0, -3), (0, 3); F: (0, -3\sqrt{10}), (0, 3\sqrt{10});$
$A: y = \frac{1}{3}x, y = -\frac{1}{3}x$

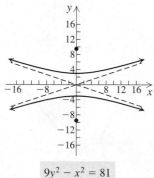

$9y^2 - x^2 = 81$

21. $C: (0, 0); V: (-\sqrt{2}, 0), (\sqrt{2}, 0); F: (-2, 0), (2, 0);$
$A: y = x, y = -x$

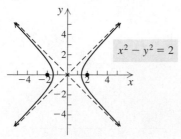

$x^2 - y^2 = 2$

23. $C: (0, 0); V: \left(0, -\frac{1}{2}\right), \left(0, \frac{1}{2}\right); F: \left(0, -\frac{\sqrt{2}}{2}\right), \left(0, \frac{\sqrt{2}}{2}\right);$
$A: y = x, y = -x$

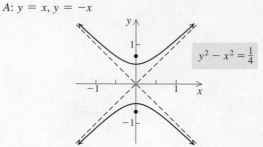

$y^2 - x^2 = \frac{1}{4}$

25. $C: (1, -2); V: (0, -2), (2, -2); F: (1 - \sqrt{2}, -2),$
$(1 + \sqrt{2}, -2); A: y = -x - 1, y = x - 3$

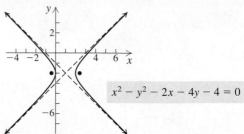

$x^2 - y^2 - 2x - 4y - 4 = 0$

27. $C: \left(\frac{1}{3}, 3\right); V: \left(-\frac{2}{3}, 3\right), \left(\frac{4}{3}, 3\right); F: \left(\frac{1}{3} - \sqrt{37}, 3\right),$
$\left(\frac{1}{3} + \sqrt{37}, 3\right); A: y = 6x + 1, y = -6x + 5$

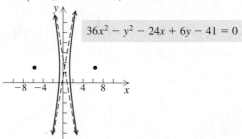

$36x^2 - y^2 - 24x + 6y - 41 = 0$

29. $C: (3, 1); V: (3, 3), (3, -1); F: (3, 1 + \sqrt{13}),$
$(3, 1 - \sqrt{13}); A: y = \frac{2}{3}x - 1, y = -\frac{2}{3}x + 3$

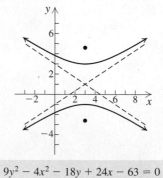

$9y^2 - 4x^2 - 18y + 24x - 63 = 0$

1. C: $(1, -2)$; V: $(2, -2)$, $(0, -2)$; F: $(1 + \sqrt{2}, -2)$, $(1 - \sqrt{2}, -2)$; A: $y = x - 3$, $y = -x - 1$

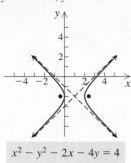

$x^2 - y^2 - 2x - 4y = 4$

3. C: $(-3, 4)$; V: $(-3, 10)$, $(-3, -2)$; F: $(-3, 4 + 6\sqrt{2})$, $(-3, 4 - 6\sqrt{2})$; A: $y = x + 7$, $y = -x + 1$

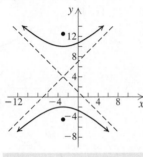

$y^2 - x^2 - 6x - 8y - 29 = 0$

35. Example 3; $\dfrac{\sqrt{5}}{1} > \dfrac{5}{4}$ **37.** $\dfrac{x^2}{9} - \dfrac{(y - 7)^2}{16} = 1$

39. $\dfrac{y^2}{25} - \dfrac{x^2}{11} = 1$ **41. (a)** Yes; **(b)** $f^{-1}(x) = \dfrac{x + 3}{2}$

42. (a) Yes; **(b)** $f^{-1}(x) = \sqrt[3]{x - 2}$ **43. (a)** Yes;

(b) $f^{-1}(x) = \dfrac{5}{x} + 1$, or $\dfrac{5 + x}{x}$ **44. (a)** Yes;

(b) $f^{-1}(x) = x^2 - 4$, $x \geq 0$ **45.** $(6, -1)$ **46.** $(1, -1)$

47. $(2, -1)$ **48.** $(-3, 4)$ **49.** $\dfrac{(y + 5)^2}{9} - (x - 3)^2 = 1$

51. C: $(-1.460, -0.957)$; V: $(-2.360, -0.957)$, $(-0.560, -0.957)$; A: $y = -1.20x - 2.70$, $y = 1.20x + 0.79$

53. $\dfrac{x^2}{345.96} - \dfrac{y^2}{22{,}154.04} = 1$

Visualizing the Graph

1. B **2.** J **3.** F **4.** I **5.** H **6.** G **7.** E **8.** D
9. C **10.** A

Exercise Set 10.4

1. (e) **3.** (c) **5.** (b) **7.** $(-4, -3)$, $(3, 4)$
9. $(0, 2)$, $(3, 0)$ **11.** $(-5, 0)$, $(4, 3)$, $(4, -3)$
13. $(3, 0)$, $(-3, 0)$ **15.** $(0, -3)$, $(4, 5)$ **17.** $(-2, 1)$
19. $(3, 4)$, $(-3, -4)$, $(4, 3)$, $(-4, -3)$
21. $\left(\dfrac{6\sqrt{21}}{7}, \dfrac{4\sqrt{35}}{7}i\right)$, $\left(\dfrac{6\sqrt{21}}{7}, -\dfrac{4\sqrt{35}}{7}i\right)$, $\left(-\dfrac{6\sqrt{21}}{7}, \dfrac{4\sqrt{35}}{7}i\right)$, $\left(-\dfrac{6\sqrt{21}}{7}, -\dfrac{4\sqrt{35}}{7}i\right)$

23. $(3, 2)$, $\left(4, \dfrac{3}{2}\right)$ **25.** $\left(\dfrac{5 + \sqrt{70}}{3}, \dfrac{-1 + \sqrt{70}}{3}\right)$, $\left(\dfrac{5 - \sqrt{70}}{3}, \dfrac{-1 - \sqrt{70}}{3}\right)$ **27.** $(\sqrt{2}, \sqrt{14})$, $(-\sqrt{2}, \sqrt{14})$, $(\sqrt{2}, -\sqrt{14})$, $(-\sqrt{2}, -\sqrt{14})$ **29.** $(1, 2)$, $(-1, -2)$, $(2, 1)$, $(-2, -1)$ **31.** $\left(\dfrac{15 + \sqrt{561}}{8}, \dfrac{11 - 3\sqrt{561}}{8}\right)$, $\left(\dfrac{15 - \sqrt{561}}{8}, \dfrac{11 + 3\sqrt{561}}{8}\right)$ **33.** $\left(\dfrac{7 - \sqrt{33}}{2}, \dfrac{7 + \sqrt{33}}{2}\right)$, $\left(\dfrac{7 + \sqrt{33}}{2}, \dfrac{7 - \sqrt{33}}{2}\right)$ **35.** $(3, 2)$, $(-3, -2)$, $(2, 3)$, $(-2, -3)$ **37.** $\left(\dfrac{5 - 9\sqrt{15}}{20}, \dfrac{-45 + 3\sqrt{15}}{20}\right)$, $\left(\dfrac{5 + 9\sqrt{15}}{20}, \dfrac{-45 - 3\sqrt{15}}{20}\right)$ **39.** $(3, -5)$, $(-1, 3)$

41. $(8, 5)$, $(-5, -8)$ **43.** $(3, 2)$, $(-3, -2)$
45. $(2, 1)$, $(-2, -1)$, $(1, 2)$, $(-1, -2)$
47. $\left(4 + \dfrac{3\sqrt{6}}{2}i, -4 + \dfrac{3\sqrt{6}}{2}i\right)$, $\left(4 - \dfrac{3\sqrt{6}}{2}i, -4 - \dfrac{3\sqrt{6}}{2}i\right)$
49. $(3, \sqrt{5})$, $(3, -\sqrt{5})$, $(\sqrt{5}, 3)$, $(-\sqrt{5}, -3)$
51. $\left(\dfrac{8\sqrt{5}}{5}i, \dfrac{3\sqrt{105}}{5}\right)$, $\left(\dfrac{8\sqrt{5}}{5}i, -\dfrac{3\sqrt{105}}{5}\right)$, $\left(-\dfrac{8\sqrt{5}}{5}i, \dfrac{3\sqrt{105}}{5}\right)$, $\left(-\dfrac{8\sqrt{5}}{5}i, -\dfrac{3\sqrt{105}}{5}\right)$
53. $(2, 1)$, $(-2, -1)$, $\left(-\sqrt{5}i, \dfrac{2\sqrt{5}i}{5}\right)$, $\left(\sqrt{5}i, -\dfrac{2\sqrt{5}i}{5}\right)$

55. True **57.** True **59.** 24 in. by 10 in. **61.** 4 in. by 5 in.
63. 30 yd by 75 yd **65.** Length: $\sqrt{3}$ m; width: 1 m
67. Length: 12 ft; width: 5 ft **69.** (b) **71.** (d) **73.** (a)
75.

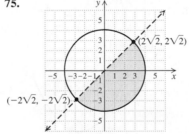

77. 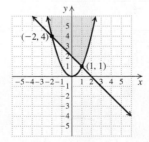 **79.**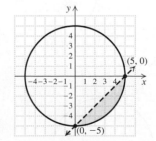

81.

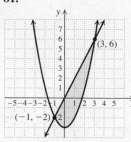

83.

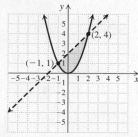

85. 2   **86.** 2.048   **87.** 81   **88.** 5

89. $(x-2)^2 + (y-3)^2 = 1$ **91.** $\dfrac{x^2}{4} + y^2 = 1$

93. There is no number x such that $\dfrac{x^2}{a^2} - \dfrac{\left(\dfrac{b}{a}x\right)^2}{b^2} = 1$, because

the left side simplifies to $\dfrac{x^2}{a^2} - \dfrac{x^2}{a^2}$, which is 0.

95. Factor: $x^3 + y^3 = (x+y)(x^2 - xy + y^2)$. We know
that $x + y = 1$, so $(x+y)^2 = x^2 + 2xy + y^2 = 1$, or
$x^2 + y^2 = 1 - 2xy$. We also know that $xy = 1$, so
$x^2 + y^2 = 1 - 2 \cdot 1 = -1$. Then $x^3 + y^3 =$
$1 \cdot (-1 - 1) = -2$.   **97.** $(2,4), (4,2)$
99. $(3,-2), (-3,2), (2,-3), (-2,3)$ **101.** $(1.564, 2.448)$,
$(0.138, 0.019)$ **103.** $(1.146, 3.146), (-1.841, 0.159)$
105. $(2.112, -0.109), (-13.041, -13.337)$
107. $(400, 1.431), (-400, 1.431), (400, -1.431)$,
$(-400, -1.431)$

Mid-Chapter Mixed Review: Chapter 10

1. True   **2.** False   **3.** True   **4.** False   **5.** (b)
6. (e)   **7.** (d)   **8.** (a)   **9.** (g)   **10.** (f)
11. (h)   **12.** (c)
13. $V: (0,0); F: (3,0);$
$D: x = -3$

14. $V: (3,2); F: (3,3);$
$D: y = 1$

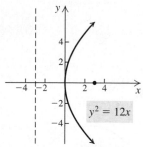

$y^2 = 12x$

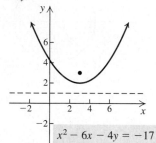

$x^2 - 6x - 4y = -17$

15. $(-2,4); 5$

16. $(3,-1); 4$

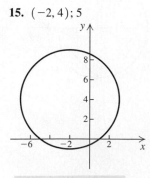

$x^2 + y^2 + 4x - 8y = 5$

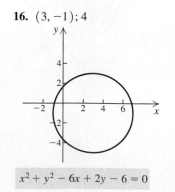

$x^2 + y^2 - 6x + 2y - 6 = 0$

17. $V: (0,-3), (0,3);$
$F: (0, -2\sqrt{2}), (0, 2\sqrt{2})$

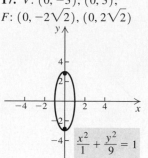

$\dfrac{x^2}{1} + \dfrac{y^2}{9} = 1$

18. $V: (1,-6), (1,4);$
$F: (1, -1 - \sqrt{21}),$
$(1, -1 + \sqrt{21})$

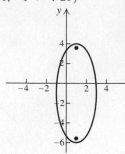

$25x^2 + 4y^2 - 50x + 8y = 71$

19. $C: (0,0); V: (0,-4), (0,4);$
$F: (0,-5), (0,5);$
$A: y = -\dfrac{4}{3}x, y = \dfrac{4}{3}x$

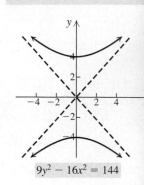

$9y^2 - 16x^2 = 144$

20. $C: (-3,2); V: (-4,2), (-2,2);$
$F: (-3 - \sqrt{5}, 2), (-3 + \sqrt{5}, 2);$
$A: y - 2 = 2(x+3), y - 2 = -2(x+3)$

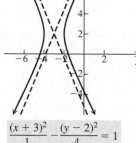

$\dfrac{(x+3)^2}{1} - \dfrac{(y-2)^2}{4} = 1$

21. $(-2,-5), (5,2)$ **22.** $(2,2), (-2,-2)$
23. $(2\sqrt{3}i, 4), (2\sqrt{3}i, -4), (-2\sqrt{3}i, 4), (-2\sqrt{3}i, -4)$
24. $\left(-\dfrac{1}{3}, \dfrac{10}{3}\right), (-1,2)$ **25.** -2 and 3

26.

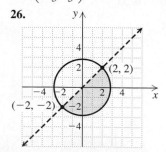

7.

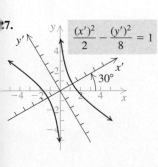

28. No; parabolas with a horizontal axis of symmetry fail the vertical-line test.   **29.** No; the center of an ellipse is not part of the graph of the ellipse. Its coordinates do not satisfy the equation of the ellipse.   **30.** No; the asymptotes of a hyperbola are not part of the graph of the hyperbola. The coordinates of points on the asymptotes do not satisfy the equation of the hyperbola.
31. Although we can always visualize the real-number solutions, we cannot visualize the imaginary-number solutions.

Exercise Set 10.5

1. $(0, -2)$ **3.** $(1, \sqrt{3})$ **5.** $(\sqrt{2}, 0)$ **7.** $(\sqrt{3}, 1)$
9. Ellipse or circle   **11.** Hyperbola   **13.** Parabola
15. Hyperbola   **17.** Ellipse or circle
19.

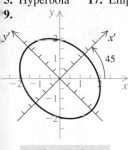

$$\frac{(x')^2}{3} + \frac{(y')^2}{5} = 1$$

21.

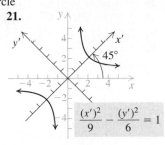

$$\frac{(x')^2}{9} - \frac{(y')^2}{6} = 1$$

23.

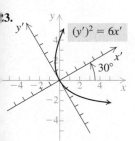

$(y')^2 = 6x'$

25.

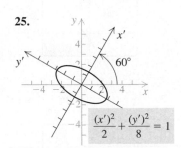

$$\frac{(x')^2}{2} + \frac{(y')^2}{8} = 1$$

27.

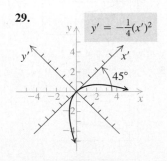

$$\frac{(x')^2}{2} - \frac{(y')^2}{8} = 1$$

29.

$y' = -\frac{1}{4}(x')^2$

31.

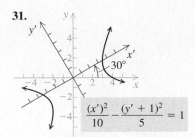

$$\frac{(x')^2}{10} - \frac{(y' + 1)^2}{5} = 1$$

33.

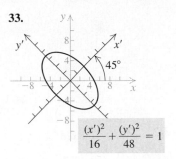

$$\frac{(x')^2}{16} + \frac{(y')^2}{48} = 1$$

35.

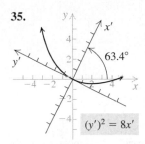

$(y')^2 = 8x'$

37.

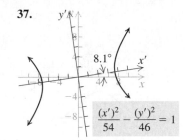

$$\frac{(x')^2}{54} - \frac{(y')^2}{46} = 1$$

39. $\dfrac{2\pi}{3}$ **40.** $-\dfrac{7\pi}{4}$ **41.** $60°$ **42.** $135°$
43. $x = x' \cos \theta - y' \sin \theta,\ y = x' \sin \theta + y' \cos \theta$
45. $A' + C' = A \cos^2 \theta + B \sin \theta \cos \theta + C \sin^2 \theta$
$\qquad\qquad\quad + A \sin^2 \theta - B \sin \theta \cos \theta + C \cos^2 \theta$
$\qquad\quad = A(\sin^2 \theta + \cos^2 \theta) + C(\sin^2 \theta + \cos^2 \theta)$
$\qquad\quad = A + C$

Exercise Set 10.6

1. (b)   **3.** (a)   **5.** (d)   **7.** (a) Parabola; (b) vertical, 1 unit to the right of the pole; (c) $\left(\frac{1}{2}, 0\right)$;
(d) $r = \dfrac{1}{1 + \cos \theta}$

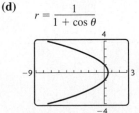

9. (a) Hyperbola; (b) horizontal, $\frac{3}{2}$ units below the pole;
(c) $\left(-3, \dfrac{\pi}{2}\right), \left(1, \dfrac{3\pi}{2}\right)$;

(d) $r = \dfrac{15}{5 - 10 \sin \theta}$

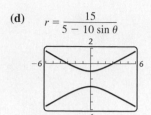

11. (a) Ellipse; **(b)** vertical, $\frac{8}{3}$ units to the left of the pole;
(c) $\left(\frac{8}{3}, 0\right), \left(\frac{8}{9}, \pi\right)$;

(d) $r = \dfrac{8}{6 - 3 \cos \theta}$

13. (a) Hyperbola; **(b)** horizontal, $\frac{4}{3}$ units above the pole;

(c) $\left(\dfrac{4}{5}, \dfrac{\pi}{2}\right), \left(-4, \dfrac{3\pi}{2}\right)$;

(d) $r = \dfrac{20}{10 + 15 \sin \theta}$

15. (a) Ellipse; **(b)** vertical, 3 units to the right of the pole;
(c) $(1, 0), (3, \pi)$;

(d) $r = \dfrac{9}{6 + 3 \cos \theta}$

17. (a) Parabola; **(b)** horizontal, $\frac{3}{2}$ units below the pole;

(c) $\left(\dfrac{3}{4}, \dfrac{3\pi}{2}\right)$;

(d) $r = \dfrac{3}{2 - 2 \sin \theta}$

19. (a) Ellipse; **(b)** vertical, 4 units to the left of the pole;
(c) $(4, 0), \left(\frac{4}{3}, \pi\right)$;

(d) $r = \dfrac{4}{2 - \cos \theta}$

21. (a) Hyperbola; **(b)** horizontal, $\frac{7}{10}$ units above the pole;
(c) $\left(\dfrac{7}{12}, \dfrac{\pi}{2}\right), \left(-\dfrac{7}{8}, \dfrac{3\pi}{2}\right)$;

(d) $r = \dfrac{7}{2 + 10 \sin \theta}$

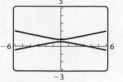

23. $y^2 + 2x - 1 = 0$ **25.** $x^2 - 3y^2 - 12y - 9 = 0$
27. $27x^2 + 36y^2 - 48x - 64 = 0$
29. $4x^2 - 5y^2 + 24y - 16 = 0$
31. $3x^2 + 4y^2 + 6x - 9 = 0$ **33.** $4x^2 - 12y - 9 = 0$
35. $3x^2 + 4y^2 - 8x - 16 = 0$

37. $4x^2 - 96y^2 + 140y - 49 = 0$ **39.** $r = \dfrac{6}{1 + 2 \sin \theta}$

41. $r = \dfrac{4}{1 + \cos \theta}$ **43.** $r = \dfrac{2}{2 - \cos \theta}$

45. $r = \dfrac{15}{4 + 3 \sin \theta}$ **47.** $r = \dfrac{8}{1 - 4 \sin \theta}$

49. $f(t) = (t - 3)^2 + 4$, or $t^2 - 6t + 13$
50. $f(2t) = (2t - 3)^2 + 4$, or $4t^2 - 12t + 13$
51. $f(t - 1) = (t - 4)^2 + 4$, or $t^2 - 8t + 20$
52. $f(t + 2) = (t - 1)^2 + 4$, or $t^2 - 2t + 5$

53. $r = \dfrac{1.5 \times 10^8}{1 + \sin \theta}$

Exercise Set 10.7

1. $x = \dfrac{1}{2}t, \; y = 6t - 7; \; -1 \le t \le 6$

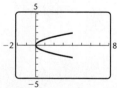

$y = 12x - 7, \; -\frac{1}{2} \le x \le 3$

3. $x = 4t^2, \; y = 2t; \; -1 \le t \le 1$

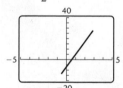

$x = y^2, \; -2 \le y \le 2$
5. $x = t^2, \; y = \sqrt{t}; \; 0 \le t \le 4$

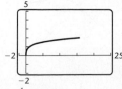

$x = y^4, \; 0 \le y \le 2$

$x = t + 3, \quad y = \dfrac{1}{t + 3}; \quad -2 \le t \le 2$

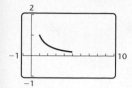

$= \dfrac{1}{x}, 1 \le x \le 5$

$x = 2t - 1, \quad y = t^2; \quad -3 \le t \le 3$

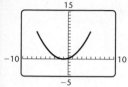

$= \dfrac{1}{4}(x + 1)^2, -7 \le x \le 5$

. $x = e^{-t}, \quad y = e^t; \quad -\infty < t < \infty$

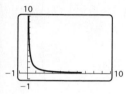

$= \dfrac{1}{x}, x > 0$

. $x = 3 \cos t, \quad y = 3 \sin t; \quad 0 \le t \le 2\pi$

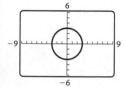

$+ y^2 = 9, -3 \le x \le 3$

. $x = \cos t, \quad y = 2 \sin t; \quad 0 \le t \le 2\pi$

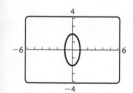

$+ \dfrac{y^2}{4} = 1, -1 \le x \le 1$

. $x = \sec t, \quad y = \cos t; \quad -\dfrac{\pi}{2} < t < \dfrac{\pi}{2}$

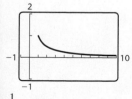

$= \dfrac{1}{x}, x \ge 1$

19. $x = 1 + 2 \cos t, \quad y = 2 + 2 \sin t; \quad 0 \le t \le 2\pi$

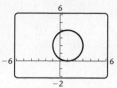

$(x - 1)^2 + (y - 2)^2 = 4, -1 \le x \le 3$

21. 0.7071   **23.** -0.2588 **25.** 0.7265   **27.** 0.5460

29. Answers may vary. $x = t, y = 4t - 3; x = \dfrac{t}{4} + 3, y = t + 9$

31. Answers may vary. $x = t, y = (t - 2)^2 - 6t; x = t + 2,$
$y = t^2 - 6t - 12$ **33. (a)** $x = 40\sqrt{3}t, y = 7 + 40t - 16t^2;$
(b) $x = 40\sqrt{3}t, \quad y = 7 + 40t - 16t^2$

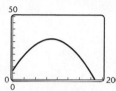

(c) 31 ft, 23 ft; **(d)** about 2.7 sec; **(e)** about 187.1 ft; **(f)** 32 ft
35. $x = 2(t - \sin t), \quad y = 2(1 - \cos t); \quad 0 \le t \le 4\pi$

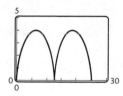

37. $x = t - \sin t, \quad y = 1 - \cos t; \quad -2\pi < t < 2\pi$

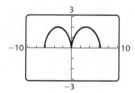

39.

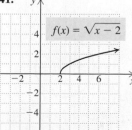

$y = x^3$

40.

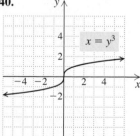

$x = y^3$

41.

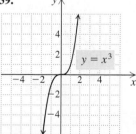

$f(x) = \sqrt{x - 2}$

42.

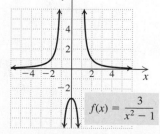

$f(x) = \dfrac{3}{x^2 - 1}$

43. $x = 3 \cos t, y = -3 \sin t$

Review Exercises: Chapter 10

1. True **2.** False **3.** False **4.** True **5.** False
6. (d) **7.** (a) **8.** (e) **9.** (g) **10.** (b) **11.** (f)
12. (h) **13.** (c) **14.** $x^2 = -6y$ **15.** $F: (-3, 0)$;
$V: (0, 0); D: x = 3$ **16.** $V: (-5, 8); F: (-5, \frac{15}{2})$;
$D: y = \frac{17}{2}$ **17.** $C: (2, -1); V: (-3, -1), (7, -1)$;
$F: (-1, -1), (5, -1)$;

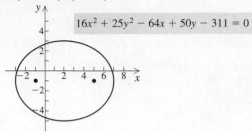

18. $\dfrac{x^2}{9} + \dfrac{y^2}{16} = 1$

19. $C: \left(-2, \dfrac{1}{4}\right); V: \left(0, \dfrac{1}{4}\right), \left(-4, \dfrac{1}{4}\right)$;

$F: \left(-2 + \sqrt{6}, \dfrac{1}{4}\right), \left(-2 - \sqrt{6}, \dfrac{1}{4}\right)$;

$A: y - \dfrac{1}{4} = \dfrac{\sqrt{2}}{2}(x + 2), y - \dfrac{1}{4} = -\dfrac{\sqrt{2}}{2}(x + 2)$

20. 0.167 ft **21.** $(-8\sqrt{2}, 8), (8\sqrt{2}, 8)$

22. $\left(3, \dfrac{\sqrt{29}}{2}\right), \left(-3, \dfrac{\sqrt{29}}{2}\right), \left(3, -\dfrac{\sqrt{29}}{2}\right)$,

$\left(-3, -\dfrac{\sqrt{29}}{2}\right)$ **23.** $(7, 4)$ **24.** $(2, 2), \left(\frac{32}{9}, -\frac{10}{9}\right)$

25. $(0, -3), (2, 1)$ **26.** $(4, 3), (4, -3), (-4, 3), (-4, -3)$
27. $(-\sqrt{3}, 0), (\sqrt{3}, 0), (-2, 1), (2, 1)$
28. $\left(-\frac{3}{5}, \frac{21}{5}\right), (3, -3)$ **29.** $(6, 8), (6, -8), (-6, 8)$,
$(-6, -8)$ **30.** $(2, 2), (-2, -2), (2\sqrt{2}, \sqrt{2})$,
$(-2\sqrt{2}, -\sqrt{2})$ **31.** 7, 4 **32.** 7 m by 12 m **33.** 4, 8
34. 32 cm, 20 cm **35.** 11 ft, 3 ft

36.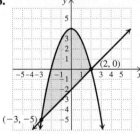

37.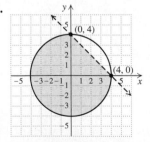

38.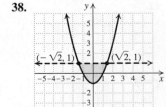

39.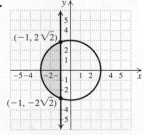

40.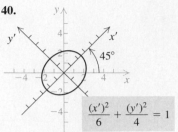

$$\frac{(x')^2}{6} + \frac{(y')^2}{4} = 1$$

41.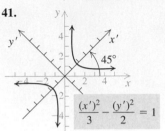

$$\frac{(x')^2}{3} - \frac{(y')^2}{2} = 1$$

42.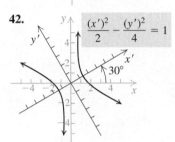

$$\frac{(x')^2}{2} - \frac{(y')^2}{4} = 1$$

43.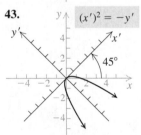

$$(x')^2 = -y'$$

44.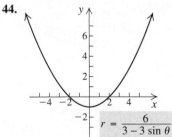

$$r = \frac{6}{3 - 3\sin\theta}$$

Horizontal directrix 2 units below the pole; vertex: $\left(1, \dfrac{3\pi}{2}\right)$

45.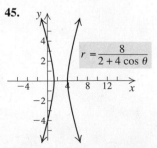

$$r = \frac{8}{2 + 4\cos\theta}$$

Vertical directrix 2 units to the right of the pole;
vertices: $\left(\frac{4}{3}, 0\right), (-4, \pi)$

46.

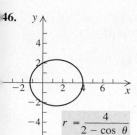

$$r = \frac{4}{2 - \cos \theta}$$

Vertical directrix 4 units to the left of the pole; vertices: $(4, 0), \left(\frac{4}{3}, \pi\right)$

47.

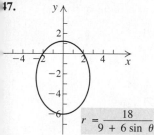

$$r = \frac{18}{9 + 6 \sin \theta}$$

Horizontal directrix 3 units above the pole; vertices: $\left(\frac{6}{5}, \frac{\pi}{2}\right), \left(6, \frac{3\pi}{2}\right)$

48. $x^2 - 4y - 4 = 0$
49. $3x^2 - y^2 - 16x + 16 = 0$
50. $3x^2 + 4y^2 - 8x - 16 = 0$
51. $9x^2 + 5y^2 + 24y - 36 = 0$
52. $r = \dfrac{1}{1 + \frac{1}{2}\cos \theta}$, or $r = \dfrac{2}{2 + \cos \theta}$ **53.** $r = \dfrac{18}{1 - 3 \sin \theta}$
54. $r = \dfrac{4}{1 - \cos \theta}$ **55.** $r = \dfrac{6}{1 + 2 \sin \theta}$

56.

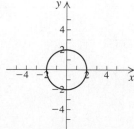

$x = t, \; y = 2 + t; -3 \le t \le 3$
$y = 2 + x, -3 \le x \le 3$

57.

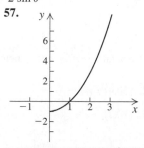

$x = \sqrt{t}, \; y = t - 1; \; 0 \le t \le 9$
$y = x^2 - 1, 0 \le x \le 3$

58.

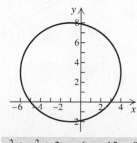

$x = 2 \cos t, \; y = 2 \sin t; \; 0 \le t \le 2\pi$
$x^2 + y^2 = 4$

59.

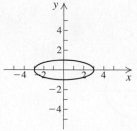

$x = 3 \sin t, \; y = \cos t; \; 0 \le t \le 2\pi$

$$\frac{x^2}{9} + y^2 = 1$$

60. Answers may vary. $x = t, y = 2t - 3; x = t + 1, \; y = 2t - 1$
61. Answers may vary. $x = t, y = t^2 + 4; x = t - 2,$
$y = t^2 - 4t + 8$ **62. (a)** $x = 75\sqrt{2}t, y = 75\sqrt{2}t - 16t^2;$
(b) 174.2 ft, 60.4 ft; **(c)** about 6.6 sec; **(d)** about 700.0 ft;
(e) about 175.8 ft   **63.** B   **64.** D   **65.** C
66. $\dfrac{8}{7}, \dfrac{7}{2}$ **67.** $(x - 2)^2 + (y - 1)^2 = 100$
68. $x^2 + \dfrac{y^2}{9} = 1$ **69.** $\dfrac{x^2}{778.41} - \dfrac{y^2}{39,221.59} = 1$
70. See the development of the formula for the standard form of a parabola that follows Fig. 1 at the beginning of Section 10.1.
71. Circles and ellipses are not functions.   **72.** The procedure for rotation of axes would be done first when $B \ne 0$. Then we would proceed as when $B = 0$.   **73.** Each graph is an ellipse. The value of e determines the location of the center and the lengths of the major and minor axes. The larger the value of e, the farther the center is from the pole and the longer the axes.

Test: Chapter 10

1. [10.3] (c)   **2.** [10.1] (b)   **3.** [10.2] (a)   **4.** [10.2] (d)
5. [10.1] $V: (0, 0)$;   **6.** [10.1] $V: (-1, -1)$;
$F: (0, 3)$; $D: y = -3$ $F: (1, -1)$; $D: x = -3$

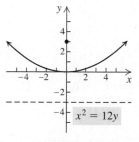
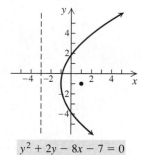

$x^2 = 12y$ $y^2 + 2y - 8x - 7 = 0$

7. [10.1] $x^2 = 8y$
8. [10.2] Center: $(-1, 3)$; radius: 5

$x^2 + y^2 + 2x - 6y - 15 = 0$

9. [10.2] C: $(0,0)$; V: $(-4,0)$, $(4,0)$; F: $(-\sqrt{7},0)$, $(\sqrt{7},0)$

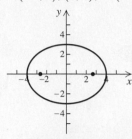

$$9x^2 + 16y^2 = 144$$

10. [10.2] C: $(-1,2)$; V: $(-1,-1)$, $(-1,5)$;
F: $(-1,2-\sqrt{5})$, $(-1,2+\sqrt{5})$

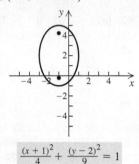

$$\frac{(x+1)^2}{4} + \frac{(y-2)^2}{9} = 1$$

11. [10.2] $\dfrac{x^2}{4} + \dfrac{y^2}{25} = 1$

12. [10.3] C: $(0,0)$; V: $(-1,0)$, $(1,0)$;
F: $(-\sqrt{5},0)$, $(\sqrt{5},0)$; A: $y=-2x$, $y=2x$

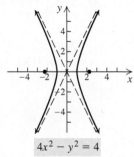

$$4x^2 - y^2 = 4$$

13. [10.3] C: $(-1,2)$; V: $(-1,0)$, $(-1,4)$; F: $(-1,2-\sqrt{13})$,
$(-1,2+\sqrt{13})$; A: $y=-\frac{2}{3}x+\frac{4}{3}$, $y=\frac{2}{3}x+\frac{8}{3}$

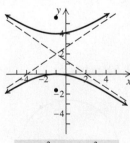

$$\frac{(y-2)^2}{4} - \frac{(x+1)^2}{9} = 1$$

14. [10.3] $y=\dfrac{\sqrt{2}}{2}x$, $y=-\dfrac{\sqrt{2}}{2}x$ **15.** [10.1] $\dfrac{27}{8}$ in.

16. [10.4] $(1,2)$, $(1,-2)$, $(-1,2)$, $(-1,-2)$

17. [10.4] $(3,-2)$, $(-2,3)$ **18.** [10.4] $(2,3)$, $(3,2)$
19. [10.4] 5 ft by 4 ft   **20.** [10.4] 60 ft by 45 ft
21. [10.4]

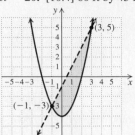

22. [10.5] After using the rotation of axes formulas with
$\theta = 45°$, we have $\dfrac{(x')^2}{9} + (y')^2 = 1$.

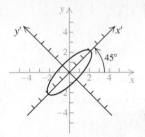

23. [10.6]

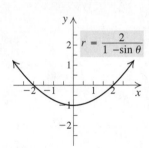

Horizontal directrix 2 units below the pole; vertex: $\left(1, \dfrac{3\pi}{2}\right)$

24. [10.6] $r = \dfrac{6}{1 + 2\cos\theta}$

25. [10.7]

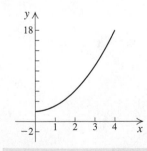

$$x = \sqrt{t},\ y = t + 2;\ 0 \le t \le 16$$

26. [10.7] $x^2 + y^2 = 9$, $-3 \le x \le 3$
27. [10.7] Answers may vary. $x = t, y = t - 5$; $x = t + 5, y = t$
28. [10.7] **(a)** $x = 125\sqrt{3}t$, $y = 10 + 125t - 16t^2$; **(b)** 119 ft,
241 ft; **(c)** about 7.9 sec; **(d)** about 1710 ft; **(e)** about 254 ft
29. [10.1] A   **30.** [10.2] $(x-3)^2 + (y+1)^2 = 8$

Chapter 11

Exercise Set 11.1

1. 3, 7, 11, 15; 39; 59   3. $2, \frac{3}{2}, \frac{4}{3}, \frac{5}{4}, \frac{10}{9}, \frac{15}{14}$
5. $0, \frac{3}{5}, \frac{4}{5}, \frac{15}{17}, \frac{99}{101}, \frac{112}{113}$ 7. $-1, 4, -9, 16; 100; -225$
9. 7, 3, 7, 3; 3; 7   11. 34   13. 225   15. $-33{,}880$ 17. 67

19.

n	u_n
1	2
2	2.25
3	2.3704
4	2.4414
5	2.4883
6	2.5216
7	2.5465
8	2.5658
9	2.5812
10	2.5937

21.

n	u_n
1	2
2	1.5538
3	1.4988
4	1.4914
5	1.4904
6	1.4902
7	1.4902
8	1.4902
9	1.4902
10	1.4902

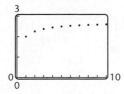
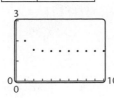

23. $2n$ 25. $(-1)^n \cdot 2 \cdot 3^{n-1}$ 27. $\dfrac{n+1}{n+2}$
29. $n(n+1)$ 31. $\log 10^{n-1}$, or $n-1$ 33. 6; 28
35. 20; 30   37. $\frac{1}{2} + \frac{1}{4} + \frac{1}{6} + \frac{1}{8} + \frac{1}{10} = \frac{137}{120}$
39. $1 + 2 + 4 + 8 + 16 + 32 + 64 = 127$
41. $\ln 7 + \ln 8 + \ln 9 + \ln 10 = \ln (7 \cdot 8 \cdot 9 \cdot 10) = \ln 5040 \approx 8.5252$
43. $\frac{1}{2} + \frac{2}{3} + \frac{3}{4} + \frac{4}{5} + \frac{5}{6} + \frac{6}{7} + \frac{7}{8} + \frac{8}{9} = \frac{15{,}551}{2520}$
45. $-1 + 1 - 1 + 1 - 1 = -1$
47. $3 - 6 + 9 - 12 + 15 - 18 + 21 - 24 = -12$
49. $2 + 1 + \frac{2}{5} + \frac{1}{5} + \frac{2}{17} + \frac{1}{13} + \frac{2}{37} = \frac{157{,}351}{40{,}885}$
51. $3 + 2 + 3 + 6 + 11 + 18 = 43$
53. $\frac{1}{2} + \frac{2}{3} + \frac{4}{5} + \frac{8}{9} + \frac{16}{17} + \frac{32}{33} + \frac{64}{65} + \frac{128}{129} + \frac{256}{257} + \frac{512}{513} + \frac{1024}{1025} \approx 9.736$ 55. $\displaystyle\sum_{k=1}^{\infty} 5k$ 57. $\displaystyle\sum_{k=1}^{6} (-1)^{k+1} 2^k$
59. $\displaystyle\sum_{k=1}^{6} (-1)^k \frac{k}{k+1}$ 61. $\displaystyle\sum_{k=2}^{n} (-1)^k k^2$ 63. $\displaystyle\sum_{k=1}^{\infty} \frac{1}{k(k+1)}$
65. $4, 1\frac{1}{4}, 1\frac{4}{5}, 1\frac{5}{9}$ 67. $6561, -81, 9i, -3\sqrt{i}$ 69. 2, 3, 5, 8
71. (a) $4150, $4305.63, $4467.09, $4634.60, $4808.40, $4988.71, $5175.79, $5369.88, $5571.25, $5780.18; (b) $8352.61
73. $16.20, $18.45, $20.70, $22.95, $25.20, $27.45, $29.70, $31.95, $34.20, $36.45   75. 1, 1, 2, 3, 5, 8, 13
77. (a) $a_n = -3.210714286n^2 + 15.76585714n + 12.76457143$; (b) $11.33   78. $(-1, -3)$ 79. Illinois: 16,200 acres; Ohio: 8200 acres   80. $(3, -2)$; 4   81. $\left(-\frac{5}{2}, 4\right); \dfrac{\sqrt{97}}{2}$
83. $i, -1, -i, 1, i; i$ 85. $\ln (1 \cdot 2 \cdot 3 \cdot \; \cdots \; \cdot n)$

Exercise Set 11.2

1. $a_1 = 3, d = 5$ 3. $a_1 = 9, d = -4$
5. $a_1 = \frac{3}{2}, d = \frac{3}{4}$ 7. $a_1 = \$316, d = -\$3$
9. $a_{12} = 46$ 11. $a_{14} = -\frac{17}{3}$ 13. $a_{10} = \$7941.62$
15. 27th   17. 46th   19. $a_1 = 5$ 21. $n = 39$

23. $a_1 = \frac{1}{3}; d = \frac{1}{2}, \frac{1}{3}, \frac{5}{6}, \frac{4}{3}, \frac{11}{6}, \frac{7}{3}$ 25. 670   27. 160,400
29. 735   31. 990   33. 1760   35. $\frac{65}{2}$ 37. $-\frac{6026}{13}$
39. 4960¢, or $49.60   41. 1320 seats   43. Yes; 32; 1600 ft
45. 3 plants; 171 plants   47. Yes; 3   48. $(2, 5)$
49. $(2, -1, 3)$ 50. $(-4, 0), (4, 0); (-\sqrt{7}, 0), (\sqrt{7}, 0)$
51. $\dfrac{x^2}{4} + \dfrac{y^2}{25} = 1$ 53. n^2 55. $a_1 = 60 - 5p - 5q$; $d = 5p + 2q - 20$ 57. $5\frac{4}{5}, 7\frac{3}{5}, 9\frac{2}{5}, 11\frac{1}{5}$

Visualizing the Graph

1. J   2. A   3. C   4. G   5. F   6. H   7. E
8. D   9. B   10. I

Exercise Set 11.3

1. 2   3. -1 5. -2 7. 0.1   9. $\dfrac{a}{2}$ 11. 128
13. 162   15. $7(5)^{40}$ 17. 3^{n-1} 19. $(-1)^{n-1}$ 21. $\dfrac{1}{x^n}$
23. 762   25. $\frac{4921}{18}$ 27. True   29. True   31. True
33. 8   35. 125   37. Does not exist   39. $\frac{2}{3}$ 41. $29\frac{38{,}569}{59{,}049}$
43. 2   45. Does not exist   47. $4545.45   49. $\frac{160}{9}$
51. $\frac{13}{99}$ 53. 9   55. $\frac{34{,}091}{9990}$ 57. $2,684,354.55
59. (a) About 297 ft; (b) 300 ft   61. $39,505.71
63. 10,485.76 in.   65. $19,694.01   67. $86,666,666,667
69. $(f \circ g)(x) = 16x^2 + 40x + 25; (g \circ f)(x) = 4x^2 + 5$
70. $(f \circ g)(x) = x^2 + x + 2; (g \circ f)(x) = x^2 - x + 3$
71. 2.209   72. $\frac{1}{16}$
73. $(4 - \sqrt{6})/(\sqrt{3} - \sqrt{2}) = 2\sqrt{3} + \sqrt{2}$, $(6\sqrt{3} - 2\sqrt{2})/(4 - \sqrt{6}) = 2\sqrt{3} + \sqrt{2}$; there exists a common ratio, $2\sqrt{3} + \sqrt{2}$; thus the sequence is geometric.
75. (a) $\frac{13}{3}, \frac{22}{3}, \frac{34}{3}, \frac{46}{3}, \frac{58}{3}$, (b) $-\frac{11}{3}, -\frac{2}{3}, \frac{10}{3}, -\frac{50}{3}, \frac{250}{3}$ or 5; 8, 12, 18, 27
77. $S_n = \dfrac{x^2(1 - (-x)^n)}{x + 1}$

Exercise Set 11.4

1. $1^2 < 1^3$, false; $2^2 < 2^3$, true; $3^2 < 3^3$, true; $4^2 < 4^3$, true; $5^2 < 5^3$, true   3. A polygon of 3 sides has $\dfrac{3(3 - 3)}{2}$ diagonals.

True; A polygon of 4 sides has $\dfrac{4(4 - 3)}{2}$ diagonals.

True; A polygon of 5 sides has $\dfrac{5(5 - 3)}{2}$ diagonals.

True; A polygon of 6 sides has $\dfrac{6(6 - 3)}{2}$ diagonals.

True; A polygon of 7 sides has $\dfrac{7(7 - 3)}{2}$ diagonals. True.

5. S_n: $2 + 4 + 6 + \cdots + 2n = n(n + 1)$
S_1: $2 = 1(1 + 1)$
S_k: $2 + 4 + 6 + \cdots + 2k = k(k + 1)$
S_{k+1}: $2 + 4 + 6 + \cdots + 2k + 2(k + 1)$
$= (k + 1)(k + 2)$
(1) *Basis step*: S_1 true by substitution.
(2) *Induction step*: Assume S_k. Deduce S_{k+1}. Starting with the left side of S_{k+1}, we have
$2 + 4 + 6 + \cdots + 2k + 2(k + 1)$
$= k(k + 1) + 2(k + 1)$ **By S_k**
$= (k + 1)(k + 2)$.   **Factoring**

7. S_n: $\quad 1 + 5 + 9 + \cdots + (4n - 3) = n(2n - 1)$
S_1: $\quad 1 = 1(2 \cdot 1 - 1)$
S_k: $\quad 1 + 5 + 9 + \cdots + (4k - 3) = k(2k - 1)$
S_{k+1}: $\quad 1 + 5 + 9 + \cdots + (4k - 3) + [4(k + 1) - 3]$
$\qquad\qquad\qquad\qquad = (k + 1)[2(k + 1) - 1]$
$\qquad\qquad\qquad\qquad = (k + 1)(2k + 1)$

(1) *Basis step*: S_1 true by substitution.
(2) *Induction step*: Assume S_k. Deduce S_{k+1}. Starting with the left side of S_{k+1}, we have
$$1 + 5 + 9 + \cdots + (4k - 3) + [4(k + 1) - 3]$$
$$= k(2k - 1) + [4(k + 1) - 3] \qquad \textbf{By } S_k$$
$$= 2k^2 - k + 4k + 4 - 3$$
$$= 2k^2 + 3k + 1$$
$$= (k + 1)(2k + 1).$$

9. S_n: $\quad 2 + 4 + 8 + \cdots + 2^n = 2(2^n - 1)$
S_1: $\quad 2 = 2(2 - 1)$
S_k: $\quad 2 + 4 + 8 + \cdots + 2^k = 2(2^k - 1)$
S_{k+1}: $\quad 2 + 4 + 8 + \cdots + 2^k + 2^{k+1} = 2(2^{k+1} - 1)$
(1) *Basis step*: S_1 is true by substitution.
(2) *Induction step*: Assume S_k. Deduce S_{k+1}. Starting with the left side of S_{k+1}, we have
$$\underbrace{2 + 4 + 8 + \cdots + 2^k} + 2^{k+1}$$
$$= 2(2^k - 1) \quad + 2^{k+1} \textbf{ By } S_k$$
$$= 2^{k+1} - 2 + 2^{k+1}$$
$$= 2 \cdot 2^{k+1} - 2$$
$$= 2(2^{k+1} - 1).$$

11. S_n: $\quad n < n + 1$
S_1: $\quad 1 < 1 + 1$
S_k: $\quad k < k + 1$
S_{k+1}: $\quad k + 1 < (k + 1) + 1$
(1) *Basis step*: Since $1 < 1 + 1$, S_1 is true.
(2) *Induction step*: Assume S_k. Deduce S_{k+1}. Now
$$k < k + 1 \qquad\qquad \textbf{By } S_k$$
$$k + 1 < k + 1 + 1 \qquad \textbf{Adding 1}$$
$$k + 1 < k + 2. \qquad\quad \textbf{Simplifying}$$

13. S_n: $\quad 2n \le 2^n$
S_1: $\quad 2 \cdot 1 \le 2^1$
S_k: $\quad 2k \le 2^k$
S_{k+1}: $\quad 2(k + 1) \le 2^{k+1}$
(1) *Basis step*: Since $2 = 2$, S_1 is true.
(2) *Induction step*: Let k be any natural number. Assume S_k. Deduce S_{k+1}.
$$2k \le 2^k \qquad\qquad \textbf{By } S_k$$
$$2 \cdot 2k \le 2 \cdot 2^k \qquad \textbf{Multiplying by 2}$$
$$4k \le 2^{k+1}$$
Since $1 \le k$, $k + 1 \le k + k$, or $k + 1 \le 2k$.
Then $2(k + 1) \le 4k$.   **Multiplying by 2**
Thus, $2(k + 1) \le 4k \le 2^{k+1}$, so $2(k + 1) \le 2^{k+1}$.

15.
S_n: $\quad \dfrac{1}{1 \cdot 2 \cdot 3} + \dfrac{1}{2 \cdot 3 \cdot 4} + \dfrac{1}{3 \cdot 4 \cdot 5} + \cdots$
$\qquad + \dfrac{1}{n(n + 1)(n + 2)} = \dfrac{n(n + 3)}{4(n + 1)(n + 2)}$

S_1: $\quad \dfrac{1}{1 \cdot 2 \cdot 3} = \dfrac{1(1 + 3)}{4(1 + 1)(1 + 2)}$

S_k: $\quad \dfrac{1}{1 \cdot 2 \cdot 3} + \dfrac{1}{2 \cdot 3 \cdot 4} + \cdots + \dfrac{1}{k(k + 1)(k + 2)}$
$\qquad\qquad = \dfrac{k(k + 3)}{4(k + 1)(k + 2)}$

S_{k+1}: $\quad \dfrac{1}{1 \cdot 2 \cdot 3} + \dfrac{1}{2 \cdot 3 \cdot 4} + \cdots + \dfrac{1}{k(k + 1)(k + 2)}$
$\qquad + \dfrac{1}{(k + 1)(k + 2)(k + 3)}$
$\qquad = \dfrac{(k + 1)(k + 1 + 3)}{4(k + 1 + 1)(k + 1 + 2)} = \dfrac{(k + 1)(k + 4)}{4(k + 2)(k + 3)}$

(1) *Basis step*: Since $\dfrac{1}{1 \cdot 2 \cdot 3} = \dfrac{1}{6}$ and
$$\dfrac{1(1 + 3)}{4(1 + 1)(1 + 2)} = \dfrac{1 \cdot 4}{4 \cdot 2 \cdot 3} = \dfrac{1}{6}, S_1 \text{ is true.}$$
(2) *Induction step*: Assume S_k. Deduce S_{k+1}.

Add $\dfrac{1}{(k + 1)(k + 2)(k + 3)}$ on both sides of S_k and simplify the right side. Only the right side is shown here.
$$\dfrac{k(k + 3)}{4(k + 1)(k + 2)} + \dfrac{1}{(k + 1)(k + 2)(k + 3)}$$
$$= \dfrac{k(k + 3)(k + 3) + 4}{4(k + 1)(k + 2)(k + 3)}$$
$$= \dfrac{k^3 + 6k^2 + 9k + 4}{4(k + 1)(k + 2)(k + 3)}$$
$$= \dfrac{(k + 1)^2(k + 4)}{4(k + 1)(k + 2)(k + 3)}$$
$$= \dfrac{(k + 1)(k + 4)}{4(k + 2)(k + 3)}$$

17.

S_n: $\quad 1 + 2 + 3 + \cdots + n = \dfrac{n(n + 1)}{2}$

S_1: $\quad 1 = \dfrac{1(1 + 1)}{2}$

S_k: $\quad 1 + 2 + 3 + \cdots + k = \dfrac{k(k + 1)}{2}$

S_{k+1}: $\quad 1 + 2 + 3 + \cdots + k + (k + 1) = \dfrac{(k + 1)(k + 2)}{2}$

(1) *Basis step*: S_1 true by substitution.
(2) *Induction step*: Assume S_k. Deduce S_{k+1}. Starting with the left side of S_{k+1}, we have
$$\underbrace{1 + 2 + 3 + \cdots + k} + (k + 1)$$
$$= \dfrac{k(k + 1)}{2} + (k + 1) \qquad\qquad \textbf{By } S_k$$
$$= \dfrac{k(k + 1) + 2(k + 1)}{2} \qquad\quad \textbf{Adding}$$
$$= \dfrac{(k + 1)(k + 2)}{2}. \qquad\qquad \textbf{Factoring}$$

19. S_n: $\quad 1^3 + 2^3 + 3^3 + \cdots + n^3 = \dfrac{n^2(n + 1)^2}{4}$

S_1: $\quad 1^3 = \dfrac{1^2(1 + 1)^2}{4}$

S_k: $\quad 1^3 + 2^3 + 3^3 + \cdots + k^3 = \dfrac{k^2(k + 1)^2}{4}$

S_{k+1}: $\quad 1^3 + 2^3 + 3^3 + \cdots + k^3 + (k + 1)^3$
$\qquad\qquad = \dfrac{(k + 1)^2[(k + 1) + 1]^2}{4}$

(1) *Basis step*: S_1: $1^3 = \dfrac{1^2(1+1)^2}{4} = 1$. True.

(2) *Induction step*: Assume S_k. Deduce S_{k+1}.

$$1^3 + 2^3 + \cdots + k^3 = \dfrac{k^2(k+1)^2}{4} \quad S_k$$

$$1^3 + 2^3 + \cdots + k^3 + (k+1)^3 = \dfrac{k^2(k+1)^2}{4} + (k+1)^3$$
$$\textbf{Adding } (k+1)^3$$
$$= \dfrac{k^2(k+1)^2 + 4(k+1)^3}{4}$$
$$= \dfrac{(k+1)^2}{4}[k^2 + 4(k+1)]$$
$$= \dfrac{(k+1)^2}{4}(k^2 + 4k + 4)$$
$$= \dfrac{(k+1)^2(k+2)^2}{4}$$

21.

S_n: $\quad 2 + 6 + 12 + \cdots + n(n+1) = \dfrac{n(n+1)(n+2)}{3}$

S_1: $\quad 1(1+1) = \dfrac{1(1+1)(1+2)}{3}$

S_k: $\quad 2 + 6 + 12 + \cdots + k(k+1) = \dfrac{k(k+1)(k+2)}{3}$

S_{k+1}:
$$2 + 6 + 12 + \cdots + k(k+1) + (k+1)[(k+1)+1]$$
$$= \dfrac{(k+1)[(k+1)+1][(k+1)+2]}{3}$$

(1) *Basis step*: S_1: $1(1+1) = \dfrac{1(1+1)(1+2)}{3}$. True.

(2) *Induction step*: Assume S_k:

$$2 + 6 + 12 + \cdots + k(k+1) = \dfrac{k(k+1)(k+2)}{3}.$$

Then $2 + 6 + 12 + \cdots + k(k+1) + (k+1)(k+1+1)$
$$= \dfrac{k(k+1)(k+2)}{3} + (k+1)(k+2)$$
$$= \dfrac{k(k+1)(k+2) + 3(k+1)(k+2)}{3}$$
$$= \dfrac{(k+1)(k+2)(k+3)}{3}$$
$$= \dfrac{(k+1)(k+1+1)(k+1+2)}{3}.$$

23. S_n: $\quad a_1 + (a_1 + d) + (a_1 + 2d) + \cdots +$
$$[a_1 + (n-1)d] = \dfrac{n}{2}[2a_1 + (n-1)d]$$

S_1: $\quad a_1 = \dfrac{1}{2}[2a_1 + (1-1)d]$

S_k: $\quad a_1 + (a_1 + d) + (a_1 + 2d) + \cdots +$
$$[a_1 + (k-1)d] = \dfrac{k}{2}[2a_1 + (k-1)d]$$

S_{k+1}: $\quad a_1 + (a_1 + d) + (a_1 + 2d) + \cdots +$
$$[a_1 + (k-1)d] + [a_1 + ((k+1)-1)d]$$
$$= \dfrac{k+1}{2}[2a_1 + ((k+1)-1)d]$$

(1) *Basis step*: Since $\frac{1}{2}[2a_1 + (1-1)d] = \frac{1}{2} \cdot 2a_1 = a_1$, S_1 is true.

(2) *Induction step*: Assume S_k. Deduce S_{k+1}. Starting with the left side of S_{k+1}, we have

$$\underbrace{a_1 + (a_1 + d) + \cdots + [a_1 + (k-1)d]} + [a_1 + kd]$$
$$= \dfrac{k}{2}[2a_1 + (k-1)d] \qquad\qquad + [a_1 + kd]$$
$$\textbf{By } S_k$$
$$= \dfrac{k[2a_1 + (k-1)d]}{2} + \dfrac{2[a_1 + kd]}{2}$$
$$= \dfrac{2ka_1 + k(k-1)d + 2a_1 + 2kd}{2}$$
$$= \dfrac{2a_1(k+1) + k(k-1)d + 2kd}{2}$$
$$= \dfrac{2a_1(k+1) + (k-1+2)kd}{2}$$
$$= \dfrac{2a_1(k+1) + (k+1)kd}{2} = \dfrac{k+1}{2}[2a_1 + kd].$$

24. $(5, 3)$

25. 1.5%: $1200; 2%: $2400; 3%: $2800

27. S_n: $\quad x + y$ is a factor of $x^{2n} - y^{2n}$.
S_1: $\quad x + y$ is a factor of $x^2 - y^2$.
S_k: $\quad x + y$ is a factor of $x^{2k} - y^{2k}$.
S_{k+1}: $\quad x + y$ is a factor of $x^{2(k+1)} - y^{2(k+1)}$.

(1) *Basis step*: S_1: $x + y$ is a factor of $x^2 - y^2$. True.
$\qquad\qquad S_2$: $x + y$ is a factor of $x^4 - y^4$. True.

(2) *Induction step*: Assume S_{k-1}: $x + y$ is a factor of $x^{2(k-1)} - y^{2(k-1)}$. Then $x^{2(k-1)} - y^{2(k-1)} = (x+y)Q(x)$ for some polynomial $Q(x)$.
Assume S_k: $x + y$ is a factor of $x^{2k} - y^{2k}$. Then $x^{2k} - y^{2k} = (x+y)P(x)$ for some polynomial $P(x)$.
$x^{2(k+1)} - y^{2(k+1)}$
$$= (x^{2k} - y^{2k})(x^2 + y^2) - (x^{2(k-1)} - y^{2(k-1)})(x^2y^2)$$
$$= (x+y)P(x)(x^2 + y^2) - (x+y)Q(x)(x^2y^2)$$
$$= (x+y)[P(x)(x^2 + y^2) - Q(x)(x^2y^2)]$$
so $x + y$ is a factor of $x^{2(k+1)} - y^{2(k+1)}$.

29. S_2: $\quad \log_a(b_1b_2) = \log_a b_1 + \log_a b_2$
S_k: $\quad \log_a(b_1b_2 \cdots b_k) = \log_a b_1 + \log_a b_2 + \cdots + \log_a b_k$
S_{k+1}: $\quad \log_a(b_1b_2 \cdots b_{k+1}) = \log_a b_1 + \log_a b_2 + \cdots + \log_a b_{k+1}$

(1) *Basis step*: S_2 is true by the properties of logarithms.
(2) *Induction step*: Let k be a natural number $k \geq 2$. Assume S_k. Deduce S_{k+1}.
$\log_a(b_1b_2 \cdots b_{k+1})$ $\qquad$ **Left side of S_{k+1}**
$$= \log_a(b_1b_2 \cdots b_k) + \log_a b_{k+1} \qquad \textbf{By } S_2$$
$$= \log_a b_1 + \log_a b_2 + \cdots + \log_a b_k + \log_a b_{k+1}$$

31. S_2: $\quad \overline{z_1 + z_2} = \bar{z}_1 + \bar{z}_2$:
$$\overline{(a + bi) + (c + di)} = \overline{(a+c) + (b+d)i}$$
$$= (a+c) - (b+d)i$$
$$\overline{(a + bi)} + \overline{(c + di)} = a - bi + c - di$$
$$= (a+c) - (b+d)i.$$
S_k: $\quad \overline{z_1 + z_2 + \cdots + z_k} = \bar{z}_1 + \bar{z}_2 + \cdots + \bar{z}_k$.
$$\overline{(z_1 + z_2 + \cdots + z_k) + z_{k+1}}$$
$$= \overline{(z_1 + z_2 + \cdots + z_k)} + \overline{z_{k+1}} \qquad \textbf{By } S_2$$
$$= \bar{z}_1 + \bar{z}_2 + \cdots + \bar{z}_k + \bar{z}_{k+1} \qquad \textbf{By } S_k$$

Mid-Chapter Mixed Review: Chapter 11

1. False **2.** True **3.** False **4.** False **5.** 8, 11, 14, 17; 32; 47 **6.** 0, −1, 2, −3; 8; −13 **7.** $a_n = 3n$

8. $a_n = (-1)^n n^2$ **9.** $1\frac{7}{8}$, or $\frac{15}{8}$ **10.** $2 + 6 + 12 + 20 + 30 = 70$ **11.** $\sum_{k=1}^{\infty}(-1)^k 4k$ **12.** 2, 6, 22, 86

13. -5 **14.** 22   **15.** 21   **16.** 696   **17.** $-\frac{1}{2}$

18. (a) 8; **(b)** $\frac{1023}{16}$, or 63.9375   **19.** $-\frac{16}{3}$ **20.** Does not exist

21. 126 plants   **22.** \$13,588.68

23. S_n: $1 + 4 + 7 + \cdots + (3n - 2) = \frac{1}{2}n(3n - 1)$
S_1: $3 \cdot 1 - 2 = \frac{1}{2} \cdot 1(3 \cdot 1 - 1)$
S_k: $1 + 4 + 7 + \cdots + (3k - 2) = \frac{1}{2}k(3k - 1)$
S_{k+1}: $1 + 4 + 7 + \cdots + (3k - 2) + [3(k + 1) - 2]$
$\qquad\qquad = \frac{1}{2}(k + 1)[3(k + 1) - 1]$
$\qquad\qquad = \frac{1}{2}(k + 1)(3k + 2)$

(1) *Basis step*: S_1: $3 \cdot 1 - 2 = \frac{1}{2} \cdot 1(3 \cdot 1 - 1)$.   **True**
(2) *Induction step*: Assume S_k:
$1 + 4 + 7 + \cdots + (3k - 2) = \frac{1}{2}k(3k - 1)$.
Then $1 + 4 + 7 + \cdots + (3k - 2) + [3(k + 1) - 2]$
$= \frac{1}{2}k(3k - 1) + [3(k + 1) - 2]$
$= \frac{3}{2}k^2 - \frac{1}{2}k + 3k + 1$
$= \frac{3}{2}k^2 + \frac{5}{2}k + 1$
$= \frac{1}{2}(3k^2 + 5k + 2)$
$= \frac{1}{2}(k + 1)(3k + 2)$.

24. The first formula can be derived from the second by substituting $a_1 + (n - 1)d$ for a_n. When the first and last terms of the sum are known, the second formula is the better one to use. If the last term is not known, the first formula allows us to compute the sum in one step without first finding a_n.

25. $1 + 2 + 3 + \cdots + 100$
$= (1 + 100) + (2 + 99) + (3 + 98) + \cdots + (50 + 51)$
$= \underbrace{101 + 101 + 101 + \cdots + 101}_{50 \text{ addends of } 101}$
$= 50 \cdot 101$
$= 5050$

A formula for the first n natural numbers is $\dfrac{n}{2}(1 + n)$.

26. Answers may vary. One possibility is given. Casey invests \$900 at 8% interest, compounded annually. How much will be in the account at the end of 40 years?   **27.** We can prove an infinite sequence of statements S_n by showing that a basis statement S_1 is true and then that for all natural numbers k, if S_k is true, then S_{k+1} is true.

Exercise Set 11.5

1. 720   **3.** 604,800   **5.** 120   **7.** 1   **9.** 3024   **11.** 120
13. 120   **15.** 1   **17.** 6,497,400   **19.** $n(n - 1)(n - 2)$
21. n **23.** $6! = 720$ **25.** $9! = 362,880$ **27.** $_9P_4 = 3024$
29. $_5P_5 = 120$; $5^5 = 3125$ **31.** $_5P_5 \cdot _4P_4 = 2880$
33. $8 \cdot 10^6 = 8,000,000$; 8 million   **35.** $\dfrac{9!}{2! \, 3! \, 4!} = 1260$
37. (a) $_6P_5 = 720$; **(b)** $6^5 = 7776$; **(c)** $1 \cdot _5P_4 = 120$;
(d) $1 \cdot 1 \cdot _4P_3 = 24$ **39. (a)** 10^5, or 100,000; **(b)** 100,000
41. (a) $10^9 = 1,000,000,000$; **(b)** yes   **42.** $\frac{9}{4}$, or 2.25
43. $-3, 2$ **44.** $\dfrac{3 \pm \sqrt{17}}{4}$
45. $-2, 1, 5$ **47.** 8   **49.** 11   **51.** $(n - 1)$ games

Exercise Set 11.6

1. 78   **3.** 78   **5.** 7   **7.** 10   **9.** 1   **11.** 15
13. 128   **15.** 270,725   **17.** 13,037,895   **19.** n

21. 1   **23.** $_{36}C_4 = 58,905$ **25.** $_{13}C_{10} = 286$

27. $\dbinom{10}{7} \cdot \dbinom{5}{3} = 1200$ **29.** $\dbinom{52}{5} = 2,598,960$

31. (a) $_{31}P_2 = 930$; **(b)** $31^2 = 961$; **(c)** $_{31}C_2 = 465$

33. $-\frac{17}{2}$ **34.** $-1, \frac{3}{2}$ **35.** $\dfrac{-5 \pm \sqrt{21}}{2}$

36. $-4, -2, 3$ **37.** $\dbinom{13}{5} = 1287$

39. $\dbinom{n}{2}$; $2\dbinom{n}{2}$ **41.** 4   **43.** 7

45. $\dbinom{n}{k - 1} + \dbinom{n}{k}$

$= \dfrac{n!}{(k - 1)!(n - k + 1)!} \cdot \dfrac{l}{k}$

$\quad + \dfrac{n!}{k!(n - k)!} \cdot \dfrac{(n - k + 1)}{(n - k + 1)!}$

$= \dfrac{n!(k + (n - k + 1))}{k!(n - k + 1)!}$

$= \dfrac{(n + 1)!}{k!(n - k + 1)!} = \dbinom{n + 1}{k}$

Exercise Set 11.7

1. $x^4 + 20x^3 + 150x^2 + 500x + 625$
3. $x^5 - 15x^4 + 90x^3 - 270x^2 + 405x - 243$
5. $x^5 - 5x^4y + 10x^3y^2 - 10x^2y^3 + 5xy^4 - y^5$
7. $15{,}625x^6 + 75{,}000x^5y + 150{,}000x^4y^2 + 160{,}000x^3y^3 + 96{,}000x^2y^4 + 30{,}720xy^5 + 4096y^6$
9. $128t^7 + 448t^5 + 672t^3 + 560t + 280t^{-1} + 84t^{-3} + 14t^{-5} + t^{-7}$ **11.** $x^{10} - 5x^8 + 10x^6 - 10x^4 + 5x^2 - 1$
13. $125 + 150\sqrt{5}t + 375t^2 + 100\sqrt{5}t^3 + 75t^4 + 6\sqrt{5}t^5 + t^6$
15. $a^9 - 18a^7 + 144a^5 - 672a^3 + 2016a - 4032a^{-1} + 5376a^{-3} - 4608a^{-5} + 2304a^{-7} - 512a^{-9}$ **17.** $140\sqrt{2}$
19. $x^{-8} + 4x^{-4} + 6 + 4x^4 + x^8$ **21.** $21a^5b^2$
23. $-252x^5y^5$ **25.** $-745{,}472a^3$ **27.** $1120x^{12}y^2$
29. $-1{,}959{,}552u^5v^{10}$ **31.** 2^7, or 128   **33.** 2^{24}, or 16,777,216   **35.** 20   **37.** $-12 + 316i$
39. $-7 - 4\sqrt{2}i$ **41.** $\sum_{k=0}^{n}\dbinom{n}{k}(-1)^k a^{n-k}b^k$, or $\sum_{k=0}^{n}\dbinom{n}{k}a^{n-k}(-b)^k$

43. $\sum_{k=1}^{n}\dbinom{n}{k}x^{n-k}h^{k-1}$ **44.** $x^2 + 2x - 2$
45. $2x^3 - 3x^2 + 2x - 3$ **46.** $4x^2 - 12x + 10$
47. $2x^2 - 1$ **49.** $3, 9, 6 \pm 3i$ **51.** $-\dfrac{35}{x^{1/6}}$
53. 2^{100} **55.** $[\log_a(xt)]^{23}$
57. (1) *Basis step*: Since $a + b = (a + b)^1$, S_1 is true.
(2) *Induction step*: Let S_k be the statement of the binomial theorem with n replaced by k and k replaced by r. Multiply both sides of S_k by $(a + b)$ to obtain
$(a + b)^{k+1}$
$= \left[a^k + \cdots + \dbinom{k}{r - 1}a^{k-(r-1)}b^{r-1}\right.$
$\quad \left. + \dbinom{k}{r}a^{k-r}b^r + \cdots + b^k\right](a + b)$
$= a^{k+1} + \cdots + \left[\dbinom{k}{r - 1} + \dbinom{k}{r}\right]a^{(k+1)-r}b^r$
$\quad + \cdots + b^{k+1}$

$$= a^{k+1} + \cdots + \binom{k+1}{r}a^{(k+1)-r}b^r + \cdots + b^{k+1}.$$

This proves S_{k+1}, assuming S_k. Hence S_n is true for $n = 1, 2, 3, \ldots$

Exercise Set 11.8

1. **(a)** 0.18, 0.24, 0.23, 0.23, 0.12; **(b)** Opinions may vary, but it seems that people tend not to pick the first or last numbers.
3. 5187 e-mails **5.** **(a)** $\frac{2}{7}$; **(b)** $\frac{5}{7}$; **(c)** 0; **(d)** 1 **7.** $\frac{1}{2}$
9. **(a)** $\frac{1}{13}$; **(b)** $\frac{2}{13}$; **(c)** $\frac{1}{4}$; **(d)** $\frac{1}{26}$ **11.** $\frac{1}{5525}$ **13.** $\frac{135}{323}$
15. $\frac{1}{108,290}$ **17.** $\frac{33}{66,640}$ **19.** **(a)** HHH, HHT, HTH, HTT, THH, THT, TTH, TTT; **(b)** $\frac{3}{8}$; **(c)** $\frac{7}{8}$; **(d)** $\frac{7}{8}$; **(e)** $\frac{3}{8}$ **21.** $\frac{9}{19}$
23. $\frac{1}{38}$ **25.** $\frac{18}{19}$ **27.** $\frac{9}{19}$ **29.** Zero
30. One-to-one **31.** Function; domain; range; domain; range
32. Zero **33.** Combination **34.** Inverse variation
35. Factor **36.** Geometric sequence
37. **(a)** $\binom{13}{2} \cdot \binom{4}{2} \cdot \binom{4}{2} \cdot \binom{44}{1}$, or 123,552; **(b)** 0.0475
39. **(a)** $13 \cdot \binom{4}{3} \cdot \binom{48}{2} - 3744$, or 54,912; **(b)** $\dfrac{54,912}{\binom{52}{5}} \approx 0.0211$

Review Exercises: Chapter 11

1. True **2.** False **3.** True **4.** False
5. $-\frac{1}{2}, \frac{4}{17}, -\frac{9}{82}, \frac{16}{257}; -\frac{121}{14,642}; -\frac{529}{279,842}$ **6.** $(-1)^{n+1}(n^2 + 1)$
7. $\frac{3}{2} - \frac{9}{8} + \frac{27}{26} - \frac{81}{80} = \frac{417}{1040}$
8.

n	u_n
1	0.3
2	2.5
3	13.5
4	68.5
5	343.5
6	1718.5
7	8593.5
8	42968.5
9	214843.5
10	1074218.5

9. $\sum_{k=1}^{7}(k^2 - 1)$ **10.** $\frac{15}{4}$ **11.** $a + 4b$ **12.** 531
13. 20,100 **14.** 11 **15.** -4 **16.** $n = 6, S_n = S_6 = -126$
17. $a_1 = 8, a_5 = \frac{1}{2}$ **18.** Does not exist **19.** $\frac{3}{11}$
20. $\frac{3}{8}$ **21.** $\frac{241}{99}$ **22.** $5\frac{4}{5}, 6\frac{3}{5}, 7\frac{2}{5}, 8\frac{1}{5}$ **23.** 167.3 ft
24. \$45,993.04 **25.** **(a)** \$7.38; **(b)** \$1365.10
26. \$88,888,888,889

27. S_n: $\quad 1 + 4 + 7 + \cdots + (3n - 2) = \dfrac{n(3n - 1)}{2}$

S_1: $\quad 1 = \dfrac{1(3 \cdot 1 - 1)}{2}$

S_k: $\quad 1 + 4 + 7 + \cdots + (3k - 2) = \dfrac{k(3k - 1)}{2}$

S_{k+1}: $1 + 4 + 7 + \cdots + (3k - 2) + [3(k + 1) - 2]$
$$= 1 + 4 + 7 + \cdots + (3k - 2) + (3k + 1)$$
$$= \dfrac{(k + 1)(3k + 2)}{2}$$

(1) *Basis step*: $\dfrac{1(3 \cdot 1 - 1)}{2} = \dfrac{2}{2} = 1$ is true.
(2) *Induction step*: Assume S_k. Add $(3k + 1)$ on both sides.
$1 + 4 + 7 + \cdots + (3k - 2) + (3k + 1)$
$$= \dfrac{k(3k - 1)}{2} + (3k + 1)$$
$$= \dfrac{k(3k - 1)}{2} + \dfrac{2(3k + 1)}{2}$$
$$= \dfrac{3k^2 - k + 6k + 2}{2}$$
$$= \dfrac{3k^2 + 5k + 2}{2}$$
$$= \dfrac{(k + 1)(3k + 2)}{2}$$

28. S_n: $\quad 1 + 3 + 3^2 + \cdots + 3^{n-1} = \dfrac{3^n - 1}{2}$

S_1: $\quad 1 = \dfrac{3^1 - 1}{2}$

S_k: $\quad 1 + 3 + 3^2 + \cdots + 3^{k-1} = \dfrac{3^k - 1}{2}$

S_{k+1}: $\quad 1 + 3 + 3^2 + \cdots + 3^{(k+1)-1} = \dfrac{3^{k+1} - 1}{2}$

(1) *Basis step*: $\dfrac{3^1 - 1}{2} = \dfrac{2}{2} = 1$ is true.
(2) *Induction step*: Assume S_k. Add 3^k on both sides.
$1 + 3 + \cdots + 3^{k-1} + 3^k$
$$= \dfrac{3^k - 1}{2} + 3^k = \dfrac{3^k - 1}{2} + 3^k \cdot \dfrac{2}{2}$$
$$= \dfrac{3 \cdot 3^k - 1}{2} = \dfrac{3^{k+1} - 1}{2}$$

29. S_n: $\quad \left(1 - \dfrac{1}{2}\right)\left(1 - \dfrac{1}{3}\right) \cdots \left(1 - \dfrac{1}{n}\right) = \dfrac{1}{n}$

S_2: $\quad \left(1 - \dfrac{1}{2}\right) = \dfrac{1}{2}$

S_k: $\quad \left(1 - \dfrac{1}{2}\right)\left(1 - \dfrac{1}{3}\right) \cdots \left(1 - \dfrac{1}{k}\right) = \dfrac{1}{k}$

S_{k+1}: $\quad \left(1 - \dfrac{1}{2}\right)\left(1 - \dfrac{1}{3}\right) \cdots \left(1 - \dfrac{1}{k}\right)\left(1 - \dfrac{1}{k + 1}\right)$
$$= \dfrac{1}{k + 1}.$$

(1) *Basis step*: S_2 is true by substitution.
(2) *Induction step*: Assume S_k. Deduce S_{k+1}. Starting with the left side of S_{k+1}, we have

$\underbrace{\left(1 - \dfrac{1}{2}\right)\left(1 - \dfrac{1}{3}\right) \cdots \left(1 - \dfrac{1}{k}\right)}\left(1 - \dfrac{1}{k + 1}\right)$
$$= \dfrac{1}{k} \cdot \left(1 - \dfrac{1}{k + 1}\right) \quad \text{By } S_k$$
$$= \dfrac{1}{k} \cdot \left(\dfrac{k + 1 - 1}{k + 1}\right)$$
$$= \dfrac{1}{k} \cdot \dfrac{k}{k + 1}$$
$$= \dfrac{1}{k + 1}. \quad \text{Simplifying}$$

30. $6! = 720$ **31.** $9 \cdot 8 \cdot 7 \cdot 6 = 3024$
32. $\binom{15}{8} = 6435$ **33.** $24 \cdot 23 \cdot 22 = 12,144$
34. $\dfrac{9}{1!4!2!2!} = 3780$ **35.** $3 \cdot 4 \cdot 3 = 36$

36. **(a)** $_6P_5 = 720$; **(b)** $6^5 = 7776$; **(c)** $1 \cdot {}_5P_4 = 120$;
(d) $_3P_2 \cdot 1 \cdot 1 \cdot 1 = 6$ **37.** 2^8, or 256
38. $m^7 + 7m^6n + 21m^5n^2 + 35m^4n^3 + 35m^3n^4 + 21m^2n^5 + 7mn^6 + n^7$
39. $x^5 - 5\sqrt{2}x^4 + 20x^3 - 20\sqrt{2}x^2 + 20x - 4\sqrt{2}$
40. $x^8 - 12x^6y + 54x^4y^2 - 108x^2y^3 + 81y^4$
41. $a^8 + 8a^6 + 28a^4 + 56a^2 + 70 + 56a^{-2} + 28a^{-4} + 8a^{-6} + a^{-8}$ **42.** $-6624 + 16{,}280i$ **43.** $220a^9x^3$
44. $-\binom{18}{11}128a^7b^{11} = -4{,}073{,}472a^7b^{11}$ **45.** $\frac{1}{12}$; 0
46. $\frac{1}{4}$ **47.** $\frac{6}{5525}$ **48.** $\frac{86}{206} \approx 0.42, \frac{97}{206} \approx 0.47, \frac{23}{206} \approx 0.11$
49. **(a)** $a_n = 0.1332631579n + 2.803157895$; **(b)** about 14.3 lb
of American cheese   **50.** B   **51.** A   **52.** D
53. **(a)** No (unless a_n is all positive or all negative); **(b)** yes;
(c) yes; **(d)** no (unless a_n is constant); **(e)** no (unless a_n is
constant); **(f)** no (unless a_n is constant)
54. $\dfrac{a_{k+1}}{a_k} = r_1, \dfrac{b_{k+1}}{b_k} = r_2$, so $\dfrac{a_{k+1}b_{k+1}}{a_kb_k} = r_1r_2$, a constant.
55. $\frac{1}{2}, -\frac{1}{6}, \frac{1}{18}$ **56.** $-2, 0, 2, 4$ **57.** $\left(\log \dfrac{x}{y}\right)^{10}$ **58.** 18
59. 36   **60.** -9 **61.** For each circular arrangement of the
numbers on a clock face, there are 12 distinguishable ordered
arrangements on a line. The number of arrangements of 12 objects
on a line is $_{12}P_{12}$, or 12!. Thus the number of circular
permutations is $\dfrac{_{12}P_{12}}{12} = \dfrac{12!}{12} = 11! = 39{,}916{,}800$. In general, for
each circular arrangement of n objects, there are n distinguishable
ordered arrangements on a line. The total number of arrangements
of n objects on a line is $_nP_n$, or $n!$. Thus the number of circular
permutations is $\dfrac{n!}{n} = \dfrac{n(n-1)!}{n} = (n-1)!$.
62. Put the following in the form of a paragraph.
First find the number of seconds in a year (365 days):
$$365 \text{ days} \cdot \frac{24 \text{ hr}}{1 \text{ day}} \cdot \frac{60 \text{ min}}{1 \text{ hr}} \cdot \frac{60 \text{ sec}}{1 \text{ min}} = 31{,}536{,}000 \text{ sec.}$$
The number of arrangements possible is 15!.
The time is $\dfrac{15!}{31{,}536{,}000} \approx 41{,}466$ years.
63. Order is considered in a combination lock.   **64.** Choosing
k objects from a set of n objects is equivalent to not choosing the
other $n - k$ objects.

Test: Chapter 11

1. [11.1] -43 **2.** [11.1] $\frac{2}{3}, \frac{3}{4}, \frac{4}{5}, \frac{5}{6}, \frac{6}{7}$
3. [11.1] $2 + 5 + 10 + 17 = 34$
4. [11.1]

n	u_n
1	0.66667
2	0.75
3	0.8
4	0.83333
5	0.85714
6	0.875
7	0.88889
8	0.9
9	0.90909
10	0.91667

5. [11.1] $\displaystyle\sum_{k=1}^{6} 4k$ **6.** [11.1] $\displaystyle\sum_{k=1}^{\infty} 2^k$ **7.** [11.1] $3, 2\frac{1}{3}, 2\frac{3}{7}, 2\frac{7}{17}$
8. [11.2] 44   **9.** [11.2] 38   **10.** [11.2] -420 **11.** [11.2] 675
12. [11.3] $\frac{5}{512}$ **13.** [11.3] 1000   **14.** [11.3] 510   **15.** [11.3] 27
16. [11.3] $\frac{56}{99}$ **17.** [11.1] \$10,000, \$8000, \$6400, \$5120, \$4096,
\$3276.80   **18.** [11.2] \$18.20   **19.** [11.3] \$74,399.77
20. [11.4]

$S_n:$ $2 + 5 + 8 + \cdots + (3n - 1) = \dfrac{n(3n+1)}{2}$

$S_1:$ $2 = \dfrac{1(3 \cdot 1 + 1)}{2}$

$S_k:$ $2 + 5 + 8 + \cdots + (3k - 1) = \dfrac{k(3k+1)}{2}$

$S_{k+1}:$ $2 + 5 + 8 + \cdots + (3k - 1) + [3(k+1) - 1]$
$$= \frac{(k+1)[3(k+1)+1]}{2}$$

(1) *Basis step*: $\dfrac{1(3 \cdot 1 + 1)}{2} = \dfrac{1 \cdot 4}{2} = 2$, so S_1 is true.

(2) *Induction step*:
$$\underbrace{2 + 5 + 8 + \cdots + (3k - 1)} + [3(k+1) - 1]$$
$$= \frac{k(3k+1)}{2} + [3k + 3 - 1] \quad \textbf{By } S_k$$
$$= \frac{3k^2}{2} + \frac{k}{2} + 3k + 2$$
$$= \frac{3k^2}{2} + \frac{7k}{2} + 2$$
$$= \frac{3k^2 + 7k + 4}{2}$$
$$= \frac{(k+1)(3k+4)}{2}$$
$$= \frac{(k+1)[3(k+1)+1]}{2}$$

21. [11.5] 3,603,600   **22.** [11.6] 352,716
23. [11.6] $\dfrac{n(n-1)(n-2)(n-3)}{24}$ **24.** [11.5] $_6P_4 = 360$
25. [11.5] **(a)** $6^4 = 1296$; **(b)** $1 \cdot {}_5P_3 = 60$
26. [11.6] $_{28}C_4 = 20{,}475$ **27.** [11.6] $_{12}C_8 \cdot {}_8C_4 = 34{,}650$
28. [11.7] $x^5 + 5x^4 + 10x^3 + 10x^2 + 5x + 1$
29. [11.7] $35x^3y^4$ **30.** [11.7] $2^9 = 512$ **31.** [11.8] $\frac{4}{7}$
32. [11.8] $\frac{48}{1001}$ **33.** [11.1] B   **34.** [11.5] 15

Just-in-Time

1. Real Numbers

1. $\frac{2}{3}, 6, -2.45, 18.\overline{4}, -11, \sqrt[3]{27}, 5\frac{1}{6}, -\frac{8}{7}, 0, \sqrt{16}$
2. $\frac{2}{3}, -2.45, 18.\overline{4}, 5\frac{1}{6}, -\frac{8}{7}$
3. $\sqrt{3}, \sqrt[6]{26}, 7.151551555\ldots, -\sqrt{35}, \sqrt[5]{3}$
4. $6, -11, \sqrt[3]{27}, 0, \sqrt{16}$ **5.** $6, \sqrt[3]{27}, 0, \sqrt{16}$
6. All of them

2. Properties of Real Numbers

1. Additive inverse property **2.** Associative property of multiplication **3.** Distributive property **4.** Commutative property of addition **5.** Multiplicative identity property **6.** Commutative property of multiplication **7.** Additive identity property **8.** Multiplicative inverse property **9.** Associative property of addition **10.** Distributive property

3. Order on the Number Line

1. False **2.** True **3.** True **4.** True **5.** False **6.** True

4. Absolute Value

1. 98 **2.** 0 **3.** 4.7 **4.** $\frac{2}{3}$ **5.** 20 **6.** 12.6 **7.** 11 **8.** $\frac{21}{8}$

5. Operations with Real Numbers

1. 19 **2.** $\frac{1}{10}$ **3.** -5 **4.** -3 **5.** -350 **6.** -5.5 **7.** 24 **8.** 10 **9.** -12.6 **10.** 20 **11.** -15 **12.** $-\frac{1}{6}$ **13.** -8 **14.** -22 **15.** $\frac{4}{5}$

6. Interval Notation

1. $[-5, 5]$ **2.** $(-3, -1]$ **3.** $(-\infty, -2]$ **4.** $(3.8, \infty)$ **5.** $(7, \infty)$ **6.** $(-2, 2)$ **7.** $(-4, 5)$ **8.** $[1.7, \infty)$ **9.** $(-5, -2]$ **10.** $\left(-\infty, \sqrt{5}\right)$

7. Integers as Exponents

1. $\frac{1}{3^6}$ **2.** $(0.2)^5$ **3.** $\frac{z^9}{w^4}$ **4.** $\frac{z^2}{y^2}$ **5.** 1 **6.** a^8 **7.** $-6x^{-4}y^4$, or $-\frac{6y^4}{x^4}$ **8.** x^{-11}, or $\frac{1}{x^{11}}$ **9.** $m^{-6}n^{-6}$, or $\frac{1}{m^6 n^6}$ **10.** t^{-20}, or $\frac{1}{t^{20}}$

8. Scientific Notation

1. 1.85×10^7 **2.** 7.86×10^{-4} **3.** 2.3×10^{-9} **4.** 8.927×10^9 **5.** 0.000000043 **6.** 5,170,000 **7.** 620,300,000,000 **8.** 0.0000294

9. Order of Operations

1. 3 **2.** 103 **3.** -235 **4.** 2048 **5.** 2 **6.** 5 **7.** 32 **8.** 44

10. Introduction to Polynomials

1. 6 **2.** 8 **3.** 4 **4.** 0 **5.** 8 **6.** Binomial **7.** Monomial **8.** Trinomial

11. Add and Subtract Polynomials

1. $9y - 4$ **2.** $-2x^2 + 6x - 2$ **3.** $3x + 2y - 2z - 3$ **4.** $2ab^2 - a^2b + 6ab + 10$ **5.** $-4x^2 + 8xy - 5y^2 + 3$

12. Multiply Polynomials

1. $-21a^6$ **2.** $y^2 + 2y - 15$ **3.** $x^2 + 9x + 18$ **4.** $2a^2 + 13a + 15$ **5.** $4x^2 + 8xy + 3y^2$ **6.** $33t^2 + 41t - 4$

13. Special Products of Binomials

1. $x^2 + 6x + 9$ **2.** $25x^2 - 30x + 9$ **3.** $4x^2 + 12xy + 9y^2$ **4.** $a^2 - 10ab + 25b^2$ **5.** $n^2 - 36$ **6.** $9y^2 - 16$

14. Factor Polynomials; The FOIL Method

1. $3(x + 6)$ **2.** $2z^2(z - 4)$ **3.** $(3x - 1)(x^2 + 6)$ **4.** $(t + 6)(t^2 - 2)$ **5.** $(w - 5)(w - 2)$ **6.** $(t + 3)(t + 5)$ **7.** $2(n - 12)(n + 2)$ **8.** $y^2(y - 2)(y - 7)$ **9.** $(2n - 7)(n + 8)$ **10.** $(2y - 3)(y + 2)$ **11.** $(b - 5t)(b - t)$ **12.** $(x^2 - 10)(x^2 + 3)$

15. Factor Polynomials: The *ac*-Method

1. $(4x + 3)(2x - 3)$ **2.** $2(5t - 3)(t + 1)$ **3.** $3(3a - 1)(2a - 5)$

16. Special Factorizations

1. $(z + 9)(z - 9)$ **2.** $(4x + 3)(4x - 3)$ **3.** $7p(q^2 + y^2)(q + y)(q - y)$ **4.** $(x + 6)^2$ **5.** $(3z - 2)^2$ **6.** $a(a + 12)^2$ **7.** $(x + 4)(x^2 - 4x + 16)$ **8.** $(m - 6)(m^2 + 6m + 36)$ **9.** $3a^2(a - 2)(a^2 + 2a + 4)$ **10.** $(t^2 + 1)(t^4 - t^2 + 1)$

17. Equation-Solving Principles

1. 10 **2.** 12 **3.** -4 **4.** 10 **5.** 2 **6.** -3 **7.** 0 **8.** 5

18. Inequality-Solving Principles

1. $[-125, \infty)$ **2.** $(-\infty, -9)$ **3.** $(-\infty, 2)$ **4.** $(-\infty, -56]$ **5.** $(-\infty, 27)$ **6.** $[11, \infty)$

19. The Principle of Zero Products

1. $0, -21$ **2.** $-7, 1$ **3.** $-\frac{3}{5}, 4$ **4.** $-\frac{5}{3}, \frac{1}{2}$ **5.** $0, 8$ **6.** $-3, 11$ **7.** $-15, 2$ **8.** $-\frac{3}{4}, \frac{4}{3}$

20. The Principle of Square Roots

1. -6 and 6, or ± 6 **2.** $-\sqrt{10}$ and $\sqrt{10}$, or $\pm \sqrt{10}$ **3.** $-\sqrt{3}$ and $\sqrt{3}$, or $\pm \sqrt{3}$ **4.** $-\sqrt{5}$ and $\sqrt{5}$, or $\pm \sqrt{5}$ **5.** -5 and 5, or ± 5 **6.** $-\sqrt{15}$ and $\sqrt{15}$, or $\pm \sqrt{15}$

21. Simplify Rational Expressions

1. The set of all real numbers except 0 and 1 **2.** The set of all real numbers except -7 and 3 **3.** $\frac{x + 2}{x - 2}$ **4.** $\frac{x - 1}{x - 3}$ **5.** $\frac{x - 3}{x}$ **6.** $\frac{2(y + 4)}{y - 1}$

22. Multiply and Divide Rational Expressions

1. 1   **2.** $m + n$ **3.** $\dfrac{4x + 1}{3x - 2}$ **4.** $\dfrac{a + 1}{a - 3}$ **5.** $\dfrac{3(x - 4)}{2(x + 4)}$

6. $\dfrac{1}{x + y}$

23. Add and Subtract Rational Expressions

1. 2   **2.** $\dfrac{2(3x^2 + 2x - 7)}{3(3x + 1)(x - 2)}$ **3.** $\dfrac{2a}{(a + 1)(a - 1)}$

4. $\dfrac{3x - 4}{(x - 2)(x - 1)}$ **5.** $\dfrac{-y + 10}{(y + 4)(y - 5)}$ **6.** $\dfrac{y}{(y - 2)(y - 3)}$

24. Simplify Complex Rational Expressions

1. $x - y$ **2.** $\dfrac{a}{a + b}$ **3.** $\dfrac{w^2 - 2w + 4}{w}$ **4.** $\dfrac{x + y}{x}$

5. $-a - b$

25. Simplify Radical Expressions

1. 21   **2.** $3y$ **3.** $a - 2$ **4.** $-3x$ **5.** $3x^2$ **6.** 2

7. $2xy \sqrt[4]{3x^2}$ **8.** $5\sqrt{21}$ **9.** $\sqrt{5y}$ **10.** $\dfrac{1}{2x}$ **11.** $x - 2$

12. $2x^2 y \sqrt{6}$ **13.** $3x\sqrt[3]{4y}$ **14.** $17\sqrt{2}$ **15.** $12\sqrt{3}$
16. $2\sqrt{2}$ **17.** $4\sqrt{5}$ **18.** $16 + 9\sqrt{3}$ **19.** -12
20. $4 + 2\sqrt{3}$

26. Rationalize Denominators

1. $\dfrac{4\sqrt{11}}{11}$ **2.** $\dfrac{\sqrt{21}}{7}$ **3.** $\dfrac{\sqrt[3]{28}}{2}$ **4.** $\dfrac{2\sqrt[3]{6}}{3}$

5. $\dfrac{3\sqrt{30} + 12}{14}$ **6.** $\sqrt{7} + \sqrt{3}$ **7.** $\dfrac{6\sqrt{m} + 6\sqrt{n}}{m - n}$

8. $\dfrac{\sqrt{3}}{3}$

27. Rational Exponents

1. $\sqrt[6]{y^5}$ **2.** $\sqrt[3]{x^2}$ **3.** 8   **4.** 128   **5.** $\frac{1}{5}$ **6.** $\frac{1}{16}$ **7.** $y^{1/3}$

8. $x^{5/2}$ **9.** $x\sqrt[6]{x}$ **10.** $(a - 2)^2$ **11.** $n\sqrt[3]{mn^2}$

28. The Pythagorean Theorem

1. 17   **2.** $\sqrt{32} \approx 5.657$ **3.** 12   **4.** 5   **5.** $\sqrt{31} \approx 5.568$

Index

Index of Applications

Photo Credits

Geometry

Plane Geometry

Rectangle
Area: $A = lw$
Perimeter: $P = 2l + 2w$

Square
Area: $A = s^2$
Perimeter: $P = 4s$

Triangle
Area: $A = \frac{1}{2}bh$

Sum of Angle Measures
$A + B + C = 180°$

Right Triangle
Pythagorean theorem
(equation):
$a^2 + b^2 = c^2$

Parallelogram
Area: $A = bh$

Trapezoid
Area: $A = \frac{1}{2}h(a + b)$

Circle
Area: $A = \pi r^2$
Circumference:
$C = \pi d = 2\pi r$

Solid Geometry

Rectangular Solid
Volume: $V = lwh$

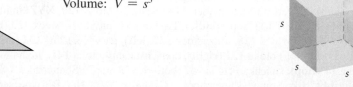

Cube
Volume: $V = s^3$

Right Circular Cylinder
Volume: $V = \pi r^2 h$
Lateral surface area:
$L = 2\pi rh$
Total surface area:
$S = 2\pi rh + 2\pi r^2$

Right Circular Cone
Volume: $V = \frac{1}{3}\pi r^2 h$
Lateral surface area:
$L = \pi rs$
Total surface area:
$S = \pi r^2 + \pi rs$
Slant height:
$s = \sqrt{r^2 + h^2}$

Sphere
Volume: $V = \frac{4}{3}\pi r^3$
Surface area: $S = 4\pi r^2$

Algebra

Properties of Real Numbers

Commutative: $a + b = b + a; \quad ab = ba$

Associative: $a + (b + c) = (a + b) + c;$
$a(bc) = (ab)c$

Additive Identity: $a + 0 = 0 + a = a$

Additive Inverse: $-a + a = a + (-a) = 0$

Multiplicative Identity: $a \cdot 1 = 1 \cdot a = a$

Multiplicative Inverse: $a \cdot \dfrac{1}{a} = \dfrac{1}{a} \cdot a = 1, a \neq 0$

Distributive: $a(b + c) = ab + ac$

Exponents and Radicals

$a^m \cdot a^n = a^{m+n}$ $\qquad \dfrac{a^m}{a^n} = a^{m-n}$

$(a^m)^n = a^{mn}$ $\qquad (ab)^m = a^m b^m$

$\left(\dfrac{a}{b}\right)^m = \dfrac{a^m}{b^m}$ $\qquad a^{-n} = \dfrac{1}{a^n}$

If n is even, $\sqrt[n]{a^n} = |a|$.

If n is odd, $\sqrt[n]{a^n} = a$.

$\sqrt[n]{a} \cdot \sqrt[n]{b} = \sqrt[n]{ab}, \quad a, b \geq 0$

$\sqrt[n]{\dfrac{a}{b}} = \dfrac{\sqrt[n]{a}}{\sqrt[n]{b}}$

$\sqrt[n]{a^m} = (\sqrt[n]{a})^m = a^{m/n}$

Special-Product Formulas

$(a + b)(a - b) = a^2 - b^2$

$(a + b)^2 = a^2 + 2ab + b^2$

$(a - b)^2 = a^2 - 2ab + b^2$

$(a + b)^3 = a^3 + 3a^2b + 3ab^2 + b^3$

$(a - b)^3 = a^3 - 3a^2b + 3ab^2 - b^3$

$(a + b)^n = \displaystyle\sum_{k=0}^{n} \binom{n}{k} a^{n-k} b^k$, where

$\binom{n}{k} = \dfrac{n!}{k!\,(n-k)!}$

$\quad = \dfrac{n(n-1)(n-2)\,\cdots\,[n-(k-1)]}{k!}$

Factoring Formulas

$a^2 - b^2 = (a + b)(a - b)$

$a^2 + 2ab + b^2 = (a + b)^2$

$a^2 - 2ab + b^2 = (a - b)^2$

$a^3 + b^3 = (a + b)(a^2 - ab + b^2)$

$a^3 - b^3 = (a - b)(a^2 + ab + b^2)$

Interval Notation

$(a, b) = \{x \mid a < x < b\}$

$[a, b] = \{x \mid a \leq x \leq b\}$

$(a, b] = \{x \mid a < x \leq b\}$

$[a, b) = \{x \mid a \leq x < b\}$

$(-\infty, a) = \{x \mid x < a\}$

$(a, \infty) = \{x \mid x > a\}$

$(-\infty, a] = \{x \mid x \leq a\}$

$[a, \infty) = \{x \mid x \geq a\}$

Absolute Value

$|a| \geq 0$

For $a > 0$,

$\quad |X| = a \to X = -a \quad \text{or} \quad X = a,$

$\quad |X| < a \to -a < X < a,$

$\quad |X| > a \to X < -a \quad \text{or} \quad X > a.$

Equation-Solving Principles

$a = b \to a + c = b + c$

$a = b \to ac = bc$

$a = b \to a^n = b^n$

$ab = 0 \leftrightarrow a = 0 \quad \text{or} \quad b = 0$

$x^2 = k \to x = \sqrt{k} \quad \text{or} \quad x = -\sqrt{k}$

Inequality-Solving Principles

$a < b \to a + c < b + c$

$a < b \text{ and } c > 0 \to ac < bc$

$a < b \text{ and } c < 0 \to ac > bc$

(Algebra continued)

Algebra *(continued)*

The Distance Formula

The distance from (x_1, y_1) to (x_2, y_2) is given by

$$d = \sqrt{(x_2 - x_1)^2 + (y_2 - y_1)^2}.$$

The Midpoint Formula

The midpoint of the line segment from (x_1, y_1) to (x_2, y_2) is given by

$$\left(\frac{x_1 + x_2}{2}, \frac{y_1 + y_2}{2} \right).$$

Formulas Involving Lines

The slope of the line containing points (x_1, y_1) and (x_2, y_2) is given by

$$m = \frac{y_2 - y_1}{x_2 - x_1}.$$

Slope–intercept equation: $\quad y = f(x) = mx + b$
Horizontal line: $\quad\quad\quad\; y = b \quad$ or $\quad f(x) = b$
Vertical line: $\quad\quad\quad\quad\; x = a$
Point–slope equation: $\quad\; y - y_1 = m(x - x_1)$

The Quadratic Formula

The solutions of $ax^2 + bx + c = 0, a \neq 0$, are given by

$$x = \frac{-b \pm \sqrt{b^2 - 4ac}}{2a}.$$

Compound Interest Formulas

Compounded n times per year: $\quad A = P\left(1 + \dfrac{i}{n}\right)^{nt}$

Compounded continuously: $\quad\quad P(t) = P_0 e^{kt}$

Properties of Exponential and Logarithmic Functions

$\log_a x = y \leftrightarrow x = a^y \quad\quad\quad\quad a^x = a^y \leftrightarrow x = y$

$\log_a MN = \log_a M + \log_a N \quad\quad \log_a M^p = p \log_a M$

$\log_a \dfrac{M}{N} = \log_a M - \log_a N$

$\log_b M = \dfrac{\log_a M}{\log_a b}$

$\log_a a = 1 \quad\quad\quad\quad\quad\quad\quad\quad \log_a 1 = 0$

$\log_a a^x = x \quad\quad\quad\quad\quad\quad\quad\; a^{\log_a x} = x$

Conic Sections

Circle: $\quad\quad (x - h)^2 + (y - k)^2 = r^2$

Ellipse: $\quad \dfrac{(x - h)^2}{(x \; a^2 h)^2} + \dfrac{(y - k)^2}{(y \; b^2 k)^2} = 1,$

$\quad\quad\quad\; \dfrac{}{b^2} + \dfrac{}{a^2} = 1$

Parabola: $\quad (x - h)^2 = 4p(y - k),$
$\quad\quad\quad\quad (y - k)^2 = 4p(x - h)$

Hyperbola: $\quad \dfrac{(x - h)^2}{a^2} - \dfrac{(y - k)^2}{b^2} = 1,$

$\quad\quad\quad\quad \dfrac{(y - k)^2}{a^2} - \dfrac{(x - h)^2}{b^2} = 1$

Arithmetic Sequences and Series

$a_1, \quad a_1 + d, \quad a_1 + 2d, \quad a_1 + 3d, \; \ldots$

$a_{n+1} = a_n + d \quad\quad\quad\quad a_n = a_1 + (n - 1)d$

$S_n = \dfrac{n}{2}(a_1 + a_n)$

Geometric Sequences and Series

$a_1, \quad a_1 r, \quad a_1 r^2, \quad a_1 r^3, \; \ldots$

$a_{n+1} = a_n r \quad\quad\quad\quad a_n = a_1 r^{n-1}$

$S_n = \dfrac{a_1(1 - r^n)}{1 - r} \quad\quad\quad S_\infty = \dfrac{a_1}{1 - r}, \; |r| < 1$

Trigonometry

Trigonometric Functions

Acute Angles

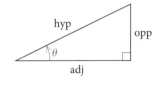

$$\sin \theta = \frac{\text{opp}}{\text{hyp}}, \quad \csc \theta = \frac{\text{hyp}}{\text{opp}},$$

$$\cos \theta = \frac{\text{adj}}{\text{hyp}}, \quad \sec \theta = \frac{\text{hyp}}{\text{adj}},$$

$$\tan \theta = \frac{\text{opp}}{\text{adj}}, \quad \cot \theta = \frac{\text{adj}}{\text{opp}}$$

Any Angle

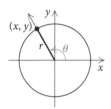

$$\sin \theta = \frac{y}{r}, \quad \csc \theta = \frac{r}{y},$$

$$\cos \theta = \frac{x}{r}, \quad \sec \theta = \frac{r}{x},$$

$$\tan \theta = \frac{y}{x}, \quad \cot \theta = \frac{x}{y}$$

Real Numbers

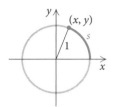

$$\sin s = y, \quad \csc s = \frac{1}{y},$$

$$\cos s = x, \quad \sec s = \frac{1}{x},$$

$$\tan s = \frac{y}{x}, \quad \cot s = \frac{x}{y}$$

Basic Trigonometric Identities

$$\sin (-x) = -\sin x,$$
$$\cos (-x) = \cos x,$$
$$\tan (-x) = -\tan x,$$

$$\tan x = \frac{\sin x}{\cos x},$$

$$\cot x = \frac{\cos x}{\sin x},$$

$$\csc x = \frac{1}{\sin x},$$

$$\sec x = \frac{1}{\cos x},$$

$$\cot x = \frac{1}{\tan x}$$

Pythagorean Identities

$$\sin^2 x + \cos^2 x = 1,$$
$$1 + \cot^2 x = \csc^2 x,$$
$$1 + \tan^2 x = \sec^2 x$$

Identities Involving $\pi/2$

$$\sin (\pi/2 - x) = \cos x,$$
$$\cos (\pi/2 - x) = \sin x, \quad \sin (x \pm \pi/2) = \pm \cos x,$$
$$\tan (\pi/2 - x) = \cot x, \quad \cos (x \pm \pi/2) = \mp \sin x$$

Sum and Difference Identities

$$\sin (u \pm v) = \sin u \cos v \pm \cos u \sin v,$$
$$\cos (u \pm v) = \cos u \cos v \mp \sin u \sin v,$$
$$\tan (u \pm v) = \frac{\tan u \pm \tan v}{1 \mp \tan u \tan v}$$

Double-Angle Identities

$$\sin 2x = 2 \sin x \cos x,$$
$$\cos 2x = \cos^2 x - \sin^2 x$$
$$= 1 - 2 \sin^2 x$$
$$= 2 \cos^2 x - 1,$$

$$\tan 2x = \frac{2 \tan x}{1 - \tan^2 x}$$

Half-Angle Identities

$$\sin \frac{x}{2} = \pm \sqrt{\frac{1 - \cos x}{2}}, \quad \cos \frac{x}{2} = \pm \sqrt{\frac{1 + \cos x}{2}},$$

$$\tan \frac{x}{2} = \pm \sqrt{\frac{1 - \cos x}{1 + \cos x}} = \frac{\sin x}{1 + \cos x} = \frac{1 - \cos x}{\sin x}$$

(Trigonometry continued)

Trigonometry *(continued)*

The Law of Sines

In any $\triangle ABC$,

$$\frac{a}{\sin A} = \frac{b}{\sin B} = \frac{c}{\sin C}.$$

The Law of Cosines

In any $\triangle ABC$,

$$a^2 = b^2 + c^2 - 2bc \cos A,$$
$$b^2 = a^2 + c^2 - 2ac \cos B,$$
$$c^2 = a^2 + b^2 - 2ab \cos C.$$

Trigonometric Function Values of Special Angles

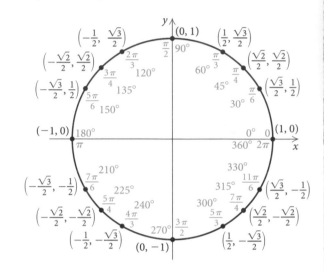

Graphs of Trigonometric Functions

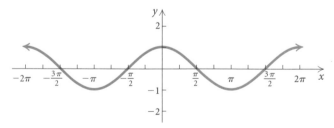

The sine function: $f(x) = \sin x$

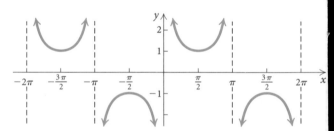

The cosecant function: $f(x) = \csc x$

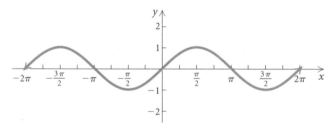

The cosine function: $f(x) = \cos x$

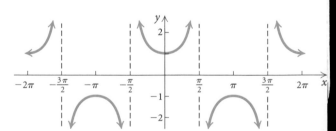

The secant function: $f(x) = \sec x$

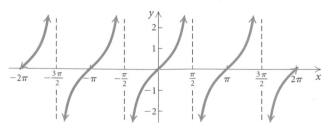

The tangent function: $f(x) = \tan x$

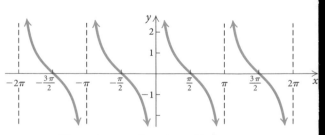

The cotangent function: $f(x) = \cot x$

A Library of Functions

Linear function

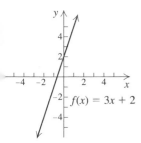

$f(x) = 3x + 2$

Linear function

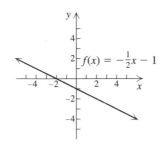

$f(x) = -\frac{1}{2}x - 1$

Constant function

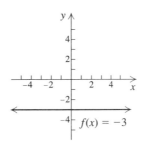

$f(x) = -3$

Absolute-value function

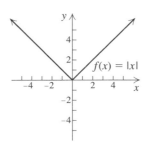

$f(x) = |x|$

Squaring function

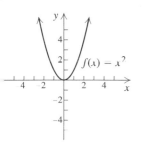

$f(x) = x^2$

Quadratic function

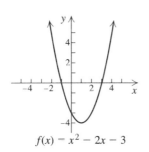

$f(x) = x^2 - 2x - 3$

Quadratic function

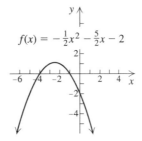

$f(x) = -\frac{1}{2}x^2 - \frac{5}{2}x - 2$

Square-root function

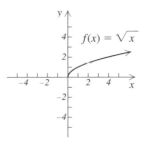

$f(x) = \sqrt{x}$

Cubing function

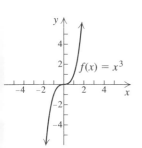

$f(x) = x^3$

Cube root function

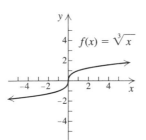

$f(x) = \sqrt[3]{x}$

Greatest integer function

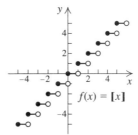

$f(x) = [x]$

Rational function

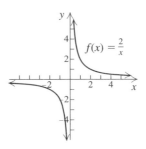

$f(x) = \frac{2}{x}$

Exponential function

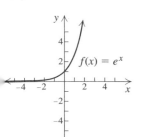

$f(x) = e^x$

Exponential function

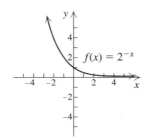

$f(x) = 2^{-x}$

Logarithmic function

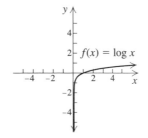

$f(x) = \log x$

Logistic function

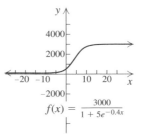

$f(x) = \dfrac{3000}{1 + 5e^{-0.4x}}$